Comprehensive Biotechnology

Comprehensive Biotechnology

Sc. & M.Sc. Students of Different Indian Universities
as per UGC Model Curriculum

K.G. RAMAWAT
Ex-Head,
Department of Botany
M.L. Sukhadia University
Udaipur,
RAJASTHAN

SHAILY GOYAL
M.Sc. Ph.D

S. CHAND
PUBLISHING
empowering minds

S. CHAND & COMPANY PVT. LTD.
(AN ISO 9001: 2008 COMPANY)
RAM NAGAR, NEW DELHI-110055

S. CHAND & COMPANY PVT. LTD.

(An ISO 9001 : 2008 Company)

Head Office: 7361, RAM NAGAR, NEW DELHI - 110 055
Phone: 23672080-81-82, 9899107446, 9911310888 Fax: 91-11-2367744
www.schandpublishing.com; e-mail: helpdesk@schandpublishing

S. CHAND PUBLISHING
empowering minds

Branches

Ahmedabad	:	Ph: 27541965, 27542369, ahmedabad@schandpublishing.com
Bengaluru	:	Ph: 22268048, 22354008, bangalore@schandpublishing.com
Bhopal	:	Ph: 4274723, 4209587, bhopal@schandpublishing.com
Chandigarh	:	Ph: 2725443, 2725446, chandigarh@schandpublishing.com
Chennai	:	Ph. 28410027, 28410058, chennai@schandpublishing.com
Coimbatore	:	Ph: 2323620, 4217136, coimbatore@schandpublishing.com (Marketing Office)
Cuttack	:	Ph: 2332580; 2332581, cuttack@schandpublishing.com
Dehradun	:	Ph: 2711101, 2710861, dehradun@schandpublishing.com
Guwahati	:	Ph: 2738811, 2735640, guwahati@schandpublishing.com
Hyderabad	:	Ph: 27550194, 27550195, hyderabad@schandpublishing.com
Jaipur	:	Ph: 2219175, 2219176, jaipur@schandpublishing.com
Jalandhar	:	Ph: 2401630, 5000630, jalandhar@schandpublishing.com
Kochi	:	Ph: 2378740, 2378207-08, cochin@schandpublishing.com
Kolkata	:	Ph: 22367459, 22373914, kolkata@schandpublishing.com
Lucknow	:	Ph: 4026791, 4065646 lucknow@schandpublishing.com
Mumbai	:	Ph: 22690881, 22610885, mumbai@schandpublishing.com
Nagpur	:	Ph: 6451311, 2720523, 2777666, nagpur@schandpublishing.com
Patna	:	Ph: 2300489, 2302100, patna@schandpublishing.com
Pune	:	Ph: 64017298, pune@schandpublishing.com
Raipur	:	Ph: 2443142, raipur@schandpublishing.com (Marketing Office)
Ranchi	:	Ph: 2361178, ranchi@schandpublishing.com
Siliguri	:	Ph: 2520750, siliguri@schandpublishing.com (Marketing Office)
Visakhapatnam	:	Ph: 2782609 visakhapatnam@schandpublishing.com (Marketing Office)

First Edition 2000
Subsequent Editions and Reprints 2000, 2004, 2008
Fourth Revised Edition 2009; Reprint 2011, 2014
Reprint 2016

ISBN : 978-81-219-1987-6 **Code :** 1003C 284

This book is for Sale in India & other SAARC Countries only.

PRINTED IN INDIA

By Vikas Publishing House Pvt. Ltd., Plot 20/4, Site-IV, Industrial Area Sahibabad, Ghaziabad-201010 and published by S. Chand & Company Pvt. Ltd., 7361, Ram Nagar, New Delhi -110 055.

Preface to the Fourth Revised Edition

"Comprehensive Biotechnology" is revised edition of the book "Plant Biotechnology". This is prepared to cater the need of students in the rapidly expanding field of Biotechnology. The word biotechnology encompasses microbial, plant and animal biotechnology. This is a very vast field and entire gamut of the field to put in a single book is a Herculean task. There are a few books already in the market then why a new book. As a researcher and teacher of biotechnology for the last 30 years, it is our humble endeavour to provide student not only latest information on this frontier area of biology but also in a language clearly understandable to the students. There are books on biotechnology which tells stories within a story and are difficult to follow in respect to progress of the information. You may need a teacher to understand such a book and to get a clearer picture.

Several new topics such as Aquaporins, Artificial intelligence, Automation in micropropagation, Biochips, Green house, Hydroponics, Inteins, Nanotechnology, Space biotechnology, Supercritical fluid extraction, etc. have been included in the book which are not available to students in any book. Besides these chapters, all updated usual topics are presented in very simple language with clear and understandable diagrams. It is the job of teacher to make the student understand the subject and direct him to find details in reference books and now days on Internet. In out efforts in this direction, a comprehensive glossary as well as internet links is provided so that students can search more information on net.

We hope students and teachers find this new book on biotechnology useful. Suggestions to improve the book are heartily welcomed. We are thankful to S. Chand & co. New Delhi to publish this book.

Udaipur

Prof. K.G. Ramawat
Dr. Shaily Goyal

Preface to the Third Revised Edition

I am happy to present the third revised edition of "Plant Biotechnology" with new chapters under the following titles : Transgenic Crops : Recent Developments, Molecular Farming, Functional Genomics, Proteomics, Microarrays, Metabolic Engineering and Intellectual Property Rights and Biotechnology. I am now sure that students will find the book more useful.

Udaipur **K. G. RAMAWAT**

Preface to the First Edition

Plant Biotechnology has been introduced as a full course in undergraduate and postgraduate classes of several Universities. Plant Biotechnology has become a very diverse subject and the literature is not available to the students and teachers alike. In recent years enormous data are generated and also compiled in the form of reference books. It is difficult for the students and teachers to have access to this information at one place in consolidated form. The exorbitant price of the reference books is also a limiting factor in developing countries in obtaining required information. Therefore, an attempt has been made to make available all the information in one book at relatively affordable low price to students and teachers.

This book has been prepared looking into the need of graduate and postgraduate students of various Universities. The subject matter has been arranged in concise, clear and simple language understandable to the students. Attempt has been made to provide basic relevant information rather than research highlights of individual cases or the exceptional cases. All the chapters are well illustrated with fine diagrams to make clear vision of the subject. The book on Plant Biotechnology is first of its kind and covers chapters such as tools and techniques of plant tissue culture, bioreactor, production of secondary metabolites, cell immobilization, impact on industry and agriculture, modern tools of genetic transformations and synthetic seed, etc., at one place. No other book is available with such an approach covering individual chapters from basic to molecular aspects of Plant Biotechnology. I am sure that this book will be an important document for the students of all the Indian Universities in the field of biotechnology.

I express my sincere thanks to Mr. Ravindra Kumar Gupta, Director, S. Chand & Company, New Delhi for his keen interest in the publication of this book.

Suggestions for further improvement of the book will always be welcomed and entertained wholeheartedly.

Udaipur –313002 **K. G. RAMAWAT**

CONTENTS

1. **Introduction to biotechnology** 1 – 12

♦ Historical developments ♦ Scope of biotechnology ♦ Industrial applications ♦ Safety concerns ♦ Questions

2. **Macromolecules** 13 – 25

♦ Carbohydrates ♦ Proteins ♦ Lipids ♦ Nucleic acids ♦ Codon, anti-codon ♦ Enzymes ♦ Questions

3. **Concept of gene** 26 – 33

♦ One gene one enzyme ♦ One gene one polypeptide ♦ Complex gene protein relationship ♦ Packaging of DNA ♦ Questions

4. **Plasmids and Vectors** 34 – 47

♦ Plasmids ♦ pBR322 ♦ Bacteriophage as cloning vector ♦ Phagemids as cloning vector ♦ Cosmids as cloning vector ♦ Artificial chromosomes (YAC, BAC, PAC, MAC, AHC) ♦ Questions

5. **Principles of recombinant technology** 48 – 55

♦ Gene cloning ♦ Selection of recombinant bacteria ♦ Transcription of foreign gene ♦ Translation of foreign gene ♦ Post transcriptional modification ♦ Expression in different host (bacteria, yeast, insect, mammalian cells, plants) ♦ Questions

6. **Gene integration and expression vectors** 56 – 65

♦ Gene expression from strong regulatable promoter (regulatable promoter, optimal gene expression and bioreactor cultivation) ♦ Expression in other microorganisms ♦ Translation expression vectors ♦ Increasing secretion ♦ Mammalian cell expression vector ♦ Questions

7. **Tools of molecular biology** 66 – 92

♦ Blotting techniques (Southern, Northern, Western, Protein transfer to a membrane, Blocking non specific binding sites, Washing primary and secondary antibodies) ♦ DNA sequencing (Sanger's method, Procedure, Automatic DNA sequencer, Slab gel sequencing system, Capillary gel electrophoresis) ♦ Cell Fractionation (Cellular fractionation, Tissue homogenization, Isolation of nuclei, Isolation of mitochondria) ♦ FISH (Fibre FISH, FISH detection) ♦ Genomic libraries (Alternative vectors, In bacteriophage, Larger inserts, YAC libraries) ♦ Nucleic acid purification (Isolation and purification of genomic DNA, Plasmid DNA isolation, Purification of plasmid DNA, Isolation of RNA, Purification of mRNA, Analysis of yield of nucleic acid) ♦ Restricion enzymes (Naming, Patterns, Mode of action, Types of restriction enzymes, Restriction enzyme as tool) ♦ Questions

8. **Polymerase chain reaction** 93 – 102

♦ Mechanism of basic PCR (Denaturation, Annealing, Extension) ♦ Comparison of PCR with gene cloning ♦ Different schemes of PCR (Standard, Inverse, Anchored, Asymmetric, Quantitative, RT, Comparison) ♦ Applications ♦ Future of PCR ♦ Questions

9. **Molecular markers** 103 – 113

♦ RFLP (Procedure, limitations) ♦ RAPD (Procedure, limitations, Difference between RAPD and RFLP) ♦ AFLP (Procedure, limitations) ♦ Microsatellites ♦ Inter-Simple Sequence Repeats ♦ Genetic mapping ♦ Questions

10. Mapping the genome 114 – 124

♦ Restriction mapping (Methodology, Direct examination of DNA) ♦ Sequence tagged site mapping ♦ Single nucleotide polymorphism ♦ Detection of polymorphic DNA in absence of sequence information ♦ Chromosome walking (map based cloning) ♦ Primer walking ♦ Chromosome jumping ♦ Questions

11. Advance Biotechnology 125 – 158

♦ Apoptosis ♦ Aquaporin ♦ Artificial Intelligence ♦ Biochips ♦ Nanosensors ♦ Biosensors ♦ Hydroponics ♦ Immunological Assays (RIA, ELISA) ♦ Nano-technology ♦ Patch Clamp Technologies ♦ Space Biotechnology ♦ Supercritical Fluid Extraction ♦ Questions

12. Introduction to Plant Tissue Culture: History, media and culture initiation 159 – 179

♦ Historical development (Haberlandt, Nobecourt, Gauthert, Morel, White, Steward, Reinert, Skoog, Cockings, Vasil, Hilderbrandt, Murashige, Maheshwari, Guha and Maheshwari) ♦ Medium and its preparation (Medium composition, medium preparation) ♦ Methods of sterilization (Explant and its sterilization, mode of action of disinfectant, equipment and medium sterilization) ♦ Callus culture ♦ Cell Suspension Culture (Batch culture, Continuous culture, Time course) ♦ Single cell culture ♦ Applications of cell culture ♦ Questions

13. Techniques for the study of cells 180 – 205

♦ Callus and cell culture (Initiation, PCV, Cell number, FW, Cell viability) ♦ Microscopy (Bright and dark field, Phase contrast, Fluorescence, Confocal, Electron microscopy) ♦ Micrometry ♦ Autoradiography ♦ Spectophotometry ♦ Centrifugation ♦ Chromatography (Paper, 2-D, TLC, Column chromatography, Adsorption, Ion exchange, Gel filtration, Affinity, GLC, HPLC) ♦ Questions

14. Cell immobilization and Synthetic seed technology 206 – 211

♦ Cell immobilization (Need for immobilization, Range of entrapment, Types of bioreactor used for immobilization of plant cells, Applications) ♦ Synthetic Seed Technology (Development of coatings, Maturation of somatic embryos, Applications, Mechanization, Economic considerations) ♦ Questions

15. Plant protoplasts: Isolation, culture and somatic hybridization 212 – 222

♦ Introduction ♦ Application of protoplasts in plant research ♦ Isolation of plant protoplasts ♦ Protoplasts culture ♦ Protoplasts fusion and somatic hybridization (Membrane properties, Fusion procedure) ♦ Regeneration from protoplasts ♦ Fusion methods (PEG, Dextran, Electrofusion) ♦ Application of protoplasts hybridization (Production of somatic hybrids and cybrids, Asymmetric hybrids, Direct DNA uptake) ♦ Identification and selection of hybrid cells ♦ Stability of hybrids ♦ Questions

16. Haploids 223 – 230

♦ Homozygous lines ♦ Haploids by different methods (Abnormal embryo development, Abnormal fertilization, Anther and Pollen culture) ♦ Questions

17. Secondary Plant Products: I Classes and production 231 – 243

♦ Primary and Secondary products ♦ Alkaloids (Defination, Occurrence, Extraction, Classification, Biological Functions) ♦ Secondary metabolite production in culture

(Alkaloids, Why tissue culture, Approaches for production, Antitumour agents, Production of steroids and saponins, Food additives, Insecticides) ♦ Questions

18. Secondary Plant Products: II Factors affecting the production in culture 244 – 260

♦ Introduction ♦ Time course study of growth and production ♦ Optimization (Effect of nutrients, Effect of plant growth regulators, Precursors, Production medium) ♦ Selection of cells (Variability in field grown plants, Producer and non producer cells, Cloning and selection, Plating technique, Selection parameters) ♦ Elicitors ♦ Hairy root culture ♦ Biotransformations (Freely suspended plant cells, Immobilized plant cells, Production of secondary metabolites by isolated plant enzymes) ♦ Questions

19. Bioreactors 261 – 272

♦ Introduction ♦ Bioreactor process ♦ Factors for growth in bioreactor (Gas-liquid mass transfer, Shear, Mixing) ♦ Types of bioreactors (Mechanically agitated bioreactor, Pneumatically agitated bioreactors, Wave bioreactor) ♦ Production of secondary metabolites (Shikonin production, Berberine production, Rosmarinic acid production, Indole alkaloids production) ♦ Organ culture in bioreactor ♦ Hairy root culture in bioreactor ♦ Commercialization ♦ Questions

20. Application of plant biotechnology in plant pathology 273 – 283

♦ Meristem culture for virus free plants (Meristem culture, Technique and media, Meristem culture and heat treatment, Mechanism of virus elimination in meristems, Effect of virus elimination, Virus indexing) ♦ In vitro culture of tumour tissue and obligate parasites (Dual cultures and Axenic cultures, Techniques and applications) ♦ Plant tumours and Habituation ♦ Plant cell and Protoplasts and Plant Virus interactions (Virus infection, Cross protection, Infection process, Study of nucleic acids, Production of secondary metabolites, Disease resistance lines) ♦ Questions

21. Micropropagation, Somatic embryogenesis and Somaclonal variation 284 – 295

♦ Introduction ♦ Organogenesis ♦ Regulation of regeneration ♦ Somatic embryogenesis (Direct and indirect somatic embryogenesis, Development, Factors affecting somatic embryogenesis, Embryogenesis as model of developmental regulation, Application of somatic embryogenesis, Advantages) ♦ Somaclonal variations (Origin and causes, Problems with somaclonal variation) ♦ Benefits of Somaclonal variation for crop improvement ♦ Questions

22. Automation in Plant propagation 296 – 311

♦ Introduction ♦ Production systems (Axillary budding, Adventitious budding, Somatic embryo development) ♦ Process of micropropagation ♦ Delivery to the growers ♦ Production cost estimates ♦ Synthetic seed technology ♦ Automation and robotics ♦ Micropropagation through bioreactors ♦ Laboratory design (General consideration for location, Design and layout of the laboratories) ♦ Greenhouse facility ♦ Packaging and shipping ♦ Establishment of micropropagation industry (Cost reduction measures, Contamination control, Washing and sterlizing operations, Transfer operations, Culture maintenance in growth rooms, Production planning, Scale of operation) ♦ Commercial aspects ♦ Questions

23. Methods of gene transfer 312 – 325

♦ DNA transfer in protoplasts (Electroporation, Chemically stimulated DNA uptake, Liposomes, Microinjection, Sonication) ♦ DNA transfer in Plant tissues (Particle gun method, Laser microbeam, Silicone Carbide fibres) ♦ Agrobacterium mediated gene transfer (Vectors, General features, Ti and Ri plasmids of

Agrobacterium, Vectors based on Ti and Ri plasmids, Cointegrative, Bianary vector, Transformation technique, Selectable markers, Reporter genes) ♦ Questions

24. **Transgenic crops** 326 – 349

♦ Insect pest control ♦ Herbicide resistant plants ♦ Virus resistant plants ♦ Abiotic stress tolerance ♦ Quality Improvement ♦ Pharmaceutical and Industrial use ♦ Economic impact ♦ Environmental impact ♦ Transplastomic plants through chloroplast transformation (Genome organization, Chloroplast vectors, stability of expressed proteins, Molecular pharming) ♦ Antisense technology ♦ Male sterility ♦ Engineering cytoplasmic male sterility via the chloroplast genome ♦ Questions

25. **Impact of Plant Biotechnology on Agriculture and Industry** 350 – 362

♦ Tissue culture: An aid to conventional breeding programme (Embryo culture, Ovary and Ovule culture, Nucellus culture, Control of fertilization, Endosperm culture) ♦ Tissue culture: A tool of novel unconventional breeding programme (Anther culture, Somatic hybridization and Genetic modification, Somaclonal variation) ♦ Propagation through tissue culture ♦ Production of disease free plants ♦ Impact on forestry (Clonal propagation, Somatic embryogenesis, Protoplasts culture and genetic engineering) ♦ Impact on horticulture ♦ Tissue Culture industry in India ♦ Scope for development ♦ Problems associated with agricultural biotechnology ♦ Commercialization-Micropropagation industry ♦ Plants produced ♦ Governmental efforts ♦ Questions

26. **Molecular Farming** 363 – 370

♦ A Brief history ♦ Edible vaccines ♦ Products on the market (Avidin, β-glucoronidase, Trypsin, Aprotinin, Collagen, Human gastric lipase, Human lactoferin, Production of Hirudin in *Brassica napus,* Production of cyclodextrins) ♦ Questions

27. **Microarrays** 371 – 375

♦ Types of microarrays (cDNA microarrays, Long-oligonucleotide microarrays, Short-oligonucleotide gene chips) ♦ Labelling and Hybridization of cDNA on to chips ♦ Applications of Microarrays (Detection of SNPs, Diagnostic chips for human diseases, Genomic mismatch scanning, DNA chips and functional genomics) ♦ Questions

28. **Metabolic Engineering** 376 – 387

♦ Strategies for metabolic engineering in plants ♦ Engineering primary metabolic pathways (Carbohydrate metabolism, Amino acid and polyamine metabolism, lipid metabolism) ♦ Engineering secondary metabolic pathways (Alkaloids, Terpenoids, Flavonoids, Lignins, Stilbene phytoalexins) ♦ Modification of stilbene synthase expression pattern causes flower colour changes and male sterility in plants ♦ Engineering novel metabolic pathways ♦ Metabolic Engineering for human health (Engineering at metabolic branch points:tocochromanol biosynthesis, Whole pathway regulation with transcription factors, Assembling multigene pathways in plants) ♦ Questions

29. **Functional Genomics** 388 – 396

♦ Differential Display, Sage and Subtrative hybridisation ♦ Microarrays (Limitations of transcriptomic approach, Limitations of proteomics) ♦ Protein Arrays, Tissue arrays and Multi-Immunoblotting antibody-based protein arrays (SELDI, Phage display, Oxyblots, Mass spectometry) ♦ Questions

30. Downstream Processing 397 – 404

♦ Steps in down stream processing (Removal of insolubles, Product isolation, Product purification, Product polishing) ♦ Industrial production system (The need for bioreactors, Bioreactor in industry, Plant cell cultures, Recombinant technology products, Growth phases and secondary metabolites, Culture conditions, Types of bioreactors, Scale-up considerations, Antibodies production) ♦ Platform technology (Resin improvement, Increased velocity, Increased capacity, Overall productivity, Selectivity and Purity, Cleaning and Sanitization) ♦ Questions

31. Animal Cell and Tissue Culture 405 – 429

♦ Introduction ♦ Advantages and disadvantages ♦ Types ♦ Laboratory facilities ♦ Culture media ♦ Substrate for cell growth ♦ Culture procedures ♦ Selection of medium ♦ Problems of cultured cells ♦ Functional Characteristics ♦ Types of cells ♦ Primary culture ♦ Cell culture and cell lines ♦ Somatic cell fusion ♦ Large scale animal cell culture ♦ Organ culture ♦ Pharmaceutical applications of animal cell culture ♦ Questions

32. Hybridoma Technology: Production of Monoclonal Antibodies 430 – 437

♦ Introduction ♦ Commercial production of monoclonal antibodies ♦ Human monoclonal antibodies ♦ Applications of monoclonal antibodies ♦ Monoclonal antibodies in transfusion research ♦ Bloodgrouping and platelet serology ♦ Recombinant monoclonal antibody technology ♦ Questions

33. Transgenic animals 438 – 451

♦ Historical background ♦ Value of transgenic animals ♦ Methods of creation of Transgenic animals (Physical, Chemical, Virus and Bacteria mediated) ♦ Applications ♦ Transgenic livestock ♦ Questions

34. Cloning animals, In vitro fertilization, Tissue Engineering 452 – 465

♦ Cloning (Cloning from different cells, Benefits and risks of cloning, Human cloning) ♦ In vitro fertilization (In vitro production of embryo, Milestones in IVM and IVF, In vitro culture of embryos, In vitro fertilization, Methods, Intracytoplasmic sperm injection) ♦ Tissue Engineering (Biomaterials, Scaffolds, Skin graft, Bone graft, Xenotransplant, Human and pig) ♦ Questions

35. Stem Cell Bioengineering 466 – 474

♦ Division of stem cells ♦ Biological role and properties of stem cells · Types of stem cells Discovery of ES cells ♦ Growing embryonic cells in the laboratory ♦ ES cell differentiation ♦ Adult stem cells and embryonic ♦ Applications (Cell replacement therapy, Cardiovascular therapy, Heamatopoiesis, Neurological disorders, Diabetes, Liver therapy) ♦ Problems (Selection of desired cells, Tumour formation, Transplant rejection) ♦ Future Perspectives ♦ Questions

36. Gene Therapy 475 – 489

♦ Introduction ♦ Diseases of genetic origin ♦ Principles ♦ Adenovirus vector system ♦ Adeno associated virus vector system ♦ Current status ♦ Limitations ♦ Ethical issues ♦ Future of gene therapy ♦ Gene silencing ♦ Genetic diseases ♦ Questions

37. Human Genome Project 490 – 496

♦ Historical ♦ Preparation of libraries ♦ Sequencing reactions ♦ Seperating trhe sequencing products ♦ Assembling the results ♦ Mapping ♦ Subclones ♦ Preparing DNA for sequencing reaction ♦ Product of sequencing ♦ Reading the sequencing working draft sequence ♦ Sanger's Institute ♦ Other organisms mapped ♦ Questions

38. Studying Proteins 497 – 512

♦ Protein purification (Importance, Source of protein, Strategies, Protocol, Examples, Evaluation, SDS –PAGE) ♦ Inteins (Distribution, Structure, Bacterial inteins- like domains) ♦ Single cell protein (Biomass and SCP, SCP production, Choice of microorganisms, Economics of SCP, *Spirulina*) ♦ Protein sequencing (Amino acid composition, Hydrolysis, Separation, Quantitative analysis, N-terminal amino acid analysis, C-terminal amino acid analysis, Edman degradation and reaction, Limitations, MS, Predicting amino sequence from nucleic acid sequence, Protein sequence analysis sites) ♦ Questions

39. Protein folding and Targeting 513 – 521

♦ Introduction ♦ Creating a functional protein ♦ Protein folding ♦ Energy to fold ♦ Process of folding ♦ Fold recognition ♦ SNARE proteins (Membrane traffic, Protein targeting, Targeting in bacteria, Targeting in Eukaryotes)♦ Questions

40. Site- directed Mutagenesis and Protein Engineering 522 – 530

♦ Introduction ♦ Cassette mutagenesis ♦ In vitro site- specific mutagenesis ♦ Site directed mutagenesis using PCR ♦ Protein engineering ♦ Enzyme engineering ♦ Protein engineering cycle ♦ Examples of biocatalysts♦ Questions

41. Protein Profiling, Protein Microarray and Proteomics 531 – 546

♦ Protein profiling (Techniques, Limitations of gels, Differential fluorescence gel electrophoresis protein profiling, ICAT profiling) ♦ Protein microarray ♦ Proteomics (Sample preparation, 2-D PAGE, Labelling, Data preprocessing, peptide identification, relative protein expression, structural proteomics) ♦ MALDI ♦ AP-MALDI ♦ Questions

42. Mass Spectrometry 547 – 557

♦ Applications ♦ Working ♦ Methods of sample ionization ♦ Electrospray ionization ♦ MALDI ♦ MS-MS ♦ Peptide sequencing summary ♦ Mass of amino acids ♦ SELDI-TOF ♦ Questions

43. Enzyme Technology 558 – 566

♦ Introduction ♦ Traditional enzyme markets ·♦ Impact of enzymes as catalyst in chemical industry ♦ Bioprocess and environment ♦ Commercialize applications in chemical industry ♦ Enzymes in other industries (Detergent, Starch, Alcohol, Textile, Feed and Food, Brewing and Cheeze, meat, fats and oils) ♦ Questions

44. Protein –Protein Interactions 567 – 579

♦ Definition ♦ Genomics ♦ Characteristics ♦ Classes of protein-protein interaction (Homo and Heterologous proteins) ♦ Protein domains in signal transduction ♦ Methods to study protein- protein interactions ♦ Characterization ♦ Biophysical techniques (Analytical ultracentrifugation, Light scattering, Fluorescence spectroscopy, Surface Palsmon resonance, Calorimetry) ♦ Databases ♦Network software ♦ Questions

45. Biotechnology in Biodiversity conservation 580 – 591

♦ Intoduction ♦ Bioresource assessment ♦ Biotechnology in ex situ conservation of biodiversity (Field gene bank, Seed banks, Pollen banks, DNA banks) ♦ In vitro conservation methods ♦ Biotechnology Vs Biodiversity ♦ Germplasm preservation: cryobiology (Need for cryopreservation, Procedure of cryoprotection and pre-treatment, Freezing methods, Freezing apparatus) ♦ Applications of cryopreservation ♦ Questions

46. **Bioremediation** 592 – 606

♦ On site versus *in situ* treatment ♦ Biostimulation Vs Bioaugmentation ♦ Phytoremediation ♦ Various phytoremediation processes ♦ Phytoextraction ♦ Phytotransformation ♦ The role of genetics ♦ Hyperaccumulators and biotic interactions ♦ Mutualism ♦ Commensalism ♦ Biocatalysts (Biocatalysis platform for green processes, Bioremediation of persistent contaminants, Microbial bioremediation, Enzymatic bioremediation, Enzymes for clean energy production, Biodiesel, Bioethanol, Biohydrogen and biofuel cells) ♦ Bioabsorption of metals ♦ Questions

Annexure 46A 607 – 611

47. **Biofertilisers and Biopesticides** 612 – 621

♦ Biofertilisers in agroecosystem ♦ Pesticides ♦ Biopesticides (Advantage of biopesticides, Genetic transformation) ♦ Microbial and antimicrobial pesticides (Bacteria as pesticides, Virus as pesticides, Fungi as biopesticides, Plant pesticides, Biochemical pesticides, Natural pesticides, Nematode biopesticides, biopesticide chemicals) ♦ Questions

48. **Biofuel** 622 – 631

♦ Biomass energy ♦ Biogas production ♦ Ethanol production for energy ♦ Hydrogen as energy source ♦ Hydrocarbon production ♦ Biodiesel ♦ Questions

49. **Biosafety Science** 632 – 643

♦ Laboratory practice and technique ♦ Safety equipment ♦ Facility design and construction ♦ Biosafety levels (Biosafety level 1, 2,3,4, Animal facilities) ♦ Clinical laboratories ♦ Importation and Interstate Shipment of certain biomedical materials ♦ Hazards and hazardous materials (Hazardous materials, Biosafety standards, Micoorganisms, Additional rules for projects involving unknown microorganisms, Additional rules for projects involving recombinant DNA (rDNA) Technologies) ♦ Classification of biological agents ♦ Working with genetically modified organisms ♦ Cartagena protocol on biosafety ♦ The World Trade Organization ♦ Horizontal gene transfer ♦ The biosafety regulatory framework in India ♦ The scope of the biosafety regulatory framework ♦ Consent-givers in the biosafety regulatory framework ♦ Useful sites for more information ♦ Questions

50. **Industrial Fermentation** 644 – 657

♦ Introduction ♦ Fermentation technology (Fermenters, Selection of microbes, Fermentation medium) ♦ Penicillin production ♦ Streptomycin ♦ Cephalosporin ♦ Vitamin B12 ♦ Riboflavin ♦ Vitamin A and C ♦ Amino acids ♦ Amylases ♦ Proteases ♦ Questions

51. **Intellectual Property Rights and Biotechnology** 658 – 664

♦ IPR of biotechnology inventions(TRIPS, Plant variety protection, GATT) ♦ US plant patents ♦ Utility patents ♦ Gene patenting ♦ Copyrights ♦ Trade secrets ♦ Questions

52. **Bioinformatics** 665 – 675

♦ Overview ♦ Sequence analysis ♦ Genome annotation ♦ Computational evolutionary biology ♦ Measuring biodiversity ♦ Analysis of gene expression ♦ Analysis of regulation ♦ Analysis of protein expression ♦ Analysis of mutation in cancer ♦ Prediction of protein structure ♦ Comparitive genomics ♦ Software tools (Biological data, BLAST, Parallel BLAST) ♦ EBI ♦ NCBI ♦ Questions

Multiple choice questions 676 – 685

Glossary 686 – 702

Bibliography 703 – 706

INTRODUCTION TO BIOTECHNOLOGY

CHAPTER 1

Biotechnology is defined as the 'application of scientific and engineering principles to the processing of material by biological agents to provide goods and services'. The Spinks Report (1980) defined biotechnology as 'the application of biological organisms, systems or processes to the manufacturing and service industries'. United States Congress's Office of Technology Assessment defined biotechnology as 'any technique that used living organisms to make or modify a product, to improve plants or animals or to develop microorganisms for specific uses'. The document focuses on the development and application of modern biotechnology based on new enabling techniques of recombinant-DNA technology, often referred to as genetic engineering.

The history of biotechnology begins with *zymotechnology*, which commenced with a focus on brewing techniques for beer. By *World War I*, however, zymotechnology would expand to tackle larger industrial issues, and the potential of industrial *fermentation* gave rise to biotechnology. The oldest biotechnological processes are found in microbial fermentations, as born out by a Babylonian tablet circa 6000 B.C., unearthed in 1881 and explaining the preparation of beer. In about 4000 B.C. leavened bread was produced with the aid of yeast. The Sumerians were able to brew as many as twenty types of beer in the third millennium B.C. In the 14th century, first vinegar manufacturing industry was established in France near Orleans. In 1680, Antony van Leeuwenhoek first observed yeast cells with his newly designed microscope. In 1857, Louis Pasteur highlighted the lactic acid fermentation by microbes. By the end of 19th century large number of industries and group of scientists were involved in the field of biotechnology and developed large scale sewage purification system employing microbes were established in Germany and France. In 1914 to 1916, Delbruck, Heyduck and Hennerberg discovered the large-scale use of yeast in food industry. In the same period, acetone, butanol and glycerin were obtained from bacteria. In 1920, Alexander Fleming discovered penicillin and large scale manufacturing of penicillin started in 1944. Table 1.1 presents chronological history of biotechnology.

Fermentation to Produce Foods Fermentation is perhaps the most ancient biotechnological discovery. Over 10,000 years ago mankind was producing wine, beer, vinegar and bread using microorganisms, primarily yeast. Yogurt was produced by lactic acid bacteria in milk and molds were used to produce cheese. These processes are still in use today for the production of modern foods. However, the cultures that are used have been purified and often genetically refined to maintain the most desirable traits and highest quality of the products.

Industrial Fermentation

In 1897, the discovery that enzymes from yeast can convert sugar to alcohol lead to industrial processes for chemicals such as butanol, acetone and glycerol. Fermentation processes are still in use today in many modern biotech organizations, often for the production of enzymes to be used in pharmaceutical processes, environmental remediation and other industrial processes.

Food Preservation

Drying, salting and freezing foods to prevent spoilage by microorganisms were practiced long before anyone really understood why they worked or even fully knew what caused the food to spoil in the first place.

Quarantines

The practice of quarantining to prevent the spread of disease was in place long before the origins of disease were known. However, it demonstrates early acceptance that illness could be passed from an infected individual to another healthy individual, who would then begin to have symptoms of the disease.

Selective Plant Breeding

Crop improvement, by selecting seeds from the most successful or healthiest plants, to obtain a new crop having the most desirable traits, is a form of early crop technology. Farmers learned that using only the seeds from the best plants would eventually enhance and strengthen the desired traits in subsequent crops. In the mid-1860's, Gregor Mendel's studies on inheritable traits of peas improved our understanding of genetic inheritance and lead to practices of cross-breeding (now known as hybridization).

I. HISTORICAL DEVELOPMENTS

Table 1.1. Historical development of biotechnology.

Year	Development
Before 6000 B.C.	Yeast employed to make wine and beer
4000 B.C.	Leavened bread produced with the aid of yeast
1521	Aztecs harvested algae from lakes as a source of food.
1670-1680	Copper mined with aid of microbes, Rio Tinto, Spain. Antony van Leeuwenhoek first observed microbes with newly designed microscope.
1869	Johann Meischer isolated DNA from the nuclei of white blood cells.
1876	Louis Pasteur identifies extraneous microbes as a cause of failed beer fermentations.
1890	Alcohol first used to fuel motors
1893	Fermentation process patented by Koch, Pasteur.
1897	Edurad Buchner discovered that enzymes extracted from yeast can convert sugar into alcohol.
1910	Thomas H. Morgan proved that genes are carried on chromosomes "Biotechnology" term coined. Large-scale sewage purification systems employing microbes were established.
1912-1914	Three important industrial chemicals- acetone, butanol and glycerol were obtained from bacteria.
1918	Germans use acetone produced by plants to make bombs.
1920	Alexander Fleming discovered penicillin. Plant hybridization.
1928	Yeast grown in large quantities for animal and glycerol.
1938	Proteins and DNA studied by x-ray crystallography.
1941	George Beadle proposed "one gene one enzyme" hypothesis.
1943-1953	Linus Pauling described sickle cell anemia calling it a molecular disease. Cortisone made in large amounts. DNA identified as the genetic material.
1944	DNA as genetic material and transformation factor in bacteria (Avery, Macleod and McCarty).
1946	Bacterial recombination discovered (Lederberg and Tatum).
1951	Lambda bacteriophage discovered (Lederberg).

1953	Double helix structure of DNA revealed (Watson and Crick).
1956	Isolation of DNA polymerase I, enzymatic synthesis of DNA (Kornberg).
1958	Semiconservative replication of DNA (Messelson and Stahl).
1960	Isolation of m-RNA.
1961	Nucleic acid hybridization (Marmur and Doty), Operon model (Jacob and Monod) and *in vitro* protein synthesis allows codon assignment (Nirenberg and Mathaei).
1962	Mining of uranium with the aid of microbes begin in Canada.
1973	Beginning of genetic engineering. Stanley Cohen produced first recombinant DNA organism. Brazilian government initiates major fuel programme to replace oil with alcohol.
1975	Hybridomas which make monoclonal antibodies were first created
1976	US National Institute of Health introduces guidelines on genetic engineering
1977	Human Growth Hormone produced by bacterial cells.
1978	Genetic engineering techniques used to produce human insulin in *E. coli* by Genentech Inc.
1979	Genentech Inc. produce human growth hormone and two kinds of interferon DNA from malignant cells transformed a strain of cultured mouse cells - new tool for analyzing cancer genes.
1980	Rank Hovis McDougall receives permission to market fungal food for human consumption in UK. The US Supreme Court rules in Diamond v. Chakrabarty that genetically engineered microorganisms can be patented.
1981	Monoclonal antibodies receive US approval for use in diagnosis.
1982	The Food and Drug Administration approves the first biotechnology therapy, a human insulin drug made by Genentech.
1983	Syntex Corporation received FDA approval for a monoclonal antibody-based diagnostic test for *Chlamydia trachomatis.*
1984	Chiron Corp. announced the first cloning and sequencing of the entire human immunodeficiency virus (HIV) genome. Stanford University received a product patent for prokaryote DNA. Charles Cantor and David Schwartz developed pulsed-field gel electrophoresis. Animal interferon's approval for protection against cattle disease.
1985	Genetically engineered plants resistant to insects, viruses, and bacteria were field tested for the first time. Axel Ullrich reported the sequencing of the human insulin receptor. Plants can be patented.
1986	Orthoclone OKT3® (Muromonab-CD3) approved for reversal of acute kidney transplant rejection. The EPA approved the release of the first genetically engineered crop, gene-altered tobacco plants.
1987	Genentech received FDA approval to market rt-PA (genetically engineered tissue plasminogen activator) to treat heart attacks. Recombivax-HB® (recombinant hepatitis B vaccine) approved.
1989	Epogen® (Epoetin alfa) a genetically engineered protein introduced, providing a means to help patients with kidney failure.
1990	GenPharm International Inc. created the first transgenic dairy cow. The cow was used to produce human milk proteins for infant formula. The Human Genome Project, the international effort to map all of the genes in the human

	body, was launched. The first successful field trial of genetically engineered cotton plants was conducted by Calgene Inc. The plants had been engineered to withstand use of the herbicide Bromoxynil.
1993	Kary Mullis won the Nobel Prize in Chemistry for inventing the technology of polymerase chain reaction
1994	The first genetically engineered food product, the Flavr Savr tomato, gained FDA approval.
1996	Dolly, the sheep was cloned.
1997	Researchers at Scotland's Roslin Institute report that they have cloned a sheep—named Dolly—from the cell of an adult ewe. Polly the first sheep cloned by nuclear transfer technology bearing a human gene appears later. A group of Oregon researchers claim to have cloned two Rhesus monkeys. Rituxan, the first antibody-based therapy for cancer (for patients with non-Hodgkin's lymphoma) was approved. A new DNA technique combines PCR, DNA chips, and computer programming providing a new tool in the search for disease-causing genes.
1998	Human embryonic stem cell lines established. University of Hawaii scientists, cloned three generations of mice from nuclei of adult ovarian cumulus cells.
2000	"Working draft" of the human genome's 3.15 billion letters was completed after a decade of research.
2003	Broad Institute was founded in Cambridge to give scientists access to the human genome project, and to understand the molecular basis of disease.
2007	Craig C. Mello, a University of Massachusetts researcher, shares the Nobel Prize with Andrew Fire of Stanford University for discovering a special kind of RNA that can shut down individual genes.

In last fifteen years progress have been made by microbiologists and genetic engineers, and we are hopeful to solve many fold problems of the present day, specially energy and food crisis to cater the need of growing population of the world. Mineral ore deposits are also becoming more scarce and expensive to recover from earth's crust. Microorganisms can be used to enhance the recovery of metals from low-grade ores and from effluents containing undesirable quantities of heavy metals or other toxins. When these technologies are applied at industrial level, they constitute bio-industry (Table 1.2).

II. SCOPE OF BIOTECHNOLOGY

Table 1.2. Principal products of bio-industry.

Technology	Examples of Products
Fermentation technology	Antibiotics, Vitamins, Enzymes, Amino acids, Citric acid, Lactic acid, Malic acid, Ethanol, Acetone, Butanol, Biogas, Biopesticides, Nucleotides, Steroids, Alkaloids, Diagnostic reagents etc.
Enzymatic Engineering	Isoglucose, Glucose syrup etc.
Genetic engineering and cell Cultures	Single cell protein, Clones, Interferon's, Vaccines, Blood Products, Monoclonal antibodies.

Genetic engineering in biotechnology stimulated hopes for both therapeutic proteins, drugs and biological organisms themselves, such as seeds, pesticides, engineered yeasts, and modified human cells for treating genetic diseases. The field of genetic engineering remains a heated topic of discussion in today's society with the advent of *gene therapy, stem cell research, cloning*, and *genetically-modified food*. Biotechnology is the applied science and has made advances in two major areas, viz., molecular biology and production of industrially important biochemicals. The scientists are now diverting themselves toward biotechnological companies; this has caused the development of many biotechnological industries. In USA alone more than 225 companies have been established and successfully working, like Biogen, Cetus, Genentech, Hybritech, etc. In world, USA, Japan, and many countries of Europe are leaders in biotechnological researchers encouraged by industrialists.

These companies are working for human welfare and opted following areas for research and development:

- Automated bioscreening for therapeutic agents.
- Bioprocessing alkenes to valuable oxides and glycols.
- Developing immobilized cell and enzyme systems for chemical process industries.
- Engineering of a series of organisms for specific industrial use.
- Genetical improvement of microorganisms for production of pharmaceutical products.
- Human gene therapy.
- Improved production of Vitamin B_{12}.
- Large-scale production of fructose from inexpensive forms of glucose.
- Manufacturing ethanol by continuous fermentation.
- Microbiological based production of human insulin and interferon's.
- Microbiologically upgradation of hydrocarbons.
- Production and development of vaccine to prevent calibacillosis.
- Production of biopesticide and biofertilizers.
- Production of diagnostic kits for toxoplasmosis identification.
- Production of monoclonal antibodies for organ transplant tissue typing.
- Production of photosynthetically efficient plants.
- Production of transgenic plants and animals.
- Production of xanthan gum in oil fields for recovery of crude mineral oils.

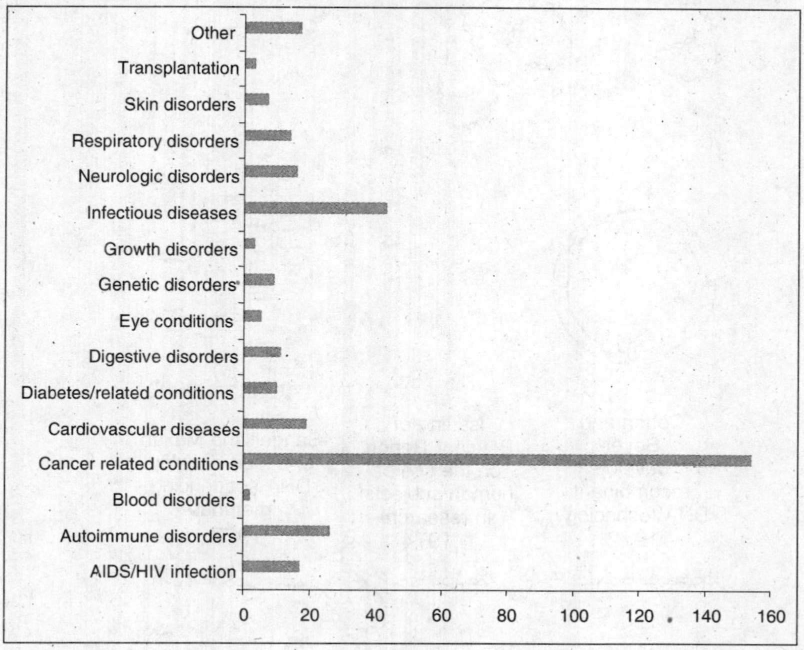

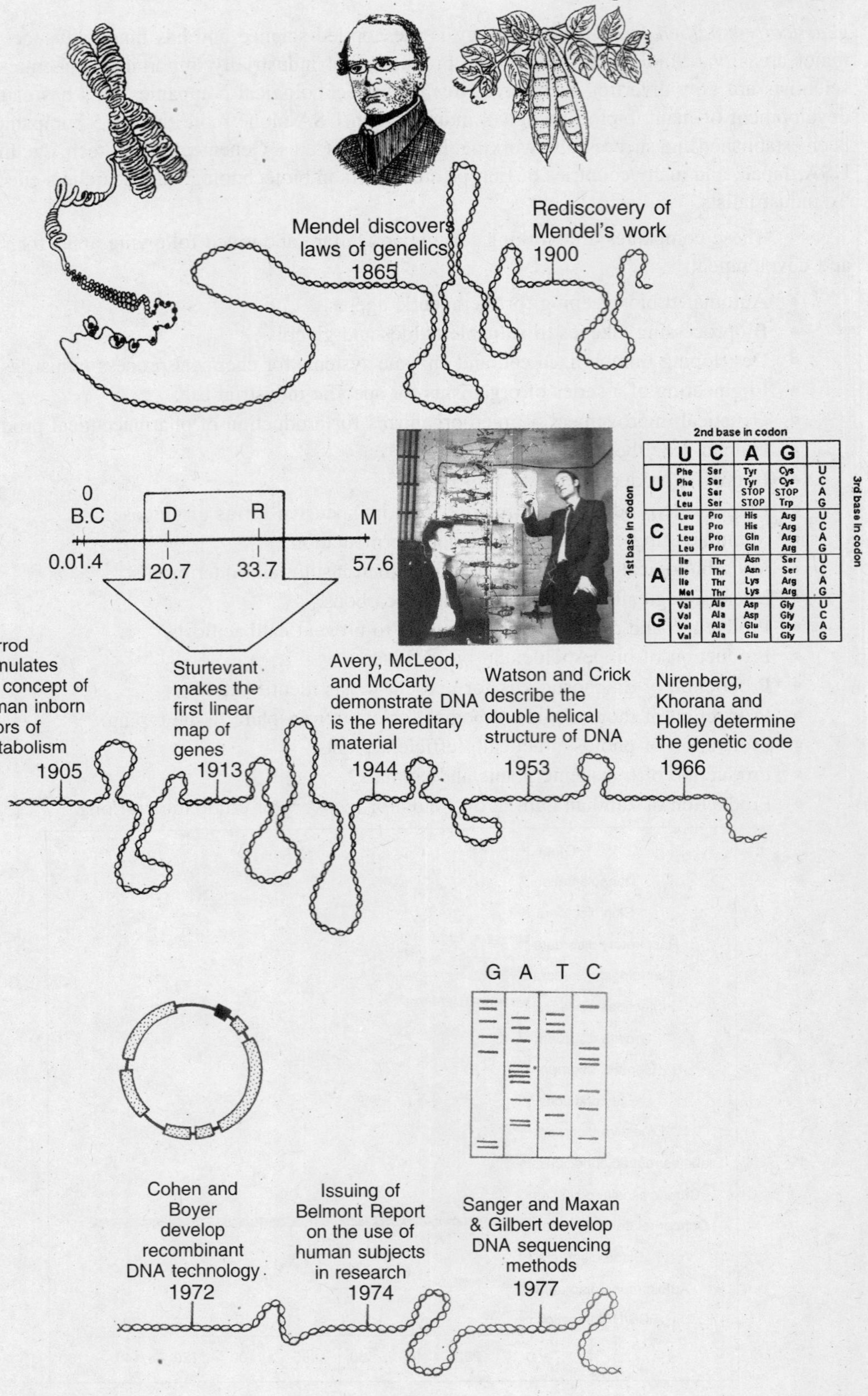
Mendel discovers laws of genelics 1865
Rediscovery of Mendel's work 1900
0
B.C
D
R
M
0.01.4
20.7
33.7
57.6
2nd base in codon
1st base in codon
3rd base in codon
U C A G
U Phe Phe Leu Leu | Ser Ser Ser Ser | Tyr Tyr STOP STOP | Cys Cys STOP Trp | U C A G
C Leu Leu Leu Leu | Pro Pro Pro Pro | His His Gln Gln | Arg Arg Arg Arg | U C A G
A Ile Ile Ile Met | Thr Thr Thr Thr | Asn Asn Lys Lys | Ser Ser Arg Arg | U C A G
G Val Val Val Val | Ala Ala Ala Ala | Asp Asp Glu Glu | Gly Gly Gly Gly | U C A G
Gorrod formulates the concept of human inborn errors of metabolism 1905
Sturtevant makes the first linear map of genes 1913
Avery, McLeod, and McCarty demonstrate DNA is the hereditary material 1944
Watson and Crick describe the double helical structure of DNA 1953
Nirenberg, Khorana and Holley determine the genetic code 1966
G A T C
Cohen and Boyer develop recombinant DNA technology 1972
Issuing of Belmont Report on the use of human subjects in research 1974
Sanger and Maxan & Gilbert develop DNA sequencing methods 1977

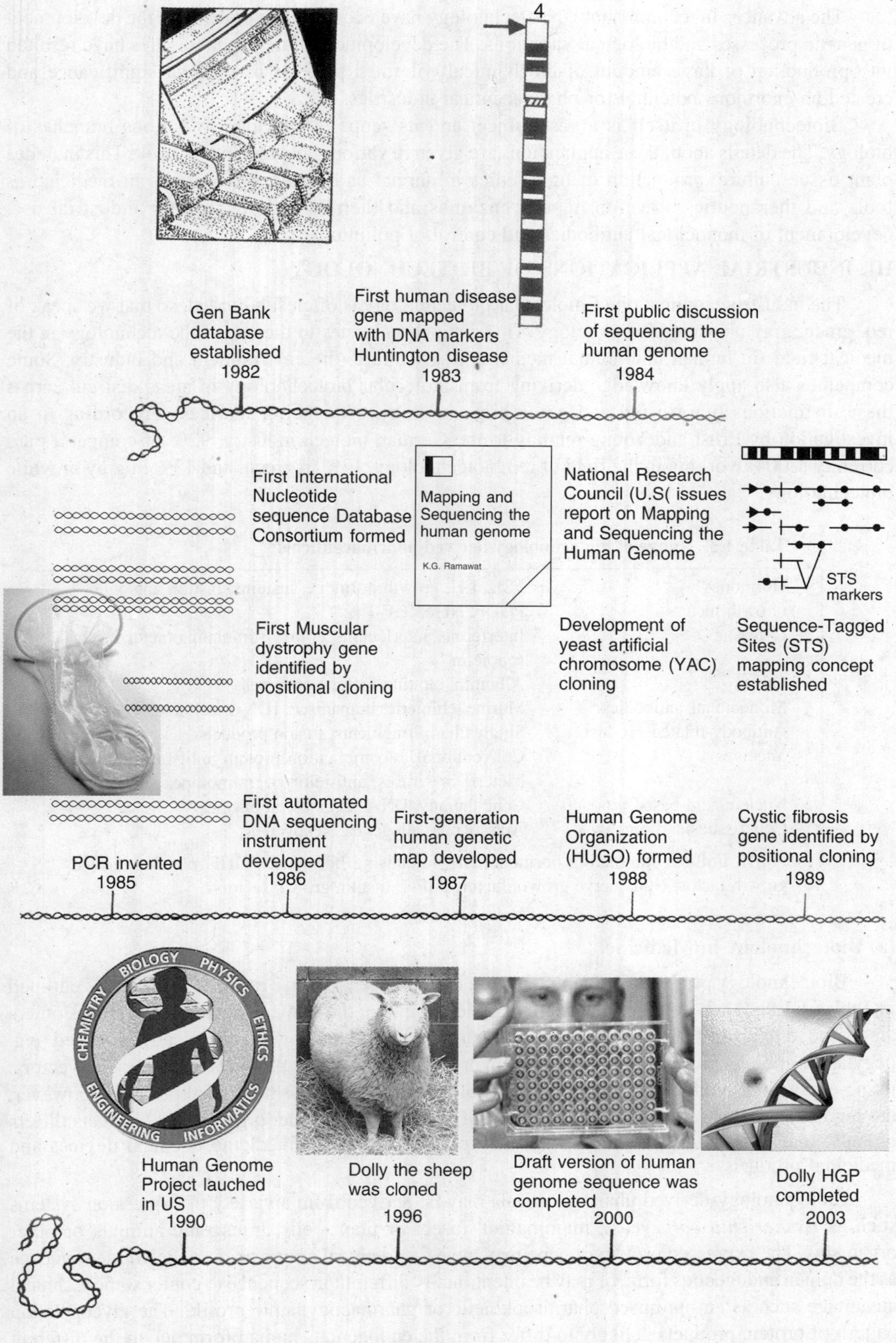
4
Gen Bank database established 1982
First human disease gene mapped with DNA markers Huntington disease 1983
First public discussion of sequencing the human genome 1984
First International Nucleotide sequence Database Consortium formed
Mapping and Sequencing the human genome
K.G. Ramawat
National Research Council (U.S(issues report on Mapping and Sequencing the Human Genome
STS markers
First Muscular dystrophy gene identified by positional cloning
Development of yeast artificial chromosome (YAC) cloning
Sequence-Tagged Sites (STS) mapping concept established
PCR invented 1985
First automated DNA sequencing instrument developed 1986
First-generation human genetic map developed 1987
Human Genome Organization (HUGO) formed 1988
Cystic fibrosis gene identified by positional cloning 1989
CHEMISTRY BIOLOGY PHYSICS ETHICS INFORMATICS ENGINEERING
Human Genome Project lauched in US 1990
Dolly the sheep was cloned 1996
Draft version of human genome sequence was completed 2000
Dolly HGP completed 2003

The advances in recombinant DNA technology have occurred in parallel with the development of genetic processes and biological variations. The development of new technologies have resulted into production of large amount of biochemically-defined proteins of medical significance and created an enormous potential for pharmaceutical industries.

Biotechnology in itself is a vast subject and its scope is extended to various branches of biology. The details about these applications are given in various chapters in this book. This includes plant tissue culture, production of transgenics in animal and plants, applications in medicine as tools and therapeutics, creation of new enzymes and their immobilization for industrial use, development of monoclonal antibodies and control of pollutions, etc.

III. INDUSTRIAL APPLICATIONS OF BIOTECHNOLOGY

The industrial application of molecular biotechnology is often subdivided, so that we speak of red, green, gray or white biotechnology. This distinction relates to the use of the technology in the medical field (in human and animal medicine), agriculture, the environment and industry. Some companies also apply knowledge deriving from molecular biotechnology in areas that cut across these distinctions (e.g. in red and green biotechnology, sequencing services). According to an investigation by Ernst and Young relating to the German biotech industry, 92% of companies are currently (2004) working in the field of red biotechnology, 13% in green, and 13% in gray or white biotechnology.

Table 1.3. Range of biotechnology derived pharmaceuticals*

Hormones	FSH, LH, growth hormone, insulins, insulin analogues
Growth factors	PGDF, NGF, IGF-1
Cytokines	Interferons, interleukins, colony stimulating factor, erythropoietin
Others	Albumin, clotting factors, enzymes
Monoclonal antibodies	Murine, chimeric, humanised, rDNA technology
Antibody-related products	Single chain, fragments, fusion products
Vaccines	Conventional, recombination protein antigen, modified bacteria or viruses, anti-idiotype, mimotopes
Nucleic acid-based products	Gene therapy, DNA vaccines, ribozymes
Cell, tissues and organs	Autologous, allogeneic xenogeneic

*FSH, Follicle stimulating hormone; LH, luteinising hormone; PGDF, platelet-derived growth factor; NGF, nerve growth factor; IGF-1, insulin growth factor-1.

(a) Biotechnology in Medicine

Biotechnology products for therapeutic use include a very diverse range of products, as outlined in Tables 1.1 and 1.2. Some products are intended to mimic the human counterpart, whereas others are intended to differ from the human counterpart and may be analogues, chemically modified (e.g. pegylated) or novel products (e.g. single chain or fragment antibody products, gene transfer vectors, tissue-engineered products). Most of these products are regulated as medicinal products; however, the regulatory status of others such as some cell therapies and tissue: organ-based products differs globally and falls within the borderline between the practice of medicine, medical devices and medicinal products.

Biotechnology-derived pharmaceuticals may be derived from a variety of expression systems such as *Escherichia coli*, yeast, mammalian, insect or plant cells, transgenic animals or other organisms. The expressed protein or gene may have the identical amino acid or nucleotide sequence as the human endogenous form, or may be intentionally different in sequence to confer some technical advantage such as an optimised pharmacokinetic or pharmacodynamic profile. The glycosylation pattern of protein products is likely to differ from the endogenous human form due to the different glycosylation preferences of the expression system used. Furthermore, intentional post-translation modifications or alterations may be made such as pegylation. It is important for the toxicologist to be aware of the nature of the product to be tested in terms of primary, secondary and tertiary structure, and any post-translational modifications such as glycosylation status, particularly as these may be altered if the manufacturing system is modified.

(b) Red Biotechnology

Within the field of red biotechnology, which deals with applications in human and animal medicine, there are various further distinctions that can be made: biopharmaceutical drug development, drug delivery, cell and gene therapies, tissue engineering/regenerative medicine, pharmacogenomics (personalized medicine), system biology, and diagnosis using molecular medicine.

(c) Biopharmaceutical Drug Development

In the field of biopharmaceutical drug development, it is the development of therapeutic human proteins by recombinant methods. (Table 1.4) for use as medicines that has the longest tradition. As mentioned above, recombinant human insulin was the first recombinant medicine in the world, produced by Genentech and brought to market in 1982. Today, recombinant human insulin has almost completely driven the other preparation of insulin (isolated from human or animal tissues) from the market.

Table 1.4. Selected examples of recombinant proteins with indication and manufacturer.

Drug	Product name	Indication	Manufacturer
Human insulin	Humulin	Diabetes mellitus type I	Eli Lilly
Somatotropin	Humatrope	Inadequate growth	Eli Lilly
Erythropoietin alpha	Erypo/Epogen	Anemia	Jansen-Cilag/Amgen
Factor VIII	Bioclate/Kogenate	Hemophilia	Centeon/Bayer
Interferon alpha 2 a	Roferon A	Cancer	Roche
Interferon beta 1 b	Betaferon	Mulitple sclerosis	Schering
Tissue plasminogen activator tPA (alteplase)	Actilyse	Thrombolytic agent	Boehringer Ingelheim

The first therapeutic antibodies, especially monoclonal antibodies, have been on the market since the late 1990s. In 2002, antibodies were (along with vaccines) the most important therapeutic class of drugs under development and there are also more recent market studies more than 100 antibodies or antibody fragments were at the clinical development stage in 2002 and research and development is being carried out on around 470 more in about 200 companies around the world (Table 1.5). Since the introduction of therapeutic antibodies onto the market, they have achieved significant turnovers, which are growing continually. The market for 2008 is estimated at a volume of US $16.7 billion (from Data-monitor, November 2003). Today, in addition to proteins, which currently play the most significant role in the biopharmaceutical field, new types of drugs based on RNA (antisense drugs, ribozymes, aptamers, Spiegelmers and RNA interference) are also being developed on the basis of advances in knowledge on molecular biotechnology (Table 1.6).

Table 1.5. Selected examples of approved monoclonal antibodies

Drug	Product name	Indication	Manufacturer
Abciximab	Reopro	Anticoagulant	Eli Lilly Centocor Europe
Trastuzumab(anti-HERA2-a)	Herceptin	Breast cancer	Roche
Adalimumab (anti-TNF-alpha)	Humira	Rheumatoid arthritis	Abbott
Infliximab (anti-TNF-alpha)	Remicade	Crohn's disease	Centocor
Alematuzumab(anti CD52)	Campth	Leukemia	Millennium & Cohn's Ilex

Table 1.6. Selected examples of therapeutic RNAs on the market or under development

Principle of action	Product name/ production stage	Indication	Company
Antisense	Vitravene/market	CMV retinitis	ISIS pharmaceuticals
Antisense	Affinitak/phase II	Cancer	ISIS pharmaceuticals
Antisense	Alicaforsen/phase III	Crohn's disease	ISIS pharmaceuticals
Antisense	AP 12009/phase II	Brain tumor	Antisense Pharma
Ribozyme	ANGIOZYME/phase II	Intestinal cancer	Sirna Therapeutics

(d) Drug delivery

Closed linked to the development of therapeutic agents are the means of achieving their targeted delivery to their site of action. These drug delivery system are mainly used for drugs whose physical and chemical characteristics make them insufficiently stable in reaching their site of action intact. They can also be used to transport drugs in a targeted way to particular sites of action (tissue specific targeting), or to overcome biological barriers such as the intestinal wall or the blood–brain barrier.

(e) Green biotechnology

Green biotechnology is the application of biotechnology processes in agriculture and food production. The main dominant forces in green biotechnology today are agro giants with a world-wide area of operation such as BASF, Bayer CropScience, Monsanto and Syngenta. They are concentrating considerable attention on molecular plant biotechnology, which is seen as a future growth factor in agroindustry. The traditional pesticide market, on the other hand has been stagnating for years.

(f) Transgenic Plants

The main emphasis in modern plant biotechnology is the production of transgenic plants. The first use of gene technology to bring about changes in plants became possible at the beginning of the 1980s, around ten years after the first experiment with bacteria. The market value of transgenic plants is estimated to be in excess of 2 billion Euros, according to the calculation of the German Federal Office for the Environment. These figures relate to transgenic crop plants, which were being grown on an area totalling about 40 million hectares world-wide in 1999 and 2000.

(g) Novel and Functional food

New types of foodstuffs with novel properties are often called functional food. Another category that is often mentioned in this context is nutraceuticals. These are foods that have a medicinal effect.

(h) Livestock breeding

Modern biotechnology is being employed commercially to introduce novel performance features in productive livestock. The transgenic specimens then display, for example, different wool characteristics for sheep, or improved milk characteristics in cattle.

(i) Grey/White Biotechnology

The terms Grey and White Biotechnology have been coined for the application of biotechnological processes in environmental and industrial production contexts. The latter is primarily focused on the production of fine chemicals, in particular technical enzymes.

(j) Technical enzymes

Modern biotechnology already dominates the technical enzymes market. They can be found as proteases, lipases, cellulases and amylases for example in modern detergents, where they serve, amongst other purposes as protein and fat solubilizers.

IV. SAFETY CONCERNS

There are a number of safety issues relating to biotechnology products that differ from those raised by low molecular weight products and need to be taken into account when designing the safety evaluation programme for a biotechnology derived pharmaceutical product (Table 1.7). The quality and consistency of the product requires careful control in terms of product identity, potency and purity because of concerns about microbiological safety, impurities arising from the manufacturing process (e.g. host-cell contaminants, endotoxin, residual DNA levels and process chemicals), and the fidelity of the protein sequence and post-translational modifications during process improvements and scale-up. The immunogenic nature of heterologous proteins, vectors, cells, tissues and process contaminants must also be considered in the design of the safety evaluation programme and appropriate monitoring for anti-product antibodies, particularly neutralising antibodies included in toxicity studies to aid interpretation of the findings. For gene transfer products, there are concerns about the distribution and persistence of vector sequences, the potential for expression of vector sequences in non-target cells, tissues and, in particular, the potential for inadvertent gonadal distribution and germ-line integration. In 1997, the Food and Drug Administration (FDA) became aware that preclinical studies from multiple clinical trial applications indicated evidence of vector DNA in animal gonadal tissues following extragonadal administration. These positive polymerase chain reaction (PCR) signals were for DNA extracts from whole gonads subsequent to vector administration. The observations involved multiple classes of vectors, formulations and routes of administration.

Table 1.7. Safety issues related to biotechnology
Microbiological safety
Adventitious agents bacterial, mycoplasma, fungal, viral, TESs, etc.
Potential for reconstitution of replication competent viral vectors for gene transfer products
Cells harbour homotropic and xenotropic viruses, e.g. retroviruses, Epstein-Barr virus, cytomegalovirus
Potential for environmental spread, e.g., vector dissemination and viral shedding
Immunological
Anti-product antibodies
Host cell proteins or other process impurities
Viral vectors for gene transfer products
Cell impurities in cell theory products
'Foreign' epitopes on cell/tissue/organ-based products
Development of immune tolerance to DNA vaccines
Pharmacological
Excessive pharmacology or unintentional receptor binding
Distribution to and exposure of non-target tissues
Altered cell phenotype, cell products, function and localisation for cell therapies
Biodistribution
Distribution and persistence of gene and cell therapics
Genetic transfer to germ cells for gene transfer products
Insertional multagenesis for gene transfer products
Vector dissemination and viral shedding
Tumourigenicity
Pharmacology or product, e.g. growth factor activity, immunomodulatory activity.
Onceogenic residual DNA in product
Insertional mutagenesis for gene transfer products
Donor-derived malignant or premalignant cells in cell therapy products
Cell culture process leading to immortalisation, transformation to maligant phenotype growth factor independence
Cellular impurities in cell therapy products
General safety concern
Physical characteristics, e.g. of the protein, viral vector
Covalently bound ligand molecules, e.g. toxin
Product formulation and excipients
Local tolerance

QUESTIONS

1. Define biotechnology. Give landmarks which helped in the development of biotechnology.
2. Discuss in brief, application of biotechnology and safety issues involved.
3. Write short notes :
   (*a*) Discovery of double helix
   (*b*) DNA sequence method
   (*c*) Use of computers in biotechnology
   (*d*) Cloning of animals
   (*e*) Human genome project
   (*f*) Discovery of PCR.

CHAPTER 2

MACROMOLECULES

The four macromolecules are proteins, lipids, carbohydrates and nucleic acids. These four groups of chemicals will come up in concepts throughout the course. Remember, these are organic molecules and so all contain carbon.

I. CARBOHYDRATES. These are defined as chains of monosaccharides (simple sugars) hooked together by glycosidic bonds. All carbohydrates have the general formula of $(CH_2O)n$ where n is variable. The monosaccharide or simple sugar can be defined as an oxygen-containing ring of carbon atoms. If the ring has six carbon atoms, the sugar is said to be a hexose sugar, e.g. glucose. If the ring has five carbons in it, it is said to be a pentose sugar, e.g. ribose and deoxyribose.

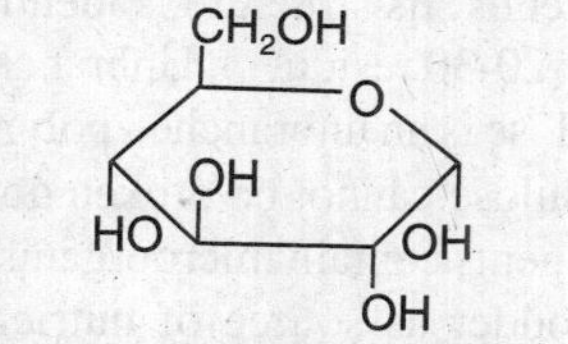

monosaccharide (glucose)

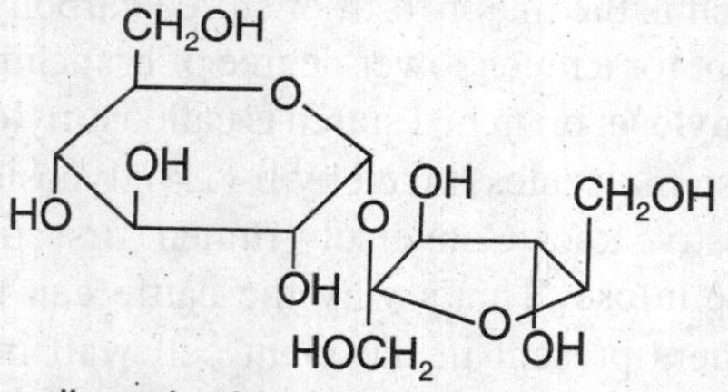

disaccharide (sucrose)

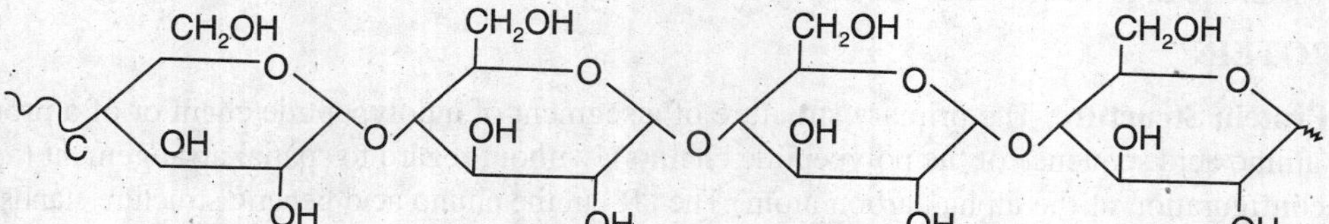

Two monosaccharides hooked together by the glycosidic bond form a disaccharide sugar. The glycosidic bond is formed by the removal of one water molecule per bond. An example of a disaccharide is sucrose; regular table sugar. To break sucrose up into glucose (dextrose) and fructose, water must be added to the bond. Therefore, the formation of polysaccharides is a dehydration synthesis as is the formation of the protein.

Monosaccharides-The monosaccharides commonly found in humans are classified according to the number of carbons they contain in their backbone structures. The major monosaccharides contain four to six carbon atoms.

Carbohydrate Classifications

Carbons	Category Name	Relevant examples
3	Triose	Glyceraldehyde, Dihydroxyacetone
4	Tetrose	Erythrose
5	Pentose	Ribose, Ribulose, Xylulose
6	Hexose	Glucose, Galactose, Mannose, Fructose
7	Heptose	Sedoheptulose
9	Nonose	Neuraminic acid also called sialic acid

Disaccharides- The linkage of two monosaccharides to form disaccharides involves a glycosidic bond. Several physiogically important disaccharides are sucrose, lactose and maltose. **Sucrose** prevalent in sugar cane and sugar beets, is composed of glucose and fructose through an α–(1,2)β–glycosidic bond. **Lactose** is found exclusively in the milk of mammals and consists of galactose and glucose in α – β-(1,4) glycosidic bond. **Maltose** the major degradation product of starch, is composed of 2 glucose monomers in an α-(1,4) glycosidic bond.

Polysaccharides-Most of the carbohydrates found in nature occur in the form of high molecular weight polymers called polysaccharides. The monomeric building blocks used to generate polysaccharides can be varied in all cases, however, the predominant monosaccharide found in polysaccharides is D-glucose.

Glycogen-Glycogen is the major form of stored carbohydrate in animals. This crucial molecule is a homopolymer of glucose in α-(1,4) linkage; it is also highly branched, with α-(1,6) branch linkages occurring every 8 –10 residues. Glycogen is a very compact structure that results from the coiling of the polymer chains.

This compactness allows large amounts of carbon energy to be stored in a small volume, with little effect on cellular osmolarity.

Starch-Starch is the major form of stored carbohydrate in plant cells. Its structure is identical to glycogen, except for a much lower degree of branching (about every 20-30 residues). Unbranched starch is called **amylose**; branched starch is called **amylopectin**. **Cellulose** is an unbranched polymer made from glucose molecules linked by β-(1,4)glycosidic bonds. Cellulose cannot be broken down in the human digestive tract. Conversely, rumen (first stomach) of ruminents contain microorganisms which produce cellulose. That's why the cattle can make use of fodder as source of nutrients. Additional polymers present in the plant cell wall include polysaccharides also called glycans, made up of cellulose fibre linked together in a diagonal fashion, pectin and lignin. Using cellulases, it is possible to digest cell walls of plant cells.

II. PROTEIN

Protein Structure. The primary structure of a segment of a polypeptide chain or of a protein is the amino acid sequence of the polypeptide chain(s), without regard to spatial arrangement (apart from configuration at the alpha-carbon atom).The "R" in the amino acid generic structure stands for the term "radical" and represents one of twenty or so different possibilities. So with 20 different "R" groups, there are twenty different amino acids in nature. All amino acids have attached to the same carbon both an amino group (NH_2) and a carboxyl group (COOH). The hydrogen from the carboxyl group actually exists in a cloud around the amino group and itself, thus creating the zwitterion. This renders the amino acid active and proteins formed from them capable of great activity. Buffers and enzymes are examples of active proteins.

The peptide bond is formed when the amino group of one amino acid and the carboxyl group of another amino acid unite with the loss of one water molecule per bond.

The commonly occurring amino acids are of 20 different kinds (a protein may contain a chain of 100 to 1000 amino acids; 26 alphabets in English making several thousands words, 20 amino acids forms different proteins) which contain the same dipolar ion group $H_3N^+.CH.COO^-$. They all have in common a central carbon atom to which are attached a hydrogen atom, an amino group (NH_2) and a carboxyl group (COOH). The central carbon atom is called the C_{alpha}-atom and is a chiral centre. All amino acids found in proteins encoded by the genome have the L-configuration at this chiral centre. This configuration can be remembered as the CORN law. Imagine looking along the H-C_{alpha} bond with the H atom closest to you.

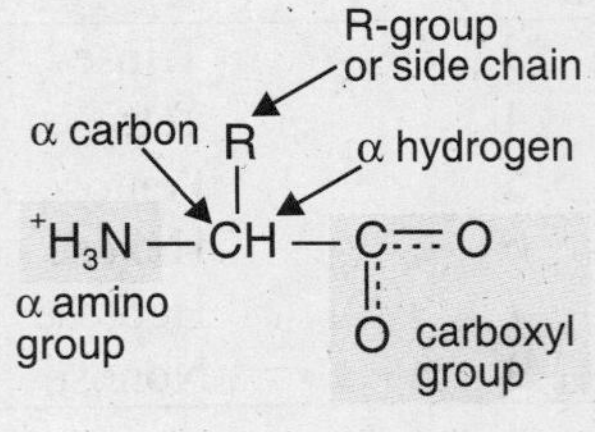

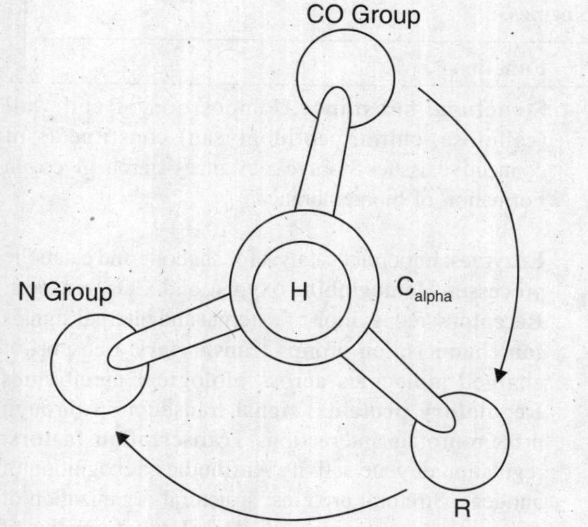

When read clockwise, the groups attached to the C_{alpha} spell the word CORN. There are 20 side chains found in proteins encoded by the genetic machinery of the cell. The side chains confer important properties on a protein such as the ability to bind ligands and catalyse biochemical reactions. They also direct the folding of the nascent polypeptide and stabilise its final conformation. Amino acids in proteins (or polypeptides) are joined together by peptide bonds. The sequence of R-groups along the chain is called the **primary structure**. Proteins can occur as primary, secondary, tertiary or quaternary structures.

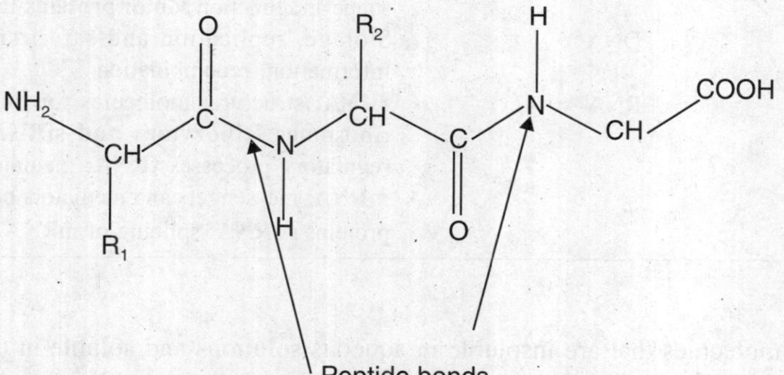

Secondary Structure. The secondary structure of a segment of polypeptide chain is the local spatial arrangement of its main-chain atoms without regard to the conformation of its side chains or to its relationship with other segments. There are three common secondary structures in proteins, namely **alpha helices, beta sheets** and **turns**.

The alpha-helix and beta-structure conformations for polypeptide chains are generally the most thermodynamically stable of the regular secondary structures. However, particular amino acid sequences of a primary structure in a protein may support regular conformations of the polypeptide chain other than alpha-helical or beta-structure.

Tertiary structure. The tertiary structure of a protein molecule, or of a subunit of a protein molecule, is the arrangement of all its atoms in space, without regard to its relationship with neighbouring molecules or subunits.

Quaternary structure. The quaternary structure of a protein molecule is the arrangement of its subunits in space and the ensemble of its intersubunit contacts and interactions, without regard to the internal geometry of the subunits. The subunits in a quaternary structure must be in noncovalent association. Haemoglobin contains four polypeptide chains ($\alpha_2\beta_2$) held together noncovalently in a specific conformation as required for its function.

Table 2.1. Compositon and function of macromolecules.

Basic building blocks	Macromolecule	Function
Simple sugar	Polysaccharide	**Structural substance**:Composition of cell wall (cellulose, chitin, peptidoglycan) constituents of connective tissues Storage substances-starch, glycogen
Phospholipids, cholesterol	Lipid	Formation of biomembranes
Amino acids	Protein	**Enzymes**: important catalyst for anabolic and catabolic processes **Hemoglobin**: oxygen and CO_2 transport. **Receptors**: recognition of external and internal signals **Ion channels, ion pumps transporters**: transport of charged molecules across biological membranes **Regulatory proteins**: signal transduction through protein-protein interactions **Transcription factors**: regulation of gene activity **Antibodies**: recognition of antigens. Strctural proteins: structural organization of supramolecular complexes **Cytoskeleton**: formation of molecular network in the cell that are important for shape and function **Motor proteins**-muscle contraction
Deoxynucleotide	DNA	Storage, replication and safe transfer of genetic information, recombination
Nucleotide	RNA	**rRNA**: structural molecules for the construction of ribosomes **Ribozymes and siRNA**: catalytic and regulatory processes **tRNA**: mediators in translation **mRNA**: messengers and mediators between genes and proteins **snRNA**: splicing of mRNA

III. Lipids

Biological molecules that are insoluble in aqueous solutions and soluble in organic solvents are classified as lipids. The lipids of physiological importance for humans have four major functions. (1) They serve as structural components of biological membranes. (2)They provide energy reserves, predominantly in the form of triacylglycerols. (3) Both lipids and lipid derivatives serve as vitamins and hormones. (4) Lipophilic bile acids aid in lipid solubilization. We shall simplify lipids by dividing them into groups as neutral lipids, phospholipids and sterols.

Neutral lipids (fats) are fatty acid chains attached to a glycerol backbone molecule. The fatty acid is a hydrocarbon chain with a terminal carboxyl group. When all four binding sites of each carbon are occupied with hydrogen atoms, the fatty acid is said to be saturated. When double bonds exist, the molecule is said to be unsaturated. If there is one fatty acid side-chain, the molecule is called a monoacylglycerol. With two fatty acid side-chains, the molecule becomes known as a diacylglycerol. With three side chains, it becomes known as a triacylglycerol. These molecules are also referred to a mono-, di- and triglycerides. The majority of body fatty acids are acquired in the diet. However, the lipid biosynthetic capacity of the body (fatty acid synthase and other fatty acid modifying enzymes) can supply the body with all the various fatty acid structures needed. Two key exceptions to this are the highly unsaturated fatty acids known as linoleic acid and linolenic acid, containing unsaturation sites beyond carbons 9 and 10. These two fatty acids cannot be synthesized from precursors in the body, and are thus considered the **essential fatty acids**; essential in the sense that they must be provided in the diet. Since plants are capable of synthesizing linoleic and linolenic acid, humans can acquire these fats by consuming a variety of plants or else by eating the meat of animals that have consumed these plant fats.

The phospholipids have a phosphate group that substitutes for one of the fatty acid side-chains in the neutral lipids. These molecules are rendered polar and are associated with cell membrane structure and permeability. (1) The steroids are the compounds containing a cyclic steriod nucleus (or ring) namely cyclopentan-perhydrophenanthrene (cPPP). Cholesterol is exclusively found in animals and is most abundant animal sterol. Cholesterol is found in animal cell membranes, regulates fluidity. (2) Cholesterol is a precursor of sex hormones (testosterone, estrogen) which are also steroids. (3) HDL (high density lipoprotein) vs. LDL (low density lipoprotein) (a) Lipoproteins are lipids linked to proteins. Since protein is denser than lipids, the higher density lipoproteins have more protein relative to lipid. (b) HDL (good cholesterol) transports fat back to the liver where it is broken down (c) LDL (bad cholesterol) transports cholesterol and fat built in the liver to our cells where it may clog arteries.

Table 2.2. Physiologically Relevant Fatty Acids

Numerical Symbol	Common Name	Structure	Comments
14:0	Myristic acid	$CH_3(CH_2)_{12}COOH$	Often found attached to the N-term of plasma membrane-associated cytoplasmic proteins
16:0	Palmitic acid	$CH_3(CH_2)_{14}COOH$	End product of mammalian fatty acid synthesis
$16:1^{\Delta 9}$	Palmitoleic acid	$CH_3(CH_2)_5C{=}C(CH_2)_7COOH$	
18:0	Stearic acid	$CH_3(CH_2)_{16}COOH$	
$18:1^{\Delta 9}$	Oleic acid	$CH_3(CH_2)_7C{=}C(CH_2)_7COOH$	
$18:2^{\Delta 9,12}$	Linoleic acid	$CH_3(CH_2)_4C{=}CCH_2C{=}C(CH_2)_7COOH$	**Essential fatty acid**
$18:3^{\Delta 9,12,15}$	Linolenic acid	$CH_3CH_2C{=}CCH_2C{=}CCH_2C{=}C(CH_2)_7$ COOH	**Essential fatty acid**
$20:4^{\Delta 5,8,11,14}$	Arachidonic acid	$CH_3(CH_2)_3(CH_2C{=}C)_4(CH_2)_3COOH$	Precursor for eicosanoid synthesis

Basic Structure of Triacylglycerides-Triacylglycerides are composed of a glycerol backbone to which 3 fatty acids are esterified

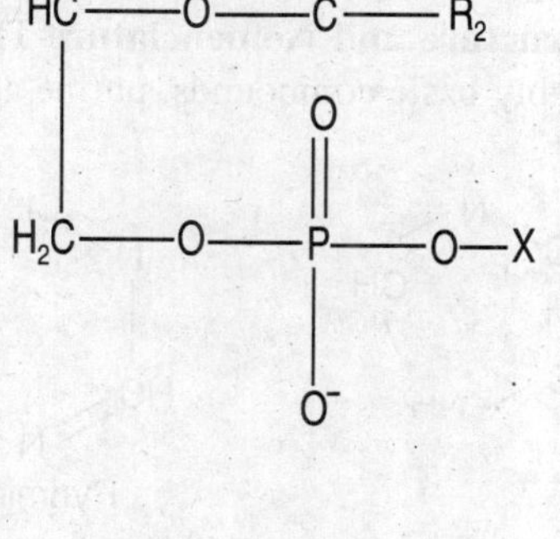

Basic Structure of Phospholipids- The basic structure of phospolipids is very similar to that of the triacylglycerides except that C-3 (sn3) of the glycerol backbone is esterified to phosphoric acid. The building block of the phospholipids is phosphatidic acid which results when the X substitution in the basic structure shown in the Figure below is a hydrogen atom. Substitutions include ethanolamine (phosphatidylethanolamine), choline (phosphatidylcholine, also called lecithins), serine (phosphatidylserine), glycerol (phosphatidylglycerol), myo-inositol (phosphatidylinositol, these compounds can have a variety in the numbers of inositol alcohols that are phosphorylated generating polyphosphatidylinositols), and phosphatidylglycerol (diphosphatidylglycerol more commonly known as cardiolipins).

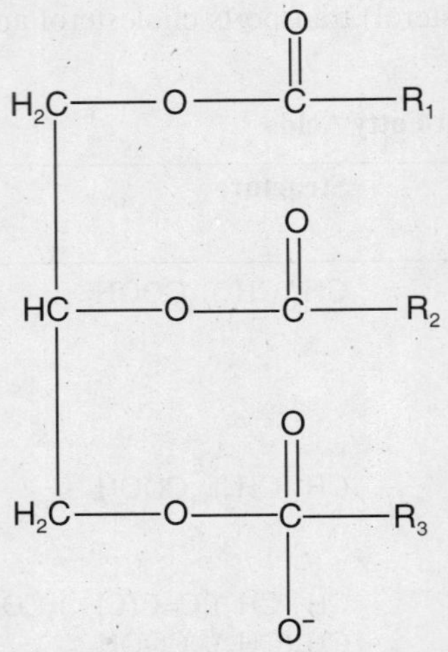

Fig. 1.0. Basic composition of a phospholipid. X can be a number of different substituents.

IV. NUCLEIC ACIDS

As a class, the nucleotides may be considered one of the most important metabolites of the cell. Nucleotides are found primarily as the monomeric units comprising the major nucleic acids of the cell, RNA and DNA. However, they also are required for numerous other important functions within the cell. These functions include:

(1) serving as energy stores for future use in phosphate transfer reactions. These reactions are predominantly carried out by ATP. (2) forming a portion of several important coenzymes such as NAD^+, $NADP^+$, FAD and coenzyme A. (3) serving as mediators of numerous important cellular processes such as second messengers in signal transduction events. The predominant second messenger is cyclic-AMP (cAMP), a cyclic derivative of AMP formed from ATP. (4) controlling numerous enzymatic reactions through allosteric effects on enzyme activity. (5) serving as activated intermediates in numerous biosynthetic reactions. These activated intermediates include S-adenosylmethionine (S-AdoMet) involved in methyl transfer reactions as well as the many sugar coupled nucleotides involved in glycogen and glycoprotein synthesis.

Nucleoside and Nucleotide Structure and Nomenclature-The nucleotides found in cells are derivatives of the heterocyclic highly basic compounds, purine and pyrimidine.

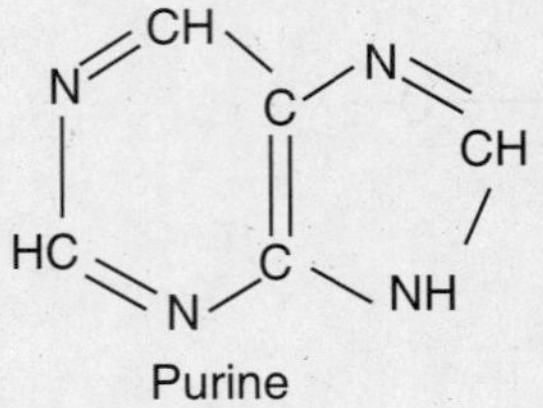

Purine

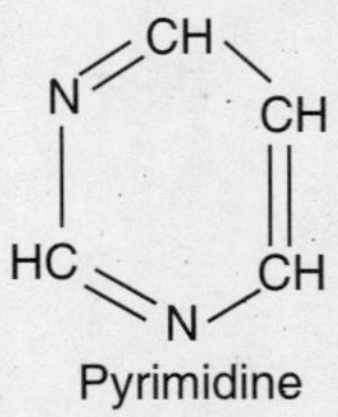

Pyrimidine

It is the chemical basicity of the nucleotides that has given them the common term "bases" as they are associated with nucleotides present in DNA and RNA. There are five major bases found in cells. The derivatives of purine are called adenine and guanine, and the derivatives of pyrimidine are called thymine, cytosine and uracil. The common abbreviations used for these five bases are, **A, G, T, C** and **U**. The sugar is a 2'-deoxy ribose which is phosphorylated at its 5' hydroxyl group. Free nucleotides contain either one, two or three phosphates representing the mono-, di-, or triphosphate form of the nucleotide, the latter being known as dATP, dGTP, dTTP, and dCTP.

The purine and pyrimidine bases in cells are linked to carbohydrate and in this form are termed, nucleosides. The rigidity and linear geometry of the H-bonds restricts base pair (bp) formation. The plane of the base pair lies perpendicular to the helix axis. The right handed B-DNA form is the physiological form of the DNA double helix as first described by Watson and Crick in 1953. The specific conformational constraints made it clear that the sequence of the bases in the polymer encodes the genetic information for the synthesis of proteins. The B-DNA is a *right handed, anti-parallel* double helix with 2nm in diameter, 10bp per turn, a twist of 36° per bp, 0.34nm (3.4Å) per base, pitch 3.4nm (34Å), base plane tilt away from being perpendicular to helix axis at 6°. The B-DNA helix as described by the above parameters represents an idealized helix. The conformation of 'real' DNA, however, deviates slightly from the B form in a sequence dependent manner as well as depending on the interaction with DNA-binding proteins. Examples of protein ? DNA interaction mechanisms will be discussed in sections 3.1 and 3.2. An important feature of the double helix are the *minor* and *major grooves,* winding along the helix surface. In those grooves, parts of the aromatic ring structures of the purines and pyrimidines are exposed to the surface of the helix. The major groove is the site where most protein-DNA interactions occur.

A measurable quantity of DNA is its contents of G+C and A+T base pairs (besides the sequence *per se*) which differs from organisms to organism (Chargaff rule) and is characteristic of an organism's genome. The G+C content in mammals varies from 39 to 46% and in bacteria from 25 to 75%. The G+C content exhibits an important correlation with the stability of the double helical conformation of DNA due to the difference in hydrogen bonding capacity of GC pairs with 3 H-bonds and AT pairs with 2 H-bonds. This has been shown through heat denaturation of the double helix into a single strand conformation. So called melting curves can be measured by increasing the temperature of the solution. At a certain temperature, called melting temperature or T_m, 50% of the DNA is found in double strand (ds) and 50% in single strand (ss) form. The melting temperature is directly proportional to the GC content.

Measuring the double strand (ds) vs. single strand (ss) content of purified DNA is done by determining the UV absorbance at a wavelength of 260nm (typical absorbance wave-length of aromatic ring structures) over a temperature range of 20 to 90° C. The resulting curve shows a

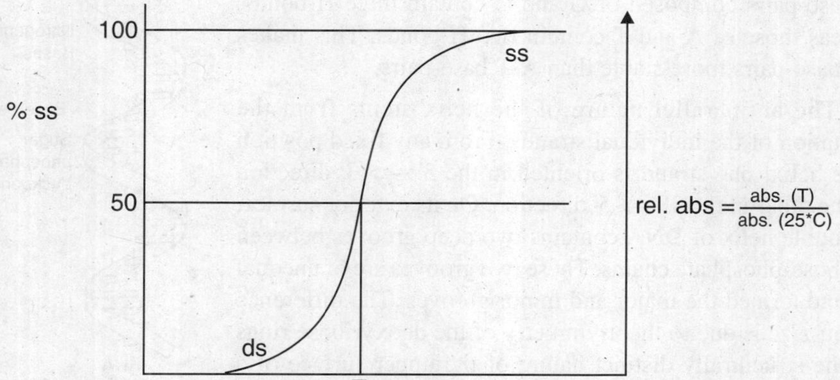

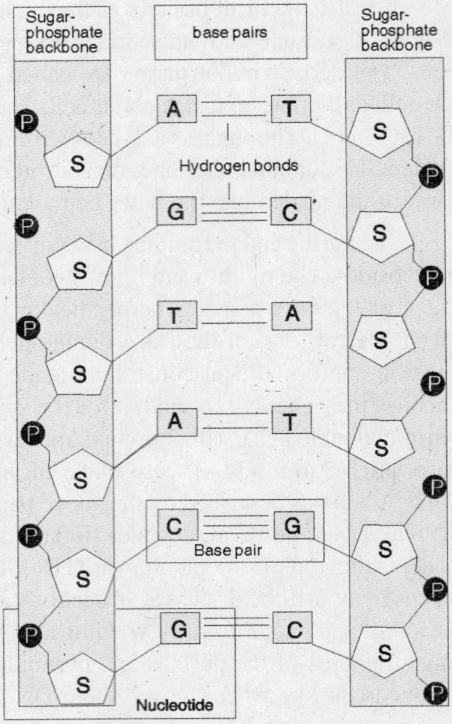

sigmoidal behavior indicating a cooperative process during denaturation and renaturation. Normally a sharp increase in absorbance indicates the transition from double to single strand form with the half point of transition being defined as the melting temperature T_m of the DNA sample in the test solution. The process is strictly reversible.

Polynucleotides-Polynucleotides are formed by the condensation of two or more nucleotides. The condensation most commonly occurs between the alcohol of a 5'-phosphate of one nucleotide and the 3'-hydroxyl of a second, with the elimination of H_2O, forming a phosphodiester bond (This determines the 3′ and 5′ end of a DNA fragment). The formation of phosphodiester bonds in DNA and RNA exhibits directionality. The primary structure of DNA and RNA (the linear arrangement of the nucleotides) proceeds in the 5' ——> 3' direction. The common representation of the primary structure of DNA or RNA molecules is to write the nucleotide sequences from left to right synonymous with the 5' ——> 3' direction as shown:

5'-pGpApTpC-3'

Structure of DNA-Utilizing X-ray diffraction data, obtained from crystals of DNA, James Watson and Francis Crick proposed a model for the structure of DNA. This model predicted that DNA would exist as a helix of two complementary antiparallel strands, wound around each other in a rightward direction and stabilized by H-bonding between bases in adjacent strands. In the Watson - Crick Model, the bases are in the interior of the helix aligned at a nearly 90 degree angle relative to the axis of the helix. Purine bases form hydrogen bonds with pyrimidines, in the crucial phenomenon of base pairing. Experimental determination has shown that, in any given molecule of DNA, the concentration of adenine (A) is equal to thymine (T) and the concentration of cytosine (C) is equal to guanine (G). This means that A will only base-pair with T, and C with G. According to this pattern, known as Watson-Crick base-pairing, the base-pairs composed of G and C contain three H-bonds, whereas those of A and T contain two H-bonds. This makes G-C base-pairs more stable than A-T base-pairs.

The antiparallel nature of the helix stems from the orientation of the individual strands. From any fixed position in the helix, one strand is oriented in the 5' —> 3' direction and the other in the 3' —> 5' direction. On its exterior surface, the double helix of DNA contains two deep grooves between the ribose-phosphate chains. These two grooves are of unequal size and termed the major and minor grooves. The difference in their size is due to the asymmetry of the deoxyribose rings and the structurally distinct nature of the upper surface of a base-pair relative to the bottom surface.

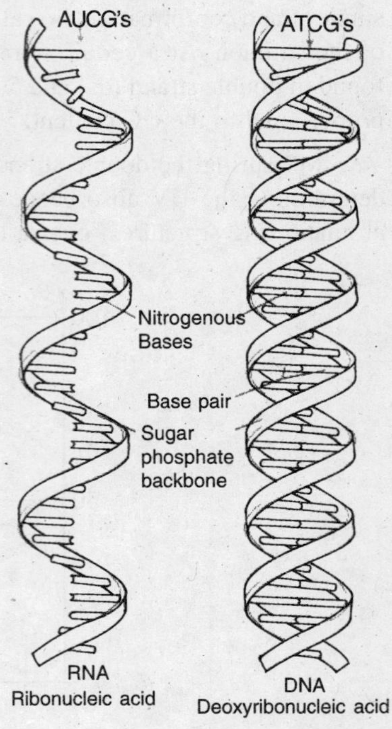

The double helix of DNA has been shown to exist in several different forms, depending upon sequence content and ionic conditions of crystal preparation. The B-form of DNA prevails under physiological conditions of low ionic strength and a high degree of hydration. Regions of the helix that are rich in pCpG dinucleotides can exist in a novel left-handed helical conformation termed Z-DNA. This conformation results from a 180 degree change in the orientation of the bases relative to that of the more common A- and B-DNA.

Genomic size of DNA

Consider the size of DNA and its structure of being a long, linear polymer. In prokaryotes, the entire genome consists of one single molecule, mostly in circular form. The total genome of *Haemophilus influenzae* has been sequenced containing more than 1.8 million base pairs. Often DNA is found to be methylated, the result of an enzymatic modification that serves as chemical protection. Methylated bases can often not be recognized by DNA binding proteins such as nucleases that hydrolyze the phosphodiester bond between nucleotides. In this way, bacteria protect their own DNA from degradation, whereas the DNA of intruders (like viruses) which is not methylated can be degraded. Methylation occurs at: N6-methyl-dA and 5-methyl-dC. Because of the large size of circular DNA, the helix is often wound up in a *super-coiled* form. The combination of supercoiling and helix conformation has an effect on the recognition of DNA binding proteins (it changes the local conformation of the B-DNA double helix).

DNA repair and mutations

During the time between the process of DNA replication and interphase, there is considerable "proofreading" of the genes. Defective areas are replaced with the correct nucleotides. The same enzymes that are used in the transcription and translation process do this. Because of this, mistakes are rare but when they occur they are called "mutations". Abnormal protein, abnormal cell function and even cell death can be the result. Mutations are also a source of the every genetic variation associated with increased survivability of an organism.

Mutations are defined as any change in the base sequence of the DNA. The most common kind of mutation is the base substitution or point mutation. Only one base is incorrect. Another kind is the frameshift mutation where base pairs have been added or deleted from the sequence. These usually result in inactive proteins. Mutations may also be spontaneous or can be caused by mutagens, which are often mitogens like benzylpyrene, aflatoxin and acridine dyes. X-rays, gamma rays and UV radiation rays are mutagens.

RNA Structure

RNA is very similar to DNA in that is made of 4 different building blocks, the ribonucleotides. The pyrimidine base thymine is modified in that it lacks a methyl group and the resulting *uracil* takes its place in base pairing. The ribose comes in its fully hydroxylated form. Together, the presence of uracil in place of thymine, and the 2'-OH in the ribose constitute the two chemical differences between RNA and DNA. RNA is composed of the four bases:

Purines:

Adenine A

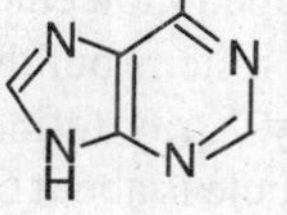

Guanine G

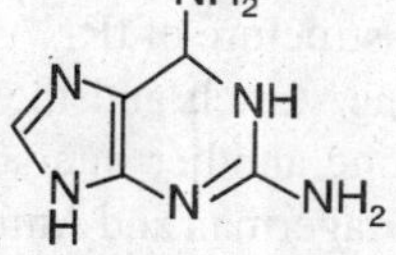

Pyrimidines:

Uracil U

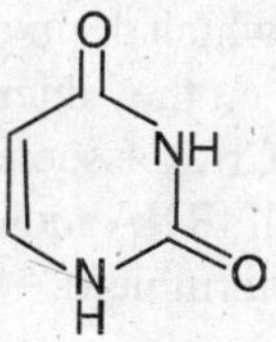

Cytosine C

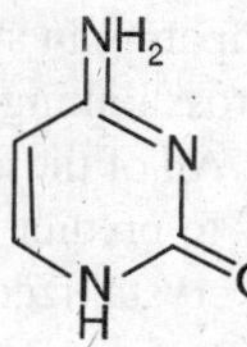

RNA differs, however, from DNA because it does not form an analogous double helical structure. RNA does, however, form base pairs with DNA resulting in a heteromeric double helix consisting of one DNA and one RNA strand. This annealing of an RNA strand to its complementary DNA strand is called hybridization and plays a crucial role in the transcription and translation of genetic sequences into protein sequences. RNA does, in contrast to DNA, form short double strand structures on itself, thereby forming so called stem and loop structures. Both DNA/RNA double helices and RNA/RNA double strands have an A-DNA like conformation, also called A-RNA or RNA-II. The helix has 11bp per turn pitch of 3.0nm (30Å).There are three major RNA species that can be distinguished both on their ability to form stem and loop structures as well as their functional role in the cell. The three types of RNA are: messenger RNA or mRNA, transfer RNA or tRNA, ribosomal RNA or rRNA.

In addition there are RNA molecules found in viruses (viral RNA) that serve as the genomic blue print that normally is encoded in DNA, and ribonucleo-proteins of diverse origin both ribosomal and non-ribosomal in nature.

Messenger RNA- mRNA is synthesized from a gene segment of DNA which ultimately contains the information on the primary sequence of amino acids in a protein to be synthesized. The genetic code as translated is for m-RNA not DNA. The **messenger RNA carries the code** into the cytoplasm where protein synthesis occurs.

Transfer RNA or tRNA-The tRNA molecules are key to the translation process of the mRNA sequence into the amino acid sequence of proteins (at least one type of tRNA for every amino acid). To be precise, the amino-acyl-tRNA-synthase proteins are the 'true' translators of the genetic code into an amino acid sequence. These synthetases *acetylate tRNA* molecules with the proper amino acid that corresponds to the anti-codon in the structure of the tRNA molecule. The anti-codon later recognizes the codon, the triple base sequence which 'codes' for the amino acid along the mRNA strand. A failure of properly acetylating the tRNA with the right amino acid results in a amino acid mutation even though the DNA sequence has not been changed. tRNA molecules are small nucleic acids of 60-95 nucleotides, mostly 76, with a molecular weight 18-20kD (kilo dalton) with the secondary structure resembling a clover leaf. Here are a few common *features shared by all tRNA molecules* found in various organisms.

(1) 5' terminus always phosphorylated
(2) 7 bp stem, may have non-Watson & Crick pairing (like GU),
3) acceptor or amino acid stem at 3' terminus, last three nucleotides CCA-3'-OH, amino acylation occurs at 3'-OH group,
(4) 3-4 bp stem and loop contains the base *dihydrouridine* (D) [D- arm]
(5) 5 bp stem and loop containing anti-codon triplet [anti-codon arm]
(6) 5 bp stem and loop contains sequence TΨ C, Ψ standing for [T- arm] pseudouridine
(7) variable arm (between anti-codon and T arm) of length 3-21 nucleotides contains numerous modified bases (up to 25%) which are all post-transcriptionally modified.

The three dimensional structure of tRNA resembles an L-shaped molecule with the D-arm and anti-codon loop building one stretch and the T-arm and acceptor stem building the other stretch being deposed by ~90° to one another interstem angle of 82° by X-ray refinement and 92° in an electron microscopy study (Hagerman and Amiri, 1996). The molecule is about 6 nm in each direction with the anti-codon to acceptor 3'-term ends being 7.6 nm apart. The diameter of both arms are about 2.0 to 2.5 nm. The structural complexity of tRNA is *reminiscent of that of a protein* with 71 out of 76 bases participating in stacking interaction (of which 42 in double helical stem structures). 9 bp interaction are cross linking the tertiary structure, i.e., they interact with bases from a different stem and loop region. All of these 9 bp are non Watson-Crick associations and are highly conserved which makes it likely to predict similar structures for all tRNA molecules (in fact, only few tRNA molecules have been crystallized and their structure determined).

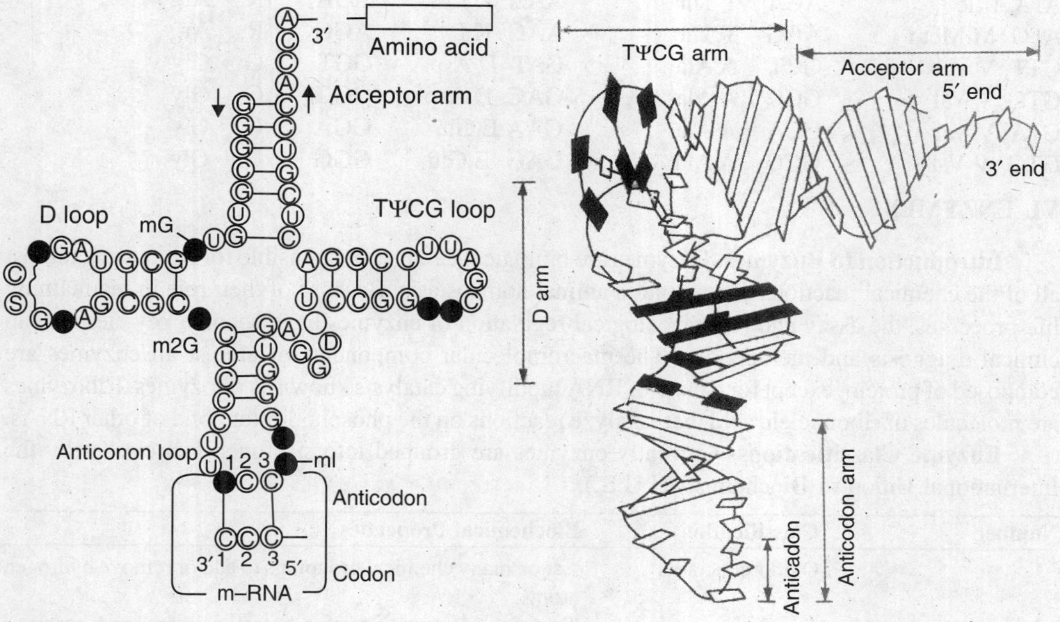

Ribosomal RNA-In the cytoplasm, **ribsomal RNA (rRNA)** and protein combine to form a nucleoprotein called a ribosome. **The ribosome serves as the site and carries the enzymes necessary for protein synthesis.**

The ribosome attaches itself to m-RNA and provides the stabilizing structure to hold all substances in position as the protein is synthesized. Several ribosomes may be attached to a single RNA at any time. In upper right corner is the 30S sub unit with mRNA and tRNA attached.

V. CODON-ANTI-CODON INTERACTION

The genetic code consists of three bases in a row, a triplet, codes for a specific amino acid. Thus a sequence of triplets in the DNA can be transcribed into a sequence of triplets in the mRNA strand. Since the DNA is a double helix formed by two complementary strands, the anti-sense strand is transcribed into mRNA resulting in the +sense on the mRNA level (with U instead of T).

As a rule, every tRNA can be covalently linked to only one type of amino acid, through the specificity of the amino acyl tRNA synthase. A specific amino acid can be linked to different tRNAs. There are as many tRNA species as codons used for translation. The codons for amino acids are show below.

TTT F Phe	TCT S Ser	TAT Y Tyr	TGT	C	Cys
TTC F Phe	TCC S Ser	TAC Y Ty	TGC	C	Cys
TTA L Leu	TCA S Ser	TAA * Ter	TGA	*	Ter
TTG L Leu i	TCG S Ser	TAG * Ter	TGG	W	Trp
CTT L Leu	CCT P Pro	CAT H His	CGT	R	Arg
CTC L Leu	CCC P Pro	CAC H Hi	CGC	R	Arg
CTA L Leu	CCA P Pro	CAA Q Gln	CGA	R	Arg
CTG L Leu i	CCG P Pro	CAG Q Gln	CGG	R	Arg
ATT I Ile	ACT T Thr	AAT N Asn	AGT	S	Ser
ATC I Ile	ACC T Thr	AAC N Asn	AGC	S	Ser

ATA	I Ile	ACA	T Thr	AAA	K Lys	AGA	R	Arg
ATG	M Met i	ACG	T Thr	AAG	K Lys	AGG	R	Arg
GTT	V Val	GCT	A Ala	GAT	D Asp	GGT	G	Gly
GTC	V Val	GCC	A Ala	GAC	D Asp	GGC	G	Gly
GTA	V Val	GCA	A Ala	GAA	E Glu	GGA	G	Gly
GTG	V Val	GCG	A Ala	GAG	E Glu	GGG	G	Gly

VI. ENZYMES

Introduction to Enzymes. Enzymes are biological catalysts responsible for supporting almost all of the chemical reactions that maintain animal homeostasis. Because of their role in maintaining life processes, the assay and pharmacological regulation of enzymes have become key elements in clinical diagnosis and therapeutics. The macromolecular components of almost all enzymes are composed of protein, except for a class of RNA modifying catalysts known as ribozymes. Ribozymes are molecules of ribonucleic acid that catalyze reactions on the phosphodiester bond of other RNAs.

Enzyme Classifications- Currently enzymes are grouped into six functional classes by the International Union of Biochemists (I.U.B.).

Number	Classification	Biochemical Properties
1.	Oxidoreductases	Act on many chemical groupings to add or remove hydrogen atoms.
2.	Transferases	Transfer functional groups between donor and acceptor molecules. Kinases are specialized transferases that regulate metabolism by transferring phosphate from ATP to other molecules.
3.	Hydrolases	Add water across a bond, hydrolyzing it.
4.	Lyases	Add water, ammonia or carbon dioxide across double bonds, or remove these elements to produce double bonds.
5.	Isomerases	Carry out many kinds of isomerization: L to D isomerizations, mutase reactions (shifts of chemical groups) and others.
6.	Ligases	Catalyze reactions in which two chemical groups are joined (or ligated) with the use of energy from ATP.

These rules give each enzyme a unique number. The I.U.B. system also specifies a textual name for each enzyme. The enzyme's name is comprised of the names of the substrate(s), the product(s) and the enzyme's functional class. Because many enzymes, such as alcohol dehydrogenase, are widely known in the scientific community by their common names, the change to I.U.B.-approved nomenclature has been slow. In everyday usage, most enzymes are still called by their common name.

Enzymes are also classified on the basis of their composition. Enzymes composed wholly of protein are known as simple enzymes in contrast to complex enzymes, which are composed of protein plus a relatively small organic molecule. Complex enzymes are also known as **holoenzymes**. In this terminology the protein component is known as the **apoenzyme**, while the non-protein component is known as the **coenzyme** or **prosthetic group**. Many prosthetic groups and coenzymes are water-soluble derivatives of vitamins.

The non-protein component of an enzyme may be as simple as a metal ion or as complex as a small non-protein organic molecule. Enzymes that require a metal in their composition are known as **metalloenzymes**.

Role of Coenzymes-The functional role of coenzymes is to act as transporters of chemical groups from one reactant to another. The chemical groups carried can be as simple as the hydride ion ($H^+ + 2e^-$) carried by NAD or the mole of hydrogen carried by FAD; or they can be even more complex than the amine ($-NH_2$) carried by pyridoxal phosphate.

Enzymes as Biological Catalysts-In cells and organisms most reactions are catalyzed by enzymes, which are regenerated during the course of a reaction. These biological catalysts are physiologically important because they speed up the rates of reactions that would otherwise be too slow to support life. Enzymes increase reaction rates, sometimes by as much as one millionfold, but more typically by about one thousand fold.

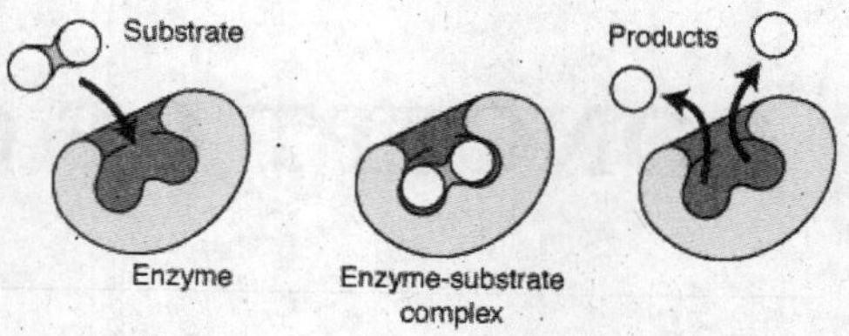

Table 2.3. Enzymes that use DNA as a substrate and are used in genetic engineering.

Enzyme	Reaction
Restriction endonuclease	Cut DNA at specific palindromic recognition sequences that are 4-6bp long
DNA polymerase I	Synthesis of the complementary DNA strand: requires a primer with a free 3'-end; important for DNA sequencing
DNA ligase	Ligates DNA strands; the enzyme forms phosphodiester bonds between phosphate residues and neighbouring 3'-ends hydroxyl group
Telomerase	Synthesizes telomere sequences at the end of chromosomes
DNA topoisomerase	Cuts DNA double strands, single – stranded or double stranded
Taq polymerase	Heat stable DNA polymerase from *Thermus aquaticus*, important for PCR
DNAse	Hydrolase that cleaves double-stranded DNA
RNAse	Hydrolase that degrades single-or double stranded RNA
RNA polymerase	Copies DNA into mRNA and rRNA
Reverse transcriptase	Copies RNA into DNA

QUESTIONS

1. What are macromolecules? Describe in brief their classes and biological role.
2. Discuss different structures formed by polypeptides ?
3. What is T_m? What is the role of base pairing in Tm ?
4. What are chemical and biological difference between DNA and RNA ?
5. Discuss different types of RNA emphasing t RNA.
6. What is codon- anticodon relationship? How amino acids are selected for addition in a polypeptide chain?

CONCEPT OF GENE

CHAPTER 3

The characteristic of every organism is governed by factors of inheritance termed as genes. As centuries have passed, the scientists have learned more and more about the nature of inheritance, and our concept of gene has undergone a remarkable evolution. The earliest studies revealed genes to be discrete factors that were retained throughout the life of an organism and then passed on to each of its progeny. Over the following century, these hereditary factors were shown to reside on chromosomes and to consist of deoxyribonucleic acid: DNA, a macromolecule with extraordinary properties. The cell's DNA is referred to in it's entirety as genome. Out of the total DNA in the genome of an organism about 25% of DNA define genes, but only 1.5% codes directly for proteins, the rest is made up of RNA genes and non coding sequences, which often either serve no function or its function is still unknown. Possibly the largest part of genome (over 50% with larger eukaryotes) is not transcribed and is partially functionless.

In this brief introduction of gene, DNA and its functioning has been described to make student understand the basic properties leading to genetic manipulation. Our aim is to make it clear, what is a gene in physical sense, a functional sense, how it makes proteins or govern a character (phenotypic like red flower colour or physiological, e.g. alkaptonurea, a genetic disease due to single defective gene, Fig. 3.1). These questions are important to understand how gene functions, what are different components (regulator, promoter, structural gene including start and stop signals) to make a gene functional as well as differentiate it from a simple DNA fragment or polynucleotide chain, what is the importance of a promoter (or its sequence) in gene expression. Then only, we can obtain a gene, make copies (cloning) and transfer to another organism (transgene) with a promoter which can regulate expression.

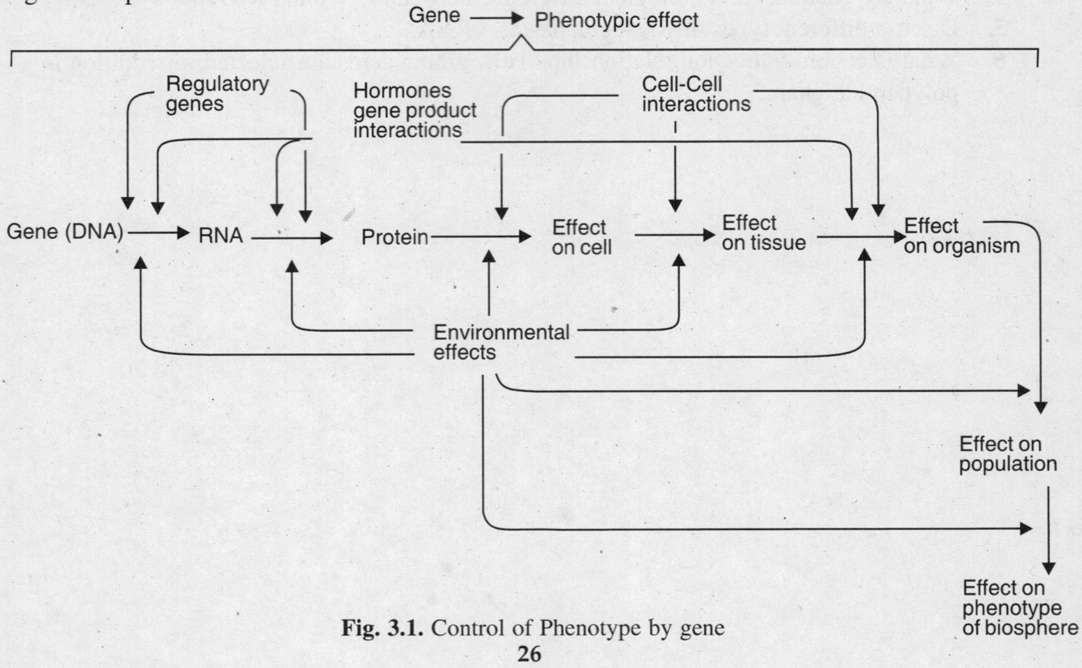

Fig. 3.1. Control of Phenotype by gene

The gene has been defined as the unit of genetic information that controls a specific aspect of the phenotype. Such a description, though accurate, does not provide a precise, unambiguous definition that can be used to identify a gene at the molecular level. At a more fundamental level, the gene has been defined as the unit of genetic information that specifies the synthesis of one polypeptide. However, it is not a very good operational definition.

Prior to 1940, genes were considered analogous to beads on a string; recombination occurred between, but not within, genes. The gene was both the basic functional unit, which controlled one phenotypic trait, and the elementary structural unit, which could not be subdivided by recombination or mutation. Oliver's 1940 report that recombination had occurred within the *lozenge* gene of *Drosophila*, stimulated both excitement and much debate about its significance. When the debate ended, the nucleotide pair had replaced the gene as the basic unit of structure, the unit of genetic material not subdivisible by mutation or recombination. In the 1960s, elegant experiments by Sydney Brenner and colleagues showed that the gene and its polypeptide product were colinear structures, with a direct correlation between the sequence of nucleotide pairs in the gene and the sequence of amino acids in the polypeptide. However, this simple concept of the gene as a continuous sequence of nucleotide pairs specifying a colinear sequence of amino acids in the polypeptide gene product was short-lived. Overlapping genes and genes within-genes were discovered in the late 1960s, and the coding sequences of eukaryotic genes were shown to be interrupted by noncoding intron sequences in the late 1970s. Thus the gene is defined as "the unit of genetic information that controls the synthesis of one polypeptide or one structural RNA molecule".

The structural gene refers to the portion that is transcribed to produce the RNA product. In the case of overlapping genes, this definition requires that some nucleotide-pair sequences be considered components of two or more genes. For those cases where exons are spliced together in various combinations to make related but different proteins, the gene may be defined as a DNA of transcription and encodes a set of closely related polypeptides, sequence that is a single unit.

I. ONE GENE-ONE ENZYME

George Beadle's collaboration with Ephrussi in studies of the synthesis of the brown eye pigment in ***Drosophila*** led him to search for the ideal organism to use in extending this work. The pink bread mold *Neurospora crassa* can grow on medium containing only (1) inorganic salts, (2) a simple sugar, and (3) one vitamin, biotin. *Neurospora* growth medium containing only these components is called "minimal medium." George Beadle and Edward Tatum reasoned that *Neurospora* must be capable of synthesizing all the other essential metabolites, such as the purines, pyrimidines, amino acids, and other vitamins, *de novo*. Furthermore, they reasoned that the biosynthesis of these growth factors must be under genetic control. If so, mutations in genes whose products are involved in the biosynthesis of essential metabolites would be expected to produce mutant strains with additional growth-factor requirements. Beadle and Tatum tested this prediction by irradiating asexual spores (conidia) of wild-type *Neurospora* with X-rays or ultraviolet radiation, and screening the clones produced by the mutagenized spores for new growth-factor requirements (Figure 3.2). In order to select strains with a mutation in only one gene, they studied only mutant strains that yielded a 1:1 mutant to wild-type progeny ratio when crossed with wild type. They identified mutants that grew on medium supplemented with all the amino acids, purines, pyrimidines, and vitamins (called "complete medium"), but could not grow on minimal medium. They analyzed the ability of these mutants to grow on medium supplemented with just amino acids, or just vitamins, and so on. For example, Beadle and Tatum identified mutant strains that grew in the presence of vitamins but could not grow in medium supplemented with amino acids or other growth factors. They next investigated the ability of these vitamin- requiring strains to grow on media supplemented with each of the vitamins separately. In this way, Beadle and Tatum demonstrated that each mutation resulted in a requirement for one growth factor. By correlating their genetic analyses with biochemical studies of the mutant strains, they demonstrated in several cases that one mutation resulted in the loss of one enzyme activity. This work, for which Beadle and Tatum received a Nobel Prize in 1958, was soon verified by similar studies of many other organisms in many laboratories. The **one gene–one enzyme** concept thus became a central part of molecular genetics.

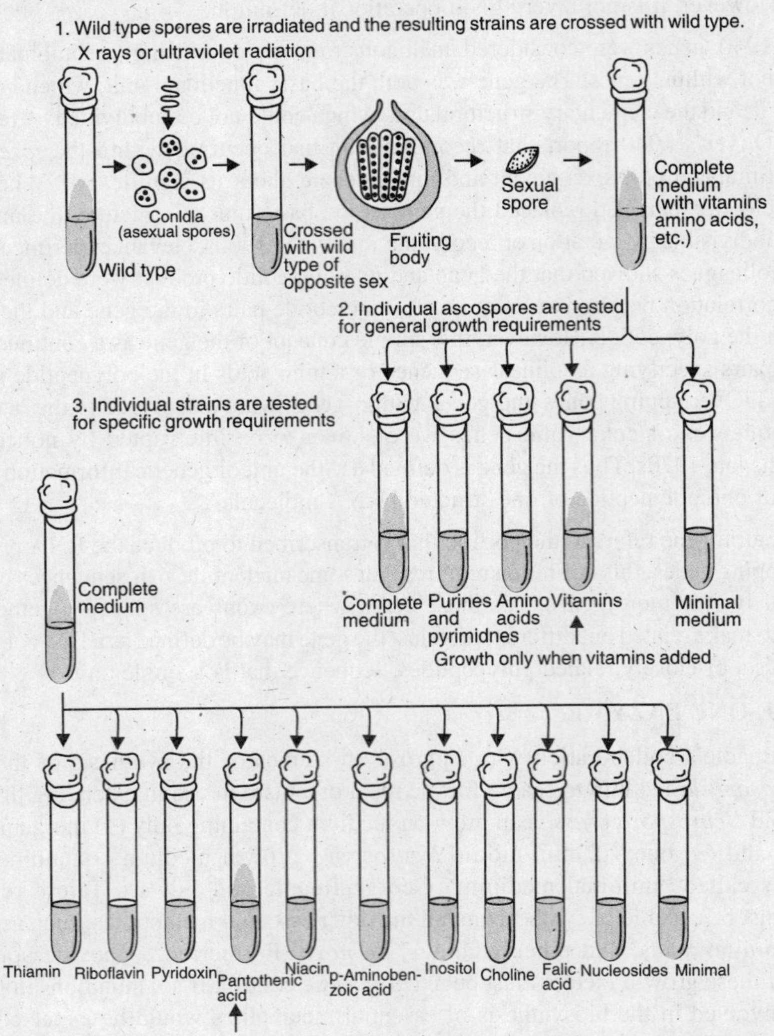

Fig. 3.2. Experiments with *Neurospora* demonstrating mutation results in gene defect

II. ONE GENE–ONE POLYPEPTIDE

Subsequent to the work of Beadle and Tatum, many enzymes and structural proteins were shown to be heteromultimeric, that is, to contain two or more different polypeptide chains, with each polypeptide encoded by a separate gene. For example, in *E. coli*, the enzyme tryptophan synthetase is a heterotetramer composed of two α polypeptides encoded by the *trpA* gene and two β polypeptides encoded by the *trpB* gene. Similarly, the hemoglobins, which transport oxygen from our lungs to all other tissues of our bodies, are tetrameric proteins that contain two α-globin chains and two β-globin chains, as well as four oxygen-binding heme groups. Other enzymes, for example, *E. coli* DNA polymerase III and RNA polymerase II, contain many different polypeptide subunits, each encoded by a separate gene. Thus, the one gene–one enzyme concept was modified to **one gene-one polypeptide**.

By the mid 1950s DNA was established as the genetic material, its structure had been analyzed by Francis Crick and James Watson (1953). Crick had stated the 'Central Dogma' of molecular biology that the linear sequence of nucleotides in a segment of a DNA molecule determines the linear sequence of nucleotides in an RNA molecule (termed as transcription), while RNA molecule in turn determines the sequence of amino acids in a protein by 'informational specificity' (termed as translation). Whole process is known as information flow from DNA to protein or central dogma.

Transcription is the process by which genetic information from DNA is transferred into RNA. DNA sequence is enzymatically copied by RNA polymerase to produce a complementary nucleotide RNA strand. One significant difference between RNA and DNA sequence is the presence of U, or uracil in RNA instead of the T, or thymine of DNA. In the case of protein-encoding DNA, transcription is the first step that ultimately leads to the translation of the genetic code, via the mRNA intermediate, into a functional peptide or protein. The stretch of DNA that is transcribed into an RNA molecule is called a *transcription unit*. A transcription unit that is translated into protein contains sequence that directs and regulates protein synthesis in addition to coding sequence that is translated into protein. Regulatory sequence that is before, or 5', of the coding sequence is called 5' untranslated (5'UTR) sequence, and sequence found following, or 3', of the coding sequence is called 3' untranslated (3'UTR) sequence. Transcription has some proofreading mechanisms, but they are fewer and less effective than the controls for copying DNA; therefore, transcription has a lower copying fidelity than DNA replication.

As in DNA replication, transcription proceeds in the 5' → 3' direction. The DNA template strand is read 3' → 5' by RNA polymerase and the new RNA strand is synthesized in the 5' → 3' direction. RNA polymerase binds to the 3' end of a gene (promoter) on the DNA template strand and travels toward the 5' end. Except for the fact that thymines in DNA are represented as uracils in RNA, the newly synthesized RNA strand will have the same sequence as the coding (non-template) strand of the DNA. For this reason, scientists usually refer to the DNA coding strand that has the same sequence as the resulting RNA when referring to the directionality of genes on DNA, not the template strand.Transcription is divided into 3 stages: *initiation*, *elongation* and *termination*.

Translation is the second process of protein synthesis. Translation occurs in the cytoplasm where the ribosomes are located. Ribosomes are made of a small and large subunit which surrounds the mRNA. In translation, messenger RNA (mRNA) is decoded to produce a specific polypeptide according to the rules specified by the genetic code. This is the process that converts an mRNA sequence into a chain of amino acids that form a protein. Translation is necessarily preceded by transcription. Translation proceeds in four phases: activation, initiation, elongation and termination (all describing the growth of the amino acid chain, or polypeptide that is the product of translation).

Though transcription and translation are explained above, it is clear from the Fig. 3.3 that in a prokaryotic system protein is direct product of gene and RNA formed in transcription is translated directly while newly formed RNA of a eukaryotic cell is processed (modified) before it is translated in to a protein. The important point to note is every gene, pro-or eukaryotic has a regulatory gene (nucleotide sequence). This promoter region is involved in regulation of translation of transcription i.e. how many times a gene should transcribe to form mRNA. This is done by consensus sequences (Consensus sequences are similar sequences present in promoter region of prokaryotes and eukaryotes and determines RNA polymerase binding) located -10, -25,-35 and -85 upstream, where RNA polymerase binds. Tight binding more transcription, loose binding less transcription. Furthermore, there are leader sequences in eukaryotics, situated very far from promoter and can regulate (by their product) down or up stream situated gene (promoter) there by further controlling the promoter.

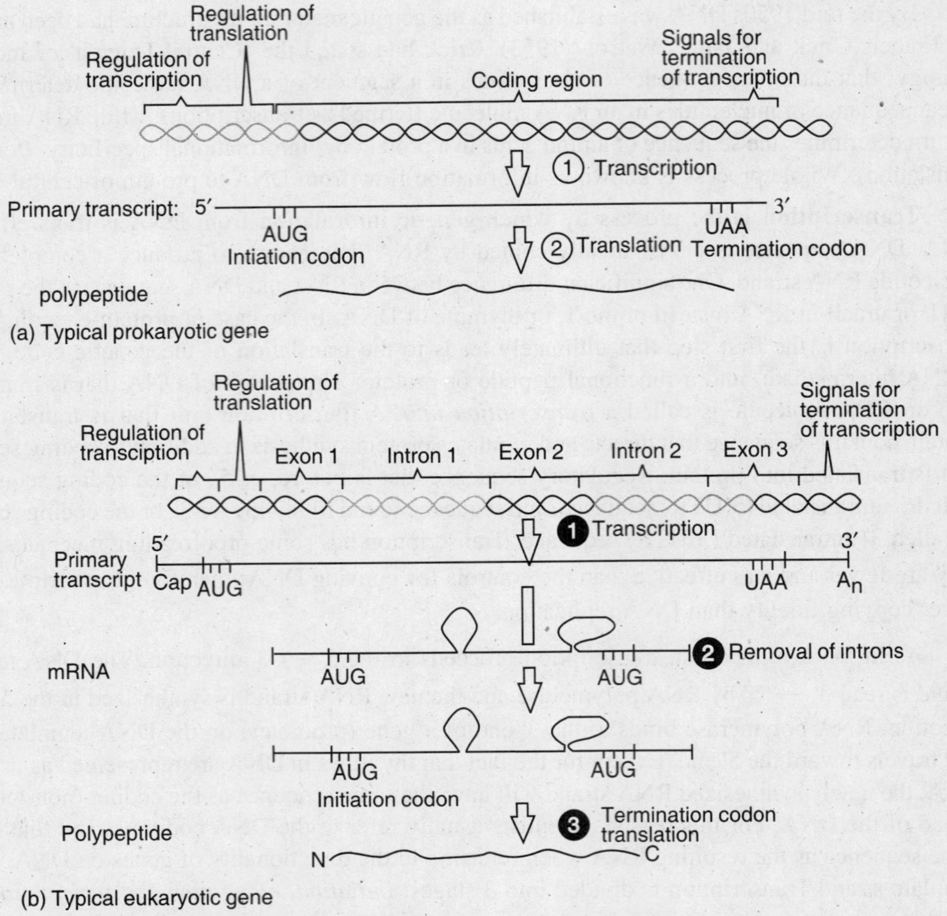

Fig. 3.3. Process of transcription and translation in prokaryotes and eukaryotes

Now the question is, when a gene should transcribe? The simple example is 'lac operon' model on which 'lactose' molecules act as 'on' and 'off' switch. Presence of lactose turned on the transcription of gene while its absence turned off the gene. That we study in molecular biology how lac/tryptophan (trp) operon function. From biotechnology point of view, if a desired gene is attached with lac or trp promoter, lactose or tryptophan is used as on /off switch to transcribe the gene of interest. Therefore, a desired gene is isolated, cloned, a promoter is selected and attached, then loaded in a Ti-plasmid, plasmid is absorbed by *Agrobacterium* and then *Agrobacterium* transfer

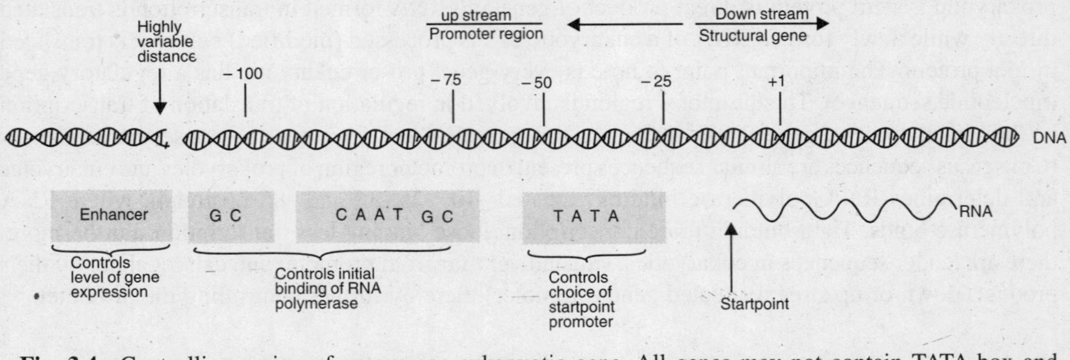

Fig. 3.4. Controlling region of an average eukanyotic gene. All genes may not contain TATA box and enhancer region is located at highly variable distance

T-DNA containing desired gene to the host cell (plant) which is in turn selected on a selection medium containing antibiotic (due to plasmid, transformed cells are resistant, untransformed not resistant and die). This gene of interest may be of prokaryotic origin (cDNA from RNA) or eukaryotic origin. In case of eukaryotic origin, it requires removal of introns. Then, a prokaryotic gene in prokaryote of prokaryotic gene in eukaryote or vice-versa is transferred. This requires modification as per organisms. These are discussed in detail in next chapters.

III. COMPLEX GENE-PROTEIN RELATIONSHIPS

Most prokaryotic genes consist of continuous sequences of nucleotide pairs, which specify colinear sequences of amino acids in the polypeptide gene products. Most eukaryotic genes are split into coding sequences (exons) and noncoding sequences (introns). However, because the spliceosomes usually excise introns from primary transcripts by *cis*-splicing mechanisms (processes that join exons from the same RNA molecule), the presence of introns in genes does not invalidate the complementation test as an operational definition of the gene. Nevertheless, in some cases, transcripts of split genes may undergo several different types of splicing, making the relationships between genes and proteins more complex than the usual one gene-one polypeptide. In other cases, expressed genes are assembled from "gene pieces" during the development of the specialized cells in which they are expressed.

Protein Isoforms

Many interrupted eukaryotic genes, such as the mammalian hemoglobin genes and the chicken ovalbumin, each encode a single polypeptide chain with a specific function. In these cases, the mRNA produced from a given gene contains all the exons of the gene joined together in the same order as they occur in the gene. However, the transcripts of some interrupted genes undergo alternate

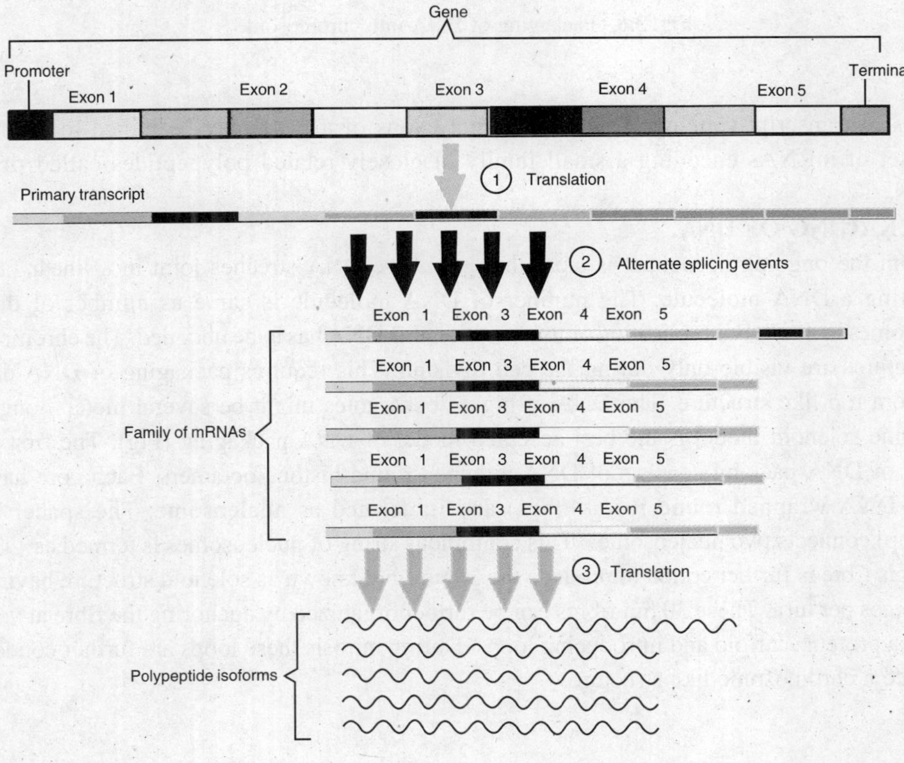

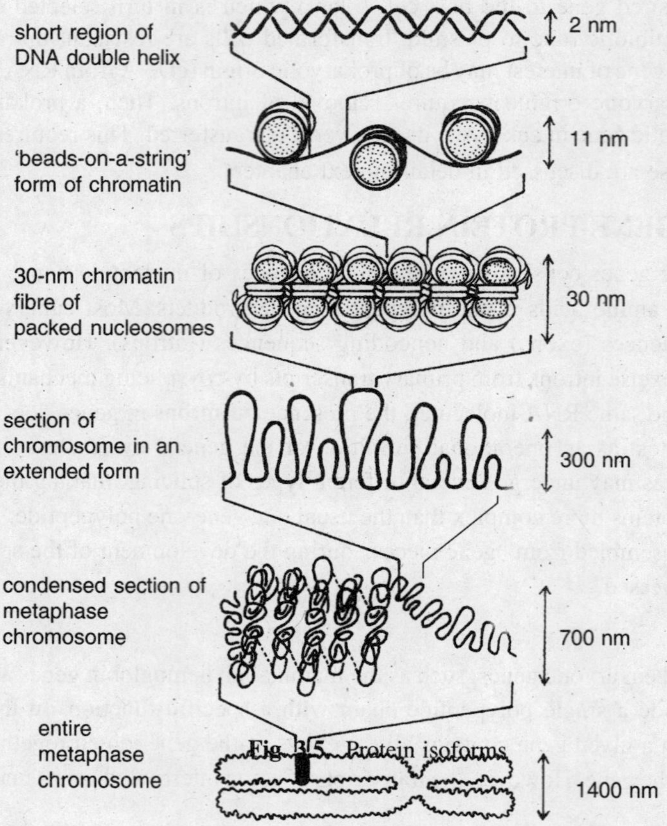

Fig. 3.6. Packaging of DNA into chromosome

pathways of transcript splicing. That is, different exons of a gene may be joined to produce a related set of mRNAs encoding a small family of closely related polypeptides called **protein isoforms.**

IV. PACKAGING OF DNA.

From the ongoing account it is clear that genes are DNA streches joint in a linear fashion constituting a DNA molecule. The number of DNA molecule is same as number of diploid chromosomes in a cell. For replication or transcription this DNA has to be uncoiled. The chromosome like structures are visible only during the cell division. This requires packaging of DNA double helix into a rod like structure. The DNA of higher eukaryotes might be several meters long. The nucleosome solenoid model is the best accepted model of DNA packaging (Fig). The first order structure of DNA packing consists of DNA wrapped round histone octamers. Each core has two turns of DNA wrapped round it (~146bps) and its termed as nucleosome. The spacer DNA (~50-60bp) connects two nucleosomes. This continuous string of nucleosomes is termed as '~10nm fibre'. This fibre is further coiled to produce the 30nm fibre known as solenoid structure having six nucleosomes per turn. These 30 nm fibres can be further organized by anchoring the fibre at various points to a protein scaffold and introducing loops. During mitosis these loops are further condensed to produce a chromosome like structure.

QUESTIONS

1. Discuss the gene as structural and functional unit of heredity.
2. Write short notes :
   (*a*) One gene one enzyme
   (*b*) One gene one polypeptide
   (*c*) Protein isoforms
   (*d*) Packaging of DNA

PLASMIDS AND VECTORS

CHAPTER 4

I. PLASMIDS

Introduction

Plasmids have a wide range of structures; they may be composed of DNA or RNA, of double or single stranded nature in circular or linear form. The size of plasmid vary from 1.5 kb (kilobase) to 1500 kb, with majority of them being circular. Large linear DNA plasmids up to 500 kb are present in streptomyces.

DNA plasmids are present in the following forms (Fig. 4.1)

1. Covalently closed circular molecule – both strands of DNA are intact.
2. Open circles – only one of the strands of DNA is intact.
3. Linear DNA molecules.
4. Supercoiled circles.

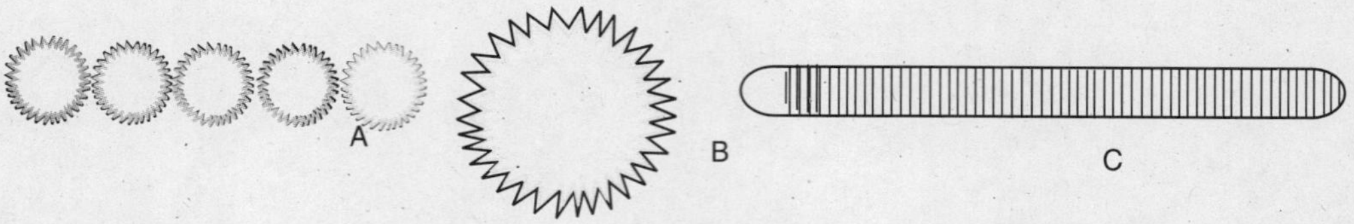

Fig. 4.1. A supercoiled; B. Circular; C. Linear plasmid

Because of different sizes and forms, they have different mobility in agarose gel electrophoresis. These are DNA molecules and hence all the enzymes of DNA synthesis and breakdown can act upon to cause changes in their forms. Mainly plasmid in bacteria exists in supercoiled form.

Definition- Plasmids are self-replicating, double stranded, extrachromosomal DNAs maintained as independent molecules by most of the bacterial genera. Lederberg in 1952 coined the term 'plasmid' to all extrachromosomal, covalently closed circular (CCC) DNA molecules. Presence or absence of plasmid from a bacterial cell makes no difference to the bacterial cell. Many types of plasmids are found in nature, in bacteria, yeast and fungi.

Classification- Plasmids are of two types :

(*a*) Conjugative or transmissible plasmids
(*b*) Non-conjugative or non-transmissible plasmids.

Conjugative plasmids can mediate their own transfer between bacteria by the process of conjugation. They possess *tra* (transfer) and *mob* (mobilizing) regions to carry on transfer function. Conjugative plasmids are found in many gram negative and some gram positive bacteria. Non-conjugative plasmids are not self-transmissible, but can be mobilized by conjugative plasmids that are present in the same cells.

Plasmids may also be classified on the basis of the number of copies found in host cells.

1. Low copy number (1-4 copies per cell)

2. Moderate copy number (10-100 copies per cell)
3. High copy number (>100 copies per cell).

Copy number is a parameter of bacterial population and may vary around mean values in the individuals of the colony. The copy number of plasmid is generally defined as the ratio of number of moles of plasmid DNA to the number of moles equivalents of chromosomal DNA. Conjugative plasmids are large with stringent control of DNA replication and are present in low copy number. Non-conjugative plasmids are small, high copy number and have relaxed control of DNA replication. Large number (high copy) of copies is produced because of relaxed control on DNA replication. High copy number plasmids are useful for expression of cloned gene to get higher yields. But when such plasmids are used as vectors, they completely utilize the host's metabolism.

Each plasmid has an origin of replication. Plasmid DNA in a bacterium may be 0.1-5% of the total DNA. Plasmids, in general, may have about 100 genes.

Plasmid Incompatibility – Some microorganisms may have as many as 8-10 different plasmids. Plasmid incompatibility is the inability of two different plasmids to co-exist in the same host cell in absence of selection pressure. Currently 30 incompatibility groups are known for *Escherichia coli* plasmids.

Importance as Cloning Vector – Plasmids have basic properties like origin of replication, self-transfer capacity and selectable markers. Due to these properties it can act as a potential vector. But plasmids also lack some important characteristics required for high quality cloning vector. The drawbacks are – (a) a small size and efficiency to transfer maximum of 15 kb of foreign DNA; (b) single restriction endonuclease site; (c) one or more selectable genetic markers. These drawbacks have been removed by genetically modifying the natural plasmid into construction of plasmid cloning vectors. The number of copies of such constructed plasmids, which is used as vector, is increased by replication. Inhibitors of protein synthesis are used to inhibit chromosomal transcription (e.g., antibiotics such as chloramphenicol or streptomycin) but plasmid replication occurs without any effect. Natural plasmids are those plasmids, which are not constructed *in-vitro* for the sole purpose of cloning, e.g., pSC 101, Col E1 and RSF 2124 are natural plasmids used for cloning.

Some plasmid derivatives, used for introducing the transposable element into cells, are known as suicide plasmids. These plasmids are incapable of replicating in the recipient cells. The replication of such plasmids is blocked due to mutation affecting replication.

Plasmid Stability – Stability of plasmid is related to certain loci in plasmid which look after faithful segregation of plasmid copies to daughter cells. These partition loci are apparently analogous to the centromere of eukaryotic chromosomes and are sites for attachment of plasmid to cell membrane to ensure distribution at cell division. Some plasmids have alternative mechanism to ensure stability. They have *cer* loci, which reverses oligomer formation of plasmid (which can cause instability and loss of plasmid).

Stability of plasmid is important when plasmid is used as vector. Genetic modifications have been used to overcome instability problem of plasmid. This is done by introducing *par* locus in high copy number plasmid vector or by inserting cloned gene into chromosome itself.

Examples of Plasmid Types – Genes located on various plasmids can express characteristics such as antibiotic resistance, antibiotic production, sugar fermentation, enterotoxin production, heavy metal resistance, hydrogen sulphide gas production, degradation of aromatic compounds and induction of plant tumours and hairy roots. Different types of plasmid showing various characteristics are presented in Table 4.1.

Table. 4.1. Plasmid types and their characteristics

Plasmid type	Examples	Characteristics
Natural	Col E1, pSC 101	Non-conjugative
	F, RK6	Conjugative
Low copy number	pBR 322	Moderate copy number plasmids are derived from this plasmid
High copy number	pUC 19, PBR 322	
Conjugative	F, R1, AK6, Ent P307	
Non-conjugative	Col E1, RSF 1030, pSC 101	
Commonly used plasmid vector	pBR 322, pAT153, pUC	Antibiotic resistance
Cryptic plasmid		No definite function
Degrative plasmid	Tol	Degrades toluene, xylene
	CAM	Degrades camphor
	OCT	Degrades octane, hexane
	SAL	Degrades salicylate
	NAH	Degrades naphthalene

Gene Cloning Vectors- An important requirement in gene cloning is a vector or a vehicle. Plasmids or viral DNA has been used for this purpose. The vehicle DNA is cleaved at a specific site by a restriction endonuclease, preferably one that produces a staggered cut in the two strands. The piece of foreign DNA which is proposed to be introduced into the vector, is also produced by the action of the same restriction endonuclease, thus generating similar ends.

Multicopy Plasmids Containing Multiple Cloning Sites- A mutation in the copy number control region of the plasmid pBR 322 has led to a class of plasmids that can exist in more than 1000 copies per cell of *E. coli*. Synthetic oligonucleotide representing sequences of various restriction endonucleases placed in succession have been introduced as multiple cloning sites (mcs). The mcs has been so designed as to from part of the β-galactosidase gene. Any insertion of a DNA fragment into any of the cloning sites would disrupt the β-galactosidase gene and can be detected by the lack of expression of the gene that can be monitored by a colour reaction (blue and white). The pUC series of plasmids, developed by J. Messing and co-workers, belong to this class. The plasmid pNEB 193, a derivation of pUC is an example.

Compatibility of Different Plasmids Among Themselves- At time it is desirable to study the *trans* acting effect of the product of a gene on another gene. In such cases it is advantageous to have one gene cloned in one plasmid vector and the other gene in a separate plasmid that can stably coexist with the first plasmid and not incompatible with it. Various plasmids have been classified into several incapability groups, implying that plasmids in the same incapability with ColE, and pMBT. Except for the F replicon all other replicons so far mentioned above are high copy number ones and all are suitable for use in *E. coli*.

Narrow Host Range Versus Broad Host Range Plasmid Vector- The replicons mentioned above have a narrow host range and these are not stable in other gram negative microorganisms like *Pseudomonas*, *Azotobacter*, *Azospirillum, Rhizobium* etc. Replication origins from broad host range plasmids like RK2, RP1, RP4 or RSF1010 have been employed for the construction of vehicles meant to be used in such organisms. These replicons, nevertheless are stable in *E. coli* and are compatible with each other and the replicons mentioned in the previous paragraph. The vector pRK404, which is derived from RK2, is an example.

II. pBR322

In the 1980s, one of the best-studied and most often used general-purpose plasmid cloning vectors was pBR322 (p=plasmid, B=F. Bolivar, R=R. Rodriguez, the scientists engineered this plasmid; 322=the numbers used by them to designate the plasmid). Plasmid pBR 322 contains 4361 bp. As shown in Fig. 4.2, pBR322 carries two antibiotic resistance genes; one provides resistance to Ampicillin (Amp^r), and the other confers resistance to tetracycline (Tet^r). This plasmid has also unique Bam HI, Hind III, and Sal I recognition sites within Tet^r gene; a unique Eco RI site that is not within any coding DNA; and an origin of replication that functions only in *E. coli*. It is maintained at a high copy number in *E. coli* and can not be readily transferred to other bacteria.

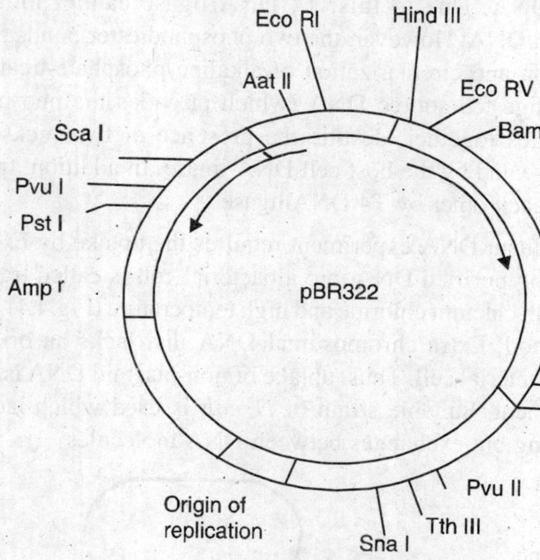

Fig. 4.2. Plasmid pBR 322

How to use pBR322 as a cloning vector ? Purified, closed, circular pBR322 molecules are cut with a restriction enzyme that lies within either of the antibiotic resistance genes and cleaves the plasmid DNA only once to create single, linear, sticky-ended DNA molecules. These linear molecules are combined with prepared target DNA from a source organism. This DNA is cut with the same restriction enzyme, which generates the same sticky ends as those on plasmid DNA. The DNA

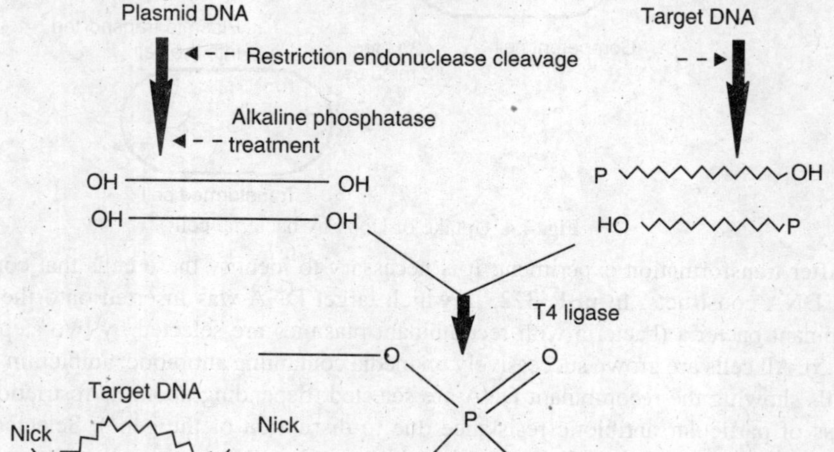

Fig. 4.3. Insertion of target DNA into plasmid and ligation

mixture is treated with T4 ligase in the presence of ATP. Under these conditions, a number of different ligated combinations are produced, including the original closed circular plasmid DNA. To reduce the amount of this particular unwanted ligation product, the cleaved plasmid DNA preparation is treated with the enzyme alkaline phosphatase to remove the 5'-phosphate groups from the linearized plasmid DNA. Due to this, T4 DNA ligase cannot join the ends of the dephosphorylated linear plasmid DNA. However, the two phosphodiester bonds that are formed by T4 DNA ligase after the ligation and circularization of alkaline phosphate-treated plasmid DNA with restriction endonuclease digested source DNA (which provides the phosphate groups), are sufficient to hold both molecules together, despite the presence of two nicks (Fig. 4.3). After transformation, these nicks are sealed by the host cell DNA ligase. In addition, fragments from the source DNA are also joined to each other by T4 DNA ligase.

The next step in a recombinant DNA experiment requires the uptake by *E. coli* of the cloned DNA. The process of introducing purified DNA into a bacterial cell is called transformation. This is carried out by treating cells with calcium chloride and high temperature (Fig.4.4) A few transformed cells are obtained by this method. Extra chromosomal DNA that lacks an origin of replication cannot be maintained within a bacterial cell. Thus, uptake of non-plasmid DNA is of no significance in a recombinant DNA experiment. Suitable strain of *E. coli* is used which lacks capabilities to destroy plasmid DNA or carrying out exchanges between DNA molecules.

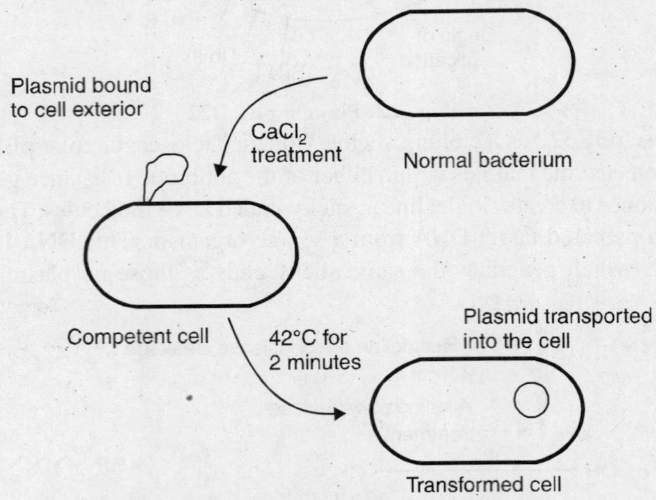

Fig. 4.4. Uptake of DNA by bacterial cell

After transformation experiment, it is necessary to identify these cells that contain plasmid-cloned DNA constructs. In pBR 322, in which target DNA was inserted onto the Bam HI site, recombinant bacteria (bacteria with recombinant plasmid) are selected by two steps as shown in (Fig. 4.5). All cells are grown successively on media containing antibiotic, ampicillin or tetracycline and cells showing the recombinant DNA are selected (depending upon the restriction enzyme site and loss of particular antibiotic resistance due to disruption of the gene). Selected recombinant bacteria are grown in bioreactor to obtain the gene product.

III. BACTERIOPHAGE AS CLONING VECTORS

The plasmid based vectors used for cloning DNA molecules generally carry up to 10 kb of inserted DNA. However, for the formation of library, it is often helpful to be able to maintain larger pieces of DNA. For this reason, *E.coli* virus (Bacteriophage, phage) lambda (λ) has been developed as a cloning vehicle. Bacteriophage λ infects *E.coli* and after injection of the viral DNA, two possibilities exist. Bacteriophage λ can enter a lytic cycle, which after 20 minutes lead to the lysis of host cells and the release of about 100 phage particles. Alternatively, the injected bacteriophage

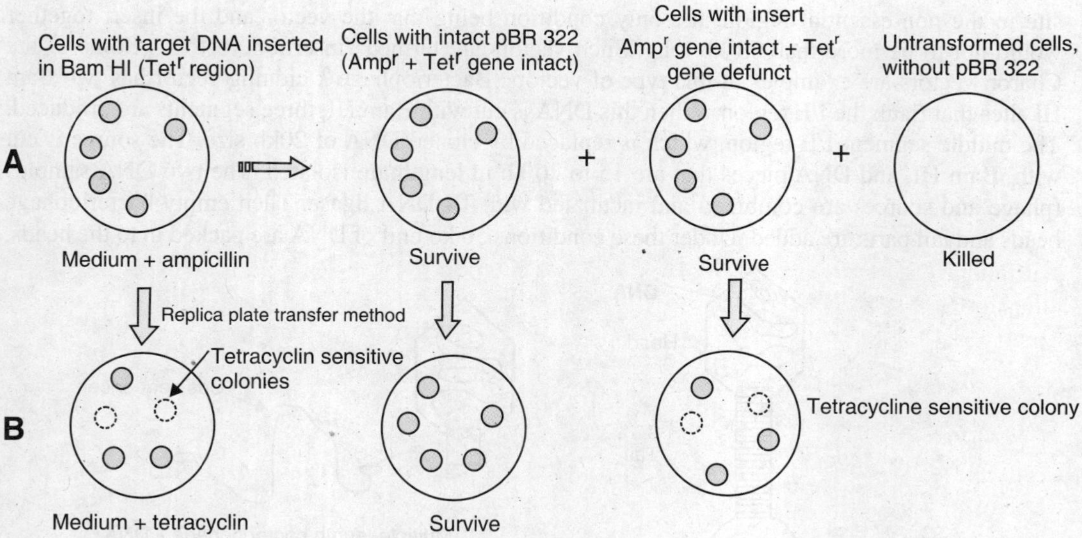

Fig. 4.5. Selection of recombinant bacterial cells containing recombinant plasmid. Colonies corresponding to tetracyclin sensitive colonies (B) are selected from original cultures (A). Growth of three types of cells obtained after transformation is also presented.

λ DNA can be integrated into the *E.coli* chromosome (DNA) as a prophage and can be maintained more or less indefinitely (Lysogeny stage). However, under conditions of nutritional or environmental stress, the integrated bacteriophage λ DNA can be excised and enter a lytic cycle. The bacteriophage λ DNA is about 50 kb in length, of which approximately 20 kb is essential for the integration/excision (I/E) events. For forming genomic libraries, this 20 kb of DNA can be replaced with 20kb of cloned DNA

Cloning Vectors Based on the Bacteriophage Lambda (λ) – Derivatives of the genome of bacterioplage lambda have been constructed to serve as cloning vectors. Transfection or transduction is used to introduce such vectors into *E.coli.* Two properties of the lambda genome make it suitable for use as a vector.

(*i*) Only about 50% of the 50 genes of lambda are essential for its replication and for lysis of the host cell. Most of these non essential genes are located together in a cluster around the middle of the genome.

(*ii*) Lambda genome is packaged inside the phage head by what is known as the 'head-full mechanism'. This means that not only there is an upper form of the amount of DNA that is packaged inside the phage head, but there is a lower limit also. Effective packaging takes place only when a minimum amount of DNA, namely 35 thousand base (kb) is present.

An infective bacteriophage λ consists of a tubular protein tail with a few tail fibres and a protein head. The production and assembly of heads and tails, and packaging of DNA are highly coordinated sequence of events.The DNA within head of a λ phage is a 50 kb linear molecule with a 12-base, single staranded extension at the 5'end. These extensions are called cohesive (cos) ends, because they contain sequences that are complementary to each other. In *E. coli*, these cos ends base-pair to form a circular DNA. DNA replication from the circular DNA creats a linear form of λ DNA that is composed of several contiguous lengths of 50kb units. Each new assembly is filled with 50kb DNA.

Inserting Type Lambda (λ) Vectors- A lambda cloning vector can therefore be constructed by deletion (*in vivo* or *in vitro* by restriction deletion) of a part of the non-essential region such that the remainder is not less than 35kb. Other mutations are also introduced such that restriction sites in the enemies regions are eliminated. A segment of foreign DNA can be cloned in an unique restriction site in the non-essential region, the only condition being that the vector and the insert together would not to be more than 53 kb long. Such vectors are termed "insertion vectors". Some of the Charon vectors are examples of this type of vectors. Bacteriophage λ cloning vector has two Bam HI sites that flank the I/E region. When this DNA is cut with Bam HI, three segments are produced. The middle segment I/E region, which is replaced by cloned DNA of 20kb size. The source is cut with Bam HI, and DNA pieces that are 15 to 20 kb in length are isolated. The two DNA samples (phage and source) are combined and incubated with T4 DNA ligase. Then empty bacteriophage heads and tail parts are added. Under these conditions 50 kb unit of DNA are packed in to the heads,

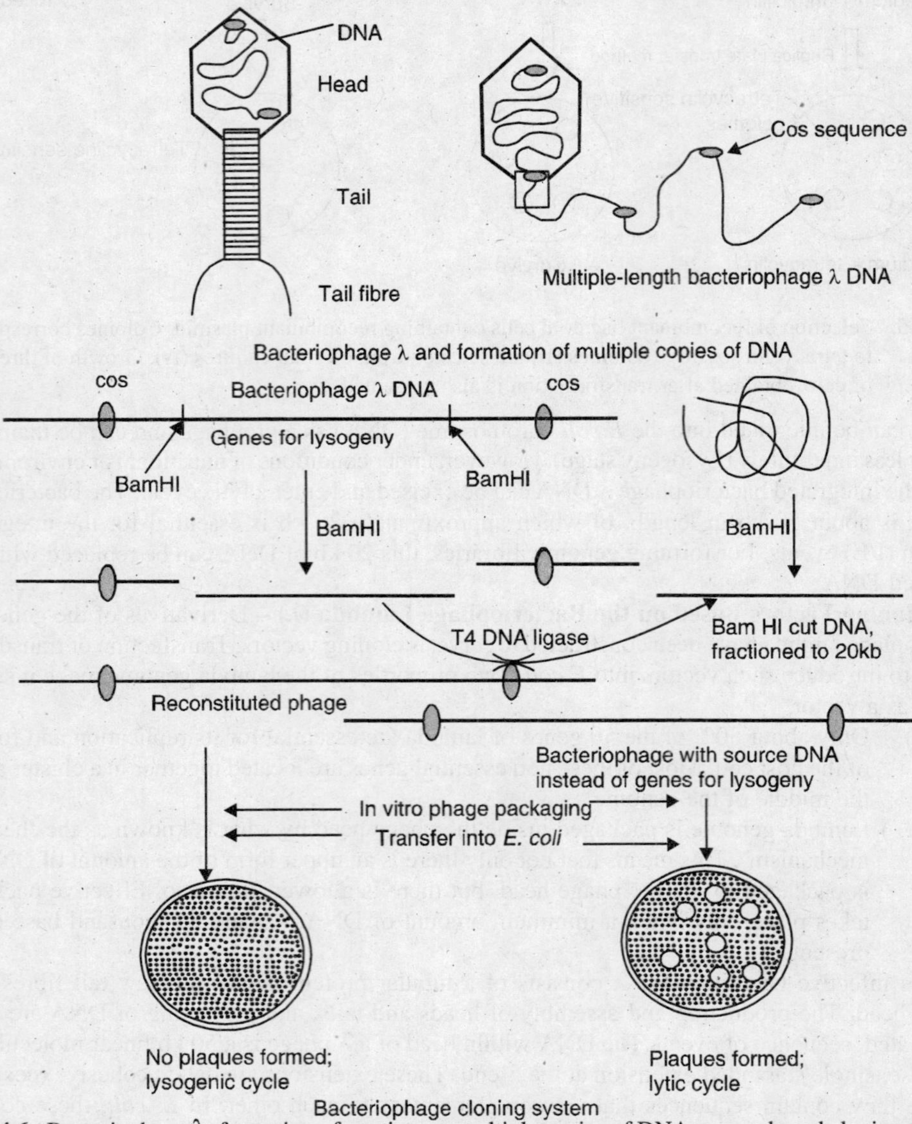

Fig. 4.6. Bacteriophage λ, formation of contiguous multiple copies of DNA, cos ends and cloning of source DNA in bacteriophage system

and infective phages are produced. Other products from ligation reaction cannot be packed, because they are either too large (>52 kb) or too small (>38 kb). Recombinant bacteriophage λ can undergo lytic cycle only in an *E.coli* strain that does not allow reconstituted phage λ (non-recombinant) with in fact I/E regions to grow. Recombinant phage is maintained by lytic cycles in fresh *E.coli* cultures (Fig. 4.6). Bacteriophage libraries can be screened by using either DNA probes or immunological assays. For this purpose, individual lytic zones are tested (compared to bacterial colonies in plasmid cloning vectors).

Substitution Type Lambda (λ) Vector- The second type of lambda vector is of the substitution type, the example being the lambda gt vectors and the EMBL vectors. These vectors have two Eco R1 sites or two Bam HI sites in the non-essential region. On digestion with Eco R1, at least three piece(s) are produced, two terminal ones containing the essential regions and the central piece(s) containing the non-essential genes. The central piece(s) is separated out by sucrose density gradient sedimentation and replacing by the foreign segment to be cloned. The limits of the size of foreign DNA that can be cloned in the lambda gt vector is 1-14 kb and in the Charon 4 vector is 8.2-22.2 kb. Such a replacement cloning vector has an advantage over the insertion vectors. The terminal pieces by themselves if joined by DNA ligase, do not make up 35 kb and hence cannot be packaged. Packaging occurs only when a segment of foreign DNA gets cloned between the two terminal pieces of the vector and hence no separate selection for recombination molecules is necessary.

Cloning Vectors Based on the Bacteriophage- Double stranded derivatives of the filamentous coliphage M13 have also found extensive use as cloning vectors. A series of such vectors (M13 mp 19 is an example) have been developed by J. Messing and co-workers. Inside the *E. coli* host, the M13 DNA exists as double stranded circles, but the released virions contain single stranded circles. The DNA fragment is cloned in double stranded M13 DNA isolated from inside the *E. coli* cell and is put back into *E. coli* by transformation. The virions containing their single stranded DNA genome are then allowed to mature and get continuously secreted out of the *E. coli* cells without killing the cells. Single stranded DNA can now be isolated from the virions in the supernatant and be directly subjected to base sequencing.

IV. PHAGEMIDS AS CLONING VECTORS

A phagemid is a hybrid of a plasmid and a filamentous coliphage that can be propagated in either form. The coliphage could be either of the three virtually identical phages, M13 fd or f1. These are male specific phages that contain single stranded circular DNA as their genome. Upon infection of *E.coli* by the bacteriophage, double stranded DNA is first formed as the replicative intermediate. Finally single stranded DNA is packaged into the virion. Both the replication origins of the plasmid and the coliphage are incorporated in the phagemid. The auxiliary replication functions necessary in *trans* for the coliphage replication are not, however, incorporate in the phagemid. Hence, replication from the coliphage origin can take place only in the presence of a helper phage. Otherwise, replication takes place from the plasmid replication origin. The pBLUESCRIPT phagemids have both Col E, (pMB 9 like) origin and the filamentous phage f1 origin. The cloning is done in any of the multiple cloning site in the double stranded circular DNA of the plasmid form of the vector. This is introduced in to *E.coli* by transformation and the synthesis of single stranded DNA from the phage f1 origin is induced by superinfection with a helper phage. The single stranded DNA formed is packaged in to the phage rods because of the presence of the phage packaging signals as well in the phagemid. Direct base sequencing can be undertaken using the single stranded DNA isolated from the virions secreted from the *E. coli* cells. Depending on the orientation of the f1 replication origin in the phagemid, either the (+) strand or the (–) strand of the phagemid is replicated in presence of the helper phage. The pEMBL phagemids are similar to the pBLUESCRIPT phagemids. Phagemids that have replication origin of pUC plasmids and of the M13 bacteriophage have also been developed and are available under the trade name of LITMUS vectors. Replication in the circular double stranded plasmid form uses the pUC replication origin. Cloning is done in the multiple cloning site located in the lacZ gene and the usual blue/white selection is available. Here also replication of the single stranded phage DNA form is induced by super infecting with the M13 helper phage.

V. COSMIDS AS CLONING VECTORS

Plasmid vectors are not suitable for cloning DNA fragments very much larger than their own size, as the transformation frequency fall beyond acceptable limits and cloned fragments or their parts very often get deleted. Takagi and co-workers observed as in early as 1976 that the presence of the cohesive end site cos λ from the bacteriophage lambda DNA in a plasmid allows it to be packaged *in vivo* into virus particles. The interesting finding was that the *in vivo* packaging mechanism would be select DNA molecules of the full size of the lambda genome (48.5 kb). Making use of this finding, cosmid vectors were first developed in 1978 by J. Collins and co-workers to facilitate cloning of larger DNA fragments in plasmids. Extracts of lambda lysogens have been successfully used for *in vitro* packaging of the lambda capsids. An example of a commonly used cosmid is pHV79 which is nothing but pBR322 containing the cohesive end site cos λ and which can accommodate up to 45 kb sized inserts. A great advantage of such a cosmid vector is that: (1) gene libraries consisting of a smaller number of clone members can span the whole genome of an organism. For example, the genome of *Escherichia coli* can be accommodated in just 120 cosmids. (2) Other advantages are that large gene can be studied intact and genetic linkage studies can be carried out at the molecular level.

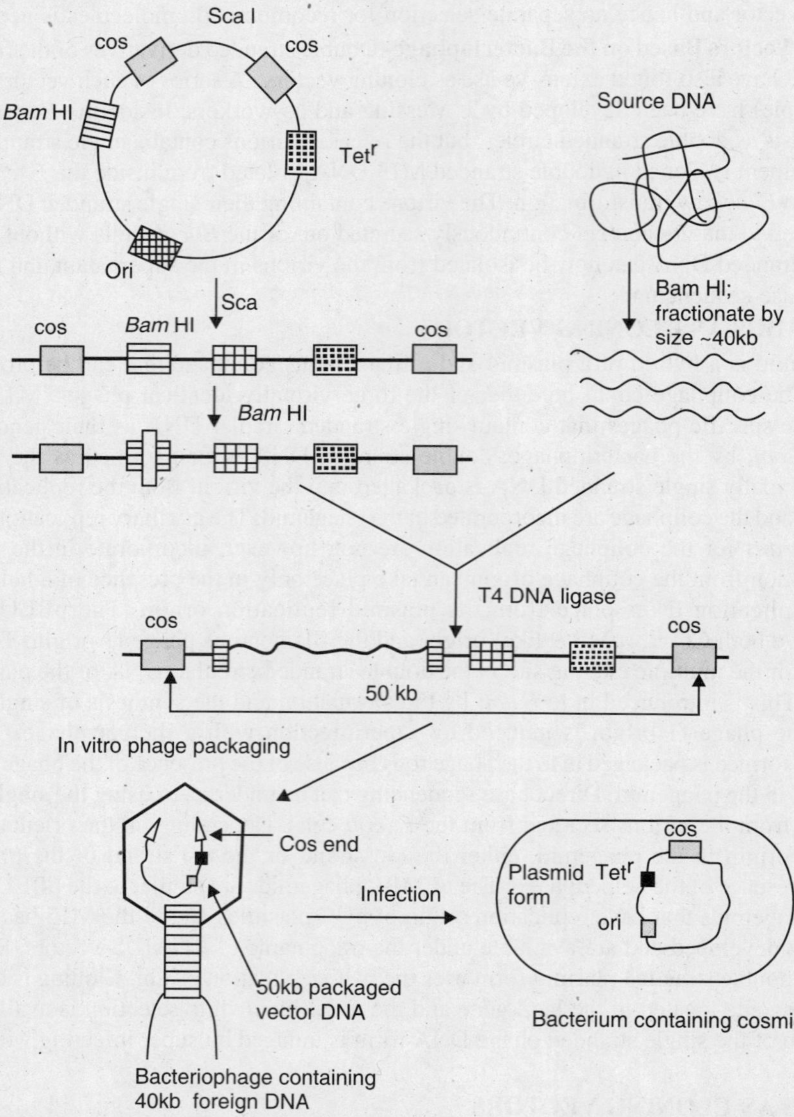

Fig. 4.7. Preparation of bacteriophage DNA for cloning of source DNA by cosmid in *E.coli*

(3) An important practical advantage of a cosmid is that, background molecules which do not have the intact and genetics linkage studies, can be carried out at the molecular level.

(4) An important practical advantage of a cosmid is that background molecules which do not have insets or have smaller inserts are eliminated during packaging. This is not possible to achieve with plasmid cloning vectors.

(5) Besides, the frequency of transformation of the lambda capsids with an *in vitro* packaging extract is much higher than the transformation frequency of plasmids.

Cosmid cloning vectors can carry 40 kb of cloned DNA and can be maintained as plasmids in *E.coli.* Cosmids combine the properties of plasmids and bacteriophage λ vectors. The commonly used cosmid pLFR-5 (6kb size) has two cos sites (cos ends) from bacteriophage λ separated by a Sca I restriction endonuclease site, a multiple cloning sequence with six unique sites (Hind III, Pst I, Sal I, Bam HI, Sma I, and EcoRI), an origin of DNA replication (ori) and a tetracycline resistance (Tet) gene. This cosmid carry about 40 kb of cloned DNA. For this vector, pieces of DNA that are approximately 40 kb in length are purified by sucrose density gradient configuration from a partial digestion of source DNA with Bam HI. The pFLR-5 DNA is cleaved first with ScaI and then with Bam HI. The two DNA samples are mixed and ligated. Some of the ligand products will have a 40kb DNA piece inserted between the two fragments that are derived from the digestion of the pLFR-5 DNA. The molecules formed by joining will be about 50 kb long, with cos sequences that are about 50 kb apart. Therefore, these DNA constructs can be successfully packaged into bacteriophage λ heads in vitro (as described above, phage λ head accommodate only 50kb DNA). After formation of complete phage, the DNA is delivered by infection into *E.coli.*

During phage packaging, cos ends are cleaved. Once inside the bacteria, the cos ends base pair and form a circular DNA molecules (Fig.4.7). This circular form is so stable, so the cloned DNA can be maintained as a plasmid-insert DNA construct because the vector contains a complete set of plasmid functions. Moreover, the tetracycline resistance gene allows colonies that carry the cosmid to grow in presence of tetracycline; non-transformed cells are sensitive to tetracycline and die.

The following are the steps for construction of a cosmid library. (i) Cleavage of the genome by partial digestion with restriction endonuclease, (ii) Sizing of the fragments by gel electrophoresis or velocity centrifugation; (iii) Cleavage of the cosmid vector and treatment with phosphate to minimize polycosmid formation; (iv) Ligation of the genomic DNA and the cosmid DNA; (v) Packaging the ligated DNA into infectious phage particles; (vi) Transduction into *E. coli*.

Table 4.2. Capacity of different cloning vectors and host organisms.

Vector	Host organism	Uptake capacity
Plasmid	E.coli	Max.10kb
Lambda phage	E.coli	Max. 25 kb
Cosmids	E.coli	35-45kb
P1 phages (PAC,P1 derived	E.coli	100-300 kb
Artificial chromosomes		
BAC	E.coli	Max. 300kb
YAC	Yeast	100-2000kb
MAC	Mammalian cells	Max.500mb

VI. ARTIFICIAL CHROMOSOMES

Efforts to clone and sequence the human genome and genomes of intensively studied organisms such as maize, *Arabidopsis thaliana, Drosophila*, and yeast require methods of cloning that gives large sized DNA fragments. Even though cosmids are excellent cloning vectors, a cosmid library of the human genome would require over 500,000 clones. Efforts to develop cloning vectors for large DNA vectors started with the extensively studied eukaryote, yeast. This organism has many of the useful attributes of *E.coli* (such as a rapid growth rate, single cells, and a relatively small genome). Yeast has a doubling time about twice that of *E.coli*, exist as a haploid and diploid, and has a genome only 2 ½ times (12mb) that of *E.coli*. Yeast undergoes mitosis and meiosis, has a nucleus, and possesses the enzymes for processing RNA after transcription. This last feature is extremely important because *E.coli* is unable to process RNA in addition, yeast can be transformed easily after treatment with the mixture of snail gut enzymes that degrade the thick, complex yeast cell wall composed of proteins and polysaccharides.

(a) Yeast Artificial Chromosomes (YAC)

Cloning of DNA fragments much larger than 45 kb became possible in 1987,when D. T. Burke and G.F. Carle developed in the laboratory of M.V. Olson an altogether new type of yeast vector,which they called yeast artificial chromosome (YAC). The development of YACs were based on the appreciation that an eukaryotic linear chromosome needs for its replication and stability, not only replication origins, but also the centromere and the telomere. The centromere sequence would attach to the mitotic spindle during cell division and help in efficient segregation of the chromosomes into the daughter cells. The telomere would in preserve the integrity of the ends of the linear chromosomes. Once these elements were provided, the vector could be replicate stably like a chromosome and could accommodate chromosomes sized inserts. Indeed, standard YACs can accommodate around 600 kb DNA inserts, while special type of YACs can accommodate upto 1400 kb DNA inserts. Though an YAC vector is meant to be propagated like a chromosome in yeast, it is a circular double stranded DNA that contains a replication origin (col E1) compatible with *E. coli* in addition to yeast replication origin or an yeast ARS element. The col E1 replication origin is useful to a yeast replication origin or an yeast ARS elements. The colE1 replication origin is useful for amplification of the vector in *E. coli*. The components of a typical YAC vector (pYAC3) are shown in Fig. 4.8. Next to the yeast replication origin is located the centomere of yeast chromosome 4 (Cen 4). The two telomere sequences are from the protozoan Tetrahymënal, which have been found to be functional in yeast. A stuffer element containing the His 3 gene of yeast is present between the two telomere sequences through two BamHI restriction endonuclease sites. Three selectable yeast marker genes, Trp 1, Ura 3 and Sup 4 are also present. The marker genes Trp1 and Ura 3 are on the two sides of the unique restriction endunuclease site SnB1, that produces blunt ends. The SnaB 1 site is used as the cloning site and is located within the Sup 4 gene. Sup 4 gene product is a tRNA that suppresses a mutation in the Ade gene of yeast resulting in a change in colour of colonies from red to white in the presence of limiting amounts of adenine in the medium. The YAC vector is digested with BamHI and SnaBI throwing the stuffer element out, disrupting the Sup4 gene and yielding two vector fragments termed as left arm and right arm. Each of the arms but at its one end a telomere sequence followed by a BamHI overhang, but a blunt flush cut at the other end. The left arm also has the Cen 4 sequence, the ARS 1, the ColE1 origin, the amp and Trp 1 markers. The right arm contains the Ura 3 marker. These two arms are blunt end ligated with the long chromosomal DNA fragment from any source, thus creating the artificial linear chromosome (Fig.4.8) among other ligation products. Ligation products are introduced into yeast cells that have mutation in the Trp1, Ura3 and Ade loci by lipofection or by fusion with yeast speroplasts. The yeast cells are allowed to regenerate the cell walls and plated on medium lacking trytophan and uracil and containing limiting amount of adenine.

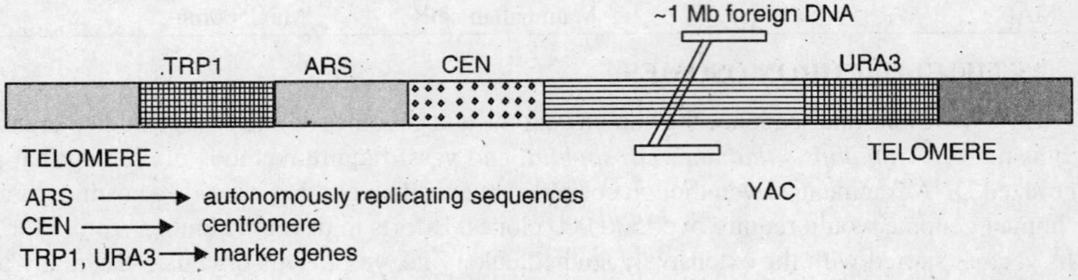

Fig. 4.8. Details of YAC

Only the yeast cells transformed with artificial chromosomes comprising both the left arm and the right arm would grow. The recombinant artificial chromosomes containing the insert would develop into white colonies. The red colonies would represent cells having the linear vector, but no insert. Cells having other ligation products like two left arms or two right arms would not grow. (Fig. 4.9).

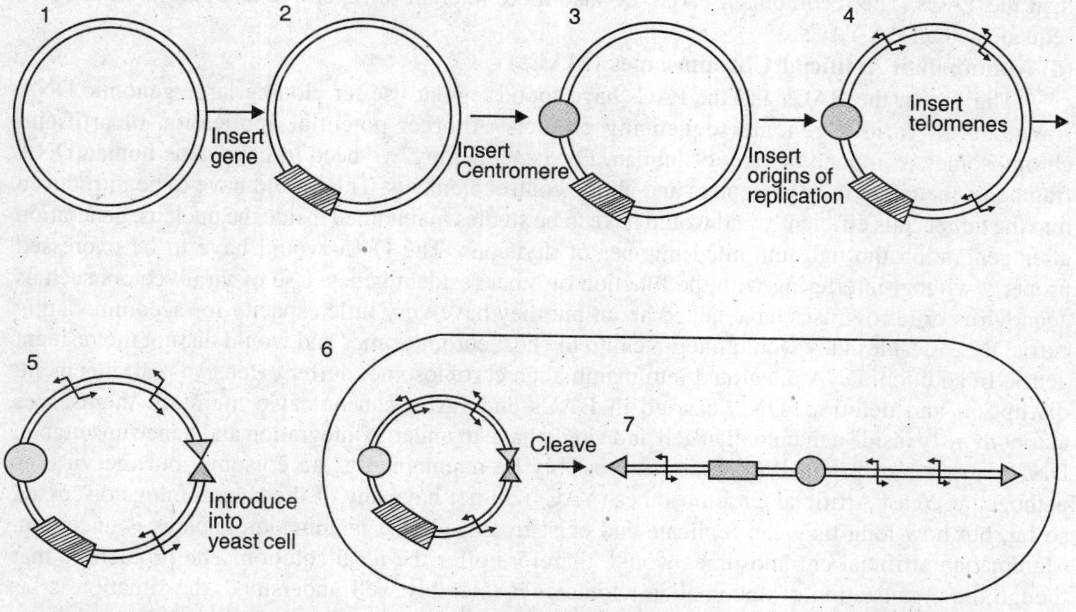

Fig. 4.9. Contraction of an artifical chromosome begins with circular piece of DNA called a plasmid

(b) Bacterial Artificial Chromosomes (BAC)

Bacterial artificial chromosomes (BAC) were developed by Mel Simmons and coworkers in the early 1990s and are based on the fertility factor (F factor) of *Escherichia coli*. The F plasmid, a ~ 100 kb circular double stranded DNA, is present in an *E. coli* cell in only 1-2 copies. The synthetic BAC vectors, which are only ~7.5 kb double stranded DNA circles contain the replication origin *oriS* and the gene *repE* of the F plasmid that are responsible for initiation and proper orientation of replication of the BAC vector. The *parA* and *parB* genes of the F plasmid ensure efficient segregation of the F factor into the daughter *E. coli* cells after its replication, are also incorporated in the BAC vector. The BAC vectors also contain multiple cloning sites (mcs), a selectable marker in the form of antibiotic resistance and colour based identification (lac Z complementation system) of recombinants carrying inserts. The naturally occurring F2 factors consist of up to 25% of the E. coli. genome integrated into the basic F factor and are very stable. This characteristics of the F factor contributes to the ability BACs to accommodate very large amount of external DNA to the extent of 300kb. The recombinant BACs have been found to exhibit a lower level of rearrangement and chimerism of the cloned DNA sequence than exhibited by YACs. The cloning of DNA in BACs is done as is done in a plasmid, by linearising the vector with a restriction endonuclease, treating with phosphatase and then ligating with the DNA fragments to be cloned. E. coli. has to be transformed by electroporation because of the large size of the recombinant BAC (Fig. 4.10).

(c) The PAC Cloning System

The PAC cloning system combines the features of the vectors derived from the bacteriophage P1 and the BAC vectors. The P1 vectors were developed by Nat Sterberg and coworkers in the early 1990s in an effort to accommodate large DNA (90-100 kb) inserts. Cloning in a P1 derived vector is similar in principle to cloning in plasmids. The large DNA fragments to be cloned are ligated to the cleaved P1 vector that contain various genetic elements of the P1 bacteriophage and a plasmid replicon/partition system. The linear recombinant molecules are then packaged in vitro into P1 particle and *E.coli* cells are transfected. A commonly used P1 vector is pAd10*sacB*ll, which contain two P1 encoded replicon, a positive selection system and an adenovirus stuffer fragment. A PAC vector has been constructed by deletion the adenovirus stuffer and inserting a pUC plasmid derivative in its place. Instead of package and transfection, the recombinant PAC vector can accommodate insert of 130-150 kb. The recombinant PACs exhibit lower rate of rearrangement and chimerism than the YACs. The recombinant PACs are also more tolerant of repetitive and palindromic DNA sequences than the YACs.

(d) Mammalian Artificial Chromosomes (MACs)

The YACs, the BACs and the PACs have found regular use for cloning large genomic DNA fragments in various genome sequencing projects. Another potential application of artificial chromosomes is in gene therapy of human. For gene therapy, we need to have gene human DNA fragments including their promoters and all the control elements. This would have to be introduced into the target cells efficiently and would have to be studies maintained inside the nucleus, generation after generation through unlimited number of divisions. The DNA would have to be expressed properly without interfering with the function of other resident genes. Use of viral vectors such as adenovirus or retroviruses have talked about but they have very little capacity for accommodating extra DNA. Besides they would integrate into the host chromosomes and would disrupt the resident genes. In addition, DNA integrated into mammalian chromosomes suffer extensive rearrangement, disruption and deletion. DNA carried in BACs and PACs can not also maintain themselves autonomously inside a mammalian cell and would have to undergo integration and hence unsuitable. DNA carried by Epstein-Barr virus can possibly be maintained as an episome, but the virus is pathogenic. Yeast Artificial Chromosomes (YACs) do not have any of the shortcomings discussed so far, but how long they can replicate and exist free inside the mammalian nucleus is uncertain. Mammalian artificial chromosomes would, therefore offer the ideal solution. The problem is that though the organization of mammalian telomeres is tolerably well understood, the situation is far from satisfactory with respect to mammalian centromeres and replication origins. Nevertheless, attempts have been made to construct MACs from "not clearly defined" large DNA fragments supposedly containing the centromere and the replication origin. Such large artificial human chromosome (aptly termed minichromosomes) if 1 mb-10mb size have been claimed to be stable for more than 100 cell generations. Satelite DNA based minichromosomes of 20 mb-30mb have also been reported. In spite of all the progress made, it is prudent to accept that MACs that can be utlilized are still in an experimental stage.

(e) Artificial human chromosomes

Researchers at the School of Medicine and Athersys, Inc. have created the first artificial human chromosomes. The synthetic chromosomes represent a breakthrough in medical research and provide scientists with a powerful new tool for the study of human genetics. Artificial chromosomes may also offer a new approach to gene therapy and the treatment of a broad range of genetic diseases. A report of the research was published in the April 2007 issue of *Nature Genetics*.

"This opens the door to a whole new avenue of research in chromosome biology and gene therapy," said Huntington F. Willard, senior author of the study and chairman of genetics at the School of Medicine and University Hospitals of Cleveland. "While it's been known since the early years of this century that chromosomes carry genes, until now the complexity and size of normal chromosomes has limited our ability to analyze their structure and function. The synthetic microchromosome system now allows us to perform detailed studies on the nature of chromosomes - essentially the next phase of the Human Genome Project which is to move from just mapping genes to actually understanding how they work and influence human disease."

Normal chromosomes are modular structures consisting not only of hundreds or thousands of genes, but also of specialized elements that are believed to be important for chromosomal stability and function. Telomeres, which consist of DNA and protein, are located at the ends of chromosomes, protecting them from damage. Centromeres are specialized regions of DNA that are essential for the proper control of chromosome distribution during cell division. Human centromeres are believed to consist of large segments of highly repetitive DNA, called alpha satellite DNA, which is thought to play a significant role in centromeric function.

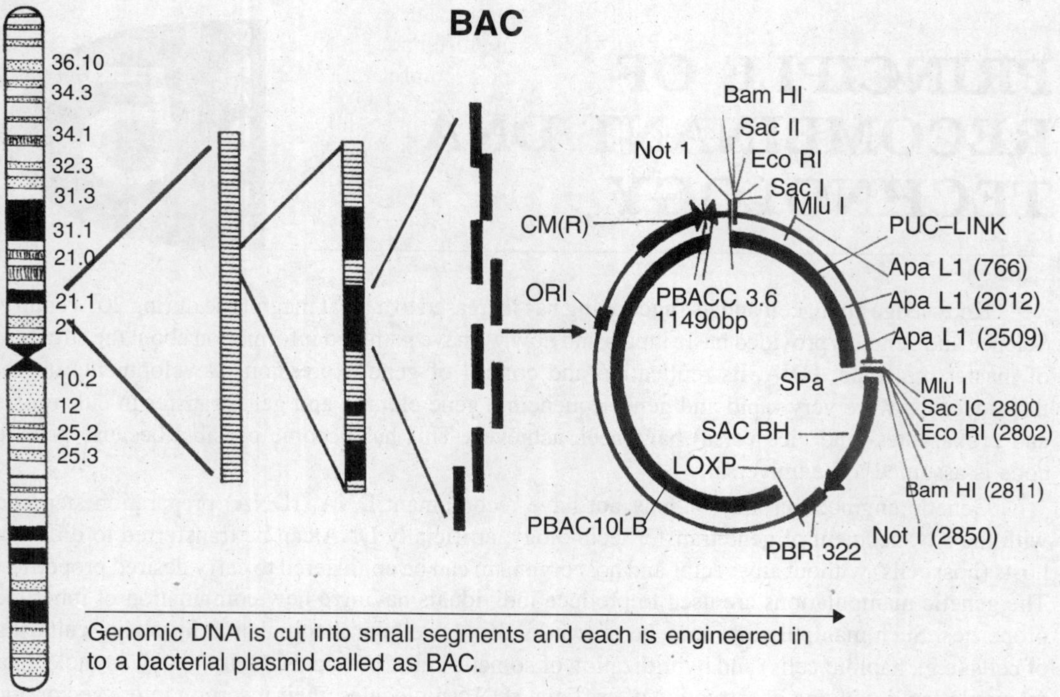

Fig. 4.10. Construction of BAC

They were successful in creation of functional centromeres and incorporation of them into artificial chromosomes, which were the critical achievements enabling the stability and normal behavior of the chromosomes throughout the cell cycle. The past attempts at producing synthetic chromosomes have failed because they lacked proper centromeres, and thus could not persist through multiple cell divisions.

In this study, the research team created artificial chromosomes from normal human material. The researchers first synthesized arrays of alpha satellite DNA, then introduced the resulting centromeric material into human cells in conjunction with telomeres and genomic DNA. Inside the cells, the independent elements assembled to form miniature chromosomes, or synthetic microchromosomes, that were structurally similar to human chromosomes, but contained less genetic material. Analysis of the newly introduced artificial chromosomes demonstrated normal centromeric activity, genetic stability, and continued gene expression through repeated rounds of the cell cycle.

Synthetic chromosomes have the potential to overcome a major stumbling block in gene therapy. The characteristic stability of synthetic chromosomes enables, for the first time, the potential long-term expression of therapeutic proteins in target tissues of patients treated using gene therapy.

The synthetic microchromosome remains independent within the host cell and functions essentially as an accessory chromosome. In contrast, most gene therapy systems currently under development utilize viral vectors, which often require the integration of the therapeutic gene into an existing chromosome and thus can result in chromosomal damage or interference with normal gene expression. Viral vectors can also induce immune responses that limit therapeutic efficacy.

QUESTIONS

1. What are plasmids ? Give characteristics of a vector required for gene transfer.
2. Write short notes :
   (*a*) Copy number in plasmid
   (*b*) Plasmid types
   (*c*) Phagemids
   (*d*) YAC
   (*e*) BAC
   (*f*) pBR 322
   (*g*) Bacteriophage
   (*h*) Mammalian artificial chromosomes
   (*i*) Artificial human chromosomes.

PRINCIPLE OF RECOMBINANT DNA TECHNOLOGY

CHAPTER 5

Knowledge about cell and its functioning has increased to a great magnitude during 20th century. Science and genetics provided basic inputs and now we have gathered information about the structure of master molecule, DNA, its replication and control of gene expression. Developments in last three decades were very rapid and gene sequencing, gene cloning and gene transfer in eukaryotes and prokaryotes (and vice versa) have been achieved. This has become possible because genetic code is assumed to be universal.

Genetic engineering may or may not have recombinant DNA (rDNA) preparation step and with the advancement of gene transfer technology, artificially DNA can be transferred to different hosts (host cells) without any vector and host organism can be engineered to carry desired properties. The genetic manipulations are used to produce individuals having a new combination of inherited properties. Such manipulations may be of two kinds: (1) cellular manipulation involving culturing of cells (e.g., haploid cells) and hybridization of somatic cells (protoplast fusion), and (2) molecular manipulation, involving construction of artificial rDNA molecules, their insertion into a vector and their establishment in a host cell or organism. The latter approach has been called "recombinant DNA (rDNA) technology". Therefore, use of term "genetic engineering" and "rDNA technology" is overlapping. However, all the manipulations involving use of constructed gene or constructed gene transfer are termed as rDNA technology.

I. GENE CLONING

Any gene to be cloned must be inserted in a cloning vector (plasmid). A foreign gene (DNA fragment) introduced (by transformation) into a bacterium cell will not be replicated with bacterium. The reason for this is that the enzyme DNA polymerase, which is responsible for copying DNA, does not initiate the process at random but at selected sites known as "origin of replication". Generally, small fragments of DNA do not possess an origin of replication. Using rDNA technology, it is possible to insert the gene into a 'cloning vector', which in turn will make copies of the fragment (inserted DNA). A cloning vector is simply a DNA molecule possessing an 'origin of replication' and which can replicate in the host cell of choice. Most commonly 'plasmids', extra chromosomal, autonomously replicating, circular DNA molecules, are used as vectors. Sometimes, viruses are used as vector for gene insertion into microorganisms, but they are better vectors for animal cells.

Cutting and insertion of desired foreign gene into the plasmid require special enzymes known as restriction endonucleases or restriction enzymes. These enzymes cut large DNA molecules into shorter fragments by cleavage at specific nucleotide sequences called 'recognition sites'. Therefore, restriction endonucleases are highly specific deoxy-ribonucleases (DNase). Both vector DNA and foreign DNA to be inserted is cut by the same restriction enzyme, generating complementary ends. Thus, ends of foreign DNA make perfect match with cut ends of vector and join to make again a circular molecule. This chimeric recombinant DNA (vector) is inserted into the host cell (bacterium) by the process of transformation. Theoretically it is possible to clone a desired gene of foreign origin into host cell (Fig. 5.1). Normally, expression of a gene is observed in the form of gene product, a protein. This involves formation of mRNA and related polypeptide. In many eukaryotic genes, there is an additional step before translation, the removal of non-coding sequences from mRNA by a process known as 'splicing'.

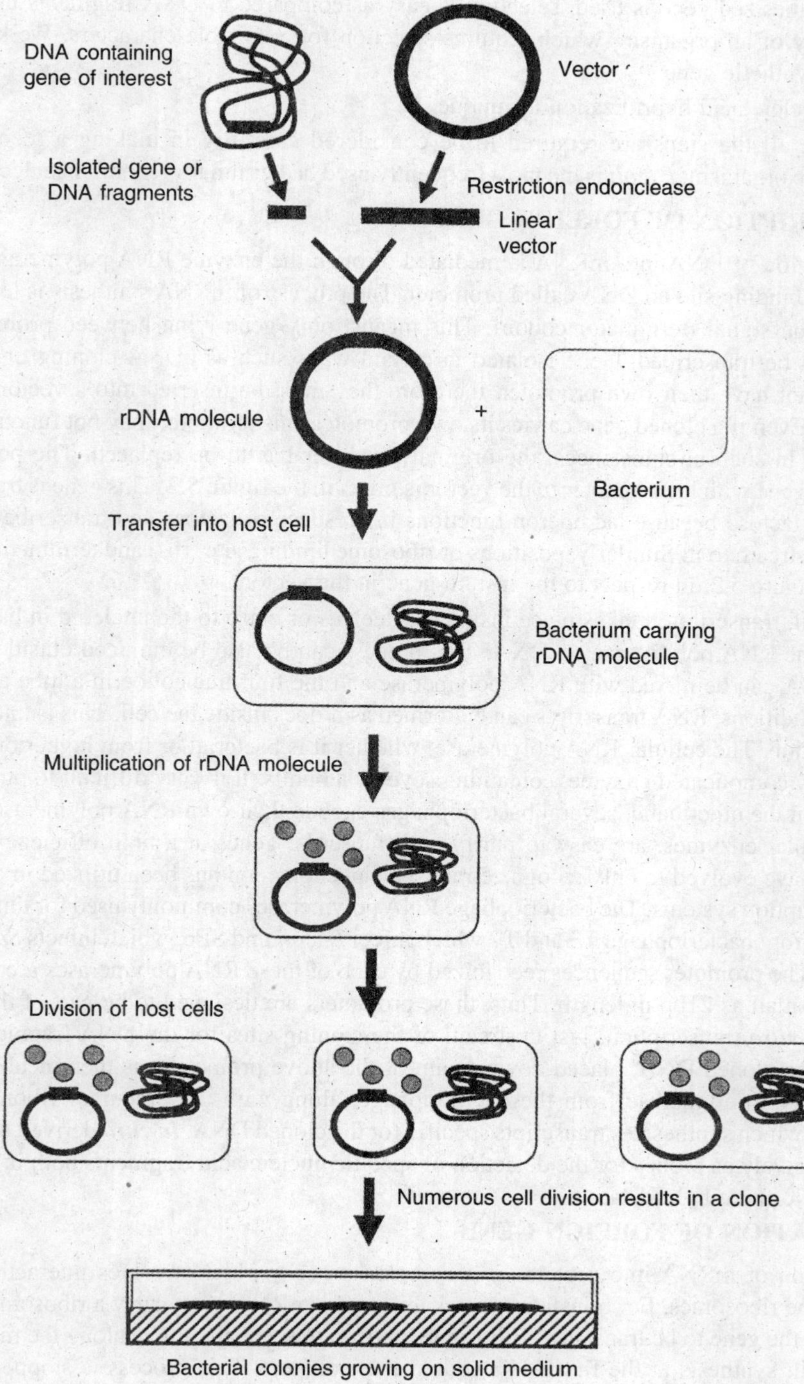

Fig. 5.1. Construction of rDNA and recombinant bacteria

II. SELECTION OF RECOMBINANT BACTERIA

Different methods of selection are used to facilitate the selection of recombinant bacteria. These are:

1. Complementation of genetic defects in nutrition; growth or no growth on a suitable selective medium.

2. If synthesized gene is used, selection is easy as compared to DNA fragments used from genomic library of an organism, which requires selection for a suitable characters. We know the properties of synthetic gene.

3. By nucleic acid hybridization technique.

Therefore, all the steps are required to be considered carefully in making a recombinant prokaryotic organism. *E. coli* is the most frequently used bacterium for recombinant technology.

III. TRANSCRIPTION OF FOREIGN GENE

Transcription of DNA into mRNA is mediated through the enzyme RNA polymerase, which recognizes the binding site on DNA called promoter. The process of mRNA synthesis is terminated by a termination signal (terminator codon). This means, only gene lying between promoter and terminator will be transcribed. Gene isolated in certain ways such as cDNA cloning or artificial synthesis, do not have their own promoter, therefore they must be inserted into a vector close to promoter site. Even if a cloned gene carries its own promoter, this promoter may not function in the new host cell. In such circumstances, the original promoter has to be replaced. The position of cloned insulin gene with lac promoter in the vector is given in the figure 5.2. This gene is transcribed in presence of lactose because lac operon functions in presence of lactose and transcribe the gene attached downstream to it. Similarly, positions of ribosome binding site (rbs) and termination codon are shown in figure 5.2, in respect to the insulin gene in the vector.

In the cell, transcription takes place inside the nucleus or close to the nucleoid in bacteria by the action of the RNA polymerase on DNA. This process can be also be mimiced outside the cell, i.e. cloned DNA can be mixed with RNA polymerase and the four nucleotide in a tube and under appropriate conditions, RNA transcripts can be formed as it does inside the cell. This is known as *in vitro* transcription. The cellular RNA polymerase, whether it is bacterial or from higher organisms, is an extremely complicated enzyme, containing several subunits. It is very difficult to purify in an active form. On the other hand, several bacteriophages encode their own RNA polymerases, which are much simpler enzymes, are easy to purify and transcribe genes at a high efficiency. This is because they have evolved to only recognize rate. This phenomenon has been utilized in designing *in vitro* transcription systems. The bacteriophage RNA polymerases commonly used for this purpose include those from bacteriophage T3 and T7 which infect *E. coli* and SP6, which infects *Salmonella typhirmurim*. The promoter sequences recognized by each of these RNA polymerases are different, but can be as small as 21bp in length. Thus, these promoters are designed to be part of the special cloning for *in vitro* transcription, just upstream of the cloning sites for the DNA fragments to be transcribed. The cloned DNA, placed downstream of the above promoters, is then incubated with the purified RNA polymerase from the bacteriophage, along with the precursor ribonucleotide triphosphates, which synthesizes transcripts specific for the cloned DNA. *In vitro* derived transcripts are used extensively as probes for the detection of specific nucleic acid fragments both in Southern as well as in Northern hybridization.

IV. TRANSLATION OF FOREIGN GENE

Translation of mRNA into proteins is a complex process which involves interaction of the mRNA with the ribosomes. For translation to take place the mRNA must carry a ribosome binding site in front of the gene to be translated. Ribosome binds to this site and move along the mRNA and initiates protein synthesis at the first AUG codon, it encounters. This process is stopped by stop codon (UAA, UAG, UGA). In case the cloned gene lacks a rbs, then it is necessary to use a vector with promoter and rbs; and the gene is inserted down stream to both these (promoter and rbs).

Translation in a cell is carried out by the ribosomes, which synthesize polypeptides by decoding the information carried by mRNA. In addition, amino-acyl tRNAs and other proteinaceous accessory factors are also utilized. Biochemically the process of translation is not yet fully characterized, which means that we do not know all the requirements for polypeptide synthesis and very few of

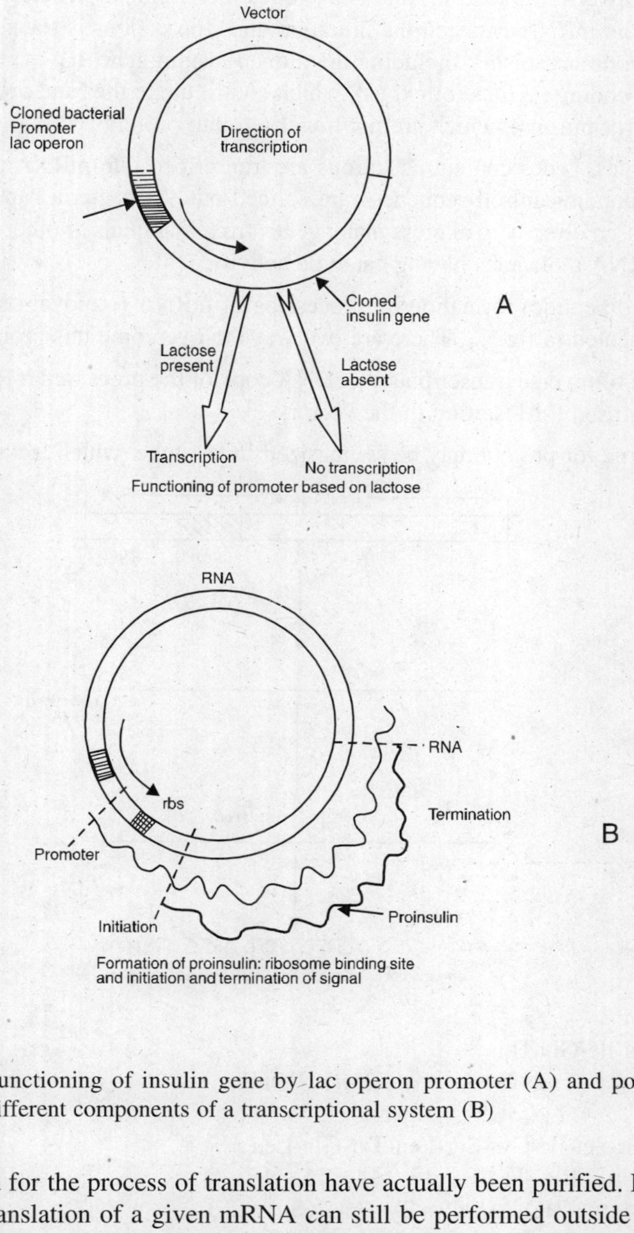

Fig. 5.2. Functioning of insulin gene by lac operon promoter (A) and position of different components of a transcriptional system (B)

the components required for the process of translation have actually been purified. Despite having the above drawbacks, translation of a given mRNA can still be performed outside the cell, if the mRNA is incubated in a mixture of components which supports translation. This is termed *in vitro* translation. The above components are generally isolated from *E. coli* cells (S-30 fraction), plants (wheat germ) or animals (rabbit reticulocytes). Because of the universality of the genetic code a given mRNA gets translated to the same extent and with the same efficiency irrespective of the *in vitro* translation system used. Since there are already many types of protein molecules in the translation mix, new protein synthesis is usually monitoring by adding radioactive amino acids in the translation mix. Thus, all newly synthesized proteins get radioactively labelled and can be easily detected by autoradiography. In many instance, a combined *in vitro* transcription and translation system is used which can produce the encoded polypeptide directly from a given cloned gene. Such systems are now commercially available.

Splicing mRNA : Bacterial and viral genes have simple structure in that all the genetic information in the mRNA between the initiation and stop codons is translated into protein. Many genes of eukaryotic organisms, including the human insulin gene, have a more complex structure. They consist of coding regions (or exons), which contribute to the final protein sequence, and non-coding regions (or introns), which are not translated into proteins.

In eukaryotes, genes containing introns are transcribed into mRNA in the usual manner, but then the corresponding intron sequences are spliced out. As bacteria can not spliced out introns, they cannot be used directly to express many genes from mammals or other eukaryotes. This is done by ribozymes-RNA molecules having catalytic activity.

Insulin (a dipeptide) formation by processing of mRNA (removal of introns) and translation of exons is presented in fig.5.3. There are two ways to overcome this problem.

1. By use of reverse transcriptase, a cDNA copy of the processed mRNA in prepared and this cDNA (gene) is used for insertion in the vector.

2. The gene for protein may be synthesized in test tube, which lacks introns.

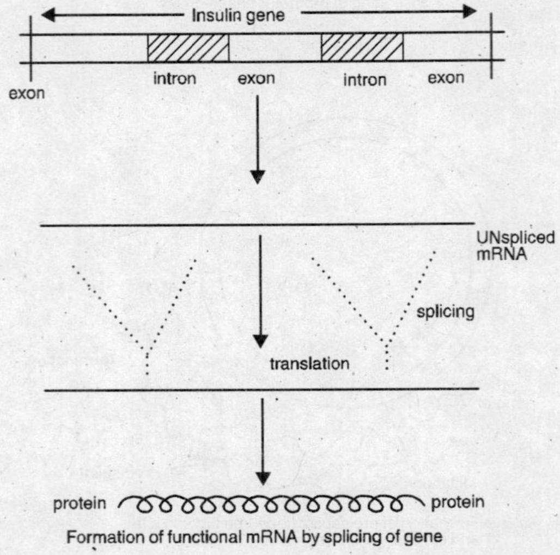

A

Gln-Glu-Val-Ile-Gly-H

Pork Insulin

5 4 3 2 1

Cys-Cys-Thr-Ser-Ile-Cys-Ser-Leu-Tyr-Gln-Leu

6 7 8 9 10 11 12 13 14 15 16

HO-Asn-Cys-Tyr-Asn-Glu

Chain A 21 20 19 18 17

Cys-Gly-Ser-His-Leu-Val-Glu-Ala-Leu-Tyr-Leu-Val-Cys

7 8 9 10 11 12 13 14 15 16 17 18 19

Leu-His-Gln-Asn-Val-Phe-H Phe-Gly-Arg-Glu-Gly

6 5 4 3 2 1 24 23 22 21 20

Chain B

Phe-Tyr-Thr-Pro-Lys-Ala-OH

25 26 27 28 29 30

B

Fig. 5.3. Processing of mRNA for insulin by splicing (A) and insulin A and B chains and their amino acid sequences (B)

V. POST-TRANSCRIPTIONAL MODIFICATION:

A number of proteins under go post-transcriptional modifications. Protein that are destined to be transported out of cell are synthesized with an extra 15-30 amino acids at the amino terminal (N-terminus). These extra amino acids are referred to as a signal sequence. The common feature of these sequences is that they have a central core of hydrophobic amino acids flanked by polar or hydrophilic residues. During passage through the membrane the signal sequence is cleaved off, making the protein active.

VI. EXPRESSION IN DIFFERENT HOSTS

Expression of desired gene in different host organisms may require modification in gene construct and promoter sequences. Problems associated with expression of eukaryotic gene in a prokaryote have to be addressed before achieving complete expression.

(a) Expression in Bacteria

Several strong and inducible promoters from bacterial systems have been well characterized and many of them have been used directly to express heterologous genes in bacterial systems. Some of the above promoters have also been used to create synthetic promoters, which are better suited for gene expression as compared to the natural ones, for example, the synthetic tac promoter utilizes the lac and trp promoters from *E. coli,* pl promoter from bacteriaophage λ and gene 10 promoter from bacteriophage T7. Since bacteria lack the capacity to process introns, the heterologous gene to be expressed should be free from introns. Such as gene is then cloned downstream of the appropriate promoter. In order to easily purify the expressed protein, the heterologous gene is usually cloned in a way, which produces a fusion protein. This is then affinity purified and cleaved with a protease to obtain the pure heterologous protein. With most bacterial expression system, it is possible to obtain the heterologous protein at a concentration of 1-2 % of the total soluble protein of the cell.

(b) Expression in Yeast

A large number of post-translational modifications generally take place in most of the eukaryotic proteins, which do not take place in bacteria. Thus, sometimes a eukaryotic expression system has definite advantages over a prokaryotic system. Yeast cells have the advantage of being amenable to the application of powerful microbial genetic selection techniques on them along with their eukaryotic nature. Several expression systems have been devised for expression heterologous genes in yeast. Yeasts generally used in the laboratory include the baker's yeast, *Saccharomyces cerevisiae* and other yeasts like *Kluvyeromyces lactis*, *Schizosaccharomyces pombe*, *Pichia pastoris* etc. In most cases, a shuttle expression vector is used, which is a plasmid capable of existing both in *E. coli* as well as yeast cells. This is possible because these plasmids are artificially created by joining a small fragments of DNA, known as the yeast 2μm fragments, into a bacterial plasmid. The above 2μm fragments required replication properties in that system. The expression of the gene takes place under the control of a cloned yeast promoter like that of glyceraldehydes phosphate dehydrogease. The plasmid also contains a yeast gene for example *leu*, which can enable the transformed *leu* yeast strain to grow on a medium lacking leucine. It turn provides a powerful method for selection of transformed yeast cell. The plasmids are also integrated into the yeast chromosome by homologous recombination which ensures stable maintenance of the heterogonous gene. Yeasts like *P. pastoris* can grow on methanol as the sole source of carbon and energy and thus can be grown very economically in large bioreactors. In addition, they can be made to secrete the heterologous protein into the medium, which makes them ideal for continuous purification from the bioreactors.

Expression in Insects. For expression of heterologous genes in insects, the baculovirus system is very popular. Baculoviruses infect a large number of insects and are characterized by the high expression of one of their genes, the polyhedron gene. The promoter of the polyhedron gene becomes active 4-5 days after infection in the insect cells and can be used to express heterologous genes in a very efficient manner. The baculovirus *Autographa californica* multipole nuclear polyhedrosis virus (AcMNPV) has been used extensively as an insect expression vector on both cultured cells as well as live larvae of the Fall Armyworm *Spodoptera frugiperda*. Expression vectors for the above are based on *E. coli* plasmids, containing a fragment of AcMNPV to provide sites for homologous recombination, the polyhedron promoter and the termination sequences. Heterologous gene to be expressed is inserted between the above promoter and terminator sequences and introduced into insect cells or larvae, along with a full AcMNPV DNA. Because the heterologous gene is now flanked by the polyhedron expression signals, it can start expressing instead of the polyhedron gene at high levels in the insect cells.

(d) Expression in Mammalian Cells

For expression heterologous genes in mammalian cells, usually vectors derived from mammalian viruses are used. These include viruses such as Simian Virus 40, polyomavirus, herpesvirus and papovavirus. One of the earliest viruses to be used as expression systems was the human papovavirus BK, which is a double stranded DNA virus, which can exist as an extra chromosomal element in cultured human cells. As is true for any vector derived from viruses, the pathogenicity genes had to be deleted, so that only the replication associated genes and those for high copy number were retained. To incorporate a strong promoter, one from mouse mammary tumour virus was incorporated to drive strong expression of transgene. The vector system also needed a selection marker, which was incorporated as neomycine resistance gene, providing resistance against the cytotoxic compound G418. Such expression vectors can be used to express heterologous proteins in cultured mammalian cells. Normally the level of expression is modest but it becomes important in case of proteins which can be correctly folded or processed only in human cells. Thus, it finds applications under special circumstances.

(e) Expression in Plants

For expression heterologous proteins in plants, the unique capacity of a soil bacterium *Agrobacterium tumefaciens* to introduce part of its resident plasmid into the plant genomic DNA has been widely utilized. The above bacteria causes a cancerous growth in the infected plant due to the introduction of this DNA fragment derived from the plasmid. The plasmid is called the Ti plasmid and the introduced fragment of DNA is called T-DNA. The tumorigenic genes, i.e., those which cause the tumour are deleted from the T-DNA and then the plasmid is incapable of causing tumours but is still capable of transferring T-DNA to the plant. Ti plasmid and T-DNA have been dealt. Any heterologous DNA can be introduced into the plant as part of the above plasmid. To ensure high expression of the heterologous gene, strong promoters derived from plants viruses, like the cauliflower mosaic virus are generally used. Several heterologous genes have been expressing plants in the recent years either to improve the field performance of crops or to impart added qualities to the edible part, like essential vitamins or even antigenic, proteins, which can serve as vaccines against common disease like cholera. The low cost of growing such plants with expression heterologous proteins, some of which can have high commercial values, has made expression of such proteins in plants a very attractive proposition. Several plants viruses can infect and multiply in their plant hosts to a very high level at a very short time. Some of these viruses, like the tobacco mosaic virus have been used to clone heterologous genes by replacing some of the non-essential viral genes. Such engineered viruses when allowed to infect plants, are capable of expressing the heterologous gene to very high levels in a matter of few weeks in the infected plants.

QUESTIONS

1. What is recombinant technology?
2. Write short notes :
   - (*a*) rDNA
   - (*b*) cDNA
   - (*c*) Introns removal
   - (*d*) Translation of recombinant gene
   - (*e*) Restriction endonuclease
   - (*f*) Transgene expression in bacteria
   - (*g*) Transgene expression in yeast
   - (*h*) Transgene expression in plants.

GENE INTEGRATION AND EXPRESSION VECTORS

CHAPTER 6

The main objective of gene cloning for biotechnological applications is the expression of the cloned gene in a selected host organism. The insertion of a gene into a cloning vector does not necessarily ensure that it will be successfully expressed. Generally, transferred gene must be expressed at high rate to produced desired protein in large quantity. This requires high expression rate, which can be achieved by highly expressive promoter. This in turn achieved by manipulating different genetic elements that control aspects of transcription, translation, protein stability, oxygen limitation and secretion from host cell. The characteristics which need to be changed for better expression are :

(1) The nature of the relevant transcription promoter and terminator sequences,
(2) The strength of ribosome binding site,
(3) The number of copies of cloned gene and whether the gene is plasmid borne or integrated into the genome of the host cell.
(4) The final cellular location of the synthesized foreign protein,
(5) The efficiency of translation in the host organism, and
(6) The intrinsic stability of the cloned gene protein within the host cell.

There is no common method to achieve maximal expression of every cloned gene. Selection of host organism is also equally important. With the available current technology, a wide range of prokaryotic and eukaryotic cells can express foreign genes, *E.coli* is the most frequently used organism for commercial production of recombinant protein. Because of vast information available on *E. coli*, proteins can be produced rapidly and inexpensively in this organism. The other system used are yeast, *Bacillus subtilis,* animal, plant and insect cells.

I. GENE EXPRESSION FROM STRONG AND REGULATABLE PROMOTER

The essential requirement for an effective gene expression system is the presence of a strong and regulatable promoter sequence upstream from a cloned gene. A strong promoter is one that has a high affinity for RNA polymerase, resultantly the adjacent downstream gene is highly transcribed. If we can regulate the promoter, we can regulate the gene transcription. Generally, 'lac operan' (lac) promoter of *E.coli* is used for expression of cloned genes. Generally a reporter gene is inserted between useful gene and promoter to notice the activity of promoter and expression of gene. It is also practiced to insert closed gene in a plasmid under the control of a continuously transcribed strong promoter. However, a high level of continuous expression may detrimental for host organism as it will exhaust the organism impairing other essential functions. The loss of plasmid during cell division is another drawback of this system.

Regulatable promoters. The most widely used strong promoters are *lac* and *trp* (tryptophan) promoters from *E.coli*. The *lac* promoter is an in vitro construct that includes (1) the -10 region (i.e. 10 nucleotide pair upstream from the site of initiation of transcription) of the *lac* promoter and -35 region of the *trp* promoter, (2) the leftward or pL promoter from bacteriophage λ and (3) the gene 10 promoter from bacteriophage T7. Each of these promoter interacts with the repressors, which provide a controllable switch for either switching on or switching off specific transcription of adjacent cloned gene.

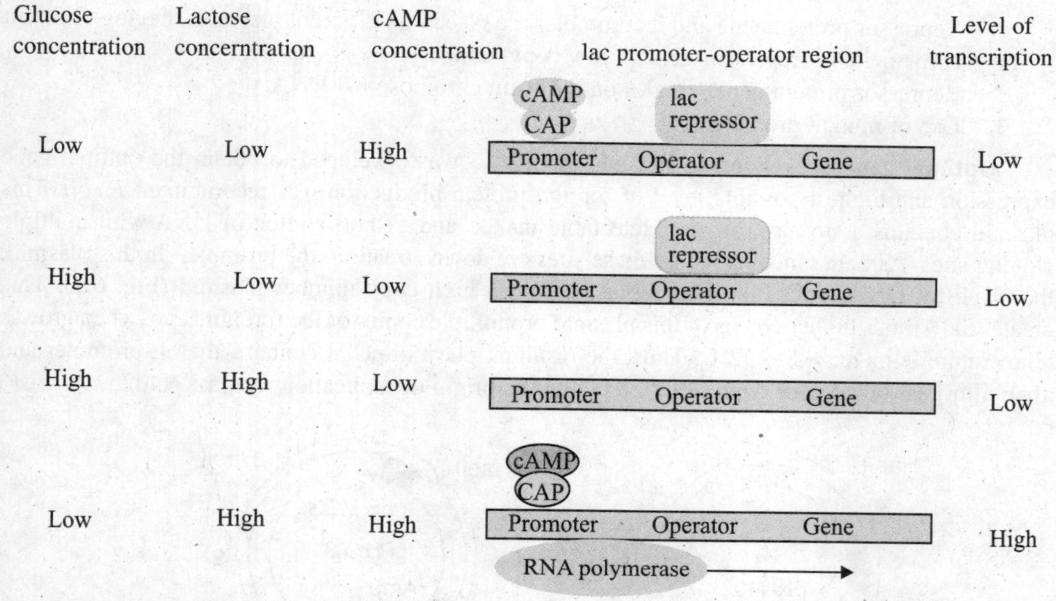

Fig. 6.1. Diagrammatic representation of the effect of the concentrations of glucose, lactose, and cAMP in the growth medium on the level of transcription from the *E.coil* lac promoter. The arrow indicates the direction of transcription.Adapted from Abeles et al., 1992, Biochemistry, p. 383, Jones & Bartlet Publishers, Boston, Mass.

Each of these promoters is recognized by *E.coli* RNA polymerase. The *lac* promoter is 'switched off' in the absence of lactose while 'switch on' in the presence of lactose in the medium. Thus , lactose acts as switch to transcribe a gene attached to *lac* promoter. This can also be performed by isopropyl-B-D thiogalactopyranoside (IPTG). Transcription from the *lac* promoter is also regulated by the binding of the catabolic activator protein (CAP) to the promoter region. In practice, *lac* UV5, (a variant of the lac promoter that contains an altered nucleotide sequence in the – 10 region and is a strong promoter than the wild-type *lac* promoter) is usually used in plasmid expression vectors.

The *trp* promoter is negatively regulated or turned off by a tryptophan-trp repressor protein complex that binds to the *trp* operator (prevents transcription). Derepression (turning on) of the trp promoter is achieved either by removing tryptophan or by adding 3-indole acrylic acid to the growth medium.

The pL promoter is controlled by the *cI* repressor protein of bactoriophage λ. In practice, a temperature-sensitive mutant of the *cI* repressor, cI857, is generally used to regulate pL directed transcription. Cells with this type of promoter are grown at 28-30°C and later on at 42°C, where thermo-sensitive *cI* repressor is inactivated and transcription take place.

The bacteriophage T7 promoter requires T7 RNA polymerase for transcription. To utilize this promoter, the T7 RNA polymerase gene is inserted in the *E.coli* chromosome – under the control of *lac* promoter. Cells containing a target gene under the control of the T7 promoter, IPTG is added to the medium. Under these conditions, the T7 RNA polymerase is produced and the closed gene is transcribed and translated.

Transcription of cloned gene depends upon the ratio of number of copies of promoter sequences and number of repressor protein molecules. There should be balance between these two, with low repressor molecules. In presence of too many repressor molecules it is difficult to induce transcription. If repressor molecules are at low number, gene may be transcribed even in the absence of inducer (known as 'leaky' promoter). Many alternatives are suggested to overcome this problem-

1. Repressor protein gene and its promoter are placed on different plasmids, having different copy member (represor gene in low copy number).
2. Repressor protein gene (single) on the main chromosomal DNA.
3. Use of mutant promoter.

Optimal gene expression – Plasmid pPLc2833 was developed to obtain the optimal gene expression and highest possible level of foreign protein production in a recombinant *E.coli.* This plasmid contains a strong promoter, selectable maker, and a short stretch of DNA with multiple cloning sites. As mentioned above all these sites are down stream to the promoter. In this plasmid, the origin of DNA replication was replaced from a high copy number plasmid (Fig. 6.2). This creates 5-10 times higher copies of this plasmid or multiple copies of the foreign gene, when growth temperature is increased to 42°C. Thus, the resultant plasmid pCP3 contains the pL promoter and ampicillin resistance gene from pPLc 2833 and the origin of replication from pKN402.

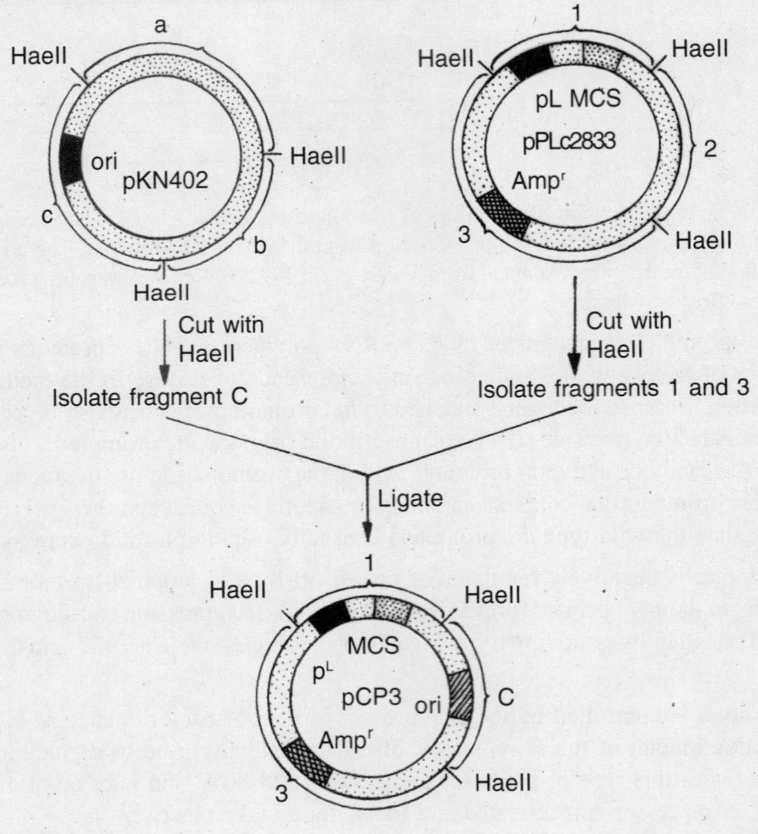

Fig. 6.2. Formation of plasmid pCP3. Restriction enzyme Hae II fragment which contains the temperature-sensitive replication origin (ori) of plasmid pKN402, was isolated and ligated to HaeII fragments 1 and 3 from plasmid pPLc2833. pPLc2833 HaeII fragment 1 contains the p^L promoter and a multiple cloning sequence (MCS). pPLc2833 HeaII fragment 3 contains the selectable marker gene for amplicillin resistance (Ampʳ).

Cells that carry the plasmid pCP3 are first grown at 28°C and then grown at 42°C. At lower temperature, *cI* repressor is functional, the pL promoter is turned off and plasmid copy number is normal. At the higher temperature, the temperature-sensitive *cI* repressor is inactivated, the pL promoter is active, and plasmid copy number increases. These properties make pCP3 an effective expression vector, producing about 20% foreign protein out of total protein.

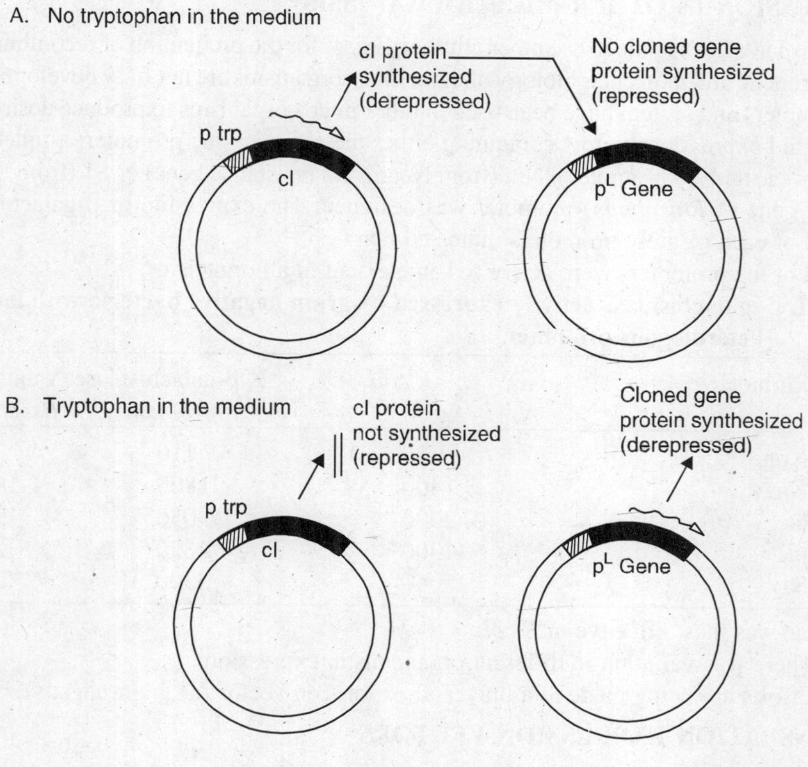

Fig. 6.3. Dual plasmid system for controlling the λ *p*Lpromoter by regulating the cI repressor with tryptophan. The tryptophan promoter (*p trp*) is inserted next to the cI gene on one plasmid, and the p*L* promoter is placed adjacent to the cloned gene (Gene) on a second plasmid. The wavy arrows denote transcription. (A) With no tryptophan in the medium, the cI gene is transcribed and translated, and the cI repressor protein binds to the pL promoter, thereby blocking the transcription of the cloned gene. B. In the presence of tryptophan, the cI gene is repressed, no cI product is made, and the cloned gene is transcribed and translated.

Bioreactor cultivation – For industrial production cultures require growth in high volume bioreactors. In small culture vessels of 1-5 litres, gene expression can be triggered either by changing the temperature or by addition of a chemical inducer. In pilot plant-sized (20-200L) and industrial-sized (>200L) bioreactors, however, a shift in temperature requires time (30-60min) and energy, both of which are costly. Similarly cost of a chemical inducer like IPTG can make the overall process uneconomical (requires in large quantity for bioreactor volume). To overcome some of the problems associated with the use of the pL promoter for large-scale fermentations, a two-plasmid system has been developed. The *cI* repressor was placed under the control of the *trp* promoter and inserted into a low-copy number plasmid Fig. 6.3. The use of a low-copy number plasmid ensures that excess *cI* repressor molecules are not produced. A second plasmid carries a cloned gene under the control of the pL promoter. The *trp* promoter is turned on in the absence of tryptophan so the cI repressor protein is synthesized and the pL promoter is turned off. In comparison to this, the *trp* promoter is turned off in the presence of tryptophan, so the cI repressor protein is not synthesized and the pL promoter is fully active. Using this two plasmid strategy, recombinant *E.coli* can be grown in a medium with tryptophan working as on/off switch. *By this method 21 and 24% of cellular protein was recombinant protein*. Thus, this system provides a potentially inexpensive means of producing proteins from recombinant microorganisms on a large-scale.

II. EXPRESSION IN OTHER MICROORGANISMS

Efforts have been made to use other microorganism for the production of recombinant proteins. However, genetic and molecular biology of other microorganisms are not fully developed. Therefore, *E.coli* promoters and system have been used in other micro-organisms to produce desired protein. A set of plasmid expression vectors containing either the *lac, tac*, (tac promoter: a functional hybrid promoter of *lac* and *trp* promoters) Nm (from Neomycin resistance gene) or S1 (from the ribosomal protein S1 gene of *Rhizobium*) promoter was designed. The expression of β-galactosidase under the control of each of these promoters indicated that –

1. All of the promoters were active to some extent in all organisms.

Table – 6.1. β- galactosidase activity expressed by gram negative bacteria with lac z gene and heterologous promoter.

Promoter	*E.coli*	β-galactosidase Δ units	
		Rhizobium	*Pseudomonas*
None	16	110	150
Nm	1400	21800	16300
lac	2000	9050	9800
tac	11300	2850	2950
S1	40	3300	3350

2. tac was most effective in *E.coli*.
3. There was variation in different organisms in expression.
4. Efforts are being made to a universal expression vector for gram-negative bacteria.

III. TRANSLATION EXPRESSION VECTORS

Transfer of a cloned gene with a regulatable, strong promoter may not necessarily sufficient for high yield of the cloned gene product. The other factors influencing the product are efficient translation and products stability. In prokaryotic cells, various mRNA are not translated with the same efficiency resulting in variation in number of protein molecules formed by them. This number may vary from a few to several hundreds.

To some extent, basis for differential translation is the presence of a translation initiation signal called a ribosome binding site (rbs) in the transcribed mRNA. A rbs is a sequence of 6-8 nucleotides in mRNA that base pair with a complementary sequence on the RNA component of the small ribosomal subunit. Stronger binding of the mRNA to rRNA results in efficient translation. Therefore, *E.coli* expression vectors have been incorporated a strong rbs. This means that heterologous pro- and eukaryotic genes can be translated in *E.coli*. However, improvement in some other conditions is also required. These are –

1. The rbs sequence must be located at precise distance from the translational start codon of the cloned gene.
2. The DNA sequence that includes the rbs through the first few codons of the gene of interest must not contain nucleotide sequences that after transcription can fold back (form intra-strand loop). This interferes with ribosome binding. Thus, cloned gene should be free from these two problems.

> A ribosome binding site (rbs) is a sequence of 6 to 8 nucleotides e.g., U A A G G A G G. At the RNA level, the translational codon is usually AUG. At the DNA level, the strand that has the ATG sequence is called the coding strand (same sequence), while the complementary strand, which acts as the template for transcription, is called noncoding strand. By convention, a start codon at the DNA level is designated as ATG. (Fig. 6.4).

A number of vectors have been developed with proper transcriptional and translational signal for the expression of cloned eukaryotic genes in *E.coli.* Such a vector has following characteristics, e.g., pkk 233-2.

- An ampicillin resistance gene as a selectable marker.
- The *tac* promoter
- The *lac* Z rbs.
- An ATG start codon located at 8 nucleotides downstream from the rbs.
- The transmission terminators T1 and T2 from bacteriophage λ.

A cloned gene is inserted at any restiriction site between rbs and the transcription terminators. After induction and transcription, mRNA of a cloned gene is efficiently translated.

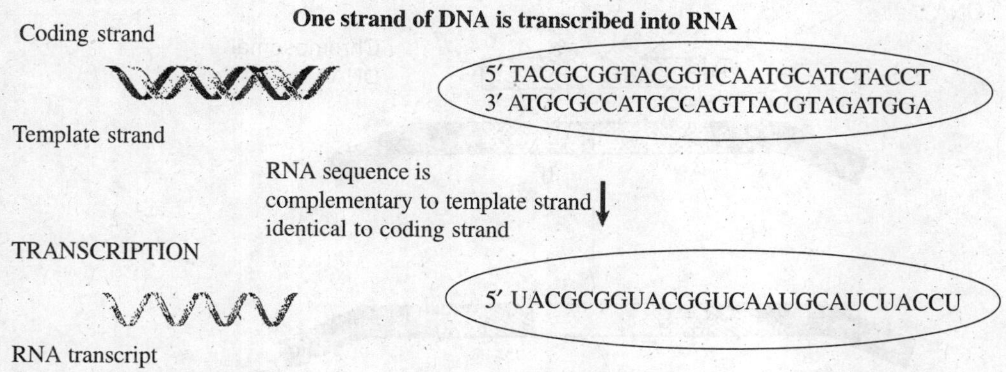

Fig. 6.4. The function of RNA polymerase is to copy one strand of duplex DNA into RNA

Under normal conditions, half-lives of protein formed ranges from a few minute to hours. This difference is due to disulfide bond formation and to the presence of certain amino acids at the N-terminus.

DNA integration into the host chromosome

Plasmid carrying a desired gene, when transferred to host cell, consumes host resources also for other genes located on it. A high copy number plasmid consumes more resources than a low copy number plasmid. Therefore, cells containing plasmid tends to loss plasmid and after a few generations, a mixture of cell population develops, with or without plasmids. Cells grow faster without plasmids, than those containing plasmid. Resultantly, after a few generations all host cells may loss plasmid. (No more plasmid, no recombinant protein production).Two approaches have been adapted to meet this problem.

1. On small-scale, plasmid-containing cells are maintained by growing cells in the presence of either an antibiotic or an essential metabolite that allow only plasmid-containing cells to grow. But addition of such compound at large-scale is very costly thus making the process uneconomical.
2. It is also important for the microorganisms that are released in the environment that they don't loss their plasmid. This can overcome by introducing the desired gene directly into host-genome. (chromosomes). This will overcome the problem of plasmid loss. When foreign DNA is part of the host chromosomal DNA, it is relatively stable for several generations and doesn't require selection agent (antibiotic).

When a cloned gene is integrated into host chromosome, it should not be placed close to an essential codings sequence. For this, the DNA sequence must be targeted to a specific non-essential site within the chromosome. To have proper protein production, a regulatable promoter can be attached to desired gene. The desired gene is integrated by homology in base pairs. About 50 bp homology is required for physical exchange (recombination) between the two DNA molecules.

For this purpose, a segment of host chromosomal DNA is selected, isolated and cloned. A desired gene is inserted with a regulatable promoter. This whole segment is transferred on host-chromosome on the basis of homology in sequence. The integration can occur as a result of a host enzyme-catalyzed double cross-over. Insertion of the entire input plasmid DNA occurs when the cloned gene is inserted next to the closed chromosomal integration site (Fig. 6.5).

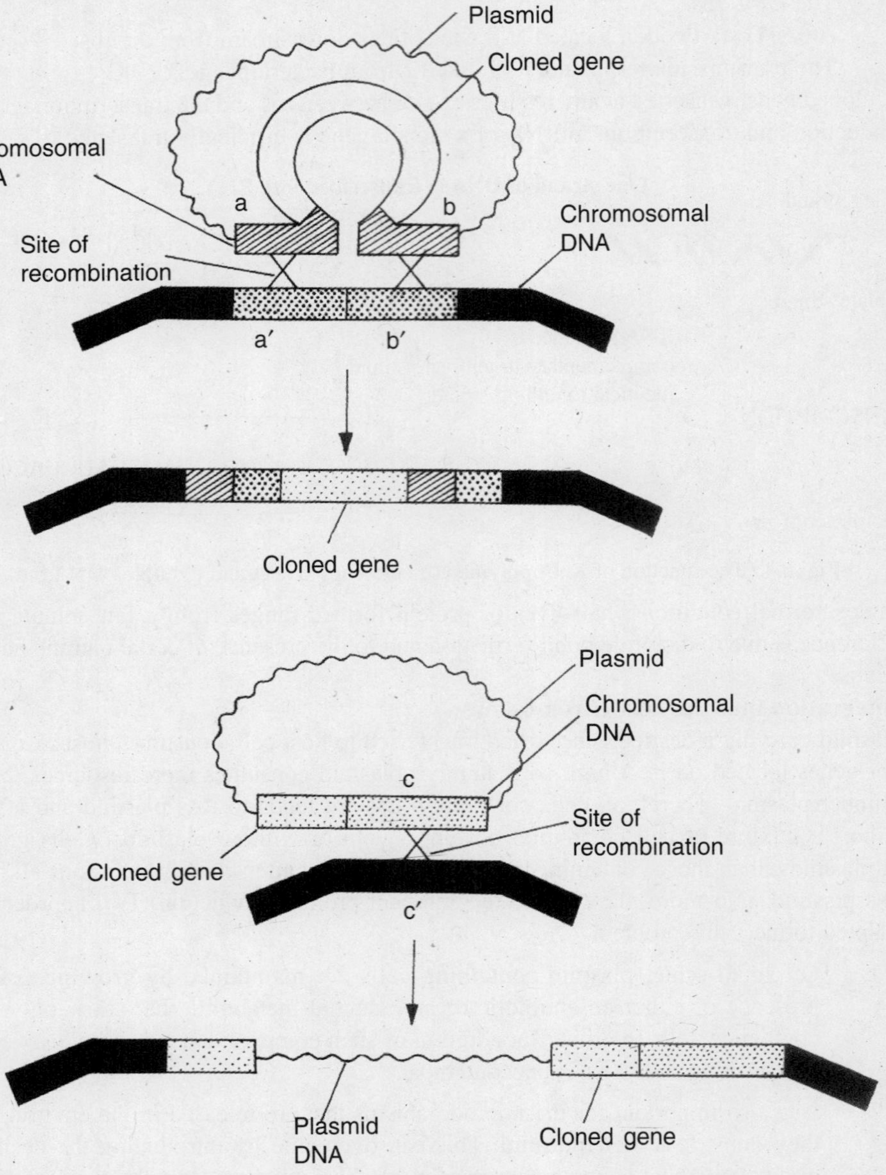

Fig. 6.5. Two ways to integrate a cloned gene into a chromosomal site. A. On a plasmid, the cloned gene has been inserted in the middle of a cloned segment of DNA (ab) from the host chromosome. Homologous DNA pairing occurs between plasmid-borne DNA regions a and b and host chromosome DNA regions a′ and b′ , respectively. A double-crossover event (X–X) results in the integration of the cloned gene. B. The cloned gene is inserted adjacent to the cloned DNA (c) from the host chromosome. Homologous DNA pairing occurs between plasmid DNA region c and host chromosome DNA region c. A single recombination event (X) within the paired c-c DNA region results in the integration of the entire plasmid, including the cloned gene.

IV. INCREASING SECRETION

It is of interest if protein produced by *E.coli* is secreted in the medium and remains stable. Protein secreted in periplasm (space between cytoplasmic membrane and cell wall), secreted proteins are easier to purify. Signal sequences located at the N-terminal of end of a peptide are responsible for (see protein targeting) secretion of proteins. Therefore, proteins can be designed (by modifying the gene) to be secreted out of bacterial cell. There are two options for releasing protein in the medium, use of gram-positive bacteria or genetically engineered gram-negative to secrete proteins (generally gram-negative do not release as having outer membrane).

Production in Eukaryotic cells

Generally, prokaryotic expression systems are used for production of recombinant proteins from closed eukaryotic cDNAs. However, this encounter some problems like instability of protein, lack of biological activity, may contains bacterial toxins (pyrogens) and post-translational modifications. Therefore, some eukaryotic systems have been developed for the production of therapeutic proteins. In eukaryote cells following changes occur in newly formed protein :

- Correct disulphide bond formation. Folding carried out by disulphide isomerase, improper folding makes protein unstable.
- Proteolytic cleavage of precursor form. Some amino acid sequences are removed to make functional protein.
- Glycosylation-addition of glucose molecule to amino acids (serine or threonine or asparginine) makes stable and imparts specific property.
- Modifications in amino acids- like phosphorylation, acetylation, sulfation, addition of fatty acids.

Prokaryotic cells are not capable of doing such modifications. Similarly, all these modifications are not possible by a single eukaryotic cell for every heterologous proteins. Therefore, for different proteins different organisms are evaluated. Similar to prokaryotic expression vectors, eukaryotic expression vector also contains :

- A selectable eukaryotic marker gene.
- A eukaryotic promoter sequence.
- Eukaryotic transcriptional and translational stop signals.
- A sequence that signals polyadenylations of the newly formed RNA.

When a plasmid is used as a vector, then it must have an origin of replication that functions in the host cell. If a vector is designed for integration into the host chromosomal DNA, then it must have a sequence that is complementary to a segment of host chromosomal DNA, to facilitate recombination. Looking to the complex nature of eukaryotic cells and vectors, shuttle vectors are designed for eukaryotic cells. These vectors have origin of replication and selectable markers both for *E. coli* and eukaryotic cell or yeast.

Introduction of foreign DNA into bacterial and yeast cells is called transformation, in animal cells transfection (In animal cells transformation is used for conversion of a normal cell into a tumour cell). Three steps are used to transform yeast cells :

1. Isolation of yeast protoplast.
2. Treatment of cells with lithium acetate.
3. By electroporation.

As compared to this transfection of cultured animal cells is achieved by :

1. Incubation of cells with DNA that has been co-precipitated with either calcium phosphate or DEAE-dextran.

2. Electroporation with purified transecting DNA. Electroporation is described in the chapter 15.

Following characteristics make the yeast an appropriate host for production of recombinant proteins:

1. Single-celled organism.
2. Physiology and genetics well-studied.
3. Can be grown in small vessels as well as large bioreactors.
4. Several promoters are known and characterized.
5. Contains a plasmid called 2μM.
6. Being eukaryotic cell, can carry out several post translational modifications.
7. Normally, it secretes only a few proteins, recombinant protein if secreted, can be purified very easily.
8. Being used in food (bread making) it is considered as safe for human consumption, and therefore, dose not require drug approval.

Three types of expression vectors are present in the yeast:

1. Episomal or plasmid vectors.
2. Integrating vectors.
3. Yeast Artificial chromosomes (YAC)

Plasmid based expression vectors aren't suitable for large scale growth of yeast. Integration of an expression vector in chromosomal DNA can be achieved, but single copy cannot produce desired quantity of protein. Details about YAC is given in the chapter 6.** , which can clone large DNA segments (100 kb). Yeast has been used for the production of hirudin (from the leach *Hirudo medicinalis*). This protein is powerful blood anti-coagulant, compared to its advantages over the conventional drug heparin.

Mammalian Cell Expression Vectors

Extrachromosomal mammalian expression vectors are helpful for study of function and regulation of mammalian genes.

These lines can also be used for production of recombinant proteins of therapeutic values. A large number of mammalian expression vectors have been designed similar to that used for eukaryotic cell. A mammalian expression vector contains a eukaryotic origin of replication from an animal virus (Simian virus 40, SV40). The promoter sequences must be of eukaryotic origin so that they can transcribe closed genes, selectable marker gene, and transcription termination sequences. Sequences for selection and propagation of a mammalian expression vector in *E. coli* are derived from *E. coli* plasmid. Bacterial gene for neomycin phosphotransferase is used to select transected mammalian cells. When grown on a medium containing geneticin, nontransformed cells are killed.

Gene for the enzyme dihydrofolate reductase (DHFR) is also used to select mammalian cells. As per this protocol, the cells should contain a defective DHFR gene, so they cannot produce any functional DHFR. The DHFR deficient cells are transected with a mammalian expression vector carrying an active DHFR gene. On a medium containing methotrexate, untransfected cells cannot grow. Many different proteins have been closed into mammalian expression vectors and expressed in host cells. A high level of production of recombinant protein can be achieved by co-ordinating the expression of a selectable marker and a cloned gene.

QUESTIONS

1. Discuss the problems associated with gene integrations and expression in different hosts.

2. Discuss the problems posed by transfer of prokaryotic gene to a eukaryotic host or vice versa, *in expression in transgenic cells.*

3. What is the importance and role of promoter in transgenic expression? How gene transcription is regulated ?
4. Write short notes :
   (*a*) Strong promoter
   (*b*) Regulatable promoter
   (*c*) Bioreactor cultivation of transgenic cells
   (*d*) Translation expression vectors
   (*e*) Secretion of recombinant protein in the medium.

TOOLS OF MOLECULAR BIOLOGY

CHAPTER 7

I. Blotting techniques

The term "blotting" refers to the transfer of biological samples from a gel to a membrane and their subsequent detection on the surface of the membrane.

(a) Southern Blotting

Southern blotting uses DNA molecule as marker (developed by E.M. Southern). Restriction enzyme digestion of genomic DNA results in so many fragments that a stained electrophoretic gel shows a smear of DNA. A probe can identify one fragment in this mixture, with the use of the technique called Southern blotting. After DNA segments are fractionated on the gel, one absorbent membrane is laid on the gel and DNA bands are transferred (blotted) on to the membrane by capillary action. These DNA bands remain in the same relative position on the membrane. The membrane is emerged in a labelled probe (a defined radiolabelled nucleic acid segment that is used to identify a specific DNA molecules bearing the complimentary sequence, produced by PCR or traditional cloning method). An autoradiogram is used to demonstrate the presence of any bands on the gel that are homologous to the probe (Fig.7.3). The cloned DNA finds wide spread application as a probe for detecting either a specific clone or a specific DNA/RNA fragment. In all these cases the ability of nucleic acids to find and bind to complementary sequences in solution is being exploited.

(b) Northern Blotting

Northern blotting uses mRNA as marker. The southern blotting can be extended to detect a specific RNA molecule from a mixture of RNAs fractioned on a gel. This technique is called Northern blotting to contrast it with the Southern's technique for DNA analysis. The fractioned RNA is blotted on to a membrane and probed in the same way as for Southern blotting. One application of Northern analysis is to determine whether specific gene is transcribed in certain tissue or under certain environmental conditions. RNAse is extracted from appropriate cell sample, then electrophoresed, blotted and probed with the cloned gene in question. A positive signal shows presence of the transcript.

(c) Western Blotting

Western blotting uses protein as marker compound. Western blotting (also called immunoblotting because an antibody is used to specifically detect its antigen) was introduced by Towbin *et al.* in 1979 and is now a routine technique for protein analysis. The specificity of the antibody-antigen interaction enables a single protein to be identified in the midst of a complex protein mixture. Western blotting is commonly used to positively identify a specific protein in a complex mixture and to obtain qualitative and semiquantitative data about that protein.

The first step in a Western blotting procedure is to separate the macromolecules using gel electrophoresis. Following electrophoresis, the separated molecules are transferred or blotted onto a second matrix, generally a nitrocellulose or polyvinylidene fluoride (PVDF) membrane. Next, the membrane is blocked to prevent any nonspecific binding of antibodies to the surface of the membrane. The transferred protein is complexed with an enzyme-labelled antibody as a probe. An appropriate substrate is then added to the enzyme and together they produce a detectable product such as a chromogenic or fluorogenic precipitate on the membrane for colorimetric or fluorimetric detection, respectively. The most sensitive detection methods use a chemiluminescent substrate that, when combined with the enzyme, produces light as a byproduct. The light output can be captured using film, a CCD camera or a phosphorimager that is designed for chemiluminescent detection. Whatever substrate is used, the intensity of the signal should correlate with the abundance of the antigen on the blotting membrane.

Detailed procedures for detection of a Western blot vary widely. One common variation involves direct vs. indirect detection as shown in Fig.7.1. With the direct detection method, the primary antibody that is used to detect an antigen on the blot is also labelled with an enzyme or fluorescent dye.

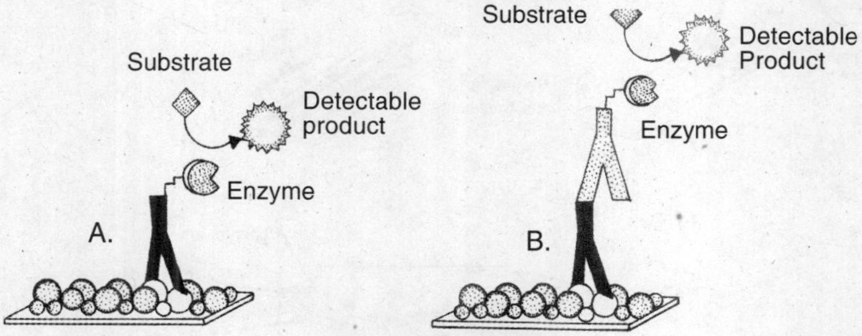

Fig. 7. 1. In the direct detection method, labelled primary antibody binds to antigen on the membrane and reacts with substrate, creating a detectable signal. **B.** In the indirect detection method, unlabeled primary antibody binds to the antigen. Then, a labelled secondary antibody binds to the primary antibody and reacts with the substrate.

In the indirect detection method, a primary antibody is added first to bind to the antigen. This is followed by a labeled secondary antibody that is directed against the primary antibody. Labels include biotin, fluorescent probes such as fluorescein or rhodamine, and enzyme conjugates such as horseradish peroxidase or alkaline phosphatase. The indirect method offers many advantages over the direct method.

(d) Protein Transfer to a Membrane

Following electrophoresis, the protein must be transferred from the electrophoresis gel to a membrane. There are a variety of methods that have been used for this process, including diffusion transfer, capillary transfer, heat-accelerated convectional transfer, vacuum blotting transfer and

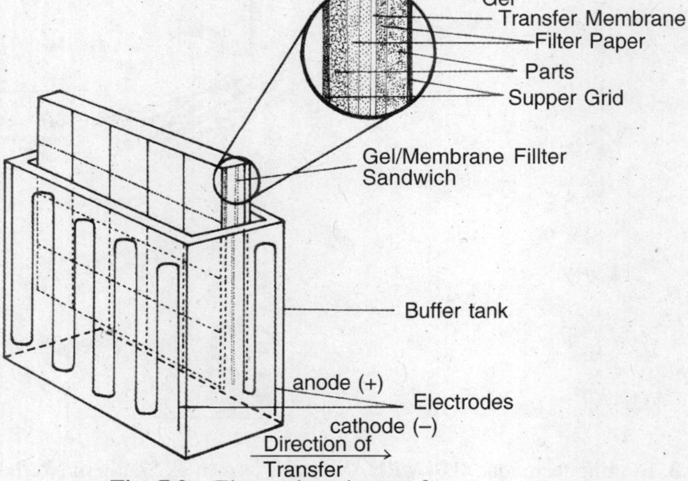

Fig. 7.2. Electrophoretic transfer

electroelution. The transfer method that is most commonly used for proteins is electroelution or electrophoretic transfer, because of its speed and transfer efficiency. This method uses the electrophoretic mobility of proteins to transfer them from the gel to the matrix. Electrophoretic transfer of proteins involves placing a protein-containing polyacrylamide gel in direct contact with a piece of nitrocellulose or other suitable, protein-binding support and "sandwiching" this between two electrodes submerged in a conducting solution (Fig. 7.2). When an electric field is applied, the proteins move out of the polyacrylamide gel and onto the surface of the membrane, where the proteins become tightly attached. The resulting membrane is a copy of the protein pattern that was found in the polyacrylamide gel itself.

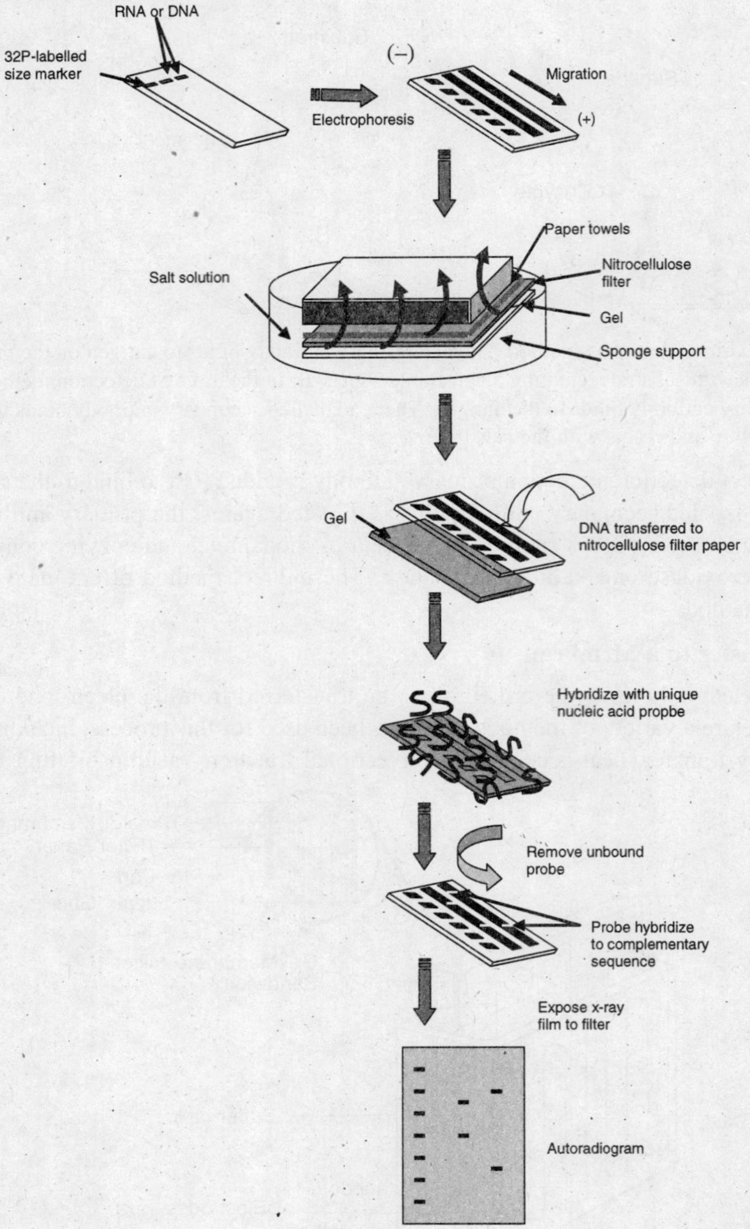

Fig. 7.3. Blotting technique (DNA/RNA is used; termed as Southern/Northern technique)

Transfer efficiency can vary dramatically among proteins, based upon the ability of a protein to migrate out of the gel and its propensity to bind to the membrane under a particular set of conditions. The efficiency of transfer depends on factors such as the composition of the gel, complete contact of the gel with the membrane, the position of the electrodes, the transfer time, size and composition of proteins, field strength and the presence of detergents. Optimal transfer of proteins is generally obtained in low ionic strength buffers and with low electrical current.

The most commonly used membranes for Western blotting (also western blot) include nitrocellulose, polyvinylidene difluoride (PVDF) and nylon membranes.

After transfer and before proceeding with the Western blot, it is often desirable to stain all proteins on the membrane with a reversible stain to check the transfer efficiency. Although the gel may be stained to determine that protein left the gel, this does not ensure efficient binding of protein on the membrane. Ponceau S stain is the most widely used reagent for staining proteins on a membrane. However, it has limited sensitivity, does not photograph well, and fades with time. MemCode Reversible Stain is a superior alternative for staining protein on nitrocellulose or PVDF membranes. MemCode Stain detects low nanogram levels of protein, is easily photographed, does not fade with time and takes less than 30 minutes to stain, photograph and erase.

Blocking Nonspecific Binding Sites

In a Western blot, it is important to block the unreacted sites on the membrane to reduce the amount of nonspecific binding of proteins during subsequent steps in the assay. A variety of blocking buffers ranging from milk or normal serum to highly purified proteins have been used to block unreacted sites on a membrane. The blocking buffer should improve the sensitivity of the assay by reducing background interference. Individual blocking buffers are not compatible with every system. For this reason, a variety of blockers in both Tris buffered saline (TBS) and phosphate buffered saline (PBS) are available. The proper choice of blocker for a given blot depends on the antigen itself and on the type of enzyme conjugate to be used. For example, with applications using an

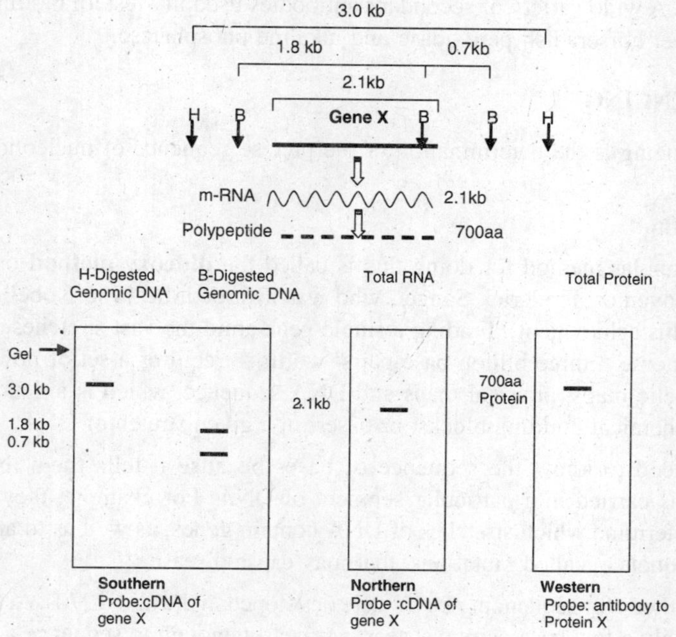

Fig. 7.4. Comparison of Southern, Northern and Western analyses of gene X.B and H are target sites for restriction enzyme B and H. Positions of bands on Southern blot cannot be compared with positions of bands on Northern and Western blots.

alkaline phosphatase conjugate, a blocking buffer in TBS should be selected, because PBS interferes with alkaline phosphatase. The ideal blocking buffer will bind to all potential sites of nonspecific interaction, eliminating background altogether without altering or obscuring the epitope for antibody binding.

A complete line of blocking buffers for Western blotting include BLOTTO, Casein, BSA, SEA BLOCK and the exclusive SuperBlock Blocking Buffers.

Washing

Like other immunoassay procedures, Western blotting consists of a series of incubations with different immunochemical reagents separated by washing steps. Washing steps are necessary to remove unbound reagents and reduce background, thereby increasing the signal:noise ratio. Insufficient washing will allow high background, while excessive washing may result in decreased sensitivity caused by elution of the antibody and/or antigen from the blot. As with other steps in performing a Western blot, a variety of buffers may be used. Occasionally, washing is performed in a physiological buffer such as Tris buffered saline (TBS) or phosphate buffered saline (PBS) without any additives. More commonly, a detergent such as 0.05% Tween 20 is added to the buffer to help remove nonspecifically bound material.

Primary and Secondary Antibodies

The choice of a primary antibody for a Western blot will depend on the antigen to be detected and what antibodies are available to that antigen. A huge number of primary antibodies are available commercially and can be identified quickly by searching sites. Both polyclonal and monoclonal antibodies work well for Western blotting. Polyclonal antibodies are less expensive and less time-consuming to produce and they often have a high affinity for the antigen. Monoclonal antibodies are valued for their specificity, purity and consistency that result in lower background. A wide variety of labelled, secondary antibodies can be used for Western blot detection. The choice of secondary antibody depends upon the species of animal in which the primary antibody was raised (the host species). A wide variety of secondary antibodies used in Western blotting are biotin, fluorescein, rhodamine, horseradish peroxidase and alkaline phosphatase.

II. DNA SEQUENCING

DNA sequencing is the determination of the precise sequence of nucleotides in a sample of DNA.

(a) Sanger's Method

The most popular method for doing this is called the **dideoxy method** or Sanger's method (named after its inventor, Frederick Sanger, who was awarded the 1980 Nobel prize in chemistry [his second] for this achievment). Finding a single gene amid the vast stretches of DNA that make up the human genome – three billion base-pairs' worth – requires a set of powerful tools. These tools include genetic maps, physical maps and DNA sequence, which is a detailed description of the order of the chemical building blocks, or bases, in a given stretch of DNA.

Scientists need to know the sequence of bases because it tells them the kind of genetic information that is carried in a particular segment of DNA. For example, they can use sequence information to determine which stretches of DNA contain genes, as well as to analyze those genes for changes in sequence, called mutations, that may cause disease.

The first methods for sequencing DNA were developed in the mid-1970s. At that time, scientists could sequence only a few base pairs per year, not nearly enough to sequence a single gene, much less the entire human genome. By the time the Human Genome Project (HGP) began in 1990, only a few laboratories had managed to sequence a mere 100,000 bases, and the cost of sequencing remained very high. Since then, technological improvements and automation have increased speed and lowered cost to the point where individual genes can be sequenced routinely, and some labs can sequence well over 100 million bases per year.

DNA is synthesized from four deoxynucleotide triphosphates. The top formula shows one of them: deoxythymidine triphosphate (dTTP). Each new nucleotide is added to the 3′ – OH group of the last nucleotide added.

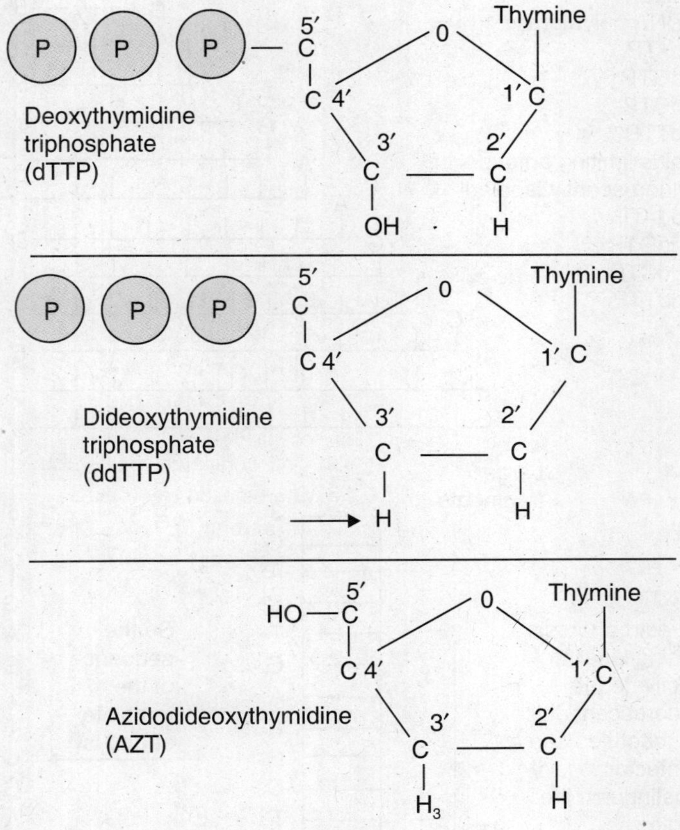

Fig. 7.5. Structure of deoxynucleotide triphosphates

The dideoxy method gets its name from the critical role played by **synthetic** nucleotides that **lack** the -OH at the 3′- carbon atom (red arrow). A dideoxynucleotide (dideoxythymidine triphosphate — ddTTP — is the one shown here) can be **added** to the growing DNA strand where **chain elongation stops** because there is no 3′- OH for the next nucleotide to be attached to. For this reason, the dideoxy method is also called the **chain termination method**. The bottom formula shows the structure of azidothymidine (AZT), a drug used to treat AIDS. AZT (which is also called **zidovudine**) is taken up by cells where it is converted into the triphosphate. The reverse transcriptase of the human immunodeficiency virus (HIV) prefers AZT triphosphate to the normal nucleotide (dTTP). Because AZT has no 3′- OH group, DNA synthesis by reverse transcriptase halts when AZT triphosphate is incorporated in the growing DNA strand. Fortunately, the DNA polymerases of the host cell prefer dTTP, so side effects from the drug are not so severe as might have been predicted.

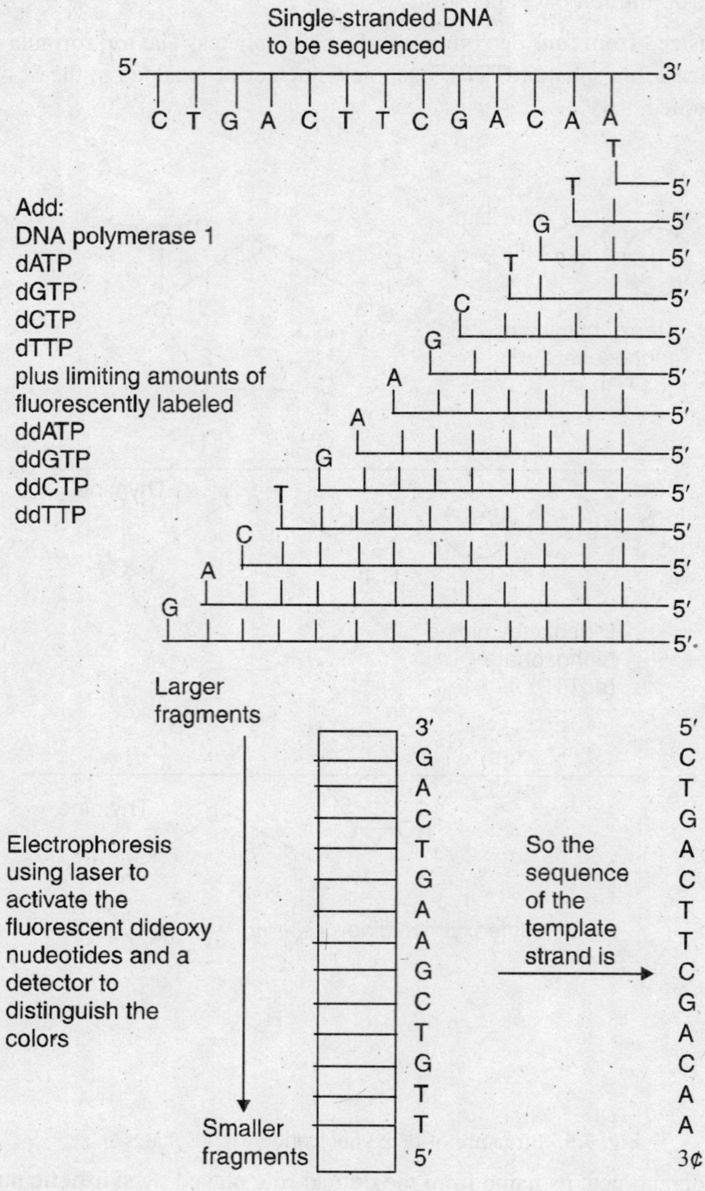

Fig. 7.6. Sanger's method of sequencing

(b) The Procedure

The DNA to be sequenced is prepared as a single strand.

This template DNA is supplied with

- a mixture of all four **normal** (deoxy) nucleotides in ample quantities
  - o dATP
  - o dGTP
  - o dCTP
  - o dTTP

- a mixture of all four **dideoxy**nucleotides, each present in limiting quantities and each labeled with a "tag" that fluoresces a different color:
    - o **ddATP**
    - o **ddGTP**
    - o **ddCTP**
    - o **ddTTP**
- DNA polymerase I

Because all four normal nucleotides are present, chain elongation proceeds normally until, by chance, DNA polymerase inserts a dideoxynucleotide instead of the normal deoxynucleotide. If the ratio of normal nucleotide to the dideoxy versions is high enough, some DNA strands will succeed in adding several hundred nucleotides before insertion of the dideoxy version halts the process.

At the end of the incubation period, the fragments are separated by length from longest to shortest. The resolution is so good that a difference of one nucleotide is enough to separate that strand from the next shorter and next longer strand. Each of the four dideoxynucleotides fluoresces a different color when illuminated by a laser beam and an automatic scanner provides a printout of the sequence.

Beginning in the late 1990s, the scientific community witnessed a remarkable climax of accomplishments related to DNA sequencing. In addition to the historic sequencing of the human genome, sequences have now been generated for the genomes of several key model organisms, including the mouse *(Mus musculus)*; the rat *(Rattus norvegicus)*; two fruit flies *(Drosophila melanogaster* and *D. pseudoobscura)*; two roundworms *(Caenorhabditis elegans* and *C. briggsae)*; yeast *(Saccharomyces cerevisiae)* and several other fungi; a malaria-carrying mosquito *(Anopheles gambiae)* along with a malaria-causing parasite *(Plasmodium falciparum)*; two sea squirts *(Ciona savignyi* and *C. intestinalis)*; a long list of microbes; and a couple of plants, including mustard weed *(Arabidopsis thaliana)* and rice *(Oryza sativa)*. Sequencing work is well underway on the honey bee *(Apis mellifera)*, and is just getting started or expected to begin soon on the chimpanzee *(Pan troglodytes)*, the cow *(Bos taurus)*, the dog *(Canis familiaris)* and the chicken *(Gallus gal)*.

Fig. 7.7. Automatic DNA sequencer

Automatic DNA sequencer

A variant of the above dideoxy-method was developed, which allowed the production of automatic sequencers. In this new approach, different fluorescent dyes are tagged either to the oligonucleotide primer (dye primers) in each of the four reaction tubes (blue for A, red for C, etc), or to each of the four ddNTPs (dye terminators) used in a single reaction tube: when four tubes are used, they are pooled. After the PCR reaction is over, the reaction mixture is subjected to separation of synthesized fragments through electrophoresis. Depending upon the electrophoretic system used, whether slab gel electrophoresis or capillary electrophoresis, following two types of automatic sequencing systems have been designed (Fig. 7.7, 7.8).

(d) Slab gel sequencing systems

These systems make use of ultrathin (75 μm) slab gels and involve running of atleast 96 lanes per gel. In these systems, automation in sample loading of sequencing gels has also been achieved, by using a plexiglass block having wells that are same distance apart as the comb teeth cut in a porous membrane that is used as a comb for drawing samples by capillary action. Each well in plexiglass block is filled with a sample (PCR dideoxy-reaction mixture), so that when the porous membrane comb is lowered onto the sample wells in the plexiglass the samples are drawn up automatically into the comb teeth by capillary action. Using this approach of employing porous combs, automated loading of up to 192, 384 or 480 samples per gel has been achieved. The porous comb with the samples is placed between the glass plates of the gel apparatus above the flat surface of the polymerized gel and the samples are driven from the comb into the gel by electrophoresis.

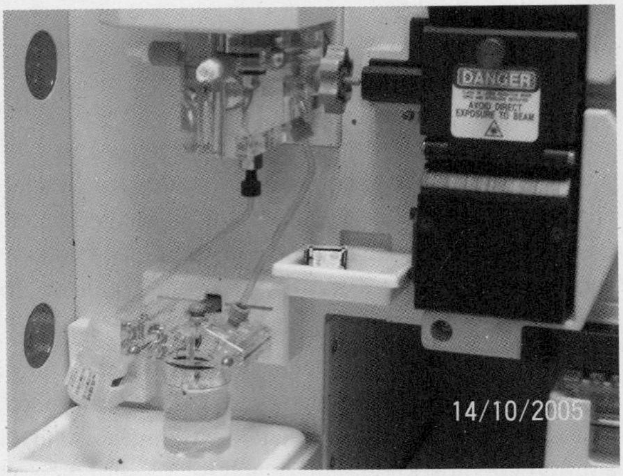

Fig. 7.8. Automated DNA sequencer details of sample loop.

(e) Capillary gel electrophoresis

In these systems, slab gel electrophoresis is replaced by capillary gel electrophoresis to analyse DNA samples. In these systems, instead of scanning DNA as it migrates through 96 lanes each in a series of 96 capillary tubes, DNA fragments pass are scanned. In the original models of the above old slab gel machines, gels must be poured and ragents frequently reloaded, interrupting the sequencing. In capillary gel sequencing systems, on the other hand, the robot moves the DNA samples and reagents through the tubes continuously, requiring attention only once a day. The system produces a steady flow of data, each signal representing one of the four DNA bases (adenine, cytosine, guanine and thymine).

DNA sequencing by PCR. DNA sequencing can also be carried out using asymmetric PCR method (see chapter on PCR).

III. CELL FRACTIONATION

Organelles are membrane-enclosed vesicles inside all eukaryotic cells that function in a variety of important cellular processes.We will isolate various subcellular fractions from rat liver by a technique termed differential centrifugation. We will then assay these fractions for the presence of specific organelles using enzyme assays.

(a) Cellular Fractionation

The arrangement of macromolecules within a cell is as important to cellular function as their catalytic activities. Cellular compartmentalization provides efficiency by bringing together related compounds that can interfere with each other (i.e. lysosomal hydrolytic enzymes). Cellular compartmentalization is accomplished in part by various subcellular organelles. The method we will use to separate the various organelles utilizes differential centrifugation to isolate components of different densities. With this technique, the heaviest or most dense organelles, nuclei pellet in less time and with less force than is required to pellet lighter organelles such as mitochondria. First, a cell homogenate is made by rupturing the cell membranes in the tissue. The homogenate is then centrifuged for a short period of time to remove cell debris and nuclei.

The supernatant is then transferred to another tube and centrifuged longer to pellet the lighter mitochondria.

For this type of fractionation, which tissue we use, and the method of homogenization are dictated by the biological system. Homogeneous cell populations from cell culture are well suited for cell fractionation. Some tissues, like those in the liver also have one cell type that predominates, so are also well suited. Most chlorophyll-free plant tissues are acceptable for preparing mitochondria, but recently-harvested plant tissues are usually required, making their use uncertain during winter months. Once a cell type is chose it is important to obtain the organelles in a biochemically active, morphologically whole state. Homogenizers are used to break open the cells without damaging the organelles. Homogenizers have a precise clearance between the glass tube and pestle, which breaks the cell membrane leaving the smaller organelle membranes intact. The homogenization buffer is a solution which often includes sucrose to partially dehydrate the organelles, keeping them intact.

No technique used to isolate organelles is perfect. It is very difficult to get pure unbroken preparations of any organelle. Techniques providing optimal isolation of one organelle may completely rupture another organelle. Thus, methods are often used to measure the contamination of one organelle fraction by another. This can be done by analyzing each organelle fraction for organelle-specific marker enzymes.

Subcellular Fraction	Relative Density
Nuclei	1
Mitochondria	2
Lysosomes	3
Microsomes	4
Cytosol	5

(b) Tissue Homogenization:

1. Transfer about one-half (~2 gm) of the tissue to a glass homogenizer on ice. Do not use a tight-fitting ground-glass homogenizer in this step since the organelle membranes may disrupt partially.
2. Add 18 ml of ice-cold Homogenization Buffer to the homogenizer. Note: the calcium in the buffer is used here to stabilize the organelle membranes.
3. Homogenize 3-5 strokes until a homogeneous homogenate is made. Keep the homogenizer on ice. Caution: Excessive grinding or heating can damage and inactivate subcellular fractions.
4. Transfer the homogenate into a plastic 50 ml tube on ice.
5. Repeat steps 5-9 (combining all homogenates) until all the minced liver tissue has been homogenized.

(c) Isolation of Nuclei:

6. Centrifuge the tube at 700 × g for 10 min. at 0ºC. The pellet contains the cell debris and nuclei. The supernatant contains all the lighter cellular fractions.
7. Without disturbing the crude nuclear pellet, carefully decant the supernatant into another tube on ice. Since the mitochondria are present in supenatant, one group member should immediately proceed to "Isolation of the Mitochondria", another member should proceed with isolation of the nuclei.

8. To continue with the nuclear isolation: Nuclear Wash :
   (*a*) Add 10 ml of ice-cold Homogenization Buffer to your crude nuclear pellet.
   (*b*) Use a teflon homogenizer to resuspend the pellet in the buffer.
   (*c*) Place two layers of cheesecloth in a funnel over a tube. Filter the 10 ml suspension through the cheesecloth.
   (*d*) Rinse the debris in the filter with another 10 ml of Homogenization Buffer collecting the filtrate in the tube. Discard the filter with its cellular debris.
   (*e*) Centrifuge the combined filtrates (20 ml) at 1500 × g for 10 min at 0ºC. The pellet contains the washed nuclear fraction. Without disturbing the nuclear pellet, discard the supernatant.
9. Add 5.0 ml of ice-cold Homogenization Buffer to the nuclear pellet.
10. Use a teflon homogenizer to resuspend the pellet in the buffer.
11. Using a pipette, aliquot the nuclear suspension into two glass test tubes.
12. Label the two glass tubes "Nuclei", and store them at - 20°C. Caution: Glass tubes should not be greater than half full, or they may crack upon freezing.

Isolation of Mitochondria

13. Obtain the supernatant from step 12 in a tube on ice.
14. Centrifuge at 5000 × g for 15 min. at 0ºC. The pellet contains the crude mitochondrial fraction. The supernatant contains microsomes, membrane fragments, ribosomes, cytoplasmic enzymes, etc.
15. Without disturbing the mitochondrial pellet, carefully decant the post-mitochondrial supernatant (and the pink partially sedimenting layer) into a graduated cylinder.
16. Pour the supernatant into a clean tube on ice.
17. With the remainder of the supernatant, Isolation of the Microsomes can be carried out.
18. To continue with mitochondrial isolation: Mitochondrial Wash:
   (*a*) Add 20 ml of ice-cold Homogenization Buffer to the crude mitochondrial pellet.
   (*b*) Using a teflon homogenizer, resuspend the pellet in the buffer.
   (*c*) Pour the suspension back into the tube, add another 20 ml of buffer, mix, and centrifuge at 234,000 × g for 10 min. at 0ºC. The pellet contains the washed mitochondria.
   (*d*) Without disturbing the mitochondrial pellet, decant and discard the supernatant.
19. Add 5.0 ml of ice-cold Homogenization Buffer to the mitochondrial pellet.
20. Using a teflon homogenizer, resuspend the pellet in the buffer.

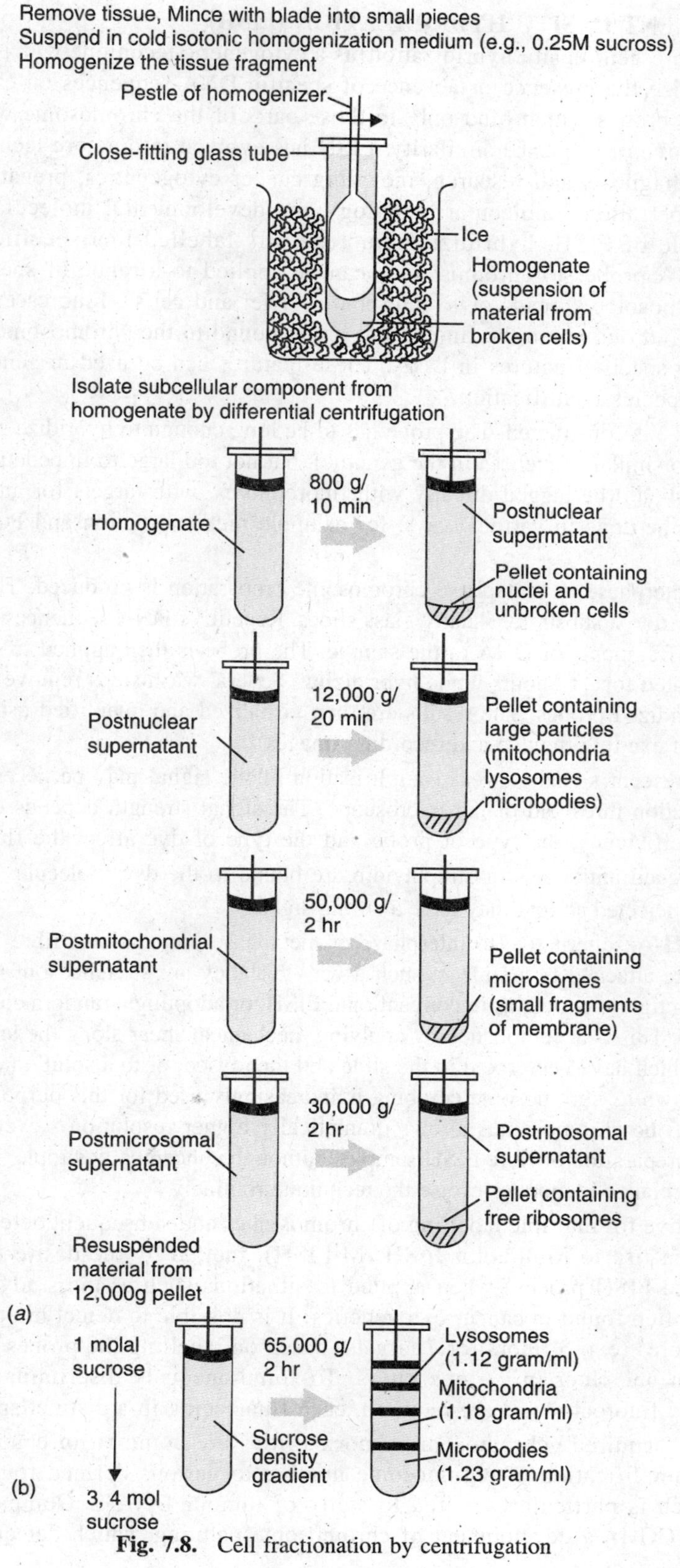

Fig. 7.8. Cell fractionation by centrifugation

IV. FLUORESCENT IN-SITU HYBRIDIZATION (FISH)

FISH (Fluorescent in situ hybridization) is a cytogenetic technique which can be used to detect and localize the presence or absence of specific DNA sequences on chromosomes. It uses fluorescent probes which bind only to those parts of the chromosome with which they show a high degree of sequence similarity. FISH has evolved into a core technique spanning many areas of diagnosis and research, including cancer cytogenetics, prenatal screening of chromosomal aberrations, molecular pathology and developmental molecular biology. The general principle of FISH, hybridizing fluorescently-labelled loci-specific, telomeric or centromeric DNA probes to chromosomes, can be applied to a range of specimens such as metaphase chromosome spreads or to interphase nuclei and cells. Fluorescence microscopy can be used to find out where the fluorescent probe bound to the chromosome. FISH is often used for finding specific features in DNA. These features can be used in genetic counseling, medicine, and species identification.

First, a probe is constructed. The probe has to be long enough to hybridize specifically to its target (and not to similar sequences in the genome), but not too large to impede the hybridization process, and it should be tagged directly with fluorophores, with targets for antibodies or with biotin. This can be done in various ways, for example nick translation and PCR using tagged nucleotides.

Then, an interphase or metaphase chromosome preparation is produced. The chromosomes are firmly attached to a substrate, usually glass slides. Repetitive DNA sequences must be blocked by adding short fragments of DNA to the sample. The probe is then applied to the chromosome DNA and incubated for ~12 hours while hybridizing. Several wash steps remove all unhybridized or partially hybridized probes. The results are then visualized and quantified using a microscope that is capable of exciting the dye and recording images.

If the fluorescent signal is weak, amplification of the signal may be necessary in order to exceed the detection threshold of the microscope. The signal strength depends on many factors; probe labeling efficiency, the type of probe and the type of dye affect the fluorescent signal. Fluorescently-tagged antibodies or streptavidin are bound to the dye molecule. These secondary components are selected so that they have a strong signal.

Fibre FISH-An alternative to interphase or metaphase preparations, fibre FISH, interphase chromosomes are attached to a slide in such a way that they are stretched out in a straight line, rather than being tightly coiled, as in conventional FISH, or adopting a random conformation, as in interphase FISH. This is accomplished by applying mechanical shear along the length of the slide; either to cells which have been fixed to the slide and then lysed, or to a solution of purified DNA. A technique known as chromosome combing is increasingly used for this purpose. The extended conformation of the chromosomes allows dramatically higher resolution - even down to a few kilobases. The preparation of fibre FISH samples, although conceptually simple, is a rather skilled art, and only specialized laboratories use the technique routinely.

The incentive for multiple labelling of chromosomes and subsequent detection in multiple colors has given rise to Multicolor FISH (M-FISH), fuelled by the desire to maximize the efficiency of the FISH process when applied to situations such as the study of the complex translocations often found in cancer cytogenetics. It is possible to detect multiple targets with fewer fluorophores (e.g. 3 fluors for 7 targets) based on labelling the probes combinatorially, with more than one fluor in various ratios. To simultaneously discriminate all 24 human chromosomes, 5 fluorochromes are required, each fluorescing in a distinct spectral region.

Complex "acquired" chromosome abnormailities are common in cancer cytogenetics, ranging from amplification of chromosome number to various defined translocations. One technique which is particularly suited to study of tumour DNA is Comparative Genomic Hybridisation (CGH), a development of chromosome painting, which detects a deviation of the ratio of two fluorochromes to indicate amplification or deletion along the chromosome strand. FISH can also be instrumental for gene mapping. Genes and DNA segments are continually being assigned to chromosome bands, but it is often necessary to determine the precise arrangement of each sequence relative to another at a more detailed level.

FISH Detection – From an instrumental standpoint, many adaptations of the FISH technique share a common need for high resolution, sharp contrast between signal and background (particularly in interphase cytogenics), and high sensitivity for achieving high signal to noise ratio from weak signals, or for determining minor comparative differences in CGH. This gives rise to the need for flexible cameras, capable of operating effectively under both "standard" and "ultra-sensititive" modes should the need arise.

Genomic in situ hybridization (GISH) is becoming an important technique to characterize parental genomes in wide hybrids and to detect introgressed segments and chromosome rearrangements. Total genomic DNA as a probe in in situ hybridization has been used to detect parental chromosomes in wheat-rye hybrids and in discriminating closely related Triliceae species. In situ hybridization using genomic probes has been demonstrated to be an effective technique for the detection of meiotic pairing between wheat-rye. However, GISH has not been used to characterize meiotic pairing in Oryza. Detection of autosyndetic and allosyndetic pairing in wide hybrids is difficult through conventional meiotic analysis.

V. GENOMIC LIBRARIES

Creation of genomic library involves cloning of whole genomic DNA, which is a huge task. Depending upon the size of genome, prokaryotic or eukaryotic, vector can be selected. This starts with isolation of DNA which is described in detail elsewhere.

Step 1) Isolation of chromosomal DNA

Step 2) Fragmentation of DNA

(a) Physical breakage - pipetting, physical agitation (vortexing), sonication, which usually leaves blunt ends.

All fragments should be blunt ended, which can be done using Klenow fragment of *E. coli* DNA polymerase I, which catalyses the 5'-to-3' addition of nucleotides (lacks 3' -to-5' exonuclease activity of intact DNA pol I) (Fig. 7.9). DNA polymerase requires a DNA template and deoxynucleotides (substrates), Mg^{2+} and a free 3'-OH on the complementary DNA strand which is being synthesised.

(b) Restriction enzyme digestion.

Advantages:

Cohesive DNA ends for efficient ligation.

Disadvantages:

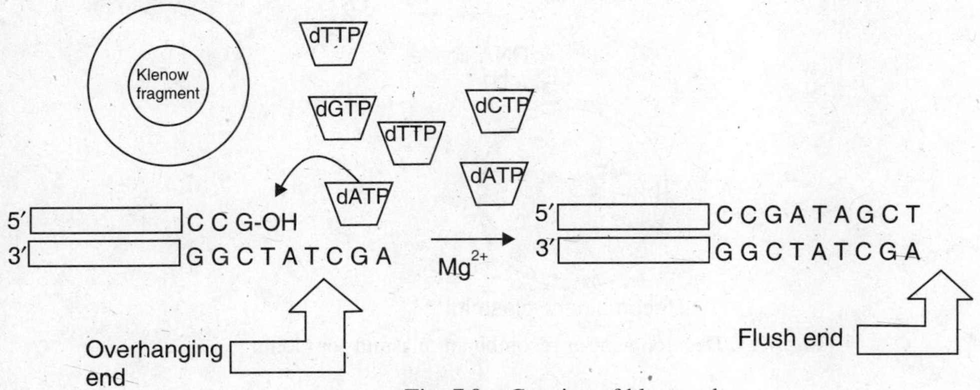

Fig. 7.9. Creation of blunt ends

- May get non-random distribution of fragment sizes, due to non-random distribution of enzyme sites.
- Enzymes may not recognise methylated DNA.
- Much eukaryotic DNA is extensively methylated.

We can vary time of digestion (or of physical treatment) to produce DNA fragments of desired size (Fig. 7.10).

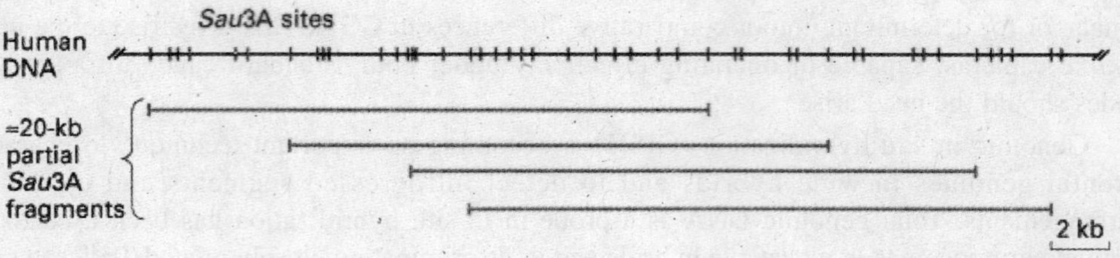

Fig. 7.10. Partial digestion for longer, overlapping DNA fragments

Step 3) Isolation and ligation of DNA fragments

- Purify fragments of required size by agarose gel electrophoresis, and elution of fragments from gel. Ligate fragments into a vector pre-cut with a restriction enzyme using DNA ligase

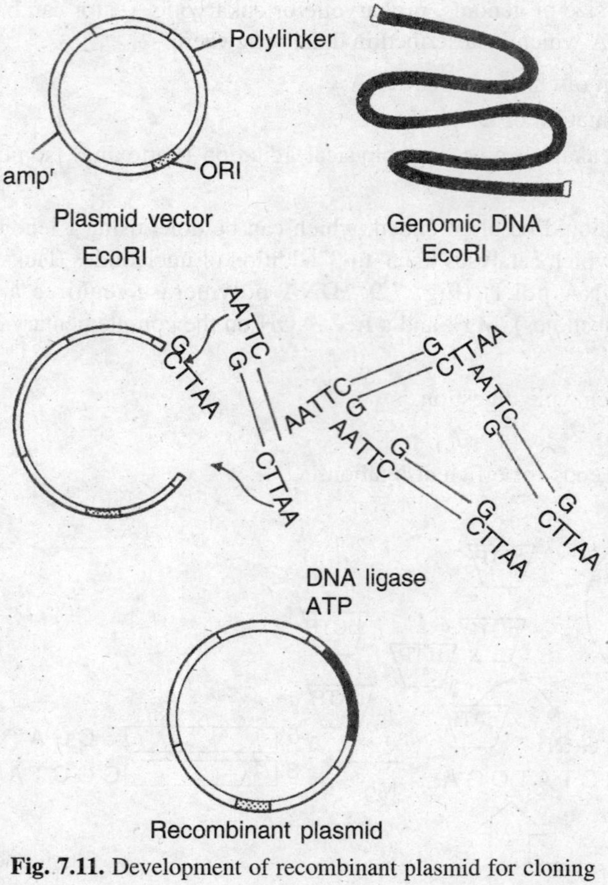

Fig. 7.11. Development of recombinant plasmid for cloning

For prokaryotes (smaller genomes) - It is viable to make gene libraries in plasmids. Plasmid inserts normally ~5-10 kb, so only need a few thousand recombinants for a representative library. For eukaryotes (larger genomes), really need vectors that can hold much larger insert DNA fragments.

(a) Alternative vector systems:

Vector type	Approximate maximum length of DNA that can be cloned into vectors
	Cloned DNA (kb)
Plasmid	20
lambda phage	25
Cosmid	45
P1 phage*	100
BAC (bacterial artificial chromosome)	300
YAC (yeast artificial chromosome	1000

Bacteriophage lambda - can perform much more efficient infection of *E. coli* than that can be achieved by plasmid transformation

*P1- *E. coli* virus capable of packaging large DNA segments

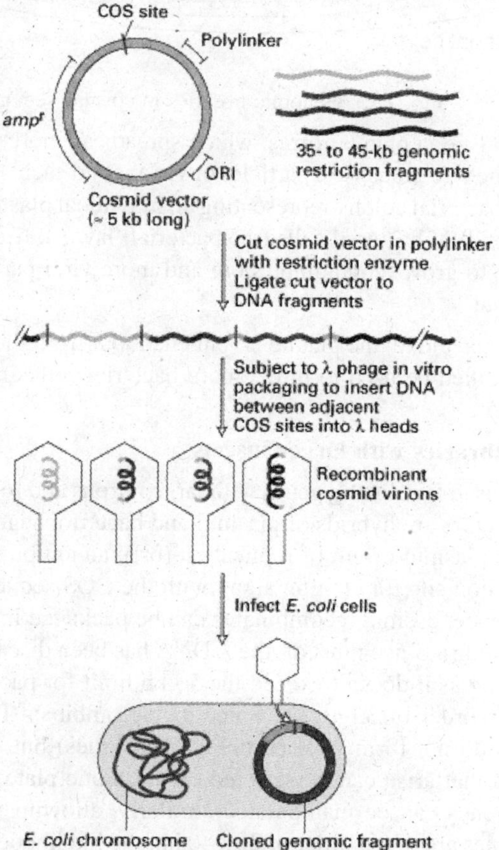

Fig. 7.12. Making a genomic library in bacteriophage lambda vector

(b) Making a genonic library in bacteriophase lambda vector

Phage lambda bind to receptors on the surface of *E. coli* and injects DNA. Infection by lambda is much more efficient than plasmid transformation - can get ~ 10^9 plaques per microgram of DNA, vs ~ 10^6 colonies per microgram of plasmid DNA.

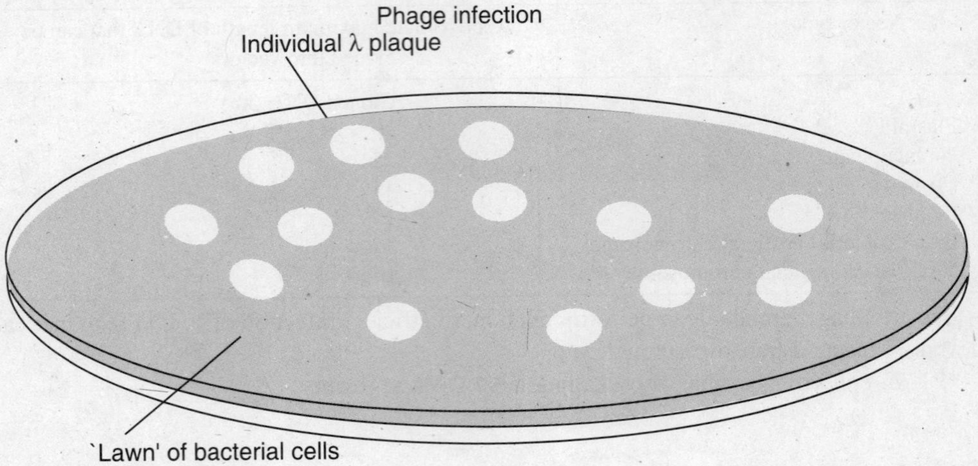

Fig. 7.13. Genomic libraries in cosmid vector

As the bacterial cells lyse, a plaque forms, which spreads as more cells become infected and lyse. The plaque contains the infectious viral particles or virions and each one represents an individual lambda clone (similar to a bacterial colony representing an individual plasmid clone). The appearance is of a clear circle in a 'cloud' of bacterial cells (the bacterial 'lawn'). If incubated for long periods, these circles will continue to grow (containing more and more viral particles) and spread until all bacterial cells are wiped out.

To pick a single lambda clone, the plaque is 'punched' out from the agar plate and stored in a holding solution. This can then be used to infect more bacterial cells and to replicate more of the recombinant lambda DNA.

(c) Vectors for making libraries with larger inserts

Cosmid libraries are used for cloning genes with large introns and for sequencing larger chunks of the genome. Cosmid vectors are hybrids of plasmid and bacteriophage lambda DNA (a small ~5 kb plasmid containing the plasmid origin of replication (ori), an antibiotic resistance gene such as amp^r and a suitable restriction site for cloning along with the COS sequence from phage λ DNA). Because of the COS sequence, cosmid recombinants can be packaged into viral particles (allowing high efficiency transformation). Since most of the λ DNA has been discarded, it can be replaced by the DNA of interest, so long as it doesn't exceed the 50 kb limit for packaging into the viral head. The insert sizes are of the order of 35-45 kb. Since the recombinant DNA does not encode any lambda proteins, cosmids do not form viral particles (or plaques) but rather forms large circular plasmids and the colonies that arise can be selected on antibiotic plates, like other plasmid DNA transformants. Cosmid clones can be manipulated similarly, allowing ease of plasmid isolation. Since many eukaryotic genes are of the order of 30 - 40 kb, the likelihood of obtaining a DNA clone containing the entire gene sequence is increased significantly when using a cosmid library.

(d) YAC libraries

YAC libraries are used for cloning very large DNA fragments (of more than 1 Mb), and are useful for cloning large genes (such as the 250 kb cystic fibrosis gene) and for creating libraries of large overlapping clones, such as for individual chromosomes isolated from organisms (chromosomal libraries). These have been used extensively for mapping genomes of complex organisms (e.g. *Homo sapiens*).

YACs are yeast artificial chromosomes and are hybrids of bacterial plasmid DNA and yeast DNA. The components required for replication/segregation of natural yeast chromosomes have been combined with *E. coli* plasmid DNA. YACs are grown in the yeast *Saccharomomyces cerevesiae* and so contain selectable markers which are suitable for the host system. Rather than antibiotic selection, yeast selectable markers enable growth of the transformant on selective media lacking specific nutrients. (Non-transformants are unable to grow). The yeast strains that are used are auxotrophic - that is, they are unable to make a specific compound.

For example, Trp1 mutants can't make tryptophan so can only grow on media supplemented with tryptophan. If the mutant strain is transformed with a YAC containing an intact Trp1 gene, then this will compensate for the inactive gene (complement) and the transfected cell is able to grow on media lacking tryptophan.

YACs are not used as extensively anymore because of inherent problems. For instance, YAC clones can contain non-contiguous segments of the genome. This means that 2 or more DNA fragments from separate parts of the genome can be integrated into an individual YAC (because they are able to support rather large inserts). A second problem is that YACs are unstable and frequently lose parts of the DNA during propagation.

BAC libraries are also used for cloning very large DNA fragments and have been particularly useful for sequencing large genomes. BACs are bacterial artificial chromosomes, and are based on a naturally occurring large bacterial plasmid, the F-factor. BAC vectors can accommodate DNA inserts up to 300 kb, still fairly respectable when needing to clone large genes or map and sequence complex genomes.

BACs have several advantages over YACs, which means they are used more extensively now. Most of the human genome has been sequenced using BAC rather than YAC clones.

VI. NUCLEIC ACID PURIFICATION

(a) Isolation and Purification of Genomic DNA

Genomic and plasmid DNA can be isolated from cells by disrupting cells and then exploiting the differences in properties between DNA, protein and other constituents.

There is no universal method for purification of genomic DNA. The method for purification of DNA depends on the nature of the source material. DNA from plants is more difficult to obtain in clonable form than from animals due to presence of secondary metabolites and polysaccharides and the extent of interfering substances present in that source. The first reported purification of chromosomal DNA was in 1944 by Avery and co-workers from the "S" cells of *Diplococcus pneumoniae* Type II. These cells contained a very large amount of capsular polysaccharide and the method was developed to purify the DNA and remove the contaminant. From about fifteen years Avery's method with some modification continued to be used for DNA solution. It was gradually realized that there was need to improve upon the yield and also the size of DNA. The deoxynucleases that were released on cell lysis degraded a large part of DNA, Marmur introduced the use of alkaline buffer (0.1 M Tris, pH 9.0) and sodium dodecyl sulphate. Both the high pH and sodium dodecyl sulphate were found to inactivate the nucleases. In addition, lysis operation to be performed at low temperature (0-2°C) to slow down the enzymes and the subsequent extraction with chilled phenol saturated with the pH 9.0 buffer, which denatured the proteins bound to DNA and consequently increased the yield of DNA. The DNA was precipitated from the aqueous layer with ethanol as was done by Avery. No protein could be detected by the Lowry test in the DNA sample, but some RNA was present. All methods subsequently developed for isolation of chromosomal DNA from bacteria, viruses and animal cells in culture have been largely derived from the method of Marmur. EDTA has been introduced in the lysis medium to inactivate nucleases which are known to be dependent on Mg++. In some methods phenol has been replaced by a mixture of equal volumes of phenol and chloroform. Since phenol treatment is thought to shear large DNA molecules, methods have been developed avoiding phenol and such methods probably yield better quality DNA at the expense of quality.

When DNA is desired to be isolated from animal or plant tissue, the above mentioned steps have to be preceded by procedure to obtain cell powder from the ground tissue. The tissue is minced finely and then disintegrated thoroughly in a Waring or equivalent blender with liquid nitrogen. This step is sometimes performed in a mortar and pestle. The powder left behind after evaporation of the liquid nitrogen is extracted with the alkaline buffer (pH 8.0 to 9.0) containing EDTA and the detergent. Subsequently the mixture is treated with protease and ribonuclease followed by phenolisation. A problem that is quite often encountered with plant tissue is the presence of a large amount of polysaccharides. Plant DNA free from polysaccharides is obtained by cesium chloride, ethidium bromide equilibrium centrifugation as described under purification of plasmid DNA.

Different types of isolation procedures

1. SDS-phenol extraction

This is the simplest and fastest method for isolation of DNA and is widely used with filamentous fungi and plant material. The technique uses SDS with phenol to denature and dissolve macerated hyphae leaving the DNA intact. DNA is then precipitated from solution with a precise volume of isopropanol.

2. SDS-proteinase K treatment

This technique is applicable to most materials and relies on proteinase K and SDS to dissolve the sample and digest the protein component without affecting the DNA. The sample is then extracted with protein denaturants (phenol, phenol-chloroform, chloroform) and the aqueous liquid containing the DNA is dialyzed and precipitated with ethanol or isopropanol in sodium or ammonium acetate solution. RNAse may or may not be used depending on different protocols used.

3. CTAB treatment

Isolation of DNA from plants is preferably achieved by treatment with CTAB. This detergent carries a positive charge which interacts with the negative charge of DNA in high salt concentration and forms a soluble complex. In subsequent steps a decrease in the salt concentration causes precipitation of DNA, leaving other compounds, especially polysaccharides, in solution. The preparation of DNA from animal tissue which is in mucopolysaccharides could also be done by this technique which yields good quality material.

(b) Plasmid DNA isolation

Plasmids are extra chromosomal double stranded circular DNA molecules present in bacteria and lower eukaryotes which exist in super helical state *in vivo*, the size of plasmid being in the range between 1kb and more than 300kb. Chromosomal DNA is much larger and though could also be circular is rendered linear by shear during cell lysis and the DNA isolation procedure. These long linear molecules invariably get entangled in the debris of lysed cells.

The early procedures of plasmid isolation were developed in the late 1960s in the laboratory of Donald R Heiliski in the University of California at San Diego. Cell lysis was achieved by using EDTA, lysozyme and mild non-ionic detergent like Triton X-100 at pH 8.0 and avoiding vigorous shaking. A centrifugal step was then incorporated to preferentially sediment the chromosomal DNA molecules entangled in the cellular debris. The supernatant solution contained the plasmids which was treated successively with ribonuclease and proteinase. Treatment with phenol, extraction with chloroform – isoamyl alcohol and precipitation of the plasmid DNA with chilled ethanol followed.

In another procedure the yield was much better increased by quickly bringing the lysozyme treated cell suspension to boil, holding at 100°C for 40 second and then immediately immersing in ice cold water. The chromosomal DNA was removed as usual by centrifugation. It may be noted here that though the hydrogen bonds between the two strands of duplex DNA are disrupted by alkali or bonds are reformed instantly on restoration of favorable pH or temperature, because of there closed circular nature. Preparations by both these methods, however, contained higher proportion of contaminating linear fragments of chromosomal DNA.

(c) Purification of plasmid DNA

1. Isopycnic Ultracentrifugation

Plasmid DNA can be purified by equilibrium centrifugation in a cesium chloride (CsCl) ethidium bromide gradient. Dyes like ethidium bromide and propidium bromide binds to duplex DNA by intercalating between the two strands. This intercalation is much more in case of linear duplex DNA, apparently because of easy access through the two open ends and least with covalent, closed circular supercoiled plasmid DNA. The bound dye caused a greater reduction in the apparent buoyant density of the linear DNA compared to that of the circular DNA and hence the plasmids could be separated from the linear chromosomal DNA by equilibrium centrifugation the CsCl density gradient, (Fig. 7.14) shows the relative positions of proteins, linear DNA or nicked plasmid DNA, supercoiled plasmid DNA and single stranded RNA. The plasmid DNA was recovered by puncturing the centrifuge tube with a hypodermic needle (not too fine) just under the plasmid band which could be visualized by ling wave length ultraviolet light. The principle behind centrifugal purification has been described.

There are many genetic engineering operations in which very pure plasmid DNA is not required. Various small columns are commercially available which can be effectively used to obtain a considerably pure plasmid DNA preparation from a crude preparation even from a cell lysate. These columns contain among other materials, silica based resin or membrane anion exchange systems, paramagnetic particles or just plain glass powder.

2. Adsorption chromatography

In chromatography – Hydroxyapatite is used which binds to the proteins and nucleic acids at low phosphate concentration. If the phosphate concentration is stepwise raised, proteins and RNA are eluted first followed by small RNA molecules i.e. plasmid DNA and finally high molecular weight, *i.e.* chromosomal DNA at the highest phosphate concentration. When clear lysate is loaded onto hydroxyapatite column, separation occurs based on above mentioned principle.

3. Preparative electrophoresis

Plasmid molecule especially in their complete superhelical CCC form can be easily separated from RNA and large chromosomal DNA by electrophoresis in agrose gel. It can yield DNA in mg quantity and can also give an idea of the molecular weight of the DNA by comparing it with the molecular weight marker.

(d) Isolation of RNA

Cells contain ribonucleases and hence inactivating this enzyme during cell lysis is crucial. Agents that effectively inactivate ribonucleases are therefore, adding cell lysis. The ionic detergents sodium dodecyl sulphate and sodium lauryl sarkosinate are known to inhibit ribonucleases and are frequently used. Reducing agents like β-merceptoethanol interferes with the tertiary structure of ribonuclease and could be effectively used for inactivating the enzyme.

Once the cell lysate is obtained, RNA is isolated through steps similar to that of DNA isolation. DNA in the cell lysate is sheared by repeated ejection through a fine hypodermic needle, proteolytic enzymes added and then the RNA is extracted with phenol or a phenol chloroform mixture. The next critical step here is to bring the pH down to pH 4.7-5.2 and also to saturate the phenol to be used for extraction with acetate buffer pH 4.7-5.2 instead of Tris buffer pH 8.0. This drives the

DNA to the interphase of phenol and water, and leaves the RNA in the water phase. The RNA is precipitated as usual from the aqueous phase by ethanol or isopropanol.

Ribosomal RNA constitutes 80-85% of total cellular RNA, transfer RNA and small nuclear RNA (in case of eukaryotes) constitute about 10-15% while messenger RNA constitutes are about 1-5%. For the most genetic engineering experiments it is the messenger RNA that is important.

(e) Purification of Messenger RNA

Bacterial messenger RNA can be separated from ribosomal and transfer RNAs by velocity sucrose density gradient centrifugation or gel electrophoresis. Method for purification of eukaryotic mRNA was developed by Philip Leder in 1972 by taking advantage of the fact that mRNA most often has a polytail. The polyA RNA could be separated from other species of RNA by affinity chromatography through a column of oligo (dT) cellulose. A polyU sepharose column has also been used by some.

(f) Analysis of Yield of Nuclei Acid

Analysis of Yield DNA

The analysis of yield of purified DNA is best performed by the diphenylamine method. The intensity of the blue color formed at acidic pH is determined by monitoring absorption at 595mμ. This assay is specific and can be used to assay DNA even in a crude DNA preparation, a quick but less sensitive and much less specific method that is quite often used to determine absorption of the nucleic acid itself at 260mμ. This assay is based on the fact that the purine and pyrimidine base present in nucleic acids absorb maximally around 260mμ. This assay would obviously be positive for both DNA and RNA, free nucleotides and nucleosides and even purine and pyrimidine based, but in a purified DNA or RNA sample the assay is reliable. Another quick but very approximate assay of double stranded DNA is by visual estimation of the ultraviolet fluorescence of DNA in a droplet elicited by ethidium bromide.

Analysis of Yield RNA

It is obvious from the preceding subsection that yield of RNA can also be determined by monitoring absorption at 260mμ as long as it is pure sample of RNA. A more specific and sensitive method is based on the monitoring of the color developed by reaction with orcinol. The RNA sample is boiled with freshly recrystallized orcinol in the presence of FeCl and HCL. The green color, developed is monitored at 655mμ. Ribose linked only to purines contribute to the color. Specificity of this reaction is not as good as the diphenyl amine reaction for DNA 2- Deoxyribose DNA methyl pentose, hexuronic acid and aldopheptoses also produce some green colour giving absorbance in that region.

RESTRICTION ENZYMES

Restriction enzymes are **DNA-cutting enzymes** found in bacteria (and harvested from them for use). Because they cut within the molecule, they are often called **restriction endonucleases**. A restriction enzyme recognizes and cuts DNA only at a particular sequence of nucleotides. For example, the bacterium *Hemophilus aegypticus* produces an enzyme named **HaeIII** that cuts DNA wherever it encounters the sequence

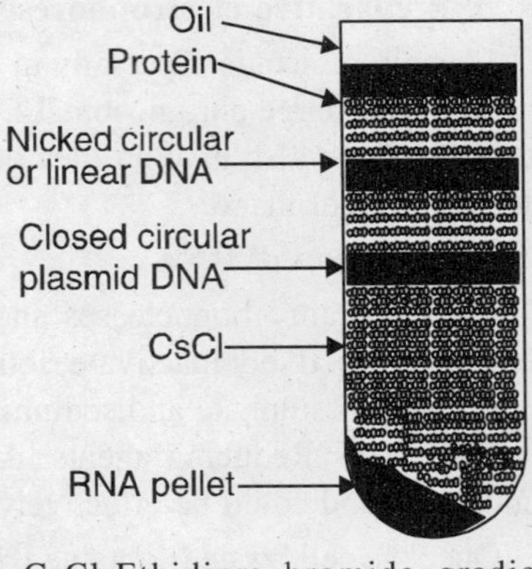

Fig. 7.14. CsCl-Ethidium bromide gradient centrifugation showing plasmid DNA in the cell lysate. DNA molecules are separated on the basis of their relative size and density

5′GGCC3'

3′CCGG5'

The cut is made between the adjacent G and C. This particular sequence occurs at 11 places in the circular DNA molecule of the virus ϕX 174. Thus, treatment of this DNA with the enzyme produces 11 fragments, each with a precise length and nucleotide sequence. These fragments can be separated from one another and the sequence of each determined. HaeIII and AluI cut straight across the double helix producing "blunt" ends. However, many restriction enzymes cut in an offset fashion. The ends of the cut have an overhanging piece of single-stranded DNA. These are called "sticky ends" because they are able to form base pairs with any DNA molecule that contains the complementary sticky end. Any other source of DNA treated with the same enzyme will produce such molecules. Mixed together, these molecules can join with each other by the base pairing between their sticky ends. The union can be made permanent by another enzyme, DNA ligase, that forms covalent bonds along the backbone of each strand. The result is a molecule of recombinant DNA (rDNA). Because recognition sequences and cleavage sites differ between restriction enzymes, the length and the exact sequence of a sticky-end "**overhang**", as well as whether it is the 5' end or the 3' end strand that overhangs, depends on which enzyme produced it. Base-pairing between overhangs with complementary sequences enables two fragments to be joined or "spliced" by a DNA ligase. A sticky-end fragment can be ligated not only to the fragment from which it was originally cleaved, but also to any other fragment with a compatible sticky end. The sticky end is also called a cohesive end or complementary end in some reference. If a restriction enzyme has a non-degenerate palindromic (the sequence on one strand reads, the same in the same direction on the complementary strand e.g. GTAATG is not a palindromic DNA sequence, but GTATAC is, GTATAC is complementary to CATATG) cleavage site, all ends that it produces are compatible. Ends produced by different enzymes may also be compatible.

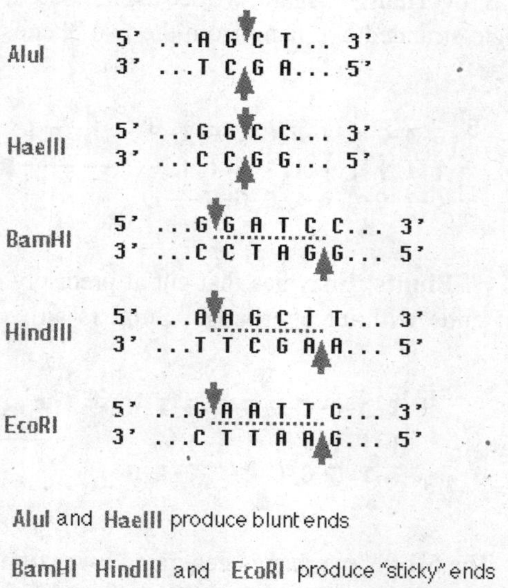

Fig. 7.15. Restriction enzymes produce blunt/sticky ends

(a) Naming

Restriction enzymes are named based on the bacteria in which they are isolated in the following manner :

E	*Escherichia*	(genus)
co	*coli*	(species)
R	RY13	(strain)
I	First identified	Order ID'd in bacterium

(b) Patterns of DNA Cutting by Restriction Enzymes

- **5' overhangs:** The enzyme cuts asymmetrically within the recognition site such that a short single-stranded segment extends from the 5' ends. BamHI cuts in this manner.

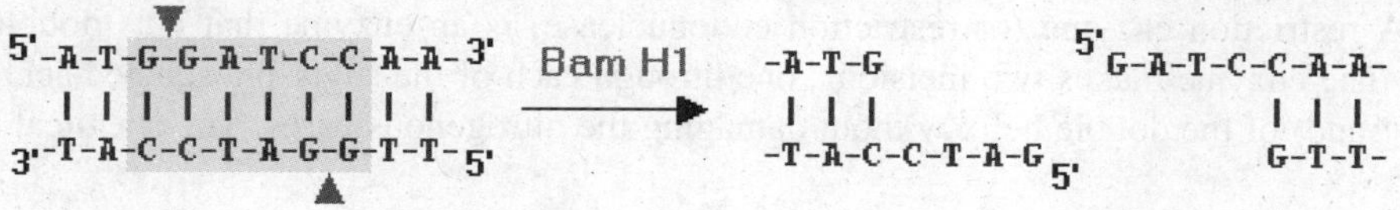

3' overhangs: Again, we see asymmetrical cutting within the recognition site, but the result is a single-stranded overhang from the two 3' ends. KpnI cuts in this manner :

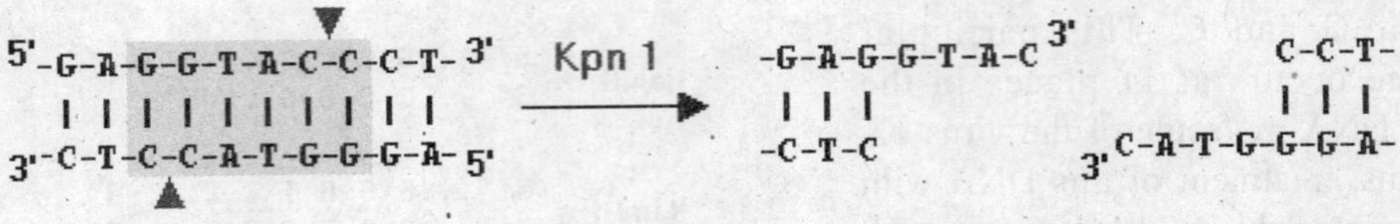

- **Blunts:** Enzymes that cut at precisely opposite sites in the two strands of DNA generate blunt ends without overhangs. SmaI is an example of an enzyme that generates blunt ends.

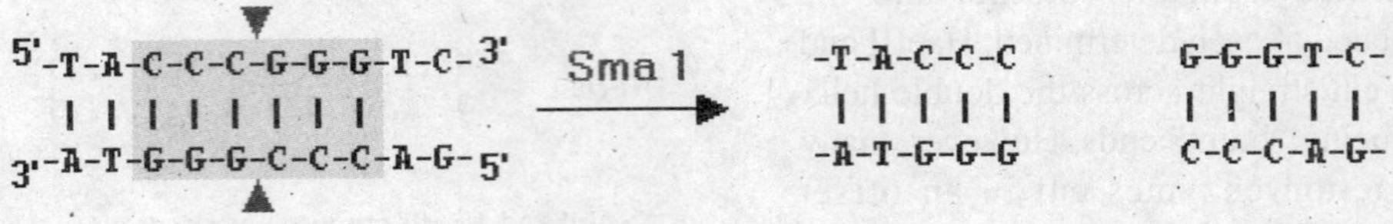

The 5' or 3' overhangs generated by enzymes that cut asymmetrically are called sticky ends or cohesive ends, because they will readily stick or anneal with their partner by base pairing

Enzyme	Source	Recognition Sequence	Cut	
EcoRI	*Escherichia coli*	5'GAATTC	5'..........G	AATTC .... 3'
		3'CTTAAG	3'..........CTTAA	G .... 5'
BamHI	*Bacillus*	5'GGATCC	5'..........G	GATCC .... 3'
	amyloliquefaciens	3'CCTAGG	3'..........CCTAG	G .... 5'
HindIII	*Haemophilus*	5'AAGCTT	5'..........A	AGCTT .... 3'
	influenzae	3'TTCGAA	3'..........TTCGA	A..........5'
TaqI	*Thermus*	5'TCGA	5'..........T	CGA ....3'
	aquaticus	3'AGCT	3'..........AGC	T ....5'
HinfI	*Haemophilus*	5'GANTC	5'..........G	ANTC ....3'
	influenzae	3'CTNAG	3'..........CTNA	G ....5'
Sau3A	*Staphylococcus*	5'GATC	5'	GATC ....3'
	aureus	3'CTAG	3'.......... CTAG	..........3'
*PovII**	*Proteus vulgaris*	5'CAGCTG	5'.......... CAG	CTG ....3'
		3'GTCGAC	3'.......... GTC	GAC .... 5'
*HaeIII**	*Haemophilus*	5'GGCC	5'.......... GG	CC ....3'
	egytius	3'CCGG	3'.......... CC	GG ....5'
*AluI**	*Arthrobacter*	5'AGCT	5'.......... AG	CT ....3'
	luteus	3'TCGA	3'.......... TC	GA .... 5'
*EcoRV**	*Escherichia coli*	5'GATATC	5'.......... GAT	ATC .... 3'
		3'CTATAG	3'.......... CTA	TAG ....5'
SalI	*Streptomyces*	5'GTCGAC	5'..........G	TCGAC ....3'
	albue	3'CAGCTG	3'.......... CAGCT	G ....5'
ScaI	*Streptomyces*	5'AGTACT	5'..........AGT	ACT .... 3'
	caespitosus	3'TCATGA	3'.......... TCA	TGA .... 5'

(c) Mode of action

A restriction enzyme (or restriction endonuclease) is an enzyme that cuts double-stranded DNA. The enzyme makes two incisions, one through each of the sugar-phosphate backbones (*i.e.*, each strand) of the double helix without damaging the nitrogenous bases. The chemical bonds that the enzymes cleave can be reformed by other enzy mes known as ligases, so that restriction fragments carved from different chromosomes or genes can be spliced together, provided their ends are complementary (more below). Many of the procedures of molecular biology and genetic engineering rely on restriction enzymes. The term *restriction* comes from the fact that these enzymes were discovered in *E. coli* strains that appeared to be restricting the infection by certain bacteriophages. Restriction enzymes therefore are believed to be a mechanism evolved by bacteria to resist viral attack and to help in the removal of viral sequences. They are part of what is called the restriction modification system.

The 1978 Nobel Prize in Medicine was awarded to Daniel Nathans, Werner Arber and Hamilton Smith for the discovery of restriction endonucleases, leading to the development of recombinant DNA technology. The first practical use of their work was the manipulation of *E. coli* bacteria to produce human insulin for diabetics. The ability to produce recombinant DNA molecules has not only revolutionized the study of genetics, but has laid the foundation for much of the biotechnology industry. The availability of human insulin (for diabetics), human factor VIII (for males with hemophilia A), and other proteins used in human therapy all were made possible by recombinant DNA.

(d) Types of restriction enzymes

Restriction enzymes are classified biochemically into three types. These are designated as Type I, Type II, Type III. A major type of Type II enzymes are sometimes referred to as Type IV enzymes.

Type I and III systems, both the methylase and restriction activities are carried out by a single large enzyme complex. Although these enzymes recognize specific DNA sequences, the sites of actual cleavage are at variable distances from these recognition sites, and can be hundreds of bases away. Both require ATP for their proper function. Type I restriction enzymes produce DNA cleavage following translocation of the DNA, which makes them important molecular motors. The cleavage of DNA appears to occur after blockage of the translocation activity (often following collision with another translocating Type R-M enzyme, but also due to other factors). These enzymes read the methylation status of their recognition sequence, compare the methylation status of two adenines within the recognition sequence, and if both adenines are unmethylated (a signal that the DNA is non-host DNA), the enzyme undergoes a conformational switch that turns the enzyme into a molecular motor and endonuclease. However, if either one of the adenines is methylated (a signal that the DNA is host DNA) then the enzyme acts as a maintenance methylase and methylates the other adenine.

In type II systems, the restriction enzyme is independent of its methylase, and cleavage occurs at very specific sites that are within or close to the recognition sequence. The vast majority of known restriction enzymes are of type II, and it is these that find the most use as laboratory tools. They produce discrete bands during gel electrophoresis, and are useful for DNA analysis and gene cloning. The first to be discovered and utilized was EcoRI, which is staggered and its recognition sequence is 5'-GAATTC-3'. Type II enzymes are further classified according to their recognition site. Most type II enzymes cut palindromic DNA sequences, while type IIa enzymes recognise non-palindromic sequences and cleave outside of the recognition site, and type IIb enzymes cut sequences twice at both sites outside the recognition sequence. Type IIs enzymes cleave the DNA at a considerable offset from the recognition sequence. The most common type II enzymes are those like HhaI, HindIII and NotI that cleave DNA within their recognition sequences

Restriction enzymes usually occur in combination with one or two modification enzymes (DNA-methyl transferases) that protect the cell's own DNA from cleavage by the restriction enzyme. Modification enzymes recognize the same DNA sequence as the restriction enzyme that they accompany, but instead of cleaving the sequence, they methylate one of the bases in each of the

DNA strands. The methyl groups protrude into the major groove of DNA at the binding site and prevent the restriction enzyme from acting upon it. Together, a restriction enzyme and its "cognate" modification enzyme(s) form a restriction-modification (R-M) system. In some R-M systems the restriction enzyme and the modification enzyme(s) are separate proteins that act independently of each other. In other systems, the two activities occur as separate subunits, or as separate domains, of a larger, combined, restriction-and-modification enzyme.

(e) Restriction enzymes as tools

Recognition sequences typically are only four to twelve nucleotides long. Because there are only so many ways to arrange the four nucleotides—A,C,G and T—into a four or eight or twelve nucleotide sequence, recognition sequences tend to "crop up" by chance in any long sequence. Furthermore, restriction enzymes specific to hundreds of distinct sequences have been identified and synthesized for sale to laboratories. As a result, potential "**restriction sites**" appear in almost any gene or chromosome. Meanwhile, the sequences of some artificial plasmids include a "**linker**" that contains dozens of restriction enzyme recognition sequences within a very short segment of DNA. So no matter the context in which a gene naturally appears, there is probably a pair of restriction enzymes that can cut it out, and which will produce ends that enable the gene to be spliced into a "plasmid".

Another use of restriction enzymes can be to find specific SNPs. If a restriction enzyme can be found such that it cuts only one possible allele of a section of DNA (that is, the alternate nucleotide of the SNP causes the restriction site to no longer exist within the section of DNA), this restriction enzyme can be used to genotype the sample without completely sequencing it. The sample is first run in a restriction digest to cut the DNA, then gel electrophoresis is performed on this digest. If the sample is homozygous for the common allele, the result will be two bands of DNA, because the cut will have occurred at the restriction site. If the sample is homozygous for the rarer allele, the sample will show only one band, because it will not have been cut. If the sample is heterozygous at that SNP, there will be three bands of DNA. This is an example of restriction mapping.

(F) Restriction mapping of DNA fragments

Restriction mapping of DNA fragments and Map construction

Restriction maps provide the basic knowledge of the physical organization of a genomic region or a whole genome with respect to the cleavage sites of restriction endonuclease. An illustrative example is the restriction map of the genome of the lambda (β) bacteriophage shown in (Fig. 7.16) Information on the map is essential before a base sequencing programme is initiated. On the other hand, the restriction map and the genetic map are complementary to each other. Proper superimposition of the restriction map and the genetic map of a genome region or the whole genome allows location of genes in absolute physical terms and eventual isolation. Some of the approaches that are most often adopted for constructing a map of the cleavage sites of restriction endonucleases are the following.

Separate digestion with individual endonuclease followed by simultaneous digestion with pairs of endonuclease

A simplified situation may be considered. Let us assume that each of the endonucleases *Bam*HI, *Bgl* II, *Eco*R I, *Hind* III and *Pst* I has a single cleavage site in a 10kb DNA fragment. Hence each endonuclease would generate two subfragments from the 10 kb DNA fragments. The results of digestions are given in (Fig. 7.17). *Bam*HI generates two fragments of 1kb and 9kb, while *Bgl* II generates two fragments of 3kb and 7kb. Either of the possibilities A or B could be correct. Simultaneous digestion with *Bam*HI and *Bgl* II, however generates three fragments of 1kb, 2kb, and 7kb. Hence, possibility A has to be the correct one and not B. Again *Hind* III generates two fragments of 4kb and 6kb. So between *Bam*H I and *Hind* III either of C or D could be correct, but since simultaneous digestion with these two endonuclease generates fragments of 1kb, 4kb, and 5kb, the possibility D has to be the correct one. The locations of the sites for *Eco*R I and *Pst* I can also be determined D has to be the correct one. The locations of the sites for EcoRI and *Pst* I can also be determined in a similar manner (E to J) and the complete restriction map determined (K) (Fig. 7.17)

Partial Digestion with Individual Restriction Endonucleases

When more than one site exists for a particular restriction endonuclease in a DNA fragment, partial digestion is often useful in locating the sites. In a partial digestion a particular DNA molecules does not suffer cleavage at each an every restriction site. Let us consider a situation when complete digestion of a 10kb DNA fragment with the endonuclease BamHI generates subfragments of 1kb, 3 kb, and 6kb, suggesting that there are two BamHI sites at locations shown in either A or B or C as shown in Fig. 7.18. Partial digestion with the same endonuclease yields DNA subfragments of 1kb, 3kb, 4kb, 6kb and 9kb. It is obvious that the possibility B would never generate on partial digestion a subfragment of 4kb, while the possibility C cannot a subfragment of 9kb. The possibility A on the other hand can generate on partial digestion, all the subfragments found. Hence, A is the correct one. The same 10 kb DNA fragment yields on complete digestion with *Bgl* II, the subfragments of 2kb, 3kb and 5 kb, suggesting that the possibilities D, E and F exist. Partial digestion with the same enzyme, however, produces the subfragments of 2 kb, 3kb, 5kb, 7kb and 8 kb. Now for possibility D can in no way produce partial digestin, subfragment of 7kb, while the possibility F would never generated a subfragment of 8kb. Again the possibility E is capable of yielding all the subfragments mentioned and thus has to be the correct one. The same kind of experimentation and argument would enable one to locate the Hind III sites. Finally double digestion with pairs of endonuclease would allow confirmation of the relative locations of the sites of all the three endonuclease as shown in Figure 7.18. The complete restriction map deduced from the observations is shown in G. Drawing of interference from this kind of experimentation can be made more certain if one end of the DNA is labelled. One could label the 5 end with ATP-$^{3}\gamma$-^{32}P using polynucleotide kinase.

Partial Digestion with an Exonuclease

This method is very useful in ordering multiple sites of a restriction endonuclease in a DNA fragment. First the DNA fragment is completely digested with the specific restriction endonuclease and the sizes of the subfragments recorded. Let us say the subfragments obtained from a 21 kb DNA fragment are of 1kb, 2kb, 3kb, 4kb, 5kb and 6kb. How does one order these subfragments? The DNA fragment is subjected to digestion with the exonuclease Bal 31, which cleaves sequentially from both ends. Samples are withdrawn periodically and Bal 31 digestion is interrupted. Each sample is then subjected to complete digestion with the restriction endonuclease and the sizes of the subfragment produced are determined. Let us say in sample 1, the subfragments obtained are of sizes 0.5 kb, 2kb, 3kb, 3.5kb 5kb and 6kb. Since the subfragments of 2kb, 3kb, 5kb and 6kb are unaffected, these must not be at the termini. So we start construction of the map, the subfragments of 1k and 4kb are at the two termni. Let us assume the sample 2 would have subfragments of 1.5 kb, 2.5 kb, 3kb, 5kb and 6kb. Now the original subfragment of 1kb is completely degraded and the subfragment that is newly affected is of 2kb. The fragments of 3kb, 5kb and 6kb are still untouched. The original subfragment of 4 kb continues to be digested (now 2.5kb). Hence we place the 2kb subfragment next to the 1 kb subfragment. The sample 3 could have subfragments of 0.5 kb, 3kb, 5kb and 5.5 kb. The subsfragments of 3 kb and 5 kb still unaffected. The original subfragment of 4 kb is still being degraded (now 0.5kb), while the degradation of the subfragment of 6kb has just been initiated (now 5,5kb). This would let us place the subfragment of 6kb next to the 2 kb subfragment. The sample 4 may have subfragments of 2.5kb, 4.5 kb and 5kb. Thus the original 4 kb subfragment is completely gone, while the original 6kb subfragment continues to be degraded (now 4.5). This time only subfragment of 3 kb has been newly affected, hence we can allot it a place next to the original 4 kb subfragment. This lets us complete the map construction, the subfragments are in the order of 1kb, 2kb, 6kb, 5kb, 3kb, and 4 kb.

QUESTIONS

1. What are restriction endonucleases? Discuss their role in biotechnology.
2. Write short notes :
   (*a*) Southern blotting
   (*b*) Northern blotting
   (*c*) Western blotting
   (*d*) Compare different blotting techniques.
   (*e*) DNA sequencing
   (*f*) Automatic DNA sequencer
   (*g*) Cell fractionation
   (*h*) FISH
   (*i*) Genomic library
   (*j*) Nucleic acid purification
   (*k*) Plasmid DNA isolation
   (*l*) Restriction enzymes.

POLYMERASE CHAIN REACTION

CHAPTER 8

A powerful technique for amplifying DNA was discovered by Kary Mullis of Cetus corporation, USA in 1985. The prestigious journal 'Science' published Polymerase Chain Reaction (PCR) as major scientific development and Taq DNA polymerase enzyme used in the technique was described as 'molecule of the year' in 1989. PCR is so powerful that from single copy of DNA molecule, millions of copies can be obtained with high accuracy, specificity and in very short time. So in few years since its invention, the polymerase chain reaction (PCR) has become a convenient and trustworthy alternative to gene cloning using vectors.

I. MECHANISM OF BASIC PCR

PCR can be defined as recurring amplification of a particular region of the genome preselected with the help of primers flanking the region of the genome to be amplified. The DNA replication involves polymerization of nucleotides, using template DNA strand with the help of enzyme DNA polymerase and suitable primers. In living cells small single stranded RNA molecule acts as primer, while in PCR short DNA molecules (which are complementary to two sites- either side of the piece of DNA to be amplified) are added as primers. Primers added are usually synthesized oligonucleotides. In PCR this reaction takes place in an 'eppendorf tube' instead of living cells.

The purpose of a PCR (Polymerase Chain Reaction) is to make a huge number of copies of a gene. This is necessary to have enough starting template for sequencing.

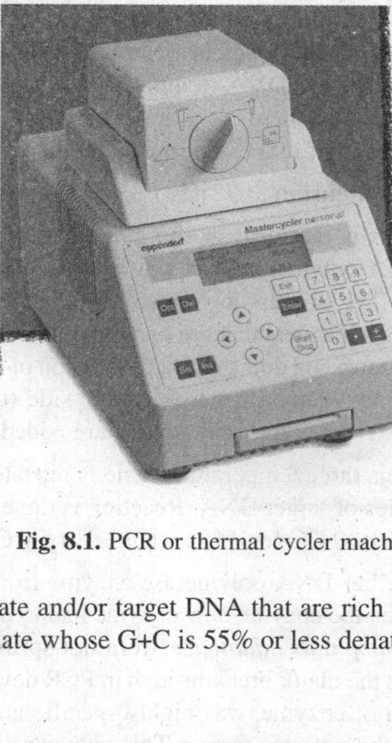

Fig. 8.1. PCR or thermal cycler machine

The cycling reactions:

There are three major steps in a PCR, which are repeated for 30 or 40 cycles in order to obtain large number of copies of desired DNA template. This is done on an automated cycler, which can heat and cool the eppendorf tubes with the reaction mixture in a very short time.

1. Denaturation at 90-98°C:

During the denaturation, the double strands of DNA separate. All enzymatic reactions stop (for example: the extension from a previous cycle). Higher temperature may be required to denature template and/or target DNA that are rich in G+C (>55%). For routine amplification of linear DNA template whose G+C is 55% or less denaturation is done at 94°C for 45 seconds.

2. Annealing at 40-60°C:

Temperature 54°C anneals two complementary primers to ends of separated single-strands of target DNA. The primers are jiggling around, caused by the Brownian motion. Ionic bonds are constantly formed and broken between the single stranded primer and the single stranded template.

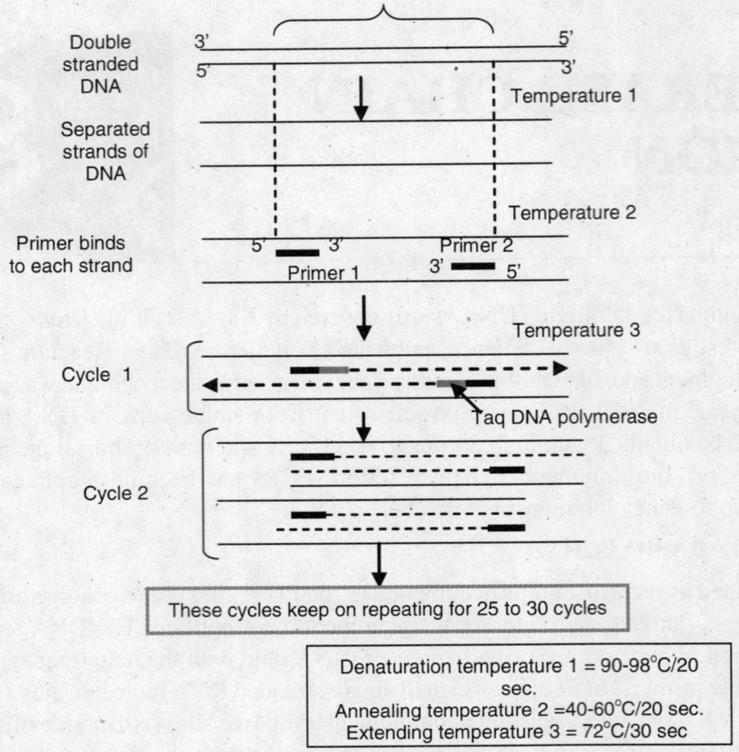

Fig. 8.2. PCR mechanism to produce DNA fragments

The more stable bonds last a little bit longer (primers that fit exactly) and on that little piece of double stranded DNA (template and primer), the polymerase can attach and starts copying the template. Once there are a few bases built in, the ionic bond is so strong between the template and the primer, that it does not break anymore.

3. Extension at 72°C :

This is the ideal working temperature for the polymerase. At this temperature thermostable Taq DNA polymerase use single stranded DNA strands of target and primers to synthesize new strands. The primers have a strong ionic attraction to the template than the forces breaking these attractions. Primers that are on positions with no exact match, get loose again (because of the higher temperature) and don't give an extension of the fragment. The bases (complementary to the template) are coupled to the primer on the 3' side (the polymerase adds dNTP's from 5' to 3', reading the template from 3' to 5' side, bases are added complementary to the template).

This three temperature cycle is repeated 25-30 times and the result is million-fold increase of the copies of target DNA. Reaction is done in DNA thermal cycler using step cycle program set to denature at 94°C for 45 sec, anneal at 54°C for 20 sec and extend at 72°C for 30 sec.

Earlier DNA polymerase enzyme from *E.coli* was used. But as it is sensitive to heat as heat destroyed the enzyme new enzyme had to be added. Use of thermostable DNA polymerase enzyme of *Thermophilus aquaticus* (from hot springs) was reported by Saiki and co-workers in 1988 and that was the major breakthrough in PCR development. It could give more yield, could avoid repeated addition of enzyme, was highly specific and longer fragments upto 10kb can be amplified instead of less than 400bp earlier. This enzyme act best at 72°C and the denaturation of 90°C does not destroy its enzymatic activity. Other thermostable DNA polymerase enzymes used today in PCR are

Pfu DNA polymerase from *Pyrococcus furiosus,* Vent polymerase from *Thermococcus litoralis* and Tbr polymerase from *T. brockianus* etc and they are found more efficiently.

II. COMPARISON OF PCR WITH GENE CLONING

PCR will ultimately replace gene cloning to a large extent except some applications of gene cloning like production of a protein in large quantities for human use. The final result of both gene cloning and PCR can be said to be a selective amplification of specific sequences. Some major point of comparison between the both are as following :

Parameter	PCR	Gene Cloning
Selection of desired DNA template	First step	Last step
Concentration of starting material	Nanogram	Microgram
Biological material required	DNA polymerase	Restriction enzymes, ligase, vector bacteria
Manipulation	In vitro	In vivo
Automation	Total	None
User skill	Not required	Necessary
Probability of error	Less	More
Labour intensive	No	Yes
Cost	Less	More
Time required for a typical experiment	3-5 hours	3-5 days

III. DIFFERENT SCHEMES OF PCR

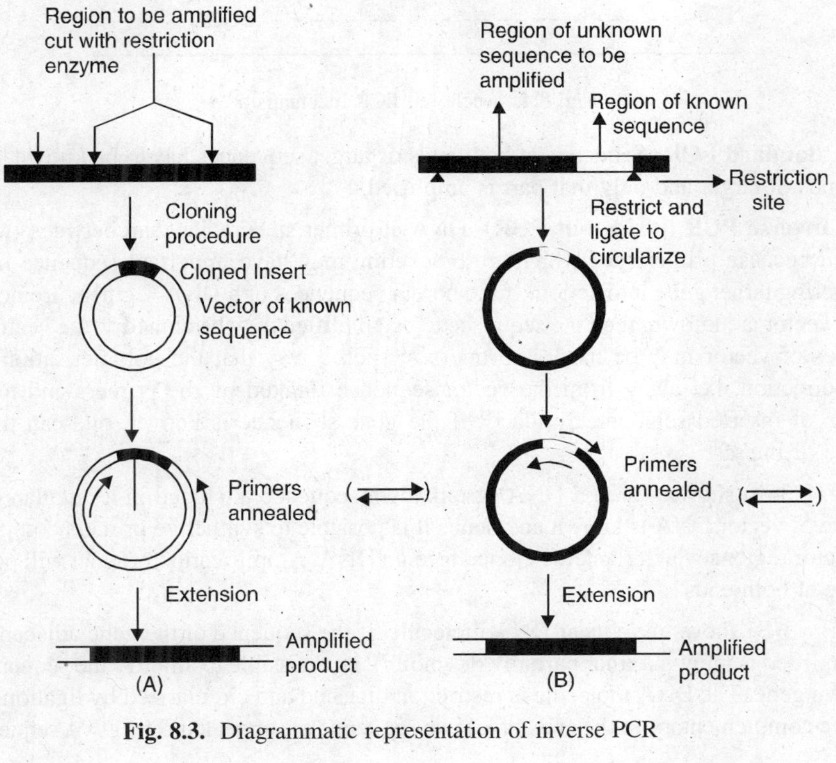

Fig. 8.3. Diagrammatic representation of inverse PCR

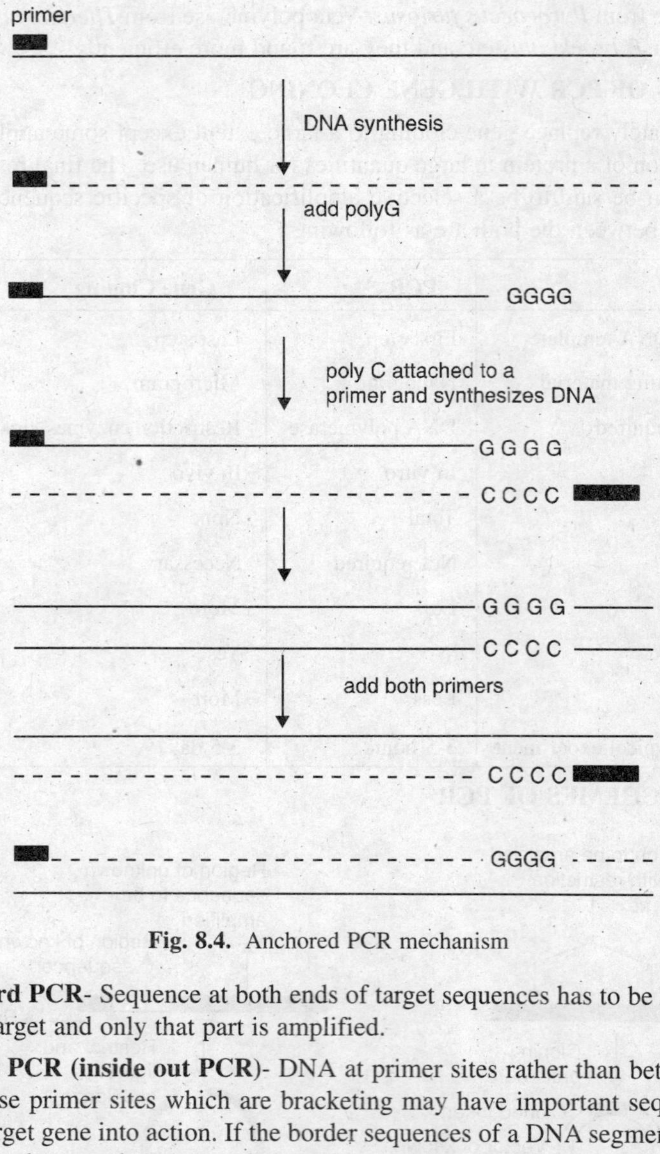

Fig. 8.4. Anchored PCR mechanism

A. Standard PCR- Sequence at both ends of target sequences has to be known. Two primers define ends of target and only that part is amplified.

B. Inverse PCR (inside out PCR)- DNA at primer sites rather than between two primers is amplified because primer sites which are bracketing may have important sequence like promoter for triggering target gene into action. If the border sequences of a DNA segment are not known and those of vector are known, then the sequence to be amplified may be cloned in the vector and border sequences of vector may be used as primers in such a way that the polymerisation proceeds in inverse direction, i.e. away from the vector sequence flanked by the primers and towards DNA sequence of inserted segment. Similarly, if the gene sequence is known, one can use its border sequence of interest.

Fig. 8.3(A) shows foreign DNA of unknown sequence is cloned in a circular vector DNA. Sequence of vector DNA is known and hence it is possible to synthesize primers complementary to those vector regions which flank the cloned foreign DNA. Amplification product will possess vector sequence at both ends.

Fig. 8.3(B) shows in a linear DNA molecule, if the sequence of the gene adjacent to the gene to be amplified is known (either partially or wholly), it is possible to amplify the sequences flanking the known gene. The DNA molecule is restriction digested and circularized by ligation.The primers which are complementary to the ends of known sequence are annealed. The DNA sequence flanking the known sequence can now be amplified. The amplifcation product will possess known sequence at both ends.

C. Anchored PCR (single sided PCR)- In the basic PCR technique and the inverse PCR, one has to use two primers representing the sequences lying at the ends of sequences to be amplified. But sometimes, we may have knowledge about sequence at only one of the two ends of the DNA sequence to be amplified. In such cases, anchored PCR may be used, which will utilize only one primer instead of two primers. In this technique, due to the use of one primer, only one strand will be copied first, after which a poly G tail will be attached at the end of the newly synthesized strand. This newly synthesized strand with poly G tail at its 3′ end will then become template for the daughter strand synthesis utilizing an anchor primer with which a poly C sequence is linked to complement with poly G of the template. In the next cycle, both the original primer and anchored primer will be used for gene amplification. Anchored PCR is developed for studying genes that encode proteins for which partial sequence are known.

D. Asymmetric PCR

Asymmetric PCR is used to preferentially amplify one strand of the original DNA more than the other. It finds use in some types of sequencing and hybridization probing where only one of the two complementary stands is required. PCR is carried out as usual, but with a great excess of the primers for the chosen strand. Due to the slow (arithmetic) amplification later in the reaction after the limiting primer has been used up, extra cycles of PCR are required. A recent modification on this process, known as **L**inear-**A**fter-**T**he-**E**xponential-PCR (LATE-PCR), uses a limiting primer with a higher melting temperature (Tm) than the excess primer to maintain reaction efficiency as the limiting primer concentration decreases mid-reaction.

E. Quantitative PCR (Q-PCR). is used to measure the quantity of a PCR product (preferably real-time). It is the method of choice to quantitatively measure starting amounts of DNA, cDNA or RNA. Q-PCR is commonly used to determine whether a DNA sequence is present in a sample and the number of its copies in the sample. The method with currently the highest level of accuracy is **Quantitative real-time PCR**. It is often confusingly known as RT-PCR (**R**eal **T**ime PCR) or RQ-PCR. QRT-PCR or RTQ-PCR are more appropriate contractions. QRT-PCR methods use fluorescent dyes, such as Sybr Green, or fluorophore-containing DNA probes, such as TaqMan, to measure the amount of amplified product in real time.

F. RT-PCR: (Reverse Transcription PCR) is a method used to amplify, isolate or identify a known sequence from a cellular or tissue RNA. The PCR is preceded by a reaction using reverse transcriptase to convert RNA to cDNA. RT-PCR is widely used in expression profiling, to determine the expression of a gene or to identify the sequence of an RNA transcript, including transcription start and termination sites and, if the genomic DNA sequence of a gene is known, to map the location of exons and introns in the gene. The 5' end of a gene (corresponding to the transcription start site) is typically identified by a RT-PCR method, named RACE-PCR, short for *Rapid Amplification of cDNA Ends.*

Real-Time Vs Traditional PCR

Real-Time chemistries allow for the detection of PCR amplification during the early phases of the reaction. Measuring the kinetics of the reaction in the early phases of PCR provides a distinct advantage over traditional PCR detection. Traditional methods use Agarose gels for detection of PCR amplification at the final phase or end-point of the PCR reaction. It is hard to differentiate between the 5-fold changes on the Agarose gel while Real-Time PCR is able to detect a two-fold change with high efficiency.

PCR Phases :

To understand why end-point PCR is limiting, it is important to understand what happens during a PCR reaction. A basic PCR run can be broken up into three phases:

Exponential: Exact doubling of product is accumulating at every cycle (assuming 100% reaction efficiency). The reaction is very specific and precise.

Linear (High Variability): The reaction components are being consumed, the reaction is slowing, and products are starting to degrade.

Plateau (End-Point: Gel detection for traditional methods): The reaction has stopped, no more products are being made and if left long enough, the PCR products will begin to degrade.

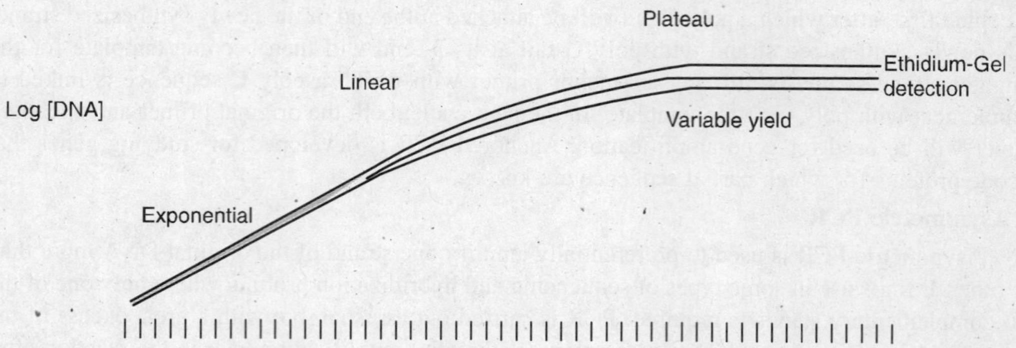

Fig. 8.5. PCR reaction rate

IV. Applications

1. PCR for screening of uncharacterized mutations

Small insertions or deletions can be simply detected by designing primers from regions closely flanking the mutation site and distinguishing the normal and mutant alleles by size on polyacrylamide or agarose gels. If the mutation changes a restriction site, mutant and normal alleles can be distinguished by amplifying across the mutant site and digesting the PCR product with relevant restriction endonuclease.

2. PCR permits rapid genotyping for polymorphic markers

Restriction site polymorphisms (RSPs) result in alleles possessing or lacking a specific restriction site. Such polymorphisms can be typed using Southern blot hybridization. A DNA probe representing the locus is hybridized against genomic DNA samples that have been digested with the appropriate restriction enzyme and size-fractionated by agarose gel electrophoresis. The resulting RFLPs have two alleles corresponding to the presence or absence of the restriction site. As a convenient alternative to RFLPs, PCR can type RSPs by simply designing primers using sequences which flank the polymorphic restriction site, amplifying from genomic DNA, then cutting the PCR product with the appropriate restriction enzyme and separating the fragments by agarose gel electrophoresis.

3. Isolation of genomic DNA

PCR allows isolation of DNA fragments from genomic DNA by selective amplification of a specific region of DNA. This use of PCR augments many methods, such as Southern and northern blotting and DNA cloning that require large amounts of DNA, representing a specific DNA region. PCR supplies these techniques with high amounts of pure DNA, enabling analysis of DNA samples even from very small amounts of starting material.

4. Amplification of DNA

Because PCR amplifies the regions of DNA that it targets, PCR can be used to analyze extremely small amounts of sample. This is often critical for forensic analysis, when only a trace amount of

DNA is available as evidence. PCR may also be used in the analysis of ancient DNA that is thousands of years old. These PCR-based techniques have been successfully used on animals, such as a forty-thousand-year-old mammoth, and also on human DNA. Viral DNA can likewise be detected by PCR.

5. Human Health and the Human Genome Project

PCR has very quickly become an essential tool for improving human health and human life. Medical research and clinical medicine are profiting from PCR mainly in two areas: detection of infectious disease organisms, and detection of variations and mutations in genes, especially human genes. The primers used need to be specific to the targeted sequences in the DNA of a virus, and the PCR can be used for diagnostic analyses or DNA sequencing of the viral genome. The high sensitivity of PCR permits virus detection soon after infection and even before the onset of disease. Such early detection may give physicians a significant lead in treatment. The amount of virus ("viral load") in a patient can also be quantified by PCR-based DNA quantitation techniques. Real-time PCR is an established tool for DNA quantification that measures the accumulation of DNA product after each round of PCR amplification.

The method is especially useful for searching out disease organisms that are difficult or impossible to culture, such as many kinds of bacteria, fungi, and viruses, because it can generate analyzable quantities of the organism's genetic material for identification. It can, for example, detect the AIDS virus sooner during the first few weeks after infection than the standard ELISA test. PCR looks directly for the virus's unique DNA, instead of the method employed by the standard test, which looks for indirect evidence that the virus is present by searching for antibodies the body has made against it.

In short, if a disorder is caused by an infectious agent, PCR can, in principle, ferret out the culprit. More than 60 PCR protocols for identifying pathogens have been described to date, and at least 10 clinical products are available for detecting the evasive organisms that cause such diseases as tuberculosis, chlamydia, viral meningitis, viral hepatitis, AIDS, and cytomegalovirus.

DNA sequencing reveals crucial variations in the nucleotides that constitute genes. These mutational changes produce disease and even death by forcing the genes to produce abnormal proteins, or sometimes no proteins at all. DNA sequencing involves first isolating and duplicating DNA segments for nucleotide analysis. Thus PCR is an essential tool for the Human Genome Project because it can quickly and easily generate an unlimited amount of any piece of DNA for this kind of study.

6. "Ancient DNA" and Evolutionary Relationships

Archaeologists have happily switched on PCR and are applying it in an amazing variety of ways. PCR is also helping sort out relationships among vanished human groups, and tracing human migrations. Studies on human brains that survived 8,000 years in a Florida sinkhole more or less intact indicate, for example, that the people who lived there were genetically different from today's Native Americans.

8. Modern Systematics, Ecological Studies, and Animal Behavior

But DNA need not be ancient to provide information about evolutionary relationships. With PCR, systematists can measure differences in DNA sequences between species directly, and if they select sequences that have changed little during evolution, between major classes of organisms. The speed and automation of the process means that scientists can easily compare dozens or even hundreds of individuals, putting their conclusions on a firmer basis.

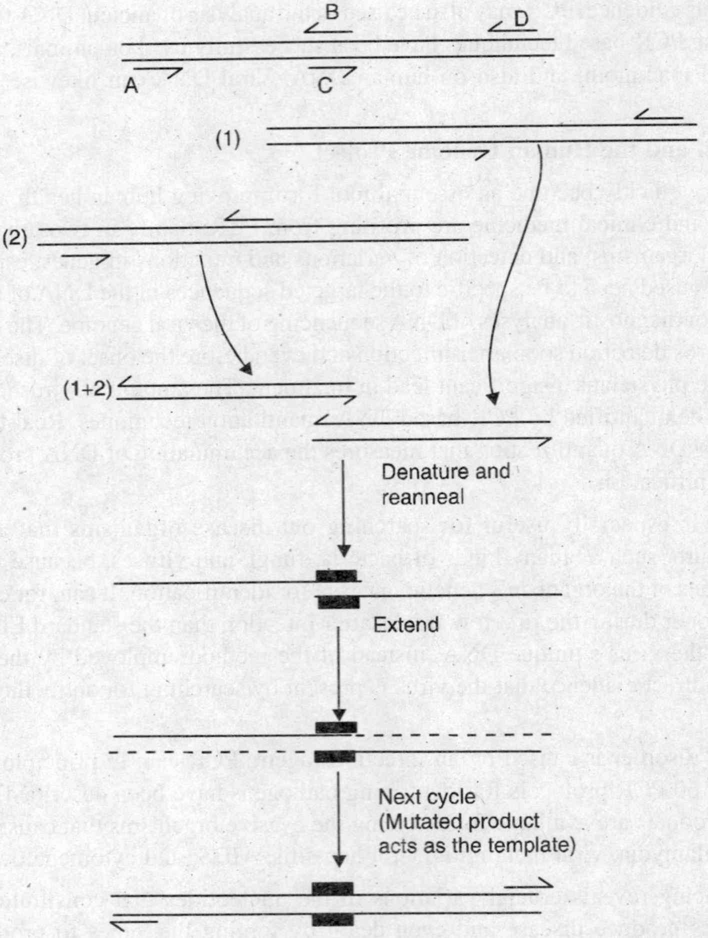

Fig. 8.6. Site directed mutagenesis by overlap extension PCR.A single stretch of DNA is amplified in two different segments by using primer pairs A, B andC.D. Internal ends of these two segments overlap because the internal primers B and C are complementary.When denatured and reannealed, the amplified segments of the two PCR reactions form base pairing at the overlap region and can be extended in the next cycle to produce the whole original sequence. The primer at the overlap region (B andC) carry alteration (s) is the base (s). Mutuations can be induced at any region of the gene by synthe sizing altered internal primers for that region.

With PCR, scientists can glean genetic information from the faintest traces of the shyest, rarest animal-urine, feces, scent marks, infinitesimal bits of hair or skin rubbed onto a tree as the elusive creature passes by. In addition to information that aids classification, individuals can be identified so as to estimate population size in a particular locale, or to determine the geographic range of a single animal, or a group of them. The technique can be adapted to similar studies of plants, for analysis of patterns of seed dispersal and the relative reproductive success of specific plants.

It is also a tool for monitoring the release of genetically engineered organisms into the environment.

9. Site-directed mutagenesis by PCR

PCR can be used to generate a mutation in DNA. In this case a primer is used which is not perfectly complimentary to the template, it has a base pair different from the template DNA. As the primers are the ends of newly-synthesized strands, should there be a mis-match during the first cycle in binding the template DNA strand, after that first round, the primer-based strand (containing the mutation) would be at equal concentration to the original template. After successive cycles, it would exponentially grow, and after 25, would outnumber the original, unmutated strand, resulting in a nearly homogeneous solution of mutated amplified fragments. The template DNA is typically eliminated by enzymatic digestion with Dpn1, a restriction enzyme which cleaves only methylated DNA. The template, which is derived from an alkaline lysis plasmid preparation and therefore is methylated, is destroyed in this step, but the mutated plasmid is preserved because it was generated *in vitro* and is unmethylated. Since the primers used are terminal, the mutations induced in the gene will also be at the ends of the gene. To overcome this limitation a method has been designed called as *overlap extension method*. Using this method it is possible to introduce mutation at will at any site within the gene of interest (Fig. 8.6). The method consists of amplifying the particular gene as two separate segments in two separate PCR reactions. This is achieved by using in both the PCR reactions one primer at the end of the gene and another primer at a sequence internal to the gene (Fig. 8.7). The internal primers used on the two reactions are complementary to each other. It is in these internal primers that a sequence variations are introduced depending upon the type of mutation desired. The two PCR products are now mixed and denatured when reannealing is allowed to take place the two overlapping strands of the PCR products anneal and their overlapping ends now act as primers. In the PCR that now will be performed, the complete gene will be synthesized bearing the mutation. The site of the internal primers can be varied within the gene depending upon the site where the mutation is to be introduced. The method is claimed to give 100% efficiency in the introducing site specific mutations.

V. THE FUTURE OF PCR

The present technology for doing PCR, about the size of a microwave oven and costing several thousand dollars, seems destined for further radical improvement. By tinkering with variables such as chemical reagents and pH, researchers have already reported success at copying larger and larger pieces of DNA, including the entire genome of HIV.

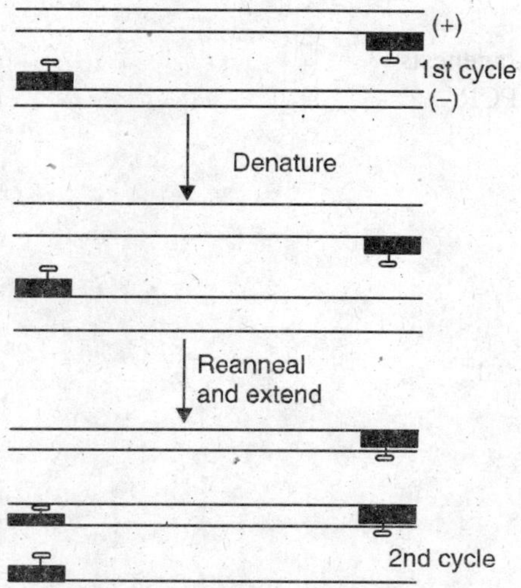

Fig. 8.7. Site directed point mutation by PCR. Basic PCR reaction is similar to that shown in figure. Here, the primers contain the required base change (▬ indicates primer with one base change at the position of the cross bar). The base change is minom and does not alter the annealing of the primer with the template. The base change is then faithfully copied into the new DNA strands in the subsequent cycles.

Extraordinary miniaturization of the hardware is also underway, as experimenters squeeze PCR onto chip-sized devices. Criss crossed with the tiniest of troughs to hold the reagents and the DNA, the chips are heated electrically and cool down much faster than the present generation of machines, so amplification is even speedier than today's swift process. Already researchers have reported using a handheld battery-powered gadget to copy pieces of DNA that contained eight different cystic fibrosis mutation sites.

While such experimental chip-based devices are not yet ready for prime time, they are hastening the day when scientists can take them on the road, and patients will be able to get on-the-spot readouts of their DNA. Before long it may be quite routine to diagnose an infectious or genetic disorder, or even detect an inherited predisposition to cancer or heart disease, right in the doctor's office.

PCR is doing for genetic material what the invention of the printing press did for written material-making copying easy, inexpensive, and accessible. In principle, PCR can reproduce the genetic material of any organism in essentially unlimited quantities, so it can be used to analyze any cells containing that material. Whether they are germs, rare medicinal plants, or human beings, eventually we can know whatever is recorded in their DNA. With simple organisms, to know their DNA will be to know almost everything about them. With complicated ones, like people, DNA is only part of the story, but a very big part. Through PCR, we will be probing the genetic past, and peering into the genetic future, for many years to come.

QUESTIONS

1. Define PCR. Discuss different variations of basic PCR technique and its applications.
2. Write short notes :
   (*a*) PCR
   (*b*) RT-PCR
   (*c*) Taq polymerase
   (*d*) Q-PCR
   (*e*) Site directed mutagenesis
   (*f*) Applications of PCR.

MOLECULAR MARKERS

CHAPTER 9

In recent years, different marker system such as Restriction Fragment Length Polymorphisms (RFLPs), Random Amplified Polymorphic DNAs (RAPDs), Sequence Tagged Sites (STS), Amplified Fragment Length Polymorphisms (AFLPs), Simple Sequence Repeats (SSRs) or microsatellites, Single Nucleotide Polymorphisms (SNPs) and others have been developed and applied to a range of crop species including cereals. The relative advantages and disadvantages of these techniques are summarised in Table 9.1.

Table 9.1. Comparison of the most common used marker systems in cereals

Feature	RELPs	RAPDs	AFLPs	SSRs	SNPs
DNA required (μg)	10	0.02	0.5 – 1.0	0.05	0.05
DNA quality	high	high	moderate	moderate	high
PCR-based	no	yes	yes	yes	yes
Number of polymorph loci analysed	1.0–3.0	1.5–50	20–100	1.0–3.0	1.0
Ease of use	not easy	easy	easy	easy	easy
Amenable to automation	low	moderate	moderate	high	high
Reproducibility	high	unreliable	high	high	high
Development cost	low	low	moderate	high	high
Cost per analysis	high	low	moderate	low	low

This information suggests that RFLP, SSR and AFLP markers are most effective in detecting polymorphism. However, given the large amountof DNA required for RFLP detection and the difficulties in automating RFLP analysis, AFLPs and SSRs are currently the most popular markers in cereals. An increasing amount of sequence information and the determination of the gene function in cereals will lead in the near future to the preferred use of new market types, such as SNPs. Application of these markers for genetic studies of cereals have been so much diverse. Main uses include :

- assessment of genetic variability and characterisation of germplasm;
- identification and fingerprinting of genotypes;
- estimation of genetic distances between population, inbreeds and breeding mater;
- detection of monogenic and qualitative trait loci (QTL);
- marker-assisted selection;
- identification of sequences of useful candidate genes, etc.

I. RESTRICTION FRAGMENT LENGTH POLYMORPHISM

RFLP (often pronounced "rif lip", as if it were a word) is a method used by molecular biologists to follow a particular sequence of DNA as it is passed on to other cells. RFLPs can be used in many different settings to accomplish different objectives. RFLPs can be used in paternity cases or criminal cases to determine the source of a DNA sample. RFLPs can be used to determine the disease status of an individual. RFLPs can be used to measure recombination rates which can lead to a genetic map with the distance between RFLP loci measured in centiMorgans.

RFLP, as a molecular marker, is specific to a single clone/restriction enzyme combination. It is a difference in homologous DNA sequences that can be detected by the presence of fragments of different lengths after digestion of the DNA samples in question with specific restriction endonucleases.

Most RFLP markers are co-dominant (both alleles in heterozygous sample will be detected) and highly locus-specific.

An RFLP probe is a labeled DNA sequence that hybridizes with one or more fragments of the digested DNA sample after they were separated by gel electrophoresis, thus revealing a unique blotting pattern characteristic to a specific genotype at a specific locus. Short, single- or low-copy genomic DNA or cDNA clones are typically used as RFLP probes.The RFLP probes are frequently used in genome mapping and in variation analysis (genotyping, forensics, paternity tests, hereditary disease diagnostics, etc.) (Fig. 9.1).

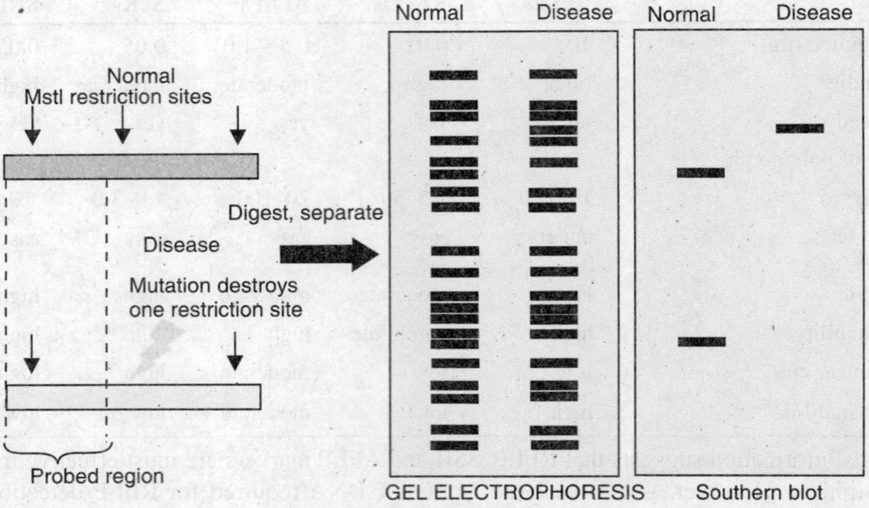

Fig. 9.1. Disease diagnosis by RFLP

Procedure

Usually, DNA from an individual specimen is first extracted and purified. Purified DNA may be amplified by polymerase chain reaction (PCR). The DNA is then cut into *restriction fragments* using suitable endonucleases, which only cut the DNA molecule where there are specific DNA sequences, termed recognition sequence or restriction sites, that are recognized by the enzymes. These sequences are specific to each enzyme, and may be either four, six, eight, ten or twelve base pairs in length. The more base pairs there are in the restriction site, the more specific it is and the lower the probability that it will find a place to be cut. The restriction fragments are then separated according to length by agarose gel electrophoresis. The resulting gel may be enhanced by Southern blotting. Alternatively, fragments may be visualised by pre-treatment or post-treatment of the agarose gel, using methods such as ethidium bromide staining or silver staining respectively.

RFLPs have provided valuable information in many areas of biology, including: screening human DNA for the presence of potentially deleterious genes. Providing evidence to establish the innocence of or a probability of the guilt of, a crime suspect by DNA "fingerprinting". The distance between the locations cut by restriction enzymes (the *restriction sites*) varies between individuals,

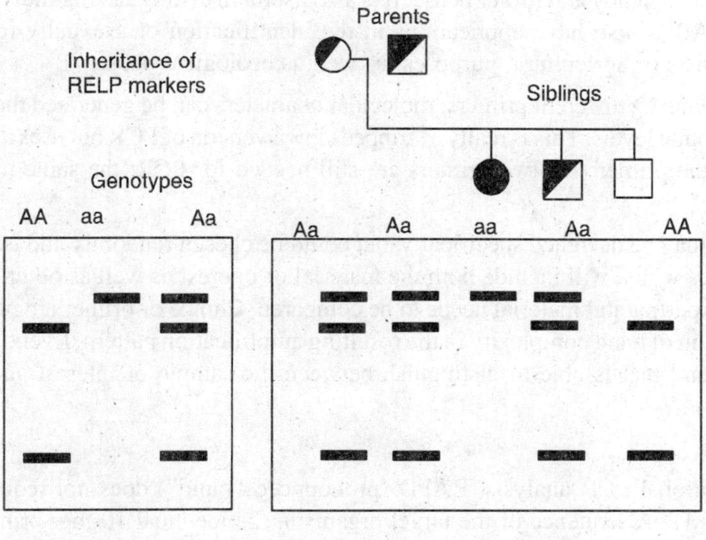

Fig. 9.2. Inheritance of RFLP markers.

due to insertions, deletions or transversions. This causes the length of the fragments to vary, and the position of certain amplicons differs between individuals (thus polymorphism). This can be used to genetically tell individuals apart. It can also show the genetic relationship between individuals (Fig. 9.2), because children inherit genetic elements from their parents. Mitochondrial DNA RFLP analyses can lead to the determination of maternal relationships. Fragments may also be used to determine relationships among and between species by comparison of the resulting haplotypes. RFLP is a technique used in marker assisted selection. Terminal Restriction Fragment Length Polymorphism (TRFLP sometimes T-RFLP) is a molecular biology technique initially developed for characterizing bacterial communities in mixed-species samples. The technique has also been applied to other groups including soil fungi.

The technique works by PCR amplification of DNA using primer pairs that have been labeled with fluorescent tags. The PCR products are then digested using RFLP enzymes and the resulting patterns visualized using a DNA sequencer. The results are analyzed either by simply counting and comparing bands or peaks in the TRFLP profile, or by matching bands from one or more TRFLP runs to a database of known species.

PCR-RFLP

Isolation of sufficient DNA for RFLP analysis is time consuming and labour intensive. However, PCR can be used to amplify very small amounts of DNA, usually in 2-3 hours, to the levels required for RFLP analysis. Therefore, more samples can be analyzed in a shorter time. An alternative name for the technique is Cleaved Amplified Polymorphic Sequence (CAPS) assay.

Limitations

1. RFLP is a multistep procedure involving restriction enzymatic cleavage, electrophoresis, southern blotting and detection of specific sequences.

2. It is a time consuming process.

II. RANDOM AMPLIFIED POLYMORPHIC DNA

This technique can be used to determine taxonomic identity, assess kinship relationships, detect interspecific gene flow, analyse hybrid speciation, and create specific probes. Advantages of RAPDs include suitability for work on anonymous genomes, applicability to work where limited

DNA is available, efficiency and low expense. It is also useful in distinguishing individuals, cultivars or accessions. RAPDs also have applications in the identification of asexually reproduced plant varieties for forensic or agricultural purposes, as well as ecological ones.

In RAPD, by using different primers, molecular characters can be generated that are diagnostic at different taxonomic levels. This is really a stripped-down version of PCR but uses a single sequence in the design of the primer (i.e. two primers are still needed for PCR: the same primer is used at either end).

The primer may be designed specifically, but could be chosen randomly and is used to amplify a series of samples which will include both the material of interest as well as other control samples with which the experimental material needs to be compared. Choice of primer length will be critical to the determination of band complexity in the resulting amplification pattern. Eventually a particular probe will be found that is able to distinguish between the sample of interest and those that are different.

Procedure

Unlike traditional PCR analysis, RAPD (pronounced "rapid") does not require any specific knowledge of the DNA sequence of the target organism: the identical 10-mer primers will or will not amplify a segment of DNA, depending on positions that are complementary to the primers' sequence. For example, no fragment is produced if primers annealed too far apart or 3' ends of the primers are not facing each other. Therefore, if a mutation has occured in the template DNA at the site that was previously complementary to the primer, a PCR product will not be produced, resulting in a different pattern of amplified DNA segments on the gel (Fig. 9.3).

RAPD is an inexpensive yet powerful typing method for many bacterial species.

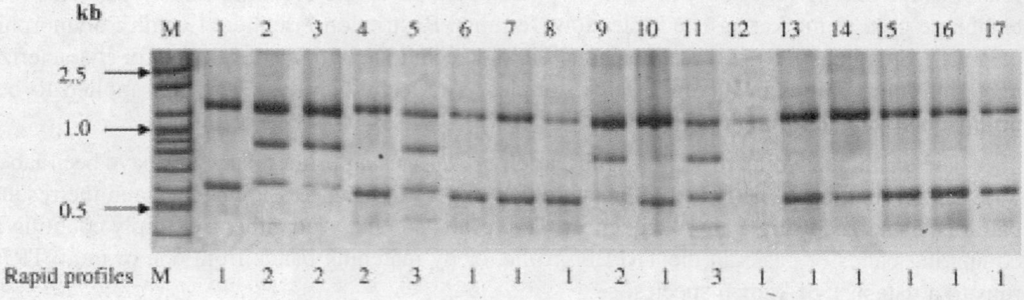

Fig. 9.3. A RAPD profile

Selecting the right sequence for the primer is very important because different sequences will produce different band patterns and possibly allow for a more specific recognition of individual strains.

RAPD amplification products can be either variable (polymorphic) or constant (non-polymorphic). In RAPD analysis of several individuals within a species, and species within a genus, constant fragments diagnostic for a genus may be identified, as well as fragments which are polymorphic between species of the genus.

RAPDs can be applied to analyse fusion of genotypes at different taxonomic levels. At the level of the individual, RAPD markers can be applied to parentage analysis, while at the population level, RAPD can detect hybrid populations, species or subspecies. The detection of genotype hybrids relies on the identification of diagnostic RAPD markers for the parental genotypes under investigation. However; RAPD markers tend to underestimate genetic distances between distantly related individuals, for example in inter-specific comparisons. It is wise to be cautious when using RAPD for taxonomic studies above the species level.

RAPDs have also begun to be used for analyses of hybridization events where allozymes have not proven to be sensitive enough for hybrid genotypes. RAPDs may also be used to assess paternity and kinship relationships in large offspring samples, to determine the actual father from a number of potential fathers. Conventional RFLP techniques are ill-suited for the analysis of paternity and estimation of reproductive success in species with large offspring clutches, because of the need to determine paternity for each individual offspring. RAPD fingerprinting provides a ready alternative for such cases. Synthetic offspring may be produced by mixing equal amounts of the DNA of the mother and the potential father. The amplification products from the synthetic offspring should ideally contain the full complement of bands that appear in any single offspring of these parents.

But certain marks may only get amplified from an offspring, but not from either of its parents. The occurrence of non-parental bands in offspring from known primate pedigrees has raised concerns in parentage determinations when single individuals are analysed. In contrast, the synthetic offspring is a complete representation of both parental genomes and will match the profile of a large sample of offspring, since in both the 'synthetic' and the real offspring clutch, allele combinations that may cause amplification artifacts are represented in equal amounts. The degree of mismatch between synthetic offspring and actual offspring clutches is indicative of mixed paternity, which can be measured by quantitative determination of mixed genome samples.

Any diagnostic RAPD marker can be eluted from the gel, reamplified, radiolabelled with P32 and then serve as an inexhaustible supply of probe in Southern hybridization. Such probes may be used to exclude the possibility of co-migration of fragments of different sequence but similar size. Other applications include generation of probes for taxonomic analysis or the quantitative estimation of the presence of a certain genome in a mixed sample by Southern analysis. RAPD probes have been used to detect RFLPs in tomato species.

Limitations of RAPD

1. Nearly all RAPD markers are dominant, i.e. it is not possible to distinguish whether a DNA segment is amplified from a locus that is heterozygous (1 copy) or homozygous (2 copies). Co-dominant RAPD markers, observed as different-sized DNA segments amplified from the same locus, are detected only rarely.
2. PCR is an enzymatic reaction, therefore the quality and concentration of template DNA, concentrations of PCR components, and the PCR cycling conditions may greatly influence the outcome. Thus, the RAPD technique is notoriously laboratory dependent and needs carefully developed laboratory protocols to be reproducible.
3. Mismatches between the primer and the template may result in the total absence of PCR product as well as merely decreased amount of the product. Thus, the RAPD results can be difficult to interpret.

Developing Locus-specific, Co-Dominant Markers from RAPDs

The polymorphic RAPD marker band is isolated from the gel. It is amplified in the PCR reaction. The PCR product is cloned and sequenced. New longer and specific primers are designed for the DNA sequence, which are called the Sequenced Characterized Amplified Region Marker (SCAR).

Table 9.2. Difference between RAPD and RFLP

RAPD	RFLP
Quantity of DNA required for analysis is small, i.e. 10-50ng.	Large quantity of purified DNA required ,i.e. 2-10ng.
Same primers with arbitary sequences can be used for different species.	Different species specific probes are required.
Fewer steps so about five times faster than RFLP	Lengthier in process
Technique comparatively less reliable	More reliable
Cannot detect allelic variants	Can detect allelic variants
1-10 loci detected	1-3 loci detected

III. AMPLIFICATION FRAGMENT LENGTH POLYMORPHISM

Amplified Fragment Length Polymorphism (AFLP) is a polymerase chain reaction (PCR) based on genetic fingerprinting technique that was developed in the early 1990's by Keygene. AFLP can be used in the fingerprinting of genomic DNA of varying origins and complexities. The amplification reaction is rigorous, versatile and robust, and appears to be quantitative. While AFLP is capable of producing very complex fingerprints (100 bands where RAPD produces 20), it is a technique that requires DNA of reasonable quality and is more experimentally demanding. AFLP uses restriction enzymes to cut genomic DNA, followed by ligation of complementary double stranded adaptors to the ends of the restriction fragments. A subset of the restriction fragments are then amplified using 2 primers complementary to the adaptor and restriction site fragments. The fragments are visualized on denaturing polyacrylamide gels either through autoradiographic or fluorescence methodologies.

Procedure

AFLP-PCR is a highly sensitive method for detecting polymorphisms in DNA.The technique was originally described by Vos and Zabeau in 1993. The procedure of this technique is divided into three steps: (Fig. 9.4).

1. Digestion of total cellular DNA with one or more restriction enzymes that cuts frequently (*Mse*I, 4 bp recognition sequence) and one that cuts less frequently (*Eco*RI, 6 bp recognition sequence). The resulting fragments are ligated to end-specific adaptor molecules.
2. Selective amplification of some of these fragments with two PCR primers that have corresponding adaptor and restriction site specific sequences.
3. Electrophoretic separation of amplicons on a gel matrix, followed by visualisation of the band pattern or from capillary electrophoresis instrument (Fig. 9.5).

In a second, "selective", PCR, using the products of the first as template, primers containing two further additional bases, chosen by the user, are used. The *Eco*RI-adaptor specific primer used bears a label (fluorescent or radioactive). Gel electrophoretic analysis reveals a pattern (fingerprint) of fragments representing about $1/4000^{th}$ of the *Eco*RI-*Mse*I fragments.AFLP's, can be codominant markers, like RFLP's. Codominance results when the polymorphism is due to sequences within the amplified region. Yet, because of the number of bands seen at one time, additional evidence is needed to establish that a set of bands result from different alleles at the same locus. If, however, the polymorphism is due to presence/absence of a priming site, the relationship is dominance. The non-priming allele will not be detected as a band. Compared to RAPD, fewer primers should be needed to screen all possible sites.

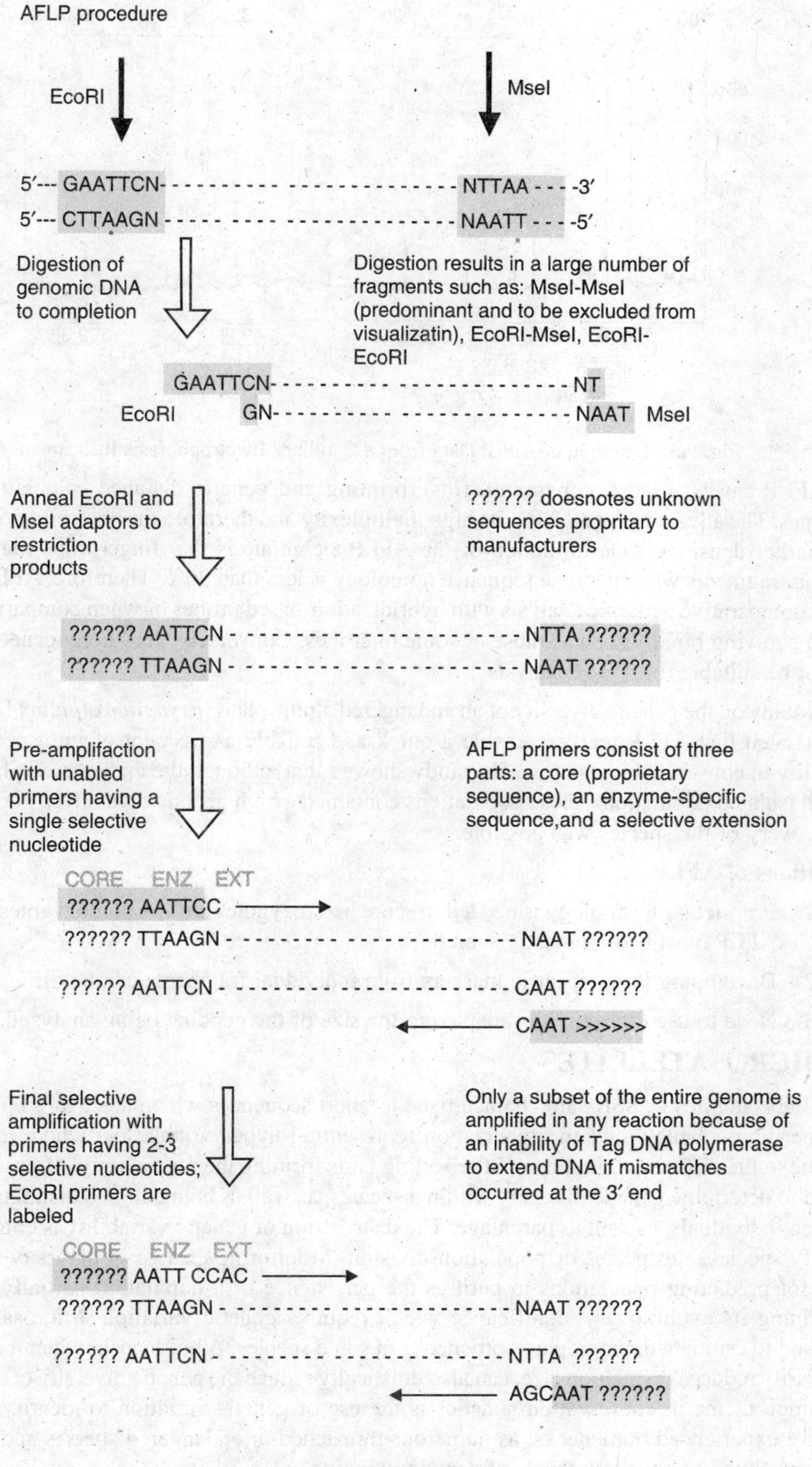

Fig. 9.4. AFLP Procedure

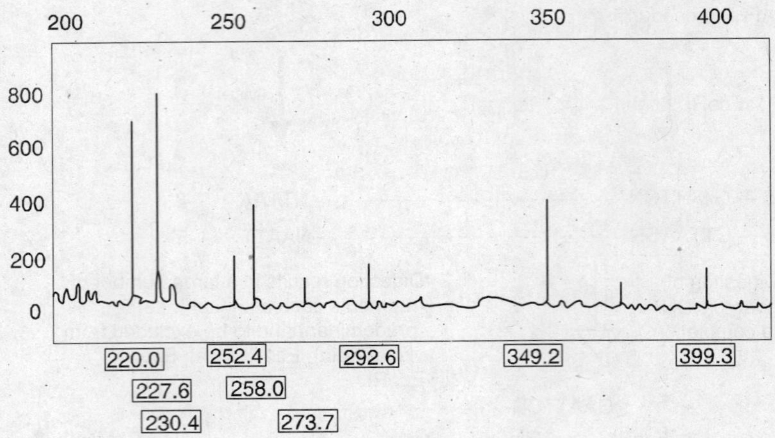

Fig. 9.5. Example of AFLP Data from a Capillary Electrophoresis Instrument

AFLP can be used for mapping, fingerprinting and genetic distance calculation between genotypes. The advantage of AFLP is its high multiplexity and therefore the possibility of generating high marker densities. One limitation of the AFLP technique is that fingerprints may share few common fragments when genome sequence homology is less than 90%. Therefore, AFLP cannot be used in comparative genomic analysis with hybridization-based probes or when comparing genomes that are evolving rapidly such as those of some microbes. Conversely, very homogeneous genomes may not be suitable for AFLP analysis.

A study on the genetic diversity of an endangered alpine plant (*Eryngium alpinum* L. (Apiaceae) demonstrated that AFLP markers enable a quick and reliable assessment of intraspecific genetic variability in conservation genetics. The study showed that although the endangered plant occurred in small isolated populations, these populations contained a high genetic diversity, a good indication that recovery of the species was possible.

Limitations of AFLP

1. Proprietary technology is needed to score heterozygotes and ++ homozygotes. Otherwise, AFLP must be dominantly scored.
2. Developing locus-specific markers from individual fragments can be difficult.
3. Need to use different kits adapted to the size of the genome being analyzed.

IV. MICROSATELLITES

Microsatellites or SSRs are characterized as short sequences which are 1 to 5 bp long. These monomer units display tandem organization representing hypervariable loci where the variations arise due to the differences in the no. of repeating units forming the repeat array. Microsatellites can be used to determine genetic diversity within a species, as well as being able to distinguish varieties and even individuals, as well as parentage. The distribution of genetic variability is commonly used to verify species, subspecies or population division. Monitoring change in diversity may also be useful for predicting populations in peril as the persistence of a population partially depends on maintaining its evolutionary significance which requires genetic variation. Microsatellites have been used to estimate demographic bottlenecks in some species. A bottleneck, when it severely and temporarily reduces population size, can also drastically reduce the genetic diversity of a population. A common theme in conservation genetics is the use of genetic variation to identify populations that have experienced bottlenecks, as numerous threatened or endangered species and populations have been found to have low levels of genetic variation.

V. INTER-SIMPLE SEQUENCE REPEATS

ISSRs can be used to assess hybridization in natural populations of plants, as a study on *Penstemon* (Scrophulariaceae) was carried out. Eight ISSR primers were used to examine patterns of hybridization and hybrid speciation in a hybrid complex involving four species, as well as allowing examination of pollen-mediated gene flow. Previous studies using allozymes, restriction-site variation of nuclear rDNA and chloroplast DNA failed to determine whether gene flow occurs among species other than *P. cenranthifolius*. The previous studies also failed to provide support for hypotheses of diploid hybrid speciation. ISSR proved to be a much more successful technique in this study, allowing all species and all DNA accessions to be differentiated.

ISSR has also been used to detect varieties and diversity in rice, revealing much more data than RFLPs. The technique allowed for dissection below the subspecies level and this gives it a good level of applicability in the study of rare or endangered plants.

ISSRs have been used in conjunction with RAPD data to determine the colonization history of Olea europaea in Macronesia, along with lineages in the species complex. The two techniques have also been utilized in examining the historical biogeography of Sea rocket (*Cakile maritima*) and Sea Holly (*Eryngium maritimum*), comparing different and only distantly related taxa of broadly similar extant distribution. The trees generated by the different methods were largely congruent topologically. Using the result, dispersal routes of the species along a linear coast line could be construed. Joint use of RAPD and ISSR has also been used to examine clonal diversity in *Calamagrostis porteri* ssp. insperata (Poaceae), a rare grass that has little or no sexual reproduction, and spreads by vegetative reproduction.

VI. GENETIC MAPPING

To calculate the genetic distance between two loci, you need to be able to observe recombination. Traditionally, this was performed by observing phenotypes but with RFLP analysis, it is possible to measure the genetic distance between two RFLP loci whether they are a part of genes or not.

Let's look at a simple example in fruit flies. Two RFLP loci with two RFLP bands possible at each locus:

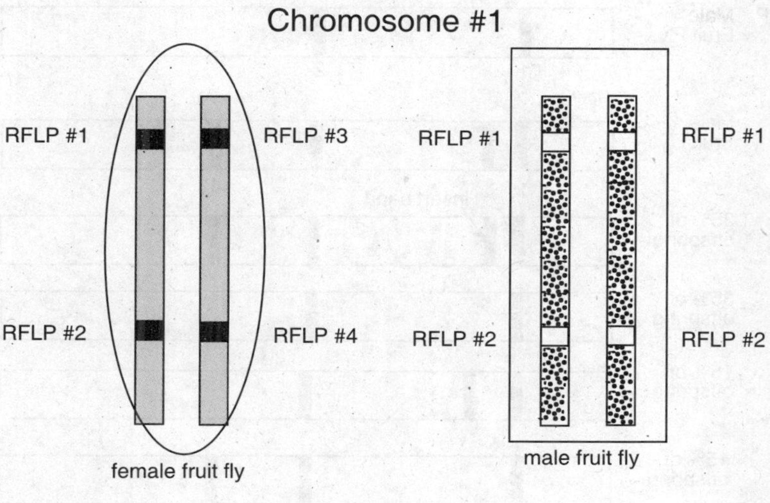

These loci are located on the same chromosome for the female (left) and the male (right). The upper locus can produce two different bands called 1 and 3. The lower locus can produce bands called 2 or 4. The male is homozygous for band 1 at the upper locus and 2 for the lower locus. The female is heterozygous at both loci. Their RFLP banding patterns can be seen on the Southern blot below:

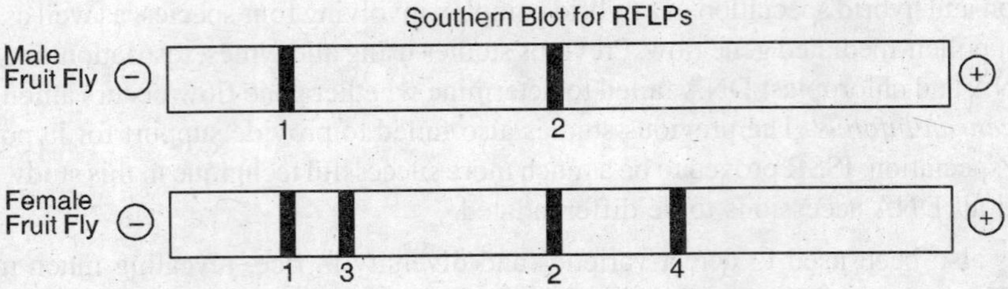

The male can only produce one type of gamete (1 and 2) but the female can produce four different gametes. Two of the possible four are called parental because they carry both RFLP bands from the same chromosome; 1 and 2 from the left chromosome or 3 and 4 from the right chromosome. The other two chromosomes are recombinant because recombination has occurred between the two loci and thus the RFLP bands are mixed so that 1 is now linked to 4 and 3 is linked to 2.

Type of Chromatid	Alleles
Parental	RFLP 1 and 2
Parental	RFLP 3 and 4
Recombinant	RFLP 1 and 4
Recombinant	RFLP 3 and 2

When these two flies mate, the frequency of the four possible progeny can be measured and from this information, the genetic distance between the two RFLP loci (upper and lower) can be determined.

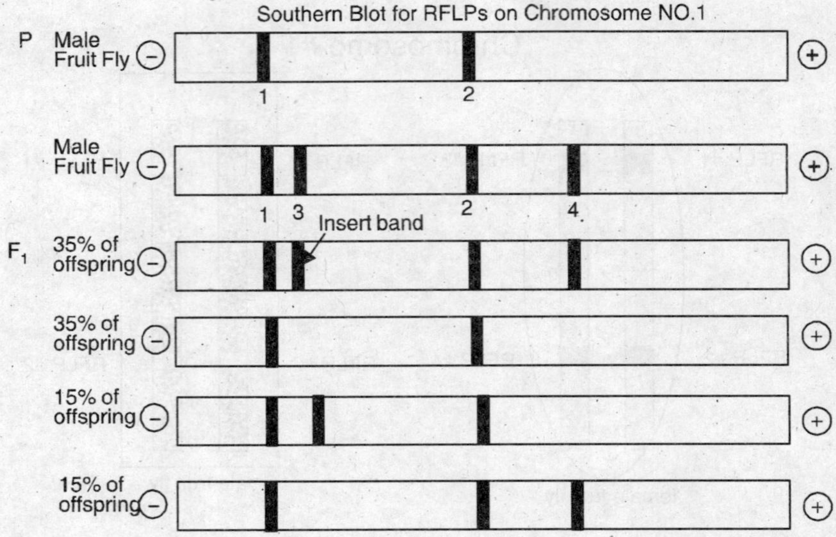

In this example, 70% of the progeny were produced from parental genotype eggs and 30% were produced by recombinant genotype eggs. Therefore, these two RFLP loci are 30 centiMorgans apart from each other.

QUESTIONS

1. What are molecular markers? Describe applications of molecular markers in solving biological problems.
2. What is RFLP? Compare this technique with RAPD.
3. Write short notes :
   (*a*) RAPD
   (*b*) RFLP
   (*c*) ISSR
   (*d*) Microsatellite
   (*e*) AFLP
   (*f*) Polymorphism.

MAPPING THE GENOME

CHAPTER 10

Gene mapping is the assignment of a gene to a locus on chromosome. Mapping relative positions of many genes and other markers results into a 'chromosome map.' There are three types of gene map (1) genetic maps (2) cytogenetic maps (3) physical maps. Genetic or linkage map assigns genes to linkage groups (groups of genes which tend to be inherited together because they are close together on chromosome). Cytogenetic map is created by correlating phenotypes with observable chromosome rearrangements and deficiencies. This type of map has a low resolution and is applicable to few species like mammals and Drosophila which display either natural or artificially inducible reproducible chromosomes banding patterns. Cytogenetic maps of simple genomes (like viruses, plasmids) are used to locate regions which differ in their physical properties. Physical or molecular map is created by ordering cloned fragments of genomic DNA. Physical map has the highest resolution and is the ultimate aim of mapping projects.

Why gene mapping?

1. Gene mapping helps us to understand and exploit the inheritance of biological traits.
2. It helps to correlate differences in phenotype with differences in chromosome structure.
3. Commercially, this helps plant and animal breeders to produce crops or herbs of improved qualities.
4. In case of human being it helps assigning genes for genetic diseases and their tracking through pedigrees by linkage to a marker (gene tracking).

I. RESTRICTION MAPPING

Genetic mapping using RFLPs as DNA markers can locate the positions of polymorphic restriction sites within a genome, but very few of the restriction sites in a genome are polymorphic, so many sites are not mapped by this technique. We increase the marker density on a genome map by using an alternative method to locate the positions of some of the non polymorphic restriction sites. This is what restriction mapping achieves, although in practice the technique has limitations that mean's it is applicable only to relatively small DNA molecules.

Methodology for restriction mapping

The simplest way to construct a restriction map is to compare the fragment sizes produced when a DNA molecule is digested with two different restriction enzymes that recognize different target sequences. An example using the restriction enzymes EcoRI and BamHI is shown in Fig. 10.1. First, the DNA molecule is digested with just one of the enzymes and the sizes of the resulting fragments measured by agarose gel electrophoresis. Next, the molecule is digested with the second enzyme and the resulting fragments again sized in an agarose gel. The results so are noble the number of restriction sites for each enzyme to be worked out, but do not allow their relative positions to be determined. Additional information is therefore obtained by cutting the DNA molecule with both enzymes together. In the example shown in Fig. 10.1, the **double restriction** enables three of the sites to be mapped. However, a problem arises with the larger EcoRI fragment because this contains two BamHI sites and there are two alternative possibilities for the map location of the outer one of these. The problem dissolved by going back to the original DNA molecule and treating

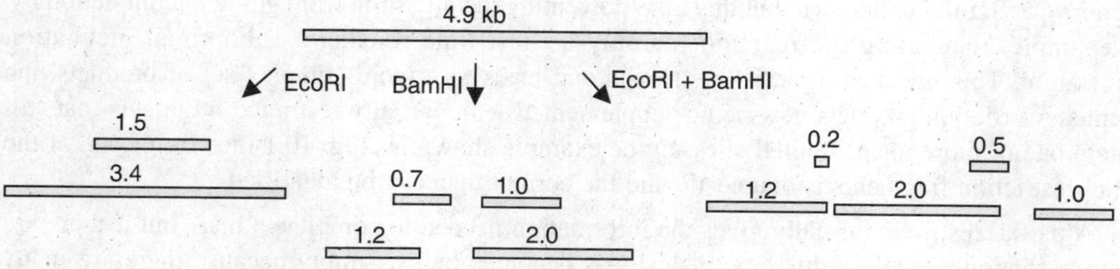

INTERPRETATION OF THE DOUBLE RESTRICTION

Fragments	Conclusions
0.2 kb, 0.5 kb	These must derive from the 0.7 kb BamHI fragment, which therefore has an internal EcoRI site :

1.0 kb	This must be a BamHI fragment with no internal EcoRI site. We can account for the 1.5 kb EcoRI fragment if we place the 1.0 kb fragment thus :

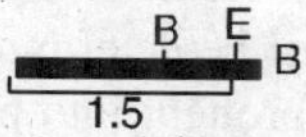

1.2 kb 2/0 kb	These must be BamHI fragments with no internal EcoRI sites. They must lie within the 3.4 kb EcoRI fragment. There are two possibilities:

PREDICTED RESULTS OF A PARTIAL BamHI RESTRICTION

If Map I is correct, then the partial restriction rpoducts will include a fragment of 1.2 + 0.7 = 1.9 kb

If Map II is correct, then the partial restriction products will include a fragment of 2.0 + 0.7 = 2.7 kb

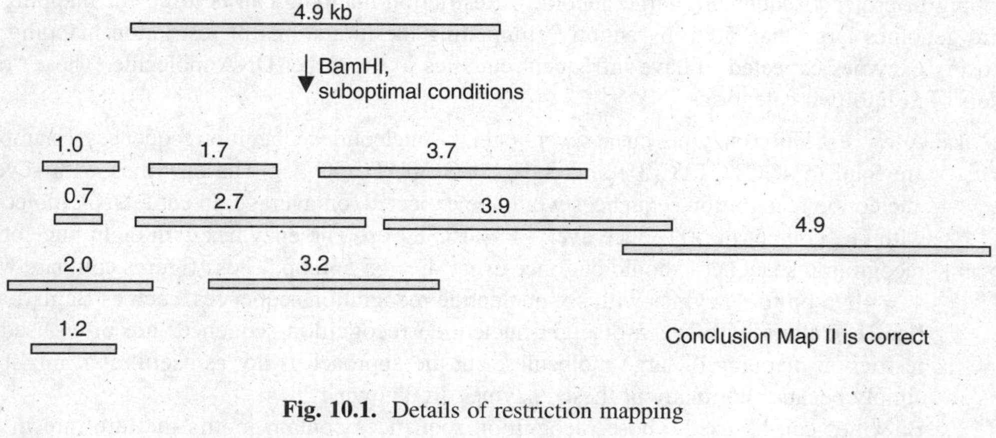

Fig. 10.1. Details of restriction mapping

it again with BamHI on its own, but this time preventing the digestion from going to completion by, for example, incubating the reaction for only a short time rousing a suboptimal incubation temperature. This is called a partial restriction and leads to a more complex set of products, the complete restriction products now being supplemented with partially restricted fragments that still contain one or more uncut BamHI sites. In the example shown in Fig. 10.1 the size of one of the partial restriction fragments is diagnostic and the correct map can be identified.

A partial restriction usually gives the information needed to complete a map, but if there are many restriction sites then this type of analysis becomes bulky, simply because there are many different fragments to consider. An alternative strategy is simpler because it enables the majority of the fragments to be ignored. This is achieved by attaching a radioactive or other type of marker to each end of the starting DNA molecule before carrying out the partial digestion. The result is that many of the partial restriction products become "indivisible" because they do not contain an en-fragment and so do not show up when the agarose gel is screened for labeled products. The sizes of the partial restriction products that are visible enable unmapped sites to be positioned relative to the ends of the starting molecule.

The scale of restriction mapping is limited by the sizes of the restriction fragments.

Restriction maps are easy to generate if there are relatively few cut sites for the enzymes being used. However, as the number of cut sites increases, so also do the numbers of single, double and partial-restriction products whose sizes must be determined and compared in order for the map to be constructed. Computer analysis can be brought into play but problems still eventually arise. A stage will be reached when a digest contains so many fragments that individual bands merge on the agarose gel, increasing the chances of one or more fragments being measured incorrectly or missed out entirely. If several fragments have similar sizes then even if they can all be identified, it may not be possible to assemble them into a clear map.

Restriction mapping is therefore more applicable to small rather than large molecules, with the upper limit for the technique depending on the frequency of the restriction sites in the molecule being mapped. In practice, if a DNA molecule is less than 50 kb in length it is usually possible to construct a clear restriction map for a selection of enzymes with six nucleotide recognition sequences. Restriction maps are equally useful after bacterial or eukaryotic genomic DNA has been cloned, if the cloned fragments are less than 50kb in length, because a detailed restriction map can then be built up as a preliminary to sequencing the cloned region. This is an important application of restriction mapping in projects sequencing large genomes. Restriction mapping can be used for mapping of entire genomes large than 50kb by slightly eliminating the limitations of restriction mapping by choosing enzymes expected to have infrequent cut sites in the target DNA molecule. These "rare cutters" fall into two categories:

1. A few restriction enzymes cut at seven- or eight-nucleotide recognition sequences. Examples are Sapl (5'-GCTCTTC-3') and SgfI (5'-GCGATCGC-3'). The enzymes with seven-nucleotide recognition sequences would be expected, on average, to cut a DNA molecule with GC content of 50% once every 4^7 = 16,384 bp. The enzymes with eight nucleotide recognition sequences should cut once every 4^8 = 65,536 bp. These figures compare with 4^6 = 4096 bp for enzymes with six-nucleotide recognition sequences, such as BamHI and EcoRI. Cutters with seven-or eight-nucleotide recognition sequences are often used in restriction mapping of large molecules, but the approach is not as useful as it might be simply because not many of these enzymes are known.
2. Enzymes can be used whose recognition sequences contain motifs that are rare in the target DNA. Genomic DNA molecules do not have random sequences and some molecules are significantly deficient in certain motifs. For example, the sequence 5'-CG-3' is rare in the genomes of vertebrates because vertebrate cells possess an enzyme that adds a methyl

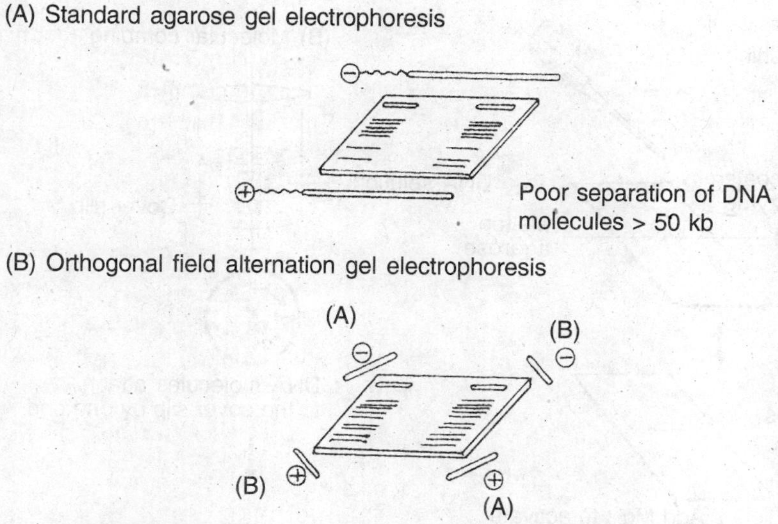

Fig. 10.2A,B. Conventional and non-conventional agarose gel electrophoresis

group to carbon 5 of the C nucleotide in this sequence. Domination of the resulting 5-methylcytosine gives thymine. The consequence is that during vertebrate evolution, many of the 5'-CG'-3 sequences that were originally in these genomes have become converted to 5'-TG-3' . Restriction enzymes that recognize a site containing 5'-CG-3' therefore cut vertebrate DNA relatively infrequently. Examples are SmaI (5'-CCCGGG-3'), which cuts human DNA once every 78 kb on average, and BssHII (5'-GCGCGC'3'), which cuts once every 390 kb.

The potential of restriction mapping is therefore increased by using rare cutters. It is still not possible to construct restriction maps of the genomes of animals and plants, but it is feasible to use the technique with large cloned fragments, and with the smaller DNA molecules of prokaryotes and lower eukaryotes such as yeast and fungi.

If a rare cutter is used then it may be necessary to employ a special type of agarose gel electrophoresis to study the resulting restriction fragments. This is because the relationship between the length of DNA molecule and its migration rate in an electrophoresis gel is not linear, the resolution decreases as the molecules get longer (Fig. 10.2A). This means that it is not possible to separate molecules more than about 50 kb in length, because all of these longer molecules run as a single, slowly migrating band in a standard agarose gel. To separate them it is necessary to replace the linear electric field used in conventional gel electrophoresis with a more complex field. An example is provided by Orthogonal Field Alternation Gel Electrophoresis (OFAGE), in which the electric field alternates between two pairs of electrodes, each positioned at an angle of 45° to the length of the gel (Figure 10.2B). The DNA molecules still move down through the gel, but each change in the field forces the molecules to realign. Shorter molecules realign more quickly than longer ones and so migrate more rapidly through the gel. The overall result is that molecules much longer than those separated by conventional gel electrophoresis can be resolved. Related techniques include CHEF (Contour Clamped Homogeneous Electric Fields) and FIGE (Field Inversion gel Electrophoresis).

Direct examination of DNA molecules for restriction sites

It is also possible to use methods other than electrophoresis to map restriction sites in DNA molecules. With the technique called optical mapping, restriction sites are directly located by looking at the cut DNA molecules with a microscope. The DNA must first be attached to a glass slide in such a way that the individual molecules become stretched out, rather than clumped together in a mass. There are various ways of doing this, including get stretching and molecular combing. To

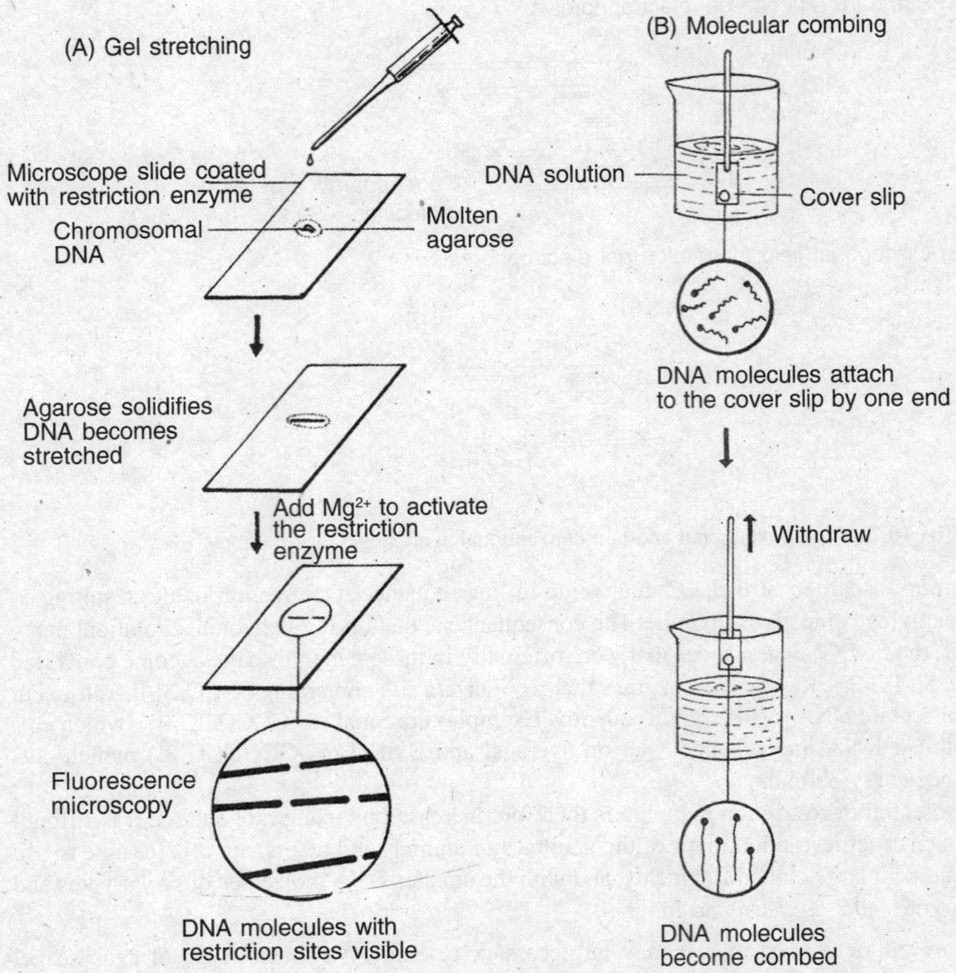

Fig. 10.3 A,B. Gel streching and molecular combing

prepare gel-stretched DNA fibers, chromosomal DNA is suspended in molten agarose and placed on a microscope slide. As the gel cools and solidifies, the DNA molecules become extended (Fig. 10.3A). In molecular combing, the DNA fibers are prepared by dipping a silicone-coated cover slip into a solution of DNA, leaving it for 5 minutes (during which time the DNA molecules attach to the cover slip by their ends), and then removing the slip at a constant speed of 0.3 mm s^{-1} (Fig. 10.3B). The force required to pull the DNA molecules through the meniscus causes them to line up. Once in the air, the surface of the cover slip dries, retaining the DNA molecules as an array of parallel fibers. After stretching or combing, the immobilized DNA molecules are treated with a restriction enzyme and then visualized by adding a fluorescent dye, such as DAPI (4,6-diamino-2-phenylindole dihydrochloride), which stains the DNA so that the fibers can be seen when the slide is examined with a high-power fluorescence microscope. The restriction sites in the extended molecules gradually become gaps as the degree of fiber extension is reduced by the natural springiness of the DNA, enabling the relative positions of the cuts to be recorded.

Optical mapping was first applied to large DNA fragments cloned in YAC and BAC vectors. More recently, the feasibility of using this technique with genomic DNA has been proverb with studies of a 1 Mb chromosome of the malaria parasite *Plasmodium falciparum* and the two chromosome and single mega plasmid of the bacterium *Deinococcus radiodurans*.

II. SEQUENCE TAGGED SITE (STS) MAPPING

To generate a detailed physical map of a large genome we need, ideally, a high-resolution mapping procedure that is rapid and not technically demanding. Restriction mapping technique that we have considered so far do not meet these requirements. Restriction mapping is rapid, easy, and provides detailed information, but it cannot be applied to large genomes. At present the most powerful physical mapping technique, and the one that has been responsible for generation of the most detailed maps of large genomes, is STS mapping. A sequence tagged site, or STS, is simply a short DNA sequence, generally between 100 bp and 500 bp in length, that is easily recognizable and occurs only once in the chromosome or genome being studied. To map a set of STSs, a collection of overlapping DNA fragments from a single chromosome or from the entire genome is needed. For example shown in Figure, a fragment collection has been prepared from a single chromosome, with each point along the chromosome represented, on average, five times in the collection. The data from which the map will be derived are obtained by determining which fragments contain which STSs. This can be done by hybridization analysis but PCR is generally used because it is quicker and has proven to be more amenable to automation. The chances of two STSs being present on the same fragment will, of course, depend on how close together they are in the genome. If they are very close then there is a good chance that will always be on the same fragment; if they are further apart then sometimes they will be on the same fragment and sometimes they will not (Fig. 3.35). The data can therefore be used to calculate the distance between two markers, in a manner analogous to the way in which map distances are determined by linkage analysis. In linkage analysis a map distance is calculated from the frequency at which crossovers occur between two markers. STS mapping is essentially the same, except that the map distance is based on the frequency at which breaks occur between two markers.

Any unique DNA sequence can be used as an STS. To qualify as an STS, a DNA sequence must satisfy two criteria. The first is that its sequence must be known, so that a PCR assay can be set up test for the presence or absence of the STS on different DNA fragments. The second requirement is that the STS must have a unique location in the chromosome ebbing studied, or in the genome as a whole if the DNA fragment set covers the entire genome. If the STS sequence occurs at more than one position then the mapping data will be ambiguous. Care must therefore be taken to ensure that STSs do not include sequences found in repetitive DNA.

These are easy criteria to satisfy and STSs can be obtained in many ways, the most common sources being expressed sequence tags (ESTs), SSLPs, and random genomic sequences.

1. **Expressed sequence tags**. These are short sequences obtained by analysis of cDNA clones. Complementary DNA is prepared by converting an mRNA preparation into double-stranded DNA. Because the mRNA in a cell is derived from protein-coding genes, the cDNAs and ESTs obtained from them represent the genes that were being expressed in the cell from which the mRNA was prepared. ESTs are looked upon as a rapid means of gaining access to the sequences of important genes, and they are valuable even if their sequences are incomplete. An EST can also be used as an STS, assuming that it comes from a unique gene and not from a member of a gene family in which all the genes have the same or very similar sequences.
2. **Simple sequence length polymorphism (SSLPs)**. SSLPs can be used as STSs in physical mapping. SSLPs that are polymorphic and have already been mapped by linkage analysis are particularly valuable as they provide a direct connection between the genetic and physical maps.
3. **Random genomic sequences.** These are obtained by sequencing random pieces of cloned genomic DNA, or simply by downloading sequences that have been deposited in the databases.

III. SINGLE NUCLEOTIDE POLYMORPHISMS (SNPs) AS FAVOURED PHYSICAL MARKER

Single nucleotide polymorphisms (SNPs, pronounced "snips") are single base-pair positions in genomic DNA at which different sequence alternatives (alleles) exist in a population (Fig. 10.4). In highly outbreed populations, such as humans, polymorphisms are considered to be SNPs only if the least abundant allele has a frequency of 1% or more. This is to distinguish SNPs from very rare mutations. In practice, the term SNP is typically used more loosely than required by the above definition. For example, single base variants in cDNAs (cSNPs) are usually classed as SNPs because most of these will reflect under lying genomic DNA variants although they could result from RNA editing. Single base-pair insertions or deletions (indels) also are considered to be SNPs by some workers. A special subset to SNPs is one where the base change alters the sensitivity of a sequence to cleavage by a restriction endonuclease. These are known as Restriction fragment length polymorphisms (RFLPs, or 'snip-SNPs'). Another important subset is one in which the two alleles can be distinguished by the presence or absence of a particular phenotype. Good examples here is sickle cell anemia caused by A > T change results in the replacement of a glutamine residue in B-globin by a valine residue but also destroys an MstII site thereby generating an RFLP.

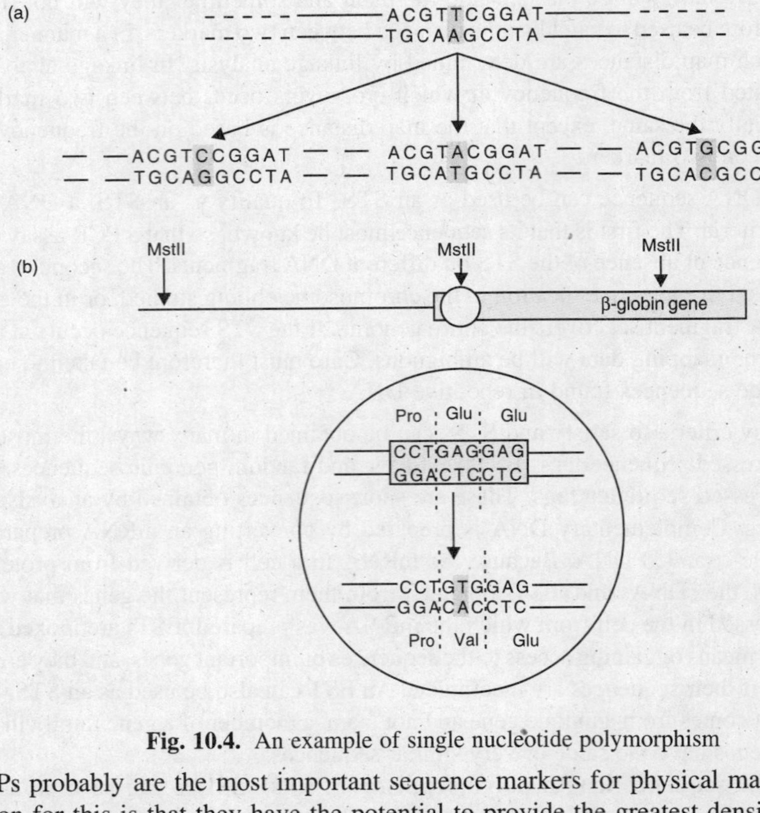

Fig. 10.4. An example of single nucleotide polymorphism

SNPs probably are the most important sequence markers for physical mapping of genomes. The reason for this is that they have the potential to provide the greatest density of markers. For example, their frequency is estimated to be 1 per 1000 base pairs which, in the case of the human genome, means 3 million in total. Originally the Human SNP Consortium planned to map 300,000 of them but the latest version of the SNP database (dbSNP) contains 10 million.

Given their importance, many different methods for detecting SNPs have been developed. SNP detection can be divided into two different activities: scanning DNA sequences for previously unknown polymorphisms to facilitate mapping, and screening (genotyping) of individuals for known polymorphisms. Scanning for new SNPs can be further divided into the global or random approach and the regional or targeted approach. The later relies almost totally on direct sequencing or on denaturing high performance liquid chromatography (dHPLC).

The technique of dHPLC takes advantage of the fact that denaturation of dsDNA is highly dependent on its sequence. Even a single nucleotide difference between two DNA molecules can alter their melting characteristics sufficiently for them to be distinguished. Several techniques for the detection of mutations are based on the differences in mobility of partially denatured molecules and include constant gradient gel electrophoresis (CGGE), denaturing gradient gel electrophoresis (DGGE), and temperature gradient gel electrophoresis (TGGE). Instead of electrophoresis as a detection step, the partially beat denatured DNA can be analyzed using HPLC matrice. The advantage of dHPLC is that analysis is rapid (~5 minutes per sample) and can be automated through the use of an autosampler.

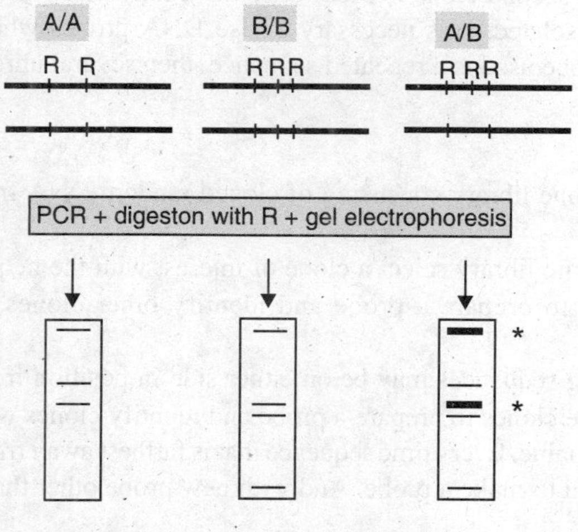

Fig. 10.5. Generation and visualization of CAPS markers

Detection of polymorphic DNA in absence of sequence information

Polymorphic DNA can be detected by amplification of the absence of the target DNA sequence information used to generate STSs. This is a simple process, distinct from the standard PCR, which is based on the amplification of genomic DNA with single primers of arbitrary nucleotide sequence. The nucleotide sequence of each primer was chosen within the constraints that the primer was nine or 10 nucleotides in length, between 50 and 80% G +C in composition and contained no palindromes. Not all the sequences amplified in this way are polymorphic but those that are RAPD are easily identified. RAPDs are widely used by plant molecular biologists to construct maps because they provide very large numbers of markers and are very easy to detect by agarose gel electrophoresis. However, they have two disadvantages- (1) the amplification of a specific sequence is sensitive to PCR conditions, including template concentration, and hence it can be difficult to correlate results obtained by different research groups. For this reason, RAPDs may be converted to STSs after isolation. (2) RAPD method cannot distinguish heterozygote from one of the homozygous genotypes.

In a different method for detecting polymorphisms, STSs are derived from genes that have already been mapped and sequenced. Where possible the primers used are chosen such that the PCR products include introns to maximize the possibility of finding polymorphisms. The primary PCR products are subjected to digestion with a panel of restriction endonucleases until a polymorphism is detected. Such markers are called cleaved amplified polymorphic sequences (CAPS). The way in which CAPS are detected is shown in Fig. 10.5.

V. CHROMOSOME WALKING

Chromosome walking is the walking along the chromosome from known site to unknown site. When a function of gene is unknown, the only way to isolate such gene is to map it genetically, and then to reach it by starting from nearest identifiable site. Chromosome walking is used as a means of finding adjacent genes (positional cloning), or parts of a gene which are missing in the original clone as well as to analyse long stretches of eukaryotic DNA. Chromosome walk are difficult for higher eukaryotic organisms, due to presence of repeated DNA sequences in the genome and flanking majority of genes. This can cause cross hybridization. Here probe used for stepping from one genomic clone to the next must be a unique sequence clone, or a subclone which has been shown to contain only a unique sequence.

A small segment of DNA from one end of the genomic clone is used as a probe to isolate clones containing this sequence and adjacent sequences encoding the next portion of the genome. The end sequence of the second clone is used to isolate a third clone and so forth until a series of overlapping clones are isolated. It is necessary to use DNA probes whose sequences are single-copy, otherwise if the probe used is a repeated sequence, then several unrelated recombinants could be identified.

Methodology

1. First obtain a gene library consisting of cloned random DNA fragments each about 20kb long.
2. From the genomic library select a clone of interest with the help of probe.
3. Use that clone to prepare a probe and identify other clones which have overlapping sequences.
4. The overlapping sequences may be on either side in position from the first clone.
5. Use any of these clones to prepare a probe and identify clones which are overlapping with it and are identifiable. Every time sequence that is farthest away from the original hybridizing sequence is used to make a probe. And with new probe other than earlier clone is selected from the library.
6. This way it is possible to repeat the above steps and to walk further along the chromosomes to reach upto target interested genes.
7. Chromosome walking can be bidirectional and restriction endonuclease mapping helps to determine the direction of successive walks (Fig.10.6).

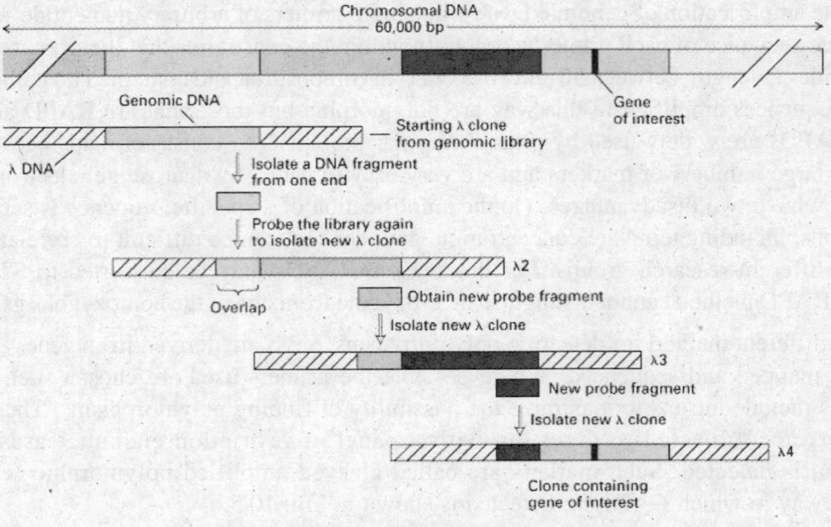

Fig. 10.6. Process of chromosome walking

Map-based Cloning: Walking

1. The strategy of map-based cloning is to find molecular markers very closely linked to the gene of interest. Those molecular markers can serve as the starting point for chromosome walking or jumping to the gene.
2. In the diagram, the walk begins with a clone containing *mkrB*. The ends of the clone (boxed) are used to probe a library. Clones from adjacent genome segments are thus identified and isolated. The distal ends of those clones are used to reprobe the library. These steps are continued until a clone contains either *mkrA* or *mkrC* sequences.

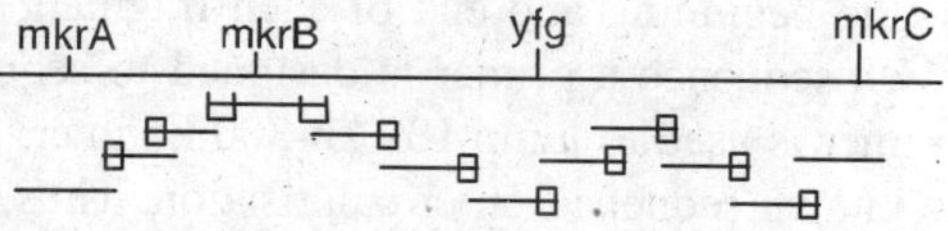

3. Clones between *mkr B* and *mkrC* must then be evaluated for the presence of *yfg*. Chromosome walking has been used in the isolation of centromere sequences, among others.
4. If the target species is a species whose genome has been completely molecularly mapped, an ordered set of YACs, PACs, BACs or cosmid clones will be available. Knowing which molecular markers are adjacent to the target gene automatically identifies the YACs and/or cosmids that need to be tested.
5. One difficulty of chromosome walking is recognizing where the gene is located between the two markers. Zoo (or garden) blots where DNAs of a variety of species have been restricted, electrophoresed and Southern blotted can be useful.

VI. PRIMER WALKING

Primer walking is a sequencing method of choice for sequencing DNA fragments between 1.3 and 7 kilobases. Such fragments are too long to be sequenced in a single sequence read using the cyclic dideoxy terminator method (also known as PCR based dideoxy sequencing or Automated Sanger method). The DNA of interest may be a plasmid insert, a PCR product or a fragment representing a gap when sequencing a genome. As the traditional chain termination method does not allow long DNA strands to be sequenced, this method works by dividing the long sequence into several consecutive short ones.

The term "primer walking" is used where the main aim is to sequence the genome. The term "chromosome walking" is used instead when we know the sequence but don't have a clone of a gene. For example, the gene for a disease may be located near a RFLP on DNA.

The fragment is first sequenced as if it were a shorter fragment — sequencing will be performed from each end using either universal primers or primers designated by the customer. This should identify the first 1000 (approx.) bases. In order to completely sequence the region of interest design and synthesis of new primers complementary to the final 20 bases of the known sequence is necessary to obtain contiguous sequence information.

Methodology

1. A primer that matches the beginning of the DNA to sequence is used to synthesize a short DNA strand adjacent to the unknown sequence, starting with the primer .
2. The new short DNA strand is sequenced using the chain termination method.
3. The end of the sequenced strand is used as a primer for the next part of the long DNA sequence.

That way, the short part of the long DNA that is sequenced keeps "walking" along the sequence. The method can be used to sequence entire chromosomes (thus, *chromosome walking*). A different method with the same purpose which becomes more popular for large-scale sequencing (e.g., the Human Genome Project) is shotgun sequencing.

VII. CHROMOSOME JUMPING

Chromosome jumping is a technique of molecular biology that is used as a tool in the physical mapping of genomes. It is related to several other tools used for the same purpose, including chromosome walking. Chromosome jumping is used to bypass regions difficult to clone, such as those containing repetitive DNA, that cannot be easily mapped by chromosome walking, and is useful in moving along a chromosome rapidly in search of a particular gene.

In chromosome jumping, the DNA of interest is identified, cut into fragments with restriction enzymes, and circularised (the beginning and end of each fragment is joined together to form a circular loop). From a known sequence a primer is designed to sequence across the circularised junction (Fig.10.7). This primer is used to jump 100 kb-300 kb intervals: a sequence 100 kb away would have come near the known sequence on circularisation. Thus, sequences not reachable by chromosome walking can be sequenced. Chromosome walking can be used from the new jump position (in either direction) to look for gene-like sequences, or additional jumps can be used to progress further along the chromosome.

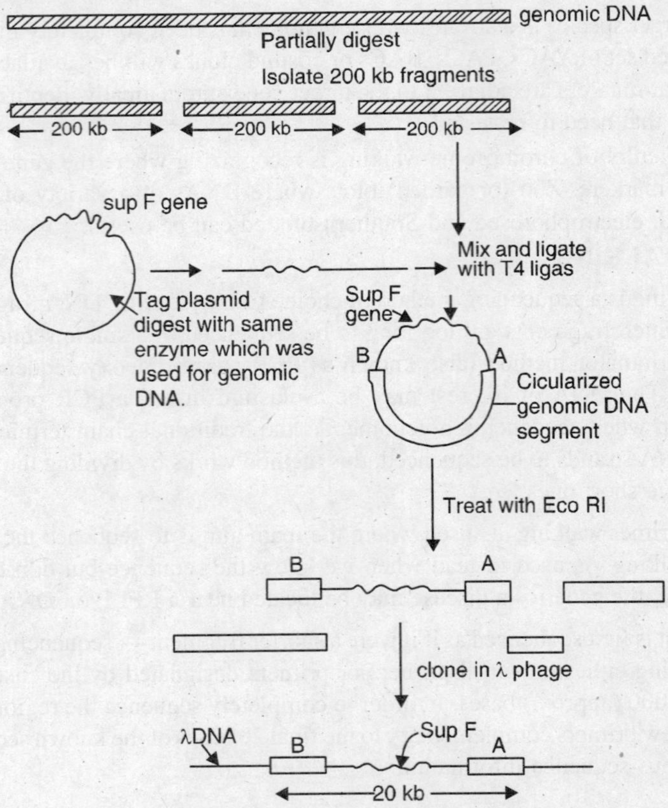

Fig. 10.7. Process of chromosome jumping

QUESTIONS

1. What are molecular markers? Describe applications of molecular markers in solving biological problems.
2. Write short notes :
   (*a*) Chromosome walking (*b*) Restriction mapping

ADVANCE BIO-TECHNOLOGY

CHAPTER 11

I. APOPTOSIS

Apoptosis was first described in the 1970s, when scientists noticed old cells dying without causing the typical burst of inflammation. Eventually, scientists realized that these cells were triggering an internal mechanism that caused them to break up so their contents could be recycled. Apoptosis is crucial to normal development. It prunes out tumor cells, immune cells that would otherwise attack the body and neurons that fail to connect with other brain cells. "It's a natural way of getting rid of old cells when new cells are being formed," says John Olney, a professor of psychiatry at Washington University in St. Louis.

Apoptosis is a quiet form of suicide. While necrotic cell death releases chemicals that can inflame nearby cells, apoptosis spares them. But because apoptotic cells quietly disappear within a day or so, programmed cell death is hard to study: You almost need an eyewitness to know it occurred.

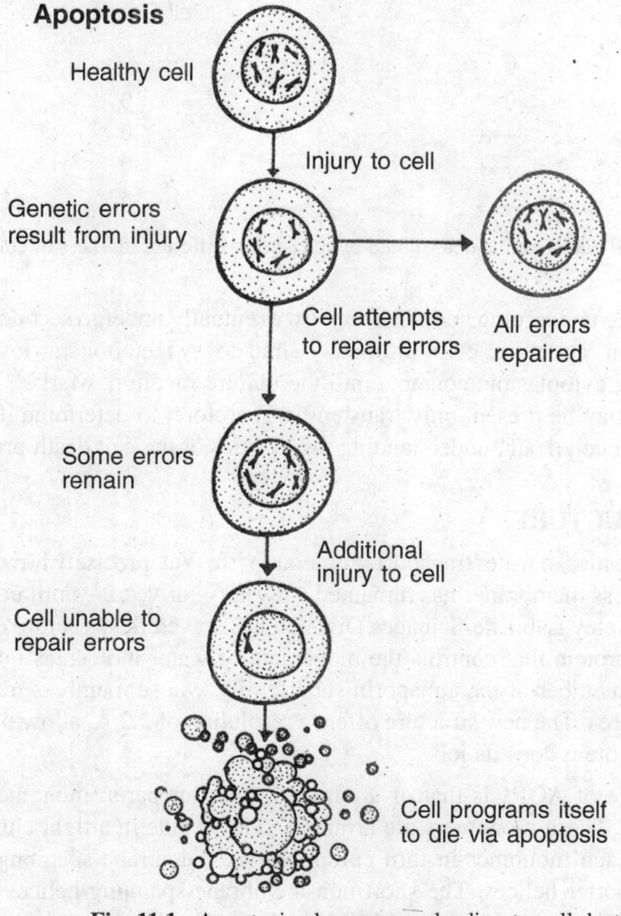

Fig. 11.1. Apoptosis phenomenon leading to cell death

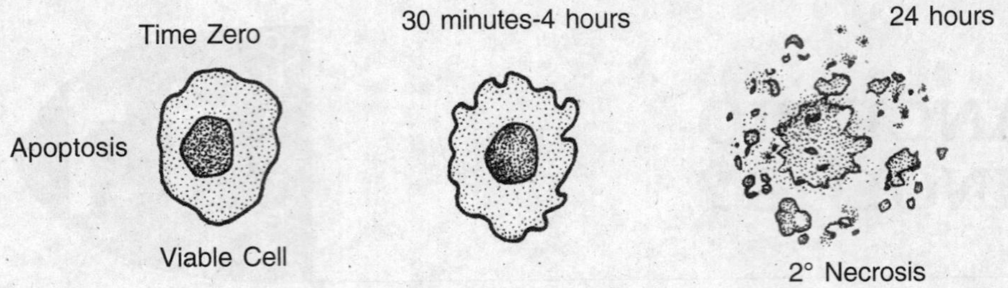

LDH Release	0	+	+++
Caspase	0	+++	++
ATP	+++	++	0
MTS	+++	++	0
Resazurin	+++	++	0

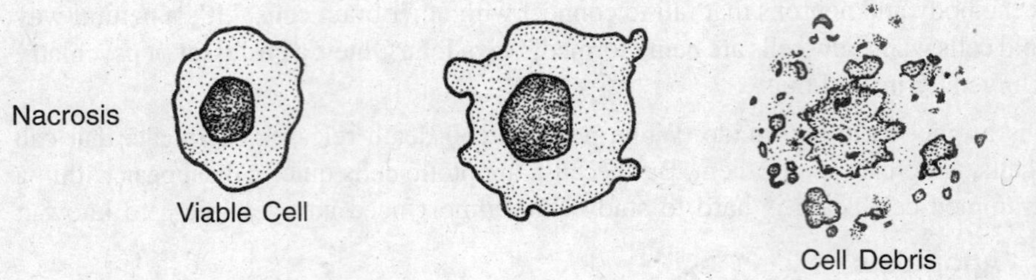

LDH Release	0	+++	++
Caspase	0	0	0
ATP	+++	0	0
MTS	+++	0	0
Resazurin	+++	0	0

Fig. 11.2. Mechanisms of cell death can be determined by measuring different markers of cell viability and apoptosis in vitro.

Cultured cells that are undergoing apoptosis in vitro eventually undergo secondary necrosis. After extended incubation, apoptotic cells ultimately shut down metabolism, lose membrane integrity and release their cytoplasmic contents into the culture medium. Markers of apoptosis such as caspase activity may be present only transiently. Therefore, to determine if apoptosis is the primary mechanism of cell death, understanding the kinetics of the cell death process in your model system is critical.

II. AQUAPORIN STRUCTURE

From aquaducts to osmosis, water transport is crucial to life. Yet, precisely how life manages the transport of water across membranes has remained a mystery for years—until now. A team of researchers from the Berkeley Lab Life Sciences Division has solved the structure of aquaporin-1 (AQP1), a membrane protein that controls the movement of water molecules into and out of mammalian cells. It is a member of the aquaporin superfamily, whose members transport water or water and glycerol or urea. The new structure offers a resolution of 2.2 Å, allowing researchers to deduce just how the protein does its job.

The overall structure of AQP1 is that of a tetramer, the four parts (monomers) of which each define a single pore. These monomers are arranged side by side in a tight cluster, with the pores running parallel. Each monomer in turn comprises six membrane-spanning helices that partially surround two shorter helices. The short non-membrane-spanning helices make up the major portion of the pore. Each pore has a dumbbell-like shape. One broad end is the cytoplasmic vestibule; the other is the extracellular vestibule. The bar of the dumbbell is the selectivity filter, which narrows to a constriction region on the extracellular end. A series of carbonyl oxygens forms a hydrophilic path across the region and through the rest of the selectivity filter. One of these oxygens, along with a histidine residue and an arginine residue, forms the hydrophilic face of the constriction region.

Opposite this face is a hydrophobic face formed by a phenylalanine residue. Three of the four residues that form the constriction region (the arginine, histidine, and phenylalanine residues) are conserved in all known water-specific aquaporins. This observation suggests that the presence of these residues can be used as a marker for identifying other water-specific aquaporins.

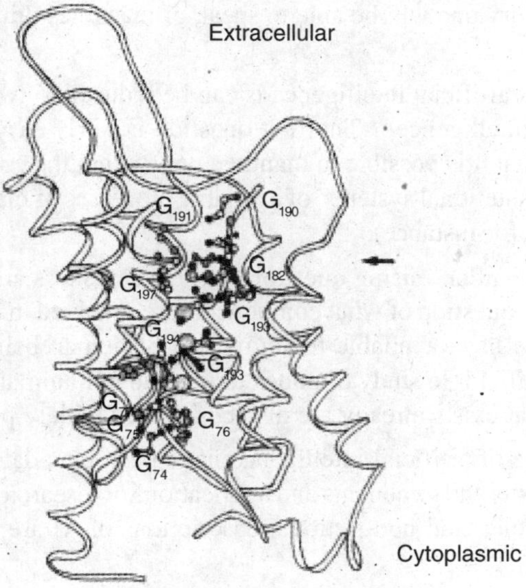

Fig.11.3. Aquaporin protein interaction with molecules

The hydrophilic path across the selectivity filter, highlighted by ball-and-stick structures of the side chains involved. The spheres are the water molecules observed in transit. The constriction region is indicated by the (Fig. 11.4) arrow.

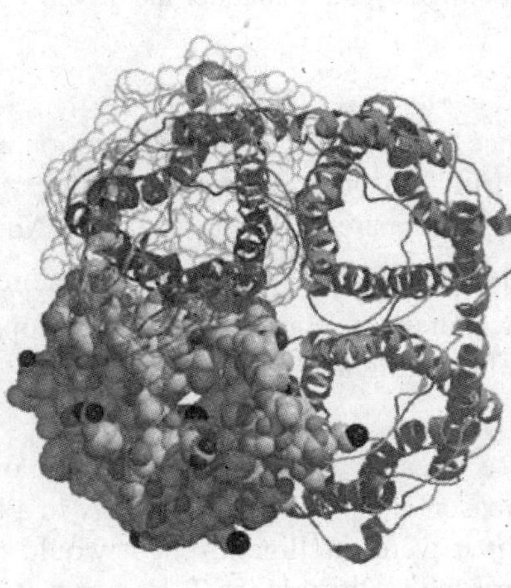

Fig. 11.4. The AQP1 tetramer viewed looking down the pores from the cytoplasmic side, normal to the membrane. One monomer of the four is represented as a solid space-filling model.

III. ARTIFICIAL INTELLIGENCE (AI)

The term artificial intelligence was first used in 1956 by John McCarthy, a computer scientist and mathematician, who also developed the LISP (LISt Processing language) programming language. LISP is particularly important in the development of artificial intelligence because it is a programming language that can support problem-solving configurations, pattern-matching rules and flexible input/output of code and data. In 1950, Turing wrote a paper for the philosophical journal *Mind* entitled "Com`puting Machinery and Intelligence," which has since become famous in the study of artificial intelligence for asking and considering the question "Can machines think." Turing presented nine objections to the idea that machines can think, in a rigorous philosophical analysis, and concludes that by the end of the century one will be able to speak of machines thinking "without expecting to be contradicted.

The question of what artificial intelligence is can be reduced to two parts: "what is the nature of artifice" and "what is intelligence"? The first question is fairly easy to answer, though it does point to the question of what it is possible to manufacture (within the constraints of certain types of system, e.g. classical computational systems, of available processes of manufacturing and of possible limits on human intellect, for instance).

The second is much harder, raising questions of consciousness and self, mind (including the unconscious mind) and the question of what components are involved in the only type of intelligence it is universally agreed we have available to study: that of human beings. Intelligent behavior in humans is complex and difficult to study or understand. Study of animals and artificial systems that are not just models of what exists already are also considered widely pertinent.

Several distinct types of artificial intelligence have been elucidated below. Also, the subject divisions, history, proponents and opponents and applications of research in the subject are described. Finally, references to fictional and non-fictional descriptions of AI are provided.

Definition

One popular and early definition of artificial intelligence research, put forth by John McCarthy at the Dartmouth Conference in 1956, is *"making a machine behave in ways that would be called intelligent if a human were so behaving."*, repeating the claim put forth by Alan Turing in "Computing machinery and intelligence" (Mind, October 1950). Another definition of **artificial intelligence** is *intelligence arising from an artificial device*. Most definitions could be categorized as concerning either systems that *think like humans*, systems that *act like humans*, systems that *think rationally* or systems that *act rationally*.

Strong artificial intelligence

Strong artificial intelligence research deals with the creation of some form of computer-based artificial intelligence that can truly reason and solve problems; a strong form of AI is said to be sentient, or self-aware. In theory, there are two types of strong AI:

- Human-like AI, in which the computer program thinks and reasons much like a human mind.
- Non-human-like AI, in which the computer program develops a totally non-human sentience, and a non-human way of thinking and reasoning.

Weak artificial intelligence

Weak artificial intelligence research deals with the creation of some form of computer-based artificial intelligence that can reason and solve problems only in a limited domain; such a machine would, in some ways, act *as if* it were intelligent, but it would not possess true intelligence or sentience. The classical test for such abilities is the Turing test.

There are several fields of weak AI, one of which is natural language. Many weak AI fields have specialised software or programming languages created for them. For example, the 'most-human' natural language chatterbot A.L.I.C.E. uses a programming language AIML that is specific to its programme and the various clones, named Alicebots.

To date, much of the work in this field has been done with computer simulations of intelligence based on predefined sets of rules. Very little progress has been made in strong AI. Depending on how one defines one's goals, a moderate amount of progress has been made in weak AI.

The term "Strong AI" was originally coined by John Searle and was applied to digital computers and other information processing machines. Searle defined strong AI: "according to strong AI, the computer is not merely a tool in the study of the mind; rather, the appropriately programmed computer really is a mind".

Experimental AI research

Artificial intelligence began as an experimental field in the 1950s with such pioneers as Allen Newell and Herbert Simon, who founded the first artificial intelligence laboratory at Carnegie-Mellon University, and McCarthy and Marvin Minsky, who founded the MIT AI Lab in 1959. They all attended the aforementioned Dartmouth College summer AI conference in 1956, which was organized by McCarthy, Minsky, Nathan Rochester of IBM and Claude Shannon.

Historically, there are two broad styles of AI research - the "neats" and "scruffies". "Neat", *classical* or *symbolic* AI research, in general, involves symbolic manipulation of abstract concepts, and is the methodology used in most expert systems. Parallel to this are the "scruffy", or "connectionist", approaches, of which neural networks are the best-known example, which try to "evolve" intelligence through building systems and then improving them through some automatic process rather than systematically designing something to complete the task.

Practical applications of AI techniques

Whilst progress towards the ultimate goal of human-like intelligence has been slow, many spin-offs have come in the process. Notable examples include the languages LISP and Prolog, which were invented for AI research but are now used for non-AI tasks. Many other useful systems have been built using technologies that at least once were active areas of AI research. Some examples include:

- **Chinook** was declared the Man-Machine World Champion in checkers (draughts) in 1994.
- **Deep Blue**, a chess-playing computer, beat Garry Kasparov in a famous match in 1997.
- **InfoTame**, a text analysis search engine developed by the KGB for automatically sorting millions of pages of communications intercepts.
- **Fuzzy logic**, a technique for reasoning under uncertainty, has been widely used in industrial control systems.
- Expert systems are being used to some extent industrially.
- **Machine translation systems** such as SYSTRAN are widely used, although results are not yet comparable with human translators.
- **Neural networks** have been used for a wide variety of tasks, from intrusion detection systems to computer games.
- **Optical character recognition systems** can translate arbitrary typewritten European script into text.
- **Handwriting recognition** is used in millions of personal digital assistants.
- **Speech recognition** is commercially available and is widely deployed.
- **Computer algebra** systems, such as Mathematica and Macsyma, are commonplace.
- **Machine vision** systems are used in many industrial applications ranging from hardware verification to security systems.

- **AI Planning methods** were used to automatically plan the deployment of US forces during Gulf War I. This task would have cost months of time and millions of dollars to perform manually, and DARPA stated that the money saved on this single application was more than their total expendiature on AI research over the last 50 years.

AI applications in biology

The post-genomic era has been characterized by two different scenarios: (1) the huge amount of available biological data sets all over the world requires suitable tools and methods both for modeling biological processes and analyzing biological sequences, (2) many new computational models and paradigms inspired and developed as metaphors of biological systems are ready to be applied in the context of computer science. Hence, the need to either develop new models or exploit and analyze the available genomes is considered a priority task for the bioinformatics research community.

There are at least 26 billion base pairs (bp) representing the various genomes available in the server of the National Center for Biotechnology Information (NCBI). Besides the human genome with about 3 billion bp, many other species have their complete genome available there. Cohen (2004) explained the needs of biologists to utilize and help interpret the vast amounts of data that are constantly being gathered in genomic research. He also pointed out the basic concepts in molecular cell biology, and outlined the nature of the existing data, and illustrated the algorithms needed to understand cell behavior. The largest known gene in the NCBI server has about 20 million base pairs and the largest protein consists of about 34, 000 amino acids. In contrast, the Protein Database (PDB) has a catalogue of only 45,000 proteins specified by their 3D structure. Bioinformatics and computational biology are concerned with the use of computation to understand biological phenomena and to acquire and exploit biological data, increasingly large-scale data. Methods from bioinformatics and computational biology are increasingly used to augment or leverage traditional laboratory and observation-based biology. These methods have become critical in biology due to recent changes in our ability and determination to acquire massive biological data sets, and due to the ubiquitous, successful biological insights that have come from the exploitation of those data. This transformation from a data-poor to a data-rich field began with DNA sequence data, but is now occurring in many other areas of biology.

DNA sequencing with artificial intelligence- The field of molecular biology is described as tailor-made for AI and machine learning approaches. This is due to the nature of AI approaches that performs well in domains where there is an immense amount of data but little theory. Since the introduction of AI to this field, numerous algorithms have been designed and applied to study different data sets. Most of these researches compare a new method with the traditional ones, affirm the effectiveness and efficiencies of their methods in particular data sets. Sequencing of DNA is among the most important tasks in molecular biology. DNA chips are considered to be a more rapid alternative to more common gel-based methods of sequencing. DNA chips commonly are made with the set of all possible probes eight nucleotides in length (octamers) generating 65,536 unique probes spaced on a 1.6 cm^2 array.

Fuzzy logic is a mathematical framework, which is compatible with poorly quantitative yet qualitatively significant data. Fuzzy logic is a natural language for linguistic modeling, thus it is consistent with the qualitative linguistic– graphical methods conventionally used to describe biological systems. Fuzzy if-then rules were also developed to describe the basic molecular properties and behaviours of DNA inside the living cell. Neural network model has also emerged as a promising AI technique in DNA sequence analysis because this approach might well embody important aspects of intelligence not captured by symbolic and statistical methods. An important direction in this work involved integrating multiple, fundamentally different AI approaches into single hybrid intelligent systems, which let each component perform the tasks for which it is best suited. Integrating symbolic knowledge into a neural network to create a knowledge-based neural network has quickly become an important hybrid-intelligence research area. Empirical observations indicate that such systems can outperform both neural network and symbolic approaches.

Bioinformatics grew out of molecular biology as it became clear that specialist skills were needed to organize and analyze the data being generated. Now molecular biology has reached a stage of development where continued progress depends on (1) combining the intelligent systems with the sheer volume of biomedical research and (2) combining the knowledge of the tiny molecular mechanisms with knowledge of the biological systems. Biological literature databases continue to grow rapidly with vital information that is important for conducting sound biomedical research. As data and information space continue to grow exponentially, the need for rapidly surveying the published literature, synthesizing, and discovering the embedded 'knowledge' is becoming critical to allow the researchers to conduct 'informed' work, avoid repetition, and generate new hypotheses. Knowledge, in this case, is defined as one-to-many and many-to-many relationships among biological entities such as gene, protein, drug, disease, etc. The knowledge discovery process basically involves identification of biological object names, reference resolution, ontology and synonym discovery, and finally extracting object–object relationships. The process of biological knowledge discovery is also evolving in terms of data and information being created continuously.

IV. BIOCHIPS

Many knowledge achievements in the life sciences have been achieved during the last decades. These technology are further improved, allowing the researchers a top-down approach, from whole organisms to the single molecule or even the single atom. This ability to characterize biological structures and observe them in real-time at the nanometric level, offers an extremely powerful tool for the understanding the phenomena occurring in the single cell. Genomics, proteomics, channelomics etc., are the recently born disciplines coping with DNA-sequencing, gene expression, immunology, regulation of cell activity, intracellular communication etc. Biochips are in most cases the technological core around which the instrumentation is built. Yet the word biochip represents a quite wide concept and stays for many different families of devices, which share the common features of being small, integrated and functionalized interfaces between the biology and the instrument. Classification of biochips can be done on the basis of their target use, their functionalization and the involved technologies.

The most important categories are:

(1) DNA-chips, for the spanning from simple glass slides with a few dozens of spots and feature size in the millimetre range, to highly sophisticated devices with extremely high density, which can analyze a whole human genome. Passive functionalization and off-chip synthesis are typical of the simplest designs, while on chip synthesis and/or active read-out are available for high-end products.

(2) Protein microarrays, allow the simultaneous analysis of hundreds of molecules extracted from small samples of tissue or body fluids by means of specific antibodies linked to the chip surface. The most important applications are found in the detection of tumours and autoimmune diseases.

(3) Patch-clamp devices, replace in automated systems the manual operation of cells pipetting. They are gaining wide acceptance in the pharmacologic research to investigate the effects of drugs on the cellular physiology, particularly on ion channel functions. Different designs are available for single cells experiments, up to large cell populations.

(4) MEA, The Micro Electrode Arrays allow the researcher to record the bioelectric activity of grown multicellular structures, like neuronal networks, or tissue slices. Most widely used are passive chips, consisting of simple electrode arrays. There is a significant trend towards the development of

active devices with integrated read-out amplifiers, stimulators, chemical sensors and the needed A/D (analogue/digital) and D/A Converters to allow a pure digital interface with the host instrumentation.

(5) Implants, This family of devices can be considered a fall-out of the previous technologies for therapeutic applications. This is gaining increasing importance due to the benefits of functional recovery and the correlated social relevance. Cochlear and retinal prosthesis are the most significant examples.

Among the applications of biochips, the most relevant is the screening of new drugs. This procedure is required and regulated by law, to determine with a high degree of reliability which are the benefits and the risks of pharmaceutical compounds. A typical screening process consists of investigation steps on the following targets: (1) single cells, (2) multicellular cultures, (3) tissue samples, (4) test animals, (5) human volunteers. The biochips can effectively help to achieve a high data throughput in the first three steps thus allowing shorter investigation times and lower cost. A special aspect in the technology of biochips is represented by the biocompatibility of the employed materials. On the one hand the biologic environment is very complex from a chemical point of view due to the presence of several different ions, enzymes, nutrients, mediators, metabolites, dissolved gases etc. These may react with the biochip and progressively degrade its performances. On the other hand the biochip can deliver substances which may influence the cell functions. Therefore, the material needs to be inert in long-term in-vivo applications like implants. Several solutions have been proposed and the most promising ones, to date, are based on polymeric materials.

Biochips are devices that can contain anywhere from tens to tens of millions of individual sensor elements (or biosensors). The sensors are packed together into a package typically the size of a microscope slide. Because so many sensors can be put into such a small area, a huge number of distinct tests can be done very rapidly.

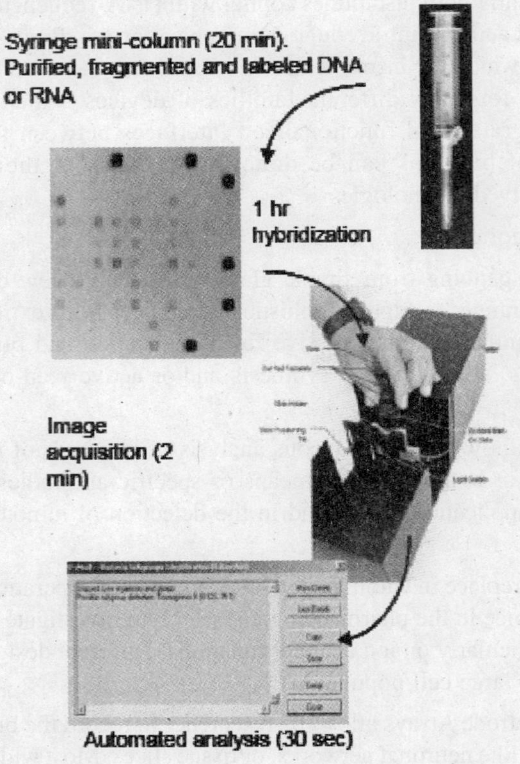

Fig. 11.5. Automated procedure for analysis using biochip

Biochips are often made using the same microfabrication technology used to make microchips. Unlike microchips, however, biochips are generally not electronic (although they can be). The key premise behind biochips is that they can do chemistry on a small scale. Each biosensor can be thought of as a "microreactor" which does chemistry designed to sense a specific analyte. Biosensors can be made to sense a wide variety of analytes, including DNA, protein, antibodies, and small biological molecules. Fluorescence is often used to indicate a sensing event. Automated microscopy systems can be used to "read" the chip, i.e. determine which sensors are fluorescing.

Biochips carry a high-density array of test sites that can analyze a single droplet of a sample for hundreds to thousands of biological agents simultaneously. The system developed at Argonne can complete such identification of infectious disease strains or biological warfare agents in less than two hours.

NANOSENSORS: The combination of nanotechnology, biology, advanced materials and photonics opens the possibility of detecting and manipulating atoms and molecules using nano-devices, which have the potential for a wide variety of medical uses at the cellular level. The nano-scale size of this new class of sensors also allows for measurements in the smallest of environments. One such environment that has evoked a great deal of interest is that of individual cells. Using these nanosensors, it is possible to probe individual chemical species and molecular signalling processes in specific locations within a cell. We have shown that insertion of a nano-biosensor into a mammalian somatic cell not only appears to have no effect on the cell membrane, but also does not affect the cell's normal function. The possibilities to monitor in vivo processes within living cells could dramatically improve our understanding of cellular function, thereby revolutionizing cell biology.

VI. BIOSENSORS

Biosensors make use of biological components in order to sense a species of interest (which by itself need not be a "biospecies"). On the other side, chemical sensors not using a biological component but placed in a biological matrix *are not biosensors* by definition. Enzymes are still indispensable tools in biosensing. An algal biosensor using alkaline phosphatase for determination of heavy metals was reported. It exploits the inhibitory action of heavy metals on the alkaline phosphatase present in the membrane of *Chlorella vulgaris* microalgae. The microalgal cells were immobilized on removable membranes placed in front of the tip of an optical fiber bundle. This section comprises sensors for biotechnological, industrial, environmental, food, pharmaceutical, medical, and related applications. Optical sensors for oxygen have been used for noninvasive analysis of dissolved oxygen in shake flasks for cell cultures. The oxygen-sensitive element is a thin, luminescent patch affixed to the inside bottom of the flask. Both intensity and decay time may be measured, the latter said to be more reliable, particularly with respect to repeated sterilization. Mass-transfer coefficients are reported as well. The sensors are fast, do not consume oxygen, and are affordable.

Biosensors are powerful tools aimed at providing selective identification of toxic chemical compounds at ultratrace levels in industrial products, chemical substances, environmental samples (e.g., air, soil, and water) or biological systems (e.g., bacteria, virus, or tissue components) for biomedical diagnosis. Combining the exquisite specificity of biological recognition probes and the excellent sensitivity of laser-based optical detection, biosensors are capable of detecting and differentiating big/chemical constituents of complex systems in order to provide unambiguous identification and acçurate quantification.

Principle

Biosensors are a combination of the biological components with electronic gadgets or transducer to make direct transduction and measurement of signals. Any change in the response of biological components, which in turn is dependent on the environment, is recorded as a change in the signal. The sample is allowed to pass through a membrane so that the selection may be exercised and the interfering molecules are retained outside the membrane. The sample then interacts with the biological sensor and forms a product which may be an electric current/charge, heat, gas or a suitable chemical. The product then passes through another membrane and reaches the transducer, which converts the biological signal into an electric signal that can be further amplified and can be either read or on digital panel or can be recorded over recorders. This allows the measurement of the concentration of the substrate without processing and thus avoiding consumption of the sample. The selectivity, sensitivity and stability of the biosensors are determined by the biological components. The latter are generally enzymes, antibodies, organelle, cells and microorganisms. The physicochemical transductors are electrodes, thermisters, transisters, and optical instruments (spectrometer, fluorimeter, etc.).

The biosensors can be used for environmental analysis: in two ways (i) the application in continuous measurement systems for monitoring and control of processes, and *(ii)* the on-field analysis by portable equipment and biosensors. Table 11.1 summarises some biosensors and relevant biocomponents for the detection and monitoring of environmental toxicants.

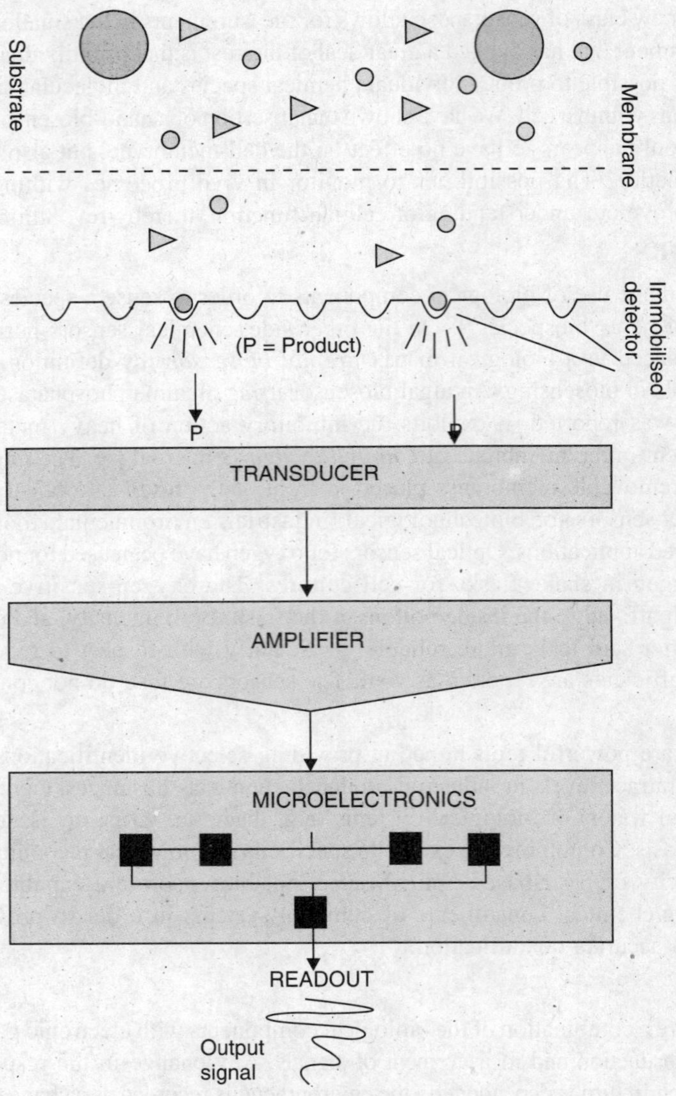

Fig. 11.6. Basic principle of a biosensor

Enzymatic Biosensors

Because of the high specificity of enzymes, enzyme biosensors are very useful for the detection of individual substances. In most of the cases, the effect is measured in terms of enzyme catalysis i.e. degradation rate of the specific substrate or production rate of the reaction product. The urease, peroxidase and tyrosinase sensors have been proved useful to detect and, in some cases, quantify the specific substrates. The tyrosine sensor (tyrosine-electrode) effectively quantifies phenol through sensing the product. In the presence of oxygen, phenol is oxidised to polymer products. The sensitivity of such sensors can be substantially increased by the measurement of a soluble intermediate reaction products, instead of quantifying the end-product. Wollenberger *et al.* (1992) developed a multienzyme system for quantification of inorganic

Table 11.1. Biological and physical components of some biosensors and their applications

Biological sensor	Transducer device (physical device)	Substance or para meter measured
Enzymes		
Glucose oxidase	Oxygen electrode	Glucose
NADH and dehydrogenase	Redox electrode	Ethanol
Creatinine iminohydrolase	NH_3 field effect transistor (FET)	Creatinine
Cholinestase	Chem FET	Nerve gas
Labelled hCG Catalase	Oxygen electrode	Human chorionic gonadotropin (hCG)
II Receptors		
ACH receptor	Conductimeter	Nerve gas
III Antibodies		
Antibody to parathion	Piezoelectric crystal	Parathion
Antibody to ϒ—globulin	Polarized light	ϒ—globulin
IV Microbial cells		
Pseudomonas fluorescens	Oxygen electrode	Glucose
Azobacteria vinelandi	NH_3 electrode	Nitrate
Methylomoans flagella	Oxygen electrode	Methane
Bacillus subtilis	Oxygen electrode	Mutagen screening
Citrolacter fluendli	pH electrode	Cephalosporin

phosphates. Such systems are effective, especially for quantification and monitoring of electro-chemically inactive reaction products. A biologically active membrane, used by Wallenberger and his co-workers, consisted of nucleoside -phosphorylase and xanthinoxidase enzyme sequence. The former catalyses inosine-phosphorylation reaction, that utilises inorganic phosphate, and subsequently the phosphorylated intermediate is converted to hypoxanthin by xanthoxidase. The rate of phosphorylation vis-a-vis hypoxanthin production is determined by the concentration of inorganic phosphate. Such a sensor has been found useful to detect inorganic phosphate within the range of 0.5-500 mmol / l. In order to increase the sensitivity of the measurement, a xanthin sensor containing co-immobilised alkaline phosphatase and nucleoside-phosphorylase was integrated. It could increase the sensitivity of the sensor, for about 20 times that of the initial sensor.

For analysis of the carbamates and organophosphates, the acetyl- or butyl-cholinesterase biosensors are highly effective. These enzyme sensors, not only measure the effect, but are also effective in determining the toxicants concentration in the medium. The simplest and easily made sensor consists of a membrane containing the cholinesterase enzyme, immobilised on the membrane and connected to an electrode, which can electrochemically oxidise the thiocholine produced by the enzyme reaction and measure the activity in terms of the O_2 consumed. Another way of measuring the activity of the enzyme, in the presence of the toxicants, is through use of the artificial substrate fluorescein diacetate (FDA) and quantification of the activity, fluorimetrically, by the measurement of the fluorescence signal from the fluorescein.

An alternative way of using the cholinesterase/cholineoxidase system is by the use of an extremely low concentration of acetylcholinesterase enzyme in the sensor, for the purpose of examining the presence of the inhibitors. Such sensors are activated after 1 min of incubation of the sensor in the toxicant containing medium and are allowed to run only for three minutes.

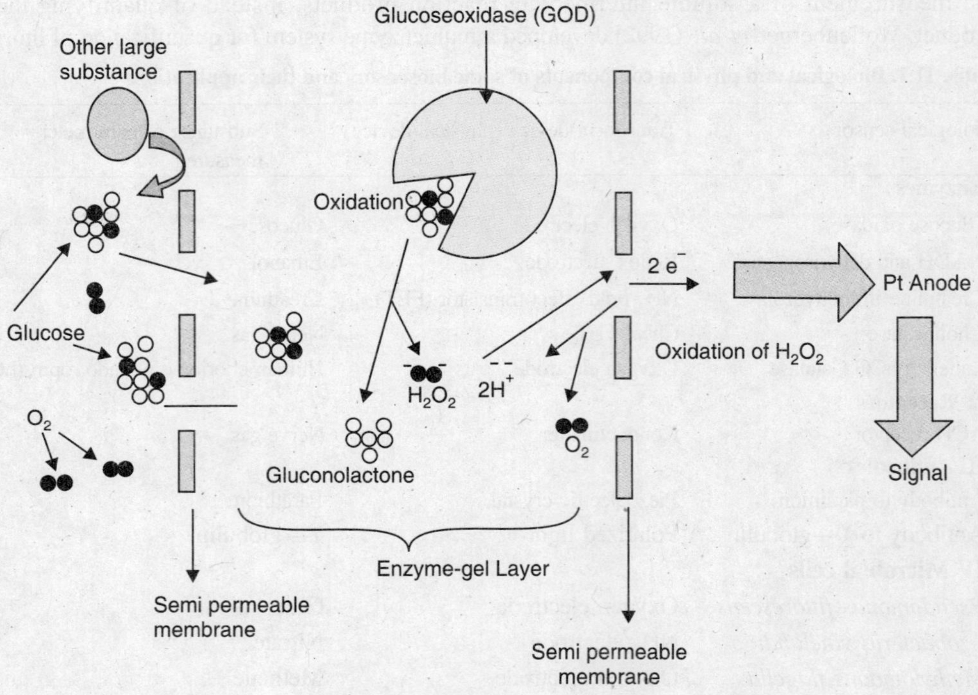

Fig. 11.7. Glucose biosensor

Microbial Biosensors

The microbial sensors cover a wide range of the substrates of the biological system and are connected with highly sensitive and reliable physical components. The microorganisms in the sensor utilise the substrate or the substrate mixture in the medium and absorb them to cause intracellular metabolism. Some of the products of such metabolic reactions are excreted or diffused out of the cells. Under aerobic conditions, most of such metabolic reactions require oxygen, and the response can be recorded by oxygen electrode, through the rate of O_2 consumption by the biosensor. Several other products like NH_4 +, CO_2, H_2S, etc. are also released from the cells which can be sensed by the sensor. The microbial activity also, often causes, change in a pH of the medium which can be measured by potentiometric electrode.

The first microbial biosensor was constructed way back in 1975, and since then, there is continuous increase in the number and sophistication of the sensors. The currently available microbial biosensors can analyse and explain over seventy biological parameters. However, because of the complexity of the biological components (microbial cell) of microbial sensors, they are less sensitive and less effective than the enzyme sensors, in terms of selectivity and detection of specific substances. However, this disadvantage of the microbial sensor is masked by its wide range of applicability and its utility to measure many biochemical and physiological reactions at the cellular level. The most important applications of the microbial biosensors are: (*i*) induction of specific transport or degradation mechanism in different nutrient media, (*ii*) metabolism of specific substrate(s) through incubation of the sensor in substrate containing medium, (*iii*) inhibition of the affected transport or degradation mechanisms, and (*iv*) influence of the substrate transport rate on specific degradation of the substrate by the sensor.

Companella and colleagues (1987) developed a microbial biosensor using *Saccharomyces cerevisiae* to measure the toxic effects of heavy metals on the cellular uptake of organic carbon. The effect was electrochemically analysed through the rate of O_2 consumption.

Because of the adaptation of microorganisms to different environmental conditions, microbes having the ability to metabolise specific environmental contaminants can be isolated and used as biological components in microbial biosensors. These sensors are especially meant for determining the biodegradability of the environmental contaminants. An example of such a sensor is the *Pseudomonas* sensor for detection and metabolism of chlorinated phenols, and the *P. diminutum*-sensor, for detection and metabolism of organ phosphorus insecticides. Similarly the cyanobacterial sensors are used for detection of herbicides, as well as for monitoring the effect of pesticides photosynthesis. Hansen (1992) prepared a cyanobacterial electrode by immobilising *Synechococcus* cells on a graphite electrode. The activity of the cyanobacterium was measured through ferricyanide conversion reaction under dark-light cycle. Fluorescence bacteria have also been used as biocomponents of the microbial biosensors and have been applied to measure the effect of chromium on luminescence.

Immunosensors

While enzyme sensors are useful to determine the effect of contaminants on enzymatic reactions and substrate metabolism, immunosensors detect the antigen-antibody reactions. Any antigen-antibody reaction can be potentially affected by environmental contaminants and, therefore, in principle, such reactions can be useful to detect the presence, as well as the concentration of the toxicant. In the last decade, numerous immunoassays have been developed and some of them are commercially available as ELISA-kit. However, the immunosensors have not yet reached the laboratories for analysing most of the environmental chemicals. The transductors include different optical instruments of different degrees of sensitivity, which can measure the decrease in antibody production and analyse the effect by comparing it with a reference value. Such sensors are usually equipped with fluorescence detection equipment and the antibodies used here are tagged with a fluorescent marker.

The construction of an electrochemical biosensor can be made by immobilising the antibodies (or antigens) in a column and connecting the biocomponent with the electronic part, which may be optical, fluorimetric or amperometric in nature. The intensity of the signal is determined by the rate of metabolism of the substrate by the enzyme. The antibodies can be marked by more than one probe that can be catalysed by more than one enzyme, and this reaction gives a more intense signal due to the production of more products. The next step in the development of electrochemical immunosensors is the spatial integration of antibody-antigen reactions through enzymatic catalysis and electrode.

The other examples of useful sensors are:

1. Oxygen gradients have been determined in engineered tissue using a fluorescent sensor spot. Optical sensor patches containing a luminescent O_2 sensitive indicator were placed on the bottom of a Petri dish, and oxygen supply in a tissue of chondrocytes was monitored over a 3-week culture period. Two-dimensional pO_2 (oxygen concentration) images across the tissue sęction were acquired over the duration of the experiment.
2. Proteins in the eluate of a preparative continuous annular chromatograph were detected with a quartz fiber-optic system that measures the intrinsic absorption of two aromatic amino

acids (Tyr, Trp) in proteins. Two types of optical multichannel detectors were developed. The first is a multichannel detector, and the second is a circular optic device. UV absorption is recorded at 280 nm. Calibration plots were established for a series of stock solutions of known concentrations of proteins.

3. Novel infrared optical probes have been introduced for process monitoring and analysis based on silver halide fibers. Compared to near-IR spectroscopy, for which quartz fiber probes can be applied, the application of previously used mid-IR fiber materials was restricted due to deficiencies with regard to their optical transmission and mechanical properties. Several flexible probes of different geometries were constructed that are suitable for process monitoring. Oil aerosols from natural pipelines can be detected with an optical sensor assembly.
4. Sensors for moisture and pH value in reinforced concrete were reported. The dye pyridinium-*N*-phenolate shows a moisture-dependent absorption. It was embedded in a polymer matrix. Increasing moisture causes a shortwave shift of the absorption. pH in concrete was measured with a fiber-optic sensor consisting of a pH indicator dye immobilized in a hydrophilic polymer matrix. The sensor system is reported to display long-term stability even in media of pH 12-13.
5. Rugged, low-cost diode laser sensors for water and temperature were described. These provide fast, accurate, and nonintrusive methods for monitoring species and temperature in combustors.
6. A commercial gas analyzer based on tunable diode laser spectroscopy has successfully been applied to control and monitoring applications in the waste incineration industry. The analyzer was optimized for process control in the industrial environment and provides real-time signals for water vapor, oxygen, and temperature.
7. An optical algal biosensor for heavy metals exploits the inhibitory effect of heavy metals on alkaline phosphatase. The enzyme is present in the external membrane of *C. vulgaris* microalgae. Oxygen in underground air can be analyzed with a plastic fiber sensor array. Oxygen, consumed in a lignite mine tailing affected by acid mine drainage formation, displays a strong concentration gradient to a depth of 6 m.
8. Fiber optics can be used to monitor the dissolution of drugs such as rifampicin and the concentration of adriamycin in rabbit blood.
9. The excitement about the possibility of monitoring glucose noininvasively via near-IR spectroscopy appears to have ceased since the number of respective articles has fallen dramatically. A novel glucose fiber probe was reported that consists of one central optical fiber around which several others were arranged in circle. Light was shone onto the skin surface through the circle fibers, and scattered light reaching the central detecting fiber was collected and transmitted to the detection system. Glucose intake experiments showed that there is good correlation between the optically determined glucose levels and the values determined with blood samples, particularly if measured at around 1600 nm.
10. Breath ammonia, resulting from *Helicobacter pylori* infection, can be analyzed with an ammonia sensor. Before urea ingestion, pylori-positive subjects had significantly lower breath ammonia levels than negative subjects, but had a significantly larger increase in breath ammonia after urea ingestion.

VII. HYDROPONICS

History. The earliest published work on growing terrestrial plants without soil was in the year 1627 in the book, Sylva Sylvarum by Sir Francis Bacon, although he died in 1626. Mineral nutrient solutions for soilless culture of plants were first perfected in the 1860s by the German botanists, Julius von Sachs and Wilhelm Knop. Growth of terrestrial plants without soil in mineral nutrient solutions was called solution culture. It quickly became a standard research and teaching technique and is still widely used today. Solution culture is now considered a type of hydroponics where there is no inert medium.

In 1929, Professor William Frederick Gericke of the University of California at Berkeley began publicly promoting that solution culture be used for agricultural crop production. He first termed it aquiculture but later found that aquaculture was already applied to culture of aquatic organisms. Gericke created a sensation by growing tomato and other plants to a remarkable size in his backyard in mineral nutrient solutions rather than soil. By analogy with the ancient Greek term for agriculture, geoponics, the science of cultivating the earth, Gericke introduced the term *hydroponics* in 1937 for the culture of plants in water (from the Greek *hydros*, water, and *ponos*, labor).

Two other plant nutritionists at the University of California were asked to research Gericke's claims. Dennis R. Hoagland and Daniel I. Arnon wrote a classic 1938 agricultural bulletin, *The Water Culture Method for Growing Plants Without Soil*. Hoagland and Arnon found that hydroponic crop yields were no better than crop yields with good quality soils. These two researchers developed several formulas for mineral nutrient solutions, known as Hoagland solutions. Modified Hoagland solutions are still used today.

In the 1960s, Allen Cooper of England developed the Nutrient Film Technique. The Land Pavilion at Walt Disney World's EPCOT Center opened in 1982 and prominently features a variety of hydroponic techniques. In recent decades, NASA has done extensive hydroponic research for their Controlled Ecological Life Support System or CELSS. Hydroponics intended to take place on mars are using LED lighting to grow in different color spectrums with much less heat.

Techniques

The two main types of hydroponics are solution culture and medium culture. Solution culture does not use a solid medium for the roots, just the nutrient solution. The three main types of solution culture are static solution culture, continuous flow solution culture and aeroponics. The medium culture method has a solid medium for the roots and is named for the type of medium, e.g. sand culture, gravel culture or rockwool culture. There are two main variations for each medium, subirrigation and top irrigation. For all techniques, most hydroponic reservoirs are now built of plastic but other materials have been used including concrete, glass, metal, vegetable solids and wood. The containers should exclude light to prevent algae growth in the nutrient solution.

Static solution culture – In static solution culture, plants are grown in containers of nutrient solution, such as glass Mason jars, plastic buckets, tubs or tanks. The solution is usually gently aerated but may be unaerated. If unaerated, the solution level is kept low enough that enough roots are above the solution so they get adequate oxygen. A hole is cut in the lid of the reservoir for each plant. There can be one to many plants per reservoir. Reservoir size can be increased as plant size increases. The nutrient solution is either changed on a schedule, such as once per week, or when the concentration drops below a certain level as determined with an electrical conductivity meter. Whenever the solution is depleted below a certain level, either water or fresh nutrient solution is added. A Mariotte's bottle can be used to automatically maintain the solution level. In raft solution culture, plants are placed in a sheet of buoyant plastic that is floated on the surface of the nutrient solution. That way, the solution level never drops below the roots.

Continuous flow solution culture- In continuous flow solution culture, the nutrient solution constantly flows past the roots. It is much harder to automate than the static solution culture because sampling and adjustments to degree and nutrient concentrations can be made in a large storage tank that serves potentially thousands of plants. A popular variation is the nutrient film technique or NFT whereby a very shallow stream of water containing all the dissolved nutrients required for plant growth is recirculated past the bare roots of plants in a watertight gully, also known as channels. Ideally, the depth of the recirculating stream should be very shallow, little more than a film of water, hence the name 'nutrient film'. The main advantage of the NFT system over other forms of hydroponics is that the plant roots are exposed to adequate supplies of water, oxygen and nutrients.

The same design characteristics apply to all conventional NFT systems. While slopes along channels of 1:100 have been recommended, in practice it is difficult to build a base for channels that is sufficiently true to enable nutrient films to flow without ponding in locally depressed areas. Consequently, it is recommended that slopes of 1:30 to 1:40 are used. This allows for minor irregularities in the surface but, even with these slopes, ponding and waterlogging may occur. The slope may be provided by the floor, or benches or racks may hold the channels and provide the required slope. Both methods are used and depend on local requirements, often determined by the site and crop requirements. As a general guide, flow rates for each gully should be 1 litre per minute. At planting, rates may be half this and the upper limit of 2L/min appears about the maximum.

Aeroponics- Aeroponics is defined as a system where roots are continuously or discontinuously in an environment saturated with fine drops (a mist or aerosol) of nutrient solution. The method requires no substrate and entails growing plants with their roots suspended in a deep air or growth chamber with the roots periodically wetted with a fine mist of atomized nutrients. Excellent aeration is the main advantage of aeroponics. Aeroponic techniques have proved very successful for propagation, but have yet to prove themselves on a commercial scale. Aeroponics is also widely used in laboratory studies of plant physiology. Aeroponic techniques have been given special attention from NASA since a mist is easier to handle than a liquid in a zero gravity environment.

Passive subirrigation

The medium generally has large air spaces, allowing ample oxygen to the roots, while capillary action delivers water and nutrients to the roots from the base of the medium. The simplest method has the container constantly sit in a shallow layer of nutrient solution or on a capillary mat saturated with nutrient solution. A variety of materials can be used for the medium: vermiculite, perlite, clay granules, rockwool, or gravel. This method requires little maintenance, requiring only occasional refilling and replacement of the nutrient solution. This keeps the medium regularly flushed with nutrient solution and air. Additional advantages of these sterile porous media are the reduction of root rotting conditions and the additional ambient humidity provided. These advantages are particularly important in the use of hydroponics for orchid cultivation.

It is important in passive subirrigation to wash out the system from time to time to remove salt accumulation. This may be checked with an electrical conductivity or TDS meter, a good average reading would be about 1500 ppm TDS. Lettuce grows well at about 800 ppm and tomatoes to 3000 ppm but both will grow reasonably well on 1500 ppm. It is important to keep the pH reading at about 6.3 to enable nutrient uptake. This is commonly employed for large display plants in public.

Top irrigation- In top irrigation, nutrient solution is periodically applied to the medium surface. This may be done manually once per day in large containers of some media, such as sand. Usually, it is automated with a pump, timer and drip irrigation tubing to deliver nutrient solution as frequently as 5 to 10 minutes every hour.

Deep water culture- The hydroponic method of plant production by means of suspending the plant roots in a solution of nutrient rich, oxygenated water. Traditional methods favor the use of plastic buckets and large containers with the plant contained in a net pot suspended from the centre of the lid and the roots suspended in the nutrient solution.

Organoponics- Organoponics is a hydroponic system converted to organic cultivation by replacing the inorganic fertilizer with compost made from sugar waste.

Diahydro- Diahydro is a natural sedimentary rock medium that consists of the fossilized remains of diatoms. Diahydro is extremely high in Silica (87-94%), an essential component for the growth of plants and strengthening of cell walls.

Expanded clay- Also known as 'Hydroton' or 'leca' (light expanded clay aggregate), trademarked names, these small, round baked spheres of clay are inert and are suitable for hydroponic systems in which all nutrients are carefully controlled in water solution. The clay pellet is also inert, pH neutral and do not contain any nutrient value.

The clay is formed into round pellets and fired in rotary kilns at 1200°C. This causes the clay to expand, like popcorn, and become porous.

Rockwool – Rockwool is probably the most widely used medium in hydroponics. Made from basalt rock it is heat-treated at high temperatures then spun back together like candy floss. It comes in lots of different forms including cubes, blocks, slabs and granulated or flock.

Fig. 11.8. Hydroton brand expanded clay pebbles

Rockwool is an excellent inert substrate for both 'free drainage' and recirculating systems. In free drainage or run-to-waste systems, the chance of disease spread is greatly lessened. Rockwool is also lightweight and self-contained, which allows plants to be grown at different densities in different stages - young plants can be grown to an advanced stage in a small area before being planted out into the main growing area, thus improving crop turnaround. Its light weight also permits setting up to be quick and inexpensive. Rockwool initially causes an increase in pH level. A pH level of 5.5-6.5 should be sufficient to create a suitable pH.

The disadvantages of rockwool are few. Although relatively inexpensive, because of its bulk, transport costs to remote regions can be prohibitive. However, the fact that it can be used several times over will reduce the growers overall costs. Before handling, gloves and long shirt sleeves should be worn to prevent minor skin irritation. This can also be lessened by wetting the rockwool before handling. When this medium is dry, care needs to be taken so as not to inhale any particles; inhaling such particles may carry a health risk.

Coir – Coco peat, also known as coir or coco, is the leftover material after the fibres have been removed from the outermost shell (bolster) of the coconut. Coco peat was a waste product from factories that used coco fibre as a raw material for making sailing ropes, chair seats and mattress fillings.

Coco is a 100% natural grow and flowering medium, which has proven its value across years and years. Coco is not only a high quality product, but also an honest and environmentally friendly product. For many years, the raw material was considered waste material, and enormous useless "Coco Mountains" appeared in the landscapes of countries like Sri Lanka and India. By developing a special biological composting process, this "waste" transformed into a high quality product. This innovation was, and still is, an important contributor to the local economy of India and Sri Lanka. This and the unique growth characteristics ensure coco is the medium of the moment and the future.

This substrate combines the tolerant, organic nature of soil with the precision of rockwool. Due to the special characteristics of the substrate the nutrient doesn't have a grow and flower variant, there is just one unique formulation for both growth and blooming phase. Due to the unique buffering capability of the coir substrate, and its sponge-like structure, the nutrients needed to ensure high yields are stored in the coco. This means that the plant itself can regulate the amount and timing of its nutrient intake. Coconut fibres have sufficient capillary action to retain enough water and nutrients. This means that the plant can go for longer periods with-out water, which could happen if a feeding pump was to break down for example. Quality coir can be used a number of times and makes an excellent soil improver after use.

Perlite- Perlite is a volcanic rock that has been superheated into very lightweight expanded glass pebbles. It is used loose or in plastic sleeves immersed in the water. It is also used in potting soil mixes to decrease soil density. Perlite has similar properties and uses to vermiculite but generally holds more air and less water. If not contained, it can float if flood and drain feeding is used.

Vermiculite. Like perlite, vermiculite is another mineral that has been superheated until it has expanded into light pebbles. Vermiculite holds more water than perlite and has a natural "wicking" property that can draw water and nutrients in a passive hydroponic system. If too much water and not enough air surrounds the plants roots, it's possible to gradually lower the medium's water-retention capability by mixing in increasing quantities of perlite.

Sand. Sand is cheap and easily available. However, it is heavy, it does not always drain well, and it must be sterilized between use.

Polystyrene packing peanuts. Polystyrene packing peanuts are inexpensive, readily available, and have excellent drainage. However, they can be too lightweight for some uses. They are mainly used in closed tube systems

Nutrient solutions. Plant nutrients are dissolved in the water used in hydroponics and are mostly in inorganic and ionic form. Primary among the dissolved cations (positively-charged ions) are Ca^{2+} (calcium), Mg^{2+} (magnesium), and K^+ (potassium); the major nutrient anions in nutrient solutions are NO_3^- (nitrate), SO_4^{2-} (sulfate), and $H_2PO_4^-$ (phosphate).

Numerous 'recipes' for hydroponic solutions are available. Many use different combinations of chemicals to reach similar total final compositions. Commonly-used chemicals for the macronutrients include potassium nitrate, calcium nitrate, potassium phosphate, and magnesium sulfate. Various micronutrients are typically added to hydroponic solutions to supply essential elements; among them are Fe (iron), Mn (manganese), Cu (copper), Zn (zinc), B (boron), Cl (chlorine), and Ni (nickel). Chelating agents are sometimes used to keep Fe soluble. Many variations of the nutrient solutions used by Arnon and Hoagland are widely used. Care is required not to allow salt concentrations to become too high, nutrients to become too depleted, or pH to wander far from the desired value.

Hydroponics fertilizers and other types of formulas for hydroponics have changed drastically during the last ten years. Many of these changes have resulted in measurably significant increases in plant growth rates, plant resistance to diseases and pests, and plant yields.

Commercial

Due to its arid climate, Israel has developed advanced hydroponic technology. They have marketed their system to Nicaragua, which uses it to produce more than one million pounds of peppers annually for sale abroad, including the United States. The largest commercial hydroponics facility in the world is Eurofresh Farms in Willcox, Arizona, which sold 125 million pounds of tomatoes in 2005. Eurofresh has 256 acres under glass and represents about a third of the commercial hydroponic greenhouse area in the U.S. Eurofresh does not consider their tomatoes organic, but they are pesticide-free. They are grown in rockwool with top irrigation.

Some commercial installations use no pesticides or herbicides, preferring integrated pest management techniques. There is often a price premium willingly paid by consumers for produce which is labeled "organic". Some states in the USA require soil as an essential to obtain organic certification. Hydroponics also saves an incredible amount of water; it uses as little as 1/20 the amount as a regular farm to produce the same amount of food.

The environment in a hydroponics greenhouse is tightly controlled for maximum efficiency and this new mindset is called Soil-less/Controlled Environment Agriculture.With this growers can make ultra-premium foods anywhere in the world, regardless of temperature and growing seasons. Growers monitor the temperature, humidity, and pH level constantly. Hydroponics have been used to enhance vegetables to provide more nutritional value. A hydroponic farmer in Virginia has developed a calcium and potassium enriched head of lettuce, scheduled to be widely available from April, 2007.

Fig. 11. 9 A, B. A view of hydropic culture and automatic irrigation

VIII. IMMUNOLOGICAL ASSAYS

I. Radioimmunoassay (RIA)

RIA is extremely sensitive method for the detection and estimation of any substance that is antigenic and that can be labeled with a radioactive isotope. Basically this method depends upon competition between labeled (known) and unlabelled (unknown) antigen for the same antibody. A known amount of labeled antigen (enzyme in this case), a known amount of specific antibody and an unknown amount of unlabelled antigen in the sample are allowed to reach equilibrium. By measuring the amount of labelled antigen bound to the antibody, a measure of the antigen in unknown sample can be obtained. By suitably modifying the concentrations of antibody and labelled antigen small amount of antigen can be detected.

In order to measure the amount of labelled antigen attached to antibody, it is necessary to separate the antibody antigen complexes from the mixture. By measuring the radioactivity, the percentage of labelled antigen bound to the antibody can be calculated. The concentration of an unknown (unlabelled) antigen can be determined by reference to a standard curve prepared from data by using varying amount of unlabelled antigen to compete.

II. ELISA

The **Enzyme-Linked Immunosorbent Assay (ELISA)** is the most common and widely used antibody application. ELISAs are designed for detecting and quantization of substances such as peptides, proteins, antibodies, hormones, haptens, drugs of abuse and their metabolites. A number of different assay formats are being used dependent on the system being investigated. The simplest format involves coating of antigen onto microtiter plates, followed by incubation with a specific antibody. The binding antibody or an appropriate secondary antibody is conjugated to an enzyme that typically catalyzes formation of a colored product. Color formation is monitored spectophotometrically and related to concentration of antigen by calibration to a standard curve. When antigen is limited, a sandwich assay format is used in which the analyte to be measured is bound between two antibodies – the capture antibody and the detection antibody. Competitive assays are often used when the antigen is small and has only one epitope, or antibody binding site. ELISAs are typically performed in 96, 384, and even 1536-well polystyrene plates and are easily robotized and adapted to high troughput screening.

The ELISA is a fundamental tool of clinical immunology, and is used as an initial screen for HIV detection. Based on the principle of antibody-antibody interaction, this test allows for easy visualization of results and can be completed without the additional concern of radioactive materials use. An HIV ELISA, sometimes called an HIV enzyme immunoassay (EIA) is the first and most basic test to determine if an individual is positive for a selected pathogen, such as HIV. The test is performed in a 8 cm x 12 cm plastic plate which contains an 8 x 12 matrix of 96 wells, each of which are about 1 cm high and 0.7 cm in diameter.

Fig. 11.10. ELISA Instrument

An ELISA plate

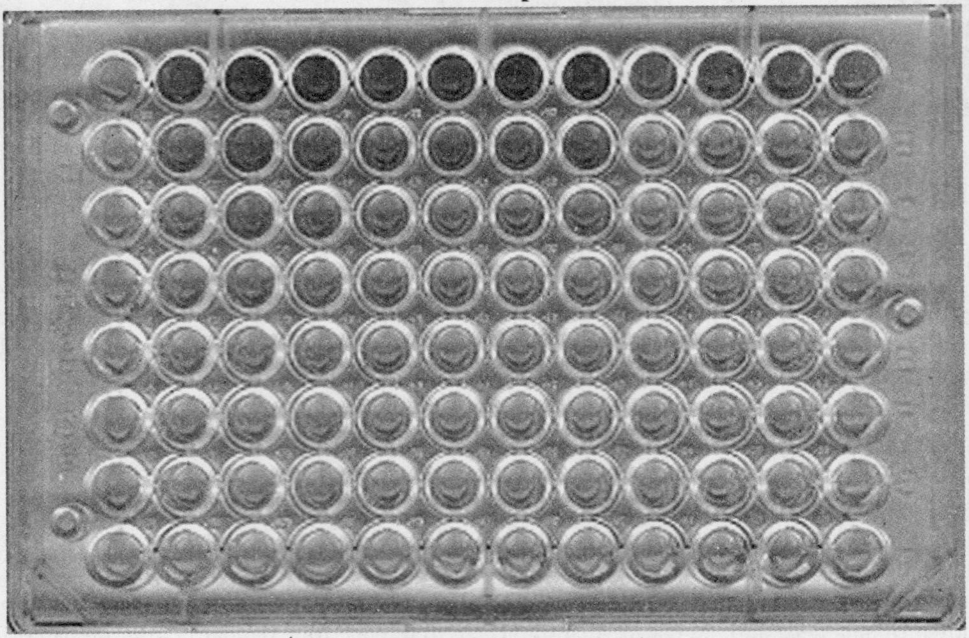

Fig. 11.11. An ELISA Multiwell Plate

The ELISA Method

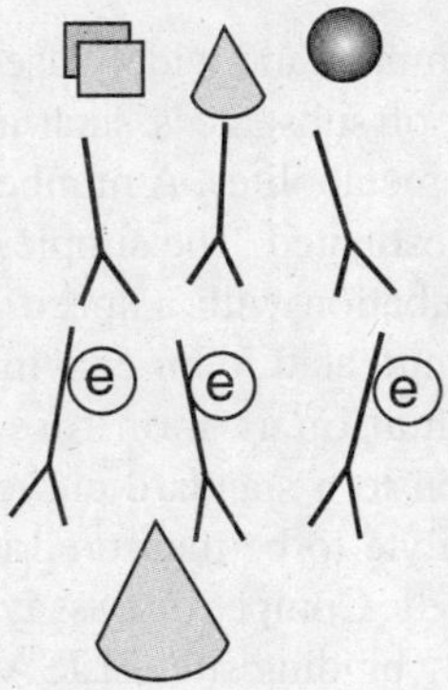

Partially purified, inactivated HIV antigens pre-coated onto an ELISA plate

Patient serum which contains antibodies. If the patient is HIV+, then this serum will contain antibodies to HIV, and those antibodies will bind to the HIV antigens on the plate.

Anti-human immunoglobulin coupled to an enzyme. This is the second antibody, and it binds to human antibodies.

Chromogen or substrate which changes color when cleaved by the enzyme attached to the second antibody.

ELISA Technique

In an ELISA test, a person's serum is diluted 400-fold and applied to a plate to which HIV antigens have been attached. If antibodies to HIV are present in the serum, they may bind to these HIV antigens. The plate is then washed to remove all other components of the serum. A specially prepared "secondary antibody" — an antibody that binds to human antibodies — is then applied to the plate, followed by another wash. This secondary antibody is chemically linked in advance to an enzyme. Thus the plate will contain enzyme in proportion to the amount of secondary antibody bound to the plate. A substrate for the enzyme is applied, and catalysis by the enzyme leads to a change in color or fluorescence. ELISA results are reported as a number; the most controversial aspect of this test is determining the "cut-off" point between a positive and negative result. (Fig. 11.12)

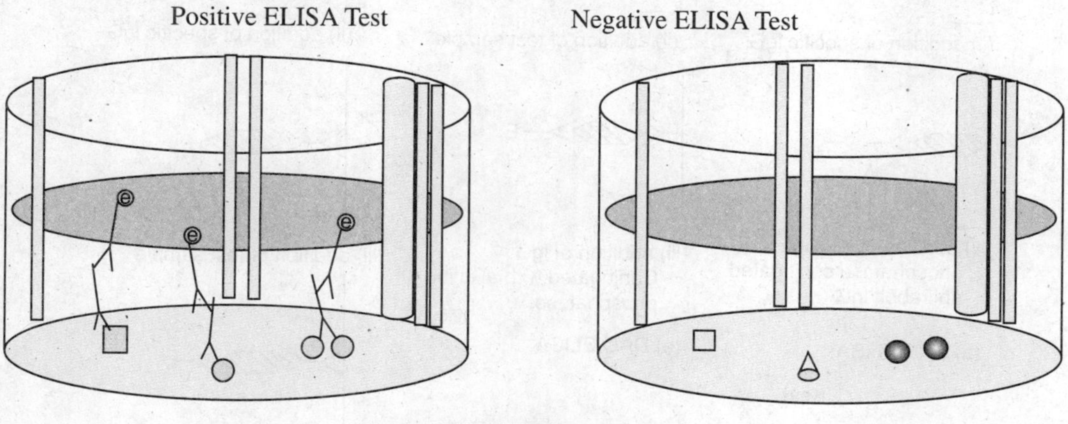

Fig. 11.12. Elisa Test

Several modifications of original ELISA are now known and are widely used for detection of viruses, fungi and bacteria etc. At least three different ELISA techniques are available for detection of plant viruses. (Fig. 11.3) In all the three techniques multiwell microplates are used.

(1) DAS-ELISA- In Double antibody sandwich (DAS)- ELISA, the wells are coated with antibody (or immunoglobulins=Ig) prepared from antisera. The test sample is then added to allow trapping of virus antigen by the coated antibody in the well. This is followed by the addition of enzyme labelled antivirus Ig, which we attach to trap antigen. A substrate for the enzyme is now added for producing colour reaction. Peroxidase (Optical Density used = 450nm) or alkaline phosphatase (Optical Density used = 405nm) is used, which allow quantitative measurement through a spectophotometric device (ELISA reader).

(2) DAC-ELISA- In this direct antigen coating (DAC) ELISA technique plant extracts prepared in a carbonate buffer are applied directly to the wells. In the second step diluted unfractioned antiserum is added. Ig attach to virus antigen is detected by the addition of enzyme-Ig conjugates. This is a simple procedure and can be completed within three hours. With a 96 well plate about 16 thousand samples can be tested in a day. This procedure is not suitable for quantitative estimation.

(3) PAC-ELISA-This protein-A coating (PAC) ELISA technique involves four steps. The wells are coated with protein-A, high dilution of unprocessed antisera is added, test samples are added and finally antisera are added. This is followed by addition of enzyme substrate for colour reaction.

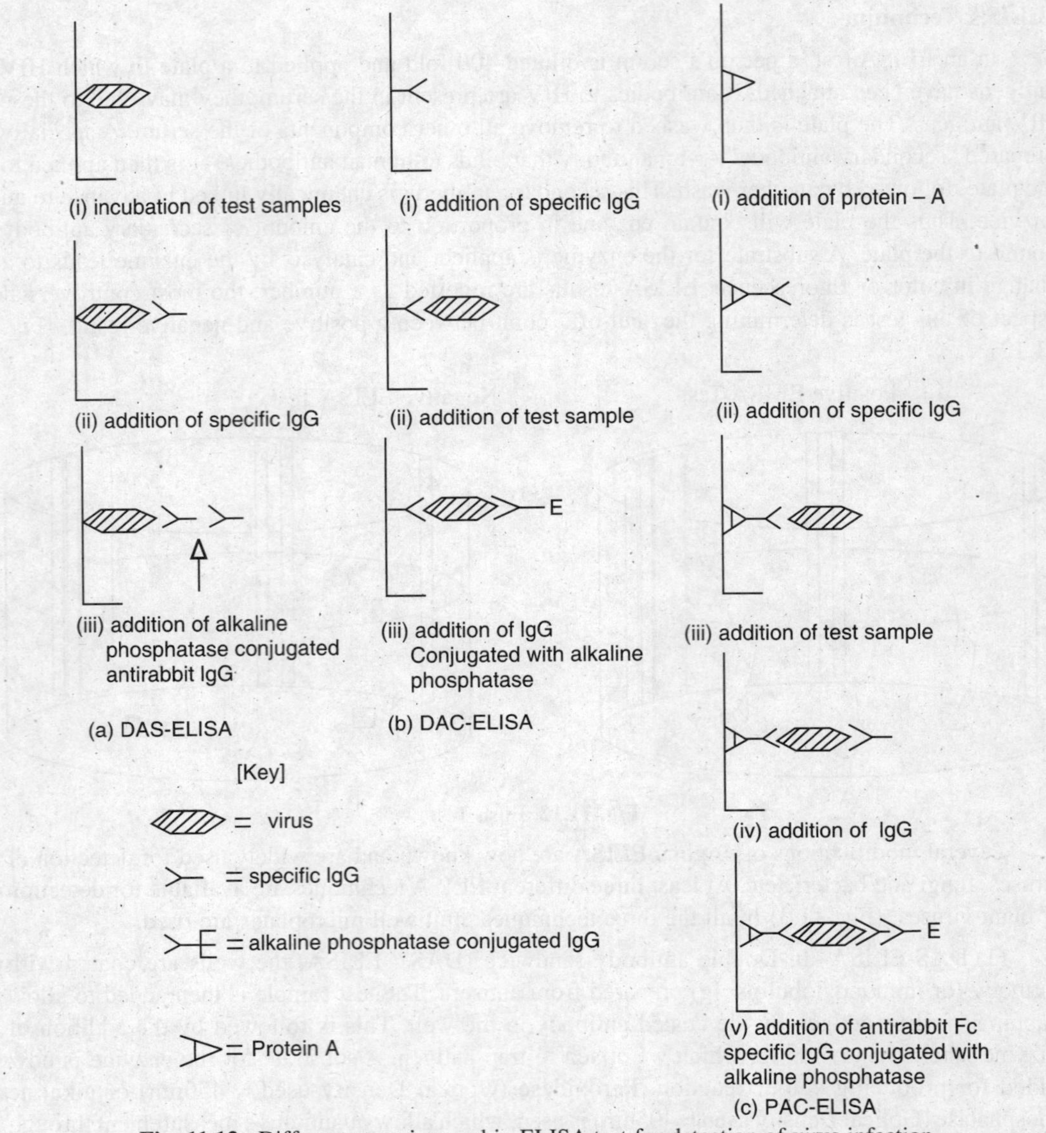

Fig. 11.13. Different steps invoved in ELISA test for detection of virus infection

IX. NANOTECHNOLOGY

1. Introduction

The emerging field of nano-scale engineering and science has generated considerable excitement recently. This field is focused on materials, devices, structures, and processes that occur on the scale of atoms or groups of atoms. The size regime is on the order of nanometers. Another common term used to describe activity and research in this field is Micro-Electro-Mechanical Systems (MEMS).

Microarray technology allows for the parallel synthesis of large numbers of nano-structured materials. These materials can then be tested using the same chip-based technology. These activities are focused on materials that could be useful for applications in the fields of electronics, energy conversion, energy storage, catalysis, and others. New approach enables the synthesis of hundreds to thousands of materials in parallel followed by testing of all of those materials at once, thus allowing for greater efficiency.

Although a metre is defined by the International Standards Organization as 'the length of the path travelled by light in vacuum during a time interval of 1/299 792 458 of a second' and a nanometre is by definition 10^{-9} of a metre, this does not help scientists to communicate the nanoscale to non-scientists. It is in human nature to relate sizes by reference to everyday objects, and the commonest definition of nanotechnology is in relation to the width of a human hair.

There is a commonly held belief that nanotechnology is a futuristic science. In the last 15 years over a dozen Nobel prizes have been awarded in nanotechnology, from the development of the scanning probe microscope (SPM), to the discovery of fullerenes. According to CMP Científica, over 600 companies are currently active in nanotechnology, from small venture capital backed start-ups to some of the world's largest corporations such as IBM and Samsung. Governments and corporations worldwide have spent over $4 billion into nanotechnology in the last year alone.

Many of the initial applications of nanotechnology are materials related, such as additives for plastics, nanocarbon particles for improved steels, coatings and improved catalysts for the petrochemical industry. All of these are technology based industries, but industries with multi-billion dollar markets.

2. The nanotechnology industry

Many of the companies working with nanotechnology are simply applying our knowledge of the nanoscale to existing industries, whether it is improved drug delivery mechanisms for the pharmaceutical industry, or producing nanoclay particles for the plastics industry. In fact nanotechnology is an enabling technology rather than an industry in its own right. While it is possible to buy a packet of nanotechnology, a gram of nanotubes for example, it would have no value. The real value of the nanotubes would be in their application within existing industry.

3. Building atom by atom

One of the defining moments in nanotechnology came in 1989 when Don Eigler used a SPM to spell out the letters IBM in xenon atoms. For the first time we could put atoms exactly where we wanted them, even if keeping them there at much above absolute zero proved to be a problem. While useful in aiding our understanding of the nanoworld, arranging atoms together one by one is unlikely to be of much use in industrial processes.

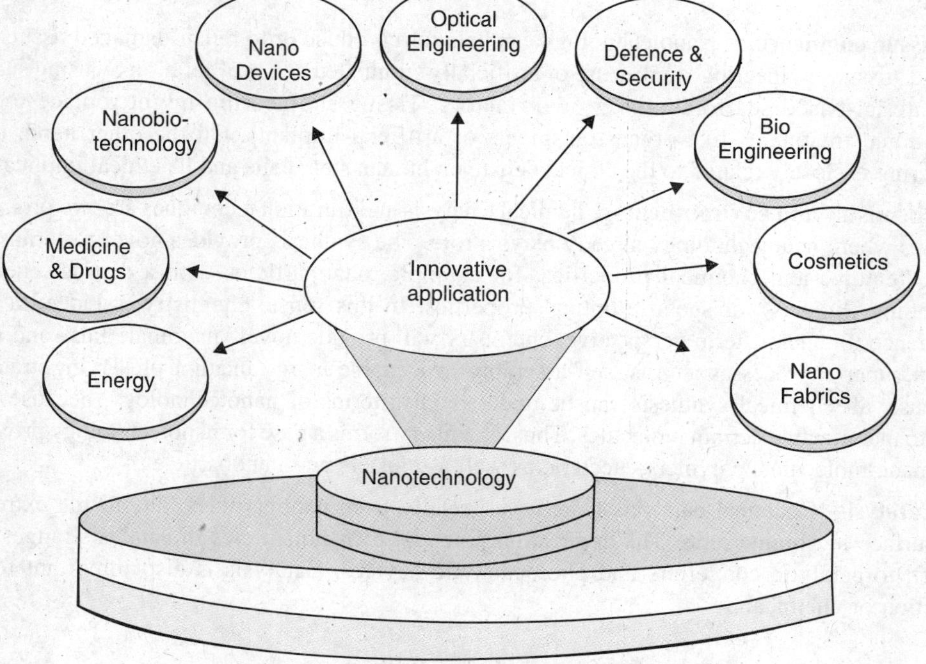

Fig. 11.14. Applications of nonotechnology

Zooming down to the microscale we still have far more complexity than we would like to attempt to replicate, with cells, cytoplasm, mitochondria, chromosomes, ribosomes and many other highly complex items of natural engineering. Moving closer to the nanoscale, we still have to deal with nucleic acids, nucleotides, peptides and proteins, none of which we fully understand, or expect to even have the computing power to understand in the near future.

4. Applications of nanotechnology

Medicine- The biological and medical research communities have exploited the unique properties of nanomaterials for various applications (e.g., contrast agents for cell imaging and therapeutics for treating cancer). Terms such as biomedical nanotechnology, bionanotechnology, and nanomedicine are used to describe this hybrid field. Functionalities can be added to nanomaterials by interfacing them with biological molecules or structures. The size of nanomaterials is similar to that of most biological molecules and structures; therefore, nanomaterials can be useful for both in vivo and in vitro biomedical research and applications. Thus far, the integration of nanomaterials with biology has led to the development of diagnostic devices, contrast agents, analytical tools, physical therapy applications, and drug delivery vehicles.

Diagnostics- Nanotechnology-on-a-chip is one more dimension of lab-on-a-chip technology. Biological tests measuring the presence or activity of selected substances become quicker, more sensitive and more flexible when certain nanoscale particles are put to work as tags or labels. Magnetic nanoparticles, bound to a suitable antibody, are used to label specific molecules, structures or microorganisms. Gold nanoparticles tagged with short segments of DNA can be used for detection of genetic sequence in a sample. Multicolor optical coding for biological assays has been achieved by embedding different-sized quantum dots into polymeric microbeads. Nanopore technology for analysis of nucleic acids converts strings of nucleotides directly into electronic signatures.

Drug delivery- The overall drug consumption and side-effects can be lowered significantly by depositing the active agent in the morbid region only and in no higher dose than needed. This highly selective approach reduces costs and human suffering. An example can be found in dendrimers and nanoporous materials. They could hold small drug molecules transporting them to the desired location.

Tissue engineering- Nanotechnology can help to reproduce or to repair damaged tissue. This so called "tissue engineering" makes use of artificially stimulated cell proliferation by using suitable nanomaterial-based scaffolds and growth factors. Tissue engineering might replace today's conventional treatments like organ transplants or artificial implants. On the other hand, tissue engineering is closely related to the ethical debate on human stem cells and its ethical implications.

Chemistry and environment – Chemical catalysis and filtration techniques are two prominent examples where nanotechnology already plays a role. The synthesis provides novel materials with tailored features and chemical properties: for example, nanoparticles with a distinct chemical surrounding (ligands), or specific optical properties. In this sense, chemistry is indeed a basic nanoscience. In a short-term perspective, chemistry will provide novel "nanomaterials" and in the long run, superior processes such as "self-assembly" will enable energy and time preserving strategies. In a sense, all chemical synthesis can be understood in terms of nanotechnology, because of its ability to manufacture certain molecules. Thus, chemistry forms a base for nanotechnology providing tailor-made molecules, polymers, etcetera, as well as clusters and nanoparticles.

Catalysis- Chemical catalysis benefits especially from nanoparticles, due to the extremely large surface to volume ratio. The application potential of nanoparticles in catalysis ranges from fuel cell to catalytic converters and photocatalytic devices. Catalysis is also important for the production of chemicals.

Filtration- A strong influence of nanochemistry on waste-water treatment, air purification and energy storage devices is to be expected. Mechanical or chemical methods can be used for effective filtration techniques. One class of filtration techniques is based on the use of membranes with suitable hole sizes, whereby the liquid is pressed through the membrane. Nanoporous membranes are suitable for a mechanical filtration with extremely small pores smaller than 10 nm ("nanofiltration") and may be composed of nanotubes. Nanofiltration is mainly used for the removal of ions or the separation of different fluids. On a larger scale, the membrane filtration technique is named ultrafiltration, which works down to between 10 and 100 nm. One important field of application for ultrafiltration is medical purposes as can be found in renal dialysis. Magnetic nanoparticles offer an effective and reliable method to remove heavy metal contaminants from waste water by making use of magnetic separation techniques. Using nanoscale particles increases the efficiency to absorb the contaminants and is comparatively inexpensive compared to traditional precipitation and filtration methods.

Reduction of energy consumption- A reduction of energy consumption can be reached by better insulation systems, by the use of more efficient lighting or combustion systems, and by use of lighter and stronger materials in the transportation sector. Currently used light bulbs only convert approximately 5% of the electrical energy into light. Nanotechnological approaches like light-emitting diodes (LEDs) or quantum caged atoms (QCAs) could lead to a strong reduction of energy consumption for illumination.

Increasing the efficiency of energy production- Today's best solar cells have layers of several different semiconductors stacked together to absorb light at different energies but they still only manage to use 40 percent of the Sun's energy. Commercially available solar cells have much lower efficiencies (15-20%). Nanotechnology could help increase the efficiency of light conversion by using nanostructures with a continuum of bandgaps. The degree of efficiency of the internal combustion engine is about 30-40% at the moment. Nanotechnology could improve combustion by designing specific catalysts with maximized surface area. Scientists have recently developed tetrad-shaped nanoparticles that, when applied to a surface, instantly transform it into a solar collector.

Displays-The production of displays with low energy consumption could be accomplished using carbon nanotubes (CNT). Carbon nanotubes can be electrically conductive and due to their small diameter of several nanometers, they can be used as field emitters with extremely high efficiency for field emission displays (FED). The principle of operation resembles that of the cathode ray tube, but on a much smaller length scale.

Consumer goods- Nanotechnology is already impacting the field of consumer goods, providing products with novel functions ranging from easy-to-clean to scratch-resistant. Modern textiles are wrinkle-resistant and stain-repellent; in the mid-term clothes will become "smart", through embedded "wearable electronics". Already in use are different nanoparticle improved products. Especially in the field of cosmetics, such novel products have a **promising potential.**

Household- The most prominent application of nanotechnology in the household is self-cleaning or "easy-to-clean" surfaces on ceramics or glasses. Nanoceramic particles have improved the smoothness and heat resistance of common household equipment such as the flat iron.

Optics-The first sunglasses using protective and antireflective ultrathin polymer coatings are on the market. For optics, nanotechnology also offers scratch resistant surface coatings based on nanocomposites. Nano-optics could allow for an increase in presicion of pupil repair and other types of laser eye surgery.

Textiles- The use of engineered nanofibers already makes clothes water- and stain-repellent or wrinkle-free Textiles with a nanotechnological finish can be washed less frequently and at lower temperatures. Nanotechnology has been used to integrate tiny carbon particles membrane and guarantee full-surface protection from electrostatic charges for the wearer. A military application could be in camouflage where nanocameras mixed with nanodisplays could create an "invisibility coat", acting like the skin of a chameleon. Many other applications have been developed by research institutions such as the Textiles Nanotechnology Laboratory at Cornell University

Cosmetics- One field of application is in sunscreens. The traditional chemical UV protection approach suffers from its poor long-term stability. A sunscreen based on mineral nanoparticles such as titanium dioxide offer several advantages. Titanium oxide nanoparticles have a comparable UV protection property as the bulk material, but lose the cosmetically undesirable whitening as the particle size is decreased.

'Mercury sponge' for removal of mercury

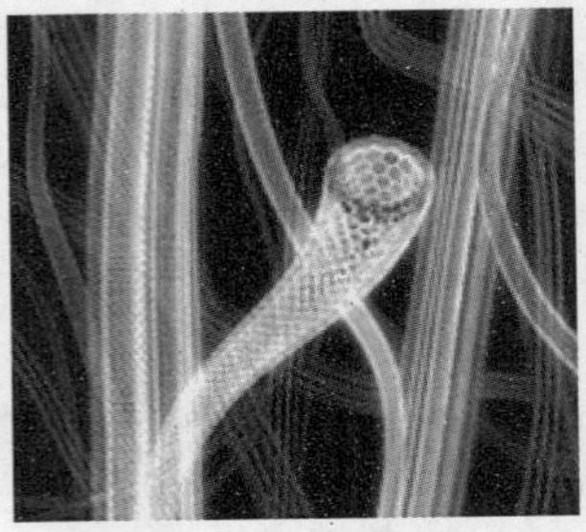

nanotubes

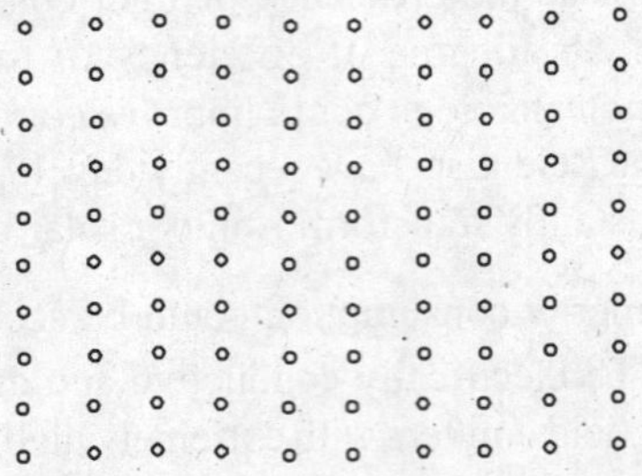

microarrays

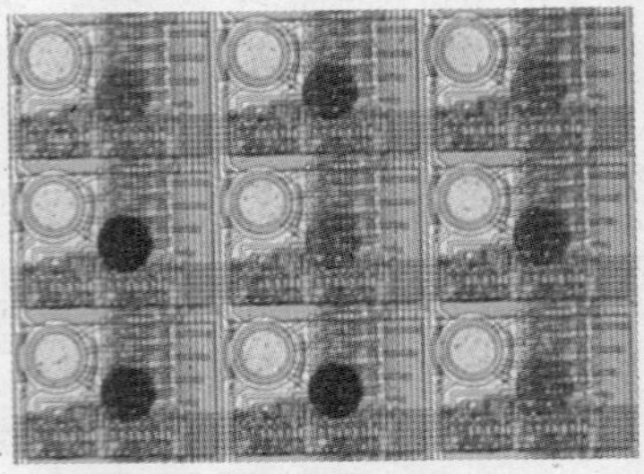

Details of microarray

Fig. 11.15. Different products of nanotechnology

Nanotechnology, like any other branch of science, is primarily concerned with understanding how nature works.

For business, nanotechnology is no different from any other technology, it will be judged on its ability to make money. This may be in the lowering of production costs by, for example, the use of more efficient or more selective catalysts in the chemicals industry, by developing new products such as novel drug delivery mechanisms or stain resistant clothing, or the creation of entirely new markets, as the understanding of polymers did for the multi-billion euro plastics industry.

In Indian context, There are only 10 companies based on nannotechnoogy. The government of India has earmarked 200 crores for research development and production of nanotechnology based materials. The principal companies involved in this direction are-Raymonds for use in fabric manufacture, Samsung in electrical hardwares, TATA, Mahindra, Nicolas Piramal and Intel in various fields like materials, drugs, and softwares. The major problem is deficiency of trained man power in this field.

X. PATCH CLAMP TECHNOLOGIES

Ion Channels

Voltage-gated and ligand-gated ion channels are well known for their role in electrical activity of nerve cells, but ion channels are found in many other places too. When a Paramecium bumps into an obstacle, mechanosensitive channels, along with calcium channels produce the action potential that switches the direction of motion of its cilia. The rate of beating of the human heart—roughly once per second—is determined by potassium channels that take about one second to close, and opposing cation channels that take about one second to open. Diabetes medications, novacaine, anti-epileptic drugs, snake and spider toxins all act on ion channels. Ion channel defects are responsible for disorders like Cystic Fibrosis, cardiac arrhythmias, and some kinds of hypertension and kidney disease.

In the nervous system, the processes underlying leaning and memory involve ion channels. Of particular interest are channels that act as "coincidence detectors," serving to signal the instances in which synaptic strength should be increased according to Hebb's rule. One such coincidence detector is the NMDA-type glutamate-activated channel, which passes current only when glutamate is present and when the membrane is depolarized. Another is the IP3 receptor, an intracellular calcium channel that is opened by the simultaneous presence of IP3 and calcium ions.

Selectivity – Some ion channels are remarkably selective. Potassium channels are orders of magnitude more permeable to potassium ions than sodium ions. Hodgkin and Keynes showed in 1955 that multiple ions pass through these channels in "single file". These sites have recently been visualized in crystal structures of potassium channels from Rod MacKinnons laboratory. Similarly, the selectivity of calcium channels for divalent over monovalent ions arises from a similar "knock-on" mechanism, as was demonstrated in the 1980s in the laboratories of Wolf Almers, Richard Tsien and Peter Hess.

Gating – Scientists are particularly interested in the way in which channels switch on and off their permeability to ions in response to electrical or chemical stimuli. This "gating" process appears to come from large mechanical motions (conformational changes) in the channel proteins. For example, Clay Armstrong demonstrated in 1962 that large ions can plug potassium channels from the intracellular side in such a way that they prevent the channel's gate from closing. Work in Rick Aldrich's lab in 1990 showed that one kind of "gate", the inactivation gate of the Shaker channel, consists of a small protein domain that acts like a tethered blocking ion. More complicated is the activation gate, which is coupled to a very sensitive "voltage sensor."

Patch-clamp recording is undoubtedly the leading method for studying the behaviour of ion channels. In 1976, Neher and Sakmann first used the cell-attached configuration of the patch-clamp technique to resolve current flow through single acetylcholine-activated cation channels in frog skeletal muscle. Subsequent improvements to the technique produced a range of methods for the high resolution recording of current flow through ion channels in both intact cells and excised membrane patches. What makes the patch-clamp technique so powerful is its ability to resolve current flow through one or more ion channels in real time. A further strength of the technique is that many different types of ion channels in a wide variety of cell types are amenable to study. The ability of polished glass patch-pipettes to form extremely high resistance electrical seals ($\geq$10G Ω) with a cell membrane underlies the success of the patchclamp technique. A high resistance seal reduces the background current noise to a level that permits the resolution of single-channel events with amplitude as small as 200 fA. Moreover, the mechanical stability of the seal between the patch-pipette and the cell membrane gives the technique its astonishing versatility. Membrane patches can be excised from cells to form either the excised inside-out or outside-out configurations. Alternatively, the membrane patch can be destroyed to establish the whole-cell configuration. The different configurations of the patch-clamp technique permit precise control of the composition of the solutions bathing either side of the cell membrane. As a result, the regulation of ion channels can be investigated simply by adding ion channel modulators into either the bath or pipette solutions while the selectivity of ion channels can determined by manipulating the ionic composition of these solutions. In 1988, Michael Gray and colleagues identified a small conductance, cAMP activated Cl-channel in the apical membrane of rat pancreatic duct cells using the cell-attached and excised inside-out configurations of the patch-clamp technique. Following the cloning, heterologous expression and functional analysis of the gene defective in cystic fibrosis, it became clear that this cAMP-activated Cl-channel first observed by Gray and colleagues was the cystic fibrosis transmembrane conductance regulator (CFTR). Since that time, the patch-clamp technique has proven to be the method of choice for investigating the behaviour of the CFTR Cl- channel. In combination with site-directed mutagenesis, it has facilitated very significant insight into the relationship between the structure and function of CFTR and the malfunction of CFTR in disease.

The patch clamp technique (Neher, 1992) is the central method in modern electrophysiology. It allows the recording of single ion-channel currents, or alternatively currents from entire small cells. It traditionally uses a glass pipette as an "electrode" which is gently applied to a cell membrane through control by a skilled operator. There is much room for improvement in the technique, especially in view of its difficulty and the large electrical capacitance introduced by the glass pipette.

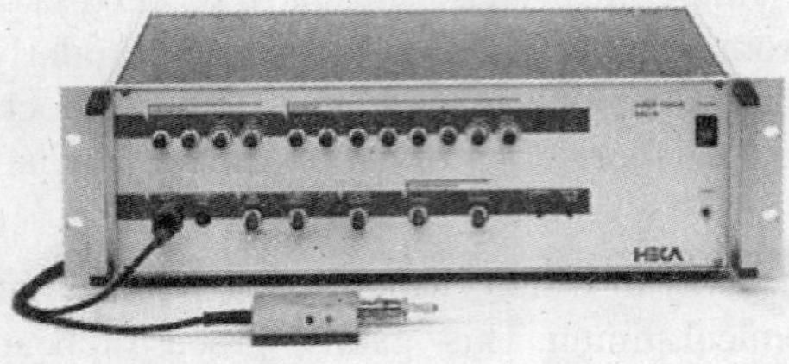

A. The instrument

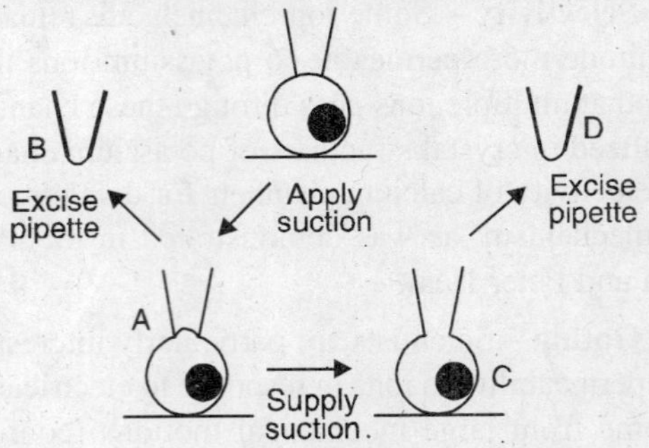

B. Configurations of the patch-clamp technique After a patch-pipette is positioned against the cell membrane, light suction is applied to the back of the patch-pipette to form a giga-ohm seal and hence, the cell-attached configuration (**A**). By pulling the patch-pipette off the cell an excised inside-out membrane patch is formed (**B**). Alternatively, further suction is applied to the back of the patch-pipette to destroy the patch of membrane and hence, establish the whole-cell configuration (**C**). By pulling the patch-pipette off the whole-cell an excised outside-out membrane patch is formed (**D**).

Fig. 11.16. The instrument (A) and patch-clamp method (B).

Cell-attached patch – The electrode remains sealed to the patch of membrane. This allows for the recording of currents through single ion channels in that patch of membrane. For ligand-gated channels or channels that are activated through the action of drug molecules, the drug of choice is usually included in the pipette solution. While the resulting channel activity can be attributed to the drug used, it is not possible to then change the drug concentration. The technique is thus limited to one point in a dose response curve per patch. Usually, the dose response is accomplished through several cells and patches. However, voltage-gated channels or channels that are activated through changes in the potential across the membrane, can be clamped at different membrane potentials using the same patch. This results in graded channel activation, and a proper I-V(current-voltage) curve can be established with only one patch.

"Inside-out" patch – After the gigaseal is formed, the electrode is quickly withdrawn from the cell, thus ripping the patch of membrane off the cell, leaving the patch of membrane attached to the electrode exposing the intracellular surface of the membrane to the external media. This is useful when an experimenter wishes to manipulate the environment affecting the inside of ion channels. For example, channels that are activated by intracellular ligands like cGMP can then be studied through a range of ligand concentrations.

Whole-cell recording or **whole-cell patch –** The electrode is left in place, but more suction is applied to rupture the portion of the cell's membrane that is inside the electrode, thus providing access to the intracellular space of the cell. The advantage of whole-cell patch clamp recording over sharp microelectrode recording is that the larger opening at the tip of the patch clamp electrode provides lower resistance and thus better electrical access to the inside of the cell. A disadvantage of this technique is that the volume of the electrode is larger than the cell, so the soluble contents of the cell's interior will slowly be replaced by the contents of the electrode. This is referred to as the electrode "dialyzing" the cell's contents. Thus, any properties of the cell that depend soluble intracellular contents will be altered. The pipette solution used usually approximates the high-potassium environment of interior of the cell. Generally speaking, there is a "grace period" at the beginning of a whole-cell recording, lasting approximately 10 minutes, when one can take measurements before the cell has been dialyzed. Whole-cell recordings involve recording currents through multiple channels at once.

The outside-out patch is obtained by pulling away from the whole-cell configuration. The membrane breaks away and folds back on itself, covering the tip of the pipette. When recording in this configuration, the pipette solution should resemble the "physiological" intracellular environment.

"Outside-out" patch: After the aforementioned whole cell patch is formed, the electrode can be slowly withdrawn from the cell, allowing a bulb of membrane to bleb out from the cell. When the electrode is pulled far enough away, this bleb will detach from the cell and reform as a ball of membrane on the end of the electrode, with the outside of the membrane being the surface of the ball. Outside-out patching gives the experimenter the opportunity to examine the properties of an ion channel when it is protected from the outside environment, but not in contact with its usual environment. In this conformation, the experimenter can perfuse the same patch with different solutions, and if the channel is activated from the extracellular face, a dose-response curve can then be studied. Single channel recordings are possible in this conformation if the bleb of membrane is small enough. This is the distinct advantage the outside-out patch variation possesses relative to the cell-attached method. However, it is more difficult to accomplish as more steps are involved in the patching process.

XI. SPACE BIOTECHNOLOGY

The Space Biotechnology Department was created in the Space Research Institute, Bulgarian Academy of Sciences in 1987 by Dr. Tania Ivanova with the basic purpose to develop an automated system for precise measurement and control of environmental parameters in order to provide optimal biotechnology for higher plant growing under microgravity. The higher plants are a key component (food source and air purification) of a Biological Life Support System (BLSS). A link of vital importance for the space crews during future long-term manned missions (first to the Moon and Mars).

During 1985, a preliminary experiment SUBSTRAT was conducted onboard SALUT-7 Orbital Station (OS) in order to choose the optimal nutritive medium for the plants. The effectiveness of the technical and technological methods and means gave us the chance to realize the first experiments with vegetables with the purpose of obtaining a vitaminous addition to the food of the space crews.

NASA funded the development of a new modification of greenhouse with optimized parameters – SVET-2 SG, launched onboard MIR OS to carry out long-term experiments (280 days in total) with different plant species in 1996-1997 as part of MIR-NASA-3,5 program. The aim of the experiments was to obtain a full life cycle plant development "from-seed-to-seed" and reproduction under microgravity.

Fig. 11.17. Growth chamber used in space

Unique results in the area of the fundamental gravitational biology have been obtained in these experiments: a new trend in the science - "Space plant growing" has been created. Healthy plants with rich biomass developed normally in the improved SVET-2 SG facility. Initially (during 1996) the heads of the "space" wheat didn't contain seeds. It was necessary to repeat the experiments on the Russian scientific program using an "Apogey" wheat variety that has a higher tolerance for the ethylene (gas in the MIR OS atmosphere that probably had caused sterility in the "Super Dwarf" wheat). Seeds were produced in February 1999 and sawed again for second crop. The achieved reproduction of *Brassica rapa* (plants grown from the "space grown" seeds were obtained) proved as well that there were no plant growth "stoppers" under microgravity and plants could be used as food for the space crews in the future BLSS.

Several synchronized ground based experiments were carried out in SVET-2 SG (in Moscow, Logan and Sofia). This allowed comparing the results obtained in the space and earth experiments and to clarify the problems of plant growing under microgravity. An international staff of scientists process, interpret and release the scientific results from the space and ground-based synchronous experiments. The plant experiments will continue in the next 21st century on the International Space Station. Space Greenhouse of a new generation, based on the Bulgarian "know-how" has been planned to work along with similar equipment of NASA and ESA.

Head start for biotechnology

Biotechnology is likely to be one of the first microgravity science payloads abroad space station. The biotechnology program has several science payloads that grow protein crystals. These are analyzed on Earth to determine the molecular structure so scientists can design drug therapies that target a specific problem with few or no side effects. It's a bit like safe-cracking at the atomic level. Most of the protein crystal growth hardware requires little of the space station's resources and crew support. They only need to be turned on, and days or months later, turned off. If crew time is available, some photo documentation may be requested.

Fig. 11.18. A view of space station to develop biotechnology

"After that, the rotating Bioreactor experiments in cell science will start on one of the utilization flights. Bioreactor is more complex and requires some crew attention since the health and growth of the cell clusters inside must be monitored, and nutrient and waste bags replaced. The NASA Bioreactor is like a rotating culture dish with a mini-life support system attached. In it, scientists can culture cells for long periods of time so they can grow in lifelike assemblies that should yield clues to how both healthy and cancerous tissues grow. From that will come new knowledge of how to improve transplants and to fight cancer.

Fig. 11.19. Rotating-horizontal bioreactor for growth of cells in space

While bioreactor research in space is relatively new, it shows great promise. For 5 days on the STS-70 mission, a bioreactor cultivated human colon cancer cells, which grew to 30 times the volume of control specimens grown on Earth. This significant result was reproduced on STS-85 Mir which grew mature structures that more closely match what are found in tumors REPLACE in humans. The space biotechnology has opened new avenues of biotechnological research to produce novel molecules as well as to provide fresh vegetables to the inhabitants of space station and possible futuristic colonies on nearby planets.

XII. SUPERCRITICAL FLUID EXTRACTION

Carbon dioxide is non-toxic, non-flammable, odorless, tasteless, inert, and inexpensive. The critical temperature of carbon dioxide is 88°F (31°C), just above room temperatures. In the past five years, research and process development activity has focused on utilizing supercritical carbon dioxide technology in processing fine chemicals, pharmaceutical intermediates, and nutraceuticals. In addition to being a solvent for extraction and fractionation (purification) of organic compounds, carbon dioxide is increasingly being utilized as a medium for reactions, as a micronizing agent in Rapid Expansion in a Supercritical Solution process (RESS), as an anti-solvent for crystallization in Gas Anti-Solvent process (GAS), and as a carrier solvent for coating and depositing materials onto or into a solid matrix. Carbon dioxide technology is one of the fastest growing new process technologies being adopted by the food, pharmaceutical and nutraceutical industries.

Fig. 11.20. A SCF extraction unit

Supercritical fluid technology will allow pharmaceutical and nutraceutical companies to develop products of standardized concentration of active ingredients. This will also produce nutraceutical and pharmaceutical products of much higher concentration (higher yields and purity) and quality (with less creation of artifacts) compared to conventional chemical engineering utilizing liquid/liquid extraction, distillation, mechanical micronization, liquid and/or gas phase reactions, etc. Supercritical carbon dioxide extraction is currently used in several food and pharmaceutical manufacturing applications. Its "greener" nature makes it a desirable option when compared with traditional organic solvent extractions.

Supercritical fluids (SCFs) are increasingly replacing the organic solvents that are used in industrial purification and recrystallization operations because of regulatory and nvironmental pressures on hydrocarbon and ozone-depleting emissions. SCF-based processes has helped to eliminate the use of hexane and methylene chloride as solvents. With increasing scrutiny of solvent residues in pharmaceuticals, medical products, and neutraceuticals, and with stricter regulations on VOC and ODC emissions, the use of SCFs is rapidly proliferating in all industrial sectors. Supercritical fluid extraction (SFE) plants are operating at throughputs of 100,000,000 lbs/yr or more in the foods industry. Coffee and tea are decaffeinated via supercritical fluid extraction and most major brewers in the US and Europe use flavors that are extracted from hops with supercritical fluids. SCF processes are being commercialized in the polymers, pharmaceuticals, specialty lubricants and fine chemicals industries. SCFs are advantageously applied to increasing product performance to levels that cannot be achieved by traditional processing technologies, and such applications for SCFs offer the potential for both technical and economic success.

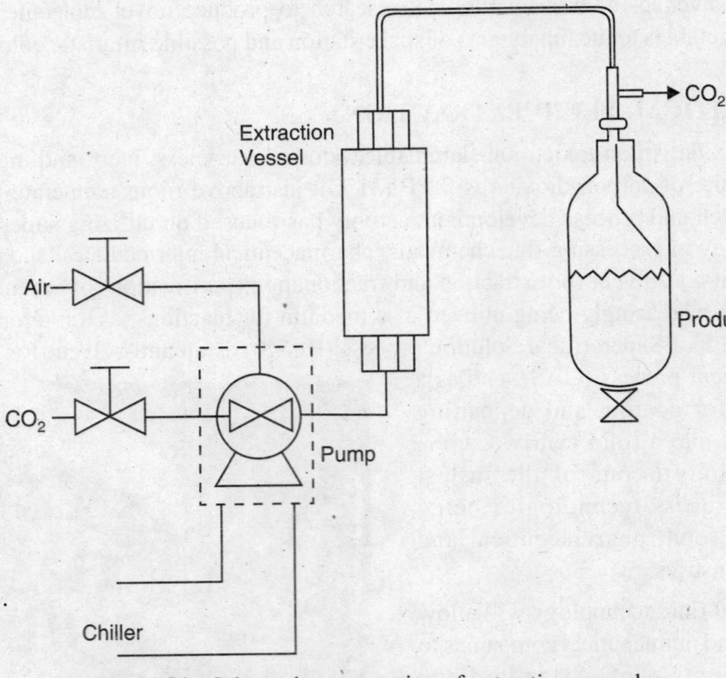

Fig. 11.21. Schematic presentation of extraction procedure

Applications in Industry

- Reactions–SC Water Oxidation–Catalysis
- Pharmaceuticals–Particle Formulation–Drug Delivery
- Extraction–Petroleum–Coffee Decaffeination–Essential Oils

Advantages for Pharmaceuticals

Carbon dioxide as a solvent has many advantages. Probably the most important advantage is that it is a Generally Regarded As Safe (GRAS) solvent that leaves no traces in the product. After extraction, the carbon dioxide is recycled and any trace carbon dioxide in the product dissipates to the atmosphere within a few hours. Also, unlike solvent extraction, the carbon dioxide is readily recycled by pressure and temperature adjustment, which is very mild and does not harm the product. Another advantage of supercritical fluid extraction is the capability of fractionating products to create co-products. Solvent extraction requires a distillation step, (in which top notes (characteristic feature) are lost and distillation notes are created), that many times alters the taste, aroma and chemical composition of the product. Also, trace quantities of residual organic solvent are usually present in the product.

Botanicals can be fractionated to produce a natural color fraction, an aroma fraction, an anti-oxidant fraction and/or a flavor fraction. This is important in producing nutraceuticals because unwanted strong flavors in certain botanicals such as garlic and rosemary can be separated from the nutraceutical components.

Supercritical Fluid Particle Sizing Technology

Particle formation by methods utilizing supercritical or sometimes subcritical carbon dioxide are subjects of great interest in the pharmaceutical and fine chemical industries. Several techniques are available to form particles utilizing supercritical carbon dioxide. If a material is highly soluble in liquid or supercritical carbon dioxide, then the process known as Rapid Expansion from a Supercritical Solution (RESS) should be considered. This is a process in which a supercritical fluid mixture is expanded into an expansion vessel through a specially designed orifice to achieve the desired mean particle size and particle size distribution. This process however is very limited in its application because very few compounds are highly soluble in supercritical carbon dioxide.

Microencapsulation

The rapid expansion of supercritical solutions (RESS) process was used to produce polymeric microparticles or microspheres loaded with pharmaceuticals for drug delivery applications. Poly (L-lactic acid) (L-PLA), naproxen, and a mixture of naproxen/L-PLA were dissolved in supercritical CO_2 and precipitated by the RESS Process. Composite particles appear as a naproxen core encapsulated in a polymer coating. Microspheres containing low molecular weight pharmaceuticals dispersed in poly (L-lactic acid) were also prepared using the PCA process. Supercritical carbon dioxide was used as the antisolvent and methylene chloride as the carrier. The drug polymer particles spherical in shape and between 0.2 and 1.0 microns in diameter as determined by scanning electron microscopy.

Production Scale SFC

Production scale SFC has been successfully used for the separation of enantiomers and fatty acid esters. Large quantities of DHA and EPA ethyl esters from fish oils are routinely separated to >95% purity on a commercial production scale SFC unit.

Partial List of Pharmaceutical Products that can be Processed by Supercritical CO_2

- Extracts of chamomile flowers for anti-inflammatory and anti-spasmodic pharmaceutically active compounds (e.g. sesquiterpene, lactone, matricin, etc.)
- Extract of calamus root as an appetite stimulant—higher yield with SFE (8.3%) when compared to steam distillation (6.4%)
- Extracts of turmeric for bile preparations—no artifacts such as tolylmethylcarbinol created in steam distillation
- Valarian as a sedative preparation—valepotriates obtained undecompossed and at high yield (>90%)
- Wormwood extract as a carminative, cholagogue and stomachic—removal of toxic â-thujone by fractional extraction from thermally unstable pharmacology active components
- RESS (micronization) of mevinolin (Lovastatin), Efrotomycin, Imipenem, Digoxin, Griseofulvin, Salicylic Acid, Stigmasterol, Testosterone, Progesterone, Cholesterol, Ketoprofen, Piroxicam, Nimesulide, and Theophyline.
- SAS, PCA, or GAS (recrystallization) of insulin, poly (l-lactic acid), chlorpheniramine maleate, indomethacin, piroxicam, thymopentine, hydrocortisone, methylprednisolone acetate, salmeterol xinafoate, lysozyme and trypsin.

- Controlled release microspheres for low molecular weight pharmaceuticals produced by Precipitation with a Compressed Antisolvent (PCA).
- Hydrogenation reactions in supercritical carbon dioxide that are a factor of 1,000 faster than conventional hydrogenation reactions with greater control over *trans* isomer formation.
- Extraction of fermentation broths producing vitamins with pharmaceutically active compounds
- Enzymatic reactions in supercritical fluids such as conversion of lipids to methyl or ethyl esters; enrichment of ibuprofen and epi-Methyl jasmonate from racemic mixtures.

Advantages of SFE

1. SCFs have solvating powers similar to liquid organic solvents, but with higher diffusivities, lower viscosity, and lower surface tension. 2. Since the solvating power can be adjusted by changing the pressure or temperature separation of analytes from solvent is fast and easy. 3. By adding modifiers to a SCF (like methanol to CO2) its polarity can be changed for having more selective separation power.. 4. In industrial processes involving food or pharmaceuticals, one does not have to worry about solvent residuals as you would if a "typical" organic solvent were used. 5. Candidate SCFs are generally cheap, simple and many are safe. Disposal costs are much less and in industrial processes, the fluids can be simple to recycle. 6. SCF technology requires sensitive process control, which is a challenge. In addition, the phase transitions of the mixture of solutes and solvents have to be measured or predicted quite accurately. Generally the phase transitions in the critical region are rather complex and difficult to measure and predict. Our research has provided much insight into this phenomenon.

QUESTION

1. Write short notes on :

(*a*) Apoptosis
(*b*) Aquaporins
(*c*) Artificial intelligence
(*d*) Biochips
(*e*) ELISA
(*f*) Nanotechnology
(*g*) Patch clamp technology
(*h*) Space biotechnology
(*i*) Supercritical fluid extraction
(*j*) What are biosensors? Discuss role of biochips in biosensors?
(*k*) What are hydroponics? Discuss their role in war and peace.

INTRODUCTION TO PLANT TISSUE CULTURE : HISTORY, MEDIA AND CULTURE INITIATION

CHAPTER 12

The technique of plant cell culture occupies a key role in the second green revolution in which gene modifications and biotechnology are being used to improve crop yield and quality. This technology has been employed to a wide range of crop plants and tree species to solve the biological problems. This state of art technology has become possible by continuous efforts of many scientists, who carried out the basic work. A brief history of this development is described below.

I. HISTORICAL DEVELOPMENTS

The history of plant tissue culture begins with the concept of cell theory given independently by Schleiden 1838 and Schwann 1839, that established the cell as a functional unit. This implies that cells are autonomous. The concept was tested experimentally by Haberlandt after 130 years, who conceived the idea of culturing plant cells. The developments in culturing of plant cells, tissues and organs are associated with the developments in our knowledge about nutritional requirements of plant cell, discovery of growth regulating factors, analytical tools and techniques, and development of microscopy.

Though, earlier attempts to grow plant cells met with failures, success was achieved in growing animal cell culture at that time. Earlier works involving plant tissue were mainly concerned with nutritional requirement of cells to make them divide and sustain growth. The work progressed rapidly after the discovery of auxin and cell division factor. Early studies also frequently used complex nutritional factors like coconut water, yeast extract, malt extract and casein hydrolysate. The developments after World War II were rapid as many inventions developed for the war purposes found their applications in biological sciences.

It was untiring efforts of Robbins, Kotte, White, Gautheret, Heller, Van Over-beek, Steward, Caplin, Miller, Nitsch, Reinert, Street, Morel, Skoog, Hildebrandt, Melcher, Cocking, La Rue and others that plant tissue culture, which was initiated as a tool has become a powerful technology in last three decades.

With the better understanding of the technique of plant tissue culture and nutritional requirements, of plant cell, it was possible to develop newer technologies by culturing plant organs (anther, ovary, ovule, petal, leaf and meristem) leading to establishment of new research lines viz., haploids, virus-free plants, *in-vitro* fertilization, embryo rescue and direct regeneration from leaf disc for genetic engineering. Subsequently, it has become possible to grow isolated epidermal cells, gland cells or even protoplasts and to regenerate from such specialized cells or individual protoplasts. Regeneration of plants and production of useful metabolites through plant biotechnology has become an industrial application. Demonstration of variability in cell cultures in relation to secondary metabolites has given the concept of morphological variations and hence 'Somaclonal variation'. The significant contributions of selected scientists are given in the following paragraphs:

Gottlieb Haberlandt: With the establishment of cell theory proposed by Schleiden & Schwann (1839), cell is considered as structural and functional unit of living organisms. In plants, totipotency of excised plant tissues and cells was established subsequently. It was Gottlieb Haberlandt who conceived the idea of culturing isolated plant cells in the nutrient solutions. Based on cell theory, he assumed that there was no limitation of divisibility, therefore, he started with isolated mesophyll cells using Knop's nutrient solution, sucrose, aspargine and peptone. Haberlandt (1902) in his famous paper described the cultivation of mesophyll cells of *Lamium purpureum* and *Eichhornia crassipes*, epidermal cells of *Ornithogalum* and hair cells of *Pulmonaria.* The cells survived 3-4 weeks but without cell division, there was increase in cell size. He conducted several experiments and carefully recorded his observations and interpreted the results. Using pieces of mature potato tubers he observed that cell division occurred in small, thin tissue almost without exception when the explant contained a vascular strand.

Gottlieb Haberlandt

If Haberlandt had used in place of highly difficult tissues, some other materials like willow or carrot, which are known to proliferate without growth substances, he would have obtained the first tissue culture. The other reason for the failure was the incomplete information at that time about nutrition of plant tissues and lack of knowledge about plant growth regulators.

P. Nobecourt

P. Nobecourt: P. Nobecourt a French plant pathologist, announced simultaneously in 1939 with White and Gautheret, the possibility of cultivating plant tissues for unlimited period. It was possible with the use of recently discovered indole acetic acid (IAA) by F. W. Went (1932). This success opened new avenues of cultivating plant cells for different studies. This success was announced a few months before the beginning of Second World War. And for six years, American & French scientists worked without knowledge of their mutual results..

Pierre Roger Gautheret: Like many others, Pierre Roger Gautheret was working earlier at the University of Sorbonne, Paris, France. Gautheret also tried to cultivate isolated cells and root tips without getting tissue cultures. Following these failures, he turned towards the tissue participating in wound healing. He first used piece of cambium cut from trees, especially *Acer pseudoplatanus, Sambucus nigra* and *Salix capraea*.. Preliminary attempts with liquid medium failed completely. Later explants were placed on the surface of medium solidified with agar. He did not expected any good results and kept those cultures in a cupboard. He was surprised to find very healthy callus on the explant after two months. He observed living turgescent dividing cells. Over the next 5 years the recognition of the importance of B vitamins in yeast extract and the auxin (IAA) allowed significant advances to be made. It was because of these factors that Gautheret could achieve the proliferation and division of cambial tissue of *Salix capraea* and *Populus alba* for several months. In 1939, Gautheret reported the propagation of carrot as the first plant tissue culture of unlimited growth. IAA stimulated the growth of undifferentiated tissue on cut surfaces of sterile explants. This tissue, termed callus, was similar in appearance to wound tissue and subsequently it was found that callus could be subcultured indefinitely. His results were immediately verified in Italy by Grelli. Finally, the possibility of cultivating plant tissues for unlimited periods was announced almost simultaneously by White (1939), Nobecourt (1939) and Gautheret (1939). In spite of difficulties encountered during the war, Gautheret was able to receive some collaborators in his laboratory and one of them was Georges Morel. Gautheret was working on fleshy organs, especially Chicory, Jerusalum artichoke and *Brassica napus* and Morel established strains of lianas, herbs and trees.

During the study of normal and tumour culture, the development of auxin non-requiring tissues were observed by French and American scientists. Gautheret discovered the reasons for this difference (Kulescha and Gautheret, 1948), the normal tissues maintained *in vitro* produces an insignificant amount of auxin while the tumour tissues produce rather important amount of this growth substance.

Later on, hyper-auxinity has been established in tumour tissues. The same was observed in case of habituated tissues not requiring exogenous supply of auxin. He proposed the term 'habituation' for such cultures.

Gautheret R.J

In 1942, Gautheret observed development of buds in tissue culture of *Ulmus* and of plantlets in Chicory tissue cultures, establishing the totipotency of the cells. The various investigations carried out by Gautheret during 30s and 40s were mainly concerned about nutrition and cell differentiation of cultivated cells. His other major contributions are establishment of hormones autonomous cultures (habituated cultures) from hormone requiring cultures and study of tumour cells. He also worked on root culture of *Helianthus tuberosus* and demonstrated that light, temperature and other factors affect the root growth. His important contributions are cited in following two monumental works.

1. Gautheret R.J. (1955). The nutrition of plant tissue cultures. *Annual Review Plant Physiol.* **6**:433-484.

2. Gautheret R.J. (1983) Plant tissue culture: A History. *Bot. Mag. Tokyo*. 96: 393-410.

Georges Morel: G. Morel was among the first to culture monocotyledonous tissues. He developed the method of meristem culture for the elimination of viruses and the micropropagation of orchids and discovered the two unique opines of crown gall tissues. These opines have become marker for analysis of transformed cultures induced by *Agrobacterium tumefaciens* and *A. rhizogenes*.

Morel, in 1950, obtained the indefinite growth of monocotyledonous tissues such as *Gladiolus*, *Iris* and lily on the medium enriched with natural extract (coconut milk, yeast extract). He also cultivated Royal Fern on this medium.

Morel and Martin (1952) developed meristem culture technique and recovered *Dahlia* shoots, free from viruses, by meristem tip culture. In 1955, they recovered virus free potato. This attained wide application of plant tissue culture to raise virus free plants in agriculture. The first industrial application was the work of George Morel who successfully multiplied tropical orchids through division of protocorms, little differentiated structures developing naturally on orchid embryos. In 1960, Morel observed the emergence of such bodies after carrying out a cymbidium shoot-tip culture. He also noted that protocorms could be sectioned into quarters and subcultured, each section regenerating a new protocorm, within a few weeks which, in turn, could be divided. The so-obtained protocorms subsequently evolved into young plantlets. This remarkable property found almost immediate commercial use since orchid producers had to cope with slow multiplications rate through cluster division. Around 1970, first commercial multiplication laboratory was established in USA for orchids.

In 1971, he studied *Agrobacterium tumefaciens*, crown gall inducing bacteria, on the dicots and isolated octopine and nopaline from the *A. tumefaciens* infected plants. These chemicals are produced only when Ti-plasmid of the bacteria integrates with the genome of the host plant cells. They are also known as opines. This investigation opened new avenue to use Ti-plasmid as a vector for gene transfer after removing opines genes from the plasmid. Morel joined the laboratory of Professor R. Gautheret in France during the World War II. He established cultures of lianas, herbs and trees. Gautheret described him as able and enthusiastic worker.

Philip R. White: P. R. White, an American scientist who worked at Rockfeller Institute, New Jersey, USA, was one of the pioneer tissue culturist of early period. In 1934 he reported for the first time successful continuous cultures of tomato root tips and obtained indefinite growth of roots. Initially he used salts of Knop's solution, sucrose and yeast extract in his medium. Later, yeast extract was replaced by three vitamins, viz. pyridoxine, thiamine and nicotinic acid. This synthetic medium has been proved to be one of the basic medium for a variety of tissue cultures. White maintained some of his root cultures till before his death in 1968.

Philip R. White

With the advice of Stanley, a famous scientist who isolated tobacco mosaic virus (TMV), White started work on the dual culture. He established cultures of tomato roots infected with TMV. Cultures of infected roots gave rise to healthy roots by root tip cultures. The success of White's experiment opened the field of root cultures, which have been explored by many workers to solve morphological, physiological and pathological problems.

In 1939, White had presented his work on tobacco tissues at a meeting of the Botanical Society of America. He reported for the first time, long term callus cultures which were established from tumour tissues of *Nicotiana glauca x N. longsdorfii*.. He was successful due to the incorporation of auxin in the medium. Similar results were also observed by Gautheret and Nobecourt in the same year by conducting independent research on carrot. White has contributed in formulating root culture medium, which is known by his name as White's medium (1943). This medium consists of essential elements, glycine, nicotinic acid, thiamine, pyridoxine, sucrose etc.

The work of White can be categorized into three aspects, continuous growth of excised tomato root tips in liquid medium, *in-vitro* cultivation of viruses on excised roots and growth of tumour tissues. On March 25, 1968, he died after suffering from acute hepatitis. His book on tissue culture was the only book on the subject.

–White P. R. (1963) *The cultivation of animal and plant cells.* Ronald Press, New York.

F.C. Steward: F.C. Steward was one of the pioneers of plant tissue culture and contributed a lot by way of understanding the requirements of plant tissue cultures and developing techniques for the different applications.

Working at Cornell University, Ithaca, USA, he had developed a School of Plant Physiologists involved in plant tissue culture work. Caplin and Steward (1948) used coconut water for the first time in plant tissue culture and obtained vigorous proliferation of carrot explants. It was also evident from this finding that the stimulating substance of coconut milk was not an auxin. Steward and co-workers were the pioneer in establishing cell cultures of single cells and clumps in liquid medium (Steward and Shantz, 1955) and developed various types of vessels for this purpose including the rotating nipple flasks (Fig12.1).

The another landmark comes in the form of somatic embryogenesis in carrot cell suspension cultures and production of complete plants (Steward and Ammirato, 1969). Steward had taken cell suspension of callus tissues derived from wild carrot embryos and plated them out on a coconut milk medium solidified with agar. From this plated cell population, thousands of embryo developed each being derived from one or a few cells of the callus tissue. This has not only opened new avenues of

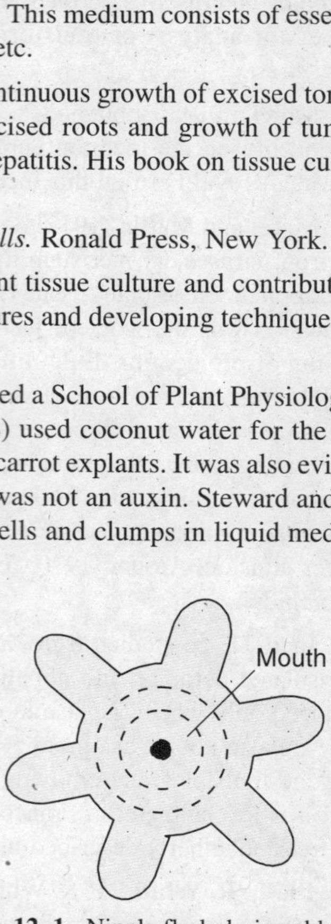

Fig. 12. 1. Nipple flask designed by Steward.

micropropagation of plant species at an enormous rapid rate but also completely established the totipotency of the cell (Fig. 12.2).

His volumes on 'Plant physiology, a treatise' were one of the most widely used reference books on plant physiology and tissue culture.

J. Reinert: Reinert, a German Scientist, was a pioneer in the field of plant culture and contemporary with Steward, Cocking and Street. In 1959, he was culturing undifferentiated parenchyma tissue of carrot root in complex medium containing sugar, salt and coconut milk. On this medium the callus had been sub-cultured repeatedly, without evident morphological change. He transferred the tissue to a synthetic medium containing an elaborate mixture of known substances, including amino acids, vitamins and hormones. The tissue became more granular and showed evidence of differentiated regions. This culture, on transfer to a medium lacking high levels of auxin, produced young plants. Histological analysis showed bipolar embryo formation in the system, which produced plants on proper medium. This had proved the development of a complete plant from mature undifferentiated parenchyma and in other words 'the totipotency' of cells.

Besides the embryogenesis in carrot, Reinert worked several years on developing technology for wide ranging applications of plant tissue culture, which includes cryopreservation of cells, regeneration of plantlets, anther culture, process of androgenesis, development of haploids, and protoplasts culture.

The book 'Plant cell, tissue and organ culture' by Reinert and Bajaj (1977) Springer Verlag, Berlin Heidelberg, was the first compilation of state of art literature on plant tissue culture.

Folke Skoog: Professor Folke Skoog was one of the leading plant physiologists of the world. Skoog was already renowned for his pioneering work with auxin, a plant growth hormone, when he joined the University of Wisconsin, Madison, U.S.A. While at Wisconsin, he discovered a major new class of plant hormones, the cytokinins, which stimulate the division of plant cells, and regulate plant growth and development. During the cultivation of tobacco pith cells, Skoog and co-workers observed that the incorporation of purine base- adenine in the culture medium enhances callus growth. Similarly, DNA extracts from yeast and Herring sperm whale also favoured callus initiation and and callus growth of tobacco pith cells. This leads to the isolation of a compound from autoclaved Herring sperm whale DNA called kinetin (6-furfuryl amino purine). Though kinetin enhances cell division in pith cells, it does not present in this form in plant cells. Therefore, they concluded that this product was formed during autoclaving of whale DNA. It is produced due to dehydration and migration of a deoxyribore moiety from the 9- position of adenine to the N^6-position (Miller et al., 1956). His work has had a profound impact on agricultural and horticultural practices around the world.

Skoog's discovery of cytokinins triggered an international flood of publications that continues to this day. His work showed that a number of cytokinins occur naturally, and that at least one of these compounds occurs in every organism tested, from bacteria to humans. With colleague Nelson J. Leonard at the University of Illinois, Skoog synthesized and tested hundreds of compounds for cytokinin activity, and established the principles that govern relationships between plant structure and activity. These discoveries are generally held to be one of the major advances made in the plant sciences during the last 50 years.

Skoog was also a pioneer in investigating how to control the formation of roots, stems and leaves from undifferentiated cells in plant tissue cultures. Tissues could be made to develop as undifferentiated masses of cells, or to become roots, stems or a combination of roots and stems, resulting in complete plants. Skoog and Miller (1957) demonstrated for the first time that a ratio of auxin and cytokinin can control the root, shoot and callus formation. Skoog's theory that plant development is controlled by hormone levels and other factors led to the realization that whole plants can be generated from cultured cells. This laid the groundwork for the production of transgenic plants and other advances in biotechnology.

In 1962 he and graduate student Toshio Murashige published a culture medium for optimal plant tissue growth that remains in use today. This is a high salt medium, which was modified by Linsmaier and Skoog (1965) and relatively low salt formulation was evolved.

Skoog

Skoog was born in Halland, Sweden on July 15, 1908. He decided to take up residence in the United States during a visit to California in 1925, and he became a U.S. citizen 10 years later. He received a bachelor of science degree in chemistry from the California Institute of Technology in 1932. He earned his Ph.D. from that institution in 1936. While at the institute, he worked closely with prominent plant physiologists Kenneth Thimann and Fritz Went (known for their work and discovery of auxin).

Between 1937-1941, Skoog was on staff at Harvard University, and from 1941-1944, at Johns Hopkins University. He served as a chemist and technical representative of the U.S. Army in Europe between 1944-1946. Skoog played an active role in a number of biological societies. He served as president of the American Society of Plant Physiologists, the Society of Developmental Biology, the American Society of General Physiologists and the International Plant Growth Substances Association. He was a member of the American Academy of Arts and Sciences, and the National Academy of Sciences. He was awarded the National Medal of Science in 1991 during a ceremony at the White House.

Cocking

Renowned plant physiologist and National Medal of Science recipient Folke Karl Skoog, professor of botany at the university of Wisconsin, Madison, U.S.A. for 32 years, died Feb. 15, 2001 after a long illness. He was 92.

E. C. Cocking: Professor Edward Cocking, a Fellow of Royal Society of London, works at the Centre for crop nitrogen fixation at Nothingham University, Nottingham, U.K. He has established an International network for research on nitrogen fixation in the worlds major non-legume crops (especially rice, wheat, maize, sorghum, and oilseed rape). This involves basic research on the interaction of crops with azorhizobia for the establishment of endophytic nitrogen fixation.

Professor Cocking and his associates have been instrumental in developing many techniques of plant cell and protoplasts culture. In 1960s, for the first time a method of isolation of protoplasts in large quantities was developed using enzymes obtained from *Myrothecium* fungus. They obtained protoplasts from root tips for *Lycopersicon esculentum* using cellulase produced by the fungus. Protoplasts isolation and their use in somatic hybridization opened new avenues of improving plant species. Subsequently, the technique found wide applications because of availability of cellulase at commercial level. At that time they started attempting fusion of legume cells containing bacteria with non-legume cells towards developing non-legume nitrogen fixing plants. They continue these efforts with the establishment of research centre for nitrogen fixation studies.

They (Bhojwani and Cocking, 1972) were also successful in isolating protoplasts from pollen grains and pollen mother cells. They used protoplasts for understanding tobacco mosaic virus infection and multiplication in plant cells. Development of technique of protoplasts isolation has many applications in cell biology, plant physiology and genetic engineering as mentioned in the chapter on protoplast culture. Besides many articles on protoplasts culture, his some of the recent publications are as follows.

1. Blackhall, NW; Jotham, JP; Azhakanandam, K; Power, JB; Lowe, KC; Cocking, EC; Davey, MR (1999): Callus initiation, maintenance and shoot induction in rice. In: Methods in Molecular Biology. Vol. 111. (Ed: Hall,RD) Humana Press Inc, Totowa NJ, 19-29.

2. Azahakanandam, K; McCabe, MS; Lowe, KC; Cocking, EC; Power, JB; Davey, MR (1999): Gene transfer into and molecular characterisation of T0, T1 and T2 transgenic rice plants mediated by *Agrobacterium tumefaciens*. Journal of Experimental Botany 50(P2.41), 20.

3. Kennedy IR and Cocking EC (1997) Biological Nitrogen Fixation: The Global Challenge and Future Needs. Position Paper of the Meeting at the Rockefeller Foundation Bellagio Conference Center, 8-12 April, 1997.

I.K. Vasil

Indra K. Vasil and Vimla Vasil: Prof. Indra K. Vasil and Dr. Vimla Vasil are products of Delhi School of Morphology headed by the then Professor P. Maheshwari. They joined the laboratory of Professor A. C. Hildebrandt at Madison, Wisconsin, as post-doctoral fellow. Vasil, V. and Hildebrandt were the first to prove the early cell theory of Schwann by regenerating a tobacco plant from a single cell derived from a suspension of cultured cells. The landmark work of Vimla Vasil on culture of single isolated cells of tobacco has opened new applications of cell cultures in genetics, morphogenesis and vegetative propagation and proved the totipotency of cell. Single cells of hybrid tobacco callus were grown in micro-chambers in the absence of any other cells, in fresh liquid medium containing coconut milk. These produced a colony of 50-75 cells in 10-25 days. Upon transfer from the micro-chambers to agar medium, cell clumps produced callus in 3 months, which ultimately produced complete plantlets (Vasil V. and Hildebrandt, 1965).

During his stay at Wisconsin, I.K. Vasil studied cell differentiation and morphogenetic behaviour in *Petroselinum hortense* and *Cichorium endivia*. He was successful in producing *C. endivia* plantlets from freely suspended cells grown *in vitro*.

Presently I.K. Vasil and V. Vasil are Professor in the University of Florida at Gainsville, USA. I.K. Vasil has contributed significantly on the cell culture, protoplasts cultures and somatic embryogenesis of cereals (monocotyledons), particularly *Pennisetum americanum, Panicum maximum, Triticum aestivum,* and *Zea mays*. He has edited a series of books on plant tissue culture and somatic cell genetics of higher plants published by Academic Press.

Monocotyledons, particularly the cereals were considered unsuitable material to regenerate via somatic embryogenesis. I.K Vasil and co-workers demonstrated that the young leaves, inflorescences or embryos are useful in initiating embryogenic cultures. The comprehensive work showed that selection and maintenance of embryogenic cultures at an early stage is important for maintaining long term embryogenic potential.

Establishment of embryogenic and non-embryogenic lines in monocotyledons and concept about these cultures were given by I.K. Vasil. Embryogenic cultures are compact, dry, amorphous and white in colour as compared to non-embryogenic cultures which are watery, dirty white to light brown in colour and soft in nature. It is necessary to isolate and maintain separately embryogenic cultures.

Riker and Hildebrandt: Professor Albert C. Hildebrandt was a professor at the University of Wisconsin, Madison, U.S.A. As a graduate student and through ensuing years as a staff member, Hildebrandt worked closely with Professor A.J. Riker, who had a keen interest in the effect of the crown gall bacterium, *Agrobacterium tumefaciens*, on the host plants and its possible relation to cancer in animals. Professor Riker was a plant pathologist and used plant tissue culture to understand the basic biology of infection of plant pathogens (Hildebrandt and Riker, 1958). As an approach to the study of crown gall disease, Hildebrandt began the study of the growth in vitro of excised plant tissue, both healthy and diseased. Prof. Hildebrandt became an internationally recognized authority on tissue culture. He was very active in the international tissue culture association. He contributed chapters to several books and manuals concerning tissue culture techniques and their uses.

The tissue culture work began with the study of the isolation and growth of tissues in culture, including the effects of nutrition and environment. Isolations were made from healthy stems, galls, apical meristems and anthers. The effects of bacteria and viruses on these callus tissues were widely studied. Tissues were induced to differentiate back to entire plants through the use of plant growth substances. This was done with a number of plant species- tobacco, sunflower, potato, geranium, gladiolus, African violets, poplar and cassava. Hildebrandt, with his students, then developed methods for isolating single cells from tissue cultures (micro-chamber, nurse tissue culture etc), growing them in liquid culture and again inducing them to differentiate back to the original plant (Sievert and Hildebrandt, 1965). A great deal of still- and cine-photomicroscopy was done on the growth of isolated sells and their relation to disease organisms.

This work on regeneration of plants from single cells (Vasil and Hildebrandt, 1965) became the basis for the eventual development of the technique for the plant transformation in the biotechnology industry. Prof. Hildebrandt also proposed for the first time the possibility of fusion of somatic cells, the somatic hybridization. Hildebrandt formulated a plant tissue culture medium known as Schenk and Hildebrandt (1971) medium.

The isolation of single cells from apical meristems and anthers was used as a means to produce virus-free plants. Hildebrandt did this for several plants, including, geranium, gladiolus and poplar. Virus-free stock was propagated and made available to commercial growers. Related to this work was the study of several virus diseases of gladiolus and geranium.

Hildebrandt was resource person for information on diseases of floricultural crops for growers of the state.

Hildebrandt was born in State college, Pennsylvania, on April 10, 1916. He obtained his Ph.D. degree in plant pathology in 1945 and joined the faculty in 1949 as assistant professor and rose to the rank of full professor in 1960. He retired in 1978. Emeritus Prof. Hildebrandt, age 84, died on January 4, 2001.

Some important citation of the group are as follows:

1. Hildebrandt AC and Riker AJ 1958 Viruses and single cell clones in plant tissue culture. Fed. Proc. 17:986-993.
2. Tamaoki T, Hildebrandt AC, Burris RH and Riker AJ 1961 Oxidation of reduced diphosphopyridine nucleotide by mitochondria from normal and crown-gall tissue cultures of tomato. Plant Physiology 36:347-351.
3. Hildebrandt AC 1977 Single cell culture, protoplasts and plant viruses. In Plant cell, tissue, and organ culture, J Reinert & YPS Bajaj, Springer-verlag, Heidelberg.

Toshio Murashige: Professor Toshio Murashige is working as Professor of plant sciences at the University of California, Riverside, U.S.A. As a graduate student he worked with Professor Skoog on nutrition of plant cells using tobacco pith cells as an experimental material. During that period plant tissue culture was in its developmental stage. Only White's medium was known. They cultured tobacco pith cells on White's medium and evaluated, one by one, inorganic and organic nutrients and formulated a medium, known as Murashige and Skoog's medium (Murashige T. and Skoog, F. 1962 A revised medium for rapid growth and bioassay with tobacco tissue cultures. Physiologia Plantarum 15:473-497). In the study, the optimal concentration of each individual element was determined empirically. This is a synthetic medium and is designed to assay the effect of organic supplements. This is the most widely used medium for plant tissue culture work. Later on, other media were developed based on this formulation.

After joining the University of California, he worked on developing micropropagation technique and established three stage micropropagation system. These are : 1.establishment of aseptic culture, 2. Multiplication of propagule and, 3. Preparation for reestablishment of plants in soil. According to him each stage has a different nutrient and light requirement. He worked on multiplication of several plant species belonging to diverse taxa e.g., ornamentals like *Dracaena, Scindapcus, Syngonium.* He also worked on process of somatic embryo formation using carrot and citrus plants and described changes at molecular levels during the process.

The following two contributions are well appreciated among several hundred articles he has published.

1. Murashige T. 1974. Plant propagation through tissue culture. Annual Review of Plant Physiology 25:134-66.

2. Murashige T. 1978 The impact of plant tissue culture on agriculture, TA Thorpe (ed) IPTCA, Calgary, Canada.

Presently he is working as emeritus scientist on problem of rejuvenation in adult tree explants. Rejuvenation of trees is evident by restored organogeneses, leaf morphology and shoot vigor, and diminishing leaf abscission, chlorosis and tissue and culture medium discoloration *in vitro*. Investigation of the underlying mechanism, using *in vitro* growing shoots, disclosed identical agarose gel electrophoretic patterns of mitochondrial DNA extracted from juvenile and rejuvenated shoots and distinct from the mitochondrial DNA of adult shoots.

Huang LC, Hsaio CK, Lius S, Liu SF, Huang BL, and Murashige T. 1992. Restoration of vigor and rooting competence in stem tissues of mature citrus by repeated grafting of shoot apices onto freshly germinated seedlings in vitro. In Vitro Cell. Dev. Biology - Plant 28P:30-32.

P. Maheshwari: The most outstanding Indian embryologist, P. Maheshwari was trained at Allahabad by Dudgeon. Maheshwari remained as expert of Botany for almost three decades. After completing the D.Sc. Thesis (1929) at Allahabad, he moved to the Agra College (1931-1937), and after serving brief tenures at Allahabad (1937-1939), Lucknow (1939) and Dacca (1939-1949), at the invitation of the Vice-Chancellor (Sir Maurice Gwyier), he joined the University of Delhi (1949-1966) and established a very productive school of embryology. The earlier work of Prof. Maheshwari was concerned with embryology, which he started in 1930 at Agra. By 1933, a flourishing school of Angiosperm Embryology was established and in coming years, this centre became internationally known. B.M. Johri, V. Puri, B. Singh and H.R. Bhargava were the prominent students trained at the above centre. Johri's work on the Alismaceae and Butomaceae done between 1933 and 1938 is among the first important piece of research in embryology in the country. Occurrence of pollen grains in the stylar canal of a typical angiosperm – *Butomopsis lancealata* by Johri (1936) was an appreciable discovery.

P. Maheshwari

P. Maheshwari worked at Haward University in 1945 and devoted most of his time to complete the book entitled, "An introduction to the embryology of angiosperms" (1950). He had also edited a few symposium volumes in 1962-63. He also established the international society of plant morphologists and started an international journal "Phytomorphology."

It was due to continuous efforts of P. Maheshwari that India acquired an international status in embryology. In 1965, he was elected fellow of the Royal Society of London.

In India, studies on tissue culture started in mid 1950 at the Department of Botany, University of Delhi under the directions of Professor P. Maheshwari. This was beginning of experimental embryology. Some faculty members who were trained abroad to acquire knowledge were – S. Narayanswami who worked with C.D. La Rue at Michigan, and B.M. Johri who worked with H.E. Street at Swansea.

Over the past four decades no other institution in India has made more significant contributions than the Department of Botany, University of Delhi, in the use of tissue culture methodologies for morphogenetic studies involving ovary, ovule, embryo, endosperm, cotyledons, and other reproductive organs. These works deal with isolation, culture, *in-vitro* fertilization, and development of plantlets from vegetative tissues. The successful *in-vitro* growth of ovary, ovule, and embryo parts in fruit set and seed development in response to exogenous growth regulators has been achieved. Emphasis was also laid on the nucellar culture for obtaining true clones of *Citrus* and the endosperm culture for the production of triploid plant in case of *Citrus, Oryza* and *Putrangiva roxburghii*.

S. Guha Mukherjee & S.C. Maheshwari: Plant tissue culture studies led to the demonstration for the first time of the development of haploids through anther and pollen culture by Sipra Guha and Maheshwari, S.C. (1964-67). This research finds many applications in crop and tree breeding programmes, being a quicker and simpler method of producing homozygous lines than conventional breeding programme. Professor S.C. Maheshwari was working in the Botany Department of Delhi University. Later on he joined the newly formed department of Plant Molecular biology of Delhi University at South Campus. He contributed significantly in the development of these departments in the University. The University Grants Commission and Department of Science and Technology have recognised the research credential of Prof. Maheshwari and granted him a centre for research on molecular biology. During his long research career he worked on various aspects of plant growth and developments like protoplasts culture, characterization of chloroplast DNA, regeneration in cereals, gene isolation and sequencing, northern analysis, transcript mapping, genomic mapping and gene delivery using electroporation in plants like spinach, mung bean, and rice towards better understanding the mechanism of gene expression during development. Presently he is working in International Centre for Genetic Engineering and Biotechnology at New Delhi.

Prof. Sipra Guha Mukherjee worked on regeneration of plants and mechanism of regeneration involving various enzymes, membrane phospholipids and second messengers at school of life sciences, J.N. University, at New Delhi (She expired in 2007).

The other centres which gradually developed as National Institutes in India are: National Chemical laboratory, Pune; Bhabha Atomic Research Centre, Bombay; Indian Agricultural Research Institute, New Delhi; Indian Institute of Science, Bangalore; National Botanical Research Institute, Lucknow; Bose Institute, Calcutta and Central Institute for Medicinal and Aromatic Plants, Lucknow. The universities at Jaipur, Jodhpur, Udaipur, Dharwar, Varanasi, Calcutta, Baroda, Pune, Hissar and Jawaharlal Nehru University, New Delhi have significantly added to our knowledge using plant biotechnology in understanding the process of growth differentiation and metabolism. The newly established department of biotechnology of the Ministry of Science and Technology has identified the key areas of research. This has encouraged the biotechnological research in India. At I.A.R.I., New Delhi, National facility for plant tissue culture repository has been established with the National Centre of Plant Genetic Resources, New Delhi.

II. MEDIUM AND ITS PREPARATION

The media used by earlier workers were based on Knop's solution. Subsequently media developed by White (1943) and Heller (1953) were used. Murashige and Skoog's medium (Murashige and Skoog, 1962) is a land mark in plant tissue culture research and is the most frequently used medium for all types of tissue culture work. Based on its composition, other media were evolved to meet the diverse experimental and species specific requirement. There are – Linsmaier and Skoog (1965), B_5 medium of Gamborg et al. (1968), Schenk and Hildebrandt (1972), Nitsch and Nitsch (1969) medium, and woody plant medium (Llyod and McCown, 1980).

The methodology of plant tissue culture has advanced to the stage, where tissues from virtually any plant species can be cultured successfully. The successful plant tissue culture depends upon the choice of nutrient medium. The cells of most plant species can be grown on completely defined media. All the media consist of mineral salts, a carbon source (generally sucrose), vitamins and growth regulators. The MS medium designed for tobacco is now used widely for various species, in callus and cell cultures.

MEDIUM COMPOSITION

The nutrient medium for most plant tissue cultures is comprised of five groups of ingredients – inorganic nutrients, carbon source, vitamins, growth regulators and organic supplements.

Inorganic Nutrients- Inorganic nutrients consist of macro- and micro-elements as their salts. Usually nutrient media contain 25 mM each of nitrate and potassium. For regular culture and cell cultures, the combined nitrogen level (nitrate and ammonium nitrogen) may reach up to 60 mM. Ammonium is essential for most cultures but in lower concentrations than that of nitrate nitrogen. A concentration of 1-3 mM of calcium, magnesium and sulphate, is always adequate. The required micronutrients include I, B, Mn, Zn, Mo, Cu, Co and Fe.

Caron Source- Glucose, fructose, maltose or sucrose (2-4%) can be used as source of energy or carbon but sucrose is the preferred source for most of the cultures. The sucrose in the medium is rapidly converted into glucose and fructose. The glucose is absorbed first followed by fructose.

Vitamins and amino acids- Thiamine, pyridoxine and nicotinic acid are commonly used as vitamins in B_5 and MS media. The former is required for most cultures while latter two promote cell growth. Amino acids and organic supplements – Amino acids serve as source of reduced nitrogen. In case of inadequate nitrogen, complex organic nitrogen supplement like casein hydrolysate (0.1-1 g/*l*) may be supplemented. Glycine is commonly used amino acid. Other organic supplements are coconut milk, yeast extract, peptone and malt extract. However, synthetic media are preferred and organic supplement of unknown chemical nature is used only when it is essential.

Table 12.1. Some of the important elements supplied to culutre medium for plant nutrition and their physiological function

Element	*Function*
Nitrogen	Component of proteins, nucleic acids and some coenzymes. Element required in greatest amount.
Potassium	Regulates osmotic potential, principal inorganic cation.
Calcium	Cell wall synthesis, member function, cell signalling.
Magnesium	Enzyme cofactor, component of chlorophyll.
Phosphorous	Component of nucleic acids, energy transfer, component of intermediates in respiration and photosynthesis.
Sulphur	Component of some amino acids (methionine, cysteine) and some cofactors.
Chlorine	Required for photosynthesis.
Iron	Electron transfer as a component of cytochromes.
Manganese	Enzyme cofactor.
Cobalt	Component of some vitamins.
Copper	Enzyme cofactor, electron –transfer reactions.
Zinc	Enzyme cofactor, chlorophyll biosynthesis.
Molybdenum	Enzyme cofactor, component of nitrate reductase.

Plant Growth Regulators (PGR)- A balanced combination of PGR is required for sustained growth. Two types of combinations are used; one for cell proliferation consisting of (preferably) 2,4- dichlorophenoxy acetic acid (2,4-D) or 1-naphthalene acetic acid (NAA) and a cytokinin (kinetin, benzyl adenosine, 2-isopentyladenosine, zeatin, thidiazuron), another for regeneration essentially containing low auxin [(NAA, IAA, Indole butyric acid (IBA)] and a cytokinin in high amount, but not 2,4-D as an auxin. 2,4-D is known to induce cell proliferation but suppresses differentiation in dicot plants. However, 2,4-D and 2,4,5-T (2,4,5-trichlorophenoxy acetic acid) are effective in inducing somatic embryogenesis in cereal (monocots) and herbaceous dicot cultures.

MEDIUM PREPARATION

A convenient approach to prepare a medium is to have stock solutions of all the nutrients in a 10x or 50x concentration. Medium is prepared by suitably diluting the appropriate amount of stock solutions for desired volume of the medium. It may be advantageous to have separate stock solutions of calcium salt and potassium iodide. All the ingredients are mixed, sugar added and pH is adjusted to 5.8-6.0 and medium is poured in the culture vessels. All the vessels are plugged with non-absorbent cotton, covered with aluminium foil and autoclaved at 121°C for 15 min. Prepared media can be stored for a few weeks before inoculation. Liquid medium for a given material is same as static medium used for callus cultures except gelling agent- agar. All the ingredients should be thoroughly mixed before dispensing in the vessels, the compositions of MS and B5 media are shown in Table 2.

Table. 12.2. Composition of Murashige & Skoog's medium and B_5 medium.

Salts	*MS*		*B_5*	
A .Macronutrient	mg/l	mM	mg/l	mM
NH_4NO_3	1650	20.6	-	-
KNO_3	1900	18.8	2500	25
$CaCl_2.2H_20$	440	3.0	150	1.0
$MgSO_4.7H_20$	**370**	**1.5**	**250**	**1.0**
KH_2PO_4	170	1.25	-	-
$(NH_4)_2SO_4$	-	-	134	1.0
$NaH_2PO_4.H_2O$	-	-	150	1.1
B. Micronutrient	mg/l	m M	mg/l	m M
KI	0.83	5.0	0.75	4.5
H_3BO_3	6.2	100	3.0	50
$MnSO_4.4H_2O$	22.3	100	-	-
$MnSO_4.H_2O$	-	-	10	60
$ZnSO_4.7H_2O$	8.6	30	2.0	7.0
$Na_2MoO_4.2H_2O$	0.25	1.0	0.25	1.0
$CuSO_4.5H_2O$	0.025	0.1	0.025	0.1
$CoCl_2.6H_2O$	0.025	0.1	0.025	0.1
$Na_2.EDTA$	37.3	100	37.3	100
$FeSO_4.7H_2O$	27.8	100	27.8	100
Sucrose (g)	30,000		20,000	
pH	5.7 – 6.0		5.5 – 5.8	
C. Vitamins				
Inositol	100		100	
Nicotinic acid	0.5		1.0	
Pyridoxin. HCl	0.5		1.0	
Thiamine. HCl	0.1		10.0	
D.Amino acid				
Glycine	2.0		-	
E. Plant Growth Regulators				
NAA/IAA/2,4,-D	0.01-10.0		0.01-10.0	
Kinetin	0.04–1 0.0		0.04– 10.0	

III. Methods of sterilization EXPLANT AND ITS STERILIZATION

Plant parts are used to initiate and establish *in-vitro* (in glass) growing cultures. In comparison to *in-vitro* grown cultures, in case of field or pot grown plants, growth is affected by various soil and environmental factors of unknown nature. Therefore, it is difficult to study the factors affecting the growth, differentiation and metabolism of the plant. Plant tissue culture provides an excellent system to study the plant growth and differentiation in totally controlled conditions. Cultures are grown in a completely defined chemical and physical environment. Thus, by changing the factors, one by one, the influence of a particular factor can be determined very easily. The field grown (*in vivo*) plants are autotrophic, i.e., they synthesize their own food by photosynthesis. In contrast to this, sugar is essential in the medium for the growth of *in vitro* cultures, i.e., they are unable to synthesize their carbohydrate requirement (even though cultures are grown in light, photosynthesis is almost negligible).

Explant type. The present knowledge permits the use of any plant part as a source of material to initiate cultures. The plant part used for this purpose is known as an 'Explant.' Nodal and inter nodal segments of stem, apical and axillary bud, leaf, leaf disc, petiole, anther, pollen, flower bud, petal, inflorescence, ovule, ovary root and even isolated epidermal peel, gland and trichome have been used as an explant.

Preparation of explant. *In vivo* material, when brought to laboratory, it is always preferable to bring material under ice/ dry ice to minimize catabolic activity. Material is washed thoroughly with tap water and liquid detergent (e.g., Tween-80), and then washed thoroughly. After proper cleaning and removal of adhering particles, material is sterilized and used. Some hard seeds (like leguminous and hard seed coat seeds) may require pretreatment (2-10 min.) with 50% sulphuric acid to break seed coat dormancy. This should be done with care and seeds should be washed thoroughly after acid treatment under running tape water for 1-2 hours. When axillary or apical buds are used, extra leaves are removed, and pre-treated with a wetting agent (quick dip in 70% ethanol or 5-10 min in detergent solution) to facilitate penetration of disinfectant. Explants are prepared in suitable size for the purpose of sterilization, inoculation and suitability to generate culture. Stem segments consist of single node or multiple nodes, or inter nodes are used while leaves can be used as whole leaf or leaf disc can be prepared.

When tuberous materials like Jerusalem artichoke (*Helianthus tuberosus*), sugar beet roots (*Beta vulagaris*)or carrot roots (*Daucus carota*) are used as an explant, large pieces/ complete tubers are sterilized and then with the help of scalpel and cork borer, tissue cylinders and discs from cylinders are prepared under aseptic conditions. The explant preparation and sterilization are carried out under laminar flow bench.

Explant sterilization

All explants have to be sterilized before transferring them on to the medium. A suitable sized explant can be sterilized by any one of the following solutions –

1. 1 % solution of sodium hypochlorite (commercial bleach).
2. 7 % saturated solution of calcium hypochlorite.
3. 1 % solution of bromine water.
4. 70% ethyl alcohol.
5. 0.2% mercuric chloride.
6. 10% hydrogen peroxide solution.
7. 1% silver nitrate solution.

Suitable sized plant material is sterilized as follows:

Clean the working area with ethanol, and start the air flow of the laminar air flow bench.

Sterilized petridishes, distilled water, scalpel, filter paper sheets (suitable sized and autoclaved) alcohol, disinfectant (mercuric chloride 0.01-0.1% aqueous solution, or 20% sodium hydrochloride), forceps, bead sterilizer or beaker or coupling jar, sprit lamp or gas burner are collected on Laminar flow bench. Put on the UV light of the bench and put it off after 30 minutes.

Bring the plant material to be sterilized on the bench and prepare pieces for sterilization.

Clean the working area and hands with alcohol, put on mask and cap, and light the sprit lamp.

Keep 3-4 petridishes in a line, add disinfectant in 1st plate and autoclaved distilled water in subsequent plates.

Place plant pieces in 1st plate and immerse the material with the help of forceps for 5-10 minutes depending upon the disinfectant used.

Transfer material from 1st to 2nd plate, rinse gently and pass to 3rd and 4th plate, one by one with thorough rinsing.

Finally, drain the distilled water or place the material in a fresh petridish or filter paper and prepare suitable sized explants (Fig-12.2).

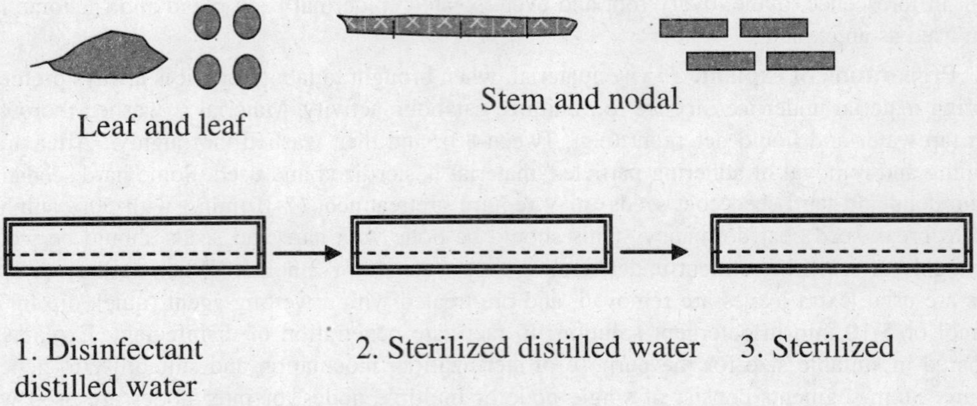

Fig. 12.2. Schematic presentation of somatic embryogenesis in carrot.

Mode of action of disinfectant

The various metallic ions can be arranged in a series of decreasing antibacterial activity. Hg^{2+} and Ag^+, are effective at less than one part per million (ppm), because of their high affinity for sulphahydryl group. Bacteria are killed by Ag containing 10^5 to 10^7 Ag^+ ions per cell. The concentration required for killing is markedly affected by inoculum size.

Chlorine was the anticeptic introduced (as chlorinated lime) by O.W. Holmes in Boston in 1835 and by Semmelweis in Vienna in 1847, to prevent transmission of puerperal sepsis by the physicians hand. Chlorine combines with water to form hypochlorous acid (HOCl) a strong oxidizing agent.

Hypochlorite solutions are (200 ppm chlorine) are used to sanitize clean surfaces in the food and the dairy industries and in restaurants, and chlorine is also added into water supply.

Equipment and medium sterilization

All the manipulations of plant tissue culture methods are carried out in aseptic conditions. All the materials used are therefore, free from microbes or sterilized. Various methods of sterilization are used depending upon the material and type of sterilization. Essentially, method for sterilization is different for living materials than tools and glass ware. Now a days many items are available as ready to use, pre-sterilized (by gamma radiation) and disposable. Use of such disposable materials have facilitated the work. In laboratory, in principle, three types of sterilization is used –

1. Dry heat
2. Wet heat
3. Filter sterilization.

1. Dry heat: Glassware, metal tools and other articles, which do not get charred by high temperature, are put in containers or wrapped in paper or thick aluminium foil and placed in dry oven and sterilized for a period of not less than three hours at a temperature of 140-160°C. Media and plasticware can not be sterilized by this method.

2. Wet heat: The most popular method of sterilization both equipment and media is autoclaving at 121°C with a pressure of 15 psi (pounds per square inch) for 15 min (1.02 kg/cm^2). Modern autoclaves are capable of providing saturated steam treatment ranging from 70-132°C, which is a pressure of up to 25 psi. All the vessels containing medium should be placed vertically and should not be filled more than 40% to their total capacity. There is a relationship between amount of medium to be sterilized and duration of autoclaving (Table-1). Larger volumes requires longer duration for autoclaving at 121°C.

Table 12.3. Sterilization period for various volumes.

ml/Container	Duration at 121°C (min)
20-50	15
75	20
250-500	25
1000	30
1500	35
2000	40

3. Filter sterilization: Filter sterilization or cold sterilization is used when a solution or medium can not be sterilized by autoclaving. It is the property of the filter (porosity 0.22 to 0.45 μm) to retain all the microorganism and make the solution free from microbes. This exclusion of microorganisms makes the solution sterilized without heating or autoclaving. Only liquids can be sterilized by this method and not the plant materials or other things. This is commonly used where thermolabile or heat sensitive chemicals like plant growth regulators (IAA, GA_3, Zeatin, Abscisic Acid), some biochemicals, enzymes for protoplasts isolation, antibiotics or nutrient solution for specific experiment are used to avoid decomposition during autoclaving. After sterilization, solution is added to the medium. In case of static medium, compound is added before solidification (cooling) after autoclaving.

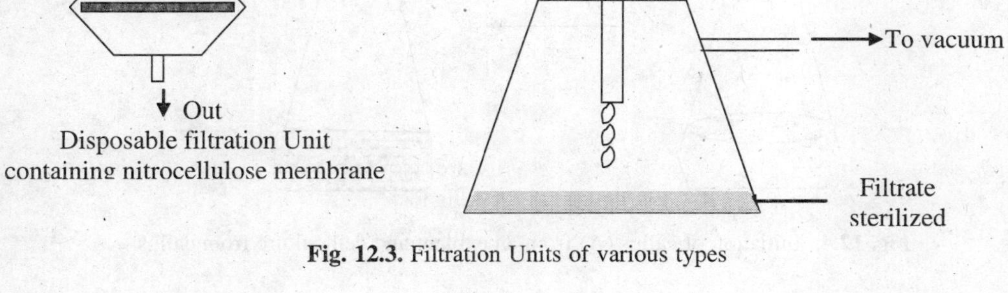

Fig. 12.3. Filtration Units of various types

Glass and membrane filters (nitro-cellulose membranes) are commercially available. Glass (sintered glass filters, Borosil) are available in different grades (G-1 to 5) and the G-5 (1-2 μm) is used for bacteriological purposes (Fig. 12.1). These filters have a porous glass disc for filtration of liquids and gases as a filter media which is non-corrosive and reusable. The grades are classified by maximum pore size which is obtained by measuring the pressure at which the first air bubble breaks away from filter under certain conditions. The pressure differential is then used to calculate the equivalent capillary diameters in microns. The desired pore size is obtained by suitably controlling the grain size, firing time and temperature and the thickness of the disc. These filters are resistant to heat, solvents and detergents.

Membrane type filters (of different diameter) and porosity (0.22 to 0.45 μm) are available as disposable or reusable filters, or as filter holders containing a filter disc (Fig. 12.3). These filters can be connected through tubing or syringe to pass solution or gases.

Filters are wrapped in aluminium foil or paper, autoclaved at 12°C for 15 min and then taken to laminar air flow bench. Solution is placed in funnel (glass type filter) or syringe (membrane type filter) and solution is allowed to pass through filter under suction (glass) or pressure (membrane). A membrane filter or glass tube, filled with cotton may be placed between vacuum flask and suction pump to avoid entry of air from the out side during operation. Sterilized filtrate is collected in flask and dispensed in the liquid or molten medium with the help of sterilized pipette fitted with cotton plug. Volume added to the medium is adjusted to arrive at a final concentration. Membrane type filters are also used at inlets and outlets of various types including air, in a bioreactor system and to draw samples.

There are a variety of wet and dry heat treatments, radiations, filtration and gas and chemical treatments available for direct sterilization of material. Gas treatments are rarely used in the laboratory. Ultraviolet light treatment of working surfaces and sterile rooms are used in the labs while gamma radiation is used in the industry for the preparation of pre-sterilized disposable plasticware.

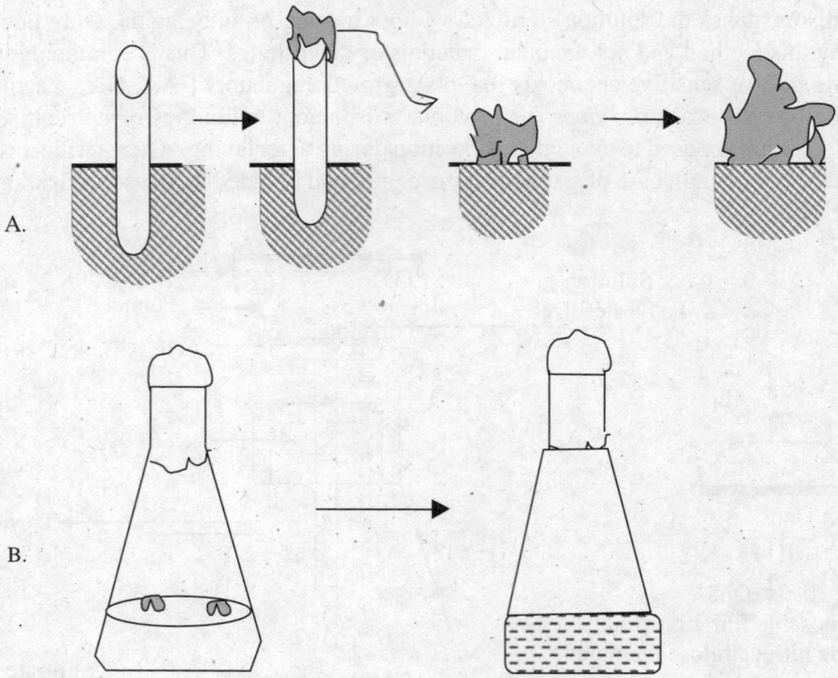

Fig. 12.4. Initiation of callus (A) from an explant and cell culture from callus

Sterilization of tools – Surgical blades and scalpels are not sterilize by dry heat because the high temperature makes the cutting edge dull. Such articles including spatula and forceps are usually immersed in 80% v/v (volume by volume)ethyl alcohol until required, and sterilized during use by frequent immersion in alcohol and flaming. Now a days, bead sterilizer (works on principle of dry heat) is available for sterilizing such tools. After an initial stabilization time of 30 minutes during which equipment attains a temperature of about 250° C, units ensure total sterilization by destruction of all microorganisms within seconds. This unit can conveniently placed on Laminar air-flow bench.

Use of fumigation (formaldehyde, sulphur etc.) have been discouraged. It is used some times to clear the laboratory or working area from microorganisms. The place is not used for 3-4 days following fumigation. Culture rooms are fumigated only after removal of cultures. Generally, wiping of working places with 80% alcohol is sufficient for regular use.

IV. CALLUS CULTURE

Cell from any plant species can be cultured aseptically on or in a nutrient medium. The cultures are initiated by planting a sterilized tissue (an explant) on an agar medium. Within 2-4 weeks, depending upon plant species, a mass of unorganized cells (callus) is produced. Such a callus can be subcultured indefinitely by transferring a small piece on to the fresh agar medium.

When a suitable sized and sterilized explant is transferred aseptically under a laminar flow bench on to an appropriate nutrient medium (for example, MS medium or B5 medium) containing appropriate combination of plant growth regulators, it produces callus (Fig. 12.4). The callus is produced from outer layers of cortical cells in a stem explant by repetitive division of the cells. These dividing cells generate pressure on the epidermis, which ultimately ruptures exposing newly formed callus. By continuous division of cells produces a mass of cells or callus on the explant. This callus is separated from the explants and transferred on to fresh medium. After attaining growth, callus is subcultures at regular interval of 3 to 4 weeks by dividing it in to pieces (250 to 500 mg each) and transferred on to fresh medium. The callus piece used to inoculate a flask is known as 'inoculum'. Inocula (plural of inoculum) of similar shape and size are used in all the subcultures and experiments (Fig. 12.5).

Callus cultures are slow growing system as compared to cell suspension cultures. Cells grow as clumps or masses in callus cultures and only lower cells are in contact with the medium whereas, cells in upper layers get their nutrients from cells in lower layers. Cells are in close association as compared to cell cultures in which all the cells are in direct contact with the medium and dissociated.

Application of callus cultures- Callus cultures are slow growing system on static medium and offers a unique system (as compared to *in-vivo* grown plants) to study the following aspects of plant metabolism and differentiation:

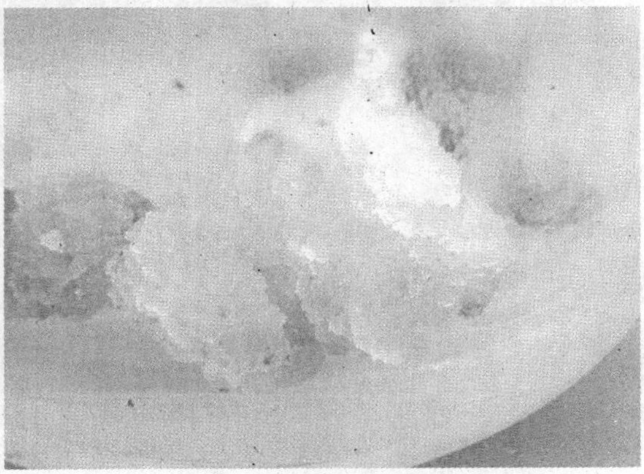

Fig. 12.5. Callus

1. Nutrition of plants.
2. Cell and organ differentiation and morphogenesis.
3. Somaclonal variation and its exploitation.
4. Developing cell suspension cultures and protoplasts cultures.
5. Genetic transformation using ballastic particle gun technology.
6. In the production of secondary metabolites and their regulation.

V. CELL SUSPENSION CULTURE

Cell suspension cultures are initiated by transferring friable callus to liquid nutrient medium. In liquid medium, plant tissues remains submerged which leads to anaerobic conditions and ultimately death of the cells. Therefore, such cultures are agitated on rotary shaker at 80-150 rpm, with an orbital diameter of 2.5-5.0 cms. Agitation serves both to aerate the cultures and to disperse the cells. Cells from the inoculum are separated during this process and a suspension of cells is produced. The division rates of suspension culture cells at the expotential phase are typically higher than callus cells, but doubling times are slow than bacterial cells and usually vary from 24-72 hrs. It is a common observation that if relatively small number cells are transferred (low inoculum density) to a new medium (either static or liquid), they may fail to divide, whereas a larger quantity of tissue transferred from the same culture may proliferate rapidly on the same medium. This observation has led to the concept of 'critical initial cell density. This is defined as the smallest inoculum per volume of medium, from which a new culture can be reproducibly grown.

There are a few conditions, which determine the critical initial density of cells. These are:

(1) The cultures physiological characteristic
(2) The length of time and conditions under which the culture was previously maintained
(3) The composition of the fresh medium.

The third point is of interest. As the isolated cells are failed to grow on fresh medium, 'conditioned medium' or 'nurse tissue' conditions are used to grow isolated cells or protoplasts. A 'conditioned medium' is the medium on which some tissues were previously grown. This makes the minor adjustment in the nutrients and chemical substances released in the medium by the callus, promotes the growth of isolated cells of protoplasts. In cell suspension cultures, cells grow as isolated single cell and cell aggregates of a few cells to a few hundred cells. Cells aggregation vary from species to species and some times it is difficult to maintain fine cell suspension culture, e.g., in members of the family Rutaceae.

Fig.12.6. Cell suspension in flasks and cells under microscope

Cell cultures are subcultured by dilution of the stock culture, 5 to 10 times v/v (volume by volume), depending upon the growth of cells. As mentioned above, growth of cell suspension cultures is always higher than callus cultures and, therefore, required rapid subculture (7-21 days) as compared to callus cultures (4-8 weeks)

Cell cultures in the liquid medium are grown as:

A. Batch cultures in shake flasks and bioreactor.

B. Continuous culture system.

Batch Culture -Whether cultures are grown in flasks (100 ml to 2 litres capacity containing 20 to 500 ml liquid medium) or in a bioreactor, system is used as a batch. Inoculated cultures are harvested after a definitive growth period (determined by growing cultures earlier and recording the growth phases during 1 to 6 weeks growth) at the end of stationary phase and cultures are harvested for a specific objective. Such cultures are grown again and again in batches for the purpose of experiments and known as batch cultures.

Continuous Culture -Continuous cultures are usually grown in abioreactor, (also known as chemostate) when inputs (nutrients, oxygen etc) and outputs (spent medium, cells) are precisely controlled and cells are grown under a defined conditions of nutrients, pH, oxygen and cell density continuously for the purpose of study. Such cultures are known as continuous cultures. This type of work is possible only after basic study of growth and metabolism in small shake flasks. In continuous culture, cells are always in an exponential phase of growth.

Time course of growth- The growth of undifferentiated callus and suspension culture can be quantified in a number of ways, but most commonly as an increase in fresh weight, dry weight or cell numbers over a time course. Growth pattern of such cultures is always forming a sigmoid curve. This curve is consist of a lag phase, in which there is a little or no cell division, an exponential phase in which maximum cell growth occurs, followed by a linear and stationary phase, in which the proportion of dividing cells gradually declines (Fig. 12.7). Each stage can be characterized by distinct structural and biochemical features but, infect whole growth cycle is a continuous physiological process. At any particular time a heterogeneous population of cells differing in morphological and physiological characters can be observed. Usually the cultures are subcultured at end of exponential phase or during stationary phase.

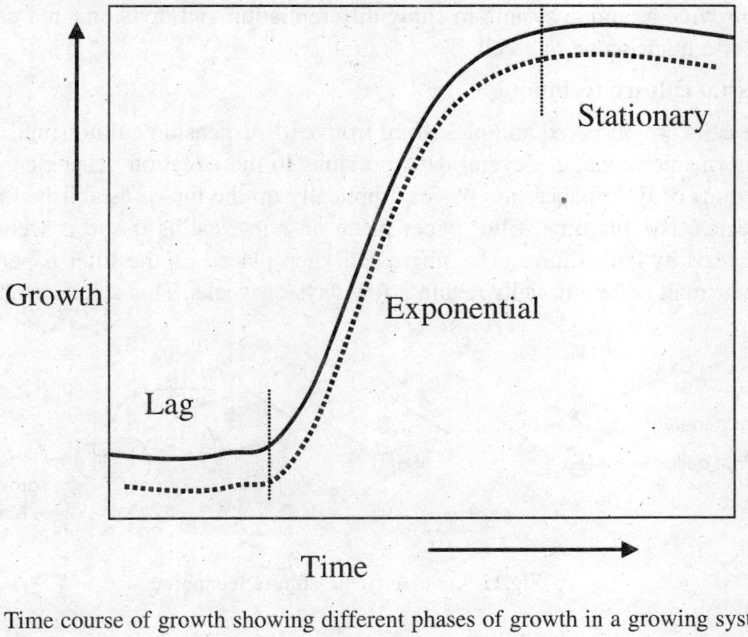

Fig.12.7. Time course of growth showing different phases of growth in a growing system.

Fresh weight of callus is determined by carefully removing the adhering agar from callus and immediately weighing the callus in a sensitive balance. Cell cultures are filtered in a butchner funnel lined with filter paper, washed with distilled water (under mild vacuum) and weighed. Dry weights are determined by drying the cells to a constant weight at 60^0C in an oven and weighing in a balance. Packed cell volume (PCV) and cell culture is determined by centrifuging it at 1000 rpm for 5 min in a graduated centrifuge tube and measuring the volume occupied by the cells out of total volume of the cell suspension. Cells are counted by a haemocytometer under compound microscope during growth period,

Comparison of callus and cell suspension cultures

Parameters	**Callus**	**Cell suspension**
Growth	Slow	Fast
Cell to cell contact	Cells in contact	Dissociated
Medium	Only lower layer is in contact with the medium.	All cells are in direct contact with the medium
Precursors	Not available to all cells.	Available to all cells.
Subculture period	Long, 4-8 weeks	Short, 7-21 days.
Accumulation of metabolites	Higher than cell suspension.	Lower that callus culture.
Scale-up in bioreactor.	Not possible	Cell suspensions are grown in bioreactor.

VI. SINGLE CELL CULTURE:

Single cell cultures are obtained from established cell suspension cultures. Cell suspension cultures are comprised of mixture of single cells and small colonies of cells. These latter strains of cells are not necessarily derived from single cells. Thus, strains desired from such cultures are not true single cell clones. Single cell clones are provided the means for pure culture studies for higher plants comparable to those used for microorganisms. Several methods are developed for obtaining single cell cultures and establishing single cell clones of higher plants. The first successful higher plants single cell isolations were made by Muir and co-workers in 1954 at Wisconsin University, Madison,USA., using nurse tissue culture technique for marigold and tobacco callus cultures. Isolated single cells provide a unique system to study differentiation and divisions in a cell by monitoring changes under a microscope in a cell.

A. Nurse tissue culture technique

Single cells are observed and picked-up from cell suspension cultures under the microscope with the help of micropipette. Several days previous to the isolation of single cells, small sterile pieces (8x8 mm) of filter papers are places aseptically on the top of established callus of same or different species. By this time, filter paper place on nurse callus tissue is wetted by liquid and nutrients released by the cultures. The single cell, then placed on the filter paper raft, may divide and produce a small colony of cells within a few days or weeks. This small colony of cells or

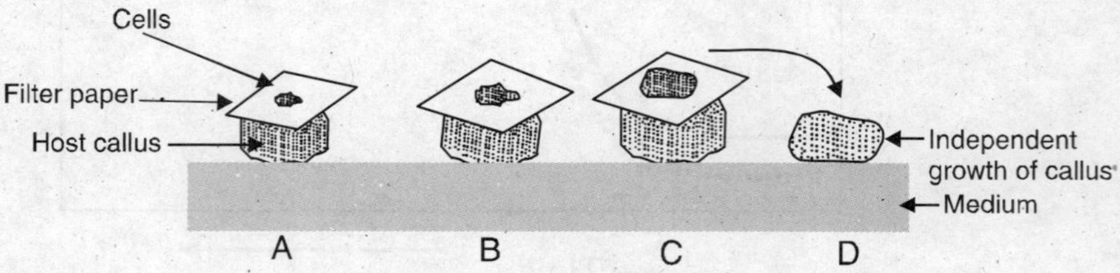

Fig.12.8. Nurse tissue culture technique

microcalli can be transferred to fresh medium, where it survives (Fig.6). A number of previously recalcitrant (difficult to grow in culture) species, notably monocotyledons such as rice and maize have been grown by such technique.

VII. APPLICATIONS OF CELL CULTURE:

1. Cell culture offers enormous opportunities in the study of single cells and group of cells.
2. In the isolation of protoplasts.
3. In cell cloning by the plating technique with or without specific treatment, e.g., mutagens, amino acid analogues etc.
4. Development of cell lines for various types of resistance e.g., salt and drought, and toxin resistant lines.
5. In scale-up technology using bioreactors of various types.
6. In providing an excellent system of micropropagation on mass-scale by somatic embryogenesis in cell culture and use of bioreactor. Such system can be used for artificial seed production.
7. In the study of nutrition.
8. In cellular differentiation, single cells constitute an excellent system to study cytodifferentiation in a cell leading to tracheiry element formation and plastids differentiation etc.
9. In secondary product formation, regulation and biosynthesis.
10. In the process of cell division and factors affecting cell divisions and related process.

QUESTIONS

1. How will you initiate cultures from an explant and measure the growth of the cultures?

2. Write short notes-

(*a*) Explant sterilization (*b*) Callus culturec. Cell culture (*d*) Cell sieving
(*e*) Microculture chamber (*f*) Nurse tissue culture (*g*) Growth curve

3. Calculate from the given data, how much of compound is required to prepare 1 litre solution?

Mol. Weight	*1 mM*	*Mol. Weight*	*5m M*
101		220	
187		175	

4. Describe preparation of Murashige and Skoog's medium. What precautions should be taken during preparation of the medium?

5. Write short notes on the followings, describing their significant contributions :

(*a*) Gautheret
(*b*) P. R. White
(*c*) J. Reinert
(*d*) G. Morel
(*e*) F. C. Steward
(*f*) I. K. Vasil
(*g*) P. Maheshwari.

CHAPTER 13

TECHNIQUES FOR THE STUDY OF CELL

I. CALLUS AND CELL CULTURE

Callus is a mass of undifferentiated cells which grows profusely on nutrient medium. Callus is obtained by growing explants on a nutrient medium containing a balanced combination of plant growth regulators and a gelling agent. Cell suspension cultures can be derived from callus cultures. Suspension cultures are a homogenous mass of cells and addition of precursors and other chemical agents is easier as there is uniform distribution.

Initiation of callus- All the procedures are to be carried out under laminar flow.

(*i*) Place the explant in the petridish.

(*ii*) Fill the dish with sterilant to be used (solution of mercuric chloride 0.02-0.1% or sodium hypochloride 5-20%) ensuring that the explants are fully submerged. Duration and concentration to be optimized, for the material.

(*iii*) Wash the explants with three changes of sterilized distilled water.

(*iv*) Use sterilized tools and transfer the explant on to the medium use any of the standard nutrient formulations (Murashige & Skoog medium, Gamborg's B_5 medium etc.) containing any auxin and cytokinin.

(*v*) Plug the tube on the flame, label it and keep in the culture room at 26^0 C.

(*vi*) Make observations regularly.

Initiation of suspension culture

(*i*) Suspension cultures can be initiated by transferring one gram of callus in 50 ml medium of the same composition except agar used for callus culture.

(*ii*) Sub cultures are made by transferring about 10% suspension culture in fresh medium.

(*iii*) The growth of cell suspension cultures can be monitored by measuring one of the following :

Packed cell volume (PCV)

A known volume of suspension is taken in a centrifuge tube and centrifuged for 5 minutes at 200 g. PCV is the volume of the pellet, which is expressed as percentage.

Cell numbers

Cells can be counted directly on hemocytometer. One volume of culture is added to two volume of aqueous chromic trioxide solution (8% w/v), heated for ten minutes at 70 ° C. After cooling, shake the cultures and count.

Fresh Weight and dry weight determinations-

Cell suspension is filtered on a Buchner funnel with Whatman filter paper under low vacuum. The cells are washed with distilled water, drained in vacuum and weighed. Dry weight is determined by drying the cells at 60 °C to constant weight or drying in a lyophilizer.

Cell viability determination using fluorescein diacetate :

Fluoresceine diacetate (FDA) is a non-fluorescing compound used to determine the cell viability. The FDA molecules pass freely across the plasma membrane, where esterase cleaves off the molecules resulting in release of fluorescein, which is retained within the cell with an intact membrane. Metabolically active cells (viable) are visualized by the excitation of fluorescein with UV light under the fluorescent microscope.

Requirement- Fluorescein diacetate, acetone, slides, cover slips, cell counter, fluorescent microscope and suitable suspension culture.

Procedure- Prepare a stock solution of FDA in acetone (0.5% w/v) and store in refrigerator until used.

(*i*) Prepare a dilute solution of stain 0.1 ml of stock solution in 5 ml culture medium. Store in ice and use within one hour.

(*ii*) Mix one drop of dilute stain with one drop of cell suspension on a slide and apply a cover slip.

(*iii*) After 2-5 min. observe with a light microscope under white light illumination to reveal all cells and turned the UV filter to reveal viable cells. All live cells fluoresce brightly.

(*iv*) For quantification of percent viability, count the total visible cells under white and UV illumination using a hand held cell counter.

[Percent cell viability = No. of fluorescing cells × 100 / Total no. of cells.]

Precautions

(*i*) Once fully developed diffuse fluorescence may appear in most cells after some time.

(*ii*) All test solutions and dilutions must be prepared in culture medium only.

II. MICROSCOPY

The microscope is an indispensable tool for cell biologists. This tool with the help of various coloring agents (dyes) widely used to study cell differentiation and metabolism, cell components and microorganisms. The first useful light microscope was developed in 1590 by Z. Jonseen and H. Jansser. The light microscopes magnify an object so that the observer can see it properly. There is a limit to the useful magnification that can be achieved in a light microscope because of the resolving power of the lenses. Several technical modifications developed over the past three or four decades have enhanced the effectiveness of the light microscope. These techniques are- bright and dark field microscopy, phase contrast, Nomarski-differential interference contrast (DIC), florescence and confocal microscopy. These techniques allow the cell biologists to examine structures in the living cells without fixing the cell or introducing any staining agent.

(a) Bright and Dark Field Microscopy

The bright field microscopy is the basic form of light microscopy. As it is named, the image of an object is dark against a brighter background. However, the only specimen that can be seen directly by bright field microscopy, are those that possess colour or affects the light passing through it. Single plant cells or microbes that don't have high concentration of molecules that absorb visible light, are invisible (details are not visible). Bright field microscopy is absolutely indispensable tool for cell biologists. This is required for routine observations of cells, cellular differentiation and pigmentation.

In dark field microscopy, oblique illumination is used. In absence of specimen, no light reaches condenser and hence a dark image is produced. However, when a specimen is present its reflective structures will cause the incident light to be reflected at all angles. Some of this light will reach the object and will appear bright. Only light that has been reflected will appear bright. Only light that has been reflected or refracted by the object is used for image formation.

According to wave theory, light is propagated from one place to another as wave travelling in a hypothetical medium. Light waves are described in terms of their amplitude, frequency, and wavelength. Amplitude is the maximum displacement of light path from the position of equilibrium. Frequency of light is the number of complete cycles occurring in a second. The property of light waves inversely associated with the frequency is the wavelength and is defined as the distance between corresponding points on a wave or distance between two successive peaks or crests. The velocity of light is about 186300 miles or $3x10^{15}$ kms per second.

A good microscope has not only good magnifying power but also good resolving power to provide finer details of the object. Thus, the basic difficulty in designing a microscope is not the magnification, but the ability of lens system to distinguish two adjacent points as distinct and separate. This ability is known as resolving power of the microscope. The resolving power of a microscope depends upon the wavelength of light and numerical aperture (N.A.). The minimum resolvable distance between two luminous points (v) is given by the following formula –

$$V = \frac{0.16\lambda}{N.A} \quad \text{Where, } \lambda = \text{wave length}$$

Thus, shorter the wavelength of light used and lower the N.A., greater is the resolving power. The limit of resolution of a microscope is approximately equal to 0.5 /N.A., which for a light microscope is approximately 200 nanometers (nm) or about the size of many bacterial cells.

The numerical aperture of a lens is dependent on the refractive index (η = the ratio of the speed of light in a given medium to the speed of light in a vacuum) of the medium filling the space between the specimen and the front of the objective lens and on the angle of the most oblique rays of light that can enter the objective lens (θ). It is given by formula [$N.A. = \eta \times \sin\theta$]. That is why immersion oil is placed between the object and oil immersion lens (100 X objective).

(b) Phase Contrast Microscopy

Dutch Physicist Frits Zernicke described it for the first time in 1934 based on Abbe's theory of image formation. It is a contrast enhancing optical technique that produces high-contrast images of transparent unstained specimens, which usually lack contrast in bright-field microscopy such as living cells and microorganisms. This involves an optical mechanism to translate minute variations in phase into corresponding changes in amplitude, which can be visualized as differences in image contrast. As the rays pass through the specimen, their velocity.

In the Zernickle Photo microscope, an annulus (phase annulus) in the condenser aperture (front focal) plane generates a hollow cone of O-order illumination that is projected into the back of focal plane of the objective (See Fig. 13.1). A matching phase ring in the objective performs 2 functions: (1) It attenuates or absorbs the non non diffracted O-order light by 70-80% and (2) It shifts its phase position by one quarter of the wavelength (λ/4) of the light. This arrangement alters the amplitude and phase relationship of diffracted versus non diffracted light wave fronts to achieve better interference, thereby enhancing contrast. For better image quality, the specimen must be thin (<5μm) to prevent excessive halation ('halo' formation due to refractive gradients passing O-order light outside the pH ring) Kohler alignment also requires centering or exact superimposition, which is accomplished through the use of a telescope lens that can focus at the back focal plane of objective.

The light microscope has evolved over the last 350 years from Antony van Leeuwenhoek's simple magnifier to the most sophisticated instruments of today. This has become symbol of research and scientific discoveries. The light microscope creates a magnified, detailed image of seemingly invisible objects or specimens, based on the principles of transmission, absorption, diffraction and refraction of light waves.

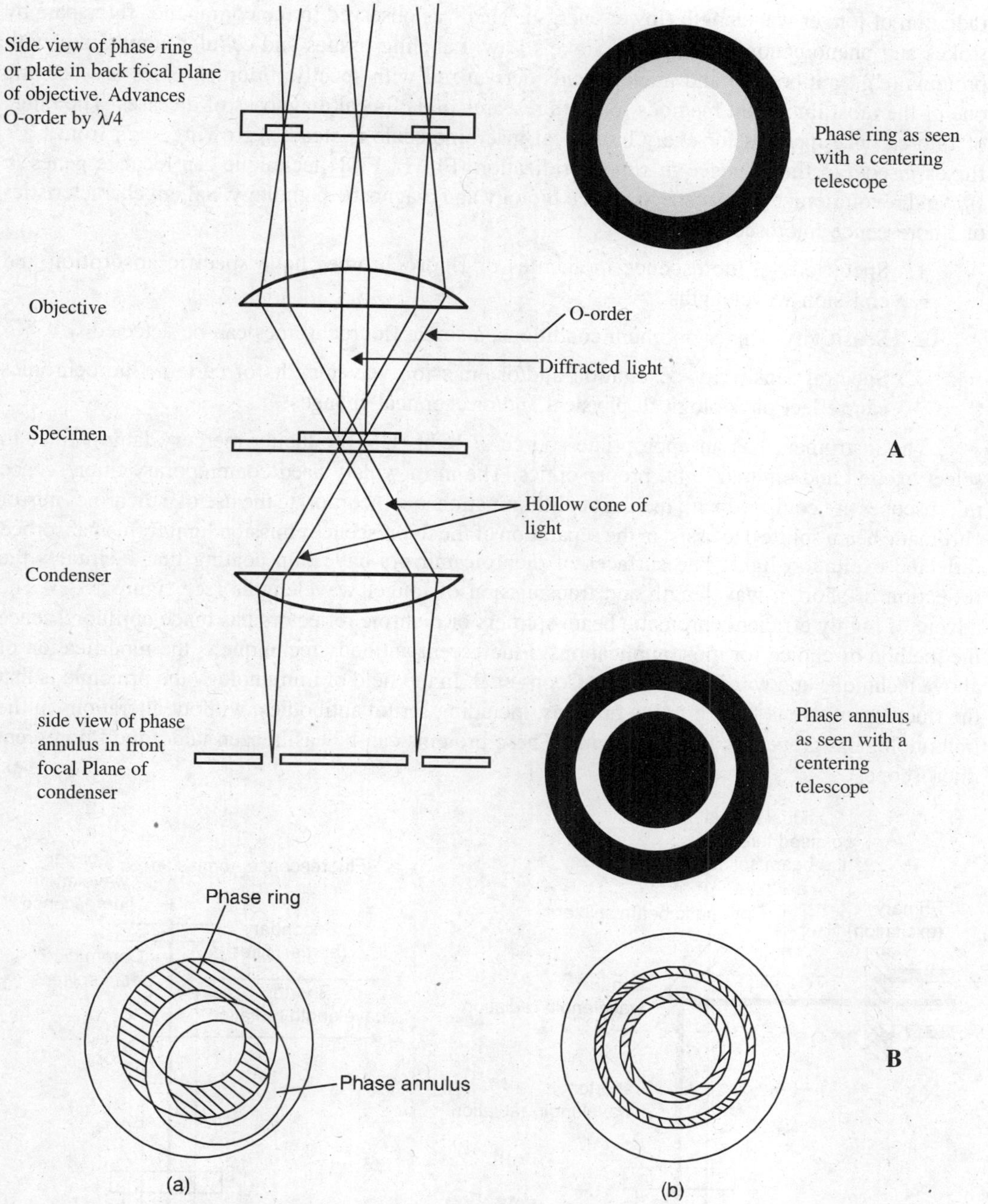

Fig. 13.1. Schematic presentation of phase contrast microscopy

(c) Fluorescence Microscopy

Several kinds of specimens absorb light radiation, become "excited", and then re-radiate or emit light. When the emission continues for some time after excitation, the process is called phosphorescence when emission stops almost instantly after excitation ($\sim 10^{-7}$ seconds), it is called "Fluorescence".

Many compounds like drugs, vitamins, and chlorophyll, display autoflourescence or primary fluorescence. These products when irradiated with light of a specific spectral region, they emit radiation of longer wavelength (lower energy). This was observed in the compound 'fluorspar' by stokes and phenomenon is known as Stoke's Law. Labeling tissues and cellular compounds with proteins (e.g., antibodies) and nucleic acids derivatized with specific fluorochromes has become one of the most important methods for both research and clinical diagnoses of diseases. The range of fluorescence markers for every biological macromolecule is steadily growing, e.g., from Ca^{+2} fluxes *in vivo* to fluorescence *in situ* hybridization (FISH). FISH technique can localize genes *in situ* and revolutionize many aspects of cell biology and diagnostic pathology. Salient characteristics of fluorescence microscopy-

1. Specificity- Fluorescence molecules or fluorochromes have specific absorption and emission wavelengths.
2. Sensitivity- Under optimum conditions, a single flourochromes can be detected
3. Spectral sensitivity- Excitation and/or emission wavelengths of certain fluorochromes can reflect physiological, physical and/or chemical changes.

The instrument has an appropriate source of light (UV producing mercury lamp), filter to select excited and emitted light, proper optics. The most widely used contemporary fluorescence microscopes are equipped with incident epiillumination and incorporate the use of a dichroic (mirror chromatic beam splitter) to assist in the separation of the fluorescence emission light from unabsorbed reflected excitation light. The surfaces of dichroic mirrors have thin coating that facilitates the reflection of shorten wavelength and transmission of longer wavelengths (see figure 13.2). The advent of highly efficient chromatic beam splitters or dichroic reflectors has made epifluorescence the method of choice for most applications. Fluorescent antibody technique is the modification of above technique and was introduced by Coons et al. In the field of immunology the principle is that the fluorescent dye is tagged to the proteins, including serum antibodies, without alterations in the immunological properties of the proteins. These proteins can be easily seen under the fluorescent microscope.

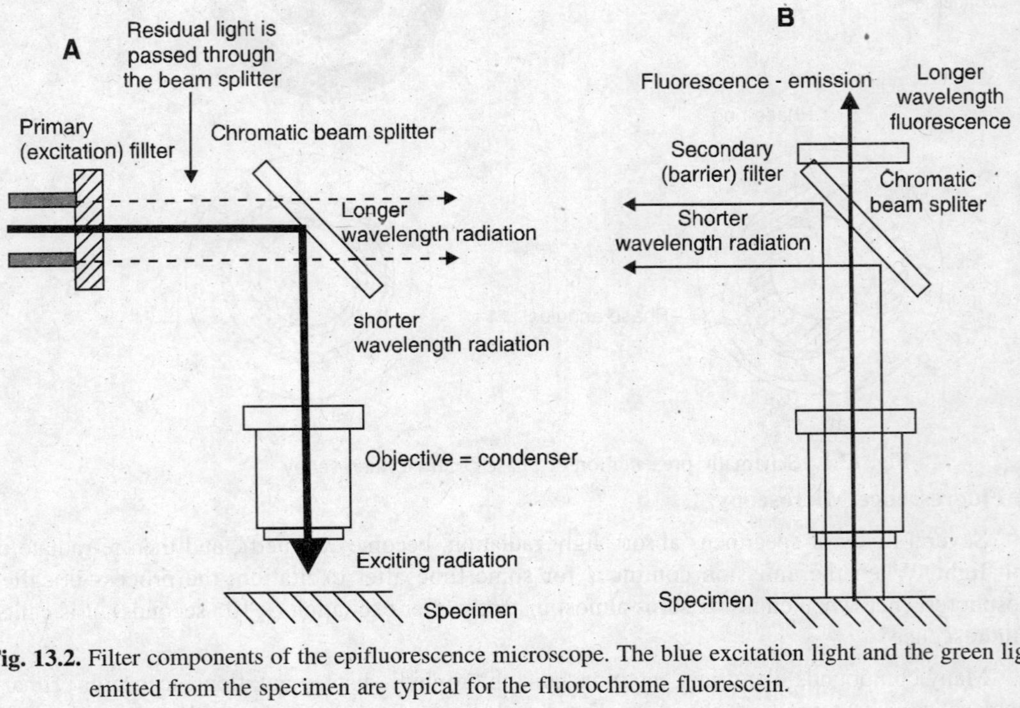

Fig. 13.2. Filter components of the epifluorescence microscope. The blue excitation light and the green light emitted from the specimen are typical for the fluorochrome fluorescein.

Fluorescence microscopes are available in market. The main object of this instrument is to transmit as much as possible, the fluorescence emitted by the object, since this fluorescence is of lens intensity. A strong source of UV light (565 nm) is used to provide light of maximum intensity to the object. Therefore, the sub-stage condensor should be made of UV transmitting glass. The objectives must have highest NA and they should be non-fluorescent.

Two-filter systems are necessary in the microscope. Primary or excited filter is placed in between the illumination source and the objective. This filter is usually blue and allows UV light to pass. Secondary barrier filter is placed in the eye piece or anywhere between eye and object. It allows only passage of light fluorescing specimen. It depends upon the type of dye used for the technique The significant advantages are –

1. The exciting radiation travels in a path opposite to that of the observed fluorescence. This minimizes the effects of the absorption of light due to specimen thickness.
2. The full-objective aperture provides excitation and collects emitted fluorescence for maximum brightness.
3. The objective lens becomes the condenser assuring perfect alignment between both.
4. Epifluorescence can be readily combined with other transmitted light techniques such as phase or DIC.

Selection of light source is governed by type of fluorochrome used. Besides this, many filter sets contain exciter filters specifically designed for the intense emission lines of mercury arc lamps at wavelength of 365, 405, 436 and 546 nm. The brightness of the fluorescence image is a function of the quantum efficiency and concentration of the fluorophone, the spectral brightness of the source, the transmission through the optical system used for both excitation and emission wavelengths, the total magnification and most significantly the numerical aperture of the objective.

Table 13.1. Comparison of various types of microscopes.

Type of microscope	Maximum useful magnification	Resolution (nm)	Remarks
Bright field	1500 X	100-200	Staining necessary
Dark field	1500 X	100-200	Staining not required, specific-men appears bright on dark background
Fluorescence	1500 X	100-200	Fluorescent dye required for non-fluorescent cells selection of cells with fluorescent SM
Phase contrast	1500 X	100-200	For living materials, staining not required
TEM	500,000-1000,000 X	1 nm	Ultra structure, protein, DNA, viruses can be observed
SEM	10,000-1000,000 X	1-10 nm	Surface structures

(d) Confocal Microscopy

Immunofluorescence is generally applied to fixed specimens, but recently there have been a number of fluorescent techniques that can be applied for living specimens. Purified proteins are derivatized with a fluorophone which are still functional. These derivatized proteins may be injected back into a cell enabling the dynamic behaviour of the protein to be observed by fluorescence microscopy. Several fluorescent indicator molecules have been developed that change their optical properties in response to particular components of their environment. Probes for Ca, ATP and membrane potential are available. These probes allow the physiological state of a living cell to be monitored optically. Although useful for many applications, fluorescence microscopy is subject to significant limitations for some purposes, e.g., thick specimens are not well resolved and produce unclear images. This problem is resolved for confocal microscopy.

A confocal imaging system consists of a laser scanning microscope with a pinhole aperture in the image plane of the objective. The confocal aperture passes light that originates from a small point (which is illuminated) and unwanted fluorescence that originates away from this point are blocked (not pass through pinhole) confocal microscopy gives very good image in thick specimens, where normal microscopy provides blurr image. In confocal microscope, most of the out-of-focus light is excluded from the final image, greatly increasing the contrast hence the visibility of fine details in the specimen (Fig. 13.3). Schematic comparison between conventional and confocal microscope is presented in the figure. In wide field conventional microscope, a condenser distributes light in a wide field compared to a fine beam of light in confocal microscope.

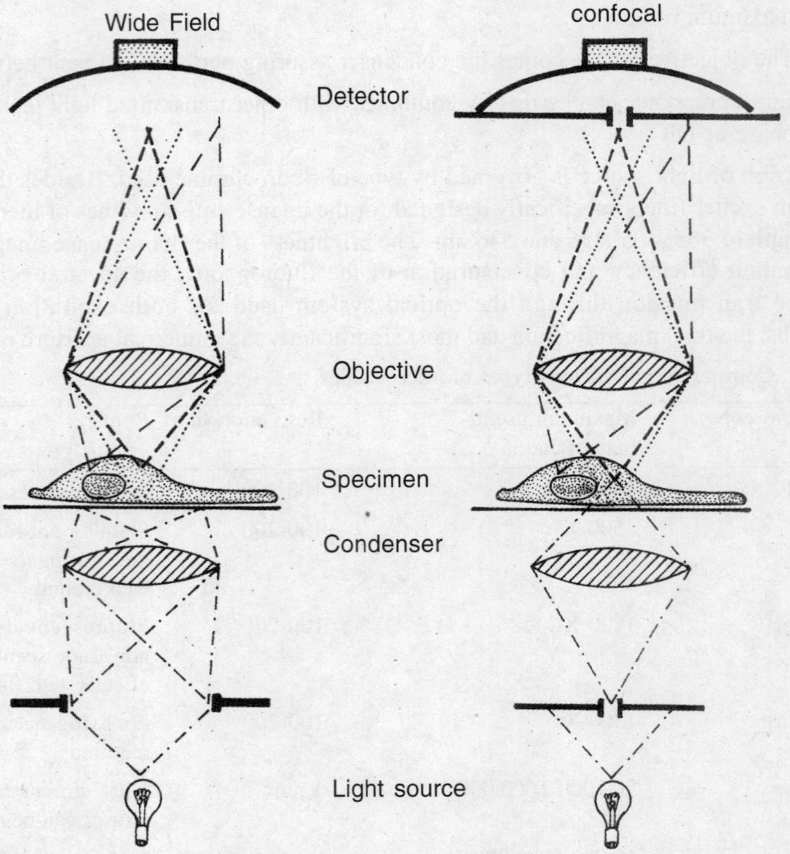

Fig. 13.3. Schematic illustration of the operating principle of the confocal microscope. Left side–a conventional, or wide-field microscope. The specimen is illuiminated over an extended region by a light source and condenser. Light rays arising from three points in the specimen are shown. The dashed lines emanate from two points in the focal plane, one centrally located (darker dashed line), the other off-axis (lighter dished lines). The thrid point is on axis but located below the plane of focus (dotted lines) – it gives a blurred image at the detector. The detector forms an image from the sum of all the simultaneously arriving light rays. Right side – a confocal microscope. Two pinhole apertures have been introduced. The upper aperture passes to the detector only the focused light rays from the on-axis in-focus point of the specimen.The lower aperture restricts the illumination so that it is focused on the point seen by the upper pinhole aperture.

Confocal microscopy offers several advantages over conventional microscopy, including controllable depth of field the elimination of image degrading out-of-focus information and the ability to collect serial optical sections from thick specimens. The key to confined approach is the use of spatial filtering to eliminate cut-of-focus light or flare in specimens that are thicker than the plane of focus.

A number of different imaging modes are used in the application of confocal microscopy to a vast variety of specimen types. They all rely on the ability of the technique to produce high resolution images, termed 'optical sections', in a sequence through relatively thick sections or whole mount specimens. Based on the optical sections as the basic image unit, data can be collected from fixed and stained specimens in single, double, triple or multiple wavelength illumination modes. These images collected with various illuminations and labeling strategies will be in register with each other. Live cell imaging and time- lapse sequences are possible and digital image processing methods applied to sequences of images allow three dimensional image of the specimen. Confocal microscopy is clinically used in the evaluation of various eye diseases.

Table 13.2. A variety of microscopic techniques exploit light properties to enhance contrast

Contrast mode	Mechanism	Comments
Bright Field	contrast depends on light absorption	usually used in conjunction with histological stains to boost contrast
Phase contrast	converts optical path differences to intensity differences	contrast proportional to local "phase dense" objects including mitochondria, lysosomes, chromosomes, nucleoli, and stress fibers
Differential interference contrast DIC)	converts rate of change of optical path across the specimen	cell and organelle edges where optical path abruptly changes stand out in relief
Dark Field	scattered light observed	produces image of cell and organelle edges
Interference reflection (IRM)	Contrast depends on interference substratum between closely spaced surfaces	used to visualize zones of cell- contact in cultured cells
Polarization	detects birefringence caused by supramolecular organization below optical resolution	used to study oriented arrays such as cytoskeleton structures (e.g., microtubules in the mitotic apparatus and stress fibres); also use dot study membranes
Fluorescence	contrast depends on absorption of light by fluorophore and its quantum yield	limited only by appropriate fluorescent probes

(e) Electron Microscopy

Electron microscopy permits a detailed study of sub-cellular organelles as its resolving power is much greater than that of the light microscope. Max Knoll and Ernst Ruska in 1931, at Technical University in Berlin, constructed electron microscope (EM). In the EM, streams of electrons are deflected by an electrostatic or electromagnetic field in the same way that a beam of light is refracted by a lens. Electron beam is generated by heating a filament in vacuum, which are accelerated by a potential and shows properties similar to light (λ = 0.005nm of electrons and 550 nm for light). Though appears to be similar there are great differences in light and electron microscopes, the principal being the image formation (Fig. 13.4.). In electron microscope, image is produced by electron scattering. Electron dispersion is a function of the thickness and molecular packing of the object and depends especially on the atomic number of the atoms in the object. The higher the atomic number, the greater is the dispersion. There are three general types of electromagnetic lenses. The one is placed between the source of illumination and the specimen. This focuses the beam of electrons on specimen and functions in a similar manner as the condenser in light microscope. The other two lens systems are on the opposite side of specimen which magnify the image in similar way as objective and ocular in light microscope. The final image is produced on a screen coated with a phosphorus compound which fluoresces upon irradiation by electrons. These images are recorded on photographic films. Air molecules in the microscope interfere with the movement of electrons. To prevent this, a high vacuum (10^{-4}-10^{-6} mm Hg) is created inside the microscope. In light microscope, magnification is largely determined by the objective and the maximum magnification of 100-120 can be obtained. This is multiplied by ocular lens by a factor of 5-15, reaching a total magnification of 500-1500x.

The resolving power of Transmission Electron Microscope (TEM) is very high. The image generated by objective can be multiplied several hundred times by projector coil, e.g. 100 objective 200 x projector coil =20,000 x. In TEM, this can be reached up to 10,000,000 x. Electron microscope has a greater depth of field as compared to light microscope. This property of EM is used to develop three dimensional images of cell organelles as ribosomes and protein structure.

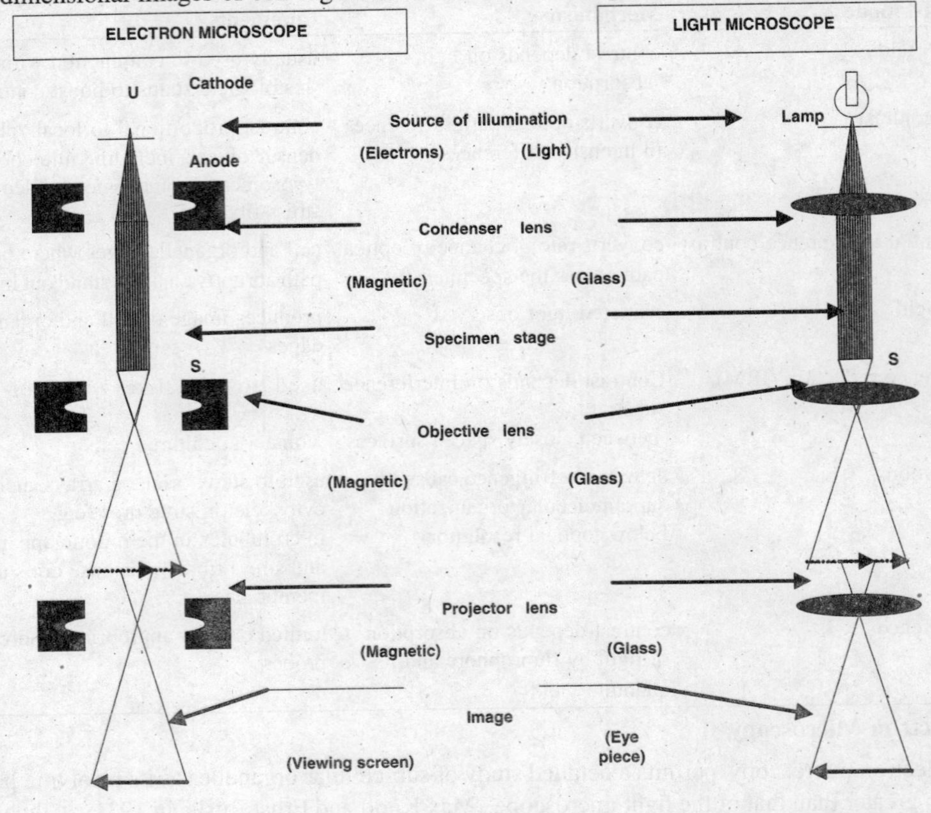

Fig. 13.4. Diagrammatic comparison of image formation in electron microscope and light microscope.

The preparation of specimen is similar to dehydration used for histology for light microscopy. The sections prepared for EM should be thin (less than 0.5µ or 500 nm), otherwise, material will appear completely dark. Such thin sections (50-300nm) are prepared with the help of ultramicrotome, fitted with glass or diamond knife. To cut sections, material is fixed in osmium tetraoxide, processed through a dehydration series and embedded in plastic or resin (araldite). Sections are placed on small grids specifically used for EM. It is difficult to handle ultra thin sections so they are placed on copper grids (400 apertures/inch square).The grids are placed in labelled boxed and can be stored till observed in EM and can be used again and again without damaging the material. Dehydrated materials are placed in vacuum chamber and vapour of heavy metals like platinum, chromium or palladium is directed from a filament of incandescent tungsten. The vaporised material is deposited on one side of the surface of elevated particles. This creates a shadow on the other side (Shadow casting). Height of the material is determined by length of shadow. In photographs such specimens appear in three dimensions, which is not possible with other techniques. Freeze fracturing (fracture of frozen material) is used to see fine details of replicated material. These are different techniques to prepare materials for sectioning and for staining to increase contrast. Special techniques are used for bacteria, viruses, proteins and nucleic acids. TEM is used to study the fine details of a cell, compare the clones, development of plastids and cell wall formation (Figs. 13. 5, 6).

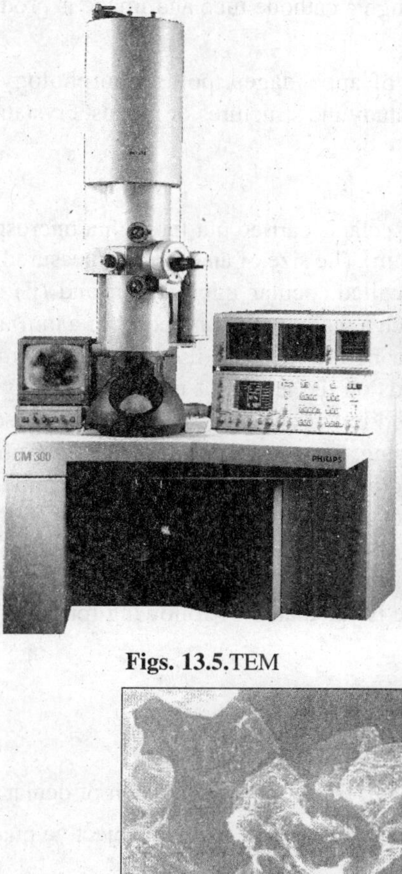

Figs. 13.5.TEM

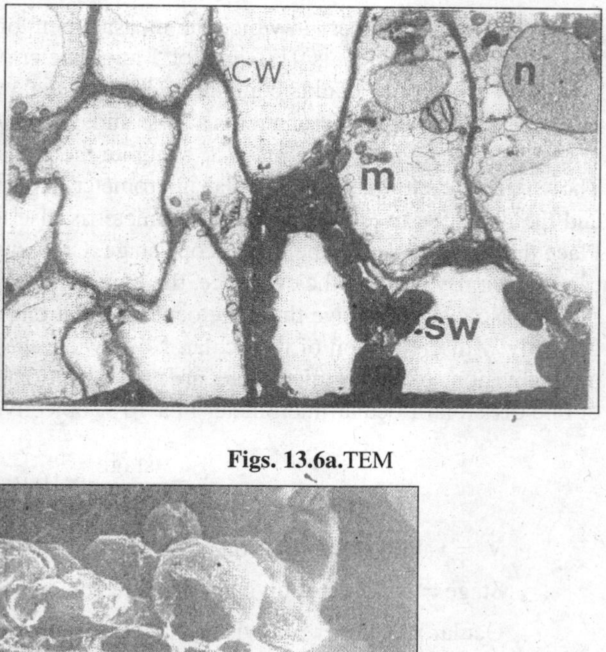

Figs. 13.6a.TEM

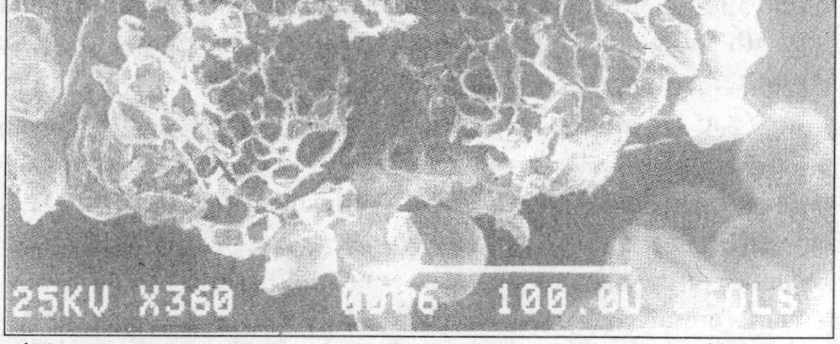

Figs. 13.6. A TEM photo showing details of cell contents (A) and a SEM photo showing details of callus cells (B). Note the shadow generated by cells. CW = cell wall, m = mitochondria, SW = Secondary Wall, n = nucleus

Electron microscope was invented in 1930 which provided detailed knowledge about bacteria, viruses, DNA and other microorganisms. The terms like prokaryote and eukaryote were validated by details of cell structure of such organisms possible only with the help of electron microscope. These technologies could be developed by joint efforts of scientists working in Physics, engineering, chemistry and molecular biology. These tools have revolutionarize our ability to explore the details of cells using microscope. Scanning electron microscope (SEM) provides surface views of whole structure of specimen (Fig.13.6). Normally, specimens are coated with a thin film of metal under vacuum. As compared to shadow casting, complete coating is essential because the scanning beam produces a charge on the uncoated biological materials which cause distortion of image. The metal coated specimen is scanned with a beam of electrons (50-100A^0) which strikes on the specimen and emits the secondary electrons. These electrons pass through a cathode tube and image is produced on the screen of Cathode Ray Tube.

In plant system, SEM is used to see the nature of appendages, pollen morphology and morphogenesis in plant tissue culture. It is also used to study the structures of metals, crystals and in forensic sciences.

III. Micrometry

Micrometry is the process in which the measurement of cells is carried out through a microscope. The unit of measurement is 0.001 mm or 1 micrometer (um).The size of an object is measured with the aid of (i) special ocular containing gradual scale called 'ocular micrometer, and (ii) stage micrometer. The stage micrometer is a glass slide on which a scale (with hundredths of a mm) is 1m printed. Stage micrometer is used to calibrate the ocular before it can be used for measurements. The value of each division of ocular micrometer depends on the length of microscope body tube and the objective used. The ocular micrometer is inserted in the eye piece (10 X) of microscope. Place the stage-ocular on the microscope stage. On focusing, lives of stage and ocular micrometers can be seen. By rotating the eye piece, the long axis of both the scales is aligned in superimposable fashion. Now slowly move the stage ocular so that one of the gradations of stage ocular coincides perfectly with graduation of the ocular. Now where the above situation is repeated, calculate how many divisions of ocular micrometer are equivalent to how many on the stage micrometer the value of one division of ocular micrometer in a 10 X objective is calculated by following formula –

$$V = \frac{\text{stage}}{\text{Oculur}} \times 10\,\mu$$

V = value of one division of ocular

Stage = Number of division of stage coincide with the number of division of ocular

Ocular = Number of division of ocular coincide with stage 10 for 10X objective place 40 in place of 10 for a 40X objective.

After calibration of the micrometer at '0' cells are placed in place of stage micrometer and a number of divisions across length and width of a cell is measured. By multiplying by value of one division of the occulur micrometer , size is calculated.

Determination of cell number in cell culture: Prepare a cell suspension in a balanced salt solution or medium. Take a haemocytometer, clean it and place the cover glass in place. Transfer small of amount of cell suspension to both chambers of haemocytometer with the help of a dropper or pipette, allowing chambers to fill by capillary action. There should not be empty space or spill, starting with one chamber of the haemocytometer, count all the cells in the 1mm centre square and four 1 mm corner squares cultured plant cells are very large as compared to pollen grains and fungal spores. Count the cells lying towards the inner side of chamber touching middle line and exclude the cells touching other chambers of protruding outside the chamber under observation. Cell aggregates can be dissociated by force of pipetting or by shaking the culture in tube or flask. Repeat observations and count cells in 10 chambers by refilling the hemocytometer.

Each square of the hemocytometer, with cover slip in place, represents a total volume of 0.1 mm^3 to 10^{-4} cm^3. Since 1 cm^3 is equivalent to 1 ml, the subsequent cell number per ml will be determined by following calculations.

[Cells per ml = the average count per square x dilution factor x 104 of cell cultures]

e.g. I : Cells/ml = 25 cells/sq $\times 10^4$ = 2.5×10^5 cells/ml

II: With dilution factor of 10 cell suspension diluted 10 times = 25 cells/sq $\times 10 \times 10^4$ = 2.5×10^6 cells/ml of medium.

IV. AUTORADIOGRAPHY

Autoradiography is a technique based on the use of radioactive isotopes. This is a common technique for the study of cell metabolism. A radioactive isotope whose presence can be traced during the entire course of an experiment is used as a tracer substance. Radioactive isotope of an element is the one that contains an unstable combination of protons and neutrons. One can obtain any type of biological molecule in which one or more commonly used isotopes are beta emitters that have a short half life, that is, they will decay in a given amount of time. The isotopes are monitored by autoradiography or by scintillation spectrometry.

The main purpose of autoradiography is to determine the location of a radioactive label in the cell, on polyacrylamide gel or on a nitrocellulose paper. A suitable radioactive material is allowed to absorbed by the cell and its path is traced on a photographic plate by extracting the metabolites at different intervals of tissue. The commonly used isotopes are: H^3, C^{14}, P^{32}, S^{35}, I1^{31} and Co^{60}.

Autoradiography is a tool to visualize biochemical events. We can know the site of synthesis of a particular metabolite in a cell and follow subsequent events in the process.

V. SPECTROPHOTOMETRY

The most commonly used method for determining the concentration of biochemical compounds is colorimetry. It uses the property of light such that when white light passes through a coloured solution, some wavelengths are absorbed more than others. Hyaline solution can be made coloured by specific reactions with suitable reagents. These reactions are generally very sensitive to determine quantities of material in the region of millimole per litre concentration. The big advantage is that complete isolation of the compound is not necessary and the constituents of a complex mixture such as blood can be determined after little treatment. The depth of colour is directly proportional to the concentration of the compound being measured, while the amount of light absorbed is proportional to the intensity of the colour and therefore, to the concentration.

Source of Radiation

A lamp is usually used as the source of radiation. The wavelength of light depends upon the quality of source lamp.

Colorimetry is an analytical device to determine the amount of an unknown substance in a solution. The device works on optical properties of the substance. Basically we measure the amount of light of a particular wavelength absorbed by that solution, and then relate solute concentration to the absorbance. A simplified schematic presentation of the instrument is as follows :

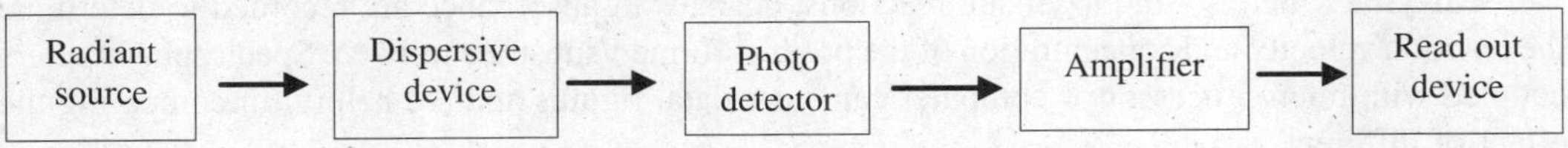

The relationship between concentration of solution and light it absorbs is given by the following laws.

Lambert's Law – When a ray of monochromatic light passes through an absorbing medium its intensity decreases exponentially as the length of the absorbing medium increases.

Beer's Law – When a ray of monochromatic light passes through an absorbing medium its intensity decreases exponentially as the concentration of the absorbing medium increases.

Transmittance – The ratio of intensities is known as the transmittance and this is usually expressed as percentage. Absorbance or extinction is known as the optical density of the substance. Absorbance increases with increase in concentration of the solution (Fig. 13.7).

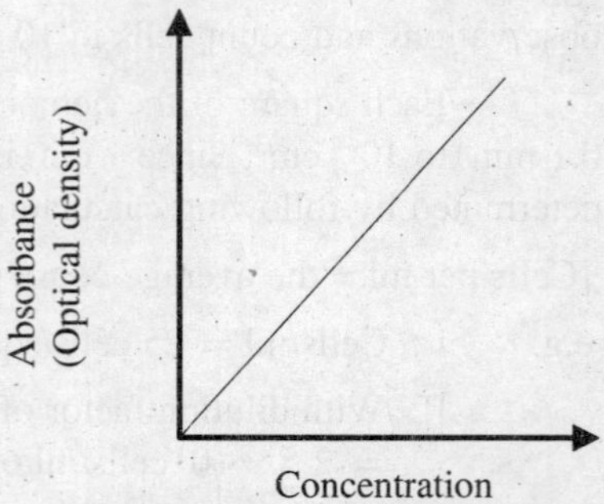

Fig. 13.7. Relationship between concentrations and absorbance

Detection of Radiation

Detection wavelength is selected on the basis of wavelength of light coming out of a solution (Fig. 13.8). Absorption of a particular wavelength of light is characteristic of a compound. Outgoing radiations are detected by photochemical detectors or vacuum phototubes. If the light intensity is very low, a photomultiplier tube is used. A phototube converts the transmitted light energy into an electric current. This electric signal is calibrated and can be produced on a chart recorder.

A better method is to split the light beam, pass one part through the sample and the other through the blank, and balance the two circuits to give zero (a double beam system). The extinction is determined from the potentiometer reading, which balances the circuit.

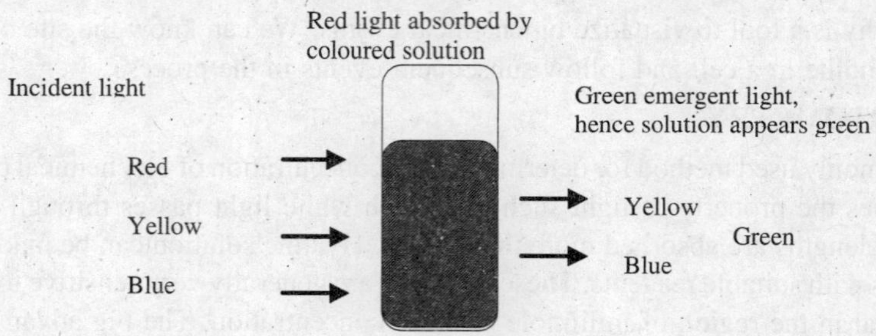

Fig. 13.8. Solution appears coloured depending upon the wavelength of the outgoing light.

Applications

Measurement of concentrations - Spectrophotometers are widely used for quantitative measurements of substance concentration in the solutions by measuring absorbance at the optimal wavelength. This is used to determine concentration in fractions obtained in column chromatography, TLC and in other solutions.

Absorption spectra - Identification and structure evaluation of various isolated pure compounds can be done by UV-Visible spectrophotometry. All the compounds have very specific absorption spectra depending on the molecular structure. Comparison of spectra helps in identification of compounds. The wavelength at which maximum absorption takes place is represented as λ_{max} of that compound in a given solvent.

Enzyme kinetics - In enzymatic reactions, changes in absorbance are recorded to determine the reaction velocity and concentration of the product formed/substrate utilized. Spectrophotometers coupled with microprocessor or computer can store data, figures and are helpful in comparing the effect of different variables.

Fluorescence Spectrometry

This procedure uses the measurement of the intensity of the fluorescence light emitted by the substance being examined in relation to that emitted by a given reference standard.

If the intensity of the fluorescence is not directly proportional to the concentration, the measurements may be effected using a calebration curve. In some cases measurement can be made with reference to a fixed standard such as a solution of unknown concentration of quinine in 0.1N sulphuric acid or of fluorescein in 0.1N sodium hydroxide. When certain chemical substances are excited electronically by the absorption of UV or visible radiation, they emit light at a longer wavelength. This phenomenon is called luminescence and depending on the life span of the excited species , two different processes could be distinguished. The first is fluorescence, where the luminescence stops within 10^{-19} to 10^{-4} sec. After the source of excitation is removed and the second is phosphorescence, where the luminescence continues for a slightly longer period of time (10^{-4} to 10 sec.).

Theory: Upon absorption of UV or visible radiation by a molecule, the electron from singlet ground state is promoted to singlet excited state. The excited species, may return to the ground state by dessipation of energy through collision or by vibrational relaxation of the excited state. The vibrational relaxed species can return to the ground state with the emission of radiation with a wavelength longer then that which originally absorbed. This radiation is referred to as fluorescence.

In order for a molecule to fluorescence, an absorbing molecular structure is required. Fluorescence may be expected to occur generally with molecules containing highly conjugated system. Atleast one c-donating group such as –NH2 or OH- should be a part of the conjugated system.

Electron with drawing groups such as –COOH- or NO2- deminish or prevent fluorescence. Fluorescence spectrometry offers detection limits lower then those of absorption spectrometry. A quantity of 1.1 microgram (u gm)/lit can be measured and linearity can be maintained up to 10,000 u gm/lit. This method is applicable in the quantitative determination of fluorescing substances.

Instrumentation : The source or irradiation of the sample is a mercury discharge or xenon lamp .. Selection of the exciting wavelength can be accomplished by means of a filter; this type is called a filter fluorometer. In a true fluorescence spectrometer two monochromaters are used, one for the excitation source and the other for analysing fluorescence emission. If the fluorescence spectrum is strong, the excitation spectrum can be determined by placing the desired solution in the instrument and evaluating the fluorescence while varying the wavelength of the excited light. The detector is a photomultiplier tube whose output is connected to a meter, a digital display or a recorder.

Applications :

1. Higher sensitivity than absorption spectrometer.
2. Applicable to fluorescent compounds.
3. It can be coupled with HPLC.

VI. CENTRIFUGATION

A centrifuge is an instrument which produces centrifugal force by rotating the samples around a central axis with the help of an electric motor. Centrifuges can be categorized as the clinical type (5-10,000 rpm), refrigerated high-speed centrifuges (10,000-20,000 rpm) and ultra - centrifuges (20,000 to 80,000 rpm). With increase in rpm, the friction of rotor with air produces so much of heat that they have to be run under refrigeration (so called refrigerated centrifuge) and both refrigeration and vacuum are used in ultracentrifuge, which runs at very high rpm. For these high speeds, even the rotor has to be made of special metal to withstand the great force.

Fig. 13.9. Different types of rotors used in centrifugation.

There are two types of rotors, angle head and swing (Fig. 13.9). In the former, the samples are kept at an angle of about 30° to the vertical axis whereas, in the latter the samples while spinning are horizontal. Simple calculations show that for the same radius, the swinging bucket method produces more gravitational force. Ultracentrifuges are of two types – analytical and preparative model.

Analytical Model

This consists of rotors and tubes, called cells. The instrument is designed to allow the operator to follow the progress of the substances in the cells, while the process of centrifugation is in progress. By estimating sedimentation velocity during the process, the molecular weight, purity etc. can be determined.

Preparative Model

This is used for purification of the components of macromolecules or other substances and all determinations are made at the end of centrifugation. The instrument has no monitoring device, while large centrifugal forces are set for a fixed time period.

Centrifugation is the most widely used technique for separation of various metabolites and also used to separate non-miscible liquids during extraction of secondary metabolites, e.g., water (aqueous), chloroform (organic) mixture. A wide variety of centrifuges are available, ranging in capacity and speed. During the process of centrifugation, solid particles experience a centrifugal force, which pulls them outwards, i.e., away from the center. The velocity with which a given solid particle moves through a liquid medium is related to angular velocity.

The principle of centrifugation is that, an object moving in a circular motion at an angular velocity is subjected to an outward force (F) through a radius of rotation (r) in cms, presented as [F = ω2r]. F frequently expressed in terms of gravitational force of the Earth, commonly referred to as RCF (Relative centrifugal force) or g by the following formula [RCF = ω2r / 980]. The operating speed of the centrifuge is expressed as revolutions per minute 'rpm', which can be converted to radians by the following formula:

$$\omega = \frac{\text{rpm}}{30}$$

$$\text{or}, RCF = \frac{\dfrac{(\pi\,\text{rpm})^2\,(\text{r})}{30^2}}{980} = \frac{980(\pi\,\text{rpm})^2\,(r)}{30^2}$$

Therefore, RCF = 1.2×10^{-5} $(\text{rpm})^2\text{r}$

Sometimes the velocity of the moving particles is expressed in the form of sedimentation coefficient (s) by the formula [V = s (ω2r)].

The sedimentation coefficient is a characteristic constant for a molecule or a particle and is a function of the size, shape and density. It is equivalent to the average velocity per unit of acceleration.

The unit Svedberg (s) is often used with reference to centrifugation and is equivalent to a sedimentation coefficient of 10^{-13}s.

Applications

Centrifugation is widely used in analytical techniques, preparation of extracts, separation of non-miscible liquid mixtures, purification of enzymes, inhibitors and removing particles (silica, etc.) from the solvents. Centrifugation with heating and vacuum (suction) is used for concentrating extraction (instrument called sample concentrator) and rapidly removing solvents. All the enzymatic work requires refrigerated centrifuges to keep the samples cool during centrifugation and to protect enzymes from inactivation by heat.

VII. CHROMATOGRAPHY

Chromatography (meaning 'coloured writing') is a technique to separate molecules on the basis of differences in size, shape, mass, charge and adsorption properties. The term chromatography was used by the Russian botanist Tswett to describe the separation of plant pigments on a column of alumina. There are different types of chromatography but they all involve interactions between these components: the mixture to be separated, a solid phase, and a solvent. The magnitude of these interactions depends upon the particular method used. Usually column chromatography is used to separate large quantities of compounds, whereas, paper chromatography (PC) or thin-layer chromatography (TLC) in one or two dimensions, is used for analytical work.

The mobile phase can be a gas or a liquid, whereas the stationary phase can only be a liquid or solid. In liquid column chromatography (LCC), separation involves a simple partitioning between two immiscible liquid phases, one stationary and the other mobile, the process is called liquid-liquid (or partition) chromatography (LLC). When physical surface forces are mainly involved in the retentive ability of the stationary phase, the process is denoted liquid-solid (or adsorption) chromatography (LSC). In ion exchange chromatography (IEC), ionic components of the sample are separated by selective exchange with counter ions of the stationary phase. Use of exclusion packing as the stationary phase brings about a classification of molecules based largely on molecular geometry and size.

A. Paper Chromatography

Principle - Cellulose in the form of paper sheets makes an ideal support medium where water is adsorbed between the cellulose fibres and forms a stationary hydrophilic phase.

The suitably concentrated mixture is spotted onto the paper, dried with a hair-drier and the chromatogram is developed by allowing the solvent to flow along the sheet. The solvent front is marked and after drying the paper, the positions of the compounds present in the mixture are visualized by a suitable staining reaction. The ratio of the distance moved by a compound to that moved by the solvent is known as the R_f value and is more or less constant for a particular compound, solvent system and paper under carefully controlled conditions of solute concentration, temperature and pH.

Sample – Generally, alcoholic extracts of plant material with or without partial purification are used for chromatography. Biological materials should be desalted before chromatography by electrolysis or electrodialysis. Excess salt results in a poor chromatogram with spreading of spots and changes in their R_f values. It can also affect the chemical reactions used to detect the compounds being separated. The sample (10-20 μl) is then applied to the paper with a micropipette or capillary.

Paper – Whatman No. 2 is the paper most frequently used for analytical purposes. Whatman No. 3 MM is a thick paper and is best employed for separating large quantities of material, the resolution is, however, inferior to Whatman No. 1. For rapid separation, Whatman Nos. 4 and 5 are convenient, although the spots are less well defined. In all cases, the flow rate is faster in the 'machine direction', which is normally noted on the box containing the paper. The paper may be impregnated with a buffer solution before use or chemically modified by acetylation. Ion exchange papers are also available commercially. For the separation of lipids and similar hydrophobic molecules, silica-impregnated papers are available commercially.

Solvent - This choice, like that of the paper, is largely empirical and will depend on the mixture investigated. If the compounds move close to the solvent front in solvent 'A' then they are too soluble, while if they are crowded around the origin in solvent 'B' then they are not sufficiently soluble. A suitable solvent for separation would, therefore, be an appropriate mixture of 'A' and 'B', so that the R_f values of the components of the mixture are spread across the length of the paper. R_f value is defined as [R_f = the distance moved by solute / the distance moved by solvent front]. This value is a constant for a particular compound under standard conditions and closely reflects the distribution coefficient for that compound. It is recommended that the developing chamber should be saturated with solvent, by using filter paper lining inside the chamber.

B. Two Dimensional Chromatography

The mixture is separated in the first solvent, which should be volatile, then after drying, the paper is turned through 90° and separation is carried out in the second solvent. After location, a map is obtained and compounds can be identified by comparing their position with a map of known compounds developed under the same conditions (Fig. 13.10).

Detection of spots- Most compounds are colourless and are visualized by specific reagents. The location reagent is applied by spraying the paper under a fumigation hood or rapidly dipping it in a solution of the reagent in a volatile solvent. Viewing under ultraviolet light is also useful since some compounds, which absorb strongly show up as dark spots against the fluorescent background of the paper. Other compounds show a characteristic fluorescence under ultraviolet light.

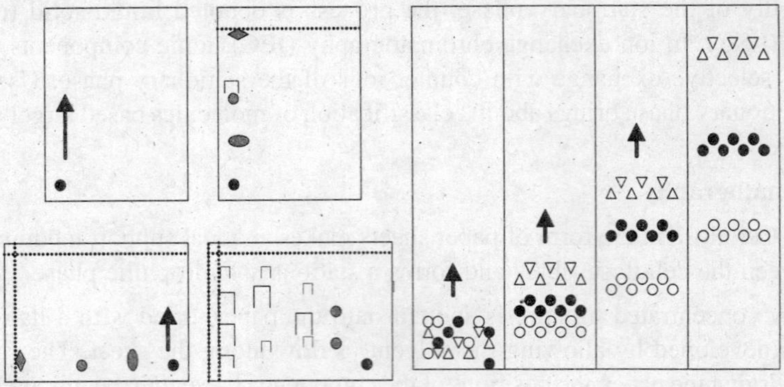

Fig. 13.10. Separation of a mixture by two-dimensional chrmatography (A) and principle behind the chromatographic separation (B).

C. Thin Layer Chromatography

Principle – Separation of compounds on a thin layer of adsorbing material is similar in many ways to paper chromatography, but has the added advantage that a variety of supporting media can be used so that separation can be by adsorption ion exchange, partition chromatography, or gel filtration depending on the nature of the medium employed. The method is very rapid compared to paper chromatography and many separations can be completed within an hour. Compounds can be detected at a lower concentration than on paper as the spots are very compact. Furthermore, separated compounds can be detected by corrosive sprays and elevated temperatures with some thin layer materials, which of course is not possible with paper.

Production of thin layer. The R_f value is affected by the thickness of the layer below 200 μm and a thickness of 250 μm is suitable for most separations. There are several good spreaders available in the market, which, can produce an even layer of required thickness by adjusting the thickness control screw. Calcium sulphate is sometimes incorporated into the adsorbent to bind the layer to the plate and, because of this, it is advisable to work rapidly once the adsorbent is mixed with water. There are now a number of prepared thin layer plates using different adsorbents on various supporting materials such as glass, plastic, and aluminium that are available commercially and these may be more convenient to use than trying to prepare plates in the laboratory.

Development - It is essential to make sure that the atmosphere of the separation chamber is fully saturated with the solvent mixture, otherwise R_f values will vary widely from tank to tank. However, horizontal chambers of high performance TLC (HPTLC) are very useful and convenient as they require less time, solvent and no stabilization time. Development of the plate is usually by the ascending technique and is very rapid

D. COLUMNS CHROMATOGRAPHY

Separation of compounds by column chromatography is one of the most widely used techniques in biochemical work and is the only technique for the separation of secondary metabolites in large quantities. It is therefore, appropriate to consider some of the general precautions to be taken when preparing and running columns before dealing with the various types of chromatography separations.

Columns are usually glass and, generally, long columns give good resolution of components but wide columns are better for dealing with large quantities of material. The essential features of chromatography columns are shown in Fig. 13.11.

Preparation of the material - A wide range of materials are used in chromatographic separations (silica, alumina, sephadex, resin) and all need to be equilibrated with the solvent before preparing the column. In addition, some form of pre-treatment is often required; for example some gel filtration materials need to be swollen, adsorbents need to be activated by heating or acid treatment, and ion exchange resins have to be obtained in the required ionized form by washing.

During the equilibration with solvent the material is allowed to settle and the fine particles remaining in suspension are removed by documentation (Fig. 13.11). If this is not carried out, the flow rate of solvent down the column will be considerably reduced due to clogging by these fine particles.

The pouring of the column. The chromatography column is packed with material by filling it about one-third full with solvent and slowly adding a slurry of the material in the solvent. This is carefully poured down a glass rod, as shown to stop air bubbles being trapped in the column. The suspension is allowed to settle and excess solvent run off. This process is repeated until the column is of the required height. The column is then washed thoroughly with solvent and the level of the liquid kept just above the surface of the material.

Application of the sample. The sample is first dissolved in the solvent or dialysed against the eluting buffer before loading it on to the column. In most class experiments the concentrated sample is carefully pipetted on to the surface and the tap opened until the top of the column is just below the level of the meniscus. The solvent reservoir is connected, and a constant head of liquid maintained at the top of the column from a pressure reservoir.

Elution - The next stage is to remove the materials from the column in order by eluting with an appropriate solvent. In displacement development, the solvent interacts more strongly with the chromatographic material than the compound on the column, than displacing the bound molecules. Large quantities of material can be separated in this way since about 50 per cent of the total column capacity is used. The separation is adequate but for better resolution of peaks elution development is preferred. In this case, no more than 10 per cent of the total capacity is loaded on to the column. The solvent interacts with the column more weakly than the solute molecules and overrides the bound molecules gradually eluting them from the column. This is probably the most commonly used means of elution and different molecules are removed from the column by changing the strength or pH of the eluting solvent in a stepwise fashion or by means of a gradient which can be linear concave or convex.

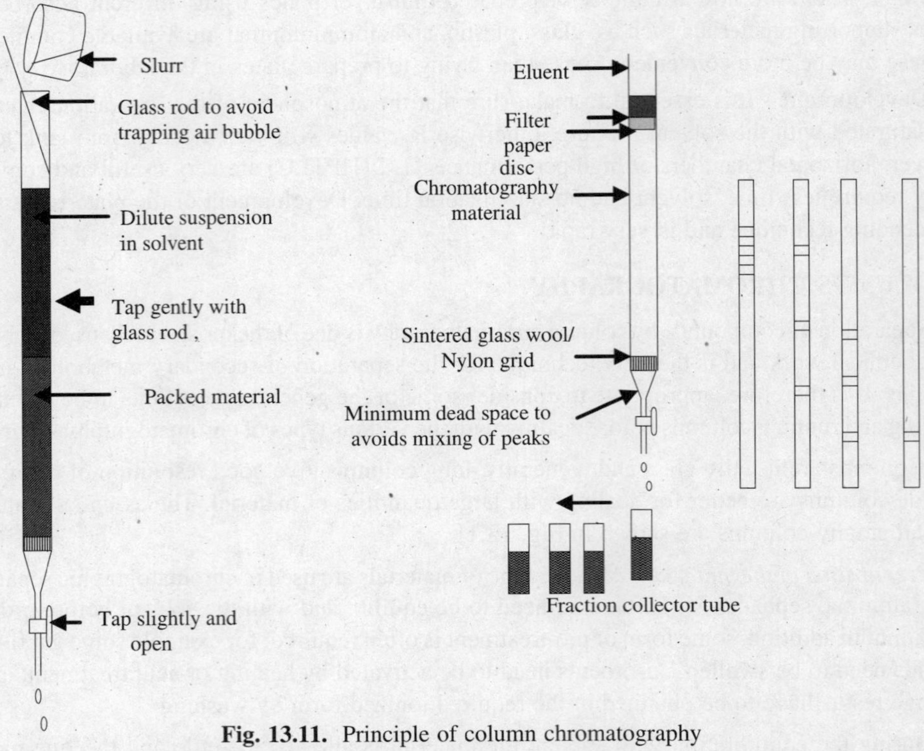

Fig. 13.11. Principle of column chromatography

The collection and analysis of fractions – The effluent from the column is collected into a series of test tubes, either manually or with a fraction collector. Each fraction is then analyzed for the presence of the compounds being examined and an elution profile prepared of the amount eluted against the effluent volume.

E. ADSORPTION CHROMATOGRAPHY

Theory- Compounds are adsorbed on to the column and an equilibrium set up between molecules bound to the column and those free in solution. The extent of the binding is governed by the charge, van der Walls' forces, dipole ineractions, hydrogen bonding and steric factors, and depends on the structure of the compounds.

F. ION EXCHANGE CHROMATOGRAPHY-

The matrix – The earliest ion exchange materials were based on synthetic resins of an aromatic nature and were suitable for the separation of inorganic ions and small molecules. However, they could not be used for the separation of large molecules such as proteins, which can not penetrate the closely linked structure of the resin and tend to be denatured by the hydrophobic matrix.

The first ion exchangers suitable for macromolecules were based on cellulose but they had low capacities as too much substitution made the cellulose soluble. Since then other materials have been developed based on dextran and an acrylamide type of polymer which have a much higher capacity.

Ionizable groups – Charged groups are attached to the matrix and the type of group defines the nature and strength of the ion exchanger. These groups may be either anionic or cationic., according to the nature of their affinity for either negative or positive ions. For example, the cation exchange materials exchange positive ions, so it is the charge carried by the exchangeable ion which decides whether a material is anionic or cationic and not the charge carried on the matrix (Fig. 13.12).

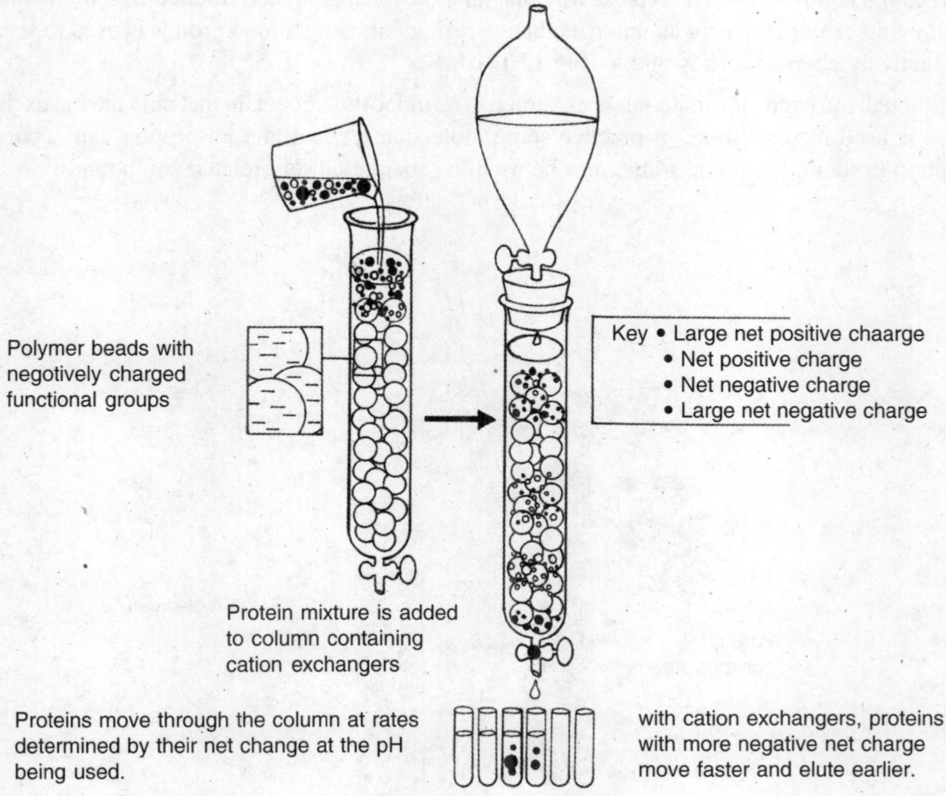

Fig. 13.12. Ion exchange chromatography

These two types can be further divided into materials that contain strongly ionized groups, such as – SO_3H and $–NR_3$, and the weakly ionized groups, such as –COOH, -OH and $–NH_2$. The strong ion exchange resins are completely ionized and exist in the charged form except at extreme pH values:

$$-SO_3H \rightleftharpoons SO_3 + H+$$

$$-NR_3OH \rightleftharpoons NR_3 + OH-$$

The weak ion exchange materials, on the other hand, contain groups whose ionization is dependent on the pH, and they can only be used at maximum capacity over a narrow pH range.

-COOH ======== COO- + H-

-NH3 ========= NH2 + H-

As a rough guide, resins containing carboxyl groups have a maximum capacity above pH 6, while those with amino groups are effective below pH 6.

The number of ionizable groups determines the capacity of the ion exchange. The total capacity is the number of ionizable groups per gram of material, whereas the available capacity is the amount of a given molecule that can bind under defined experimental conditions. In the case of some materials, large molecules may be unable to penetrate the matrix and can only react with charged groups on the surface. In this case, the available capacity will be considerably less than the total capacity.

Ion exchange equilibria – The typical way that an ion exchange material functions is illustrated in the following example, where an anion exchange resin containing amino groups is used to separate two negatively charged ions X and Y (Fig.13.12).

Although ion exchange materials are claimed to be mono-functional, in that only the ion exchange process is used in separation, in practice some molecular sieving and adsorption can occur. The adsorption is small, but it can sometimes be used to separate closely related compounds.

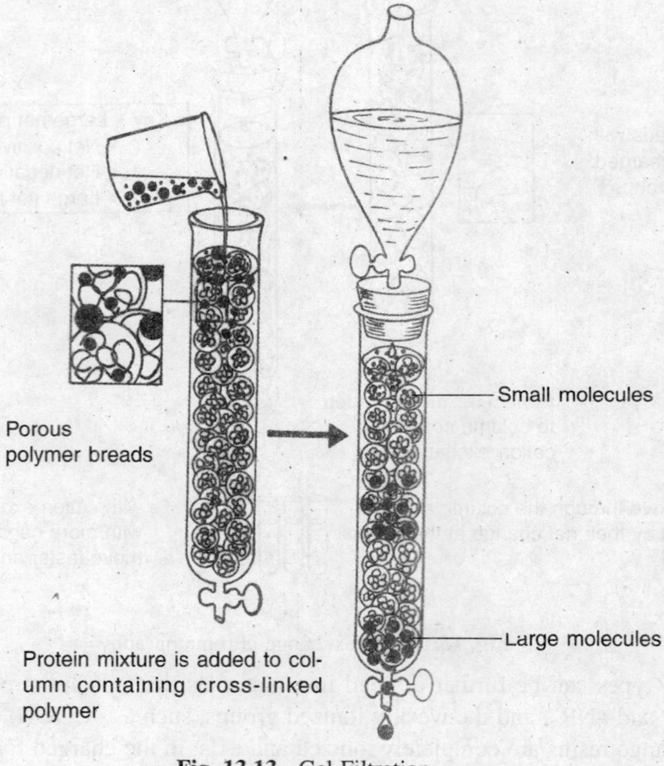

Fig. 13.13. Gel Filtration

Preparation of material. Ion exchange materials are first allowed to swell in the buffer and the fines removed as described earlier.

The ion exchange material is then obtained in the required ionic form by washing with the appropriate solution. For example, the H+ form of a cation exchange resin is obtained by washing the material with hydrochloric and then water until the washings are neutral. Similar the Na+ form is prepared by washing the resin with sodium chloride or sodium hydroxide then water as above.

The final stage before preparing the column is to equilibrate the material by stirring with the eluting buffer.

Gel filtration

Molecules can also be separated on the basis of differences in their size by passing them down a column containing swollen particles of a gel. Small molecules can enter the gel but larger molecules are excluded from the cross-linked network (Fig. 13.13).

This means that the accessible volume of solvent is very much less for molecules totally excluded from the gel than for small molecules, which are free to penetrate. The separation of molecules by gel filtration is illustrated in the diagram. First the mixture of large and small molecules is placed on top of the column. As they pass down the column, the small molecules diffuse into the gel and follow a longer path than the large molecules, which are completely excluded from the gel particles. Eventually, complete separation occurs, with the large molecules leaving the column first and the smaller ones last.

Materials- Pharmacia manufacture cross-linked dextrans (Sephadex) and agarose (Sepharose) and cross-linked agarose (Sepharose CL) and Sephacryl, while Bio-Red Laboratories make polyacrylamide (Biogel P), agarose (Biogel A), porous glass (Bio-Glas), and polystyrene (Bio-Beads). The degree of cross-linking etc. is carefully controlled to give a range of products able to fractionate molecules over a limited size frange. The useful fractionation range of molecules is only approximate since separation depends on the shape and, to a minor extent, the charge, as well as the size of the molecules. Sephadex G25 and G50 are made in several particle sizes, fine, coarse, medium and superfine grades. The fine particles are suitable for most laboratory work requiring high resolution, while the coarse material is convenient for preparative work with large columns since this gives flow rate.

Sepharose is stable from pH 4 to pH 10 and can be used over the temperature range 0-30°C, and Sepharose CL is stable over pH 3-14 and can be used up to temperature of 70°C. These more porous gels are used to fractionate very large molecules such as nucleic acids and viruses.

After use the columns can be washed in water and stored in the cold room for quite a time provided a trace of preservative such as phenol or chloroform is added to prevent bacterial growth.

H. AFFINITY CHROMATOGRAPHY

Principle -Affinity chromatography is a type of adsorption chromatography in which there is a high degree of specificity in the interaction between the adsorbent and the compound to be separated. In the case of enzymes, the ligand attached to the adsorbent is usually a powerful inhibitor with a high affinity constant which will bind only one enzyme in a complex mixture of proteins.

Affinity material - The ligand is covalently attached to a supporting medium so that the chromatographic material can be designed for a specific purification task. The matrix has to be macroporous to allow large molecules access to the binding sites and needs to have good flow properties. It has to be devoid of non-specific adsorption sites but must contain functional groups to which ligands can be attached. There are a number of commercially available materials that fit these criteria including cross-linked Sepharose (CL-Sepharose) available from Pharmacia.

A spacer arm is nearly always inserted between the matrix and the ligand so that large molecules can gain access to the binding sites. If this spacer arm is too short, then steric hindrance can still occur while if it is too long there is an increased risk of non-specific adsorption, particularly of hydrophobic compounds. In practice, a spacer arm of 2-10 C atoms has been found to be optimal.

Purification by affinity chromatography - The principle of affinity chromatography is relatively simple to understand and this is illustrated in Fig. 13.14. which shows the purification of an enzyme from a mixture of other proteins. However, like most things the pracitice is not quite so straight forward and technical problems arise such as unwanted side reactions during the synthesis and attachment of the ligand to the side arm and matrix. This tends to increase the non-specific adsorption of the affinity material with a consequent lowering of the specificity. In other cases the binding of the compound to the ligand can be so strong that is becomes quite difficult to remove it from the affinity column.

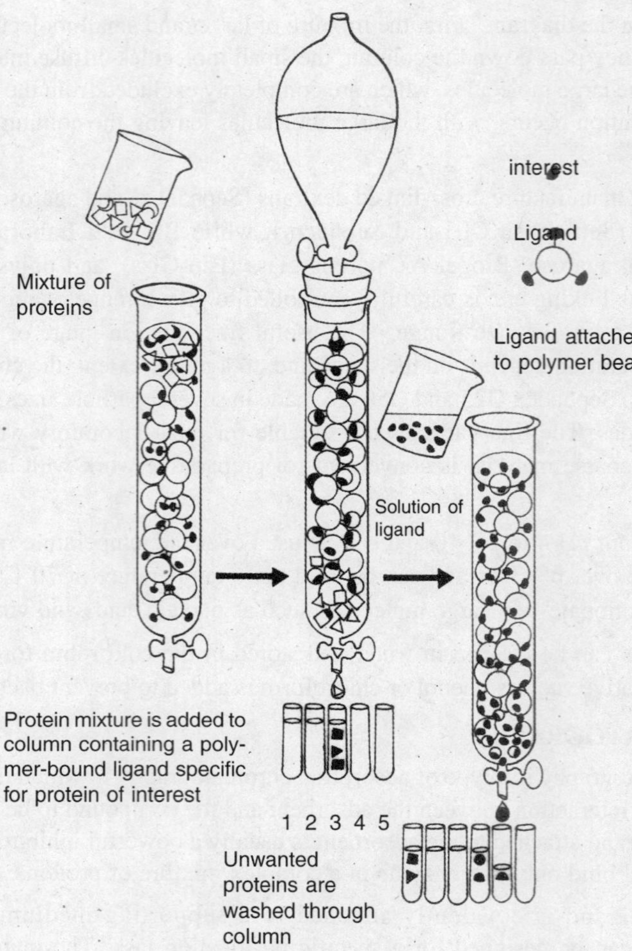

Fig. 13.14. Affinity chromotography

However, in spite of these and other technical problems, separation by affinity chromatography can lead to a very high degree of purification in a single step.

I. Gas Liquid Chromatography

A separation by partitioning solutes between a mobile gas phase and a stationary liquid phase held on a solid support is called Gas Liquid Chromatography (GLC). A GLC has following components, gas cylinders and regulators (Nitrogen or Helium gas, Hydrogen fuel, air fuel), oven with column, injector unit, detector system, recorder and control panel.

Columns

Two basic types of columns are used, the packed (stainless steel, copper or glass columns packed with inert material coated with a liquid phase) and the capillary columns with open tubular structure (capillary's inner wall is coated with the liquid stationary phase). Capillary columns are narrow and long (0.35 mm X 50-150 m) as compared to packed columns (1.6-9.5 mm x 3 m).

Sample Injection System- As a rule of chromatography, sample should be small, and concised . In GLC, it should be in vapour form introduced in very short time (at once) and should be highly reproducible, otherwise great variation within the sample will appear between different replicates.

Small liquid samples (1-10ml) are injected by a micro syringe through a self-sealing silicone rubber septum onto a heated compartment. Here, sample is vaporised and transported into column by the carries gas, GLC is a very sensitive system, therefore, sample has to be diluted suitably to detection and recording limits of the instrument. Sample maybe injected in the heating compartment directly, or through a sample loop or even directly in the (capillary) column.

Ovens – The column ovens evenly heated and programmable chamber reaching a temperature upto 450oC. It facilitate vaporization of solute in successive steps and better separation of the solute components.

Detectors – Detectors are sensing devices to detect the amount of individual components leaving the column. Various types of detectors are available for GLC and they vary in their sensitivity to detect small sample concentration. Thermal conductivity detector is the least sensitive while electron capture detector is the most sensitive. The commonly used detector is the hydrogen flame ionization detector. The signals generated by a detector are sent to recorder to produce chart.

Regulation of gas flow – In a GLC, a sample (liquid or gaseous) is injected into a heating block, where it is vaporized and carried to column by the carrier gas. The solutes are adsorbed by the stationary phase and then desorbed by fresh flow of carrier gas (mobile phase) in the column towards outlet. Different solutes will move with different speed in the column depending upon their adsorption and partion ratio. The solutes pass the column in increasing order of their partition ratios and enters in a detect, where they are detected by different detectors (flame ionization, thermal conductivity, electron capture, photo ionization and flame photometric detector). Different detectors have different sensitivity, therefore, used according to requirement of the sample. In case a recorder is used, the signals appear on the chat as a plot of time versus the composition of carrier gas stream. The time for emergence of a peak is characteristic for each component, the peak area (on the chart) is proportional to the concentration of the component in the mixture (sample).

For a GLC quantification, the volatile nature of sample is a prerequisite but availability of high column temperature (up to 450°C) and pyrolytic techniques increased the application of the system. Test conditions for the metabolites have to standardized using authentic samples. A concentration gradient of authentic sample is used to determine the quantity of the component is unknown mixture as a general practice in all the chromatographic work.

J. High performance liquid chromatography (HPLC)

Among the most difficult problems a researcher faces in plant analysis is the isolation, purification and identification of compounds in mixture (explant extract). Some of the separation problems are solved by TLC, paper chromatography and ion exchange chromatography. However, for many biologically active compounds these techniques are inadequate. The development of high performance liquid chromatography (HPLC) has made possible the rapid analysis of non-volatile, ionic, thermally labile compounds that were previously difficult to separate. Molecular constituents of the cell can be determined with high sensitivity, speed, accuracy and resolution. The major requirement of the technique is the solubility of solutes in an appropriate mobile phase, and it can handle very small amounts, of the solutes.

HPLC is liquid chromatography and belongs to the category of column chromatography in which the mobile phase is a liquid. This mobile phase if forced under high pressure up

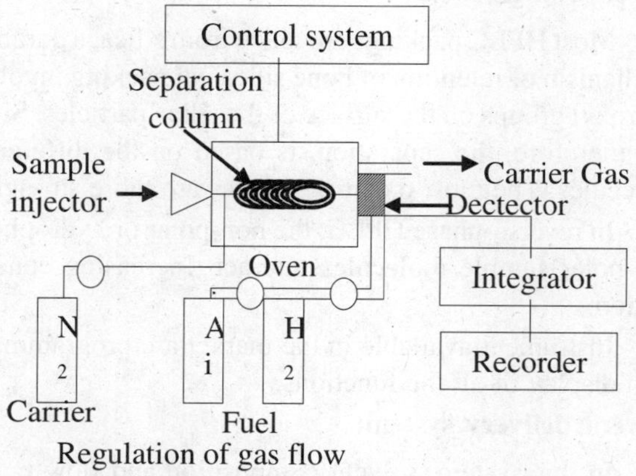

Fig. 13.15. Schematic GLC

to 6000 psi, into the column (c/f column chromatography – without such presume). Most of the HPLC operations are done by normal phase chromatography, bonded phase chromatography and reverse phase chromatography.

The role of pore size in HPLC is a crucial one liquid column packings do not have a discrete pore size though packing pore size is usually represented by a single number. The inter-relationship between pore size, pore volume, silica density, and surface area and their individual and combined effects on a given separation number is complex. Column is the very heart of the system, hence it has to be chosen with great care.

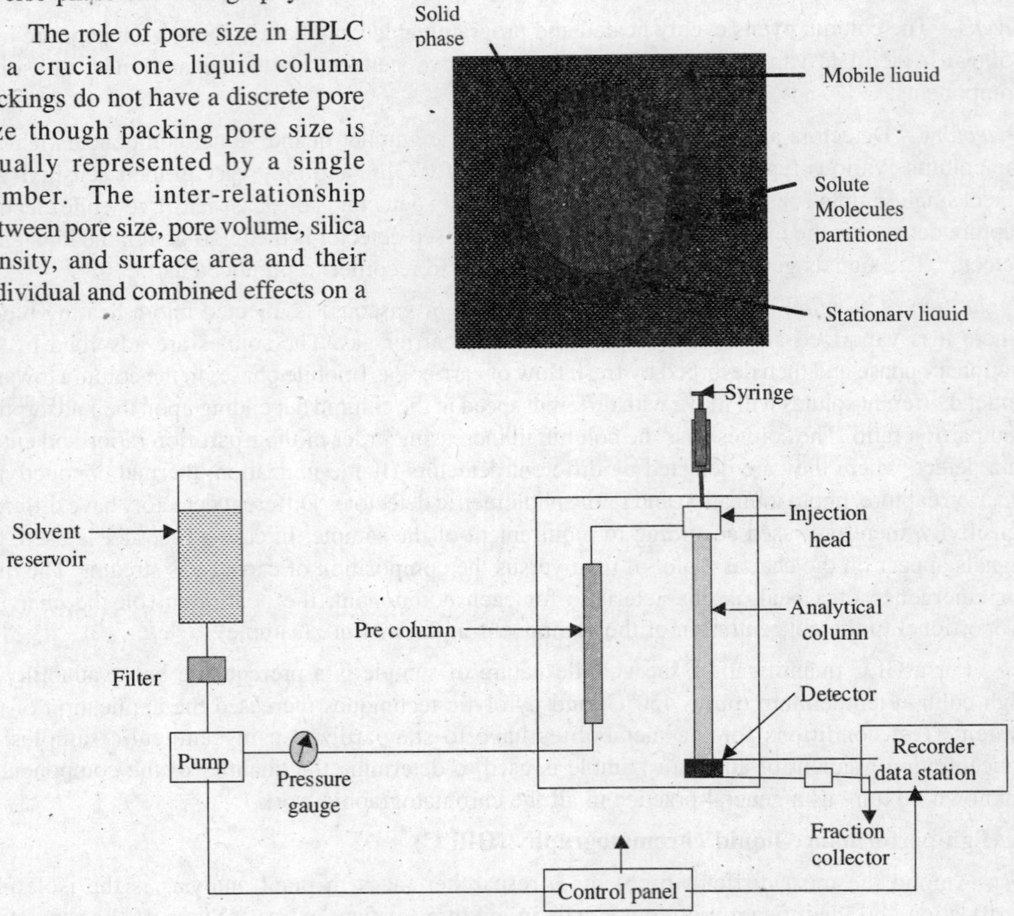

Fig. 13.16. Mechanism of HPLC (A) and schematic HPLC system (B)

Most HPLC packings are made from silica, a durable and economical adsorbent. The functional mechanism of retention of bone silica gel packing involves interactions of the sample with the polar hydroxyl groups on the surface of the silica particles. Since the silica surface is very polar, retention, and therefore, the separation, is based on the difference in the polar functionalities of different molecules. The most p;olar molecules are more strongly retained or absorbed to the silica surface.

In reverse-phase HPLC, the non-polar or hydrophobic column packing is used, with which the non-polar sample molecules interact. Increasing concentration of organic solvents are used for elution.

Instrument available in the market are programmable, co-ordinated and sophisticated devices with display of all the functions.

Solvent delivery system:

In this system, solvent composition and flow rate are programmable (as compared to GLC, temperature) for better separation of the components of a mixture. This function is performed by the pump, solvent of some composition (isocrartic) can be used with the help; of one pump. Pump should work precisely over a wide flow range with very high pressure (up to 5000 psi) and resistant to solvents. Various types of pumps are available in the market like displacement type, reciprocating type and pressure vessel type.

For gradient analysis, a second reservoir and pump and a gradient controller are required. This is an electronic device which synchronices the operation of the two pumps to provide a mobile phase mixture of the desired concentration.

Sample injection is done by using a rotary value and loop (20-2000 μl) injector,. By changing the position of this value, solute is transferred in the column. Pre-column is an optimal column, and used to protect the analytical column by saturating the mobile phase.

The analytical column in which the actual separation takes place, is a stainless steel tube (25 cms x 2-4.6 mm) containing stationary phase. The material used to pack the column are of two types; superficially plorous or pellicular and totally porous. Silica gel is the most commonly used for microparticulate column packing.

Detectors- There is no universal detector. Various types of detectors are used depending upon the nature and amount of compounds to be measured. Detectors respond to the physical property of the solute which is not exhibited by pure mobile phase. Ultraviolet/visible absorption, fluorescence and electrochemical detectors are widely used detectors.

Applications

HPLC has wide applications. It is used in food technology, beverages, pharmaceutics and biotechnology. It is equally used in quantification of biomolecules, routine quality control work and for separation of compounds from mixture. It is very useful, efficient, quick and reliable for large number of complex sample mixtures like alkaloids, steroids and phenolics.

QUESTIONS

1. Write short notes on the following :

(*a*) Affinity chromatography
(*b*) Bright field microscope
(*c*) Centrifugation
(*d*) Column chromatography
(*e*) Confocal microscope
(*f*) Fluorescence microscopy
(*g*) Fluorescence spectrophotometer
(*h*) Gel filtration
(*i*) GLC
(*j*) HPLC
(*k*) Paper chromatography
(*l*) SEM
(*m*) Spectrophotometry
(*n*) TEM
(*o*) Thin Layer chromatography.

CELL IMMOBILIZATION AND SYNTHETIC SEED TECHNOLOGY

CHAPTER 14

I. Cell Immobilization

Immobilization of plant cells and organs is a relatively new development in the techniques of plant tissue culture used for the production of secondary metabolites and development of synthetic seeds. To achieve industrial production of secondary plant products, it is necessary to evolve newer methods to reduce the cost and make the plant system comparable to microbial system in the production and cost. Immobilization of plant cells, protoplasts or embryos (also enzymes and microbes) is achieved by binding these materials on to or within a solid support.

(a) Need for immobilization

Plant cells are characterized by large size, sensitivity to shear (see chapter on bioreactor for details) and oxygen concentration, and need cell to cell contact for metabolite production. The secondary metabolite production is triggered by short period of stress in culture (e.g., depletion in the nutrients during the stationary phase). Immobilization can over-come many of the above mentioned limiting factors of suspension cultures with the distinct advantages of easier separation of biocatalyst (immobilized cells) from the product. Immobilized cells are useful tools for the bio-transformation of high value products. However, the applications of cell immobilization are limited to selected cell lines.

(b) Range of entrapment

Three types of entrapment systems are used:

A. Gel formation by ionic linking on a charged polymer (co-valent linkage to a polymer),
B. Gel formation by cooling off a heated polymer (entrapment within a polymer),
C. Gel formation by chemical reactions (cross linking, radical polymerization).

Table 14.1. Some polymers used for entrapment of cells.

Polymer	Principle of Gel formation	Gel-forming procedure
Synthetic Polyacrylamide	Polymerization	Addition of initiators to a solution of monomers
Epoxy resin	Polycondensation	Epoxy and polyamine precursors are cured.
Polyurethane	Polycondensation	Reaction of polyisocyanates with water
Agar, agarose	Thermal gelation + Ionotropic stabilization	Cooling of heated polymer solution
Alginate	Ionotropic gelation	Dripping into a calcium chloride solution.
Chitosan	Ionotropic gelation	Dripping into a polyphosphate solution
Proteins Collagen	Neutralization	Neutralization of an acidic solution
Gelatin	Thermal gelation	Cooling of heated polymer solution
Fibrin	Enzymatic gelation	Enzymatic conversion of fibrinogen of fibrin.

Agarose, alginate, carrageenan, urethane polymers and phenylene oxide have been used as solid matrix for entrapment for the production of many pharmaceutical compounds and synthetic seeds (Table 14.1).

The most widely used technique for the immobilization of cells with preserved viability has been their entrapment in alginate or carrageenan. Advantage in using these polymers is the simplicity with which spherical particles can be obtained by dripping a polymer-cell (or embryo) suspension into a medium containing appropriate cation.

This process was reported for the first time for cells of *Catharanthus roseus, Morinda citrifolia* and *Digitalis lanata* by Brodelius et al., (1979, 1980). The different compounds used for entrapment and brief procedures are as follows.

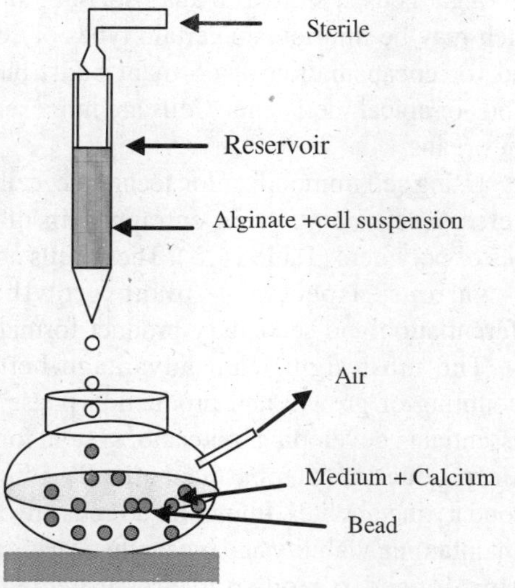

Fig.14.1. Process of preparation alginate beads

a. Alginate- A polysaccharide composed of glucuronic acid and mannuronic acid, and thus, contains one carboxyl group on each sugar moeity. In the presence of calcium or other multivalent cations, alginate forms a gel due to formation of coionic bridges. This gel can be solubilized by the addition of a calcium-complexing agent such as phosphate ion, citrate, or ethylene diamine tetra acetic acid (EDTA 1-2%).

b. Carrageenan Kappa-carrageenan is a polysulfonated polysaccharide that forms a relatively strong gel in the presence of potassium ions. Carrageenan (3% w/v) in 0.9% sodium chloride is dissolved by heating and then sterilized by autoclaving. Cells are mixed in this sterilized carrageenan and dropped in a medium containing 0.3 M KCl solution, where they form beads.

c. Agarose - Agarose has the advantage that no counter-ion is required for gel stability. Therefore, immobilization can be carried out in any medium. There are several types of agarose preparations available in the market but low-gelling temperature type is preferred. This is a complex polysaccharide. A solution of 3% agarose is prepared in medium and autoclaved. The plant cells (2g) are suspended in agarose (8g) at 35 °C and mixed. The solidified gel is passed through a screen to obtain small particles. The small blocks are transferred in the medium.

To generate beads, warm cell-agarose liquid is dispersed in paraffin oil by magnetic stirrer. When droplet of appropriate size is formed, the mixture is cooled on an ice bath under continuous stirring until the polymer is solidified. The mixture is then centrifuged and oil is removed. After washing with sterilized distilled water, beads are suspended in the medium.

Table 14.2. Immobilization of plant cells for various uses.

Method	Plant species
Alginate	*Apium graveolens, Brassica oleracea, Cannabis sativa, Capsicum annuum, Catharanthus roseus, Coffea arabica, Datura innoxia, Digitalis lanata, Glycirrhiza achinata, Haplopappus gracilis, Ipomoea sp. Papaver somniferum, Vicia faba.*
Agarose	*Brassica napa, Crepis capillaris, Hyoscyamus muticus, Lycopersicon peruvianum, Petunia hybrida, Solanum tuberosum.*
Chitosan	*Amaranthus tricolor, Asclepias syriaca.*
Carageenan	*Lavandula vera*
Hollow fibre	*Glycine max*
Polyurethane	*Beta vulgaris, Capsicum frutescens , Cinchona pubescens, Daucus carota, Jasminium sp.*

Agar is used similar to agarose, but it has relatively high (around 45 °C) melting temperature, which may be injurious to certain types of cells. High melting temperature agarose or agar can be used for encapsulation of dormant buds, bulblets, bulbils or apical meristems. Cells are more sensitive than organs.

Using cell immobilization technique, cells from diverse species have been entrapped in different types of polymers (Table 14.2). These cells are used for various aspects of plant growth, cell differentiation and secondary product formation.

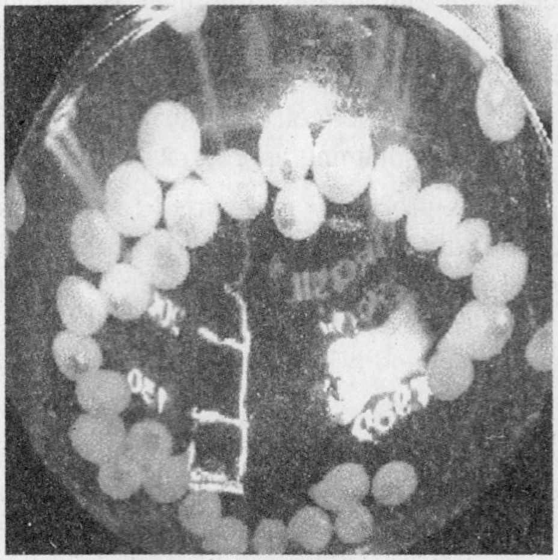

Fig. 14.2. Encapsulated somatic embryos

The most significant advantage being the uncoupling of growth and production phase which is essential in developing bioreactor system for large-scale production of pharmaceutically important secondary metabolites. Immobilized cells are capable of maintaining viability and biosynthetic capacity for the prolonged periods and are protected from mechanical stresses. In case product is released in the medium, it facilitates product recovery and down-stream processing (isolation and purification of the desired compound).

Types of bioreactors used for immobilization of plant cells

The following types of bioreactors are used for plant cell immobilization:

A. Packed bed reactors

In this reactor, the cells can be immobilized either on the surface of or throughout the support, the fluid containing substrate flows past the support. When cells are immobilized through the support, the packed bed can accommodate a large number of cells per reactor volumes.

B. Well mixed reactors

The advantages of a well mixed reactor for immobilized cells are that gaseous substrates can be directly sparged (passed and mixed through bubbling), the temperature and pH can be carefully controlled, and mass transfer rates can be improved over packed bed reactors. The suspension culture with complete cell recycle is essentially a well mixed "immobilized" cell reactor.

C. Fluidized bed reactors

Such reactors utilize the energy of the flowing fluid (liquid and/or gas) to suspend the particles. Small-immobilized particles are used as they require less energy. The mass transfer advantages of these small particles is one of the major benefits of the fluidized bed reactor. To obtain adequate conversions of substrate to product, the fluid must often be retained in the reactor for an extended period. During this period biocatalyst converts the substrate in the product.

D. Membrane type reactors

The most commonly considered membrane reactors are the hollow fibre and spiral wound reactors. In hollow fibre reactor, the cells are retained either within the tubes or in outer region. The spiral wound membrane reactor is a flat plate reactor rolled into a cylindrical shape.

Applications

A. Enhanced production of secondary metabolites

The initial interest in the use of plant cells in an immobilized state stemmed mainly for its bio-processing advantage and the potential reuse of biomass. Experimental evidence indicates that immobilization can have a dramatic impact on cellular physiology and secondary metabolism. Lindsey

(1985) demonstrated that the process of immobilization reduces the rate of cell division, protein synthesis and these effects are conducive for increasing the yield of secondary metabolites. The another consequence of plant cell immobilization is to reduce the production of cell wall material, which contain a substantial amount of bound phenolic compounds. This will cause increased availability of precursors for secondary metabolism (Table 14.3).

Table 14.3. Production of secondary metabolites in immobilized cells:

Species	Product	Increase (x folds)	Type of immoibilization
Capsicum frutescens	*Capsaicin*	>100	*Foam*
Capsicum annuum	*Capsaicin*	>100	*Gel*
Coffea arabica	*Methyl xanthines*	13	*Gel*
Catharanthus roseus	*Ajmalicine*	35	*Gel*
Mucuna pruriens	*L-DOPA*	*No new synthesis*	*Gel.*

B. Biotransformations

Hydroxylation of cardiac glycosides has proved an interesting application of immobilized plant cells. Bioconversion of β-methyl digitoxin into β-methyl digoxin has been achieved using ***Digitalis lanata*** immobilized cell cultures up to 70 days.

Table 14.4. Selected one-step bioconversions by immobilized cells. Further details are discussed in the text.

Cell culture species	Reaction type	Precursor	Product	Matrix and remarks
Papaver somniferum	Reduction	Codeinone	Codeine	Alginate and PUR; in both cases stereospecific bioconversion, product release.
Digitalis lanata	Hydroxylation	β-methyl-digoxitoxin	β-methyl-digoxin	Alginate; very long bioconversion period (160 d)
Mentha species	Reduction	(-)-menthone	(+)-neomenthol	PAAH; stereospecific reduction, product release.
Daucus carota	Hydroxylation	Digitoxigenin	Periplogenin	Alginate; five times reuse possible
Mucuna pruriens	Hydroxylation	L-tyrosine	L-DOPA	alginate, pectinate, agarose, gelatine; product release

II. Synthetic seed technology

Synthetic seeds offer a potential technology towards use of somatic embryogenesis for large-scale propagation of plants through automations. The micropropagation industry started with orchids and ferns and is still largely restricted to horticulture, plantation and ornamental plants. During the past decades production of bulb and tuber plants, such as lilies, potato, gladiolus, and cassava, has become quite common. Large–scale propagation requires planting material of suitable age, shape and of germination percentage. Somatic embryos are suitable materials for this purpose as large number of somatic embryos of same age can be produced in a bioreactor. Technology is developed to encapsulate them to produce artificial or synthetic seeds. Encapsulated embryos are protected against desiccation and mechanical injury by the gel (Fig. 14.2). It was Murashige (1977) who conceived the idea of developing synthetic seeds. The first constructions of synthetic somatic embryos were achieved by Kitto et al, (1982) and single embryo encapsulation of hydrated alfalfa somatic embryos was obtained by Redenbaugh et al (1984). The necessary components of synthetic seeds depend on whether they are hydrated or desiccated.

Four types of synthetic seeds have been proposed on the basis of embryos and its encapsulation.

(*a*) Uncoated, desiccated somatic embryos e.g. for orchard grass.
(*b*) Coated, desiccated somatic embryos e.g. carrot.
(*c*) Encapsulated (coated), hydrated somatic embryos e.g. alfalfa.
(*d*) Uncoated hydrated embryos (in a fluid drilling gel) e.g. carrot.

Development of coatings

Different materials are used for coating the somatic embryos such as agarose, sodium alginate, polyoxyethylene.

Attempts were made to citrus embryos with various compounds to develop a suitable synthetic seed coat. Of the eight compounds tested, poly-oxyethylene was found to have the best film-forming ability, to readily dissolve in water, and to have a non-damaging effect on the embryos. Consequently, polyoxyethylene was chosen for further research on creating synthetic seeds using celery somatic embryos, producing multiple embryo synthetic seeds.

Alfalfa and celery somatic embryos were used for developing hydrated synthetic seeds using various hydrogels, particularly sodium alginate. Hydrated synthetic seeds have many uses in propagation, e.g., reduced cost of vegetatively propagated elite lies, direct delivery of tissue culture plants to field, and genetic uniformity of plants. This technology is still in research phase and has not yet been commercialized. This technique has been used for following species for *in-vitro* and in soil sowing (Table 14.5).

Table 14.5. Plants produced from synthetic seeds sown in vitro and in soil.

In-vitro	In soil
Apium graveolens, Brassica sp, Daucus carota, Gossypium hirsutum, Lactuca sativa, Medicago sativa, Oryza sativa, Picea abies, and Zea mays	*Apium graveolens, Daucus carota and Medicago sativa*

Maturation of somatic embryos

Application of synthetic seed technology is limited by the quality of somatic embryos because our knowledge of somatic embryogenesis is insufficient; much of the synthetic seed research is focussed on maturation of somatic embryos.

(a) Pretreatment. Yields of alfalfa somatic embryos were significantly improved when amino acids such as proline, alanine, arginine or glutamine were added to the callus maintenance medium. Up to 100 times more somatic embryos were produced.

Organic acids applied during the last subculture of callus growth (pretreatment) were found to increase the number of embryos formed, conversion of embryos and the amount of seed specific proteins. Although the organic acid pretreatment significantly reduced callus growth during the last subculture, the total number of plants produced was almost double the control treatment.

(b) Maturation. Embryo maturation with ABA was essential for obtaining high soil conversion. ABA matured alfalfa somatic embryos had an average of 50% soil conversion. Optimum concentration of ABA for maturation of alfalfa embryos, was 5 μM.

ABA was also critical for soil conversion of carrot and celery somatic embryos with 1-5 μM ABA giving the highest conversion frequency.

(c) Storage – After ABA maturation hydrated, non-encapsulated alfalfa somatic embryos can be stored at 4°C for over 2 months without loss of conversion ability.

(d) Conversion – Green house conversion of somatic embryos depends on quality of somatic embryos and watering system. Watering should be perfect and adequate, neither high, nor dry.

(e) Problems with synthetic seeds –There is low gas exchange in encapsulated somatic embryos and conversion declines with increase in gel percentage. In a recent report, steep oxygen gradients were found single calcium alginate beads and up to 70% of the bead being free of oxygen at the end of logarithmic growth phase. One potential solution to this problem is to increase oxygen supply. In an attempt to increase oxygen supply, perfluorochemical (PFC) oils have been used as oxygen carriers and can be mixed with gel to increase oxygen supply. It is proposed that new polymer, besides alginate and polyoxyethylene, should be tested – e.g., Tullanox, Cab-o-gil, etc.

Applications

(*a*) For development of plants for breeding purposes, e.g. alfalfa.
(*b*) In forestry for transplanting improved/selected material reduce the cost of breading, e.g., used in Eastern White pine, white spruce, black spruce, red spruce, Norway spruce, Japanese larch, European larch and hybrid larch.

Mechanization

1. Bioreactor development – For horticultural crops millions of somatic embryos and for agronomic crops millions to billions somatic embryos are required for large-scale commercialization. This can be met only with the development of bioreactor technology. This technology has to be improved as somatic embryos on agar plates are found better with better conversion frequencies than in liquid cultures. This technology has been used in alfalfa, celery, carrot and lilies.
2. Encapsulation mechanization – In mechanization main emphasis has been given on mechanical sowing of planting material or encapsulated seeds. More efforts and improvement is required for mechanical encapsulation to accommodate various types of somatic embryos. A method to encapsulate somatic embryogenesis is presented in simplified and schematic manner in fig. 14.1. Somatic embryos move in a vibrating bowl, fall one by one and encapsulated in sodium alginate gel. At various points sensors have been incorporated to monitor the embryo movement, air flow and gel formation.

Economic considerations

Cost of synthetic seeds of alfalfa has been estimated to \$ 0.00026 based on development of efficient automation and bioreactor system. However, in a research laboratory, it came to about 0.068 US \$. This is due to difference between the cost calculations based on the research unit and the production unit. If the lower cost is achieved by improving the technology, it will be commercially viable.

QUESTIONS

1. What is a synthetic seed? Describe various methods of synthetic seed production and its implication.
2. With suitable examples describe the technique and agents used for cell immobilization.
3. Write short notes on the following :
(*a*) Synthetic seeds
(*b*) Encapsulation
(*c*) Desiccated embryos
(*d*) Cell immobilization
(*e*) Gels for immobilization
(*f*) Immobilized cell bioreactors
(*g*) Immobilization for production of secondary metabolites production.

PLANT PROTOPLASTS : ISOLATION, CULTURE & SOMATIC HYBRIDIZATION

CHAPTER 15

I. INTRODUCTION

Plasma membrane of plant cells is surrounded by cellulose wall and adjacent cells are joined together by a thick pectin rich matrix. Separation of plant cells and removal of the cell wall experimentally by an enzymatic process, results in the production of isolated, nacked plant cells termed 'protoplasts'. Protoplasts thus can be called as a functional individual cell with plasma membrane as outermost layer and the only barrier between environment and the cell contents.

The real start of plant protoplasts research started in 1960's with the enzymatic removal of cell wall by E.C. Cocking at the University of Nottingham, England. It became a widely used technique with the availability of commercially produced enzymes. Takabe and coworkers using commercially available enzymes isolated protoplast in 1968 and subsequently achieved regeneration of whole tobacco plant from mesophyll protoplasts in 1971. Carlson's and coworkers were successful in obtaining the first interspecific somatic hybrids by protoplast fusion between *Nicotiana glauca* and *N. longsdorffii*. During these experiments, polyethylene glycol (PEG) was established as a potent fusogenic agent. Subsequently, the improved fusion methods have resulted in the production of several somatic hybrids including the first intergeneric hybrid (called pomato) between potato and tomato by Melchers and coworkers in 1978. The first interspecific hybrid between protoplasts of normal *Nicotiana sylvistris* and x-ray irradiated protoplast of male sterile *N. tabacum* was obtained by Zeleer and coworkers in 1978. Next year in 1979, stimulation of protoplasts fusion by electrical stimulus and gene transfer through protoplast fusion were demonstrated. Zimmerman and Scheurich (1981) improved the electric fusion method for protoplasts isolation. The 1980s witnessed many protoplasts based works particularly those reporting novel protoplast to plant systems for genetic manipulation. During the 1990s protoplast based technologies for gene transfer were overshadowed by *Agrobacterium* and Biolistics® mediated gene delivery to plants. However, public antagonism (especially in Europe) to rDNA technology renewed interest in exploiting protoplast in somatic hybridization, cybridization, protoclonal variation study, proteomix and metabolomics.

II. APPLICATIONS OF PROTOPLASTS IN PLANT RESEARCH

Plant cells are normally connected to each other via many plasmodesmata in multicellular tissues. Plant protoplasts cultures provide an excellent system to obtain single cells of higher organisms and study its functioning in a controlled environment. Protoplasts isolation provides millions of cells which is comparable to microbial system and can be used in many ways.

(*a*) Isolation of mutants, developed spontaneously or through mutagens, is easier in single cell derived colony.

(*b*) Single cell cloning can easily be performed with protoplasts.

(*c*) Somatic cell fusion can only be performed with protoplasts.

(*d*) Genetic transformations through DNA or organelles uptake can be achieved using protoplasts.

(*e*) Complete plants of single cell origin can be obtained, as plant cells are totipotent.

(*f*) It is a unique experimental system to study *de novo* cell wall formation.

(*g*) Protoplasts are very useful objects of membrane studies e.g. membrane transport or uptake processes and membrane structures.

(*h*) Protoplasts are also excellent experimental material for ultrastructural studies e.g., in the study of microtubules formation during cell division.

(*i*). Protoplasts are very suitable for isolation of cell organelles and chromosomes because of absence of thick cell wall barrier.

III. ISOLATION OF PLANT PROTOPLASTS

Protoplasts are isolated by two methods, mechanical and enzymatic. Mechanical isolation method of protoplasts is no more practically used. It is only a historical method. This involves cutting of plasmolysed cells (in fact cell walls) with a sharp blade to release protoplasts.

Isolation of protoplasts

This involves four steps –

(*a*) Sterilization of leaves

(*b*) Peeling off the lower epidermis

(*c*) Incubation in enzyme solution and

(*d*) Isolation and cleaning of the protoplasts.

Healthy leaves are obtained from green house grown plants and leaves are sterilized as described earlier (sterilization of explant). After sterilization lower epidermis is removed with the help of fine forceps and the stripped leaves are cut into small pieces. Flaccid leaves facilitate peeling. *In vitro* grown cultures are aseptic material and therefore, used directly for the isolation of protoplasts. Actively growing cultures in the exponential phase are good material for the isolation of protoplasts.

Usually one gm material is used to obtain protoplast and conditions are optimized empirically by changing enzyme concentration and composition, concentration of osmoticum, duration of incubation and types of tissues.

Procedure

Enzymes – Different enzyme preparations are available in the market but the idea is to combine one middle lamella dissolving and one cell wall digesting enzymes in proper composition to achieve maximum protoplasts release from one gm material. Following enzymes are used –

1. Macerozymes R-10
2. Cellulase – Onozuka R-10
3. Hemicellulase
4. Pectinase
5. Drieselase

A combination of these enzymes in a concentration of 0.5-2% is used. In many cases, only macerozyme and cellulase are sufficient to obtain protoplasts in significant number. The enzyme solution (pH 5.5) is prepared in 10-15% sorbitol or mannitol containing small amount of $CaCl_2$ (7mM) for membrane stability. This solution is sterilized through a membrane filter (cold sterilization) and leaf or callus tissues are placed in it. Petriplates containing tissue and enzyme mixture are sealed with parafilm and incubated for 4-12 hours (sometimes 0.5 to 20 hrs) on a rocking shaker at 24-26°C. After incubation, solution is filtered through a wire or nylon mesh (50-100 mm) to remove debris (undigested cells, tissues, broken cells etc.), transferred into screw capped small centrifuge tubes (sterilized), and centrifuged at 100 g. The protoplasts formed a pellet while the debris in the supernatant is carefully removed. Fresh sterilized sorbitol solution (no enzyme) is added to tube and centrifuged. By repeating the process 2 to 3 times, protoplasts are cleaned (debris is removed).

If 20% sucrose solution is used, protoplasts will float and debris will settle during centrifugation at 200g for 1 min. Floating protoplasts are carefully removed with the help of sterilized pipette and bulked together for further use. Protoplasts are counted by haemocytometer and then diluted to proper strength (number per ml) in the culture medium containing osmoticum Fig. 15.1.

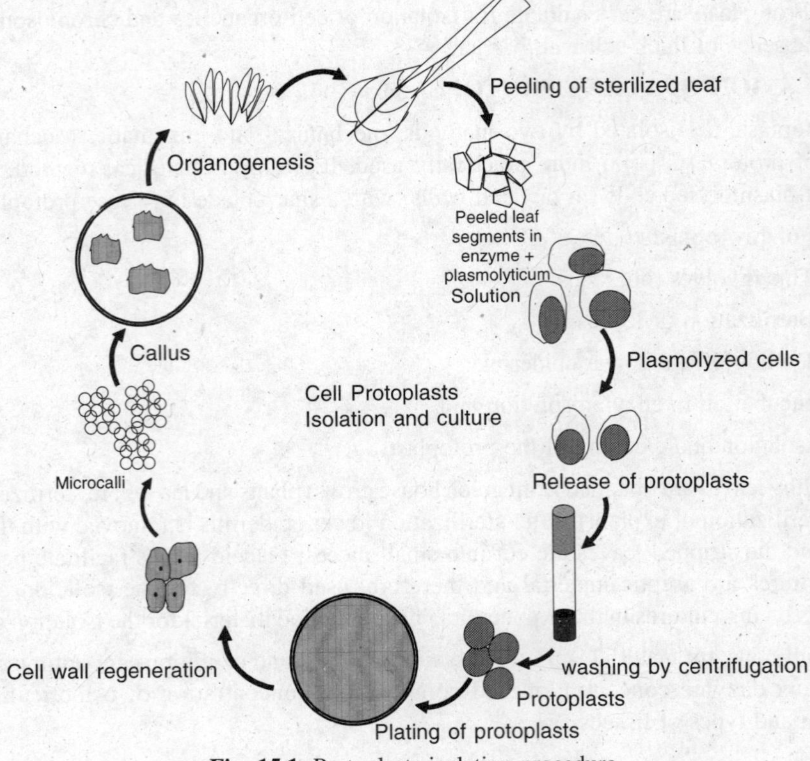

Fig. 15.1. Protoplasts isolation procedure

IV. PROTOPLASTS CULTURE

(i) Osmotic pressure – Isolated protoplasts require somatic protection in the culture medium until they generate a strong wall. Osmolarity in the medium is adjusted to the same level as in the enzyme and washing solution. Prolonged culture at high osmotic pressure can result in browning of the cultures and inhibition of callus growth. The osmolarity of the medium is gradually decreased with cell wall formation and cell divisions. Abrupt decrease in osmolarity may cause bursting of protoplasts and cells and also affects the growth of cells.

(ii) Nutritional requirement – Generally the basic constituents in protoplast culture medium are similar to those used for cell cultures viz. MS or B_5 medium. Some modifications can have beneficial effects on protoplasts survival and colony formation. Usually carbon, nitrogen, vitamins and plant growth regulators are suitably modified. This can be very well illustrated by ammonium ion concentration. The presence of ammonium in the culture medium was found to be toxic to protoplasts of potato, several fruit tree species and some species of composite. It is possible to grow protoplasts at low densities by enriching the medium with vitamins and complex organic substances. The medium used for callus cultures can be used for protoplasts culture of the same species.

(iii) Growth regulators - All the culture media contain a combination of an auxin and a cytokinin to stimulate growth and division. Only a few protoplasts systems can be cultured without an exogenous supply of growth regulators, e.g., *Citrus* and *Petunia* tumour protoplasts. Usually the concentration and combination of growth regulators used for cell culture can be used for protoplasts cultures. However, it is recommended to change growth regulator supply after the first few divisions, as the phytohormone requirement of protoplasts and protoplasts derived cells may be very different.

(iv) Environmental factors. High light intensities often inhibit protoplasts division, therefore, protoplasts are cultured in dim light or darkness. Protoplasts are generally cultured at 24-26°C. A short heat treatment at 45°C for 5 min improved the protoplasts division in wheat, rice and *Pennisetum.* Similarly, low voltage electric pulse treatment also stimulated the division of protoplasts in *Trifolium.* Protoplasts from most species grow well at pH 5.5 to 5.8.

(v) Density. Successful culture of plant protoplasts often depends on the appropriate medium volume per cell ratio (density). The optimal protoplasts density usually has to be experimentally determined. The usual density ranges from 0.5 to 2×10^5 protoplasts per ml of culture medium. Protoplasts of some species, however, prefer much higher plating density ($1\text{-}2 \times 10^6$ cells per ml). In practice, the density of protoplasts derived cells is serially diluted during culture with simultaneous lowering of the osmoticum in the medium.

V. PROTOPLASTS FUSION AND SOMATIC HYBRIDIZATION

In nature, gametic fusion forms zygot with a new combination of genes. Diploid nature is restored by haploid gametic fusion. Sexual crossing is one of the major sources of genetic variability. Incompatibility at various levels, before fertilization and after fertilization, restricts recombination between distantly related species and species specific characters are maintained. A limited success has been achieved in obtaining interspecific and intergeneric hybrids by conventional breeding programmes. These hybrids are useful in further breeding programmes and in the understanding of cytological and genetical behaviour of chromosomes in these hybrids. However, it is not possible to obtain hybrids by conventional method in all the desired plant materials. Protoplasts fusion provides a novel approach overcoming sexual incompatibility and obtained somatic hybrids. Fusion of somatic cells (2n) and production of hybrids is known as somatic cell fusion or somatic hybridization.

A. Membrane properties

Plant membranes are flexible, asymmetric structure of membrane proteins integrated into phospholipid molecules. The fluid mosaic model with lipid bilayer is the established structure of intact membranes. The process of protoplasts fusion requires the direct contact between membrane surfaces and integration of membranes of two different cells. The adhesion of protoplasts depends on surface charge. According to electrophoretic studies negatively charged outer membrane surface. This charge varies from –10 mV to –50 mV and can significantly be reduced by Ca^{2+}. Positively charged protoplasts can be obtained in the presence of poly-L-lysin, poly-L-ornithrine or synthetic phospholipids. After establishment of protoplast – protoplast contact, membrane fusion results by structural modifications and can be enhanced by fusogenic agents. Protoplast fusion is a non-specific event and hence fusion can be induced between any two protoplasts types. Due to these fusion properties of the membranes, protoplast fusion has been achieved between dicot and dicot, dicot and monocots and plant and insect cells (*Drosophila*), amphibian cells and mammalian cells. Protoplast fusion between haploid (gametic) and diploid (somatic) cells has been achieved which clearly demonstrate that fusion is independent of cell type.

B. Fusion Procedure

Protoplasts of both types are mixed in an equal proportion attaining a density of 5×10^4 to 2×10^5 protoplasts per ml. Protoplasts density is usually determined by haemocytometer.

There are several chemical and physical agents which enhance fusion between protoplasts, known as fusogen or fusion agent, e.g., sodium nitrate, polyethylene glycol of different molecular weight (PEG 1500, PEG 4000, PEG 6000), dextran sulphate and gelatin etc. High calcium ion concentration and pH of the incubation mixture have profound effect on fusion percentage. Fusion is stimulated under electric charge as membranes are negatively charged.

Isolated protoplasts are suspended in an aggregation mixture (5.5% sodium nitrate /20-40% PEG in 10% sucrose solution), kept at 35°C for 5 minutes and then centrifuged at 200 g for 5 min to obtain dense pellet. The tube with pellet of protoplasts is once again transferred to a water bath maintained at 30°C for 30 minutes during which most of the protoplasts undergo fusion. The protoplasts are gently washed and transferred to liquid medium or plated. The fusion and fusion products are observed under inverted microscope.

VI. REGENERATION FROM PROTOPLASTS

Protoplasts form cells by regenerating cell wall and subsequently form microcalli by cell divisions. Thus, regeneration from protoplasts is through callus formation. Both, organogenesis or embryogenesis can be obtained as from callus cultures of the given species. In some cases, protoplasts can directly form somatic embryos, e.g., in *Medicago*, *Brassica juncea* and *Citrus.* Regeneration from protoplasts and mode of regeneration are dependent upon the species, genotype, and all the factors governing regeneration in callus or cell cultures. The successful regeneration from protoplasts has been reported in several species. A list of selected species is presented in Table 15.1.

Table 15.1. Plant species regenerated from protoplasts.

Species	Plant source	Regeneration
Cereals. *Oryza sativa* var. japonica	Embryogenic cell suspension	Plant
Zea mays	ECS	Sterile plant
Triticum aestivum	***ECS***	***Plantlet***
Hordeum vulgare	ECS	Albinoplant
Vegetables		
Brassica oleracea var. stalica	Hypocotyl	Plant
Cucumis sativus	***Cotyledon and leaf***	***Plant***
Asparagus officinalis	***Embryogenic callus***	***Plant***
Capsicum annuum	Leaf	Fertile plant
Tuber and root crops *Ipomoea batatas*	***Petiole and stem***	***Plant***
Beta vulgaris	Leaf	Plant
Oil crops		
Brassica napus var. oleifera	Leaf	Plant
Helianthus annuus	***Hypocotyl***	***Fertile plant***
Legumes		
Glycine max	***Immature cotyledons***	***Fertile plant***
Woody trees		
Larix eurolepsis	***Embryogenic cell suspension***	***Plant***
Picea glauca	Embryogenic cell suspension	Plantlet
Populus sp.	Leaf	Plant
Prunus avium	***Cell suspension (root)***	***Plant***
Coffea cansphora	Somatic embryo	Plantlet
Ornamentals		
Rosa sp.	Embryogenic cell suspension	Plantlet
Pelargonium sp.	Cell suspension (leaf, petiole)	Plantlet
Chrysanthemum sp.	Cell suspension (leaf)	Plant

VII. Fusion Methods. While several methods were proposed to induce protoplasts fusion, earliest success was achieved using sodium nitrate (Carlson et al, 1972). Plant protoplasts have uniformly negatively charged membranes. Divalent cations, especially Ca^{2+}, modify the electrophoretic mobility of protoplasts. Aggregation mediated by neutralization of plasmalemma negative charges by 50 mM Ca^{2+} mixed with 0.4M mannitol in a solution of high pH (10.5) has led to the genuine cytoplasmic fusions of tobacco protoplasts and *Petunia* protoplasts. The other fusogens are as follows.

(i) PEG – Various fusogenic agents have been used for getting protoplasts fusion. The discovery of the powerful aggregation effect of polyethylene glycol (Kao and Michayluck, 1974) provided the awaited tool. A few minutes after being immersed in a 20-40% (w/v) solution of PEG (1500 to 6000 mw) virtually all protoplasts exhibit adhesion. However, actual fusion occurs upon dilution of PEG. A combination of PEG induced adhesion followed by elution with PEG, high Ca^{2+}, high pH was even more reliable. Actual role of PEG in the process of membrane fusion remains yet to be clearly determined. Addition of dimethyl sulphoxide (DMSO) with PEG has stimulatory effect on protoplast fusion. It has been suggested that bridge formation mediated by Ca^{2+} and PEG through its slightly negative polarity might be the cause of PEG induced fusion of protoplasts.

(ii) Dextran – Similarly, fusogenic properties of dextran and of polyvinyl alcohol suggest that weak non-ionic surfactant properties be involved. It seems perturbation of negative charges of the protoplast plasmalemma seems essential. This is illustrated by fusion induced by a positively charged phospholipid and above all electric field induced fusion.

(iii) Electrofusion – In electrofusion method of Zimmerman (1982) protoplasts are brought in close contact by a non-uniform alternating field between two electrodes, then fusion is initiated by field pulse of high voltage (500 V – 3 KV, cm-1) for a very short time (10-100 micro seconds). This is the most widely used apparatus (Fig15.2) and method in the studies on membrane fusion and for the production of somatic hybrid colonies or plants. Electrofusion is a biophysical process that involves two steps. In the first step the protoplasts are exposed to alternating electric field (0.5-1.5 MHz) that generates dipoles through dielectrophoresis. In this non-uniform field, the protoplasts acts as dipoles, move in the direction of the increasing field as a function of the field strength, cell size and the mobility of charge within the protoplast and in the surrounding solution. As a result of mutual dielectrophoresis in non-uniform as well as uniform field a protoplasts clump, and bind together to form the chain. The second step for electrofusion is the application of one or more short (10-100 μs) direct current (DC, 103 KV/cm) pulses that cause reversible membrane breakdown. This results in pore formation in the membranes touching to each other. This way contacting membranes can be fused and this fusion opens the way for hybrid cell formation.

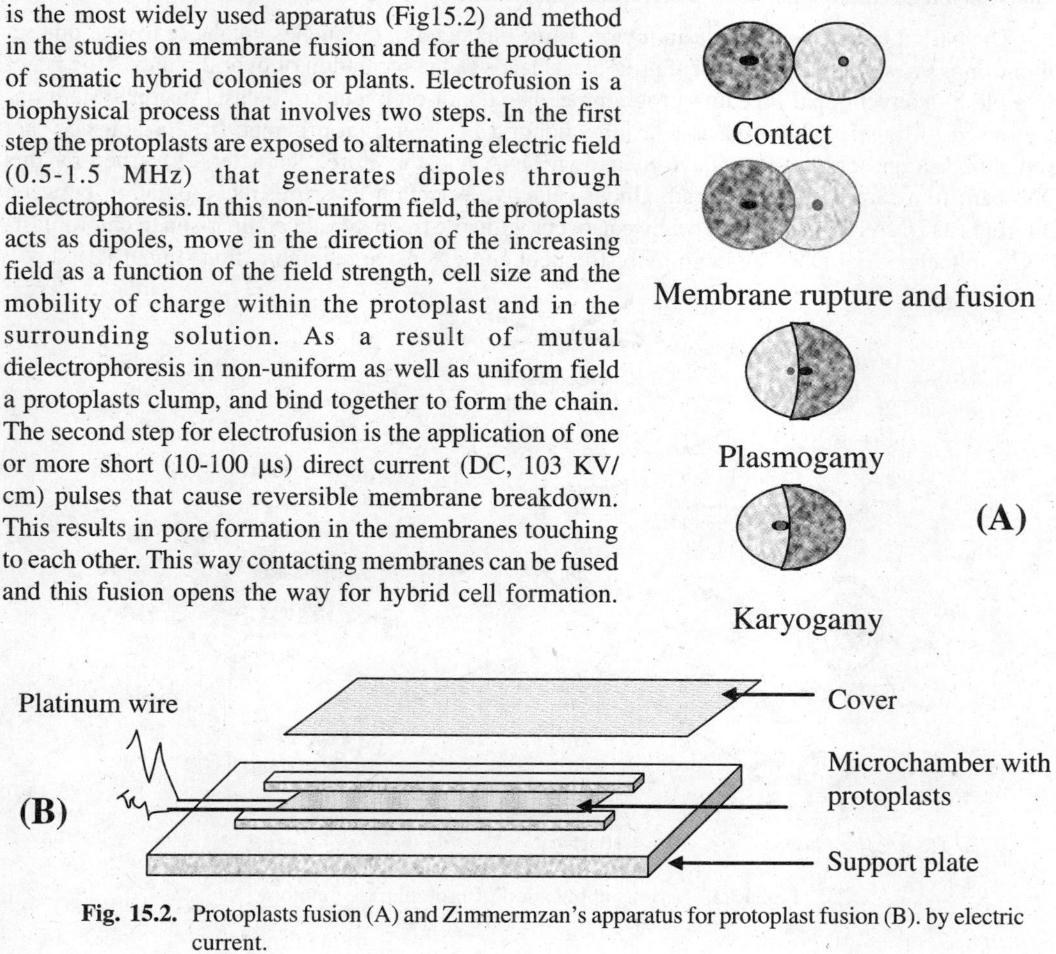

Fig. 15.2. Protoplasts fusion (A) and Zimmermzan's apparatus for protoplast fusion (B). by electric current.

Generally several cycles of AC and DC current are given to increase the fusion frequency.

The fusion chamber is a small apparatus having two parallel running platinum wires for passing the current (Fig.15.2). Protoplasts suspension is placed in this microchamber having platinum electrodes (wires). This can be placed on inverted microscope and fusion process can be observed simultaneously (Fig.15.2). PEG fusion efficiency is given as the best 5-30% heterospecific fusion products under optimal conditions. The electric fusion method appears very promising, as its efficiency is very high.

VIII. APPLICATIONS OF PROTOPLASTS HYBRIDIZATION (somatic cell hybridization)

There are three major applications –

(*a*) Production of somatic hybrids and cybrids
(*b*) Asymmetric hybrids
(*c*) Direct DNA uptake

(a) Hybrids and cybrids production

Protoplasts fusion technique during 1970s established that parasexual hybrids can be produced by this method. Intra and inter-specific hybrids have been produced by this method in several Solanaceous plants like *Datura, Petunia* and *Nicotiana* species. By this method complete hybrid plants of intermediate characters were produced. Mostly amphidiploid plants were obtained.

The partial transfer of cytoplasm (cytoplasmic characters), organelles uptake or loss of one set of chromosomes of one of the parent protoplasts leads to the formation of cybrid plants. This is not possible by conventional breeding programme. The donor-recipient protoplasts fusion system has been used to transfer chloroplasts and mitochondria in several crop plants. It is established that mitochondria and chloroplasts have their own DNA and therefore, characters governed by this DNA are transferred and expressed. Under selective selection pressure, recombination between chloroplasts DNAs from different origin can occur with low frequency. In comparison to chloroplasts DNA, mitochondrial DNA undergo more frequent and extensive recombination (Fig. 15.3).

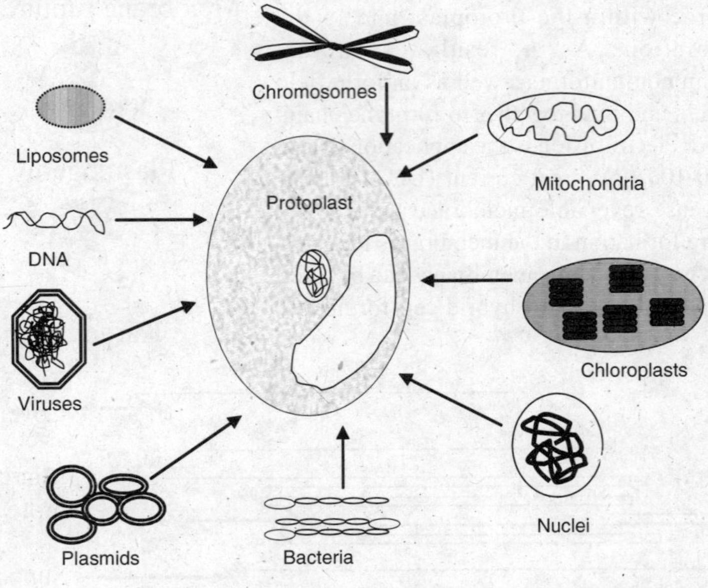

Fig. 15.3. Various application of protoplast technology

(b) Asymmetric hybrids Production

Because of physico-chemical nature of fusion, hybrid cells could be produced between distantly related taxa irrespective of their phylogenetic relationship. In majority of cases, these wide cell hybrids (hybrid of distantly related parents) showed instability in their chromosome constitution and to regenerate plants, e.g., in hybrid cells obtained in fusion between soybean x *Nicotiana glauca*, soybean x *N. tabacum*, carrot x tobacco, carrot x rice, *Duboisia* x tobacco and soybean x rice etc.

In some intergeneric fusion combinations, hybrid plants with additive chromosome number were regenerated e.g., after fusion between potato and tomato, *Arabidopsis* and *Brassica*, *Datura innoxia* and *Atropa balladonna*, *Lycopersicon esculentum* and *Solanum lycopersicoides*, *Eruca sativa* and *Brassica juncea* and *Lycopersicon esculentum* and *Solanum nigrum*. In some of the cases, two parental diploid genome coexisted in the cybrid cells (amphidiploid). However, morphological abnormalities and sterility occurred in several of these intergeneric hybrids.

(c) Direct DNA and macromolecules uptake

Cell membranes are semi-solid structures as per the fluid mosaic model and fluidity of membrane is physiological characteristic. Transient aqueous pores in the bilayer portion can make the membrane permeable for macromolecules. The formation of pores and repair of membrane pores (sealing) opens the new approach for introduction of nucleic acid molecules into cytoplasm and the nucleus. The reversible, non-destructive permealization of membranes can be achieved by both the chemical treatments and action of short duration high electric pulses. Uptake of DNA molecules into a variety of plant protoplasts was successfully promoted by application of PEG treatment or electroporation.

Plant protoplasts have the remarkable property to take up particles and macromolecules, like polystrene, latex particles, ferritin, proteins, DNA, viruses, bacteria or even whole chloroplasts and nuclei can be transplanted. Mayo and Cocking (1969) reported that the uptake is by pinocytosis, and demonstrated this by differential staining effects of phosphotungustic acid on the plasmalemma.

(i) DNA uptake – The discovery that *Diplococcus pneumoniae* transforms itself by taking up exogenous DNA into its genomes has made genetic engineering a routine task in bacteria. New genes can be synthesized and DNA-RNA hybrids can be obtained.

DNA can now be incorporated into the higher plant cells protoplasts. Doy et al (1972-73) transformed haploid tomato cell cultures by phage lambda (bacterial virus). Transformed cells can utilize galactose and lactose as carbon source for their growth. This is one of the approaches for obtaining transgenic plants (Table-2). The transient expression of transferred gene in a few days permits the use of this technique for DNA uptake studies. It is evident from the data in the table that various marker genes have been used for transient expression. Production of transgenics requires the stable integration of the introduced DNA molecules into the genome of recipient species used as protoplasts source. The use of various reporter genes as chloramphenicol acetyl transferase (CAT) and β-glucoronidase (GUS) also help in the identification of transformed cells (colonies). A reporter gene produces phenotypic expression (e.g. colour) to confirm the transfer of gene. This way, integretion levels can be detected early in the transgenic cells.

(ii) Virus uptake – E.C. Cocking reported that tomato fruit protoplasts take up TMV particles by pinocytosis. This system has been used to demonstrate mechanism of infection and nucleic protein replication.

(iii) Bacterial uptake and nitrogen fixing – Nitrogen fixing bacteria *Rhizobium japonicum* can be introduced into the pea leaf protoplasts during enzymatic digestion of the cell wall. This uptake occurs by invaginations of the plasmalemma during plasmolysis. Similarly, Davey and coworkers obtained protoplasts from root nodules cells packed with bacteria and try to fuse it with non-legume protoplasts. All these attempts to introduce nitrogen fixing genes/properties to non-legumes were not succesful.

(iv) Chloroplasts transplantation – Carlson incorporated the non-mutant tobacco chloroplasts into the cytoplasm of an albino protoplasts. Whole plant was regenerated later on. This may lead to improvement of plants for efficient photosynthesis.

(v) Transplantation of nuclei – After smaller particles, nuclei were tried and they could be incorporated through plasmolemma without bursting it. Potrykus and colleagues transplanted nuclei from *Petunia hybrida* into protoplasts of *Zea mays*, *Petunia hybrida* and *Nicotiana glauca*. Further development of this technology led to the development of cloned sheep and monkey by nuclear transplantation.

Table 15.2. Transgenic plants produced through DNA uptake into protoplasts.

Species	Promoter/gene	Method	Regeneration
Oryza sativa	Promoter and the first intron of the alcohol dehydrogenase gene of maize/-glycuronidase gene.	Polyethylene glycol	Plant
	Cauliflower mosaic virus 35S gene promoter/neomycin phosphotransferase gene.	Electroporation	Plant
	-do-	--	-
	Cauliflower mosaic virus 35S gene promoter (35S) hygromycin phosphot ransferase gene (H+P)		
Zea mays	Promoter and the first intron of the alcohol dehydrogenase gene of maize/β-glucovonidase gene.	Electroporation	Sterile plant
Citrus sp.	Mannopine synthase gene promoter of Agrobacterium tumefaciens (chloramphenicol acetyl transferase gene	PEG PEG	Plant Plant
Solanum tuberosum	Nopaline synthase gene of A. tumefaciens/ neomycin phosphotransferase gene (neo.)		
Lactuca sativa	35S/neo	Electroporation	Fertile plant
Arabidopsis thaliana	35S/hph	PEG	Fertile plant

IX. IDENTIFICATION AND SELECTION OF HYBRID CELLS

Following fusion treatment, the protoplasts in liquid culture medium regenerate cell walls and undergo mitosis resulting in a mixed population of parental cells, homokaryotic fusion products and heterokaryotic fusion products or hybrids. Hybrid cells must be distinguished from the other cells present. Heterokaryotic recognition techniques as well as colonies selection methods are thus highly desirable. For this purpose, several methods are available. These are dependent on –

(*a*) Physical properties of fused cells.

(*b*) Biological properties of fused cells.

(*c*) Biological properties of colonies derived from fused cells. Following methods have been adopted for various types of protoplasts.

(*a*) Fusion partners such as non-green protoplasts (albino protoplasts from tissue culture) and leaf derived chloroplastic protoplast allow microscopic visual detection of heterokaryotes. A few hours or a few days after fusion products are pipetted and cultivated individually. Because visual techniques applied are tedious, other non-visual methods have been sought.

1. By centrifugation in density gradient, enriched bands of fusion products can be recovered at an intermediate density.

2. Vital staining, specially using fluorescent dyes is another possibility to detect heterokaryotes providing a means of automatic separation with cell sorter.

3. Controlled fusion by electric field should provide a possible way to cultivate only true heterokaryotes.

4. A method adapted from animal cells to plant protoplasts allows the recovery of fusion products by complementation between inactivated protoplasts by different metabolic inhibitors. Cybrids as well as hybrids selection *in vitro* should be even more effective by use of complementary markers selectable *in vitro*.

5. One of the principal features to characterize somatic hybrid is the chromosome complement. Chromosome number and morphology may be compared to the patterns displayed by the fusion parents to determine whether or not presumptive hybrids are true hybrids.

6. Supportive evidences for hybridity can be obtained from an electron microscopic morphometric determination of chromatic texture in hybrid nuclei. This technique was used for the verification of *Arabidopsis-Brassica* somatic hybrids and it may prove useful for identifying other interspecific and intergeneric hybrids.

7. Morphological characters of the hybrid plants and parents, are used to ascertain the hybrid nature of the regenerated hybrids.

8. Isoenzymes-proteins ultimately manifest the diversity laid down in the genetic programme of each individual. Isoenzymes (multiple molecular forms of an enzyme exhibiting identical and similar catalytic properties) are frequently used for this purpose. Isoenzymes used for hybrid identification include esterase, peroxidase, amylases, phosphatase, alcohol dehydrogenase, lactate dehydrogenase, aspartate and aminotransferase etc.

9. More recent techniques are – RFLP (Restriction Fragment Length Polymorphism), use of reporter gene and RAPD (Random Amplified Polymorphic DNA) for looking into genetic variability and transformations in the plants regenerated after hybridization or DNA uptake studies.

X. Stability of hybrids

Unfortunately somatic hybrids of plants display chromosomic instabilities. Even for sexually compatible species, regenerated somatic hybrids exhibit quite variable chromosome numbers relative to normal parental additions. Nevertheless, it has been shown in several cases that hybrids are capable of morphogenesis, although resulting plants are abnormal and sterile.

Table 15.3. Transfer of useful agronomic traits by protoplasts technology.

Species	Agronomic trait transferred
Brassica oleracea	C3-C4 photosynthesis intermediate trait
Raphanus sativus	C3-C4 photosynthesis intermediate trait
Citrus reticulata	Virus resistant trait
Citrus sinensis	*Phytophthora* and tristeza virus tolerance
Solanum melongena	Bacterial and fungal wilt resistance
Solanum tuberosum	resistance to *Phytophthora*

Protoplasts cultures have their own advantages and disadvantages. The problems faced by scientists in recovering hybrids, regeneration of hybrids, sexually viable hybrids etc. are the main difficulties and disadvantages with the technique. Therefore, other methods of transferring genetic material have been developed. Therefore, protoplast culture is being used selectively for specific purposes for gene transfer.

QUESTIONS

1. Describe the technique of protoplasts isolation , culture and regeneration with the help of suitable diagram.
2. Describe the protoplast technology for crop improvement.
3. What is somatic cell hybridization? Describe the procedure application and stability of somatic cell hybrids
4. Write short notes on the following :
   (*a*) Protoplast culture
   (*b*) Fusion agents
   (*c*) Somatic hybridization
   (*d*) Selection of hybrid cells
   (*e*) Application of somatic hybridization.

HAPLOIDS

CHAPTER 16

Among seed-propagated species, the creation of a new variety necessarily requires several inbred sexual generations to obtain homogeneity. Usually, about 12 inbred generations are required.

The new varieties can be registered in official catalogues and marketed only if they exhibit the following characteristics:

- they have to be of higher performance than the official controls;
- they must be identified and different from former varieties;
- they have to be homogeneous and stable.

This is a long process via breeding programmes, never perfect because a complete homozygosity cannot be reached, and the number of selected characters is limited.

The discovery of viable haploid plants and doubling of their chromosomes meant that homozygous plants could be obtained in one generation. It should be clear that in an angiosperm plant, microspores and embryo sac are haploid structures, and any resultant product obtained from these would be haploid. A brief reproduction in angiosperm is given in Fig. 16.1 showing haploid stages. Fertilization of egg, fusion with one male gamete, results in zygote (2n) formation. Therefore, haploid cells are used (before fertilization) to obtain haploid plants, but dissection and isolation of egg is difficult as compared to pollen grains.

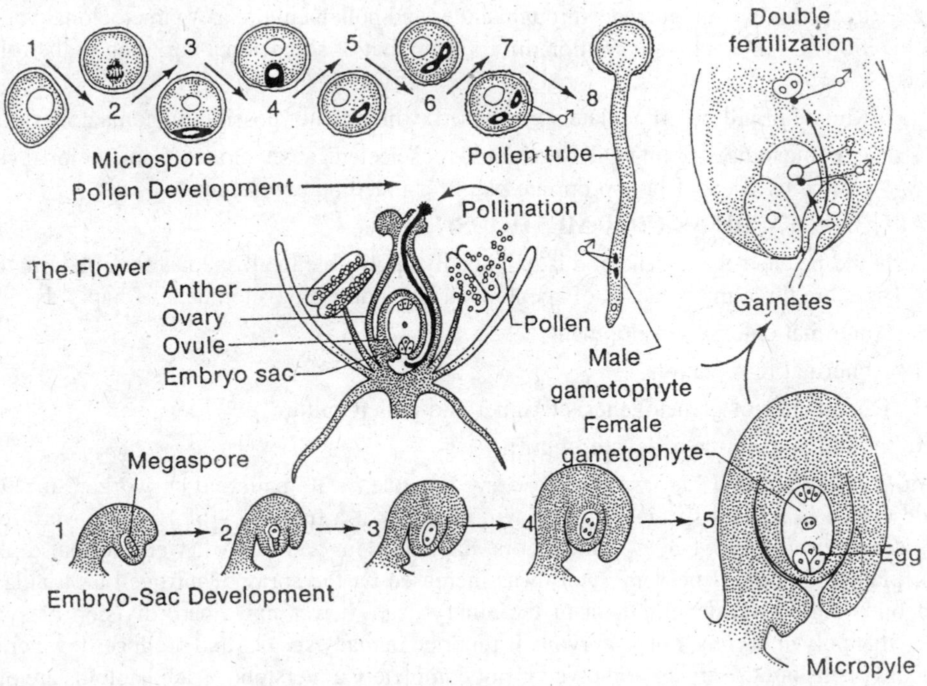

Fig. 16.1. Development of microspore and megaspore in an angiosperm, haploid and diploid stages, and fertilization.

Haploid (plant): Plant, which has gametic chromosome number (n).

Double-haploid (D.H. plant): Originally haploid plant which has recovered the normal chromosome number of the species spontaneously via endomitosis or after colchicine treatment.

Dihaploid (plant): Haploid plant from a tetraploid species.

Parthenogenesis or Gynogenesis: Spontaneous or induced production of a maternal haploid plant from female gametophyte.

Androgenesis: *Sensus stricto*= Production of haploid plant from the male gametic nucleus inside the embryo sac.

Sensus largo= Production of a haploid plant from *in vitro* culture of the male gametophyte.

Gametophyte: Tissues or cells involved in gametic cell formation through meiosis event (germ cells).

I. HOMOZYGOUS LINES

The chromosome number of haploid plants is readily doubled by using colchicine; the plants so obtained are called Doubled Haploids (DH). DH plants are completely homozygous and can be produced in one to two years. Homozygous lines are regularly produced during breeding of both self and cross- pollinated crops, usually in six years. Thus DH lines are economical in terms of time, labour and space. These homozygous lines are used in breeding programmes of rice, wheat, barley, *Brassica* sps, tobacco, potato and rye etc. in Japan, China, Canada and Europe. These breeding programs resulted in many commercial varieties.

Gametoclonal Variation:

The term somaclonal is widely used and adopted for tissue culture generated variation in the regenerants. The term is also used to describe variation developed during:

1. Protoplast culture as "protoclonal variation " and
2. Regeneration from gametes through anther and pollen culture as "gametoclonal variation".

The utilization of gametoclonal variation differs from that of somaclonal variation in the following respects:

(a) Mutation will be immediately expressed, which is not possible in somaclones.

(b) Chromosome doubling will be required for selected gametoclones. Gametoclonal variation can be used to select appropriate plant for breeding programme.

II. HAPLOIDS BY DIFFERENT METHODS:

It should be clear that no method is universally applicable to all species for obtaining haploid plants. There are three main causes of spontaneous or induced development of haploids:

1. Abnormal embryo development
2. Abnormal fertilization
3. Disturbance of gametogenesis (Auther and pollen culture)

1. Abnormal embryo development:

(a) Polyembryony: This was first observed in rice by Ramaih and co-workers in 1933 and later described in several species. In rice twin plantlets at a frequency of 10^{-2}-10^{-5} were recorded during seed germination. 1-30% contains one haploid plant and the rest were normal diploid. In pepper plants, the rate of polyembryony was increased by the spray of nitrogen protoxide on the flower buds during the development of the embryo sac. It was explained, division of synergids induced that polyembryony. Polyembryony is reported in many species and produces low percentage of haploid plants but the process involved is not completely understood. Such haploids are obtained in cocoa and coffee. The rate of haploid formation was too low to be used as a tool for crop improvement.

(b) Chromosome elimination: A large number of barley haploids can be obtained by interspecific hybridization between *Hordeum vulgare* (cultivated species) and *H. bulbosum* (wild species). Whatever the direction of cross, the haploids recovered *H. vulgare* genome. Cytological investigation demonstrated that fertilization is normal but the chromosomes of *H. bulbosum* are eliminated during the first mitosis of the embryo (Fig. 16.2). These embryos cannot continue their complete growth on the mother plant because of the lack of albumen. They have to be grown up in tissue culture conditions (known as embryo rescue). It is suggested that asynchrony in the mitotic cycles of the two species and/or aberration in the fixation of *H. bulbosum* chromosome to the protein of the achromatic spindle of the hybrid is the cause of loss of chromosomes. The other example of obtaining haploid by this method is wheat. When wheat is crossed, either with maize or *H. bulbosum*, haploid wheat embryos are obtained. After embryo rescue, haploids are diplodized with colchicines and homozygous plants are used for wheat breeding.

2. Abnormal Fertilization: *In situ* parthenogenesis and androgenesis are achieved when the nucleus of only one gamete, female (parthenogenesis or male (androgenesis) is involved in the embryo formation. In both the situations the cytoplasmic organelles are those of female gametophyte. This has been shown using nucleic markers linked to the cytoplasm, e.g. male sterility trait in maize and in Petunia. The nucleus of the male gametophyte may be treated with chemical or physical agents to prevent the formation of two functional generative nuclei, e.g. use of toluiedene blue in Poplar (Populus).

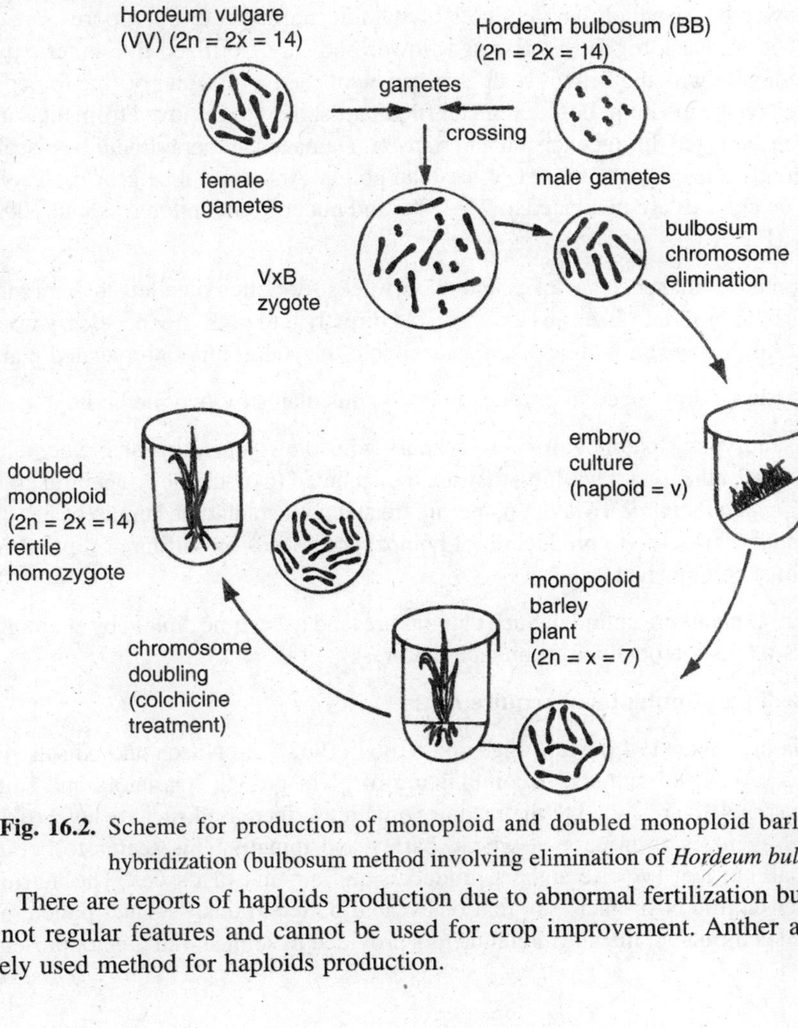

Fig. 16.2. Scheme for production of monoploid and doubled monoploid barley plants using wide hybridization (bulbosum method involving elimination of *Hordeum bulbosum* chromosomes).

There are reports of haploids production due to abnormal fertilization but these observations are not regular features and cannot be used for crop improvement. Anther and pollen culture is widely used method for haploids production.

3. Anther and Pollen culture: Haploid plants have the gametophytic number of chromosomes, a single set (n) of chromosomes in the sporophyte. These are of great significance for plant improvement, useful in mutations and for the production of homozygous plants. Conventional breeding programmes are tedious and are not very efficient for the production of homozygous lines or haploids. These methods are not very successful in cross-pollinated crops for obtaining homozygous plants. Use of anther and pollen cultures for the induction of haploids by Guha and Maheshwari in 1964 at Delhi University opened new avenues for obtaining homozygous plants in short duration and in large number for breeding programmes.

(a) Anther culture. In 1953 Tulecke, for the first time observed that mature pollen grains of a gymnosperm *Ginkgo biloba* can be induce to proliferate in culture to form haploid callus. This work demonstrated that haploid callus can be produced from highly differentiated cells like pollen instead of normal process of pollen germination. It was not untill 1964 that Sipra Guha and Satish C. Maheshwari reported the direct development of embryoids from mirospore of *Datura innoxia* by the culture of excised anther. Latter Bourgin and Nitsch in France (1967) obtained complete haploid plants of *Nicotiana tabacum*. Since then intrest in the production of haploids has become a major area of research interest and haploid cultures have been produced in a large number of species of diverse taxa.

(b) Procedure. Plant tissue culture technique for obtaining haploid cultures is simple and efficient. Closed flower buds, which have anthers containing uninucleate micropores, are most suitable for the induction of androgenosis. Excised flower buds are sterilized like other explants and cut open longitudinally with the help of scalpel or can be opened with a pair of fine forceps and anthers are transferred on the medium. Before transfer filaments should be removed from the anthers. Anthers should not be damaged during excision and transfer. Damaged anthers should be discarded, as they often tend to produce callus from parts other than pollen. Anthers can be grown on solid or in liquid medium. The cultures are incubated at 24-27 °C and under illumination of about 2000 lux for 12-16 hours/day (Fig. 16.3).

Depending upon the plant species it lakes about 1-8 weeks for pollen plantlets to emerge from the anthers. Plantlets of appropriate size can be transferred directly into pots after necessary washing and to remove agar. All the care should be taken as desirable for young micropropagated plants.

Haploid plants can be diplodized to produce homozygous plants by two methods:

(a) By treating the young plantlets for 24-48 hours with 0.5% aqueous colchicine solution while still attached to the anther. After washing they are transplanted (Colchicine is a neutral alkaloid obtained from *Colchicum antumnale*). Mature plants are treated with colchicine lanolin paste (0.4%) to obtain diploid shoots. This leads to production of homozygous seeds by selfing or diploid callus can be obtained from such branched.

(b) Haploid stem explants are cultured. Such cultures are tand to become diploid by endomitosis. Diploid homozygous plants are obtained from such cultures.

(c) Culture media and nutritional requirements

Basal media of White (1943) Murashige and Skoog (1962) and Nitsch and Nitsch (1969) are used for cultures of excised anthers. A combination of plant growth regulators and suitable concentration of sugar (2-4%) are added to the media. Sometimes, increased sucrose level (6-12%) is beneficial for androgenesis as in case of wheat, barley and tomato. This seems to be act as osmoticum rather than nutrient because anthers produce small amount of tissues. The nutritional requirement of anther culture is simpler than that of isolated pollen grains. Isolated pollen grains are highly specialized cells and all the nutritional factors provided by anther wall are not present for isolated pollen grains. Therefore, pollen culture may require additional nitrogen, particularly in the form of amino acids (organic reduced nitrogen source). L-glutamine is the most frequently used amino acid for haploid production from anther culture of potato, *Brassica napus* and *Triticum aestivum*. Today the culture of isolated micropore is possible for many crop plants like tobacco, rice, rapeseed, barley, wheat, potato and Maize in case of a *Nicotiana tobacum* variety incorporation of 2% charcoal in the medium increase the response of anthers from 41 to 91%.

(d) Androgenesis

There are several pathways of androgenesis which leads to the formation of haploids either directly by embryogenesis or indirectly through callus formation (Fig. 16.1). From from breeding point of view direct embryogenesis is preferred over indirect embryogenesis involving callus formation which may produce variation. There are several hundred plant species in which haploid callus or complete plantles have been produced. A few selected species are listed in the table 16.1.

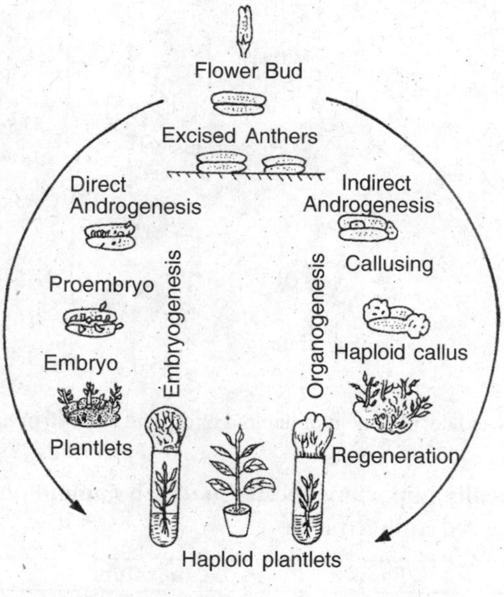

Fig. 16.3. Process of anther culture and androgenesis leading to haploid plants production.

In nature, pollen mother cells (P.M.C) give rise to pollen tetrads through meiotic division. These microspores are released in the atmosphere. The first mitotic division in the microspores produces a large vegetative cell and small generative cell. The former remains quiscent while the later divides to form two male nuclei or sperms. In culture although androgenesis can be induced in anthers in the tetrad stage at the binuclear pollen stage. However, microspores just before or at the time of first mitosis, are most suitable at the induction of androgenesis. Some of the modes of androgenesis are presented in (Fig. 16.4) on the basis of results obtained with various species, notably *Datura*, tobacco and barley. Androgenesis may be produced by pollen grains through mitosis producing equal or unequal nuclei. In all cases, they may give rise somatic embryos directly or may produce callus. Direct androgenesis, behaviour of microspores as zygote leading to embryo formation, has been observed in *Atropa belladonna, Datura innoxia* and *Nicotiana tabacum*. Indirect androgenesis is the most common method (96%) of producing haploids this type of development (indirect androgenesis) may be due to disturbed polarity and / or complex nutrient medium.

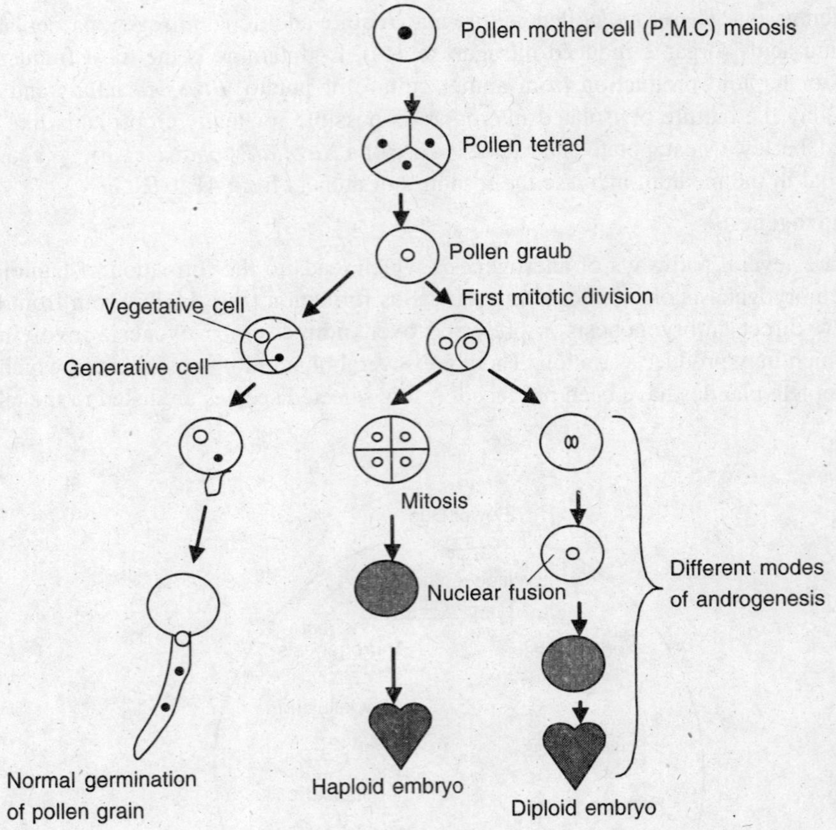

Fig. 16.4. Microsporogenesis and development of haploid or diploid embryo or normal germination of pollen grain.

Table 16.1. Some economically important species in which haploid plants and calli have been obtained by excised anther/pollen.

Plant species	Family	Inoculum	Response
Asparagus officinalis	**Liliaceae**	**Anther**	**Callus, plant**
Brassica oleracea x B. alboglabra	Cruciferae	Pollen	Callus
B. napus	Cruciferae	Anther	Callus, embryos
Capsicum annuum	Solanaceae	Anther	Callus, embryos
Coffea arabica	Rubiaceae	Anther	Callus, embryos
Hordeum vulgare	Graminae	Anther	Callus, plants.
Lycopersicum esculentum	Solanaceae	Anther	Callus, plants
Nicotiana tabacum	Solanaceae	Anther, pollen	Embryos, plants
Oryza sativa	Graminae	Anther	Callus ,embryos, plants
Petunia hybrida	Solanaceae	Anther	Callus, embryos, plants.
Secale cereale	Graminae	Anther	Callus, embryos, plants
Solanum melongena	Solanaceae	Anther	Callus, plants
S. tuberosum	Solanaceae	Anther	Embryos, plants.
Triticum aestivum	Graminae	Anther	Callus, embryos, plants.
Zea mays	Graminae	Pollen	Embryos

(e) Stability of haploid cultures

Like plant tissue cultures of diploid plants, cultures of haploid origin are difficult to maintain with genetic stability. Somaclonal variation occurs during long-term culture as well as, cultures tend to become diploid. There is continuous increase in the proportion of diploid cells in the haploid culture. Use of specific chemicals like para-fluoro phenylalanine has been tried to enhance haploid calllus growth and maintenance of haploid nature. Therefore, haploid cultures are heterogeneous population of haploid, diploid and polyploid cells and there is a continuous decrease in percentage of haploid cells in this mixture with increase in culture period. Such unstability has been reported in *Atropa belladonna, Nicotiana tabacum* and *Azadiracta indica.*

(f) Utilization of haploids in agriculture and tree breeding

Haploids are useful for two reasons: (1) the presence of one set of chromosome facilitates the isolation of mutants, and (2) isogenic diploids can be obtained by chromosome diplodization. Isogenic lines are defined as individuals that posses the same genotype irrespective of their homo or heterozygous nature. In classical plant breeding this means that vegetatively propagated clones are isogenic and lines of inbreeders become isogenic after repeated self fertilization. In outbreeders, isogeny is achieved in hybrid varieties. The creation of the homozygous lines from which hybrids are produced, normally requires special breeding techniques and take long time (several years).

By anther culture technique, haploids can be obtained in a few weeks and by doubling their chromosome number homozygous diploids can be produced in a single generation. These fertile homozygous plants can be used for producing the inbred lines required for hybridization programme. The advantage with these quickly produced homozygous diploid plants is that they are stable and do not lose further desirable characters.

A. Haploid production

Production of haploids and their importance in plant breeding have been established. Attention has been focused on how to improve the frequency of haploid formation and their best utilization for the improvement of plants. There are several alternative methods for haploid production. These methods have been tried by different workers in different species viz., Parthenogenetic development of egg cells into ovule, chromosome deletion, delayed pollination, temperature shocks and regeneration from the callus have all been tried. Haploids in *Triticum dicocum* and *Triticum persicum* have been produced by x-rayed pollens. However, there is no comparison of these methods with androgenesis in tissue culture. Later is the best convenient and rapid method for production of large number of haploids in many taxa.

B. True breeding lines

The usefulness of haploids in obtaining true breeding diploid lines by the application of chromosome doubling technique, finds its practical utility in the production of new varieties of crops by anther culture of monosomic tobacco plants, it has been possible to isolate nullihaploids, and regeneration of nullisomic would serve as a tool for studying linkage relationships.

C. Virus free plants

Selection of disease resistant geranium plants free from virus infection has been obtained through anther culture. Haploid cell lines have provided a basis for recovering defined biochemical and temperature sensitive mutants as there is direct impression of affected gene.

D. Experimental embryology

In plant breeding, haploids could be exploited to advantage in studies concerned with experimental embryogenesis, plant breeding and cytogenetics, because of the case with which homozygous cultivars could be obtained in shorter times following diplodization by colchicine. Several double haploid lines have been selected for improved yield, early flowering and increased disease resistant.

E. Mutation breeding

To induce mutation, young plantlets emerging from anthers are subjected to gamma radiation (1500 – 3000 rads) and following selection mutants are easily selected as all changes in haploid genome are directly expressed. Similarly Chemical mutagens and x-rays can also be used. Therefore, haploid plants are very important to produce and detect mutants as all changes are directly expressed. By diplodization stable diploid are produced. These are of great economic importance in case of cereals, particularly, self-incompatible plants like rye. By conventional breeding it is just not possible to obtain homozygous lines.

F. Vegetative propagation crops

In vegetatively propogated crops, like potato, haploids are very useful for breeding programme. They are difficult to produce by conventional breeding programme. Fertile diploid and tetraploids are produced in potato by anther culture and tetraploids were also used for breeding.

G. Long duration crops

In long duration crops, like coffee, haploids and homozygous diploids are of great importance to find out resistant plants and once resistant plants propagated, they show stable resistance as they are homozygous for that character.

H. Tree breeding

Long term haploid callus cultures were first initiated from *Taxus brevifolia* pollen in 1955 in the lab of professor C.D. La Rue. Application of haploidy in the improvement of woody species is being attempted but progress has been slow. Endrogenic haploids have been obtained in following genera *Albizia*, *Populus*, *Camellia*, *Hevea*, *Prunus*, *Ulmus*.

Later on Tulecke (1953) established haploid cultures of *Ginkgo biloba* in same laboratory but no regeneration was achieved. Trees are difficult to breed as they have long vegetative phase. Therefore, haploid production through anthers and pollen culture provide an excellent system to produce homozygous/ haploid plants.

QUESTIONS

1. Write short notes on the following :
   (*a*) Haploids
   (*b*) Anther culture
   (*c*) Homozygous lines
   (*d*) Androgenesis.

2. Describe the technique and applications of anther and pollen culture for the production of homozygous lines.

3. Discuss the role of haploids in crop improvement.

SECONDARY PLANT PRODUCTS

CHAPTER 17

I. CLASSES AND PRODUCTION (PRIMARY & SECONDARY PRODUCTS)

A plant cell produces two types of metabolites: primary metabolites involved directly in growth and metabolism, viz., carbohydrates, lipids and proteins, and secondary metabolites considered as end products of primary metabolism and are in general not involved in metabolic activity, viz., alkaloids, phenolics, essential oils, steroids, lignins, tannins etc. Primary metabolites are produced as a result of photosynthesis and these products are further involved in the cell component synthesis.

In general, primary metabolites obtained from higher plants for commercial use are high-volume, low-value bulk chemicals. They are primarily used as industrial raw materials, foods or food additives, examples: vegetable oils, fatty acids (used for making soaps and detergents), and carbohydrates (sucrose, starch, pectin and cellulose). These materials cost Indian Rs. 15-150 per kg (or US $ 0.5 to 4 per kg) and are readily available in large quantities. However, some primary metabolites such as myoinositol and β-carotene are expensive because their extraction, isolation and purification are difficult.

Secondary metabolites are compounds biosynthetically derived from the primary metabolites but more limited in occurrence in the Plant Kingdom, and may be restricted to a particular taxonomic group (genus, species or family). As mentioned above, secondary metabolites are mostly accumulated by plant cells in smaller quantities than primary metabolites. These secondary metabolites are synthesized in specialized cells at particular developmental stages making their extraction and purification difficult (compared to the primary product produced by the whole plant or organ).

The medicinal plants rich in secondary plant products are termed as 'medicinal' or 'officinal' plants. These secondary metabolites or products exert in general a profound physiological effect on the mammalian system and, thus are known as active principles of plant. The physiological effect of these active principles is used for curing ailments and therefore, these are drugs of plant origin or natural drugs. The use of crude drugs of plant origin (unpurified preparations of active principles, plant extract or sometimes powdered plant material) is used in the Indian system of medicine or 'Ayurveda'. A large number of drugs of plant origin are still used in Western medicine (Table-17.1). Though some of the drugs are obtained by synthesis for commercial uses, others are still obtained from natural sources.

Table 17.1. Plant derived drugs used in Western medicine.

Ajmalicine	Digoxin	Papain	Reserpine
Atropine	Ephedrine	Papaverine	Scopolanine
Caffeine	Hyoscyamine	Pseudoephedrine	Sennosides
Codein	L-Dopa	Quinidine	Vincaleukoblastine
Colchicine	Morphine	Quinine	

Besides secondary plant products, there are several examples of primary metabolites exerting strong physiological effects. In this category proteins are principal compounds having as diverse functions as blood agglutinants from fabaceae, hormones (e.g. insulin), various snake venom poisons, ricin from *Ricinus communis*, and abrine, and precatorine from *Abru9+9s precatorius*. Other examples of primary metabolites exerting a strong physiological effect include antibiotics, vaccines and several polysaccharides acting as hormone or elicitor.

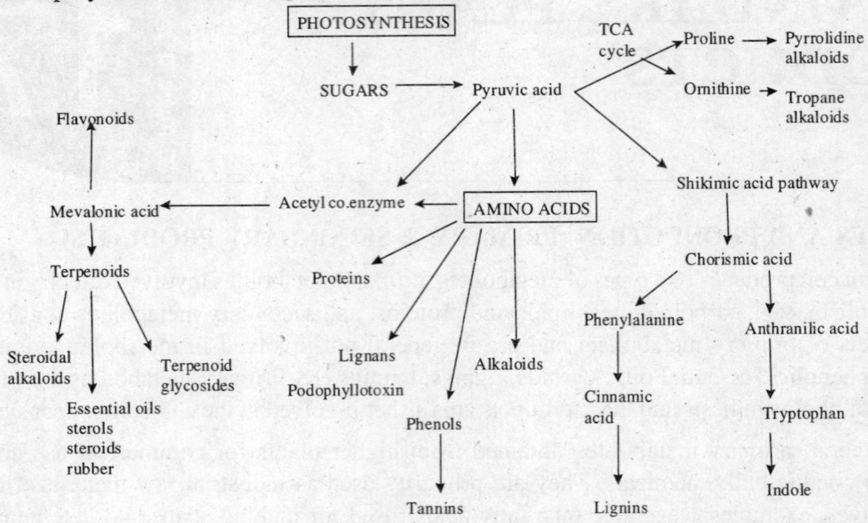

Fig. 17.1. Synthesis of major classes of secondary metabolities from primary metabodities.

Synthesis of various classes of secondary metabolites from primary metabolites is presented in a schematic form in Fig.17.1. All the necessary carbon skeletons are derived from carbohydrates synthesized from photosynthesis. The other major primary products are amino acids. Acetyl co-enzyme and mevalonic acid play a key role in the synthesis of various terpenoids, while the shikimic acid pathway is envolved in the synthesis of lignins and indole alkaloids.

II. ALKALOIDS

Alkaloids have been known to man for several centuries. Morphine, an alkaloid of latex of the opium poppy was isolated by F.W. Serturner in 1806. Latter other alkali-like active principles were isolated and identified, e.g., narcotine in 1817 by Robiquet, emetine in 1817 by Pelletier and Magendie and so on. The term 'alkaloid' was coined by W. Meibner, a German pharmacist, meaning 'alkali like'. Latter it was demonstrated that the alkalinity was due to the presence of a basic nitrogen atom. The first alkaloid to be synthesized was coniine in 1886 by Ladenburg, which had already been isolated in 1827.

(a) Definition – The alkaloids are defined as 'basic nitrogenous plant products, mostly optically active and possessing nitrogen heterocycles as their structural unit, with a pronounced physiological action'. Pelletier (1982) suggested following definition for alkaloids: 'an alkaloid is a cyclic organic compound containing nitrogen in a negative oxidation state which is of limited distribution among living organisms'. Examples of true alkaloids according to definition are: morphine (*Papaver somniferum*), the first alkaloid isolated, quinine (*Cinchona* species), coniine (poison hemlock, the first alkaloid synthesized, from *Conium maculatum*; this plant alkaloid was given to Socrates in 400 B.C.) and reserpine (*Rauwolfia serpentina).*

(b) Occurrence of alkaloids- Around 5000 alkaloids of all types have been known to occur in 15% of all land plants and in more than 150 families. The important families are: Apocynaceae, Papaveraceae, Papilionaceae, Ranunculaceae, Rubiaceae, Rutaceae and Solanaceae and less common lower plants and fungi (ergot alkaloids).

In plants, alkaloids generally exist as salts of organic acids such as acetic acid, oxalic, citric, malic, lactic, tartaric, tannic acid etc. Some weak basic alkaloids (nicotine etc.) occur free in nature. A few alkaloids also occur as glycosides of sugar (e.g., glucose, rhamnose and galactose) e.g.

alkaloids of the solanum group (solanine) as amides (piperine) and as esters (atropine, cocaine) of organic acids.

(c) Extraction of alkaloids – Alkaloids are soluble in water and can be extracted with alcohol. Powdered drug is extracted by: 1) refluxing (1-3 h), 2) using a Soxhlet extractor (3-4 h) and 3) in cool solvent with agitation (overnight) using methanol or ethanol. The filtrate is evaporated to dryness and residue is mixed with 1% sulphuric acid. This aqueous phase is extracted (separated) with organic non-polar solvent (e.g., chloroform) in a separating funnel to remove non-alkaloidal compounds. The organic phase is discarded and the aqueous phase is made basic (pH 9-10) by adding a few ml ammonia solution. The basic aqueous phase is again extracted with organic non-polar solvent (e.g., chloroform), whereby alkaloids, which are basic in nature, are transferred to the organic phase.

After evaporation of the organic phase, the suitably diluted residue is used directly for Thin Layer Chromatography (TLC), Gas Liquid Chromatography (GLC), High Performance Liquid Chromatography (HPLC) or after necessary purification, if required. Usually 100 mg to 1 g dried material is sufficient and if alkaloids are fluorescent in Ultra-Violet (UV) light (e.g., ajmalicine in *Catharanthus roseus*), small samples are sufficient to work with TLC and spectrofluorimetry. For non-fluorescent compounds (e.g., *Cocculus* and *Papaver* alkaloids) TLC plates with silica gel containing fluorescent indicator (silica gel -G-60 F_{254}) can be used.

Various solvents are used for the separation of different classes of alkaloids and closely related products. Fluorescent and non-fluorescent alkaloids can be visualized by using TLC plates with or without fluorescent indicator. Plates can be sprayed with alkaloid-reactive reagents like Draggendorff's reagent (General reagent), Salkowski's reagent (indole group) and CAS (Cerric Ammonium Sulphate) reagent for indole alkaloids etc.

Quantification of alkaloids is achieved by colorimetry, TLC densitometer, Optical Density in UV by spectrophotometer, fluorescence in spectrofluorimeter, and separation and quantification automated like GLC and HPLC. Alkaloids are isolated and purified by column chromatography and preparative TLC (thick TLC plates, 2 mm thick silica gel- G-60) in large quantity required for identification and pharmacology works.

(d) Classifications of alkaloids. There are several ways to classify the alkaloids, e.g., taxonomic (say solanaceous, papilionaceous alkaloids), pharmacological (say analgesic, cardio active alkaloids) and chemical classification. Since some families contain alkaloids of several types or the same alkaloids can produce various pharmacological effects in different systems, these classifications were confusing. Therefore classification based on chemical structure is universally acceptable. In the following pages text has been arranged on the basis of chemical structure.

Phenylalkylamines- This class of alkaloids is synthesized from phenylalanine, an aromatic amino acid. These are ephedrine, pseudoephedrine, (*Ephedra* species), taxine (*Taxus)* and hordenine (*Hordeum vulgare*). Several species of *Ephedra* (Gnetaceae) contain the alkaloidal amine 'ephedrine'. The ephedrines are L-ephedrine, d-pseudoephedrine, *p*-N-methyl ephedrine, p-nor-ephedrine, d-nor pseudoephedrine and d-N-methyl pseudoephedrine. Other plants known to contain ephedrine or its derivatives are *Catha edulis* and *Taxus baccata*.

The alkaloid ephedrine and pseudoephedrine are largely used as antispasmodic and circulatory stimulants. They were used in China for at least 2000 years before being introduced into Western medicine in 1924. Ephedrine is largely used as substitute for epinephrine against bronchial asthma of allergic and reflexive types. It is also used orally and locally in patients suffering from hay fever, urticaria and other allergic reactions. E. foliata grows in the Thar desert contains traces of pseudoephedrine, E. geraradiana, a cold desert species growing in western Himalayas contains 2% alkaloids.

Pyrrolidines, Piperidines and Pyridines. Piperidine alkaloids such as coniceine, coniine and N-methyl coniine are present in *Conium maculatum*, but not detected in the cultures of this plant. Lobeline and other piperidine alkaloids are present in *Lobelia inflata* tissue cultures and intact plants. The alkaloids belonging to these groups are known to have divergent physiological activities.

Apart from tobacco alkaloids, nicotinic acid and its derivatives are the major pyridine alkaloid present in plants. The most commonly occurring compound is trigonelline (N-methyl-nicotinic acid) present in *Trigonella foenum-graecum*. Plants and cultures of *Nicotiana tabacum* contain nicotine, anatabine, anabasine, myosmine and nicotelline. Nicotine has been a major target of study among them.

Tropane alkaloids. The alkaloids hyoscyamine, atropine and hyoscine (scopolamine) are found principally in plants of the family Solanaceae and categorized as anticholinergics. More than 30 alkaloids are known to be present in *Datura. Datura stramonium* and *D. innoxia* are the main source of hyoscyamine and scopolamine, respectively. Other species known to contain tropane alkaloids are *Dubosia* hybrids, *Hyoscyamus niger* (henbane), *H. muticus, D. myoporoides, D. leichhardtii,* and *Atropa belladona* (deadly nightshade). Cocaine in coca (*Erythroxylon coca*) was the first local anaesthetic to be discovered. Leaves of a few species of *Erythroxylon* indigenous to Peru and Bolivia, contain 0.6 –1.8% cocaine. The leaves of the plant have been used for centuries by the natives to increase endurance and to promote a sense of well-being. Cocaine was isolated in 1859 by A. Niemann for its Central Nervous System (CNS) stimulatory activity, which can lead to dependence liability, hence cocaine has been used as drug of. These alkaloids are synthesized from tropic acid.

Quinolizidine and pyrrolizidine- Quinolizidine alkaloids are common natural products of many Fabaceae and commonly called as lupin alkaloids because of their presence in all species of the genus *Lupinus*. This group of alkaloids is synthesized from lysine via cadaverine, e.g. lupanine, sparteine etc.

Isoquinoline-type alkaloids show strong pharmacological activities like those of morphinane-, protoberberine- and benzophenanthridine-type alkaloids; they are widely distributed in the plant kingdom, mainly in Papaveraceae, Berbidaceae, Ranunculaceae and Menispermaceae.

Papaver alkaloids require a special mention in isoquinoline alkaloids. The opium poppy, *Papaver somniferum*, is one of the man's oldest cultivated plants. The therapeutic use of latex obtained from unripe capsules of poppy was recorded by Theophrastus in the third century B.C. Discorides (A.D.77) described the curative properties of the opium poppy and presented various uses for both latex and extracts of whole plants. Purified alkaloids are used in modern medicine for treatment of pain, cough and diarrhoea. Opium is the dried cytoplasm of a specialized internal secretory system, the laticifer. Morphine content increases in the morning with decrease in codeine and thebaine content in the latex. So capsules are incised in the morning to obtain morphine-rich latex. When the green unripe capsule (fruit) is cut, milky latex oozes out, which turns dark brown on drying and is collected as raw opium? This is a labour intensive manual process and requires strict vigilance and control to check illegal drug transport. Alkaloids are purified in laboratories (factories controlled by the Government).

More than 40 alkaloids have been identified from *Papaver somniferum* including 25 from latex. However, from the medicinal point of view, benzyl isoquinolines (papaverine and narcotine) and phenanthrines (morphanians: morphine and codeine) are important. The demand for legal opium is estimated to be more than 1000 tons per annum. In India, cultivation and alkaloid isolation is a business completely controlled by the Government. World trade of opium is governed by the United Nations Opium Conference Protocol (1953).

Morphine was the first alkaloid isolated in pure form. Synthesis of morphine takes place from tyrosine via DOPA (Dihydroxy phenylalanine), nor-landanosoline to papaverine or reticuline. Reticuline is converted into thebaine, thebaine in to codeine and then morphine as the end product. Thus codeine is an intermediate product, not present in large quantities in the latex.

Quinoline alkaloids of *Cinchona*. Extracts obtained from the bark of *Cinchona* species have been of great therapeutic value. The antimalarial activity of these alkaloids has been known for many centuries. The main compound found to be responsible for this activity is quinine. With the discovery of new and more potent synthetic antimalaria drugs, use of quinine was decreased. But with the development of resistance in the malarial parasite (*Plasmodium Falciparum*) towards these to synthetic drugs, interest in the natural drug has been renewed. The alkaloid has been prepared by total synthesis but these procedures are too complex and not commercially viable. Quinidine, another major cinchona alkaloid, is used for the treatment of cardiac arrhythmia. It is also effective against malaria *para*site. Besides their pharmaceutical use, cinchona alkaloids are used frequently in the food and soft drink industry because of their bitter taste. The principal species are *C. ledgeriana, C. pubescens (*syn. C. succirubra) and *C. officinalis.*

Cinchona trees have been cultivated in plantations for more than 130 years for production of cinchona bark. Though the plant is native to certain parts of South America it is presently cultivated in India, Java and Indonesia. When trees are 7 to 12 years old, the bark of the tree is harvested; at this stage of maturity the alkaloid content is approximately 12-15%. In all 35 alkaloids are found in bark. The important ones are quinine, quinidine, cinchonine, and cinchonidine, and their dehydro derivatives (Fig. 2.9). Besides these, several indole alkaloids (aricine, cinchonamine, quinamine) are also present in the leaves and bark.

Indole alkaloids. The monoterpene indole alkaloids represent a large and diverse group of plant products, mainly present in Loganiaceae, Apocynaceae and Rubiaceae. The monoterpene indole alkaloids are formally derived from a unit of tryptamine and a C_9/C_{10} unit of terpenoid origin (secologanin). These alkaloids are classified in to three groups- Corynanthe-, Aspidosperma- and Iboga- type on the basis of carbon skeleton. Examples of Corynanthe type alkaloids are ellipticine, reserpilline and other alkaloids of this type in *Ochrosia elliptica, reserpine and* ajmaline in *Rauwolfia serpentina* and *Catharanthus roseus,* and cinchonamine, quinamine, aricine in *Cinchona*. Conoflorine, tubotiwine are Iboga type alkaloids present in *Tabernanthe iboga*. Tabersonine, ichnericine and minovincine are Aspidosperma alkaloids present in *Voacanga africana.*

Large number of indole alkaloids produced by *Catharanthus roseus* (synonym *Vinca roseus*) have been identified. Several of these have been found to be valuable agents in the treatment of hypertension and others of cancerous growths. In particular, vincristine and vinblastine, two dimeric indole alkaloids are used for the treatment of leukaemia and Hodgkin's disease. The other indole alkaloids of *C. roseus* are ajmalicine, serpentine, lochneridine, tabersonine and vindoline.

Purine alkaloids – Purine alkaloids are widely distributed within the plant kingdom and have been detected in at least 90 species belonging to 30 genera. Their occurrence, however, is limited to dicots. Caffeine and threobromine, methylated derivatives of xanthine, are generally the main purine alkaloids and are regularly accompanied in low concentrations by the two methylxanthines- theophylline and paraxanthine as well as by methylated uric acids such as theacrine, methylliberine and liberine.

Since purine alkaloids are present in tea and coffee are widely consumed in human diet across the continents. Plant species from different families are made in to a pleasant stimulant, e.g., coffee (*Coffea arabica, C. robusta*), tea (*Camellia sinenesis*), Cocoa (*Theobroma cacao*), mate (*Ilex paraguayiensis)* and cola *(Cola nitida).*

Tropolone alkaloids – The neutral alkaloid colchicine present in the bulbs of *Colchicum autumnale* and tubers of Superb Glory (*Gloriosa superba*), Liliaceae is an example of a tropolone-type alkaloid. Colchicine is synthesized from phenylalanine via cinnamic acid and sinapic acid.

Other important secondary metabolites categorized as alkaloids are acridone alkaloids present in plants of the family Rutaceae and steroidal alkaloids present in plants of the family Solanaceae.

(e) Biological functions of alkaloids:

The biological functions of alkaloids within the plants are not clearly understood but it is clear that they are not produced in plants for a single function, but for many functions. The following functions have been observed in different plant species. Alkaloids are considered as:

- Reserve substances to supply nitrogen, but very little evidence is available about this function.
- End products of the detoxification mechanism, otherwise their accumulation in plants might cause damage to the plants, e.g., in tobacco, 10% of carbon metabolism is directed to synthesize nicotine biosynthesis. Thus, it is an energy expensive process.
- Poisonous substances to protect the plant itself from insects and animals. Nicotine has insecticidal properties. Sheep avoid grazing lupin plants with high alkaloid content. Some cacti repel fruit fly but *Drosophila pachea* is resistant and breeds on cactus.
- Plant stimulants or regulators, e.g., alkaloids inhibit rye and oat seedling growth. Colchicine inhibits cell division.
- Reservoirs for protein synthesis.
- Excretory products of plant.
- Inhibition of enzymatic activity by alkaloids is also known.

III. SECONDARY METABOLITE PRODUCTION IN CULTURE:

Plant cell cultures have great potential for the production of secondary metabolites. In recent years, considerable success has been achieved in increasing the secondary metabolites using cell suspension cultures in several plant species. Plant cells grown in culture have potential to produce and accumulate chemicals similar to the parent plant from which they were derived. There are numerous reports describing the production of diverse secondary metabolites, viz., anthocyanins, alkaloids, carotenoids, flavones, coumarins, naphthaquinones, saponins, sesquiterpenes, steroidal alkaloids, sterols, tannins, terpenoids and several others. It is not possible here to describe the production of each and every class of compounds. Details of some of the classes of secondary metabolites produced in callus and cell cultures are presented. How to produce such compounds and factors govering their production is presented in preceeding chapter 17.

(a) Alkaloids

Alkaloids have been known to man for several centuries and are used for human welfare. Among various groups of secondary metabolites, alkaloids are the most extensively investigated compounds. Of all known natural products, about 20% (i.e., about 16,000) are classified as alkaloids. The biological activity of several of these has been investigated but only 30 alkaloids are produced at the commercial level. Most are medicines but some are used as flavouring, poison, and model compounds for pharmacological activity. These alkaloids can be classified as novel chemicals, as their world production is limited, e.g., quinine and quinidine have a yearly production of 300-500 kg, while compounds such as vincristine and vinblastine are produced in a few kilogram range. If we compare this system to cane sugar production from sugar cane (plant based primary product), biomass utilization for extraction of alkaloids is very small. For aforesaid quantities of alkaloids, 5,000-10,000 tons of *Cinchona* bark, 200-300 tons of *Catharanthus roseus* roots are required for quinine and vincristine, respectively. The price of these compounds is exorbitant, which why secondary metabolites are termed low-volume- high-value products.

Plants have been used since ancient times to cure various ailments. With the help of the modern tools of chemistry, active principles of many age-old drugs have been identified and plant explorations have lead to the addition new plant species and compounds to the existing list of medicinal plants. Though many drugs are of synthetic origin, several compounds are still obtained from plants (Table 17.1). The demand for these plants is so high that most of them are presently cultivated (except *Camptotheca, Ephedra* and *Ochrosia*) to obtain sufficient biomass for extraction of drugs. To meet the demand of an increasing population and to save natural resources, alternative methods including chemical synthesis are always being looked into to increase the yield of the active principle.

(b) Why tissue culture

These pharmacologically active novel compounds are extracted from plants. In plant systems, they accumulate in leaves (nicotine in *Nicotiana*), roots (ajmalicine in *Catharanthus roseus*), bark (quinine in *Cinchona*) or in the whole plant (ephedrine in *Ephedra*). Sometimes these products are produced in specialized differentiated tissues such as resin in resin ducts and latex in laticifers. Except the herbaceous cultivated plants (e.g., *Papaver somniferum*), most of the secondary metabolites are accumulated after a certain age or maturity of the plant. In the case of tree or shrub species, e.g., *Cinchona, Rauwolfia, Camptotheca, Ochrosia*, etc., plants attain maturity in a few years before they accumulate the active principle in high amounts. It is difficult to increase the area under plantation for a particular species and growth of plants takes it own time. To meet the ever-increasing demand (e.g., vincristine) the natural resources are not sufficient. The world political scenario may also affect the supply of a particular raw material. To overcome all these hurdles, the industry requires alternative methods of assured supply of uniform material throughout the year. Harvesting of plants (except cultivated species) from natural forest resources is not only difficult, but also makes them endangered species; e.g., *Ephedra gerardiana* and several other Himalayan plants.

Table 17.2. Important alkaloids of plant origin and their pharmacological activity.

Alkaloid	Source	Pharmacological property
Ajmalicine	*Catharanthus roseus*	hypotensive
Atropine	*Atropa belladonna*	anticholinergic
Berberine	*Berberis* spp., *Coptis japonica*	antispasmodic, antiprotozoal
Codeine	*Papaver somniferum*	sedative, analgesic
Colchicine	*Colchicum autumnale*	antimitotic
Caffeine	*Coffea arabica, Camellia sinensis*	stimulant
Camptothecine	*Camptotheca acuminata*	antitumour
Emetine	*Cephaelis ipecacuanha*	emetic
Ellipticine	*Ochrosia elliptica*	antitumour
Ephedrine	*Ephedra gerardiana*	spasmolytic
Morphine	*Papaver somniferum*	analgesic, sedative
Papaverine	*Papaver somniferum*	spasmolytic
Quinine, quinidine	*Cinchona ledgeriana*	antimalaria
Reserpine	*Rauwolfia serpentina*	hypotensive
Vinblastine, vincristine	*Catharanthus roseus*	anticancer

When plant material is not available throughout the year in a quantity sufficient for industrial production and chemical synthesis is not possible, particularly for large complex molecules, biotechnological methods offer an excellent alternative. But before implementing this approach, cost of the product and its demand should justify production by biotechnological means.

Production of various alkaloids in cultures, their yield and approximate price are given in Table 17.2. Demand and cost of colchicine is not very high, but both criteria are very high for vincristine. Alkaloids such as ephedrine are obtained by chemical synthesis, a cheap method, while nicotine is obtained from field-grown plants (again a cheap source); therefore, their production through plant tissue culture is not commercially viable.

Table 17.2. Alkaloids produced in culture and their market value.

Alkaloid	Culture type	Yield in culture	Price US $/g
Ajmalicine	Cell suspension	0.2 g/l	37.0
Ajmaline	Cell suspension	0.04 g/l	10.0
Camptothecine	Cell suspension	0.00025% D.W.	480.0
Colchicine	Callus culture	0.0006% D.W.	50.0
Ellipticine	Cell suspension	0.005% D.W.	2630.0
Emetine	Root culture	0.3-0.5% D.W.	26.30
Morphine	Cell suspension	0.25 g/l	340.0
Quinine	Cell suspension	Traces	0.50
Reserpine	Cell suspension	0.002 g/l	8.30
Scopolamine	Hairy root	0.08 g/l (*Duboisia*) 0.4% D.W.(*Hyoscyamus*)	17.30
Vincamine	Cell suspension	3.3 g/l	19.70
Vinblastine	Shoot culture	Traces	10530.00
Vincristine	Shoot culture	Traces	25200.00

D.W., Dry weight basis.

(c) Approaches for production

Production of alkaloid from *in vitro* culture requires basic information about botany, phytochemistry and importance of alkaloids (in pharmaceutics or otherwise) before commencing work. After collection of proper information about the material, cultures are raised as static or cell suspension cultures. Callus cultures are slow growing systems and accumulate more as compared to fast-growing cell suspension cultures. But cell suspension cultures have obvious advantages of growing at a large-scale in a bioreactor essentially required for ultimate industrial production of the compound. By manipulation of medium components growth is optimized. It is well established that cultures produce very low amounts of secondary metabolites (alkaloids) as compared to the intact plant. Therefore, production has to be increased to make the system productive and ultimately commercially viable. Some of the extensively studied alkaloids produced in tissue culture are listed in Table 17.3.

Optimization of alkaloid production is done using physical factors (light, temperature, vessel type), nutrients (carbon source, nitrogen source - nitrate versus ammonium nitrogen, organic (reduced) nitrogen, precursor molecules, phosphate, etc.), plant growth regulators (auxin type and concentration, cytokinin) and perhaps complex organic supplements (casein hydrolysate, coconut milk, yeast extract, etc.). It has been established from a large number of reports that fast-growing cultures accumulate alkaloids in low amounts during exponential phase of growth and in high amounts during the stationary phase. During this phase nutrients are exhausted and primary metabolism comes to a halt and the stored pool of primary metabolites is diverted to the synthesis of secondary products.

Zenk *et al.* (1977) used a two step culture system: medium for optimal growth (growth or maintenance medium) and medium for production of alkaloids (production medium). In the latter medium, growth is practically arrested by manipulating the nutrients, viz., high sucrose (4-10%) and low phosphate, 2,4-D and nitrogen. On transfer of growing cells to the production medium, product yield is increased several-fold.

The other strategy is to manipulate the plant component. Cultures are mixture of producer and non-producer cells and selection of high alkaloid producing cells/aggregates enhance the product yield. But these high-yielding cultures are not stable and repetitive clonal selection is required to maintain high yields of alkaloids. Other strategies include shoot culture, root culture, transformed root culture and single cell (protoplasts) selection on the basis of fluorescence of the alkaloid. When all the conditions of product yield are standardized in batch cultures, large-scale cultivation and product yield can be attempted using bioreactor.

More recent approaches seek to know the mechanism of transport and accumulation, enzymes involved in the synthesis of alkaloids, identification of gene/s and transfer of such genes (recombinant technology) for expression in other eukaryotes or prokaryotes. These are strategies to improve the production of alkaloids in particular and secondary metabolites in general.

Table 17.3. Alkaloid production in cell and callus cultures

Plant species	**Alkaloid**
Atropa belladonna	Atropine
Camellia sinensis	Caffeine, theobromine
Catharanthus roseus	Ajmalicine, serpentine, vincristine, vinblastine
Cinchona succirubra, C. ledgeriana, C. officinalis	Quinine, quinidine
Cocculus pendulus, C. hirsutus	Cocculine, coccutrine
Coffea arabica	Caffeine, theobromine
Coptis japonica	berberine
Datura stramonium, D.innoxia	Scopolamine, hyoscyamine
Ephedra sinica, E. gerardiana	Ephedrine, pseudoephedrine
Hyoscyamus muticus, H. niger	Scopolamine, hyoscyamine
Nicotiana tabacum	Nicotine,
Ochrosia elliptica	Ellipticine, 9-methoxy ellipticine
Papaver somniferum	Papaverine, morphine, codeine
Rauwolfia serpentina	Serpentine, ajmalicine
Ruta graveolens	Platydesminium, ribalinium, rutalinium
Thalictrum minus	berberine

(d) Antitumour agents – Plants have been used in the treatment of cancer for over 3500 years. But it is only since 1959 that a concentrated systematic effort has been made to screen crude plant extract for their inhibitory activity against tumour systems. Under the programme sponsored by National Cancer Institute, USA, around 2,00,000 plant extracts from 2,500 genera of plants have been screened.

The number of medicines used in chemotherapy is around 20. These comprise the synthetic drugs and drugs obtained from plants. A large number of plant products have been shown to possess cytotoxic or antineoplastic properties.

Lomax and Narayanan (1988) of the National Cancer Institute (NCI), USA have presented a list of potent antitumour agents of interest to the Institute, which are at a developmental stage and the clinical trial level. Information about the new compounds of cancer therapy can be obtained from NCI, USA. The institute also helps in evaluating the chemicals against experimental tumours. Plant tissue culture provides an excellent system to produce such compounds in controlled conditions and to study their production, regulation, biosynthesis and bio-transformation. Unavailability of pure compounds in sufficiently large quantities is a limiting factor in their evaluation on animal systems. The drugs are cleared for human consumption only after extensive clinical trials to evaluate their efficacy and side effects.

Table. 17.4. Atitumour agents produced in culture

Plant species	Compound
Baccharis megapotamica	Baccharin
Brucea antidysenterica	Bruceantine
Camptotheca acuminata	Camptothecine
Catharanthus roseus	Vincristine, vinblastine
Cephalotaxus harringtonia	Harringtonine, homoharringtonine
Fagara zanthoxyloides	Fagaronine
Heliotropium indicum	Indicine=N-oxide
Maytenus bucchananii, Putterlickia verrucosa	Maytansine
Ochrosia elliptica	Ellipticine,9-methoxy-E
Podophyllum peltatum, P. hexandrum	Podophyllotoxin
Taxus brevifolia, T. baccata	Taxol
Thalictrum dasycarpum	Thalicarpine
Tripterygium wilfordii	Tripdiolide,tryptolide
Withania somnifera	Withanolide-A, withaferine

Antitumour compounds produced using various techniques of plant tissue culture are presented in Table 17.4. As stated earlier, production of secondary metabolites in the cultures is very low compared to the parent plant of the same species (Table 17.5). Therefore, it is desirable to apply various factors to optimize the product yield and use special techniques for obtaining high-yields. These techniques were helpful in increasing the production of ellipticine, podophyllotoxin and tripdiolide in the cultures of *Ochrosia elliptica, Podophyllum peltatum* and *Tripterygium wilfordii*, respectively.

Table 17.5. Anti-tumour compounds: production *in-vitro* and *in-vivo*

Anti-tumour compound	*In-vivo* plant% dry weight	*In-vitro* culture% dry weight
Brucentine	1.0×10^{-2}	5.8×10^{-5}
Camptothecine	5×10^{-3}	2.5×10^{-4}
Ellipticine	3.2×10^{-5}	2.7×10^{-4}
Maytansine	2.0×10^{-5}	5.0×10^{-7}
Podophyllotoxin	6.4×10^{-1}	7.1×10^{-1}
Taxol	5.0×10^{-1}	3.0×10^{-2}
Taxane	2.0×10^{-1} - 10^{-2}	$4\text{-}8 \times 10^{-3}$
Tripdiolide	1.0×10^{-3}	1.0×10^{-2}
Vinblastine, Vincristine	5.0×10^{-3}	5.0×10^{-4}

Among the anticancer agents of natural origin, vincristine, vinblastine, podophyllotoxins and taxol are currently used clinically. All these products are in great demand, production is low, difficult to isolate and purify and hence very high cost forced the world scientific community to search for alternative means of these drugs production. If newer methods are available it will not only reduce the cost of drug but also relieved the pressure on plants which have become endangered because of over-exploitation, e.g., *Taxus* and *Podophyllum.* Therefore, efforts are being to develop efficient protocols for large scale multiplication of endangered medicinal plants like *Podophyllum* and *Taxus* and toprotect the habitat to conserve the plants *in situ*. The conservation programmes in different countries are part of global efforts to save medicinal and other plant species from being extinct.

(e) Production of steroids and saponins: Steroids are natural compounds distributed in microbes, plants and animals. They are classified in to sterols, saponins, cardiac glycosides, and steroidal alkaloids. In animals, fungi, algae and higher plants, there is a common pathway from which steroids are synthesized starting from acetate, mevalonate to squalene epoxide. From this point the biosynthetic pathway diverges considerably. In animals, it is converted to cholesterol, in fungi to ergosterol whereas in plants it is converted to phytosterols by step-wise demethylation of lanosterol and cycloartenol.

In view of therapeutic and industrial applications, steroids constitute a group of secondary metabolites of significant importance, of which cardiac glycosides obtained from Digitalis cannot be replaced by any synthetic substitutes. Sapogenins such as diosgenin and hecogenin are valuable starting material for the synthesis of steroid hormones (contraceptive, anabolic anti-inflammatory agents). Certain steroids such as withanolides and ecdysones, are recognized as cytostatic, insecticidal and anti-inflammatory agents. Most of the steroids are produced from natural resources or through marginal cultivation (*Digitalis*, *Glycyrrhiza*). There is considerable demand and scope for the production of these compounds through alternative biotechnological methods and plant tissue and cell culture holds promise towards achieving these objectives. The promising plant species grown in culure for the production of saponins and steroids are listed in Table 17.6. *Panax, Digitalis* and *Withania* are the most widely used plants of International importance from commercial and therapeutic points of view.

(f) Food additives : There is an upsurge of interest to produce natural food additives in view of increasing consumer preference and gradual phasing out of synthetic compounds due to biosafety reasons. Several countries are moving towards total adoption of natural food additives. Despite the high demand for natural additives, there are problems in these compounds through conventional cultivation methods for several reasons, such as agroclimatic conditions, seasonal specificity and diseases. Therefore, there is need to apply biotechnological methods of production, whereby plant cell, tissue and organ culture offers an alternative method of producing food additives. The compounds produced through plant cell and tissue culture methods are listed in Table 17.7. It is interesting to note that stigma culture of Saffron, *Crocus sativus* produced yellow coloured callus and on drying, this formed a yellow-orange- coloured saffron like fragrant powder.

Table 17.7. Steroids and saponins produced through tissue cultures.

Plant species	Product formation
a. Saponins	
Aesculus hippocastamum	Aescin
Agave sisalana	Hecogenin
Dioscorea deltoidea	Diosgenin
Glycyrrhiza glabra	Glycyrrhizin
Panax ginseng	Ginseng
b. Cardiac Glycosides	
Digitalis lanata, D. purpurea	Digoxin, Digitoxin

Strophanthus sps.	***Quabain***
Urginea maritima	Proscilariddin
c. Other steroids	
Holarrhena antidysenterica	Sitosterol, stignasterol, cholesterol
Solanum xanthocarpum	Solasodine,
Withania somnifera	Withanolides

Insecticides:

Commercial insecticides of plant origin include pyrethrins, rotenoids, nicotine, neemix, azatin and quassin. Natural insecticides of plant origin are effective against a wide range of insects, many of which cannot be successfully controlled by synthetic insecticides. Limonoids are also insecticidal compounds, founds mainly among members of the families Meliaceae and Rutaceae. Unlike synthetic insecticides, natural compounds are relatively non-hazardous to human beings and other mammals. Plant insecticides generally exert their effect by interfering with the physiology of insects, affecting their nervous system, hampering their normal development and acting as antifeedents. Insecticides produced through in vitro method are presented in Table 17.8. More and more products are investigated for their production through alternative biotechnological methods. Technology is upgraded for the existing products of proven commercial value for their production at large-scale in bioreactorss.

Table 17.7. Production of food additives by plant tissue culture.

Plant species	Product
a. Colours	
Daucus carota, Euphorbia milli, Vitis vinifera,	AnthocyaninAnthocyaninAnthocyanin
Beta vulgaris	Betalaines
Crocus sativus	Crocin , crocetin
b. Flavours	
Allium cepa	Onioin flavour
Capsicum annuumC. frutescens	Capsicum, capsaicin
Crocus sativus	Safranal
Vanilla planifolia	Vanilla, vanillin
c. Sweetner	
Stevia rebaudiana	Stevioside
Thaumatococcus danielli	Thaumatin

Table 17 .8. Insecticides of plant origin.

Plant species	Insecticide
Azadirachta indica	Azadirachtin
Derris elliptica, Tephrosis vogelii, Lonchocarpus utilis	Rotenoids
Chrysanthemum cinerariaefolium, Tagetes erecta	Pyrethrins
Nicotiana rustica, N. tabacum	Nicotine
Quassia amara	Quassin

QUESTIONS

1. What are secondary plant products? Describe the pharmaceuticals, insecticides and food additives produced through biotechnological methods.
2. What are alkaloids? Describe the methods of production in culture with suitable examples.
3. Write short notes: a) alkaloids, b) Saponins and steroids, c) Food additives, d) Antitumour agents, e) Insecticides of plant origin.
4. Describe the secondary metabolites of natural origin with special reference to alkaloids.
5. Write short notes on the following : A. True alkaloids B. Primary v/s secondary metabolites C. Classification of alkaloids D. Biological functions of alkaloids E. sapogenins F. Resin and essential oils G. Shikimic acid pathway.
6. Describe in brief classification, biosynthesis and major classes of alkaloids.

SECONDARY PLANT PRODUCTS

CHAPTER 18

FACTORS AFFECTING THE PRODUCTION IN CULTURE

I. INTRODUCTION

Some plants are the source of several natural drugs and hence these plants are termed as the medicinal plants. Essentially plant cells contain primary and secondary metabolites. Primary metabolites are substances widely distributed in nature, occurring in one form or another in virtually all organisms. In higher plants, these compounds accumulate in storage organs or seeds and are required for physiological development because of their role in basic cell metabolism.

The production of secondary metabolites is controlled by genes and plant cell cultures produce the same product in culture as produced by the intact plant. But in most cases, production has remained far lower than the intact plant barring a few examples where it exceeded the parent plant. Therefore, to enhance production of the active principle, various approaches have been applied.

Various factors affecting the production of secondary metabolites in plant tissues grown in culture are presented in Fig. 18.1. Initially, growth and production of secondary metabolites are optimized by manipulating the physico-chemical factors followed by selection of high-productive cells. Afterwards, use of the production medium and specific techniques such as elicitation, hairy roots and immobilization are applied. When cultures are considered suitable for commercial production, cultures are grown in a bioreactor. In this chapter, factors affecting optimization and selection are presented. The special techniques are presented in respective chapters.

There are two approaches to enhance product synthesis and accumulation. The most widely applied empirical approach is optimization wherein medium and environmental factors of the cultures are manipulated. The second approach is to select suitable cells/tissues under normal or selective conditions. The most recent method is to combine both the approaches, i.e., selection of cells/tissues on the optimal medium condition for high product yield. For large- scale production of a compound, stable high-producing cell lines of plants of interest have to be obtained. Therefore, optimization of culture conditions and production of secondary metabolites in high amounts are prerequisites for industrial production. For this elicitation and hairy root cultures are also used with production system.

II. TIME COURSE STUDY OF GROWTH AND PRODUCTION:

With the initiation of research on secondary product formation, it is of paramount importance to record the growth and secondary product formation in the stable cultures during the incubation period untill tissues are subcultured. This observation provides insight about into the relationship between growth and secondary product synthesis and accumulation in the cultures.

This is helpful in designing experiments for optimization, harvesting of tissues in a particular day for secondary product analysis and modulating the growth and transferring cells into the induction medium . Growth curve of a stabilized culture is always sigmoid (Fig. 18.2). It is evident from the figure that the growth curve has three phases: a) lag, when there is almost no growth, cells adjust to the new medium after subculture, b) exponential phase of growth when primary metabolism increases and tissue proliferates rapidly with the consumption of medium nutrients; and c) log or stationary phase when primary metabolism and cell proliferation come to a halt as nutrients in the medium are

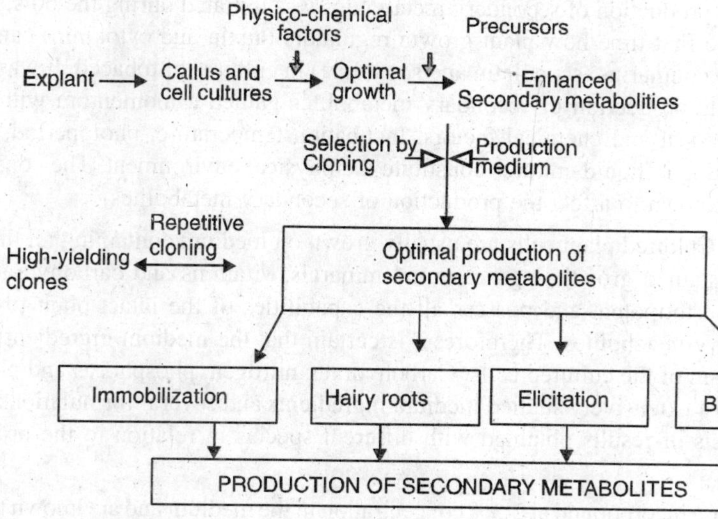

Fig. 18.1. Various approaches used for the optimal production of secondary metabolites

exhausted (nutrients become limiting factors). There may be a decline in dry weight of the cells as cells utilise stored reserve material. If we look at secondary metabolites synthesis during these phases, it is evident that synthesis is almost lowest during the lag phase (secondary metabolites accumulation of previous passage get diluted further due to cell proliferation) and the early exponential phase. It is at the end of exponential phase that increased synthesis of secondary metabolites is evident as primary metabolites start converting into secondary metabolites. It is during the log phase that there is no more cell growth and all primary metabolites are diverted into the synthesis of secondary metabolites. Therefore, the highest amounts of secondary metabolites are accumulated in the stationary phase cultures. It is at this stage, that if cells are not harvested, there may also be degradation of secondary metabolites with primary metabolites. At this, stage both nutrient medium and cells, are exhausted of carbohydrates and other nutrients. When these observations are in hand, experiments are designed to optimize the yield of secondary metabolites.

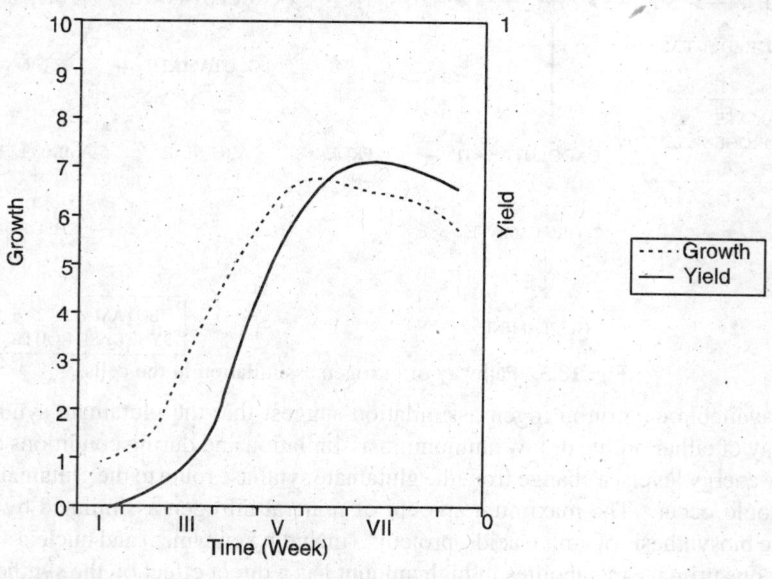

Fig.18.2. Growth and yield of secondary metabolites during time course showing sigmoid curves.

III. OPTIMIZATION :

Research on the production of secondary metabolites was initiated during the 60's. In 1961, it was described for the first time how plant growth regulators (auxin and cytokinin) can modulate the production of two coumarins (scopolatin and scopoline) in cultures of tobacco. It was only after 1970 that work on the production of secondary metabolites gained a momentum with increased knowledge about physical and chemical factors. Incubation temperature, photoperiod, the pH of the medium and aeration in liquid cultures constitute the physical environment. The constituents of the environment are known to affect the production of secondary metabolites.

Effect of Nutrients- Cultured plant cells are usually grown on medium containing all the elements required for their sustained growth, e.g., essential minerals, vitamins and carbohydrate sources. Plant cell cultures are totipotent and possess all the capabilities of the intact plant to synthesize primary and secondary metabolites. Therefore, it is certain that the medium ingredients affect the growth and metabolism of the cultured cells. Carbohydrate, nitrogen, phosphorus and plant growth regulators are the most extensively studied medium ingredients and several formulations have been suggested on the basis of results obtained with different species in relation to the production of secondary metabolites

Carbohydrates are incorporated at 2-3% concentration in the medium and are known to influence the production of phytochemicals. In *Catharanthus roseus* cultures, alkaloid content fluctuated with sucrose concentration in the medium, usually increasing with increased sucrose concentration (4-10%).

A mixture of nitrate and ammonium is used in all the standard media as a source of nitrogen. Many plant cells may not be able to tolerate high amounts of ammonium used in these media. That is why different nutrient formulations have been devised to suit the requirement of different species. It has been established that the composition of cells vary on the media supplemented with nitrate or ammonium. Nitrogen metabolism is presented in simplified scheme as follows (Fig. 18.3).

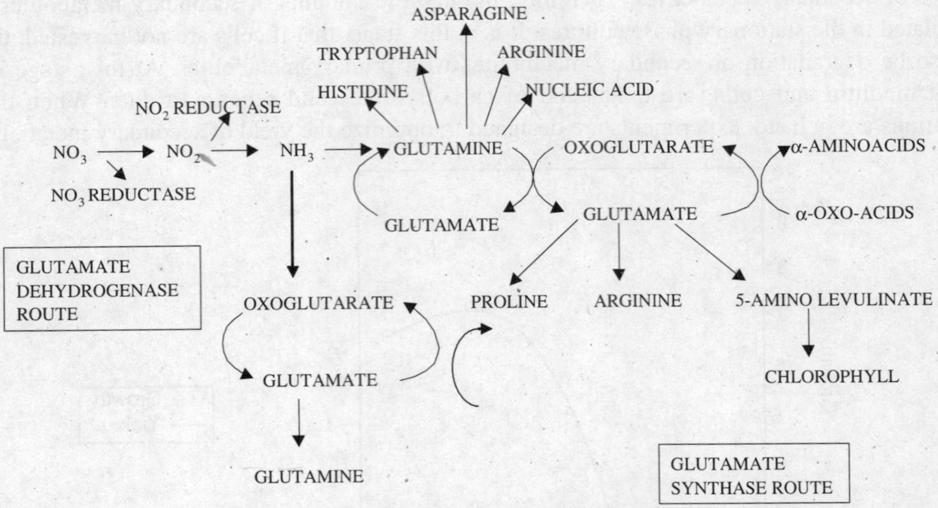

Fig. 18.3. Pathway of nitrogen assimilation in the cells.

The available data on nitrogen assimilation suggest that the glutamate synthase route is the key pathway of either nitrate or low ammonium assimilation and during conditions of high ammonia load or low energy levels, a change from the glutamate synthase route to the glutamate dehydrogenase pathway could occur. The maximum amount of mineral nitrogen assimilated by cultured cells is used for the biosynthesis of amino acids, proteins, (including enzymes) and nucleic acids. Availability of nitrogenous primary metabolites in high amount has a direct effect on the synthesis of secondary metabolites. Though diverse results in relation to secondary metabolites by varying quality and quantity of nitrogen in the medium are obtained, high ammonium ion concentrations are inhibitory to secondary metabolites and lowering or removal of ammonium nitrogen results in increased secondary metabolites.

Involvement of inorganic phosphate in metabolic regulation is well established in photosynthesis and respiration (glycolysis) and is essential to malic acid and phospholipid synthesis. It seems quite logical that altered phosphate levels in the growth media may profoundly affect the biosynthesis of phytochemicals by cultured plant cells. Many secondary products are synthesized through phosphorylated intermediates, e.g., terpens, terpenoids and phenylpropanoids, which subsequently release the phosphate. Thus, the phosphate cleaving steps must occur in the synthesis of such compounds. In both eukaryotic and prokaryotic systems secondary pathways are often inhibited by Pi levels which appear to be optimal for growth and low Pi concentrations are often beneficial for an active secondary metabolism. Therefore, Pi levels are reduced in the production medium designed for high yields of secondary metabolites.

Effect of Plant Growth Regulators- Effect of growth regulators on cultured plant cells is manifested in growth, metabolism and differentiation. Concentration and ratio of growth regulators directly govern the differentiation. But the effect of plant growth regulators is a complex response and yet to be clearly understood. A large number of reports describe the effect of growth regulators on secondary metabolite levels of cultured cells. Two types of growth regulators are required by plant cells, namely auxins and cytokinins, (indole acetic acid, indolebutyric acid, 1-nqphthaleneacetic acid, 2,4-dichlorophenoxyacetic acid, 2,4,5-trichlorophenoxyacetic acid, benzyladenine, kinetin, 2-isopentenyladenosine, zeatin, thidiazuron). Plant growth regulators do not react with intermediate compounds of biosynthetic pathways but appear to change cytoplasmic conditions of product formation to higher or lower levels. The growth regulator effect is not compound specific, production of all secondary metabolites is affected by growth regulators.

Plant growth regulator types and product formation have been well studied. Nicotine production in *Nicotiana tabacum* cv. Bright yellow was among the few secondary metabolites studied in great detail for the first time. In fact, tobacco cultures were used as an experimental material for many first studies in tissue cultures. There are several reports in the literature stating that by reducing the concentration of 2,4-D in the medium or replacing it with another auxin, enhances the accumulation of secondary metabolites, e.g., alkaloids in the cultures of tobacco *Ephedra* and pigment (shikonin) in the cultures of *Lithospermum erythrorhizon*. Among the auxins, 2,4-D has a very marked effect in suppressing differentiation and secondary metabolites in the tissues of *Ochrosia elliptica, Ruta graveolens* and *Ephedra gerardiana*.

Similarly, accumulation of indole alkaloid in cell cultures of *Catharanthus roseus* was suppressed as long as cultures were grown in medium containing 2,4-D, while yields of alkaloid were high in IAA supplemented medium. Similar, results were also obtained by growing cultures of *Datura innoxia, Hyoscyamus niger, Atropa belladonna, Papaver rhoeas* and *Papaver somniferum* in the presence or absence of 2,4-D. It may be generalised to certain extent from the data that higher concentrations of auxins suppress and lower concentrations permit the production of metabolites in cell cultures.

It is worth mentioning that the concentration and combinations of plant growth regulators modulate growth of the tissues and the production of secondary metabolites. Increased cytokinin levels in the medium also affect the cellular differentiation and metabolite production associated with such differentiation is expressed/enhanced in the cultures. A similar effect of cytokinin on alkaloid metabolism was observed in 2,4-D-dependent cell line; omitting 2,4-D from the culture medium for one transfer induced alkaloid accumulation, adding a cytokinin to the 2,4-D-free medium further increased the accumulation. Therefore cytokinin has a positive role in increasing alkaloid production and its effectiveness increased in the absence of 2,4-D.

Generally, a balanced combination of cytokinin to auxin is used and therefore results should be interpreted in terms of a combined effect. As both cell multiplication and cell differentiation have to occur in cell cultures when metabolites production is to be increased, a better method is to use two step culture medium condition; promotion of cell growth on a nutrient medium with high levels of 2,4-D followed by transfer of cells to medium with low level of auxin (2,4-D) or devoid of auxin. This has led to development of concept for the maintenance (growth) medium and the production medium known as two stage culture system.

Precursors - Precursors are molecules which are directly incorporated into secondary metabolites, but perhaps with some structural changes. When such precursor molecules are fed to cultures they incorporate them into biosynthetic pathways of secondary metabolite. Information about enhancement of alkaloid biosynthesis by feeding of precursor amino acids suggests that the amino acid not only acts as a precursor but also an inductor and, secondly, the level of amino acid in the free pool is increased. It is presumed that in the latter condition, the pool of amino acid previous to amino acid feeding might be too low to get metabolised into alkaloid. There are examples of precursors significantly enhancing the production of secondary metabolites (Table 18.1).

Table 18.1. Effect of added precursors on secondary product level in the tissues.

Species	Precursor	Metabolite	Stimulation
Ruta graveolens	4-OH-2-Quinoline	Dictamine	Trace - 0.6% DW
Cinchona ledgeriana	Tryptophan	Quinoleines	Trace - 0.9% DW
Lithospermum erythrorhizon	Phenylalanine	Shikonin	37 μg - 126 μg/g FW
Ephedra gerardiana	Phenylalanine	Ephedrine	0.17% - 0.5% DW
Capsicum frutiscens	Vanillylanine + isocarpic acid	Capsaicine	Trace - 10 mg
Catharanthus roseus	Tryptamine + Secologanin	Ajmalicine	Trace - 0.6 mg/l

There are two important points to be noted that - (*i*) different class of compounds have different precursor molecule as they are synthesized by different biochemical pathways and are structurally and functionally different molecules, (*ii*) there may be several precursors of a compound produced at different stages of biochemical synthesis. This shows that secondary metabolites have many steps, many precursors, many enzymes and many genes involved in the synthesis. Thus, the reaction (biosynthesis) may be stopped at any step and may affect the synthesis and accumulation of secondary products. These intermediate precursors are also directed sometimes in the synthesis of primary metabolites, thus affecting the concentrations of secondary products. The ultimate content of a cell thus depend upon internal catabolism, extracellular degradation and availability of precursors and enzymes of biosynthetic pathway.

Production medium . *It has been concluded from the results obtained from the several above-mentioned studies on optimization of secondary product formation in cultured cells that -*

1. Higher concentration of auxin in the medium, particularly 2,4-D suppresses secondary metabolites (also suppress cellular differentiation).

2. Higher concentration of phosphate in the medium causes cell growth and lower concentration enhances secondary metabolite level.

3. Lower carbohydrate level (sucrose) favours cell proliferation while higher concentration arrests cell growth and increases secondary product formation.

4. In certain cases, higher nitrogen level in the medium enhances cell proliferation while low concentration increases secondary product formation.

5. Increased synthesis of secondary products occurs during the stationary phase of cultures when primary metabolism and cell proliferation come to a halt.

On the basis of these conclusions, a secondary metabolite production or induction medium was devised by Zenk et al. in 1977 in which the above conditions were combined. Cells grown on maintenance medium (optimal for growth) proliferate rapidly and such cultures are then transferred to induction or production medium (optimal for secondary metabolites) in which growth is arrested or cells enter in a stationary phase of growth. Such induction medium contains the same constituents but with low or zero levels of phosphate, nitrogen (not always) and auxin (2,4-D), and very high sucrose concentration (6-10%). Results of increased production of secondary metabolites in some cultures grown in production medium are shown in Table 18.2.

Table 18.2. Effect of growth of cells in production medium on secondary product synthesis.

Species	Metabolite	Increased production
Catharanthus roseus	Ajmalicine	0.02 - 0.6 mg/l
C. roseus	Phenols	4.9 - 17.2 mg/l
Eschscholtzia californica	Benzophenanthridines	13 - 146 mg/l
Peganum harmala	β carolines	7.4 - 17 mg/g DW
Vitis vinifera	Anthocyanins	Trace - 1200 mg/l

Therefore, if during the exponential phase of growth, cells in optimal growth medium (maintenance medium) are transferred into production medium, growth comes to a halt and carbohydrate and other nutrients are available. So primary metabolites are rapidly diverted to synthesis of secondary metabolites instead of cell growth, thereby enhancing the secondary product synthesis.

IV. SELECTION OF CELLS

An *in vitro* growing cell system has two components : the cell and its environment. In the above paragraphs, the details of manipulating physical and chemical environment of the cell are presented. Here, we discuss how selection procedures are helpful in increasing the yield of cultures. Before producing secondary metabolites at the industrial/commercial level, it is a prerequisite to achieve optimal yield of secondary metabolites through optimization and select the cells for maximal yield. Though the production of a secondary metabolite is a genetically controlled phenomenon, cells can be selected for high yield of secondary metabolites from a heterogeneous population to improve the overall quality of the cultures, in relation to the production of active principle. Before starting the selection procedures, it is necessary that heterogeneous nature of the cultures be established and a sensitive method of analysis be available to analyze a large number of clones. The 'stability' of such selected clones is of paramount importance for developing further method to achieve industrial production.

(a) Variability in field grown plants:

Field-grown plants, particularly the cross- pollinated plants, are heterozygous and express different phenotypic and physiological characteristics. It is because of the heterozygocity that the difference is also expressed in terms of secondary metabolite content. Variation in the alkaloid yield is well-documented in such plants as *Catharanthus roseus* (Apocynaceae, cross pollinated crop) and *Lupinus polyphyllus* (Leguminosae, self-pollinated crop).

In a population of field grown *C. roseus* plants, a complete spectrum from very low to very high alkaloid (ajmalicine and serpentine)-producing plants was recorded and very high alkaloid producing plants were found scattered in the population. Similarly, in lupin fields, animals avoid grazing plants with high alkaloid contents. From these two examples, it is clear that there are plants with high alkaloid (secondary metabolites) levels and if cultures are initiated from such plants optimization and selection procedures can generate very high alkaloid-yielding clones. Similar variation in alkaloid yield has also been observed in the fields of *Nicotiana* and *Hyoscyamus* species.

(b) Producer/non-producer cells :

In the plant itself, certain cells, tissues or organs accumulate more secondary products than others. This is evident from the presence of a high level of alkaloid containing idioblasts distributed in leaf epidermis of *Catharanthus roseus* or gland cells in citrus, or glandular trichomes or hairs present in many species. When an explant is transferred onto a medium, it grows by division of cells producing an undifferentiated mass of cells- called callus. Callus is derived, mostly, from parenchyma cells present in that explant. When sufficient callus is produced by an explant, it is separated and subcultured onto a fresh medium. In this growing system certain cells divide rapidly while others are slow in division. In cell suspension cultures plant cells grow as free cells or cell aggregates. The size of the cell aggregate depends on the inherent genetic make up of the species as well as the growth cycle. We know that rapidly dividing cells accumulate less compared to stationary phase cultures. In fast-growing cultures, certain cells accumulate more secondary metabolite compared to others because cells differ in growth. Therefore, all the cultures (cell populations) are mixtures of cell containing high secondary metabolites (producer) and low secondary metabolites (non-producers). This has been observed in cultures of several species. In the case of pigmented cells, demarcation between high and low pigment containing cells can be made very easily by the naked eye e.g., anthocyanin production in carrot and grape cells. In species in which, secondary metabolites constitute a fluorescent compound, it is very common to visualize and differentiate between the producer and non-producer cells; the producer cells are highly fluorescent under ultraviolet (UV) light. Cells with a higher amount of secondary metabolites fluoresce more intensely compared to other cells, e.g., in the case of *Catharanthus roseus*, cells with tryptamine fluoresce yellow while cells with ajmalicine fluoresce blue. Other cells do not fluoresce at all or fluoresce very faintly. Since *C. roseus* cultures are grown in darkness, they lack chlorophyll; otherwise chlorophyll fluoresces red under UV light and may interfere with alkaloid fluorescence. This way cells plated on a petridish can be marked, selected and separated mechanically to get high alkaloid-yielding cell lines. Thus, all the cultures are a mixture of producer and non-producer cells and analysis of such cultures gives an average value of secondary metabolites. From such cultures, if producer or high product forming cells are separated from non-producer cells and grown separately, they give rise to high product-yielding clones. The population of cells can be compared with a population of field-grown plants, which also differ in metabolite level.

(c) Cloning and selection

It is important to know the cyclic changes in the endogenous cellular concentrations of secondary metabolites for effective selection of high-producing cell lines. The powerful influence of exogenously supplied plant growth regulators, sucrose concentrations and phosphate is well established. Lastly, a balance between the growth of cells and the rate of synthesis of secondary metabolites determines the cell content.

Increase in secondary metabolite contents during the stationary phase indicates that fast-growing cultures during the exponential phase do not accumulate secondary metabolites in high amounts. This has been demonstrated in *Morinda citrifolia* in which anthraquinone concentration increased from 36 mg to 215 mg/gm dry weight during the growth cycle. It is imperative that slow growing cultures accumulate more compared to fast-growing cultures. Proper selections should be made to isolate cells which are fast growing and simultaneously synthesizing secondary metabolites as well. Such cultures are advantageous for scale-up technology.

It is clear from figure 18.4 that a very high percentage of clones contain low alkaloid content and a clones containing high-alkaloid level were absent in non-selected cells. Whereas, it was possible to obtain certain clones with high to very high serpentine containing clones after selection of cells through plating. These high alkaloid-producing cells were present in a population of cells and analysis of cells provided average alkaloid contents. The technique can also used to select

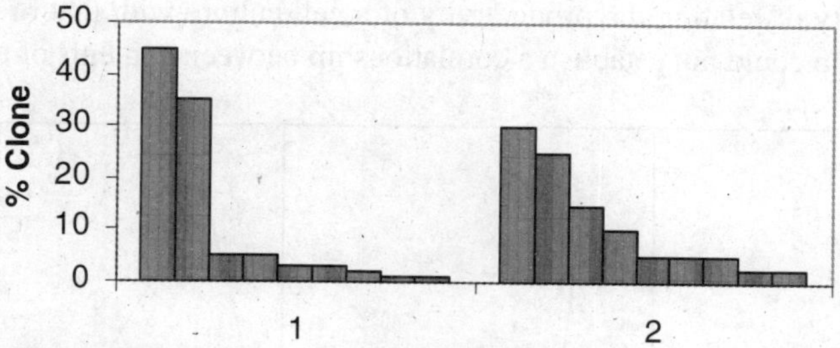

Fig. 18.4. Frequency distribution of serpentine content in selected (2) and non-selected (1) clones of *Catharanthus roseus*. Note the high percentage of serpentine rich clones in selected cultures.

clones on a selection medium (containing or devoid of a specific compound) or on normal growth medium. *Catharanthus roseus* cells plated on a medium devoid of the auxin, 2,4-D, produced clones differing in their ajmalicine and serpentine contents (Fig.18.5). On such selection medium high-yielding clones can be obtained, multiplied and used for further work. These experiments demonstrated the importance of plating and selection procedures to obtain high-alkaloid yielding clones

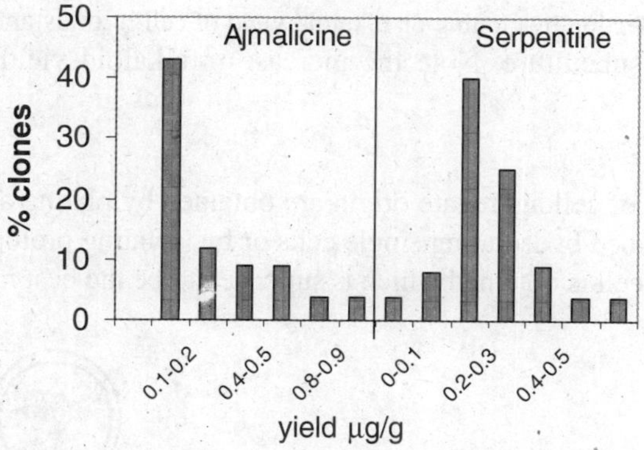

Fig.18.5. Ajmalicine and serpentine production in clones of *C. roseus* grown from cell suspension cultures obtained in a medium with 2,4-D.

High alkaloid-producing clones become a heterogeneous population of producer and non-producers cells after a few generations of cell cycles. This time period may vary from species to species. Therefore, to maintain a high level of secondary metabolites, it is necessary to repeat cloning and separate the producer cells again. Repetitive secondary and tertiary cloning have been used to maintain high-alkaloid levels in *Catharanthus roseus, Nicotiana, tabacum* and *Lithospermum erythrorhizon*. Figure18.6 demonstrates the effect of repetitive cloning on alkaloid yield *of Catharanthus roseus*. Drastic increase in alkaloid level is due to selection and subsequent selection in the selected cell lines, while decrease in the alkaloid level in the clones is due to development of heterogeneous population of producer and non-producer cells after subsequent subcultures. Therefore, repetitive cloning is required to maintain high levels of secondary metabolites in the clones required for technological development at the industrial level. This technique has been used in *Catharanthus roseus, Papaver somniferum* and other species to maintain the alkaloid yield.

It is advantageous to select high-metabolite producing single cells by single cell analysis and sorting technique. This method is quicker and saves time and high-metabolite producing cells as it is a non-destructive method. Individual analysis of single cells from suspension cultures of *Anchusa officinalis* for rosmarinic acid content by microspectrophotometry showed that cell content varies very widely. By combining non-destructive microspectrophotometric analysis with single-cell cloning, cells of known productivity were used to derive true clonal cell suspensions. This enabled the possibility of relating the productivity of a cell culture with that of its original mother cell. This investigation could not establish a corelationship between contents of the clones to that of the mother cell.

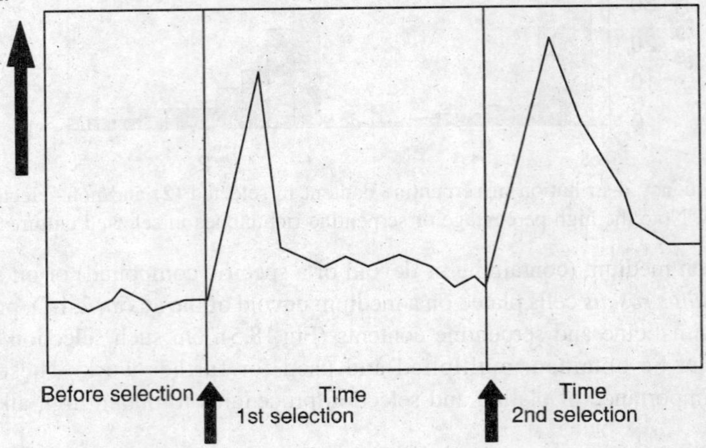

Fig 18.6. Effect of repetitive selection by cloning on alkaloid yield of cell clones and their performance during subsequent subculture. Note the increase in alkaloid yield immediately after selection

(d) Plating Technique

Clones of single cells or of cell aggregate origin are obtained by plating appropriate cultures. Single-cell clones can be obtained by isolating single cells or by isolating protoplasts. Therefore, in both cases a product of single cell is obtained which is supposed to be the best method for obtaining single-cell product.

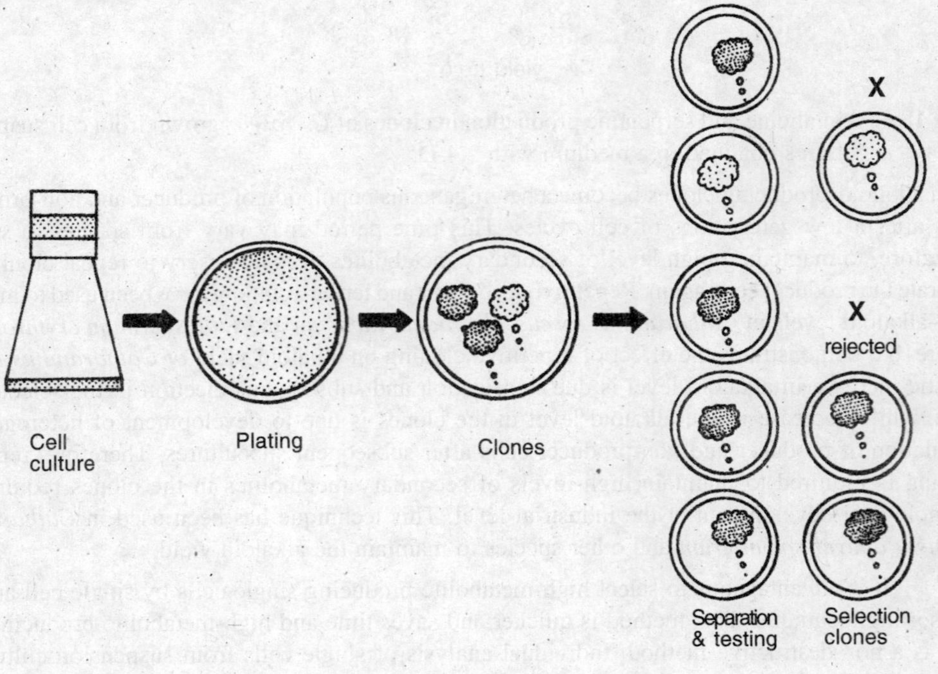

Fig- 18.7. Plating of cell suspension in the plates and selection of clones.

This way a definite selection of producer cells can be made. Plant cells grow in suspension in units ranging from single cells to clusters of more than 1000 cells. Such cultures can be made uniform in cell size by filtration through screens of appropriate size (Sigma chemical Co., USA). A screen of appropriate size is selected on the basis of size of the cells in the suspension. A uniform and defined, 1 ml suspension is spread over the surface of a 25 ml medium in 100x15 mm plastic (presterilized) petri dishes. The concentration of cells in the suspension determines the amount of inoculum. Generally, 100 mg fresh weight is reasonable to conduct plating. Single cells, smaller or larger aggregates can be used for plating and selection, depending on the growth of species in the cell cultures. Obtaining a suspension of single cells for plating has been usually achieved by filtration through a nylon net (150-250 mm). In practice, it is impossible to get a pure single cell fraction and mostly inoculum consists of single cells and 3-4 cell aggregates. It is assumed that the aggregates are derived from single cells. By obtaining protoplasts, true single-cell clones can be obtained. Optimization of culture conditions is a prerequisite for inducing cell diversions in single-cell cultures. (Fig. 18.7)

Chemically defined cell culture media may not support cell division at low cell densities of less than 9000-15,000 cells per ml, a critical inoculum density. However, in order to achieve single cell clones, cells must be plated at low densities in order to prevent overlapping of the growing colonies. Use of conditioned medium in varying proportion with fresh medium has been suggested to achieve growth of single cells. A conditioned medium is one used for the growth of cells previously. Once colonies are developed, they are grown separately by regular sub cultures of whole colony on fresh medium as deemed necessary for that species. When sufficient callus is produced (usually after 4-6 subcultures), half the callus is used for subculture and half for analysis of secondary metabolites. Selected colonies (clones) are grown and their growth and secondary metabolite contents are discarded. Quicker analytical methods are helpful in early selection of clones.

(e) Selection parameters

Clones can be selected on the basis of visual (morphological) characters, growth characteristics and physiological parameters (protein banding, secondary metabolite production). If colonies are coloured (pigmentation: anthocyanins, chlorophyll: presence-absence) or their fluorescence properties evident under ultraviolet light (secondary products), it is much easier to select producer colonies. This procedure has been effectively used in selection of high-ajmalicine producing cells of *Catharanthus roseus* and pigmented anthocyanin and shikonin producing cells in *Vitis vinifera* and *Lithospermum erythrorhizon* cell cultures, respectively.

Once the clone is grown in size, sufficient to divide and subculture, part of the clone is used for subculture while the other part is retained for qualitative/quantitative secondary compound determination. Once all clones are quantified for their useful metabolite contents, then high secondary metabolite-containing clones are selected and the poorly productive may be discarded or screened for new product/s. Any of the following criteria is used to select the clones depending on the nature of secondary metabolites.

Visible markers: Coloured pigments and compounds fluorescent under UV light may be used as visible markers for selection of clones.

Invisible compounds as markers: for such compounds a quick analytic system sensitive enough to detect minor quantities of secondary metabolites is required to evaluate the clones. These are:

- chemical tests like Dragendorff's reagent for alkaloids or Libermann-Buchard test for steroids etc.
- Thin-layer chromatography of crude samples and use of specific reagent as used for cultures of that particular species.

More sensitive methods such as GLC, HPLC, RIA may also be used which requires partial purification of samples before analysis is carried out.

V. ELICITORS

Use of special techniques such as elicitation and hairy roots cultures not only provide a method to enhance the production of secondary metabolites but also a method to clearly understand the mechanism under lying the production of secondary metabolites. Plant cells produce secondary metabolites when treated with elicitors. This phenomenon is used to understand the gene expression and gene involved in the production of secondary metabolites.

Strictly speaking elicitors are compounds of biological origin involved in plant microbe interaction. In the context of product accumulation by plant cell cultures, elicitors are mediator compounds of microbial stress (biotic elicitors) or are stress agents like UV light, alkalinity, osmotic pressure, or heavy metal ions (abiotic elicitors). Molecules, which stimulate secondary metabolism leading to the induction of stress metabolites, are called elicitors and those derived from fungi may be referred to as fungal elicitors.

The discovery of 'elicitation response' in 1972 by Keen and co-workers demonstrated that enhanced production of phytoalexins may also be achieved by treating plant cells with extracts prepared from pathogenic fungi in the absence of live organisms. Although alkaloids have not generally fallen into the phytoalexin category, in the mid 1980s a number of investigators demonstrated that the production of certain alkaloids in plant cell cultures could be induced by treating cells with elicitors derived from a variety of pathogenic fungi. It has been demonstrated repeatedly that fungal-induced stress of normal, intact plant tissues leads to the induction and accumulation of phytoalexins. The ability of cultures plant cells to produce certain metabolites in response to molecules isolated from fungi appears to be a general phenomenon, and must be correlated with physiological stress imposed on the cells.

Chemically defined elicitors (actinomycin-D, arachidonic acid sodium salt, chitosan, nigeran, poly-L-lysine and ribonuclease-A) or fungal preparations (whole extract or cell wall fraction) from common pathogenic fungi (*Pythium, Alternaria, Helminthosporium, Fusarium, Colletotrichum, Sclerotinia* etc) are used. Fungus is grown on fresh culture medium from stored stock cultures, homogenized and autoclaved at 12°C for 20 minutes and suitably diluted fungal preparations or chemicals are added in to medium to evaluate the elicitation effect.

It is well established that plant produces phytoalexins in varying concentrations when challenged by pathogenic fungi. These compounds are fungal antagonists. Though the production of phytoalexins has been demonstrated in several species, production of plant- specific secondary metabolites, other than phytoalexins, is a different manifestation. The works involving the production of plant-specific secondary products in elicitor-challenged cultures met with mixed results. The data obtained by challenging plant cells with fungal elicitors are presented in Table 18.3.

A common feature observed after treating cell suspension cultures with a fungal elicitor is the vigorous browning of the cells as well as the culture medium. It has been confirmed that elicitor treated *Catharanthus roseus* cells released phenolics in the culture medium which were independent of alkaloid accumulation. From the results obtained with co-cultivation system using tobacco cells derived from different genotypes and *Verticillium alboatrum,* it was concluded that the host defence mechanism is based upon induction of PAL activity resulting in overall increase in the phenolics profile in resistant genotypes.

Induction of *in vitro* accumulation of phytoalexins has been observed by treating plants or cultured plant cells with elicitors. Elicitor treatment has resulted in an increase in amount and activity of mRNAs encoding enzymes of phytoalexin biosynthesis in cell cultures of soybean (*Glycine max*), French bean (*Phaseolus vulgaris*) and parsley (*Petroselinum crispum*). Exposure of plant cells to the elicitors requires a minimal period of contact to induce the physiological response. It was further concluded that elicitors rapidly decharged the plasmalemma potential of the cells resulting in differential uptake.

During investigation on the regulation of benzophenanthridine alkaloid biosynthesis in suspension-cultured cells of *Sanguinarina canadensis* (blood root), the production of sanguinarine and chelerythrine was significantly enhanced by a variety of media manipulations, e.g., by phytohormone concentration, medium nutrient composition or by the addition of abiotic elicitors.

VI. HAIRY ROOT CULTURE

In the production of hairy root cultures, the explant material is inoculated with a suspension of *Agrobacterium rhizogenes*. This culture is generated by growing bacteria in yeast maltose broth (YMB) medium for two days at 25°C with rotary shaking, pelleting by centrifugation ($5x10^3$ rpm, 20 min.) and resuspending the bacteria in YMB medium to form thick suspension (approximately 10^{10} viable bacteria per ml). Transformation may be induced on aseptic plants grown from seed, or on detached leaves, leaf discs, petioles or stem segments from greenhouse plants followed by sterilization of the excised tissues. In some species a profusion of root may appear directly at the site of inoculation, but in others a callus will form initially and roots emerge subsequently from it. In either case, hairy roots normally appear within one to four weeks. The susceptibility of species to infection is very variable. Addition of acetosyringone, the compound produced during the wounding response of plants, activates the Vir genes of *Agrobacterium* adding plasmid T-DNA transfer. An alternative approach to transformation is to use cocultivation of plant protoplasts with *A. rhizogenes*. Cultures may be cleared of bacteria by several passages in media containing 200 mg/l cephalosporin or 500 mg/l ampicillin. The infection of plants with *A. rhizogenes* causes one or both of two pieces of T-DNA (Tt and Tg) to be inserted into the plant genome. Integration alters the auxin metabolism of transformed tissues in such a way that the hairy root phenotype is expressed and amino acid metabolism is modified in such a way that specific metabolites such as opines are produced. Both TL-DNA and TR-DNA have rhizogenic functions but in most species the TL-DNA appears to be more important in determining the hairy root phenotype. Although synthesis in opines (manopine or agropine) is the fine indication that roots are transformed, the expression of the opine genes in hairy root tissue may be unstable over time. Thus, detection of the T-DNA by Southern hybridization is necessary to confirm that the cultured root tissue has transformed. Hairy roots are characterized by a high degree of lateral branching, a profusion of root hairs and an absence of geotropism. Hairy root cultures do not require conditioning of the medium. A feature of hairy root systems of paramount importance for their commercial exploitation is their stable, high-level production of secondary metabolites. Although the productivity of untransformed root cultures may be similarly exploited, establishment and maintenance of such cultures is difficult and the auxin supplement needed for optimal growth can depress productivity. The transformed hairy root cultures utilized in the production of secondary metabolites are listed in Table-18.4. It is important to note that several products are synthesized in roots (e.g., nicotine) and in such cases hairy root cultures offer a unique fast-growing organized culture system.

Table 18.3. A variety of physiological responses produced under chemically defined and fungal elicitor stress.

Plant species	Fungus	Preparation	Secondary metabolite	Effect
*Apium graveolens**	*Fusarium* sp.	—	Furanocoumarins	Increase
Catharanthus roseus	*Pythium*	Homogenate	Indole alkaloid and related enzymes	Increase
Catharanthus roseus	*P. aphanidermatum*	Homogenate	Indole alkaloid and phenolics (TDC,PAL)	—
Daucus carota	*P. aphanidermatum*	Elicitor	Anthocyanins & enzymes	Increase

Glycine max	*Phytophthora* megasperma	Elicitor Chitosan	Phytoalexin Isoflavonoids	Production production
Papaver somniferum	*Botrytis* sp.	Mycelium Actinomycin	Sanguinarine Codeine	26 times increase Production
Petroselinum hortense	*Alternaria sp.*	Cell wall	Bergapten	Lowering cytoplasmic phosphate
Petroselinum hortense	*Alternaria* sp.	*Cell wall*	*PAL, mRNA*	Inhibition of uptake *of phosphate and amino acid*
Phaseolus vulgaris	*Alternaria* sp.	Poly-L-lysine RNAse	Phaseollin Phaseollin	Production Production
Sanguinaria canadensis	*Penicillium expansum*	Glucan	Benzophenanthridine alkaloids	Role of calcium in signal transduction
Solamum tuberosum	*Phytophathora infestans*	Elicitor	Phenolic & PAL activity	Increase in enzyme activity

Table 18.4. Development of hairy root cultures in an attempt to regulate the production of secondary metabolite.

Family	Species	Secondary products
Solanaceae	Atropa belladonna	Atropine
	Datura stramonium	Hyoscyamine
	Hyoscyamus muticus	Hyoscyamine
	Nicotiana rustica	Nicotine, anatabine
	N. tabacum	Nicotine, anatabine
	Scopolia japonica	Hyoscyamine
	Solanum laciniatum	Steroidal alkaloids
Apocynaceae	*Catharanthus roseus*	Ajmalicine, serpentine, catharanthine, vindolinine
Chenopodiaceae	*Beta vulgaris*	Betacyanin, betaxanthine
Boraginaceae	*Lithospermum erythrorhizon*	Shikonin
Asteraceae	*Tagetes patula*	Thiophenes
Rubiaceae	*Cinchona ledgeriana*	Quinoline alkaloids
Lamiaceae	*Mentha piperita*	Monoterpenes
	M. vulgaris	Monoterpenes

VII. BIOTRANSFORMATIONS

A bioconversion can be defined as the conversion of one chemical into another, *i.e.*, of a precursor (or substrate) into a final product, using a cell suspension acting as biocatalyst (Fig. 18.8). The biocatalyst can be microorganisms, plant or animal cells, either growing or in a quiescent state, or an extract from such cells or a purified enzyme. The biocatalyst may be free, in solution, immobilized or on solid support or entrapped in a matrix (To understand these processes, see chapter on cell immmobilization). In bioconversion by whole cells or extract a single enzyme or several enzymes may be involved. By means of recombinant DNA technology genes encoding relevant enzymes can be introduced in the host cells.

$$\underset{\textbf{(Substrate)}}{\textbf{A}} \xrightarrow{\text{Cell/enzyme}} \underset{\textbf{(Product)}}{\textbf{B}}$$

Fig.18.8. Bioconversion of substrate into product in presence of biocatalyst.

Each individual cell contains many enzymes, which can display different catalytic properties depending on the conditions to which they are exposed. A plant cell contains 800-1000 different enzymes belonging to primary and secondary metabolism. Enzymes has two properties similar to catalysts; the ability to speed up a reaction, and to remain unchanged at the end of the reaction, they 'recycle' in the reaction. The principle for this process is that enzymes efficiently bind to the transition state of a precursor. Consequently, the free energy of activation for the reaction is strongly reduced and the bioconversion facilitated. The reaction is speeded up without changing the equilibrium. Once the product is formed it is released and the active site becomes available for further biocatalysis.

Since each enzyme has its own specific site, enzymes are selective towards their precursors (substrate specificity). However, a number of enzymes with broad specificity are known, e.g., phenoloxidases, lipases, and esterases. Most biocatalyst are active at room temperature ensuring stability of the precursor and the product.

(a) Freely suspended plant cells:

Freely suspended cells form the most simple biocatalytic system, as precursors can be supplied directly to the cultures. Often batch-grown cultures consisting of undifferentiated plant cells are used. The only barriers a precursor has to pass are the cell wall and cell membrane. A wide range of potentially valuable compounds has been produced by adding precursors to various culture species. To produce one particular compound the best way is to use plant cells that can perform a one-step bioconversion. Unfortunately, often the precursor undergoes more than one bioconversion resulting in complex mixtures of (unknown) products.

Nevertheless, a number of one-step bioconversions by freely suspended cells have been described. The bioconversion rates have been calculated in terms of mg product formed per liter culture per day (mg $l^{-1}d^{-1}$). This way of calculation enables comparison of bioconversion rates of freely suspended with those obtained by immobilized cells. The bioconversion of papaverine (an opium alkaloid with spasmolytic properties) has been studied by Dorisse et al. (1988) using cell suspensions of *Glycyrrhiza glabra.* After 21 days, 31.5% of this precursor was hydroxylated, mainly into papaverinol. The bioconversion rate of 3.7 mg $l^{-1}d^{-1}$ was rather low. Addition of gitoxigenin (belonging to the cardenolides with action on the insufficient heart) to a culture of *Daucus carota* yielded a single bioconversion product, 5 b - hydroxygitoxigenin (Table 18.5). The precursor was added at 0 h or at 72 h after inoculation of the cells. In the first case 50% of the precursor was completely hydroxylated into 5 –hydroxygitoxigenin after 51h, in the latter case this level was already reached after 24h, corresponding to a high bioconversion rate of 50 mg $l^{-1}d^{-1}$.

Besides hydroxylation, glycosylation is one of the most interesting bioconversions. Glycosylation, particularly gl<u>u</u>cosylation (glucose coupling), occurs readily in plant cells, but only on a limited scale in micro-organisms and the organic chemist has problems with sugar coupling as well. Consequently many glucosyltransferases with different substrate (precursor) specificities for several hydroxyl bearing compounds have been investigated. Examples are glucosylations by cells of *Ruta graveolens, Perilla frutescens, Catharanthus roseus, Lithospermum erythrorhizon, Mallotus japonicus* and *Eucalyptus perriniana.* Glucosylations by cells of *Glycyrrhiza echinata, Aconitum japonicum, Coffea arabica, Dioscoreophyllum cumminsii* and *Nicotiana tabacum* have been performed as well. Examples of other glycosylation reactions are the fructosylation of ergot alkaloid (ergotamine is used against migraine, ergometrine provokes uterus contraction) by cells of *Claviceps purpurea* (a fungus of class ascomycetes). Other examples are listed in Table 18.6.

Podophyllotoxin from *Podophyllum* resin (podophyllin is highly active against several cancer cell lines but its general toxicity is too high to allow clinical application. Chemical modification of the naturally occurring lignan has led to the clinically applied anticancer drugs teniposide and etoposide. The organic synthesis of podophyllotoxin-like lignans is difficult, particularly with regard to their chiral centres and, therefore, alternative routes to lignans are needed. More than 30 lignans are known to exhibit powerful cytotoxic action. Lignans are clearly candidates for plant cell biotechnological production. For the formation of podophyllotoxin and 5-methoxypodophyllotoxin by cultures of *Podophyllum hexandrum* and *Linum flavum* respectively, a series of bioconversions in the cells are required.

(b) Immobilized plant cells

Immobilization may have benefits with respect to the production of secondary metabolites. Note the difference between the words immobilized and entrapped to immobilize means to make immobile or immovable, to entrap means to catch in. Thus, immobilized cells can be entrapped cells, but also cells adsorbed into support material. By entrapment the cells become protected against shear damage in bioreactors. The entrapped cells can be used over a prolonged period, high concentrations of biomass are possible, principally giving high bioconversions of precursor, and the method simplifies recovery of the cell mass and products. An immobilization method selected for plant cells should not negatively affect the cells, must be carried out easily under aseptic conditions, should enable operating for long periods, and for large scale applications in particular, should be low in cost.

Immobilized plant cells may have higher production rates as compared with the freely suspended cells under the same bioconversion conditions. Entrapment of cells may result in a kind of micro-environment, resembling that of the organized tissue in the intact plant. Entrapment of plant cells is one of the means to create non-growth conditions under which the production of secondary metabolites may be improved.

Some successfully applied, characteristic one-step bioconversions reported in literature are summarized in Table 18.7. Alginate-entrapped cells of *Papaver somniferum* were able to reduce stereospecifically condeinone into codeine at a low rate of 3.8 mg $l^{-1}d^{-1}$, while more than 80% of the product was released into the medium during an incubation period of three days. The freely suspended cells showed a lower bioconversion capacity. Alginate-entrapped cells of *Digitalis lanata* were able to hydroxylate b-methyldigitoxin into methyldigoxin at a moderate rate of 9.0 mg $l^{-1}d^{-1}$, but in contrast with the other systems mentioned here, the bioconversion period was very long, namely more than 60 days. The freely suspended cells possessed 2-fold higher hydroxylation capacities for a much shorter period.

Alginate is a very suitable matrix for the entrapment of plant cells. Alginate entrapped cells of *Mucuna pruriens* showed the highest bioconversion rate for L-DOPA production when compared to pectinate-, agarose- and gelatine-entrapped cells.

When enzyme preparations are chosen for bioconversion purposes, some criteria have to be met. It is necessary that sufficient amounts of active enzyme can be isolated easily from cell cultures and that they remain stable.

(c) Production of secondary metabolites by isolated plant enzymes:

The small numbers of isolated plant enzymes that have been described in the literature, usually catalyze one-step reactions. Strictosidine synthase has been isolated from cell suspension cultures of *Catharanthus roseus* followed by its characterization. The enzyme forms strictosidine, the central intermediate of the indole alkaloid biosynthesis. By the stereospecific condensation of tryptamine and secologanin, the 3a (S)- isomer is exclusively formed. Digitoxin 12 b- hydroxylase, a Cytochrome P-450-dependent mono-oxygenase present in the microsomes in cell cultures of *Digitalis lanata*, has been isolated and characterized (Table 18.8). The enzyme is responsible for the hydroxylation of methyldigitoxin into methyldigoxin, a reaction of pharmaceutical interest. The microsomes were isolated followed by entrapment in 2 % calcium alginate and 70% of the enzyme activity was retained. The biocatalyst was applied in a continuous flow system, in which the methyldigoxin production continued for more than 20h. The product was not isolated during the process, which might explain the relatively low bioconversion rate of 3.1μ kat kg^{-1}. Product inhibition, a common phenomena in enzyme catalysis, is likely to have occurred.

Table.18.5. Production rates of plant cell cultures biosynthesizing high and low amounts of secondary metabolites.

Cell culture species	Product	Production rate($mg\,l^{-1}\,d^{-1}$)
Anchusa officinalis	Rosmarinic acid	400
Mucuna pruriens	L-DOPA	200
Linum flavum	Coniferin	127
Lithospermum erythrorhizon	Shikonin	100
Artemisia annua	Artemisinin	2.0
Campthoteca accuminata	Camptothecine	1.9
Podophyllum hexandrum	Podophyllotoxin	1.3

Table 18.6. Selected one-step bioconversions by freely suspended cells.

Cell culture species	Reaction type	Precursor	Product	Remarks
Glycyrrhiza glabra	*Hydroxylation*	*Papaverine*	*Papaverinol*	*Long period of bioconversion (21 d)*
Daucus carota	*Hydroxylation*	*Gitoxigenin*	*5-hydroxyg-itoxigenin*	*Precursor added after 72 h of cultivation.*
Mentha species	*Reduction*	*(-) menthone*	*(+)-neomenthol*	*Stereospecific bioconversion*
Rauwolfia seperpentina	*Glucosylation*	*Hydroquinone*	*Arbutin*	*Precursor added continuously*
Coffea arabica	*Glycosylation*	*Vanillin*	*vanillin—D-glucoside*	*High bioconversion percentage *85%)*
Choisya ternata	*Hydroxilation*	*ellipticine*	*5-formyl ellipticine*	*In static medium*

Table 18.7. Selected one-step bioconversions by immobilized cells.

Cell culture species	Reaction type	Precursor	Product	Matrix and remarks
Papaver somniferum	*Reduction*	*Codeinone*	*Codeine*	*Alginate and PUR; in both cases stereospecific bioconversion, product release.*
Digitalis lanata	Hydroxylation	b-methyl-digoxitoxin	b-methyl-digoxin	Alginate; very long bioconversion period (160 d)
Mentha species	Reduction	(-)-menthone	(+)-neomenthol	PAAH; stereospecific reduction, product release.
Daucus carota	Hydroxylation	Digitoxigenin	Periplogenin	Alginate; five times reuse possible
Mucuna pruriens	Hydroxylation	L-tyrosine	L-DOPA	alginate, pectinate, agarose,gelatine; product release

Table 18.8. Selected one-step bioconversions by enzyme preparations.

Enzyme	Reaction type	Precursor	Product	Remarks
Strictosidine synthase	Condensation	Tryptamine + secologanine	Strictosidin	Stereospecific condensation, adsorbed enzyme with long half life (60 d)
Digitoxin 12 –' hydroxylase	Hydroxylation	Methyl-digitoxin	Methyl-digoxin	Continuous flow system for more than 20 h.
O-glucosyl transferase	Glucosylation	Hyoscyamine	6- hydroxy-hyoscyamine	Only S-forms of precursors are converted, also epoxylating activity
Phenoloxidase	Hydroxylation	L-tyrosine	L-DOPA	Continuous flow system for more than 2 d, very high activity.

QUESTIONS

1. What is biotransformation? Describe different systems used to obtain products of interest.
2. Write short notes on the following :
   (*a*) Immobilized enzymes
   (*b*) Biotransformation
   (*c*) Heterologous expression and
   (*d*) Reaction types in biotransformations
3. Write short notes on the following : (*a*) Hairy root Cultures (*b*) Ri plasmid (*c*) Elicitors.
4. Describe the various factors affecting the production of secondary metabolites.
5. Write short notes on the following :
   (*a*) Optimization (*b*) Selection
   (*c*) Cloning (*d*) Production medium (*e*) Precursor
6. How will you improve upon production of secondary metabolites in a given cell culture system ?
7. What are prerequisites for th production of useful metabolites in a bioreactor system?
8. why recurrent selection of clones is required for stable production of secondary metabolites?

BIOREACTORS

CHAPTER 19

I. INTRODUCTION

Plant cell and callus cultures have been extensively used in the past three decades to explore the possibility of producing useful secondary metabolites through biotechnological methods. These empirical approaches generated enormous data about the plant species producing secondary metabolites in culture, physico-chemical factors like light, temperature, plant growth regulators and nutrients affecting the production and lastly, selection of cells with high yield of secondary metabolites. However, all the cultures investigated were not commercially viable systems. Therefore, two way approach was made: 1) to generate the basic know-how for the production of secondary metabolites, and 2) to develop technology for specific product using plants like *Catharanthus roseus, Berberis, Coptis* and *Panax* etc. as a model system. Investigations were focused to develop technology for these plants and also to better understand the process as a whole.

Plant cell in the agitated liquid medium produces secondary metabolites characteristics of the parent plant, a process governed by several genes (i.e., genetically controlled) and hence it is a multi-steps reaction. To produce secondary metabolites at commercial level there are certain pre-requisites like- high demand, high product cost, and availability of uniform raw material without interruption. Technology should be cost effective otherwise the production will not be commercially viable. Only a few selected compounds can fulfil these criteria because of high production cost. It is imperative that the production cost can be reduced by increasing the yield of secondary metabolites by improving the bioreactor technology to reduce the capital cost of inputs. Serious attempts are being made to increase the yield by manipulating the biosynthetic pathways, removing the limiting factors barriers, and incorporating the techniques of genetic engineering. Here, we are presenting the details of bioreactor system and various cultures used to produce secondary metabolites in these bioreactors, towards developing technology for the industrial-level production.

Application of plant cell tissue or organ culture for the production of secondary metabolites was originated in 1947, when James first reported the occurrence of alkaloids in meristem culture of solanaceous plants. Scale-up (process at large-scale as compared to small-scale process in laboratory) problems are common in all kinds of engineering sciences. Various approaches have been developed to solve these problems, ranging from the fundamental solutions to the processes where 'trial and error' and 'rule of thumb' are used. The classical processes of the fermentation industry have been developed along the time with trial and error method. Bioreactors are advantageous then shake flasks as they provide better control on the system (i.e., pH, dissolved gas concentration, cell growth, etc) there are examples of cultivation of plant cells in large fermenter demonstrating feasibility of the technology for the scale-up production of secondary metabolites. Tulecke and Nickel (1959) successfully developed a 10L system in a simple carboy for the cultivation of plant cells. Noguchi and co-workers (1977) cultivated *Nicotiana tabacum* cell suspensions in a 20,000 L stirred tank reactor while Schiel and Berlin (1987) studied *Catharanthus roseus* cells in a 5,000L stirred tank reactor. The first commercial process used *Lithospermum erythrorhizon* cell suspensions grown in stirred tank reactors of 200 and 750L. Various published results show that plant cell cultures can be grown in the microbial type bioreactors, thus suggesting the use of existing fermentation facilities and conventional equipment for industrial production of plant products using the cell culture technology. In the contrary to above, Archanbault *et al.* (1998) suggested that specifically designed and operated bioreactors might be more appropriate for plant cell cultures. It is worth mentioning that different types of bioreactor designs and processes are currently in use for the production of secondary metabolites (Table 19.1).

Table.19.1. Some examples of large-scale production of plant cells.

Plant species	Capacity in litres	Compound produced
Catharanthus roseus	85	Serpentine
Coleus blumei	450	Rosmarinic acid
Lithospermum erythrorhozon	750	Shikonin
Nicotiana tabacum	20,000	Biomass
Panax ginseng	200-20,000	Saponins

II. BIOREACTOR PROCESS

Bioreactor is a large culture vessel made up of glass for use at laboratory-scale (up to 10L) but large-scale bioreactors are made up of stainless steel. Laboratory-scale bioreactors can be designed and fabricated in the laboratory or can be purchased as a functional unit (Fig.19.1). The commercially available systems are sophisticated equipment fitted with microprocessor control unit (or a computer) to control the pH, dissolved oxygen, gas flow rate, agitation speed, nutrients, temperature inside the vessel and cell density for the optimal growth and/or the production. All the units have a culture vessel and a control unit. The details of a typical stirred type bioreactor are given in the Fig.19.2. There are different types of stirrers available to suit the requirement for tissue shear pressure and effective agitation. The modern bioreactors are fitted with various sensors like pH, temperature, foaming and nutrient concentrations. The control unit records the changes in the composition of the medium and then transfer the desired chemical/nutrient to adjust the medium according to set parameters. Samples are taken, at time intervals to study the time course of the growth, with the help of sample tubes (Fig.19.3) fitted with sterilized membrane filter and syringe to create a suction. For sterilization, small reactors are autoclaved while commercial scale reactors are sterilised *in-situ* by passing steam at appropriate pressure. Due care should be taken to sterilize the system according to its volume. Various kinds of bioreactors have been designed depending upon the requirements and the systems, which are discussed later in this chapter. However, all the bioreactors require a cooling unit (cryobath with water circulating pump) to circulate cold water (being a tropical country, in temperate countries, tape water is quite cold and can serve the purpose), a membrane type air pressure pump to supply oil-free air at a desired pressure, an autoclave of desired dimensions to sterilise the unit, disposable or autoclavable air filters, and autoclavable tubings to connect the system besides general facility for tissue and cell culture. A sufficient amount of stock of cell culture is required to inoculate the bioreactor.

This system is used to develop technology for the industrial production of useful secondary metabolites or for the production of large number of plantlets using somatic embryogenesis. Secondary products produced in plant cells are either released into the medium or accumulated in the cells. Depending upon the tissue type, either the spent medium or the biomass is harvested after suitable incubation period for the extraction of compounds. Moreover, if growth and production occur simultaneously (*i.e.*, growth associated production), a process mode which supports growth of cells over an extended period should be chosen. But, if the product synthesis follows a period of rapid growth (*i.e.*, non-growth associated production), then an appropriate process mode is chosen which is capable of maintaining cultures in a non- or slow- growing rate and retain its productivity. Depending upon these factors for the accumulation / release of secondary products, different process modes are used viz., batch culture, fed batch culture, rapid feed batch (semi-continuous) cultures, two-stage batch culture, continuous chemostat type cultures and continuous profusion culture. It is imperative that a bioreactor is used when basic studies related to optimization of product yield have been completed. In most cases, a growth medium and a production medium are used to obtain maximum yield of the secondary metabolite.

III. FACTORS FOR GROWTH IN BIOREACTOR

A. Gas-liquid Mass Transfer

Maintaining a constant oxygen mass transfer coefficient ($K_L a$) is a basis of scale-up of many bio-processes. The oxygen transfer requirements of cultured plant cells are low as compared to that of bacterial cultures because unlike microorganisms, plant cells have lower respiration rates. For instance, if cells which respires at a rate of 0.2 mM g^{-1} h^{-1} are to be grown to 10 gl^{-1} without allowing the dissolved

Fig.19.1. A stirred tank bioreactor with its control system.

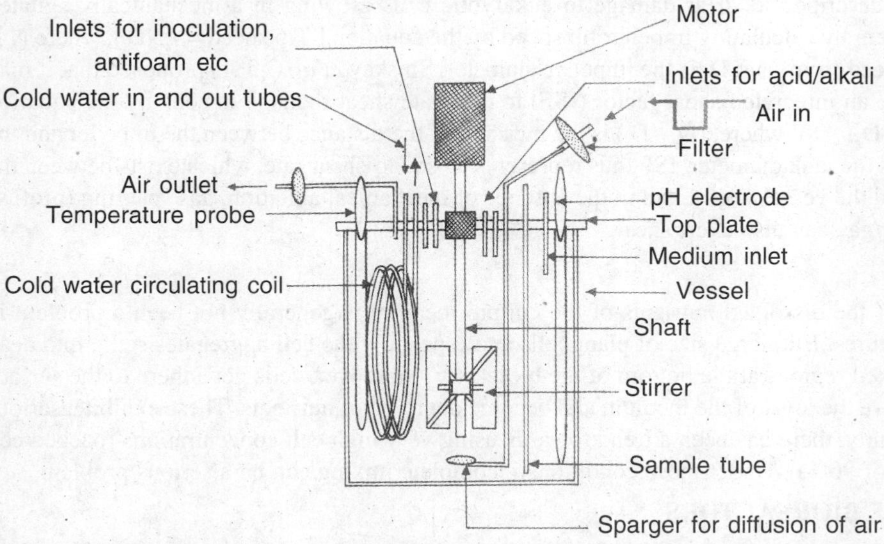

Fig.19.2. A schematic presentation of various parts of a stirred type bioreactor.

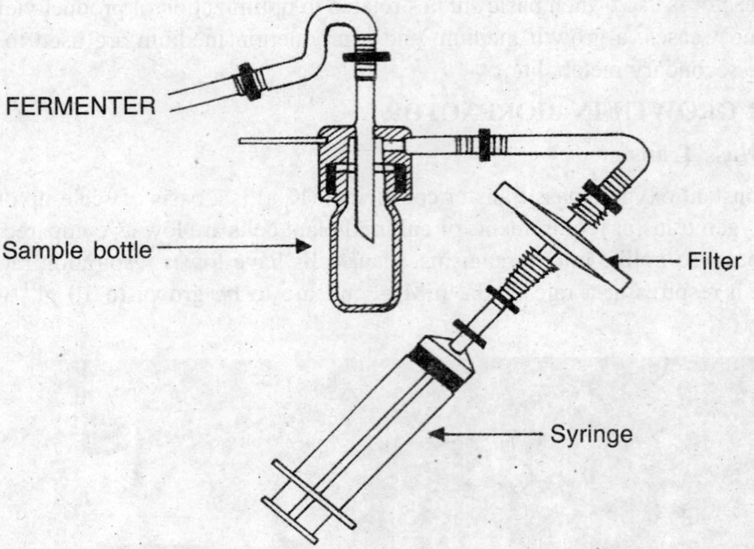

Fig.19.3. Arrangement for taking small sample from bioreactor using a filter and syringe.

oxygen concentration to fall below 20% of saturation. Plant cell culture in bioreactors typically requires K_La values between 10-30 h^{-1}. Operation of bioreactor at higher K_La values results in poor cell growth or production of secondary metabolites. This may be due to either increased shear associated with high K_La conditions or due to enhanced CO_2 stripping from the medium. To maintain a constant dissolved oxygen level, bioreactors should be equipped with a dissolved oxygen probe to adjust aeration and agitation rate.

B. Shear

Plant cells are shear sensitive. Shear refers to forces exerted on the surface of a body in a direction parallel to surface. This is contrasted to normal forces, which are exerted on a surface but, perpendicular to the surface. Mathematically, shear can be described by the equation [$\tau = n\gamma$], where t is the shear stress (a force per unit area), g is the shear rate (a change in velocity across a distance), and h is a viscosity (a coefficient which describes the resistance to flow). Earlier, efforts were made to describe the shear damage to eukaryotic cells growing in a mechanically agitated bioreactor system by calculating impeller tip speed by the equation [Tip speed = πND_i], where N is the impeller speed (rpm) and D_i is the impeller diameter. Sinskey *et al.* (1981) proposed that it may be better to use an integrated shear factor (ISF) to correlate shear damage to cells by the equation [$ISF = \pi ND_i /(D_t - D_i)$] where, $(D_t - D_i)$ is the measure of the distance between the impeller and the tank wall. D_t is the tank diameter. ISF thus represents a pseudo-shear rate, which exists between the impeller tip and the vessel wall. Even in the absence of mechanical agitation, gas sparging (profuse bubbling of air/gas)can also exert shear.

C. Mixing

Mixing of the dissolved nutrients of the culture medium is generally not been a problem in suspension cultures. But, large size of plant cells and especially the cell aggregates settle into dead zones or unmixed regions at the bottom of the bioreactor. Moreover, cells get adhere to the surface of the tank above the level of the medium and become deprived of nutrients. These exhibit a serious problem. Recently, there has been a trend towards using very high cell concentrations [packed cell volume (PCV of 90%)]. Under these conditions incomplete mixing can be a serious problem.

IV. TYPES OF BIOREACTORS

Depending upon the mode of agitation, bioreactors can be basically classified into following two types:

A. Mechanically agitated bioreactors

The mechanically agitated bioreactors are most commonly used for large-scale culture of plant, animal and microbial cells. In this type of bioreactor, the medium is agitated with the help of a mechanically driven impeller. Various types of impellers are in use, depending upon the requirement. Two traditional types of impeller used in this system are shown in Fig.19.4.

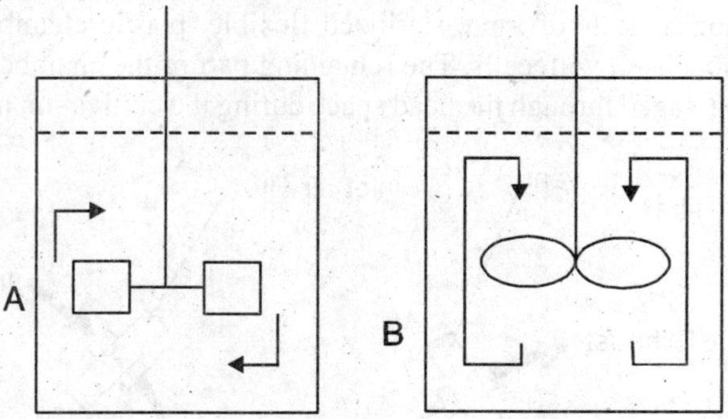

Fig.19.4. Types of impellers: A) Flat-blade turbine impeller and B) Marine propeller.

Flat-blade turbine impeller (Fig.19.4A) with high speed is generally used in bacterial culture. High agitation breaks the incoming air into small bubbles. Since plant cells cannot tolerate high shear conditions and mixing of air may be a more serious problem with plant cell cultures, an alternate impeller, capable of inducing low shear have been used. Marine propeller impeller (Fig. 19.4B) is better suited for low shear mixing. It provides axial mixing of the medium. On the other hand, flat-blade turbine promotes radial mixing. For large-scale systems, neither of these two types of impeller is used. Alternatively, low-shear impellers (e.g., paddle and helical types) have been shown to more useful for plant cell cultivation. Since low agitation is insufficient to break incoming gas into small bubbles, incoming gas stream is dispersed as fine bubbles using an appropriate gas distributor. Enrichment of incoming gas with oxygen is beneficial.

B. Pneumatically agitated bioreactors

Pneumatically agitated bioreactors are of two types viz., the bubble column and air-lift (Fig. 19.5). These bioreactors are tall and thin as compare to mechanical agitation type reactors. Typically, the height-to-diameter ratio in pneumatically agitated bioreactors is high. In bubble columns, air is bubbled at the base of the column and medium is agitated with this. In air-lift bioreactors, gas is fer.

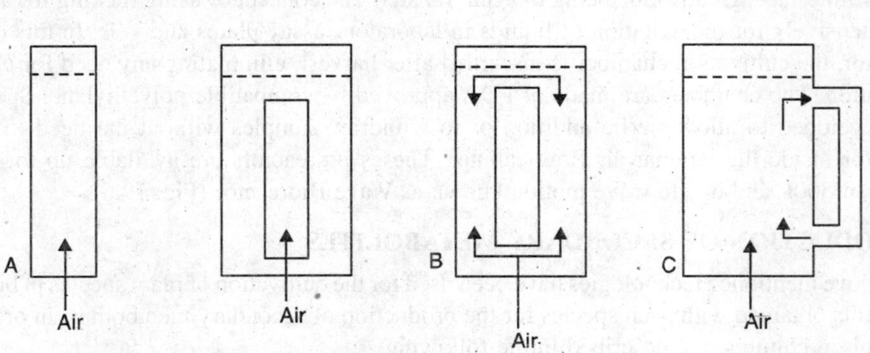

Fig.19.5. Pneumatically agitated bioreactors; A) Bubble column, B) Draft tube air-lift and C) External loop air-lift.

sparged in the riser section and after the gas is disengages at the top of the column, the medium then flows downward in the down-corner section. These two sections may be separated using a baffle, a concentric cylinder, or an external loop. Circulation in the air-lift bioreactor promotes better mixing and therefore, has advantages in suspending cells and clumps more uniformly. But, in large-scale air-lift bioreactors, oxygen transfer rate is low in the down-corner section. It should be emphasized that the performance of an air-lift bioreactor is strongly dependent upon the geometry of the system. Ratio of the cross-section area of the riser section and down-corner sections is of special importance because this ratio affects mixing and oxygen transfer.

C. Wave bioreactor

Wave bioreactor consists of a pre-sterilized flexible, plastic chamber that is partially filled with medium and inoculated with cells. The remaining part of the chamber is inflated with air. The air is continuously passaged through the head space during the cultivation mixing and mass transfers

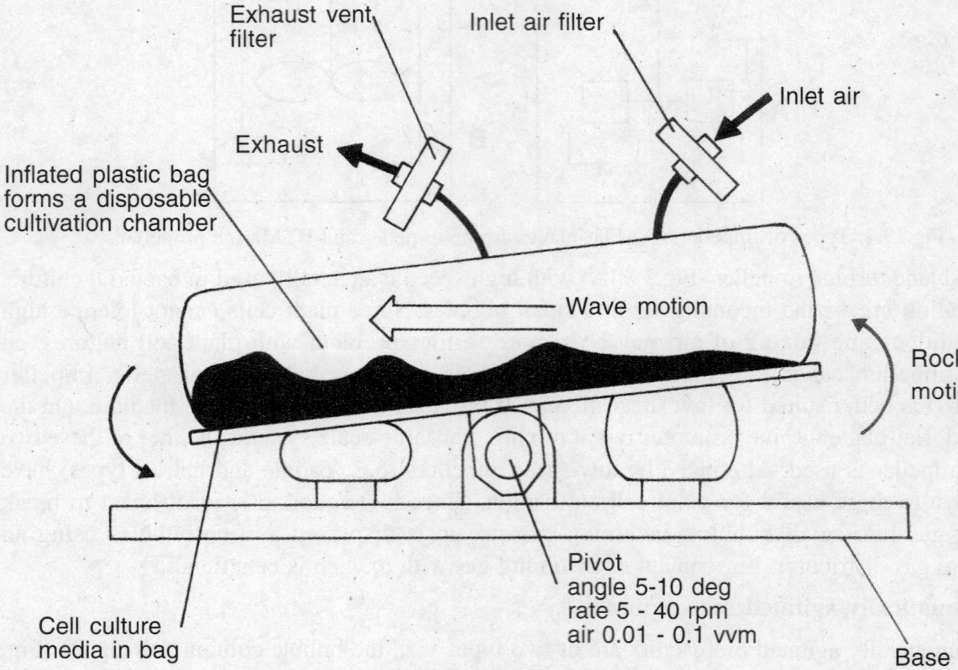

Fig. 19.6. Wave bioreactor

are achieved by rocking the chamber back and forth. This rocking motion generates waves at the liquid air interface, greatly enhancing oxygen transfer. The concept of using rocking for agitation is used extensively for the agitation of liquids in laboratory assay plates and gels. In this disposable bioreactor, the cultivation chamber is discarded after harvest, eliminating any need for cleaning or sterilization. The chambers are made of FDA approved biocompatible polyethylene. Special ports were developed to allow sterile addition or to withdraw samples without the need to place the bioreactor inside the laminar air flow cabinet. Theses bioreactors are available up to a working volume of 500L and due to wave motion known as Wave Bioreactor (Fig. 19.6)

V. PRODUCTION OF SECONDARY METABOLITES

Above mentioned technologies have been used for the cultivation of plant species in bioreactors. The results obtained with such species for the production of secondary metabolites, in organised or disorganised cultures, are described in the following.

A. Shikonin Production

Shikonin is used as a remedy for various skin ailments and as a dye for skin and cosmetics. Shikonin is synthesized in cells of *Lithospermum erythrorhizon* from the shikimic acid and isopropanoid pathways as shown in Fig. 19.7. Commercial production of shikonin was stimulated by its value and problems encountered in meeting the demand from the natural resources. The first process for the commercial production of natural plant product by the cell suspension cultures was developed in Japan for the production of the naphthoquinone shikonin. Through visual inspection it was possible to identify the high-producing cultures.

Since the growth of cells was low in the medium (designated as M-9) which supports high levels of shikonin production, a two-stage culture process was developed. The cells are first grown in a growth promoting medium (designated as MG-5) for 9 days, filtered and pumped to a second vessel in which the M-9 production medium is added (Fig. 19.8). The cells are cultured for 14 additional days, after which they are recovered by filtration, and the shikonin is extracted from the cells. Skikonin levels at that time were reported to be 4g l^{-1}. They reported the effect of oxygen on shikonin production. Growth and production were enhanced by increased K_La for values upto $18h^{-1}$. In a fermenter with a paddle-type impeller, shikonin production was enhanced by increased K_La upto 10 h^{-1}. Paddle-type impeller favours cell growth however, shikonin production was reduced as compared to that in shake-flask cultured cells. High speed of the impeller further reduced shikonin production due to mechanical injury of the cells. Efforts to use air-lift type bioreactor were frustrated by the 'bubbling-up' of the cells and the subsequent adherence of the cells to the tank wall. These authors then adopted a rotary drum type tank which revolved slowly, thus washing cells from the wall (Fig. 19.9). This system has been scaled-up to 1000L without loss in shikonin production.

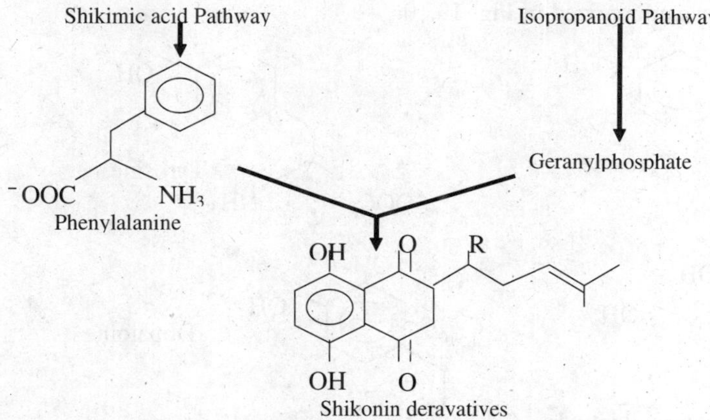

Fig. 19.7. Pathway of shikonin and its derivatives.

Skihonin derivatives are a hydrophobic compound and exist on the surface of the cells as oil particles. Organic solvents such as ethanol or ether are commercially used for the extraction of the derivatives from the dried roots of *L. erythrorhizon*. The derivatives from the dried cultured cells are extracted with ethanol.

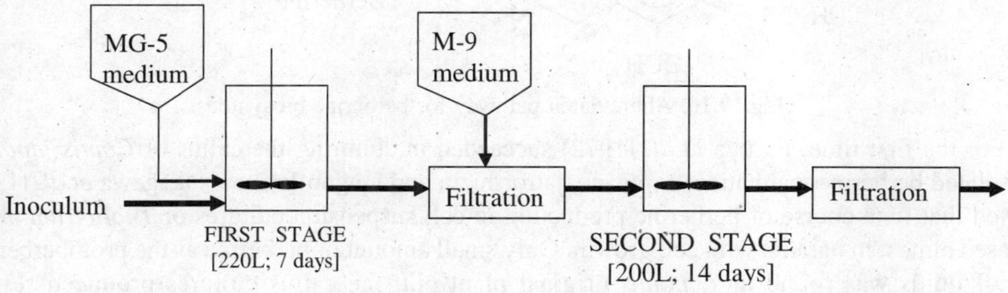

Fig.19.8. Schematic diagram of process of commercial production of shikonin.

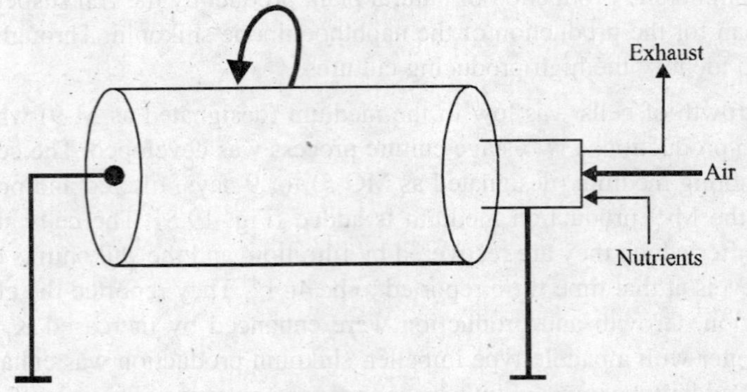

Fig.19.9. Rotary-drum bioreactor for the production of shikonin (Tanaka et al. 1983).

B. Berberine production

Berberine alkaloids are widely distributed in plants belonging to families berberidaceae, ranunculaceae, menispermaceae and rutaceae. Production of berberine has been scaled-up to the 40 bioreactor by Mitsui Petrochemicals Industries in Japan. Berberine is derived from two molecules of tyrosine as summerized in Fig. 19.10.

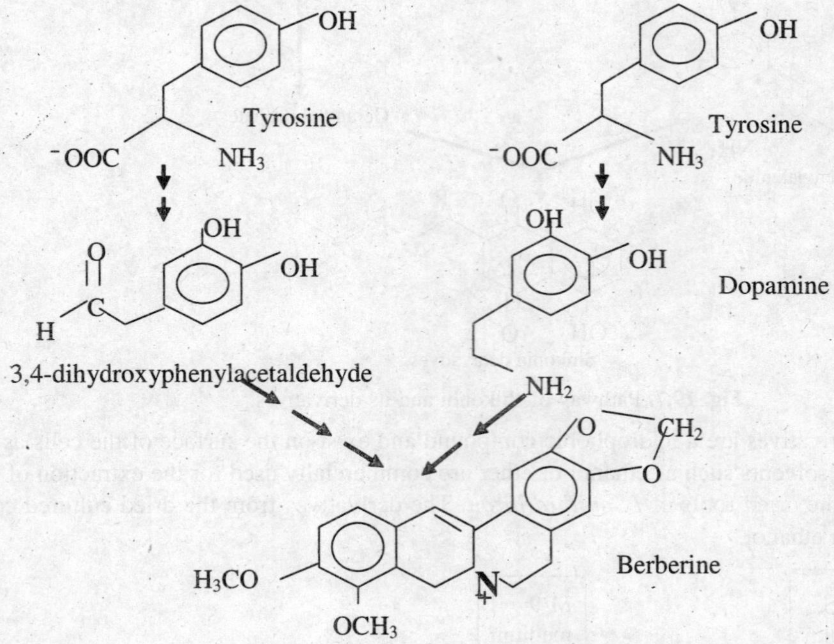

Fig.19.10. Abbreviated pathway for berberine biosynthesis.

For the first time, Furuya *et al.* (1972) succeeded in culturing the callus of *Coptis japonica* and isolated berberine, palmatine, coptisine, jatrorrhizin and magnoflorine. Nakagawa *et al.* (1984) reported that time course of berberine production in cell suspension cultures of *Thalictrum minus* increased almost in parallel with cell growth. Only small amount of berberine as the protoberberine-type alkaloids was found in *T. minus* original plant but, the callus cultures produced several protoberberine-type alkaloids, moreover, berberine was the main product. Berberine production in the cultured cells of *Thalictrum* was about 350 times higher than that in field grown plants. In contrast with cell cultures of *T. minus, Coptis japonica* and *Berberis aristata* cells in culture accumulated most of the protoberberine alkaloid in the cells and released only a small amount of alkaloids into the medium, when partial autolysis of senescent cells occurred. Kobayashi *et al.* (1988) developed an immobilized cell bioreactor for the production of berberine. Mitsui Petrochemical Industries has developed a technique for the commercial production of berberine using scaled-up system of 4000L vessel.

C. Rosmarinic acid production

Rosmarinic acid (a-o-caffeoyl-3,4-dihydroxypharyllacetic acid) has been most commonly found within the families lamiaceae and borigenaceae, and seems largely restricted to the tubiflorae. Rosmarinic acid is produced from the two distant precursors, phenylalanine and tyrosine (Fig. 19.11). Although most of the rosmarinic acid was reported to be produced during the period of rapid growth, production only occurred in *Coleus blumei* cell cultures when detectable levels of (PAL) was present. Oxidized rosmarinic acid displayed anti-thyrotropic activities in tests with human thyroid membrane preperations and the pure compound has been shown to effectively suppress the complement-dependent components of endotoxin shock in rabbits .

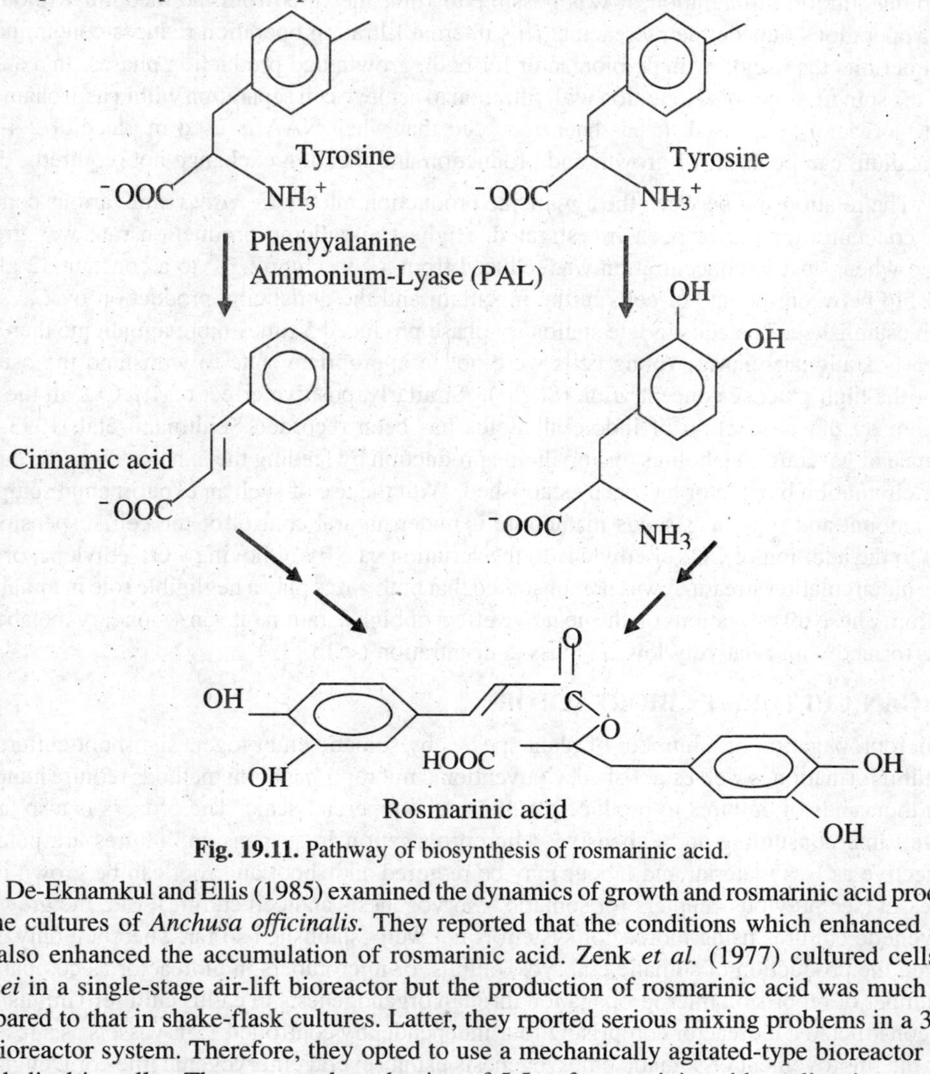

Fig. 19.11. Pathway of biosynthesis of rosmarinic acid.

De-Eknamkul and Ellis (1985) examined the dynamics of growth and rosmarinic acid production by the cultures of *Anchusa officinalis.* They reported that the conditions which enhanced growth and also enhanced the accumulation of rosmarinic acid. Zenk *et al.* (1977) cultured cells of *C. blumei* in a single-stage air-lift bioreactor but the production of rosmarinic acid was much low as compared to that in shake-flask cultures. Latter, they rported serious mixing problems in a 32L air-lift bioreactor system. Therefore, they opted to use a mechanically agitated-type bioreactor system with helical impeller. They reported production of 5.5g of rosmarinic acid per liter in a two-stage batch culture system. Firstly, cells were grown for 8 days in a glucose fed batch system and after suitable growth, cells were filtered out and pumped into a second culture vessel in which they were diluted with a sucrose solution to initiate production of rosmarinic acid. Incorporation of the precursors of rosmarinic acid biosynthesis inhibits both, growth and production.

D. Indole alkaloids production

A number of indole alkaloids are produced by *in vitro* cultures of *Catharanthus roseus* including anti-tumour agents. Since *C. roseus* cells enlarge as they age, they may become more shear sensitive, and this increased shear sensitivity may explains low production of secondary metabolites. Due to this shear sensitivity of cells, several workers have adopted pneumatically agitated type bioreactors for culturing *C. roseus* cells. Both, external loop and draft tube airlifts as well as bubble column systems have been used. In contrast, others used a mechanically agitated bioreactor with flat blade turbine impellers, and the alkaloid production in this type of bioreactor was less than that observed in shake flasks even when same inoculum was used in both the systems.

Since the medium which favours growth of cells do not favours production of the alkaloids in *C. roseus* cells, a two-stage culture system was used similar to that for the production of shikonin. By coupling filtration to aeration, it was possible to filter the cells from the medium without the need for operations outside the bioreactor. This internal filtration operation reduces contamination risk and permits the use of a single bioreactor for both, growth and production phases. In a similar system, the spin filter-coupled agitation with filtration to achieve cell separation within a mechanically agitated bioreactor was used. It has been recorded that when NAA is used in place of 2,4-D, a single medium can be used for growth and production and medium exchange not required.

The relationship between the ajmalicine production rate by *C. roseus* and various constant glucose concentrations have been investigated. Highest ajmalicine production rate was greatly enhanced when glucose concentration was reduced from a constant 57 gl^{-1} to a constant 32 gl^{-1}. A relationship between the age of cells in the inoculum and the ajmalicine production by *C. roseus* has been established. The cells in late stationary phase produced 5 times more ajmalicine than cells in the early stationary phase. Young cells were not in appropriate state to withstand the osmotic shock of the high glucose concentration ($80gl^{-1}$). Similarly, positive effect of 4% CO_2 in the inlet gas stream on the production of indole alkaloids has been recorded Schltmann etal. 1993. The importance of gaseous metabolites on ajmalicine production by feeding the exhaust air to the culture in an air circulation bioreactor has been established. With the use of such an experimental setup, the timing, amount and type of gaseous metabolite is under natural control of the cell suspension, in contrast to the addition of CO_2 or ethylene to the aeration gas. By removing CO_2, ethylene, or both from the air circulation stream, it was demonstrated that both gases play a negligible role in ajmalicine production. These investigations on the negative effect of high aeration rate on secondary metabolites were performed with relatively low biomass concentration ($< 15\ gl^{-1}$).

VI. ORGAN CULTURE IN BIOREACTOR

Micropropagation of a number of plant species by somatic embryogenesis, shoot culture and microbulbil formation is well established. Conventional micropropagation methods require hundreds or even thousands of cultures to produce plantlets at commercial-scale. The process is also labour intensive, time consuming and expensive. Micropropagation in suspension cultures are potential cost-effective as less material and labour may be required and shoot and root can be grown in the same vessel (see previous chapters for somatic embryogenesis and cell culture). But, the growth of organogenetic cultures using bioreactor system is not well established so far. There are only a few reports on the production of somatic embryos, bulbils, or microtubers in bioreactor as compared to large number of reports on micropropagation through organogenesis in callus culture. Timmis *et al.* (1998) constructed a bioreactor comprising four independently controlled 1.5L vessels as a research tool for the improvement of somatic embryogenesis using several embryogenic lines of Douglas-fir (*Pseudotsuga menziesii)*.Bapat and colleagues at Bhabha Atomic Research Centre, Bombay have successfully produced large number of somatic embryos of *Santalum album* (sandal wood) using a stirred-type bioreactor system of one litre. Similarly, Bamboo shoots are produced in an air lift bioreactor at National Chemical Laboratory, Poona. Production of *Lillium longifolium* bulblets, *Gentiana* shoot culture and *Artemisia* shoot culture are other examples of organ cultures raised in bioreactor.

The culture of differentiated plant tissues is strongly affected by the four main factors in all bioreactors : moisture, temperature, soluble nutrients and gases. Light is also a critical factor required for some cultures. Assuring efficient gas exchange is a particular challenge, especially in densely packed reactors containing more than 50% biomass volume. Nutrient mist technology, or aeroponics, is one approach to alleviate this problem. Inventions in the design of nutrient mist reactors, includes the use of acoustic windows, have made the technology simple and cost effective. Other then angiosperms, a few fern species has also been cultured in bioreactors like e.g., gametophytes of *Pteridium aquilinum* and *Anemia phyllitidis* in 8L air-lift type bioreactor.

VII. HAIRY ROOT CULTURE IN BIOREACTOR

Hairy root cultures, as an alternative approach to produce secondary metabolites *in vitro*, have received attention during the last decade. It has been claimed that hairy root cultures have a stable and high biosynthetic capacity for the secondary metabolite production as compared to unorganised cultures. Furthermore, hairy roots can be easily retained within the bioreactor, facilitating continuous operation. An additional feature of hairy roots is the low inoculum density required for growth, as compared to cell suspension cultures and immobilized cell systems where high density is required. During incubation, the structured nature of hairy root cultures leads to formation of interconnected, non-homogeneous material unevenly distributed throughout the bioreactor, which made it necessary to develop novel process strategies and modified bioreactors. Furthermore, the tendency of hairy roots to grow in a large cluster results in mass transfer limitations inside the cluster and in inefficient exploitation of the reactor volume. Scale-up of processes using hairy roots will be hampered by these technological drawbacks.

A large number of investigations have been carried out on the growth of roots and production of secondary metabolite using bioreactor systems. *Catharanthus roseus* roots were grown in been grown in 20L stirred-tank-reactor and the production of indole alkaloids in such cultures was reported. The other works on root culture includes effect of reactor types on hairy root growth of *C. roseus*, cultivation of hairy roots of *Datura stamonium* in 14 L stirred tank bioreactor, the production of thiophene from hairy root cultures of *Tagetes*, the production of anthraquinone pigments by hairy roots of *Rubia* and Carrot hairy roots in 1.5L stirred tank bioreactor. In 1990, a 500L bioreactor was developed, which could be operated as nutrient mist reactor as well as with submerged hairy root cultures. Unfortunately, the scale-up of hairy root culture is still troublesome.

Despite of technological disadvantages, hairy roots can be grown at a relatively high biomass densities in small bioreactors. Biomass yield of *D. stramonium* cultures grown in 1.5L bubble column reactor was achieved up to $40gl^{-1}$.

IX. COMMERCIALIZATION

Commercial-scale plant cell culture for the production of secondary metabolites has acquired much attention in Japan. Shikonin, a natural dye, is produced by Mitsui Petrochemcal Industries. Berberine is sold exclusively in Asia and can be produced commercially in a continuous-flow bioreactor in which cells are retained by a membrane and harvested periodically. This system has been operated at a scale-up to 4000 L. Production of ginseng roots as organized tissues (tuberous-roots) is being exploited commercially. In Japan, Ushiyama (1989) successfully cultured ginseng roots in 20,000 L bioreactor and efforts are being made to exploit this system for the production of ginsenosides. The production of a natural vanilla from cell culture is being developed by Escagenetics in USA. Elicitation is apparently a key component of the production strategy. A 300 L air-lift type bioreactor system has been developed by Vipont Research Laboratories, U.S.A. for the production of sanguinarine from *Papaver somniferum*. Colgate has signed a formal arrangement with Vipont, which suggests strong interest in completing the development of commercialization of this project. A new start-up company, Phyton Catalytic, U.S.A. has been founded to make commercial scale levels of taxol and other plant products. In Europe, several processes have seriously considered for the commercialization of natural plant products but, ultimately put on hold or abandoned. However, a 75,000 L reactor facility has been established near Hamburg in Germany for use and exploitation of secondary metabolites production.

In addition to these efforts to produce secondary metabolites, whole plants are being produced. Levin *et al.* (1988) reported that Plant Biotechnology Industries in Israel uses a bioreactor as an important component of a micropropagation work that has already been successfully applied to several species. A collaborative venture between Kirin Brewery and the US Firm Plant Genetics has produced a large-scale lettuce and cereal crops from immobilization of somatic embryos.

Technology for the production of useful secondary metabolites has come to a threshold level. A final push is required to make this technology widely applicable for the production of pharmaceuticals, dyes and fragrances. demand for natural products for fragrances, dyes and pharmaceuticals (anti-AIDS, anti-aging, anti cancer etc.), technology should be ready to meet the demand in coming decades.

QUESTIONS

1. What is a bioreactor? Describe various types and the factors governing the cell growth in a bioreactor.
2. What is the effect of air flow and mixing on cell growth in a bioreactor?
3. Write short notes on the following :
   (*a*) Bioreactor
   (*b*) Immobilized cell bioreactor
   (*c*) Stirred Tank bioreactor
   (*d*) Air-Lift Bioreactor
   (*e*) Wave bioreactor
   (*f*) Organ culture in bioreactor
   (*g*) Secondary metabolites production in bioreactor.

APPLICATIONS OF PLANT BIOTECHNOLOGY IN PLANT PATHOLOGY

CHAPTER 20

I. MERISTEM CULTURE FOR VIRUS FREE PLANTS

Development of virus free plants is one of the most significant applications of plant biotechnology. The technique has applications in diverse areas such as rapid clonal multiplication of vegetatively propagated crop plants, virus elimination and germplasm preservation. From a plant pathologist's point of view, control of viral diseases can be classified under the following headings:

(*a*) Eliminating the sources of virus infection.

(*b*) Avoiding the disease carrying vectors.

(*c*) Directly attacking the vectors.

(*d*) Breeding resistant varieties.

(*e*) Curing virus infected plants.

(*f*) Adopting special methods of propagation like meristem culture.

Virus is a intracellular, obligate parasite and its multiplication is so intimately associated with normal metabolic process in plants that selectively interfering with it is not possible at all. Moreover, such methods can not eliminate virus from a whole plant. Above mentioned all the methods except meristem culture are conventional. Plant pathological approaches and once the control pressure is removed, virus population is increased to its early concentration. Fortunately, the majority of known viruses are not transmitted by seeds and seeds of infected plant usually develop into healthy plant. However, seedlings are useless for maintenance of clones, hybrids and varieties because they do not produce plants identical to parents. Seeds are produced by gametic recombination (sexual reproduction) and thus they produce offspring different from parents. This not required in plantation and horticultural crops. Once systemically infected, the disease is transmitted from one vegetative generation to the next in vegetatively propagated crops. By this process of multiplication and perpetuation of virus, all the varieties of various crops become completely infected. It is presumed that most of the clonally propagated crops are infected by one or more viruses. Some of them may be latent and not detectable by their symptoms.

Virus free plants are free from virus for which they are tested or produced experimentally but they may contain undetectable (latent) viruses for which they are not tests. Therefore, another term ' virus tested' has been suggested in place of virus free plants to give a clear idea about the virus for which the plant is tested.

There are four stages in therapeutic method for virus elimination. These are:

(*a*) Identification of virus (es) present in the clone.

(*b*) Therapy.

(*c*) Testing of treated plants.

(*d*) Propagation and continued testing of cure plants under conditions that avoid reinfection.

The most common and widely used method of virus elimination for planting material is heat treatment. Keeping the sugarcane cutting at 50-52° C in water for 30 minutes can eliminate several viruses. This technique has been used in several plant species like potato, tomato, apple strawberry and raspberries. Conventional techniques of plant pathology are used but with the discovery of new biotechnological methods, this has become a routine method to eliminate virus. The biotechnological approach of meristem culture has been widely employed to raise virus free plants.

a. MERISTEM CULTURE

Limasset & Cornnet (1949) observed that in systematically infected plants, virus concentration decreased as they approached the apical meristem. In apical meristem, no virus was detectable in 50% of the material tested. On the basis of this observation, Morel and Martin (1952) proposed that it might be possible to isolate the apical meristem of a systematically infected plant *in vitro* to obtain a virus free plant. They were successful in confirming this hypothesis and obtain virus free *Dahalia* plants through meristem culture of the infected plants. Since then, this technique has been used for obtaining virus free plants in a number of genera.

The meristem is a dome of actively dividing cells, about 0.1mm in diameter and 0.25 mm long. As rule meristem of both main shoot and axillary buds are used for obtaining virus free plants.

b. TECHNIQUE AND MEDIA

Shoots of all angiosperms and gymnosperms grow by virtue of their apical meristems. The apical meristem is usually a dome of tissue located at the tip of shoot and measures approximately 0.1mm in diameter and 0.2-0.3mm in length. From plant anatomy point of view, this mass is separated as tunica, corpus, central mother cells, flank meristem and rib meristem. The apical meristsms are first formed during embryo development and they remain (except in dormant buds) in an active state of division throughout the vegetative phase of the plant. The totipotency of the plants cells form the basis of meristem culture technique.

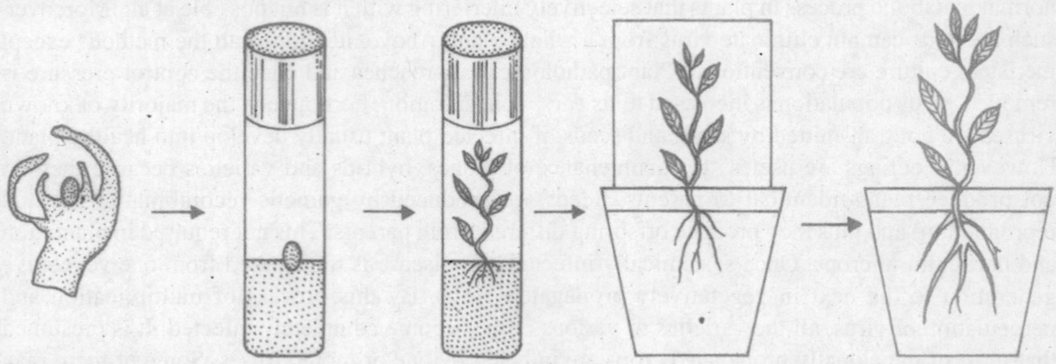

Fig.20.1. Schematic presentation of isolation and meristem culture and regeneration of virus free plantlets.

After proper sterilization of shoot buds, (apical or axillary) apical meristem (0.3 to 0.5mm) is removed with the help of fine forceps and scalpel under a stereo binocular microscope. Before removing the meristem, apical dome is cleared from leaves. Isolated meristem is immediately transferred on to the surface of solid medium and incubated in the culture room (Fig. 20.1). The medium composition and culture conditions are usually the same as used for the material under investigation. Usually low salt and high vitamins are considered favourable for the growth of isolated meristem. There is no simple rule for the success of meristem tip culture. The *in vitro* culture of meristem and shoot tip involves several phases such as initiation of culture and establishment of explant, growth and differentiation, proliferation of shoots and finally plantlet formation by rooting of shoots. Each phase is influenced by several factors as discussed in different chapters in this book. Therefore, any technique involving micropropagation through explant culture has all the factors of micropropagation influencing the growth and multiplication. All the conditions of nutrients, plant growth regulators and meristem have to be decided empirically to obtain success. Virus free plants obtained in different species through this technique are presented in the table-1.

It may be concluded from the results obtained with these species that all the factors which influence the shoot growth and proliferation, also influence the meristem culture, viz., apical or lateral position of the meristem as in *Chrysanthemum cinerariaefolium,* concentrations and composition of plant growth regulators, pre-condition of the plants from which meristems are obtained, pH of the medium, and addition of gibberellic acid. In some cases, use of filter paper with liquid medium was found better for growing meristem as compared to static medium, e.g. in rhubarb.

c. MERISTEM CULTURE AND HEAT TREATMENT

It has been observed that without heat therapy, percent age of virus-free plants produced through meristem culture is 10 W. Therefore, Pre-treatment of plantlets with heat therapy, increases the chance of obtaining more virus-free plants. In this technique, the temperature and its duration for the treatment of apical buds has to be determined experimentally. Neither the treatment should kill the plant material (excessive treatment) nor it should be ineffective (less treatment). This will depend upon the precondition of plant, type of virus infection and its population, and treatment itself. Experiments with C*hrysanthemum* cv. Blanche portevine supreme showed increase in virus-free plants from 9 to 90% by heat treatment of the plant material at 10 or 30 days. Treatment at older age of 40 and 60 days did not increase the percentage of meristems producing virus-free plants, bud adversely affected the regeneration of meristem.

Heat treatment is also helpful in eliminating certain viruses which are not eliminated by normal meristem culture, e.g., in potato infected with PVX, PVS. In such crops, heat treatment improves the efficiency of the technique.

Usually heat treatment is given to the plant before excision of meristem. However, in case of *Nicotiana rustica*, excised meristems grown in static or liquid medium were heated at 32 °C and elimination of cucumber mosaic virus and alfalfa mosaic virus was achieved as compared to cultures grown at 22° C, which remained infected. TMV was not eradicated by this method. Thus, heat treatment to the mother plant or excised meristem is beneficial in improving the efficiency of the method to obtain virus-free plants.

Table 20.1. Effect of use of meristem culture for virus elimination.

Plant species	Virus eliminated
Allium sativum (garlic)	Mosaic virus
Ananas sativus (pineapple)	Mosaic virus
Brassica oleracea (cauliflower)	Turnip mosaic virus, cauliflower mosaic virus
B. oleracea var. gemmifera (brussels sprouts)	Turnip mosaic virus, cauliflower mosaic virus
Caladium hortulanum	Dasheen mosaic virus
Colocasia esculenta (Taro)	Dasheen mosaic virus
Chrysanthemum sps.	Virus B.
Cymbidium sps. (Dahlia sps.)	Cymbidium mosaic, Dahlia mosaic virus, Complex of viruses
Dianthus barbatus	Ring spot virus, mottle virus, vein mottle virus, latent virus.
Dianthus caryophyllus (Carnation)	Ring spot virus, mottle virus, vein mottle virus, latent virus, streak virus, mottle virus.
Fragaria sps.	Vein banding, crinkle, yellow virus complex, latent C virus, yellow edge (virus2) vein chlorisis (virus4), crinkle (virus3). Pallidosis virus, middle yellow edge virus.
Freesia sps.	Freesia mosaic virus, Phaseolus virus2

Gladiolus sps.	Unidentified viruses
Hyacinthus sps.	Hyacinth mosaic virus
Ipomoea batata (sweet potato)	Internal cork virus, Rugos mosaic virusFeathery mottle virus
Iris sps.	Iris mosaic virus
Musa sps. (Banana)	Cucumber mosaic virus
Narcissus tazetta(Nargis)	Arabis mosaic virusNarcissus degeneration virus
Pelargonium sps.(Geranium)	Tomato ring spot virus, cucumber mosaic virus, tomato ring spot
Petunia sps.	TMV
Saccharum officinarum (Sugarcane)	Mosaic
Solanum tuberosum (potato)	Potato virus-X, virus S, Potato virus-Y and Potato virus-A, Spindle tuber leaf role virus.

(d) Mechanism of virus elimination in meristems

Several reasons have been provided for the absence of virus particles in the meristems by different workers based on experimental evidences obtained by them. No reason is conclusive and applicable to all the plant materials or tested for different crop plants. Therefore, none of these is considered as general rule or the exact mechanism operative by which virus particles are eliminated from the growing meristem. The various mechanism proposed for virus elimination from the apical meristems are as follows.

(*a*) During 1950s, it was presumed that virus inhibiting effect of plant growth regulators might be responsible for virus-free apical meristem as apical meristem has always high concentration of plant growth regulators. This hypothesis has never been proved.

(*b*) WU and Co-workers (1960) suggested that rapid cell division is the cause of virus free cells. Normal nucleo-proteins are synthesised first and viral nucleoprotein is synthesised during cell elongation. So there is a competition between cell division and viral multiplication as evidenced by experiments with tobacco callus.

(*c*) It has been demonstrated on the basis of experiments with carnation that carnation mottle virus containing meristems on transfer to nutrient medium become free of virus after 30-40 hrs. It was concluded that contact with nutrient was responsible for virus elimination from the meristems grown *in vitro*.

At present, therefore, there is no general explanation available for the fact that some viruses are and others are not eliminated in meristem culture.

e. EFFECT OF VIRUS ELIMINATION

Several workers have used meristem culture for the production of virus-free plants in many plant species (Table 20.1). But this process does not impart resistance in the species. Therefore, virus-free plants will become diseased when exposed to the virus. The duration in which virus reinfect the plant will depend upon the agricultural practices and epidemiology of virus. Following effects can be noticed in virus-free plants raised through meristem culture:

(*a*) It will be not surprising that new pathological (virus) symptoms are produced by virus free plants. This may be due to the fact that the removal of major virus provides the host free for latent or less infectious virus to produce symptoms.

(*b*) Increased growth and vigour of plants, e.g., as observed in *Narcissus tozetta,* rhubarb, cauliflower, *Pelargonium, Chrysanthemum.*

(*c*) Improvement in rooting of stem cutting, e.g., in *Pelargonium.*

(*d*) Increased flower size, e.g., *Chrysanthemum, Narcissus.*

(*e*) In potato, healthy tubers produce higher yield of potato tuber than infected tubers.

(f) Virus Indexing

The availability of a rapid, effective and sensetive procedure for virus indexing is of significant importance for success of a programme designed at production of virus-free plants. All the plant materials should be certified for virus free status for international exchange. Usually biological assays were used for virus indexing. These methods were slow and not useful for large number of plants produced through biotechnogical methods. Latter on, transmission electron microscopy, serology or a combination of both were developed for virus detection in plant tissues. Among various serological techniques, the enzyme linked immunosorbent assay (ELISA) has been extensively used for virus indexing. With the availability of recombinant DNA technology, nucleic acid hybridization techniques have also been applied for detection of viruses.

Meristem culture is also used extensively in germplasm preservation of endangered and crop species.

II. IN-VITRO CULTURE OF TUMOUR TISSUES AND OBLIGATE PARASITES

Plant biotechnological methods have been employed for solving the various plant pathological problems (Fig. 20.2). With the advancement in the knowledge about growing plant cells, more and more applications were developed. Growth of plant tumour cells and autonomous growth of plant cells is one of the oldest aspects of the study of plant cells. Similar to plant cells, in-vitro growth of obligate parasites (pathogenic forms of fungi and nematodes that grow only on specific living host) was a challenge. Plant tissues grown in culture solved the problem of growing and maintaining such parasites for better understanding the host parasite relationship, establishing axenic culture of the fungi and evolving methods of disease control. Similar is the application in the study of plant viruses (also an intra-cellular, obligate parasite) and development of meristem culture for virus-free plants. The phenomenon of production of phytoalexins by host cells in response to attack by pathogenic fungi has been well studied only in cultures of plant cells. Now this subject has attained great significance in the production of useful secondary metabolites and used extensively in understanding the mechanism of secondary metabolites production using signals generated by elicitors for the expression genes. The application of plant biotechnology in plant pathology includes study of host parasite interaction and mechanism of infection, life cycle of nematodes and evaluation of plant species for disease resistance reaction *in vitro*.

(a) Dual cultures and Axenic cultures

Plant tissue cultures provide an excellent system to grow and maintain the obligate parasites in cultures for plant pathological research which is otherwise not possible to maintain throughout the year. Since obligate parasites (downy mildew fungi, white rusts, powdery mildews, smuts and chytridiales) can only be grown on living hosts, cultures of host tissues which can be maintained for several years, provide a living host system to grow such fungi. A schematic presentation is given in the figure-1 to obtain dual cultures. The obligate parasitic fungi grown as dual cultures are given in the following.

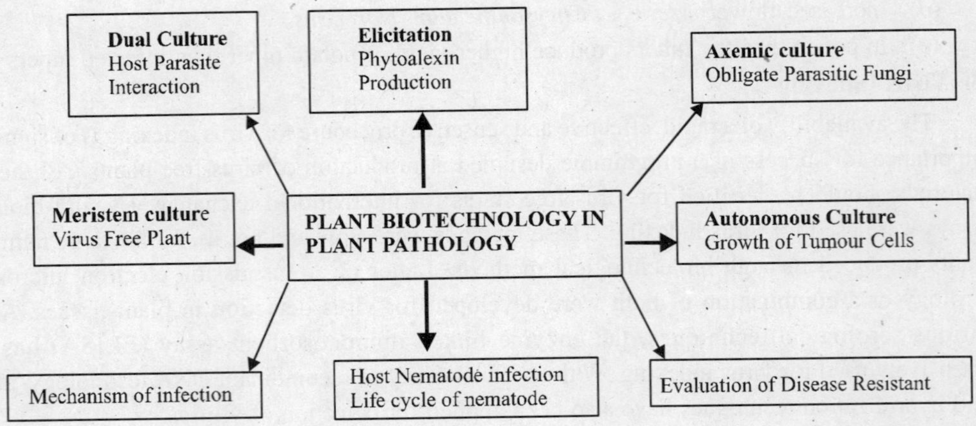

Fig-20.2.. Role of plant biotechnology in solving plant pathological problems.

a. Several of the downy mildew parasitic fungi that have been grown as dual culture –

(*i*) *Peronospora farinosa* on sugarbeet (*Beta vulgaris*).

(*ii*) *P. parasitica* on cabbage or turnip (*Brassica* spp).

(*iii*) *Phytophthora infestans* on potato (*Solanum tuberosum*).

(*iv*) *Sclerospora graminicola* on pearl-millet (*Pennisetum americanum*).

(*v*) *S. sorghii* on *Sorghum vulgare*.

(*vi*) *S. sacchari* on sugarcane (*Saccharum officinarum*).

(*vii*) *Plasmopara halstedii* on subflower (*Helianthus annuus*).

(*viii*) *Bremia lactuceae* on lettuce cultivars.

(*b*) White rust causal agent – *Albugo* on *Evolvulus*.

(*c*) Powdery mildews (*Erysiphales* on *Helianthus annus; Brassica* spp; *Vitis* and others).

(*d*) Smuts (Ustilaginales) on *Silene alba*.

(*e*) Rusts (Uredinales) on cereals, *Cronartium ribicola* on white pine.

(*f*) *Chytridales* on *Solanum tuberosum*.

(*g*) Plasmodiophorales on *Brassica* spp.

(b) Technique and applications

Dual cultures are established by transferring healthy explants of the host plant and callus is obtained. Then a surface sterilized explant of diseased host is placed on the callus itself. The parasitic fungus infects and grows on the living callus cells of the host. In subsequent subcultures, diseased explant is removed and dual-cultures are established. In the other approach, as in case of *Sclerospora graminicola* infected ear heads showing witches-broom symptoms, diseased explants obtained from highly diseased ear heads of pearl millet, are transferred on the nutritive medium. Both the host cells and parasitic fungus grow simultaneous on the medium. These ways dual cultures can be obtained and used for various purposes (Fig. 20.3). During several subcultures, fungus is maintained on host cells but slowly and slowly, fungi starts growing on the medium without host cells. This way pure cultures of fungi can be obtained, known as 'axenic culture'. Other methods of obtaining dual cultures are also available, by surface sterilization of leaf explants, as in case grapevine (*Vitis vinifera*), and transferring them to appropriate medium. In culture, fungus sporulates on leaf surface and ripe sporangia are isolated in distilled water. The liberated zoospores from the sporangia are used to inoculate callus of grape vine. This technique has been used for several other downy mildew fungi. In slightly modified method, infected cotyledons are used to inoculate the host callus. In all the methods, regular subcultures of dual cultures are required to maintain the cultures. Similar procedures have been adopted for obtaining contamination-free uredial inocula of rust fungi. Such axenic cultures provide new avenues to study the physiology of fungi otherwise, not possible by any means. Therefore, dual cultures are useful in studying following aspects of host-parasite relationships. As such host cultures, fungus and dual cultures, all the three systems are available for various studies.

(*i*) **Process of infection**: In nature it is difficult to study the process of infection but under controlled conditions, both parasite and host are available and by inoculating cultures or formed organs, process of infection can be very well studied in laboratory.

(*ii*) **Host-parasite interaction**: This is an important aspect of disease cycle and symptoms. Structural and physiological changes associated with onset of infection to progress of disease can be easily studied in-vitro by dual culture system.

(*iii*) **Disease control**: Evaluation of various chemical agents for the control of parasitic fungi have become easy. These methods are cost and time effective with the availability of axenic cultures of such obligate fungi, which otherwise seasonal, influenced by environmental factors and laborious and costly. By simply incorporation of chemicals in the medium factors controlling and modulating the growth can be determined very easily.

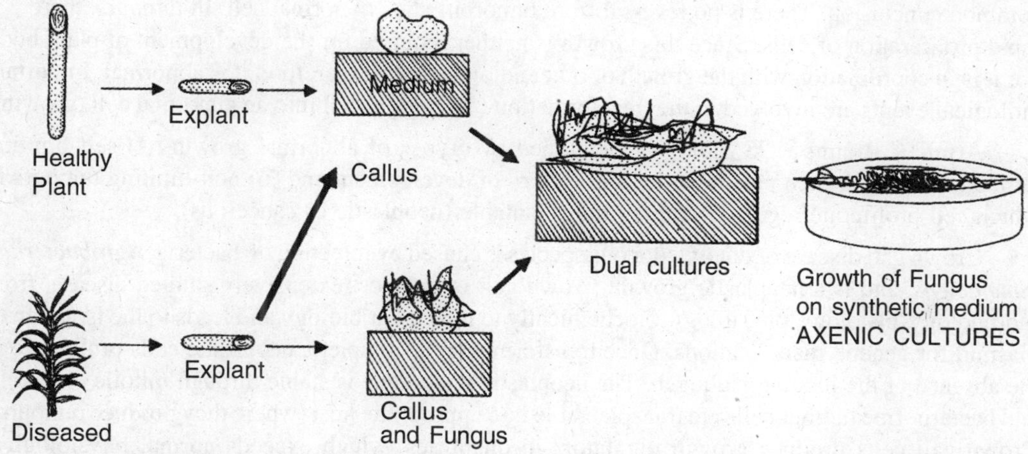

Fig. 20.3. Schematic presentation of development of dual cultures and axenic cultures of obligate parasites

(*iv*) **Life cycle**: Life cycle of fungi and nematodes can easily be observed by establishing host cultures and inoculating them with the parasitic organism. The pathogen completes its life cycle in the flasks as host cells are available. This is a very important aspect of disease cycle and understanding the life-cycle of causal agent provides many information about developing disease control methods. By this method living nematodes are maintained and used for pathological studies e.g. root knot nematodes.

(*v*) **Perenation:** Mode of perenation of parasitic organism can be observed in culture. Changed environmental conditions can affect the life-cycle of the pathogen and disease cycle as well as reoccurrence of the disease. This is only possible by culturing the pathogen.

(*vi*) **Axenic cultures:** Axenic cultures of the parasitic fungi provide the opportunity, which not possible in nature, to study fungal physiology, control measures, life cycle, designing of new chemicals, toxin production, if any and material for all other purposes to other scientists. Axenic cultures have been raised in *Selerospora graminicola* and *Peronospora farinosa*.

(*vii*) **Resistant lines:** Obtaining disease free callus (resistant cell lines or plants) is possible by dual culture or growing cells with the fungal toxin/extract.

All these factors which affect the growth of normal host callus also affect the growth of diseased explant. In most of the infections by downy mildews, abnormal growth is associated with disease development. This is result of hyperauxinity and hyper cytokinin like activity of fungal metabolites. Plant growth regulators and rich nutrients like coconut water, casein hydrolysate and yeast-extract, affect the growth of host cells and parasitic fungi. Obtaining healthy plants or cell strains from diseased cultures is only possible with dual culture system.

The pathogenecity of the fungus maintained in dual culture or as axenic culture requires periodic testing by exposing the experimental plant material. Field trials should be carried out in controlled fields with proper variety and season. Inoculum should consists of proper stage of fungi required for natural infection. This requires pathological training or help from a pathologist and care should be taken for not spreading the disease to other fields.

III. PLANT TUMOURS AND HABITUATION

Plant tumours are unique examples of complex interaction and mutual adaptation between plants and tumour inducing agents. Among the various types of abnormal growths, galls or tumours, hypertrophies, malformations and witches-brooms are common. Neoplastic transformation leads to an apparently stable alteration in cellular phenotype. Once a tumour cell, always a tumour cell is a common conclusion. There is no reversal from tumorous state to normal cell. In tumours, there is a rapid proliferation of cells. Since this growth is neither required for the development of plant body nor it is in coordination with the growth of other adjoining cells, therefore, it is abnormal. Important biological events are involved in the transformation of a normal cell into an abnormal cell (growth).

Armic C.Braun, a US biologist mentioned two types of abnormal growth; (I) self-limiting tumours (benign) which possess a definite pattern of development and (ii) non-limiting tumours in which cell proliferate regularly and are transplantable (neoplastic or cancerous).

Crown gall disease of dicotyledonous species is caused by infection of bacteria, *Agrobacterium tumefaciens*, and is a neoplastic growth. Crown gall is the most extensively studied disease, from morphology, histology, physiology, biochemically to molecular biology and leads to the isolation of plasmid for genetic manipulations. Once transformation is complete, neoplastic cells proliferate in the absence of the inciting pathogen. The neoplastic phenotype is stable through mitotic divisions, and bacteria-free tumour cells are transplantable into appropriate host (where they produce tumours). Crown-gall cells produce growth regulators in quantities, which exceeds normal physiological requirements and as a result, they are able to grow *in vitro* on media not supplemented by growth hormones.

Crown gall tumour cells synthesize amino acid metabolites known as opines. Opines of the octopine group are synthesized by the reductive condensation of basic amino acids with pyruvate, viz., octapine, lysopine, histopine and octopinic acid. Opines of the nopaline group are synthesized by the condensation of α-ketoglutarate with amino acids; they are nopaline and nopalinic acid. *A. tumefaciens* utilizes opines produced by host tumour cells as carbon and nitrogen sources.

Oncogenic strains of *A. tumefaciens* possess a large plasmid (90-150x10^6 daltons) known as the tumour inducing (Ti) plasmid. Transformation is associated with and accomplished by transfer of a stable, replicating portion of Ti plasmid DNA to the plant cell. In morphological terms, tumours induced by *A. tumefaciens* are of two types: unorganized tumours and teratomas (means monster tumour/growth). While scattered vascular elements may differentiate within unorganized tumours, they do not produce recognizable organs. Teratomas on the other hands, are tumours in which transferred cells have a capacity for organogenesis. Majority of teratomas are induced by nopaline strains.

In vitro cultivation of tissues provides a system in which many variables can be controlled enabling the assessment of excised organs, tissues and cells. This forms an important tool to study the fundamental process of conversion of normal cell into an abnormal (cancerous) cell and comparison of metabolism of the both types of cells.

Crown gall has been extensively studied with tissue culture methods. Continued proliferation of bacteria free crown gall was described as early as in 1942 by P.R. White and A.C. Braun. Tumour cells from marigold, paris-daisy, sunflower and tobacco were extensively studied in culture for their nutritional requirements.

The majority of plant cells require our exogenous source of plant growth regulators to grow in culture. Crown gall cells, in comparison to normal cells, have no such requirement. Gautheret (1955) found that normal plant cells grown in culture often became hormone-independent spontaneously. This phenomenon is known as 'habituation'. Cells grown for prolonged periods in culture may become autonomous for cytokinin, auxin, or both growth regulators. Habituation is genetically stable, since it persists over many mitotic generations.

Clones of habituated cells were often more cytokinin autonomous than the parent cultures from which they were derived. In a sense this change toward greater autonomy is similar to the phenomenon of tumour progression in animals.

Habituated cultures can be used for following studies for comparison of metabolism between normal and transformed cells.

1. Primary metabolism concerning transformation.
2. Plant growth regulators production, oxidation and protection mechanism.
3. Production and control mechanism of secondary plant products.
4. Study of effect of extracellular supply of plant growth regulators on cell growth, differentiation and metabolism.
5. Different gene expression in normal and transformed cells.

Isolated tissues growing *in vitro* appeared important in the study of insect induced plant galls tissues, specially those obtained from woody plants. A large number of single cell clones have been isolated from established callus cultures that originated from normal, crown gall, insect gall and virus infected tissues. Single cells were isolated from complex crown gall tissues to examine the recovery of gall cells from disease.

Effect of carbohydrates, nitrogenous compounds, auxins, vitamins and nucleic acids on *Phylloxera* gall, and on normal grape stem were studied in great detail by Professor Arya during his stay at Wisconsin in the Laboratory of Prof. A.C. Hildebrandt. It has been concluded from the above cited and other studies on tumour cells that the transformation of normal cell to that of a tumour involves altered synthesis of carbohydrates, proteins, plant growth regulators and nucleic acids.

IV. PLANT CELLS AND PROTOPLASTS, AND PLANT-VIRUS INTERACTIONS

(*a*) Virus infection

Viruses are intracellular obligate parasites and thus require host cell for their growth and multiplication. Plant protoplasts offer a unique system to study host-parasite interaction in the controlled conditions.

For the establishment of infection and virus cultures, concentrated virus suspension (with sufficient high purity and infectivity) and freshly isolated suspension of host plant protoplasts are co-cultivated in presence of polycations such as poly-L-ornithine, poly-L-arginine or polymin P. The virus in first mixed with polycation and then mixed with protoplasts. This procedure has been adopted for a large number of host-parasite interaction studies, e.g., DNA virus (cauliflower mosaic virus) on turnip and tobacco mosaic virus (TMV) on tobacco.

The tobacco mosaic disease of plants induced by TMV has been studied with isolated cells , tissue and protoplasts cultures of tobacco and other species. Tomato and callus infected with TMV *in vitro* showed that infected tissues could be cultured for unlimited periods *in vitro*. Such cultures raised from infected tobacco stem were maintained for 20 years in the laboratory for G. Morel without loss of virus from the culture (through regular subcultures). However, cultures artificially inoculated (infected) with TMV, often have become virus-free after a few subcultures. The TMV activity and nutrients of the medium affect the host-parasite growth in the culture. These results showed the possibility of culturing virus on host tissues grown *in vitro* and its use in the study of host-parasite interaction, infection process and virus multiplication. There are limitations with the use of protoplasts for virus infection study. These are freshly isolated host protoplasts are required each time, which is a delicate system to handle for injury, and application of results obtained with protoplasts to intact plant is not possible directly because the two systems (protoplasts and intact plant) are very different.

B. Cross protection

Phenomenon of cross-protection by viruses (infection by mild or non-virulent strain cause development of resistance against more virulent strain) is known in plants and animals. Infection of protoplasts with two or more viruses is of immense importance to understand this phenomenon. Tobacco protoplasts were infected with cowpea chlorotic mottle virus (CCMV) and broom mosaic virus (BMV).This provides resistance against more virulent TMV.

C. Infection process

Use of host protoplasts from resistant and succeptible cultivars provided insight about process of infection. It appears that the resultant cultivars or non-host cells do not have specific virus attachment sites on cell surface. TMV could cause infection of cowpea leaves, which was undetactable, but replicated well in the protoplasts of this plant. This suggests that non-host immunity might be due to some barrier to infection at the cell wall level or due to inhibition of viruses to multiply in protoplasts of non-host species. Protoplasts from resistant plants appeared to show resistance by lesser number of infected cells and lesser virus multiplication as compared to susceptible host cells.

D. Study of nucleic acids

A protoplast suspension is inoculated with a purified virus preparation or with a viral nucleic acid or viral nucleoprotein, that would facilitate studies on the function of individual RNA species of some viruses with multiparasite genomes. Use of cauliflower mosaic virus (double stranded DNA) and Gamini virus (unique viron morphology in having single stranded DNA) are being used to study virus replication and their possible role as vector for gene transfer using cell culture methods.

Therefore, it is apparent that plant cells and protoplasts cultures offer several possibilities to study plant pathological problems. Use of cell cultures in virus infection requires injury to facilitate infection and thus a limiting factor in the study of viruses using cell cultures.

E. Production of Secondary metabolites

Possibility of production of useful secondary metabolites through fungal-elicitors treatment is attracting the scientists and technique is used in the gene expression for understanding the production of secondary metabolites.

F. Disease resistant lines

Similarly, production of resistant cell lines are not discussed in detail, but always a possibility to obtain resistant cell lines against pathogen or its toxin. Infection process in resistant and susceptible varieties gives idea about genes involved in the mechanism and subsequently, identification and isolation of such genes will be of immense help in developing resistant cultivars. Therefore, plant biotechnological methods are helpful in solving fundamental and applied problems of plant pathology.

QUESTIONS

1. Discuss the role of plant biotechnology in solving plant pathological problems.
2. Write short notes :
   - A. Dual cultures
   - B. Axenic cultures
   - C. Applications of biotechnology in pathology
   - D. Habituation
   - E. Neoplastic growth
   - F. Teratoma
   - G. Use of protoplasts in virus study.
3. What is meristem culture? Discuss it's application in the production of virus free plants.

MICROPROPAGATION, SOMATIC EMBRYOGENESIS & SOMACLONAL VARIATION

CHAPTER 21

I. INTRODUCTION

Potentiality of a plant cell to regenerate the entire organism (plant) is termed as 'totipotency'. This potentiality has been exploited through the culture of protoplasts, cells, tissues and organs *in vitro*. In cultured materials it has been possible to study such processes as differentiation of a parenchyma cell into tracheid (cytodifferentiation), organ formation (organogenesis) and somatic embryo formation (somatic embryogenesis). Morphogenesis or the origin of form, can be examined description, physiological changes, biochemical changes or even at molecular level through manipulation of physico-chemical (physical and chemical) environment of the cell by Besides the study of fundamental process of differentiation, the capacity of cell to form organs and embryos can be exploited to regenerate plantlets for clonal (clone is vegetatively obtained group of plants from an individual) propagation. The basic concept mentioned above led to the establishment of diverse multiplication techniques. The genetic stability of yielded copies (vegetatively produced plantlets) is likely to vary according to the procedure chosen.

II. ORGANOGENESIS

This is the process by which cells and tissues are forced to undergo changes which lead to the production of a unipolar structure, namely a shoot or root primordium, whose vascular system is often connected to the parent tissue. This system is commonly produced in callus cultures, but can be produced directly from the explant. Organogenesis from callus and explants became possible because of discovery of auxin by Went and Thiman (1939) and subsequently kinetin from DNA hydrolysate by Skoog in 1955. The earliest reports on controlled organogenesis *in vitro* were by White 1939 who obtained shoots on callus of a tobacco hybrid and by Nobecourt 1939, who observed root formation in carrot callus. The finding of White was confirmed and extended by Skoog 1944, who showed that auxin could stimulate rooting and inhibit shoot formation. Further studies of Skoog and co-workers conclusively established that a balanced combination of auxin and cytokinin controls the root and shoot formation (Skoog and Miller, 1957). Studies with tobacco callus showed that a high ratio of auxin to cytokinin in the medium favoured root formation, a reverse ratio having high cytokinin favours shoot formation and that intermediate ratios promoted callus formation. As an outcome of this approach, several hundred-plant species have been reported to form shoot and/or root *in vitro*.

The earliest report on controlled somatic embryogenesis *in vitro* was with carrot reported simultaneously in 1958 by J. Reinert and F.C. Steward. The process occurs naturally in a wide range of species from both reproductive and somatic tissues. Somatic embryos can be formed on callus, in cell suspensions and protoplasts cultures, or directly from cells of organized structures, such as stem segments or zygotic embryo.

There are several advantages of plantlet regeneration through plant biotechnological methods using organogenesis or embryogenesis, as compared to conventional methods of propagation. The advantages include the efficiency of process (the formation of plantlet in fewer steps, simultaneously with reduction in labor, time and cost), the potential for the production of much higher number of plantlets and the morphological and cytological uniformity of the plantlets. Today about 150 species from angiosperms and gymnosperms have been reported to produce somatic embryos in culture.

The first plantlet formation *in vitro* was reported as early as mid 1940s; Ernest Ball (1946) reported in *Tropaeolum* and *Lupinus*. It was only during 1960s that Morel reported plantlets formation in orchids, which became commercially viable programme. Today this use of tissue culture technology is its widest application, with over 300 commercial operations. The plant multiplied through tissue culture includes ornamental food crops, vegetables, spices, fruits and nut crops, medicinal plants and forest trees. There are four ways by which micropropagation can be achieved (Fig. 21.1).

Somatic embryogenesis leads to the production of bipolar structure containing a root/shoot axis, with a closed independent vascular system. Somatic embryogenesis can be produced directly from the explant (termed as direct embryogenesis) or can be produced via callus formation from explant (indirect embryogenesis).

Axillary budding ensures genetic stability, the production of plants from axillary buds has proved to be the most reliable method of in-vitro propagation. Yielding about 90% of the current production it is now by far the most widely used multiplication technique. Shoots and subsequently plantlets formation from axillary bud (pre-existing meristem) ensures genetic fidelity and termed as '**clonal propagation**' (clone constitutes vegetatively derived individuals from a plant, e.g. by cuttings, explants). The plantlets produced by this method constitute the clone of the parent plant from which the were derived.

Adventitious budding in response to external stimulus, pre-existing meristem as well as meristematic cells of the explant produce shoot buds. Shoot buds originating from other then existing meristems are known as adventitious buds such as those produced by stem cortex, petiole. In highly proliferating explants, adventitious shoots are produced giving a gregarious appearance to the explant

Morphogenesis is defined as 'origin of form'. Such morphological changes in the callus lead to the *de-novo* organ formation from undifferentiated tissues are termed as **organogenesis.** Organs (shoot root or both) formation is common phenomenon in callus cultures. Large number herbaceous plants are produced by this method. The regenerants may not be true to parent plant as intermediate callus formation has occurred. This is comparable to indirect somatic embryogenesis in the stability for genetic characters of the regenerated plants.

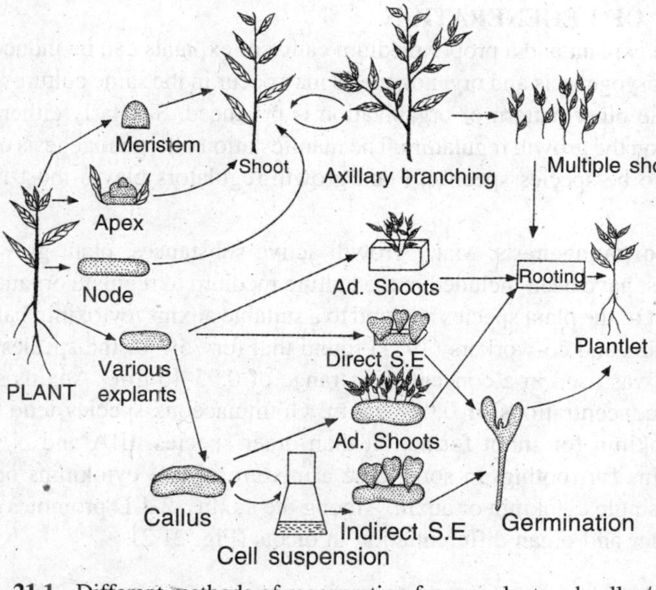

Fig. 21.1. Different methods of regeneration from explant and callus/cell cultures.

Table 21.1. Selected examples of regeneration from different explants and cultures.

- **Root culture**
- *Aegle marmelos*
- *Albizzia lebbeck*
- *Dalbergia sissoo*
- **Tree species from callus cultures**
- *Betula pendula*
- *Citrus grandis*
- *Citrus sinensis*
- *Populus deltoides*
- *Tecomella undulata*
- **Stem culture**
- *Eucalyptus camaldulensis*
- *Psidum guajava*
- *Syzygium cuminii*
- *Rose hybrida*
- *Camellia sinensis*
- *Tecomella undulata*
- *Ziziphus mauritiana*
- **Endosperm culture**
- *Dendrophthoe falcata*
- **Bud culture**
- *Atropa belladonna*
- *Chrysanthemum monifolium*
- *Dalbergia latifolia*
- *Morus indica*
- *Tectona grandis*
- *Terminalia bellerica*
- **Leaf culture**
- Curculigo orchioides
- *Dioscorea floribunda*
- *Rauwolfia serpentina*
- *Petunia hybrida*
- *Solanum melongena*
- *Artemisia annua*
- *Saccharum officinarum*
- **Flower culture**
- *Rannunculus scleratus*
- *Tagetes erecta*
- *Utricularia inflexa*
- **Embryo culture**
- *Hordeum vulgare*
- *Coccus nucifera*
- *Arachis hypogaea*
- *Costus speciosus*
- *Podophyllum hexandrum*
- **Seed and seedling callus**
- *Capsicum* spp.
- *Cajanus cajan*
- *Vigna mungo*
- *Coptis teeta*
- *Brassica juncea*
- *Azadirachta indica*
- *Carthamus tinctorium*
- **Inflorescence culture**
- *Pennisetum americanum*
- *Sorghum almum*
- *Triticum aestivum*
- *Triticale*
- *Zea mays*
- *Musa* sp.
- *Brassica oleracea var. botrytis*

III. REGULATION OF REGENERATION

Using a suitable explant and a proper medium callus or explants can be induced to form organ or embryos. Both embryogenesis and organogenesis may occur in the same cultures e.g. in pumpkin. But usually one or the other pattern of organization is produced. Similarly either root or shoot is formed depending upon the growth regulators. The manifestation of organogenesis or embryogenesis in the culture tends to be species specific. Plant growth regulators play a most important role in these processes.

A. Control of organogenesis. Many growth active substances, plant growth regulators as well as other nutrients, have been included in the culture medium to regulate organogenesis *in vitro*. As stated earlier, most of the plant species respond to a suitable auxins /cytokinin balance by forming shoots and roots. Evans and co-workers (1981) found that for 75% of the species forming shoots, either kinetin or BA was used in a concentration range of 0.05-46 μm. Auxins such as IAA and NAA were used in concentrations of 0.06-27 μm. Graminaceous species tend to have a lower requirement for cytokinin for shoot formation than other species. IBA and NAA are the most commonly used auxins for rooting. In some case a mixture of two cytokinins or two auxins has proven superior to a single cytokinin or auxin. Among the auxins, 2,4-D promotes cell proliferation and suppresses cellular and organ differentiation in dicots (Fig. 21.2).

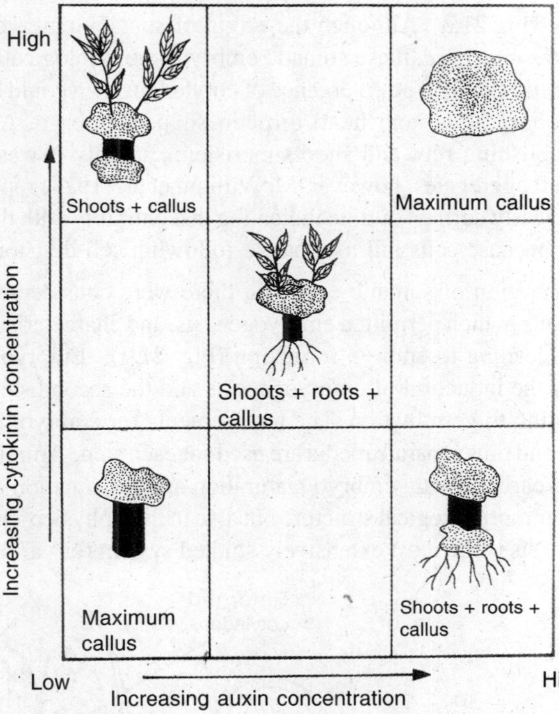

Fig. 21.2. Effect on a combination of auxin and cytokinin ratio

Table 21.2. Plant genera regenerated and micropropagated through tissue culture.

A. Ornamentals	B. Wood species	C. Vegetables, crop and other species
Anthurium	Araucaria	Actidinia
Bromeliads	*Betula*	*Allium*
Cordyline	*Coffea*	*Apium*
Cymbidium	*Eucalyptus*	*Asparagus*
Dianthus	*Hevea*	*Beta*
Draceana	*Malus*	*Brassica*
Syngonium	*Prunus*	Cardamon
Chrysanthemum	*Cryptomeria*	Musa
Gladiolus	*Populus*	*Trifolium*
Gerbera	*Pyrus*	
Pelargonium	*Rose*	
Tulipa	Salix	
Orchids (many)	*Santalum*	
Rhododendron	*Sequeoia*	
Saintpaulia	*Tectona*	
Spathiphyllum	*Vitis*	

IV. SOMATIC EMBRYOGENESIS

Plant cells are totipotent and can produce whole new plants under favourable conditions of nutrients and plant growth regulators. Steward and co-workers in U.S.A. and Reneirt in Germany almost simultaneously reported for the first time somatic embryo formation in carrot cell suspension cultures in 1958-59. These somatic embryos (called somatic embryos because originate from somatic cells instead of gametic fusion which forms zygotic embryo) were similar to zygotic embryos in development and structure (Fig.21.3). Although the origin of somatic embryos produced in carrot cell suspension cultures was uncertain, these somatic embryos morphologically developed through the globular, heart and torpedo stages. The emergence of cotyledons then could be used to distinguish between mature cotyledonary stage and heart/torpedo shaped embryo. At torpedo stage cell differentiation occurs establishing root and shoots meristem. Initially it was observed that these embryo originated from cell aggregates, however, McWilliam et al. (1974) reported that the "young embryoids have always a clearly defined outline indicating discontinuity with the surrounding cells". The aggregates are found because cells fail to separate following cell divisions.

After the initial observation of somatic embryos, there were considerable efforts devoted to define the medium conditions, which permitted embryogenesis, and characterising the morphological and developmental events leading to embryo formation (Fig. 21.4). Embryogenesis is a two step process. The first stage is the induction of embryogenesis and the second step is the development of embryo, ultimately leading to germination. The requirements for embryo induction and embryo development are different, and thus separate media are used for each step. Ammirato (1987) described four stages viz., induction, early growth, embryo maturation and germination of conversion. These stages not only differ in their morphological structure but also in their physico-chemical requirements. Carrot somatic embryogenesis is the most extensively studied system for various aspects of somatic embryogenesis.

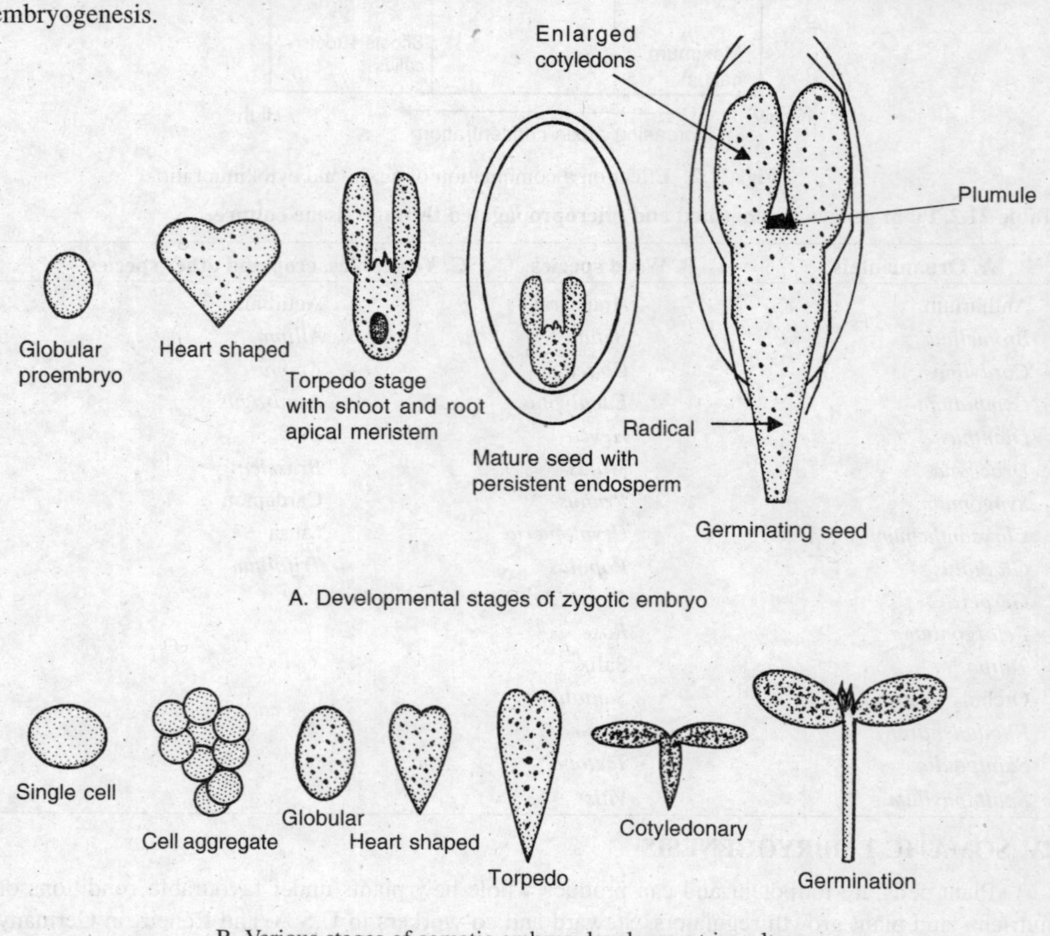

B. Various stages of somatic embryo development in cultures

Fig. 21.3. Development of zygotic and somatic embroys (A) and various stages of somatic embryo development (B).

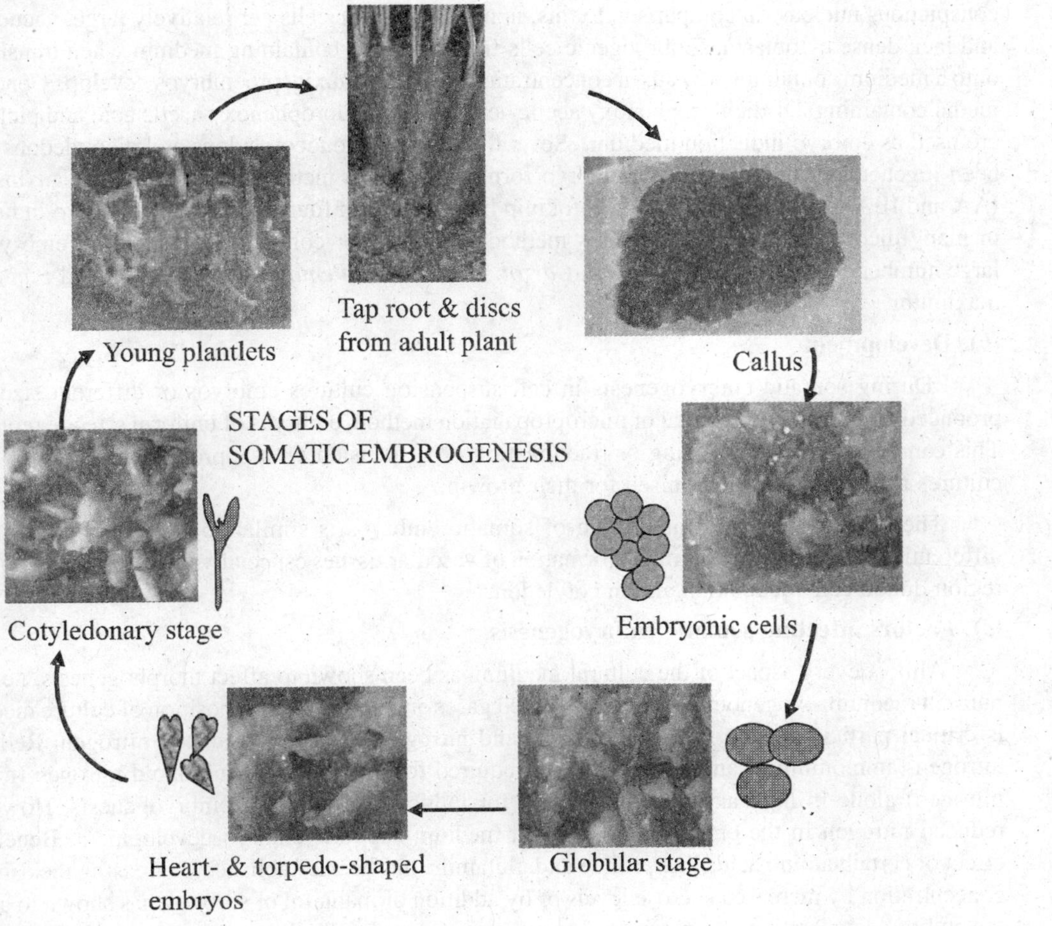

Fig. 21.4. Somatic embryogenesis in higher plants

(a) Direct and indirect somatic embryogenesis

From place of origin point of view embryogenesis is of two types; direct embryogenesis and indirect embryogenesis.

Juvenile explants like hypocotyls, cotyledons and young immature zygotic embryos are best materials to initiate embryogenic cultures. When embryogenesis occurs directly on the explants without production of callus it is known as 'direct embryogenesis'. When explants produce callus and the callus forms embryos then it is known as 'indirect embryogenesis'. Usually an exogenously supplied auxin is required in appropriate concentration for the induction of somatic embryogenesis from callus or explant. However, in some cases like carrot, exogenous supply of auxin may not be required for embryogenesis to occur. It is proposed that in case of direct embryogenesis cells of explanted tissues are already determined for embryogenic development and termed as pre-embryogenic determined cells (PEDC's). In case of indirect embryogenesis cells require redetermination through a period in culture and termed as induced embryogenic determined cells (IEDC's). It is postulated that this phenomenon is determined by epigenetic factors.

Embryogenesis occurs from a single cell or from a group of cells. The earliest cell divisions in embryogenetically determined cells follow various patterns, but finally produce embryos of similar shape. Embryogenic cells are small, isodiametric in shape, filled with dense cytoplasm and have a conspicuous nucleus. In comparison to this, non-embryogenic cells are relatively large, vacuolated and lack dense cytoplasm. Embryogenic cells from an auxin-containing medium when transferred onto a medium containing low auxin concentration or without auxin, proembryos develop. Generally media containing 2,4-dichlorophenoxy acetic acid, 2,4,5-trichlorophenoxy acetic acid and picloram are used as embryo induction medium. Several species of monocotyledons and dicotyledons have been regenerated through somatic embryo formation by this method. However, other auxins like IAA and IBA in combination with a cytokinin have also been found suitable for embryo induction in many dicotyledonous species. This method is useful for continuously obtaining embryos in large number e.g., in *Atropa belladona*, *Carrot*, *Ranunculus*, *Pennisetum purpurium* and *Pannicum* maximum.

(*b*) Development

During somatic embryogenesis in cell suspension cultures embryos of different sizes are produced. For any experimental or micropropagation method, embryos of uniform size are required. This can be achieved by sieving or fractionation of suspension with appropriate sieve size. Such cultures may be fully synchronised for their growth.

The development and maturation of somatic embryos is similar to zygotic embryos. Cell differentiation is most evident in the formation of vascular tissues especially visible in the hypocotyl region due to cell vacuolation and in cotyledon.

(*c*) Factors affecting somatic embryogenesis

Almost every aspect of the cultural mediumhas been shown to affect morphogenesis, i.e., the nutrient medium, exogenous hormones, evolved gases and light. The composition of culture medium is critical particularly the levels of sucrose and nitrogen, especially reduced nitrogen. Reduced nitrogen (ammonium or amino acids) is not required for the induction and oxoid nitrogen (nitrate nitrogen) alone in high amount is sufficient for induction of somatic embryogenesis. However, reduced nitrogen in the embryo development medium supports embryo development. Beneficial effect of certain amino acids like proline and glutamine has been established. Increasing the osmotic concentration by increased sucrose levels or by addition of mannitol or sorbitol, has shown to affect the embryo development. As mentioned above, auxins, especially the synthetic auxin 2,4-D, appear to be required for embryo induction but adversely affect embryo development. Cytokinins, except zeatin, suppress embryogeneisis. Ethylene suppresses embryogenesis, abscisic acid (ABA) suppresses abnormal development of embryos, whereas, gibberellic acid and zeatin produce in an increased frequency of abnormal embryo development. ABA imparts dormancy and helps in the formation of cotyledonary stage somatic embryo.

In liquid cultures, dissolved oxygen can play a critical role in differentiation. Carrot cells produce somatic embryos below 16% dissolved oxygen while above 16% oxygen levels, roots are produced.

(*d*) Embryogenesis as a model of developmental regulation

With the establishment of somatic embryogenesis in synchronous cell culture, the system is used as a model to study organ differentiation and regulation.

(*a*) Callus specific and embryo specific proteins have been identified. It was observed that embryo specific genes were expressed long before the morphologically visible differentiation.

(*b*) A glycoprotein was released into the medium when embryogenic carrot cell aggregates were transferred into a 2, 4-D medium and embryo development was initiated. A second glycoprotein was released into the medium from non-embryogenic tissues.

(*c*) Rapid increases in the rates of protein and RNA synthesis after transfer embryogenically competent carrot cells to an auxin free medium were recorded.

(*d*) Polyamine synthesis and its role in embryogenesis suggested embryogenic cultures had higher level of polyamines, inhibitors of polyamines biosynthesis could suppress embryogenesis whereas the addition of the polyamine spermidine could reverse this inhibition and levels of enzymes for polyamine synthesis were higher under embryogenic conditions.

(e) Application of somatic embryogenesis

(*a*) Somatic embryogenesis provides potential plantlets in the form of somatic seeds. These are available in enormous number once the method is established. These somatic embryos can be used for the production of synthetic seeds for direct sowing in the field. Synthetic seeds are produced by encapsulation in agarose or calcium alginate as described latter in this book (Chapter 14).

(*b*) Somatic embryo provides organized culture system comparable to intact plant in sterile liquid cultures. Such cultures produce organ specific or differentiation related compounds in higher amounts compared to cell culture of that species, e.g., as studied in *Digitalis* and *Theobroma cacao.*

(f) Advantages

Somatic embryo is a versatile technique for micropropagation of plant species. A large number of herbaceous dicots and monocots have been regenerated through somatic embryogenesis (Table 21.3). However, woody trees are still difficult materials to regenerate via somatic embryogenesis. This method of micropropagation offers several advantages over organogenesis and clonal propagation through explants.

(*a*) Rapid multiplication through production of somatic embryogenesis in cell cultures and use of bioreactors for scale-up technology.

(*b*) Presence of bipolar structures in the same unit (presence of both root and shoot) avoids the rooting step required in organogenesis.

(*c*) Somatic embryos grow individually making the system easy to manipulate (to sub-culture) and develop scaling-up methods.

(*d*) It is possible to induce dormancy and store the culture for long duration.

(*e*) Possibilities of encapsulation and other methods of packing and direct delivery system can be employed.

(*f*) Provides an important resource for the analysis of molecular and biochemical events that occur during induction and maturation of embryo.

(*g*) Isolation of specific storage protein is possible.

(h) It shortens the breeding cycle of deciduous trees and increases the germination of hybrid embryos where delayed germination of seeds is a significant handicap in rooting of the plants of horticultural interest.

Therefore, somatic embryogenesis offers several advantages over organogenesis and is used for large number of crop plants for rapid multiplication.

Table 21.3. Somatic embryogenesis reported in tissue culture.

DICOTYLEDONS	MONOCOTYLEDONS
Albizzia lebbeck	*Allium sativum*
Atropa belladonna	*Asparagus officinalis*
Brassica oleracea	*Chlorophytum*
Carica papaya	*Avena sativa*
Citrus spp.	*Bambusa arundinacea*
Coffea arabica	*Hordeum vulgare*

Dancus carota	*Oryza sativa*
Solanum melangena	*Panicum maxicum*
Theobroma cacao	*Pennisetum americanum*
Vitis vinifera	*Saccharum officinarum*
	Secale cereale
	Sorghum bicolor
	Zea mays

V. SOMACLONAL VARIATION

The growth of plant cells *in-vitro* and their regeneration into whole plants is an asexual process, involving only mitotic division of the cell. Therefore, it was expected that the process will produce genetically uniform plants or clonal multiplication is possible through callus regeneration. This expectation has formed the basis for micropropagation work and provided a technical basis for genetic manipulation in plants using callus. The occurrence of uncontrolled variation during the callus regeneration was unexpected and undesirable.

Micropropagation has been established as most widely used application of plant tissue culture almost and a large numbers of plants were regenerated from explants, callus and cell cultures and lastly from protoplasts cultures. In these regenerated plants, chromosomal instability and phenotypic variations were observed. As changes in agronomic trails, were observed in crop plants, interest in exploiting this variation through plant tissue culture for the improvement in crop plants increased. In 1981, Larkin and Scowcroft, named the phenomenon of variation found in plants regenerated from cell cultures as "somaclonal variation". The term is now universally adopted for variation found or created through use of somatic cells and similar terms like 'protoclonal ' and 'gametoclonal' are used for describing variation from protoplasts and anther (pollen) culture, respectively.

(a) Origin and causes

Plant cells are totipotant. It is possible to obtain regeneration from single cells or protoplasts. But in this process, a cell divides and redivides several hundred times to produce callus and subsequently organs. During this process of division, several internal and external factors influence the cell. This leads to creation of variations in cell culture, called somaclonal variation. This is the reason that, direct shoot formation from explant through existing meristems and meristem culture are used for clonal propagation and considered that no genetic variation occur in such system. In comparison to this, callus and cell cultures, organogenesis from cultures and indirect somatic embryogenesis are sources of somaclonal variations. Somaclonal variation is common in long term cultures. The followings are considered causes of somaclonol variations.

In culture, plant cells divide under the influence of plant growth regulators. A balanced combination of plant growth regulators is required to sustain growth and division. Abnormalities in cell division can occur such as multi-polar spindle formation, bridges and fragments, laggards, micronuclei and chromosome fragmentation. These errors result in numerical and structural chromosome changes in regenerated plants. Growth regulators, temperature, light, osmolarity and agitation rate of the culture medium are all known to affect the cell cycle *in-vitro* in plants. Therefore, it may be concluded that inadequate control of the cell cycle *in-vitro* is one of the causes of somaclonal variation. It has been proposed that the various mutational events are directly or indirectly related to the modification of DNA, specially DNA hypo/hyper methylation. It is clear that base modifications and chromatic structural changes are responsible physiological and phenotypic changes associated with somaclonal variation.

(b) Problems with Somaclonal variation

A. Variation is cultivar dependent- Variation observed in one cultivar may not be present in other cultivars. Therefore, it is unpredicable to obtain similar variation in related cultivars and the same technique cannot be applied in all the cultivars for their improvement. Secondly, when variation is observed in small experimental material, it is again not certain that the same variation will be reproduced at large-scale cultivation. Due to these reasons, it is difficult to exploit the variation for crop varieties.

B. Frequencies of change vary -There are no standard parameters for measuring somaclonal variation. Some uses very sensitive molecular methods (RFLP, RAPD's) while others may be morphological characters (leaf size, branching etc.). When different factors like genotype, ploidy, time in culture, tissue source and plant growth regulators affect the variation, its manifestation and magnitude in different or even same cultivars will be different when tested by different workers. Therefore, it is not possible to use the data produced by earlier experiments to apply and use somaclonal variation.

(c) Many changes are undesirable -It is just not possible to have desirable changes through somaclonal variation as it is not under direct control. There may be change in several characters simultaneously along with desirable and non-desirable.

(d) Some changes are unstable-Depression in vigour and yield are common in regenerated plants. Variation in chromosome number and structure are common cause of unstability in growth, fertility and vigour. This unstability is not required for crops and large-scale cultivation through somaclonal variation. In fact, three types of variations have been recognised :

(*a*) Heritable stable variation.
(*b*) Heritable unstable variation.
(*c*) Non-heritable (epigenetic) variation.

Heritable stable variation results from changes in the information content of the genome whereas, heritable unstable variation results from heritable changes in gene activity. Epigenetic variation has been referred to as transient phenotypic modifications which result from the effect of plant growth regulators.

E. Many changes are not novel -Much of the variation observed in the somaclones can be produced by conventional methods like wide crossing (cross or hybridization between distantly related or unrelated materials) or through mutation breeding (use of radiation or chemical mutagenic agents). In fact, it is a matter of expression of a gene (or may be epigenetic), but character exists in the plant material. It is not a new combination of characters. Therefore, conventional breeders are opposing these techniques for use in crop plants.

F. Characters of interest may not change-Since changes in characters are not produced at will it is difficult to obtain desirable characters by somaclonal variation.

VI. BENEFITS OF SOMACLONAL VARIATION FOR CROP IMPROVEMENT

An understanding of the causes and origin of somaclonal variation is crucial for its realistic, application to crop improvement. With the first information about somaclonal variation, it was considered as a new source of creating variation. The initial results were obtained without proper understanding somaclonal variation, therefore, these results could not be utilized for further research. It has become clear that somaclonal variation is not an alternative to conventional breeding programme. However, it holds certain promises and results are coming for its utilization. The benefits of somaclonal variation are as follows:

A. A rapid source of variation is available

Somaclonal variation through plant tissue cultures offers a system of extensive variation and this variation can be created rapidly as compared to conventional breeding programme. Even within a species, the range of changes can be broad. This aspect of somaclonal variation still retains promise particularly in asexually propagated crops where alternative breeding approaches are limited (because due to lack of seed production, hybridization programmes can not be practiced). In such crop, variation can be created by explant culture and this variation can then be used for selection of desired clones.

B. Some changes occur at high frequency

This is the benefit of somaclonal variation that this system of producing changes in characters at high frequency can be used as source of useful mutants. For example, numerous somaclones with altered patterns of seed storage proteins have been identified in plants regenerated from cultured immature embryos of cereals. High frequencies of chromosome doubling have been exploited as means of obtaining polyploid plants. Similarly, the high incidence of chromosome re-arrangement (especially translocations) has been exploited as a means of achieving introgression in interspecific and intergeneric hybrids.

C. Agronomic trait can change

Many changes in agronomic trait have been observed in somaclonal variants. Useful variation observed in somaclones include changes in disease response, heading (tillering) and maturation dates, grain yield, protein content and grain quality. Changes in characters related to environmental stresses like cold, frost, salinity, water logging and freezing have also been observed in the somaclones of crop plants. This technique is used to select plants for fungal toxins, salts or herbicides by culturing cells with these selection pressures (by incorporating in the medium) and selecting resistant lines. Plants selected for various characters through selection pressure are presented in table-1.

D. Novel variants can arise

Initially somaclonal variation was considered as new source of variation and even novel changes were that the changes, which are not possible. by conventional breeding could be produced by somaclonal variation.

E. Improved plants through somaclones

Improved plants have been produced in many crop varieties by exploitation of somaclonal variation, similar to that produced by conventional breeding programme. It takes about 5 years to grow and test the improved material in different localities. There are a few varieties already released in the market developed through somaclonal variation, e.g., ultra-crisp celery and sweeter carrots produced by DNA plant technology in the USA; tomato with higher solid content, altered pigment; potato with disease resistance, improved cooking quality, and variants of strawberry, lettuce and maize. 'Somaclonal snowstorm' as its name suggests, is a new variety of *Paulownia tomentosa* derived by somaclonal variation by Marcotrigiano and Jagnnathan in 1988.

Table 21.4. Use of Chromosome variation in the production of useful lines.

Chromosome variation	Examples of use
Chromosome doubling for production of polyploids	Potato Sugar beet Tobacco *Brassica*
Structural chromosome variation for hybrid introgression in inter-specific hybrids.	*Lolium multiflorum x Lolium perenne Hordeum vulgare x Hordeum jubatum Hordeum vulgare x Hordeum bulbosum Oryza sativa* x *Oryza latifolia*
Numerical and structural chromosome variation for production of amphidiploids and introgression in intergeneric hybrids.	*Triticum crassum x Hordeum vulgare Triticum aestivum x Secale cereale Elymus canadensis x Secale cereale Triticum tauschii x Secale cereale Hordeum vulgare x Triticum aestivum Lolium multiflorum x Festuca pratensis Triticum aestivum x Agropyron sp. Saccharum officinarum x Sclerotachyafuchsa*

QUESTIONS

1. Write short notes on the following :

(*a*) Organogenesis

(*b*) Somatic embryogenesis

(*c*) Regulation of embryogenesis.

2. Describe the various methods of achieving regeneration *in vitro* and the factors affecting regeneration.

3. Describe the micropropagation methods for the multiplication of flowering plants.

4. Write short notes on the following :

(*a*) Stages of embryo development

(*b*) Bud breaking

(*c*) Advantages and disadvantages of micropropagation.

5. What do you understand by somaclonal variation? Discuss its importance in crop improvement.

6. What is somatic embryogenesis? Describe various stages of somatic embryogenesis and factors influencing somatic embryos formation with the help of suitable diagrams.

AUTOMATION IN PLANT PROPAGATION

CHAPTER 22

I. INTRODUCTION

Large scale clonal propagation of plants (micropropagation) is the first major application of plant biotechnology. This industry has been developed on 3 major discoveries-

1. Meristem culture for virus free plant.
2. Protocorn formation in shoot tip cultures of orchids.
3. Role of cytokinins in shoot morphogenesis.

The successful application of these principles to the in vitro multiplication of plants by micropropagation has been a key factor in the development of this technology in to an important world wide industry which produces 250-300 million units annually.

The micropropagation industry started with orchids and ferns and is largely restricted to horticultural plants, such as *Anthurium, Catt leya, Gerbera, Nephrolepsis, Saint paulia, Spathiphyllum and Syngenium.*

During last 15 years production of many bulb and tuber plants such as lilies, potato, and cassava has become quite common. Most importantly many vegetables (cauliflower, leek, garlic and onion, lettuce, asparagus, cucumber), fruits (Strawberry, kiwi fruit, raspberry, banana, datepalm) and a few spice and medicinal species (cardamom, pyrethrum) are now being propagated in vitro in large quantities. Rapid progress has also been made in the propagation of woody species like apple, plum, peach, cherry, poplar, eucalyptus, teak, sandalwood and many conifers. The industry is at a critical stage and requires reduction of production cost and identification by high value high volume nature.

II. PRODUCTION SYSTEMS

Plant tissue culture refers to growing and multiplication of cells, tissues and organs on defined solid or liquid media under aseptic and controlled environment. Plant tissue culture technology is being widely used for large-scale plant multiplication. The commercial technology is primarily based on micropropagation, in which rapid proliferation is achieved from tiny stem cuttings, axillary buds, and to a limited extent from somatic embryos, cell clumps in suspension cultures and bioreactors.

A. Axillary budding - Development of shoots from pre-existing meristems on nodal region ensures genetic stability of the regenerants and termed as axillary budding. This is the most reliable method of *in vitro* propagation producing about 90% of the current production of the micropropagated plants. Propagation of long (life) cycle plants such as forest and fruit trees is multiplied mainly through this method. This is also the principal method of propagation of plants at industrial level where high standards of homogeneity are to be maintained e.g., strawberry, rose, potato, artichoke, asparagus, pineapple and banana etc. In this method the variation developed during multiplication does not exceed the percentage of variation found in the natural population prepared by conventional methods.

B. Adventitious budding – *De novo* formation of adventitious buds (not from pre-existing meristems) may occur directly from the tissues of the explant e.g. from leaf petiole or root segments, from internodal region of stem explant. This method offers the required level of genetic stability in the regenerated plants except for chimeral plants such as *Saintpaulia* and *Begonia*. In some cases adventitious shoots form indirectly from callus, e.g. in *Ficus lyrata* and *Anthurium andreanum*. Since calli are unorganised structures, they are prone to nuclear unstability during rapid cell divisions. Therefore, chances of genetic unstability increase in such cases where intermediate callus is produced.

C. Somatic embryo development – Somatic embryos occur either directly on explants or more frequently in callus cultures (indirect somatic embryogenesis). Indirect embryogenesis involves callus or cell cultures and hence will lead to the production of somaclonal variation in the regenerated plant population. This method does not ensure genetic stability. However, somatic embryogenesis in cell cultures is very useful for developing scale-up technology using bioreactor and in the production of synthetic seeds through encapsulation. This is not possible with any other method. Therefore, this method can be used when very high standards of genetic stability are not required.

Thus, in practice, most micropropagation is achieved by maintaining organized tissues by the multiplication of meristems and axillary buds.

Advantages of biotechnology - Plant tissue culture offers several advantages over conventional propagation methods, vegetative and sexual, for large-scale micropropagation. These are –

1. Shoot multiplication can be achieved in small space because miniature plantlets are produced. Each flask can accommodate hundreds of plantlets. In case of somatic embryogenesis, this number is enormous producing 20,000 to 50,000 embryos per litre of medium (small space).
2. Propagation is carried out under sterile conditions. No damage is caused due to insects and diseases, and plantlets produced are free from microbes (pathogens).
3. In case virus-free material is used (even through virus elimination by meristem culture), a large number of virus-free plants can be obtained (virus-free plants).
4. Plant tissue culture is carried out under defined conditions of environmental, nutritional and tissue system, therefore, it is a highly reproducible system under the defined set of conditions (controlled conditions reproducibility).
5. This production is unaffected by seasonal variations as uniform conditions are maintained (no seasonal effect).
6. No care is required between two subcultures as compared to conventional vegetative propagation systems like watering, weeding (less care).
7. Small glass house space is required because of miniature size of plantlets (small glass house space).
8. Mother plant or genotype of stock plant can be stored and maintained *in-vitro* without damage to environmental factors and stock plants.
9. The plants which are difficult to propagate vegetatively by conventional methods can be propagated by this method (difficult to root plants).
10. It is possible to mechanize whole process of vegetative propagation for large-scale plantations (mechanization).
11. Miniature storage organs (tubers, corm, tuberous roots) can be produced for genotype storage and subsequent plantations (Germplasm storage).
12. Being sterile, transport across countries is permissible without difficulties (transport across countries does not require phytosanitory regulations).

Disadvantages - Though, there are several advantages, there are certain disadvantages with the system. These are –

1. Micropropagation methods through use of tissue culture involve capital intensive expensive materials like autoclave, laminar air flow bench, controlled culture rooms etc. (capital intensive).
2. This is a technically skilled work, knowledge about material, techniques and decision making (during subculture and multiplication of propagule) are required in the personnel (skilled type work).
3. Contamination is a serious threat and cause severe damage to material and add substantively lot to the cost of production, affects time schedule delivery of the material (contamination).
4. Specific conditions of micropropagation, rooting, and hardening may be required. Therefore, each material requires separate research methods (No uniform standard method) and multiplication can be commercialized where protocols are available.
5. Small delicate plantlets are produced, which take longer initial time to grow.
6. Genetic stability is doubtful in certain methods.
7. It is a capital intensive industry, if plants are produced in small number, they cost too much, otherwise also cost is a major factor for the production and sale of tissue culture raised plants (high cost of plantlets).
8. Only those plants which have consumer preference can be produced by the industry.

III. PROCESS OF MICROPROPAGATION

The process of plant micro-propagation aims to produce clones (true copies of a plant in large numbers). The process is usually divided into the following stages:

Stage 0- pre-propagation step or selection and pre-treatment of suitable plants.

Stage I - initiation of explants - surface sterilization, establishment of mother explants.

Stage II - subculture for multiplication/proliferation of explants.

Stage III – shooting and rooting of the explants.

Stage IV - weaning/hardening.

These stages are universally applicable in large-scale multiplication of plants. The individual plant species, varieties and clones require specific modification of the growth media, weaning and hardening conditions. A rule of the thumb is to propagate plants under conditions as natural or similar to those in which the plants will be ultimately grown *ex-vitro*. For example, if a chrysanthemum variety is to be grown under long day-length for flower production, it is better to multiply the material under long-day length at stages III and IV. There is a wide option to undertake production of plant material up to a limited number of stages. For example, many commercial tissue culture companies undertake production up to Stage III, and leave the remaining stages to others.

Pre-propagation stage

The pre-propagation stage (also called stage 0) requires proper maintenance of the mother plants in the greenhouse under disease and insect-free conditions with minimal dust.

Stage I

This stage refers to the inoculation of the explants on sterile medium to initiate aseptic culture. Initiation of explants is the very first step in micropropagation. A good clean explant, once established in an aseptic condition, can be multiplied several times; hence, explant initiation in an aseptic condition should be regarded as a critical step in micropropagation. More than often, explants fail to establish and grow, not due to the lack of a suitable medium but because of contamination. The explants are transferred to *in vitro* environment, free from microbial contaminants. In herbaceous plants e.g. potato, chrysanthemum, carnation, streptocarpus, strawberry, and African violet; the explant sources are meristems, apical and axillary buds, young seedlings, developing young leaves and petioles, and unopened floral buds.

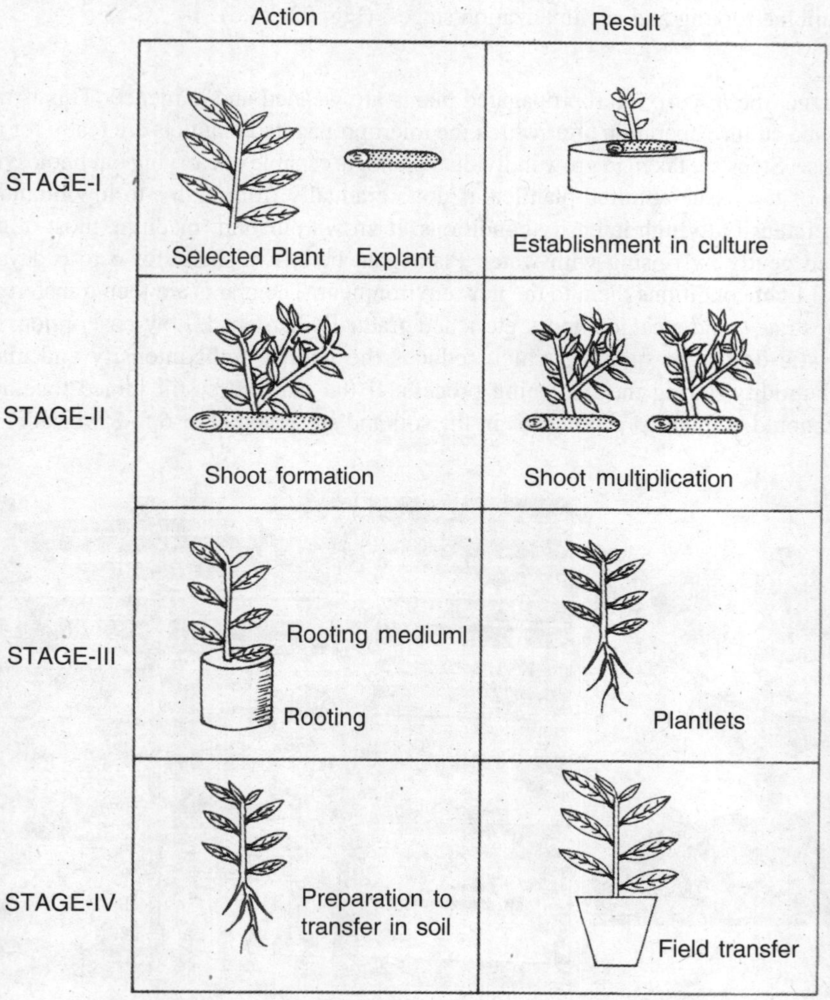

Fig.22.1. Micropropagation method- stages to obtain *in vitro* multiplication

Stage II

Stage II is the propagation phase in which the explants are cultured on the appropriate media for multiplication of shoots. The primary goal is to achieve propagation without losing the genetic stability. Repeated culture of axillary and adventitious shoots, cutting with nodes, somatic embryos and other organs from Stage I leads to multiplication of propagules in large numbers. The propagules produced at this stage can be further used for multiplication by their repeated culture. Sometimes it is necessary to subculture the *in vitro* derived shoots onto different media for elongation.

Stage III

The *in vitro* shoots obtained at Stage II are rooted to produce complete plants. If the proliferated material consists of bud-like structures (e.g. orchids) or clumps of shoots (banana, pineapple), they should be separated after rooting and not before. Many plants (e.g. banana, pineapple, roses, potato, chrysanthemum, strawberry, mint, several grasses and many more) can be rooted on half-strength-

MS (Murashige and Skoog, 1962) medium without any growth-regulators. Good sturdy well-rooted plants are essential for high survival during weaning and later transfer to soil. This stage is labour intensive and expensive. The process of *in vitro* rooting has been estimated to account for approximately 35–75% of the total cost of production (Debergh and Maene, 1981). Efforts should be made to combine rooting and acclimatization stages. (Fig. 22.2 a.d).

Stage IV

At this stage, the *in vitro* micropropagated plants are weaned and hardened. This is the final stage of the tissue culture operation after which the micropropagated plantlets are ready for transfer to the greenhouse. Steps are taken to grow individual plantlets capable of carrying out photosynthesis. The hardening of the tissue-cultured plantlets is done gradually from high to low humidity and from low light intensity to high intensity conditions. If grown on solid medium, most of the agar can be removed gently by rinsing with water. Plants can be left in shade for 3 to 6 days where diffused natural light conditions them to the new environment. The plants are then transferred to an appropriate substrate (sand, peat, compost, etc.), and gradually hardened. Low-cost options include the use of plastic domes or tunnels, which reduces the natural light intensity and maintains high relative humidity during the hardening process. If the plants are still joined together after rooting, these should be planted as bunches in the soil and separated after 6 to 8 weeks of growth (Fig. 22.2 a-d).

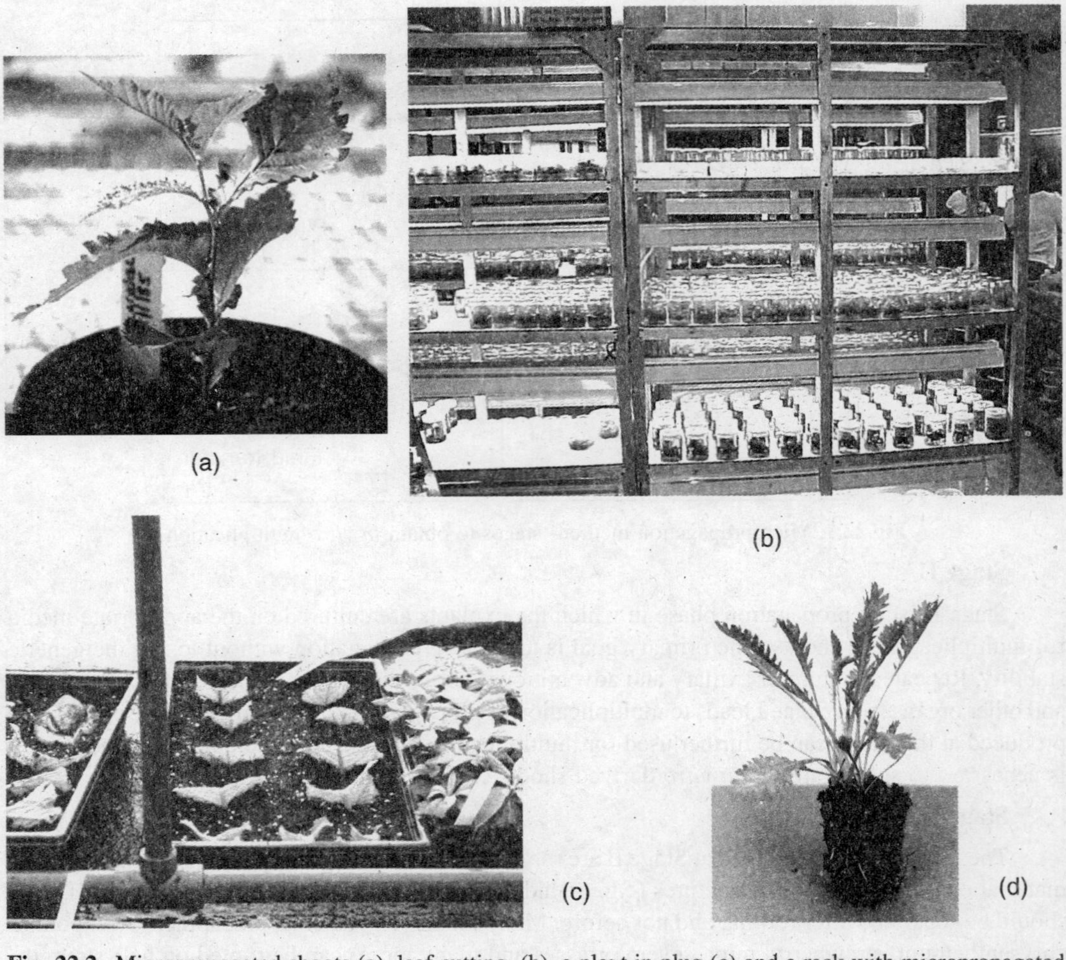

Fig. 22.2. Micropropagated shoots (a), leaf cutting, (b), a plant in plug (c) and a rack with micropropagated plants (d).

IV. DELIVERY TO THE GROWERS

The delivery of the rooted and hardened small micropropagated plants to growers and market requires extra care (Fig. 22.3). In some cases, plant losses can occur during shipment and handling by growers. This is particularly true when the plants are not fully hardened and rooted or not grown for sufficient duration after transfer to soil. Growers should be given clear instructions how to handle the material provided. Apart from the economic loss, poor survival of planted material erodes the confidence of growers in the technology. The transfer of individual plants to soil in black plastic or polythene bags is widely used as a low-cost option to provide fully-grown banana plants directly to farmers in many developing countries.

V. PRODUCTION COST ESTIMATES

Micropropagation based industry requires proper calculations of land space, water and electric resources, trained manpower and a very well programmed production line. It is very important that company should make profit to become commercially viable. In Indian context, selection of industrially backward area ensures many incentives and subsidies. Land and manpower are the two major cost inputs, besides, selection of climatically favourable locality reduces electricity and irrigation charges. Usually production cost may be calculated on the basis of number of plants per square foot. However, it will also vary from crop to crop and factors like number of shoots per cluster multiplication rates, losses due to contamination, also affect the cost calculations, labour is the major cost component in the production of tissue culture plants.

Cost component should also include expenses incurred on plant maintained in green house, including labour and irrigation. Calculation can also be divided on the basis of inputs provided at each stage of micropropagation. Percentage of cost component for laboratory is given in the table. This does not include cost of infrastructure created. (Takes 22.1–2).

Table 22.1.

Relative laboratory production cost components

Cost components		Percentage
Direct labour	34	64
Indirect labour (media preparation, initiation etc)	15	
Supervision	15	
Utilities	10	
Depreciation	12	
Other production costs	14	
Total	100	

Table 22.2.

Relative green house production cost components

Cost components		Percentage
Planting direct labour	11	50
Indirect labour (irrigation, maintenance, sorting, dispatch)	30	
Supervision	9	
Utilities	8	
Depreciation	12	
Other production costs	30	
Total	100	

VI. SYNTHETIC SEED TECHNOLOGY

The concept of production and utilization of synthetic seeds (somatic embryo as substitutes for true seeds) was first suggested by Murashige in 1977 (Cf. Bajaj, 1995; Cyr, 2000). Synthetic seeds can be produced either as coated or non-coated, desiccated somatic embryos or as embryos encapsulated in hydrated gel (usually calcium alginate) (Redenbaugh *et al.*, 1993). Successful utilization of synthetic seeds as propagules of choice requires an efficient and reproducible production system and a high percentage of post-planting conversion into vigorous plants. Artificial coats and gel capsules containing nutrients, pesticides and beneficial organisms have long been thought as substitutes for seed coat and endosperm (Bajaj, 1995). However, this technology is still in the developmental stage, and currently cannot compete with the other methods of commercial plant propagation. Encapsulation of somatic embryos in a suitable gel is known as synthetic seeds. The details of synthetic seed formation is given in the chapter 14.

Fig. 22.3. Plant trays for transport

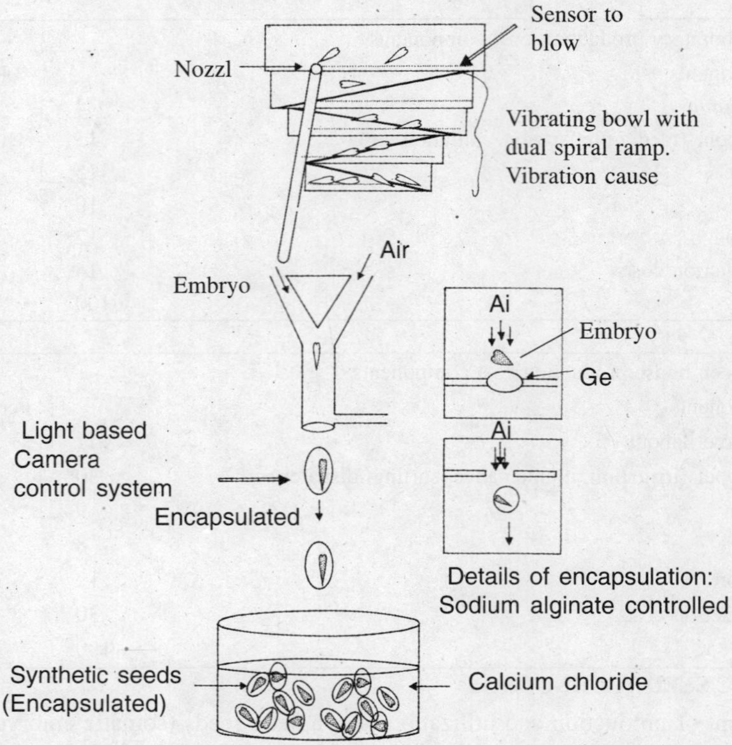

Fig. 22.4. Schematic presentation of automation unit for the production of synthetic seeds (after Plant Genetics Inc., USA)

Here we will discuss how these synthetic seeds can be used for direct sowing in the fields by mechanization. For horticultural crops, tens of thousands to millions of somatic embryos will be needed for large scale commercialization. For agronomic crops, millions to billions of embryos are required. Current technology using shake flasks and bioreactors are not evolved to produce such huge crops. Major difficulty is conversion rate (percent development of plants out of cotyledonary stage embryos) in liquid cultures.

A sophisticated encapsulation machine was constructed by the university of California, Agricultural Engineering Department, in collaboration with plant Genetics Inc. USA for encapsulating seeds and somatic embryos in calcium alginate beads. Machine has a vibratory bowl to move the somatic embryos to one of two nozzles. The movement of alginate into the nozzles is controlled by a positive displacement, adjustable stroke and air pressure driven pump. Seed singulation on the vibratory bowl is controlled by a panel box on each unit. It is a microprocessor based system, can be controlled by one person. Each nozzle is calibrated to produce approximately one encapsulated embryos per second. The machine is designed for embryos similar to the size of tomato seeds Fig. 22.4.

VII. Automation and robotics

Replacement of the people who do micropropagation work in laminar flow hoods with a machine of any kind is neither technologically simple nor economically feasible. The fundamental fact is that there is no comparison of human eye-hand brain combination. Micropropagation requires elimination of contaminated plants, selection of desired sized plants separation of shoots from a propagule and lift and transfer of shoots. This requires on the spot judgement, handling of delicate material, and transportation. This kind of sophistication and judgement need high technology and cost inputs. This has become technologically achievable by artificial intelligence. Proto-type machines

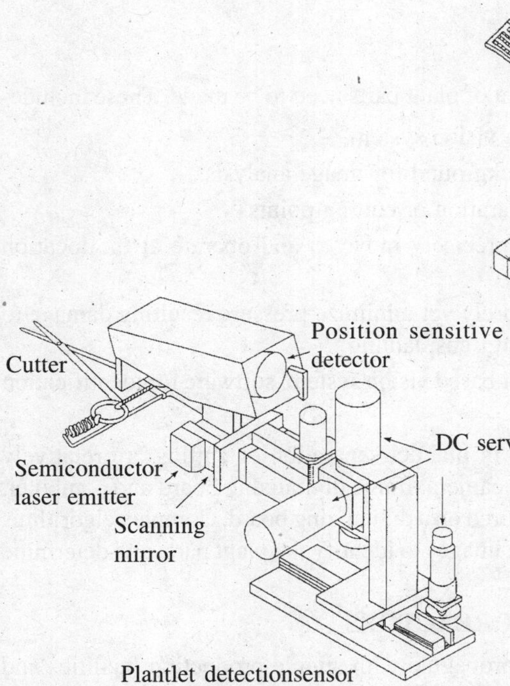

Fig. 22.5. Plant detection sensor for robotics

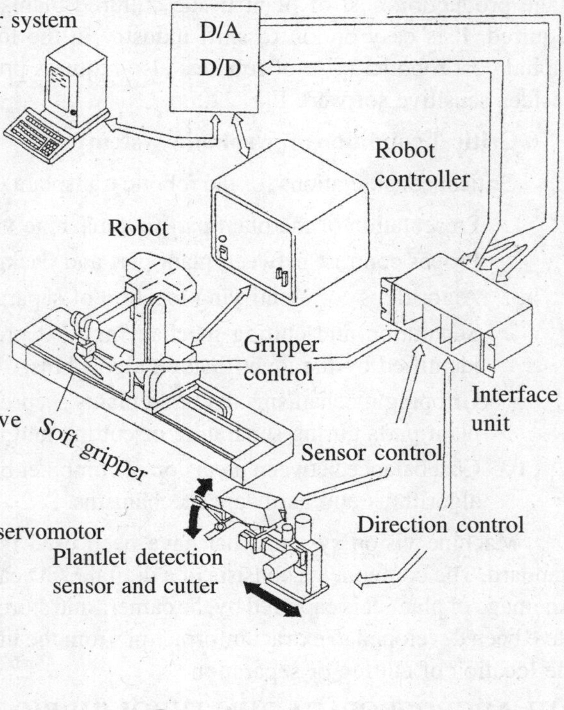

Fig. 22.6. System structure of the prototype robot system.

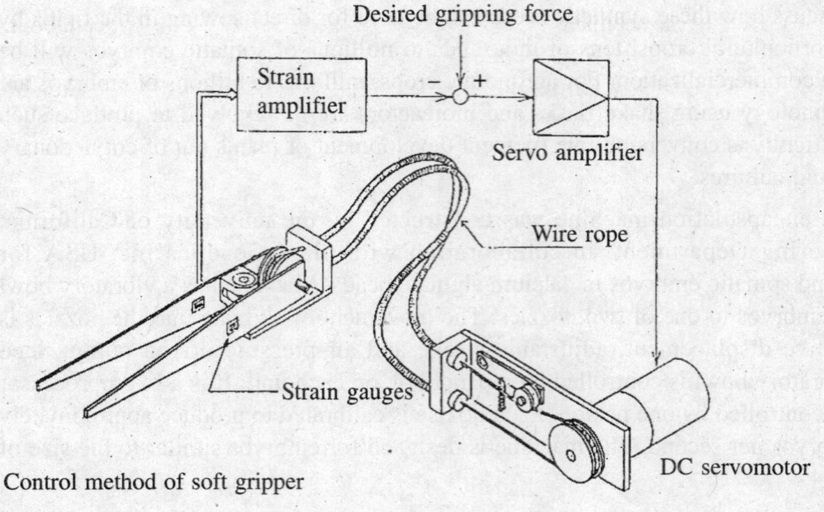

Fig. 22.7. A schematic presentation of robotic system and control method for soft gripper

were developed by Japanese scientist for cutting and handling of shoot tips using microprocessor based technology. This robot has a cutting tool and a soft gripper for handling the plants. This robot cut and transfers one plantlet in 20 seconds. Later on, an improved version was developed which was equipped with Motorola's 32 bit microcomputer. This has improved cutting, gripper and sensing tool. These robots are not developed for plant tissue culture work; therefore, improvement for plant tissue culture work requires significant economic inputs for development of the machine. At this stage production cost of plant tissue cultured plants is very high and at least 50% reduction is required. It is easy option to shift industry in the low-labour cost zones rather than using high capital cost robotics. Use of robotics also requires proper architecture of the plant and propogule, besides sensitive software Fig. 22.5,6,7.

Critical conditions for robotic system

Critical considerations for the robotic transplant of plant parts need to be made. These include-

1. Presentation of the plant part to a machine vision system.
2. Proper contrast between plant part and background for image analysis.
3. Algorithms for identifying location of separation or cutting points.
4. Separation and cutting mechanisms that precisely move to and operate at the location identified by the identification algorithms.
5. Gripping mechanisms that hold plants securely yet minimize pressure resulting damage to plant parts during separation or cutting and transplanting.
6. Cost balance between precision of computer based vision system, software for identification algorithms and hardware mechanisms.

Machine vision systems that have been used in plantlet separation or cutting are relatively standard. The components consists of a digital CCD camera, frame digitalizing board and computer. An image of plantlet is captured by the camera and stored on a digitalizing board. Complex algorithms have been developed to extract information from the images to identify the plant parts and determine the location of cutting or separation.

VIII. MICROPROPAGATION THROUGH BIOREACTORS

Many attempts for large scale production of propagule with simple production facilities and techniques have been made including robotics, photo autotropic cultures, bioreactor techniques etc. Among them bioreactor technique seems to be the most promising because it can reduce labour, and the cost of production. This is required for the commercialization of mass propagation of plants. Bioreactors are described in chapter 19. A bubble column bioreactor was used for the first time by Motomu Akita (Japan) in 1981 to propagate *Begonia*. Use of bioreactor enhances the productivity and the efficiency of plant propagation, The bioreactor offers following advantages

1. Large number of plants in a batch, possibility of scale-up.
2. Single harvest, no multiple vessels.
3. Improved growth rate.
4. Axillary branching proliferation is high (loss of apical dominance because of movement in vessel).

However, there are some drawbacks like hyperhydracity, plantlet size variation and microbial contamination. Following species have been grown in bioreactor for micropropagation-

Shoots: *Atropa belladonna, Begonia, Chrysanthemum, Dianthus, Fragania, Nicotiana, Petunia, Primula, Stevia rebandiana* (upto 500L vessel).

Bulbs: *Fritillaria, Hyaciinthus, Lilium*

Corns: *Caladium, colocasia, Pinellia*

Tubers: *Solanum tuberosum*

Adventitious buds: *Atropa belladona*

IX. LABORATORY DESIGN

(a) General consideration for location

A convenient location for a small laboratory can be a room or part of the basement of a house, a garage, a remodeled office or a room in the header house. The minimum area required for media preparation, transfer and primary growth shelves is about 14 m^2. Walls should be installed to partition different areas. Before setting up a commercial micropropagation unit, it is essential to check out the area keeping in mind the climate, and access to water, electricity, transportation, and infrastructure for supplies. A temperate climate is usually better suited to tissue culture ventures. This greatly reduces the cost of cooling required to maintain the temperature for optimum growth of the cultures. The availability of electricity and water is of utmost importance, and should be taken into consideration while choosing the location of the facility. For example in India, of the 76 commercial tissue culture units, nearly 52 are located in and around the cities of Bangalore and Pune, where the climate is moderate. Hence, cooling is required only during certain periods of the year. However, in cities like Delhi, which have extremes of climate, tissue culture facilities require both heating and cooling. In a facility, which produces five million plants, the electricity cost per thousand plants is around US $0.30 in Bangalore and Pune; the same is about US $0.80 in Delhi. Disruption of power and water supply causes major breakdown in the smooth running of tissue culture units. Poor quality water adds to the cost of media. Location consideration should also include a check with local authorities about zoning and building permits before construction begins. The buildings should be located away from sources of contamination such as a gravel driveway, parking lot, soil mixing area, shipping dock, pesticide storage, and dust and chemicals from fields. Sanitation of the area is important in selecting location. The units should be located in areas where insect populations are minimal and air has low dust and pollen counts. In the case of export-oriented units, these should be near an international airport, which will reduce the time lag between packaging and shipment. This is critical to assure timely delivery of quality tissue culture products. For new ventures, the size of the facility should be kept small until the market acceptance is ensured.

(b) Design and lay out of the laboratories

Plant tissue culture laboratories have specific design requirements. Careful initial planning is, therefore, a prerequisite for successful running of a facility. The location and design of the laboratories should take into account isolation from foot traffic, control of contamination from adjacent rooms, thermostatically controlled heating and cooling, water supply and drains for a sink, adequate electrical service, provisions for a fan and intake blower for ventilation, and good lighting. Large sized facilities are frequently built free standing. Although more expensive to build, the added isolation from adjacent activities keeps the laboratory clean. Prefabricated buildings make convenient low-cost laboratories. They are readily available in various sizes in many countries. Prefabricated buildings assembled on site can also be used. A single span building allows for a flexible arrangement of walls for dividing into convenient sized rooms. The floor should be of concrete or capable of carrying 170.5 kg/m^2 (50 pounds per sq. foot). Walls and ceiling should be insulated to at least R-15, and covered inside with water resistant material. Windows, if desired, may be placed wherever convenient in the media preparation and glassware washing rooms. The heating system should be capable of maintaining room temperature at 200C during the coldest part of winter. A minimum of 2cm pipes should be used for water supply. Connection to a septic system or sanitary sewer should be provided. Air conditioning requirements should carefully estimated. Electrical service capacity for equipment, lights and future expansion should be calculated. For safety reasons, the electrical installation should be carried out professionally. Most electrical wiring will require 220 Volts, and autoclaves 230/250 Volts. The areas such as the media preparation room, inoculation room and growth chambers should be isolated as 'clean zones' (Broome, 1986). The office, storage area, staff centre and packaging rooms can be maintained under ordinary conditions. The working areas must be demarcated according to the activities involved in the facility. Cleanliness is the major consideration when designing a plant tissue culture laboratory to minimise contamination. A positive pressure module should be installed to circumvent air intake from outside. Routine cleaning and aseptic procedures can decrease contamination losses to less than 1%. An enclosed entrance should precede the laboratories, and sticky mats should be placed to collect dirt from shoes.

X. GREENHOUSE FACILITY

A critical stage in plant tissue culture is the interim phase between the laboratory and field conditions. *In vitro* derived plants need to be gradually hardened to field conditions. Plant hardening is usually carried out under greenhouse that ensures high survival of the tissue-cultured plants in the field. There are three types of greenhouses: Ground to ground, Gable, and Quonset type. The most commonly used greenhouse is the Quonset type. It contains movable or fixed benches with hardening tunnels on them. The size of the greenhouse must be based on the scale of production. Greenhouse glazing can be of glass or fibre glass. Polyethylene films or sheets of polycarbonate or acrylic can also be used. Air inflated double polyethylene covering is the most economic (Table 22.1). Appropriate light, shading and blackout systems can be achieved with supplementary lighting. Drip irrigation systems, misting and fogging can be installed as needed. Greenhouses erected in warm climates should have fan-assisted drip pad cooling especially during summer (Broome, 1986).

Greenhouses in colder climates need to be heated. Floor and bench systems can be used for heating and cooling the air. Low cost plastic pipes can be used to circulate warm air, which are adequate and cost effective (Fig. 22.8).

XI. PACKAGING AND SHIPPING

A separate area should be designated for packaging in a commercial tissue culture unit. Packaging materials such as cardboard cartons and labels should be stored in this area. The type of packaging of a particular plant depends greatly on the temperature zones through which the consignment has to pass from the point of shipment to its destination. For example, if the consignment of tissue-cultured plants is passing anywhere within temperate zones, it is only necessary to take care of the frost conditions and pack accordingly. But if the consignment is passing from a tropical

Fig. 22.8. Glass house facility : external and internal view showing crop of micropropagated plants.

country to a temperate country, it becomes necessary to take care of the temperature zones in different places through which the consignment passes. Thus, the type of packaging of tissue-cultured products varies with plant and destination. Widely used packaging materials are made of plastics, wood, glass, paper, textile fabrics such as made of cotton, jute, and synthetic. Some of the products from the above materials are kraft, kraft liner, carton board, straw board, corrugated paper and paperboard, paperboard polyethylene (PE), polypropylene (PP), polystyrene (including expansible polystyrene, PS & EPS), polyethylene terephthalate (PET), polyvinyl chloride (PVC). For example, in a package of 22 kg, about 40,000 lily plantlets can be packed. However, in the same parcel, only 20,000 *Spathiphyllum* plantlets or 10,000 *Syngonium* plantlets can be packed because of the higher respiration rates in these species. The heat build-up is much more in *Spathiphyllum* and *Syngonium* plants than the dormant lilies. Similarly, only 8000 *Gerbera* or 7000 *Cordyline* plants can be packed in the same package. Cooling material such as ice, dry ice and other material take up much of the packaging space. Before loading and shipping, the packed items should be properly counted and rechecked. The cartons containing the cultures for shipping to the customer should be properly labelled with the names and addresses of the consignor and the consignee, and the details of the commodity (storage temperature, handling, etc.). Adequate care must be taken while packaging large consignments, so that there is no disturbance or damage during transit. To prevent any sort of delay, ensure that the consignment is accompanied by documents such as invoice, packinglist, import permit, phytosanitary certificate and Generalized System of Preferences (GSP).

In conventional Agronomical practices, the crops are being grown / cultivated in the open field under natural conditions where the crops are more susceptible to sudden changes in climate i.e. temperature, humidity, light intensity, photo period and other conditions due to which the quality, yield of a particular crop can get affected and may be decreased. Green Houses are available in different sizes and constructed as per customer requirement. The sizes vary from as small as 100 Sq. M to 10,000 Sq M and more. The degree of sophistication also varies from a simple polyhouse with polyethylene sheet covering to highly sophisticated, fully automated systems with Poly carbonate sheet roofing (double walled), PAR lightings, boom irrigation, rolling benches and full scale computerized (fully automated) systems.

XII. ESTABLISHMENT OF MICROPROPAGATION INDUSTRY

The building of a tissue culture facility includes the cost of land, construction, electrical installation and plumbing. If the available funds are limited, well-designed laboratories can be established by modifying existing structures. For example, a two or three room house or a trailer with appropriate modifications can be converted into a medium sized micropropagation facility (Fig. 22.8). Depending on the production and storage capacity, a facility can be small- less 100,000 plants, medium- 100,000 to 500,000 plants, or large-scale- 500,000 to 2,000,000 plants per annum. The initial investment required for setting up a low cost, medium scale tissue culture laboratory (ca. 195 m2) in India is given in Table 22.3. Land is sold at a premium, especially near large cities. In India, the land and building costs work out to about US $62500. A house, ca. 195 m^2, can be converted into a tissue culture facility with a capacity of 200,000 plants. The rental charges for such a structure would be around US $5000 per year. The initial investment would thus include only plumbing and electrical work, reducing the capital cost to US $2750. Thus, in large cities, it would be more economic to rent than build the premises (Table 22.3). If the hardened tissue-cultured plants were the end products, an additional investment of US $20,000 would be required to erect a greenhouse. If the production is limited to-in-agar and ex-agar products (rooted or non-rooted micro-cuttings), greenhouse would not be required.

Table.22.3. Cost of setting up a tissue culture facility in India

Requirement Cost (US $)	Rented building	Purchased building
Infrastructure	2750	62,437
Equipment	13885	13885
Accessories	4193	4193
Furniture	7333	7333
Miscellaneous	521	521
Total	28,682	88,369

(a) Cost reduction measures

The major drawback of plants produced through plant tissue culture is their high cost. It is necessary to reduce the production cost so as to produce more and more species. Proper attention should be made to optimize the production protocol by manipulating physical and chemical parameters. Similarly, elimination is a growth stage, e.g. direct rooting eliminates the rooting stage, or reduction in green house period. Reduction in explant to plant period as well as number of plants produced has a direct effect on cost. As shown in the table, labour component is 60-70% of the inputs. Automation in medium preparation, use of robotics in separation of shoots, and in green house technology will reduce the production cost. More and more industries are shifting to the Asian countries because of low labour cost and producing millions of plants of banana, sugarcane, cardamom and other horticultural crops.

(b) Contamination control

The loss of cultures increases the cost of production. Contamination in cultures is caused from natural contaminants such as dust, air-borne particles, bacterial and fungal spores, fibres, and hair. Man-made contamination occurs mainly from body, clothing, and from faulty procedures in the laboratory. It has been estimated that each operator generates a minimum of 1 to 5 million particles (bigger than 0.5ìM diameter) per minute. Modern, high quality clean rooms have environment control systems that minimise contamination. Wearing laboratory coats (which are mandatory in Class 100 clean rooms) reduces contamination from clothing, skin and hair.

(c) Washing and sterilising operations

Costly dishwashing machines for cleaning the culture containers should be replaced by manual washing if labour is relatively inexpensive. The washed culture containers can be dried in sun instead of costly hot air ovens. In small-scale laboratories, the autoclaves can be replaced by large sized pressure cookers, which are much cheaper. Instead of having one or two huge horizontal autoclaves, which generate hot air pockets in the sterilising rooms, it is better to have more of smaller vertical autoclaves, which keep the air cool. It is a normal practice to use costly aluminium foil for wrapping instruments for sterilisation. This can be replaced by stainless steel containers, which are autoclavable. Savings can also be made in media and containers.

(d) Transfer operations

Instead of Petri dishes as stage for manipulation of culture, stainless steel plates, ceramic tiles and brown wrapping paper can be used; all of these can be autoclaved. Ethanol for hand and workbench sterilization can be replaced by industrial spirit.

(e) Culture maintenance in growth rooms

Plant cultures can be maintained in rooms with air conditioners and tube lights instead of highly priced plant growth chambers. The conventional method of downward illumination can be replaced by sidewise lighting systems, which not only reduces the number of lights but also provides more uniform illumination to the cultures. In the tropical and Mediterranean regions, the electrical lighting systems can be replaced by sunlight. The tissue culture laboratories in Cuba produce millions of tissue-cultured sugarcane, pineapple and banana plants using natural light.

(f) Production planning

Most plants are seasonal in demand. To meet the requirements of extremely large number of plants, commercial production has to be backed by well defined working procedures and monitoring performance of the operation. Commercial laboratories should produce a range of plants for different seasons to maximise the use of the facilities throughout the year. This lowers the unit cost of plant production. Large-scale micropropagation is a labour-intensive process, and therefore organising the availability of personnel is quite important. To reduce organisational problems in big units, the management structure must be well planned. Supporting information systems such as inventory control, production scheduling, space utilisation and daily targets should be well defined. To produce quality products on large-scale, there should be good co-ordination among the technicians, supporting staff, supervisors and the researchers. The job description and reporting system should be very clearly stated.

(g) Scale of operation

The physical components required for the functioning of a commercial tissue culture laboratory can be either scaled up or down according to the interests of the propagator. The correct design of a laboratory, big or small, will help maintain asepsis, thereby increasing the efficiency of the unit and achieving a high standard of work. A micropropagation company in India converted a three-room apartment into a medium-sized tissue culture laboratory. Plant such as *Spathiphyllum*, *Syngonium*, *Ficu*s, hosta, calla lily, gerbera, and cordyline, were produced on a commercial scale. Delivery of the tissue-cultured plants to the tune of 2.5 million per year has been made from this unit by adopting various low cost alternatives. Similar tissue culture facilities, aptly called 'bio-factories', in villages of Cuba produce up to one million banana plants and 2.5 million sugarcane plants annually.

XIII. COMMERCIAL ASPECTS

There are now many commercial companies, in India, third world and developed countries, producing millions of plantlets through micropropagation. A few selected genera micropropagated commercially are enlisted in Table 22.4. In the Netherlands alone, where there is an efficient pot plants and glass house industry, there are ten companies each producing ten million plants per year. Similar micropropagation companies exist everywhere and some are specializing in only one type of plant e.g., for ferns or gerberas.

In micropropagation companies facilities exists for simultaneous inoculation of large volume of plant material. For this purpose, large media rooms with precisely controlled medium formulations are used, large numbers of laminar flow benches and large growth chambers exist. Inoculation chambers and growth rooms are clean areas and sterility is strictly maintained. The secondary multiplication of easy to root flowering plants may be carried out *in vivo* by rooting of the stem cuttings. This reduces the cost and save time in multiplication.

Large glass houses and green houses are essential component of micropropagation industry. Hardening and acclimatization of delicate *in vitro* raised plantlets is carried out in these glass houses. Now-a-days, chamber made of polycarbonate and polypropylene sheets are used for creating large working place. These houses are fitted with mist and fog generating units with cyclic auto-regulation. Light is provided through proper light sources.

Table 22.4. Plant genera micropropagated at large scale

A. Ornamentals	B. Wood species	C. Vegetables, crop and other species
Anthurium	***Araucaria***	***Actidinia***
Bromeliads	*Betula*	*Allium*
Cordyline	*Coffea*	*Apium*
Chrysanthemum	***Cryptomeria***	***Arachis***
Cymbidium	*Eucalyptus*	*Asparagus*
Dianthus	*Hevea*	*Beta*
Draceana	*Malus*	*Brassica*
Ferns (many)	*Pinus*	Cardamon
Freezia	***Populus***	***Cicer***
Fuchsia	*Prunus*	*Festuca*
Grbera	*Pyrus*	*Fragaria*
	Ribes	*Glycine*
Gladiolus	Rose	*Musa*
Hemerocallis	Rubus	*Phaseolus*
Hyacinth	*Salix*	*Rheum*
Iris	*Santalum*	*Solanum*
Narcissus	*Sequeoia*	*Trifolium*
Orchids (many)	*Tectona*	
Pelargonium	***Vitis***	
Phlox		
Rhododendron		
Saintpaulia		
Saxifraga		
Spathiphyllum		
Syngonium		
Tulipa		

QUESTIONS

1. Describe the micropropagation methods for the multiplication of flowering plants.
2. Discuss the role of plant tissue culture based micropropagation industry in developing countries.
3. Write short notes:
   (a) Automation in micropropagation
   (b) Use of robotics in micropropagation
   (*c*) Synthetic seed for mass cropping
   (*d*) Measures to reduce cost in micropropagation industry
   (*e*) Cost calculation for tissue culture plants
   (*f*) Establishment of tissue culture industry
   (*g*) Use of bioreactor for mass multiplication
   (*h*) Green house maintenance.

METHODS OF GENE TRANSFER

CHAPTER 23

The first transgenic plant was produced via *Agrobacterium* mediated modified transformation of *Nicotiana tabacum* protoplasts by Horsch and co-workers in 1984. Since then several dozen plant species have been genetically engineered using different techniques. Simultaneous development of other techniques such as selectable markers facilitated the development in genetic engineering for obtaining transformed plants. But this technique is not suitable for monocotyledon plants as they are not natural host of *Agrobacterium* (There is evidence that limited gene transfer is possible in monocots by this system). Therefore, other methods of direct gene transfer have been developed for use with monocots and other species. These can be categorized on the basis of the use of protoplasts or cell and tissue as the target materials. The isolation and purification of protolasts have been presented in Chapter-10. Freshly isolated protoplasts are used for genetic transformation.

I. DNA TRANSFER IN PROTOPLASTS

(*i*) Electroporation
(*ii*) Chemically stimulated DNA uptake by protoplasts
(*iii*) Liposomes
(*iv*) Microinjection
(*v*) Sonication

II. DNA TRANSFER IN PLANT TISSUES

(*i*) Acceleration of DNA coated microparticles
(*ii*) Laser microbeam
(*iii*) Silicon carbide fibres

Direct uptake of DNA by isolated protoplasts is a genotype dependent response. Regeneration from protoplasts is not common in all the species. Due to this, production of fertile transgenic plants has remained difficult in most of the cereal species. However, transgenic fertile plants have been produced in *Sorghum vulgare, Oryza sativa* and *Hordeum vulgare*. Direct delivery of free DNA molecules into plant protoplasts by physical (electroporation and micro injection) and chemical (polyethylene glycol) methods have been developed to facilitate DNA delivery across the plasma membrane.

III. AGROBACTERIUM MEDIATED GENE TRANSFER

I. DNA TRANSFER IN PROTOPLASTS

A. Electroporation

This method is based on the use of the short electrical pulses of high field strength. Electroporation causes the uptake of DNA into protoplasts by temporary permeabilization of the plasma membrane to macromolecules. Protoplasts and foreign DNA are placed in a buffer between two electrodes and a high intensity electric current is passed, the alternating current of about 1 MHz is applied to align the protoplast by dielectrophoresis. Once aligned, fusion is induced by applying one or more rectangular direct current pulses (1-3kV /cm, 10-100 μs), then the alternating field is reapplied briefly to maintain close membrane contact for fusion (Fig. 23.1). Electric field damages membranes and creates pores in membranes. DNA diffuses through these pores immediately after the electric field is applied, until the pores are resealed. Technique is optimized by using appropriate electric field strength (defined as the applied voltage divides by the distance between two electrodes). The optimum field strength is dependent on the followings –

1. The pulse length of electric current
2. Composition and temperature of the buffer solution
3. Concentration of foreign DNA in the suspension
4. Protoplasts density, and
5. Size of the protoplasts.

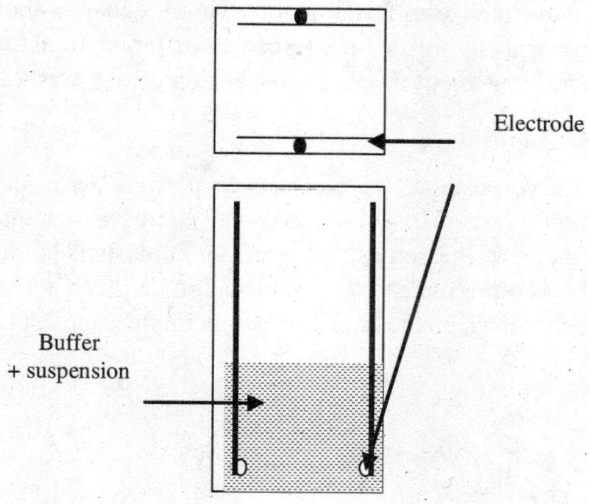

Fig. 23.1. Top (above) and side view (below) of glass cell with electrodes used for electroporation.

It has been demonstrated that the removal of pectin from the plant wall increases the amount of DNA which can be introduced by electroporation. Tobacco mosaic virus was introduced in tobacco protoplasts by this method. Electroporation has been used successfully for transient and stable transformation of protoplasts from a wide range of species. Plating efficiency (i.e. number of colonies recovered out of number of cells transferred on plates) of electroporated protoplasts grown on selection medium (containing selective marker) can be as high as 0.5%. The highest plant transformation efficiencies have been reported for tobacco, with 0.2% of electroporated leaf mesophyll protoplasts giving rise to transgenic calli. Low transformation efficiency is common in cereals, e.g. in rice 0.002% efficiency was recorded.

B. Chemically Stimulated DNA Uptake

Direct uptake of DNA by protoplasts is stimulated by polyethylene glycol (PEG) and PEG is the most widely used chemical for this purpose. PEG mediated transformation involves mixing of freshly isolated protoplasts with DNA and immediately adding 15-20% PEG dissolved in a buffer containing divalent cations. This mixture is incubated for 30 minutes, protoplasts are washed and then plated in Petri plates for culture and growth. The optimization of transformation frequencies by this method include factors that follows.

1. PEG concentration in the mixture
2. Composition and concentration of salts used
3. The pH of the solution
4. Concentration of the foreign DNA
5. Size and form (linear, supercoiled) of the DNA molecules used
6. Culture and selection techniques used for protoplasts.

PEG mediated transformation is generally preferred over electroporation for stable transformation of monocot protoplasts due to relatively higher survival rates after treatment.

PEG also stimulates the uptake of liposomes and improves the efficiency of electroporation. PEG causes precipitation of ionic macromolecules like DNA and stimulates their uptake by endocytosis. PEG mediated DNA uptake typically transforms 0.1 to 0.4% of the total protoplasts treated. Production of transgenic plants depends upon the regeneration competance of the transformed protoplasts. In case of *Petunia*, 40% transformed calli derived from mesophyll protoplasts could be induced to form fertile plants. This is equal to about 0.1% transformation efficiency of the treated protoplasts. In different species different transformation efficiency has been observed, e.g., in embryogenic protoplasts suspension of rice was 0.0004% and, soyabean and tobacco was 0.7-1%.

C. Liposomes

Liposomes have also been used as a carrier for the introduction of nucleic acid into plant protoplasts. Liposomes are small lipid sacs containing plasmids and are prepared artificially. The fusion of liposomes with plant protoplasts is stimulated by chemicals such as PEG (endocytosis). Liposomes mediated transformation has been achieved by including positively charged agents such as cations in the transformation mixture or using the cationic liposome preparation (Fig. 23.2).

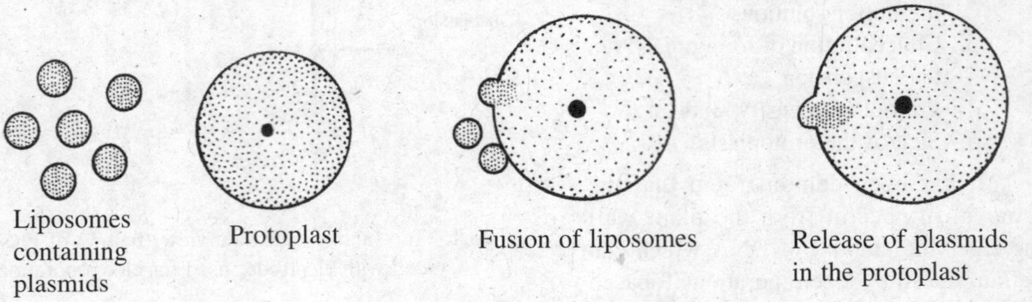

Fig. 23.2. Liposomes mediated gene transfer.

Other chemical agents like polycation Polybrene® or lipofectin have also been used for both transient and stable transformation for maize protoplast. Cationic liposome and polycation mediated DNA delivery are new protoplasts transformation methods and are considered better than other methods of transformation. There are several advantages for the use of this technique.

1. Protection of DNA/RNA from nuclease digestion.
2. They have low cell toxicity.
3. Encapsulation of nucleic acids makes them more stable during storage.
4. High degree of reproducibility.
5. This method is applicable to wide range of plant cell types.

D. Microinjection

Delivery of nucleic acids to protoplasts or intact cells via microinjection is a labour intensive procedure that requires special capillary needles, pumps, micromanipulators, inverted microscope and other equipment. However, injection into the nucleus or cytoplasm is possible and cells can be cultured individually to produce callus or plants. In this way selection of transformants by drug resistance or marker genes may be avoided. This method involves skill of the worker to insert needle into the cytoplasm or in the nucleus. The basic technique is similar to that used for animal cell microinjection. In order to microinject protoplasts or other plant cells, the cells need to be immobilized (Fig. 23.3). The cells are immobilized by –

1. The use of a holding pipette which holds the cells by vacuum.
2. Attachment of cells to poly-L-lysin coated cover slips.
3. Embedding the cells in agarose, agar or sodium alginate.

Glass micropipette are prepared to have openings of about 0.3 μM in diameter and are inserted into plant cell cytoplasm and nuclei with the aid of a micro manipulators device. A syringe like device is used for the controlled delivery of volume (10^{-11} – 10^{-4} μ*l*) into the plant cell. Most plant cells are injected while keeping inside microdroplets (2-50 μ*l*) of medium using a chamber which is sterile, vibration free and permits temperature and humidity regulation. A maximum of 100-200 cells per hour can be microinjected by this method.

The recovery of transformants is dependent upon the regeneration ability of the microinjected cells. Different methods have been used to grow injured (microinjected) single cells

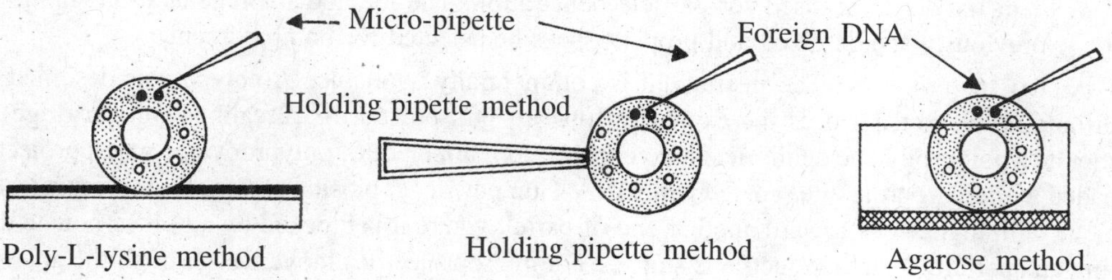

Fig. 23.3. Various methods of immobilizing the cell and microinjection

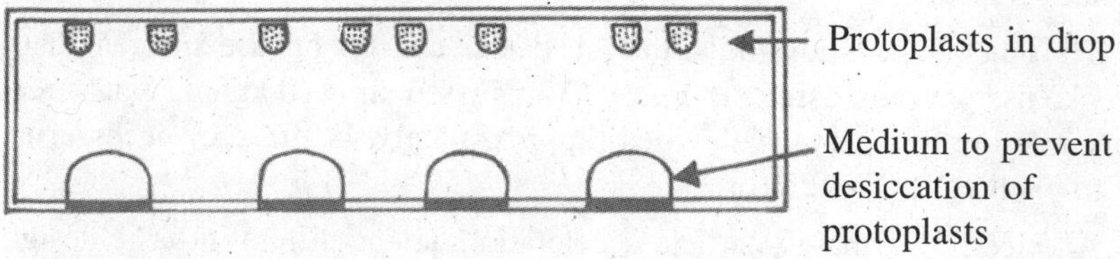

Fig. 23.4. Hanging drop culture method

or protoplasts. Hanging droplets, covered under thin layer of agar or agarose, and micro culture have been used (Fig. 23.4). Attempts have been made to inject linear, or super coiled DNA, in cytoplasm or in nucleus. Nuclear injections are found better for transformations.

E. Sonication

Mild sonication (20 KHz ultra-sound) has been used to facilitate the uptake and transient expression of a chloramphenicol acetyltransferase (CAT) gene in protoplasts of sugar beet (*Beta vulgaris*) and tobacco. This method was found superior than electroporation method used for the same material. Plating efficiency was also similar to untreated cells. However, transgenic plant production using this technique has not been reported so far.

II. DNA TRANSFER TO INTACT TISSUE

A. Acceleration of DNA Loaded Microparticles (Particle Gun or Biolistic® Method)

This is latest technology to transfer DNA into intact tissues. Several devices are developed using different methods. All to achieve the transfer of micro-sized particles (microprojectiles) coated with DNA to penetrate the cells. In this procedure, micron size tungsten or gold particles are accelerated in a gun barrel to velocities sufficient for non-lethal penetration of cell walls and membranes. Klein and co-workers in 1987 developed and used for the first time the particle gun to transfer chimeric DNA and viral RNA molecules into intact onion cells. Tungsten acted as a carrier of nucleic acids because it was available in micro-size balls, non-toxic to cells, and dense enough (high density) for rapid penetration of target material.

Microprojectile mediated transformation is a mechanical method of introducing DNA in to any plant species. This method can be successfully used where plasmids or protoplasts mediated transformation can not be used. An acceleration device is used to propel particles (micro projectiles) carrying plasmid DNA, is called by various names based on machine or technique used to accelerate the particles such as 'particle gun technology', 'biolistic method', 'DNA bombardment', 'particle acceleration of DNA method' and 'electric discharge particle acceleration method'.

This is a quick method of stable transformation and testing a gene for cell and organelle specific expression, this technique has three components –

1. The basic equipment to generate particle acceleration.
2. Metal particles coated with precipitated DNA (desired gene).
3. Plant tissues to be used for particle penetration. The method for regeneration should be previously standardized and proper tissues be selected for bombardment.

(*a*) **Instrument :** The instrument is commercially available. Prototype was designed by Klein and co-workers. It uses the explosive force of gun powder (0.22 caliber gun cartridge) to accelerate a polypropylene cylindrical macroprojectile. Thin piece of polypropylene macroprojectile is loaded with microprojectiles coated with DNA. Gun powder explosion forces thin macroprojectile to move with high speed toward another end of barrel, where it is blocked by a polycarbonate disc having an aperture. Macroprojectile is stopped but microprojectiles move fast through the aperture towards tissue placed in the same direction. For each transfer 50 μg tungsten is accelerated upto 2000 ft per second in a partial vacuum. With this speed, particles reach upto lower layers of cells in target tissues (Fig. 23.5).

The other devices are similar in basic design concept but use different methods to accelerate particles like use of compressed air or gas. Compressed air (130 kg/cm^2 pressure) has been used to accelerate microprojectiles at velocities (approximately 440 m/sec) necessary to achieve DNA delivery to plant cells (Fig. 23.6).

An electric discharge particle acceleration device differs in basic design from the above described devices. In this device, a high voltage discharge (14 KV current) delivered to a small water droplet which quickly vaporizes and releases energy to propel DNA coated gold spheres into target cells (Fig. 23.7). In similar way to above devices, a DNA carrier is attracted (accelerated) due to potential differences, stopped in-between by a screen, DNA coated particles cross the screen and fly towards target tissue and deliver the DNA into cells.

(*b*) **DNA coating** : This is a sophisticated technology and requires precise preparation of DNA coated gold or tungsten particles. The particles should have following properties.

1. High density (19 g/cm^2 or greater) to ensure proper acceleration and penetration through cell walls.

2. Size (0.5-5 μm) should match with size of the cells. Large sized spheres can be used with large cell size.

3. Gold is costlier than tungsten, but it does not oxidize like tungsten.

DNA is precipitated onto the particles prior to bombardment. Most commonly, $CaCl_2$ and spermidine are used to precipitate plasmid DNA (the desired gene attached to plasmid is suitable for integration into plant genome) onto tungsten particles. Ethanol is also used for precipitation of the DNA on gold particles.

(*c*) **Plant tissue :** Plant tissue used for transformation should be competent to regenerate. Mostly embryogenic tissue is an ideal material for transformation (e.g., in corn, cotton, soyabean, papaya and wheat) but transformed plants have been obtained after leaf bombardment in tobacco, and stem bombardment in cranberry. Normally, reporter gene and selectable marker genes are used to isolate and select transformed cells/plantlets.

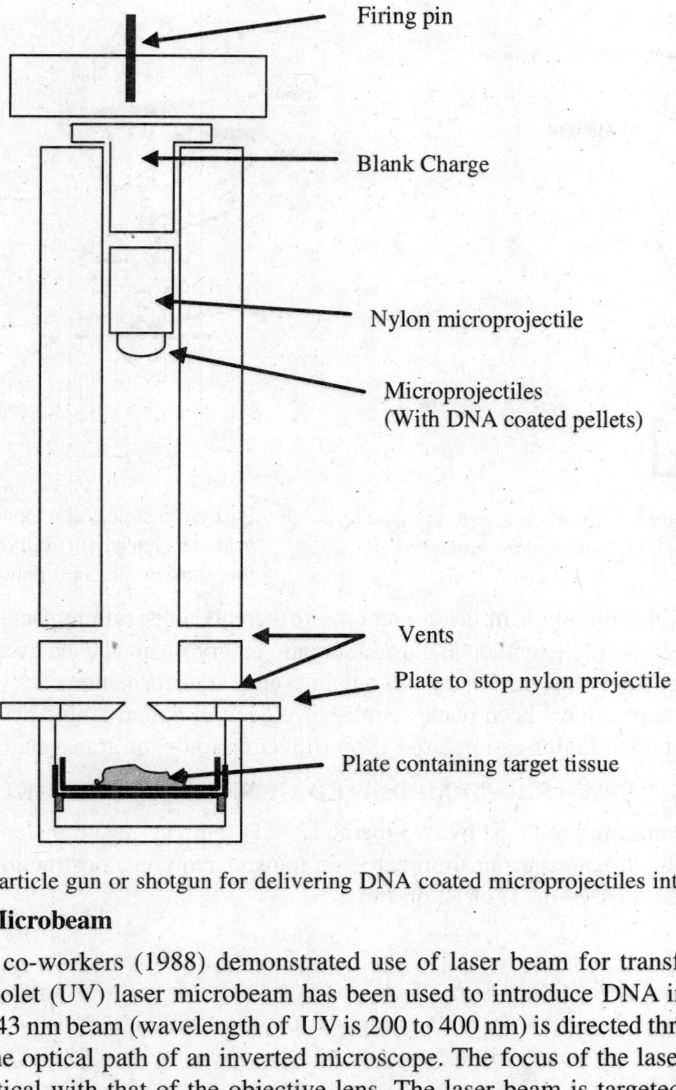

Fig. 23 5. Particle gun or shotgun for delivering DNA coated microprojectiles into plant cells.

B. Laser Microbeam

Weber and co-workers (1988) demonstrated use of laser beam for transformation of plant cells. An ultraviolet (UV) laser microbeam has been used to introduce DNA into plant cells and chloroplasts. A 343 nm beam (wavelength of UV is 200 to 400 nm) is directed through an adjustable attenuator into the optical path of an inverted microscope. The focus of the laser beam is adjusted so that it is identical with that of the objective lens. The laser beam is targeted by focusing on a specimen in the microscope. This laser beam can then make holes in any part of cell which is in focus. Laser micropuncture of the cell wall and plasma membrane allows uptake (entry) of plasmid DNA into cells. *Brassica napus* (rape seed) cells and microspores have been used for transformation by this technique. This technique has also been used to transfer genes into isolated chloroplasts and chloroplasts of intact protoplasts. 20% transformation was achieved by this method but fertile plants are yet to be produced by this method.

C. Silicone Carbide Fibres

Microinjection and electroporation methods have also been used for transfer of DNA using intact plant cells and tissues. Similarly, vortexing plasmid DNA and plant cells with silicon carbide fibre (0.6 μm in diameter and 10-80 μm in length) produced transformed cells at low frequency. Under vortex (vigorous shaking by vibration), silicone fibres penetrate cells and create fine holes permitting entry of DNA.

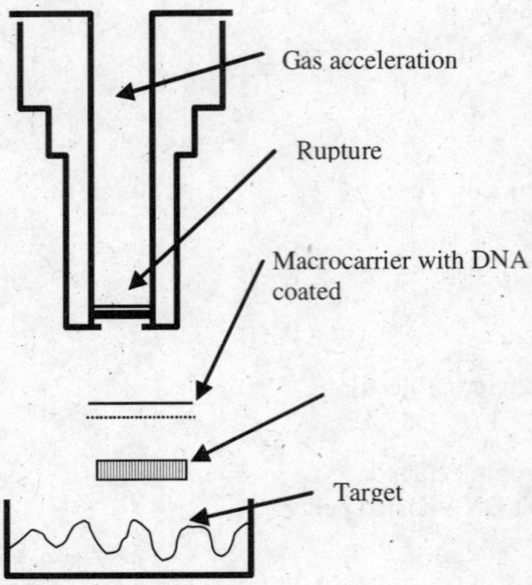

Fig. 23.6. Compressed air particle acceleration device for delivering DNA coated microprojectiles into plant cells.

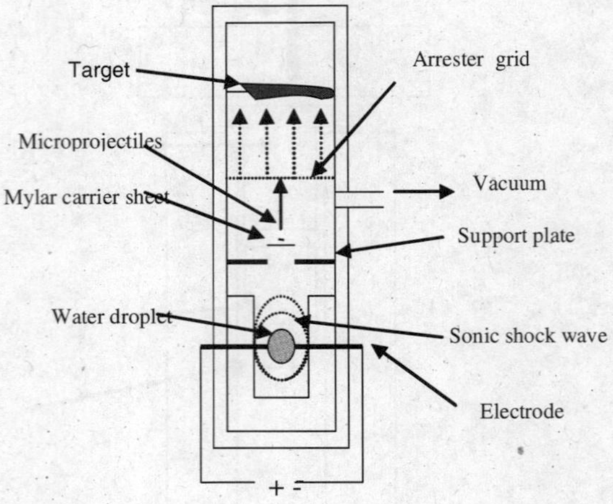

Fig. 23.7. Particle acceleration device based on high voltage charge for delivering DNA coated microprojectiles into plant cells.

DNA uptake by imbibition in dried embryos of cereals and legume species has also been reported. This is a very simple method and dried somatic embryos can also be used for this method. Only transient gene expression has been observed and stable transformations are yet to be achieved by this method. Attempts have been made to introduce gene in pollen grains by microinjection in the anthers. These pollen grains can be used for fertilization to obtain transgenic plants.

III. AGROBACTERIUM MEDIATED GENE TRANSFER METHOD

Transgenic plants are produced by two methods – (1) Indirect gene transfer using plasmid as a vector and (2) direct gene transfer. Indirect gene transfer requires construction of a vector for carrying foreign genes based on Ti or Ri plasmids.

A. Vectors

Vectors are the carrier DNAs into which 'foreign' DNAs or genes of interest are inserted to make a recombinant DNA (rDNA). Vectors along with this 'foreign' DNA (i.e., rDNA) are then introduced into appropriate host cell. Vectors are of two types-cloning vectors [used for obtaining millions of copies (cloning) of DNA segment] and expression vectors [used for expression of cloned gene to produce the product (protein)].

Plasmids are most commonly used vectors in gene cloning work. Isolation and purification of plasmids is a very routine experimental procedure. Several methods are available for isolation and purification of plasmid. The most critical stage in the process is lysis of the cell wall which is just sufficient for isolation of plasmid DNA without contamination by chromosomal DNA. Clear lysate contains plasmid. Alkali-SDS lysis and rapid boiling procedures are commonly used methods for plasmid isolation.

B. General Features

Most vectors carry marker genes which allow recognition e.g., antibiotic resistance – selectable markers, e.g. npt II (kanamycin resistance). Other features include (i) multiple unique restriction sites – a synthetic poly linker and (ii) bacterial origin of replication e.g. (Col E1).The problem is that a vector having these properties may not be easy to transfer to plant system. Therefore, *Agrobacterium* Ti plasmid is preferred because of (1) wide host range and (2) presence of T-DNA border sequence.

C. Ti and Ri Plasmids of *Agrobacterium*

Crown gall disease of dicotyledonous species is caused by infection of the bacteria, *Agrobacterium tumefaciens*, and is a neoplastic growth. Crown gall is the most extensively studied disease and leads to the isolation of plasmid for genetic manipulations. Neoplastic transformation leads to an apparently stable alteration in cellular phenotype.

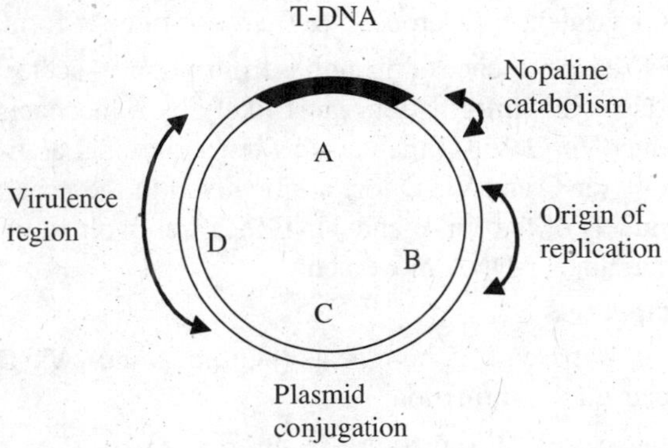

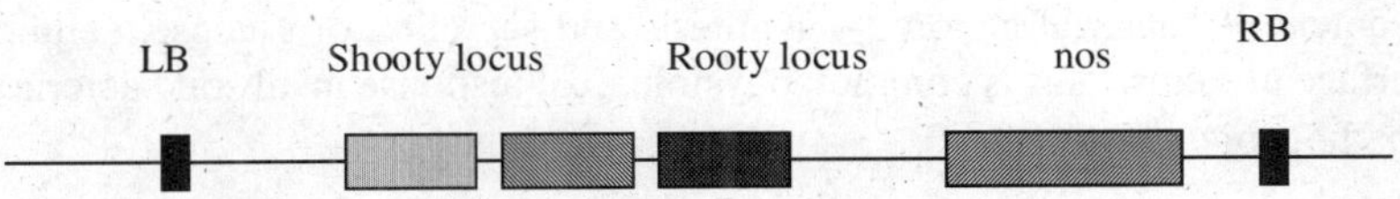

Details of T-DNA

Fig. 23.8. Structure of Ti plasmid and T- DNA.

A. tumefaciens utilize opines produced by host (tumour) cells as carbon and nitrogen sources. Details are given in Chapter 22.

Oncogenic strains of *A. tumefaciens* possess a large plasmid (90-150x10^6 daltons) known as the tumour inducing (Ti) plasmid. Transformation is associated with and accomplished by transfer of a stable, replicating portion of Ti plasmid DNA to the plant cell.

Although the precise mechanism of T-DNA transfer is not understood, molecular analysis of T-DNA integrated in host chromosomes has shown the presence of 25 nucleotides directly repeated sequences flanking the boundaries of the integration sites (left boundary and right boundary, LB, RB). Genes that are to be introduced in the plant cells must be inserted between these left and right border sequences, or just adjacent to one border of T-DNA.

(*i*) **Structure of Ti plasmid :** Ti-plasmids have four distinct regions (Fig. 23.8)

A – T-DNA, main transfer DNA responsible for tumour formation.

B – Responsible for replication

C – Responsible for conjugation

D – responsible for virulence (vir region, vir-genes responsible for virulence; mutation in this region may lead to non-virulence). Important region required for transfer of T-DNA.

(*ii*) **Structure of T-DNA :** Transfer – DNA is transferred and integrated into host genome during the infection. Infection brings about physiological and morphological changes in the tissue due to expression of genes located on T- DNA. These are :

– Onc region – shooty, rooty genes, IAA and cytokinin production (Fig. 23.8)
– OS region – opines – unusual amino acids, octopine and nopaline. Octopine synthase and nopaline synthase (nos) coded by T-DNA. So depending upon synthesis of amino acids plasmid is known as a) Octopine type – Ti plasmid, or b)Nopaline type – Ti plasmid. These

amino acids are used only by bacteria as source of carbon and nitrogen and these genes are located outside T-DNA.

– Other important point of interest is 25bp flanking border sequences (Lb,Rb),

Vir region – If vir region and T-DNA are physically separated but present on two plasmids in the same bacteria, transfer of DNA takes place. This is an important property for vector. It consists of 36 kbp, and six operons;Vir- A, B, C, D, E & G. Except A & G, others are polycystronic (Fig. 23.9). Operons ABDG are for virulence. Operons C & E are for tumour formation.

Vir-A – chemoreceptor – sense presence of phenolics from wound – acetosyringone, B-hydroxy aceto syringone. Vir A tranduces this information, most likely by a mechanism involving protein phosphorylation, to the product Vir-G. Vir-G then acts as transcriptional activator of itself and the other Vir loci. The products of Vir-C and Vir-D loci are involved in the generation and processing of the T-DNA copy. The products of the Vir-B and Vir-E loci are involved in forming most of the structural components that facilitate T-DNA movement.

(*iii*) T-DNA: Transfer process

- Nicking between 3rd and 4th base of 25 bp repeats (bottom strand). Vir-D operon encodes for an endonuclease that cause nick formation.
- Initiation of DNA synthesis in 5'-3' direction – as per Fig. 23.10.
- Involvement of bacterial genome – synthesis and secretion of glucose, cellulose, fibrils, and cell surface proteins. This is common physiological response in all soil bacteria and is involved with pathogenic characters.

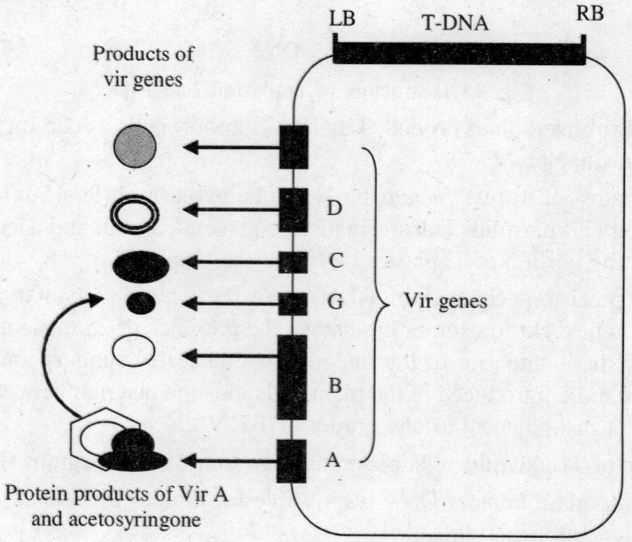

Fig 23.9. Vir genes and products of vir genes.

The generation of the T-strand is the first step in the complex process of *Agrobacterium*-mediated plant cell transformation. Following its formation, this DNA must transverse the bacterial cell membrane, the bacterial cell wall, the plant cell wall and plant cell and nuclear membranes. Once inside the nucleus, the T-DNA must finally integrate stably into the plant cell genome (Fig. 23.10). During this entire transit process, the T-DNA strand also must avoid degra-dation by nuclease. The T-DNA exists as a DNA protein complex. The T-DNA complex protects it and mediates its travel. The final step in the genetic transformation of plant cell is integration of T-DNA copy, presumably T-strand, into plant cell DNA. It has been argued that the T-strand might be converted to a double stranded (ds) DNA prior to integration.

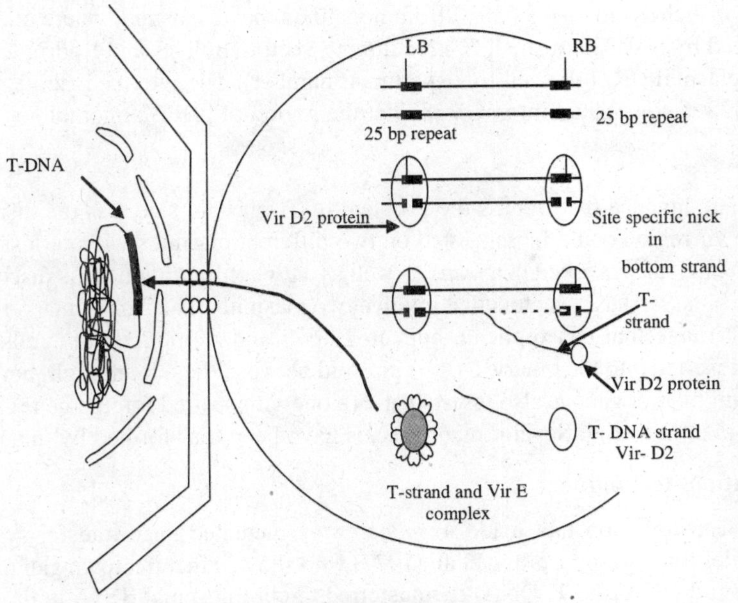

Fig. 23.10. T-DNA transfer process.

Thus by placing foreign genes into T-DNA region of Ti -plasmid, it is possible to clone (make copies) the introduced genes with the multiplication of plasmid residing inside the bacteria (self replication of plasmid makes millions of copies) which is grown on a medium and with the multiplication of bacterial population, residing plasmid is also multiplied by this method. It is possible to exploit the natural ability of *Agrobacterium* to transfer new DNA into the plant genome.

D. Vectors Based on Ti and Ri Plasmids

The Ti or Ri-plasmid can not be used directly. There are limitations for direct use of these plasmids. These are (i) Large size of vector make it difficult to manipulate (ii) Absence of unique restriction enzymes sites and (iii) Tumour induction. Therefore, vectors are designed with useful characteristics. This involves-removal of tumour induction property or disarming the plasmid. This is achieved by replacing of tumour induction genes in T-DNA by selectable markers such as npt-II (kanamycine). Promoters and polyadenylation signal isolated from octopine and nopaline synthase genes were used for expression of selectable markers. As there is no excess production of plant hormones, whole plants transformed with such disarmed *Agrobacterium* strains can be produced and detected by the production of opines. When a selectable marker gene (kanamycin resistant) is introduced, transformed cells can be selected by their ability to grow on media containing the selective antibiotic. Untransformed cells will not survive on this medium.

Other promoters - *onc* – CaMV35S, CaMV19S isolated from cauliflower mosaic virus have also been used. Therefore, it is concluded that T-DNA and Vir genes are two essential components of a vector.

E. Cointegrative Vectors

Cointegrative vectors recombine, via DNA homology, with an intermediate cloning vector, which is used for manipulation and cloning of the gene in *E. coli*. *Agrobacterium* containing cointegrative vector and *E. coli* containing intermediate cloning vector are allowed to undergo conjugation, but the intermediate vector can not replicate in *Agrobacterium* so it has to transfer the marker genes as well as the DNA segment to the resident Ti plasmid (cointegrative vector) through recombination in the region of DNA homology.

Example of such vector – pGV3850 from nopaline type Ti plasmid, where almost all T-DNA has been replaced by pBR322, a small *E. coli* cloning vector (Bolivar and Rodriguez prepared and hence name – plasmid BR, followed by experiment number). The intermediate vector (pGV1103) based on pBR322 is conjugated into pGV3850 at the region of pBR322 homology.

F. Binary Vector

A significant advance that bypass the problem of Ti-plasmid size was the discovery that the T-DNA and the vir region could be separated on two different plasmids without loss of the T-DNA transfer capacity, i.e., they worked in a *trans* as well as a *cis* configuration. This discovery led to the development of binary T-DNA vectors that involve two plasmids. The small binary T-DNA plasmid has a wide host range that can replicate both in *E. coli* and *Agrobacterium* cells. The desired foreign gene is inserted into the binary T-DNA plasmid between the left and right border sequences. A selectable plant marker gene is also inserted and (alongwith desired foreign gene) allow selection of transformed plant material. Several plant species have been transformed by this method.

G. Transformation Technique

The critical information that made *Agrobacterium* mediated gene transfer systems possible came from the elegant work by Chilton et al. (1977) who showed that in crown gall disease a region of bacterial plasmid DNA (the T- DNA) is transferred to chromosomal DNA in the plant nucleus, stably maintained there, and expressed in the absence of the bacterium. The first transgenic *Nicotiana tabacum* plant was produced by Horsh and co-workers in 1984 using *Agrobacterium.*

Agrobacterium mediated gene transfer methods are being developed for a wide range of dicotyledonous plants and gymnosperm species. The gene tagging approach employs.

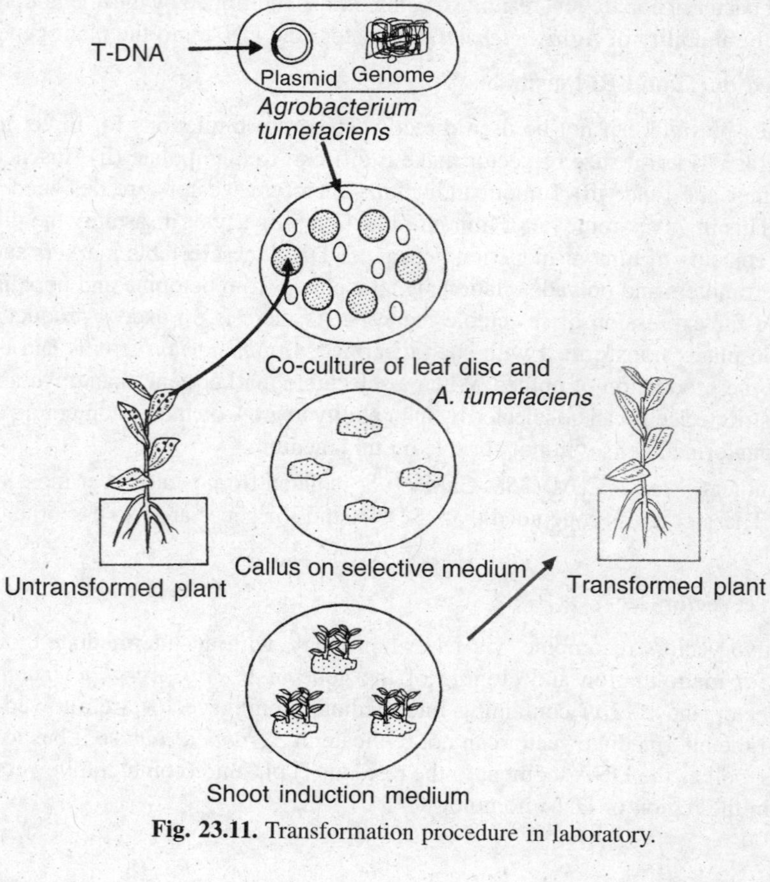

Fig. 23.11. Transformation procedure in laboratory.

T-DNA containing either a reporter or strong transcriptional enhancer transform a large number of cells. Normally bacteria are incubated with plant cells (few hours to few days) during that period T-DNA transfer takes place. The cells are then washed and treated with antibiotics to remove the bacteria. The cells are then cultured in the presence of the selectable agent, and transformed shoots are regenerated and characterised. (Fig. 23.11)

Plasmids of *Agrobacterium* have been used as vector for transfer of foreign DNA into a number of dicot species. However, seed legumes are still not amenable, exceptions *Glycine max* and monocotyledons (cereals) can not be used as *Agrobacterium* is host specific and do not infect monocots. But with an exception of *Asparagus* (which has been transformed with this technique). There is no cambial activity and wound healing process in monocots, therefore, acetosyringone is not produced; this results in failure of vir genes to recognise by chemoreception.

(*i*) Pre-requisites for agroinfection-

1. Production of acetosyringone.
2. Bacteria have access to actively dividing cells (DNA replication) such as meristems, fresh protoplasts, dedifferentiated tissues.
3. Regeneration in transformed tissues should be possible.

(*ii*) Explants for cocultivation- following materials can be used :

- Protoplasts
- Cell suspension cultures
- Callus \ thin cell layers – epidermis, tissue slice, organ section (leaf disc, section of roots, stem or flower).
- Wounded and inoculated whole plant.

H. Selectable Markers

Selectable markers are usually required for efficient recovery of transgenic cells and plants. After gene transfer, transformed cells are few in number compared to untransformed cells. It is not possible to separate transformed and non-transformed cells by any physical method. A selectable marker gene incorporated with the desired gene helps the growth of transformed cells on a nutrient medium containing corresponding selective agent. The availability of multiple selectable markers is useful in developing efficient transformation methods for diverse species as well as for the introduction of multiple novel traits (characters) through successful transformation and regeneration of transformed plants from such cells (Table 23.1).

Selective agents differ in their toxicity to different plant species. The different developmental states of the plant cells or tissues give different response to the selectable marker. The cells will react differently than whole plant or organ. Therefore, it is necessary to use correct concentration of selective agent for the transformed cells of a given species to select the cells and to inhibit the growth of untransformed cells. This can be illustrated by following example. Transgenic plants of *Lycopersicon esculentum*, *Brassica napus* and *Lactuca sativa* can be selected on low kanamycin concentration (15-100 mg/*l*), while other plants like *Beta vulgaris* requires high kanamycin concentration (400 mg/*l*) as selection agent.

Table. 23.1. Selectable markers used in transformation.

Marker gene	*Enzyme encoded*	*Resistance against*
Antibiotics		
Npt-II	Neomycin phosphotransferase	Kanamycin, G418 neomycin
Npt or *aph IV*	Hygromycin phosphotransferase	Hygromycin
Aad A	Aminoglycoside-3'-adenyl transferase	Streptomycin

Herbicides		
Bar	Phosphinothricine acetyltransferase	Phosphinothricin
Aro A	5-enolpyruvyl shikimate-3-phosphate synthase	Glyphosphate
Other modified *als*	Acetohydroxy acid synthase (or acetolactate synthase)	
Chlorsulfuron imidazolanones		
DHFR	Dihydrofolate reductase	Methotrexate

Herbicides are generally more toxic to plant cells than antibiotics. This is due to their specific mode of action in plant cells as well as their efficient uptake and translocation within the plant tissues. Herbicides and other highly toxic compounds may require delayed application in order to ensure that the transformed cells have produced sufficient quantities of enzyme responsible for the protection to such compound.

Callus and cell cultures are better systems than organized explants to achieve transformations and selection of transformed cells. Explants may give rise to organs from untransformed cells which may escape selective agent. This is not possible with isolated cells and only transformed cells will be able to grow on the selective agent.

I. Reporter Genes

In all transformation experiments, regeneration of non-transformed plants is also observed alongwith transformed plants. If these non-transformed plants can be detected at an early stage, a lot of time, labour and analyses can be saved. There are several methods to screen and identify the transformed plants, but usually these methods are laborious and time consuming, e.g. southern blotting technique to compare DNA sequences requires large amount of tissue (DNA) for comparison. The other methods use selective agent. In both these methods, prolonged growth and subcultures are involved before selection can be made.

The use of reporter gene eliminates these drawbacks and transformed plants can be recognised easily. A reporter gene is a coding unit whose product is easily assayed (such as GUS whose product can easily be detected by histochemical assay). This gene may be connected to any promoter of interest so that expression of the gene can be used to assay promoter function. In fact, the reporter gene describes the transfer and expression of other promoter. Two genes that are now widely used for this purpose are those coding for B-glucoronidase (GUS) and luciferase (LUC). The coding regions of these non-plant genes have been fused to plant promoters and polyadenylation sequences such that they give high level of expression in plant cells (Table 23.2). The anthocyanin regulatory genes can serve as a unique reporter system in maize and some other cereals. Transformed cells accumulate reddish purple anthocyanin pigments. The anthocyanin genes are extremely sensitive marker used in *Arabidopsis* and *Nicotiana* transgenic plants.

Table.23.2. Reporter genes, assay and identification method in transferred plants.

Reporter genes used for	Substrate and assay	Identification
Chloramphenicol acetyl transferase (CAT)	^{14}C chloramphenicol + acetyl Co-A, TLC separation	Detection of acetyl chloram-phenicol by autoradiography
B-glucoronidase (GUS)	Glucoronides (PNPG, X-GLUC, NAG, REG)	Fluorescence detection colorime-tric, fluorimetric and histochemical
B-galactosidase (Lac Z)	β-glactoside (x-gal)	Colour of colony
Luciferase (LUC)	Decanal and $FMNH_2$ATP + O_2 + luciferin	Bioluminiscence (exposure of x-ray films)
Octapine synthase	Arginine pyruvate + NADH	Electrophoresis
Nopaline synthase	Arginine + ketoglutaric acid+NADH	

Without selection agent, transformed plants are selected with the help of scoreable gene or reporter gene.

Examples of Transgenic Plants- Herbaceous Dicot - Tobacco, *Petunia hybrida*, tomato, potato, egg plant, *Arabidopsis thaliana,* lettuce, *Apium graveolens* (celary), sunflower, flax, rape oil seed, cauliflower, cabbage, cotton, soyabean, pea, chicory, carrot, licorice, sweet potato, kiwi, papaya, grape, rose, *Chrysanthemum*, etc.

Woody Dicot - *Populus, Malus, Pyrus communus, Azadirachta indica*

Monocot – *Asparagus, Secale cereale, Oryza sativa, Triticum aestivum, Zea mays, Avena sativa.*

Gymnosperm – *Picea glauca.*

QUESTIONS

1. Describe the various methods of direct gene transfer giving suitable examples.

2. Write short notes on the following :

a. Transgenic plants
b. Particle acceleration
c. DNA uptake by protoplasts
d. Transgenics for disease resistance
e. Microinjection
f. Electroporation
g. Lasers and Liposomes for transformations.

3. Write short notes on the following :

a. Agrobacterium
b. Ti plasmid
c. Vector
d. Binary vector
e. Plasmid copy number
f. T-DNA
g. Vir gene
f. T-DNA transfer
i. Selectable marker
j. Reporter gene.

4. What are plasmids? Describe the plasmid mediated gene transfer process.

TRANSGENIC CROPS

CHAPTER 24

Since the bringing of agriculture, approximately 10,000 years ago, humans have profoundly modified the shape and form of most cultivated species. These changes were largely made in pursuit of better adaptation and increased productivity within the context of a cultivated field in contrast to the natural environment. Though the selection of naturally occurring mutants presenting novel and useful phenotypes, countless generations of farmers practiced the earliest and simplest form of genetic modifications. It is only in the last century that a science based approach to the genetic modifications of most cultivated plants was developed largely through the work of plant breeders and biotechnologists.

The methodology and tools available to achieve the hasten increases in productivity or changes in various quality traits have increased dramatically in number and technical sophistication over the last 50-100 years. Sexual crosses between carefully selected parents still play a major role in all present day breeding programme. However, constant efforts are made to increase useful genetic diversity and its exploitation for increasing productivity. Such increases have been created through the use of wide crosses (inter-specific and inter-genetic) use of in-vitro techniques such as embryo rescue to make use of already existing variations. Other approach to transfer useful traits from one plant to another (by over coming the species specific barrier) through use of somatic cells (e.g. protoplast fusion) rather than reproductive cells. various mutagens (irradiation, chemicals, insertional sequences) have also been used extensively to accelerate the identification of desired mutants for breeding or gene discovery. The development of the ability to transfer a gene or genes from one species to another via transgenic is only the most recent addition to a long list of methods which aim to broader the available genetic diversity in a given plant species.

Maize is a good example of continuous efforts for improvement of a crop species. Its ancestral progenitor, teosinte, is phenotypically quite different from modern maize in many important respects. These differences were so striking to the botanists that until relatively recently these two sub-species of *Zea mays* were not even thought to be close relatives. From an agronomic view-point, teosinte is hardly an attractive crop. Its yield is estimated at 100kg/ha whereas modern maize hybrids can yield more than 10 tons /ha or 100 times more. However, the two plants are compatible and can be crossed easily.

Previous methods of genetic improvement relied mostly on the shuffling (or addition in the case of some translocations) of a relatively large number of genes (hundred to thousand) from the same or related plant species. In contrast, nuclear transgenesis involves the addition of a small number of genes (usually less than five, typically two) into the nucleus. Transgensis is thus results in quantitative changes that are quite small in comparison to what occurs through "conventional" methods of genetic improvement. This is not to minimize the potential consequences of this genetic transformation, however as a single gene can have a large impact (be it beneficial or deleterious).

The introduced genes usually serve two distinct purposes, one of the transferred genes (transgenes) is a selectable marker, meaning that it confers on the transformed cells a distinctive phenotype that can be used to select the few genetically modified (GM) cells among a population of untransformed cells. These marker genes typically confer resistance to selective agent, often an antibiotic. The other transgenesis is meant to confer a desirable phenotype (a character) e.g., agronomic (herbicide, pest/stress resistance) or related to food quality (self-life, taste, nutritional value).

A second novel feature in transgenic plants is that the introduced genes can originate from almost any living organism or can even be entirely synthesized *de novo*. Thus, transgenesis offers the possibility of introducing relatively 'foreign' DNA into a plant cell.

No matter what the method used to transform plant, such as biological vector (*Agrobacterium tumefaciens*) or a physical device (particle gun), there is no means of controlling either the number of copies, or the location of the introduced material. Therefore, transgenic lines obtained with the same genetic construct need not be identical, either because the expression of transgene(s) differ due to copy number or due to position effect (not all the chromosomal loci are equally favourable to gene expression).

Among the cultivated transgenic crop plants, herbicide tolerance has consistently been the dominant trait followed by Bt based insect resistance, herbicide tolerant soybean and canola, and Bt maize and cotton constitute the four major genetically modified (GM) crops. The two commercialised transgenic vegetable crop are tomato with delayed fruit ripening and potato with insect and virus resistance.

Transgenic crops for –

1. Insect pest control
2. Herbicide resistance
3. virus resistance
4. Abiotic stress tolerance
5. Quality improvement
6. Pharmaceutical and industrial use.
7. Economic impact
8. Environmental impact
9. Transplastomic plants

1. Insect pest control

The most talked insect pest control through use of transgene is Bt gene (from *Bacillus thuringiensis*) producing toxin.

The adoption of insect resistant transgenic crops have been increasing annually at double digit rates since the commercial release of first generation maize and cotton expressing a single modified *B. thuringiensis* toxin (Bt) nine years ago. Studies have shown that these Bt crops can be successfully deployed in agriculture, which has lead to a decrease in pesticide usage, and that they are environmental friendly. However, sustainability and durability of pest resistance continued to be discussed. Now, scientists are developing second and third generation insect resistant transgenic plants and examine the proposed models for longevity of such resistance.

First generation transgenic plants – Transgenic plants constaining only marker genes, which are useful in the development of transformation systems. **Second generation transgenic** plants – Transgenic plants containing, in addition to the selectable marker, one or two transgenic encoding simple agronomic traits (such as pest and herbicide resistance) **Third generation transgenic plants.** – Transgenic plants that contain multiple transgenes targeting multiple pests and disease, often in a temporal or spatial manner. These might also express additional value added or agronomic traits.

By using a variety of technique, it has become possible to transform plants with foreign genes. Expression of foreign genes implants makes it possible to produce a very wide range of new plants varieties. Transgenic plants have been developed to be resistant to a range of environmental stresses, including insects, viruses, herbicides, pathogens and salt stress to have flower with modified colour to a modified nutritional content including modifications in amino acids, lipids, discolouration and sweetness.

Bt gene and toxin (*Bacillus thuringiensis*)-Several species of bacteria produce protein in abundance. When insect larvae ingest these bacteria with their food, protein kill larvae. The most widely studied of these bacteria is *Bacillus thuringiensis* or Bt in short. This species lives all over the world. When these bacteria form space, they also form a large crystal like structure, in the bacterial cytoplasm, that is made out of protein. This bacterium comprises a number of different strains and subspecies, each of which different protein (toxin) that can kill certain specific insects

Insecticidal toxins from some strains of Bt.

Strain	Toxin class	Protoxin size (KDa)	Target species
Berliner	Cry I	130-140	Lepidoptera
Kurstaki KTO, HD-I	Cry I	130-140	Lepidoptera
Kurstaki HD- I	Cry II	71	LepidopteraDiptera
Tenebrionis (sandiego)	Cry III	66-73	Coleopters
Israelensis	Cry IV	68	Diptera

One of the proteins in the crystal like structure is called the Bt-protoxin. When insect larvae eat the bacterial cells along with leaves, the spores and the crystalline like structure containing the proteins are released in the larval gut, where the digestive enzymes cleave the protoxin producing an active toxin.

Protoxin activated within its gut by the combination of alkaline pH (7.5- 8.0) and specific digestive proteases. The toxin binds to the membrane of the epithelium cells of the gut, inserts itself into that membrane and creates an ion channel through which other molecules (e.g. ATP) can freely pass. Punctured by many holes, the gut cell cannot survive long, so the insect larvae starve for lack of nutrition and ultimately die. Because conversion of the protoxin to the active toxin requires both alkaline pH and the presence of specific proteases, such conditions are not present in mammals and hence they are safe from the protoxin.

No significant role for the bacterium has been attributed to the parasporal crystal structure. The parasporal crystal usually consists mainly of protein (~95%) and small amount of carbohydrate (~5%). The crystal protein can generally be dissociated by mild alkali treatment into subunits. The insecticidal toxins of *B. thuringiensis* strains can be grouped into four major classes: Cry-I, Cry II, Cry III and Cry IV. This is based on insecticidal activity against various insect. These toxins are further classified in sub classes and sub groups according to DNA sequence of the toxin gene, e.g. Cry I gene has six sub classes (Cry I A to F) and Cry I has subgroup (Cry IA a to c)

As a result of co-evolution between insects and their pathogens, there is host specificity between Bt-toxin and the membranes of the gut cells. The Bt toxin of a particular Bt. Strain will bind to the gut of Lepidoptera larvae, or only some species of Lepidoptera, but not to other. When toxin does not bind, there is no effect on the cells that line the gut, and the larvae do not die. Thus some Bt toxin will kill lepidoptera (butter flies and moths), other coleopteran (beetles and weevils) and others diptera (mosquitos). For the biological control of insect pests, approximately 1.3 x108 to 2.6x108 spores per sq foot of the target area is applied. Administration of the spores is timed to coincide with the peak of the larval population of the target organism.

Btare Subspecies kurustaki contains a protoxin gene on one of seven different plasmids that are approximately 2.0, 7.4, 7.8, 8.2, 14.4, 45 and 71kb in length. Protoxin is 130 KDa, therefore, not present on small plasmids. This gene has been transferred to other bacteria to kill mosquito larvae as well as gene has been modified to produce toxin during vegetative phase of bacterial growth rather than only during sporulation. Thus it is possible to produce toxin continuously in fermentor by growing bacteria. It has also been attempted to increase the host range.

This Bt toxin has been used in several ways to control the insects. A relatively simple way is to grow the Bt. bacteria, dry them out, and prepare the heat killed and dried bacteria in such a way that they can be sprayed or dusted on crops. These preparations are initially highly effective, but the Bt protoxin is not stable after product is sprayed on plants. The Bt protoxin crystals are released from the bacteria and protoxin quickly disappears from the plants.

Scientists at Mycogen, a biotechnology company in San Diego, California (USA), introduced a Bt gene in a different bacterium (*Pseudomonas fluorescens*). These bacteria can readily be grown in large fermentors, killed and than formulated as a spray. With this bacterium the protoxin crystal remain in the bacterial cells, and as a result they are stable even after they have been sprayed on the plants. The spraying Bt toxin works well with insect larvae that live on the surfaces of leaves, but would be less effective with insect larvae that live in soil or larvae living inside the plants. To control there insects, scientists have transferred Bt gene using particle gun transfer system in cotton, tomato, tobacco, potatoes, and other crop plants. Transgenic plants produced containing Bt gene are listed in Table.

Table 24.1. Plants with insecticidal gene derived from *B. thuringiensis*.

Company		Transgenic crop
Monsanto	-	Potatos, cotton, tomato, corn
Calgene	-	Cotton, tobacco, potato,
Ciba-geigy	-	Tobacco, corn
Agrigenetics	-	Canola (rape seed)
Campbell Institute	-	Tomato
Rohm & Hass	-	Tobacco

Constitutive or tissue specific expression – Although constitutive expression of insecticidal transgene products has provided high levels of resistance in crop plants, tissue-specific or inducible expression might be desirable under some circumstances. Because the epidermal cells are first to be attached by insects, defence genes expressed under epidermal cell-specific promoters (e.g. CEF6, an enzyme of curricular wax production) might be useful. Phloem –feeding insects can be targeted using the root phloem specific promoter AAP 3, the phloem specific pumpkin promoter PPZ and the rice sucrose synthase RSS promoter. Progress is being made with chemically inducible promoters, including those induced by ethanol, tetracyclin, copper, glucocorticosteriod hormones and steroidal and non-steroidal ecdysone agonists.

Transgenic plants with multiple resistance genes -The simultaneous introduction of three genes expressing insecticidal proteins, Cry1Ac, Cry2A and Gna, into indica rice to control three major pests, rice leaf folder, yellow stem borer and the brown plant hopper has been achieved. The Bt genes target the leaf folder and the stem borer, and the Gna gene target the plant hopper. Triple transgenic plants were more resistant compared with their binary counterparts. Comparison of three different transgenic Bt cotton populations containing either the single Cry1Ac or Cry 2Ab gene or both genes, for fruit penetration and damage revealed that transgenic cotton containing two Bt genes performed better.

Cry toxins domain change - Most activated cry toxins share a common three-domain structure. It has been shown that hybrid cry toxins exhibited substantially enhanced toxicity or host range. Enhanced efficacy of Bt cry proteins was achieved by creating fusions between domain III of cry 1 Ac and domains I and II of various other cry 1 proteins.

A hybrid *B. thuringiensis* gene was constructed using a synthetic and truncated cry 1 Ba gene with cry IIa gene, encoding domain II. Transgenic potato plants expressing this hybrid toxin were resistant to several insect pests, including potato tuber moth and European corn borer.

Plant derived lectins –The ability of the mannose – specific snowdrop lectin (GNA, agglutinin) to serve as a carrier protein to deliver insecticidal peptides and proteins to the haemolymph of lepidopteran larvae was demonstrated by expressing GNA-allatostatin and GNA-SFI-1 fusions and using the purified fusion protein in artificial diets against the tomato moth. SFI1 is an insecticidal venom neurotoxin from a spider. Whereas the two individual components of the toxins showed no oral toxicity, the fusion proteins were insecticidal to lepidopteran larvae.

Unmodified lectins have been shown to be insecticidal towards sap-sucking insects outside the host range of Bt ; e.g., garlic leaf lectin expressed in transgenic tobacco plants substantially decreased the percentage survival of peach potato aphids feeding on these plants compared with control.

Engineering plants with fusion protein combining cry1AC with the galactose binding domain of the non-toxic ricin B-chain provides the toxin with additional binding domains, thus increasing the potential number of interactions at the molecular level in target insects. Transgenic rice and maize plants engineered to express the fusion protein were significantly more toxic than those containing Bt gene alone.

Second generation new transgenic insect pest resistant plant are being produced using new sources of novel proteins such as vip (vegetative insecticidal proteins) produced by *B. thuringiensis* during vegetative growth, and proteins produced by symbiotic bacteria living with nematodes. The technology is being continuously improved.

2) Herbicide resistant plants

Certain herbicides can be used as pre emergence herbicides to kill weeds before the crops are planted. If the crop plants are resistant to these chemical then they can be used with the crop plant (post emergency). By understanding the mechanism of action of these herbicides and development of resistance by certain bacteria to such chemical can provide clone for developing herbicide tolerant plants. Some plants or bacteria are resistant because they have an enzyme that detoxifies the herbicides. In other words they possess a gene for this action. Transfer of this gene to a crop plant should provide the crop plant by same action or mechanism. Some plants or bacteria become resistant to herbicide because of mutation in the target enzyme (or gene) and because of this change they are no more sensitive to herbicide or are not damaged by herbicides. The enzyme can work in presence of herbicide. Therefore the detoxifying mechanism or change in affected enzyme can make the organism herbicide tolerant.

Glyphosate (a herbicide) act by inhibiting one of the enzymes that is necessary for the synthesis of amino acids in the chloroplast. Glyphosate initially produced and marketed by Monsanto under the trade name Roundup® is a widely used non selective herbicide it effectively kills 76 of the world's 78 worst weed species.

Scientist at Monsanto isolated a gene for an enzyme involved in amino acid biosynthesis enzyme EPSP-synthase (5 enol pyruvinyl) shikimate 3-phosphate synthase) from resistant *E. coli* bacteria. They modified the gene in such a way that it could be expressed in plants, and then transferred it to plants e.g. tobacco, tomato and soybean. Expression of bacterial gene in plants required a control region that would direct the expression at the gene in the plant (because bacterial control regions do not work in plants). In addition to this the gene had to be modified in such a ways that the enzyme, which is synthesized in cytoplasm, would be transported to chloroplast. This is important that when gene of prokaryotic origin is used the product should be transported to right cellular compartment in the plant. This should not affect the quantity of quality of yield. The gene has been successfully transferred in soybean where the plants showed resistance with out change in yield.

Phosphinothricin is a herbicide that acts by inhibiting another enzyme necessary for amino acid biosynthesis (glutamine synthetase) and nitrogen metabolism. This enzyme converts ammonia to glutamate. Inhibiting the activity of this enzyme leads to rapid accumulation of ammonia within the plant cell. Higher concentration of ammonia are toxic to the cell. Phosphinothricin produced and marketed by Hoechst AG under the trade name Basta®, is also a very effective non selective herbicide. This product is related to an antibiotic that is also a herbicide, produced by the fungus *Streptomyces hygroscopicus*. Scientists at Plant Genetic Systems, Belgium obtained a gene from this fungus that encodes an enzyme that converts phosphinothricin to a non-herbicidal derivative by combing it with cell metabolite. This gene, known as bar-gene, has been transferred to tobacco and potato, where it is expressed showing herbicide tolerance in these plants. The yield performance of the plants remained unchanged.

Herbicides are simply chemical compounds that kill or inhibit the growth of plants without deleterious effects on animals. Herbicides usually inhibit processes that are unique to plants, e.g. photosynthesis. Mostly herbicides act as inhibitors of essential enzyme reactions. Any change which can reduce the inhibitory effect of herbicide, can reduce the inhibitory effect of herbicide, and can provide increased herbicide tolerance.

Glyphosate acts by inhibiting the enzyme 5 enol pyruvinyl shikimate 3 phosphate synthase (EPSP synthase), an essential enzyme in the biosynthesis of the aromatic amino acid, tyrosine, phenylalanine and tryptophan. These are essential components in the diets of higher animals. Therefore higher animals do not contain EPSP synthase, and are not affected by glyphosate.

Glyphosate does inhibit the EPSP synthase of micro organisms as well as those of plants. Selection of organisms is made on inhibitory concentration of herbicides by growing them in presence of herbicide. This way researches isolated glyphosate tolerant mutant of *Salmonella typhimurion*, *Aerobacter acrogens,* and *Escherichia coli*. In bacteria, EPSP synthase is encoded by the aero A gene. When aeroA genes (with plant promoter and adenylation signals) were transferred in plants, the transgenic plants showed increased tolerance to glyphosate. In plants, aromatic amino acids are synthesized in chloroplasts, but gene for EPSP is localized in nucleus. Therefore, a protein is attached to EPSP synthase, which translocate the EPSP synthase into chloroplast, where the protein is removed by clearage. It has been shown that the petunia transit peptide targets the *E. coli* aeroA gene product into tobacco chloroplasts and will impart glyphosate tolerance.

In another method, glyphosate-tolerant plants have also been produced by using an EPSP synthesis cDNA isolated from a glyphosate tolerant petunia cell culture line. Such lines can be selected by growing cells on medium containing increasing concentration of selection factor, e.g. glyphosate. In the cell line, tolerance resulted from amplification (an increase in copy number) of the EPSP synthesis gene, resulting in over production of EPSP synthase in these cells.

The EPSP synthase cDNA isolated from the cell line was joined to the CaMV 35s promoter and to the Ti nos. polyadenylation signal. The strong CaMV 35s promoter (35s +EPSP synthase + nos) gene was introduced into petunia plants on a Ti vectors, the transgenic developed were tolerant to the four times higher concentration which kills control plants.

The other examples of herbicides resistance plants are given in the table 24.2. It is evident from these examples that a resistant factor is developed on the basis of mode of action of herbicide, by modifying or over producing the target product.

Table 24.2. Gene- based herbicide resistance in plants

Herbicide	Mode of development of herbicide resistance
Triazines	Resistance is due to an alteration in the psbA gene, which codes for the target of this herbicide, chloroplast protein D1.
Sulphonylureas	Genes encoding resistant version of the enzyme acetolactate synth--etase have been introduced into poplar, canola, flax, and rice.
Glyphosate	Resistance is from overproduction of EPSPS, the target of this herbicide.
Bromoxynil	Resistance to this photosystem II inhibitor has been created by transforming tobacco and cotton plants with a bacterial nitrilase gene, which encodes an enzyme that degrades this herbicide.
Phenoxy carboxylic acid (e.g. 2,4-D and 2, 4, 5–T)	Resistant cotton and tobacco plants have been created by transformation with the rfdA gene from Alcaligenes, which encodes a dioxygenase that degrades this herbicide.

Gluphosinate	over 200 different plants have been transformed with either the bar (Phosphinothricin) gene from *Streptomyces hygroscopicus* or the pat gene from *S. viridochromogenes*. The phosphinothricin acetyltransferase that these genes encode, detoxifies this herbicide.
Cyanamide	Resistant tobacco plants were produced when cyanamide hydrates gene from the fungus *Myrothecium verrucarla* was introduced. The enzyme encoded by this gene converts cyanamide to urea.

Canola (*Brassica napus*) cultivars engineered to tolerate the application of broad-spectrum herbicides have been developed both via transgenic and mutagenesis. This type of canola has been adopted rapidly by the Canadian farmers. The proportion of farmers growing transgenic herbicide-resistant canola has increased from 7% in 1995 to 80% in 2000. The use of transgenic herbicide – tolerant canola varieties had increased net return by 32%, had reduced pesticide use by 6000 tones and fuel consumption by 31 million litres. There are undoubtedly very real benefits both for the farmers and for the environment.

3) Virus resistant plants

Plants viruses often cause considerable damage and significantly reduce yield. Breeding for disease resistance is the best method to protect plants from viral and other infections. Recently scientists have used the techniques of genetic engineering to develop virus resistant transgenic plants. There methods used immunization with viral coat protein genes, other viral genes, or viral gene antisense sequence to confer resistance.

Potato is one of the most important food crops after cereals and pulses. It is very difficult to improve potato through breeding techniques as it is a tetraploid. Most of the cultivars are susceptible to various diseases caused by fungi, nematodes, and virus. Potatoes are vegetatively propagated. Therefore, seed material (tubers) for planting must be virus free. Potatoes suffer from three important virus diseases called Photo virus X, (PVX, PVY) and potato leaf-roll virus.

The phenomenon of cross protection or immunization of plant is not clearly understood. It is similar to immunization of human being for bacterial disease. When a plant is inoculated with a form of the virus, when virus infects, plant cell start synthesizing coat proteins instead of its own proteins. This cross protection is in some way related to the synthesis of the coat protein by the plant cell. In 1986, Roger Beachy and colleagues at Washington University introduced the gene that encodes that coat protein of TMV into tobacco plants, resultantly each and every cell of the transgenic plant start producing coat protein. These plants showed considerable resistance to infection by TMV. The virus was unable to multiply in the cells already containing some coat proteins. Therefore, the number of virus particles per cell remained low in transgenic plants as compared to normal control plants.

Scientists at Mogen International in the Netherlands, used the same approach to make potatoes resistant to PVX. The gene encoding the coat protein of PVX was introduced into two cultivars. The transgenics showed 100 times less virus particles as compared to control plants after two weeks of inoculation. The yield performance of most cultivars was same but potatoes produced were elongated. The viral coat protein gene approach has been used to transfer tolerance to a number of transgenic plants for a number of different crops. Although complete protection is not usually achieved, but high levels of virus resistance have been reported. Moreover, a coat protein gene from one virus sometimes provides tolerance to a number of unrelated viruses.

In both eukaryotes and prokaryotes, an RNA molecule that is complementary to a normal gene transcript (mRNA) is called antisense RNA. The mRNA, being translatable, is considered to be a sense RNA. The presence of antisense RNA can decrease the synthesis of the gene product by forming a duplex molecules with the normal sense mRNA. Thereby, preventing it from being translated. The antisense RNA-mRNA duplex is also rapidly degraded, a response that diminishes the amount of that particular mRNA in the cell. Therefore, in principle it should be possible to prevent plant viruses from replicating and subsequently damaging plant tissues by creating transgenic plants that synthesize antisense RNA that is complementary to virus coat protein mRNA.

The Ti binary vector system was used to transfer both protein producing sense and antisense RNA producing cDNA sequence to separate tobacco cells, from which transgenic plants were regenerated. The transgenic tobacco plants that expressed the cucumber mosaic virus (CuMV) coat protein were produced from viral particle accumulation and did not show symptoms of viral infection, irrespectively of whether the inoculum of the challenge virus was high or low. However, the transgenic tobacco plants expressing the CuMV coat protein antisense RNA were protected only when the concentration of the challenge virus in the inoculums was low. Therefore, this approach is not successful when virus infection is high.

Table 24.3. Virus resistant transgenic plants developed that contain cloned viral coat protein (gene)

Plant species	Virus that provided the coat protein gene
Nicotiana benthamians	Plum pox virus, watermelon mosai virus 2
Papaya, tobacco	Papaya ring spot virus
Potato	Potato virus (PVX, PVY, PVS)
Rice	Rice stripe virus
Tobacco	Soybean mosaic virus, tobacco streak virus, Tobacco spotted wilt virus, PVX
Tobacco, alfalfa, tomato	Alfalfa mosaic virus
Tobacco, cucumber	Cucumber mosaic virus
Tomato	Tomato mosaic virus

4. Abiotic stress tolerance

Abiotic stresses such as drought, salinity and extreme temperatures cause significant losses of crop productivity and quality. Development of crops with an inherent capacity to withstand abiotic stress would help stabilize the crop production and significantly contribute to food and nutritional security in developing countries.

Transcriptome engineering or over expression of a master switch gene (such as stress sensors, protein kinases or transcription factors) that regulate several target genes coding for osmolyte biosynthesis enzymes, antioxidant enzymes and stress protein (such as late embryogenesis abundant proteins) is emerging as an important tool to combat abiotic stress. Stress-induced transcription factors such as c-repeat binding protein (CBF) or dehydration responsive element binding proteins regulate the expression of many genes for compatible osmolyte biosynthesis and oxidative stress management. Over expression or stress responsive promoter driven expression of CBF3 gene in transgenic *Arabidopsis* provided protection against multiple environmental stress such as cold, salt and drought. Components of the *Arabidopsis* CFF pathway are conserved in *B. napus*, wheat, rye, and tomato. Transgenic tomato with CBF1 gene and a CAM35s promoter showed significant chilling tolerance.

Transgenic tomato plants expressing *Arabidopsis thaliana* CBF1 gene, showed enhanced tolerance to oxidative stress, as CBF1 over expression induced a high level of expression of a catalase gene in these transgenic tomato plants.

5. Quality Improvement

The goal of plant biotechnology is not confined to improvements of crop plants for agronomic traits and significant efforts are also being made to improve the nutritional content and organoleptic qualities such as taste and aroma in fruits and vegetables.

Nutritional improvement -Plant produces various compounds such as storage proteins, vitamin, flavonoids, carotenoids that perform vital functions for plants and also have nutritional importance for human beings. Vegetables are sources of minerals, proteins, micronutrients, vitamins, antioxidants, phytosterols and dietary fibre. However, some of the vegetables are deficient in essential amino acids such as methionine and lysine. The amino acid content can be modified or enhanced by expression of synthetic protein, over expression of homologous or heterogonous proteins, modifying the amino acid sequence of the protein or through metabolic engineering.

Potato is an important food crop, the nutritive value of potato protein is diminished due to deficiency in essential amino acids lysine, tyrosine and the sulphur containing amino acids methionine and cysteine. To improve the nutritive value of potato an *Amaranthus* seed albumin gene AmA1 has been expressed in transgenic potato tubers. This protein is non-allergenic and rich in all essential amino acids corresponding with WHO standards for human diet requirements. Similarly, a 292 bp artificial gene (asp-1) encoding a storage protein composed of essential amino acids was introduced in sweet potato. One of the transgenic lines showed a four fold increase in protein as compared to that of storage roots of control plants.

Carotenoids, such as β-carotene and lycopene, give the fruit its characteristic colour. Carotenoids are good antioxidants and are precursors of vitamin A. These are synthesized through the isoprenoid biosynthetic pathway. Provitamin content of tomato was increased by transferring a bacterial gene encoding for the phytoene - desaturase enzyme that converts phytoene to lycopene into transgenic tomato. These transgenic plants produced three-fold more β-carotene content than that of control plants. Similarly, a six-fold increase in carotenoid content and two to three fold increase in α-tocopherol content was achieved in transgenic potato plants by antisense technology.

Another group of metabolites exploited for its antioxidant property are the flavonoids. These are a diverse group of polyphenolic secondary metabolites, which impart colour to the fruits. Flavonoids are present only in tomato peel. A transgenic approach has been used to increase the flavonoid content by over-expression of either the enzymes involved in flavonoids biosynthesis or transcription factors that regulate the genes of this pathway. Transgenic tomato plants expressing petunia *CHI-A* gene encoding chalcone isomerase showed significant increase in flavonoids content. Similarly, a 10-fold increase in flavonoid content has been achieved by ectopic expression of the maize transcription factors LC and C1 in transgenic tomato.

Golden rice-with Pro-vitamin A –According to the World Health Organization (WHO), vitamin A deficiency (VAD) is the leading causes of preventable blindness in children. For children, a lack of vitamin A causes severe visual impairments and blindness and significantly increases the risk of severe illness and even death from common infections such as diarrhoea and measles.

The genes from daffodil and one from the bacterium *Erwinia uredovora* were inserted in the rice genome. These three genes produce the enzymes necessary to convert GGDP to pro vitamin-A. The inserted genes are controlled by specific promoters such that the enzymes and the provitamin-A are only produced in the rice endosperm.

Provitamin –A is not produced by traditional rice varieties. However, geranylgeranyl diphosphate (GGDP), a compound naturally present in immature rice endosperm, with the help of several enzymes not normally found in rice can be used to produce provitamin-A.

Through the work of two European scientist, Dr. Ingo Potrykus of the Swiss Federal Institute of Technology in Zurich and Dr. Peter Beyer of the University of Freiburg in Germany, rice plants were developed containing two daffodil genes and one bacterial gene that carry out the four steps required for the production of beta-carotene in rice endosperm. Endosperm is the nutritive tissue surrounding the embryo of a seed and makes up the majority of the rice grain that we eat. The resulting plants appear normal, except after milling (to remove the brown bran), their grain is a golden yellow colour due to the presence of provitamin-A.

When golden rice is ingested, the human body splits the provitamin-A to make vitamin A. Detailed information can only be obtained once the golden rice trait is transferred to local varieties and produced in quantities sufficient to support necessary field experiments. According to Swiss scientist Potrykus, the intent of golden rice is to supplement the diet with vitamin A, it not provide 100% of the Recommended Daily Allowance (RDA). Potrykus maintains the goal of golden rice having a beneficial effect on vitamin A-deficient people is realistic with experimental golden rice lines possibly already in the 20-40% RDA range.

Golden rice is the result of an effort to develop rice varieties that produce provitamin-A (beta-carotene) as a means of alleviating vitamin A (retinol) deficiencies in the diets of poor and disadvantaged people in developing countries. Because traditional rice verities do not produce vitamin-A, transgenic technologies were required.

Improvement of aroma -The armoa of fruits, vegetables and flowers are mixtures of volatile metabolites such as alcohols, phenols, ethers, aldehydes, ketones etc. Some of the short-chain aldehydes and alcohols are derived form lipid components by the action of lipases, hydroperoxide lipases and alcohol dehydrogenases. When yeast Δ–9 desaturase gene was transferred in tomato plants, changes in certain flavour compounds such as cis-3-hexenol, 1-hexanol, hexanal and cis-3-hexenal was recorded.

Linalool, an acyclic monoterpene alcohol, markedly influences the flavour of tomatoes. Linalool imparts a sweet, floral alcoholic note to fresh tomatoes. Hence linalool levels were altered by engineering the S-linalool synthase (LIS) gene from *Clarkia breweri* in tomato plants. The expression of S-linalool synthase enzyme, which catalyses the formation of linalool, resulted in elevated levels of linalool in the transgenic fruit.

Seedless vegetables- Browning and loss of flavour are two problems associated with potato. Transgenic potato have been generated in which browning is overcome by antisense inhibition of polyphenol oxidase. Cystathionine gamma synthase (CGS) is a key enzyme regulating methionine biosynthesis in plants. To increase the level of soluble methionine in potato, *Arabidopsis thaliana* CGS cDNA was introduced under transcriptional control of the cauliflower mosiac virus 35s promoter into potato. Increase in 2.4 – to 4.4 fold increase in methional level in transgenic potato tubers was recorded.

The seedless nature of parthenocarpic (development of fruit without fertilization) fruits increases consumer acceptance, makes processing of vegetables easier, and also improves the quality of vegetables, e.g., brinjal (where seeds are associated with bitter substances). Parthenocarpy has been shown to be regulated by auxins. Hence, efforts have been made to increase the auxin production or the sensitivity of ovary to auxins, towards inducing parthenocarpy. Expression of *iaaM* gene driven by the ovule specific promoter *defH9* has been shown to confer parthenocarpy to transgenic tomato and eggplant.

In another approach, the *Agrobacterium rhizogenes* derived gene *rol B* has been used for the induction of parthenocarpy in tomato. Transgenic tomato plants transformed with the *rol B* under the control of ovary and young fruit specific promoter *TPRP-F1* developed parthenocarpic fruits.

6. Pharmaceutical and industrial use

The ability to transfer gene across different plant species and kingdoms through genetic engineering is being exploited in term of biofarming. Biofarming refers to production of proteins and biomolecules in transgenic plants at agricultural scale. The proteins mainly include antigens, antibodies, enzymes that are of immense importance in therapeutics, pharmaceutical and industrial applications. Though many of these proteins are being made in bacterial, fungal or animal systems, plants are now being preferred for manufacturing these proteins. The use of plants as biofactories is attributed to many factors; 1) plants offer cost effective and environmentally safe production of proteins as they use low cost inputs such as light, water and minerals, 2) plants allow mass production, 3) suitable for production of eukaryotic proteins which many require post-translational modifications, oligomerization etc, and 4) plants are not pathogenic to human beings. The feasibility of vegetables as plant factories is very well illustrated in the form of edible vaccines, plantibodies (plant derived antibodies) and plant derived recombinant enzymes. The details of these products are given in the chapter on biofarming (chapter –26).

Industrial applications -Use of plants as biofactories for producing industrial enzymes, chemicals and raw materials can be carried out in two ways : 1) introduce foreign protein gene and produce in it high amounts, and ii) modify the plant metabolic pathways to generate sufficient end products, by product or a novel biomolecule. The first category involves production of enzymes such as cellulases, amylases (required in large quantities) and foreign proteins such as silk protein or sweet protein 'monellin'.

Cellulases find wide applications in food ingredients, beverage and textile industries. Endoglucanase (E1), a cellular gene from *Acidothermus cellulotyticus* with leaf specific promoter and RbcS-2A signal peptide were transferred in potatoes which resulted in E1 protein accumulation (2.6% of total leaf soluble protein). Human milk proteins such as lactoferin and casein have also been expressed in transgenic potatoes. These can be used as supplement in human infant formulas to enhance nutrition, digestibility and antimicrobial properties.

The remarkable mechanical property of spider silk finds several industrial applications. However, obtaining spider silk in large quantities from natural sources is very difficult. Hence efforts have been made to obtain large quantities of spider silk by over-expressing *Nephila claviceps* dragline protein gene in tobacco and potato. Dragline protein accumulation of up to 2% of the soluble protein has been achieved through this gene in transgenic potato lines.

A sweet protein monellin from *Dioscoreophyllum cumminsii* plant has been expressed in lettuce and tomato. Monellin is 100,00 times sweeter than sucrose. Due to high calorific value and dental caries – inducing properties, sucrose had been replaced by artificial sweeteners. Artificial sweeteners have low caloric value. However, these artificial sweeteners also have limitations in terms of taste and are suspected to be carcinogenic. An alternative has been developed. Palatinose is a low calorific structural isomer of sucrose with physio-chemical properties similar to that of sucrose. For converting sucrose into palatinose, an enzyme sucrose isomerase from *Erwinia rhapontici* was cloned under the control of tuber specific promoters and transferred to potato. Transgenic potato plants accumulated 1.7 to16 μmol/g of palatinose in tubers. This gene can also modify the taste of several fruits.

Manipulation of starch quality and quantity is the most common example of metabolic engineering. Starch has a very wide spectrum of application starting from food to fermentation, pharmaceutical, plastic, paper, textile and many other industries. Due to its varied use, natural starch is modified chemically or enzymatically to suit industrial use. The chemical modification processes are costly and byproduct causes environmental deterioration. Hence, increasing the starch production and tailoring the plant cell to produce starch derivatives will be cost effective and eco-friendly. Increase in starch content has been achieved in transgenic potato over-expressing barley endosprem ADP-glucose pyrophosphorylase. Transgenic increase in the adenylate content by over expression of adenylate kinase in potato 2 to 4 fold increase in concentration of several amino acids.

The endogenous pathways have been manipulated to redirect carbon flow towards the production of novel biomolecules. Fructans are polymers of fructose and represent alternative storage carbohydrate in several plant species. Potatoes, which are normally non-fructan producing plants, have been successfully engineered for producing fructans as novel carbohydrates. The potato plants were modified by introducing either the microbial fuctosyltransferase gene from *Bacillus substilis* (Sac B) or *Streptococcus mutans* (ftf) or plant gene 1-SST (sucrose : sucrose 1-fructosyl transferease) and 1-FFT (Fructan : Fructan 1- fructosyltransferase) from globe artichoke (*Cynara scolymus*). The fructan content in the transgenic potato plants expressing microbial fructosyltransferase genes varies from 1 to 7% of dry weigh in microtubers. Similarly, transgenic potatoes carrying plant gene 1-SST and 1-FFT showed inulin (fructan) content up to 5% of the dry weight in transgenic tubers. However, fructan produced from microbial fructosyl transferase gene was of considerable higher molecular weight than that produced by plant genes 1-SST and 1-FFT.

7. Economic impact

In the USA, the six biotechnology-derived crops planted in 2003 (canola, corn, cotton, papaya, squash and soybean) produced an additional 2.4 million tonnes of food and fibre and increased farm income by US $ 1.9 billion. This biotechnology – derived crops also reduced the use of pesticides by 21,000 tonnes.

Several GM rice varieties have been entered and passed field and environmental release trails, and four varieties entered preproduction trials in farmer's fields in 2001. Farm surveys of randomly selected household cultivating insect-resistant GM rice varieties indicate that GM rice cultivation is beneficial to small and poor farm households because crop yields are higher and pesticide use is reduced. Reduced use of pesticide also contributed towards improved health as compared to non-GM rice.

For cotton, key documented benefits are a 70% reduction in insecticide applications in Bt cotton fields in India, resulting in a saving up to US $ 30 per hectare in insecticide costs, and an 80-87% increase in harvested cotton yield; a dramatic reduction in pesticide application in Bt cotton fields has also been reported in china, this also resulted in substantial reduction in pesticide pollution from 22% to 47%.

Transgenic corn was introduced in the USA in 2003. In this case, synthetic variant of wild-type cry 3Bb1 gene from *B. thuringiensis kumamotoeniss* was use. This gene product – protein has 8 times more insecticidal property and provided resistance to corn plant against larvae feeding on roots. The gene was codon optimised for optional expression in monocot plants, and was expressed under the control of the root-enhanced 4AS1 promoter. The gene was introduced into corn cell culture using particle bombardment.

Terminator gene -One potential use of transgenic technology is to allow seed producers to realize profits from their investments in new product development. Seed companies have preferred to invest heavily in developing new varieties of crops such as corn for which the farmers typically purchases new seeds each years. Two biotech protection methods, dubbed 'Terminator' and 'Traitor' by opponents, may allow companies to increase profits on their cultivars.

Terminator, officially named as "Technology protection system" (TPS), incorporates a trait that kills developing plant embryos, so seed cannot be saved and replanted in subsequent years.

Traitor, officially known as "Trait-specific genetic use restriction technology" or T-Gurt, incorporates a control mechanism that requires yearly application of a preparatory chemical to activate desirable traits in the crop. The farmer can save and replant seeds, but cannot gain the benefits of the controlled traits unless he pays for the activating chemical each year. Both methods avoid the difficulties associated with enforcing 'no replanting' agreements. Because TPS and T-Gurt plants would be transgenic, their commercial use will require approval by the government. Scientist from agricultural research service (USDA) and Delta and Pine Land Company jointly developed this technology in 1998.

The technology protection system (TPS) inserts half a dozen sequences into the DNA of the parent plant that is designed for protection. These DNA sequences are arranged into a system that kills seeds at a prearranged time in their development. The system can be left inactive while the seed company grows several generations of seeds for sale. The system is switched on by soaking the seeds in a special chemical before the seeds are delivered to the farmer for planting.

The special chemical triggers a slow cascade of events that lead eventually to the death of progeny seeds developed on the protected plant. For the purpose of preventing replanting, the progeny seeds should be killed only after they have completed production of all commercially valuable products such as oil. Therefore, the system is designed to take effect only after the crop has grown to maturity in the field and the progeny seeds are nearly ripe.

8. Environmental impact

The use of Bt gene containing crops has been the most hotly debated issues regarding GM crops. Two different concerns have been broadly raised regarding such engineered insecticide resistance. The first concerns the broader impact of the presence of such insecticidal proteins on other organisms coming in contact with the transgenic crop. The second centres on the possibility of the target insets developing resistance to the insecticidal protein.

The possibility of detrimental environmental impacts of Bt corn become headline news in 1999. A paper was published suggesting that the presence of this insecticidal protein in the pollen of transgenic corn was detrimental to the larvae of the Monarch butterfly (*Danaus plexippus*). This was a laboratory based study and not a field study, even though sparked a controversy about use of transgenic crops and its impact on ecosystem. Later on, based on field studies by American Universities, the issue was settled in 2001. This episode illustrates the fact that the first generation of transgenic crops is largely lacking mechanisms to target gene expression to precise cell organ or cell types. Rather, the introduced transgenes are typically expressed constitutively in all the cells of the plant. In case of Bt corn, the presence of insecticidal cry protein in the pollen, where it serves no useful purpose as the insect attacks the stem of the plant, raised environmental concerns without any reason. If the expression is controlled, particularly in open-pollinated crop like corn, the incidental damages to the environment can be minimize.

The second concern expressed regarding Bt gene was that insect would develop resistance to the insecticidal protein. This would not only make the transgenic crop worthless but might also the usefulness of Bt spray. One of the few tools available to organic farmers in the fight against insect pests. The seed industry is encouraging farmers to keep some area for non-GM crops, where insect can multiply, and also cross with resistant insect, if any, generating susceptible progeny. This will delay the true breeding resistant strain of insect.

9. Transplastomic plants through Chloroplast transformation

Chloroplasts are a class of plastids - organelles in plant cells – apparently derived from a cyanobacteria (blue-green bacteria) ancestor that once lived symbiotically inside the plant cell. Chloroplasts contain chlorophyll and are found in the shoots and leaves of green plants, while colourless plastids are found in the roots and other coloured plastids are found in fruit. The number of plastids in each cell is variable, and each plastid contains multiple copies of its own genome, typically 50 to 100. Many plastid genomes have been sequenced. They resemble bacterial genomes in many respects; though features normally found in muticellular organisms, such as interrupted genes and RNA editing are also present. The chloroplast genome codes for the transcription and translation machinery of the chloroplast plus numerous structural proteins. But the vast majority of the chloroplast proteins are encoded in the plant nucleus and imported into the chloroplast after synthesis.

Stable transformation of the chloroplast – putting foreign genes into the chloroplast genome - was first achieved in the single cell green alga *Chlamydomonas reinhardtii* in 1988, soon to be followed by tobacco plant, and more recently, *Arabidopsis thaliana*. Several biotech companies, including Monsanto, Rhone-Poulenc, Novartis, American Cyanamid, Calgene, Pioneer Hybrid have initiated major programmes on chloroplast transformation since the late 1990s.

Chloroplast transformation has been appreciated at least as far back as 1998 as a means of "containing" transgenes; that is, preventing them from transferring to non-GM crops or wild relatives through pollen, and hence preventing the creation of transgenic herbicide tolerant weeds. The theory is that chloroplasts are inherited exclusively through the female line.

Joe Cummins has exposed the fallacy of this claim. He pointed out that tobacco pollen does transfer chloroplast transgenes under selection with a herbicide-like drug tentoxin. It is well known that chloroplasts are mainly inherited through pollen in conifers, and major crops such as alfalfa inherit chloroplasts from both pollen and egg. There is also occasional biparental inheritance of chloroplast genes in rice, and cultivars of peas vary in the presence of chloroplast DNA in pollen. These cases, he emphasized, are just a few examples from a large literature showing that chloroplasts are inherited through pollen, pollen and egg, or selectively influenced by stress to transmit chloroplast genes through pollen where maternal transmission is usual.

Chloroplast transformation involves homologous recombination. This not only minimises the insertion of unnecessary DNA that accompanies transformation of the nuclear genome, but also allows precise targeting of inserted genes, thereby also avoiding the uncontrollable, unpredictable rearrangements and deletions of transgene DNA as well as host genome DNA at the site of insertion that characterises nuclear transformation. In practice, the inserted transgene has short DNA sequence tails added at each end, the tails are homologous to sequences on the chloroplast target gene, which thus initiate homologous recombination. Once the transgene is inserted into the chloroplast chromosome, the target gene is disrupted. The disruption of the target gene is expected to alter the growth and metabolism of the plant.

Leaf discs are bombarded with plasmid constructs containing a selectable antibiotic resistance marker physically linked to the gene of interest, flanked by DNA for inserting into the correct site of the chloroplast genome. The antibiotic resistance marker most frequently used is the *aadA* gene encoding resistance for spectinomycin and streptomycin, driven by the promoter of the chloroplast encoded 16S rRNA gene.

This transformation procedure applied to tobacco, *Arabidopsis* or oil seed rape, generates plants in which all the chloroplast genomes are uniformly transformed (a condition referred to as homoplasmic), despite the fact that tobacco leaf cells may contain 100 chloroplasts, each containing 100 copies of the chloroplast genome.

Genome organization

The chloroplast genome typically consists of basic units of double-stranded DNA of 120 to 220 kb arranged in monomeric or multimeric circles as well as in linear molecules. The chloroplast genome generally has a highly conserved organization, with most land plant genomes having two identical copies of a 20- to 30-kb inverted repeat region (IRA and IRB) separating a large single copy (LSC) region and a small single copy (SSC) region. Plastid transformation is typically based on DNA delivery by the biolistic process or occasionally by polyethylene glycol (PEG) treatment of protoplasts. This is followed by transgene integration into the chloroplast genome via homologous recombination facilitated by a RecA-type system between the plastid-targeting sequences of the transformation vector and the targeted region of the plastid genome. Chloroplast transformation vectors are thus designed with homologous flanking sequences on either side of the transgene cassette to facilitate double recombination. Targeting sequences have no special properties other than that they are homologous to the chosen target site and are generally about 1 kb in size. Both flanking sequences are essential for homologous recombination. Transformation is accomplished by integration of the transgene into a few genome copies, followed by 25 to 30 cell divisions under selection pressure to eliminate untransformed plastids, thereby achieving a homogeneous population of plastid genomes. If the transgene is targeted into the IR region, integration in one IR is followed by the phenomenon of copy correction that duplicates the introduced transgene into the other IR as well.

Chloroplast vectors. Similar to introduction of new genes in chromosomes (nucleus), it is also possible to modify the plastid genome by introducing novel genes in the plastid's single chromosome (DNA). These plants are referred to as "transplatomic". As with nuclear transformation, various technical approaches are available.

One of the most distinctive features of plastid transformation is that the novel DNA integrates into the plastic genome via homologous recombination. The genetic construct to be introduced is flanked by regions of DNA (usually > 400 bp) that are homologous to a portion of the plastid genome. This means that the site of integration is precisely known. A practical consequence is that this reduces the number of independent transformants that need to be obtained, as there is no variation in expression due to position (compared to nuclear transformation, many chromosomes or positions, variations in transformants).

A second feature of tansplastomic plants is the extremely high number of copies of the transgene. Indeed, there are numerous copies of the genome in each plastid and many plastids per cell, it is estimated that a transplastomic plant can carry as many as 10,000 copies of the transgene per cell. Such high copy numbers may allow extreme high levels of expression of a foreign protein. In the case of transplastomic plants expressing the cry2 Aa2 gene from *B. thuringienis*, this provided a very high level of protection against insects, including some strains resistant to normal cry protein.

The proposal of a "universal vector" containing the *trnA* and *trnI* genes from the IR region of the tobacco chloroplast genome as flanking sequences for homologous recombination to transform several other plant species (of unknown genome sequence) was suggested several years ago. This concept was based on the high conservation of this intergenic spacer region among the higher plant chloroplast genomes. Vectors designed for transformation of the tobacco plastid genome have been successfully used for potato and tomato plastid transformation, because the homologous flanking sequences present in these vectors showed adequate homology to the corresponding sequences of potato and tomato plastid DNA but the efficiency of transformation is significantly lower than tobacco.

Stability of expressed proteins

Newly synthesized proteins are highly susceptible to proteases and require protection from chloroplast proteases. One such approach used the CRY chaperone (encoded by the *orf*2 gene) to fold the insecticidal protein, Cry2Aa2, into cuboidal crystals. The crystal structure protected the foreign proteins from degradation, thereby increasing protein accumulation over 128-fold (from 0.36% to 46.1% of total soluble protein). Similarly, when the human serum albumin coding sequence was regulated by the chloroplast *psbA* 5' and 3' UTRs (untranslated regions) in the light, protein expression increased 500-fold, resulting in the formation of protective inclusion bodies. A 3- to 10-fold reduction in HSA protein expression was seen when leaves were harvested in the dark . This illustrated the power of regulatory sequences during illumination and protection from proteases when their access is limited.

Molecular pharming

There are currently 37 patents for molecular pharming by chloroplast transformation listed. The first commercial exploitation of chloroplast transformation for molecular pharming is likely to be in *Chlamydomonas reinhardtii* ("GM pharmaceuticals in common green alga").

Several crop chloroplast genomes have been transformed via organogenesis (cauliflower [*Brassica oleracea*], cabbage [*Brassica capitata*], lettuce [*Lactuca sativa*], oilseed rape [*Brassica napus*], petunia [*Petunia hybrida*], poplar [*Populus* spp.], potato [*Solanum tuberosum*], tobacco [*Nicotiana tabacum*], and tomato [*Solanum lycopersicum*]) or embryogenesis (carrot [*Daucus carota*], cotton [*Gossypium hirsutum*], rice [*Oryza sativa*], and soybean [*Glycine max*]), and maternal inheritance of transgenes has been observed. Chloroplastderived biopharmaceutical proteins, including insulin, interferons (IFNs), and somatotropin (ST), have been evaluated by in vitro studies. Human INFa2b transplastomic plants have been evaluated in field studies. Chloroplast-derived vaccine antigens against bacterial (cholera, tetanus, anthrax, plague, and Lyme disease), viral (canine parvovirus [CPV] and rotavirus), and protozoan (amoeba) pathogens have been evaluated by immune responses, neutralizing antibodies, and pathogen or toxin challenge in animals. Chloroplasts have been used as bioreactors for production of biopolymers, amino acids, and industrial enzymes.

Chloroplast transformation to produce GM pharmaceuticals entails specific risks that are associated with its advantages.

1. The high level of transgene expression that can be achieved increases the hazards of environmental contamination and inadvertent exposure of human subjects, domestic livestock and wild life.
2. The high copy number of transgenes increases the hazards of horizontal gene transfer to bacteria and viruses, with the potential of creating dangerous pathogens and spreading antibiotic resistance marker genes. It is now known that DNA persists in all environments, and transformation by direct uptake of DNA is a major route of horizontal gene transfer among bacteria.

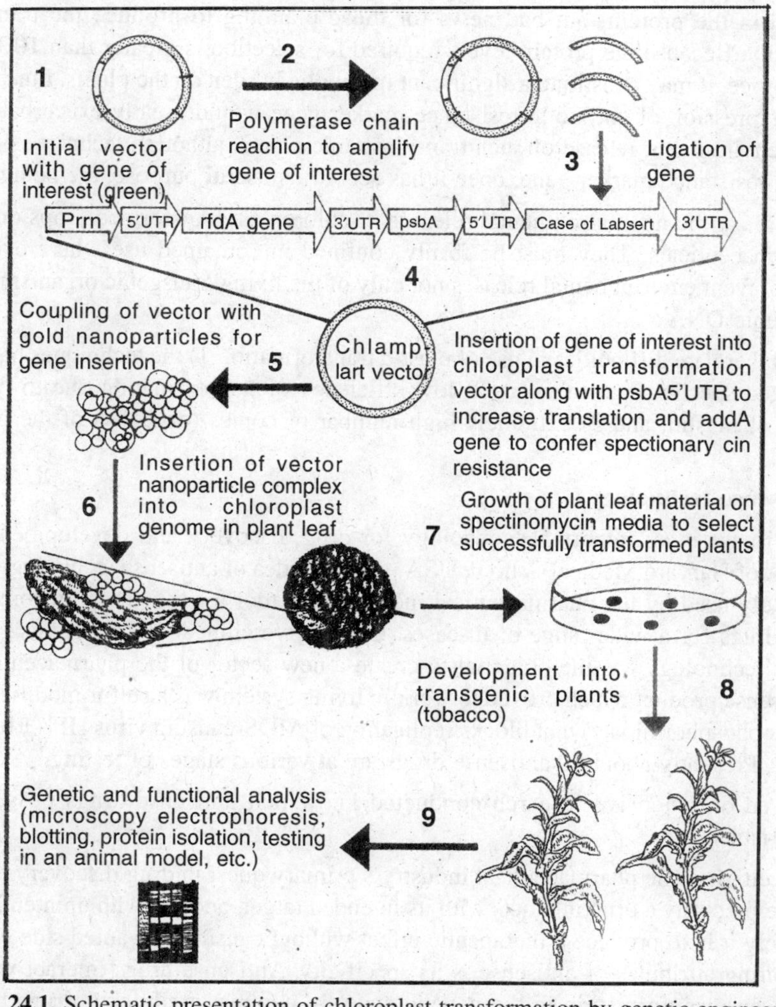

Fig. 24.1. Schematic presentation of chloroplast transformation by genetic engineering

3. The close similarities (homologies) between plastid and bacterial genomes is expected to greatly increase the frequency of horizontal gene transfer, up to a billion-fold. Furthermore, the horizontal transfer of non-homologous DNA occurs at relatively high frequencies when a homologous DNA 'anchor sequence' is present, which can be as short as 99bp.
4. There are at least 87 species of naturally transformable bacteria in the soil.
5. The disruption of the target gene in transformation results in changes in the growth and metabolism of the plant that may pose risks to health and the environment.

Another advantage of chloroplast transformation is that foreign genes can be over-expressed, due to the high gene copy number, up to 100 000 compared with single-copy nuclear genes. And there does not seem to be gene-silencing and other instability that plague nuclear transformation. The gene product is retained inside the chloroplasts or can in principle be targeted to a specific compartment in the chloroplast.

Another problem is to get high level of protein expression, even though the gene copy number is high. In chloroplasts, post-transcriptional processes determine the levels of proteins expressed, depending on translational signals.

High protein accumulation of transgene product in the tomato fruit is good news for those interested in the protein, but bad news for those planning to produce the transplastomic crop successfully. Because the protein levels required for selection is greater than 10% of total soluble protein in rice, it may constitute a significant metabolic burden on the plants. Furthermore, the high level of expression of antibiotic resistance marker gene would greatly exacerbate public concern over the environment release of such transplastomic plants, although techniques for removing the antibiotic resistance marker gene, once it has served its useful purpose, are being developed.

There can be no environmental releases of chloroplast transformed crops or algae producing GM pharmaceuticals. They must be firmly confined in contained use where every precaution is taken to prevent environmental releases not only of the living transgenic organism or cells, but also of transgenic DNA.

Still there are difficulties in use of plastid transformation;1) the techniques has been successful in a limited species, e.g., tobacco; 2) low efficiency of transformation due to double membrane structure of plastid; and 3) extremely high number of copies (of DNA) of the plastic genome in each cell.

10. Antisense technology

Antisense is an important technology for drug discovery and development. Professor Paul Zamecnik of Harvard Medical School, USA, gave the idea of antisense technology in late seventies. It is broadly used by the pharmaceutical industry as a tool for functional genomics and as highly specific drugs for a wide range of diseases. Cumulative clinical and preclinical data suggest that antisense technology has the potential to create a new sector of the pharmaceutical industry. The first antisense product which proved to work in living system was a sulfur modified oligonucleotide (called as phosphorthioate) that blocks replication of AIDS-causing virus HIV-1 by targeting a viral gene gag. Presently about 50 antisense drugs are at various stages of testing.

Based on extensive research conducted know that antisense drugs possess the following characteristics:

Specificity-The pharmaceutical industry's primary quest in drug discovery is drug specificity. The more precisely a drug interacts with its intended target, and not with unintended molecules, the more likely it is to produce a therapeutic effect without causing unwanted side effects. One of the most alluring attributes of antisense is its specificity. Antisense drugs interact with their intended target based on information in the genetic code. In contrast, traditional drugs bind based on the shape of proteins and charge interactions, creating more opportunity for unwanted interactions, and, often undesirable side effects. Based on their specificity, antisense drugs have the potential to be safer than other types of drugs.

Broadly Applicable- The nature of the antisense receptor, RNA, makes virtually any molecular target accessible to antisense drugs. There are many highly desirable molecular targets for drug discovery that are considered "undruggable" by traditional small molecule technology. These targets are often members of families of closely related proteins that are too similar in structure for traditional drugs to distinguish amongst, the biological function of the protein is unknown, or it is difficult to develop an assay for drug screening. Antisense drugs discriminate between targets based on their genetic sequence, so the universe of potential targets is significantly greater.

Rational Design- Antisense drug discovery is more rational than any other type of drug discovery because:

1) The rules for creating these drugs are known, as they bind by hybridization to the target RNA. 2) The chemistry is constant, with only the order of the drug's nucleotides changing to make the drug target specific. 3) The distribution and metabolism of antisense drugs are very similar from drug to drug, resulting in a common and often times, predictable, safety profile across antisense drugs.

Efficiency- The benefit of a more rational approach to drug discovery is efficiency. The efficiency of antisense results in shortened timelines to identify lead drug candidates in the discovery phase of development. The efficiency of antisense technology is also realized in the area of manufacturing.

Several antisense drugs are now in late stage clinical development with one antisense drug developed by Isis, Vitravene•, marketed in the US and Europe. At the same time, the explosion of genetic information provided by the completion of the Human Genome Project (HGP) provides a rich resource of information on target genes for the design of antisense drugs. Each of our genes is a set of instructions for, and control of the manufacture inside the cell of a unique protein. Some proteins form the cell structure, while others are enzymes that carry out the functions of the cell or hormones interacting with the external environment of the cell.

Most pharmaceutical drugs available today interact with one or more proteins. They act to enhance or inhibit the action of a protein or mimic its role thereby bringing about the specified therapeutic effect. To date the pharmaceutical drugs on the market target a total of only about 500 different proteins. Although not all proteins will be suitable targets for therapeutic intervention, there is clearly enormous scope for drug discovery and new therapies in this post genomic era, considering that the human genome codes for at least 30,000 different proteins.

Antisense v/s conventional drugs

The approximately 30,000 genes in our human genome can be transcribed into about 85,000 different mRNA, each used in the cell as a template to synthesise a different protein. Conventional pharmaceutical drugs (small chemicals), peptides, or proteins (for example, hormones), and antibodies (which are very large proteins) typically bind to the target protein directly to treat a disease. Antisense drugs are designed to bind to the mRNA of a target protein, inhibiting the protein production process. The completion of the sequencing and initial analysis of the human genome through the HGP provides a resource for the design of antisense drugs without requiring the complex and time consuming analysis of the structure of the target protein which is required for conventional (small molecule) drugs.

How antisense works

The antisense concept was first conceived in 1978 and has been through various technical refinements. Antisense compounds are designed to have the right nucleotide sequence to bind

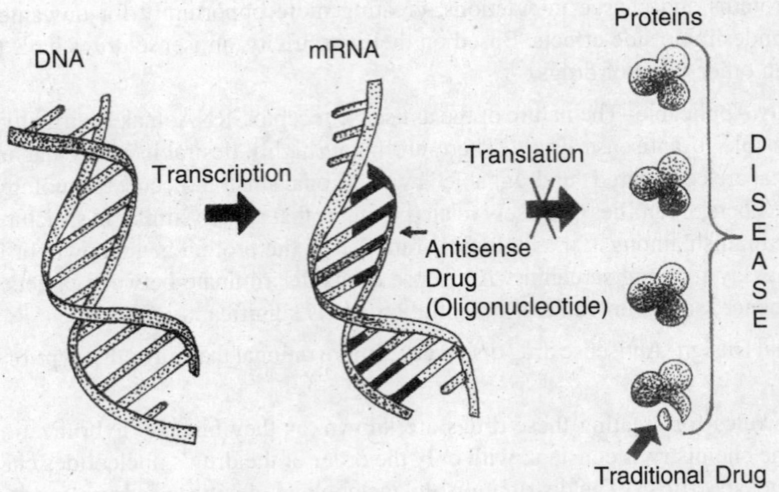

Fig. 24.2. Principle of antisense technology

specifically to and interfere with its associated mRNA, the instructions for the production of a particular protein. To create antisense drugs, special chemically stabilized nucleotides are synthetically linked together in short chains of about 12-30 nucleotides (called oligonucleotides). Each antisense drug is designed with the right complementary genetic code to bind to a specific sequence of nucleotides in its mRNA target to form a short area of double strands.

This double stranded region can inhibit the production of protein by a number of mechanisms. They include stopping the ribosome from reading the message, or by leading to the destruction of the mRNA by an enzyme already in the cells called RNase H. Rnase H present both inside and outside of the nucleus, destroys any such double-stranded nucleotides. The rapid development of antisense technology offers almost unlimited scope for the development of new and highly specific therapeutics. Antisense Therapeutics therefore is well positioned to play a significant role in the progression of antisense technology for drug development in human diseases (Fig. 24.2).

Antisense refers to short DNA or RNA sequences, termed oligonucleotides, which are designed to be complementary to a specific gene sequence. The goal is to alter specific gene expression resulting from the binding of the antisense oligonucleotide to a unique gene sequence.

In principle, antisense technology is supposed to prevent protein production from a targeted gene. The exact mechanism by which this occurs remains uncertain. Proposed mechanisms include triplex formation, blocking RNA splicing, preventing transport of the mRNA antisense complex into the cytoplasm, increasing RNA degradation, or blocking the initiation of translation. Initially, cellular nucleases dramatically reduce the effectiveness of antisense oligonucleotides by rapidly degrading these molecules after administration. These obstacles can be overcome in applications utilizing synthetic oligonucleotides by altering the nature of the phosphodiester bond by replacing an oxygen with sulfur. Such modified oligonucleotides are termed phosphorothionates. Delivery of antisense oligonucleotides into target cells or the cell nucleus has been problematic. The variety of viral and non-viral delivery systems previously discussed are currently being explored to overcome this obstacle. Animals treated with antisense oligonucleotides have had significant side effects, some of which have been lethal. Currently, the most problematic aspect associated with antisense technology revolves around the specificity of their action. In some cases, non-specific antisense sequences, in other words, sequences which do not bind to the targeted gene or RNA, have prevented gene expression to the same degree as their sequence-specific antisense counterparts. This has led to considerable complication in data interpretation and requires detailed and careful data analysis before thinking of clinical trials. Since antisense technology focuses on preventing gene expression, it has been most widely applied to cancer gene therapy.

Evolution of Antisense Technology

The first researchers who worked on antisense technology saw the possibilities of new methods of drug design. However, that promise has not yet been fully realized. First-generation drugs have demanded daily, intravenous dosing, and their lower specificity for their target mRNA keeps them from being as potent as is desirable. New developments in the chemistry of antisense give further hope to developing potent, long-lasting antisense therapies that are more conveniently administered.

The first antisense constructs studied were based on the natural backbone structure of DNA. These constructs had an extremely high affinity for their target sequences. However, structures based on naturally occurring DNA are subject to swift degradation by nucleases and proved to be unsuitable for therapeutic uses.

In an effort to make antisense technology more amenable to therapeutic use, scientists attempted to alter the backbone of the antisense molecules. The first successful backbone chemistry modified the phospho diester structure in the DNA backbone to a phospho-rothioate structure, which increased the construct's stability and made the technology more amenable to therapeutic use. However, these constructs still had plasma half lives of only hours, and the modified DNA backbone decreased the affinity of the antisense molecule to its target mRNA sequence.

To increase the stability of the antisense compound, a second-generation chemistry was developed, adding the 2'MOE (2'-methoxyethyl) modification to the oligonucelotide backbone. A further improvement incorporates both first-generation and second-generation chemistries into "gapmers": antisense sequences that have their ends modified with 2'MOE chemistry while retaining at their centers first-generation chemistry. The 2'MOE modified ends protect the construct from degradation, giving significantly improved half life and binding affinity, while the first generation center permits enzymatic degradation of the mRNA/antisense complex. Gapmer second-generation chemistry drugs have increased half-lives of 5- to 10-fold over first-generation chemistry and increased affinity to target mRNA.

The improved affinity of second-generation drugs is primarily attributable to their design and composition. Second-generation drugs are composed of both RNA-like and DNA-like nucleotides, while first-generation drugs are entirely DNA-like. Because RNA hybridizes more tightly to RNA than to DNA, the second-generation drugs have a greater affinity for their RNA targets and, therefore, greater potency. With increased potency, second-generation drugs are more active at lower doses.

Additionally, second-generation chemistry with gapmer design significantly slows degradation of the drugs by protecting the drug from destructive nucleases. The resulting slower clearance from the body allows for less frequent dosing, and in the future may allow for oral delivery, a useful feature for long-term use given added patient convenience.

Agricultural applications-

1. Slow ripening tomato- The enzyme polygalacturonase causes softening of tissues by degrading pectin in fruit wall. Using antisense technology, an antisense gene of polygalacturonase was added making the gene ineffective. The fruits remain hard for longer period of time.

2. Male sterility- Flavonoids are required for normal pollen development and function. Incorporation of antisense gene for chalcone synthase (key enzyme for flavonoid biosynthesis) resulted in non-functional pollen or male sterile lines in Petunia (with white flowers also). Addition of flavonoids on stigma or mixing with pollen restores the viability of pollen and fertilization takes place.

3. A seed specific protein promoter (napin or ACP promoter) was attached to ACP-desaturase coding gene. ACP saturase (ACP = acyl carrier protein) catalyzes the conversion of stearoyl-ACP

to oleoyl-ACP. This is first desaturation step in fatty acid biosynthesis in seeds. Due to decreased activity of this enzyme, oil with high percentage of saturated fatty acids (oleic acid) has been obtained, which can be used in place of cocoa butter. This technique has also demonstrated the use of antisense RNA technology to produce oil with desired composition. Gene was obtained from *Brassica campestris* and transgenic *B. rapa* and *B. napus* were obtained by antisense technology.

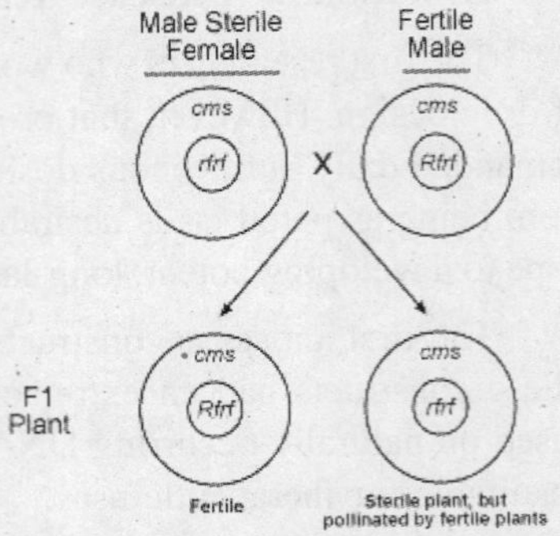

Fig. 24.3. A cross between male sterile female and fertile male plant.

MALE STERILITY

Male sterility is defined as the failure of plants to produce functional anthers, pollen, or male gametes. The first documentation of male sterility came in 1763 when Kölreuter observed anther abortion within species and specific hybrids. Cytoplasmic male sterility as the name indicates is under extra nuclear genetic control. They show non-Mendelian inheritance and are under the regulation of cytoplasmic factors. It is more prevalent than female sterility, probably because, the male sporophyte and gametophyte are less protected from the environment than the ovule and embryo sac. Male sterility is easy to detect because a large number of pollen are produced and are easily studied. Male sterility is also easily assayed through staining techniques (carmine, lactophenol or iodine); while detection of female sterility requires crossing. Male sterility has propagation potential in nature since it can still set seed and is important for crop breeding, while female sterility does not. Male sterility can be aroused spontaneously via mutations in nuclear and/or cytoplasmic genes.

Among the two types of male sterility- genic and cytoplasmic- cytoplasmic male sterility is caused by the extranuclear genome (mitochondria or chloroplast) and show maternal inheritance. Manifestation of male sterility in these may be either entirely controlled by cytoplamsic factors (cytoplasmic) or by the interaction between cytoplasmic and nuclear factors.

When nuclear genes for fertility restoration (Rf) are available for CMS system in any crop, it is called as Cytoplasmic Genetic Male Sterility (CGMS). This type of male sterility system is common in many plant species across plant kingdom. The sterility is manifested by the influence of both nuclear and cytoplasmic genes. There are commonly two types of cytoplasms, N (normal) and S (sterile). There are also restorers of fertility (Rf) genes, which are distinct from genetic male sterility genes. The Rf genes do not have any expression of their own unless the sterile cytoplasm is present. Rf genes are required to restore fertility in S cytoplasm which causes sterility. Thus a combination of N cytoplasm with rfrf and S cytoplasm with Rf- produces fertiles; while S cytoplasm with rfrf produces only male steriles. Another feature of these systems is that Rf mutations (i.e., mutations to rf or no fertility restoration) are frequent, so N cytoplasm with Rfrf is best for stable fertility.

Because of the convenience to control the sterility expression by manipulating the gene – cytoplasm combinations in any selected genotype, cytoplasmic genetic male sterility systems are widely exploited in crop plants for hybrid breeding. Incorporation of male sterility evades the need for emasculation in cross pollinated species, thus encouraging cross breeding producing only hybrid seeds under natural conditions.

History of Cytoplasmic Male sterility. Male-sterility-inducing cytoplasms have been known for over 100 years. In 1921, Bateson and Gairdner reported that male sterility in flax was inherited from the female parent. Chittenden and Pellow observed in 1927 that male sterility in flax was due to an interaction between the cytoplasm and nucleus. In 1943, Jones and Clarke established that male sterility in onion is conditioned by the interaction of the male-sterile (S) cytoplasm with the homozygous recessive genotype at a single male-fertility restoration locus in the nucleus. The authors also described the technique used today to exploit cytoplasmic-genic male sterility (CMS) for the production of hybrid seeds. CMS inbred lines have been widely used for hybrid production of many crops. The first application of organelle biotechnology was the role played by cytoplasmic male sterility in hybrid seed production, a major contribution towards the "Green Revolution."

Cytoplasmic male sterility in hybrid breeding – Hybrid production requires a female plant in which no viable male gametes are borne. Emasculation is done to make a plant devoid of pollen so that it is made female. Another simple way to establish a female line for hybrid seed production is to identify or create a line that is unable to produce viable pollen. This male sterile line is therefore unable to self-pollinate, and seed formation is dependent upon pollen from the male line.

Cytoplasmic male sterility (CMS) is used in hybrid seed production. In this case, the sterility is transmitted only through the female and all progeny will be sterile. This is not a problem for crops such as onions or carrots where the commodity harvested from the F1 generation is produced during vegetative growth. These CMS lines must be maintained by repeated crossing to a sister line (known as the maintainer line) that is genetically identical except that it possesses normal cytoplasm and is therefore male fertile. In genic cytoplasmic male sterility restoration of fertility is done using restorer lines carrying nuclear restorer genes in crops. The male sterile line is maintained by crossing with a maintainer line which has the same genome as that of the MS line but carrying normal fertile cytoplasm.

ENGINEERING CYTOPLASMIC MALE STERILITY VIA THE CHLOROPLAST GENOME

Naturally-occurring cytoplasmic male sterility (CMS) has been known for over 100 years. CMS systems are used to produce commercial F1 hybrid lines. Ruiz and Daniell (2005) recently reported the first engineered cytoplasmic male sterility system in plants. They studied the effect of light regulation of the phaA gene coding for β-ketothiolase engineered via the chloroplast genome. The phaA gene was efficiently expressed in all tissue types examined, including leaves, flowers, and anthers. The transgenic lines were normal except for the male sterile phenotype, lacking pollen. Scanning electron microscopy revealed a collapsed morphology of the pollen grains. Transgenic lines showed an accelerated pattern of anther development, affecting their maturation, and resulted in aberrant tissue patterns. Abnormal thickening of the outer wall, enlarged endothecium, and vacuolation decreased the inner space of the locules, affected pollen grain, and resulted in the irregular shape or collapsed phenotype. Reversibility of the male sterile phenotype was observed under continuous illumination, resulting in viable pollen and a copious amount of seeds. This study offers a new tool for transgene containment for both nuclear and organelle genomes and provides an expedient mechanism for F1 hybrid seed production.

Advantages of the chloroplast transformation system. A chloroplast genetic engineering approach offers a number of attractive advantages, including high-level transgene expression, multi-gene engineering in a single transformation event, transgene containment via maternal inheritance, lack of gene silencing, position effect due to site-specific transgene integration, and lack of pleiotropic effects due to sub-cellular compartmentalization of transgene products. Genetically engineered cytoplasmic male sterility via the chloroplast genome may be used for the safe integration of foreign genes via the nuclear genome and in those rare cases in which plastid genomes are paternally or biparentally transmitted.

Reversibility of male sterility. To test whether depletion of the acetyl-CoA pool destined for de novo fatty acid biosynthesis in the chloroplast by β-ketothiolase is the cause of the male sterility phenotype, a continuous light could revert male fertility of the chloroplast transgenic lines. Recent reports have shown light-dependent regulation of the key enzyme in fatty acid biosynthesis. Therefore, acetyl-CoA carboxylase (ACCase) male sterile transgenic lines were exposed to continuous light for a period of 10 days. It was observed that from 20 flowers produced during the 10 days of illumination by the two transgenic plants, four flowers produced pollen, normal length anther filaments, fruit capsules, and seeds. The seeds recovered from the reverted male fertility study were able to germinate in the selection medium, confirming that these seedlings were transgenic and contained the transgene; wild type tobacco seedlings were bleached. These findings support the hypothesis that an increase in ACCase activity outcompetes the removal of acetyl-CoA by β-ketothiolase.

The worldwide use of CMS to produce competitive hybrid cultivars has been achieved. Major investments of time and resources are required to backcross a male-sterility-inducing cytoplasm into elite lines. These generations of backcrossing could be avoided by transformation of an organellar genome of the elite male-fertile inbred line to produce female inbred lines for hybrid seed production. Because the male-fertile parental and male-sterile transformed lines would be developed from the same inbred line, they should be highly uniform and possess the same nuclear genotype, excluding mutations and residual heterozygosity. Therefore, the male-fertile parental line becomes the maintainer line to seed-propagate the newly transformed male-sterile line. A few generations of seed increases would produce a CMS-maintainer pair for hybrid seed production. An additional advantage of organellar transformation would be the diversification of CMS sources used in commercial hybrid-seed production. Transformation of the chloroplast genome would allow breeders to introduce different male-sterility-inducing factors into superior inbred lines. Introduction of a male-sterility inducing transgene into one of the organellar genomes of a higher plant would be a major breakthrough in the production of male-sterile inbred lines. This technique would be of great potential importance in the production of hybrid crops by avoiding generations of backcrossing, an

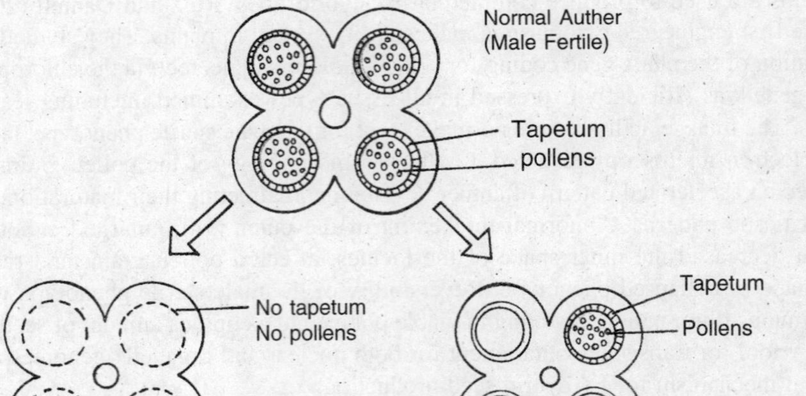

Fig. 24.4. Male sterile plants by barnase gene.

approach especially advantageous for crop plants with longer generation times. Moreover, transgenes that are engineered into annual crops could be introgressed into wild crops, persist in the environment, and thereby have negative ecological consequences. Therefore, it may be necessary to engineer a male sterility system that is 100% effective. For vegetable, fruit, or forage crops, restoration of male fertility in the hybrid is not necessary.

By barnase gene

Several pollination-control mechanisms based on genetic engineering of nuclear male sterility and its restoration have also been described in literature and have emerged as tangible options for development of male sterile/ restorer lines. Mariani and co-workers in 1990 demonstrated induction of male sterility in transgenic tobacco and Brassica napus plants by expression of a ribonuclease gene (barnase) from *Bacillus amyloliquefaciens* in tapetal tissues of anthers using a tobacco tapetum-specific promoter, TA29. The barnase nuclease is an extremely potent cellular toxin.

In normal anther, tapetum surrounds the pollen mother cells, which in turn form the pollen grains. The tapetum also provides nourishment to the developing pollen grains.Barnase gene is fused with TA29 tapetum specific promoter and transferred in rape seed. The resulting plants are male sterile because of destruction of tapetum due to expression of barnase gene. This toxicity was lethal for tapetum and pollen grains. Fertility restoration makes use of another protein, called barstar. This protein is also produced by the same bacterium, *Bacillus amyloliquefaciens.* Barstar product is an intracellular protein that protects against the lethal effect of barnase gene by forming a stable, enzymatically inactive complex in the cytoplasm. Hence, barstar is the bacterial cell's self-defence against its own barnase gene. The plants transformed with chimeric TA29-barstar gene are healthy and fertile. As the barstar gene is coupled with tissue specific TA29 promoter, little protein is produced in the tapetum cells. However, when male-sterile TA29-barnase plants are crossed with those carrying TA29-barstar, the resulting F1 hybrids having both barstar and barnase genes are male fertile. The reason of restoration of male fertility is that the barstar protein combines with the barnase nuclease and makes it ineffective. However, several of these lines showed instability of the male sterile phenotype in subsequent generations.

For commercial crops following features are required :

1. Efficient fertility restoration system for seed production.
2. Easy maintenance of male sterility.
3. Eradication of male fertile plants with ease.
4. Beneficial traits should not be affected.
5. Stability of male sterile lines.

QUESTIONS

1. What are transgenic crops ? Give an account of recent developments about transgenic crops for insect pests resistance and nutritional quality.
2. Write short notes on : (*a*) Golden rice, (*b*) Terminator gene, (*c*) Transplastomic plants, (*d*) industrial applications of transgenic crops, (*e*) Bt gene, (*f*) cry protein, (*g*) herbicide resistance, (*h*) virus resistance.
3. Discuss the importance of transgenic crops for developing countries

IMPACT OF PLANT BIO-TECHNOLOGY ON AGRI-CULTURE & INDUSTRY

CHAPTER 25

Plant organ, tissue and cell culture procedures have developed rapidly in the last half-century since the pioneering efforts of Gautheret, White and Nobecourt. The potential application of the methods of tissue culture are of special significance in crop improvement since conventional methods involve several difficulties, including heterozygosity and a long span between successive generations, hence many investigators are devising methods whereby tissue culture could be fully exploited to improve crop varieties.

The role of tissue culture in crop improvement could be identified in four areas: (a) as an aid to conventional breeding programme; (b) as a tool of unconventional breeding programme; (c) in clonal propagation, and (d) in obtaining disease-free plants.

Before launching a large-scale programme of crop improvement by tissue culture, it should be ensured that the method is economically viable. Monocotyledonous plant material has never been a favourable system for tissue culture though most crop plants are monocots. In recent years success has been achieved in growing graminaceous crops in cultures using root, *Triticum aestivum*; shoot primordia, *Sorghum bicolor;* mesocotyl segments, *Panicum miliaceum*; internodes of leaf segments, *Saccharum officinarum*; anther culture *Oryza sativa* and endosperm *Lolium perenne.* Obtaining embryogenic cultures and regeneration from protoplasts obtained from such cultures are the final steps to achieve crop improvement through plant biotechnological methods. Regeneration from protoplast has been achieved in *Pennisetum americanum, Oryza sativa, Triticum duram, Zea mays* and several others. The ultimate objective of all these methods is to improve plant characters for consumer preference, quality and marketability.

I. TISSUE CULTURE: AN AID TO CONVENTIONAL BREEDING PROGRAMME

In a conventional breeding programme, a plant breeder comes across several obstacles. How plant breeder has advantageously used the techniques of tissue culture to overcome such obstacles is described in the following paragraphs.

A. Embryo Culture

Interspecific crosses may fail because of several reasons, but when the development of embryo is arrested owing to the degeneration of the endosperm, or when the embryo aborts at an early stage of development, embryo culture is the only technique to recover hybrid plants. Incidentally, embryo has been the first plant part successfully cultured in isolation. It was in 1904 that Hanning could rear seedlings from cruciferous embryo excised from immature pods. Laibach (1925) for the first time, employed embryo culture for obtaining hybrids from a cross between *Linum perenne* and *L. austriacum.* More recently, a number of hybrids have been successfully raised through embryo culture : *Hordeum vulgare* X *Secale cereale, H. vulgare* X *Agropyron repens, H. vulgare* X *Triticum* sp. inter-specific hybrids in *Abelmoschus* and *Secale.*

Transplantation experiments have also helped in rearing hybrid embryos to maturity e.g. barley endosperm was used as the nurse tissue for growing hybrid embryos of *Hordeun vulgare* X *Agropyron repens* and *H. vulgare* X *Triticum aestivum.* During recent years, the technique of embryo culture is also being employed for the recovery of haploid plants by the elimination of a set of chromosomes. This has been quite successful in a few interspecific and intergeneric crosses. *Hordeum bulbosum* has provided a system whereby haploid plants of *H. vulgare* and *Triticum aestivum* can be obtained by selective somatic elimination of either a genome or genomes. When *H. vulgare* is crossed to *H. bulbosum*, the caryopsis (the fruits) develop normally only up to 10 days and then degenerate. Scientists cultured 7- to 8- day old fruits on B_5 medium and after one week, cultured the embryos from such fruits on synthetic medium. Haploid *H. vulgare* seedlings were recovered. Similarly, when *H. bulbosum* is crossed to *T. aestivum*, the fruits exhibit vigorous development initially but after 14-18 days the embryos abort. Hybrid embryos, when excised and cultured on suitable nutrient medium, developed satisfactorily and produced normal seedlings. Embryo culture has helped in overcoming self-sterility of seeds, especially of crop plants propagated vegetatively, when the seeds do not germinate in nature. In a wild relative of commercial banana, *Musa balbisiana* for example, the seeds do not germinate under natural conditions. However, if the embryos are excised and grown on a simple culture medium of mineral salts, seedlings are readily obtained. Likewise, tuber crop *Colocasia esculenta* propagates only vegetatively and the seeds do not germinate in nature. This natural sterility barrier could be overcome by resorting to culture of embryos.

In many crosses the culture of embryo before it aborts has led to the recovery of hybrid seedlings. In cotton the culture of hybrid embryo has given encouraging results for obtaining viable seedlings in crosses between *Gossypium arboreum* (2n=26) and *G. hirsutum* (4n=52).

In vitro zygotic embryo culture continues as a means of rescuing hybrid embryos during interspecific gene transfer. Parker and Michaels (1986) employed this technology to recover hybrids of *Phaseolus vulgaris* and *P. acutifolius*. *P. acutifolius* has improved tolerance to common bean bacterial blight (*Xanthomonas compestris*) than *P. vulgaris* (common field bean) but hybrid and early backcross embryos from crosses between these species frequently abort. Embryos rescue assisted the transfer of improved bacterial blight tolerance to economically important field beans. In another case, embryo rescue technique has been utilized in the transfer of cytoplasmically- inherited herbicide tolerence from *Brassica napus* to *B. oleracea*.

The orchid industry owes a great deal to the technique of embryo culture. The culture of orchid embryos was initiated at the Singapore Botanic Gardens in 1928. To date, they have succeeded in producing 1,431 hybrids, and 141 selfed (plants raised through self pollination) species. Callus induction from embryos of orchids can be very useful in obtaining more hybrids.

B. Ovary and Ovule Culture

Embryos at very early stages of development have not been amenable to culture, primarily because of the problems involved in excision and the complex nutrient requirements. Where such a problem exists, it is logical to culture either the entire ovary or the ovule. This facilitates the availability of maternal tissue and eliminates the cumbersome problem of dissecting very young embryos and selecting a very exacting medium. The culture of pollinated ovaries and ovule *in vitro* has been successful in a number of plants.

The credit for developing and firmly establishing the technique of ovary culture goes to J.P. Nitsch (1951), who experimented with the ovaries of *Phaseolus vulgaris, Cucumis anguria, Fragaria chilensis, F. virginiana,* and *Lycopersicon esculentum*. Ovaries excised several days after pollination showed satisfactory growth on a simple medium. However, the size of the fruit development *in vitro* was smaller than that of field grown fruits. Later on ovaries of tomato, wheat, *Pisum* and *Phlox* were grown in culture.

Beasley (1971) observed good growth of fibres of cotton (*Gossypium hirsutum*) ovules, excised 2 days post-anthesis, and grown on liquid medium. The unfertilized ovules were found to shrivel and turn brown on the basal medium. However, the addition of GA_3 or IAA (0.5 to 50 μM) resulted in larger ovules and the production of fibres. When fertilized ovules were cultured in GA_3 medium, a marked stimulation in fibre production was observed. IAA showed a tendency towards maximum fibre units at 5.0 μM. Kinetin along with IAA inhibited fibre development. Fibre development in cultured ovules was similar to that *in vivo*.

Cotton breeders have long been interested in producing hybrids between *Gossypium arboreum* and *G. hirsutum* in order to combine several desirable characters. The crosses between these species are often unsuccessful because of the abortion of hybrid endosperm and embryo at early stages of development. The technique of culturing *in vitro* offers an alternative to overcome such handicaps in making the crosses successful. Ovules and embryos of the hybrids can be cultured on suitable media to obtain viable plants.

Unripe seeds of *Cattleya,* an orchid, germinate in light. Protocorms are formed after 6 weeks. When 5-month old seedlings were transferred to Heller's liquid medium with Na-FeEDTA (salt of ethylene diamine tetracetic acid; see its preparation in the chapter on medium preparation), additional protocorms or seedlings developed after 10 weeks. The medium thus helps in reducing the time as well as in increasing the number of plantlets.

C. Nucellus Culture

The culture of nucellar tissue has also yielded encouraging results and is now being employed in the improvement of *Citrus* crop. Rangaswamy (1958) evolved a novel technique of studying nucellar polyembryony. He excised the nucellus from pollinated ovules of *Citrus japonica* (*C. microcarpa*) and cultured on different media. The micropylar-half of the nucellus proliferated on white's medium containing casein hydrolysate and callus differentiated into pseudo bulbils, which eventually developed into plantlets.

During recent years, nucellus culture has been attempted in many vegetable and fruit crops towards clonal propagation of selected materials, e.g., *Aegle marmelos, Cucumis melo, Luffa cylindrica, Trichosanthes anguina, Vitis vinifera.* Nucellar culture also helps in the elimination of viruses, e.g., in *Citrus* species.

D. Control of Fertilization

In his attempts to evolve a desirable variety, the plant breeder is often confronted with several barriers. The common ones are: (a) disharmony in the flowering periods of the parents, (b) short life span of pollen, (c) failure of pollen germination, (d) slow growth of pollen tube, (e) inhibition or bursting of pollen tube in style, and (f) failure of male gametes to cause fertilization. Using the techniques of tissue culture, embryologists have successfully evolved methods for overcoming these barriers to crossability. Test-tube pollination and fertilization, first employed at the University of Delhi, have now been shown to have wider implications. With the success of intra-ovarian pollination by Kanta (1960), investigations were extended to demonstrate the feasibility of the technique in pollinating and fertilizing excised pistils and ovules *in vitro*. For successful pollination and fertilization *in vitro*, the following considerations are important: (a) culture of pollen grains and ovules at the right stage of development, (b) a suitable nutrient medium, and (c) appropriate cultural conditions. Success has been achieved in *Nicotiana tabacum*, where the entire pistil is cultured and stigma pollinated artificially. Consequent to such test-tube pollination, pollen germination, growth of pollen tubes, entry of tubes into ovules, fertilization, development of embryo and endosperm, viable seed-setting and seed germination take place *in vitro*. This technique has been used for cross pollination in *N. rustica* and *Petunia violacea.* In another method, of ovule from unpollinated ovaries are aseptically excised and cultured on a suitable nutrient medium. Pollen from freshly dehisced anthers is then dusted the next day on cultured ovules. The pollen grains germinate, and growth of pollen tube and fertilization are brought about followed by normal development of endosperm, embryo and seed. This has already been successfully achieved in *Argemone mexicana*, *Eschscholzia californica* and *Papaver somniferum*.

E. Endosperm Culture

Endosperm, usually a triploid tissue in angiosperms, can be suitable explant for raising triploids of cereals. As early as 1949, LaRue reported the proliferation of immature endosperm of maize. In

Oryza, induction of callus was observed at the cellular stage of endosperm containing starch grains. Endosperm culture to obtain triploid plants have been used in *Putranjiva roxburghii, Nigella damascena, Petroselinum crispum, Citrus*, apple and *Vitis vinifera*. The successful culture of endosperm tissue has opened up new avenues for plant breeders who can apply this technique to economically important crop plants, especially where conventional breeding methods prove futile.

II. TISSUE CULTURE: A TOOL OF NOVEL UNCONVENTIONAL BREEDING PROGRAMME

A. Anther Culture

In vitro gametic embryogenesis has become an efficient means of producing haploids in a growing number of species since the success of Guha and Maheswari (1964) with cultured anthers of *Datura*. Success in anther culture and more recently in isolated microspore culture have stimulated interest in this technology for plant breeding and applications in both monocots and dicot species of plants. The utility of haploid production in the rapid production of homozygous breeding lines from highly heterozygous parents is well recognized including reduced time requirements per cycle of plant breeding due to immediate phenotypic expression of the recessive traits in haploid derived homozygous plants.

For successful induction of androgenic haploids, the anthers should be cultured just before, or during or immediately following microspore mitosis. Most of the cultured anthers require an inductive phase, which depends on exposure to certain conditions of light and temperature for a particular period. A definite stimulatory effect of high temperature on embryoid formation has been shown in tobacco and *Datura innoxia*.

Of the 12 species of *Nicotiana* tested, J.P. Nitsch (1969) reported the production of haploids in anther cultures of 5 species (*N. alata, N. glutinosa., N. rustica, N. sylvestris*, and *N. tabacum*) on a chemically defined medium with or without the addition of vitamins and IAA.

Anthers of *Oryza sativa* grow very well when either 2,4-D or NAA is added to Murashige and Skoog's medium. The nutritional experiments indicate that anthers cultured at the microspore stage have a similar requirement than anthers cultured at meiotic stages.

Anther culture undoubtedly offers a unique tool for obtaining haploid plants. Already, a number of economically important crop plants have yielded encouraging results: *Aegilops*, rice, *Brassica* hybrid, *B. napus, Capsicum annuum, Coffea arabica, Hordeum vulgare, Lycopersicon esculentum, Secale cereale, Solanum melongena, S. tuberosum, Triticum aestivum,* and *Vitis vinifera.* In China, the programme of large-scale production of haploid plants through anther culture is reported to be progressing well and almost every crop is reported to be under test for obtaining haploids through anther culture. More details about haploid plants are given in Chapter 16.

B. Somatic Hybridization and Genetic Modification

Hybridization, involving both sexual fertilization and a variety of protoplast-protoplast and cytoplast-protoplast fusion have been successfully demonstrated *in vitro* in a few crop species. 'Test-tube' fertilization has been employed to help overcome pre- and post-fertilization cross-incompatibilities, while parasexual fusion have been employed to produce interspecific and intergenomic hybrids, and recombine cytoplasmically inherited traits. Utilization of *in vitro* hybridization and fusion has been of a limited value in crop improvement, till date. Most applications have been related to synthesis and resynthesis of amphiploids or transfer of cytoplasmic organelles encoded traits (e.g. triazine herbicide tolerance, cold tolerance) into otherwise potentially useful crop germplasm, almost exclusively in Brassicaceae and Solanaceae.

Plant protoplasts cultures offer a potential system to achieve somatic hybridization and to create genotypically and cytoplasmically novel hybrids (hybrids and cybrids; see Chapter-10). Plant protoplasts also offer a system for genetic modification to attain the objective of crop improvement.

This could be possible due to the remarkable property of plant protoplasts for the uptake of macromolecules such as proteins, DNA, ferritin, latex, polystrene, virus, bacteria and even isolated chloroplasts and whole nuclei.

Several investigators have reported interspecific and intergeneric protoplasts fusion in *Petunia, Atropa belladonna, Vicia, Glycine, Bromus,* carrot, lily, oat x maize, carrot x tobacco, *Torenia fournieri x T. flava* and *Brassica chinensis X B. oleracea*. Melchers et al. (1978) reported regeneration of somatic hybrid plants of potato and tomato from fused protoplasts.

Cells with novel phenotype can be produced by genetic engineering sometimes by combining the features of several different organisms. Numbers of medicinally or commercially important molecules are routinely produced in genetically engineered cells. Altering the genotypes of plants is an important application of recombinant DNA technology. By such means, it is possible to transfer genes from one plant species to another. These include genes affecting yield, hardiness or disease resistance. The first engineered recombinant plant of commercial value was developed in 1985. An economically important herbicide is glyphosate, a weed killer that inhibits a particular essential enzyme in many plants. The target gene of glyphosate is also present in the bacterium, *Salmonella typhimurium*. Mutagenesis and growth of *Salmonella* in the presence of glyphosate obtained a resistant form of gene. Then the gene was cloned in *E. coli* and recloned in the T-DNA of *Agrobacterium*. Transformation of plants with the glyphosate resistant gene has yielded varieties of maize, cotton and tobacco that are resistant to herbicides. Thus the fields of these crops can be sprayed with glyphosate at any stage of growth of the crop. The weeds are killed but crops are unharmed. The bacterium, *Bacillis thuringiensis* produces a protein that is lethal to the black cutworm but the bacterium does not normally grow in association with the plants that are damaged by the worm. However, the gene coding for the lethal protein has been introduced in the soil bacterium *Pseudomanas fluoriscens*, which lives in association with maize and soyabean roots. Inoculation of soil with engineered *P. fluoriscens* helps control the black cutworm and reduced crop damages.

Genetic recombinations leading to alterations in the production of specific secondary compounds provide better yielding crop plants. Such modifications have been used for the production of pyrethrins in *Chrysanthemum cinerariefolium* and altered flower colour in carnation.

C. Somaclonal Variation

In vitro cell and tissue cultures of plant give rise to genetic variation spontaneously. While spontaneous variation is not desired during propagule multiplication, it has been useful in providing some genetic variants among crop species. The observed frequencies of such somaclonal variants vary widely and are probably related to both time and culture and the observer's ability to detect genetic variants. Several point mutations have been observed, the majority of analysed spontaneous genetic variants from somatic cultures appears to have resulted from induction of aneuploidy, loss or interchanges or intra-chromosomal segment duplication, deletions or structural rearrangements. Better cultivars have been produced in *Citronella* grass for higher biomass and oil contents.

Spontaneous or mutagen induced genetic variation in somatic cell culture coupled with *in vitro* selection techniques have been effective in isolating desired novel genetic variants with cellular level expression while in a heterozygous condition. Haploid cell cultures were reported to have higher frequencies of detectable variants than diploid cultures.

Among the numerous reports of *in vitro* variants, several may provide genetic variation useful for crop improvement. Modified tolerance to specific herbicide, altered composition, and modified pest / stress tolerance if not associated with deleterious pleiotrophic effects (the production by a single gene, two or more apparently unrelated effects) will likely to emerge in commercial crop varieties in future.

III. PROPAGATION THROUGH TISSUE CULTURE

Clonal propagation

Clonal propagation by vegetative methods is a practice followed since man started cultivating plants. The main objective of clonal propagation has been to reproduce plants of selected, desirable qualities uniformly and in bulk. The traditional propagation methods, requires long duration, whereas tissue culture helps in rapid plant multiplication. A selected examples of clonal multiplication of trees and horticultural plants are as follows: *Aegle marmelos, Tecomella undulata, Musa* sp., citrus, peach, prunus, *Malus, Prunica granatum* etc.(for details on this topic see the Chapters 21 and 22). Improved cultivars developed by biotechnological methods are then clonally multiplied to replace inferior cultivated varieties, e.g. as in *Mentha arvensis, M. piperata, M. spicata* and several other aromatic plants.

IV. PRODUCTION OF DISEASE-FREE PLANTS

A majority of the commercial crop plants propagated vegetatively through tubers, bulbs, cuttings and grafting which may contain systemic bacteria, fungi and viruses. Such infections affect the yield and quality of the crop, unless it is readily detected and the plants made free from infection. The traditional method of eliminating viruses by heat treatment is applicable only to a few varieties. Generally, the apical meristem of a virus diseased plant contains very little virus, or is altogether free. Hence, plants obtained from apical meristem are usually free of viruses. There are a number of instances where meristem and/or floral bud culture has resulted in elimination of the virus and restoration of healthy plants ,e.g., *Solanum tuberosum, Saccharum officinarum, Petunia* spp, *Allium sativum, Ananas comosus, Brassica oleracea, Musa* sp. and *Pisum sativum.*

The task of crop improvement by conventional methods is quite laborious and time-consuming, and poses many obstacles in the way of a plant breeder. The techniques of tissue culture offer a potential method to resolve these difficulties, and in many instances to shorten the time taken in crop improvement programmes. One of the important applications of tissue culture has been in clonal propagation, and in obtaining disease-free plants. These unimproved traits are responsible for severe losses of crop yield and low market price of the produce. Therefore, biotechnological methods are extensively used to improve these characters.

V. IMPACT ON FORESTRY

Forest resources are disappearing at an alarming rate. An important component of forest ecosystem is trees, which provide food, fuel, construction and industrial products. In addition, trees are reorganized as the critical elements in maintaining stability in the world's atmosphere. Erosion of genetic variability is also a matter of concern now-a-days. Therefore, both *in situ* and *ex situ* management is necessary for biodiversity, conservation and consequently for tree crop improvement.

Tree improvement involoves managing genetic resources, and includes conservation, selection and breeding and propagation. Biotechnology has the potential to accelerate the genetic improvement of forest yield.

The traditional methods for propagation of trees are by seeds and rooted cuttings or grafting. Micropropagation offers several advantages over conventional vegetative propagation. viz., small area requirement, large number of plant formation in short duration, cloning of selected material, easy to transport etc. Rapid progress has been made in tree tissue culture after production of first true plantlet via organogenesis from *Populus tremuloides* by Winton (1970). Conifers are soft wood plants used in pulp and paper industry. Sommer and co-workers (1975) produced complete plantlets from *Pinus palustris* for the first time.

The ability to regenerate complete plants from individual protoplasts, cells, or tissue is a necessary requirement that must be established prior to the application of most biotechnological methods for crop improvement. The methods employed in micropropagation and improvement of tree species are-

A. Clonal propagation
B. Organogenesis and embryogenesis
C. Protoplasts culture and genetic engineering

A. Clonal propagation

There are two main objectives for developing clonal techniques for forest trees. Firstly, viable large scale clonal systems can be an asset for selected trees and secondly, reliable regeneration protocols are necessary for further genetic manipulations. Some of the requirements for a large scale system are simplicity, reproducibility, cost effectiveness and adaptability to nursery practices.

Asexual multiplication of both hard woods and softwoods using tissue culture methods can be achieved by axillary bud breaking, production of adventitious buds and somatic embryogenesis. The first two approaches lead to the development of a bipolar embryo. A large number of forest trees have been multiplied by these methods (Table 25.1). Axillary bud break is common in angiosperm while adventitious shoots formation is used in gymnosperm for multiplication. Clonal multiplication is the most frequently used method is tree propagation.

Table 25.1. Multiplication of forest tree species by different methods

A. By clonal multiplication	B. By somatic embryogenesis
Populus tremula	*Abies balsamea*
Sterculia urenes	*Biota orientalis*
Tectona grandis	*Cupressus orizonica*
Prosopis cineraria	*Dalbergia sisso*
Tecomella undulata	*Dalbergia latifolia*
C. Protoplasts culture & genetic engineering	
Eucalyptus grandis	*Picea glauca*
Populus	*Pinus sp.*
Juglans	*Santalum album*
Pithoclobium dulce	*Commiphora wightii*

B. Somatic embryogenesis

To date, shoot cultures are the most successful method of propagation for hardwood trees. However, somatic embryogenesis is alternative method providing large number of plantlets at low cost, the potential for generating artificial seeds and possibility of automation. Suspension cultures can provide embryogenis protoplasts which potentially can be used for genetic engineering. Somatic –embryogenic tissue can be generated directly of treated explants or obtained indirectly by manipulating non –embryogenic callus in vitro. In confers, this is the principal method of propagation. The examples of forest tree species propagated by somatic embryogenesis are –*Larix decidua, Picea abies* and others (Table 25.1).

C. Protoplast culture and genetic engineering

Protoplast culture provides an alternative method of improving plants, and is one of the few systems which can utilize somaclonal variation, somatic hybridization, cybridization and genetic engineering. Protoplasts have been isolated from both tropical and temperate species of hard woods, e.g., sandal wood, *Eucalyptus grandis*, *E. urophylla*, *E. saligna*. Transformation of forest species has been successful with only a few genera like *Populus* and *Juglans* mainly using the *Agrobacterium* vector system.

One of the major problems preventing a wider use of micropropagation technology is the difficulty in using explants from mature trees. Success has been achieved in culturing juvenile explants like excised zygotic embryos and seedling parts, and by the time trees are old enough for evaluation, they are often recalcitrant (not responding) in culture. Therefore, tree propagation for the most plant is carried out with unproven (juvenile) material. Contamination and secretion of phenolics are the other problems associated with the tree species culture. Lastly, the cost of regenerating plantlets is very high.

VI. IMPACT ON HORTICULTURE

In vitro culture techniques have had numerous applications to fruit crops beginning nearly 60 years ago with embryo rescue techniques for stone fruits. Later on, the method has been applied successfully to produce commercially acceptable early-ripening peach and nectarine cultivars. The methods have been adapted to other crops as well, e.g. in breeding programmes to produce both early-ripening and seedless grapes.

Historically, the second application of in vitro culture methods to fruit crops was to eliminate disease causing viruses from strawberries. Meristem-tip culture has since become an integral part of virus- indexing progrmmes for a number of fruit crops, usually in conjunction with thermotherapy. In some cases, it has been necessary to micrograft the meristem tip into an in vitro-grown seedling as is done for citrus.

Within the past 25 years, micropropagation of fruit crops has become an important application of in vitro technology. Strawberry was the first fruit crop for which the method was developed. Now many fruit crops are being micropropagated commercially. More recent uses of in vitro culture emphasize application for genetic improvement of fruit crops. These application include production of hybrid plants from fused protoplasts, somaclonal variation and mutagen-driven changes in regenerated plants, haploids plants from anther culture and transfer of specific genes via *Agrobacterium*-mediated transformation.

Table 25.2. Micropropagation of fruit crops

A *Temperate fruit crops*	B. *Tropical fruit crops*
Apple (*Malus*)	Kiwifruit (*Actidinia chinensis*)
Pear (*Pyrus communis*)	Pineapple (*Ananas cosmosus*)
Peach (*Prunus persica*)	Cashew (*Anacardium occidentale*)
Cherry, sweet (*Prunus avinum*)	Papaya (*Carcica papaya*)
Cherry, sour (*Prunus cerasus*)	Lime (*Citrus aurantifolia*)
Strawberry (*Fragaria*)	Lemon (*Citrus limon*)
Black berry (*Rubus sp.*)	Mulberry (*Morus alba*)
Rasp berry (*Rubus sp.*)	Banana (*Musa sp.*)
Grape (*Vitis vinifera*)	Date palm (*Phoenix dactylifera*)
	Guava (*Psidium guajava*)
	Pomegranate (*Prunica granatum*)

Following methods are used to improve and multiply the horticultural crop plants. Details about these methods are given in respected chapters, therefore not presented here.

1. Micropropagation
2. Virus eliminaton
3. Genetic improvement
4. Germplasm conservation
5. Haploid production

The above mentioned methods have been employed for fruit crops (Table 25.2) plantation crops and vegetable crops. Examples of plantation crops are –Pineapple, palms, banana, cacao (*Theobroma cacao*), coffee (*Coffea sp*), tea (*Camellia sp.*) rubber (*Hevea brasiliensis*), Vanilla (*Vanilla planifolia*), clove (*Syzygium sp.*) ginger (*Zingiber officinale*), turmeric (*Curcuma longa*), cardamom (*Elleteria cardamomum*). Some of these plants are produced in millions by plant tissue culture industry. In vegetable crops, major emphasis has been given on – cabbage (*Brassica oleracea*), mustard (B. *campestris*), artichoke (*Helianthus tuberosus*), garlic (*Allium sativum*), onion (*A. cepa*), tomato, carrot, melon (Cucumis melon), cucumber (*Cumis sativus*) pepper (*Capsicum annum*), and watermelon (*Citrullus vulgaris*).

Micropropagation of ornamental plants (foliage and flower plants) has become a major plant tissue culture based industry on developed and developing countries. Detail of this and examples are given in other chapters.

VII. TISSUE CULTURE INDUSTRY IN INDIA

India is an agricultural country and its economy is largely dependent on agriculture. Therefore, technological developments in the agriculture sector have a direct impact on country's economy. The abundance of genetic base (biodiversity), diverse agroclimatic zones and highly qualified man-power offer a potential scope for technological advances in agricultural biotechnology.

India has been traditionally a leading exporter of primary agricultural produce such as tea, coffee, rubber, cardamon, cotton and several spices, flavouring, aromatic and medicinal plants. This status is declining due to low productivity, non-competitive prices and sometimes, poor quality of the material. Therefore, our urgent need to improve plant cultivars for quality and quantity is required using biotechnological methods.

Most of the agricultural produce in India is produced as rain fed crops. The shortage of healthy seeds/planting material and lack of disease resistant clones have often been affecting the agricultural economy. Large scale production of elite clones through micropropagation or somatic embryogenesis, production of disease free clones and improvement of varieties through cellular, molecular and conventional breeding methods hold the promise of improving some of the problems of Indian agriculture, horticulture and forestry.

VIII. SCOPE FOR DEVELOPMENT

A. Diverse Agroclimatic Regions

India has rich resources of land spread over diverse agroclimatic zones that have the potential to produce a wide variety of agricultural and commercially high value crops. It has large biodiversity, germplasm of rare ornamentals and medicinal plants, most of them native to this continent. Therefore, India offers many advantages over other countries in the world.

B. Forest Wealth

The native forests all over the country are well known for their richness and variety in foliage and ornamental plants. These plants can be propagated by tissue culture method and introduced in the market.

C. Different Climatic Zones

In India all the climatic zones, from temperate to tropical are available. Therefore, any plant growing in any climate can be multiplied and grown in the country. On the same time, cool climatic zones are less expensive for micropropagation work as extensive air conditioning for plant growth is not required.

D. Scientific Talent

In India, the scientific talent and the operational skills of the people is at par with the best in the world. The salaries are low in India. This makes it an ideal place for production facility and that is why, some companies have been shifted to India from USA and Europe. It is worth mentioning here that anther and pollen culture for haploids and *in-vitro* fertilization like plant biotechnologies were discovered in India. Embryo culture and rescue of hybrid embryo is of quite significance in this aspect.

Propagation of plants through tissue culture has become an important and popular technique to multiply crops that are otherwise difficult to propagate conventionally by seeds and/or vegetative methods. Differentiated cells or organs are manipulated to give rise to multiple copies of the parent plant under optimum aseptic environmental conditions and appropriate stimuli. It offers several advantages over conventional propagation methods such as rapid multiplication of valuable genotypes, rapid improvement of crop plants, production of disease and virus free plants, production throughout the year possible, germplasm conservation and easy transport and transfer (international) of material (*in vitro* material is contamination free). There are several methods developed in last 3 decades for vegetative multiplication of plant species involving axillary budding, adventitious budding, somatic embryogenesis and organogenesis from callus cultures. In fact, the genetic stability of the plant copies (population) produced is likely to vary according to the procedure chosen. However, these methods have certain limitations.

IX. PROBLEMS ASSOCIATED WITH AGRICULTURAL BIOTECHNOLOGY

Development and use of any technology is dependable on infra-structural facility and simultaneous development of other related technologies. Agricultural biotechnology requires infrastructure like electricity, air-conditioned rapid transport system, skilled labour and related biotechnology like innovative and novel packaging material, plastic ware and glass appliances etc. Followings are major hurdles -

(*a*) Industry is a capital intensive (requires elaborate establishment money).

(*b*) Seasonal activity with peak demand; delivery of plant material at one time requires large facility which may be under utilized at other times.

(*c*) High production cost applicable to selected crop plants with established protocols for multiplication.

(*d*) Management of personnel and organization of work schedule in chain oriented manner.

(*e*) Contamination problem adds a big cost factor.

(*f*) Somaclonal variation at large multiplication rate.

In context of India, on export scene, lack of variety of packing material at affordable price, high air freights, unsuitable air connections, lengthy phytosanitory, custom and export documentation formalities make this non-traditional, highly perishable export commodity less attractive as compared to potential it has. However, recently the government of India has set up suitable cargo-handling cold storage facility at New Delhi and Bangalore for export of biotech products.

X. COMMERCIALIZATION-MICROPROPAGATION INDUSTRY

Commercialization of agricultural biotechnology began in India in real sense in 1987 when various private sector companies adopted this technology as a means of increasing the productivity of their primary produces or as a part of their diversification programme. This includes A.V. Thomas group of companies, Indo-American Hybrid seeds (IAHS), Hindustan Lever, Tata Tea, Unicorn Biotech, Nath seeds, RP Goenka Enterprises, India Tobacco company, Maharashtra Hybrid Seeds Company, Hindustan Agrigenetics Ltd. and some others.

The first professionally managed commercial plant tissue culture unit in the country was set up in 1987-88 by A.V. Thomas, at Cochin in export processing zone, and remained one of the largest in Asia and only one of its kind in India (producing 190 million plants in 1996), until IAHS began production in its Bangalore based unit. IAHS has 48 Laminar flow work stations, hundred growth rooms (each can accommodate 400,000 culture at one time) and a green house of 30,000 sq. ft. (2790 sq m) with most advanced equipment such as automatic shade system, fogging, blackout and irrigation systems, and movable benches for hardening of tissue culture plants.

Subsequently, advancement in commercialization of plant tissue culture and acceptance of plantlets raised through tissue culture have led to continued exponential growth within the industry in terms of number of new units as well as number of plants produced by these units (Table 25.3). Most of these units are set up with foreign know how. Most of our commercial units prefer to seek the entire technology package on a turn-key basis. The major foreign collaborators (mostly with buy back arrangements) are M/S Cultiss, Holland; M/S Green Teek, Holland; M/S Microplants Ltd., UK; Centre de Research Agronomique, Belgium; Planex, Australia; and Agrobio, France etc.

Table 25.3. Plant Micropropagation Industry in India.

State	No. of Industries	Annual production capacity in million
Maharashtra	25	More than 31
Karnataka	9	30.1
West Bengal	9	5
Andhra Pradesh	6	16.25
Tamil Nadu	4	26.35
Kerala	4	25.35
Haryana	3	13.25
Gujarat	3	11.25
U.P.	3	25.0
Punjab	1	1.7
H.P.	1	5
Orissa	1	Not known
Rajasthan	1	Not known

Indo-American Hybrid Seeds (IAHS), A.V. Thomas & Co., Harrisons Malayam Ltd., Godrej Plant Biotech Ltd. and SPIC have established ultramodern micropropagation facilities. IAHS and A.V. Thomas have exported millions of tissue cultured plants to Europe and released several lakhs banana and cardamon plantlets to farmers in domestic market. A.V. Thomas have developed a cardamon variety with yield of 250 kg/hectare as compared to normal 70 kg/hectare and can be cropped in two years instead of usual three years. A.V. Thomas have also developed high glucovanillin producing and early flowering vanilla plants through tissue culture. Cost effective micropropagation system has also been developed in tea and coffee.

IAHS has won several contracts from the Netherlands, UK, and Denmark to develop new plant varieties by somatic hybridization, selection of somaclonal variants, endosperm culture, *in-vitro* pollination and embryo rescue method in combination with conventional plant breeding. IAHS is providing service for screening plants for 40 viruses, 2 viroids and several bacterial, fungal and mycoplasmal diseases for 19 ornamental and 26 horticultural and agricultural crops. This company had exported cut flowers worth Rs. 14 million to Holland, Denmark and UK in 1991.

Beena Nursery (P) Ltd., Trivandrum offers high quality and disease free plantlets of orchids suitable for the cut flower industry. Hindustan Lever Ltd, Mumbai is involved in clonal multiplication of high yielding cultivars of plantation crops such as cardamon through meristem culture, virus-free microtubers of potato. It has also designed a bioreactor to scale-up the microtuber production. Godrej plant Biotech Ltd. (earlier known as Unicorn Biotech), multiply trees and plants by apical and axillary meristems and somatic embryogenesis. Micropropagation protocols for 2 cultivars of strawberry suitable for cultivation in tropical plains and conventional hilly regions have been developed. The company sold over 2 lakhs plants to farmers in Maharashtra, Gujarat and Andhra Pradesh during 1992-94.

Micropropagated banana, *Chrysanthemum*, rose, carnation and foliage ornamental plants are not only being produced at large scale but also being exported. Dhampur Sugar Mills, Dhampur and EID Parry, Chennai are involved in production of virus-free sugarcane through tissue culture.

SPIC science foundation, Chennai have introduced sex-typed female papaya plants in commercial tissue culture market for the first time in India. SPIC Agrobiotech centre is working on coffee, banana, potato, sunflower, rose, gerbera, orchids and tree species (*Callistemon, Simarouba*, jack fruit, mango and date palm).

National Agricultural and Scientific Research foundation, Calcutta is engaged in large scale production of *Musa*, *Spathiphyllum, Anthurium, Elettaria* and orchids. It has already developed protocols for tea, jojoba, lilies, gladiolus, gerbera, *Ficus* and *Philodendron*. In all there are 75 units of small, medium and large capacity, capable of producing 190 million plants per annum. The major 4 units are producing 15-25 million plants per annum. The others are capable of producing plants from 1-10 million per annum.

XI. PLANTS PRODUCED

Indian units are currently producing fruit crops, forest trees, ornamental (foliage and flowering plants), vegetative crops and plantation crops (Table 25.4). The major emphasis is on the production of ornamentals.

Table 25.4: Plant species currently micropropagated at commercial level in India.

Fruit crops – Banana, Citrus, Grapes, Papaya, Pomegranate, Strawberry, Pineapple, Cherry apple, Peach, Datepalm, Mango, Apple.

Forest Trees – Callistemon, *Eucalyptus*, Teak, Kadamb, Jack fruit, *Simarouba*, Salvadora, Bamboo.

Ornamentals – Foliage *Aglaonema*, *Calathea*, *Cordyline*, *Diffenbachia*, *Ficus*, *Philodendron*, *Spathyphyllum*, *Syngonium*.

Ornamental – Flowering plants *Alstroemeria, Anthurium,* Callalily, Carnation, *Chrysanthemum,* Gerbera, *Impatiens,* Rose, *Dracaena, Gladiolus, Iris,* Orchids, Lily, Dendrobium.

Vegetative crops – Potato, *Asparagus, Brassica napus, B. Juncea,* tomato, cauliflower, cabbage.

Plantation crops – Cardamon, Sugarcane, Vanilla, Coffee, Ginger, Turmeric, Tea, Black pepper.

XII. GOVERNMENTAL EFFORTS

The Indian Government has identified micro propagation of plants through tissue culture as an industrial activity under Industrial (Development and Research) act, of 1951 which was made effective in 1991. The units engaged in floriculture can avail duty free imports under export oriented units (EOU)/ export processing zone (EPZ) scheme even if they sell 50% in domestic market and export the remaining 50%. The import duty on live trees, plants, bulbs, roots, cut flowers as well as ornamental and foliage plants has been reduced from 55% to 10%. Import duty on seed development machinery for soil preparation, greenhouses, pre-cooling unit and refrigerated transport units has been reduced to 25%. A subsidy on air-freight to a maximum of 25% has been allowed for export of cut flowers and tissue culture plants.

Government of India has formed National Board of Biotechnology in 1982 to promote research and development work in the area of Biotechnology in the country. Since then several well defined programmes of National priorities began in the country through National Institutes and Universities. Later on, Department of Biotechnology (DBT) was set up in February 1986 for better functioning under the Ministry of Science and Technology. This department is planning, promoting and coordinating biotechnological programmes in the country Fig. 25.1.

DBT has set up task forces in various biotechnological areas. The research projects funded by DBT are in various research themes like micropropagation for fuel, fodder and biomass, medicinal and endangered plant propagation, protoplasts culture and genetic engineering, and crop improvement. Plant species have been identified for Research and Development programmes. Two pilot plant for micropropagation have been set up at National Chemical Laboratory, Poona and Tata Energy Research Institute, Delhi.

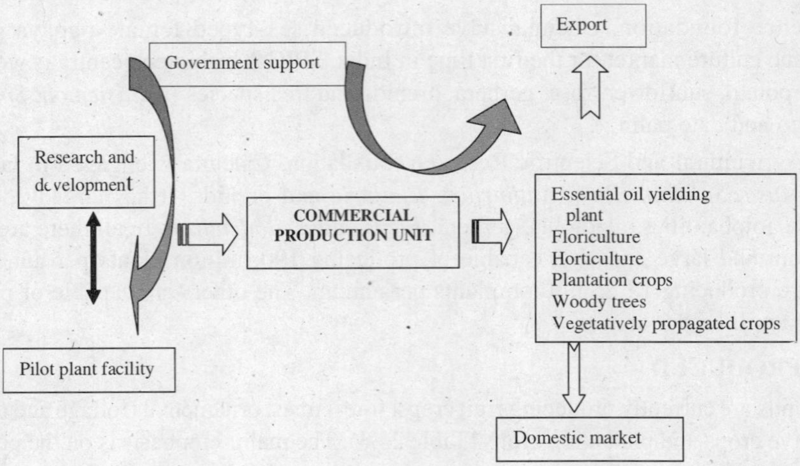

Fig. 25.1. A schematic presentation showing the government's role in supporting biotechnological industry and production of various crops.

Government of India has liberalized rules for import of basic materials, provides tax exemptions and subsidies on air freight of plants raised through tissue culture, cold storage facilities have been created at Delhi, Mumbai and Bangalore for storage of consignment awaiting shipment. Tissue culture consignments are given quarantine clearance within 6 hours. Due to these reasons, Indian export of floriculture products increased from Rs. 738 lakhs in 1990-91 to Rs. 1139 lakhs in 1991-92.

QUESTIONS

1. Discuss the role of plant biotechnology in agriculture.
2. Describe the impact of plant biotechnology on crop improvement.
3. Write short notes on the followings :
   (*a*) Anther culture
   (*b*) Ovary and ovule culture
   (*c*) Embryo rescue
   (*d*) *In vitro* fertilization
   (*e*) Genetic modifications.
4. Describe the tissue culture based industry in India and discuss the role of the federal government in supporting the Industry.
5. Write short notes on the following:
   (*a*) Micropropagation Industry
   (*b*) Department of Biotechnology, New Delhi
   (*c*) Governmental aid to biotechnology
   (*d*) Production Units.

MOLECULAR FARMING

CHAPTER 26

A revolution in plant biology has occurred over the last 15 years as tools of plant biotechnology and molecular biology were integrated and employed for plant genetic engineering. It has now become possible to introduce foreign genes into crop plants and study the expression of transgenes in selected plant tissues and several useful genes including those for insect resistance, disease resistance, herbicide tolerance, food quality have been the choice for genetic manipulation. Some other interesting characters are increased, starch and sucrose levels, modifications and synthesis of desired fatty acids, manipulations of flower colour and synthesis of plastics.

With the development of technology for the isolation of genes of prokaryotic and eukaryotic origin, high value compounds can now be produced in plants. This ability to produce and isolate high-value production transgenic plant systems is popularly known as "molecular farming" or "biomanufacturing of speciality products". Studies in this direction have began exploring the production of proteins of pharmaceutical and industrial value, e.g., human serum albumin, antibodies, blood factors, neuropathies, anti-genic proteins and vaccines.

Plants are source of many products such as food, clothing, building material, medicines etc. owing low cost of growing plants on large area, ease in scale-up, availability of natural storage organs, and the knowledge based harvest, transport and processing systems. Therefore there have been a greater interest in employing plants for the production of high value products. Against chemical synthesis, plants have additional features like the elimination of purification steps, when the compound of interest is used as food or feed supplement.

It is evident from Table 26.1 that several products are produced by transgenic plants and many more will be produced in future. Once the recombinant proteins are synthesized in appreciable levels in transgenic plants, the plant systems should serve as the bio manufacturing devices. The only current limitation in this direction is the low levels of the accumulation of the recombinant proteins.

I. A BRIEF HISTORY

Since the first transgenic was produced in 1983, possibility of producing protein developed. The first report of human antibodies produced in plants was by Dr. During (1988) and was expanded to include secretary antibodies by Hiatt et. al (1989). The first report of a protein being produced in plants for the specific purpose of extraction, purification, and scale of that protein was by Hood et. al (1997). These workers provided method for the production of avidin, an egg protein. Aprotinin, one of the first molecularly farmed pharmaceutical proteins to be produced in plants in 1999, may soon be used on patients for wound closure.

Edible fruits and vegetables are good choice to develop transgenic oral vaccines. Potato, tomato, banana, grapes are the examples of plant species grown all over the world and particularly in developing countries, where cheap vaccines are required the most. This reduces the cost of purification and down stream processing and transportation. Transgenic plants have been produced in the following other edible plants and species can be suitably modified to desired vaccine production: - apple, asparagus, cabbage, carrot, cauliflower, cucumber, egg plant, papaya, pea etc.

There are currently two methods of protein production from plants :-

(*i*) Stable integration of foreign DNA into plant genome introduced either by genetic transformation : *Agrobacterium* mediated or directly by using micro projectile bombardment and,

(*ii*) Transient expression of candidate DNA using viral vectors. The stable integration is advantageous because it passes in subsequent generations of large number of transgenic plants, either by vegetative or sexual means and also the possibility to introduce more than one gene for possible multi-component vaccine production. To produce sufficient amount of vaccines by recombinant cell culture technology, fermentation and purification are required which are very expensive. If the antigens are expressed in edible tissues of transgenic plants, it will become a cost effective production and delivery system. This is referred to as "edible plant vaccine technology". In addition tissue or organ specific expression of foreign antigens is possible by using tissue specific promoters.

Classes of proteins within molecular family –

These can be categorized in four broad areas –

(*i*) **Industrial proteins-enzymes –** This group includes hydrolyses-both glycosidases and proteases. Oxido-reductase enzyme such as laccase, a fungal enzyme used in fibre bleaching and bioglue of wood products, represents a separate class of industrial enzymes. Enzymes used in ethanol production from biomass are candidates for molecular farming. All these products will usually be required in large quantities and must therefore, be produced very inexpensively. Regulatory hurdles as they exist today may be a major hindrance to the molecular farming of these products since very large acreage will be required. Therefore, rules for cultivation and production of such industrial molecules should be simplified.

(*ii*) **Therapeutics –** This group includes all proteins used directly as pharmaceuticals along with those proteins used in the making of pharmaceuticals. The list of such proteins is long, growing and includes such products as thrombin and collagen (therapeutics) and trypsin and aprotinin (intermediates). Practically speaking, only those proteins with high value will be considered as candidates for molecular farming. Products in this category must be manufactured under stringent cGMP (current good manufacturing practices) producers and be of high purity.

(*iii*) **Monoclonal antibodies –** This group includes all antibody forms (IgA, IgG, IgM, secreatroy IgA etc.) and antibody fragments (Fv). They can be produced in plants in both glycosylated and non-glycosylated forms. These plant-derived monoclonal antibodies (plant bodies) have the potential of alleviating the serious production bottleneck that exists as dozens of new antibodies products attempt to reach the market place. Examples of plant-derived monoclonal antibodies in products development include a-caries for prevention of dental decay and a-herpes for prevention of herpes transmission.

(*iv*) **Antigens for edible vaccines –** Specific protein antigens can be produced in plants that will induce a humeral immune response when cater by an animal or human. Protection studies have shown good efficacy when these edible (or oral) vaccines have been used. In some cases, protection has actually been better with the edible vaccine than with the commercially available vaccine. Examples of edible vaccines developed are given below.

II. EDIBLE VACCINES

Vaccines are the most important and cost effective sources for fighting infection diseases. Every year, an estimated 17 million people die of infectious diseases which include 7 million children. Although there have been opportunities for the production of cell cultures and recombinant vaccines, there is an increasing demand and the current production facilities are inadequate to supply the vaccines on a large-scale at an affordable price for the people living in developing countries. In recent years, considerable progress has been made in producing functionally active proteins, peptide of medical importance in transgenic plants. The expression of subunit antigens of infectious microorganisms in transgenic plants and their subsequent immunogenic properties led to the production of edible vaccines.

Production of therapeutic proteins in Plants

Until recently small organic molecules synthesized by microbes or by organic chemistry were used as therapeutic agent. These include most antibiotics, analgesics and other pharmaceuticals. Now attention has been focussed on large and complex protein molecules as therapeutic agents. The structure and functionality of a protein is determined by its sequence of amino acids, which in turn determines its three dimensional structure. Complex proteins undergo further processing suhc as the addition of phosphate groups (phosphorylation) or carbohydrate molecules (glycosylation) which modify the protein's functions. As we know proteins are direct gene product. While short peptide chains (less than 30 amino acids) can be synthesized chemically larger proteins are produced by living cells.The DNA that encodes the instructions for producing the desired protein in inserted into cell,and as the cells grow they synthesize the protein. This protein is isolated and purified from the cells.Since 1982, more than 95 therapeutic proteins have been licensed for production using bacterial fungal and mammalian cells grown in sterile cultures. However about 50% of the total cost of production is in extraction and purification of proteins which is required in all the systems.Like mamalian cells, plant production systems have the advantage over microbial system of being able to produce active forms of complex proteins with appropriate post translational modifications.

Antibiotics – Transgenic plants have been used for the production of antibodies for dential caries, rheumatoid arthritis, cholera, *E. coli* diarrhoea, malaria certain canceres, Norwalk virus, HIV, Rhinovirus, influenza, hepatitis B virus and herpes simplex virus.

Vaccines – Plant derived accines have been produced against *Vibrio cholerae*, enterotoxigenic E. coli, Norwalk, hepatitis B virus, rabies virus, human cytomegalovirus, rota virus and respiratory syncytial virus F. Insulin expression in plant resulted in production of a useful vaccine against insulin dependent autoimmune diabetes. Plant virus particles expressing multiple antigens from various pathogens have been sueful as vaccines against pulmonary infections of *Pseudomonas*, malaria, HIV, hepatitis B virus and respiratory syncytial virus F.

Table 26.1. Production of biopharmaceuticals in transgenic plants.

Application	Plant	Protein
I. Anticoagulants		
Protein C pathway	Tobacco	Human protein C (serum protease)
Indirect thrombin inhibitors	Tobacco, oilseed rape, Ethiopian mustar	Human Hirudin variant 2
II. Recombinant hormones/protein		
Neutropenia	Tobacco	Human granulocyte-macrophage
Anaemia	Tobacco	Colony-stimulating factor
Antihyper analgesic by opiate activity	Thale cress, oilseed	Human enkephalins
Wound repair/control of cell proliferation	Tobacco	Human epidermis growth factor
Hepatitis C and B	Rice, turnip	Human interferon-α

Treatment		
Liver cirrhosis	Potato, tobacco	Human serum albumin
Blood subunits	Tobacco	Human hemoglobin
Collagen	Tobacco	Human homotrimeric collagen I
III. Protein/peptide inhibitors		
Cystic fibrosis, liver disease, and hemorrhage	Rice	Human a-1 antitrypsin
Trypsin inhibitor for transplantation surgery	Maize	Human aprotinin
Hypertension	Tobacco/tomato	Angiotensin-1-converting enzyme
HIV Therapies	Nicotiana bethamiana	A-trichosanthin from TMV-U1 subgenomic coat protein
IV Recombinant enzymes		
Gaucher's disease	Tobacco	Glucocerebrosidase
V Neutraceuticals		
Provitamin A deficiency	Rice	Daffodil phytoene synthase

The modern biotechnological tools demonstrated the feasibility of using a genetically engineered food as an inexpensive oral vaccine production and delivery system for diarrhoea disease. Recently clinical trials have been conducted for heat labile enterotoxin (LT-B vaccine against cholera) from *E. coli* in the form of edible vaccine. (Table 26.2.)

Table 26.2. The production of vaccines in transgenic plants.

Application	Plant	Protein
Hepatitis B	Tobacco	Recombinant HbsAg
	Tobacco	Murine hepatitis epitope
Dental caries	Tobacco	*Streptococcus mutans* surface protein SpaA
Autoimmune diabetes	Potato	*Vibrio cholera* toxin B subunit-human insulin fusion
	Potato	Glutamic acid decarboxylase
Cholera and *E. coli* diarrhea	Tobacco/ Potato	*E. coli* heat-labile enterotoxin LT-B
Cholera (oral vaccine)	Potato Rabies	*V. cholerae* toxin CtoxA and CtoxB subunit
Mucosal vaccines not requiring adjuvants	Cowpea	D2 peptide of fibronectin-binding protein B of *Staphylococcus aureus*
Diarrhoea due to Norwalk virus	Tobacco/Potato	Coat protein of Norwalk virus
Rabies	Tobacco/Spinach	Rabies virus glycoprotien
HIV	Tobacco/Bean	HIV epitope (gp120)
Rhinovirus	Blackeyad bean	Human rhinovirus epitope (HR14)
Foot and mouth	Blackeyad bean	Foot and mouth virus epitope (VP2)
Resistance of milk, dog and cat against virus	Blackeyad bean	Milk enteritis virus(VP2)
Malaria	Tobacco	Malaria B-cell epitope
Influenza	Tobacco	Hemagglutin
Cancer	Tobacco	c-Myc

In the developing countries, diarrhoea disease is a leading cause of death, especially among children and travellers. Travellers who visit these tropical areas are victim of diarrhoeal diseases because they are frequently exposed to bacterial contaminations in food, water and common places. Enterotoxigenic *E. coli*, which produce a heat Labile (LT) and heat stable (ST) enterotoxin, are the most common causes of travellers diarrhoea throughout the world. LT is comprised of six sub-units. LT-A is an enzymatically active protein which enters the epithelial cells of the gut and initiates cellular metabolic changes that lead to loss of water from cells.

LT-B has five identical enzymatically inactive proteins, which form a pentamer that binds to GM_1 gangliosides in the membranes epithelial cells. Binding of LT-B initiates transport of the active subunit inside the cells resulting in diarrhoea and any interference with binding will block the action of toxin. LT-B elicits oral immune response when given orally without any symptoms of disease.

Tobacco plants containing LT-B bacterial gene accumulated LT-B toxin and this LT-B was similar to that produced by bacteria. When mice were orally inoculated with tobacco derived LT-B, both serum and mucosal antibodies were induced. In another experiment; potato plants were transformed to produce LT-B and upon feeding the transgenic tubers directly to mice, serum antibodies were induced. Potato and production of serum and mucosal antibodies was confirmed.

Norwalk Virus

Norwalk virus is the causative agent of acute epidemic gastroenteritis in humans. Recent advances in cloning the Norwalk virus genome and expression of the capsid protein in insect cell cultures leave facilitated the study of the virus and the development of candidate vaccines for oral immunization. Norwalk virus capsid protein (NVCP) gene has been transferred and expressed in tobacco leaves and potato tubers. Partially purified antigen from tobacco or potato tubers is used for vaccination.

Hepatitis-B surface antigen (HbsAg)

Hepatitis is the single most important cause of viremia in humans and currently there are about 300 million carriers all over the world. The world-wide problem of infection and its association with chronic liver disease has necessitated the development of an effective vaccine. In many parts of the developing world, the expense of immunization programme limits the usage of the currently available serum or yeast could offer as a relatively low-cost method. The transfer of hepatitis-B surface antigen gene in tobacco, expression of recombinant gene in tobacco followed by partial purification of protein from the plant. When this protein was injected into mice, it elicited antibody response similar to that obtained with yeast derived commercially available vaccine. This is clear that gene product obtained from two different organisms has same property and transgenic plants can be used as source of antibodies.

III. PRODUCTS ON THE MARKET

1. Avidin

Avidin is a glycoprotein found in avian, reptilian and amphibian egg white. It is used primarily as a diagnostic reagent. The protein is composed of four identical submits, each 128 amino acids long. The usual source of commercial quantities of avidin is from chicken egg white but the resultant product is relatively expensive due to the cost of maintaining live animals. Avidin gene has been cloned and transferred in corn. The avidin gene sequence had been optimized for expression in corn and corn has produced avidin similar to chicken egg white. The resultant avidin had properties almost identical to those of avidin from chicken egg white. Both the corn-derived and the chicken egg white derived avidin were glycosylated. Corn-derived avidin expression levels over 20% total soluble protein have been observed. Technology has been developed to purify this protein and the product is sold by Sigma-Aldrich Company.

2. β-glucoronidase

GUS (B-glucoronidase) is a homotetrameric hydrolase (68 kDa/unit) that cleaves β-linked terminal glucoronic acids in mono and oligo-saccharides and phenols. GUS is widely used as a visual marker in transgenic plant research. It was first reported to be produced commercially in transgenic corn, where its properties were compared with GUS extracted from *Escherichia coli*, the original source of the protein. The sequence of two genes and most properties were similar, with the exception of the most N-terminal amino acids. It was presumed that removal of the terminal amino acid in plants was a result of the N-terminal rule in action as a result of the change in sequence. Corn-derived GUS is also marketed by Sigma-Aldrich.

3. Trypsin

Although both plant-derived GUS and avidin are protein products that have been commercially produced for last several years, their markets have been quite small. These are conceptual products showing feasibility of producing a foreign protein in a crop plant. Maize-derived trypsin, a more recent introduction, has a significant market potential. Trypsin is a protease that is used in a variety of commercial applications including the processing of some bio-pharmaceuticals. The availability of maize-derived bovine trypsin helps to supply the growing market of animal-free reagents. This market is being exploded by a desire by pharmaceutical companies to eliminate animal-source materials and reduce the fears of contamination of products by mammalian viruses and prions.

Trypsin is an aggressive proteolytic enzyme. Although trypsin has been expressed in a variety of recombinant systems, more of these systems has been demonstrated to be commercially viable on a large-scale.

Expression of this enzyme at commercial levels in maize was possible only by expressing the enzyme in an inactive zymogen form. Although zymogen gene was put into plants, the active enzyme was recovered in extracts from maize seeds. Zymograms and Western blots of the purified trypsin suggested that almost all of the trypsinogen was converted to trypsin in the seed, either prior to, or immediately upon extraction of flour from the seeds.

The purified maize-derived tyrpsin had a specific activity of 175 unit/mg protein, which was comparable to that of bovive pancreas-derived trypsin (1664/mg protein). The maize-derived trypsin was similar to bovine trypsin.

Table 26.3. Molecular farming products likely to hit the market in near future

Product	*Company*	*Use*
Trypsin	*Prodigene*	*Pharmaceutical intermediate*
GUS	*Prodigene*	*Diagnostic reagent*
Avidin	*Prodigene*	*Immunological reagent*
Aprotinin	*Prodigene, Large Scopale Biology*	*Reduce SIR and bleeding, promote would closure.*
Collagen	*Prodigene, Medicago, Meristem Therapeutics*	*Gel caps, skin sealent, sear treatment*
Lipase	*Mesistem Therapeutics*	*Exocrivie pancreatic insufficiency, cystic fibrosis*
Lacto ferin	*Ventria, Meristem Therapeutics*	*Natural defense protein against infections, Iron repository*
Lysozyme	*Ventria*	*Antiviral, anti-bacterial, anti-fungal*
Brazzein	*Prodigene*	*Natural protein sweetner*
α-carries MAb	*Planet Biotechnology*	*Prevention of dental carries*
α -Herper MAb	*Epicyte Pharmaceutical*	*Prevention of herpes transmission*

Aprotinin

Aprotinin is a protein that inhibits serine proteases (including trypsin, chymotrypsin and pepsin). This activity is known to modulate and lesser the systemic inflammatory response (SIR) associated with cardiopulmonary bypass surgery. The use of aprotinin results into a decreased need for blood transfusions, reduced bleeding and decreased re-exploration for bleeding (Table 26.3).

Current material supplied to the market is native aprotinin extracted from bovine lungs. With growing concern regarding bovine spongiform enceplalopathy (BSE) and other animal pathogens, a high market price in surgery and wound healing, and limitations on natural sources of aprotinin, a non-animal derived, lower cost and reliable source adds significant value to the user. Aprotinin can produced in crop plants without any limitations and plants are not affected by animal viruses and prions.

Recombinant bovine aprotinin from transgenic corn seed was produced in 1999 - using particle bombardment technique.

Collagen

Collagen is a structural protein currently obtained from hooves and connective tissue of animals. Vast quantities of collagen are consumed throughout the world in the form of gelatin. Based on their structural roles and compatibility within the body, collagen and gelatin are commonly used biomaterials in the medical, pharmaceutical and cosmetic industries. Collagen is used commercially in the areas of arterial sealants, bone grafts, corneal implants, drug delivery, tissue engineering (artificial skin and cartilage), and as a vascoelastic supplement. Gelatin is used in the stabilization and delivery of vaccines and drugs, in capsules and soft gels, nutraceuticals, and as plasma expanders.

The first report of human collagen produced in plants was published in 2000. The cDNAs for complete human pro a 1(I) collagen chain was inserted into tobacco using *Agrobacterium* and *nptz* as the selectable marker gene. The resultant protein was organized into a triple helix, which was somewhat surprising since plant do not contain the specific post-translational machinery needed for collagen assembly.

Human gastric lipase

Certain humans lack gastric lipase (a protein), the condition is known as exocrine pancreatic insufficiency (PI). The person cannot digest food lipids in the absence of lipase. The disease mainly affects patients who have cystic fibrosis or who suffer from pancreatic disease. The current supply of gastric lipase is from porcine pancreatic tissue. There are several companies (table) attempting to produce lipase in transgenic tobacco.

Human Lactoferin

Human lactoferin is a natural defence iron-binding milk protein that supposedly possesses antibacterial, anti-fungal, anti-viral, anti-neoplastic and anti-inflammatory properties. Each natural lactoferin molecule has two iron-binding domains, which reversely bind iron. Recombinant human lactoferin has been produced in *Aspergillus* sp and tobacco. Attempts are being made to produce lactoferin at commercial level in rice and maize seeds.

Production of Hirudin in *Brassica napus*

Hirudin is a naturally occurring anti coagulant protein produced in the salivary glands of medicinal leeches (*Hirudo medicinalis*) and secreted to facilitate feeding. Since its discovery almost 50 years ago, it has been extensively studied. Hirudin possesses a number of desirable properties which advocate its use as an therapeutic pharmaceutical. It is extremely specific and potent inhibitor of thrombin, the last enzyme in the blood agglutination cascade. It is also rapidly cleared from the body, exhibits low toxicity and has low immunogenicity. A small number of closely related isoforms of hirudin have been isolated all of which show strict conservation for six cysteine residues. These residues participate in the formation of three disulfide bridges whose precise pairing is necessary for protein activity. Although the native protein is sulphated at the Tyr-63 position, recombinant non-sulphated hirudin exhibits significant activity.

Functional hirudin has been produced in both bacterial and yeast systems. Looking to high demand (several lacs kilogram of protein), it is a good candidate for production in a plant based system. After tobacco, *Brassica napus* can be transformed easily with *Agrobacterium*. Values obtained for the specific activity of *B. napus*-derived hirudin are equivalent to those reported for recombinant hirudin produced in yeast systems.

Production of Cyclodextrins

Cyclodextrins are cyclic oligosaccharides containing six, seven or eight glucose molecules in α, β and γ - cyclodextrins (CDs), respectively. CDs can be obtained from starch by the action of cyclodextrin glucosyl transferees (CG Tase) enzyme, and are useful for pharmaceutical delivery systems, flavour and odour enhancement and for removal of undesirable compounds (e.g. caffeine) from foods. Cyclodextrins can be regarded as torus-like molecules. As a consequence of chain conformation of the sugar units, all secondary hydroxyl groups (at C_2, C_3) are located at the inner side of the tourus, while all the primary hydroxyl groups at C_6 are located on the outer side. As a result the external faces are hydrophilic, making the cyclodextrin water-soluble. In contrast, the cavities of CDs are hydrophobic, since they are lined by hydrogen atoms of C_3 and C_5, and ether-like oxygen's. These matrices allow complexation with a variety of hydrophobic compounds like – drugs, oils, aromas etc.

Tissue specific expression of CG Tase in potato tuber and its targeting to plastids for action on stored starch was achieved through the use of a chimeric gene construct consisting of the followings :

(*i*) A patatin promoter for tuber specific expression,

(*ii*) A DNA sequence encoding transit peptide of the small submit (SSU) of ribulose biphosphate carboxylase,

(*iii*) CG Tase structural gene from Klebsiella and,

(*iv*) 3' region of the gene encoding nopaline synthase (nos.)

Therefore, the technology is ready for biofarming of bioactive molecules using crop plants.

QUESTIONS

1. What do you understand by molecular farming ? Give suitable examples where molecular farming has been successful.
2. Write notes : (*a*) Edible vaccines (*b*) Avidin (*c*) Trypsin (*d*) Cyclodextrins.

MICROARRAYS

CHAPTER 27

Documenting gene expression on a genome- wide scale is the analysis of transcriptome. The transcriptome is the complete set of transcripts and their relative levels of expression in a particular cell or tissue type under defined conditions. Several technologies have been developed for parallel analysis of the expression of thousands of genes, among which cDNA microarrays and oligonucleotide arrays are the best known. These methods can provide differences in expression levels in two different tissues but unable to give data on the absolute levels of expression. A third method called Serial Analysis of Gene Expression (SAGE) relies on counting of sequences tags to estimate absolute transcripts levels. This method is not suitable for large number of samples.

The basic procedure of microarray analysis is to deposit a very small amount of DNA corresponding to each one of a collection of thousands of genes ("the probes") on to a few square centimetre array surface, and then to challenge these probes by hybridization to a "target" of mRNA that has been labelled with a fluorescent dye. The amount of target that sticks to each spot of probe is proportional to the abundance of the transcripts in the sample, and is detected as the intensity of fluorescent signal. A change in abundance is measured as an increase or decrease in the signal, relative either to the signal from a control reference sample or to the signals from the other probes on the array. Estimates of the absolute quantity of each transcript in each sample are only approximate, so most inferences involve comparison of expression levels between treatments.

I. TYPE OF MICROARRAYS

There are currently three types of platform: cDNA, long oligonucleotide microarrays and short oligonucleotide gene chips.

1. cDNA microarrays

The technology of cDNA microarrays is a conceptually simple method for monitoring the relative levels of expression of thousands of genes simultaneously. Polymerase Chain Reaction (PCR) amplified cDNA fragments (Expression Sequence Tags: ESTs) are spotted at high density (10-50 spots per mm^2) onto a microscope slide or filter paper and probed against a fluorescently labelled target. The platform is very flexible, since users print their own microarrays and are free to add or subtract clones as they refine their collection of ESTs. Since cDNA libraries are tissue specific, multiple libraries must be sampled to produce an unbiased set of clones that is representative of genes expressed under different conditions.

The most expensive and time-consuming step in cDNA microarray analysis is the amplification of EST set. While each spot on a microarray should contain just long of DNA, it is delivered as few nanolitres of 5mg/ml DNA solution. PCR amplified DNA fragments have been found to yield much stronger signals than concentrated plasmid DNA. PCR amplification is performed in a multi-well plate and the product is precipitated and re-suspended in a special spotting solution. Clones in the 1-2 kb range are used.

Once amplified, fragments may be spotted on to either a coated glass microscope slide or a nitrocellulose or nylon membrane. Membranes are most suited to applications where radioactivity is used to label cDNA, while glass only supports fluorescence-based detection. Printing is performed with a robot that picks up samples of DNA from a multi-well plate and deposits aliquots sequentially onto a field of 100 slides. The spacing between spot centres is specified from 120-250μm according to density required. The entire microarray usually covers an area 2.5x 5.0 cm. However, shorter grids can be printed if sample size is small.

Commonly used print heads have from 4 to 32 individual printing pins spaced several millimetres apart. The pins pick up DNA solution from nearby wells on the master plate. At a rate of a few deposits per second, it takes up to 2 days for the robots to prepare 100 microarrays containing at least 5000 clones. Since only a few micro-litres of PCR products are used in each print run, each amplified batch of clones is sufficient for multiple print runs and hence at least 1000 slides (Fig. 27.1).

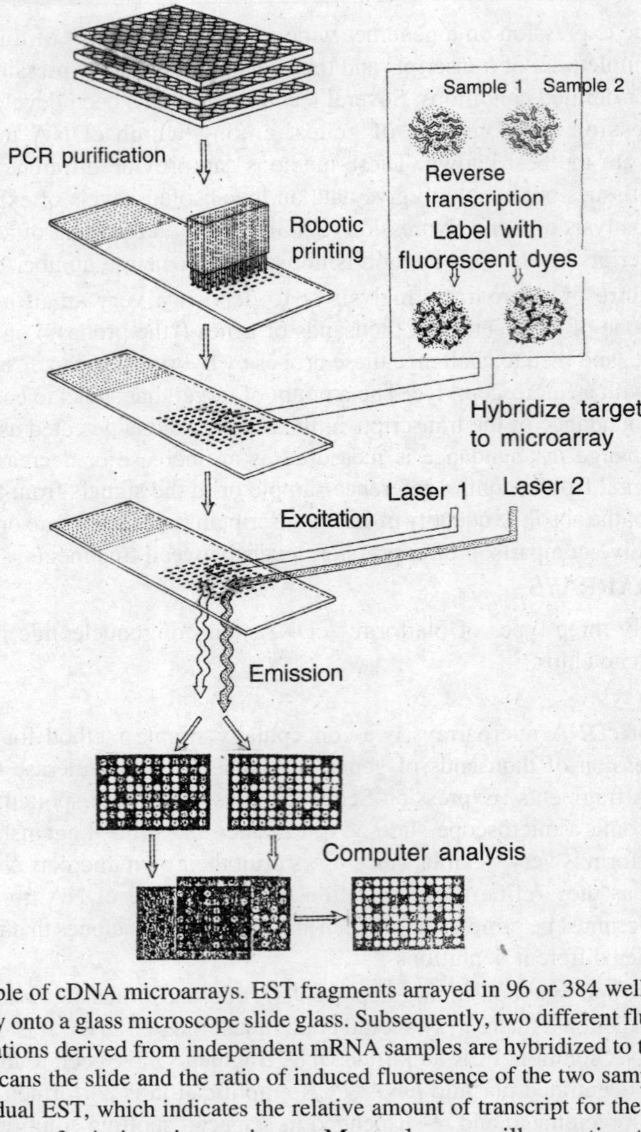

Fig. 27.1. Principle of cDNA microarrays. EST fragments arrayed in 96 or 384 well plates are spotted at high density onto a glass microscope slide glass. Subsequently, two different fluorescently labeled cDNA populations derived from independent mRNA samples are hybridized to the array. After washing, a laser scans the slide and the ratio of induced fluoresençe of the two samples is calculated for each individual EST, which indicates the relative amount of transcript for the EST in the samples.

Several types of printing pins are in use. Most rely on capillary action to transfer solution from the pin to the slide. This method has the advantage of controlling spot size by adjusting the duration of time the pin touches the slide. The smaller the size of spot, higher the density of microarrays. The pins are expensive and fragile and are also prone to clogging. An alternative method is the pin and loop system, in which the DNA is picked up as a meniscus in a small loop, and a pin of a given diameter literally stamps the solution onto the slide. An ink-jet printer promises highly reproducible and efficient spot deposition (Fig. 27.2)

2. Long- oligonucleotide microarrays

Long –oligonucleotides in the range of 50-70 bases have gained in popularity as probes since the late 1990s. The main differences relative to cDNA arrays are that time consuming PCR amplification of thousands of clones is not required (instead oligonucleotides are purchased from a company), and a single strand of DNA of uniform length is deposited thereby increasing greater control over hybridization specificity. Once the strand of interest has been identified in a cDNA or genomic sequence, freely available algorithms (e.g. oligoarray) can be used to design long oligonucleotides with the desired base composition, which promotes hybridization at the same temperature for all probes. Checks are also performed to eliminate the potential for cross-hybridization, and in general over 90% of genes yield suitable probes.

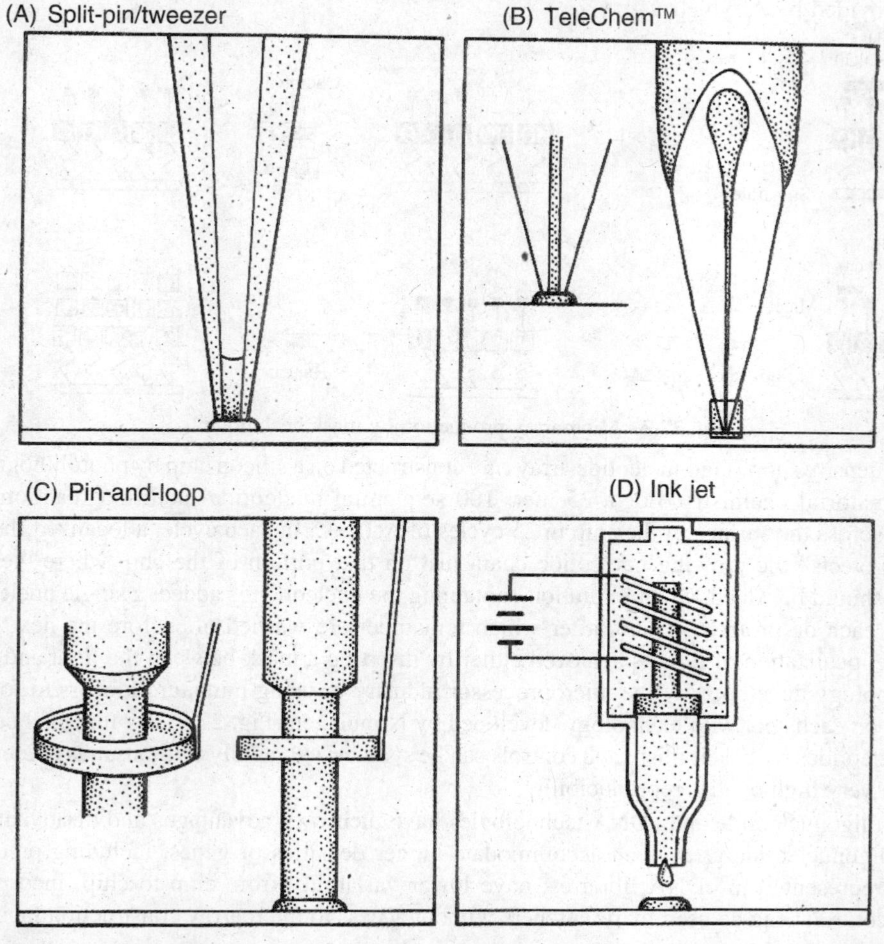

Fig. 27.2. Types of printing pins. (A) Tweezer or split-pin designs transfer low nanoliter amounts of DNA to the array by capillary action as the tip strikes the solid surface. (B) TeleChem™ tips and pins apply small droplets by contact between the pin and substrate. (C) The pin-and-loop design picks up the DNA in a small loop, and a pin stamps solution on a slide at a uniform density. (D) Ink jets spray picoliter droplets of liquid under pressure.

Oligonucleotide can be synthesized *in situ* on the slide surface. Agilent Technologies will synthesize arrays containing up to 44,000 long oligonucleotides (60-mers) using a combination of ink-jet printing to deliver the appropriate nucleotide in 60 consecutive steps and phosphoramitide chemistry to ensure the incorporation of a single base in each step.

3. Short-oligonucleotide gene chips

This approach uses short oligonucleotide microarrays, originally developed and marketed under the trademark Affymetrix Genechip®. The unit of hybridization here is a series of 25-mer oligonucleotides designed by a computer algorithm to represent known or predicted open reading frames. Each gene is represented by between 10 and 20 different oligonucleotides to control for variation, in hybridization efficiency due to factors such as GC content. The possibility of cross-hybridization with similar short sequences in transcripts other than the one being probed is controlled for by including a mismatch control adjacent to each oligonucleotide that has a single base change at the centre of oligonucleotide. With Affymetrix chips, the level of expression of each gene is calculated as an 'average' of the differences between the perfect match and mismatch using a software provided by Affymetrix.

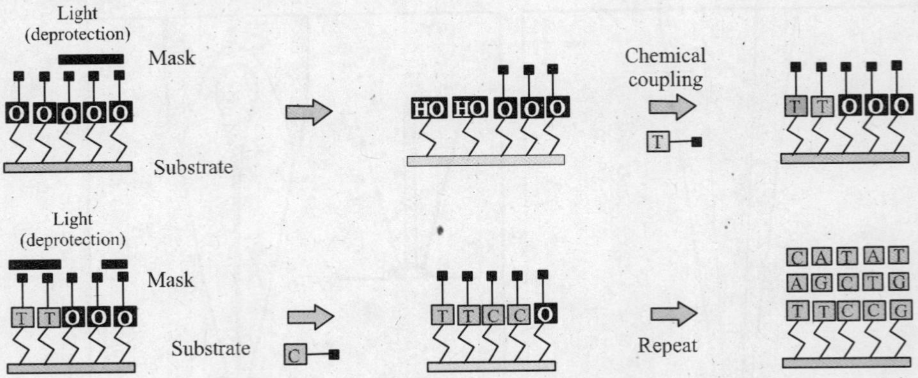

Fig. 27.2. Microarray production by mask bechnique

High-density short oligonucleotide arrays are constructed on a silicon chip by photolithography and combinatorial chemistry. For a 25 mer, 100 sequential nucleotide- addition reactions are performed across the surface of the chip in 25 cycles of A,T,G,C. In each cycle, a localized flash of light "de protects" the growing nucleotide chain just on that portion of the chip where the next nucleotide should be added, when a solution containing the nucleotide is added, a single nucleotide adds on to each de protected chain, after which remainder are washed off before the next cycle begins. The localization of light is achieved either by inserting a mask between the light and chip, using technology developed by the microprocessor industry or using miniature mirrors to focus a laser beam on each spot with technology developed by Nimblegen (Fig. 27.3). By this method lacs of oligonucleotides with their mismatch controls can be synthesized rapidly on thousands of identical chips, with very high quality reproducibility.

Both oligonucleotide and cDNA technologies have their own advantages and disadvantages. In brief, oligonucleotide arrays can accommodate higher densities of genes, including predicted genes not represented in cDNA libraries; have lower variability from chip to chip; incorporate mismatch controls; can be used by researchers without access to microarray construction facilities; and lend themselves to data comparison across research groups.

As compared to this, cDNA methods can be applied to any organism, irrespective of the status of genome sequencing efforts; are considerably cheaper once the platform has been developed; and are more flexible in design. The cDNA technology uses kilobases rather than tens of bases for hybridization; this avoids cross-hybridization artifacts.

II. LABELLING AND HYBRIDIZATION OF cDNA ON TO CHIPS

Once the DNA chips with microarray are produced, the next step is hybridization of sample DNA to these DNA chips. Mixtures of DNA or RNA isolated from biological systems (test materials) are labelled enzymatically by incorporating nucleotides bearing reporter tags and hybridized to microarrays. Hybridization reactions yields heteroduplexes (hybridized double helix) between DNA with fluorescent probe and the complementary target sequences on the chip. Since the each target sequence is chemically homogenous and occupies a known position, the identity and quantity of each component in unknown sample can be determined by measuring the intensity of the fluorescence of each position on the microarray.

Hybridization is performed in special humidified chambers. Visualizing the hybridized target on a microarray can be performed by laser-induced fluorescence imaging using either a confocal detector or a CCD (close circuit digital) camera. This allows collection of quantitative data at a fast speed.

III. APPLICATIONS OF MICROARRAYS

Earlier it was believed that microarray can replace Sanger's dideoxy method of sequencing. This was described as the method of sequencing by hybridization (SBH). However, this technique could not be utilize successfully since uniform hybridization signals are not available for a large number of oligonucleotides in parallel and due to sequence dependent variability in heteroduplex formation.

1. Detection of single nucleotide polymorphism (SNPs)

Identification of single base mismatches in the known DNA sequences of a genome can be achieved resequencing DNA segments using SBH. In this case, one need to distinguish a perfect match from a single base mismatch and precise sequencing is not required.

2. Diagnostic chips for human diseases

Now information about known alleles of a gene for particular character is available. Using this information chips can be prepared to detect these polymorphism of that gene e.g., diagnostic chips have been prepared to detect mutant alleles in cystic fibrosis, cancer susceptible gene, and beta-globin. For cystic fibrosis, one microarray containing 428 variations was designed to detect mutations in exon-11 of this gene. Another microarray contained 1480 features for detection of known deletions, insertions, or base substitutions was designed. DNA obtained from patients with already characterized mutant DNA gene, give positive results. This way disease can be diagnosed.

3. Genomic mismatch scanning

Basically this technique utilizes basic property of DNA *i.e.* Complementary sequences make duplex. Homologous segments are identified by the formation of heteroduplexes that are free of any mismatches.

4. DNA chips and functional genomics

(a) Transcript monitoring

Measurement of level of transcripts for thousands of genes (expression of a genome at particular stage) in parallel is one of the most wide spread uses of DNA chip technology. Both cDNA and oligonucleotide microarray work well for estimating the level of transcripts. This way two or more samples can be compared, for comparison purpose (*i*) a limited number of genes involved in a biological activity may be selected, e.g. comparison of two different clones, one drug producing another non producing; (*ii*) clones from a library may be used as has been done in human genome for monitoring 1000 genes; (*iii*) a complete expressed sequence content of an organism is used e.g. as in yeast.

(b) DNA microarrays and proteomics

DNA sequencing by microarray can be used to derive information about proteins. Proteins are gene products. Hybrid proteins are used to activate the transcription of reporter genes. This work involves sequencing of hybrid system.

QUESTIONS

1. What are microarrays ? How do oligonucleotide and cDNA based DNA chips differ? Discuss the different methods for the production of microarrays.
2. How do DNA chips and microarray help in variety of molecular biology and biotechnology applications.

METABOLIC ENGINEERING

CHAPTER 28

Metabolic engineering in plants involves the modification of endogenous pathways to increase flux towards particular desirable molecules. In some cases the aim is to enhance the production of a natural product, whereas in others it is to synthesize a novel compound or macro-molecule. Over the past few years, significant advances have been made in metabolic engineering through the application of genomics and proteomics technologies to elucidate and characterize metabolic pathways in a complete manner, rather than on a step-by-step basis. This has been supported by direct studies at the level of the metabolome (defined as a set metabolites synthesized by an organism constitute its metabolome). Further work has been carried out on the modelling of metabolic pathways on a genomic scale. Such studies have shown that metabolic pathways are controlled at multiple levels and any form of small change can have wide-ranging effects at the whole -system level. As a result, it has been realized that the manipulation of single genes is of only limited value in metabolic engineering, and attention has shifted towards more complex and sophisticated strategies in which several steps in a given pathway are modified simultaneously to achieve optimal flux. In this chapter, we look at recent examples of single and multiple gene metabolic engineering in primary and secondary metabolism that demonstrate how this new knowledge of metabolic systems is being applied in plants.

I. STRATEGIES FOR METABOLIC ENGINEERING IN PLANTS

There are three basic goals of metabolic engineering in plants: the production of more of a specific desired compound, the production of a novel compound (i.e. a molecule that is produced in nature, but not usually in the host plant), or a completely novel compound. Strategies for achieving these goals include the engineering of single steps in a pathway to increase or decrease metabolic flux to target compounds, to block competitive pathways or to introduce short cuts that divert metabolic flux in a particular way (Figure 28.1a). However, this approach has only limited value, because the effects of modulating single enzymatic steps are often absorbed by the system in an attempt to restore homeostasis. Targeting multiple steps in the same pathway could help to control metabolic flux in a more predictable manner. This might involve upregulating several consecutive enzymes in a pathway, upregulating enzymes in one pathway while suppressing those in another competing pathway, or using regulatory genes to establish multipoint control over one or more pathways in the cell (Figure 28.1b). Examples are provided below to show how such strategies are applied to manipulate the pro-duction of different classes of compound.

II. ENGINEERING PRIMARY METABOLIC PATHWAYS

(a) Carbohydrate metabolism

Carbohydrate metabolism in plants centres on the rever-sible conversion of sugars, the direct products of photo-synthesis, into storage and structural carbohydrates such as starch and cellulose. Starch is a storage carbohydrate that accumulates transiently in leaves and stably in seeds, tubers and roots. It is a staple source of dietary carbohy-drate for animals and also has a large number of industrial uses. Metabolic engineering has been used in an attempt to increase starch yields and to modify the properties of starch by changing the relative proportions of its two components, amylose and amylopectin. There has also been some interest in producing novel starches using- bacterial enzymes to change the nature and frequency of branching. Recent examples of metabolic engineering to increase the quantity of starch include over expressing the potato sucrose transporters in transgenic pea seeds to accelerate sugar uptake and boosting the adenylate pool in potato tubers, resulting in a 60% increase in the level of starch in combination with an increased tuber yield. Smidansky and colleagues have shown that wheat and rice plants transformed with a modified form of the maize gene for ADP-glucose pyrophosphorylase (*shrunken2*) utilize their spare capacity for starch biosynthesis and produce 40% and 20% more seed biomass, respectively, than wildtype controls. Many of the enzymes that control starch synthesis and branching have been modulated in potatoes to produce highamylose starch (e.g. for improved frying or for industrial use as gelling agents and thickeners) and highamylopectin starch (e.g. for improved freeze thaw characteristics or for improved paper quality or adhesive manufacture). More recently, similar experiments have been carried out in other starchproducing crops. For example, a lowamylose rice grain has been produced by expressing an antisense *Waxy* gene and rice grains with modified amylopectin have been produced by anti-sense inhibition of the gene encoding isoamylase. Several novel strategies for starch metabolic engineering have been explored, including the use of starch-binding domains to target recombinant enzymes to starch granules during starch biosynthesis.

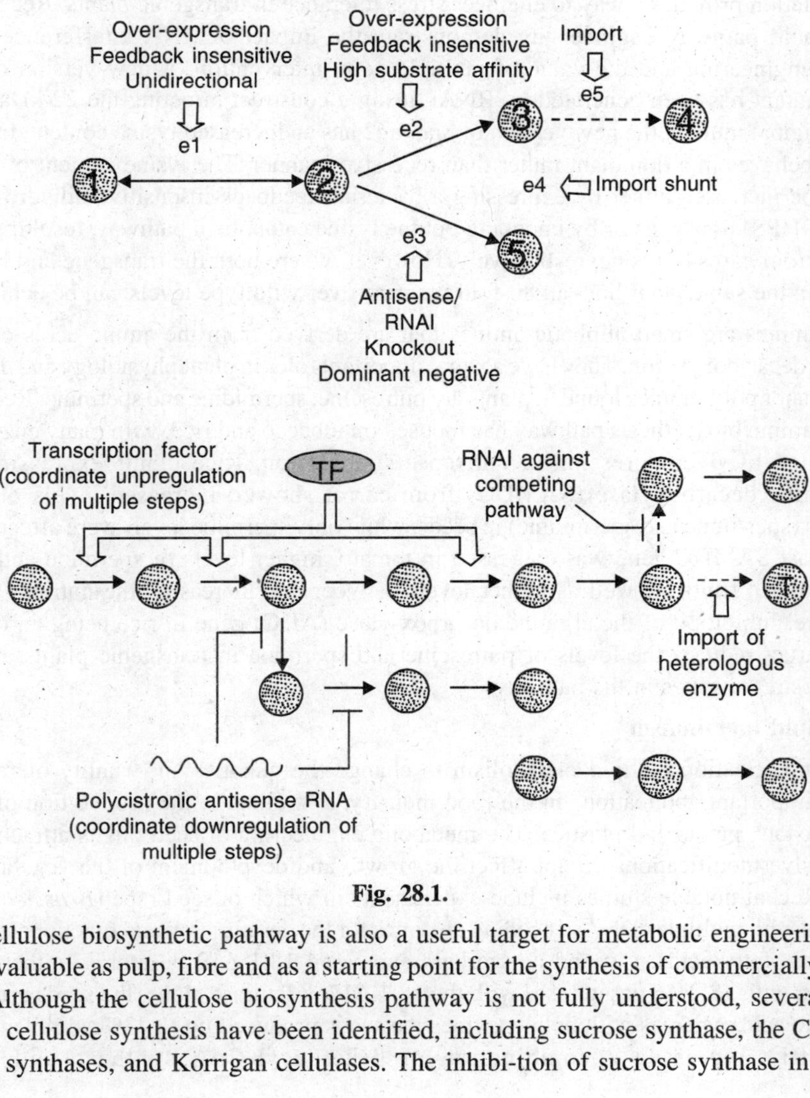

Fig. 28.1.

The cellulose biosynthetic pathway is also a useful target for metabolic engineering because cellulose is valuable as pulp, fibre and as a starting point for the synthesis of commercially important polymers. Although the cellulose biosynthesis pathway is not fully understood, several enzymes required for cellulose synthesis have been identified, including sucrose synthase, the CesA family of cellulose synthases, and Korrigan cellulases. The inhibi-tion of sucrose synthase in cotton has been shown recently to inhibit fibre formation, whereas antisense and dominant negative experiments involving CesA genes in *Arabidopsis* have also demonstrated roles for several of the CesA proteins in primary and secondary cell-wall formation.

The fructans are another group of storage carbohydrates that are important targets for metabolic engineering; as they have a similar texture to fats and can substitute for them in many foods, yet contain far fewer calories. At the present time the only commercially available fructan is inulin from chicory, but bacterial and plant enzymes for fructan biosynthesis have been introduced into several crops to increase fructan levels to levels suitable for extraction. Another use envisaged for fructans is the protection of plants from abiotic stress. This applica-tion is shared with sugars such as trehalose, as well as other small-molecule compounds like glycine betaine, mannitol and sorbitol, all of which have been subject to metabolic engineering to produce improved plant varieties.

(b) Amino acid and polyamine metabolism

Amino acid synthesis pathways have been targets for metabolic engineering predominantly to increase the abundance of essential amino acids (lysine, threonine, methionine and tryptophan) in food and feed crops. Other amino acids, such as proline, are important in plant stress responses and their modulation provides a way to engineer stress tolerance in transgenic plants. Recent examples of amino acid pathway engineer-ing demonstrate the impact of RNA interference (RNAi) on metabolic engineering and the value of multiple point intervention. A new version of the maize *opaque2* mutant has been generated by RNAi, using a construct targeting the 22 kDa zein genes. Like the original mutant, the new version of *opaque2* has an increased lysine content, but unlike the original it behaves in a dominant rather than recessive manner. The lysine content of *Arabidopsis* seeds can be increased either by expressing a bacterial, feedback-insensitive dihydro-dipicolinate synthase (DHPS) transgene or by knocking out the lysine catabolism pathway, resulting in twelve - fold or fivefold gains in lysine, respectively. However, where both the transgene and knockout are combined in the same plant line, an 80-fold increase over wild type levels can be achieved.

Polyamines are small aliphatic amines that are derived from the amino acids ornithine and arginine by decar-boxylation. They have several important roles in plant physiology and development. The three major poly-amines found in plants are putrescine, spermidine and spermine. Recent analysis of the polyamine biosynthesis pathway has focused on tobacco and rice, with many enzymatic steps being studied by overexpression or antisense suppression. Rice plants expressing S-adenosylmethionine decarboxylase (SAMDC) from carrot showed increased levels of the higher polyamines (sper-mine and spermidine) in seeds, while only spermine levels were affected in leaves. When a yeast SAMDC gene was expressed in tomato, higher levels of spermine and spermidine were detected in fruits, as well as higher levels of lycopene, increasing the nutrient value of the tomato juice. Inhibition of the arginine decarbox-ylase (ADC) gene in rice using an oat antisense ADC construct reduces the levels of putrescine and spermine in transgenic plants, but does not effect downstream genes in the pathway.

(c) Lipid metabolism

The modification of lipid metabolism to change the quantity and quality of fatty acids in plants has important applications in the food industry, as well as in the produc-tion of detergents, fuels, lubricants, paints and plastics. The metabolic engineering of seed oils is attractive, because even extensive modifications do not affect the growth and development of the vegetative parts of the plant. Recent notable studies include experiments in which oilseed rape (*Brassica napus*) was transformed with multiple cDNAs encoding enzymes for the desaturation and functional modification of long-chain fatty acids. Liu and co workers expressed 18:1 Δ12 desaturase either alone or in combination with 18:2 Δ6 desaturase. With the 18:1 Δ12 desaturase alone, the seeds produced up to 46% linoleic acid, whereas with both enzymes the seeds produced up to 43% γ-linolenic acid. Han et al. (2001) expressed a combination of β-ketoacyl-CoA synthase and 22:1 acyl-

CoA:lysophosphatidic acid acyl transferase in oilseed rape, yielding 60% erucic acid. More recently, the α-linolenic acid content of rice grains has been increased tenfold by expressing the soybean FAD3 gene encoding Ω3 fatty acid desaturase.

III. ENGINEERING SECONDARY METABOLIC PATHWAYS

(a) Alkaloids

The alkaloids are perhaps the largest group of secondary products synthesized by plants and the pathways, enzymes and regulatory genes involved in their synthesis have been extensively reviewed (see chapter 17). Several distinct types of alkaloids can be distinguished, including the isoquinoline, indole and tropane classes. It has been demonstrated that antisense suppression of the berberine bridge enzyme in poppy cells, reduced the amount of benzophenanthridine alkaloids but increased the levels of several amino acids. There have been many attempts to modulate the indole alkaloid biosynthetic pathway, which produces important pharmaceutical compounds such as the anticancer drugs vinblastine and vincristine.

(b) Terpenoids

Terpenoids (isoprenoids) are structurally diverse polymers constructed from the basic building block isoprene. They are classified according to the number of isoprene units they contain, which increase by fives. Thus, hemiterpenes contain five isoprene units, monoterpenes contain 10, sesquiterpenes contain 15 and diterpenes contain 20. Carotenoids, another important group of plant secondary metabolites, are terpenoids with 40 isoprene units. The largest polymers contain thousands of units and include natural products such as rubber. Despite their diversity, all terpenoids are synthesized from the common precursors dimethylallyl pyrophosphate and isopentyl pyrophosphate. This occurs through two distinct pathways, the mevalo-nate-independent pathway and the mevalonate path-way, both of which have been targets for metabolic engineering.

Thus far, there have been few examples of monoterpenoid metabolic engineering. In one study, petunia was transformed with the (*S*)-limonene synthase gene from *Clarkia breweri* resulting in the production of linalool, which repels aphids. In another, the same gene was introduced into tomato, under the control of a fruitspecific promoter, resulting in the accumulation of linalool in tomato fruits.

Much more attention has been paid to the carotenoid metabolic pathway, which produces pigments with roles in light harvesting and photoprotection, and which form vital components of human and animal diets. In plants, the first step in carotenoid biosynthesis is the coupling of two geranyl phosphate molecules to yield geranylgeranyl pyrophosphate (GGPP). This is converted into phytoene by the enzyme phytoene synthase and then to β-carotene by the enzyme phytoene desaturase. Further desaturation results in the formation of lycopene, which is converted either into δ-carotene (provitamin A) by the enzyme β–cyclase, or into δ-carotene and then δ-carotene by consecutive enzymatic steps catalyzed by ε-cyclase and β-cyclase.

The principal example of carotenoid metabolic engineer-ing in plants is of course the synthesis of β-carotene in rice endosperm (Golden Rice), which normally accumulates GGPP but lacks the subsequent enzymes in the pathway. Since publication of the original reports, further 'Golden Rice' lines have been generated containing only two foreign enzymes (the daffodil phytoene synthase and a bacterial phytoene desaturase (see chapter-24). Hoa and colleagues (2003) reported the production of indica Golden Rice varieties that are amenable to deregulation and also assure freedom to operate in developing countries due to the removal of constraints brought about by intellectual property and regulatory issues.

Transgenic tomatoes have also been described expressing phytoene desaturase or phytoene synthase. In the former case, there was no overall increase in carotenoid content, although l3-carotene levels were three times higher than normal; in the latter case, the total carotenoid content was up to fourfold normal levels.

(c) Flavonoids

The flavonoids are phenolic compounds derived ultimately from phenylalanine, and thus share a common metabolic source with lignins and lignans. The first committed step in flavonoid biosynthesis is the conversion of the precursor 4-coumaroyl-CoA into chalcone by the enzyme chalcone synthase. Chalcone is then derivative in a series of enzymatic steps to produce a variety of molecules that act as pigments, defence chemicals (phytoalexins) and regulatory molecules. Flavonoids have been a favourite target for metabolic engineering since the early 1990s, because the result of different modifications can be determined through alterations in flower colour. As for other pathways, the number of experiments involving multiple gene transfer strategies is beginning to increase. A recent example is the stacking of dihydroflavonol 4-reductase and anthocyanidin synthase transgenes in forsythia by sequential transformation. Increasing the levels of flavonoids in food plants can provide health benefits, as these molecules often have antioxidant activity. This has been demonstrated in the case of tomatoes by over expressing the petunia gene for chalcone isomerase, leading to an 80-fold increase in the flavonoid content of the tomato peel and a corresponding 20-fold increase in the flavonoid level in tomato paste. In addition, chalcone synthase and flavonol synthase transgenes were found to act synergistically to upregulate flavonol biosynthesis significantly in the flesh of tomato fruits.

(d) Lignins

Lignins, like flavonoids, are derived from phenylalanine and can be classed as phenylpropanoid compounds. They are complex polymers composed of caffeic, ferulic and sinapic acid monomers joined together and to the cell wall. We will mention here a few notable examples that demonstrate contemporary metabolic engineering approaches. One interesting example is the use of the LIM regulatory protein from tobacco to control several of the metabolic enzymes involved in the lignification pathway coordinately. Similarly, two Myb-like transcription factors have been isolated

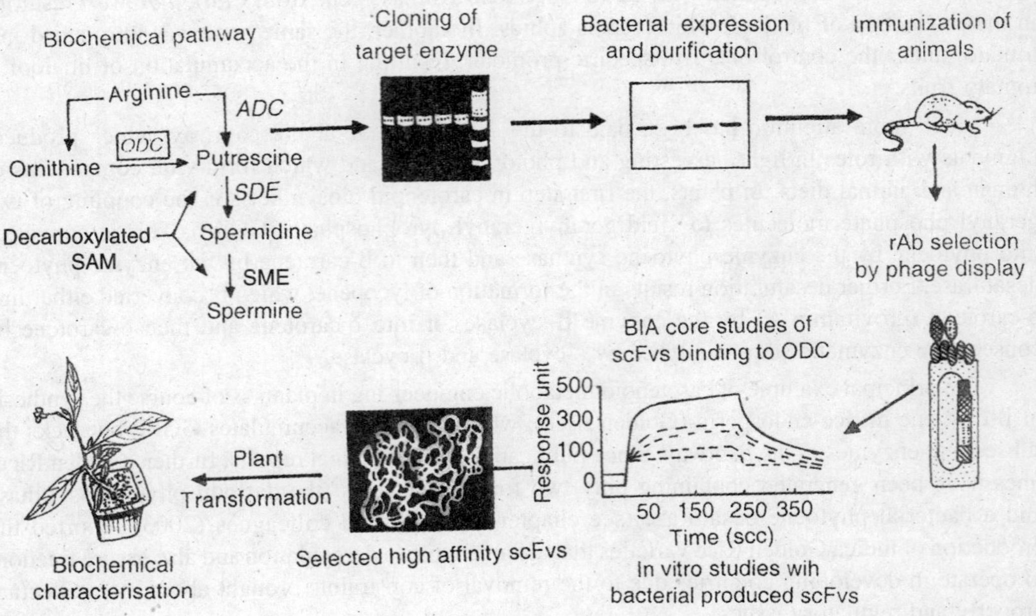

Fig. 28.2. Schematic overview of the anti-body metabolic engineering strategy, using immuomodulation of polyamine pathway by ectopic expression of an anti-ODC-scFv as an example. ADC : arginine decarboxylase, ODC : ormithine decarboxylase, rAb: recombinant antibody, SAM: S-adenosylmethionine, SDE : spermidine synthase, SME: spermine synthase.

(e) Stilbene phytoalexins

The strategy of engineering pathogen resistance by the introduction of novel phytoalexins is to date limited by the availability of only a few biosynthetic genes. In most cases such genes are members of small gene families whose individual transfer may not lead to the level of expression required for enhanced disease resistance in the foreign plant. This is particularly so, because due to association with their original promoters expression patterns differ among gene family members with respect to developmental control, tissue-specificity, inducibility and promoter strength. So far, stilbene synthase (STS) genes have been studied most intensively. STS have been isolated from several unrelated species such as peanut, grapevine and pine. In most cases the biosynthesis of stilbenes with phytoalexin properties is induced by pathogens and under environmental stress conditions like ozone and UV-light. Stilbene formation specifically requires the sole presence of the key enzyme STS, which can be further classified according to its-substrate preference. The STS enzymes from peanut and grapevine uses malonyl-CoA and 4-coumaroyl-CoA as substrates to form the trihydroxystilbene phytoalexin resveratrol. Tobacco plant does not have a resveratrol synthase, though the precursor molecules, which are also used by chalcone synthase, the key enzyme of flavonoid biosynthesis, are present throughout the plant kingdom. Tobacco plants carrying a STS gene from peanut showed an inducible *de novo* formation of the phytoalexin resveratrol. Moreover, transfer of two stilbene synthase genes from grapevine, which were expressed under the control of their natural promoters, resulted in increased disease resistance of the transgenic tobacco plants. However, the amount of STS mRNA accumulated in transgenic tobacco after induction was only 5% or less compared to the amount found in grapevine.

Transfer of the whole STS gene family of grapevine to a foreign plant could surmount this limited expression but is yet impracticable. Therefore attempt was made to optimise the expression of single STS genes by using heterologous promoters and enhancer elements, which promises strong transcriptional activity. The first approach was to increase the expression of a grapevine STS gene in tobacco by the insertion of transcriptional enhancer elements, yet maintaining the natural regulation defined by the STS promoter. It is known that upstream sequences of the *35S* RNA promoter from Cauliflower Mosaic Virus (CaMV) are able to stimulate transcription of heterologous promoters, although their properties do not strictly meet the definition of enhancers as being active in an orientation independent manner at distances greater than 1 kb up or downstream of the transcription start site. The scientists successfully used a tetramer of the -343 to -90 region, which had been successfully employed as a transcriptional enhancer in T-DNA tagging experiments. Insertion of the tetrameric enhancer 1.5 kb upstream of the transcription start site of the STS gene (VE5+) led to a significant increase in STS mRNA accumulation without altering the kinetics of induction. This enhancement was striking in transgenic plants derived from *Agrobacterium tumefaciens*-infected leaves (about 14-fold) and in shoots regenerated from protoplasts after direct gene transfer (about 4-fold). The enhancing property was lower when the tetrameric enhancer was inserted in the inverse orientation, supporting earlier observations.

The tetrameric enhancer element was, however, less active when it was inserted downstream of the STS coding region at a distance of 1.8 kb from the stop codon. Here only a 5-fold stimulation was achieved confirming previous observations and, in contrast to the upstream position, the activity was independent of the enhancer orientation. Insertion of the 1.3 kb enhancer fragment into the intron of the STS gene, turned out to have a negative impact on transcription efficiency, presumably due to interference with the RNA processing. In biological trials conducted in the greenhouse VE5+ enhancer plants were reproducibly less susceptible towards *Botrytis cinerea* than control plants. In comparison to line Vst1 expressing the unmodified grapevine STS gene, line VE5+ exhibited an additional attenuation of disease incidents of approximately 20%. Thus, the level of disease resistance based on the inducible synthesis of an additional phytoalexin can be influenced positively by modulation of the corresponding promoter strength.

IV. MODIFICATION OF STILBENE SYNTHASE EXPRESSION PATTERN CAUSES FLOWER COLOUR CHANGES AND MALE STERILITY IN PLANTS

In another strategy, the effects of constitutive stilbene synthesis were studied. A rather common approach to enhance the expression of a foreign gene makes use of the 35S RNA promoter, which ensures constitutively high rates of transcription. In the case of phytoalexin genes the permanent synthesis of the antimicrobial compound was possibly undesirable and could negatively impact metabolic profiles in the transgenic plant. To study the effects of permanent STS expression in plants, the *Vst1* gene from grapevine was expressed under control of the cauliflower 35S promoter strengthened by the duplication of its enhancer region. The effects of high foreign phytoalexin gene expression were exemplified in tobacco and petunia. Interestingly, very high constitutive STS expression has a dramatic impact on flower colour and pollen development in tobacco plants. A very high constitutive level of STS expression caused white and male sterile flowers, while the remaining organs did not reveal any changes.

V. ENGINEERING NOVEL METABOLIC PATHWAYS

Although metabolic engineering is often used to modify or extend existing pathways in the host plant, there are some examples where multi gene engineering has been used to introduce completely novel pathways and thus to produce completely alien products. In one such experiment two multifunctional cytochrome P450 enzymes and a uridine diphosphate glucose (UDPG)-glucosyltransferase were transferred from sorghum (*Sorghum bicolor*) into *A. thaliana*, resulting in the production of the cyanogenic glucoside dhurrin (which confers resistance to the flea beetle (*Phyllotreta nemorum*). These investigators also generated transgenic plants carrying one or both of the cytochrome P450 enzymes, but lacking the UDPG-glucosyltransferase, and demon-strated the accumulation of appropriate metabolic intermediates.

Another example of this general approach is the production of polyhydroxyalkanoate (PHA) polymers in transgenic plants. PHAs are polyesters of hydroxyacids synthesized by a range of bacteria, which can be used as biodegradable plastics. The best-studied PHA, poly (3--hydroxybutyrate) (PHB), can be made to accumulate in the plastids of transgenic *A. thaliana* plants at levels exceeding 40% dry shoot weight. This is achieved through the introduction of three bacterial genes, encoding the enzymes 3-ketothiolase, acetacetyl-CoA reductase and PHA synthase. PHB is a brittle plastic with few potential uses. The copolymer poly (3-hydroxybuty-rate-co-3-hydroxyvalerate) is much more useful, and is marketed under the trade name Biopol. This has been produced in *A. thaliana* and rapeseed through the coordinated expression of four bacterial genes: the three mentioned above plus *Escherichia coli* threonine deaminase. Although high-level accumulation of these plastics is generally detrimental to plant growth (by draining the pools of acetyl and propionyl-CoA), both plastics have recently been produced at relatively high levels in alfalfa with no adverse effects.

IV. METABOLIC ENGINEERING FOR HUMAN HEALTH

The goals of plant metabolic engineering are to produce valuable compounds in an economically attractive platform, to increase the resistance of a plant to various stresses or to increase the seed yield of a crop plant. Early attempts at plant metabolic engineering led to several successes in flux redirection, particularly in the production of novel plant oils through manipulation of lipid metabolism. Most of these attempts relied on the expression or silencing of a single gene, using well established gene expression technology to manipulate pathways that were relatively well understood. This approach is especially effective when the engineered step is at a potential metabolic branch point. Expressing a heterologous enzyme at this step can create novel substrates for existing enzymes, leading to the creation of a new branch of the pathway and formation of a novel product. Producing tocotrienols in a plant that normally produces only tocopherols by the expression of a single, heterologous gene encoding homogentisate geranylgeranyltransferase (HGGT) is a good example of this strategy.

Several technical advances have recently been made in simultaneously manipulating multiple steps in plant metabolic pathways and in constructing novel, multienzyme pathways in plant tissues. In the former case, novel or cryptic aspects of a pathway have often been revealed, as in the re-engineering of plant tocochromanol biosynthesis. In the latter case, transcription factors can be used to modulate entire pathways or even to induce the expression of pathways in organs where they do not normally occur, such as the induction of proanthocyanidin biosynthesis in the leaves of forage crops. Advances in gene expression technology and genomics have recently led to the successful engineering of heterologous, multigene pathways in plants, such as microbial long-chain polyunsaturated fatty acid (LCPUFA) biosynthetic pathways reconstituted in the seeds of crop plants. One of the great promises of biotechnology is the cost effective production of nutritional compounds that have, the potential to significantly improve human health.

a. Engineering at Metabolic Branch Points: Tocochromanol Biosynthesis

Tocochromanols constitute the vitamin E family of lipophilic antioxidants found in the tissues of photosynthetic organisms. They are an essential part of the human diet, protecting the body against the effects of free radicals, reactive oxygen species and lipid oxidation. Vitamin E is an essential component of the human diet, with a recommended daily allowance of 15-20 mg, and higher doses, taken as dietary supplements, have been associated with cardiovascular and immune benefits, as well as a reduction in the risk for some cancers.

Tocochromanols probably play a similar protective role against free radicals in the plant tissues in which they are synthesized because the tocochromanol content of plants changes significantly during plant growth and development (when the oxidative load is high) and in response to abiotic stress. This change in content is a result of the changes in the expression of genes encoding tocochromanol pathway enzymes.

Tocochromanols consist of an aromatic head group with an isoprenoid side chain that can be saturated (tocopherols) or unsaturated (tocotrienols). Tocopherols and tocotrienols can be further divided into α-, β-, γ and δ-forms, depending on the number and position of methyl groups on the aromatic head group. The first committed reaction in tocochromanol biosynthesis is also the first branch point in the pathway. The prenylation of homogentisic acid (HGA) with phytyldiphosphate (PDP) to form 2-methyl-6-phytylbenzoquinol (MPBQ) leads to tocopherol synthesis, whereas the prenylation of GA with geranylgeranyldiphosphate (GGDP) to form 2-methyl-6-geranylgeranylbenzoquinol (MGGBQ) leads to tocotrienols. The second branch point of this pathway occurs in the subsequent reactions because the prenylated HGA is the substrate in both pathways for one of two enzymes, a tocopherol cyclase (VTE 1), which leads ultimately to the formation of δ- and β-tocochromanols, or a methyltransferase (VTE3), which is part of a pathway leading to the formation of g-and a-tocochromanols. Although the steps in these reactions have been known for over 30 years, it is only in the past few years that the genes controlling the various steps have been cloned, the control steps identified and the pathway manipulated in transgenic plants.

Seeds of most monocots and some dicots contain tocotrienol and tocopherol in almost equal proportions, whereas the predominant form of tocochromanols in dicots are tocopherols. One of the interesting observations that has arisen from the study of mutants and transgenic plants is that the cyclases and methyltransferases involved in the conversion of prenylated HGA to tocopherols or tocotrienols are common to both pathways. Although this possibility had been raised in early *in vitro* studies of purified enzymes, it was recently confirmed in transgenic *Arabidopsis* leaves, which normally only make a trace amount of tocotrienols. Expression of a barley HGGT resulted in the formation of tocotrienols by the action of the endogenous cyclases and methyltransferases from the tocopherol biosynthetic pathway.

This example of metabolic engineering at a metabolic branch point led to several other interesting conclusions. First, it was apparent that an organism's ability to synthesize tocopherols or tocotrienols depended upon the presence of either homogentisic acid prenyltransferase (VTE2) or HGGT and thus, for the first time, provided a genetic basis for tocotrienol synthesis in monocots and tocopherol synthesis in dicots. Second, the total amount of tocochromanols synthesized by the transgenic plants was about 10-15-fold greater than that of controls, with all of the extra tocochromanol content coming from toco-trienols. Clearly, the HGGT had a strong preference for GGOP over POP. Furthermore, these results showed that the HGA pool was not limiting tocopherol synthesis under normal conditions because there was enough HGA to make additional tocotrienol in the transgenic line and this was in concurrence with studies showing that VTEZ, not its substrate supply, was rate limiting under normal flux conditions.

In humans, the highest vitamin E activity is seen with α tocopherol, although it is suspected that α-tocotrienol would have similar if not quite as potent biological activity. In all the above studies, where the total tocochromanol content was increased by inducing tocotrienol biosynthesis, the predominant forms of tocotrienol accumulated were γ- and δ-tocotrienols. Indeed, when tocotrienol biosynthesis was induced by increasing the HGA pool through the use of the *TYRA* gene, almost all the tocochromanol observed was δ-tocotrienol. This form of tocotrienol is produced by the direct cyclization of MGGBQ by VTE1. MGGBQ is also the substrate for a methyl tansferase, VTE3, which converts MGGBQ to 2,3-dimethyl-6-geranylgeranyl benzoquinol (DMGGBQ).

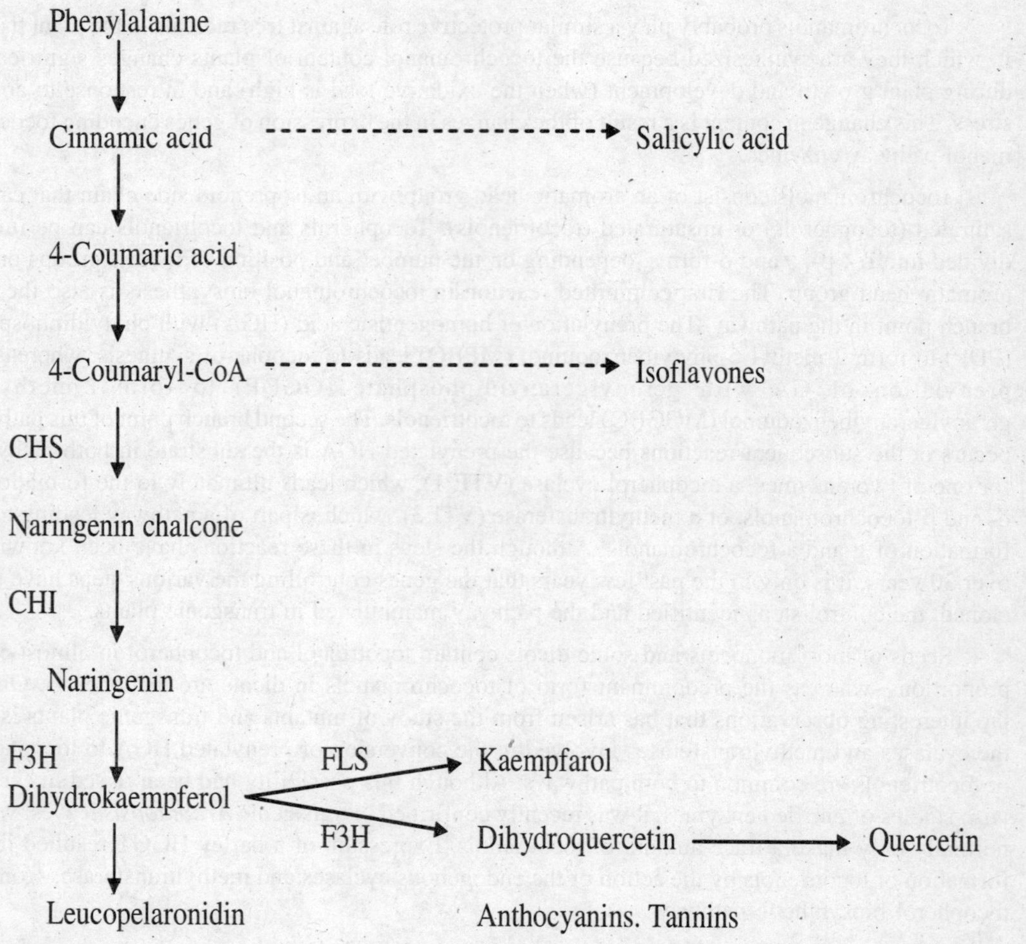

Fig. 28.3. Phenyl propanoid metabolism in plants

DMGGBQ then undergoes cyclization catalyzed by VTE1 to form -tocotrienol and this, in turn, is methylated by 'γ-methyltransferase (VTE4) to produce the biologically active α-form. The same enzymes act on MPBQ to produce the equivalent methylation patterns in the tocopherol branch.

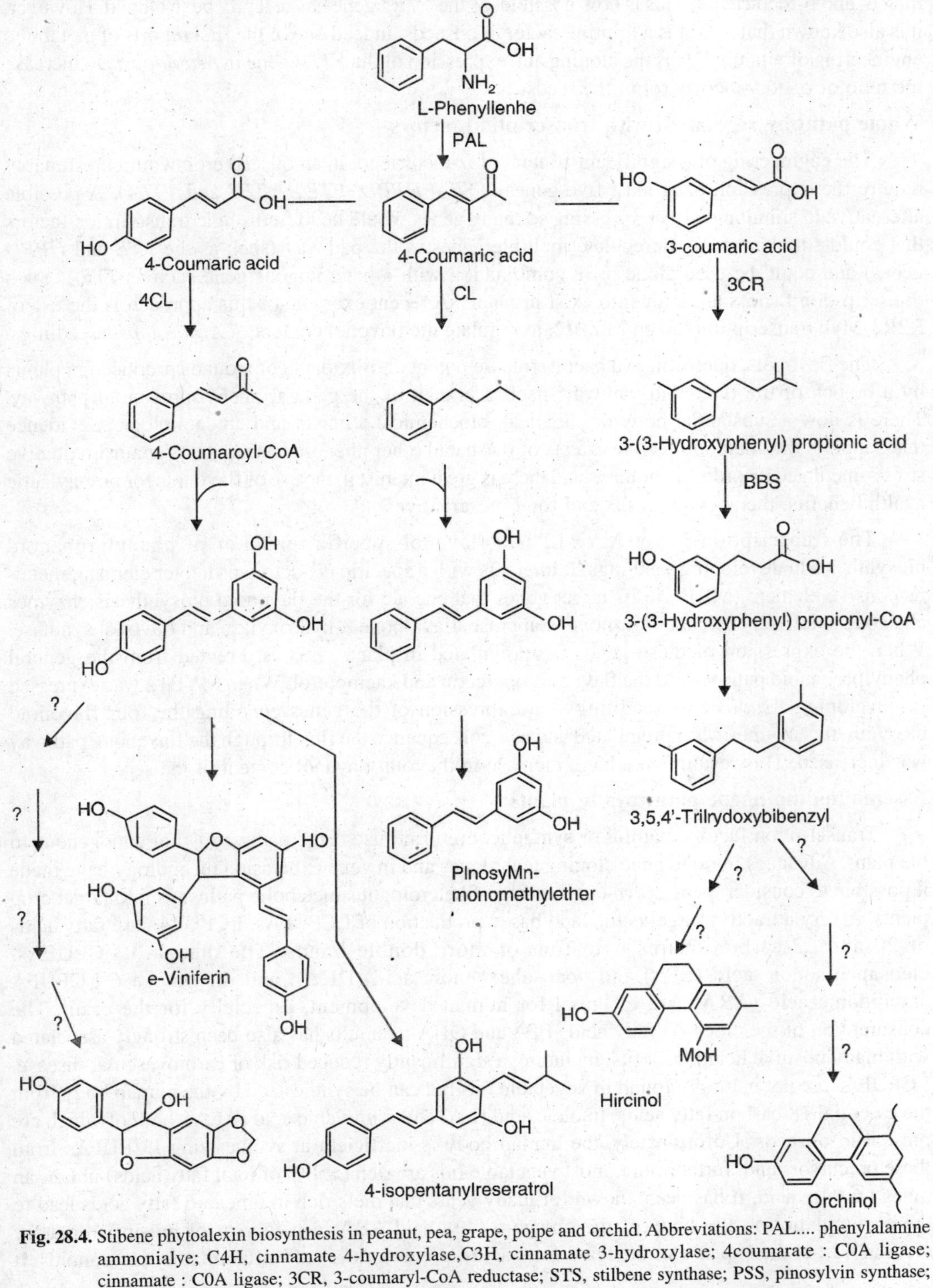

Fig. 28.4. Stibene phytoalexin biosynthesis in peanut, pea, grape, poipe and orchid. Abbreviations: PAL.... phenylalamine ammonialye; C4H, cinnamate 4-hydroxylase,C3H, cinnamate 3-hydroxylase; 4coumarate : COA ligase; cinnamate : COA ligase; 3CR, 3-coumaryl-CoA reductase; STS, stilbene synthase; PSS, pinosylvin synthase; 3CL, cinnamate 3-hydroxylase; PSMT, pinosylvin monomethyltransfe; BBS, bibenzyl synthase.

The results from the transgenic oilseeds producing δ-tocotrienol would suggest that successful engineering of α-tocotrienol requires over expression of the methyltrans-ferase, VTE3, at the second branch-point of the tocochromanol pathway to increase flux towards γ- and a-tocotrienol rather than δ- and β-tocotrienol. This is now possible, as the *VTE3* gene has recently been cloned. However, it is also known that VTE4 is a limiting factor in oilseeds, indeed one of the first reports of metabolic engineering of vitamin E was the cloning and expression of the *VTE4* gene in *Arabidopsis* to increase the ratio of α- to γ-tocopherol in the seeds.

Whole pathway regulation with transcription factors

The engineering of a significant content of α-tocotrienol in an oilseed crop would presumably require the expression of at least five genes *(TYRA, HPPD, VTE2, VTE3* and *VTE4)*. A possible alternative to simultaneously expressing so many genes would be to find single transcription factors that could simultaneously upregulate multiple genes in the pathway (such as the *VTE* and *HPPD* genes) and could be used alone or in combination with a heterologous gene such as *TYRA*. Such transcrip-tion factors are known to exist in plants. A recent example of this approach is the use of R2R3-Myb transcription factor, MYBI2, to regulate the flavonol content of *Arabi-dopsis* seedlings.

The flavonols, quercetin and kaemferol, are potent anti-oxidant compounds produced in plants by a branch of the flavonoid pathway, itself a branch of the general phenyl-propanoid pathway. There is now a substantial body of chemical, biochemical, clinical and epidemiological evidence that supports the chemoprotective effects of these and other phenolic antioxidants against oxidative stress--mediated disorders in humans and there is great interest in the use of flavonols for prophylactic health benefits, therapeutic agents and food preservatives.

The transcription factor MYB12 is a flavonol-specific regulator of phenylpropanoid biosynthesis in developing seedlings. It interacts with a specific *cis*-acting promoter element, metal-responsive element (MRE), in its target genes that encode for the flavonoid biosynthesis enzymes chalcone synthase, chalcone flavanone isomerase, flava-none 3-hydroxylase and flavonol synthase. When the expression of these genes is upregulated in plants, flux is directed from the general phenylpropanoid path-way to the flavonols quercetin and kaempferol. When MYB12 was expressed in developing *Arabidopsis* seed-lings the expression of the genes encoding the four flavonoid biosynthetic enzymes was upregulated and as a consequence the flux through the flavanone pathway was increased. This resulted in a large increase in the total flavanol content of the seeds.

Assembling multigene pathways in plants

Transcription factors, natural or synthetic, are of no use if the pathway is not endogenous to the plant. Advances in rapid gene cloning techniques and in gene expression technology have made it possible to consider the *de novo* engineering of heterologous metabolic pathways into target crop plants. A very attractive target is the land based production of LCPUFAs. LCPUFAs are fatty acids of 20 and 22-carbon chains with four or more double bonds. The omega 3 LCPUFAs, eicosapentaenoic acid (EPA) and docosahexaenoic acid (DHA), and the omega 6 LCPUFA arachidonic acid (ARA), are essential for human development, especially for the brain. The consumption of the omega-3 fatty acids EPA and DHA by adults has also been strongly associated with many positive health benefits, including a significantly reduced risk of cardiovascular disease. LCPUFAs are not naturally found in vegetable oils but can be synthesized by the human body from the essential 18-carbon fatty acids, linoleic acid (LA) and a-linolenic acid (ALA), both of which are present in oilseeds. Unfortunately, the human body is inefficient at synthesizing LCPUFAs from these precursors and, furthermore, most vegetable oils are rich (>50% of total fatty acids) in LA, an omega-6 fatty acid. It has been known for many years that diets rich in omega-6 fatty acids lead to the accumulation in the body of the omega-6 fatty acid ARA, a precursor of pro-inflammatory eicosanoids, at the expense of omega-3 fatty acids, precursors of anti-inflammatory eicosanoids. It has also been observed that supplementation of omega-3 LCPU-FAs in the diet can alleviate this imbalance.

The current major source of omega-3 LCPUFAs is the human diet is fish and fish-oil products. Most of the omega-3 LCPUFAs in fish originate from their diet of marine microorganisms or plankton, which are primary producers of LCPUFAs in the food chain. The genes responsible have been isolated from microorganisms and transferred in oil seed plants. The total omega-3 LCPUFAs content in the seeds of transgenic linseed, Brassica and soybean, have been reported to comprise 1.5 to 20% of the total seed fatty acids. These examples demonstrate the significant advances in developing desired crops by metabolic engineering methods.

QUESTIONS

1. What is metabolic engineering? How it helps in obtaining desired product and plant?
2. Give an account of various metabolic engineering approaches for the production of useful metabolites.
3. Write notes : (*a*) omega-3 LCPUFAs (*b*) Stilbene gene (*c*) Fructans.

FUNCTIONAL GENOMICS

CHAPTER 29

Recent technological advances enabling the analysis of hundred to thousands of mRNAs and protein simultaneously offer the possibility of not only identifying individual components but also provides insight about the signalling pathways by using an integrated approach.

The general term for transcriptomics and proteomics is 'functional genomics'. The core facility for this work has been established both in academic and industrial settings. Transcriptomics is defined as the genome-wide study of mRNA expression levels. Transcriptomic approach includes qRT-PCR, SAGE, subtractive hybridization, differential display and the use of microrrays. Proteomics is defined as the study of the full set of proteins encoded by a genome. Its foundation is the invention of two-dimensional (2D) gel electrophoresis. Following the separation of protein extracts, different types of mass spectrometers are available for protein identification (Fig. 29.1).

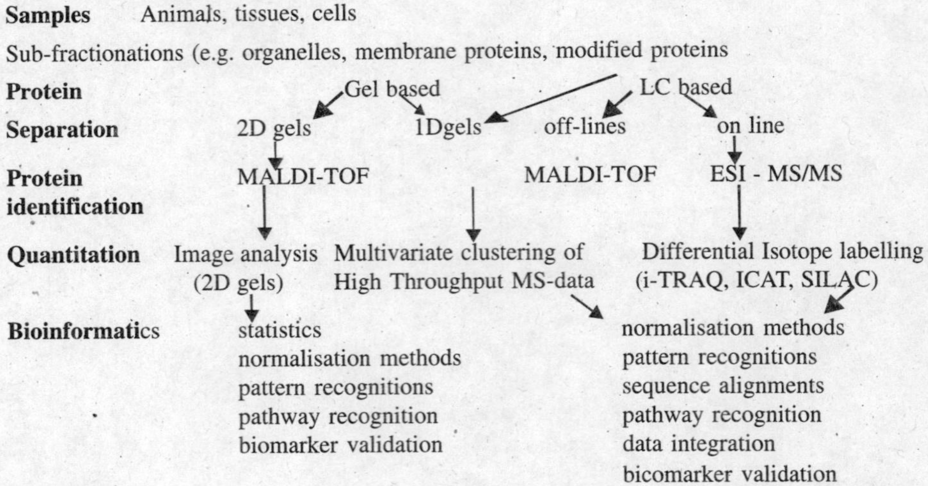

Fig. 29.1. Schematic presentation of tools and tehmiques of functional genomics

The human genome is estimated to contain 20-25,000 genes, half of which are expressed in the brain. By conventional methods such as northern blot analyses, it is only possible to determine the expression of one or a few genes where as in a particular tissue, the microarray technology offers the unique chance to analyse hundred to thousand transcripts simultaneously. Historically, microarray technology has been followed by substrative hybridisation and differential display.

I. DIFFERENTIAL DISPLAY, SAGE & SUBTRATIVE HYBRIDISATION

Northern blot analysis using RNA derived from tissue extracts correlates the various transcriptional events to a particular group of cells within t-000000000000000he tissue of interest. *In situ* hybridisation allows gene expression profiling of cellular level but both these approaches depend on the prior knowledge of defined gene sequences. This is also true for microarrays, whereas techniques such as differential display, serial analysis of gene expression (SAGE) and subtractive hybridisation are capable of identifying novel genes.

Differential display was developed in 1992. Similar to subtractive hybridisation, it required only small amounts of RNA and is capable of detecting known and novel genes. But in comparison to subtractive hybridisation it can compare more than two mRNA populations. The major disadvantages of differential display are the high number of false positives and the low sensitivity to detect rare species. In differential display, following reverse transcription using oligo (dt) primers with one or two additional specific nucleotides, the cDNA sequences are amplified by PCR, and the resultant products are separated on polyacrylamide gels. The bonds can then be visualized, e.g. by auto radiography. If a band is present in one lane and absent in the other (i.e. band present in one sample and not the other), this band is considered to represent an expressed gene. This band is excised and amplified by PCR. The resultant amplification product is used to screen a library to obtain a full-length cDNA as differential display (similar to subtractive hybridisation) tends to amplify 3' end selectively.

SAGE is based on two principles (Fig. 29.2); small square tags contain sufficient information to identify a transcript and are analysed in a serial fashion due to concentration of the short sequences tags, allowing sequencing to multiple tags within single clone. In a first step, polyadenylated mRNA is extracted, followed by synthesis of cDNAs using a biotinyated oligo (dt) primer, which is then cleaved using a 4 bp recognition site restriction endonuclease. The cut cDNA is isolated with streptavidin beads and ligated to linkers that contain a type IIS restriction site, which confers cleavage at a defined distance up to 20 bp away from the recognition site, cleavage releases short cDNA tags from the beads, and by virtue of linker, these tags are ligated, forming long concatemers. These contcatemers are introduced into a vector and sequenced, problem of tag specificity can be overcome by increasing the tag length to 18 bp.

Subtractive hybridisation is used to isolate genes that are upregulated in one cell type or tissue compared to another. The starting material is generally cDNA obtained by the poly (A) reverse transcription polymerase chain reaction (Poly (A) RT-PCR). Following first stand cDNA synthesis for a limited time using a poly (dt) primer, PCR is used to amplify the second strand, again using a poly (dt) primer resulting in a cDNA population of a defined size range (usually between 100 and 500 bp). This strand can be further amplified, if required to identify differentially expressed transcripts in two samples, denatured cDNA derived from the two samples is hybridised to each other, with one sample (the driver cDNA) being biotinylated and in excess compared to the other (the tracer cDNA). By using several cycles of hybridisation and removal by streptravidin precipitation, in the end those genes are left which are either unique or massively upregulated in the tracer population. This technique has been designated to identify rare mRNAs. In general it is reliable to isolate genes, which are enriched at least 5-10 folds in the tracer pool compared to the driver, since its invention, the technique of subtractive hybridisation has been further developed and is now combined with differential display.

Common to all these techniques is that the genes, which are identified as being differentially expressed, need to be validated by northern blot hybridisation or qRT-PCR.

Quantitative RT – PCR, Taq Man and light cycler

Since its introduction in 1980s the PCR reaction has been modified and optimised for an increasing number of applications. Initially the focus was on the amplification of specific nucleic acids templates into quantities amenable to identification and experimental manipulation. Then the PCR technology was developed further to accurately quantify the amount of a specific nucleic acid template in a complex sample. Quantitative PCR (qPCR) rather than simply analysing the amount of the final product, measures the accumulation of PCR products during the course of a reaction (hence named real-time qPCR, qRT-PCR). Under appropriate conditions the number of PCR cycles that is required for the accumulations of a specific amount of product (during the exponential phase of reaction) is a reflection of the relative amount of nucleic acid template present in the sample

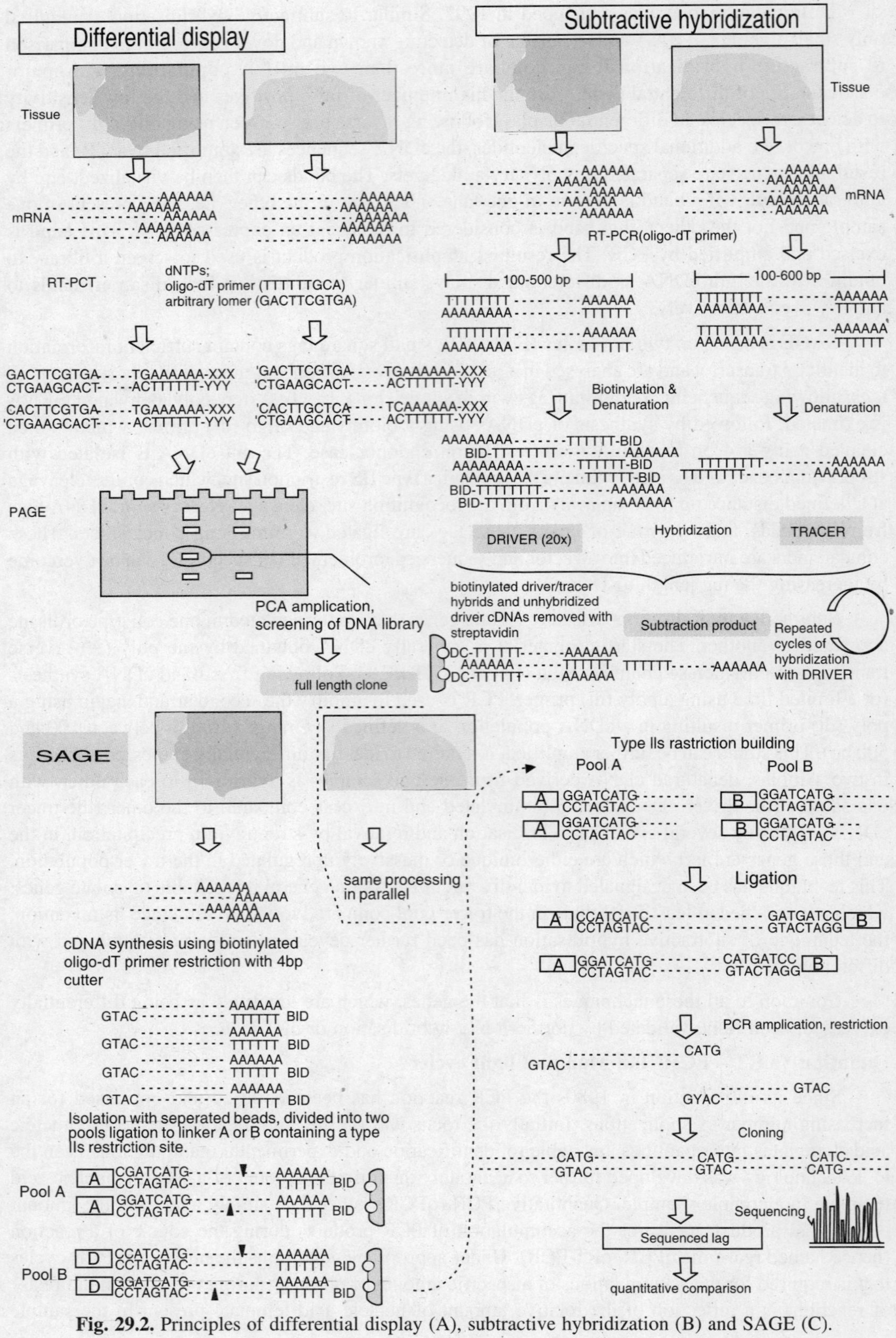

Fig. 29.2. Principles of differential display (A), subtractive hybridization (B) and SAGE (C).

Table 29.1. Software available in Function Genomics

Name	Functions	Application range	Internet links
EASE	Category representation analysis and annotation tool	Proteomics and genomics	http://genomebiology.com/2003/4/6/P4
Spotfire	Multipurpose tool for analyzing and visualising data	Multipurpose	http://www.spotfire.com
Genespring	Expression raw data analysis clustering and functional categories	Genomics	http://www.sigenetics.com
DMT	Expression raw data analysis and statistical analysis	Genomics	http://www.affymetrix.com
MAS	Advanced statistical analysis	Genomics	http://www.affymetrix.com
NetAffx	Affymetrix probes linked to public and proprietary databases	Genomics	http://www.netaffx.com
dChip	Expression raw data analysis, clustering and statistical analysis	Genomics	http://www.dchip.org
TXTGate	Profiling gene groups with text-based information	Proteomics and genomics	http://www.genomebiology.com/2004/5/6
Ingenuity	Pathway interaction	Proteomics and genomics	http://www.ingenuity.com
CLUSTER	Pure clustering tool	Genomics	http://rana.standford.edu
Tree View	Pure clustering tool	Genomics	http://www.rana.stanford.edu
GeneData Expressionist	Expression raw data analysis, statistics, and clustering promoter analysis	Large scale proteomics and genomics	http://www.gendata.com
Informax Xpression NTI	Expression raw data analysis, clustering and statistical analysis	Proteomics and genomics	http://www.invitrogen.com
Rosetta Resolver Gene Expression Data Analysis System	Statistics and clustering algorithms	Large scale proteomics and genomics	http://www.rosettabio.com
Lion Bioscience array SCOUT	Expression raw data analysis: clustering, statistics, annotation	Large scale proteomics and genomics	http:/www.lion-bioscience.com
In Vitrogen Corporation ResGen Pathways	Expression raw data analysis: clustering, stats, annotation and pathway analysis	Large scale genomics	http://www.lion-bioscience.com
Biomax Gene Expression Suite	Clustering tool with link to metabolic pathways	Large scale genomics	http://www.biomax.de
GO ToolBox	Category representation analysis and annotations	Genomics	http://bioinfo.cnio.es/docus courses/FatiGO/pathways.
		resgen.com	

under analysis. With qRT-PCR, samples can be analysed within a short period of time and different types of transcripts can be analysed in parallel. The mainly used commercial systems are the TaqMan (Applied Biosciences) and light cycler (Roche) both of which use fluorescence to quantify the amplification of the transcript of interest during the exponential phase of the PCR reaction. The length of the amplified transcript of interest (the so-called 'amplicon') is typically less than 100 bp.

The TaqMan qRT-PCR system uses two non-fluorescent primers combined with a fluorescent probe, which contains both the fluorochrome and a quencher (so called primer/probe approach). During PCR amplification, the polymerase separates the fluorochrome from the quencher causing a fluorescent signal proportional to the number of amplified transcripts during the cycle

The light cycler uses a 'hybridisation probe'. Here the three essential components are two labelled oligonucleotides and the amplification product. One oligonucleotide carries a fluoresceine label at its 3' end whereas the other carries a second label at its 5' end. The sequences of the two oligonucleoties are selected such that they hybridise to the amplified in a head-to-tail arrangement allowing Fluorescence Resonance Energy Transfer (FRET).

A third approach, which is compatible with both the TaqMan and the light cycler is the use of SYBR green a fluorescents dye. Here a fluorescent signal is produced when the dye was bound to double-stranded DNA. Which is formed during PCR amplification, compared to only a weak signal when not bound. Fluorescence is measured during polymerisation following primer annealing.

II. MICROARRAYS

Several factors accelerated the development and widespread application of the gene expression array or microarray technology. These are decoding to the genome of an increasing number of organisms, the development of fluorochrome's miniaturization and automation of the techniques.

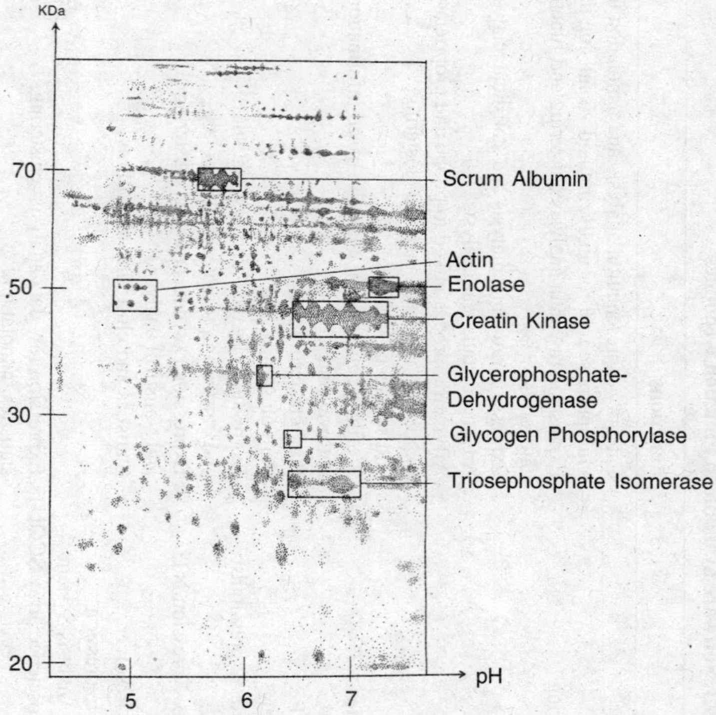

Fig. 3. 2D-SDS page of proteins

The term microarray includes a variety of distinct methodological and practical approaches, they are all high-density arrays of oligonucleotide or cDNA sequences coupled onto solid support, commonly used is a derivatized glass microscope slide or silicon wafer. The basic principle that complementary sequences of DNA can be used to 'probe' immobilized sequences of DNA molecules. This basic principle of complementary sequences hybridisation is used for long time in molecular biology e.g. in Southern blot techniques.

The modern application of microarrays differs from classic blotting procedure in that the labelled pool of nucleic acid molecules is isolated from the tissue of interest and hybridised to a large number of immobilized DNA molecules arrayed in specific locations on a solid surface. As such each sequences is individually spotted onto the support, and by building up thousands of such spots, an array can be made to cover the entire known expressed gene content of an organism. By hybridising the sample to all the sequences present on the array simultaneously, highly parallel (or high-through put) analysis of expressed genes can be performed. RNA is extracted, labelled with florescent dye and hybridised to the sequences on the array. The signal detected from each spot is being in principle proportional to the amount of that specific RNA in the original sample. By analysing each spot, a picture not only of which genes are expressed but also the level of expression can be obtained.

In general, there are two types of arraci, "soligonucleotide arrays where defined oligonucletides are spotted or directly synthesized onto the support and ii) cDNA arrays where selected cDNAs or PCR products are spotted. The array fragments can include expressed genes obtained from various sources, including those previously identified in public database, or from expressed sequences tags (ESTs).

ESTs are randomly selected clones constructed from RNA derived from a specific tissue or cell type. Several software's are available (Table - 29.1) to normalize, measure and interpret microarray data. It is important to keep in mind that generally any functional information obtained is dependent on the database use. Here most commonly used database are GO biological processes, GO molecular function, GO cellular components, Interpro, KEGG pathway, molecular localization, RIKEN and Gen MAPP pathways.

In addition to profiling gene expression, the microarray technology can be used for genotyping and to detect sequences polymorphism. This involves examining single nucleotide polymorphism (SNPs). These are single base differences in DNA sequences that occur between individuals in a population. As these polymorphisms can affect the properties of the encoded proteins (such as activities of enzymes), this information may ultimately be useful for tailoring medicine.

(a) Limitations of transcriptomic approach -

The way microarray functions, a principal limitation of transcriptomic approach is that many functions of a cell are executed at the translational and post-translational level (and not at transcriptional level). However, these are reflected by changes in gene expression for down stream targets of signal transduction pathways (such as genes that are targets on by transcription factors). Then it should be kept in mind that mRNAs are unstable molecules, which are programmed for enzymatic degradation with considerable variation in half-lives. This means that mRNAs with extremely short half-lives will be difficult to extract in reproducible quantities. Another issue is alternative splicing.

In addition to theoretical limitation, technical limitations also need to be considered; since on a microarray chip up to 40,000 transcripts can be analysed simultaneously, gene expressions changes need to be relatively robust to be detected and often a fold cut off is used as a measure of significant changes in expression. Therefore modest changes in gene expression can be overlooked, as transcriptional differences are often subtle. Therefore it is important to get the noise out the gene arrays (i.e. a clear distinction between false and true differences). Furthermore, the use of different platforms doesn't necessarily yield same results.

Proteomics

Measuring gene expression at protein level is potentially more informative than the corresponding measurement at the mRNA level. Protein as major catalysts of biological function contain several different aspects of information that collectively give more actual information as compared to mRNA based information. This information includes the abundance of activity, state of post translational modification, sub-cellular localization and association with each other. By definition the proteome is the full complement of protein produced by a particular genome of a cell, tissue or species at a given time. It represents a higher complexity than the transcriptome, and it displays a higher degree of dynamics due to post translational modifications with a genome in humans of an estimated 20-25, 000 genes, there are probably more than 500,000 different proteins in humans due to alternative splicing and post translational modifications.

The term proteomics essentially refers to the large-scale analysis of many proteins simultaneously. While this could include all the proteins encoded by the genome at any given time (i.e. the proteome), it more commonly refers to specific subsets of proteins found in a specific region or cell type under specific conditions. The majority of proteomic techniques involves some form of the separations procedure in order to isolate individual proteins (such as 2D PAGE and HPLC), ionisation of peptides, followed by detection and identification by mass spectrometry (MS). Details of all these techniques have been described in the chapter on proteomics. Here, very brief description is given as required as a part of functional genomics.

2D gel electrophoresis -

Provides the opportunity to determine quantities and qualitative differences in protein composition of different samples (2D – PAGE- polyacrylamide gel electrophoresis).

Isoelectric focusing and second dimension separation –

In 2D-PAGE, proteins are first separated according to charge and then in the second dimensions by mass, this method is in principle capable of resolving up to 10,000 proteins in a single gel.

In the first phase of 2D – PAGE proteins are separated by isoelectric focusing using immobilized pH gradient (IPG) in a thin gel strip. Proteins migrate through the IPG strip until their net charge is neutral (the isoeelctric point, *pI*). Following separation in the first dimension, proteins are separated in the second direction (90^0) using electrophoresis in acrylamide gel containing sodium dodecyl sulphate (SDS), where SDS impacts a net negative charge, allowing protein separation by mass (SDS-PAGE) Fig. 29.3.

Once proteins are separated, they can be visualized using a variety of stains. Silver stains are highly sensitive and detect as little as 100 pg proteins. Other dyes are SYPRO- a fluorescent dye ; coomassie brilliant blue dye.

(b) Limitations of proteomics

2D – PAGE is not suitable for highly acidic or basic proteins as these are difficult to solubilize and do not separate well in the isoelectric focusing process. Due to the aqueous nature of the system, hydrophobic membrane proteins are also difficult to solubilize and separate. High molecular weight proteins have difficulties entering the IPG strips. One SDS-PAGE system is not adapted to resolve both high and low molecular weight proteins at the same time. Additional limitations are –

1. Proteins of less abundance constitute the major protein proportion of a cell, which are difficult to quantify from gel.

2. Many spots contain more than one protein, and

3. Due to post translational modification and processing, same proteins migrate to different positions in gel, thus complicating quantification.

Recent improvement in 2D gel analysis include prior prefractionation to reduce the complexity or the sample, the use of fluorescent dyes with enhanced detection sensitivity and use of IPG strips with enhanced separation by using overlapping narrow pH gradients. These have improved resolution and solved certain old problems.

III. PROTEIN ARRAYS, TISSUE ARRAYS, AND MULTI-IMMUNOBLOTTING ANTIBODY – BASED PROTEIN ARRAYS

Based on experience gained with gene based microarrays protein microarrays is developed recently to simultaneously detect and quantify proteins and peptides, molecules usually antibodies are immobilized at a high density on a solid support such as a pre-treated glass microscope slides. When a protein extract is added each individual antibody captures its target protein and by arraying hundred or thousands of species antibodies on a single slide this method allows large –scale and high –through put analysis. A serious problem with antibody arrays however, is antibody specificity. Similar to gene arrays, protein arrays are probed by direct labelling of the protein sample with fluorescent dyes such as Cy3 or Cy5, and the abundance of a protein is related to the fluorescent intensity of the particular spot on the arrays.

(a) SELDI

In addition to coupling of antibodies to solid supports a range of chemistries has been developed to selectively enrich different types of proteins or peptides and captures these on to a solid phase chromatographic surface. This technique is called SELDI (surface –enhanced laser desorption/ionisation) and is similar to MALDI .

The sample can be applied to a cationic affinity surface and washed with an appropriate buffer such that only proteins and peptides with affinity for the support are retained. SELDI chips are available with gallium coatings, which have a hight affinity for phosphate containing proteins. By using SELDI targets with a different affinity matrix, it is possible to analyse different subsets of proteins.

(B) PHAGE DISPLAY

In phage display, proteins or peptides are displayed on the surface of a bacteriophage, in ease of antibodies, phage display is a method to evaluate a huge range of potentially useful synthetic antibodies and then produce large quantities of the selected ones. Phage libraries comprising collections of these antibody-covered phages typically each contain a billion different antibodies, a number comparable to that in the human immune system.

(C) OXYBLOTS

2D PAGE is flexible since the gel format can be altered for example mini gels have been used to identify oxidized proteins. Following 2D PAGE, proteins are transferred to a members for western blotting, and the proteins are visualized using fluorescent dyes followed by probing with an antibody with an antibody oxidized protein groups. This method is capable of identifying a hundred or more proteins on these so- called oxyblots and reveals differences in, e.g., protein levels in a diseased as compared to a normal tissue.

(D) MASS SPECTROMETRY

Mass spectrometry (MS) is the preferred method for the identification of proteins following separation by 2D-PAGE and HPLC. It forms an essential tool in proteomic analysis, by providing structural information such as peptide mass and amino acid sequences as well as information on protein modifications.

Fifteen years ago, the two major techniques were developed matrix assisted laser desorption ionisation (MALDI) time of flight (TOF) and electrospray ionization (ESI) MS. The development of these relatively non-destructive, soft ionisation methods made it possible to convert proteins into volatile ions. Usually for peptide mass fingerprint (PMF) analysis, the protein to be analysed is purified and digested enzymatically (by trypsin) to cleave the protein at specific bonds giving a reproducible pattern of digestion. The details of this technique is given separately.

QUESTIONS

1. What do you understand by functional genomics ? Give various steps involved in the study of two different samples.
2. Short notes : (a) transcriptomics (b) Proteomics (c) MALDI-TOF (d) 2D-PAGE (e) SELDI (f) Microarray.
3. What is differential display? Discuss its role in biomedicine and agriculture.

DOWNSTREAM PROCESSING

CHAPTER 30

Downstream processing refers to the recovery and purification of biosynthetic products, particularly pharmaceuticals, from natural sources such as animal or plant tissue or fermentation broth, including the recycling of salvageable components and the proper treatment and disposal of waste. It is an essential step in the manufacture of pharmaceuticals such as antibiotics, hormones (e.g. insulin and human growth hormone), antibodies (e.g. infliximab and abciximab) and vaccines; antibodies and enzymes used in diagnostics; industrial enzymes; and natural products including alkaloids, fragrance and flavor compounds. Downstream processing is usually considered a specialized field in biochemical engineering, itself a specialization within chemical engineering, though many of the key technologies were developed by chemists and biologists for laboratory-scale separation of biological products.

Downstream processing and analytical bioseparation both refer to the separation or purification of biological products, but at different scales of operation and for different purposes. Downstream processing implies manufacture of a purified product fit for a specific use, generally in marketable quantities, while analytical bioseparation refers to purification for the sole purpose of measuring a component or components of a mixture, and may deal with sample sizes as small as a single cell. Generally we deal with gram quantities of purified proteins and natural products. Sometimes even difficult to isolate and purify gram quantities of natural products from a complex plant extract. Scaling up of this process to industrail level, efficient recovery of the final product and cost beneficiation are part of downstream processing.

I. STEPS IN DOWNSTREAM PROCESSING

A widely recognized categorization of downstream processing operations divides them into four groups which are applied in order to bring a product from its natural state as a component of a tissue, cell or fermentation broth through progressive improvements in purity and concentration.

(a) Removal of insolubles is the first step and involves the capture of the product as a solute in a particulate-free liquid, for example the separation of cells, cell debris or other particulate matter from fermentation broth containing an antibiotic. Typical operations to achieve this are filtration, centrifugation, sedimentation, flocculation, electro-precipitation, and gravity settling. Additional operations such as grinding, homogenization, or leaching, required to recover products from solid sources such as plant and animal tissues, are usually included in this group. (Fig. 30.1)

(b) Product Isolation is the removal of those components whose properties vary markedly from that of the desired product. For most products, water is the chief impurity and isolation steps are designed to remove most of it, reducing the volume of material to be handled and concentrating the product. Solvent extraction, adsorption, ultrafiltration, and precipitation are some of the unit operations involved.

(c) Product Purification is done to separate those contaminants that resemble the product very closely in physical and chemical properties. Consequently steps in this stage are expensive to carry out and require sensitive and sophisticated equipment. This stage contributes a significant fraction of the entire downstream processing expenditure. Examples of operations include affinity, size exclusion and reversed phase chromatography, crystallization, and fractional precipitation.

Different levels of purifications may be achieved depending upon the ultimate use, e.g. enzyme phytase for animal feed is not purified, enzyme for industrial use is partially purified while ezyme for analytical use is highly purified. Details about various types of chromatographic techniques for protein purification is given in the chapter on techniques of cell biology.

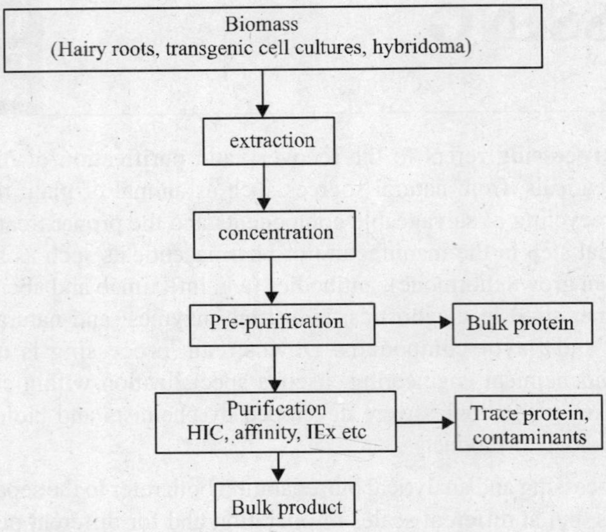

Fig. 30.1. Steps in downstream processing to obtain product (Hydrophobic interaction chromatography, Ionexchange chromatography)

(d) Product Polishing describes the final processing steps which end with packaging of the product in a form that is stable, easily transportable and convenient. Crystallization, desiccation, lyophilization and spray drying are typical unit operations. Depending on the product and its intended use, polishing may also include operations to sterilize the product and remove or deactivate trace contaminants which might compromise product safety. Such operations might include the removal of viruses or depyrogenation.

A few product recovery methods may be considered to combine two or more stages. For example, expanded bed adsorption accomplishes removal of insolubles and product isolation in a single step. Affinity chromatography often isolates and purifies in a single step. Downstream processing may cost up to 50-70% of product cost. The major issue concerned are- recovery, uniformity and ourity and quality control.

II. INDUSTRIAL PRODUCTION SYSTEM

In general industrial production involves use of bioreactor for cultivation of plant, animal and bacterial cells. The biomass so generated, may be of recombinant cells, is processed to recover the compound of interest; may be a primary product like protein, a recombinant protein, a growth hormone or an alkaloid. As mentioned above, we need to scale up the process from lab scale to industrial scale.

The need for bioreactors- The discoveries of Louis Pasteur soon lead to the development of bioreactors for the large scale culture of bacteria and yeasts. Small cultures originally were performed in dishes which could not create appreciable quantities of fermentation products (such as acetone, ethyl alcohol). These products had value if they could be produced in large volumes. To achieve this goal, the development of bioreactors was required

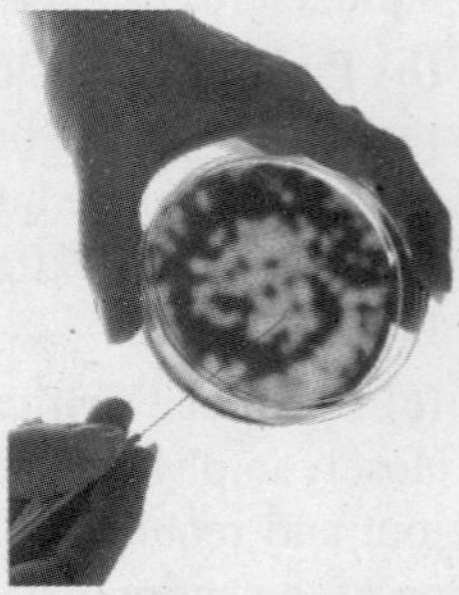

Fig. 30.2. Bacterial culture in dish

(beginning in the early 1900s) that could culture at least a few litres of cells and media. Eventually, bioreactors would be able to ferment tens of thousands of litres at a time.

Bioreactors in industry- In the first half of the 20th century, ethyl alcohol and butyl alcohol were the most important industrial fermentation products in the world. Such products were vital to the synthesis of synthetic rubber, especially during World War I. This required the development of bioreactors that could culture thousands of litres. By the 1960's, the fermentation of many simple organic products diminished as chemical synthesis (mostly from petroleum) provided a cheaper alternative.

Laboratory scale bioreactor

Industrial bioreactor of large capacity

Fig. 30.3. Bioreactors for sc ale-up production

Bioreactors and bacterial fermentation have played a vital role in the production of many medicinal products including the majority of antibiotics. The first large scale manufacture of antibiotics occurred during World War II with the production of penicillin. It was discovered early on that the most efficient production of penicillin occurred through fermentation processes rather than using direct chemical synthesis. Since this time many other antibiotics have been characterized and mass-produced using fermentation (Fig. 30.3).

(*c*) **Plant cell cultures**- Plant secondary metabolites are important for plant interactions with their environments. These metabolites give plants important properties such as antibiotic, antifungal, antiviral, UV protection, and pest deterrents. These chemicals have also been used by man for thousands of years as medicines. Today plants are responsible for the production of over 30,000 types of chemicals including pharmaceuticals, aromas, pigments, cosmetics, nutraceutics, and other fine chemicals. It was discovered early that metabolites could be accumulated in higher quantities using plant cell cultures as compared with whole plant cultivation. However, since plant cell cultures are low yield compared to their bacterial counterparts, they are primarily used only for high-value metabolites.

Recombinant technology products- The biosynthetic routes and mechanisms describing the production of plant secondary metabolites are often very complex and can involve dozens of enzymes. Through the study of plant metabolism a new discipline called metabolic engineering has emerged. The goal of metabolic engineering is to improve cellular activity by manipulating enzymatic, transport and regulatory functions of a cell. This is accomplished through the use of recombinant DNA technology. Production of novel proteins including eukaryotic proteins like insulin in prokaryotes or in eukaryotes like yeast or insect cell falls under this category.

Growth phases and secondary metabolites. The growth of a bacterial culture can be represented by a curve that consists of four stages or phases:

- Lag phase - growth and reproduction are just beginning,
- Log phase - reproduction is occurring at an exponential rate,
- Stationary phase - environmental surroundings and food supply cannot support any more exponential growth, therefore, growth slows down.
- Death phase - when all of the nutrients have been exhausted, the population dies off

The production of secondary metabolites (such as antibiotics) often occurs during the stationary growth phase (or idiophase) of a bacterial culture. Secondary metabolites are non-growth related and seem to play no obvious role in cell maintenance. The production of secondary metabolites largely concerns the first growth phases of a bacterial culture. This production is considered a two-stage process, with a trophophase (growth phase) and an idiophase (production phase), which usually occurs when culture growth has slowed or ceased. The major aim for mass production of secondary metabolites is to maximize the biomass in a short trophophase, while optimizing the conditions for high, sustained idiophase production.

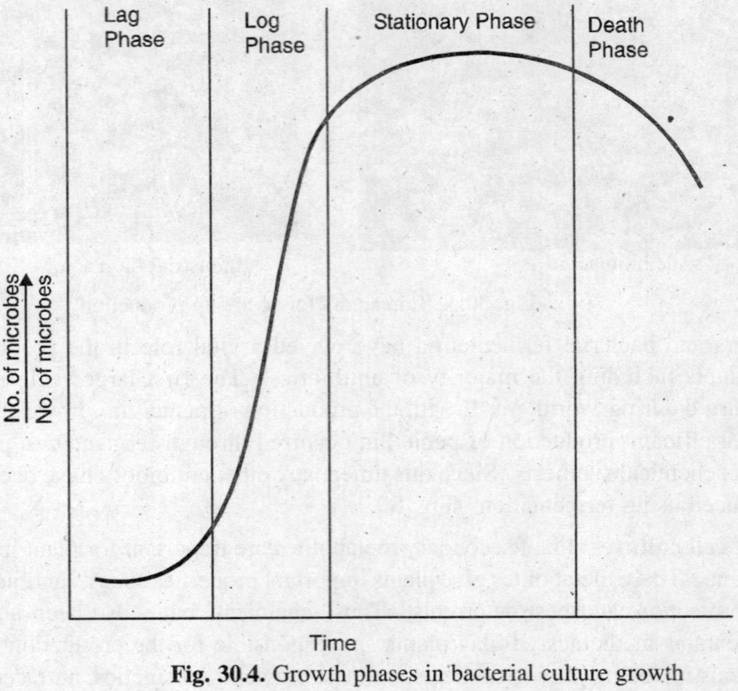

Fig. 30.4. Growth phases in bacterial culture growth

Culture conditions

A variety of parameters are important to influence the production of desired secondary metabolites in bioreactor cultures.

These include:

- Substrate
- Slow or fast utilization
- Nitrogen requirements
- Amino acid supplements
- Temperature
- pH
- Oxygen concentration
- Pressure

Types of bioreactors

Several types of bioreactors enjoy use in both research and industry. Here are some of the most common:

- Stirred Tank - mechanical mixing and air-sparged liquid media
- Airlift - uses air sparging to pneumatically mix the media
- Combinations and variations of stirred tank and airlift reactors
- Membrane - two permeable membranes to deliver nutrients and export wastes and products from bacteria in a centre tube

Bacterial and plant cultures can be grown in bioreactors using several modes of operation including: batch, semi-batch, or continuous. Details of bioreactors and processes are given in the chapter on bioreactor.

Scale-up considerations

This process is the important step for transferring bench-top fermentations to mass production. Scale-up would be very simple if all parameters affecting bacteria remained the same. Numerous empirical and semi-empirical relationships are often used to correlate variables such as shear rates and oxygen mass transfer with physical parameters such as impeller speed and reactor dimensions. One of the most important and often overlooked factors in scale-up is power consumption. When scaling-up a process, it is important to factor in the power costs of operating large fermenters.

Antibodies Production

Antibody-based therapeutics are expected to continue to be a major source of new therapies for the next decade. However, monoclonal antibodies (MAbs) are among the world's most expensive drugs. Pressure from healthcare providers is driving the need to lower the costs of manufacturing MAbs by as much as an order of magnitude, from \$1,000 per gram to \$100 per gram.

Table. 30.1. Upstream productivity increases from 1994 to 2004

Mammalian cell line	*E.coli*	*Pichia pastoris*	
1994	<1-2g/l	2-3g/l	3g/l
2004	4.7g/l	9g/l	15g/l

So far, the most significant improvement is the increasing titers from concentrations of low milligram- to multigram per liter. This increased upstream productivity has subsequently created a demand on downstream processes to provide enhanced capacity and speed. In addition, because only a small proportion of biotherapeutics that enter clinical trials make it to market, companies tend to populate the candidate pipeline quickly, meaning more processes require development in shorter timeframes.

To solve these multiple demands, several solutions have been developed, including, among others, the use of platform technologies and chromatography resin improvements.

PLATFORM TECHNOLOGIES

A downstream platform technology is based on experience with purifying a class of products with similar properties. With more than 500 MAbs currently in preclinical studies, monoclonal antibodies represent the greatest number of biotechnology products to which similar purification processes can be applied. In particular, the unique selectivity of Protein A has enabled the development of platform technologies for the purification of monoclonal antibodies.

Platform technologies facilitate rapid and economical process development and scale-up, potentially allowing evaluation of a larger number of product candidates, speeding market entry, and even reducing the validation effort. Familiarity with a process also can result in more robust processes and better technology transfer from development to manufacturing. Platform technologies create predictable activities and durations that result in a generic timeline.

Platform technologies can also reduce validation costs. The use of platform technologies, combined with the precision and sensitivity of quantitative Polymer Chain Reaction (Q-PCR) has helped the evaluation of the efficiency of generic and matrix viral clearance. Anion exchange steps in the platform are used as polishing steps—i.e., to remove impurities such as DNA, host cell proteins, and adventitious viruses. Recovery of recombinant protein from transgenic plants is presented in the table 30.2.

Table 30.2. Downstream processing of recombinant proteins from transgenic plant system

Transgenic system	Recombinant protein	Purification steps	Yield, purity
corn	trypsin	**Clarification** centrifugation **Purification** - trypsin inhibitor affinity chromatography - cation exchange chromatography	25% yield >98% pure
tobacco	Monoclonal Ab (IgG)	**Clarification** - basket centrifugation - tubular centrifugation **Purification** - expanded-bed protein A affinity chromatography	60% yield >90% oyre
rice	Human lysozyme	**Clarification** - three-step filtration **Purification** - cation exchange chromatography - size exclusion chromatography	60% yield. Single band on SDS-PAGE

Because anion exchange chromatography is commonly used as part of the platform purification process for monoclonal antibodies, a great deal of knowledge has been gained in this area. Most monoclonal antibodies have basic isoelectric points, typically greater than 8, so they do not bind under neutral pH and low conductivity conditions. Experience with fast-flow agarose anion exchangers combined with Q-PCR analysis has shown that bracketing and generic approaches demonstrate efficient removal from the process of a nonenveloped virus, SV40.

Another potential improvement in validation is minimizing resin lifetime studies for Protein A and anion exchange chromatography. Studies with anion exchangers in flow-through mode have demonstrated that surrogate measurements such as DNA clearance, back-pressure, and band spreading can predict when columns will no longer provide sufficient virus clearance.

(a) RESIN IMPROVEMENTS

Today some chromatography resins can achieve binding capacities greater than 250 grams of protein per liter. Fast mass transfer rates enable the use of more than 80% of the capacity after only one minute, and flow velocities can now exceed 1,000 cm/h. For ion exchangers, the most commonly used resin, these improvements have been achieved through a greater understanding of resin structure and surface chemistry, that enables bead design with appropriate mechanical properties, porosity, and distribution of charged functional groups.

Binding capacity can influence process economics in several ways. High capacity can reduce column size requirements, which in turn can reduce space requirements and water consumption for buffers and cleaning solutions. Binding kinetics also influence raw material costs per gram of product, though this area is rather complex. As an example, resin A has a high total capacity, slow kinetics, and low maximum flow velocity compared to resin B. Resin B has low total capacity, fast kinetics, and a high maximum flow velocity. At very short residence times, resin B gave higher productivity than resin A; but when residence time was optimized for resin A, it provided a more economical solution. The example also points out the importance of process development for each individual biotherapeutic.

(b) INCREASED VELOCITY

Improvements in resin design can be illustrated by comparing a classic fast-flow, agarose-based resin with a recently introduced resin designed to improve productivity. The increased flow velocity that can be achieved with this second-generation anion exchanger is demonstrated in an industrial column with a 20 cm bed height and a 1 meter internal diameter (i.d.). A constraint of the industrial production environment is the need to work with moderate pressures to achieve adequate flow rates. Realistically, very large columns for capture applications should not need to handle more than 1–2 bar to avoid weight and handling issues. Designing an ion exchanger for capture means striking a balance between bead properties that give high capacity and high resolution, and properties that allow high flow rates at low back-pressures. The packed bed of the new resin enables a linear flow rate of more than 700 cm/h at a back-pressure of less than 3 bar. High flow rates can improve turnaround time in downstream processing by decreasing washing, cleaning, and re-equilibration times. In most cases, sample loading and elution times can also be reduced, though this reduction depends on the properties of the target molecule and impurities. These time savings can result in the production of more batches to meet the quantity demands for high-dose biopharmaceuticals.

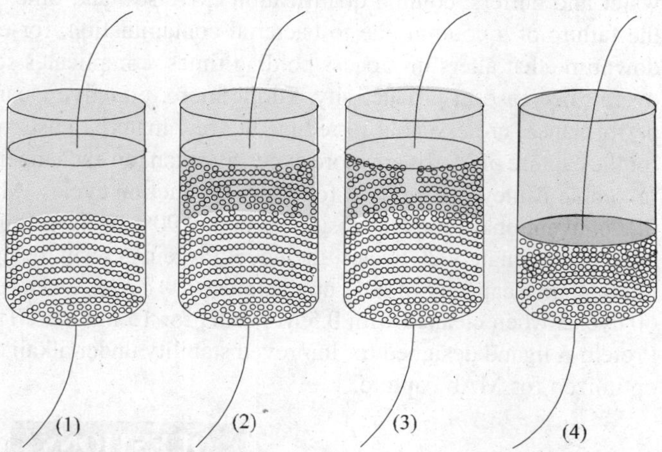

Fig. 30.5. Changes in separation bed results in faster separation

(c) INCREASED CAPACITY

High flow velocity alone does not make a productive process. Binding capacity at high speeds, however, is essential for capture steps. The second-generation resin provides significantly higher dynamic binding capacities compared to classical fast-flow agarose anion exchangers. A high binding capacity of 80–120 g/L IgG (at residence times of 2–6 minutes and typical pH and conductivities) could be obtained with a novel, recently introduced cation exchanger belonging to the same high-flow agarose generation.

(d) OVERALL PRODUCTIVITY

By combining speed and binding capacity, overall productivity is enhanced, which compares high-scale productivity for the capture of a target protein from an *E. coli* homogenate on three different resins. This study demonstrated that it is feasible to purify more than 100 kg of target protein product in 24 hours. Reducing the steps in downstream processing has significant effect on cost reduction.

(e) SELECTIVITY AND PURITY

Selectivity plays a capital role in platform technologies because it reduces the number of purification steps, thereby shortening overall production times and increasing recovery. Specific binding of the target molecule to a ligand-substituted resin enhances productivity by rapidly removing the bulk of the impurities from the feedstream and concentrating the product in one step.

(f) CLEANING AND SANITIZATION

The potential for carryover of product and impurities from one run to another must be addressed for packed columns, and the cleaning protocol must be shown to remove residual materials. Many cleaning and sanitization protocols use NaOH. Another factor is the stability of resins, columns, and associated equipment in sufficiently harsh conditions for cleaning and sanitization. In addition, column packing at large scale is time-consuming and costly; maintaining a packed column is essential for economical production of consistent product. Repacking columns requires large quantities of water and buffers, column qualification exercises, and time. If reserve columns are not available, the failure of a column due to bacterial contamination, for example, could result in a production downtime that alters in-process holding times, complicates scheduling in the plant, and creates an overall decrease in productivity. Furthermore, effective cleaning and sanitization, with preserved performance, are essential to reduce the raw material costs of resins. Protein A-based resins used for the capture of MAbs are more expensive than ion exchangers, but their use becomes economically favorable if they can be used for many production cycles. Although the dynamic binding capacity of a conventional Protein A resin was below 90% of the original capacity for a polyclonal antibody after approximately 20 cycles, the novel Protein A resin, designed to tolerate NaOH, was observed to maintain a capacity of 90% up to almost 200 cycles when 0.1 M NaOH was used, and up to about 60 cycles when cleaned with 0.5 M NaOH for 15 minutes. The resin uses a genetically engineered Protein A ligand designed for improved stability under alkali conditions. Other properties were also optimized for MAb capture.

QUESTIONS

1. What is downstream processing?. How these technologies are effective in cost reduction?
2. Write short notes:
   (*a*) Steps in downstream processing
   (*b*) Downstream processing
   (*c*) Downstream processing of recombinant protein
   (*d*) Laboratory and industrial bioreactors
   (*e*) Scale-up technology
   (*f*) Chromatographic methods
   (*g*) Platform technology
   (*h*) Cost reduction measures in downstream process.

ANIMAL CELL AND ISSUE CULTURE

CHAPTER 31

I. INTRODUCTION

The foundation of animal cell and tissue culture were laid at the beginning of the present century, when cell could be shown to survive and divide in-vitro (Jolly, 1903). The true beginning of animal cell and tissue culture was however, made by Ross Harrison (1907), by using frog as a source of tissue. Alenis Carrel (1912) used tissue and embryo extract as culture media. For about 50 years, mainly tissue explants rather than cell was used for culture techniques, although later after 1950s, mainly dispersed cell cultures were utilized. As in plants, the terms 'cell culture', 'tissue culture' and 'organ culture' are commonly used for animal material also. However, for animals if the explant maintains its structure and function in culture, it is described as an 'organotypic culture' and if cells in culture reassociate to create a three dimensional structure, irrespective of the tissue from which it was derived, it is described as histolytic culture.

II. ADVANTAGES AND DISADVANTAGES OF TISSUE CULTURE

There are several advantages and disadvantages of animal tissue culture; therefore, this technique should not be used indiscriminately.

The advantages of tissue culture includes:

1. Physico-chemical environments: control of pH, temperature, osmolarity, dissolved gases.
2. Physiological conditions: control of hormone and nutrient concentrations in medium.
3. Cell line homogeneity: availability of selective media, cloning.
4. Characterization: cytology and immunostaining are easily performed.
5. Preservation: can be stored in liquid nitrogen.
6. Replicates and variability: quantitation is easy.
7. Reagent saving: reduced volumes, direct access, lower cost.
8. Control of CXT: ability to define dose, concentration and time.
9. Mechanization: available with micro-titration and robotics.
10. Reduction of animal use: cytotoxicity and screening of pharmaceuticals, cosmetics etc.

The disadvantages and limitations of tissue culture are:

1. Expertise is necessary: handling, chemical contamination, microbial contamination, cross contamination.
2. Environmental control: workplace, incubation, pH controls, containment and disposable of biohazards.
3. Quantity and Cost: capital equipment, consumables, medium, and serum.
4. Genetic instability: heterogeneity, variability.
5. Phenotypic instability: de-differentiation, adaptation, and selection.
6. Identification of cell type: expression of markers, histology, cytology geometry and microenvironment.

III. TYPES OF TISSUE CULTURES

- Organ culture implies that the architecture characteristics of the tissue in-vivo are retained, at least in part in the culture. Towards this end, the tissue is cultured at the liquid-gas interface (on a soft, grid or gel), which favours the retention of a spherical or three-dimensional shape.
- In primary explants culture, a fragment of tissue is placed at a glass (or plastic)- liquid interface, where, following attachment, migration is promoted in the plane of the solid substrate.
- Cell culture implies that the tissue, or out growth from the primary explants is dispersed (mechanically or enzymatically) into a cell suspension, which may then be cultured as an adherent monolayer on a solid substrate or as a suspension in the culture medium.

The properties of different types of cultures are summarized as:

1. Organ cultures tend to retain the differentiated properties of the tissue from which culture was derived.
2. The organ cultures do not grow rapidly, cannot be propagated easily, poorer reproducibility of the sample than is achieved with cell culture. Quantitation is more difficult.
3. Organ cultures do not retain specific histological interactions and it is difficult to reproduce the characteristics of the tissue.
4. Cell cultures may be derived from primary explants or dispersed cell suspensions. The propagation of cell lines is feasible.
5. The cell line obtained from primary culture may be characterized and the characteristics will apply for most of its finite life span. The derivation of continuous cell live usually implies a phenotypic change or transformation.
6. Cultured cell lines are more representative of precursor cell compartment in-vivo than of fully differentiated cells, as normally most differentiated cell do not divide.
7. Cell cultures have many advantages in quantitation, characterization and replicate sampling but lack the potential for cell-cell interaction and cell-matrix interaction afforded by organ cultures.
8. Organotypic culture has given new prospectus for the study of cell interaction among discrete, defined populations of homogeneous and potentially genetically and phenotypically defined cells.

IV. LABORATORY FACILITIES FOR TISSUE CULTURE

The laboratory facilities for animal culture should allow for the following:

(*i*) Sterile handling
(*ii*) Incubation
(*iii*) Preparation of glassware, media and tissue.
(*iv*) Wash-up
(*v*) Sterilization
(*vi*) Storage

The clean area for handling should be available at one end of the room; wash up and sterilization at the other end, with preparation, incubation and storage in between.

V. CULTURE MEDIA

Culture medium is one single most important factor for culturing cells and tissues. It provides optimum conditions of factors like pH, osmotic pressure etc. and the chemical constituents, which the cell or tissues are incapable of synthesizing.

The natural media used to promote cell growth include coagula such as plasma clots, biological fluids such as serum and tissue extracts of which embryo extract is the most common. Although natural media are very useful and convenient for a wide range of uses, they suffer from disadvantages of poor reproducibility. For this reason synthetic media have been designed. A balanced salt solution (BSS) is composed of inorganic salts and may include sodium bicarbonate and in some cases glucose.

BSS forms the basis of many complete media and examples include Earle's BSS, Hank's BSS, Eagles MEM with Hanks's salts, Eagle's MEM with Earle's salts. BSS is also used as a diluent for concentrates of amino acids and vitamins to make complete media, as a washing or dissection medium and for short incubations upto 4 hour usually with glucose. These recipes are often modified; for example by omitting glucose or phenol red from Hank's BSS or by leaving out Ca^{+2} or Mg^{2+} ions from Dulbecco's PBS. PBS without Ca^{2+} and Mg^{2+} is known as PBS solution A. (PBSA). The compositions of commonly employed BSS's are given in Table 31.1.

Table 30.1. Constitutions of different balanced salt solutions.

Components	**Earle's balanced salt solution; g/1**	**Dulbecco's phosphate buffered saline (solution A) (PBSA);g/1**	**Hanks' balanced salt solution g/1**	**Spinner salt solution (Eagle);g/1**
Inorganic salts				
$CaCl_2$ (anhyd)	0.02	-	0.14	-
KCl	0.04	0.20	0.40	0.40
KH_2PO_4	-	0.20	0.06	-
$MgCl_2$ 6H2O	-	-	0.10	-
$MgSO_4$ $7H_2O$	0.20	-	0.10	0.20
NaCl	6.68	8.00	8.00	6.80
$NaHCO_3$	2.20	8.00	8.00	6.80
Na_2HPO4 $7H_2O$	-	2.16	0.09	-
NaH_2PO_4 $7H_2O$	0.14	-	-	1.40
Other components				
D-glucose	1.00	-	1.00	1.00
Phenol red	0.01	-	0.01	0.01

Complete media range in complexity from relatively simple Eagle's MEM (Eagle, 1959), which contains essential amino acids, vitamins and salts to complex media such as Medium 19 (M199) (Morgan et al., 1950), CMRL 1066 (Parker et al, 1957), MB752/1 (Waymouth, 1959), RPMI 1640 (Moore et al., 1967), and F12 (Ham, 1965) and a wide range of serum free formulations. The complex media contain a large number of different amino acids including non-essentials amino acids and additional vitamins and are often supplemented with extra metabolites (e.g. nucleosides, tricarboxylic acid cycle intermediates and lipids) and minerals. Common constituents of commonly used media with serum for animal cell and tissue culture are listed in Table 31.2.

The use of serum in culture media has several disadvantages, which led to development of many serum free media. Some of the disadvantages include:

(*i*) Serum varies from batch to batch and deteriorates within one year.

(*ii*) Serum may stimulate undesirable growth and may even inhibit growth in some cases.

(*iii*) Changing serum batches requires extensive testing to ensure that the replacement is as close as possible to the previous batch.

(*iv*) If more than one cell type is used each type may require a different batch of serum, so that a number of bathes may need to be maintained and co-culturing may be difficult.

(*v*) The demand of serum is increasing and it will probably exceed the supply.

(*vi*) Physiological variability is another factor, limiting the use or serum in culture media.

(*vii*) The presence of serum can create a major obstacle to purification in recovering cell products and may limit the pharmaceutical acceptance.

(*viii*) Serum is frequently contaminated with viruses, many of which may be harmful to cell culture.

(*ix*) Serum increases the cost of the medium.

Table. 31.2. Different commonly used media (with serum), for animal cell and tissue culture and their constituents (all quantities in mg/l).

Component	Eagle's MEM	Dulbecco's	Ham's F12	CMRL 1066	RPMI 1640	199	L15	Fischer's	Way mouth MB 752
Amino acids									
L-alanine	-	-	8.90	25.0	-	25.0	225	-	-
L-arginine (free base)	-	-	-	-	200	-	500	-	-
L-arginine-HCl	126	84.0	211	70.0	-	70.0	-	15.0	75.0
L-asparagine	-	-	-	-	50.0	-	-	-	-
L-asparagine-H2O	-	-	15.0	-	-	-	260	11.4	-
L-aspartic acid	-	-	13.3	30.0	20.0	30.0	-	-	60.0
L-cysteine (free base)	-	-	-	-	-	-	120	-	61.0
L-cystine	24	48.0	-	20.0	50.0	-	-	-	15.0
L-cystine 2Na	-	-	-	-	-	23.7	-	23.7	-
L-cysteine HCl.H2O	-	-	35.1	260	-	0.0987	-	-	-
L-glutamic acid	-	-	14.7	75.0	20.0	66.8	-	-	150
L-glutamine	292	584	146	100	300	100	300	200	350
Glycine	-	30.0	7.50	50.0	10.0	50.0	200	-	50.0
L-histidine (freebase)	-	-	-	-	15.0	-	250	-	128
L-histidine HCl.H2O	42.0	42.0	21.0	20.0	-	21.9	-	81.1	-
L-hydroxy proline	-	-	-	10.0	20.0	10.0	-	-	-
L-isoleucine	520	105	3.94	20.0	50.0	20.0	125	75.0	-
L-leudcine	520	105	13.1	60.0	50.0	60.0	125	30.0	-
L-lysine HCl	73.1	146	36.5	70.0	40.0	70.0	93	50.0	240
L-methionine	15.0	30.0	4.48	15.0	15.0	15.0	75.0	100	50.0
L-phenylalanine	33.0	66.0	4.96	25.0	15.0	25.0	125	67.0	50.0
L-proline	-	-	34.5	40.0	20.0	40.0	-	-	50.0
L-serine	-	42.0	10.5	25.0	30.0	25.0	200	15.0	-
L-threonine	48.0	95.0	11.9	30.0	20.0	30.0	300	40.10	75.0
L-tryptophan	10.0	16.0	2.04	10.0	5.0	10.0	20.0	10.0	40.0
L-tyrosine	36.0	72.0	5.40	40.0	20.0	-			40.0
L-valine	47.0	94.0	11.7	25.0	20.0	25.0	100	70.0	65.0

Vitamins									
L-ascorbic acid	-	-	-	50.0	-	0.050	-	-	17.5
Biotin	-	-	0.0073	0.010	0.200	0.010	-	0.010	0.02
D-Ca-pantothenate	1.00	4.00	0.480	0.010	0.250	0.010	1.00	0.500	1.00
Calcieferol	-	-	-	-	-	0.100	-	-	-
Choline chloride	1.00	4.00	14.0	0.500	3.00	0.500	1.00	1.50	250
Folic acid	1.00	4.00	1.30	0.010	1.00	0.010	1.00	10.00	0.40
i-inositol	2.00	7.20	18.0	0.050	35.0	0.050	2.00	1.50	1.00
Nicotinamide	1.00	4.00	0.04	0.025	1.00	0.025	1.00	0.50	1.00
Pyridoxal HCl	1.00	4.00	0.062	0.025	-	0.025	-	0.50	-
Riboflavin	0.10	0.40	0.038	0.010	0.20	0.010	-	0.50	1.00
Thiamin HCl	1.00	4.00	0.34	0.010	1.00	0.010	-	1.00	10.0
Vitamin $B1_2$	-	-	1.36	-	0.005	-	-	-	0.20
Pyridoxine HCl	-	-	0.062	0.025	1.00	0.025	-	-	1.00
Cholesterol	-	-	-	0.200	-	-	-	-	-
Para-amino-benzoic acid	-	-	-	0.050	1.00	0.050	-	-	-
Menaphthone sodium bisulphate. $3H_2O$	-	-	-	-	-	0.019	-	-	-
Dl-a tocopherol	-	-	-	-	-	-	0.10	-	-
PO_4. 2Na									
Inorganic salts									
$CaCl_2$ (anhyd.)	200	200	-	200	-	-	-	-	-
$CaCl_2$ $2H_2O$	-	-	44.0	-	-	186	186	91.0	120
$Fe(NO_3)_3$. $9H_2O$	-	0.10	-	-	-	0.10	-	-	-
KCl	400	400	224	400	400	400	400	400	400
KH_2PO_4	-	-	-	-	-	-	60.0	60.0	-
$MgCl_2.6H_2O$	-	-	122	-	-	-	-	-	-
$MgSO4.7H_2O$	200	200	-	200	100	200	400	121	200
NaCl	6,800	6,400	7,599	6,799	6,000	8,000	8,000	8,000	6,100
$NaHCO_3$	2,200	3,700	1,176	2,200	2,200	350	-	1,125	2,200
$Na_2H_2PO_4.H_2O$	140	125	-	140	-	-	-	78.0	-
Na_2HPO_4 (anhyd)	-	-	-	-	-	47.5	190	60.0	-
$Na_2HPO_4.7H_2O$	-	-	268	-	1,512	-	-	-	-
$CuSO_4$. $5H_4O$	-	-	0.00249	-	-	-	-	-	-
$FeSO_4.7H_2O$	-	-	0.834	-	-	-	-	-	-
$ZnSO_4.7H_2O$	-	-	0.863	-	-	-	-	-	-
$CaNO_3.4H_2O$	-	-	-	-	100	-	-	-	-
Other components									
D-glucose	1.000	4,500	1,802	1,000	2,000	1,000	-	1,000	5,000
D-galactose	-	-	-	-	-	-	9,000	-	-
Lipoic acid	-	-	0.21	-	-	-	-	-	-
Phenol red	10.0	15.0	12.0	20.0	5.00	17.0	10.0	5.00	10.0
Sodium pyruvate	-	110	110	-	-	-	550	-	-
Hypoxanthine	-	-	4.10	-	-	0.30	-	-	-
Linoleic acid	-	0.084	-	-	-	-	-	-	25.00
Putrescine .2HCl	-	-	0.161	-	-	-	-	-	-
Thymidine	-	-	0.73	10.0	-	-	-	-	-

Cocarboxylase	-	-	-	1.00	-	-	-	-	-
Coenzyme A	-	-	-	2.50	-	-	-	-	-
Deoxycytidine HCl	-	-	-	10.0	-	-	-	-	-
Deoxyguanosine	-	-	-	10.0	-	-	-	-	-
Diphosphopyridine nucleotide.	$4H_2O$	-	-	-	7.00	-	-	-	-
Ethanol for solubilzing lipid components	-	-	-	16.0	-	-	-	-	-
Flavine adenine dinucleotide	–	–	-	1.00	-	-	-	-	-
Glutathione	-	-	-	10.0	1.00	0.05	-	-	-
5-methyl-deoxy-cytidine	-	-	-	0.10	-	-	-	-	-
Sodium acetate. $3H_2O$	-	-	-	83.0	-	-	-	-	-
Sodium glucuronate. H_2O	-	-	-	4.20	-	-	-	-	-
Triphosphopyridine nucleotide	-	-	-	1.00	-	-	-	-	-
Tween 80	-	-	-	5.00	-	5.00	-	-	-
Uridine triphosphate. $4H_2O$	-	-	-	1.00	-	-	-	-	-
Adenine. SO_4	-	-	-	-	-	10.0	-	-	-
5'AMP	-	-	-	-	-	0.20	-	-	-
ATP-2Na	-	-	-	-	-	10.0	-	-	-
Cholesterol	-	-	-	-	-	0.20	-	-	-
2-Deoxyribose	-	-	-	-	-	0.50	-	-	-
Guanine HCl	-	-	-	-	-	0.30	-	-	-
D-ribose	-	-	-	-	-	0.50	-	-	-
Na acetate	-	-	-	-	-	36.7	-	-	-
Thymine	-	-	-	-	-	0.30	-	-	-
Uracil	-	-	-	-	-	0.30	-	-	-
Xanthine	-	-	-	-	-	0.30	-	-	-
CO_2 (gas phase)	5%	10%	5% (pH 7.0) 2%(pH 7.4)	5%	5%	Ambient	Ambient	5%	5% (pH 7.0) 2% (pH 7.4)

The major advantage of serum free media is the ability to make a medium selective for a particular cell type. Hormones that have been used to replace serum include growth hormone (somatotropin), 50 ng/ml, insulin at 1-10 U/ml, which enhances plating efficiency in a number of different cell types and hydrocortisone, which improves the cloning efficiency of glia and fibroplasts. Various combinations of estrogen, androgen or progesterone with hydrocortisone and prolactin at around 10 ng can be shown to be necessary for the maintenance of mammary epithelium. The composition of different serum free media is listed in Table 31.3. These media were designed to grow various types of cell cultures e.g. MCDB110 for human lung fibroblasts, MCDB202 for chick embryofibroblasts, Iscove's for lymphoids cells and LHC-9 for bronchial epithelium.

Antibiotics were originally introduced into the culture media to reduce the frequency of contamination. The addition of antibiotics into the culture media has a number of disadvantages such as development of antibiotic resistant organisms, hide low level. Cryptic contaminants and mycoplasma infections, antimetabolic effects etc. It is strongly recommended that their use be restricted to primary culture, large-scale labour intensive experiments with a high cost of consumables. A number of antibiotics used in tissue culture are moderately effective in controlling bacterial infections (Table 31.4).

Table. 31.3. Examples of serum-free media (basal formulae*) (modified from Freshney, 1987)

Component	MCDB110	MCDB202	Iscove's	LHC
Amino acids				
(All as L-enantiomers)				
Alanine	1.0F4	1.0Fe4	1.0E-4	2.8E-4
Arglnine HCl	1.0F3	3.0F-4	4.0E-4	2.0E-3
Asparagine	1.0F4	1.0F-3	1.9E-4	1.0E-4
Aspartic acid	1.0F4	1.0F-4	2.3E-4	3.0E-5
Cysteine-HCl	5.0E-5	2.0F4	-	2.4E-4
Cystine	-	2.0E-4	2.9E-4	-
Glutaminc acid	1.0E-4	1.0E-4	5.1E-4	1.0E-4
Glutamine	2.5E-3	4.0E-3	4.0E-3	6.0E-3
Glycine	3.0E-4	1.0E-4	4.0E-4	1.0E-4
Histidine HCl	1.0E-4	1.0E-4	2.0E-4	1.6E-4
Isoleucine	3.0E-5	1.0E-4	8.0E-4	3.0E-5
Leucine	1.0E-4	3.0E-4	8.0E-4	1.0E-3
Lysine HCl	2.0E-4	2.0E-4	8.0E-4	2.0E-4
Methionine	3.0E-5	3.0E-5	8.0E-4	2.0E-4
Phyenylalanine	3.0E-5	3.0E-5	4.0E-4	6.0E-5
Proline	3.0F-4	5.0E-5	3.5E-4	3.0E-4
Serine	1.0E-4	3.0E-4	4.0E-4	1.2E-3
Threonine	1.0E-4	3.0E-4	8.0E-4	2.0E-4
Tryptophan	1.0E-5	3.0F-5	7.8E-5	3.0E-5
Tyrosine	3.0E-5	5.0E-5	4.6E-4	3.0E-5
Valine	1.0E-4	3.0E-4	8.0E-4	6.0E-4
Vitamins				
d-Biotin	3.0E-4	3.0F-8	5.3E-8	6.0E-8
Folic acid	-	-	9.1E-6	1.8E-6
Folinic acid	1.0E-5	1.0E-8	-	-
DL-a-lipoic acid	1.0E-8	1.0E-8	-	1.0E-6
Niacinamide	5.0E-5	5.0E-5	3.3E-5	3.0E-7
D-pantothenate ½ Ca	1.0E-6	1.0E-6	1.7E-5	1.0E-6
Pyridoxal	-	-	2.0E-5	-
Pyridoxine-HCl	3.0E-7	3.0E-7	-	3.0E-7
Riboflavin	3.0E-7	3.0E-7	1.1E-6	1.0E-7

Thiamin HCl	1.0E-6	1.0E-6	1.2E-5	1.0E-6
Vitamine B12	1.0E-7	1.0E-7	9.6E-9	3.0E-7
Other organic constituents				
Acetate	-		-	3.7E-3
Adenine	1.0E-5	1.0E-6	-	1.8F-4
Choline chloride	1.0E-4	1.0E-4	2.9E-5	2.0E-4
D-glucose	4.0E-3	8.0E-3	2.5E-2	6.0E-3
Inositol	1.0E-4	1.0E-4	4.0E-5	1.0E-4
Linoelic acid	-	2.0E-7	-	-
Putrescine-2HCl	1.0E-9	1.0E-9	-	1.0E-6
Na Pyruvate	1.0E-3	5.0E-4	1.0E-3	5.0E-4
Thymidine	3.0E-7	3.0E-7	-	3.0E-6
Major Inorganic salts				
CaCl2	1.0E-3	2.0E-3	1.5E-3	1.1E-4
KCl	5.0E-3	3.0E-3	4.4E-3	1.5E-3
KNO2	-	-	7.5E-3	-
MgCl2	-	-	-	2.2E-2
MgSO4	1.0E-3	1.5E-3	8.1E-4	-
NaCl	1.1E-1	1.2E-1	7.7E-2	1.0E-1
Na2HPO4	3.0E-3	5.0E-4	1.0E-3	2.0E-3
Trace elements				
CuSO4	1.0E-9	1.0E-9	-	1.0E-8
FeSO4	5.0E-6	5.0E-6	-	5.4E-4
H2SO4	3.0E-8	3.0E-8	1.0E-7	3.0E-8
MnSO4	1.0E-9	5.0E-10	-	1.0E-9
Na2SiO3	5.0E-7	5.0E-7	-	5.0E-7
(NH4) 6 Mo7O24	1.0E-9	1.0E-9	-	1.0E-9
NH4VO3	5.0E-9	5.0E-9	-	5.0E-9
SaCl2	5.0E-10	5.0E-12	-	5.0E-10
ZnSO4	5.0E-7	1.0E-7	-	4.8E-7
Buffers and Indicators				
Hepes	3.0E-2	3.0E-2	2.5E-2	2.3E-2
NaHCO3	-	-	3.6E-2	1.2E-2
Phenol red	3.3E-6	3.3E-6	4.0E-5	3.3E-6
CO2	2%	2%	10%	5%

Computer style notation used concentration e.g. 3.0, E-2 = 30 mM. Sufficient NaOH is added to give pH 7.3 – 7.4 at 37°C in correct gas phase.

Table 31.4. Antibiotics used in tissue culture.

S.No.	Antibiotic	Concentration (μg/ml) Working	Concentration (μg/ml) Cytotonic	Activity against
1.	Amphotericin B	2.5	30	Yeasts, Fungi
2.	Ampicillin	2.5		Bacteria (Gram+ve, Gram-ve)
3.	Ciprofloxacin	100		Mycoplasma
4.	Erythromycin	50	300	Mycoplasma
5.	Gentamycin	50	>300	Bacteria (Gram +ve, Gram-ve) Mycoplasma
6.	Kanamycin	100	10,000	Bacteria (Gram +ve, Gram-ve); Mycoplasma
7.	Neomycin	50	3000	Bacteria
8.	Nystatin	50		
9.	Penicillin-G	100 U/ml	10,000 U/ml	Bacteria (Gram+ve)
10.	Polymixin B	50	1000	Bacteria (Gram-ve)
11.	Streptomycin Sulfate	100	20,000	Bacteria (Gram+ve, Gram-ve)
12.	Tetracyclin	10	35	Bacteria (Gram+ve, Gram-ve)
13.	Tylosin	10	300	Mycoplasma

VI. SUBSTRATE FOR CELL GROWTH

The majority of vertebrate cells cultured in vitro have been grown as monolayers on an artificial substrate. Spontaneous growth in suspension is restricted to hemopoietic cell lines, rodent ascites tumor cell lines and a few other selected cell lines. With these exceptions, most cell need to spread out on a substrate is order to proliferate. Inadequate spreading will inhibit cell proliferation. Cell sown to require attachment for growth are said to be anchorage dependent.

However cells, which have undergone transformation, frequently become anchorage independent and can grow in suspension. (Note: In microbiology and plant cell culture, transformation means uptake of new genetic material but in animal cell culture, particularly mammalian cell culture transformation means phenotypic modification of cells in culture, without involving the uptake of new genetic material, which is described as transfection to distinguish from transformation. Whenever, a substrate is needed, it may be adhesive (glass, plastic, palladium etc) or a metallic surface; these surfaces may sometimes be variously modified) or non-adhesive (agar, agarose, methocel).

(i) Glass as a substrate: Glass is most commonly used substrate, in the form of slides, cover ships, flasks and test tubes, which may or may not be variously modified. It is cheap, easily washed, without losing its growth supporting properties, can be readily sterilized by dry heat or moist heat and is optically clear for microscopic examination of cultures.

(ii) Disposable plastic as substrate: Polystyrene is often used as the most common and cheapest substrate, to be discarded after single use. This plastic is supplied sterile and is not autoclavable. However, it is hydrophobic and needs to be treated by gamma irradiation or some chemical or with electric are to produce a charged surface, which thus becomes wettable. Other plastics used for tissue culture include polyvinyl chloride (PVC), polycarbonate, polytetra fluoroethylene or teflon (PTFE), thermanox (TPX) etc. PTFE films are available as disposable dishes or as membranes to be incorporated in an autoclavable resuable culture vessel. Substrates involving polystyrene, sephadex and polyacrylamide are also available in the form of beads for propagation of anchroage dependent cells in suspension.

(iii) Palladium and metallic surface as substrates: For 90% of animal cell and tissue cultures, glass or plastic is employed as a substrate. For instance palladium deposited on agarose (using electron microscopy shadowing equipment) as a substrate for growth of fibroblasts and gila. Stainless steel discs and other metallic surfaces were also used.

VII. CULTURE PROCEDURES

Culture procedures involve preparation and sterilization of glassware's, equipment, reagents and media. The apparatus and liquids meant for tissue culture should be reserved for that purpose above and washing of glassware's should also be separate, since the requirement of washing for tissue culture is higher than that of general glassware.

(i) Preparation and sterilization of apparatus: Sterilization procedures are designed not just to kill replicating microorganisms but also to eliminate the most resistant spores. Moist heat is more effective than dry heat. For dry heat a minimum temperature of 160°C maintained for 1 hour is required. Moist heat (for fluids and perishable items) should be maintained at 121°C for 15-20 min.

(ii) Preparation and sterilization of reagents and media: Most reagents and media can be sterilized either by moist heat i.e. by autoclaving (if heat stable) or by membrane filtration (if heat liable). During autoclaving the container may be kept sealed (borosilicate or polycarbonate) or the caps may be left stack (soda glass) to avoid breakage. Media and reagents supplied on-line to large-scale culture vessels and industrial or semi-industrial bioreactors can be sterilized on-line by short duration ultra high temperature treatment. Filtration through 0.1-0.2 μm microporous filters is the method of choice for sterilizing heat liable solutions. Culture media are also available in autoclavable form.

Media Preparation

Media can be purchased as prepared liquid media, made up in the laboratory from dried powders containing most of the components of the nutrient mixtures, or prepared in the laboratory from individual stocks of the individual components or groups of the components. Medium components deteriorate with time, and do so faster in solution. Some necessary components break down and are lost, others create toxic breakdown products or oxidize to toxic components. While it varies from cell to cell and with serum-supplemented or serum-free media, it is found that 2 weeks is a safe storage time for serum-free media, or longer if serum is added when the medium is prepared. Outdated medium can be used for washing cells or preparing tissues for primary culture. Some prepared liquid media can be frozen. Those that form a precipitate when thawed should not be frozen. In any case, it is always safe to store the prepared powdered medium and make liquid medium in the laboratory on a regular basis.

When preparing primary cultures, an antibiotic may be added to the wash medium during the initial stages of tissue handling. This may be done by preparing a 1000-fold concentrated stock solution of an antibiotic such as gentamycin (in phosphate-buffered saline) and adding it directly to the wash medium.

(iii) Preparation of animal material: It is essential that all sort of contamination should be avoided during experimental manipulations by collecting tissue aseptically or sterilizing it after removal. Embryo or other contents of uterus are free from contamination while other adult tissues are contaminated with bacteria. The embryo should be removed carefully under aseptic conditions eliminating the possibility of contamination. The adult tissue may be washed with a solution having high concentration mixture of antibiotic. The mixture is a balanced salt solution containing 100 units of penicillin and 0.5 mg of treptomycin or neomycin per ml. Future handling of tissue should be carefully done under fully aseptic conditions, which can be created either through preparing an aseptic room or through the use of laminar flow cabinet or hood. The use of hoods has greatly facilitated the handling of material under aseptic conditions (Fig. 31.1).

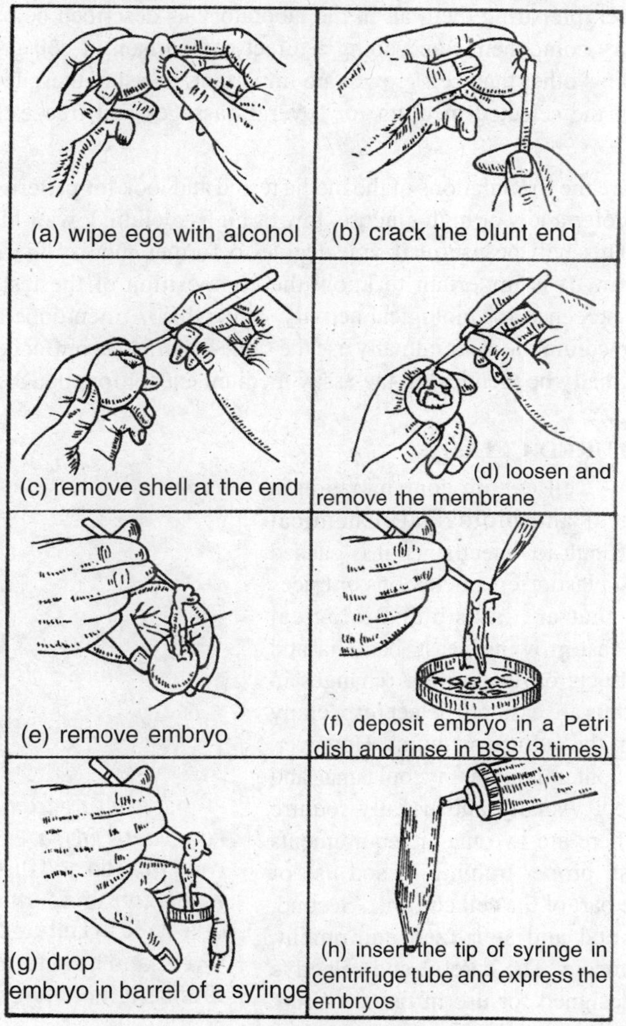

Fig. 31.1. Procedure of preparing embryo extract

VIII. SELECTION OF APPROPRIATE MEDIUM

If a new cell line is brought into the laboratory, determine what medium is recommended for its growth. This information can be obtained from the same source as the cells. If the recommended medium is incompatible with the CO_2 settings on the incubator used for other cells grown in the laboratory, or is not commonly prepared in the laboratory, one might wish to change the growth medium. It is best to initially grow the cells in their original medium and compare this with the more convenient medium after a passage or two in each. If the growth rate and morphology of the cells look the same, then a medium switch can be made. However, keep in mind when trying to repeat published data that cells grown in a different medium may respond differently in some parameters measured. For example, estrogen-requiring cells previously grown in F12 medium and then switched to DME medium may have a diminished response to exogenously added estrogens in this medium because of the weak estrogenic activity of the phenol red, which is present at a much higher concentration in the DME medium.

If one wishes to grow a primary cell in culture and no published data exist on growing that cell type *in vitro,* or one wishes to grow the cells in a different manner (e.g., with defined supplements rather than serum), it is best to screen several of the commercially available media before deciding on the one that is best for that particular use. This can be done by obtaining five to ten candidate media powders from a supplier, preparing them all in the laboratory as described below using the same water and supplementary components, and doing a direct comparison of cell growth in the different conditions. If endpoints other than cell growth are important, measure these too in each of the media. Carry the cells in the selected medium for several passages and freeze them in this medium for future use.

It is interesting to compare the formulations of the media tested and look for patterns of whether the cells of interest seem to prefer a low or high glucose, low or high calcium, low or high levels of amino acids, and so forth. This will be useful if you decide to further modify the medium, as described in the section below. It is important to know the composition of the test medium if conditioned medium is to be screened for biological activity. Since this is often done at 10¨C50% conditioned medium, the test medium should minimally use the same bicarbonate buffer concentration as the assay medium and optimally be identical to the assay medium except for changes introduced by the conditioning cells.

IX. PROBLEMS OF CULTURED CELLS

Avoiding contamination- Cell culture contamination is of two main types: chemical and biological. Chemical contamination is the most difficult to detect since it is caused by agents, such as endotoxins, plasticizers, metal ions or traces of chemical disinfectants, that are invisible. Biological contaminants in the form of fast growing yeast, bacteria and fungi usually have visible effects on the culture (changes in medium turbidity or pH) and thus are easier to detect (especially if antibiotics are omitted from the culture medium). However, two other forms of biological contamination, mycoplasmas and viruses, are not easy to detect visually and usually require special detection methods. There are two major requirements to avoid contamination. First, proper training in and use of good aseptic technique on the part of the cell culturist. Second, properly designed, maintained and sterilized equipment, plasticware, glassware, and media. The careful and selective (limited) use of antibiotics designed for use in tissue culture can also help avoid culture loss due to biological contamination.

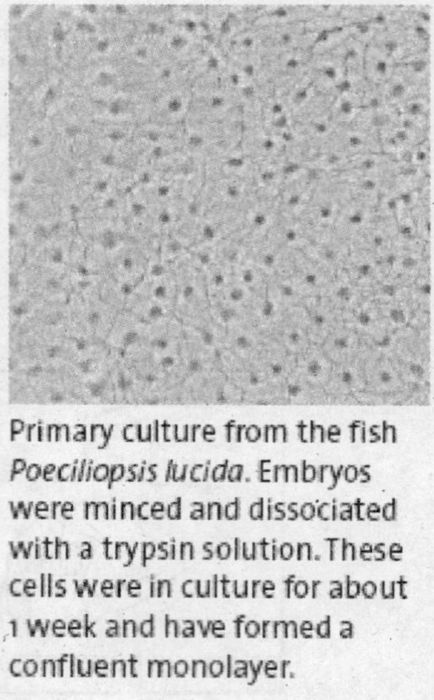

Primary culture from the fish *Poeciliopsis lucida*. Embryos were minced and dissociated with a trypsin solution. These cells were in culture for about 1 week and have formed a confluent monolayer.

X. FUNCTIONAL CHARACTERISTICS

The characteristics of cultured cells result from both their origin (liver, heart, etc.) and how well they adapt to the culture conditions. Biochemical markers can be used to determine if cells are still carrying on specialized functions that they performed in vivo (e.g., liver cells secreting albumin). Morphological or ultrastructural markers can also be examined (e.g., beating heart cells). Frequently, these characteristics are either lost or changed as a result of being placed in an artificial environment. Some cell lines will eventually stop dividing and show signs of aging. These lines are called Finite. Other lines are, or become immortal; these can continue to divide indefinitely and are called continuous cell lines. When a "normal" finite cell line becomes immortal, it has undergone a fundamental irreversible change or "transformation". This can occur spontaneously or be brought about intentionally using drugs, radiation or viruses. Transformed cells are usually easier and faster growing, may often have extra or abnormal chromosomes and frequently can be grown in suspension. Cells that have the normal number of chromosomes are called Diploid cells; those that have other than the normal number are Aneuploid. If the cells form tumors when they are injected into animals, they are considered to be Neoplastically Transformed.

XI. TYPES OF CELLS

Cultured cells are usually described based on their morphology (shape and appearance) or their functional characteristics. There are three basic morphologies:

1. Epithelial-like: cells that are attached to a substrate and appear flattened and polygonal in shape.
2. Lymphoblast-like: cells that do not attach normally to a substrate but remain in suspension with a spherical shape.
3. Fibroblast-like: cells that are attached to a substrate and appear elongated and bipolar, frequently forming swirls in heavy cultures.

XII. PRIMARY CULTURE

When cells are surgically removed from an organism and placed into a suitable culture environment, they will attach, divide and grow. This is called a primary culture. There are two basic methods for doing this. First, for Explant Cultures, small pieces of tissue are attached to a glass or treated plastic culture vessel and bathed in culture medium. After a few days, individual cells will move from the tissue explant out onto the culture vessel surface or substrate where they will begin to divide and grow. The second, more widely used method, speeds up this process by adding digesting (proteolytic) enzymes, such as trypsin or collagenase, to the tissue fragments to dissolve the cement holding the cells together. This creates a suspension of single cells that are then placed into culture vessels containing culture medium and allowed to grow and divide. This method is called Enzymatic Dissociation.

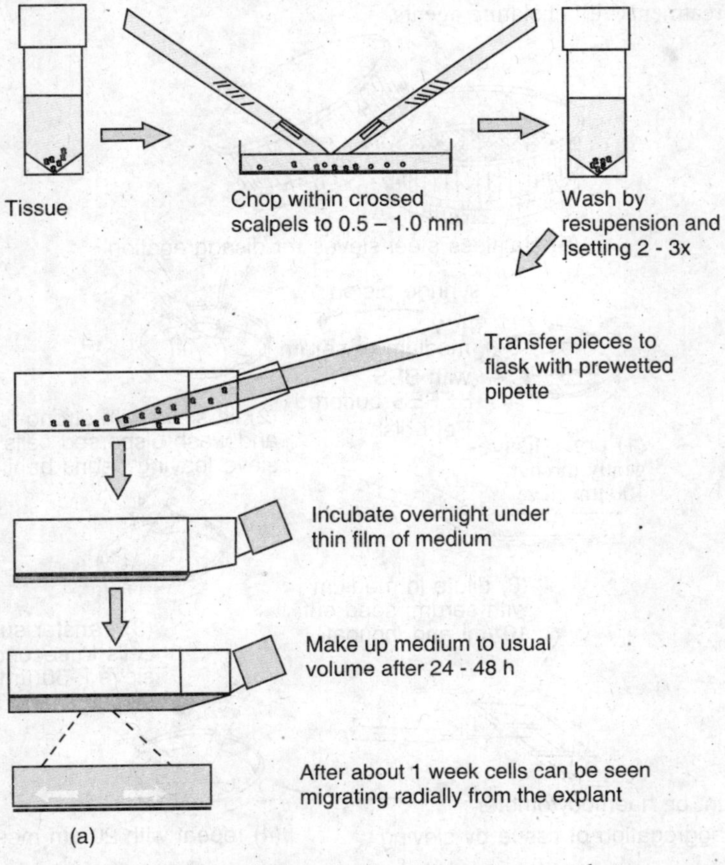

Fig. 31.2. Stages in dissection and seeding primary explants

A primary culture is that stage of the culture following isolation of the cells but before the first subculture. There are three stages to consider:

(*i*) Isolation of the tissue.

(*ii*) Dissection and/or disaggregation

(*iii*) Culture following seeding into the culture vessel.

Following isolation a primary cell culture may be obtained either by allowing cells to migrate out from the fragments of tissue adhering to a suitable substrate or by disaggregating the tissue mechanically or enzymatically to produce a suspension of cells, some of which will ultimately attach to the substrate. It appears to be essential for most normal untransformed cells-with the exception of hematopoietic cells-to attach to a flat surface in order to survive and proliferate with maximum efficiency. Transformed cells, on the other hand, particularly cells from transplantable animal tumors, are often able to proliferate in suspension (Fig. 31.2).

A. (i) Isolation of tissue: For tissue isolation sterilize the site with 70% alcohol and remove the tissue aseptically. The isolated tissue may be stored in a refrigerator before transferring it to balanced salt solution (BSS) or to a suitable culture medium because viable cells may be recovered from the chilled tissue several days after explanation. Different protocols are available for isolation of tissue like mouse embryo, hen's egg, human biopsy material etc.

B. (ii) Disaggregation: It is achieved by anyone of the following three methods-

(*a*) Mechanical disaggregation (Physical disruption)

(*b*) Enzymatic digestion

(*c*) Treatment with chelating agents.

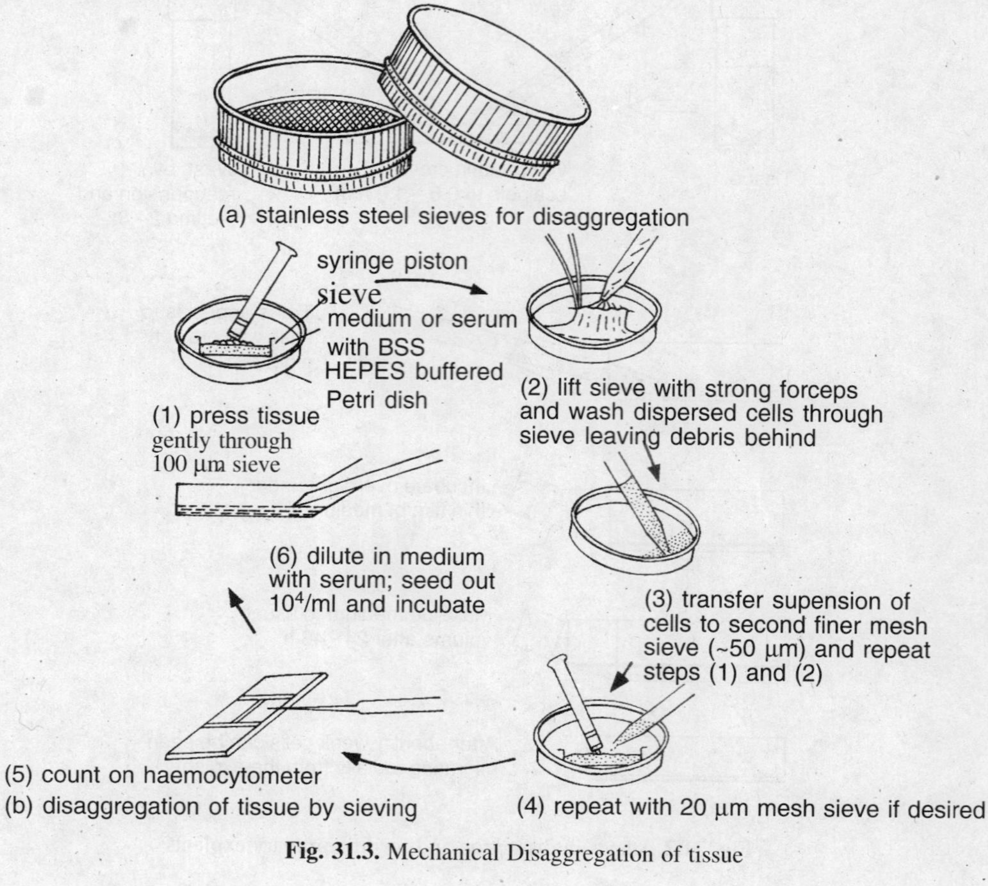

Fig. 31.3. Mechanical Disaggregation of tissue

(a) Mechanical disaggregation: The mechanical disaggregation is force-based technique in which the tissue is carefully sliced and the cells, which spill out, are collected. In this technique the cells are either pressed through the sieves of gradually reduced mesh or forced through a syringe and needle or repeatedly pipetted. Although the method may cause mechanical damage, the cell suspension is more quickly obtained than in the enzyme disaggregation. When the availability of tissue is no limitation and the efficiency of yield unimportant mechanical disaggregation may be used to obtain good yield of cells in a shorter time, but at the expense of very much more tissue. However, mechanical disaggregation provides good yield of cells in a shorter time (Fig. 31.3).

(b) Enzymatic disaggregation: The enzymes used most frequently are crude preparations of trypsin, collagenase, elastase, haluronidase, DNAse, pronase, dispase, alone or in various combinations. Trypsin and pronase give the most complete disaggregation, but may damage the cells. Collagenase and dispase, on the other hand, give incomplete disaggregation but are less harmful. Haluronidase can be used in conjunction with collagenase to digest the intracellular matrix, and DNAse is employed to disperse DNA released from lysed cells, as it tends to impair proteolysis and promote reaggregation.

Disaggregation by trypsin (for embryonic tissue) is termed as trypsinization and may be of two types: (1) Warm trypsinization and (2.) Cold trypsinization. Crude trypsin is the most common enzyme used for disaggregation. It can be tolerated by a variety of cells and is effective for many tissues. Its residual activity is neutralized by the serum of the medium or by a trypsin inhibitor (e.g. soybean trypsin inhibitor) in case of a serum free medium. (1) In warm trypsinization it is important to minimize the exposure of cells to active trypsin in order to preserve maximum viability thus, the cells are exposed to the warm enzyme (37°C) for a minimum period, the dissociated cells are collected every half an hour and the trypsin is removed by centrifugation after 3-4 hour required for complete disaggregation. This technique is useful for the disaggregation of large amounts of tissue in a relatively short time, particularly for whole mouse embryo, chick embryo etc. It does not work as well with adult tissue in which there is a lot of fibrous connective tissue, and mechanical agitation can be damaging to some of the more sensitive cell types, such as epithelium. (2) The cold trypsin method usually gives a higher yield of viable cells, with improved survival after 24 hrs culture and preserves more different cell types than the warm method. Cold trypsinization involves soaking of tissue in trypsin at 4°C to allow penetration of enzyme, followed by incubation at 36.5°C for a shorter period. The cold trypsin method is also convenient, as no stirring or centrifugation is required and the incubation at 4°C may be done overnight (Fig. 31.4).

(3) Other enzymatic procedures- Some times the disaggregation by trypsin may be damaging for epithelial cells or it becomes ineffective for fibrous tissues. The extracellular matrix contains collagen particularly in connective tissue and muscle, therefore collagenase has proved effective for disaggregation of several normal and malignant tissues. It is of great benefit when the tissue is either too fibrous or too sensitive to allow the successful use of trypsin. Often crude collagen is used. The process is gentle and requires no mechanical agitation or special equipment. With more than 1g of tissue, however it becomes tedious at the dissection stage and can be expensive because of the amount of collagenase required. The discrete clusters of epithelial cells produced by disaggregation in collagenase and by the cold trypsin method can be selected under a dissection microscope and transferred to individual wells on a microtitration plate, alone or with irradiated or mitomycin C treated feeder cells (Fig. 31.5).

(c) Treatment with chelating agents: Cell –cell adhesion in tissues is mediated by a variety of homotypic interacting glycopeptides (cell adhesion molecules CAMs), some of which are calcium dependent and hence are sensitive to chelating agents such as EDTA or EGTA. Chelating agents are used for the production of cell suspensions from established cultures. Epithelium tissues, which need Ca^{2+} and Mg^{2+} ions for its integrity, can be treated with chelating agents; e.g. include citrate and ethylene-diamine- tetra-acetic acid (EDTA).

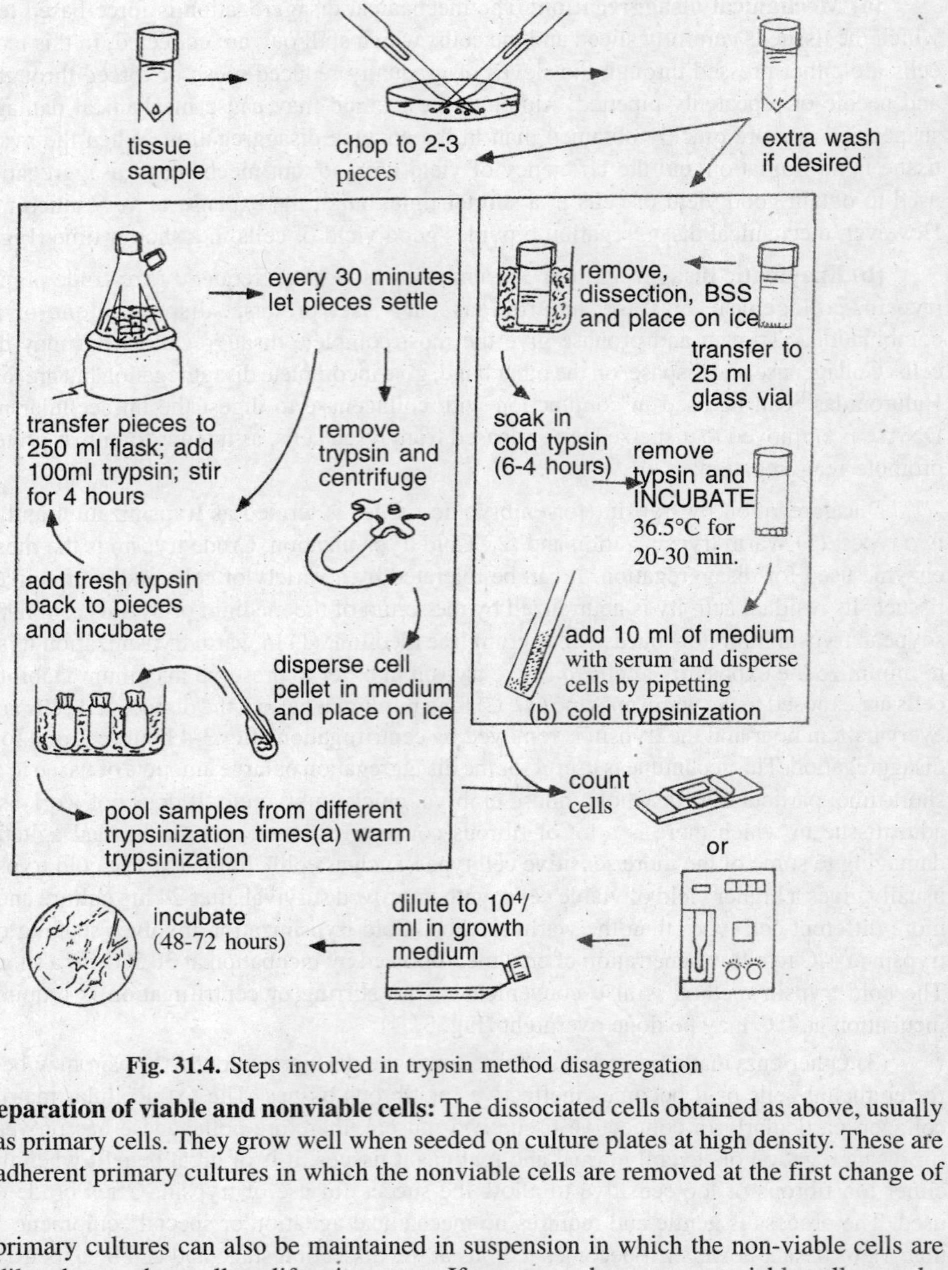

Fig. 31.4. Steps involved in trypsin method disaggregation

(d) Separation of viable and nonviable cells: The dissociated cells obtained as above, usually described as primary cells. They grow well when seeded on culture plates at high density. These are known as adherent primary cultures in which the nonviable cells are removed at the first change of medium.

The primary cultures can also be maintained in suspension in which the non-viable cells are gradually diluted out, when cell proliferation starts. If necessary, however, nonviable cells may be removed by centrifuging the cells on a mixture of Ficoll and sodium metrizoate. The dead cells will form a pellet at the bottom of the tube.

(e) Subculturing. When the cells in the primary culture vessel have grown and filled up all of the available culture substrate, they must be subcultured to give them room for continued growth. This is usually done by removing them as gently as possible from the substrate with enzymes. These are similar to the enzymes used in obtaining the primary culture and are used to break the protein bonds attaching the cells to the substrate. Some cell lines can be harvested by gently scraping the cells off the bottom of the culture vessel. Once released, the cell suspension can then be subdivided and placed into new culture vessels. Once a surplus of cells is available, they can be treated with suitable cryoprotective agents, such as dimethylsulfoxide (DMSO) or glycerol, carefully frozen and then stored at cryogenic temperatures (below -130°C) until they are needed. The regrowth of the subcultured cell line follows the standard growth pattern. A lag period after seeding is followed by a period of exponential growth, called the log phase. When the cell density (cells /cm^2 substrate) reaches a level such that all of the available substrate is occupied, or when the cell concentration (cells/mL medium) exceeds the capacity of the medium, growth ceases or is greatly reduced. The cells should be taken between the middle of the log phase and the time before which they entered the plateau phase of a previous subculture. The cells should not be subcultured while they are in the lag phase.

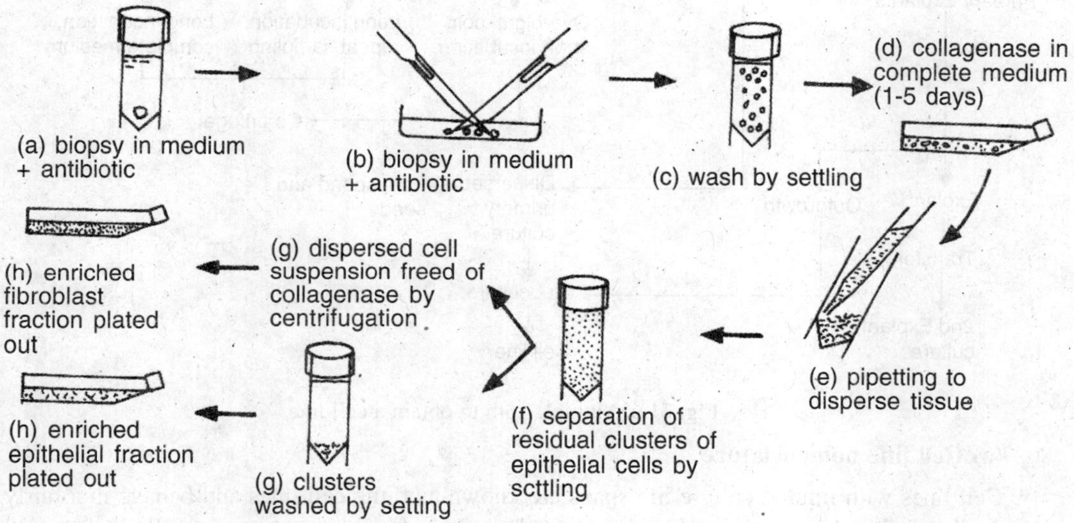

Fig. 31.5. Steps involved in disaggregation by collagenase

XIII. CELL CULTURE AND CELL LINES

Two basic culture systems are used for growing cells. These are based primarily upon the ability of the cells to either grow attached to a glass or treated plastic substrate (Monolayer Culture Sytems) or floating free in the culture medium (Suspension Culture Systems). Monolayer cultures are usually grown in tissue culture treated dishes, T-flasks, roller bottles, or multiple well plates, the choice being based on the number of cells needed, the nature of the culture environment, cost and personal preference. Suspension cultures are usually grown either:

1. In magnetically rotated spinner flasks or shaken Erlenmeyer flasks where the cells are kept actively suspended in the medium;

2. In stationary culture vessels such as T-flasks and bottles where, although the cells are not kept agitated, they are unable to attach firmly to the substrate. Many cell lines, especially those derived from normal tissues, are considered to be Anchorage-Dependent, that is, they can only grow when attached to a suitable substrate. Some cell lines that are no longer considered normal (frequently designated as Transformed Cells) are frequently able to grow either attached to a substrate or floating free in suspension; they are Anchorage-Independent. In addition, some normal cells, such as those found in the blood, do not normally attach to substrates and always grow in suspension.

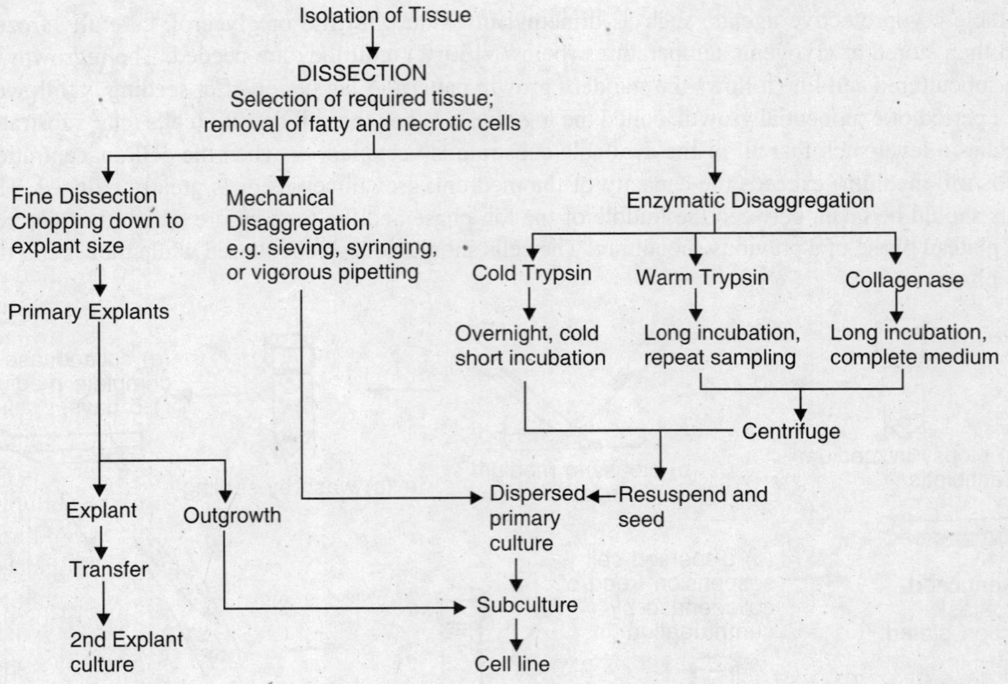

Fig. 31.6. Multiple path to obtain a cell line

(a) Cell line nomenclature

Cell lines with limited culture life spans are known as finite cell lines and behave in a fairly reproducible fashion. They grow through a limited number of cell generations, usually between 20 and 80 cell population doublings, before extinction. Cell lines should be given a code or designation (e.g. normal human brain (NHB)) a cell strain or cell line number (if several cell lines were derived from the same source e.g. NHB1, NHB2 etc) and if cloned a clone number NHB2-1, NHB2-2 etc. For finite cell lines, the number of population doublings should be estimated and indicated after a forward slash e.g. NHB2/2, and increases by one for a split ratio of 1:2 (e.g. NHB2/2, NHB2/3 etc), by two for a split ratio of 1:4 (e.g. NHB2/2, NHB2/4 etc.), and so on.

When referenced in reports it is helpful to prefix the cell line designation with a code indicating the laboratory in which it was derived e.g. NCI for National Cancer Institute.

Fig. 31.5. Properties of finite and continuous cell lines

Properties	Finite	Continuous (transformed)
Transformation	Normal	Immortal, tumorigenic
Ploidy	Euploid, diploid	Aneuploid, heteroploid
Anchorage dependent	Yes	No
Contact inhibition	Yes	No
Density limitation of cell growth	Yes	Reduced or lost
Mode of growth	Monolayer	Monolayer or suspension
Maintenance	Cyclic	Steady state possible
Serum requirement	High	Low
Cloning efficiency	Low	High
Markers	Tissue specific	Chromosomal, enzymic, antigenic

Special functions (virus, susceptibility, differentiation)	May be retained	Often lost
Growth rate	Slow	Rapid
Yield	Low	High
Control parameters	Generation no.; tissue –specific markers	Stain characteristics

Once a primary culture is sub-cultured (passaged or transferred), it becomes known as a cell line. This term implies the presence of several cell lineages of either similar or distinct phenotypes. Among these cell lineages, if a particular cell lineage has specific properties, which are identified in the bulk of the cell of that culture, it is described then as a cell strain. If a cell line transforms in vitro, it give rise to a continuous cell line and if selected or cloned and characterized, it is known as continuous cell strain.

Most primary cultures or continuous cell lines grow as monolayers. These cultures will need a periodic change of medium, whether or not the cells are proliferating. Intervals between medium changes and between subcultures vary from one cell line to another, depending upon the rate of growth and metabolism; rapidly growing transformed cell lines such as Hela, are usually sub cultured once per week, and the medium should be changed four days later. More slowly growing cell lines are required to be subcultured every two, three or even four weeks and the medium should be changed weekly between subcultures. The factors that indicate the needs for replacement of culture medium are drop in pH, cell concentration, cell type and morphological deterioration. Cells, which grow continuously in suspension, either because they are non-adhesive (e.g. many leukemia's and murine ascite tumors) or because they have been kept in suspension mechanically, may be subcultured like bacteria or yeast. No enzyme treatment is required in this case and the subculturing is done by dilution of cell density, rather than by replacement of the medium.

Subculture need

The need to subculture a monolayer is determined by the following criteria (1) Density of culture- Normal and transformed cells should be subcultured as they reach confluence. Though the transformed cells will continue to proliferate beyond confluence but after about two doublings they will start deteriorating. (2) Exhaustion medium- exhaustion of media usually indicates that medium requires replacement, fall in the pH is the main indicator of need to subculture. (3) Last subculture- Routine subculture should be performed so that reproducible behavior is achieved and monitored.to reduce or increase the seeding density is based on the confluence. (4) Other procedures- When cells are required for other proposes other than the routine maintenance then this should be done at the time of subculture Fig. 31.7.

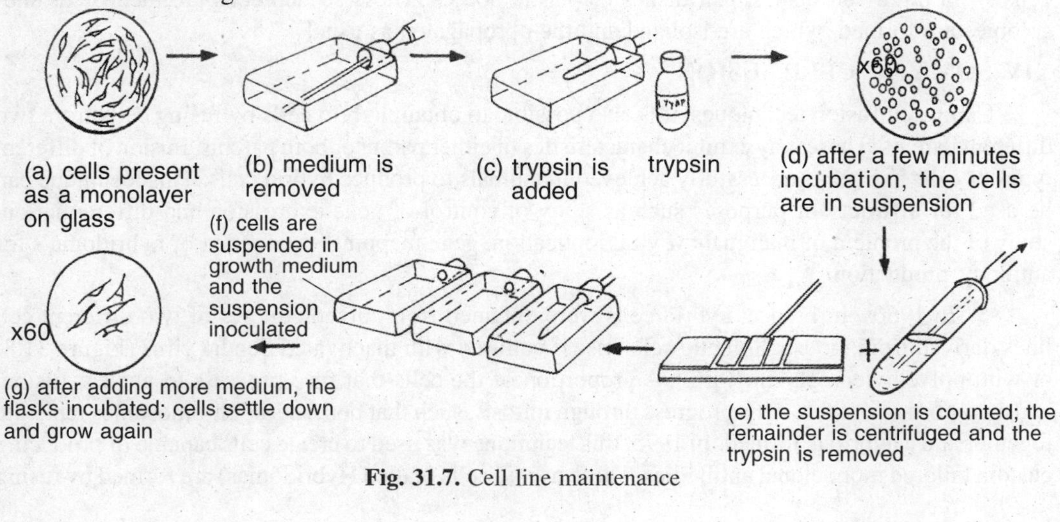

Fig. 31.7. Cell line maintenance

(c) Primary cell cultures vs. cell lines Two types of cell culture should be distinguished: **1.** Primary cultures, which are obtained directly from an animal and can keep the differentiated state for a short period (days to weeks). Functionally differentiated primary cell cultures have a limited life span, and although maintenance of the differentiated properties has been improved by additives to the culture medium, components of the extracellular matrix or by different forms of co-culture (see Figures 2 and 3), cell specific functions will eventually decline. **2.** Permanent cultures (e.g. HeLa, 3T3, MDCK), which have an unlimited proliferation capacity and which originated from embryos, tumors or transformed cells.

Cells can proliferate and/or differentiate, both with different limitations, depending on the cell type studied (e.g. neurons). Numerous publications provide protocols for the isolation of different cell types, their culture conditions, and for the evaluation of the degree of differentiation. Much attention is presently given to stem cells (see our News section). Primitive pluripotent embryonic stem cells derived from post-mortem foetal tissue have been used in cell therapy studies to replace diseased tissues. Until recently it was difficult to isolate and culture stem cells from adult tissues other than bone marrow. However, the cell therapy field is rapidly expanding, and it appears that many tissues harbour adult stem cells that possess a much greater plasticity than previously thought. Adult stem cells may obviously also be used as a model system, and it is to be expected that they will receive more attention also in the veterinary sciences.

(d) Cloning of cell lines

From heterogeneous cultures, pure cell lines can be obtained by cloning but since many cultures have each of finite life span, the cloning will be successful only in isolating variants from continuous cell lines. These variants may include biochemical mutants and cell strains with marker chromosomes cloning may be carried out in monolayer culture, using petridishes, multiwell plates or flasks. Micromanipulation is the only conclusive method for determining genuine clonality of a colony. Cloning can also be carried out in suspension by seeding cells into a gel, such as agar or a viscous solution, such as methocel. The stability of the gel or viscosity of the methocel ensures that daughter cell do not break away from the colony as it forms. Hematopoietic cells are usually cloned in suspensions. A continuous cell line generally have high plating efficiency in monolayer and is suspension due to their transformed status, while normal cells, which may have a moderately high cloning efficiency in monolayer, have a very low cloning efficiency in suspension.

Dilution cloning proposed by Puck and Marcus (1955) is most widely used method. Trypsinized individual cells are used for preparing suspensions and from the suspension, cell are seeded at low density, in multiwell dishes, petridishes or plastic bottles. These seeded cells are incubated, until colonies are formed, which are isolated and then propagated as usual.

XIV. SOMATIC CELL FUSION

Using cell fusion techniques, it is also possible to obtain hybrid cells by fusing cells from two different parents. These may exhibit characteristics of either parent or both parents. Fusion of different types of cells has been successfully achieved in animals to produce hybrid cells. This technique can be used for a variety of purposes such as study of control of gene expression and differentiation, study of the problem of malignancy, viral replication, gene mapping, production of hybridoma's for antibody production.

In 1960s, in France, hybrid cells were obtained from mixed cultures of two different cell lines derived from mouse. Somatic cells fuse if cultured with inactivated Sendai virus (Figure 31.8) or with polyethylene glycol (PEG). A proportion of the cells that fuse progress to nuclear fusion, and a proportion of these cells progress through mitosis, such that both sets of chromosomes replicate together and a hybrid is formed. In 1975, this technique was used to create cells capable of producing custom tailored monoclonal antibodies. These hybrid cells (called Hybridomas) are formed by fusing two different but related cells. The first is a spleen-derived lymphocyte that is capable of producing the desired antibody. The second is a rapidly dividing myeloma cell (a type of cancer cell) that has the machinery for making antibodies but is not programmed to produce any antibody. The resulting hybridomas can produce large quantities of the desired antibody. These antibodies, called Monoclonal Antibodies due to their purity, have many important clinical, diagnostic, and industrial applications with a yearly value of well over a billion dollars.

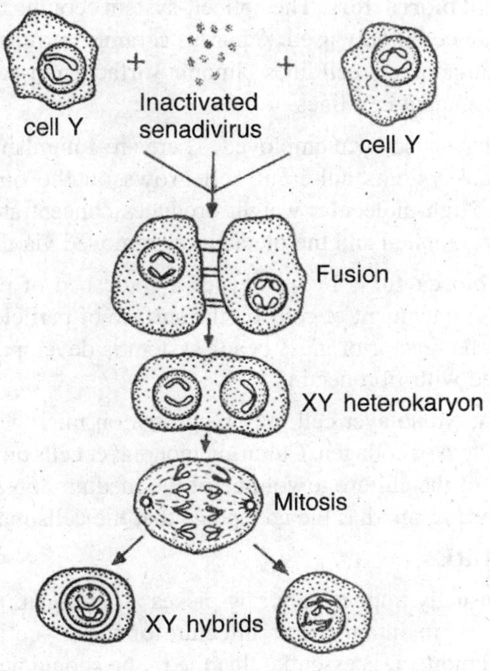

Fig. 31.8. Various steps of production of somatic cell hybrids using Sendai virus.

XV. LARGE-SCALE ANIMAL CELL CULTURE

The name fermenter derives from microbiological culture systems designed originally for bacteria and yeast and was used initially for laboratory equipment of around 50-100 l capacity, but the increase in scale in line with development in the biotechnology industry. Coupled with a greater diversity in design to cope with monolayer cultures as well as suspension, has led to the introduction of the name bioreactor. A wide variety of reactor types have been used for large-scale up of animal cell culture.

1. Process for cells in suspension culture

(i) Stirred tank bioreactors: Cells in suspension can be grown in stirred tank reactors. It is important to control the means of agitation of animal cell culture because animal cells are devoid of cell wall and fragile. The stirring speed should be between 30 and 100 rpm. Impeller fitted at the end of mechanical driveshafts is designed to allow vertical as well as horizontal liquid flow. Temperature is controlled by a thermo circulator which pumps heated water or air around and outer jacket or through coiled pipes within the culture. It may also be noted that sparging with CO2 and air is required to maintain adequate gas exchange. Antifoam or Pluronic F68 (0.01-0.1%) should be included when the serum concentration is above 2%. In the absence of serum, it may be necessary to increase the viscosity of the medium with (1-2%) carboxy methylcellulose.

(ii) Airlift fermenters: It consists of two cocentric cylinders, with the inner cylinder being shorter at both and than the outer, thereby creating an outer and inner chamber. The bottom of the inner chamber carries a sintered steel ring through which 5% CO2 in air is bubbled. The bubbles rise, carrying the cell suspension with them. CO2 is vented from the top and the displacement ensures the return of the cell suspension down the outer chamber.

2. Scale-up in monolayer

For anchorage-dependent monolayer cultures, it is necessary to increase the surface area of the substrate in proportion to the number of cells and the volume of the medium.

(i) Capillary system bioreactors: The opticell-system ceramic cylinders or bundles of hollow semi-permeable fibers are commonly used. A porous ceramic permits growth on inner walls of the channels, suitable for non-adherent cell lines. Smooth surface, non-porous, permits attachment and growth of anchorage dependent cell lines.

Hollow fibers have also been employed. There are a number of systems based on hollow-fiber perfusion. In such systems adherent cell grows on the outer surface of the perfused microcapillary bundles. High-molecular weight produces concentrate in the outer space with the cells, while nutrients are supplied and the metabolites removed via the inner space.

(ii) Packed bed bioreactors: In such types a static bed of particulate material is held in containers and it permits attachment of cells on the surface of particles. The medium is allowed to flow through the bed. A bioreactor of glass beads system is developed for the growth of cells; the glass beads are precoated with fibronectin.

(iii) Microcarriers: Monolayer cell may be grown on microbeads (90-300 μm in diameter) made of plastic, glass, gelatin or collagen. Culturing monolayer cells on microbeads gives a maximum ratio of the surface area of the culture to volume of the medium, upto 90,000 cm^2/l, depending on the size and density of beads, and has the advantage that the cells may be treated as a suspension.

XVI. ORGAN CULTURE

Organ culture usually implies culturing pieces of an organ (not necessarily whole or gan) in vitro. Its objective is to maintain the architecture of the tissue and direct it towards normal development. In this technique, it is essential that the tissue should never be disrupted or damaged. Media used for growing organ cultures are generally the same as those used for tissue culture. The techniques of organ culture can be divided into (i) those employing a solid medium and (ii) those employing liquid medium.

Embryonic organ culture is easier than to normal organ from adult animals. Embryo organs can be cultured, using any of the following three techniques:

(i) Organ cultures on plasma clots: A plasma clot is prepared by mixing 15 drops of plasma with five drops of embryo extract in a watch glass. This watch glass is placed on a pad of cotton wool in a petridish; cotton wool is kept moist to prevent excessive evaporation from the dish. A small carefully dissected piece of tissue is placed on top of the plasma clot in the watch glass. This technique has now been modified, and a raft of lens paper or rayon net is used on which the tissue is placed. Transfer of the tissue can then be achieved by moving the raft easily. Excess fluid is removed and the net with the tissue placed again on fresh pool of medium.

(ii) Organ cultures on agar: Media solidified with agar are also used for organ culture and these media consist of 7 parts: 1% agar in BSS, 3 parts thick embryo extract and 3 parts horse serum. Defined media with or without serum are also used with agar. The medium with agar provides mechanical support for organ culture and does not liquefy. Embryonic organs generally grow well on this medium but tumors from adults will not survive.

(iii) Organ cultures on fluid media: Organ cultures also make use of liquid media. In this method, often a raft made of lens paper or cellulose acetate net or perforated metal gauze is used.

The culture of organs or parts from adult animal is more difficult due to their greater requirement of O2. A variety of adult organs (e.g. liver) have been cultured using special media (e.g. T8; Paul,

1975) with a special apparatus (Towell's Type II culture chamber). Since serum was found to be toxic; serum free media were used, and the special apparatus permitted use of 95% O2.

(iv) Whole embryo culture: The effect of metabolic inhibitors on embryonic development was studied by Spratt (1956) using embryo cultures. In this techniques old embryos were used and the embryo development could be followed for another 24-48 hours in vitro before the embryo dies. It involves the following steps:

(*a*) Prepare a suitable defined medium and added to sterile watch glasses, placed on moist absorbent cotton wool pads in Petri dishes.

(*b*) Incubate hen's eggs at 38°C for 40-42 hours to provide about a dozen embryos.

(*c*) The shell is wiped with alcohol and broken into a sterile evaporating dish containing 50 ml chick saline or BSS.

(*d*) A circular cut is made (using scissors) into the vitelline membrane, around the blastoderm and the latter is transferred to a petridish containing BSS.

(*e*) The adherent vitelline membrane is removed with the aid of forceps and the embryo is examined under the microscope to determine the stage of development.

(*f*) The blastoderm is transferred to the top of the medium in the watch glass prepared as above (a).

(*g*) The blastoderm is spread on agar gel (ventral side down) and the excess BSS is removed.

(*h*) Culture is incubated at 37.5°C.

XVII. PHARMACEUTICAL APPLICATIONS OF ANIMAL CELL CULTURE

1. Source of valuable products: Animal cell products are mainly high molecular weight proteins and examples include enzymes, hormones, vaccines, monoclonal antibodies, interferon's etc. Some important enzymes obtained from animal cell cultures include asparaginase, collagenase, hyaluronidase, pepsin, trypsin, renin, tyrosine hydroxylase and urokinase. Hormones obtained from cultured cells include chorionic hormone, luteinizing hormone, follicle-stimulating hormone and erythropoetin. Some important products obtained from animal cell lines are listed in Table 31.6.

The most important commercial utilization of animal cell cultures at present is still the multiplication of viruses for the manufacturing of vaccines. There are several vaccines including poliovaccine, bovine leukemia virus (BLV) vaccines, rabies vaccines, foot and mouth disease (FMD) vaccines, which are produced at commercial scale using cell cultures. Today the antiviral action of interferons is well understood and interferon therapy is about to become routine in viral diseases like herpes zoaster and virus encephalitis. The bulk of L-interferon is being obtained by a process from bufy coats (leukocyte pellets), which are available as a waste product from blood banks. The leukocytes can be brought into suspension and induced to secrete interferon with new castle-disease or sendai virus. Human â-interferon can be produced by cultivation and induction of human primary fibroblast cells.

Production of monoclonal antibodies with the help of cultured cells has also achieved significance in industry, due to the increasing utility of these antibodies in in-vitro diagnosis, in-vivo imaging, therapies in humans and animals, industrial applications like immuno purification etc. Hybridoma cells used for production of antibodies can be grown in airlift fermenter or stirred fermenter. These cells may also be immobilized and may then involve perfusion type systems. After fermentation, cells and debris are removed by centrifugation and supernatant is used for isolation of monoclonal antibodies. In recent years recombinant DNA techniques are being utilized for the production of monoclonal antibodies.

A number of other proteins have also been produced using cultured cells; examples include tissue-type plasminogen activator (t-PA), factor VIII, factor IX, protein C, human growth hormone and ballikrein. Clinical trails have shown that recombinant t-PA is safe and effective for dissolving blood clots in patients with heart diseases and thrombotic disorders. A number of antigens that are used in viral vaccines are also being synthesized using cell cultures.

Table 31.6: Some animal cell lines and the products obtained from them.

S.No.	Cell Line	Product
1.	Human leucocytes	Interferon
2.	Human tumor	Angiogenic factor
3.	Human kidney	Urokinase
4.	Human kidney cell (293)	Human Protein (HPC)
5.	Dog kidney	Canine distemper vaccine
6.	Cow kidney	Foot and Mouth Disease vaccine (FMD)
7.	Chick embryo fluid	Vaccines for influenza, measles and mumps
8.	Duck embryo fluid	Vaccines for rabies and rubella.

2. Identification and study of hereditary metabolic disorders: The precise nature of many of the genetic diseases can be determined only by demonstrating the metabolic abnormality, commonly the absence of an enzyme in tissue. Disorders such as absence of catalase in acetalasia, alpha-D-galactase-1-phosphate, UDP glucose uridyl transferase in galactosaemia or the failure to decarboxylate branched chain beta-acid in maple syrup urine disease can be demonstrated in primary fibro blast cultures. In some instances, the heterozygous locus, a deficiency of the transferase enzyme per cell can be measured. Some human genetic disorders which can be studied in this way are: glycogen storage diseases, hemocystinuria, Hunter's syndrome, Hurier's syndrome, I-cell disease, Niemann-pick disease, orotio aciduria, Refsum's disease, xeroderma pigmentosa etc.

3. Karyological studies: The karyotype of a population of cells or an organism is the catalogue of the chromosomes of a typical or an average cell. While karyo typing is a useful technique in experimental biology, for example for identifying the species of origin of cell-lines, its most important application in medicine is in the early diagnosis of some congenital abnormalities. Most of these are abnormalities resulting from the inheritance of a unbalanced complement of chromosomes, such as Klinefelter's syndrome (XXY), Turner's syndrome (XO) or Down's syndrome (mongolism) in which chromosome number 21 is represented by 3 instead of 2. Some other anomalies such as translocations or deletions can also be recognized.

4. Tissue culture in virology: The ability of tissue cultures to support the growth of viruses and to reveal their presence by lesions, which are in some cases specific, has been applied in virology. This approach has been applied to study host-parasite relationships, detection and identification of viruses as well as for production of viruses for Vaccine manufacture. Of very great important in recent years have been the development of vaccines against poliomyelitis which are made from the viruses grown in tissue cultures. In addition to viruses other intracellular parasites, which have been grown in tissue culture and studied, include Rickettsiae, mycobacteria (tubercle and leprosy bacilli), lepraemurium, parasitic protozoans etc.

5. Tissue culture in cancer research: Tissue culture techniques have been very useful in understanding the problem of cancer. In includes the development of quantitative assay procedures for tumor viruses stimulated by the exploitation of cell culture methods by virologists. Since Rous demonstrated the infective nature of the fowl sarcoma, many other infective cancer-producing agents have been recognized.

The development and standardisation of species-specific primary cell cultures will likely become more important in veterinary research. We can expect to get new insights while exploring and modulating metabolism and function at the cellular level. The diversity of the physiological and pathological features of companion and farm animals is guaranteed to reveal the equivalent on the cellular level. Perhaps the best feature of primary cell cultures is the possibility to modulate the metabolic and regulatory pathways of cells of interest and to delineate the physiological effects of various compounds and drugs in a controlled way. The refinement of this experimental tool can be expected to accelerate species-specific research. Furthermore, cell cultures can be employed as screening systems in various fields of veterinary science, thereby reducing the need for live animals.

QUESTIONS

1. Discuss balanced salt formulation for animal cell culture. Compare animal cell culture with plant cell culture.
2. What is primary cell culture and cell lines. Discuss applications of animal cell cultures.
3. How primary animal cell cultures are initiated? Describe various approaches for establishment of cell cultures?
4. Write short notes on :

   (*a*) Organ culture
   (*b*) Application of animal cell culture
   (*c*) Monolayers
   (*d*) Cloning in animal cell culture
   (*e*) Somatic cell fusion
   (*f*) Primary cell culture
   (*g*) Balanced salt solutions.

HYBRIDOMA TECHNOLOGY : PRODUCTION OF MONO-CLONAL ANTIBODIES

CHAPTER 32

I. INTRODUCTION

Antibodies are glycoprotein molecules present in the serum and are produced in response to antigens. Antibodies produced in the blood are made by a large number of different lymphocytes (B cells). Each B cell makes a unique antibody, so the antibodies that recognize any particular antigen are a mixture of molecules. This mixture is called 'a' polyclonal antibody; an antibody preparation that reacts with only one antigen, but which nevertheless derives from many different 'clones' of B cells. Antibodies may be classified under five major classes, which are IgG, IgA, IgM, IgE and IgD. The antibodies produced in response to an antigen are heterogeneous in nature. The heterogeneity of antibodies increases immune production in-vivo often reduces the efficacy of an antiserum for various in-vitro uses. Removal of antibodies with unwanted specification from a polyclonal antibody preparation is a time consuming task involving adsorption techniques. These methods often result in the loss of much of the desired antibody. An alternative approach is to generate pure (monospecific) clones of plasma cells in vitro from which monoclonal antibody (MAb) with a single antigeic specificity can be isolated. Kholer and Milstein in 1975 were the first to report on the production of monoclonal antibodies.

Over 25 years ago, Kholer and Milstein described their outstanding work on the fusion of immune spleen cells with myeloma cells to produce hybridomas that secrete monoclonal antibodies. This hybridoma technology, for which these authors were awarded a Nobel Prize, has infiltrated virtually all areas of biology and medicine and has set the scene for impressive advances in the fields of cell biology and immunodiagnostics. During the past few years, refinements in technology and the production of a wide range of monoclonal antibodies MAbs, which bind to proteins, carbohydrates, nucleic acids, and haptens, have certainly broadened the applications of this exciting technology. Increasingly, attention has been focused on the application of Mabs to research and clinical diagnostics as well as to therapy. Blood transfusion and transfusion medicine in particular have benefited enormously from the development of hybridoma technology, and this has resulted in the generation of a vast range of Mabs to human blood group antigens. Some of those have proved invaluable as blood grouping reagents and worldwide have virtually replaced polyclonal ABO and Rh D reagents for routine immunohematology testing. The development of Mabs to blood group antigens has also contributed significantly to the study of the structure and functions of these antigens and to their value as diagnostic and pro- phylactic reagents

Hybridomas are the hybrid cells of myeloma (cancer) cells with antibody producing cells (lymphocytes) from an immunized donor (animal). The hybrid cell or hybridoma resulting from this fusion has the ability to multiply rapidly and indefinitely in vitro and to produce an antibody of predetermined specificity, known as monoclonal antibody. Such antibodies are highly specific in their binding capacity for the corresponding antigen. The production technique of monoclonal antibodies involves the following steps:

(i) Immunization: A mouse is usually immunized with the target antigen. This is done by injecting the antigen mixed with an adjuvant, intradermally or subcutaneously at multiple sites repeatedly at different times to stimulate immune response.

(ii) Splenectomy: When the concentration of antibodies is found to be optimal, the animal is sacrificed and the spleen is removed from the mouse. The spleen, which contains a large number of plasma cells, is dissociated into single splenocytes by mechanical or enzymatic method.

(iii) Cell fusion: The lymphocytes are fused with an immortal cell line (myeloma cell). The fusion is carefully carried under controlled conditions by including a fusing agent polyethylene glycol (PEG) and 30-40% w/v dimethylsulphoxide (DMSO) in a concentrated suspension of lymphocyte-myeloma cells. Fusion occurs within a few minutes after exposure to PEG, the plasma membranes of cells in contact to fuse to form heterokaryons of B-cells and myeloma cells. Majority of myeloma cells may not be fused to spleen lymphocytes and infused cells have to be destroyed. This is accomplished by culturing the cells in the presence of aminopterin or amethopetrin drugs which inhibit folic acid metabolism and therefore, kill the myeloma cell deficient in HGPRT enzyme (hypoxanthine guanine phosphoribosyl transferase), which are mutational variants of parent myeloma and produced for purposes of selecting the hybrid myeloma cells in the background of HGPRT minus cells.

(iv) Selection of clones: The fused cells are grown in HAT (hypoxanthine/aminopterin thymine) medium. The HAT medium inhibits the growth of $HGPRT^-$– myeloma cells. Only $HGPRT^+$ fused hybridoma capable of continuous growth will survive. Unfused defective HGPRT lymphocytes in the cell population do not survive.

(v) Cloning: The fused cells are put, at very low concentrations, into the wells of multiwell plate, hence each well with a cell will only have one cell in it. The cells that grow in each well will be a clone, ensuring to have a pure cell line and this cell line is termed as hybridoma.

(vi) Production of Antibody:

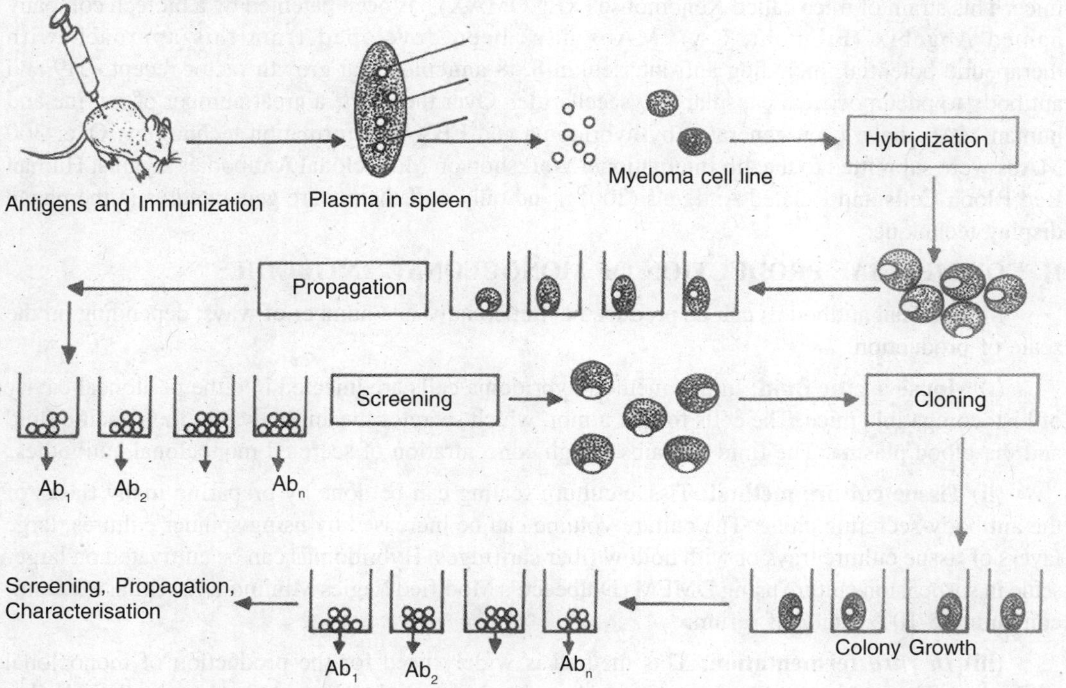

Fig. 32.1. Monoclonal antibodies production

Adopting the conventional approach of Kohler and Milstein for generating murine antibodies of predefined specificity by somatic cell hybridization (Fig 32.1), many MAbs with specificities to red cell blood group antigens in systems such as ABO, LE, MNS, P, and FY have been produced.

However, it was soon recognized that mice did not respond to some clinically important polymorphic antigens of human red cells, notably the D antigen of the RH system. This led to the development of a new approach in generating human MAbs from immortalized antiboy-secreting human B lymphoblastoid cell lines (LCLs). Transformation of human B lymphocytes with Epstein-Barr" virus (EBV) has been widely used to immortalise in vitro antibody secreting B cells. Specific antibody-secreting cells of interest were enriched/concentrated by this procedure and then cloned into MAbs producing cell lines; this appears to be a suitable approach to developing IgG-type human MAbs. Genetic engineering techniques have been used to convert existing routine MAbs into mousehuman chimeric antibodies with mouse (Fab) variable regions joined to human (Fc) constant region and into humanized antibodies, which have the murine complementarity-determining region (CDR) loops grafted onto the framework regions of a human immunoglobulin gene. The effects of these changes have increased the suitability of MAbs for human use by prolonging the half-life of antibodies, reducing their immunogenicity, and enhancing their capacity to recruit complement and effector cells. Although chimeric and humanized antibodies have advantages over murine antibodies for therapeutic use in humans, full human antibodies would be safer and more efficient. In the 1990s, a new approach was introduced that involved the insertion of human immunoglobulin genes into mouse embryonic stem (ES) cells. This process involved the inactivation of routine IgH and IgK loci by deleting the J~ and CK genes by homologous recombination in mouse ES cells. DNA fragments encoding parts of the human IgH and IgK loci were intergrated into the ES cells by fusion with yeast spheroplasts containing antibody DNA inserted into yeast chromosomes. By extensive breeding, crossing, and backcrossing with Ig knockout mice, strains of animals were obtained that produced only human IgH and IgK antibodies.

By conventional hybridoma technology, humanized Mabs were generated from transgenic mice. This strain of mice called Xenomouse (XENOMAX) has been patented by a biotech company named Abgebix (Fremont, CA). MAbs have been developed from this approach with therapeutic'potential, including anti-interleukin-8, t8 antiepidermal growth factor receptor, t9 and antibody to pneumococcal capsular polysaccharide. Over the years, a great number of murine and human MAbs have been generated by hybridoma and EBV transformation technology. Over 300 MAbs were submitted to the 4th International Workshop on Monoclonal Antibodies Against Human Red Blood Cells and Related Antigens (2001), and only 4 of them were generated usng the phage display technique.

II. COMMERCIAL PRODUCTION OF MONOCLONAL ANTIBODIES

Monoclonal antibodies can be produced commercially in a number of ways, depending on the scale of production:

(i) **Mouse ascitic fluid**: In this method hybridoma cells are injected into the peritoneal cavity of histocompatible mice. The cells form a tumor, which secretes the antibody into their ascitic fluid and the blood plasma. The fluid contains a high concentration of secreted monoclonal antibodies.

(ii) Tissue culture method: Tissue culture scaling can be done by preparing many flasks of the antibody-secreting clone. The culture volume can be increased by using spinner cultures, large layers of tissue culture trays or with hollow fiber cartridges. Hybridomas can be cultivated on large-scale in suspension culture using DMEM (Dulbecco's Modified Eagles Minimum Essential Medium) containing 5-10% fetal calf serum.

(iii) *In vitro* fermentation: This method is widely used for the production of monoclonal antibodies. The traditional bioreactor technology has been used to grow hybridoma cells in bulk.

(iv) Immobilized cell reactors: These have been used to make monoclonal antibodies on a scale of a few grams and the most popular is hollow fiber reactor. The beads of calcium alginate (3-5 mm) coated with cationic poly amino acid such as polylysine are commonly employed to encapsulate the hybridomas. The beads are treated with chelating agents to re-liquefy the alginate. The homogeneous suspension of cells was retained within the semipermeable coat of polylysine membrane, which selectively retains antibodies within the capsule, permitting at the same time the diffusion of metabolites. Cells in microcapsules, suspended in the growth medium in a stirred tank bioreactor to have high cell densities. Monoclonal antibodies accumulate in the microcapsules to high concentrations. The antibodies can be extracted from the separated capsule by homogenization.

III. HUMAN MONOCLONAL ANTIBODIES

The antibody combines only the amino acids responsible for making the antigen binding site (the hypervariable regions) of a mouse (or rat) antibody with the rest of a human antibody molecule thus replacing its own hypervariable regions. Daclizumab, Vitaxin, Mylotarg®, Herceptin, and Xolair® are examples. The most common method of generation of human hybridomas secreting human monoclonal antibodies (hMAbs) involves formation of interspecies hybrid heteromyeloma fusion partner. This fusion partner can be obtained from fusion of mouse and human myeloma cells. Such a heteromyeloma when fused with human lymphocytes, synthesizes more stable 'hybrid hybridomas' of mouse human hybrid. Transformation can be performed by making use of Epstein Barr Virus (EBV) with specificity for human B-lymphocytes, which resulted into production of lymphoblastoid cells. The antibody secreting lymphoblastoid cells can be isolated by anti-body screening and subsequent cloning can be performed. Human myeloma cell fusion is another method for production of human hybridomas. The success in fusing mouse-x human cells using this technique is limited.

IV. APPLICATIONS OF MONOCLONAL ANTIBODIES

Monoclonal antibodies have tremendous applicability in bio-medical research and in diagnosis and management of human diseases. Therefore the applications of monoclonal antibodies can be classified as:

(*i*) Diagnostic uses
(*ii*) Therapeutic applications
(*iii*) Investigational and analytical applications
(*iv*) Drug targeting
(*v*) Other applications.

(i) Diagnostic uses: Monoclonal antibody based diagnostic reagents include products for detecting pregnancy, diagnosis of infectious protozoan, bacterial and viral pathogens, monitoring therapeutic drug levels, detecting heart damage; detecting diabetes and detecting tumor cells. They are also used to detect allergies as well as for monitoring the presence or appearance of malignant cells after surgical or radio-therapeutic treatments. Diagnostic imaging with radiolabelled monoclonal antibodies is known as immuno- scintiography. Diagnostic imaging is useful for detection of cardiovascular diseases, sites of bacterial infections and cancer. Radioisotope commonly used in imaging with MAbs is listed in Table 32.1.

Table 32.1. Radioisotopes used in imaging.

Isotope	Half life	Gamma emission (KeV)	Disease
^{99m}Tc	6 hours	140	Brain tumors
^{111}In	2.8 days	173	Myocardial infraction
^{125}I	60 days	35	Sexually transmitted diseases
^{131}I	8 days	364	Breast cancer

Monoclonal antibodies have found an important place in cell surface diagnosis. Specific monoclonal antibodies against all cell surface antigenic determinants enable one to select antibodies which are specific only to a given cell line. It has been shown that specific monoclonal antibodies against various T cell antigens have brought substantial changes in T-cell population in variety of diseases such as systemic lupus erythroematosus, multiple sclerosis, rheumatoid arthritis etc. Hybridoma technologies have been employed for the identification of new and previously unknown cell surface differentiation antigens. It is now possible to study cell-cell interactions. Cell sub-populations belonging to liver, kidney, central nervous system, sperm cells, macrophage and monocytes have been separated using labeled antibodies and re-isolating as pure cells.

Monoclonal antibodies can be used in the detection of tumors, which shed tumor specific antigens into the blood. Tumor markers used in MAb imaging include carcinoembryonic antigen, alpha fetoprotein, ferritin, gangliosides, epidermal growth factor receptor, transfer receptor, tumor associated cell surface antigen, prostatic acid phosphate, human chorionic gonadotrophin and oncogenic products. Anti-ABD and Anti-HLA MAbs have been developed by immunizing the mouse with human cells are used in blood grouping. Monoclonal antibodies against membrane glycophorin A of human erythrocytes at different sites as well MAbs against surface molecules of other erythrocytes have been developed. The human leukocytes in some immunological phenomena like cytotoxic reactions, cellular co-operation, graft rejection etc. can be detected using antisera from multiparous women. But such antisera are not available against rare HLA-antigens, so hybridoma technique can be used to elicit MAbs against rare HLA-antigens.

(ii) Therapeutic applications: Radioisotope immuno conjugates are monoclonal antibody conjugates that deliver cytotoxic doses of radioactivity to the target cells. B emitters are able to penetrate multiple cell diameters in thickness. This technique can be used for the destruction of tumor cells. The limitation is that it causes damage to surrounding non-target cells. Another approach is to couple the antibodies to cell-destroying drugs or toxins. The antibodies serve to concentrate the harmful agents in the tumour. Target specific immuno-drug conjugates of daunomycin has been prepared and found to be potentially cytocidal. Toxins include, ricin, abrin, gelonin, pokeweed antiviral protein etc. Chemotherapeutic agents include methotrexate, doxorubicin, chlorambucil, vinca alkaloids and mitomycin C.

Monoclonal antibodies could also neutralize the action of lymphocytes responsible for the rejection of grafts and destroy the autoantibodies produced in autoimmune diseases. The first monoclonal antibody licensed for human use was orthrochrome OKT3 and has shown promising results in the management of immune related graft rejection events. OKT3 is also useful as prophylactic during post transplantation period. The humanized versions of OKT3 show promise in inhibiting the autoimmune destruction of beta cells in Type 1 diabetes mellitus. Infliximab (Remicade®) binds to tumor necrosis factor-alpha (TNF-$\propto$) and shows promise against some inflammatory diseases such as rheumatoid arthritis. Omalizumab (Xolair®) binds to IgE thus preventing IgE from binding to mast cells; showing promise against allergic asthma. Daclizumab (Zenapax®) binds to part of the IL-2 receptor produced at the surface of activated T cells and can be used to prevent acute rejection of transplanted kidneys. It has also showed promise against T-cell lymphoma. Herceptin® (trastuzumab) binds HER2, a receptor for epidermal growth factor (EGF) that is found on some tumor cells (some breast cancers, lymphomas); it is the only monoclonal antibody so far that seems to be effective against solid tumors. Vitaxin binds to a vascular integrin (alpha-v/beta-3) found on the blood vessels of tumors but not on the blood vessels supplying normal tissues. In Phase II clinical trials, Vitaxin has shown some promise in shrinking solid tumors without harmful side effects. Bevacizumab (Avastin®) binds to vascular endothelial growth factor (VEGF) preventing it from binding to its receptor; approved by the US FDA in February 2004 for the treatment of colorectal cancers. Abciximab (ReoPro®), inhibits the clumping of platelets by binding the receptors on their surface that normally are linked by fibrinogen. It is helpful in preventing reclogging of the coronary arteries in patients who have undergone angioplasty.

Monoclonal antibodies can be used for serogenetic classification of infections microorganisms, protozoan's and metazoan parasites. MAbs against endotoxin are being evaluated as aids in antibiotic therapy of bacteria and sepsis. The role of hybridoma technology for production of monoclonal antibodies is gaining importance in immunoparasitology and in parasite immunodiagnosis.

(iii) Investigational and analytical applications: Monoclonal antibodies have been useful in determining the structure of cell membranes. The purification of complex mixtures or substances such as proteins, enzymes, hormones etc. could also be carried out with monoclonal antibodies. Monoclonal antibodies have the potential to replace or supplement conventional antibodies in a series of assays. A number of antibodies have the requisite specificity and sensitivity for use in radio assays.

(iv) Drug targeting: The use of monoclonal antibodies to target drugs to specific cells types is a promising approach. Monoclonal antibodies to tumor-associated antigens have great potential for the specific targeting of radioactivity and anti-neoplastic agents to tumors. Selecting in drug targeting can be achieved by the attachment of monoclonal antibodies can be fixed on nanoparticles by direct adsorption or via spacer molecule (protein A) or by covalent linkage. The in-vitro studies performed suggest the specificity of these immuno-nanoparticles to tumour cells. Magnetic nanoparticles coated with artificial polypeptides can be conjugated to an antibody specific to carcinoembryonic antigen. Recent development in liposome technology makes it possible to explore therapeutic applications involving site-specific delivery using monoclonal antibodies. Anti-target monoclonal antibody conjugated with liposome's with specific avidity directed to carbohydrate containing antigens have been investigated to delivery drugs like idoxyuridine and acyclovir in the treatment of herpes simplex virus (HSV) injected cell lines.

(v) Other Applications: Monoclonal antibodies are being used directly in serotherapy as well as in the preparation of very specific vaccines active against viral strains. The potential of monoclonal antibodies as tools in enzyme genetics is developing fast as a discipline of immunogenetics. Hence by using a panel of monoclonal antibodies it is now possible to distinguish genetically closely related enzymes. Some other applications include assignment of genes encoding particular enzymes to specific chromosomes or regions of chromosomes in somatic hybrid studies. Monoclonal antibody based assay will soon replace conventional tissue section immuno florescence assays. Such assay will be very useful in forensic sciences.

V. MONOCLONAL ANTIBODIES IN TRANSFUSION RESEARCH

Red Cell Surface Antigens With Clinical Significance- MAbs have been used to study high frequency blood group antigens and null phenotypes. Many of these antigens are associated with clinically/immunologically important proteins and red cell membrane structure and function.

The Rh blood group system- The D antigen is the most clinically significant of over 50 Rh antigens that have been defined. The D antigen is one of the most immunogenic of human red cell antigens in spite of its relatively low number of antigen sites per cell. The D antigen is also known to consist of a mosaic of epitopes. With the aid of polyclonal antibodies, the number of D antigen sites has been estimated at between 10,000 and 40,000 depending on the phenotype and zygosity.

The Fy blood group system- Fy a and Fy b are the most important antigens in the FY system in transfusion medicine because both anti-Fy a and anti-Fy b have caused hemolytic transfusion reactions, and anti-Fy a has been implicated in hemolytic disease of the newborn (HDN) (although much less commonly than anti-D). More than 95% of some West Africans and about 68% African Americans have the Fy(a-b-) phenotype. It has been well known that a high proportion of African American and African blacks is resistant to malaria. Miller and co-workers were the first to discover that Fy(a-b-) red cells were resistant to *Plasmodium knowlesi* (a strain of malarial parasite) invasion in vitro. The resistance to invasion by *Plasmodium vivax* was later confirmed in red cells of Fy (a-b-) individuals.

VI. BLOODGROUPING AND PLATELET SEROLOGY

When the hybridoma techniques were used to generate MAbs for use as ABO typing reagents, their value, when compared with polyclonal antibodies, was questioned. Since then Mabs to red cell antigens have been extensively evaluated for use as routine and reference diagnostic blood grouping and phenotyping reagents. The high specificity of MAbs has provided evidence of multiple epitopes in some antigens, which could result in the incorrect typing of an individual's red cells.

Platelet Serology- Although it has not been possible to raise MAbs directed to human platelet antigens (HPA) by conventional hybridoma techniques, recombinant, human monoclonal, single-chain phage fragments with specificity for HPA-la was first isolated by Griffin and Ouwehand in 1995. Since then, phage display technology has made a significant impact in the field of fetomaternal alloimmune thrombocytopoenia. A whole blood HPA-la phenotyping assay has been developed using an IgG1 construct of the recombinant anti-HPA-la together with a murine MAb to IIb/IIIa in an enzyme-linked immunosorbent assay-based technique. The use of MAbs in the monoclonal antibodyspecific immobilization of platelet antigens (MAIPA) assay 43 has made a significant contribution to human platelet antibody identification, particularly for the resolution of HLA and HPA alloantibody mixes and for the detection of weak antibodies.

VII. RECOMBINANT MONOCLONAL ANTIBODY TECHNOLOGY

Recombinant monoclonal antibody technology comprises a group of molecular approaches useful for the cloning and production of antibodies without requiring the use of conventional tissue culture (hybridoma) methods. Although the latter methods have led to great advances in the development of murine antibody-based diagnostic and therapeutic reagents. Antibody phage display, in contradistinction to conventional tissue culture methods, seeks to immortalize immunoglobulin (Ig) genes rather than the cells from which they were derived. The hallmark of the approach is the establishment of a physical link between an antibody's phenotype (*i.e.*, that portion of the antibody protein which binds antigen, i.e., the Fab domain) and its genotype (*i.e.*, the heavy- and light-chain DNA that encodes that particular Fab molecule). cDNA is prepared from the mRNA of the rearranged Ig genes expressed by donor B-cells, and heavy- and light-chain gene segments are amplified by the polymerase chain reaction (PCR). The heavy and light chains are cloned into one of a number of available plasmid expression vectors, resulting in a library of over 108 Fab DNAs. Co-transfection of these plasmids with M13 helper phage into *E. coli* results in phagemid particles which carry Fab DNA on the inside and Fab protein on the outside by fusion to a phage coat protein Fig. 32.2. In a process called panning, such 'Fab/phage display' libraries are incubated with antigen to bind Fab/phage bearing specificities of interest. Unbound Fab/phage are washed away and the bound Fab/phage are eluted and allowed to re-infect *E. coli* cultures for further propagation. After two to four rounds of panning, essentially all Fab/phage are specific to the antigen of interest. Phage particles bearing expressed mAb on their tips can be used directly in diagnostic assays (e.g., blood group typing) or can be genetically manipulated to express soluble antibody molecules (free of phage) that can be used in conventional (e.g., indirect Coombs type) assays. Because Fab/phage display is a molecular method, it offers certain advantages over hybridoma techniques. First, it does not rely on the immortalization of human donor lymphocytes, thus avoiding certain technical issues noted above. In point of fact, since the method does not depend on the availability of species-specific cell fusion partners, phage display can be used to isolate mAbs from any animal for which one can design appropriate sets of PCR primers for Ig gene amplification (e.g., rabbit, chicken, monkey, camel, shark, mouse, etc.). Since it is an RNA-based technology, the method allows access to any B-cell compartment in which there is immunoglobulin (Ig) mRNA – resting B cells, activated B cells, as well as plasma cells (the latter of which are not amenable to hybridoma formation). In addition, the isotype of the antibodies that are desired canconstant region PCR primer(s) used when making the phage display library. Because the screening for positive clones takes place in one step

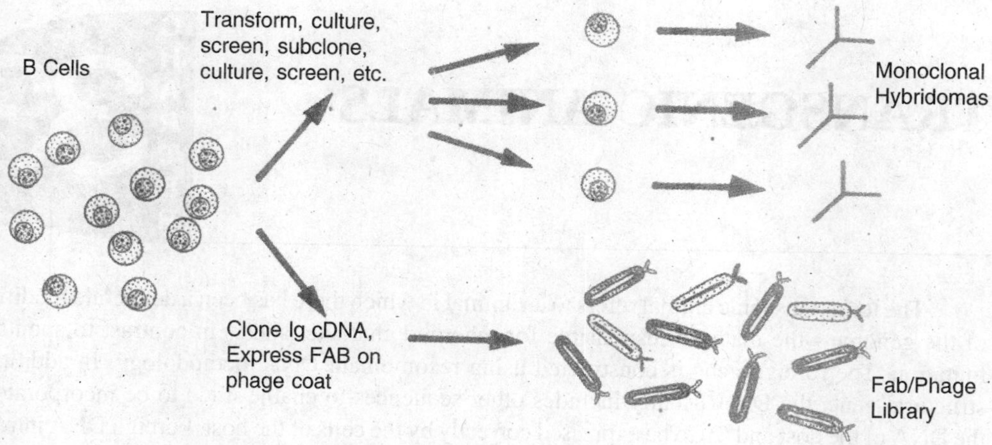

Fig. 32.2. Production of recombinant antibodies

(a 'panning'), millions of clones are evaluated during each round of selection, which requires only a few hours instead of the days, weeks and months required for the individual screening and subcloning of transformed B-cells plated in hundreds or thousands of microplate wells. With experience, a Fab/phage library can be constructed in less than a month and large numbers of antigen-specific mAbs can be obtained within the few weeks which follow. The clones are indefinitely stable and capable of self-replication when supplies become low. They can be stored as bacterial stocks, phage stocks, or as plasmid DNA either frozen or spotted on a piece of filter paper.

QUESTIONS

1. What are monoclonal antibodies ? Discuss the principle utilized and method followed in the preparation of monoclonal antibodies.
2. How recombinant monoclonal antibodies are produced? How they are advantageous over conventional monoclonal antibodies?
3. Discuss the various applications of MAbs in human welfare.

TRANSGENIC ANIMALS

CHAPTER 33

The term transgenic animal refers to an animal in which there has been a deliberate modification of the genome - the material responsible for inherited characteristics - in contrast to spontaneous mutation. The foreign gene is constructed using recombinant DNA methodology. In addition to a structural gene, the DNA usually includes other sequences to enable it (1) to be incorporated into the DNA of the host and (2) to be expressed correctly by the cells of the host. Foreign DNA introduced into the animal, using recombinant DNA technology must be transmitted through the germ line so that every cell, including germ cells, of the animal contains the same modified genetic material. Much of the research in this area is centred on developing transgenic mice. The commonly used protocols includes introducing a cloned gene into a fertilized egg by microinjection, implanting the treated fertilized egg into a receptive female, and then testing the offspring to determine whether any of them have the input genes in their cells. By using these experimental strategies, transgenic versions of mice, cattle, sheep, goats, pigs, birds and fish has been generated. It is hoped that transgenic can be used to enhance existing genetic properties of livestock and provide the genetic basis for novel features. In addition, it is anticipated that the mammary glands-especially those of cows, sheep and goats will be used as a biological factory for cloned gene products that can be readily purified in large quantities for milk.

I. HISTORICAL BACKGROUND:

Prior to the development of molecular genetics, the only way of studying the regulation and function of mammalian genes was through the observation of inherited characteristics or spontaneous mutations. Long before Mendel and any molecular genetic knowledge, selective breeding was a common practice among farmers for the enhancement of chosen traits, e.g., increased milk production. During the 1970s, the first chimeric mice were produced. The cells of two different embryos of different strains were combined together at an early stage of development (eight cells) to form a single embryo that subsequently developed into a chimeric adult, exhibiting characteristics of each strain. The mutual contributions of developmental biology and genetic engineering permitted rapid development of the techniques for the creation of transgenic animals. The potential of genetic engineering to improve agricultural animals has been much discussed since Palmiter and colleagues first demonstrated that the phenotype of a mouse could be changed by the addition of a transgene in 1982. DNA microinjection, the first technique to prove successful in mammals, was first applied to mice and then to various other species such as rats, rabbits, sheep, pigs, birds, and fish. Two other main techniques were then developed: those of retrovirus-mediated transgenesis and embryonic stem (ES) cell-mediated gene transfer. Since 1981, when the term transgenic was first used by J.W. Gordon and F.H. Ruddle, there has been rapid development in the use of genetically engineered animals as investigators have found an increasing number of applications for the technology.

II. VALUE OF TRANSGENIC ANIMALS

The creation of transgenic animals is resulting in a shift from the use of higher order species to lower order species, and is also affecting the numbers of animals used. This shift in the patterns of animal use is being monitored by the CCAC through the use of the Animal Use Data Form. An example of the replacement of higher species by lower species is the possibility to develop disease models in mice rather than using dogs or non-human primates. In the long term, a reduction in the number of animals used, for example to study human diseases, is possible due to a greater specificity of the transgenic models developed. On the other hand, the success of the method has led to using its potential for investigating a wider range of diseases and conditions. The actual use of some species may be increased, in addition to the numbers of animals which are sacrificed as donors during the creation process. The potential of the technology has also made it possible to consider employing cattle, swine, sheep and goats as processing units to manufacture proteins or as organ donors.

The complex interactive processes of living mammals are not reproducible *in vitro*. However, transgenic animals provide a means of evaluating genetic modifications in terms of anatomical and physiological changes in a complex system. Transgenic models are more precise in comparison to traditional animal models, for example the oncomouse with its increased susceptibility to tumor development enables results for carcinogenicity studies to be obtained within a shorter time-frame, thus reducing the course of tumor development in experimentally affected animals. However, models are not strict equivalents, so as with any other system care must be taken in drawing conclusions from the data.

Transgenic animal systems combine the virtues of cell culture and congenic breeding strategies while avoiding the negative aspects of each system. Using transgenic techniques, a characterized genetic sequence may be evaluated within the specific genomic background of the whole animal. Therefore, transgenic animals may be utilized to study the regulation of a specific genetic sequence in a realistic fashion. A representative, but non-inclusive, list of purposes for which transgenic animals have been used indicates the wide ranging application of this biotechnology-

(1) In medical research, transgenic animals are used to identify the functions of specific factors in complex homeostatic systems through over- or under-expression of a modified gene (the inserted transgene).
(2) In toxicology: as responsive test animals (detection of toxicants).
(3) In mammalian developmental genetics.
(4) In molecular biology, the analysis of the regulation of gene expression makes use of the evaluation of a specific genetic change at the level of the whole animal.
(5) In the pharmaceutical industry, targeted production of pharmaceutical proteins, drug production and product efficacy testing.
(6) In biotechnology: as producers of specific proteins.
(7) Genetically engineered hormones to increase milk yield, meat production; genetic engineering of livestock and in aquaculture affecting modification of animal physiology and/or anatomy; cloning procedures to reproduce specific blood lines.
(8) Developing animals specially created for use in xenografting.

III. METHODS OF CREATION OF TRANSGENIC ANIMALS

For practical reasons, i.e., their small size and low cost of housing in comparison to that for larger vertebrates, their short generation time, and their fairly well defined genetics, mice have become the main species used in the field of transgenics.

The four principal methods used for the creation of transgenic animals are- (1) physical transfection, (2) chemical transfection (3) virus (retrovirus) mediated gene transfer (transduction) and (4) DNA packaged inside a bacterium (bactofection).

(1) Physical transfection-This method involves the direct microinjection of a chosen gene construct (a single gene or a combination of genes) from another member of the same species or from a different species, into the pronucleus of a fertilized ovum. It is one of the first methods that proved to be effective in mammals. The introduced DNA may lead to the over- or under-expression of certain genes or to the expression of genes entirely new to the animal species. The insertion of

DNA is, however, a random process, and there is a high probability that the introduced gene will not insert itself into a site on the host DNA that will permit its expression. The manipulated fertilized ovum is transferred into the oviduct of a recipient female or foster mother that has been induced to act as a recipient by mating with a vasectomised male. A major advantage of this method is its applicability to a wide variety of species. Other transfection methods includes particle bombardment, ultrasound and electroporation. These methods are described in chapter on transgenic plant development.

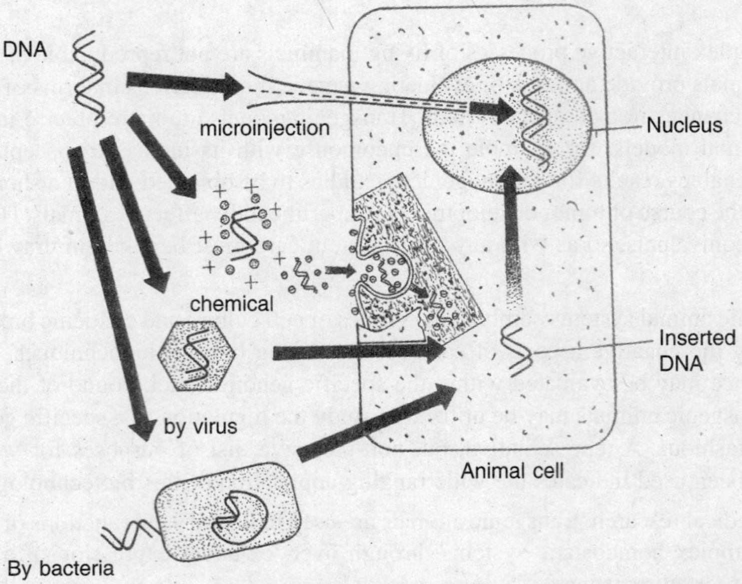

Fig. 33.1. Transfection methods

(2) Chemical transfection- There are several chemical transfection techniques for animal cells but all are based on similar principles. The calcium phosphate mediated DNA uptake involves the formation of a co-precipitate which is taken up by endocytosis. The ability of mammalian cells to take up exogenous DNA from the cultured medium was first reported in 1962. Formation of a fine DNA and calcium phosphate co-precipitate (should be prepared fresh) facilitates DNA uptake by endocytosis. Some of the DNA fragments which enter the cell may reach the nucleus and integrated. Expression of such genes confers the transfection. The transformation frequency of calcium precipitate method is generally low (1-2%), therefore use of soluble complexes (polyplexes) or liposomes and lipoplexes (fusogenic) phospholipid are used to package DNA inside these vehicles. The desired gene is transferred in plasmid and it may be used directly for chemical transfection or inserted in a bacterium for delivery into a mammalian cell. Yeast cells with the cell wall removed (spheroplast) therefore have been used to introduce YAC DNA into mouse, using liposomes and embryonic stem cells for the production of YAC transgenic mice.

(3) Retrovirus-mediated gene transfer- To increase the probability of expression, gene transfer is mediated by means of a carrier or *vector*, generally a virus or a plasmid. Retroviruses are commonly used as vectors to transfer genetic material into the cell, taking advantage of their ability to infect host cells in this way. Offspring derived from this method are chimeric, i.e., not all cells carry the retrovirus. Transmission of the transgene is possible only if the retrovirus integrates into some of the germ cells.

For any of these techniques the success rate in terms of live birth of animals containing the transgene is extremely low. Providing that the genetic manipulation does not lead to abortion, the result is a first generation (F1) of animals that need to be tested for the expression of the transgene.

Depending on the technique used, the F1 generation may result in chimeras. When the transgene has integrated into the germ cells, the so-called germ line chimeras are then inbred for 10 to 20 generations until homozygous transgenic animals are obtained and the transgene is present in every cell. At this stage embryos carrying the transgene can be frozen and stored for subsequent implantation.

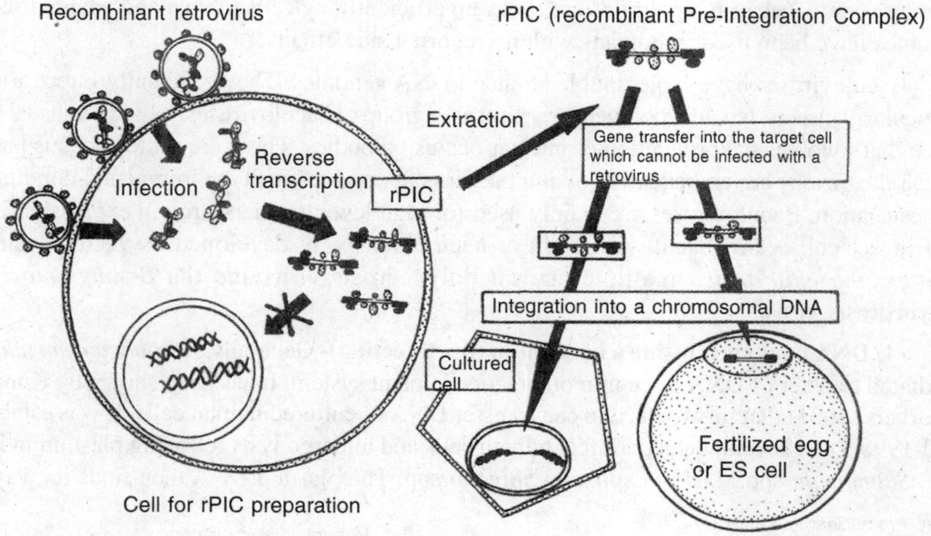

Fig. 33.2. Principles of gene integration by rPIC method.

There have been numerous reports and applications of transgenic plants in agriculture, mainly to benefit the producer. However, the realization of genetically engineered livestock has been much slower. The production of transgenic animals has focused mainly on producing models (e.g. the mouse) for basic and medical research. In terms of commercially important livestock species, work has revolved around specialized non-agricultural purposes such as pharmaceutical production and xenotransplantation and, to a lesser extent, applied agricultural purposes to improve animal production traits and animal-food products. In this case, one of the most important production animals, the dairy cow, was given enhanced resistance to a common and often devastating infection of the mammary gland, a potential benefit to both the producer and the animal's well-being.

Virus vector – Viruses have a natural ability to adsorb to the surface of host cells and infect. This property can be exploited to deliver rDNA into animal cells. The viral system is efficient in transfer, expressed well and replicate rapidly in the host cells. Several classes of viral vectors have been developed for use in human gene therapy and at least eight have been used in clinical trials. The general properties of viral vector are-

1. Transgene may be incorporated into viral vectors as additional gene or as replacement to certain genes of viral genome by ligation or homology recombination. If virus can propagate independently it is called helper independent. If essential viral genes are replaced by transgene then virus need a replication gene in *trans* position (another virus similar to binary vectors) and virus is called 'helper-dependent'.

2. It is necessary to prevent replication as well as recombination of helper virus.

Icosahedral viruses such as adenovirus and retroviruses package their genome into preformed capsid, their volume is fixed with defined amount of DNA can be packaged. Rod shaped baculoviruses form the capsid around the genome, so there are no such size constraints. There is no ideal virus, each has his own advantages or disadvantages.

Adenoviruses are DNA viruses with a linear, double stranded genome of approx 36 kb. They are used in gene transfer because of they show some advantageous features like stability, a high capacity for foreign DNA, a wide host range that includes non dividing cells and the ability to produce high titer stock (up to 10^{11} pfu/ml). They are suitable for transient expression in dividing cells because they do not integrate efficiently into the genome. Adenoviruses are also used as gene therapy vectors because the virions are taken up efficiently by cells *in vivo* and adenovirus derived vaccines have been used in humans with no reported side effects.

Baculoviruses have large double stranded DNA genomes. They efficiently infect arthropods, particularly insects. Nuclear polyhedrosis viruses, a group of baculoviruses, have an unusual infection cycle that involves the production of nuclear occlusion bodies. These are proteinaceous particles in which the virions are embedded allowing the virus to survive harsh environmental conditions such as desiccation. Baculoviruses are mainly used for high level transient protein expression in insects and insect cells. Two baculoviruses have been extensively developed as vectors, namely the *Autographa caloformica* multiple nuclear polyhedrosis virus and the *Bombyx mori* nuclear polyhedrosis virus.

(4) **DNA packaged inside a bacterium (bactofection)-** Generally, *Agrobacterium tumefaciens* mediated transfer of DNA is a common practice in plant system. It has been shown by Kunik and co –workers (2001) that this bacterium can transfer DNA in cultured human cells. It was established in mid-1990's that several bacteria infect human cells and undergo lysis releasing plasmid in host cells e.g., *Salmonella* spp., *Listeria* spp. and *Shigella* spp. The plasmid DNA then finds its way to the

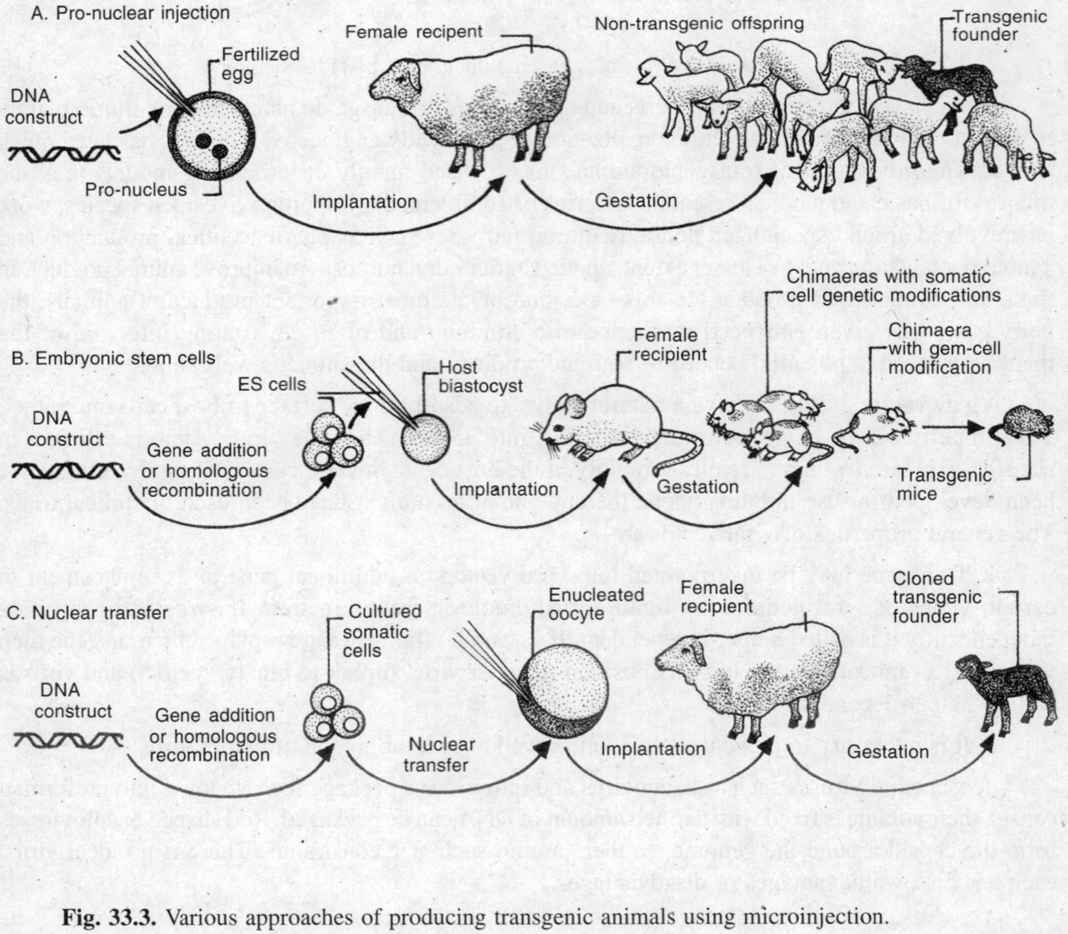

Fig. 33.3. Various approaches of producing transgenic animals using microinjection.

host cell nucleus, where it is integrated in the genome and expressed. In comparison to this *A. tumefaciens* transfer DNA by conjugation [the transfer of DNA through a pilus (plural pilli)]. This pilus is formed by bacterial cell. When live bacteria are used, it is necessary that the bacteria are attenuated. This is because the gene transfer system uses the natural ability of bacteria to infect eukaryotic cells. The bacteria may multiply and destroy host cells.

IV. APPLICATION POSSIBILITIES FOR GENE TRANSFER

In recent years, several application possibilities for gene transfer in domestic animals have been discussed. Until now it has been possible to influence only traits that are based on a single gene or on a limited number of genes. There is only a very limited number of traits of interest to breeders, which are based on a single gene.

In these authors' well-known experiments, growth, one of the classical quantitative traits in animal breeding, was changed to become a quasi-qualitative trait through the transfer of a single growth hormone gene which was related to a feedback independent regulation mechanism. Similar consequences can be expected by application of somatotropins in dairy cattle.

As far as gene transfer in cattle is concerned, there are realistic prospects that it will be possible to influence positively different production traits. Traditional selection programmes using conventional breeding techniques have achieved important results and will continue to do so. However, it seems preferable to concentrate the very expensive and complex technique of gene transfer to fields which until now could only be improved with limited success through conventional breeding programmes, such as breeding for increased disease resistance.

Table 33.1. Various markers used in transgenic animals

Marker	Product	Selection
Ada	Adenosine deaminase	Xyl-A (9-β-D-xylofurenosyl adenosine) and 2'-deoxycoformycin
Tk	Thymidine kinase	Thynidine and aminopterin to block de novo dTTP synthesis selected on HAT medium
Ble	Glycopeptide binding protein	Confers resistance to glycopeptides, antibiotics bleomycin, pleomycin, pheomycin, Zeocin
His D	Histidinol dehydrogenase	Confers resistance to histidinol
Hpt	Hygromycin	Confers resistance to hygromycin
Npt II	Neomycin phosphotransferase	Confers resistance to kanamycin, neomycin
Pac	Puromycin N-acetyl transferase	Confers resistance to puromycin
trpB	Tryptophan synthase	Confers resistance to indole

V. TRANSGENIC LIVESTOCK

1. History of transgenic livestock for agricultural purposes

The production of transgenic animals for agricultural purposes has focused on improving growth or milk production in farm animals to benefit the producer. The first reports of transgenic livestock centered around introducing growth-promoting genes into pigs and sheep. Although transgenic animals were produced, these initial efforts were not completely successful as animals expressing transgenes encoding growth hormone manifested several health problems– most likely due to the systemic nature of transgene expression. Recent work in the pig has corrected some of these health issues with the muscle-specific expression of an IGF-1 transgene. In another approach to increase growth in swine, transgenic pigs have been generated that overexpress bovine α-lactalbumin in their milk. Levels of α-lactalbumin in milk are positively correlated with milk production and increased levels of α-lactalbumin in milk led to a significant increase in litter weight gain. Transgenic pigs have also been generated to address the environmental issue of manure-based phosphorous pollution. As swine are not capable of breaking down dietary phosphorous, transgenic pigs expressing a bacterial phytase gene in their saliva have up to a 75% reduction in fecal phosphorous.

2. Mammary gland-specific transgenic livestock-The modification of milk composition by genetic engineering has been frequently discussed. Milk proteins are made exclusively in the mammary gland and only during lactation. By using the promoter and regulatory components of a gene encoding a milk protein, it is possible to target the expression of a transgene to the mammary gland. Many groups have been working to take advantage of this natural system to produce important human pharmaceuticals in the milk of dairy animals. In contrast to the degree of work done to use the mammary gland as a bioreactor and recover transgene products for other uses, work in the area of modifying the milk itself has not been as extensive. Work of this type was first modeled in the mouse, where several studies demonstrated that the composition and functionality of milk could be changed by expression of a specific transgene in the mammary gland. Transgenic Jersey cattle designed to express lysostaphin in the mammary gland, were generated by nuclear transfer. Lysostaphin is a peptidoglycan hydrolase, of bacterial origin, capable of cleaving specific bonds in the cell wall of *Staphylococcus aureus*. This is relevant to the dairy cow as *S. aureus* is the most common cause of mastitis, a bacterial infection of the mammary gland. Confirming earlier work in mouse models, lysostaphin transgenic dairy cattle were more resistant to *S. aureus* infection than controls. When the mammary glands of three transgenic cows were presented with an intramammary challenge of *S. aureus*, 14% of the infused quarters became infected compared with 71% in the nontransgenic controls. These results have important implications in terms of animal health, milk production and dairy farming.

Mastitis is a very important issue for the dairy industry. If left untreated, the presence of bacteria in the udder can lead to changes in the milk and overall health of the animal. Such changes include decreased milk production and increased somatic cell count resulting in subsequent economic losses due to: the dumping of milk; culling and animal replacement costs and increased veterinary and labour charges to treat the infection. Mastitis is the most costly disease in the dairy industry, with over 1.7 billion dollars a year in losses in the US alone. It is also the most common reason for antibiotic use in dairy cattle and the most frequent cause of antibiotic residues in milk. In addition, current antibiotic treatments are not totally effective and have been implicated in causing antibiotic resistance.

Growth -Attempts are being made to increase growth and body composition of animals through the transfer of these genes which are responsible for growth-hormone regulation. Experiments with the respective proteohormones in growing animals have shown that such effects can be achieved. These gene transfer experiments have been undertaken particularly in pigs and in sheep.

Disease resistance -Only a very limited number of genes is known that are able to influence the resistance of domestic animals to diseases. A model for such work is the influenza resistance caused by the Mx-gene. The authors have made attempts to produce influenza-resistant pigs through the transfer of three Mx-gene constructs.

Quality of animal products- The improvement of the quality or composition of animal products through the transfer of respective gene constructs could provide new prospects for animal production. A model proposed by Mercier has the reduction of the lactose content of milk as its objective. In the milk of transgenic sheep and cattle carrying a lactose gene, combined with an udder specific promoter, lactose is split into galactose and glucose. Such milk could then also be consumed by the large percentage of the world population suffering from lactose intolerance. Through gene transfer, researchers have tried to promote synthesis of cysteine in the mammal organism and to influence positively wool growth.

Gene farming- A suitable combination of tissue-specific promoters and the transfer of these genes into domestic animals may lead to efficient and biologically reliable production of proteins. In particular, efforts have been made to use animals as bioconversion systems.

Isolation and characterization of genes of interest for breeding purposes. The results of basic research in molecular biology are an excellent foundation for practical work on the genome of domestic animals.

Isolation and characterization of suitable regulation elements. The provision of suitable promoters and enhancers is even more difficult than the search for structure genes. For convenient use in breeding programmes, it is of the utmost importance to produce gene constructs that allow the transferred constructs to be expressed in the right tissue at the right time, as well as in the right amount.

3. Transgenic animal in pharmaceuticals

A transgenic animal for pharmaceutical production should (1) produce the desired drug at high levels without endangering its own health and (2) pass its ability to produce the drug at high levels to its offspring. The current strategy to achieve these objectives is to couple the DNA gene for the protein drug with a DNA signal directing production in the mammary gland. The new gene, while present in every cell of the animal, functions only in the mammary gland so the protein drug is made only in the milk. Since the mammary gland and milk are essentially "outside" the main life support systems of the animal, there is virtually no danger of disease or harm to the animal in making the "foreign" protein drug. After the DNA gene for the protein drug has been coupled with the mammary directing signal, this DNA is injected into fertilized cow, sheep, goat, or mouse embryos with the aid of a very fine needle, a tool called a micromanipulator, and a microscope. The injected embryos are then implanted into recipient surrogate mothers where, hopefully, they survive and are born normally.

Success in creating a transgenic animal that can produce the drug is far from guaranteed. About 10 to 30 percent of mouse embryos produce transgenics, but less than 5 percent of goats, sheep, or cows do. Production of the drug is measured during lactation after the animal is raised to maturity and bred. Because of the long time periods involved and low success rates, developing transgenic animals is currently very expensive, as the dollar amounts in Table 32.2 indicates.

Table 33.2. Current market price of the drug and supply produced by one animal

Drug	Animal	Value/Animal/Yr* ($)
AAT	Sheep	15,000
Factor VIII	Sheep	37,000
Factor IX	Sheep	20,000
Haemoglobin	Pig	3,000
lactoferin	Cow	20,000
CFTR	Sheep, mouse	75,000
Human protein C	Pig	1,000,000

Table 33.3. Proteins with therapeutic and industrial value that have been produced (but not commercialized) in the milk of transgenic animals include:

Protein	Animal	Use
Antithrombin III	Goat	Reduce the amount of blood needed in some surgeries
Factor VIII, Factor IX	Goat, Pig, Sheep	Treatment of hemophilia
CFTR	Sheep	Treatment of cystic fibrosis
Lactoferrin	Cow	Natural antibiotic and used in coronary surgery
Alpha-1-antitrypsin	Sheep	Treatment of cystic fibrosis and emphysema
Lysostaphin	Cow	An anti-bacterial compound that prevents mastitis in cows
Spider silk protein	Goat	Production of ultra-strong, lightweight medical and industrial materials

Although most protein drugs are made in milk, a notable exception is human hemoglobin that is being made in swine blood to provide a blood substitute for human transfusions. Because hemoglobin is naturally a blood protein, it is likely to be one of few exceptions to the usual method of production in milk.

Furthermore, the economics of blood production are less favorable, because to recover human hemoglobin, the animal producing it must be slaughtered

4. Transgenic Animal Model Development

(a) The Transgenic Mouse

It is designed to support investigators doing biology of aging research by creating mice that have been genetically altered by either inserting a new gene or removing a normal gene. This method has become one of the most exciting approaches of discovering the functions and interactions of genes in mammals. At the University Of Washington, Nathan Shock Center, this transgenic technology is used to develop new animal models for studying genetic mechanisms of the aging process.

During the previous year, transgenic mouse production has focused on constructs with enhanced defense against free radical injury in aging (e.g., catalase, superoxide dismutase, glutathione S-transferase), Werner Syndrome, adult onset diabetes, Alzheimer's Disease, thrombospondin, and rheumatoid arthritis in aging. Almost 4000 embryos, mainly of the C57BL/6 ingred strain, have been transferred, 498 pups analyzed and at least 40 contained the integrated construct. In addition, this core concentrated an appreciable portion of effort into embryonal stem line (ES) methodologies for generation of knockouts and targeted ES transgenics. This included work to generate mouse models of Werner's Syndrome, models for study of presenillin genes related to Alzheimer's Disease and study of models of thrombospondin in aging. In the past year, a total of 396 embryos were transferred, and 79 pups were born, of which 37 were chimeric. The isolation of mammalian genes is of utmost importance to the biology and medicine of aging because of the contributions these studies can make to the understanding of physiology and development. Techniques for introducing foreign genes into the mouse germ line provide novel approaches for modeling human genetic and chronic degenerative diseases.

Since the initial report in 1980 describing transgenic mice, methods for the direct microinjection of DNA into the pronuclei of fertilized embryos have become established. Foreign genes can be incorporated into somatic germline tissues, with expression of these elements in the progeny of founder mice. The creation of "transgenic" animals that make a specified gene product presents a spectrum of opportunities for basic studies in molecular pathogenesis and pre-clinical investigations applicable to a wide variety of medical problems of aging.

An additional gene transfer technology developed in the 1980's involved the use of stem cells from the early embryo, so-called embryonic stem (ES) cells. The capacity of ES cells to undergo differentiation makes them useful for investigating the effects of genetic modifications of either the gain of function or loss of function. These pluripotent genetically modified ES cells can then be used to make mice with deleted genes (gene knockout) or targeted mutagenesis of genes thought to be involved in the aging process. It is also possible to develop lines of transgenic mice carrying very large DNA constructs (>600 kb) transfected into ES cells. The University of Washington Aging Program provides a focused level of expertise and resources to enhance and facilitate the development of transgenic animal models using ES cell transfer technology.

(b) Two methods of producing transgenic mice are widely used :

(1) transforming **embryonic stem cells** (**ES** cells) growing in tissue culture with the desired DNA (2) injecting the desired gene into the **pronucleus** of a fertilized mouse egg.

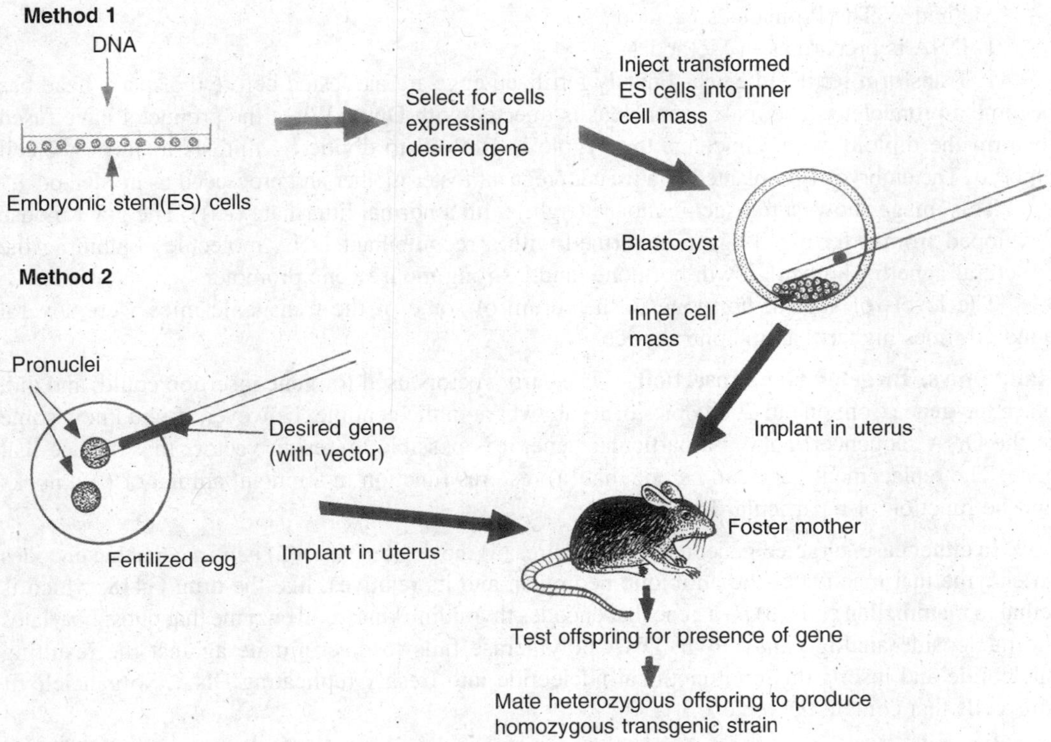

Fig. 33.4. Method of producing transgenic mice

Method 1- The Embryonic Stem Cell Method

Embryonic stem cells (**ES** cells) are harvested from the **inner cell mass** (ICM) of mouse blastocysts. They can be grown in culture and retain their full potential to produce all the cells of the mature animal, including its gametes.

1. Make your DNA-Using recombinant DNA methods, build molecules of DNA containing the structural gene you desire (e.g., the insulin gene), vector DNA to enable the molecules to be inserted into host DNA molecules, promoter and enhancer sequences to enable the gene to be expressed by host cells

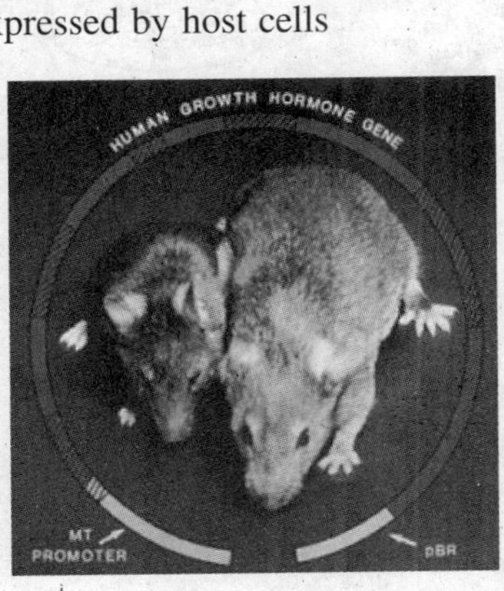

2. Transform ES cells in culture- Expose the cultured cells to the DNA so that some will incorporate it.

3. Select for successfully transformed cells.

4. Inject these cells into the inner cell mass (ICM) of mouse blastocysts.

5. Embryo transfer- Prepare a pseudopregnant mouse (by mating a female mouse with a vasectomized male). The stimulus of mating elicits the hormonal changes needed to make her uterus receptive. Transfer the embryos into her uterus.

6. Test her offspring - Remove a small piece of tissue from the tail and examine its DNA for the desired gene. No more than 10–20% will have it, and they will be heterozygous for the gene.

7. Establish a transgenic strain - Mate two heterozygous mice and screen their offspring for the 1:4 that will be homozygous for the transgene. Mating these will find the transgenic strain (Fig. 33.4).

Method 2 -The Pronucleus Method

1. DNA is prepard as in Method 1

2. Transform fertilized eggs - Freshly fertilized eggs are harvested before the sperm head has become a pronucleus. The male pronucleus is injected with DNA. When the pronuclei have fused to form the diploid zygote nucleus, the zygote is allowed to divide by mitosis to form a 2-cell embryo. The embryos is implanted in a pseudopregnant foster mother and proceeded as in Method 1.

This image shows a transgenic mouse (right) with a normal littermate (left). The giant mouse developed from a fertilized egg transformed with a recombinant DNA molecule containing the structural gene for human growth hormone and a strong mouse gene promoter

The levels of growth hormone in the serum of some of the transgenic mice were several hundred times higher than in control mice.

Random vs. Targeted Gene Insertion – The early vectors used for gene insertion could, and did, place the gene (from one to 200 copies of it) anywhere in the genome. However, if you know some of the DNA sequence flanking a particular gene, it is possible to design vectors that replace that gene. The replacement gene can be one that (a) restores function in a mutant animal or (b) knocks out the function of a particular locus.

In either case, targeted gene insertion requires (a) the desired gene (2) *neo*^{r}, a gene that encodes an enzyme that inactivates the antibiotic neomycin and its relatives, like the drug G418, which is lethal to mammalian cells (3) *tk*, a gene that encodes thymidine kinase, an enzyme that phosphorylates the nucleoside analog gancyclovir. DNA polymerase fails to discriminate against the resulting nucleotide and inserts this nonfunctional nucleotide into freshly-replicating DNA. So ganciclovir kills cells that contain the *tk* gene.

Knockout mice are valuable tools for discovering the function(s) of genes for which mutant strains were not previously available. Two generalizations have emerged from examining knockout mice: Knockout mice are often surprisingly unaffected by their deficiency. Many genes turn out not to be indispensable. The mouse genome appears to have sufficient redundancy to compensate for a single missing pair of alleles. Most genes are **pleiotropic**. They are expressed in different tissues in different ways and at different times in development.

The Cre/loxP System – One of the bacteriophages that infects E. coli, called P1, produces an enzyme — designated Cre — that cuts its DNA into lengths suitable for packaging into fresh virus particles. Cre cuts the viral DNA wherever it encounters a pair of sequences designated *loxP*. All the DNA between the two *loxP* sites is removed and the remaining DNA ligated together again (so the enzyme is a recombinase).

Using "Method 1" (above), mice can be made transgenic for the gene encoding Cre attached to a promoter that will be activated only when it is bound by the same transcription factors that turn on the other genes required for the unique function(s) of that type of cell; a "target" gene, the one whose function is to be studied, flanked by *loxP* sequences.

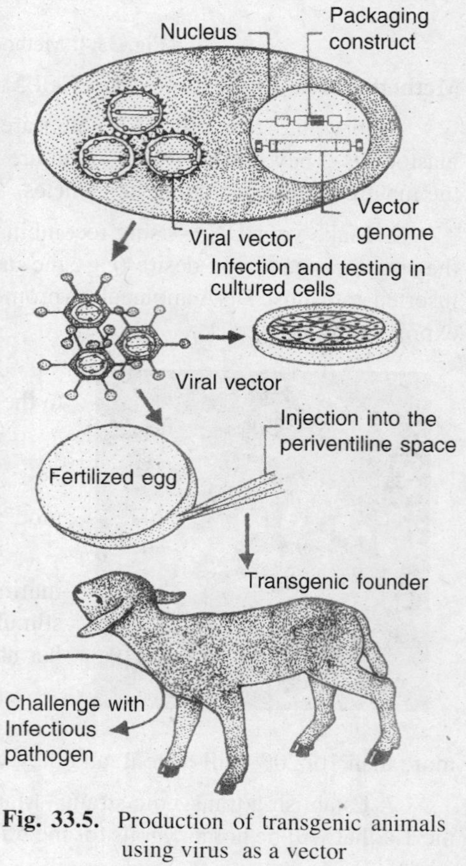

Fig. 33.5. Production of transgenic animals using virus as a vector

In the adult animal, those cells that receive signals (e.g., the arrival of a hormone or cytokine) to turn on production of the transcription factors needed to activate the promoters of the genes whose products are needed by that particular kind of cell will also turn on transcription of the Cre gene. Its protein will then remove the "target" gene under study. All other cells will lack the transcription factors needed to bind to the Cre promoter (and/or any enhancers) so the target gene remains intact.

The result: a mouse with a particular gene knocked out in only certain cells. The Cre/*loxP* system can also be used to remove DNA sequences that block gene transcription. In such a "knockin" mouse, the "target" gene is turned on in only certain cells.

5. Gene transfer in various other species

A review of the literature shows that to date the only successful method to produce transgenic rabbits, pigs, sheep, goats and cattle is DNA-microinjection technique into the pronucleus.

Rabbits- Rabbits are used as experimental models in gene transfer experiments. In 1985, the successful production of transgenic rabbits was reported, for the first time, and included the growth hormone construct MT-hGH. The rate of degeneration of rabbit zygotes caused by injection was below 10 percent. The pre-implantation development capacity of injected zygotes is significantly lower compared with control embryos.

Pigs- Pig zygotes must be centrifuged to show the pronucleus. Fifty percent of the centrifuged non-injected zygotes develop in vivo up to the morula or blastocyst stage. After microinjection, 10-20 percent development to various stages of embryonic development occurs. Of the injected zygotes, 5.6 percent to 11 percent developed and led to the birth of piglets. The integration rate in pigs is approximately 10 percent. Growth-hormone constructs used in initial experiments led to an expression rate of 50 percent.

The production of transgenic F_1 offspring is possible. In the authors' own experiments, inheritance of the transgene could be proved in two out of five animals.

It is not necessary to centrifuge sheep embryos to make the pronucleus visible. According to "Nomarski optics", 80 percent of the pronuclei can be located if a microscope with interference contrast is available. The capacity for in vivo development of sheep zygotes with injection (26 percent) and without (10 percent) is half that of pig embryos after similar treatment. Seven days after in vivo culture of non-treated and non-in vitro cultured sheep zygotes, Rexroad and Wall in1987 observed a development rate of 86 percent. An in vitro culture of five hours' duration reduced this development rate to 65 percent, and after the injection of a buffer solution a reduction to 42 percent was observed. Nineteen percent developed to the 32-cell stage after injection of DNA solution.

Sheep and Goats- Until recently, the transgenes introduced into sheep inserted randomly in the genome and often worked poorly. However, in July 2000, success at inserting a transgene into a specific gene locus was reported. The gene was the human gene for alpha1-antitrypsin, and two of the animals expressed large quantities of the human protein in their milk.

(1) It was done as sheep fibroblasts (connective tissue cells) growing in tissue culture were treated with a vector that contained these segments of DNA (2) regions homologous to the sheep COL1A1 gene. This gene encodes Type 1 collagen. (Its absence in humans causes the inherited disease osteogenesis imperfecta.) This locus was chosen because fibroblasts secrete large amounts of collagen and thus one would expect the gene to be easily accessible in the chromatin. A neomycin-resistance gene to aid in isolating those cells that successfully incorporated the vector. The human gene encoding alpha1-antitrypsin. Some people inherit two non- or poorly-functioning genes for this protein. Its resulting low level or absence produces the disease Alpha1-Antitrypsin Deficiency (A1AD or Alpha1). The main symptoms are damage to the lungs (and sometimes to the liver). (1)

Promoter sites from the beta-lactoglobulin gene. These promote hormone-driven gene expression in milk-producing cells. (2) Binding sites for ribosomes for efficient translation of the mRNAs.

Successfully-transformed cells were then (a) fused with enucleated sheep eggs and (b) implanted in the uterus of a ewe (female sheep). (c) Several embryos survived until their birth, and two young lambs have now lived over a year. (d) When treated with hormones, these two lambs secreted milk containing large amounts of alpha1-antitrypsin (650 µg/ml; 50 times higher than previous results using random insertion of the transgene).

The work on transgenic milk production is expensive requiring large facilities for purifying the protein from sheep's milk. Purification is important because even when 99.9% pure, human patients can develop antibodies against the tiny amounts of sheep proteins that remain.

GTC Biotherapeutics, won preliminary approval to market a human protein, antithrombin, in Europe. Their protein, the first made in a transgenic animal to receive regulatory approval for human therapy and was secreted in the milk of transgenic goats (Fig. 33.5).

Chickens-Chickens have several advantages over other farm animals. (1)grow faster than sheep and goats and large numbers can be grown in close quarters (2) synthesize several grams of protein in the "white" of their eggs.

Two methods have succeeded in producing chickens carrying and expressing foreign genes.

(1) Infecting embryos with a viral vector carrying the human gene for a therapeutic protein and promoter sequences that will respond to the signals for making proteins (e.g. lysozyme) in egg white.

(2)Transforming rooster sperm with a human gene and the appropriate promoters and checking for any transgenic offspring.

Preliminary results from both methods indicate that it may be possible for chickens to produce as much as 0.1 g of human protein in each egg that they lay. Not only should this cost less than producing therapeutic proteins in culture vessels, but chickens will probably add the correct sugars to glycosylated proteins something that *E. coli* cannot do.

Transgenic Fish- Aquatic animals are being engineered to increase aquaculture production, for medical and industrial research, and for ornamental reasons. Genes inserted to promote disease resistance may allow transgenic fish to absorb higher levels of toxic substances, including heavy metals. In turn, consumers of these fish may be ingesting higher amounts of substances such as mercury and selenium. Transgenic fish that have genes from species such as peanuts or shellfish that are common causes of allergic reactions in humans may prompt allergic reactions in an unsuspecting consumer. Transgenic species may behave much like invasive species when interacting with the natural environment. They may compete with native species for resources and pose a threat to the genetic diversity of native populations, especially when genetic modifications such as a rapid growth rate offer advantages over slower-developing native species. Despite industry assurances that transgenic fish would be unable to naturally reproduce or significantly threaten the environment, some scientists are far more doubtful.

The sample bill included in this package addresses these concerns by banning the importation, transportation, possession, spawning, incubation, cultivation, or release of aquatic transgenic animals except under a permit.

Fluorescent cat- Recently South Korean scientist produced transgenic white Turkish angora cats to glow red under ultraviolet light these cats contain a fluorescent gene for flu protein and expressed under skin. Subsequently they produced a number of cloned cats from the skin cells of transformed mother cat. They proposed that such cat could be beneficial in diagonistic of genetic diseases and also showed a way to produce endangered animal by cloning.

QUESTIONS

1. What do you understand by transgenic animals ? How they are produced? What are the utilities of such transgenic animals ?
2. Discuss different methods to produce transgenic animals ?
3. What is the utility of transgenic animals ?
4. Write short notes on :
   (*a*) Retrovirus
   (*b*) Bactofection
   (*c*) Transgenic for mastitis
   (*d*) Lactoferin
   (*e*) Transgenic mouse
   (*f*) Transgenic fish.

CLONING ANIMALS, IN VITRO FERTILIZATION, TISSUE ENGINEERING

CHAPTER 34

I. CLONING

Advances in biotechnology have allowed scientists to make genetically identical copies or clones of animals. Duplication of an organism's genome occurs naturally when identical twins are born or when a plant is grown from a cutting of another plant. However, the world really took notice of cloning in 1997 when a group of Scottish researchers announced the birth of Dolly the sheep, which had been cloned using a single cell from an adult sheep. Dolly had only one "parent;" her nuclear genome was exactly like her "mother's" instead of being a combination of two parents. Therefore, Dolly could generally be thought of as her mother's identical twin. Complex biotechnological procedures have enabled scientists to successfully clone mice, sheep, cows and other mammals. The technology is still at early stages and currently, one in three cloned animals is born abnormally large or with other developmental problems. Embryos contain two copies of each gene one from each parent. It is thought that about 60 genes are 'imprinted' with instructions to switch one copy on or off to allow for normal growth and development. If this doesn't happen correctly and both copies are switched on, or both copies are switched off, it results in problems in growth and development, both prenatal and postnatal. We do not understand this imprinting process in cloned embryos. The closest scientists have come to cloning a non-human primate occurred in October 2004. Biologists successfully transferred cloned monkey embryos into monkey mothers. None of the resulting pregnancies lasted more than a month.

(a) Cloning from different cells. Cloning can be performed using adult or embryonic cells cultured for varying periods. This can be categorised as follows-

Cloning from embryonic cells - Dr. Ian Wilmot working at Roslin Institute, Edenburg, UK, used embryonic cell nucleus to transfer into the egg. When embryonic cells are used for cloning the animals obtained are only clonal to each other but not clonal to the animal to which the donor cells were derived.

Cloning from adult and fetal cells- Adult cells are obtained from somatic cells of an adult animal or fetus.

Cloning from short term cultured cells- Animal cells are though grown in culture but are neither totipotent nor can be grown infinite. Therefore, fresh cultures are initiated from embryonic cells and used for experimental purposes. Dolly, Molly and Polly were produced using short term cultured cells at Roslin Institute for this purpose cells were taken either from udder (mammary glands) of a female sheep or from the foetal fibroblasts (common cells found in connective tissues).

Cloning from long term cultured cells- Instead of using short term cultured cells long term cultured skin fibroblasts derived from seventeen years old bull were used as donor cells and six cloned calves were obtained. These experiments showed the potentiality of cells to regenerate and prolong cultures does not affect the cloning competence of adult somatic donor cells. Besides this, short term cultured cells do not allow targeted gene transfer while use of adult donor cells result in rapid aging of the cloned animal (adult donor cells have shortened telomeres which is the characteristic of aging).

Dolly, the first animal to be cloned, was created using the technique of somatic cell nuclear transfer (SCNT). To do this, cells are taken from the animal that is going to be cloned. In the case of

Dolly the sheep, a cell was taken from her udder. These are normal body cells – somatic cells and the nucleus is removed. The nucleus contains all of the genetic material to make the animal and so, is termed the donor cell. Egg cells are used for cloning because of their ability to grow rapidly. The egg cell's nucleus is removed and the nucleus from the donor cell is inserted in its place. The egg is then stimulated to grow by numerous stimulants which activate the reconstructed embryo to divide and grow. The division of the egg cell follows the same process that would occur if the egg was fertilised by sperm during natural reproduction. The cell division continues for 5 days until a blastomere forms. A blastomere is a ball of nearly 100 cells all with the same genetic material as the donor. Once a cloned embryo reaches the blastomere stage of development, it can be implanted into a uterus of a female to create a whole organism. This is called reproductive cloning..When Dolly was born, she was the only lamb born from 277 attempts. She was a clone of the sheep whose udder cell was used. Blastomeres can also be used as a source of stem cells.

Somatic cell nuclear transfer

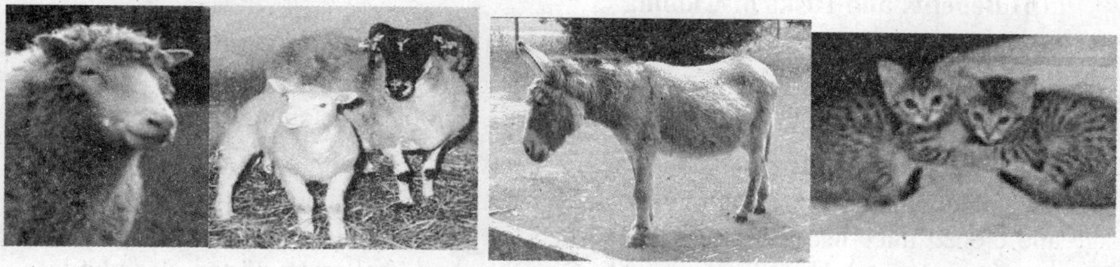

Fig. 34.1. Cloned animals

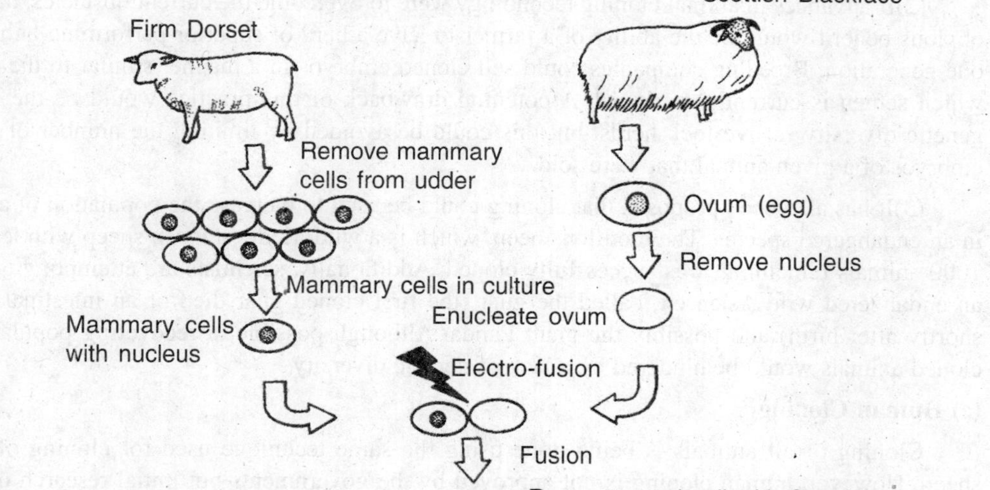

Fig. 34.2. Creation of Dolly Sheep by nuclear transfer cloning

Embryo splitting – Another cloning technique is embryo splitting. Using microsurgery (surgery conducted under a microscope), an embryo is split while it still consists of only a few cells. Genetically identical individuals develop from each portion in the same way as identical twins are formed in nature. This technique has been used to successfully create cloned embryos and cloned animals.

Chromatin transfer – There are a number of problems associated with nuclear transfer - the method used to clone Dolly and almost all cloned animals since then. When the nucleus is transferred to a new egg cell, the egg reprograms the incoming nucleus to allow it to go back to its undifferentiated state. Because it has come from an adult cell, it no longer needs to produce the proteins, hormones and other molecules associated with it being an embryo and growing to produce all the different tissues in a whole body. Incomplete reprogramming of the donor cell is thought to be a leading factor in the low success rate of animal cloning. Chromatin transfer is a new cloning technique aimed at reducing these problems. It involves treating the cell of the animal to be cloned to remove molecules associated with cell differentiation before the nucleus is removed.

(b) Benefits and Risks of Cloning

Researchers have cloned other mammals including cows, goats, pigs, and mice. However, the overall low rate of successful cloning and frequent occurrence of developmental abnormalities in cloned animals demonstrate the need for further research before cloning will be practical.

(1)It has been reported that cloned animals may exhibit health problems throughout their life. Cloned animals may age prematurely as Dolly was diagnosed with arthritis at a seemingly young age and cloned mice had a shorter than normal life span. Additionally, it was demonstrated that cloned mice were both larger in size and heavier than a control group of non-cloned mice.

(2)If advances in animal cloning technology were to overcome the current obstacles, the most obvious benefit would be the ability of a farmer to have a herd of superior performing animals in one generation. Breeding companies could sell cloned embryos in a manner similar to the way in which semen is currently marketed. A potential drawback of this practice would be the loss of genetic diversity in livestock herds, but this could be avoided by limiting the number of cloned embryos of a given animal that were sold.

(3)It has also been proposed that cloning could be used to increase the population of animals in an endangered species. The mouflon sheep, which is a wild Mediterranean sheep with less than 1000 animals remaining, was successfully cloned. Additionally, scientists are attempting to clone an endangered wild Asian ox, called the guar (the first cloned guar died of an intestinal illness shortly after birth) and possibly the giant Panda. Although possible, a recovering population of cloned animals would be hindered by a lack of genetic diversity.

(c) Human Cloning

Cloning of all animals is being done using the same technique used for cloning of Dolly sheep. However, human cloning is not approved by the governments but initial research on stem cells have been permitted.

The Benefits of Human Cloning – There are many ways in which in which human cloning is expected to benefit mankind. Below is a list that is far from complete.

1. **Human cloning technology could be used to reverse heart attacks-** Scientists believe that they may be able to treat heart attack victims by cloning their healthy heart cells and injecting them into the areas of the heart that have been damaged. Heart disease is the number one killer in the United States and several other industrialized countries.

2. **Infertility-** With cloning, infertile couples could have children. Despite getting a fair amount of publicity in the news current treatments for infertility, in terms of percentages, are not very successful. One estimate is that current infertility treatments are less than 10 percent successful.

Couples go through physically and emotionally painful procedures for a small chance of having children. Many couples run out of time and money without successfully having children. Human cloning could make it possible for many more infertile couples to have children than ever before possible.

3. Plastic, reconstructive, and cosmetic surgery. With the new technology, instead of using materials foreign to the body for cosmetic procedures, doctors will be able to manufacture bone, fat, connective tissue, or cartilage that matches the patients tissues exactly. Victims of terrible accidents that deform the face should now be able to have their features repaired with new, safer, technology. Limbs for amputees may be able to be regenerated.

4. Down's syndrome- Those women at high risk for Down's syndrome can avoid that risk by cloning.

5. Organ failure- We may be able to clone livers or kidneys for transplants.

6. Leukemia- We should be able to clone the bone marrow for children and adults suffering from leukemia. This is expected to be one of the first benefits to come from cloning technology.

7. Testing for genetic disease. Cloning technology can be used to test for and perhaps cure genetic diseases.

(d) The current law on human cloning

Different countries have formulated laws for the cloning. The United Nations General Assembly on December 12, 2001 began elaborating an international convention against the reproductive cloning of human beings.

European Union-The European Convention on Human Rights and Biomedicine prohibits human cloning in one of its additional protocols, but this protocol has been ratified only by Greece, Spain and Portugal. The Charter of Fundamental Rights of the European Union explicitly prohibits reproductive human cloning, though the Charter currently carries no legal standing. The proposed Treaty of Lisbon would, if ratified, make the charter legally binding for the institutions of the European Union.

U.S.-In 1998, 2001, and 2003 the U.S. House of Representatives voted whether to ban all human cloning, both reproductive and therapeutic. Each time, divisions in the Senate over therapeutic cloning prevented either competing proposal (a ban on both forms or reproductive cloning only) from passing. Some American states ban both forms of cloning, while some others outlaw only reproductive cloning.

Current regulations prohibit federal funding for research into human cloning, which effectively prevents such research from occurring in public institutions and private institutions such as universities which receive federal funding. However, there are currently no federal laws in the United States which ban cloning completely, and any such laws would raise difficult Constitutional questions similar to the issues raised by abortion.

(e) Ethical issues

Human cloning is a matter for the medical profession's attention since it would involve medical procedures and technology, and it may result in the creation of new genetic and psychological conditions that would require professional care. Therefore, the medical profession must evaluate the ethics of human cloning, and in particular, the potential role of physicians in the practice. The Council's purpose here is to consider whether physicians should participate in human cloning, not to determine whether it should be legal or illegal.

The Council on Ethical and Judicial Affairs offers the following report to assess the ethical uncertainties involved in human cloning. It will address what are currently perceived to be the most widely discussed applications of human cloning, and it will lay the groundwork for future reports.

Issues involving embryo research, stem cell research, embryo splitting, embryo twinning, and embryo donor organisms will be addressed in future reports. A scientific analysis of cloning technology can be found in a companion report issued by the Council on Scientific Affairs.

(f) Existing Limits on Human Cloning

Coverage of advances in cloning has described the prospects of manufacturing armies of programmed killers, duplicating sports stars or academic geniuses, and recreating deceased loved ones. Based on the intrinsic limitations of human cloning technology, some widely mentioned undesirable applications of cloning are impossible, and others, which may be possible technically, are clearly prohibited by existing law, public policy, and professional ethical standards. In order to clarify the many misconceptions about human cloning, physicians should help educate the public about the intrinsic technical limits of human cloning as well as the ethical and legal protections that should prevent abuses of human cloning. The term "cloning" may suggest that one organism is the exact replica of another. Human clones would be identical in so far as they would have the same nuclear genes as the donor. However, as observed in natural monozygotic twins, having identical genes does not result in two indistinguishable individuals. A clone must because of the different environment and circumstances in which he or she creates his or her life story be a different person from the person from whom he or she was cloned. Although human cloning may be thought of as a sort of "delayed twinning," twins may be more similar than clones since most twins are conceived and nurtured in the same environment *in utero* and often during childhood.

Since environment has a profound influence on development, human clones likely would be different in terms of personality and other characteristics. Because cloning would not produce exact replicas, several applications of human cloning are illogical. In particular, human cloning would not be a solution to terminal illness or mortality. Children are already thought of as a way to "soften the blow of mortality," and clones may be seen as a more powerful approach since there is no sharing or mixing of genomes. The possibility of having one's life to live over again, or of getting back a lost child, might be attractive. But the clone would not be the same person as the cloned individual. The fact remains that the person does die and cannot be replaced.

The same reasoning applies to recreating sports stars, dictators, and geniuses— genetics does not wholly define a person. Cloning may allow the persistence of certain genotypes and derived phenotypic traits, but it does not provide individual immortality or replication. A clone of a sports star will not necessarily be a superb athlete, and even if he or she did possess keen athletic ability, he or she would not be identical to the cloned sports star. All the religions do not permit the experiments with human body including cloning and techniques like invitro fertilization and transgenics. Therefore, a rational development of technology might be acceptable to the society.

II. IN-VITRO FERTILIZATION

The 21st century seems set to see a revolution in the application of biotechnological procedures to farm animals. This revolution has been heralded by the production of transgenic mice by Brinster and colleagues, followed rapidly by the production of transgenic cattle, sheep and pigs, the development of practicable sperm sexing in cattle and the dramatic production of the sheep, Dolly, cloned from a somatic cell by Campbell and co-workers in 1996 and Wilmut and co-workers 1997 with all the possibilities that offers for the production of genetically modified animals. However, the successful development and application of this and related technologies is critically dependent on a whole range of basic reproductive technologies such as in vitro maturation of oocytes, in vitro fertilization, in vitro culture of embryos and in vitro culture of sperm precursors. Without significant improvements in these reproductive technologies, application of developments in cloning and the production of transgenic farm animals will remain limited and extremely costly.

(a) In-vitro Production of Embryos

In the provision of *in vitro* produced (IVP) embryos, the processes of *in vitro* maturation of oocytes (IVM) and of *in vitro* fertilization (IVF) are almost inextricably linked.

(b) Milestones in IVM and IVF

Investigators of mammalian oogenesis usually refer to oocyte maturation as the set of processes occurring from the germinal vesicle stage to completion of the second meiotic division with the formation of the first polar body. Milestones in the study of *in vitro* maturation of oocytes include the discovery by Pincus and Enzmann in 1935 that release of immature rabbit oocytes from the inhibitory influence of the follicle allowed them to resume maturation in culture and the demonstration by Minato and Toyoda in Japan in 1983 and by Schroeder and Eppig in 1984 in the USA that in vitro matured mouse oocytes fertilized in vitro could result in live offspring. Other milestones include the finding that 39°C was the optimal temperature for maturation of cattle oocytes, the birth of live calves from *in vitro* matured and fertilized cattle oocytes and the birth of a live mouse pup after complete development from primordial oocyte to *in vitro* fertilized embryo and transfer to a host mother. This last work, which was carried out in Eppig's laboratory, involved a two step procedure for maturation of the oocytes in which the ovaries of new born mice were first cultured for a number of days, the ovaries then dissociated with enzyme, the oocyte–granulosa cell complexes matured *in vitro*, fertilized and cultured *in vitro* to the 2-cell and blastocyst stages. The live pup was born after transfer of a 2-cell embryo to the oviduct.

Milestones in IVF include the discovery of the need for sperm capacitation in rabbits by Austin (1951) and Chang (1951), the definitive proof of in vitro fertilization in rabbits by Chang (1959), the birth of the first human *in vitro* fertilized baby in 1978 as a result of the work of Edwards, Steptoe and co-workers, the birth of the first *in vitro* fertilized calf in Brackett's laboratory (1982), the discovery that 39 °C was the optimal temperature for bovine IVF and the development of improved methods for *in vitro* capacitation of bull sperm such as the use of heparin.

(c) In vitro culture of embryos

Milestones in the development of in vitro culture of embryos include the first attempt at mammalian embryo culture by Brachet (1912) using coagulated blood plasma to culture 5–7 day rabbit blastocysts in glass dishes, the cinematographic studies of Lewis and Gregory (1929) on cleaving rabbit embryos, and the discovery by Hammondin 1949 that 8-cell mouse embryos could be cultured to blastocysts in a simple salt solution supplemented with hen egg white and yolk. This was followed by the absolutely crucial discovery of Whitten in 1957 that the addition of lactate as an energy source allowed culture of 2-cell mouse embryos to blastocysts in a simple Krebs Ringer bicarbonate solution supplemented with serum albumin, and the demonstration by McLaren and Biggers (1958) that embryos cultured to the blastocyst stage in this simple medium developed to live young after transfer to host mothers. This discovery of Whitten's, which was subsequently extended by the work of Brinster, Biggers, Whitten and their collaborators, laid the basis for the subsequent huge expansion of research into preimplantation development in the mouse and other species and without this discovery embryo biotechnology would not exist. Later milestones in embryo culture include the finding in Foote's laboratory that, unlike the mouse, amino acids were essential for blastocyst formation in the rabbit and the cleavage and development to blastocysts of *in vitro* fertilized human embryos. A major milestone for farm animal embryo culture was the discovery in First's laboratory in 1983 of the importance of using a temperature equal to the animal's core body temperature (about 39 °C for cattle) for in vitro maturation and fertilization of cattle oocytes.

A further very significant factor in the development of embryo culture in farm animals was the finding that embryos tended to develop better in gas phases with oxygen concentrations lower than 20%, especially when cultured in the absence of somatic cell support.

(d) In vitro fertilization

In vitro fertilization (IVF) is a procedure in which eggs (ova) from a woman's ovary are removed, they are fertilized with sperm in a laboratory procedure, and then the fertilized egg (embryo) is returned to the woman's uterus. IVF is one of several assisted reproductive techniques (ART) used to help infertile couples to conceive a child. If after one year of having sexual intercourse without the use of birth control a woman is unable to get pregnant, infertility is suspected. Some of the reasons for infertility are damaged or blocked fallopian tubes, hormonal imbalance, or endometriosis in the woman. In the man, low sperm count or poor quality sperm can cause infertility.

IVF is one of several possible methods to increase the chances for an infertile couple to become pregnant. Its use depends on the reason for infertility. IVF may be an option if there is a blockage in the fallopian tube or endometriosis in the woman, or low sperm count or poor quality sperm in the man. There are other possible treatments for these conditions, such as surgery for blocked tubes or endometriosis, which may be attempted before IVF. IVF will not work for a woman who is incapable of ovulating or with a man who is not able to produce at least a few healthy sperm. IVF has been used successfully since 1978, when the first child to be conceived by this method was born in England. Over the past 20 years, thousands of couples have used this method of ART or similar procedures to conceive.

(e) Methods

In vitro fertilization is a procedure in which the joining of egg and sperm takes place outside of a woman's body. A woman may be given fertility drugs before this procedure so that several eggs mature in the ovaries at the same time. The mature eggs (ova) are removed from the woman's ovaries using a long, thin needle. The physician has access to the ovaries using one of two possible procedures. One involves inserting the needle through the vagina (transvaginally); the physician guides the needle to the location in the ovaries with the help of an ultrasound machine. In the other procedure, called laparoscopy, a small thin tube with a viewing lens is inserted through an incision in the navel. This allows the physician to see on a video monitor inside the uterus to locate the ovaries. The naked oocyte is identified under the microscope and is handled with great care. The oocyte are incubated for 5-10 hours after oocyte recovery depending upon the estimated maturity of the follicles and oocytes. The following four culture media have been tried for oocyte culture and IVF: modified Ham's F10 medium, Earl's solution, modified Whitten's culture medium, Whittingham's T6 medium. The culture medium is prepared weekly and tested by culturing mouse embryos from 2-cell stage to blastocyst stage. The use of the same medium for fertilization and embryo culture has actually simplified the fertilization and culture techniques. Once the eggs are removed, they are mixed with sperm in a laboratory dish or test tube. (This is the origin of the term "test tube baby.") The eggs are monitored for several days. Once there is evidence that fertilization has occurred and the cells have begun to divide, they are then returned to the woman's uterus.

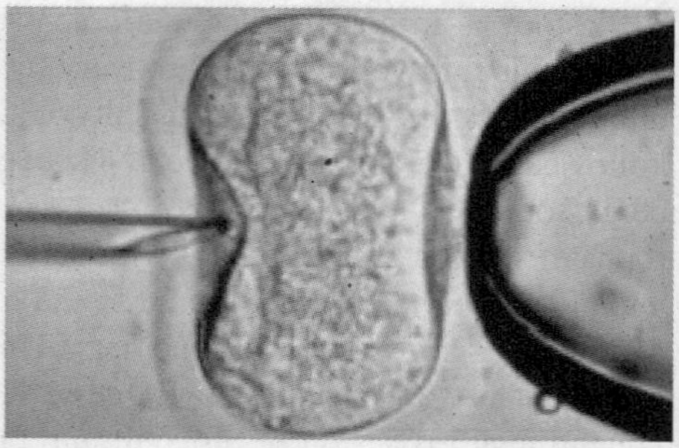

Fig. 34.3. A microscopic image of a needle injecting sperm cells directly into a human egg. At right is a pipette used to hold the ovum steady. This is carried out with the help of micromanipulator fitted with an inverted microscope.

Once a woman is determined to be a good candidate for in vitro fertilization, she will generally be given fertility drugs to stimulate ovulation and the development of multiple eggs. The maturation of the eggs is then monitored with ultrasound tests and frequent blood tests.

A microscopic image of a needle injecting sperm cells directly into a human egg. At right is a pipette used to hold the ovum steady. This is carried out with the help of micromanipulator fitted with an inverted microscope.

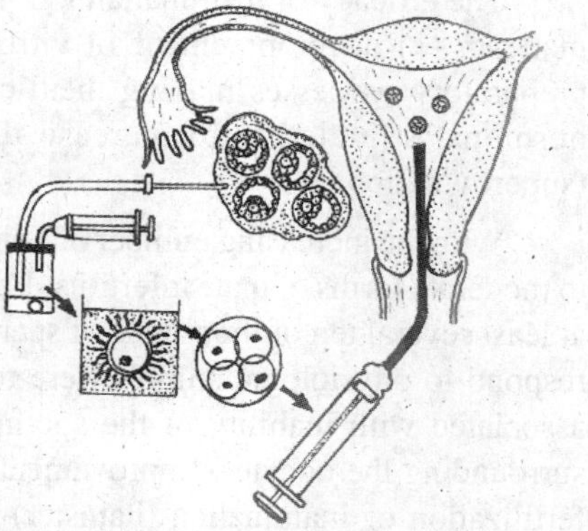

Fig. 34.4. For in vitro fertilization, hormones are administered to the patient, and then eggs are harvested from her ovaries. The eggs are fertilized by sperm donated by the father. Once the cells begin to divide, one or more embryos are placed into the woman's uterus to develop.

The screening procedures and treatments for infertility can become a long, expensive, and, sometimes, disappointing process. Each IVF attempt takes at least an entire menstrual cycle and can cost \$5,000–10,000, which may or may not be covered by health insurance. The anxiety of dealing with infertility can challenge both individuals and their relationship. The added stress and expense of multiple clinic visits, testing, treatments, and surgical procedures can become overwhelming. Couples may want to receive counseling and support through the process. After the IVF procedure is performed, the woman can resume normal activities. A pregnancy test can be done approximately 12–14 days after the procedure to determine if it was successful.

The risks associated with in vitro fertilization include the possibility of multiple pregnancy (since several embryos may be implanted) and ectopic pregnancy (an embryo that implants in the fallopian tube or in the abdominal cavity outside the uterus). There is a slight risk of ovarian rupture, bleeding, infections, and complications of anesthesia. If the procedure is successful and pregnancy is achieved, the pregnancy carries the same risks as any pregnancy achieved without assisted technology.

For in vitro fertilization, hormones are administered to the patient, and then eggs are harvested from her ovaries. The eggs are fertilized by sperm donated by the father. Once the cells begin to divide, one or more embryos are placed into the woman's uterus to develop.

(f) Other alternative methods

Other types of assisted reproductive technologies might be used to achieve pregnancy.

1. A procedure called intracytoplasmic sperm injection (ICSI) utilizes a manipulation technique that is performed using a microscope to inject a single sperm into each egg. The fertilized eggs are then be returned to the uterus, as in IVF.
2. In gamete intrafallopian tube transfer (GIFT), the eggs and sperm are mixed in a narrow tube, and then deposited in the fallopian tube, where fertilization normally takes place.
3. Another variation on IVF is zygote intrafallopian tube transfer (ZIFT). As in IVF, the fertilization of the eggs occurs in a laboratory dish. And, similar to GIFT, the embryos are placed in the fallopian tube, rather than in the uterus as with IVF.

(g) Intracytoplasmic sperm injection

Sexual reproduction is mediated by specialized cells, the male (spermatozoon) and the female (oocyte) gametes (haploid stage). During fertilization, spermatozoon and oocyte unite to form a diploid zygote, which divide by mitotic division to form a multicelluar organism. In most invertebrates and nonmammalian vertebrates, adult animals deposit gametes outside their body, where fertilization subsequently takes place. In comparison to this, in mammals, fertilization occurs within the female genital tract. Hence in vitro fertilization (IVF) and early embryonic development can be achieved relatively easily in invertebrates and nonmammalian vertebrates. In these animals this is termed as extracorporeal mode of natural fertilization. Recreation in vitro of the local environment of the female genital tract required for mammalian IVF, is a much more difficult task.

The efficacy of mammalian IVF remained low until the late 1950s, when improvements were made in -1) development of in vitro culture technique that allowed spermatozoa to undergo preparatory processes making them capable of fertilization (sperm capacitation), and 2) use of hormonal stimulation to increase the number of mature oocyte recovery from the ovaries (superovulation).

With the increasing number of human IVF attempts, indications of IVF treatment were extended to moderate forms of male infertility. Earlier IVF was considered only for female infertility. However, at least several ten of thousands of spermatozoa with adequate motility, morphology and capacity to respond to physiologic stimuli were required for successful IVF.Because most IVF failures were associated with inability of the sperm to penetrate the zona pellucida (a rigid glycoprotein coat surrounding the oocyte). Improvement in this process can further help the IVF. The first report on fertilization of mammalian (hamster) oocyte by intracytoplasmic sperm injection (ICSI) showed the way. Substantial improvement in cell micromanipulation technique in the 1980 raised the possibility of use of this technique for humans. The technique was perfected by a group of scientists at Free university of Brussels. Knowledge obtained through this human application has contributed significantly to the understanding the mechanisms that governs mammalian oocyte activation and applied to a number of other species.

III. TISSUE ENGINEERING

There are three principal therapeutic strategies for treating diseased or injured tissues in patients: (*i*) implantation of freshly isolated or cultured cells; (*ii*) implantation of tissues assembled in vitro from cells and scaffolds; and (*iii*) in situ tissue regeneration. For cellular implantation, individual cells or small cellular aggregates from the patient or a donor are either injected into the damaged tissue directly or are combined with a degradable scaffold in vitro and then implanted. For tissue implantation, a complete 3D tissue is grown in vitro using patient or donor cells and a scaffold, and then is implanted once it has reached "maturity." For in situ regeneration, a scaffold implanted directly into the injured tissue stimulates the body's own cells to promote local tissue repair. Sources of cells for implantation include autologous cells from the patient, allogeneic cells from a human donor who is not immunologically identical to the patient, and xenogeneic cells from a different species. Each category may be further delineated in terms of whether the cells are adult or embryonic stem cells (capable of both self renewal and differentiation into a variety of cell lineages), or a mixture of differentiated cells at different stages of maturation (including rare stem and progenitor cells). Some approaches use cell mixtures, whereas others rely on separation enrichment of stem cells. Although the prospect of using xenogeneic cells for tissue repair remains controversial because of the potential for transmitting animal pathogens to humans, xenogeneic cells could perhaps temporarily support an ailing tissue until either a human donor organ becomes available for transplant, or the tissue repairs itself. For example, pig liver cells (hepatocytes) grown in extracorporeal bioreactors are being tested clinically to see whether they can support patients with liver failure until a liver transplant can be performed.

To engineer living tissues in vitro, cultured cells are coaxed to grow on bioactive degradable scaffolds that provide the physical and chemical cues to guide their differentiation and assembly into three-dimensional (3D) tissues. The assembly of cells into tissues is a highly orchestrated set of events that requires time scales ranging from seconds to weeks and dimensions ranging from 0.0001 to 10 cm. Coaxing cells to form tissues in a reliable manner is the typical engineering design problem that must be accomplished under the classical engineering constraints of reliability, cost, government regulation, and societal acceptance. The successful large-scale production of engineered tissues requires an adequate source of healthy expandable cells, the optimization of scaffolds, and the creation of bioreactors, which mimic the environment of the body and that are amenable to scale-up. Additional challenges include the preservation of the product so that it has a long shelf-life and the successful use of various approaches to prevent tissue rejection.

(a) Biomaterials

A crucial mainstay of tissue engineering is the biomaterial from which scaffolds are fashioned. Many biomaterials direct the growth of cells in culture. However, tissue regeneration in vivo involving the guided growth of nerve, bone, blood vessels, or corneal epithelia across critical injury sites requires that cells receive more specific instructions. In vivo, the cells that repair and regenerate damaged tissues are bombarded with molecular signals, both from the "hostile" wound site and from healthy surrounding tissues. The ideal biomaterial for a scaffold would selectively interact with the specific adhesion and growth factor receptors expressed by target cells in surrounding tissues required for repair of damaged tissue. The scaffold could guide migration of these target cells into the injury site and stimulate their growth and differentiation, finally degrading in response to matrix remodeling enzymes released by the cells as tissue repair progresses. The discovery of adhesion domains in fibronectin and other extracellular matrix glycoproteins containing the amino acid sequence Arg-Gly-Asp (RGD) has enabled the design of synthetic materials that can modulate cell adhesion. Yet the process is not as simple as identifying an adhesion peptide and incorporating it into a biodegradable material. Cell motility is an adhesion-dependent process required for cell migration, angiogenesis, and regrowth of severed nerve ends, among many other physiological events. An engineering model that incorporates the biophysics of how cells bind to extracellular matrix adhesion molecules and how they contract predicts that increasing the number of adhesion contacts between cells and the extracellular matrix may not always be advantageous. If too few adhesive ligands (such as RGD) are vailable, cells cannot get a strong enough grip to enable them to move, but if there are too many ligands, cells adhere so firmly that they remain stuck in place. Thus, intermediate adhesion is required for optimal cell migration. In vivo, bone scaffolds coated with adhesion proteins containing the RGD motif promote maximal tissue ingrowth only at intermediate values of ligand surface density likewise, only at an intermediate density do adhesion proteins on scaffolds induce neural progenitor cells to extend neurites, a prerequisite for nerve regeneration. Cells are also responsive to the nanoscale spatial organization of RGD peptides—such peptides more effectively induce cell adhesion and migration when they are clustered rather than random. The ability of fibrin scaffolds modified with the peptide ligand L1Ig6, which binds to the cell surface adhesion receptor integrin avb3, to promote angiogenesis is influenced by the supramolecular organization of the fibrin. Design principles are emerging for modulating the interactions of cells with growth factors.

(c) Scaffolds

Scaffolds are porous, degradable structures fabricated from either natural materials (collagen, fibrin) or synthetic polymers (polyglycolide, polylactide, polylactide coglycolide). They can be sponge like sheets, gels, or highly complex structures with intricate pores and channels fabricated using new materials-processing technologies. Virtually all scaffolds used in tissue engineering are intended to degrade slowly after implantation in the patient and be replaced by new tissue. Many epithelial and connective tissues have a simple macroscopic architecture consisting of a number of thin layers. Bladder, intestine, and blood vessels are composed of a layer of smooth muscle sandwiched between a layer of collagenous vascularized support matrix and an epithelial lining. Such structures can be built by seeding the different cell types for each layer sequentially on degradable scaffolds made from synthetic fibers of polyglycolide or its derivatives that are 10 to 20 micrometers in diameter. Polyglycolide, unlike polylactide, does not dissolve in solvents such as chloroform. Thus, 3D polyglycolide scaffolds can be sculpted by dipping them in a solution of polylactide dissolved in chloroform and shaping the wet fabric on a mold. When the chloroform evaporates, the polylactide serves as a solid glue to hold the fabric in the desired shape. A scaffold made in this way in the shape of a bladder and seeded with urinary epithelial cells and smooth muscle cells has been implanted into dogs. This artificial bladder acquired near-normal function.

New methods are being developed to process fabrics for more demanding scaffolding applications. For example, using solid free-form fabrication techniques, complex 3D polymer structures have been built from a series of very thin 2D layers, starting with a computer model of the desired shape derived from an MRI (magnetic resonance imaging) or computerized tomography image of the patient's original tissue. Scaffolds can also be designed to release growth factors that induce cellular differentiation and tissue growth in vitro, or cell migration into the wound site in vivo. Scaffolds containing small, degradable polymer beads that release nerve growth factor improve the viability of fetal neural cells transplanted into rat brain. The fragile nature of proteins has motivated design of scaffolds that release naked plasmid.

Controlling the diffusion rates of genes and proteins from scaffolds so that they are in the physiological range is the next challenge. New bioactive materials, such as those that covalently incorporate growth factors and other molecules that regulate cell behavior, offer alternatives for enhancing scaffold performance.

(b) Skin graft

History of the Procedure: In its most basic sense, skin grafting is the transplanting of skin and, occasionally, other underlying tissue types to another location of the body. The technique of skin harvesting and transplantation was initially described approximately 2500-3000 years ago with the Hindu Tilemaker Caste, in which skin grafting was used to reconstruct noses that were amputated as a means of judicial punishment. More modern uses of skin grafting were described in the mid-to-late 19th century, including Reverdin's use of the pinch graft in 1869, Ollier's and Thiersch's uses of the split-thickness graft in 1872 and 1886, respectively and Wolfe's and Krause's use of the full-thickness graft in 1875 and 1893 respectively. Today, skin grafting is commonly used in dermatologic surgery.

Skin grafting is only one means of reconstructing a defect in the skin, regardless of the cause of the defect. Most commonly, it is used for reconstruction after the surgical removal of cutaneous malignancies; however, skin grafts are also used to cover chronic nonhealing cutaneous ulcers, to replace tissue lost in full-thickness burns, and to restore hair to areas of alopecia. As skin cancer continues to be prevalent, more defects will be produced in the surgical treatment of these tumors, and as a consequence, individuals who perform cutaneous reconstructive surgery should acquire a thorough understanding of the techniques of skin grafting.

Generally, skin grafting is used when, in the opinion of the reconstructive surgeon, other methods of reconstruction such as primary closure, second-intention healing, or local skin flaps are inappropriate, unavailable, or would produce a suboptimal result. Skin grafts are divided into 2 major categories: full-thickness skin grafts (FTSGs) and split-thickness skin grafts (STSGs). STSGs may be subdivided into thin (0.008- to 0.012-mm), medium (0.012- to 0.018-mm), and thick (0.018- to 0.030-mm) grafts. Although the use of skin grafts is indicated primarily in reconstruction after surgical excision of malignancies, skin grafting has several other applications and considerations.

Allogeneic cells have been used successfully to treat skin ulcers, diabetes, and liver disease. Patients with diabetic or venous skin ulcers have been treated with two FDA approved living skin products engineered in the lab. One product is composed of neonatal dermal fibroblasts obtained from human foreskins. The neonatal fibroblasts are expanded in culture and seeded onto a thin scaffold composed of the polymer polylactide coglycolide (originally developed for use in surgical sutures) which breaks down gradually in the presence of water. The cells on their scaffold are cultured in custom-designed bioreactors for several weeks until they form a tissue similar to the inner dermal layer of skin. This neo-dermis is then frozen for shipment to physicians. The second skin product has both dermal and epidermal layers. It is composed of dermal fibroblasts in a collagen solution that forms a gel when heated to body temperature; the gel is coated with several layers of human epidermal cells (keratinocytes). After transfer to the patient, this skin product is at least partially replaced by host skin cells as healing progresses. The dermal fibroblasts in the skin products naturally secrete extracellular matrix proteins and are able to respond to growth-regulatory molecules secreted by the host. These skin products can persist up to 6 months after implantation.

Bone graft

There is an FDA-approved autologous cell product for the repair of articular cartilage. A small piece of cartilage is removed from the healthy section of a patient's injured knee. Cartilage cells (chondrocytes) are isolated, expanded in culture, and are then implanted at the injury site. In a variation on this approach, mesenchymal stem cells have been harvested from patient bone marrow, expanded in culture, and then induced to differentiate into cells that can help to repair damaged bone, cartilage, tendon, or ligament. Stem cells hold great promise for treating damaged tissue where the source of cells for repair is extremely limited or not readily accessible. Embryonic stem (ES) cells are attractive because they can be expanded in an undifferentiated state in vitro and can be induced to form many different cell types. But we still need better markers to identify stem cells and their progeny, better ways to expand them in culture, and more research to see whether there is an immunological barrier in implanting stem cells derived from allogeneic donors. The addition of bone marrow to engineered bone grafts improves healing of bone defects, concentrating and selecting for the marrow stem cells that form bone improves healing still further *Tissue architecture.*

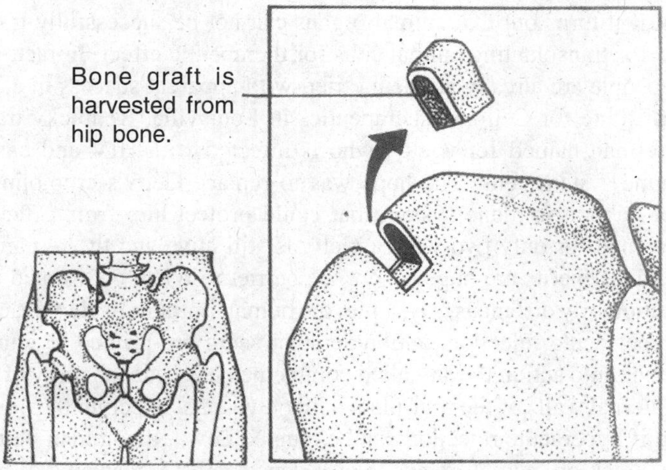

Fig. 34.5. Bone graft

The correct molecular and macroscopic architecture of cartilage, blood vessels, bone, and other tissue is essential for proper tissue function. Connective tissue cells grown on 3D scaffolds in vitro secrete biochemically appropriate extracellular matrix molecules yet fail to acquire the appropriate tissue architecture. The answer may lie in providing appropriate physiological stresses during engineering of the tissue in vitro. The first generation of bioreactors for culturing cells was designed simply to pump nutrient liquid (culture medium) through the assembling tissue. The next wave of bioreactors designed for growing blood vessels and cartilage subjected the nascent tissue to compression, shear stresses, and even pulsatile flow of culture medium. Such stresses profoundly improve the mechanical properties of engineered vessels, cartilage and cardiac muscle. In a promising result for the 500,000 patients who need heart bypass surgery, an artificial artery engineered in the laboratory under pulsatile conditions using endothelial cells seeded on a polymer scaffold developed a burst strength of 2000 mmHg compared with 300 mmHg for unstressed tissues (although physiological responses to vasoactive factors were not completely normal). A typical saphenous vein graft used surgically has a burst pressure of 1000 mmHg, normal systolic pressure exerted on blood vessels in vivo is about 120 mmHg. A critical issue for engineering 3D tissues in vitro is scale-up for clinical use. Hundreds or thousands of tissues must be grown and cryopreserved under sterile conditions.

If cells to be engineered into tissues for implantation must first be derived from the patient, then a separate culture system will be required for each patient, posing formidable regulatory challenges and high costs. Thus, there is much interest in the in situ growth of tissue from injected cells where mechanical stresses are applied naturally.

(e) Xenotransplants (Organ transplant). When surgeon Joseph Murray performed the world's first successful human organ transplant in 1954 (a kidney transplant between identical twins) he had no idea what he was beginning. Today, organ transplants no longer make news, about 20,000 Americans each year receive life-saving transplants of hearts, kidneys, livers, or lungs, from people who have signed organ-donor cards or whose relatives approve the donation. To address the shortage of human organs, many scientists and several biotechnology companies have been working to use organs from animals. The procedure is called "xenotransplantation" (from the Greek "xeno" meaning "stranger"; the "x" is pronounced like a "z" as in Xerox). Even if it turns out that animal organs can not be successfully transplanted, researchers also have ideas for transplanting animal cells for therapeutic effect. In fact, transplants of living animal cells into people are already are being tried-with apparent success in early experiments. The, director of the Institute for Cellular Therapeutics in Louisville, Kentucky transplanted baboon bone-marrow into a man named Jeff Getty, who is infected with HIV and has AIDS. Bone marrow produces immune system cells. The hope was to replace Getty's crumbling immune system with an HIV-proof baboon immune system that could protect him from infection. Although the baboon cells functioned for only two weeks, Getty is still alive and the researchers learned a great deal. If the potential benefits are huge, so are the barriers. When confronted with an organ from an animal as evolutionarily distant as, say, a pig, the human immune system reacts violently. In a response known as hyper acute rejection, antibodies that seem pre-primed to attack tissues from another species summon into action the so-called complement cascade, an array of proteins in the blood that attacks the internal walls of the transplant's blood vessels, rejecting the organ within hours or even minutes. Even if hyperacute rejection can be tamped down, the human body mounts a more vigorous long-term attack on animal organs than it does against transplants of human organs. More blood cells, primarily B lymphocytes and natural killer cells, join the attack on the foreign tissue. Today, physicians can suppress many immune responses with drugs such as cyclosporine, FK506, and prednisone. These drugs are used in human-to-human transplants, known as allotransplants. In xenotransplants, heavier doses are required, and the patient's immune defenses against infectious organisms may be crippled. The easiest way to deal with immune-system rejection of xenotransplants would be to sidestep them-to use organs from the animal that is the closest possible to human beings. That, of course, is the chimpanzee, whose genome is more than 98 percent identical with the human genome.

(f) Human and pig.

Strange as it may sound, the animal that is emerging as the most likely source of transplantable organs is the pig. Pigs' organs are the right size. The animals are highly domesticated, they have large litters, and they grow quickly to maturity. They can be raised in sterile environments, which would reduce the likelihood of transmission of at least some pig diseases to humans. Many researchers, however, still worry about viruses that are unknown or that have become part of the animals' genome and cannot be dislodged. Unfortunately, pig organs have molecular characteristics that make the human immune system react violently. But there may be ways around that, and researchers are exploring different approaches.Despite the obstacles, some xenotransplantation experiments involving humans are going on today-although not whole-organ transplants. By the end of 1998, says Amy Patterson, a scientist with the NIH Office of Recombinant DNA Activities, more than 200 people in the United States had received xenografts of animal cells or tissues. These experiments included implanting fetal pig neurons into the brains of people with Parkinson's disease, and using plastic-wrapped pig liver cells to cleanse the blood of people with liverfailure, keeping them alive until a human donor liver can be found.

To treat insulin-dependent diabetes, scientists at Neocrin Co. in Irvine, Calif., and VivoRx Inc. in Santa Monica, Calif., have devised ways to wrap pig islets-cells of the pancreas that produce insulin-in a plastic material that shields the islets from immune-system cells and antibodies but permits nutrients to enter and insulin to exit. These "encapsulated" islets have functioned successfully in laboratory animal.

Researchers at Circe Biomedical Inc. in Lexington, Mass., have used encapsulated pig liver cells (hepatocytes) as temporary measures to extend the lives of people with liver failure until a human liver transplant can be found. Scientists at Nextran Inc. in Princeton, N.J., have used whole transgenic pig livers as similar temporary "bridges." The patient's blood is cleansed as it passes through the specially packaged hepatocytes or through the transgenic liver, which are maintained outside the patient's body. Scientists at CytoTherapeutics Inc. in Providence, have used encapsulated calves' cells that produce natural pain-blocking substances relieve the pain of cancer patients. The bovine adrenal chromaffin cells produce several opium-like chemicals that relieve pain (catecholamines, encephalins, and opioid peptides, or proteins).

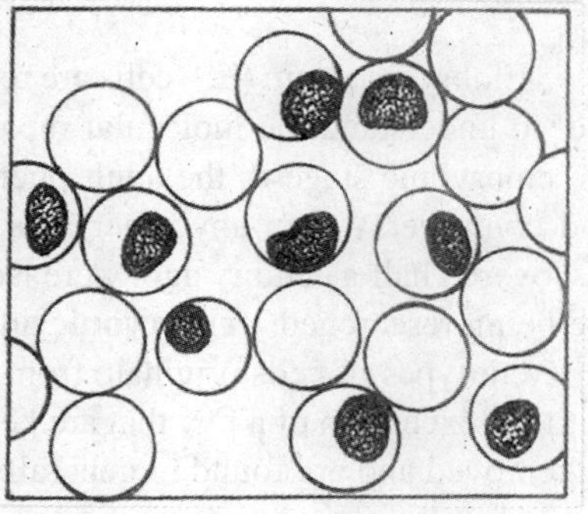

Fig. 34.6. Islet cell masses encapsulated in a protective polymer

Other researchers are trying to modify a sugar molecule that appears on cell surfaces in most mammals-but not in humans and their close primate relatives. This molecule, galactose alpha-(1-3) galactose, is apparently the target for the "xenoreactive antibodies" that all adult humans have, says Jeffrey Platt, of the Mayo Clinic in Rochester, Minn. So researchers are trying to insert into pigs a human gene that will replace the pig molecule with a human sugar residue, fucosyl transferase. Whether the transgenic pigs will function normally after this switch is not yet known.

QUESTIONS

1. Describe methods to produce animal clone. Discuss its applications in medicine.
2. Write brief notes :
   (*a*) In vitro fertilization
   (*b*) Intracytoplasmic sperm injection
   (*c*) Biomaterials
   (*d*) Xenotransplants
   (*e*) Tissue engineering
   (*f*) Bone graft
   (*g*) Skin graft.

STEM CELL BIOENGINEERING

CHAPTER 35

Embryonic stem (ES) cells are revolutionizing the field of developmental biology as a potential tool to understand the molecular mechanisms occurring during the process of differentiation from the embryonic stage to the adult phenotype. Stem cells have generated more excitement, scrutiny and controversy than any other area of recent scientific study. The first stem cells, which were discovered half a century ago, were isolated from blood. There are two main types of stem cells that are being researched – **embryonic** and **adult stem cells**. Embryonic stem cells are what all of our different types of cells originate from. Embryonic stem cells disappear before the organism is born with the exception of a few that are kept in the testis and ovaries as germ cells. These germ cells can be removed and are found to generally act in the same manner as embryonic stem cells (Fig. 35.1).

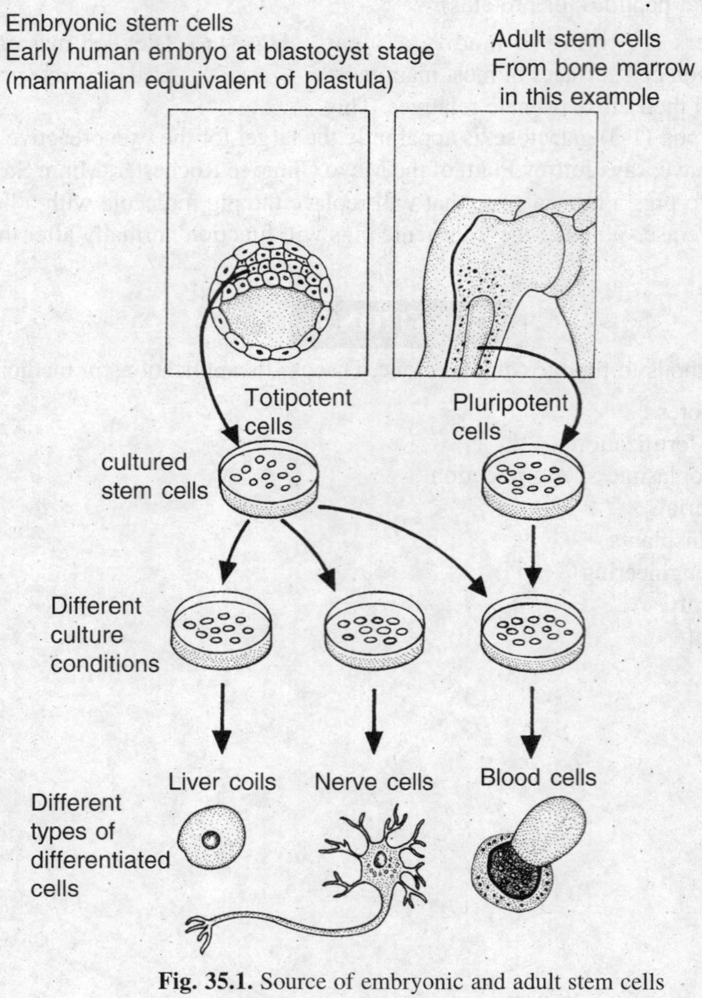

Fig. 35.1. Source of embryonic and adult stem cells

I. DIVISION OF STEM CELLS

Stem cells undergo following two type of divisions: (i) Symmetric cell division to expand their number during development, and (ii) asymmetric cell division to self renew themselves and to give rise to a more differentiated progeny. In asymmetric cell division for generating differentiated cells, two different strategies may be followed. In invariant symmetric cell division, each stem cell gives rise to one daughter stem cell and another daughter cell that undergoes differentiation. In populational asymmetric cell division, on the other hand, a stem cell divides to give two daughter cells which may be either both stem cells or committed progenitor cells, or else like invariant asymmetric division, one daughter cell may be stem cell (self renewal) and the other a progenitor (due to differentiation).

II. BIOLOGICAL ROLE AND PROPERTIES OF STEM CELLS

Stem cells differ from other kinds of cells in the body. All stem cells, regardless of their source, have three general properties: they are capable of dividing and renewing themselves for long periods; they are unspecialized; they can choose to become one of the many different types of cells present in the body based on signals from their environments.

1. Stem cells are unspecialized

One of the fundamental properties of stem cells is that they do not perform specialized functions. A stem cell cannot pump blood through the body (like a heart muscle cell) or supply oxygen to the cells of the body (like a red blood cell). However, unspecialized stem cells can give rise to specialized cells including heart muscle cells, blood cells, or nerve cells. They do this by coordinating their gene expression in an elaborate and complex pattern spanning many generations of cells.

2. Stem cells are self- renewing

Most specialized cells in our body (such as muscle cells, blood cells, or nerve cells) do not replicate themselves. The supply of these cells is maintained by stem cells, which replicate many times and then differentiate into the specialized cells that are needed. In this way, cells are continuously replenished as they die. This is called homeostasis.

3. Stem cells can give rise to specialized cells

Differentiation is the process whereby stem cells give rise to specialized cells. Scientists are just beginning to understand the signals that trigger stem cell differentiation. These signals include chemicals secreted by other cells, physical contact with neighboring cells, and contact with molecules in the microenvironment. Specific sets of signals be identified as responsible for the promotion of differentiation into specific cell types. The development of stem cell-based technologies will depend on our ability to reproducibly drive the differentiation of stem cells into specific tissue lineages. This issue is of particular importance for the development of ES cell-based technologies, where the differentiation of certain subsets of cells and tissues is uncommon. The generation of ES cell-derived heart muscle cells is one example where significant improvements in tissue-specific differentiation are needed. Under default mechanisms, only approximately 1% of ES cells form cardiac tissue.

III. TYPES OF STEM CELLS

The stem cells can be broadly classified into (i) adult stem cells (derived from mature organs in the adult individual), (ii) fetal stem cells (derived from the developing foetus) and (iii) embryonic stem cells (derived from the totipotent cells of early embryo). Depending upon the source of their origin, stem cells from the foetus can be classified in the following types: (i) haematopoietic stem cells (HSCs): these are found in bone marrow and are a continuous source of progenitors of red cells, platelets, monocytes, granulocytes and lymphocytes; (ii) mesenchymal stem cells or marrow stromal cells(MSCs) : these are also found in bone marrow and can be differentiated in culture into osteoblasts; (iii) neural stem cells: these are found in the central nervous system(CNS) and have the potential to differentiate into neurons, astrocytes and oligodendrocytes.

Embryonic stem (ES) cells

Embryonic stem (ES) cells are derived from the embryo when it exists as a blastocyst. They have the ability to develop into all the different cell types found in the body. Actually, the resulting single cell formed by fertilized egg, begins to divide and multiply at a rate much faster than that observed in somatic cells. Scientists refer to these cells as **totipotent** stem cells. These primordial embryonic cells have the potential to grow into a complete mammal. Within days of fertilization, these new and dividing cells form a hollow sphere, called a blastocyst. Stem cells arising in the inner mass of the blastocyst are called the ES cells. The ES cells are considered **pluripotent** - they can divide indefinitely and blossom into all the various tissue types of the human body. After roughly 14 days, ES cells divide to give rise to what will eventually develop into the spine. At this stage, the stem cells within the embryo are considered **multipotent**. These stem cells can grow into some tissues, but not all tissues. Those destined to become bone or blood, for example, may not be able to form stomach or skin. By definition, stem cells have two important characteristics that distinguish them from other types of cells. First, they are unspecialized cells that renew themselves for long periods of time through cell division of at least one daughter cell. Secondly, under certain physiological or experimental conditions, they can be induced to differentiate. This means that they can divide into cells with special functions, such as the beating cells of the heart muscle or the insulin-producing cells of the pancreas Fig. 35.2.

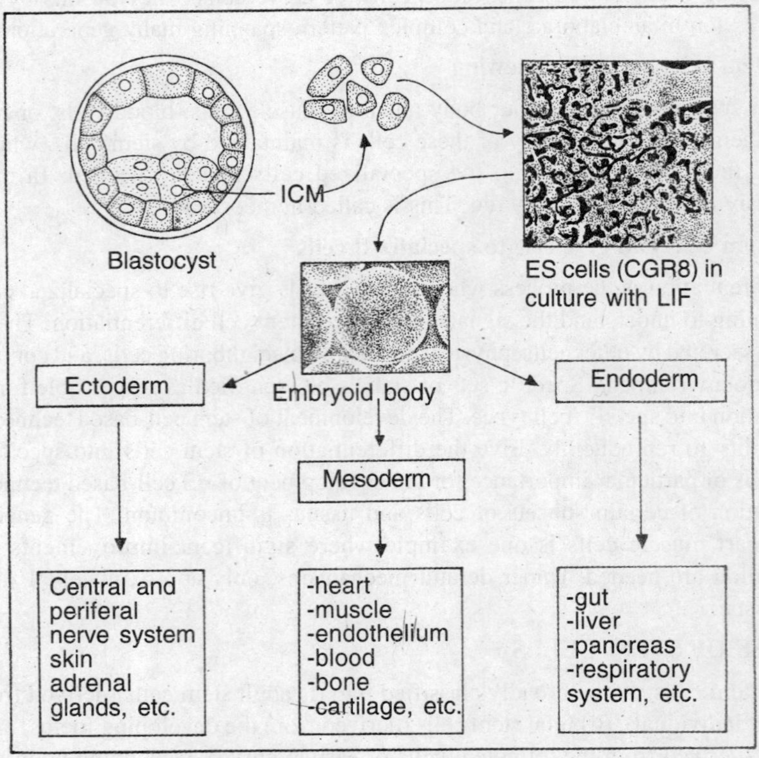

Fig. 35.2. Overview of the derivation of EC cells and their pluripotency

IV. DISCOVERY OF ES CELLS

The work that laid the foundation for ES cell discovery was the study of teratocarcinomas, complex tumors containing a mix of specialized cell types as well as a population of unspecialized cells. These unspecialized cells are called embryonal carcinoma (EC) cells. The latter were shown to be pluripotent and could give rise to various cell types both *in vitro* and *in vivo*. It was therefore natural to consider using these cells for therapeutic purposes. However, EC cells never seemed ideal for this purpose because they had an abnormal number of chromosomes and originated from tumors. Careful study of the induction of teratocarcinomas in experimental animals, as well as an understanding of the biology of EC cells and early embryos, led scientists to the discovery of ES cells. In 1981, ES cells were isolated for the first time in mice and it was a major breakthrough in the field of developmental biology.

Two fundamental questions about stem cells that relate to their long-term self-renewal is crucial to our ability to successfully grow these cells in laboratory and in turn use them for various tissue engineering and cellular therapies. The first question deals with why embryonic stem cells can proliferate for extended periods of time in the laboratory without adopting a specialized fate, while most adult stem cells cannot. The second issue addresses which factors in living organisms normally regulate stem cell self-renewal and differentiation. Discovering the answers to these questions may make it possible to understand how cell proliferation is regulated during normal embryonic development or during the abnormal cell division that ultimately leads to cancer.

V. GROWING EMBRYONIC STEM CELLS IN THE LABORATORY

The process of growing cells in the laboratory is known as cell culture. Mouse (or human) embryonic stem cells are isolated by transferring the inner cell mass (ICM) of the blastocyst into a plastic laboratory culture dish containing medium, which provides nutrients and necessary growth factors. The cells divide and spread over the surface of the dish. In the case of human ES cells, the inner surface of the culture dish is typically coated with mouse embryonic skin cells, which are non-dividing and secrete a host of growth factors. This is known as a feeder layer and is necessary for the prevention of ES cell differentiation. The feeder layer provides a sticky surface to which the inner cell mass cells can attach. Mouse ES cells, however, can be maintained in culture without the need of a feeder layer. In this case, the culture medium is supplemented with high concentrations of a protein known as leukemia inhibitory factor (LIF), which prevents differentiation (Fig. 35.3). Over the next several days, the cells of the inner cell mass grow and divide and spread on the culture dish. When the cell concentration is high enough, they are removed gently and plated into several fresh culture dishes. This process of replating the cells is repeated for many months, and is called subculturing. After six months or more, the original 20 to 30 cells of the inner cell mass yield many millions of embryonic stem cells. When embryonic stem cells have proliferated in cell culture for six or more months without differentiating, are pluripotent, and appear to be genetically normal, they are referred to as an ES cell line. Once ES cell lines are established, batches of them can be frozen and kept for further experiments or even shared with other laboratories for further culture and experimentation.

VI. ES CELL DIFFERENTIATION

As long as mouse ES cells in culture are grown in the presence of LIF or a feeder layer, they can remain undifferentiated (unspecialized). If LIF is removed from the culture medium, cells will begin to clump together to form 3-dimensional structures called embryoid bodies (EBs). The cells then begin to differentiate spontaneously. Allowing spontaneous differentiation, a good way to determine if a culture of embryonic stem cells is healthy. It is not an efficient way to produce cultures of a specific cell type. In order to generate a large number of cells of a specific type, scientists use the process of directed differentiation. This is done by changing the chemical composition of the culture medium, altering the surface of the culture dish, or modifying the cells by inserting specific genes. Nineteen different tissues have been created through the use of chemical signaling and physical stimuli in mouse embryonic stem cells. Some of these include cells of the heart, skin, brain, bone, and pancreas (Fig. 35.3).

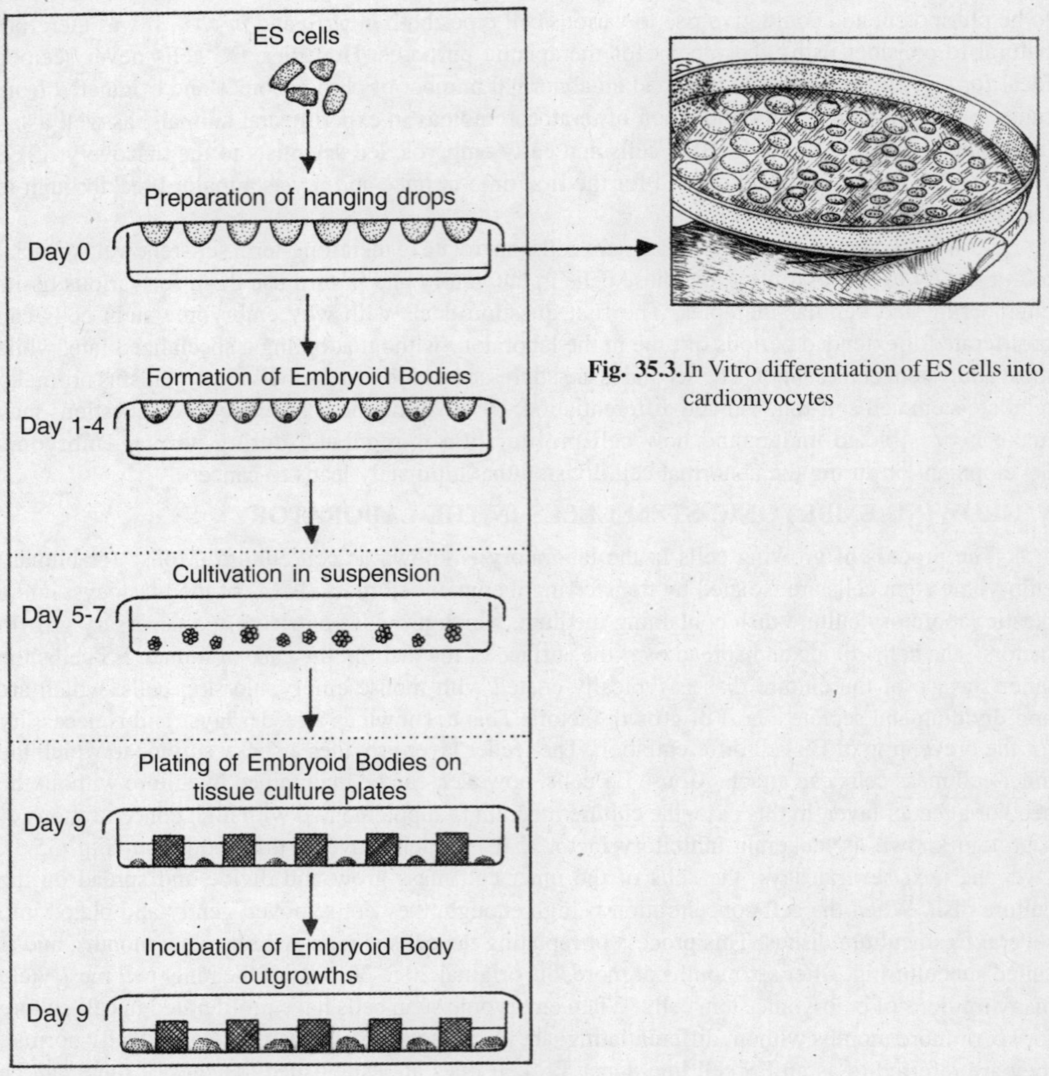

Fig. 35.3. In Vitro differentiation of ES cells into cardiomyocytes

VII. ADULT STEM CELLS

Many adult tissues (such as the bone marrow, brain and gut) contain stem cells. Like ES cells, adult stem cells can make identical copies of themselves for long periods of time (self-renewal). At the same time, they can give rise to mature cell types that have characteristic shapes and specialized functions. These cells are **unipotent** and will differentiate into the specific type required to replenish the cells of the tissue to which these stem cells belong e.g. skin, mammalian brain etc. Stem cells typically generate an intermediate cell types, before they achieve their fully differentiated state. The intermediate cell is called a **progenitor cells**. Progenitor cells are partly differentiated cells in the sense that they are committed to a particular cell lineage and, upon division give rise to differentiated cells. Similarly, **precursor cells** are those which are further restricted in their potential and represent the penultimate stage in the developmental pathway of any specialized cell type. Of all the adult stem cells identified so far, haematopoeitic stem cells (HSCs) are the best characterized. When stem cells divide, some of the progeny cells differentiate into cells of specific types, while others remain as stem cells. For example, intestinal stem cells continuously regenerate the lining of the gut, skin stem cells make skin and haematopoietic stem cells (from bone marrow) give rise to a range of blood cells. Thus, stem cells repair everyday wear and tear in different organs of our body.

Until recently, adult stem cells were considered to be irreversibly committed to specific lineages of differentiation. HSCs, for example, normally give rise to all types of blood cells such as red blood cells, white blood cells and platelets. It was previously believed that HSCs could not give rise to any cells of a different tissue. However, a number of experiments over the last several years have raised the possibility that stem cells from one tissue may be able to give rise to cell types of a completely different tissue. This phenomenon is known as 'stem cell plasticity'. Examples of such plasticity include bone marrow stem cells becoming neurons, or pancreatic islet cells that are capable of producing insulin. Exploring the possibility of using adult stem cells for cellular therapies has recently become an active area of research.

VIII. ADULT STEM CELLS AND EMBRYONIC STEM CELLS

Adult stem cells are more limited. Instead of being able to convert into any type of cell, they are more specialized (pluripotent) and can only transform themselves into a few different types of cells. In the body, adult stem cells replace dying red blood cells, immune system cells and many other cells thought the owner's life. While embryonic stem cells are better suited for creating a wide range of organs and tissues, research in adult stem cells is also needed. Adult stem cells may also be easier to manage, since they so not tend to differentiate spontaneously, like the ESCs do. Using embryonic cells to treat a patient may result in an immune response, since the cells would not actually be from the patient. If adult stem cells from a sick patient can be removed, cultured, altered and reinserted, there would be virtually no risk of having an immune system response. Embryonic cells also have an ethical stigma. Research on embryonic cells was forbidden until President Bush's decision on August 9, 2001 to allow research on embryonic cell cultures that already existed.

IX. POTENTIAL THERAPEUTIC APPLICATIONS OF STEM CELLS

The ultimate objective of stem cell bioengineering is to be able to understand and possibly control stem cell differentiation and lineage commitment in vitro. One potential application is the generation of different types of neurons for the treatment of Alzheimer's disease, spinal cord injuries, or Parkinson's disease. The production of heart muscle cells for heart attack survivors may also be possible. The generation of insulin-secreting pancreatic islet cells for the treatment of type-1 diabetes, and even the generation of hair follicle stem cells for the treatment of certain types of baldness, have been considered. Stem cells could also be useful for a number of tissue engineering applications such as the production of complete organs including livers, kidneys, eyes, hearts, or even parts of the brain. This represents a considerably greater challenge, beyond the generation of specialized cell types, and will require considerable time and effort to develop. Other areas that would benefit from a better understanding and control of stem cell proliferation in vitro are drug testing, cancer research, and fundamental research on embryonic development.

Human ES cells could potentially be used for all of the above applications. Human ES cells are obtained from aborted fetuses or fertilized eggs. This has come under ethical scrutiny since use of these procedures requires serious moral consideration by society. A possible way to avoid this issue would be to use stem cells isolated from adult tissues. If, as recent studies have suggested, the plasticity of stem cells is greater than previously imagined, then it may be feasible to use adult stem cells as an initial cell population for therapeutic modality. Bone marrow, peripheral blood or umbilical cord blood HSCs could be ideal for such procedures. They are relatively easy to harvest and robust assays exist to quantify HSCs and distinguish them from more differentiated progenitors. Although much remains to be learned about the factors controlling their self-renewal and differentiation, the state of knowledge about HSCs is considerably more advanced than for other stem cell systems.

(a) Cell replacement therapy

Degenerative disorders like myocardial infarction, Parkinson's disease, diabetes are due to the progressive (acute/chronic) loss of functional cells due to disease or injury to the cells or ageing. Replacing the worn out or injured cells by functional cells to restore the normal function of the tissues or organs is the underlying principle of CRT, otherwise also called as regenerative medicine.

(b) Cardiovascular therapy

Several cardiovascular diseases are associated with myocardial cell death resulting from ischemic heart disease, viral infections or immunopathological conditions. The only current treatment for the patients with end stage heart failure is the cardiac transplantation. Often, this is limited by the availability of a perfectly matched organ. The demand for organ supply is increasing steadily, necessitating the vital need for the development of new therapeutic options. In the last decade, several animal studies demonstrated that transplantation of isolated cardiomyocytes might offer an alternative approach for the treatment of cardiac degenerative disorders. Several animal studies demonstrate that engraftment of cardiac myocytes into the adult heart can be achieved successfully.

The emerging results with ES cells have the potential to revolutionize medicine and could find a cure for the cardiac degenerative disorders obviating the need for the allogeneic organ transplantation. Several factors alone or in combination have been shown to enrich cardiac differentiation such as hepatocyte growth factor (HGF), epidermal growth factor (EGF), basic fibroblast growth factor (bFGF), transforming growth factor b1 (TGFb1), platelet derived growth factor (PDGF), sphingosine-1-phosphate, retinoic acid, 5-azacytidine, vitamin C, and overexpression of GATA-4. The so called CRT where the damaged cardiac cells will be replaced by cells derived from the ES cells to restore the normal function of the heart could be a boon for heart patients in near future.

(c) Hematopoiesis

Hematopoiesis in ES cell aggregates (ESCAs) develops spontaneously in the absence of exogenous cytokines demonstrating that the microenvironment of the developing ESCAs is enough to supply the necessary inductive signals for commitment of mesodermal precursors to hematopoietic cell fate, and also to provide the differentiation signals for blood cell maturation. The pattern of hematopoiesis in the ESCAs mimics yolk sac development and is dominated by primitive erythroid cells that express embryonic forms of hemoglobin (bH1).

(d) Neurological disorders

Neuroectodermal differentiation is also relatively easy to achieve from murine ES cells. High yields of neurons, astrocytes and oligodendrocyte like cells have been achieved with several different optimised protocols. These protocols have since been further simplified to remove the ESCAs stage and differentiate murine and human ES cells into neuroectodermal lineages in monolayer. This is the only somatic cell type for which this is known to be possible. Preliminary implantation studies of ES derived neurons into mouse models have reported that these cells are functionally active, integrate into the brain and in one case have corrected the phenotype of a neurodegenerative disease.

(e) Diabetes

Cell replacement therapy via islet transplantation through stem cells has gained more popularity due to the recent upcoming results in the animal model studies. Early investigations demonstrate that ES cells are the promising sources for alleviating the pathological conditions of diabetic patients. It has been reported that the insulin-expressing cells that had spontaneously differentiated within embryoid bodies were selected by an approach where an introduced antibiotic resistance marker under the control of the insulin promoter was made use of.

(f) Liver therapy

Hepatocytes or hepatocyte-like cells derived from ES cells can be a promising approach in the therapy of liver diseases. Several entities like chronic viral hepatitis, alcoholic liver disease, haemochromatosis etc. result in liver cirrhosis and finally organ failure making a liver transplantation inevitable. Differentiated ES cells could be used either for direct transplantation into the diseased liver or for the generation of bioartificial liver (BAL) devices to bypass situations of acute liver failure until an organ for transplantation is available, for example acute fulminant viral hepatitis or severe intoxications. Studies have shown the differentiation of murine and human ES cells into hepatocyte-like cells and results have also been achieved on the transplantation of differentiated murine ES cell into the liver. Especially BAL devices seem to be promising for the next future. Main difficulties of CRT like tumor formation or problematic engraftment of the cells are of no relevance in this therapeutic approach.

X. PROBLEMS ENCOUNTERED WITH USE OF ES CELLS

(a) Selection of desired cell phenotype

The routinely used differentiation protocols give rise to a mixture of different cell populations. To harvest a single population of cells of our interest, several strategies have been developed. Magnetic bead tagged antibodies directed against the unique cell surface marker of the desired cell population can be directed and the desired population can be harvested by running through a magnetic column where the magnet tagged antibodies bound with the desired cell population will be retained, allowing other cell population washed out. Fluorescence labelled antibodies raised against unique cell surface markers of the desired cell population can be lifted up by the fluorescence -activated cell sorting (FACS). Some protocols employ selective culture conditions that promote the growth of one particular cell type at the expense of the other. ES cells may be genetically engineered to have selection markers. Expression of selection markers (antibiotics resistant genes) makes the desired population resistant to that particular antibiotic when applied to kill the other cell population. These antibiotics block the endogenous translation or transcription machineries if cells do not express the particular enzyme/ protein which antagonize the blocking activity of these antibiotics. The drawback of this approach is that the genetically manipulated ES cell derivatives might be immunologically rejected since they express the foreign proteins.

(b) Tumor formation

Implantation of undifferentiated ES cells leads to formation of benign teratoma in the recipients. This demands a pure population of terminally differentiated cell phenotype. Negative selection of Oct4 expressing cells might be a solution for this issue. New strategies and methodologies need to be developed to isolate the terminally differentiated cells.

(c) Transplant rejection

ES cell derived cellular grafts might be rejected due to immunogenicity of the transplanted cells. ES cells can be easily manipulated genetically to overcome the problem of immune rejection so that lifelong pharmocologic immunosuppression will not be needed. One attractive approach is that the MHC genes in ES cells can be replaced with host MHC genes so that the host immune system will recognize the implanted cells as 'self'. This approach is practically possible with help of homologous recombination techniques. One interesting approach in this regard is the application of nuclear transplantation technology in ES cells. In this approach, the nucleus from a normal somatic cell of the recipient, say from a skin biopsy is extracted and is then injected into an enucleated oocyte. The cytoplasm of this oocyte has the required potential to reprogram the differentiated nucleus injected and re-establishes an embryonic gene expression pattern in the chromatin of the somatic cell nucleus. The blastocyst formed from this oocyte would be the source for the derivation of new ES cell lines which would be genetically matched for each nuclear gene of the recipient. In this case, whole MHC regions in ES cells and other immune molecules will be identical to that of the patient, except minor molecules derived from mitochondrial genes. The immune rejection in this case will be the least of all the above-mentioned approaches.

XI. FUTURE PERSPECTIVES

To serve as a versatile renewable source of healthier cells for the use in CRT, the problems mentioned above should be overcome. That is, new methodologies and protocols should be developed for the derivation of every single cell lineage population of body tissues/organs. New strategies should be developed to harvest only the terminally differentiated phenotype of the cell or the phenotype of the cell which serves as nearly an ideal substitute to be used for CRT. This population of cells should be free from the tumourigenic population or from the population that may lead to severe complications during the post transplantation period. The problem of the immunogenic of ES cell derived cellular transplant in the host should be overcome by new approaches then ES cells will prove to be the potential versatile source of the cellular transplants to be used in CRT to treat a number of degenerative disorders.

Utilization and practical application of ES cells in cell replacement therapy are still in their infancy and need further exploration and extensive investigations, experimental confirmations and clinical trials before they will be made available as an ideal cell substitute for the treatment of a number of degenerative disorders. Once first successful clinical studies will be achieved and evaluated, ES-based cell replacement therapy will revolutionize medicine in the near future offering therapeutically alternatives for treatment of severe degenerative disorders.

QUESTIONS

1. What are stem cells ? Describe different types of stem cells and their source.
2. Distinguish between embryonic stem cells and adult stem cells and discuss their attributes and relative utility in treating various ailments.
3. Describe the types of divisions in stem cells. What mechanism is involved in each case and what purpose does each of these cell divisions serve?
4. How stem cells can be proved beneficial in the cure of various degenerative diseases?
5. How ES cells can be grown in laboratory and what factors govern their differentiation in to a particular cell type ?

GENE THERAPY

CHAPTER 36

I. INTRODUCTION

Genes, who are carried on chromosomes, are the basic physical and functional units of heredity. Genes are specific sequences of bases that encode instructions on how to make proteins. Although genes get a lot of attention, it's the proteins that perform most life functions and even make up the majority of cellular structures. When genes are altered so that the encoded proteins are unable to carry out their normal functions, genetic disorders can result.

Gene therapy is a technique for correcting defective genes responsible for disease development. Researchers may use one of several approaches for correcting faulty genes:

(*i*) A normal gene may be inserted into a nonspecific location within the genome to replace a nonfunctional gene. This approach is most common.

(*ii*) An abnormal gene could be swapped for a normal gene through homologous recombination.

(*iii*) The abnormal gene could be repaired through selective reverse mutation, which returns the gene to its normal function.

(*iv*) The regulation (the degree to which a gene is turned on or off) of a particular gene could be altered.

II. DISEASES OF GENETIC ORIGIN

Most of us do not suffer any harmful effects from our defective genes because we carry two copies of nearly all genes, one derived from our mother and the other from our father. The only exceptions to this rule are the genes found on the male sex chromosomes. Males have one X and one Y chromosome, the former from the mother and the latter from the father, so each cell has only one copy of the genes on these chromosomes. In the majority of cases, one normal gene is sufficient to avoid all the symptoms of disease. If the potentially harmful gene is recessive, then its normal counterpart will carry out all the tasks assigned to both. Only if we inherit both copies of the same recessive gene from our parents, we will develop a disease.

On the other hand, if the gene is dominant, it alone can produce the disease, even if its counterpart is normal. Clearly only the children of a parent with the disease can be affected, and then on average only half the children will be affected. Huntington's chorea, a severe disease of the nervous system, which becomes apparent only in adulthood, is an example of a dominant genetic disease.

Finally, there are the X chromosome-linked genetic diseases. As males have only one copy of the genes from this chromosome, there are no others available to fulfill the defective gene's function. Examples of such diseases are Duchenne muscular dystrophy and, perhaps most well known of all, hemophilia.

Queen Victoria was a carrier of the defective gene responsible for hemophilia, and through her it was transmitted to the royal families of Russia, Spain, and Prussia. Minor cuts and bruises, which would do little harm to most people, can prove fatal to hemophiliacs, who lack the proteins (Factors VIII and IX) involved in the clotting of blood, which are coded for by the defective genes.

Sadly, before these proteins were made available through genetic engineering, hemophiliacs were treated with proteins isolated from human blood. Some of this blood was contaminated with the AIDS virus, and has resulted in tragic consequences for many hemophiliacs. Use of genetically engineered proteins in therapeutic applications, rather than blood products, will avoid these problems in the future.

Not all defective genes necessarily produce detrimental effects, since the environment in which the gene operates is also of importance. A classic example of a genetic disease having a beneficial effect on survival is illustrated by the relationship between sickle-cell anemia and malaria. Only individuals having two copies of the sickle-cell gene, which produces a defective blood protein, suffer from the disease. Those with one sickle-cell gene and one normal gene are unaffected and, more importantly, are able to resist infection by malarial parasites. The clear advantage, in this case, of having one defective gene explains why this gene is common in populations in those areas of the world where malaria is endemic.

III. BASIC PRINCIPLES OF GENE THERAPY

Much attention has been focused on the so-called genetic metabolic diseases in which a defective gene causes an enzyme to be either absent or ineffective in catalyzing a particular metabolic reaction effectively. A potential approach to the treatment of genetic disorders in man is gene therapy. This is a technique whereby a working gene replaces the absent or faulty gene, so that the body can make the correct enzyme or protein and consequently eliminate the root cause of the disease.

The most likely candidates for future gene therapy trials will be rare diseases such as Lesch-Nyhan syndrome, a distressing disease in which the patients are unable to manufacture a particular enzyme. This leads to a bizarre impulse for self-mutilation, including very severe biting of the lips and fingers. The normal version of the defective gene in this disease has now been cloned.

If gene therapy does become practicable, the biggest impact would be on the treatment of diseases where the normal gene needs to be introduced into only one organ. One such disease is phenylketonuria (PKU). PKU affects about one in 12,000 white children, and if not treated early can result in severe mental retardation. The disease is caused by a defect in a gene producing a liver enzyme. If detected early enough, the child can be placed on a special diet for their first few years, but this is very unpleasant and can lead to many problems within the family.

Before treatment for a genetic disease can begin, an accurate diagnosis of the genetic defect needs to be made. It is here that biotechnology is also likely to have a great impact in the near future. Genetic engineering research has produced a powerful tool for pinpointing specific diseases rapidly and accurately. Short pieces of DNA called DNA probes can be designed to stick very specifically to certain other pieces of DNA. The technique relies upon the fact that complementary pieces of DNA stick together. DNA probes are more specific and have the potential to be more sensitive than conventional diagnostic methods, and it should be possible in the near future to distinguish between defective genes and their normal counterparts, an important development.

The transfer of DNA is usually achieved either by mixing the DNA with a substance that enhances its uptake or by packaging it inside a disabled virus (a virus that can carry DNA into a cell but cannot cause a disease). First removing cells from the patient and then re-introducing those that have been appropriately modified may carry out Gene therapy. Alternatively, the direct introduction of DNA into the patient's body may be the only effective strategy.

There are several types of gene therapy :

(i) **Gene augmentation therapy:** This is appropriate for the treatment of inherited disorders caused by the loss of a functional gene product. The aim is to add a functional copy of the lost gene back into the genome and express it at sufficient levels to replace the missing protein. It is only suitable if the pathogenic effects of the disease are reversible.

(ii) Gene inhibition therapy: This is suitable for the treatment of infectious diseases, cancer and inherited disorders caused by inappropriate gene activity. The aim is to introduce a gene whose product inhibits the expression of the pathogenic gene or interferes with the activity of its product.

(iii) Killing of specific cells: This is suitable for diseases such as cancer that can be cured by eliminating certain populations of cells. The aim is to express within such cells a suicide gene, whose product is toxic. One approach is the expression of an enzyme that converts a harmless prodrug into a highly toxic molecule. Another is the expression of a protein that makes the cells vulnerable to attack by the immune system. It is very important to ensure that suicide genes are appropriately targeted; otherwise the therapy would result in widespread cell death.

(iv) Somatic and germ line gene therapy: There is an important distinction between somatic gene therapy (DNA transfer to our normal body tissue) and germ line gene therapy (DNA transfer to cells that produce eggs or sperm). The distinction is that the results of any somatic gene therapy are restricted to the actual patient and are not passed on to his or her children.

In most gene therapy studies, a "normal" gene is inserted into the genome to replace an "abnormal," disease-causing gene. A carrier molecule called a vector must be used to deliver the therapeutic gene to the patient's target cells. Currently, the most common vector is a virus that has been genetically altered to carry normal human DNA. Viruses have evolved a way of encapsulating and delivering their genes to human cells in a pathogenic manner. Scientists have tried to take advantage of this capability and manipulate the virus genome to remove disease-causing genes and insert therapeutic genes.

Target cells such as the patient's liver or lung cells are infected with the viral vector. The vector then unloads its genetic material containing the therapeutic human gene into the target cell. The generation of a functional protein product from the therapeutic gene restores the target cell to a normal state.

Some of the different types of viruses used as gene therapy vectors:

IV. ADENOVIRUS VECTOR SYSTEM

Adenoviruses infect a wide range of non dividing human cells and have been used extensively as live vaccines against respiratory infections and gastroenteritis without side effects. These features make adenovirus a likely prospect for delivering genes to target cells. For use as a vector for gene therapy, a cell line that produces the adenoviral E1 gene products, which are essential for adenoviral replication, is cotransfected with two portions of the adenovirus genome (Fig. 36.1). One of these segments, which is maintained in Escherichia coli as a plasmid, carries a therapeutic gene in place of the E1 region and is flanked by adenovirus DNA sequences. The E1 region is located near the 5' end of the adenovirus genome. The second transfected DNA segment is an adenovirus DNA molecule that lacks part of its 5'end , including the E1 region, but shares a region of adenoviral DNA with the plasmid that carries the therapeutic gene. A recombination event in the region of overlap between the two transfected pieces of DNA reconstitutes a full length adenovirus genome that contains the therapeutic gene and lacks the E1 region. The E1 gene products that are supplied by the host cell initiate virus production; virus particles are assembled and then released after cell lysis. The DNA cloning capacity of this adenovirus vector system is about 7.5 kb. If recombination does not occur, both transfected DNA molecules are too small to be packaged properly. Moreover, recombination between the E1 DNA region in the genome of the host cell and the recombinant adenovirus DNA to form a replication –competent virus is extremely unlikely.

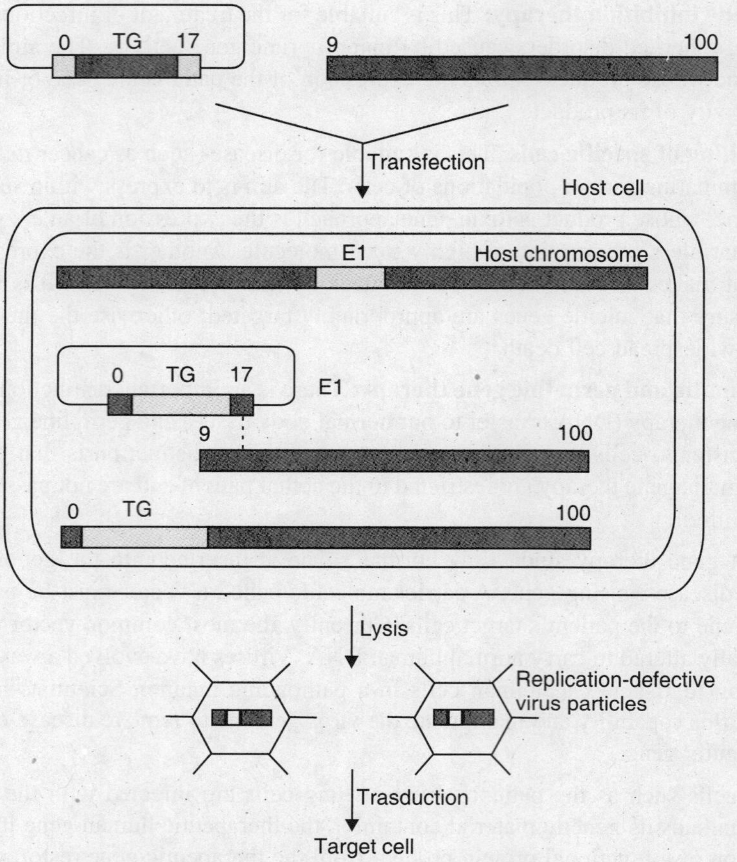

Fig. 36.1. Adenovirus vector system. A plasmid with a therapeutic gene (TG) that is inserted into a segment of the adenovirus genome extending from map unit 0 to 17 and a linear adenvirus DNA molecule that consists of map units 9 to 100 are both introduced into a host cell that carries an integrated and functional adenovirus E1 gene (E1). The entire adenovirus genome is 100 map units. A recombination event (dotted line) between the shared DNA regions of the plasmid and linear adenovirus DNA molecule creates a DNA molecule that is equivalent to a full-length adenovirus genome. The recombinant DNA molecule with the therapeutic gene is packaged and released after cell lysis. The recombinant virus particles are replication defective. Plasmid DNA which is part of the final construct does not impact on the packaging of the recombinant DNA molecule (not shown).

After infection of a target cell with a recombinant adenovirus, the DNA is passed into the cell nucleus, where the therapeutic gene is expressed. The recombinant DNA construct does not integrate into a chromosome, and consequently, it does not persist for long periods. Therefore, adenovirus based gene therapy requires periodic administration with additional recombinant viruses.

Adenovirus gene delivery systems have been used in gene therapy trials for treating cystic fibrosis.

V. ADENO ASSOCIATED VIRUS VECTOR SYSTEM

Adeno- associated virus is a small, non pathogenic, single-stranded human DNA virus (4.7kb) that can integrate into a specific site on chromosome 19. Productive infection by adeno-associated virus depends on proteins from another virus (helper virus) such as adenovirus, hence the name adeno-associated virus. After the adeno-associated virus enters the nucleus, the polymerase of a host cell converts the adeno-associated virus genome to a souble-stranded DNA that is then transcribed.

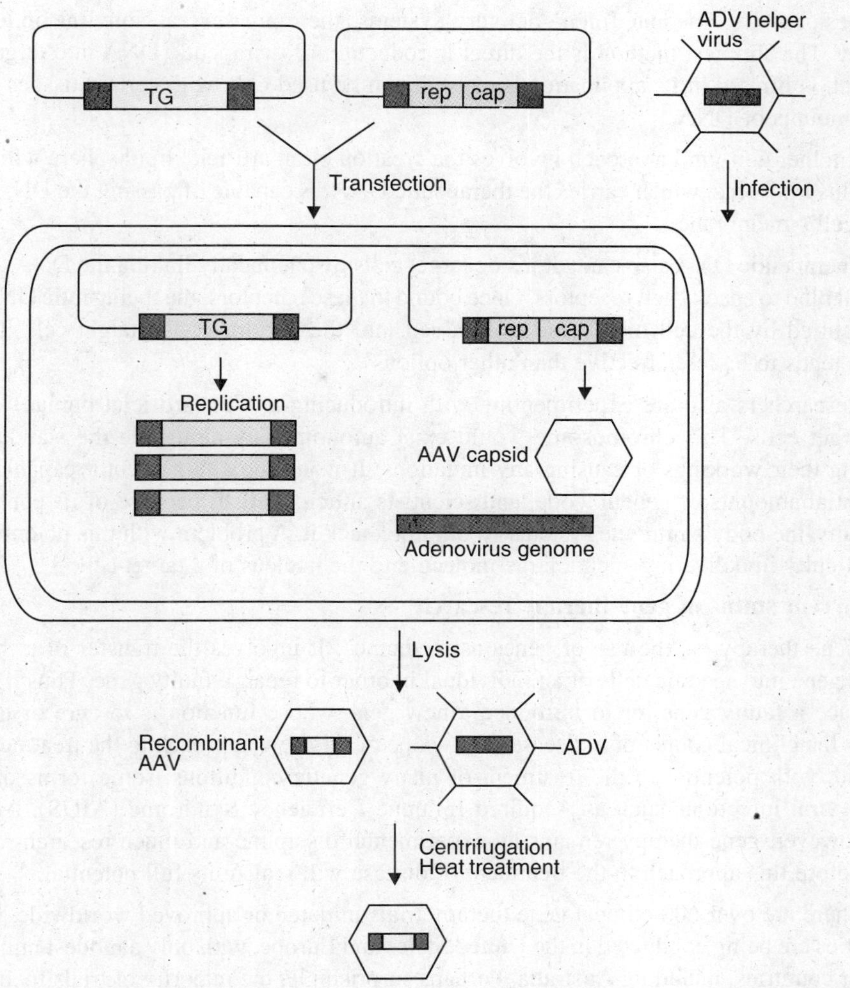

Fig. 36.2. Adeno-associated virus (AAV) vecotr system. Two plasmids, one containing a therapeutic gene (TG) thatis flanked by the inverted terminal repeats (ITR) of AAV and the other with the AAV genes for replication (rep) and capsid formation (cap) under the control of a promoter (p) and polyadenylation sequence (pa), are contrasfected into a host cell that is infected with a helper adenovirus (ADV). Both recombinant AAV and ADV particles are released after cell lysis. The recombinant AAV can be separated from ADV by centrifugation, and any remaining ADV is inactivated by heat treatment.

Its absence of pathogenicity makes adeno-associated virus a good candidate as a vector for the delivery of therapeutic genes. Recombinant adeno associated virus is generated by cotransfection of two plasmids into a host cell that has been infected with adenovirus (helper virus) (Fig. 36.2). One of the plasmids carries a therapeutic gene flanked by the inverse terminal repeat sequences (~125bp each) from the adeno-associated virus. The second plasmid contain the two genes (rep and cap) of the adeno-associated virus genome that are responsible for replication of the adeno-associated virus capsid, respectively. After lysis of infected host cells, recombinant adeno-associated virus particles are purified from adenovirus by centrifugation and dialysis. Any residual adenovirus (helper virus) in the sample is killed by heat treatment. In the absence of the *rep* gene, the recombinant adeno-associated virus DNA is unlikely to integrate into chromosome 19. On the other hand, without any adeno-associated virus genes, the recombinant vector will not evoke an immune response.

Besides virus-mediated gene-delivery systems, there are several nonviral options for gene delivery. The simplest method is the direct introduction of therapeutic DNA into target cells. This approach is limited in its application because it can be used only with certain tissues and requires large amounts of DNA.

Another non-viral approach involves the creation of an artificial lipid sphere with an aqueous core. This liposome, which carries the therapeutic DNA, is capable of passing the DNA through the target cell's membrane.

Therapeutic DNA also can get inside target cells by chemically linking the DNA to a molecule that will bind to special cell receptors. Once bound to these receptors, the therapeutic DNA constructs are engulfed by the cell membrane and passed into the interior of the target cell. This delivery system tends to be less effective than other options.

Researchers also are experimenting with introducing a 47th (artificial human) chromosome into target cells. This chromosome would exist autonomously alongside the standard 46 —not affecting their workings or causing any mutations. It would be a large vector capable of carrying substantial amounts of genetic code, and scientists anticipate that, because of its construction and autonomy, the body's immune systems would not attack it. A problem with this potential method is the difficulty in delivering such a large molecule to the nucleus of a target cell.

VI. Current status of gene therapy research

Gene therapy is "the use of genes as medicine". It involves the transfer of a therapeutic or correct gene into specific cells of an individual in order to repair a faulty gene. Thus it may be used to replace a faulty gene, or to introduce a new gene whose function is to cure or to favourably modify the clinical course of a disease. The scope of this new approach to the treatment of disease is broad, with potential in the treatment of many genetic conditions, some forms of cancer and certain viral infections such as Acquired Immune Deficiency Syndrome (AIDS). At the present time however, gene therapy remains an experimental discipline and much research remains to be done before this approach to the treatment of disease will realize its full potential.

There are over 600 clinical gene therapy trials initiated or approved worldwide. Of these, the majorities are being conducted in the United States and Europe, with only a modest number initiated in other countries, including Australia. Perhaps surprisingly, the majority of trials focus on treating acquired diseases such as cancer and AIDS, although an increasing number of inherited conditions are being targeted. A form of immune deficiency called adenosine deaminase (ADA) deficiency was the first disease to be treated with a gene therapy approach in humans in the early 1990's, and was also the first condition for which therapeutic gene transfer into stem cells has been attempted in the clinical arena.

Another from of immune deficiency is due to a mutation in a gene located on the X chromosome and is called Severe Combined Immune Deficiency (SCID). This "X-linked condition" only affects boys (see Genetics Fact Sheet 8 for an explanation of the inheritance pattern of genes located on the X chromosome). The use of gene therapy in the treatment of this condition by a French research group Cavazzana-Calvo M *et al.* (2000), gene therapy of human severe combined immunodeficiency (SCID-X1) disease. It was hailed as the first example of a genetic condition being successfully treated by gene therapy and is a milestone in medical history. Seven out of 10 infants treated to date have restored immune function but 2 of the children treated initially have developed a form of leukemia. The leukemia in these two patients was caused when the virus used to deliver the therapeutic gene activated a cancer-causing gene. After the first boy developed leukemia in October 2002, and the second in January 2003, the clinical trials of the gene therapy being conducted in a number of countries were halted. These have now been resumed but only for patients with no other treatment options. Work is continuing to make the therapy as safe as possible.

Table 36.1. Current therapies for 10 single-gene human disorders

Gene product	Disease and symptoms	Frequency	Current therapy	Prognosis
Adenosine deaminse	Severe combined immunodeficiency Loss of T and B cells	1:1,000,000	Bone marrow transplant; adenosine deaminase replacement	Without therapy : fatal by 2 years of age With therapy: clinical improvement
LDL receptor	Familial hypercholesterolemia Elevated blood serum cholesterol level, coronary artery disease	1:500 (hecterozygotes)	Diet, drugs, liver transplant	Some clinical improvement
Glucocerebrosidase	Gaucher disease Accumulation of glucocerebroside in macrophages causing liver, spleen, and bone damage.	1:2500 (Jewish populations); rare in non-Jewish populations	Symptomatic treatment: removal of spleen, antibiotics, repairing bone damage, bone marrow transplant, enzyme replacement	Some clinical improvement
Blood clotting factor VIII	Hemophillia A Altered plasma protein that causes defective blood clotting, chronic internal bleeding into joints, excessive bleeding after wounding	1:10,000 (males)	Concentrates of factor VIII by transfusion	Extended life expectancy requires continual treatment, risk of viral infection from transfusions
Cystic fibrosis transmembrane regulator	Cystic fibrosis Multisystem disease pancreatic insufficiency in some cases, intestinal blockage, blocked airways of the lungs	1:2,500 (Caucasians)	Antibiotics, physical clearing of the lungs, upgrading diet	Fatal by late 20s
Ornithin transcarbamylase	Hyperammonemia Urea cycle defect, ammonia accumulation arginine deficiency Early form; within 72 h of birth, lethargy, vomiting, coma, death, and if survival, irreversible brain, damage Late form: vomitting, lethargy, seizures.	1:40,000	Restricted protein diet, agrinine-supplemented diet, drugs, liver transplant	Late form: good Early (severe) form: symptoms diminished
Dystrophin	Duchenne muscular dystrophy Progressive muscle wasting	1:7,500 (males)	Only supportive treatments: good nutrition, and in respiratory function, confinement to a wheelchair	Fatal by early 20s
β-Globin	Sickle-cell disease Chronic anemia, multisystem disease, damage to spleen, heart, kidneys, liver, and brain; Heterozygotes have a mild form of the disease	1:500 (heterozygotes in populations of black African origin; less frequejnt in other populations)	Blood transfusions, drugs, analgesics, bone marrow transplant	Symptoms diminished; treatments suboptimal

The US Food and Drug Administration (FDA) have not yet approved any human gene therapy product for sale. Current gene therapy is experimental and has not proven very successful in clinical trials. Little progress has been made since the first gene therapy clinical trial began in 1990. In 1999, gene therapy suffered a major setback with the death of 18-year-old Jesse Gelsinger. Jesse was participating in a gene therapy trial for ornithine transcarboxylase deficiency (OTCD). He died from multiple organ failures 4 days after starting the treatment. His death is believed to have been triggered by a severe immune response to the adenovirus carrier.

Another major blow came in January 2003, when the FDA placed a temporary halt on all gene therapy trials using retroviral vectors in blood stem cells. FDA took this action after it learned that a second child treated in a French gene therapy trial had developed a leukemia-like condition. Both this child and another who had developed a similar condition in August 2002 had been successfully treated by gene therapy for X-linked severe combined immunodeficiency disease (X-SCID), also known as "bubble baby syndrome."

FDA's Biological Response Modifiers Advisory Committee (BRMAC) met at the end of February 2003 to discuss possible measures that could allow a number of retroviral gene therapy trials for treatment of life-threatening diseases to proceed with appropriate safeguards. FDA has yet to make a decision based on the discussions and advice of the BRMAC meeting.

The most active research being done in gene therapy for children has been for genetic disorders such as cystic fibrosis. Other gene therapy trials involve children with severe immunodeficiencies, such as adenosine deaminase (ADA) deficiency (a rare genetic disease that makes children prone to serious infection), and those with familial hypercholesterolemia (extremely high levels of serum cholesterol).

Advances in understanding and manipulating genes have set the stage for scientists to alter a person's genetic material to fight or prevent disease. Gene therapy is an experimental treatment that involves introducing genetic material (DNA or RNA) into a person's cells to fight disease. Gene therapy is being studied in clinical trials (research studies with people) for many different types of cancer and for other diseases. It is not currently available outside a clinical trial.

Researchers are studying several ways to treat cancer using gene therapy. Some approaches target healthy cells to enhance their ability to fight cancer. Other approaches target cancer cells, to destroy them or prevent their growth. Some gene therapy techniques under study are described below:

- In one approach, researchers replace missing or altered genes with healthy genes. Because some missing or altered genes (e.g., p53) may cause cancer, substituting "working" copies of these genes may be used to treat cancer.

- Researchers are also studying ways to improve a patient's immune response to cancer. In this approach, gene therapy is used to stimulate the body's natural ability to attack cancer cells. In one method under investigation, researchers take a small blood sample from a patient and insert genes that will cause each cell to produce a protein called a T-cell receptor (TCR). The genes are transferred into the patient's white blood cells (called T lymphocytes) and are then given back to the patient. In the body, the white blood cells produce TCRs, which attach to the outer surface of the white blood cells. The TCRs then recognize and attach to certain molecules found on the surface of the tumor cells. Finally, the TCRs activate the white blood cells to attack and kill the tumor cells.

- Scientists are investigating the insertion of genes into cancer cells to make them more sensitive to chemotherapy, radiation therapy, or other treatments. In other studies, researchers remove healthy blood-forming stem cells from the body, insert a gene that makes these cells more resistant to the side effects of high doses of anticancer drugs, and then inject the cells back into the patient.

• In another approach, researchers introduce "suicide genes" into a patient's cancer cells. A pro-drug (an inactive form of a toxic drug) is then given to the patient. The pro-drug is activated in cancer cells containing these "suicide genes," which leads to the destruction of those cancer cells.

• Other research is focused on the use of gene therapy to prevent cancer cells from developing new blood vessels (angiogenesis).

Some of the recent developments in gene therapy research include:

(*i*) University of California, Los Angeles, research team gets genes into the brain using liposomes coated in a polymer called polyethylene glycol (PEG). The transfer of genes into the brain is a significant achievement because viral vectors are too big to get across the "blood-brain barrier." This method has potential for treating Parkinson's disease.

(*ii*) RNA interference or gene silencing may be a new way to treat Huntington's. Short pieces of double-stranded RNA (short, interfering RNAs or siRNAs) are used by cells to degrade RNA of a particular sequence. If a siRNA is designed to match the RNA copied from a faulty gene, then the abnormal protein product of that gene will not be produced.

(*iii*) New gene therapy approach repairs errors in messenger RNA derived from defective genes. Technique has potential to treat the blood disorder thalassaemia, cystic fibrosis, and some cancers.

(*iv*) Gene therapy for treating children with X-SCID (sever combined immunodeficiency) or the "bubble boy" disease is stopped in France when the treatment causes leukemia in one of the patients.

(*v*) Researchers at Case Western Reserve University and Copernicus Therapeutics are able to create tiny liposomes 25 nanometers across that can carry therapeutic DNA through pores in the nuclear membrane.

(*vi*) Sickle cell is successfully treated in mice.

VII. LIMITATIONS OF GENE THERAPY

(i) Short-lived nature of gene therapy: Before gene therapy can become a permanent cure for any condition, the therapeutic DNA introduced into target cells must remain functional and the cells containing the therapeutic DNA must be long-lived and stable. Problems with integrating therapeutic DNA into the genome and the rapidly dividing nature of many cells prevent gene therapy from achieving any long-term benefits. Patients will have to undergo multiple rounds of gene therapy.

(ii) Immune response: Anytime a foreign object is introduced into human tissues, the immune system is designed to attack the invader. The risk of stimulating the immune system in a way that reduces gene therapy effectiveness is always a potential risk. Furthermore, the immune system's enhanced response to invaders it has seen before makes it difficult for gene therapy to be repeated in patients.

(iii) Problems with viral vectors: Viruses, while the carrier of choice in most gene therapy studies, present a variety of potential problems to the patient —toxicity, immune and inflammatory responses, and gene control and targeting issues. In addition, there is always the fear that the viral vector, once inside the patient, may recover its ability to cause disease.

(iv) Multigene disorders: Conditions or disorders that arise from mutations in a single gene are the best candidates for gene therapy. Unfortunately, some the most commonly occurring disorders, such as heart disease, high blood pressure, Alzheimer's disease, arthritis, and diabetes, are caused by the combined effects of variations in many genes. Multigene or multifactorial disorders such as these would be especially difficult to treat effectively using gene therapy.

VIII. ETHICAL CONSIDERATIONS FOR GENE THERAPY

While the body has many billions of cells, only a very small proportion of these cells are involved in reproduction, the process by which our genes are handed on to future generations. In males these cells are located in the testes and in females, in the ovaries. These special reproductive cells are called "germ cells". All other cells in the body, irrespective of whether they are brain, lung, skin or bone cells, are known as "somatic cells". In gene therapy, *only* somatic cells are targeted for treatment. So any changes to the genes of a person by gene therapy will only impact on the cells of their body and cannot be passed on to their children. Changes to the somatic cells cannot be passed on to future generations (inherited). Gene therapy treats the individual and has no impact on future generations. The possible genetic manipulation of the egg or sperm cells (germ cells) remains the subject of intense ethical and philosophical discussion. The strong consensus view at present is that the risks of germ-line manipulation far exceed any potential benefit and should not be attempted.

Gene therapy does have risks and limitations. The viruses and other agents used to deliver the "good" genes can affect more than the cells for which they're intended. If a gene is added to DNA, it could be put in the wrong place, which could potentially cause cancer or other damage. Genes can also be "over-expressed," meaning they can drive the production of so much of a protein that they can be harmful. Another risk is that a virus introduced into one person could be transmitted to other people or into the environment.

Gene therapy trials with some children present an ethical question. We don't test children to see if they are carriers of genetic diseases, because at this point, there's nothing we can do about it. If a child knows he has a problem gene, there's no guarantee he will be affected by it, but he has to live with the knowledge for the rest of his life. For example, the gene for some types of breast cancer has been identified. But if we test a young girl to see if she's a carrier for it, what will she do with that information?

No therapy, established or experimental, is without some associated risk and the potential benefits of new treatments must always be balanced against such risks. The experience with gene therapy for the immunodeficiency condition SCID as described above illustrates the need for this therapy to be conducted as part of clinical trials. Safety will appropriately remain an important consideration as the field of gene therapy evolves.

The principle objections are summarized as:

- The technology is imperfect. The effects of gene transfer are unpredictable and, even if the target disease was cured, further defects could be introduced into the embryo.
- Denial of human rights. Individuals resulting from germ line gene therapy would have no say in whether their genetic material should have been modified.
- Potential abuse. Germ line gene therapy could be used not only to eliminate disease, but also to enhance favourable characteristics and suppress unfavourable ones. On a small scale, this would result in a generation of 'designer children', with traits chosen by their parents. On a large scale, gene therapy could result in eugenics - manipulation of the genetic properties of a population.

Some Questions to Considered by the society

- What is normal and what is a disability or disorder, and who decides?
- Are disabilities diseases? Do they need to be cured or prevented?
- Does searching for a cure demean the lives of individuals presently affected by disabilities?
- Is somatic gene therapy (which is done in the adult cells of persons known to have the disease) more or less ethical than germline gene therapy (which is done in egg and sperm cells and prevents the trait from being passed on to further generations)? In cases of somatic gene therapy, the procedure may have to be repeated in future generations.

- Preliminary attempts at gene therapy are exorbitantly expensive. Who will have access to these therapies? Who will pay for their use?

IX. FUTURE OF GENE THERAPY

To cure genetic diseases, scientists must first determine which gene or set of genes causes each disease. The Human Genome Project and other international efforts have recently completed the initial work of sequencing and mapping virtually all of the 25,000 to 35,000 genes in the human cell. This research will provide new strategies to diagnose, treat, cure, and possibly prevent human diseases. Although this information will help scientists determine the genetic basis of many diseases, it will be a long time before diseases actually can be treated through gene therapy. "The Human Genome Project is just a start". It's going to locate genes for us, but it's not going to tell us what these genes do. That will be the next step. Once we have that information, we'll be able to take advantage of that knowledge to provide treatment and/or cures."

Gene therapy's potential to revolutionize medicine in the future is exciting, and its expectations for curing and preventing childhood diseases is encouraging. One day it may be possible to treat an unborn child for a genetic disease even before symptoms appear.

Scientists are hoping the mapping of the human genome will lead the way toward cures for many diseases and that the successes of current clinical trials will create new opportunities and challenges. For now, however, it's a wait-and-see situation, calling for cautious optimism.

GENE SILENCING

Interruption or suppression of the expression of a gene at transcriptional or translational levels is termed as gene silencing. Scientists have been working on strategies to selectively turn off specific genes in diseased tissues for the past thirty years to get rid of the disease. Three major landmarks in direction are:

- 1980's – use of antisense oligodeoxynucleotides (ODNs)
- 1990's – discovery of ribozymes
- 2000's - RNA interference (RNAi)

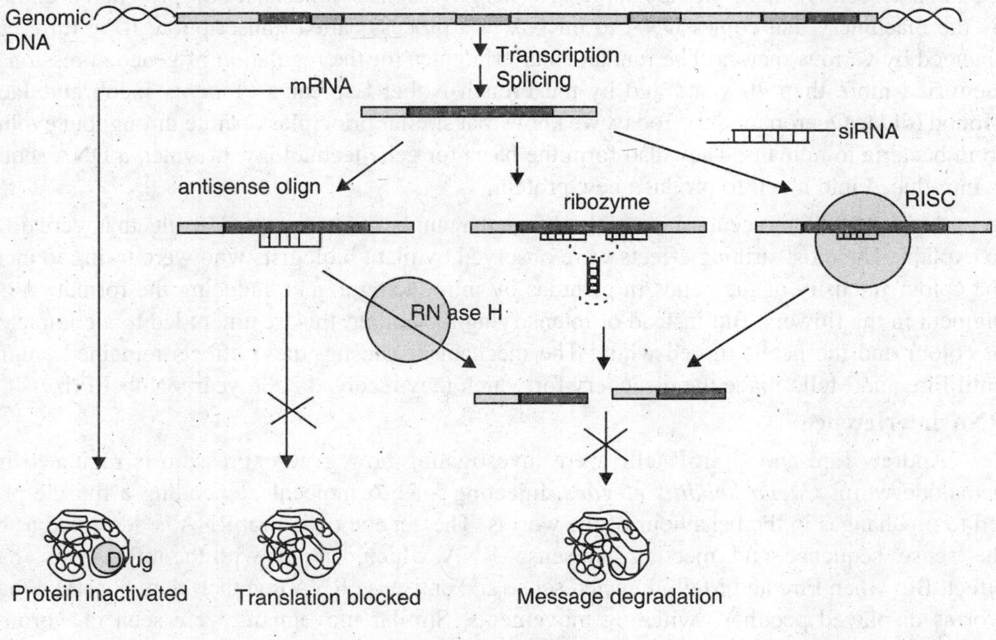

Fig. 36.3. Different ways of gene silencing.

The first evidence that dsRNA could lead to gene silencing came from work in the nematode *Caenorhabditis elegans*. Drs. Guo and Kemphues were attempting to use antisense RNA to shut down expression of the *par-1* gene in order to assess its function. As expected, injection of the antisense RNA disrupted expression of *par-1*, but quizzically, injection of the sense-strand control did the same.

Nobel Laureates of 2006 have discovered a fundamental mechanism for controlling the flow of genetic information. Our genome operates by sending instructions for the manufacture of proteins from DNA in the nucleus of the cell to the protein synthesizing machinery in the cytoplasm. These instructions are conveyed by messenger RNA (mRNA). In 1998, the American scientists Andrew Fire and Craig Mello published their discovery of a mechanism that can degrade mRNA from a specific gene. This mechanism, RNA interference, is activated when RNA molecules occur as double-stranded pairs in the cell. Double-stranded RNA activates biochemical machinery which degrades those mRNA molecules that carry a genetic code identical to that of the double-stranded RNA. When such mRNA molecules disappear, the corresponding gene is silenced and no protein of the encoded type is made.

RNA interference occurs in plants, animals, and humans. It is of great importance for the regulation of gene expression, participates in defense against viral infections, and keeps jumping genes under control. RNA interference is already being widely used in basic science as a method to study the function of genes and it may lead to novel therapies in the future.

Central Dogma

The genetic code in DNA determines how proteins are built. The instructions contained in the DNA are copied to mRNA and subsequently used to synthesize proteins. This flow of genetic information from DNA via mRNA to protein has been termed the central dogma of molecular biology by the British Nobel Laureate Francis Crick. Proteins are involved in all processes of life, for instance as enzymes digesting our food, receptors receiving signals in the brain, and as antibodies defending us against bacteria.

Our genome consists of approximately 30,000 genes. However, only a fraction of them are used in each cell. Which genes are expressed (i.e. govern the synthesis of new proteins) is controlled by the machinery that copies DNA to mRNA in a process called transcription. It, in turn, can be changed by various factors. The fundamental principles for the regulation of gene expression were identified more than 40 years ago by the French Nobel Laureates François Jacob and Jacques Monod (of lac Operon model). Today, we know that similar principles operate throughout evolution, from bacteria to humans. They also form the basis for gene technology, in which a DNA sequence is introduced into a cell to produce new protein.

Around 1990, molecular biologists obtained a number of unexpected results that were difficult to explain. The most striking effects were observed by plant biologists who were trying to increase the colour intensity of the petals in petunias by introducing a gene inducing the formation of red pigment in the flowers. But instead of intensifying the colour, this treatment led to a complete loss of colour and the petals turned white. The mechanism causing these effects remained enigmatic until Fire and Mello made the discovery for which they received 2006 year's Nobel Prize.

RNA interference

Andrew Fire and Craig Mello were investigating how gene expression is regulated in the nematode worm *Caenorhabditis elegans*. Injecting mRNA molecules encoding a muscle protein led to no changes in the behaviour of the worms. The genetic code in mRNA is described as being the 'sense' sequence, and injecting 'antisense' RNA, which can pair with the mRNA, also had no effect. But when Fire and Mello injected sense and antisense RNA together, they observed that the worms displayed peculiar, twitching movements. Similar movements were seen in worms that completely lacked a functioning gene for the muscle protein. What had happened?

When sense and antisense RNA molecules meet, they bind to each other and form double-stranded RNA. Could it be that such a double-stranded RNA molecule silences the gene carrying the same code as this particular RNA? Fire and Mello tested this hypothesis by injecting double-stranded RNA molecules containing the genetic codes for several other worm proteins. In every experiment, injection of double-stranded RNA carrying a genetic code led to silencing of the gene containing that particular code. The protein encoded by that gene was no longer formed.

After a series of simple but elegant experiments, Fire and Mello deduced that double-stranded RNA can silence genes, that this RNA interference is specific for the gene whose code matches that of the injected RNA molecule, and that RNA interference can spread between cells and even be inherited. It was enough to inject tiny amounts of double-stranded RNA to achieve an effect, and Fire and Mello therefore proposed that RNA interference (now commonly abbreviated to RNAi) is a catalytic process.

Mechanism

The components of the RNAi machinery were identified during the following years. Double-stranded RNA binds to a protein complex, Dicer, which cleaves it into fragments. Another protein complex, RISC, binds these fragments. One of the RNA strands is eliminated but the other remains bound to the RISC complex and serves as a probe to detect mRNA molecules. When an mRNA molecule can pair with the RNA fragment on RISC, it is bound to the RISC complex, cleaved and degraded. The gene served by this particular mRNA has been silenced.

RNA interference — a defense against viruses and jumping genes, RNA interference is important in the defense against viruses, particularly in lower organisms. Many viruses have a genetic code that contains double-stranded RNA. When such a virus infects a cell, it injects its RNA molecule, which immediately binds to Dicer. The RISC complex is activated, viral RNA is degraded, and the cell survives the infection. In addition to this defense, higher organisms such as man have developed an efficient immune defense involving antibodies, killer cells, and interferons (Fig. 36.4).

Jumping genes, also known as transposons, are DNA sequences that can move around in the genome. They are present in all organisms and can cause damage if they end up in the wrong place. Many transposons operate by copying their DNA to RNA, which is then reverse-transcribed back to DNA and inserted at another site in the genome. Part of this RNA molecule is often double-stranded and can be targeted by RNA interference. In this way, RNA interference protects the genome against transposons.

Table 36.2. Comparison of different gene silencing strategies.

Agent	Mechanism	Result
Most drugs	Bind to target protein	Protein inhibition
RNase H-independent ODNs	Hybridize to target mRNA	Inhibition of translation of the target protein
RNase H-dependent ODNs	Hybridize to target mRNA	Degradation of the mRNA by RNase H
Ribozymes and DNA enzymes	Catalyze cleavage of target mRNA	Degradation of the mRNA
siRNA	Hybridize to target mRNA by its antisense strand and guide it into endoribonuclease enzyme complex (RISC)	Degradation of the mRNA

RNA interference regulates gene expression

RNA interference is used to regulate gene expression in the cells of humans as well as worms. Hundreds of genes in our genome encode small RNA molecules called microRNAs. They contain pieces of the code of other genes. Such a microRNA molecule can form a double-stranded structure and activate the RNA interference machinery to block protein synthesis. The expression of that particular gene is silenced. We now understand that genetic regulation by microRNAs plays an important role in the development of the organism and the control of cellular functions.

Applications

RNA interference opens up exciting possibilities for use in gene technology. Double-stranded RNA molecules have been designed to activate the silencing of specific genes in humans, animals or plants. Such silencing RNA molecules are introduced into the cell and activate the RNA interference machinery to break down mRNA with an identical code.

1. This method has already become an important research tool in biology and biomedicine. For instance, a gene causing high blood cholesterol levels was recently shown to be silenced by treating animals with silencing RNA. Plans are underway to develop silencing RNA as a treatment for virus infections, cardiovascular diseases, cancer, endocrine disorders and several other conditions.

2. Post-transcriptional gene silencing (PTGS), which was initially considered a bizarre phenomenon limited to petunias and a few other plant species, is now one of the hottest topics in molecular biology. In the last few years, it has become clear that PTGS occurs in both plants and animals and has roles in viral defense and transposon silencing mechanisms.

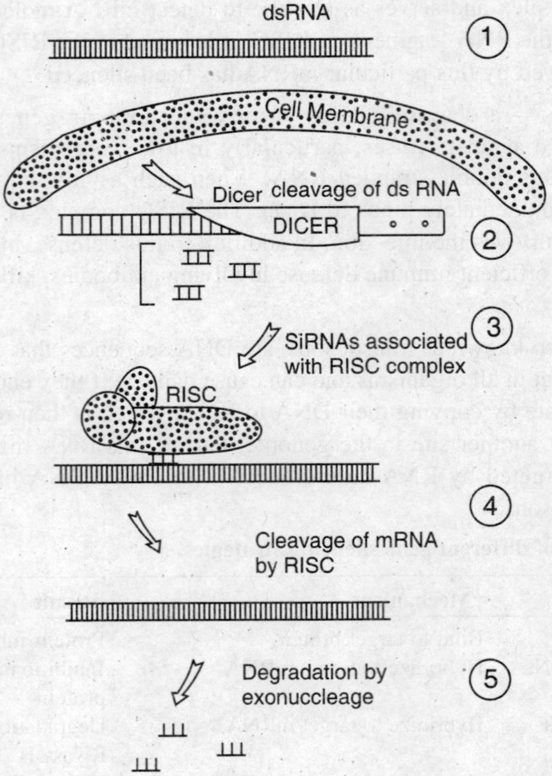

Fig. 36.4. Mechanism of action.

GENETIC DISEASES

One of the most important threats for human's health is, undoubtedly, the **genetic diseases**. It is a disorder caused by genetic factors and especially abnormalities in a human's genetic material (genome). There are four main types of **genetic disorders**. Of course, some of these changes in genome can cause interesting advantages in specific environments (Darwinian Fitness). But there is no doubt that all these abnormalities (disorders) bring destructive results to a living being in the present environment. The four types of (human) Genetic diseases are:

1. Single-gene/monogenic Genetic Diseases:

In this category the starting point is a mutation/change in one gene. The next question is how a change in the sequence of a single gene can cause severe disorders. Genes code for proteins which are some of the most important tools for the living beings, and also take place in the structures of the cells. The results of a mutation that happens in a part of gene that codes for a functional part of a protein are unwelcome. The protein is no more functional and as a result, many severe consequences take place. Almost 6000 single gene disorders are known and it is estimated that 1 of 200 newborns face a single gene genetic disorder. Some of these are **sickle cell anaemia**, **cystic fibrosis**, **Aicardi Syndrome**, **Huntington's disease**.

2. Multifactorial/Polygonic Genetic Diseases:

The second type of human genetic diseases is caused by mutations in more than one genes. The environment combines with these mutations in order these diseases to appear. We can easily conclude that polygenic disorders are more complicated than the previous type (single gene diseases). These abnormalities are also difficult to analyze, because there are many factors that researchers should take into consideration in order to reach to some useful conclusions. Many well known chronic diseases are Multifactorial Genetic Diseases. Everybody knows **Alzheimer, diabetes, obesity** and **arthritis**. Besides many cancer types are caused by multi mutations.

3. Chromosomal Genetic Diseases:

Chromosomes are big DNA molecules composed from genes. The chromosomes are located in the cell nucleus. Abnormalities in the structure, number (and not only) of the chromosomes can cause some of the most dangerous genetic disorders. This type of disorders seems to be much easier to observe because they are, sometimes, detected by examination with microscope. **Down syndrome** is the most well known disease caused by chromosomal abnormalities. In this disorder there is a third copy of chromosome 21 (there are two copies of each chromosome in the cells of healthy humans). Chromosomal diseases can be also caused by segments and joins of parts of chromosomes.

4. Mitochondrial Genetic Diseases:

It is not a common situation. Mitochondrial DNA is a DNA molecule found in the mitochondria (out of the nucleous) – a necessary organelle for cellular respiration. Mutations in the mitochondrial DNA can also cause undesirable abnormalities. Mitochondrial diseases are a group of disorders produced by defects in the oxidative phosphorylation system (Oxphos system), the final pathway of the mitochondrial energetic metabolism, resulting in a deficiency of the biosynthesis of ATP. Part of the polypeptide subunits involved in the Oxphos system are codified by the mitochondrial DNA. In the last years, mutations in this genetic system have been described and associated to well defined clinical syndromes. The clinical features of these disorders are very heterogeneous affecting, in most cases, to different organs and tissues and their correct diagnosis require precise clinical, morphological, biochemical and genetic data. The peculiar genetic characteristics of the mitochondrial DNA (maternal inheritance, polyplasmia and mitotic segregation) give to these disorders very distinctive properties.

QUESTIONS

1. What is gene therapy? How it is useful for human disease cure ?
2. What are different approaches of gene therapy
3. Write short notes :
   (*a*) Genetic diseases (*b*) Adeno vector
   (*c*) Future of gene therapy (*d*) Ethics in gene therapy
   (*e*) Gene silencing (*f*) Genetic diseases of humans
   (*e*) RNA interferences.

HUMAN GENOME PROJECT

CHAPTER 37

Begun formally in 1990, the U.S. Human Genome Project was a 13-year effort coordinated by the U.S. Department of Energy and the National Institutes of Health. The project originally was planned to last 15 years, but rapid technological advances accelerated the completion date to 2003. Project goals were to

- *identify* all the approximately 20,000-25,000 genes in human DNA,
- *determine* the sequences of the 3 billion chemical base pairs that make up human DNA,
- *store* this information in databases,
- *improve* tools for data analysis,
- *transfer* related technologies to the private sector, and
- *address* the ethical, legal, and social issues (ELSI) that may arise from the project.

To help achieve these goals, researchers also studied the genetic makeup of several nonhuman organisms. These include the common human gut bacterium *Escherichia coli*, the fruit fly, and the laboratory mouse.

A unique aspect of the U.S. Human Genome Project is that it was the first large scientific undertaking to address potential ELSI implications arising from project data.

Another important feature of the project was the federal government's long-standing dedication to the transfer of technology to the private sector. By licensing technologies to private companies and awarding grants for innovative research, the project catalyzed the multibillion-dollar U.S. biotechnology industry and fostered the development of new medical applications.

I. HISTORICAL

Rapid progress in genome science and a glimpse into its potential applications have spurred observers to predict that biology will be the foremost science of the 21st century. Technology and resources generated by the Human Genome Project and other genomics research are already having a major impact on research across the life sciences. The potential for commercial development of genomics research presents U.S. industry with a wealth of opportunities, and sales of DNA-based products and technologies in the biotechnology industry are projected to exceed $45 billion by 2009.

Some current and potential applications of genome research include

- Molecular medicine
- Energy sources and environmental applications
- Risk assessment
- Bioarchaeology, anthropology, evolution, and human migration
- DNA forensics (identification)
- Agriculture, livestock breeding, and bioprocessing

Studying the human genome - the complete set of human genes - is a way of studying fundamental details about us. The three billion letters of the human genome are written using the four-letter alphabet of DNA. The DNA is divided among 23 pairs of chromosomes that are found in each of the trillions of cells in our bodies. In 2003, The Human Genome Project produced a complete representative sequence of the human genome. Of course, people are not identical, and DNA sequences do differ subtly between individuals. Currently, a number of separate projects are charting sequence variations found in human populations.

The representative sequence is a composite from several people who donated blood samples. Originally, close to 100 people volunteered to give a sample of their blood. Each person provided their informed consent, affirming that they agreed to the study of their DNA. No names were attached to the blood samples and ultimately scientists used only a few of them. These measures ensured that the DNA sequences remained anonymous; not even the donors knew whether their samples were actually used or not.

The main goal of The Human Genome Project was to read, letter by letter, the three billion bases of human DNA. Before starting to sequence the human genome, scientists built maps of the chromosomes and developed and refined techniques for analyzing DNA. With the tools in place, project scientists began large-scale DNA sequencing in 1999. In just one year, they had amassed sequence data covering more than 80 percent of the genome.

The human genome is a massive text. If the three billion letters (or bases) of the genome were printed in telephone books, they would require a stack of books nearly as tall as the Washington monument.

To accurately determine the sequence of every base in the genome, scientists needed to read the three billion bases not just once, but at least six to ten times. Individual sequencing reactions could only reveal the order of a few hundred bases of DNA at a time - amounting to a fraction of a page. This meant that to place in order all of the DNA bases, it was necessary to produce many thousands of overlapping segments of DNA sequence.

Human genome project 1990's onward developments

1990-1994

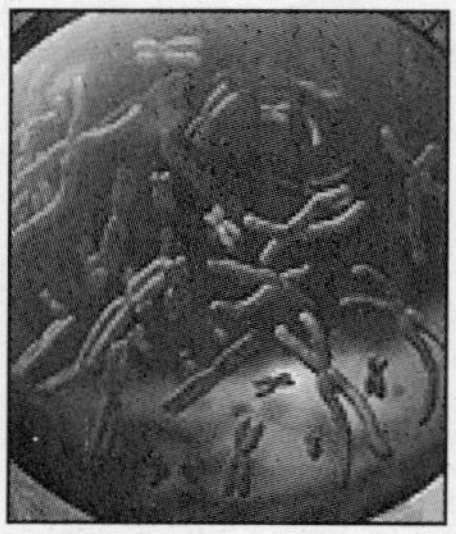

1990: Launch of the Human Genome Project

1990: ELSI Founded

1990: Research on BACs

1991: ESTs, Fragments of Genes

1992: Second-generation Genetic Map of Human Genome

1992: Data Release Guidelines Established

1993: NEW HGP Five-year Plan

1994: FLAVR SAVR Tomato

1994: Detailed Human Genetic Map : Microbial Genome Project

1995-1996

1995: Ban on Genetic Discrimination in Workplace

1995: Two Microbial Genomes Sequenced

1995: Physical Map of Human Genome Completed

1996: International Strategy Meeting on Human Genome Sequencing

1996: Mouse Genetic Map Completed

1996: Yeast Genome Sequenced

1996: Archaea Genome Sequenced

1996: Health Insurance Discrimination Banned

1996: 280,000 Expressed Sequence Tags (ESTs)

1996: Human Gene Map Created

1996: Human DNA Sequence Begins

1997-1999

1997: Bermuda Meeting Affirms Principle of Data Release
1997: E. coli Genome Sequenced
1997: Recommendations on Genetic Testing
1998: Private Company Announces Sequencing Plan
1998: M. Tuberculosis Bacterium Sequenced
1998: Committee on Genetic Testing
1998: HGP Map Includes 30,000 Human Genes
1998: New HGP Goals for 2003
1998: SNP Initiative Begins
1998: Genome of Roundworm C. elegans Sequenced
1999: Full-scale Human Genome Sequencing
1999: Chromosome 22

2000 - 2001

2000: Free Access to Genomic Information
2000: Chromosome 21
2000: Working Draft
2000: Drosophila and Arabidopsis genomes sequenced
2000: Executive Order Bans Genetic Descrimination in the Federal Workplace
2000: Yeast Interactome Published
2000: Fly Model of Parkinson's Disease Reported
2001: First Draft of the Human Genome Sequence Released
2001: RNAi Shuts Off Mammalian Genes
2001: FDA Approves Genetics-based Drug to Treat Leukemia

2002 -2003

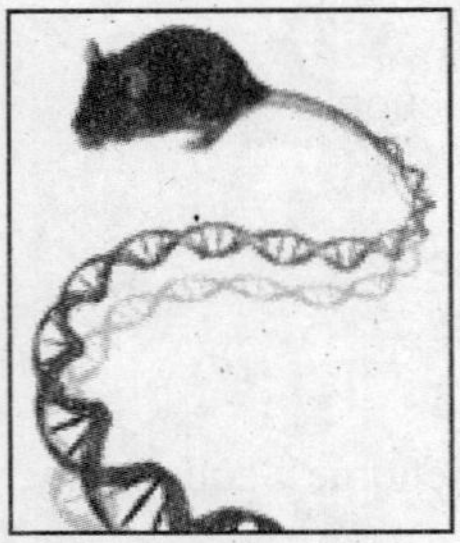

2002: Mouse Genome Sequenced
2002: Researchers Find Genetic Variation Associated with Prostate Cancer
2002: Rice Genome Sequenced
2002: The International HapMap Project is Announced
2002: The Genomes to Life Program is Launched
2002: Researchers Identify Gene Linked to Bipolar Disorder
2003: Human Genome Project Completed
2003: Fiftieth Anniversary of Watson and Crick's Description of the Double Helix
2003: The First National DNA Day Celebrated
2003: ENCODE Program Begins
2003: Premature Aging Gene Identified
2004 - The Future
2004: Rat and Chicken Genomes Sequenced
2004: FDA Approves First Microarray
2004: Refined Analysis of Complete Human Genome Sequence
2004: Surgeon General Stresses Importance of Family History
2005: Chimpanzee Genomes Sequenced

2004: The Future
2005: HapMap Project Completed
2005: Trypanosomatid Genomes Sequenced
2005: Dog Genomes Sequenced
2006: The Cancer Genome Atlas (TCGA) Project Started
2006: Second Non-human Primate Genome is Sequenced
2006: Initiatives to Establish the Genetic and Environmental Causes of Common Diseases Launched

II. PREPARATION OF LIBRARIES

Clone libraries offered the same advantage of real libraries: orderly access to information. In most clone libraries, the DNA fragments were stored in *E. Coli*. These are bacteria that normally live in our intestines. Each *E. Coli* cell stored a single segment of human DNA and represented a single book of the library. Clone libraries allowed each human fragment to be tracked and easily copied.

E. coli cells containing fragments of human DNA, or any other type of DNA, can be stored in freezers indefinitely. When scientists need to retrieve DNA from the library, they simply revive the cells by bringing them back up to 37 degrees Centigrade - gut temperature.

The *E. coli* cells act as copiers, producing many copies of the human DNA sequence that they contain. To prepare to sequence DNA, a clone of cells containing the same bit of human DNA is released into a rich, warm broth. The cells are shaken vigorously to provide them with air. This causes them to divide rapidly - about once every half hour. After incubating for just a single night, one third of a teaspoon of broth contains billions of *E. coli* cells and so, billions of copies of the particular fragment of human DNA they contained.

III. SEQUENCING REACTIONS

A DNA sequencing reaction includes four main ingredients, "Template" DNA copied by the *E. coli*; free bases, the building blocks of DNA that come in 4 types; short pieces of DNA called "primers"; and DNA polymerase, the enzyme that copies DNA.

The chemical reaction that makes DNA in a test tube is similar to what happens in a living cell: both rely on DNA polymerase and, in both cases, DNA strands have a head end, which is called the 5' end, and a tail end, which is called the 3' end. A DNA strand can grow only from its 3' end.

Making DNA in cells and sequencing DNA in test tubes both depend on complementary base pairing. The building blocks on opposite strands of DNA pair specifically - a C always pairs with a G, and an A always pairs with a T.

The primer sequence binds to its complementary sequence on the template DNA.

Free bases that match the template sequence can attach to the new strand's growing (3') end.

Among the free bases in the solution are a few that have a fluorescent dye attached to them. When a dye-bearing base attaches to the growing strand, it stops the new DNA strand from growing any further. A different colored dye is attached to each of the four kinds of bases.

Human genome is sequenced by two approaches, one that used Sanger's method of sequencing and the another that known as 'shotgun method' developed by Dr. J.C. Venter from the Institute for genomic research. Later on, Dr Venter joined Celera genomic, a private company working on genomic sequencing. The company has elaborated automatic sequencing facility to sequence human genome and genome of other organisms at very low cost as compared to public research laboratories.

IV. SEPARATING THE SEQUENCING PRODUCTS

The DNA molecules produced during the sequencing reaction are separated from each other by a process called electrophoresis. DNA molecules are negatively charged. The sequencing machine sets up an electric field; all the DNA moves through a porous gel toward the positive electrode. The gel acts like a sieve; shorter DNA fragments move more quickly through the holes of the gel than do larger DNA fragments.

V. ASSEMBLING THE RESULTS

A computer program integrates the data from individual sequencing reactions. It can spot where DNA fragments overlap and order them as they originally were on the chromosome.

Many overlapping sequences reads are needed to generate the uninterrupted sequence of the original stretch of DNA. During the Human Genome Project, every base pair of DNA was sequenced an average of nine times. Some stretches of DNA were easy to read and needed to be sequenced little less often, while other stretches were more difficult to read and had to be sequenced more often.

During the Human Genome Project scientists ran more than 50 million sequencing reactions. Some 2000 scientists from more than two dozen labs around the world, worked on the project.

VI. MAPPING

To begin the project, researchers built maps of the human genome. They identified thousands of DNA sequence landmarks that helped them navigate across the chromosomes.

Developing genome maps was necessary preparation for DNA sequencing. These same maps also served to orient geneticists who were hunting for disease genes.

With enough landmarks in place, project scientists created "libraries" of clones that spanned the genome. Each clone contained a manageably small fragment of human DNA that was stored in bacteria. Scientists used the landmarks to tell them what part of the human genome each fragment came from.

This clone-by-clone approach made it possible to double check the location of each DNA sequence. It also allowed participating laboratories from around the world to carve up the genome and coordinate their work.

VII. SUBCLONES

The clone libraries were prepared using bacterial artificial chromosomes, or BACs. Each BAC clone contained 100,000 to 200,000 bases of DNA sequence. The large BAC clones were used to establish the order of the DNA sequences. To sequence the DNA, smaller-sized clones were needed. Project scientists cut the large BAC clones into smaller fragments of about 2,000 bases. These smaller fragments were typically stored in viruses called phage that can infect *E. coli cells*.

VIII. PREPARING DNA FOR SEQUENCING REACTIONS

The next morning, the *E. coli* cells are broken open to release their DNA. The human DNA is separated from the cell debris and washed clean.

Now there are enough copies of the human DNA fragment to set up a sequencing reaction.

Area	HGP Goal	Standard Achieved	Date Achieved
DNA Sequence	95% of gene-containing part of human sequence finished to 99.99% accuracy	99% of gene-containing part of human sequence finished to 99.99% accuracy	April 2003
Capacity and Cost of Finished Sequence	Sequence 500 Mb/year at < $0.25 per finished base	Sequence >1,400Mb/ year at <$0.09 per finished base	November 2002
Human Sequence Variation	100,000 mapped human SNPs	3.7 million mapped human SNPs	February 2003

IX. PRODUCTS OF SEQUENCING REACTIONS

A completed sequencing reaction contains an array of colored DNA fragments. The shortest fragments correspond to the length of the primer plus one dye-colored base. The longest fragments are usually between 500 and 800 bases long, depending on when the sequencing reaction ran out of steam.

The products of sequencing reactions are fed into an automated sequencing machine. Automated sequencers have become increasingly sophisticated during the past decade. They can run more samples, process them more quickly, and are easier to operate.

X. READING THE SEQUENCING PRODUCTS

As each DNA fragment reaches the end of the gel, a laser excites its fluorescent dye. A camera detects the color of the emitted light and passes that information to a computer. One by one, the machine records the colors of the DNA fragments that pass through the gel.

A single sequencing reaction can reveal the order of several hundred DNA bases.

XI. WORKING DRAFT SEQUENCE

Whenever a stretch of DNA that spanned 2,000 or more bases was assembled, it was placed into public databases within 24 hours. Anyone with access to the Internet could see and analyze the sequence data.

After sequencing the 3 billion letters in the human genome an average of nine times, the Human Genome Project had released DNA sequence for 99 percent of the genome. This finished sequence was 99.99 percent accurate. The project had completed all of its goals ahead of schedule and under budget.

Sanger Institute

The Wellcome Trust Sanger Institute is one of the world's leading genomics centres. Its mission is to further the knowledge of genomes, particularly through large-scale sequencing and analysis. The institute is making a major contribution to the Human Genome Project by sequencing one billion base-pairs of the human genome, one-third of the total.

Although the Sanger Institute is not a commercial organisation, it is under considerable pressure to research and sequence genomes quickly so that the information is available for purposes such as fighting disease. In conjunction with the institute's continued expansion, this leads to a constant demand for more and better computers, which need more space to house them. Recently, it embarked on the construction of a brand-new data centre, which when completed in 2005 will be one of Europe's largest. In the meantime, the infrastructure had to be moved to a temporary location.

Fig 37.1. Human karyotype

The relocation had to be completed without disruption to normal operations. To achieve this, the Sanger Institute put in place a duplicate infrastructure that ensured services would continue in the event of a problem. It also engaged HP Services to implement the move within a carefully planned window of downtime.

Together, the two companies devised a low-cost, low-risk approach to the relocation. HP's business-critical team was involved in the project together with its on-site managed services team, which had extensive experience of the Sanger Institute infrastructure. At the outset, HP invested time in acquiring a full understanding of the requirements of the project and the risks involved. The services provided included business-critical environmental reviews (pre-implementation and post-implementation), insurance, transportation, and installation by a team of engineers. HP assumed responsibility for dealing with any equipment faults during the move and for ensuring that the systems were fully operational afterwards.

The relocation, which involved moving 407 HP AlphaServer systems and 12 HP storage systems, was an outstanding success. No scheduled downtime window was exceeded. In fact, the total amount of downtime amounted to only one-third of the anticipated 48 hours.

Other organism mapped with HGP Map Viewer - genome annotation updates:

Species	Build	Map Viewer Release
Nasonia vitripennis (jewel wasp)	1.1	September 12, 2007
Ostreococcus 'lucimarinus'	2.0	August 9, 2007
Caenorhabditis elegtans (nematode)	WS170	July 18, 2007
Equus caballus (domestic horse)	1.1	July 11, 2007
Ornithorhynchus anatinus (platypus)	1.1	July 11, 2007
Mus musculus (mouse)	37.1	July 5, 2007
Monodelphis domesticus (opossum)	MonDom5	March 8, 2007
Danio rerio (zebrafish)	Zv6	February 27, 2007
Populus trichocarpa (black cottonwood)	1.1	January 12, 2007
Bos taurus (cow)	3.1	January 3, 2007
Gallus gallus (chicken)	2.1	November 30, 2006
Drosophila melanogaster (fruit fly)	5.1	November 21, 2006
Dictyostelium discoideum	2.1	November 2, 2006
Strongylocentrotus purpuratus (sea urchin)	2.1	October 18, 2006
Pan troglodytes (chimpanzee)	2.1	October 5, 2006
Homo sapiens (human)	36.2	September 14, 2006
Apis mellifera (bee)	Amel_4.0	August 11, 2006
Rattus norvegicus (rat)	RGSC v3.4	July 6, 2006
Macaca mulatta (rhesus macaque)	1.1	June 23, 2006
Tribolium castaneum (red flour beetle)	1.1	April 18, 2006
Dictyostelium discoideum	1.1	November 22, 2005
Arabidopsis thaliana (mouse-ear cress)	TAIR6.0	November 21, 2005
Canis familiaris (dog)	2.1	September 8, 2005
Anopheles gambiae (mosquito)	2.2	June 30, 2005

QUESTIONS

1. What is a genome? Why human genome project is so important? Describe other organisms involved in deciphering genome.
2. Write short notes on :
   (*a*) Sanger Institute
   (*b*) Sanger's contribution
   (*c*) Human genome project
   (*d*) Application of HGP
   (*e*) Ethical issue of HGP
   (*f*) Methods used to decipher Human Genome.

STUDYING PROTEINS

CHAPTER 38

I. PROTEIN PURIFICATION

The study of proteins and their function is central to understanding both cells and organisms. Following are a few of the reasons why proteins are important in biology:

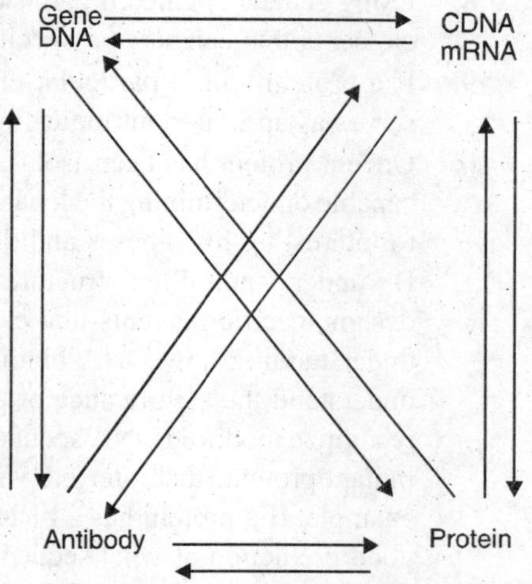

- they serve as catalysts that maintain metabolic processes in the cell,
- they serve as structural elements both within and outside the cell,
- they are signals secreted by one cell or deposited in the extracellular matrix that are recognized by other cells,
- they are receptors that convey information about the extracellular medium to the cell,
- they serve as intracellular signalling components that mediate the effects of receptors,
- they are key components of the machinery that determines which genes are expressed and whether mRNAs are translated into proteins,
- they are involved in manipulation of DNA and RNA through processes such as: DNA replication, DNA recombination, RNA splicing or editing.

(a) Importance

1. By purifying a protein it can be clearly established that a particular biological activity (enzymatic activity, signaling capacity, etc.) actually resides in a unique protein.
2. Purified proteins serve as extremely valuable biochemical reagents. It is remarkably valuable to be able to obtain things like purified growth factors or hormones, proteases, DNA polymerases, reverse transcriptases, ligases, phosphatases, or antibodies that recognize a particular epitope of interest.
3. Once a protein is purified, it is possible to study its enzymology, understand its affinity for particular substrates, or dissect its ability to catalyze enzymatic reactions. Such approaches have allowed us to understand how biological molecules can act as catalysts in metabolic processes or as transducers that will convert chemical energy into ionic gradients or mechanic forces.
4. The availability of purified proteins allows bio-organic chemists to modify specific residues to help understand how these residues confer particular structures or allow the protein to operate as catalyst.
5. Purified proteins can be sequenced either by Edmund degradation, which obtains sequence information from the N-terminal sequence of a protein or a peptide derived from a protein.

This approach is rapidly being replaced by mass spectroscopy. Sequence information can be used to design probes to isolate cDNAs, and this information can be used to deduce the primary sequence of the entire protein. MALDI-TOF (Matrix Assisted Laser Desorption Ionization Time of Flight) is a technology that can provide much information about a protein, even when only small quantities are available.

6. Partially purified proteins can often be identified on the basis of fragments generated by ionizing the proteins or peptides derived from them and performing mass spectroscopy.
7. Traditional structure-function studies have become remarkably more powerful because of the ability to isolate and purify not only conventional enzymes but also mutant forms of those proteins.
8. Using genetic engineering, it is possible to create novel proteins or combination of protein elements that can serve a specific function.
9. If a protein with a particular enzymatic activity or signaling capacity can be purified it serves as an important indication of how biological systems work.
10. Once a protein has been isolated, it is possible to produce reagents (antibodies) that are capable of determining the location of that protein in vivo which can give important support to interesting hypotheses and disprove incorrect speculations.
11. By understanding the structure-function relationships of proteins, it is often possible to design specific reagents that can be used to test the function of a protein in vivo. Thus, understanding how DNA-binding proteins recognize DNA sequences allows us: (1) to understand the significance of particular DNA mutations, (2) to design proteins that can recognize modified DNA sequences or recognize a known DNA sequence and (3) to design mutant proteins that interfere with the activity of proteins normally present in the cell. For example, if a protein has a bHLH motif (basic Helix Loop Helix) it is possible to make a good prediction of what sequences interact with DNA and what mutations might prevent binding of the protein to DNA or which sequences might compete with an endogenous proteins for DNA binding.
12. Finally, the increasing power of x-ray diffraction and 2-D NMR to determine protein structure requires the availability of larger quantities of purified proteins. Understanding 3-D protein structure has begun to allow us to understand how protein folding is controlled and it has provided a remarkable insight into the way the proteins act in vivo. The ability to visualize an ion channel, even at low resolution, must forever modify the way a scientist understands ion flow through the membranes.
13. Understanding how particular amino acid residues are involved in protein function, especially when it is combined with knowledge of the 3-D structure of the protein, helps in understanding how particular sequences in proteins are involved in biological functions.
14. Purification of a protein can also help purify a nucleic acid of interest. For example, if the hnRNA, snRNA proteins are purified, they can be co-purified with associated RNA, and bound RNA can be amplified

(b) Sources of protein- Second, protein purification requires a reasonable source of the protein. It is not difficult to appreciate that it is easier to purify a protein from a rich source than from a poor source. Furthermore, a rich source is likely to produce a greater yield of the protein of interest. The choice of what starting material should be tried will frequently determine the success or failure of a purification

Given a reasonable assay, and a source of material, it should be possible to purify any protein given enough time, money, and patience. Certainly it is possible that the task may prove difficult because the protein of interest may be extremely unstable. It may be extremely sensitive to proteolysis or it may be easily denatured, the attention to detail will frequently allow these problems to be overcome. Frequently, the problem of identifying a rich source of protein can even be obviated by substituting an artificial source for the protein. If the cDNA encoding a protein can be obtained the protein can usually be expressed at high levels giving an experimentalist an immediate advantage. Of course, in many cases, the gene for a protein of interest is not yet available.

(c) The strategy of protein purification- To purify a protein one must begin with the starting material and fractionate it using any one of a large number of physical or biochemical approaches. The initial material can be separated into fractions using centrifugation, precipitation with salt, binding to ionic columns (Ion-exchange chromatography), velocity sedimentation, equilibrium density sedimentation, binding to affinity columns, separation by sizing columns (see gel filtration chromatography), fractionation by isoelectric focusing, its ability to associate with known ligands or binding partners, or any other method available to the experimentalist. Once such a fractionation is accomplished, the activity of interest should be found in some fractions but not others. Assuming that the activity is not destroyed by the fractionation process (an assumption that is almost always less than completely correct), the fractionation serves two purposes. First it removes contaminating material and, second, it enriches the fraction containing the activity of interest. The enriched fraction can then be subjected yet to another round of fractionation and the process can be repeated. If the amount of activity can be accurately determined (assayed) and the total amount of the material (usually protein) present in the fraction can be determined. The specific activity of the preparation (i.e., the activity present in the fraction divided by the total amount of the material in the fraction) can be determined in each stage of the preparation. In a well-designed purification procedure, each fractionation step should result in the removal of the contamination material (proteins, nucleic acids, lipids, and anything else) and a progressive increase in specific activity.

(*d*) **Protocol for purification**

1. Selection of tissue source.
2. Solubilize the enzyme to obtain a homogeneous solution.

Cell Lysis (mechanical breakage of cells)

Detergents (to reduce hydrophobic interactions) Salts (to reduce electrostatic interactions)

3. Stabilize the enzyme at all purification steps.

Temperature (Most proteins are more stable at lower (4°C to -85°C) temperatures.)

Protease inhibitors and antimicrobials prevent peptide bond cleavage.

Ligands (Some proteins are more stable with a specific ligand bound.)

Salts (Proteins with highly charged surfaces may be unstable at low ionic strength.)

Metal ions (Some proteins are more stable with bound metal ions.)

4. Separate the components

1. Precipitation

(*i*) Salts–Aggregation at low salt due to unscreened charged-charged interactions Precipitation at high salt due to change in water structure, e.g. ammonium sulfate precipitation.

(*ii*) Organic Solvents - effects on water structure and hydrophobic interactions

2. Column chromatography: General Features. In most cases chromatography is performed in long glass tubes filled with a matrix or resin (particle size similar to a fine sand) that is completely immersed in a buffered salt solution. The mixture of proteins is added to the top of this column and buffer is allowed to flow through the column. As the buffer flows through the column the mixture of proteins is drawn down through the column and interacts with the matrix or resin. The actual mode of separation depends on the nature of the resin. In the case of gel filtration the proteins do not interact with the resin. In the case of other forms of chromatography the protein does stick to the resin and must be eluted by reducing the strength of the interaction between the protein and the resin.

Usually several different chromatographic steps are performed. There is no set order, but a typical order might be cation exchange, gel filtration, and then anion exchange.

(c) Examples of Chromatography

(i) Gel filtration

The matrix contains pores that allow smaller molecules to enter but excludes larger molecules. Thus larger molecules spend less time in the matrix and elute first. The size of the protein can be estimated from the elution time or volume.

If the total column volume is V_t and V_x is the volume occupied by the resin, then the volume available to the solvent is V_0:

$$V_t = V_x + V_0$$

The elution volume, V_e, is the volume of solvent required to elute the protein. The molecular weight of the protein can be obtained from the following formula:

$$V_e / V_0 = a^*\log(MW) + b$$

Where "*a*" and "*b*" are constants obtained by calibration of the gel filtration column. Note that the native molecular weight is obtained, e.g. Hemoglobin would give a measured molecular weight of 64 KDa.

(ii) Ion exchange

The protein sticks to these resins by electrostatic interactions. Resins contain either negative (cation exchange resins) or positive (anion exchange resins) charges. In very general terms a protein will stick to a cation exchange resin below its pI and to an anion exchange resin above its pI. However, what really matters is the local charge distribution on the surface of the protein (i.e. patches of residues with similar charges). Elution of proteins from ion exchange resins involves either a change in pH that results in a change in the charge of the protein or by increasing the salt concentration. The latter method provides additional ions than compete with the protein for binding sites on the resin.

(iii) Affinity

The resin contains a specific group (i.e. ligand or antibody) that causes the protein of interest to bind to the resin. In the first example (ligand affinity) elution can be accomplished by the addition of excess ligand. Note that this is analogous to the ion exchange column. In the case of antibody based affinity columns the protein-antibody binding must be weakened by changes of the pH and/or the salt concentration.

(iv) Adsorption

Adsorption based on the surface properties of the protein. The most common example is hydrophobic interaction chromatography. In this case

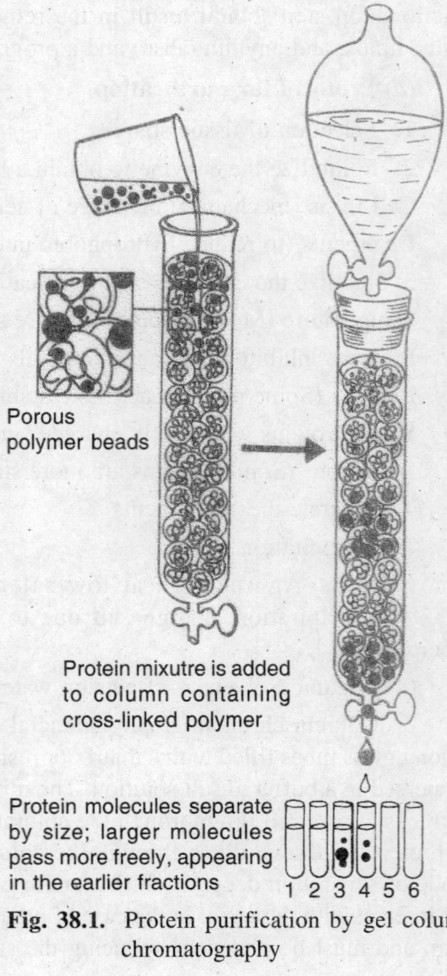

Fig. 38.1. Protein purification by gel column chromatography

the resin is coated with short-chain alkanes that makes it hydrophobic. Proteins will bind to the resin if they have hydrophobic patches on the surface. Elution is usually accomplished by changing the solvent polarity (i.e. addition of alcohol, etc.).

(f) Evaluation of Purity

(*i*) No increase in specific activity with subsequent purification steps.

(*ii*) Single polypeptide species by:

Mass Spectrometry
Gel Filtration/Electrophoresis
Identification of amino-terminal amino acid residue

Electrophoresis. Application of an electric field, E, to a particle with a charge (q) will generate a force resulting in constant velocity (v). The velocity of a charged particle is directly proportional to the strength of the applied field (E).

What happens when an electric field is applied to a solution of charged proteins?

- Negatively charged molecules (anions) migrate to the + terminal (anode). In electrophoresis, "*Anions migrate to the anode.*"
- Positively charged molecules (cations) migrate to the - terminal(cathode)
- Neutral molecules don't move.

SDS gel electrophoresis (SDS-PAGE) – Most electrophoretic separations (including IEF) are performed in gel matrices to avoid the problem of convectional mixing. These gels can either be composed of agarose (a carbohydrate polymer from seaweed) or polyacrylamide (a chemically crosslinked polymer). In addition to preventing thermal mixing, these gels also act as size exclusion media (similar to gel filtration). Therefore, the electrophoretic mobility of proteins in gels will depend on both the size as well as the charge-to-mass ratio. Therefore, to separate proteins according to size it is necessary to give them to same charge-to-mass ratio. This can be accomplished by denaturing the protein in SDS (sodium dodecyl sulfate, shown below).

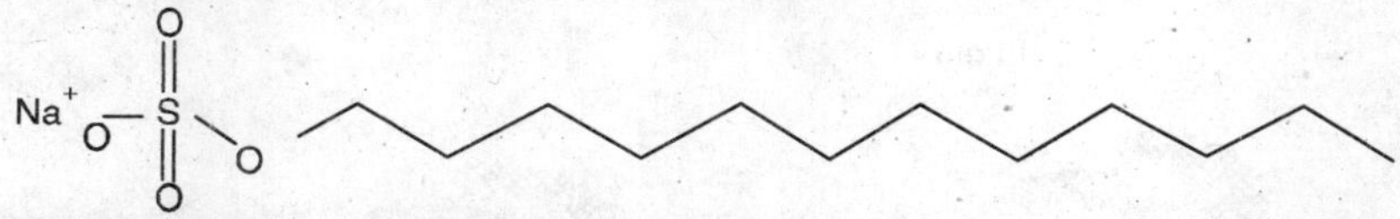

The SDS binds to most proteins in a uniform manner giving all proteins the same charge-to-mass ratio.

The electrophoretic mobility is defined as the distance migrated in a certain time period. Thus the mobility is directly proportional to the velocity. The effect of the gel matrix is to produce the following dependence of the molecular weight on the mobility.

$$\log(MW) = c*\text{Mobility} + d$$

Where "c" is a proportionality constant that depends on the gel properties. Each gel has a region where the above equation holds and a plot of log (MW) vs. Mobility will be linear. The determination of molecular weight requires the calibration of the SDS gel using proteins of known molecular weight. Note the similarity between the above equation and that used for gel filtration. For example, if hemoglobin were run as a standard, it would result in a band on the gel at a mobility corresponding to MW = 16KDa, i.e. its monomer molecular weight.

Detection in gel

1. Coomassie brilliant blue (blue dye that binds to proteins)
2. Fluorescamine (fluorescent compound that reacts with proteins)
3. Autoradiography (Detection of radioactivity in proteins)

4. Immunoblots (Use of an antibody against the protein to localize it, similar to an ELISA assay).

SDS polyacrylamide gels provide a simple, easily accessible, analytical tool that can be easily used to determine what proteins are present in a mixture. It is important because:

- this technique dissociates multimeric proteins and separates peptides by molecular weight (see dictionary entry on *SDS gels*)
- it allows estimation of the molecular weight of each peptide
- it is sensitive (about 0.1 micrograms can be detected with staining by *Coomassie* Brilliant Blue and 'silver staining' is about 100 fold more sensitive)
- it is sensitive to protein modification (glycosylation & phosphorylation frequently change mobility)
- it can resolve and allow visualization of 50 or more proteins
- it can be combined with isotopic labelling to study synthesis and turnover of proteins
- it can be modified to follow antigens (see western blotting) .
- it is fast (can be done in part of a day)
- it is easy to learn and is done in the same way for all protein mixtures
- of course, it denatures proteins so biological activity is usually (but not always) lost.

II. INTEINS

Inteins are selfish DNA elements inserted in-frame and translated together with their host proteins. This precursor protein undergoes an autocatalytic protein splicing reaction resulting in two products: the host protein and the intein. Protein splicing was shown to occur in heterologous organisms, in different in-vitro systems and with various natural and engineered host flanks. Hence, protein splicing is probably independent of specific host cell factors. Most inteins have an endonuclease domain inserted in the protein splicing domain. The endonuclease activity of these inteins can mediate the specific transposition of their gene into unoccupied integration sites of intein-less homologs (homing).

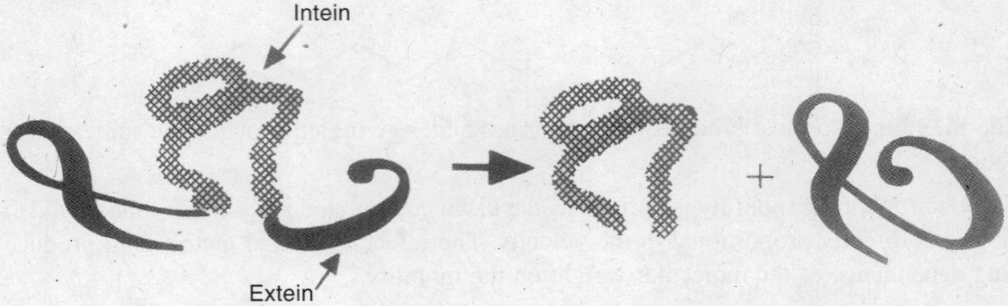

Protein-splicing scheme. A precursor protein is shown on the right, with intein protein-splicing domain shown in dotted and the host protein flanks (exteins) shown in black. The intein protein-splicing domain autocatalyzes its excision and the ligation of its two flanks.

(a) Distribution

Inteins have a diverse and sporadic distribution across species and proteins. Inteins are found in all three domains of life: Archaea, Bacteria, and Eukaryotes. However their distribution is sporadic in species and in hosts. Some species have no inteins, some just one and *Methanococcus jannaschii* has nineteen. For species with completely sequenced genomes like *E.coli*, *M.jannaschii* and *S. cerevisiae* we know the total number of inteins in the strain sequenced.

One major group of organisms where inteins are not known in is multicellular , eukaryotes both metazoa and plants. The multicellular red alga *Porphyra* does contain an intein but in its chloroplast genome. The reasons for this absence are not clear. Inteins may yet be found in these organisms and only turn out to be scarcer or perhaps difficult to detect. It is interesting to note that some intein-containing organisms, such as *Mycobacteria tuberculosis* and the CIV virus, are intracellular pathogens of metazoa.

Inteins are currently known in more then 50 types of proteins with diverse in functions. These proteins include metabolic enzymes, DNA and RNA polymerases, proteases, ribonucleotide reductases, and vacuolar-type ATPases. Inteins integration points also vary in structure and function. Their only apparent common feature is being in highly conserved protein motifs.

Intein protein domain family is part of the Hint superfamily, termed after the characteristic structure fold first identified in Hedgehogs and Intein protein domains. Four characterized Hint domain families are currently known: Hog Hints, inteins, and two types of Bacterial intein-like (BIL) domains. Together with sharing the same structure fold and common sequence features, Hint domains have similar biochemical activities. The domains post-translationally process the proteins in which they are present by protein-splicing, self-cleavage or ligation activities. This site mainly introduces inteins and some their sister families. It explores the relation between the activities of these domains, their sequence motifs, and protein structure. A database of inteins is maintained at New England Biolabs. Intein registry, publications, sequence search and information on their mechanism can be found there.

(b) Structure

Intein proteins contain a number of conserved sequence motifs (blocks). The motifs can be grouped in three domains according to their location and inferred function. Intein structures show that the inteins protein-splicing and endonuclease active sites are formed from conserved motifs. The intein's domain organization, deduced by sequence analysis, exactly corresponds to the structural domains.

The protein-splicing N-terminal (N) domain spans about 100-150 aa. Mutations in this domain affect the first step in protein splicing, an N-S/O acyl shift in the peptide bond connecting the N terminus of the intein and the N-terminal flank (intein flanks are termed exteins). The N domain motifs are similar in sequence and function to motifs found at the Hint domain of Hog regions found in the Hedgehog animal developmental proteins and a few related nematode protein families. Hog Hint domains self-process their precursor proteins, cleaving themselves off the N-terminal parts of the proteins. Cleavage was shown to utilize a cholesterol molecule that is consequently covalently attached to the cleavage site in the N-terminal part modulating the activity of that part. The relation between inteins and Hog Hint domains was first determined by their similarity in sequence and function. Determination of protein structures from both families provided final proof for the common origin of inteins and Hog Hint domains. The intein protein-splicing C-terminal (C) domain is composed of the two adjacent motifs in the C-terminal 25-40 aa (including the conserved aa immediately C' to the intein). Residues in these motifs are necessary for the catalyzing the next steps of protein splicing: the branch formation and its resolution. Most (but not all) inteins also include a central endonuclease (EN) domain. The EN domain is usually of the LAGLIDADG (dodecapeptide) homing endonucleases type. Intein LAGLIDADG EN domains are characterized by 4 motifs that probably form the endonuclease active site. An intein from the cyanobacteria *Synechocystis* species PCC6803 (Ssp gyrB) has a different type of endonuclease domain. In this intein the endonuclease domain contains an HNH motif. This motif is found in various homing and other endonucleases. The endonuclease domain is optional in inteins. Mutations in it affect the intein endonuclease activity but not the protein splicing activity, some inteins are missing this domain, and inteins were shown to protein splice with this domain removed.

Functional inteins with no EN domain (minimal inteins), the relation of the protein splicing domain to other Hint domains and the presence of different EN domains in inteins all indicate that the primeval inteins had no EN domains. Different EN domains, perhaps from homing endonucleases

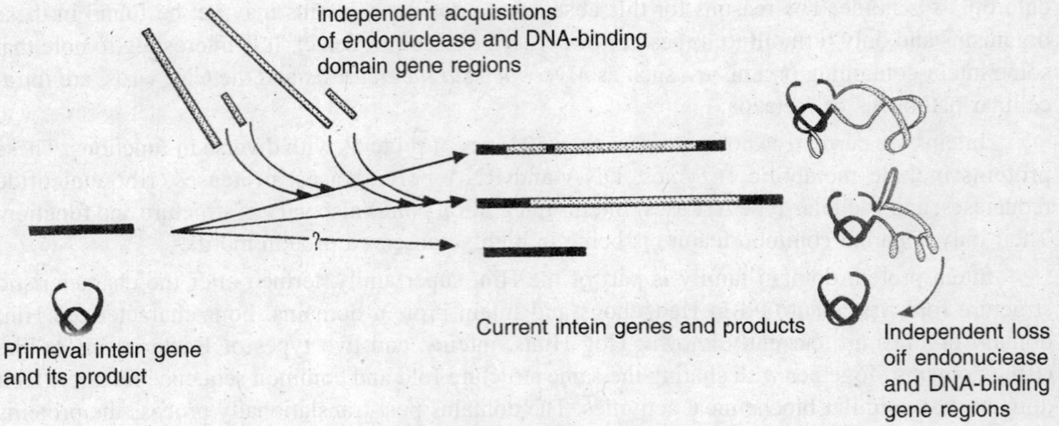

Fig. 38.2. Functional inteins

and DNA binding domains invaded intein genes to form the typical present day intein. Some present day minimal inteins clearly lost their EN domain (such as Mxe gyrA) and some maybe never acquired one.

Some protein families, such as ribonucleotide reductases and archaeal DNA polymerase type B, are more prone to contain inteins. These proteins contain inteins in different organisms and in different integration sites. Some of the ribonucleotide reductases and most of the DNA polymerases with inteins contain more than one intein.

Bacterial intein-like domains (BILs)

BILs are auto-catalytic protein domains. BILs are ~140 amino acids long, and have characteristic sequence motifs of the Hint (Hog/ Intein) domain. These domains are highly modular, being integrated in variable positions of non-conserved protein hosts, some of which are extracellular. Similarly to other types of Hint domains, BIL domains may contribute to their host variability through several post-translational biological activities like protein-splicing, self-cleavage and ligation. BILs may also be involved in maturation and conjugation of poly-Ubiqutin-like precursor proteins.

Most BIL domains are present in diverse bacterial species, yet recently they were also found in two ciliates. BILs are divided into two distinct families (**A- and B-type**) according to their unique sequence motifs. A-type BIL domains catalyze protein-splicing, and of N'- or C'-terminal cleavages, and B-type BIL domains can autocleave at either their N', C' or both ends. Several models are proposed for the biochemistry of BIL domains, based on inteins protein-splicing mechanisms. Yet, unlike inteins, BILs protein-splicing did not depend on Cys, Ser or Thr residue to be directly C-terminal to the BIL domain. BIL domains are distinct from inteins in having additional unique sequence motifs, not being integrated in highly conserved sites of essential proteins and in not including endonuclease domains. They are also distinct from Hog protein Hint domains by: (*i*) lacking conserved motifs characteristic to those domains; (*ii*) occurring in bacteria; and (*iii*) being flanked by protein domains unlike those found in all known Hog proteins.

III. SINGLE CELL PROTEIN

Single cell protein has the potential to be developed into a very large source of supplemental protein that could be used in livestock feeding. In some regions single cell protein could become the principal protein source that is used for domestic livestock, depending upon the population growth and the availability of plant feed protein sources. This could develop because microbes can be used to ferment some of the vast amounts of waste materials, such as straws; wood and wood processing wastes; food, cannery and food processing wastes; and residues from alcohol production or from human and animal excreta. Producing and harvesting microbial proteins is not without costs, unfortunately. In nearly all instances where a high rate of production would be achieved, the single cell protein will be found in rather dilute solutions, usually less than 5 % solids. Methods available for concentrating include filtration, precipitation, coagulation, centrifugation, and the use of semi-permeable membranes. These de-watering methods require equipment that is quite expensive and would not be suitable for most small-scale operations. Removal of the amount of water necessary to stabilize the material for storage, in most instances, is not currently economical. Single cell protein must be dried to about 10 % moisture, or condensed and acidified to prevent spoilage from occurring, or fed shortly after being produced.

(a) Biomass and Single Cell Protein

Microbial cells are produced for two main applications, (a) as a source of protein for animal or human food, (Single Cell Protein) or (b) for use as a commercial inoculum in food fermentations and for agriculture and waste treatment. As a commodity, SCP must be competitive with commercial animal and plant proteins, in terms of process and nutritional value and must conform to human and animal food safety requirements. Productivity, yield and selling price are the major factors affecting the economics of SCP production. Microbial inoculants, which are used as a process aid, generally have a higher value. In this case, the objective of the production process is to optimize yield of viable cells of defined biological activity with good shelf life characteristics. *Saccharomyces cerevisiae,* is categorised primarily as a microbial inoculant. Inactive dried brewer's or baker's yeast is also used as a dietary source of vitamins and trace minerals in specific medical conditions. Considerable amounts of yeast extract are produced from bakers yeast as a source of flavour and vitamins.

(b) Single Cell Protein Production

Product Safety and Quality- SCP has applications in animal feed, human food and as functional protein concentrates. Some bacterial SCP have amino acid profiles similar to animal/ plant protein. Yeast, fungal and soya bean proteins tend to be deficient in methionine. Ingestion of RNA from non-conventional sources should be limited to 50g per day. Ingestion of purine compounds arising from RNA breakdown, leads to increased plasma levels of uric acid, which can cause gout and kidney stones. High content of nucleic acids causes no problems to animals since uric acid is converted to allantoin which is readily excreted in urine. Nucleic acid removal is not necessary from animal feeds but is from human foods. A temperature hold at 64°C inactivates fungal proteases and allows RNA-ases to hydrolyse RNA with release of nucleotides from cell to culture broth. A 30 min stand at 64°C reduces intracellular RNA levels in *Fusarium graminearum* from 80mg/g to 2mg/g.

Production-Large scale fermenters are required. High biomass productivity requires high oxygen transfer rates which promotes high respiration rates which in turn increase metabolic heat production and the need for an efficient cooling system. In order to maximise fermentation productivity it is essential to operate continuous fermentation processes. Different processes have adopted different fermenter designs with respect to process requirements.

A wide range of substrates can be used to grow microbial proteins. Single cell protein has a good potential for a supplemental protein source for feeding livestock. Various materials can be used as a substrate for producing single cell protein (whey, orange peel residue, sweet orange residue, sugarcane bagasse, paper mill waste, rice husks, wheat straw residue, cassava waste, sugar beet pulp, coconut waste, yam waste, banana pulp, mango waste, grape waste, sweet potato).

Substrates like bananas have been shown to be excellent substrates that can be used to produce single cell protein. Bananas contain 0.95 % Cell Protein (CP), but when subjected to microbial (*Aspergillus niger* or *Aspergillus foetidus*) fermentation the CP level increased to 21.5 %, and up to 20 % could be used in broiler diets. Some microorganisms may produce mycotoxins. Single cell protein seems to be utilized better by ruminant animals. Single cell protein was a suitable supplemental protein source for lactating dairy goats.Milk production and milk production efficiency was increased when single cell protein replaced groundnut meal in lactating goat diets.

Single cell protein was found to be able to replace up to 55 % of the fish meal and soybean meal in swine diets, while still maintaining satisfactory performance. Single cell protein (yeast) was found to normal growth in rats, but inferior to casein.

Candida on n-paraffin- Mechanically agitated fully baffled fermenters with turbine mixers. Reactor feed consists of paraffin, gaseous ammonia and other salts. Oxygen requirements per unit biomass produced by aerobic micro-organisms grown on n-hexadecane is 2.5 times higher than that required for growth on glucose and amounts to 2.2 kg O_2 per kg biomass. Heat produced was 25000 kJ/kg biomass. Fermenters require substantial agitation. Because of the insoluble nature of alkanes, they exist in agitated fermenter as suspensions of alkane drops 1-100mm in diameter. Cell recovery is by centrifugation, producing 15% dry solids, evaporation to 25% dry solids and spray drying.

***Methylophilus methylotrophus* from methanol.**

Pressure recycle fermentor. A combination of an airlift and loop reactor consisting of an airlift column, a down-flow tube with heat removal and a gas-release space. The nitrogen source is ammonia gas and pH is controlled between 6 and 7. Cell specific growth rate is approximately 0.5h-1 and cell yield of 0.5 g/g. Methanol is oxidised via dehydrogenation to formaldehyde which can either be assimilated for conversion to cell mass or further oxidised to CO2 with concomitant energy production. Cells are recovered by agglomeration followed by centrifugation, flash dried and ground.

***Kluyveromyces marxianus* from whey.**

Whey which contains about 5% lactose, 0.8% protein and 0.2-0.6% lactic acid, is used as a substrate. Biomass production requires an aerobic fermentation whereas aeration is minimal for ethanol production. For feed grade biomass, the entire fermentation contents, containing yeast, residual whey proteins, minerals and lactic acid may be recovered. For preparation of food grade material, cells are harvested by centrifugation, washed and dried. Cell yield is 0.45-0.55 g/g based on lactose consumed.

***Fusarium graminearum* from glucose.**

Medium components include food grade glucose syrup, gaseous ammonia, salts, and biotin.

Fermentation pH is controlled at 6 by gaseous ammonia addition, fed into the air inlet stream. Cell concentrations are 15-20g/L and a specific growth rate of up to 0.2h-1 is achieved. Following cyclone separation and an RNA reduction step, cells are recovered by rotary vacuum filtration and formulated into a range of products.

Potential Substrates for SCP Production : Sulphite waste liquor

Candida utilis has been produced as a protein supplement by fermentation of sulphite waste liquor in Germany during both world wars. More recently a Finnish company developed a fungal SCP production process, the Pekilo Process, to grow *Paecilomyces varioti* using sulphite waste liquor.

Cellulose

Cellulose from natural sources and waste wood is an attractive starting material for SCP production because of its abundance. The association of cellulose with lignin in wood makes it somewhat intractable to microbial degradation. Thermal or chemical pretreatment, used in combination with enzymatic hydrolysis, is usually required. Systems using cellulolytic organisms appear to have promise, but economic viability has yet to be achieved.

Whey

Whole milk whey or deproteinised whey is a carbohydrate source, which creates disposal problems. (High BOD) Problems associated with whey for SCP production are usually insufficient substrate, seasonal supply variations and its high water content (>90%) which makes transport prohibitively expensive. While most organisms do not grow on lactose as a carbon source, strains of the yeast *Kluyveromyces marxianus* readily grow on lactose.

Starch

The symba process was developed in Sweden to produce SCP from potato starch using two yeast strains. *Saccharomyces fibuligera* produces the enzyme necessary for starch degradation enabling co-growth of *Candida utilis*. This process is known as simultaneous saccharification and fermentation.

Glucose

Food grade glucose was the substrate chosen by RHM for production of fungal SCP using *Fusarium graminearum.* The strategy adopted was to take advantage of mycelial fibre content to produce a range of high added value products including meat analogues for human consumption.

Methane/Methanol

Methane was initially considered as a SCP raw material because, as a gas product, purification problems after fermentation would be minimal. Disadvantages associated with methane-based processes are related to: (a) the greater oxygen requirements necessary to fully oxidise methane compared with paraffins, (b) the low solubility of methane in water and (c) the requirement that the fermentation plant be flame proof as methane-oxygen mixtures are highly explosive. Methane is however easily converted to methanol which requires less oxygen, less fermenter cooling, ishighly water soluble and has minimal explosion risks. ICI, which manufactures bulk methanol, chose this substrate for backterial SCP production for animal feed using the trade name Pruteen. The company designed a non-mechanical "pressure cycle fermenter" which uses air for both agitation and aeration in the worlds largest single aerobic fermenter of 3000m3 capacity. The process which produces 50-60,000 tonnes SCP per year, using the organism *Methylophilus methylotrophus,* was commissioned in 1979-1980, but has suffered from dramatic increases in methanol prices. The economic difficulties encountered by ICI with animal feed processes lead to a joint venture with RHM to produce *Fusarium* SCP in the ICI plant.

(c) Choice of Microorganism

The key criteria used in selecting suitable strains for SCP production should consider the following: The substrates to be used as carbon energy and nitrogen source and the need for nutrient supplementation. High specific growth rates, productivity and yields on a given substrate. pH and temperature tolerance. Aeration requirements and foaming characteristics. Growth morphology in the reactor. Safety and acceptability – non-pathogenic, absence of toxins. Ease of recovery. Protein, RNA and nutritional composition of the product. Structural properties of the final product. In general, fungi have the capacity to degrade a wider range of complex plant materials, particularly plant polysaccharides. They can tolerate low pH which contributes to reducing fermenter infections. Growth of fungi as short, highly branched filaments rather than in pellets is essential in order to optimize growth rate. Bacteria, in general, have faster growth rates than fungi and grow at higher temperatures, thereby reducing fermenter cooling requirements. Bacterial and yeast fermentations are easier to aerate. In contrast to fungi, which are easily recovered by filtration, bacteria and yeast require the use of sedimentation techniques and centrifugation. Bacteria, in general produce a more favourable protein composition than yeast or fungi. Protein content in bacterial can range from 60-65% whereas fungi selected for biomass production and yeast have protein contents in the range of 33-45%. However, associated with the higher bacterial protein levels is a much higher level of nutritionally undesirable RNA content of 15-25%. Microorganisms involved in SCP production must be safe and acceptable for use in food. Organisms should be stable genetically so that the strain with optimal biochemical and physiological characteristics may be maintained in the process through many hundreds of generations.

(d) Economics of SCP Production

The initial reasoning by companies such as BP and ICI to enter SCP production was to produce, at low cost, high value SCP from petroleum, for addition to animal feed.The intention was to replace imported protein additives such as soyabean meal. Factors which contributed to the failure of hydrocarbon SCP to make a major commercial impact included the 1973 dramatic oil price increases, which raised feedstock and energy costs and the lower price increases achieved by agricultural products such as soyabean. When one considers that crude oil prices increased by a factor of 6 in 1973, and that the cost of substrate for SCP processes represents 40-60% of total manufacturing costs, the negative impact on hydrocarbon based SCP processes may be understood. Agricultural crops, the major competitor to SCP for animal feed, manifest a remarkable ability to respond to market forces and maintain price stability. By producing human protein supplements, the end value of the product can be appreciated, however the production process has to utilise more conventional and costly substrates for production (glucose). These products take advantage of the high fibre content in certain fungi to produce a product that is a good meat analogue. This product is also low in sodium and fat. The product Quorn is a fungal protein.

(e) Spirulina

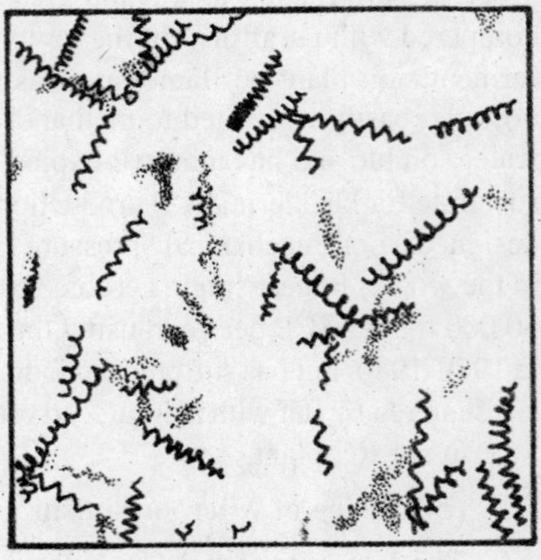

There are family species of microalgae such as blue-green algae *Spirulina* (*Arthrospira platensis*) and *Aphanizomenon,* and green algae *Chlorella* and *Dunaliella.* This edible algae, spirulina, has a long history of safe human consumption and over 30 years of safety testing. Special farms where spirulina is cultivated under controlled conditions do not allow the growth of other contaminant blue-green algae, as in lakes and waterways.

It contains compounds like phycocyanin, polysaccharides and sulfolipids that enhance the immune system. Spirulina contains the most remarkable concentration of functional nutrients ever known in any food, plant, grain or herb. Several years ago, the National Cancer Institute announced sulfolipids from blue-green algae like spirulina were remarkably active in test tube experiments against the AIDS virus.

In 1993-95, research showed natural polysaccha-rides in spirulina increased T-cell counts, strengthened the immune system and raised disease resistance in chickens, fish and mice. The animal feed industry is embracing spirulina as a new probiotic to replace overused antiobiotic drugs in animal feeds. In 1994, a Russian patent was awarded for spirulina as a medicine for reducing allergic reactions from radiation in the Children of Chernobyl. This tiny aquatic plant offers 60% all-vegetable protein, essential vitamins and phytonutrients such as the antioxidant beta carotene, the rare essential fatty acid GLA, sulfolipids, glycolipids and polysaccharides. Its deep green color comes from its rainbow of natural pigments - chlorophyll (green), phycocyanin (blue) and carotenoids (orange) - that harvest the sun's energy. Easy-to-digest so nutrients are absorbed quickly.

The healthiest food

The good news is, many essential nutrients recommended by experts to help protect our bodies are concentrated in Spirulina. It contains the most powerful combination of nutrients ever known in any grain, herb or food.

- **World's highest beta carotene food reduces long term health risks.** Spirulina beta carotene is ten times more concentrated than carrots. So even if you don't eat the recommended 4 to 9 servings of fruits and vegetables every day (most people eat only 1-2, including french fries), get your natural beta carotene insurance from spirulina to help support your body's defenses.
- **60% easy-to-digest vegetable protein without the fat and cholesterol of meat.** People are eating less meat and dairy protein because they want to lower fat, cholesterol, and chemicals in their diet. Spirulina is the highest protein food with all the essential amino acids and has only a few calories to keep your waistline where you want it.
- **A rare essential fatty acid is a key to health.** Gamma-linolenic acid (GLA) in mother's milk helps develop healthy babies. Studies show nutritional deficiencies can block GLA production in your body, so a good dietary source of GLA can be important. Spirulina is the only other whole food with GLA.
- **Iron for women and children's health.** Iron is essential to build a strong system, yet is the most common mineral deficiency. Spirulina is rich in iron, magnesium and trace minerals, and is easier to absorb than iron supplements.
- **High in Vitamin B-12 and B Complex.** Spirulina is the highest source of B-12, essential for healthy nerves and tissue, especially for vegetarians.
- **Unusual phytonutrients for health and cleansing.** Scientists are discovering the benefits of polysaccharides, sulfolipids & glycolipids, and the rainbow of natural pigments that give spirulina a deep green color. Green (chlorophyll), blue (phycocyanin) and orange (carotenoids) colors collect the sun's energy and power growth. Chlorophyll is a natural cleanser and is often referred to as nature's green magic.

IV. PROTEIN SEQUENCING

Proteins are found in every cell and are essential to every biological process, protein structure is very complex: determining a protein's structure involves first **protein sequencing** - determining the amino acid sequences of its constituent peptides; and also determining what conformation it adopts and whether it is complexed with any non-peptide molecules. Discovering the structures and functions of proteins in living organisms is an important tool for understanding cellular processes, and allows drugs that target specific metabolic pathways to be invented more easily.

The two major direct methods of protein sequencing are mass spectrometry and the Edman degradation reaction. It is also possible to generate an amino acid sequence from the DNA or mRNA sequence encoding the protein, if this is known. However, there are a number of other reactions which can be used to gain more limited information about protein sequences and can be used as preliminaries to the aforementioned methods of sequencing.

(a) Determining amino acid composition

It is often desirable to know the unordered amino acid composition of a protein prior to attempting to find the ordered sequence, as this knowledge can be used to facilitate the discovery of errors in the sequencing process or to distinguish between ambiguous results. Knowledge of the frequency of certain amino acids may also be used to choose which protease to use for digestion of the protein. A generalised method for doing this is as follows:

1. Hydrolyse a known quantity of protein into its constituent amino acids.
2. Separate the amino acids in some way.
3. Determine the respective quantities of the amino acids.

(b) Hydrolysis

Hydrolysis is done by heating a sample of the protein in 6 Molar hydrochloric acid to 100-110^{0}C for 24 hours or longer. Proteins with many bulky hydrophobic groups may require longer heating periods. However, these conditions are so vigorous that some amino acids (serine, threonine, tyrosine, tryptophan, glutamine and cystine) are degraded. To circumvent this problem, Biochemistry Online suggests heating separate samples for different times, analysing each resulting solution, and extrapolating back to zero hydrolysis time. Rastall suggests a variety of reagents to prevent or reduce degradation - thiol reagents or phenol to protect tryptophan and tyrosine from attack by chlorine, and pre-oxidising cysteine. He also suggests measuring the quantity of ammonia evolved to determine the extent of amide hydrolysis.

(c) Separation

The amino acids can be separated by Ion-exchange chromatography or hydrophobic interaction chromatography. An example of the former is given by using sulfonated polystyrene as a matrix, adding the amino acids in acid solution and passing a buffer of steadily increasing pH through the column. Amino acids will be eluted when the pH reaches their respective isoelectric points. The latter technique may be employed through the use of reversed phase chromatography. Many commercially available C8 and C18 silica columns have demonstrated successful separation of amino acids in solution in less than 40 minutes through the use of an optimised elution gradient.

(d) Quantitative analysis

Once the amino acids have been separated, their respective quantities are determined by adding a reagent that will form a coloured derivative. If the amounts of amino acids are in excess of 10 nmol, ninhydrin can be used for this - it gives a yellow colour when reacted with proline, and a vivid blue with other amino acids. The concentration of amino acid is proportional to the absorbance of the resulting solution. With very small quantities, down to 10 pmol, fluorescamine can be used as a marker: this forms a fluorescent derivative on reacting with an amino acid.

(e) N-terminal amino acid analysis

Determining which amino acid forms the N-terminus of a peptide chain is useful for two reasons: to aid the ordering of individual peptide fragments' sequences into a whole chain, and because the first round of Edman degradation is often contaminated by impurities and therefore does not give an accurate determination of the N-terminal amino acid. A generalised method for N-terminal amino acid analysis follows:

1. React the peptide with a reagent which will selectively label the terminal amino acid.
2. Hydrolyse the protein.
3. Determine the amino acid by chromatography and comparison with standards.

There are many different reagents which can be used to label terminal amino acids. They all react with amine groups and will therefore also bind to amine groups in the side chains of amino acids such as lysine - for this reason it is necessary to be careful in interpreting chromatograms to ensure that the right spot is chosen. Two of the more common reagents are Sanger's reagent (2,4-dinitrofluorobenzene) and dansyl derivatives such as dansyl chloride. Phenylisothiocyanate, the reagent for the Edman degradation, can also be used. The same questions apply here as in the determination of amino acid composition, with the exception that no stain is needed, as the reagents produce coloured derivatives and only qualitative analysis is required, so the amino acid does not have to be eluted from the chromatography column, just compared with a standard. Another consideration to take into account is that, since any amine groups will have reacted with the labelling reagent, ion exchange chromatography cannot be used, and thin layer chromatography or high pressure liquid chromatography should be used instead.

(f) C-terminal amino acid analysis

The number of methods available for C-terminal amino acid analysis is much smaller than the number of available methods of N-terminal analysis. The most common method is to add carboxypeptidases to a solution of the protein, take samples at regular intervals, and determine the terminal amino acid by analysing a plot of amino acid concentrations against time.

(g) Edman degradation

The Edman degradation is a very important reaction for protein sequencing, because it allows the ordered amino acid composition of a protein to be discovered. Automated Edman sequencers are now in widespread use, and are able to sequence peptides up to approximately 50 amino acids long. A reaction scheme for sequencing a protein by the Edman degradation follows - some of the steps are elaborated on subsequently.

1. Break any disulfide bridges in the protein by oxidising with performic acid.
2. Separate and purify the individual chains of the protein complex, if there are more than one.
3. Determine the amino acid composition of each chain.
4. Determine the terminal amino acids of each chain.
5. Break each chain into fragments under 50 amino acids long.
6. Separate and purify the fragments.
7. Determine the sequence of each fragment.
8. Repeat with a different pattern of cleavage.
9. Construct the sequence of the overall protein.

Digestion into peptide fragments Peptides longer than about 50-70 amino acids long cannot be sequenced reliably by the Edman degradation. Because of this, long protein chains need to be broken up into small fragments which can then be sequenced individually. Digestion is done either by endopeptidases such as trypsin or pepsin or by chemical reagents such as cyanogen bromide. Different enzymes give different cleavage patterns, and the overlap between fragments can be used to construct an overall sequence.

(h) The Edman degradation reaction

The peptide to be sequenced is adsorbed onto a solid surface - one common substrate is glass fibre coated with polybrene, a cationic polymer. The Edman reagent, phenylisothiocyanate (PTC), is added to the adsorbed peptide, together with a mildly basic buffer solution of 12% trimethylamine. This reacts with the amine group of the N-terminal amino acid.

The terminal amino acid derivative can then be selectively detached by the addition of anhydrous acid. The derivative then isomerises to give a substituted phenylthiohydantoin which can be washed off and identified by chromatography, and the cycle can be repeated. The efficiency of each step is about 98%, which allows about 50 amino acids to be reliably determined.

(i) Limitations of the Edman degradation

Because the Edman degradation proceeds from the N-terminus of the protein, it will not work if the N-terminal amino acid has been chemically modified or if it is concealed within the body of the protein. It also requires the use of either guesswork or a separate procedure to determine the positions of disulfide bridges.

(j) Mass spectrometry

The other major direct method by which the sequence of a protein can be determined is mass spectrometry. This method has been gaining popularity in recent years as new techniques and increasing computing power have facilitated it. Mass spectrometry can, in principle, sequence any size of protein, but the problem becomes computationally more difficult as the size increases. Peptides are also easier to prepare for mass spectrometry than whole proteins, because they are more soluble. One method of delivering the peptides to the spectrometer is electrospray ionization, which won the Nobel Prize in chemistry in 2002. The protein is digested by an endoprotease, and the resulting solution is passed through a high pressure liquid chromatography column. At the end of this column, the solution is sprayed out of a narrow nozzle charged to a high positive potential into the mass spectrometer. The charge on the droplets causes them to fragment until only single ions remain. The peptides are then fragmented and the mass-charge ratios of the fragments measured. (It is possible to detect which peaks correspond to multiply charged fragments, because these will have auxiliary peaks corresponding to other isotopes - the distance between these other peaks is inversely proportional to the charge on the fragment). The mass spectrum is analysed by computer and often compared against a database of previously sequenced proteins in order to determine the sequences of the fragments. This process is then repeated with a different digestion enzyme, and the overlaps in the sequences used to construct a sequence for the protein.

(k) Predicting protein sequence from DNA/RNA sequences

The amino acid sequence of a protein can also be determined indirectly from the mRNA or, in organisms that do not have introns (e.g. prokaryotes), the DNA that codes for the protein. If the sequence of the gene is already known, then this is all very easy. However, it is rare that the DNA sequence of a newly isolated protein will be known, and so if this method is to be used, it has to be found in some way. One way that this can be done is to sequence a short section, perhaps 15 amino acids long, of the protein by one of the above methods, and then use this sequence to generate a complementary marker for the protein's RNA. This can then be used to isolate the mRNA coding for the protein, which can then be replicated in a polymerase chain reaction to yield a significant amount of DNA, which can then be sequenced relatively easily. The amino acid sequence of the protein can then be deduced from this. However, it is necessary to take into account the possibility of amino acids being removed after the mRNA has been translated.

(l) Protein Sequence Analysis sites

Pestfind :

Find PEST (proline, glutamic acid, serine and threonin) sequences from a single protein sequence (D. Mathog, M. Rechsteiner)

PLSEARCH :

search protein sequences for similarity to AACC patterns

Prose:

search Prosite patterns in a sequence (Schuerer)

EMBOSS:

- fuzzpro: Protein pattern search.
- fuzztran: Protein pattern search after translation
- patmatdb: Search a protein sequence database with a motif
- patmatmotifs: Search a motif database with a protein sequence

QUESTIONS

1. What are proteins? How they are purified ?
2. What are single cell proteins ? How they are produced ?
3. How proteins are target to organelles or sites ?
4. How proteins are sequenced ?
5. Write short notes on :
   (*a*) Protein sequencing (*b*) Sephadex column
   (*c*) Single cell protein (*d*) Spirulina.

PROTEIN FOLDING AND TARGETING

CHAPTER 39

I. INTRODUCTION

Proteins are the biological workhorses that carry out vital functions in every cell. To carry out their task, proteins must fold into a complex three-dimensional structure, but how and which shape a protein should attain is determined protein itself.

Of all the molecules found in living organisms, proteins are the most important. They are used to support the skeleton, control senses, move muscles, digest food, defend against infections and process emotions. Proteins come in all shapes and sizes. They can be round (like haemoglobin), long (like collagen), strong (like spectrin c which protects erythrocytes (the cells that carry oxygen from the lungs to our tissues) from the powerful shearing forces they are exposed to), or elastic (like titin, which controls muscle stretching and contraction). The name protein is derived from the Greek word *prôtos*, meaning 'primary' or 'first rank of importance? ? and with good reason. These are the most abundant component within a cell'. More than half the dry weight of a cell is made up of proteins and they have a range of indispensable roles; for example, enzymes, the biocatalysts that carry out crucial biochemical reactions in every cell that would otherwise be too slow to sustain life.

What is remarkable is that the more than 100,000 proteins in our bodies are produced from a set of only 20 building blocks, known as amino acids. All amino acids have the same basic structure an amino group, a carboxyl group and a hydrogen atom, but differ due to the presence of a side-chain. This side-chain varies dramatically between amino acids, from a simple hydrogen atom in the amino acid glycine to a complex structure found in tryptophan. Depending on the nature of the side-chain, an amino acid can be hydrophilic (water-attracting) or hydrophobic (water-repelling), acidic or basic; and it is this diversity in side-chain properties that gives each protein its specific character.

II. CREATING A FUNCTIONAL PROTEIN

The sequence of amino acids in a protein defines its primary structure. The blueprint for each amino acid is laid down by sets of three letters known as base triplets that are found in the coding regions of genes. These base triplets are recognized by ribosomes, the protein building sites of the cell, which create and successively join the amino acids together. This is a remarkably quick process: a protein of 300 amino acids will be made in little more than a minute (Fig. 39.1).

The result is a linear chain of amino acids, but this only becomes a functional protein when it folds into its three-dimensional (tertiary structure) form. This occurs through an intermediate form, known as secondary structure, the most common of which are the rod-like α helix and the plate-like α-pleated sheet. These secondary structures are formed by a small number of amino acids that are close together, which then, in turn, interact, fold and coil to produce the tertiary structure that contains its functional regions (called domains).

Ribonuclease can be denatured by adding certain chemicals or by heat. The disulphide bridges break and other forces of attraction between amino acids disappear, which makes the enzyme collapse into a tangled, useless ball. The ribonuclease then folds back to its natural functional state on its

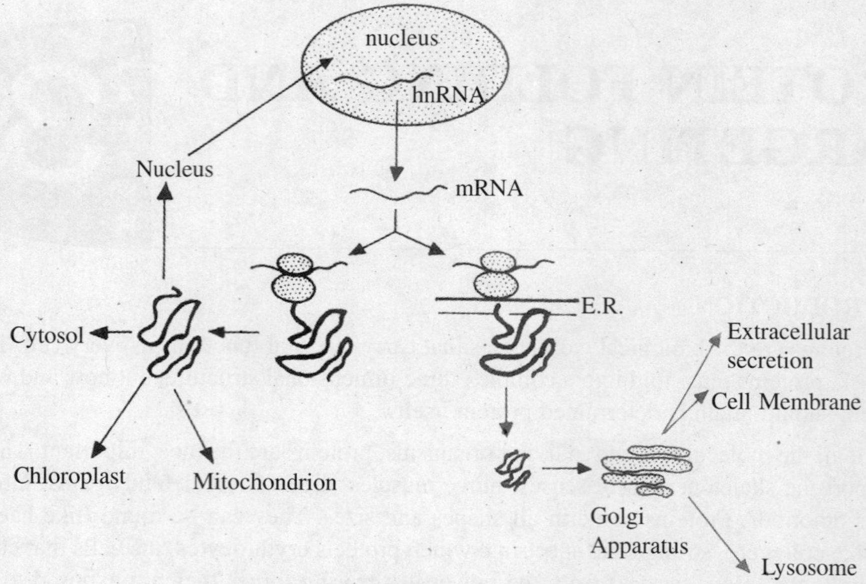

Fig. 39.1. Formation of protein from DNA

own. So, Anfinsen concluded that the amino-acid sequence determines the shape of a protein, a finding for which Anfinsen received the Nobel Prize in Chemistry in 1972 (Fig. 39.2).

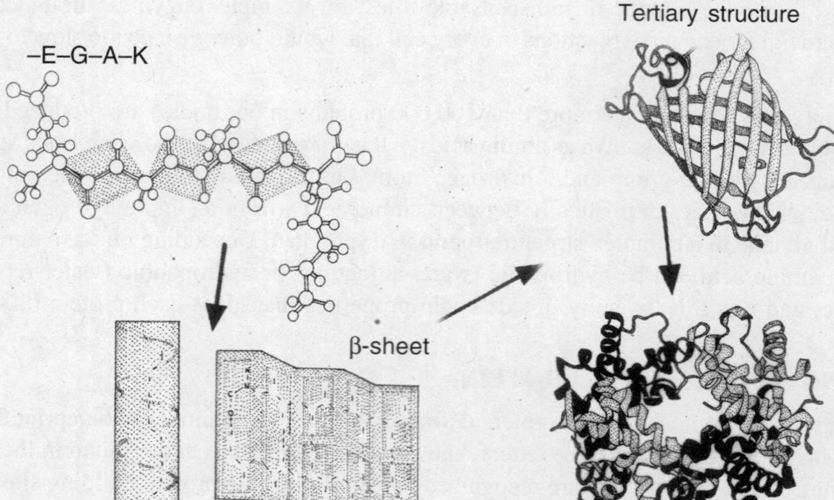

Fig. 39.2. Protein forming primary, secondary, tertiary and quaternary structures

Although it is possible to deduce the primary structure of a protein from a gene/s sequence, its tertiary structure cannot be determined (although it should become possible to make predictions when more tertiary sequences are submitted to databases). It can only be determined by complex experimental analyses and, at present, this information is only known for about 10% of proteins. It is therefore not yet known how an amino-acid chain folds into its tertiary structure in the short time scale (fractions of a second) that occurs in the cell. So, there is a huge gap in our knowledge of how we move from protein sequence to function in living organisms: the line of sight from the genetic blueprint for a protein to its biological function is blocked by the impenetrable jungle of protein folding.

III. PROTEIN FOLDING.

One of the most important results in understanding the process of protein folding was a thought-provoking experiment that was carried out by Christian Anfinsen and colleagues in the early 1960s. They investigated a protein called ribonuclease, which they isolated from the pancreatic tissue of cattle. This enzyme, made up of 124 amino acids, cleaves any ribonucleic acid (RNA) that could be harmful to the cell, such as truncated RNA that would not make a fully operational protein. It briefly binds RNA in a binding site and requires several sulphur-containing amino-acid cysteine residues in the protein, which form bonds with each other (called disulphide bridges) and hold the protein structure together.

IV. ENERGY TO FOLD

As with all processes in nature, protein folding also needs energy, the process has to obey the laws of thermodynamics (Fig. 39.3). A protein always folds so that it achieves the lowest possible energy just as we always try to adopt the most comfortable position, in which we need to move about least, when going to sleep. It is thought that this is achieved by using an energy gradient or funnel along the path from the random tangle to the folded protein. Alan Fersht of Cambridge University used the following analogy to illustrate this model: if you blindfold a golfer and let him hit the ball in any direction he likes, the probability that he will hole the ball is almost infinitesimal. The same is true of a protein finding the right form by chance. However, if all parts of the golf course slope toward the hole, which is at the lowest point in the area, even a blindfolded golfer has a good chance of finding the hole. So, fixed reaction pathways are not necessary, as each protein seeks out its natural shape through a funnel of declining energy; it can take many folding routes and still reach its target of the completed tertiary structure.

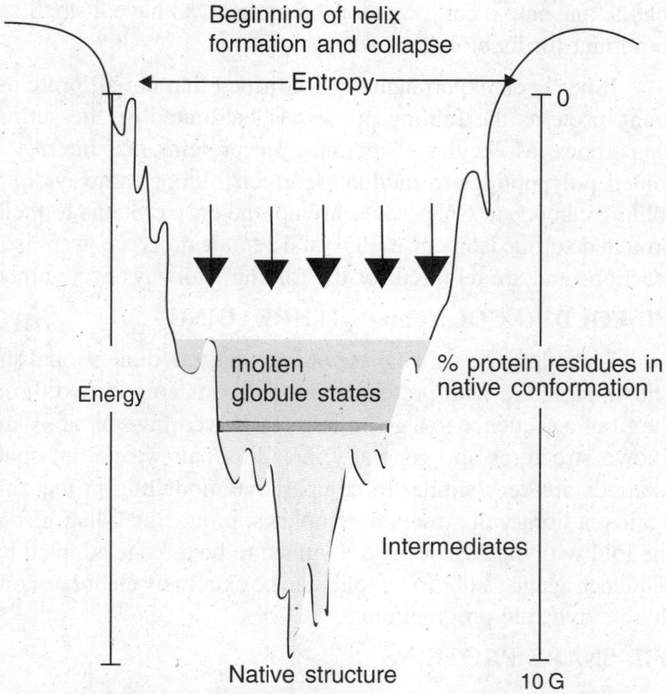

Fig. 39.3. Schematic presentation of energy innovolement in protein folding

V. THE PROCESS OF FOLDING

The folding pathway of a large polypeptide chain is very complicated, and not all the principles that guide the process have been worked out. However, many plausible models have attempted to describe protein folding. One model views folding as a hierarchical process where local secondary structures form first. Under this model, á helices and â sheets form first, with longer range interactions between helices and sheets forming super-secondary structures later. This process continues until the entire polypeptide folds. An alternative model describes folding as a spontaneous collapse of the polypeptide into a compact state. This collapsed state is known as a molten globule. It may be that the actual folding process of proteins incorporates features of both models. Instead of following a single pathway, a population of peptide molecules may take a variety of routes. Thermodynamically, the folding process can be viewed as a kind of free-energy funnel, where the unfolded states are characterized by a high degree of conformational entropy and relatively high free energy. In a trivialized definition, entropy is a measure of chaos, a measure of all different conformational states that the protein can be in. Obviously, there is more chaos in the protein in its unfolded state. On the other hand high free energy is a measure of unstableness, which is higher in a protein's unfolded state. Therefore, as folding proceeds, the narrowing of the funnel represents a decrease in the number of conformational states present. At the bottom of the funnel, also known as the global minimum, an ensemble of folding intermediates are reduced to a single conformation. It is important to realize that although we often describe the free energy funnel as having *one* global minimum - that is, one native conformation - a protein can have a small set of native conformations, each one important for its biological function(s).

It has been experimentally confirmed that not all proteins fold spontaneously in the cell. For many proteins the folding process is facilitated by the action of specialized proteins known as chaperones. Molecular chaperones are proteins that interact with partially folded or improperly folded polypeptides to facilitate correct folding pathways of provide microenvironments so that folding can occur. Chaperones are not the only proteins to facilitate protein folding. Two enzymes, protein disulfide isomerase (PDI) and peptide prolyl cis-trans isomerase (PPI), catalyze isomerization reactions and are required for the folding pathways of a number of proteins.

VI. FOLD RECOGNITION (THREADING)

Threading uses a database of known three-dimensional structures to match sequences without known structure with protein folds. This is accomplished through a scoring function that assesses the fit of a sequence to a given fold. These scoring functions are usually derived from a database of known structures and generally include a pair-wise atom contact and solvation terms. Threading methods are very similar to comparative modelling in that threading compares a target sequence against a library of structural templates, producing a list of scores. The scores are then ranked and the fold with the best score is assumed to be the one adopted by the sequence. The methods to fit a sequence against a library of folds can be extremely elaborate computationally, such as those involving double dynamic programming.

VII. SNARE PROTEINS

Eukaryotic cells contain membrane-bound compartments that are connected by trafficking of vesicular intermediates. To maintain compartmental organization, proper targeting of transport vesicles is required which involves two different molecular interactions.

(a) Membrane traffic

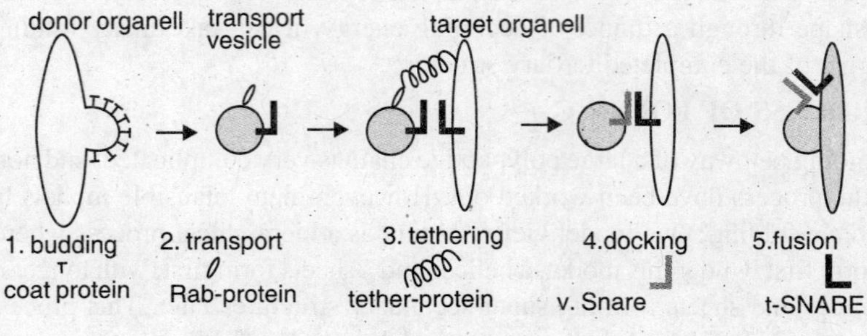

Fig. 39.4. Membrane trafficking of protein

The first selective recognition of vesicle and target membrane takes place during docking. In this reaction specific rab/ypt small GTPases and tethering factors on opposite membranes interact. Docking then provides a signal for complex formation of specific evolutionary conserved transmembrane proteins that reside on vesicles (v-SNAREs) and target membranes (t-SNAREs). The acronyme SNARE means: soluble N-ethylmaleimide-sensitive factor attachment protein (SNAP) receptor. Twenty one SNARE proteins have been found in yeast and over 35 in mammals so far. They are localized to specific organelles and take part in one or in a few membrane transport steps.

Conserved domains of v- and t-SNAREs close to the membrane anchor bind each other. Complex formation is accompanied by changes in the protein structure which probably drives membrane fusion. Recent structural studies suggest that SNARE complexes have a common structure: four conserved SNARE domains of about 50 amino acid residues form a parallel four-helix bundle. The helices interact through hydrophobic amino acid residues except in the center of these helices. Here an arginine (R) residue of a v-SNARE interacts with three glutamine (Q) residues from the other helices. These amino acids are very conserved leading to a reclassification of SNAREs as R- and Q-SNAREs. All t-SNAREs have a glutamine, while some v-SNAREs contain an arginine and others a glutamine in this 0 layer. Q-SNAREs can be subdivided into three different groups. According to an emerging model all SNARE complexes are formed by one helix from each of the four groups.

(b) Protein Targeting

Proteins that must be targeted within the cell must be intercepted early during synthesis so that this can happen correctly. As a protein is being synthesized, decisions must be taken about sending it to the correct location in the cell where it will be required. The information for doing this resides in the nascent protein sequence itself. Once the protein has reached its final destination, this information may be removed by proteolytic processing.

(c) Targeting in Bacteria

In bacterial cells, the targeting decision is relatively straightforward: is the protein destined to be an intracellular protein or an extracellular one? However, even here nuances are possible. In a gram-negative bacterium, such as *E. coli,* there must be some way of knowing whether a protein is destined to go to the cell membrane, the periplasmic space, or the outer membrane.

Secreted proteins contain a **signal sequence**. This is a short (6 - 30) stretch of hydrophobic amino acids, flanked on the N-terminal side by one or more positively charged amino acids such as lysine or arginine, and containing neutral amino acids with short side-chains (such as glycine or alanine) at the cleavage site. Some examples are:

Table 39.1. Examples of signal sequences

***E. coli* Signal Sequences from the SwissProt Database**

Entry	Description	Sequence
P31550	Thiamin-binding periplasmic protein precursor	MSAPAVAVTAPVFA
P29679	Acyl-coa thioesterase i precursor (protease i)	MMNFNNVFRWHLPFLFLVLLTFRAAA
P19935	Tolb protein precursor	MKQALRVAFGFLILWASVLHA
P10904	Glycerol-3-phosphate-binding periplasmic protein precursor	MKPLHYTASALALGLALMGNAQA
P37387	D-xylose-binding periplasmic protein precursor	MKIKNILLTLCTSLLLTNVAAHA

As proteins with signal sequences are synthesized, they are bound by the SecB protein. This prevents the protein from folding. SecB delivers the protein to the cell membrane where is is secreted through a pore formed by the SecE and SecY proteins. Secretion is driven by the SecA ATPase.

After the protein has been secreted, the signal sequence is removed by a membrane bound leader peptidase.

(*d*) **Targeting in Eukaryotes**

In eukaryotic cells, the situation is more complex. Extracellular proteins can be targeted for secretion or to the cell membrane, or to one of the many internal organelles such as the lysosome. Intracellular proteins can be targeted for the cyoplasm, to the nucleus or to special organelles such as the mitochondrion or the chloroplast.

The Signal Sequence hypothesis was first enunciated by **Gunther Blöbel** who was awarded the Nobel Prize in Medicine in 1999 for his work.

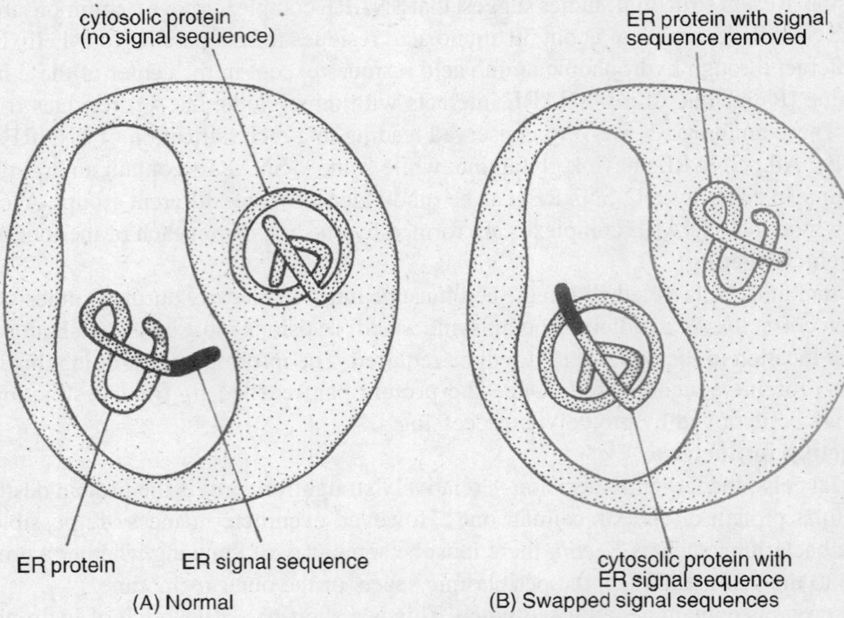

Fig. 39.5. Role of signal sequencs in transport of protein

Protein secretion in eukaryotic cells also involves a signal sequence. As in bacteria, the signal sequence is a short (6 - 30) stretch of hydrophobic amino acids, flanked on the N-terminal side by one or more positively charged amino acids such as lysine or arginine, and containing neutral amino acids with short side-chains (such as glycine or alanine) at the cleavage site. Signal sequences are recognized by Signal Recognition Particles (SRPs) which are ribonucleoprotein particles containing a stable 305 nt 7S RNA and 6 polypeptides. The 7S RNA has 2 domains - the Alu domain (so called because it is related to the Alu family of repeating sequences) and the S domain. The 7S RNA has

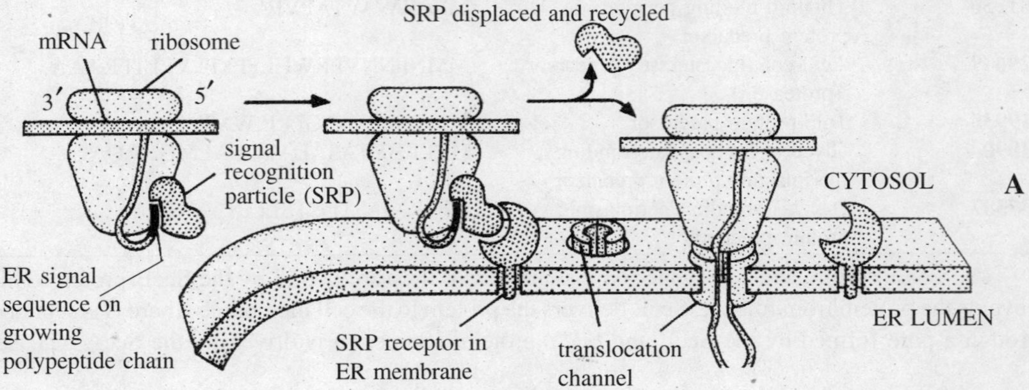

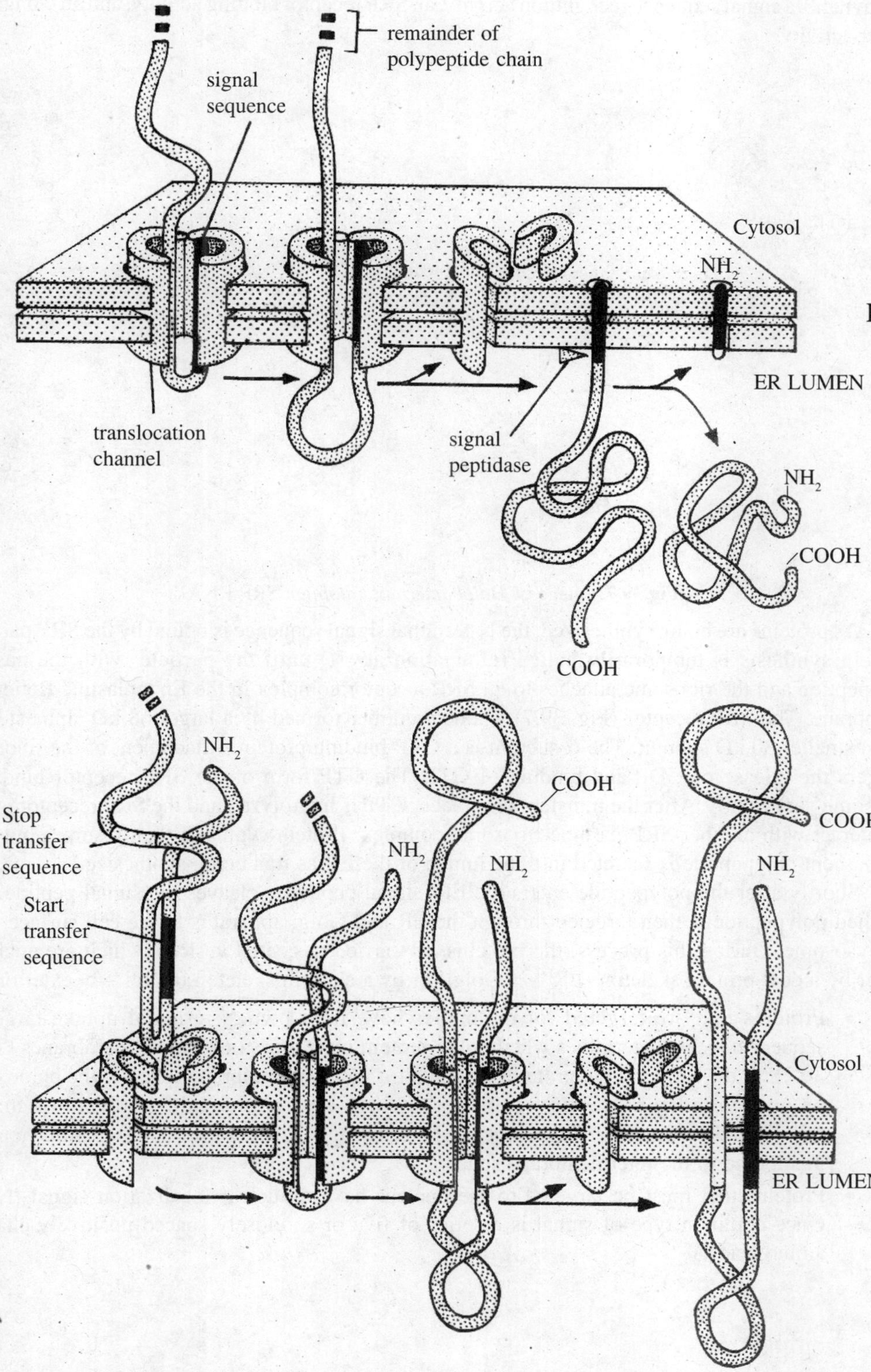

Fig. 39.6. Synthesis of polypeptides (A), and transport across the membrane (B, C) involving signal sequences for start and stop signals.

3 activities - a signal sequence recognition activity, an SRP receptor binding activity, and an elongation arrest activity.

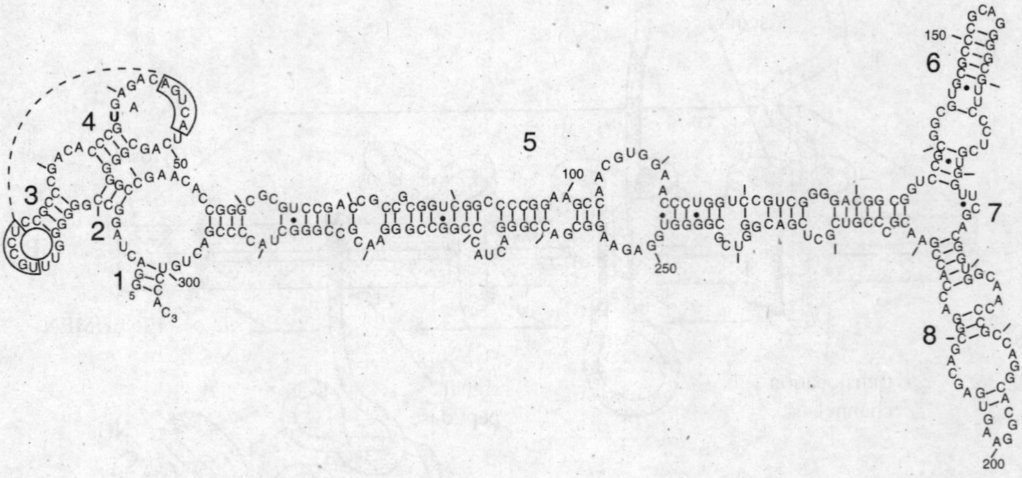

Fig.39.7. Image of *Halobacterium halobium* SRP RNA

As proteins are being synthesized, the N-terminal signal sequence is bound by the SRP particle. Protein synthesis is temporarily halted (elongation arrest) until the particle- with the nascent polypeptide and the ribosome attaches to an SRP receptor complex in the Endoplasmic Reticulum membrane. The SRP receptor (Fig. 39.7) is a heterodimer formed by a larger 68 kD alpha subunit and a smaller 30 kD subunit. The α subunit is a GTP-binding protein. Attachment of the ribosome triggers the release of GDP and binding of GTP. The GTP form of the SRP receptor binds the ribosome very tightly. After the transfer takes place, GTP is hydrolyzed and the SRP receptor is free to interact with another SRP particle-ribosome complex. Protein synthesis then resumes but now the nascent polypeptide is secreted into the lumen of the ER as it is being synthesized.

Shortly after the polypeptide enters the ER, signal peptidase cleaves the signal peptide. The finished polypeptide is then targeted through the ER and Golgi apparatus to the cell surface or to the lysosome. During this process, the proteins are carried in coated vesicles which are enclosed either by a coat protein skeleton (ER —> Golgi) or by a clathrin skeleton (Golgi —> exterior).

- Proteins that are targeted for organelles have their own N-terminal uptake-targeting sequence(s) that determines whether the protein must cross one or two membranes (Table 39.2). In the former case, proteins destined for the intermembrane space of the chloroplast or mitochondrion are first transported into the stroma (chloroplast) or matrix (mitochondrion) and then re-transported back through the inner chloroplast or mitochondrial membrane to the intermembrane space.
- Proteins that must be targeted to the nucleus have a nuclear localization signal (NLS). Once common type of signal is a series of five or so closely spaced positively charged amino acids.

Table 39.2. Location of signals in protein targeted to various organelles.

Organelle	Location of Signal in the Protein(Usually)	Nature of Signal
Endoplasmic Reticulum	C-terminus	**KDEL** or **HDEL** (in yeast)
Mitochondrion	N-terminal	3-5 nonconsecutive Arg or Lys; often contains Ser & Thr; never has acidic amino acids; no consensus sequence.
Chloroplast	N-terminal	Generally rich in Ser, Thr, and small hydrophobic amino acids; no acidic amino acids
Nucleus	Internal	Single cluster of 5 basic amino acids (e.g. **KKKRK** in the SV40 T antigen), or 2 smaller clusters separated by 10 amino acids
Lysosome	Internal	Covalent attachment of Mannose-6-phosphate
Peroxisome	C-terminus	**SKF** tripeptide at C-terminus

QUESTIONS

1. Discuss phenomenon and importance of protein folding.
2. Write short note :
   (*a*) Protein folding
   (*b*) Protein targeting
   (*c*) Transport across membrane
   (*d*) Signal sequences.

SITE-DIRECTED MUTAGENESIS & PROTEIN ENGINEERING

CHAPTER 40

I. INTRODUCTION

Site-directed mutagenesis is a molecular biology technique in which a mutation is created at a defined site in a DNA molecule, usually plasmid. In general, site-directed mutagenesis requires that the wild-type gene sequence be known. This technique is also known as site-specific mutagenesis or Oligonucleotide-directed mutagenesis. The principle of site-directed mutagenesis is that a mismatched oligonucleotide is extended, incorporating the "mutation" into a strand of DNA that can be cloned.

An important milestone in modern molecular biology came in 1978 with the first description of site-directed mutagenesis using oligonucleotides for *in vitro* synthesis of mutant DNA. Michael Smith, its pioneer, shared the Nobel Prize in Chemistry in October 1993 with Kary B. Mullis who developed PCR. In 1987 Kunkel and co-workers introduced an improvement to the technique that eliminated the need for phenotypic selection. The plasmid to be mutated would be transformed into an *E. coli* deficient in two genes, dUTPase & uracil deglycosidase. The former would prevent the breakdown of dUTP, a nucleotide that replaces dTTP in RNA, resulting in an abundance of the molecule, and deficiency in the latter would prevent the removal of dUTP from newly synthesized DNA. As the double-mutant E. coli replicates the up-taken plasmid, its enzymatic machinery incorporates the dUTP, resulting in a distinguishable copy. This copy is then extracted and incubated with an oligonucleotide containing the desired mutation, which attaches by base pair hydrogen bonding to the complementary wild-type gene sequence, as well as the Klenow enzyme, dNTPs, and DNA ligase. The reaction essentially replicates the dUTP-containing plasmid using as primer the oligonucleotide, giving a nearly identical copy. The essential differences being that the copy contains dTTP rather than dUTP, as well as the desired mutation. When the chimeric double-stranded plasmid, containing the dUTP, unmutated strand and the dTTP, mutated strand, is inserted into a *normal*, wild-type E. coli, the dUTP-containing strand is broken down, whereas the mutation-containing strand is replicated (Fig. 40.1).

II. CASSETTE MUTAGENESIS

This type of mutagenesis involves the cleavage by a restriction enzyme (RE) at a site in the plasmid and subsequent ligation of an oligonucleotide containing the mutation in the gene of interest to the plasmid. Usually the RE that cuts at the plasmid and the oligonucleotide is the same, permitting sticky ends of the plasmid and insert to ligate to one another.

III. IN VITRO SITE-SPECIFIC MUTAGENESIS

Mutagenesis is a fundamentally important DNA technology which seeks to change the base sequence of DNA and test its effect on gene or DNA function. The mutagenesis can be conducted *in vivo* (in studies of model organisms, or cultured cells) or *in vitro* and the mutagenesis can be directed to a specific site in a pre-determined way (site-directed mutagenesis), or can be random. In the case of *in vivo* mutagenesis, for example, gene targeting offers exquisite site-directed mutagenesis within living cells, while exposure of male mice to high levels of a powerful mutagen such as ethyl nitrosurea (ENU) and subsequent mating of the mice offers a form of random mutagenesis which can be important in generating new mutants.

***In vitro* mutagenesis** can involve essentially random approaches to mutagenesis, which may be valuable in producing libraries of new mutants. In addition, if a gene has been cloned and a functional assay of the product is available, it is also very useful to be able to employ a form of *in vitro* mutagenesis which results in alteration of a specific amino acid or small component of the gene product in a predetermined way.

Oligonucleotide mismatch mutagenesis

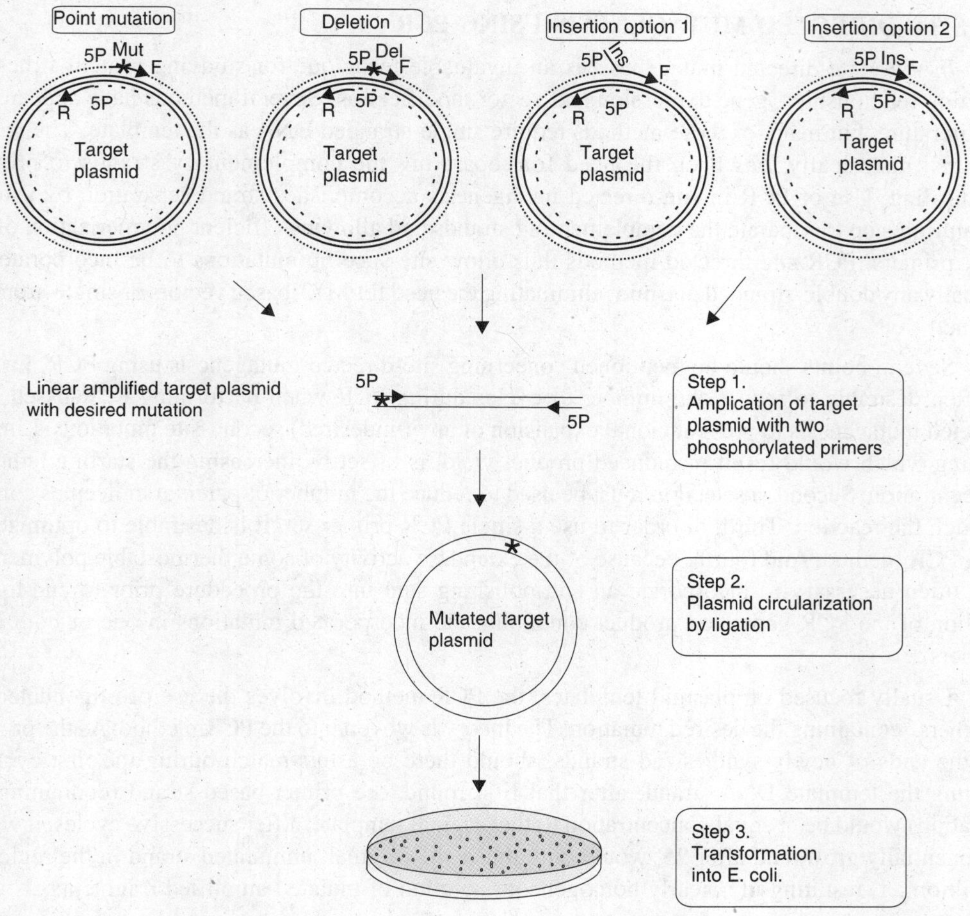

Fig. 40.1. Various approaches to ereat changes in plasmid

Many *in vitro* assays of gene function wish to gain information on the importance of individual amino acids in the encoded polypeptide. This may be relevant when attempting to assess whether a particular missense mutation found in a known disease gene is pathogenic, or just generally in trying to evaluate the contribution of a specific amino acid to the biological function of a protein. A popular general approach involves cloning the gene or cDNA into an M13 or phagemid vector which permits recovery of single-stranded recombinant DNA. A mutagenic oligonucleotide primer is then designed whose sequence is perfectly complementary to the gene sequence in the region to be mutated, but with a single difference: at the intended mutation site it bears a base that is complementary to the desired mutant nucleotide rather than the original. The mutagenic oligonucleotide is then allowed to prime new DNA synthesis to create a complementary full-length sequence containing the desired mutation. The newly formed heteroduplex is used to transform cells, and the desired mutant genes can be identified by screening for the mutation.

Other small-scale mutations can also be introduced in addition to single nucleotide substitutions. For example, it is possible to introduce a three-nucleotide deletion that will result in removal of a single amino acid from the encoded polypeptide, or an insertion that adds a new amino acid. Provided the mutagenic oligonucleotide is long enough, it will be able to bind specifically to the gene template even if there is a considerable central mismatch. Still larger mutations can be introduced by using cassette mutagenesis in which case a specific region of the original sequence of the original gene is deleted and replaced by oligonucleotide cassettes

III. SITE-DIRECTED MUTAGENESIS USING PCR

In vitro site-directed mutagenesis is an invaluable technique for studying protein structure-function relationships, gene expression and vector modification. Several methods have appeared in the literature, but many of these methods require single-stranded DNA as the template. The reason for this, historically, has been the need for separating the complementary strands to prevent reannealing. Use of PCR in site-directed mutagenesis accomplishes strand separation by using a denaturing step to separate the complementing strands and allowing efficient polymerization of the PCR primers. PCR site-directed methods thus allow site-specific mutations to be incorporated in virtually any double-stranded plasmid; eliminating the need for M13-based vectors or single-stranded rescue.

Several points should be mentioned concerning site-directed mutagenesis using PCR. First, it is often desirable to reduce the number of cycles during PCR when performing PCR-based site-directed mutagenesis to prevent clonal expansion of any (undesired) second-site mutations. Limited cycling which would result in reduced product yield, is offset by increasing the starting template concentration. Second, a selection must be used to reduce the number of parental molecules coming through the reaction. Third, in order to use a single PCR primer set, it is desirable to optimize the long PCR method. And fourth, because of the extendase activity of some thermostable polymerases it is often necessary to incorporate an end-polishing step into the procedure prior to end-to-end ligation of the PCR-generated product containing the incorporated mutations in one or both PCR primers.

Usually focused on plasmid templates, the PCR method involves the use of oligonucleotide "primers" containing the desired mutation. The theory is woven into the PCR method: As the primers are the ends of newly-synthesized strands, should there be a mis-match during the first cycle in binding the template DNA strand, after that first round, the primer-based strand (containing the mutation) would be at equal concentration to the original template. After successive cycles, it would exponentially grow, and after 25, would outnumber the original, unmutated strand in the region of 8 million : 1, resulting in a nearly homogeneous solution of mutated amplified fragments.

Despite the fact that the PCR product is overwhelmingly composed of mutated plasmid, the template DNA is typically eliminated by enzymatic digestion with Dpn1, a restriction enzyme which cleaves only methylated DNA. The template, which is derived from an alkaline lysis plasmid preparation and therefore is methylated, is destroyed in this step, but the mutated plasmid is preserved because it was generated *in vitro* and is unmethylated as a result.

IV. PROTEIN ENGINEERING

Protein engineering is the application of science, mathematics, and economics to the process of developing useful or valuable proteins. It is a young discipline, with much research currently taking place into the understanding of protein folding and protein recognition for protein design principles.

There are two general strategies for protein engineering. The first is known as *rational design*, in which the scientist uses detailed knowledge of the structure and function of the protein to make desired changes. This has the advantage of being generally inexpensive and easy, since site-directed mutagenesis techniques are well-developed. However, there is a major drawback in that detailed structural knowledge of a protein is often unavailable, and even when it is available, it can be extremely difficult to predict the effects of various mutations.

Computational protein design algorithms seek to identify amino acid sequences that have low energies for target structures. While the sequence-conformation space that needs to be searched is large, the most challenging requirement for computational protein design is a fast, yet accurate, energy function that can distinguish optimal sequences from similar suboptimal ones. Using computational methods, a protein with a novel fold has been designed, as well as sensors for unnatural molecules.

The second strategy is known as directed evolution. This is where random mutagenesis is applied to a protein, and a selection regime is used to pick out variants that have the desired qualities. Further rounds of mutation and selection are then applied. This method mimics natural evolution and generally produces superior results to rational design. An additional technique known as DNA shuffling mixes and matches pieces of successful variants in order to produce better results. This process mimics recombination that occurs naturally during sexual reproduction. The great advantage of directed evolution techniques is that they require no prior structural knowledge of a protein, nor it is necessary to be able to predict what effect a given mutation will have. Indeed, the results of directed evolution experiments are often surprising in that desired changes are often caused by mutations that no one would have expected. The drawback is that they require high-throughput, which is not feasible for all proteins. Large amounts of recombinant DNA must be mutated and the products screened for desired qualities. The sheer number of variants often requires expensive robotic equipment to automate the process. Furthermore, not all desired activities can be easily screened for.

Rational design and directed evolution techniques are not mutually exclusive; good researchers will often apply both. In the future, more detailed knowledge of protein structure and function, as well as advancements in high-throughput technology, will greatly expand the capabilities of protein engineering.

V. ENZYME ENGINEERING

A most exciting development over the last few years is the application genetic engineering techniques to enzyme technology. A full description this burgeoning science is beyond the scope of this text but some suitable references are given at the end of this chapter. There are a number of properties which may be improved or altered by genetic engineering including the yield and kinetics of the enzyme, the ease of downstream processing and various safety aspects. Enzymes from dangerous or unapproved microorganisms and from slow growing or limited plant or animal tissue may be cloned into safe high-production microorganisms. In the future, enzymes may be redesigned to fit more appropriately into industrial processes; for example, making glucose isomerase less susceptible to inhibition by the Ca^{2+} present in the starch saccharification processing stream.

Protein engineering techniques have been available for over two decades beginning with the development of methods for genetic engineering. Since that time, the engineering of enzymes has advanced rapidly along with a revolution in the range and efficiency of new techniques and strategies for designing and evolving proteins in the laboratory. While recent advances in high-throughput screening techniques are permitting larger libraries of enzymes to be screened more rapidly, a combination of new genetic tools and computational methods are enabling the efficient application of random mutagenesis targeted to areas of enzyme structures that are more likely to elicit the desired enhancement of their biocatalytic properties. Meanwhile the rational design of enzyme properties, in particular, by computational design is showing great potential. This review article summarises recent and important patents relating to the engineering of enzymes for biocatalysis.

The amount of enzyme produced by a microorganism may be increased by increasing the number of gene copies that code for it. This principle has been used to increase the activity of penicillin-G-amidase in *Escherichia coli*. The cellular DNA from a producing strain is selectively cleaved by the restriction endonuclease HindIII. This hydrolyses the DNA at relatively rare sites containing the 5'-AAGCTT-3' base sequence to give identical 'staggered' ends.

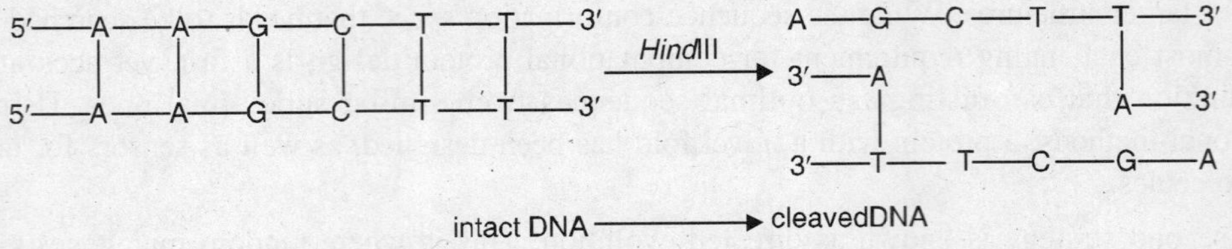

The total DNA is cleaved into about 10000 fragments, only one of which contains the required genetic information. These fragments are individual cloned into a cosmid vector and thereby returned to *E. coli*. These colonies containing the active gene are identified by their inhibition of a 6-amino-penicillanic acid-sensitive organism. Such colonies are isolated and the penicillin-G-amidase gene transferred on to pBR322 plasmids and recloned back into *E. coli*. The engineered cells, aided by the plasmid amplification at around 50 copies per cell, produce penicillin-G-amidase constitutively and in considerably higher quantities than does the fully induced parental strain. Such increased yields are economically relevant not just for the increased volumetric productivity but also because of reduced downstream processing costs, the resulting crude enzyme being that much purer.

VI. PROTEIN ENGINEERING CYCLE

The process starts with the isolation and characterisation of the required enzyme. This information is analysed together with the database of known and putative structural effects of amino acid substitutions to produce a possible improved structure. This factitious enzyme is constructed by site-directed mutagenesis, isolated and characterised. The results, successful or unsuccessful, are added to the database, and the process repeated until the required result is obtained.

New enzyme structures may be designed and produced in order to improve on existing enzymes or create new activities. An outline of the process of protein engineering is shown in Figure 8.1. Such factitious enzymes are produced by site-directed mutagenesis. Unfortunately from a practical point of view, much of the research effort in protein engineering has gone into studies concerning the structure and activity of enzymes chosen for their theoretical importance or ease of preparation rather than industrial relevance. This emphasis is likely to change in the future.

As indicated by the method used for site-directed mutagenesis, the preferred pathway for creating new enzymes is by the stepwise substitution of only one or two amino acid residues out of the total protein structure. Although a large database of sequence-structure correlations is available, and growing rapidly together with the necessary software, it is presently insufficient accurately to predict three-dimensional changes as a result of such substitutions. The main problem is assessing the long-range effects, including solvent interactions, on the new structure. As the many reported results would attest, the science is at a stage where it can explain the structural consequences of amino acid substitutions after they have been determined but cannot accurately predict them. Protein engineering, therefore, is presently rather a hit or miss process which may be used with only little realistic likelihood of immediate success. Apparently quite small sequence changes may give rise to large conformational alterations and even affect the rate-determining step in the enzymic catalysis. However it is reasonable to suppose that, given a sufficiently detailed database plus suitable software, the relative probability of success will increase over the coming years and the products of protein engineering will make a major impact on enzyme technology.

Much protein engineering has been directed at subtilisin (from *Bacillus amyloliquefaciens*), the principal enzyme in the detergent enzyme preparation, Alcalase. This has been aimed at the improvement of its activity in detergents by stabilising it at even higher temperatures, pH and oxidant strength. Most of the attempted improvements have concerned alterations to:

1. the P_1 cleft, which holds the amino acid on the carbonyl side of the targeted peptide bond;
2. the oxyanion hole (principally Asn_{155}), which stabilises the tetrahedral intermediate;
3. the neighbourhood of the catalytic histidyl residue (His_{64}), which has a general base role; and
4. the methionine residue (Met_{222}) which causes subtilisin's lability to oxidation.

It has been found that the effect of a substitution in the P_1 cleft on the relative specific activity between substrates may be fairly accurately predicted even though predictions of the absolute effects of such changes are less successful. Many substitutions, particularly for the glycine residue at the bottom of the P_1 cleft (Gly_{166}), have been found to increase the specificity of the enzyme for particular peptide links whilst reducing it for others. These effects are achieved mainly by corresponding changes in the K_m rather than the V_{max}. Increases in relative specificity may be useful for some applications. They should not be thought of as the usual result of engineering enzymes, however, as native subtilisin is unusual in being fairly non-specific in its actions, possessing a large hydrophobic binding site which may be made more specific relatively easily (e.g. by reducing its size). The inactivation of subtilisin in bleaching solutions coincides with the conversion of Met_{222} to its sulfoxide, the consequential increase in volume occluding the oxyanion hole. Substitution of this methionine by serine or alanine produces mutants that are relatively stable, although possessing somewhat reduced activity.

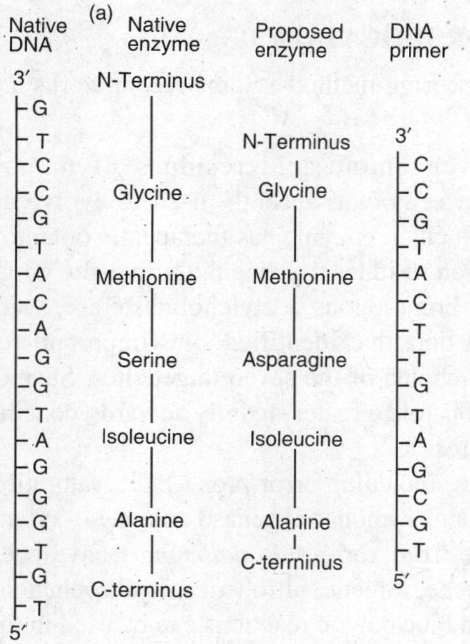

Fig. 40.2 (*a*)

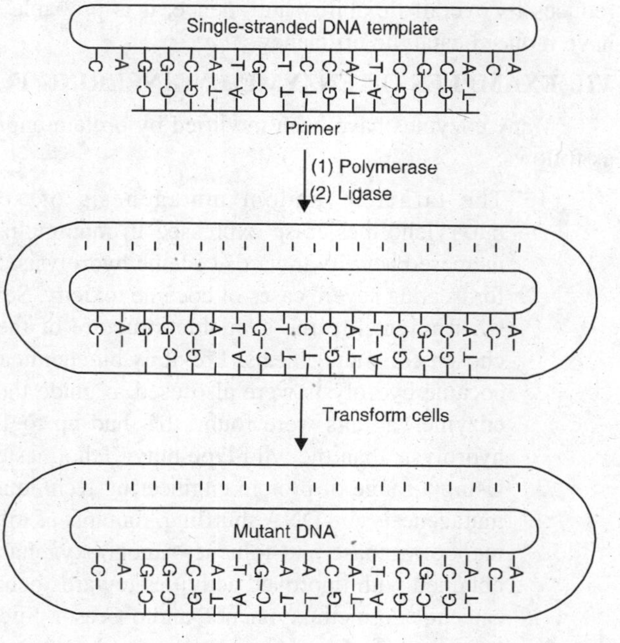

Fig. 40.2 (*b*)

An outline of the process of site-directed mutagenesis, using a hypothetical example. (a) The primary structure of the enzyme is derived from the DNA sequence. A putative enzyme primary structure is proposed with an asparagine residue replacing the serine present in the native enzyme. A short piece of DNA (the primer), complementary to a section of the gene apart from the base mismatch, is synthesised. (b) The oligonucleotide primer is annealed to a single-stranded copy of the gene and is extended with enzymes and nucleotide triphosphates to give a double-stranded gene. On reproduction, the gene gives rise to both mutant and wild-type clones. The mutant DNA may be identified by hybridisation with radioactively labelled oligonucleotides of complementary structure (Figure 40.2).

An example of the unpredictable nature of protein engineering is given by trypsin, which has an active site closely related to that of subtilisin. Substitution of the negatively charged aspartic acid residue at the bottom of its P_1 cleft (Asp_{189}), which is used for binding the basic side-chains of lysine or arginine, by positively charged lysine gives the predictable result of abolishing the activity against its normal substrates but unpredictably also gives no activity against substrates where these basic residues are replaced by aspartic acid or glutamic acid.

Considerable effort has been spent on engineering more thermophilic enzymes. It has been found that thermophilic enzymes are generally only 20-30 kJ more stable than their mesophilic counterparts. This may be achieved by the addition of just a few extra hydrogen bonds, an internal salt link or extra internal hydrophobic residues, giving a slightly more hydrophobic core. All of these changes are small enough to be achieved by protein engineering. To ensure a more predictable outcome, the secondary structure of the enzyme must be conserved and this generally restricts changes in the exterior surface of the enzyme. Suitable for exterior substitutions for increasing thermostability have been found to be aspartate ® glutamate, lysine ® glutamine, valine ® threonine, serine ® asparagine, isoleucine ® threonine, asparagine ® aspartate and lysine ® arginine. Such substitutions have a fair probability of success. Where allowable, small increases in the interior hydrophobicity for example by substituting interior glycine or serine residues by alanine may also increase the thermostability. It should be recognised that making an enzyme more thermostable reduces its overall flexibility and, hence, it is probable that the factitious enzyme produced will have reduced catalytic efficiency.

VII. EXAMPLES OF ENZYME ENGINEERING FOR BIOCATALYSIS

Many enzymes have been modified by protein engineering methods. Some recent works are as follows:

1. The targeted random mutagenesis of seven amino-acid residues of human butyrylcholinesterase expressed in mammalian cells, was recently used to evolve an increased activity towards cocaine hydrolysis. Such an enzyme has therapeutic potential for treating severe cases of cocaine toxicity. Seven residues that line the active-site gorge when aligned to the available structure of the homologous acetylcholinesterase, were chosen for mutagenesis. Previous biochemical data that identified sites important for cocaine hydrolysis were also used to guide the choice of the seven target sites. Several enzyme variants were found that had up to 100-fold greater activity towards cocaine hydrolysis than the wild-type butyrylcholinesterase.
2. Using a range of protein engineering techniques, including error-prone PCR, saturation mutagenesis and DNA shuffling, mutants of toluene 4-monooxygenase, toluene-o-xylene monooxygenase and toluene-4-monooxygenase from various *Pseudomonas* have been obtained with improved activities towards benzene, toluene, nitrobenzene, nitrophenols, catechols, o-methoxyphenol, and o-Cresol. Such biocatalytic reactions can be used in the synthesis of phenol, catechol, nitrophenols, nitrocatechols, 3-methoxy-catechol, 1,2,3-trihydroxybenzene, methoxyhydroquinone, methyl-hydroquinone, 4-methylresorcinol, methylhydroquinone, and pyrogallol.
3. B12-dependent dehydratases with improved reaction kinetics have been obtained using error-prone PCR and oligonucleotide-directed mutagenesis to target the DhaB1 gene, encoding the a-subunit of glycerol dehydratase. Such enzymes are useful in the production of glycerol and 1,3-propanediol.

4. Nitrilases can be used in biocatalytic processes to convert nitriles to carboxylic acids. Recently, error-prone PCR has been used to generate mutants of an *Acidovorax facilisnitrilase* having improved activity towards 3-hydroxynitriles. Isolated enzyme variants were capable of up to 1.9-fold improved activities when used to convert 3-hydroxynitrile to 3-hydroxycarboxylic acid, 3-hydroxyvaleronitrile to 3-hyd-roxyvaleric acid, or 3-hydroxybutyronitrile to 3-hydroxybutyric acid.
5. In another example highlighting the general applicability of error-prone PCR, two carotenoid biosynthesis genes were simultaneously mutated randomly to improve the production of astaxanthin from beta-carotene. Targeting of the two enzymes, CrtO ketolase and CrtZ hydroxylase, simultaneously resulted in mutations within both genes and a net increase in the astaxanthin biosynthesis yield of up to 20% above that obtained with the wild-type enzyme pathway.
6. Error-prone PCR has also been used to improve the activity of 5'-xanthylic acid (XMP) aminase, which is potentially useful for the biocatalytic synthesis of 5'-guanylic acid (GMP) from XMP in microbial fermentations. Six improved mutants were found, with between two and six accumulated mutations, and the best mutant having 3.8-fold greater activity resulting from three mutations.
7. Xylanases are useful enzymes added to animal feed to aid the digestion of feed grains containing xylan. The growing need to sterilize the feed grain requires a xylanase enzyme that can retain activity after heat treatment, as well as being active at physiological digestive conditions (pH 3.5-5.0, 40°C). A recent patent has combined an engineered disulphide bond, previously introduced into a homologous xylanase enzyme, and the Q162H mutation previously reported for xylanases. Previously, the disulphide bond was shown to only marginally increase the xylanase thermostability, whereas the Q162H mutation had no effect on thermostability. However, the two mutations combined together result in a xylanase enzyme with a thermal denaturation mid- point increase of 14°C, and which also retains 40% of the original activity after a 30 minute treatment at 70°C, and at least 30% activity at pH3.5-6, and 40-60°C.
8. A similar product, 5'-inosinic acid (IMP), is synthesized in an industrial process that uses mutant acid phosphatases from *E. blattae*, for which the phosphomonoesterase activity has been reduced to less than 40% of the wild-type enzyme due to the mutations G74D or I153T. Decreased esterase activity allows the mutant enzymes to synthesise IMP from inosine.
9. Another enzyme that can be used as a supplement to animal feeds is phytase. This enzyme is useful during digestion for releasing inorganic phosphate from phytate, which contains over 70% of the phosphate in plant material. A variant of the *E. coli* pH 2.5 acid phosphatase has been shown, rather unexpectedly, to exhibit increased phytase activity upon digestion of the enzyme with a mixture of trypsin and pepsin.
10. Alpha-amylase is an enzyme that is commonly used for the degradation of starch during textile and paper desizing, and also as a component of household detergents. Several examples of engineered alpha-amylases have been recently patented, including variants with improved activity and/or thermostability; solvent stability, and resistance to multimerisation.
11. Polysaccharide lyases have potential uses in polysaccharide sequencing and degradation. A recent patent describes rationally designed variants of the polysaccharide lyase, chondroitinase B, and also potential uses in the inhibition of anticoagulation, angiogenesis and maternal malarial infection.

12. Active-site residues involved in catalysis and substrate binding were targeted by site-directed mutagenesis. While mutagenesis of the catalytic residues were damaging to enzyme activity, the mutation of a substrate binding arginine (R364) to alanine resulted in an altered product profile to that of wild-type enzyme after the digestion of dermatan sulfate.

QUESTIONS

1. How site directed mutagenesis is used to engineer proteins ?

2. What is PCR? How it is used in site directed mutagenesis ?

PROTEIN PROFILING, PROTEIN MICROARRAYS AND PROTEOMICS

CHAPTER 41

I. Protein Profiling.

Protein profiling, an independent, emerging sub-specialty of proteomics, is poised to provide unprecedented insight into biological events. Protein profiling is defined here as the quantitative assessment of protein expression levels. As profiling evolves, the term will increasingly refer to the study of multiple proteins, protein forms, or protein families, almost always comparing two different states (diseased vs. healthy, treated vs. untreated, experimental vs. control).

As this approach is perfected, protein expression profiling will make its mark in high-value research markets, including the elucidation of protein-protein interactions and signaling pathways, finding biomarkers for drug discovery and development, serum profiling to identify patient populations that respond to various treatments, and, eventually, medical diagnostics. Where proteomics is the study of a proteome in a given circumstance, including how proteins interact and function, protein profiling refers more specifically to which proteins are expressed, under which circumstances and times, and in which tissues. In human being proteome is very complex; 30,000 genes coding for between ten and thirty times as many proteins, with concentration ranges varying by 1012.

(a) Techniques. Many high-throughput protein identification and characterization methods developed for proteomics have been applied to protein profiling. These include gel electrophoresis, mass spectrometry, and protein microarrays. Unlike DNA sequencing, where capillary electrophoresis changed everything by allowing high-throughput sequencing, no equivalent exists for proteins. As profiling efforts will rely on multiple techniques, so will each depend on the perfection of numerous underlying technologies. Protein microarrays are a case in point. High-throughput antibody generation technologies, artificial molecular recognition surfaces of target proteins, phage-display libraries of antibody fragments, and proteome arrays containing full-length, functional proteins from a library of expression clones are new areas in protein study.

(b) Limitations of Gels. Although 2-D gel electrophoresis will never be replaced in protein characterization studies, (Fig. 41.1.) it suffers from a few drawbacks. Gels are time-consuming, imprecise, and their results are difficult to reproduce. Gels work well for soluble, high molecular weight proteins, but not so well for low molecular weight or relatively insoluble species. Bands on gels often contain several proteins, and great skill is required to assure uniform measurement across gels.

On the plus side, 2-D gels offer a window to visualizing a thousand proteins or more in one analysis and are easily interfaced to mass spectrometry (MS). To varying degrees, MS provides the same advantages for profiling experiments as for proteomics in general: high sensitivity and accuracy, mass and sequence information that is difficult to obtain otherwise, and reasonably high throughput.

For differential expression measurements, MS experiments are either isotopically label based or label free. In labeling experiments, the experimental and control proteins are generated and tagged independently, one with a "heavy" ^{13}C-hydrocarbon chain (or other stable isotope) and the other with a lighter version of the same label (typically with fewer ^{13}C atoms). The two samples are then combined, digested, and analyzed using LC-MS or LC-MALDI MS/MS. This technique labels at the protein level.

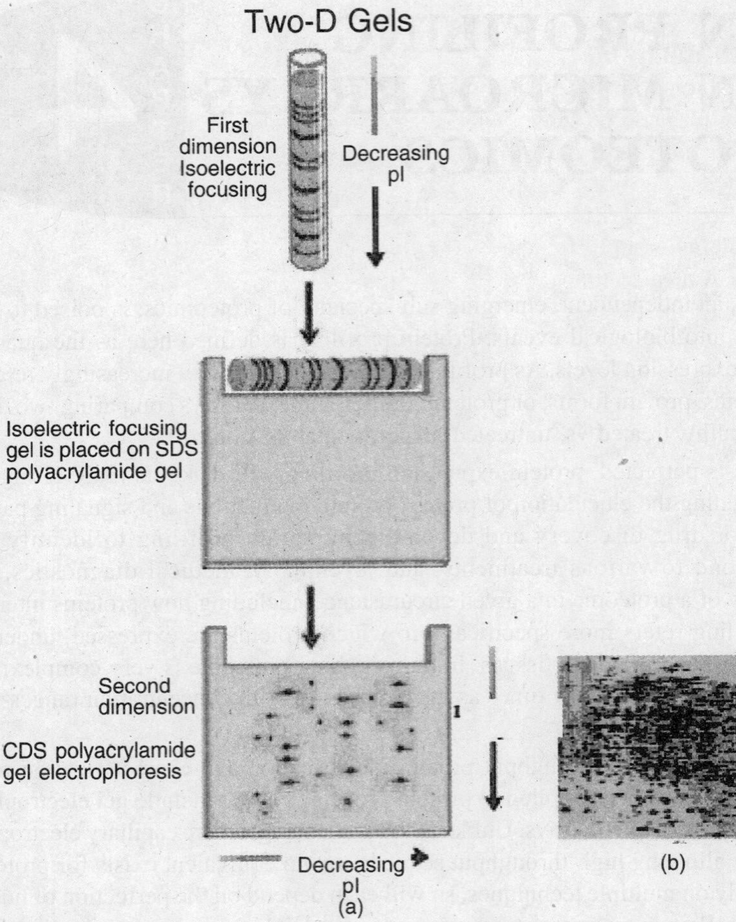

Fig. 41.1. Schematic presentation of 2-D gel electrophoresis

Coupled with MS, label-free protein profiling provides a reasonably precise mass analysis and relative abundance for each peptide. Results show 10-30% variability, about the same as with isotopic labelling.

(c) Differential fluorescence gel electrophoresis proteome profiling -Two dimensional difference gel electrophoresis (DIGE) utilizes mass- and charge-matched, spectrally resolvable fluorescent dyes (Cy3 and Cy5) to label two different protein samples in vitro prior to 2-D electrophoresis. Compared to conventional 2D gel electrophoresis, DIGE has the major advantage that both the control and experimental sample are run in the same gel. These samples are then imaged separately but because they were run in the same gel, the images can be perfectly overlaid without "warping". This reduces the number of gels that must be run to make statistically valid comparisons and raises the confidence with which protein changes between samples can be detected and quantified. Use of a third dye (Cy2) permits an internal standard to be created by pooling an equal aliquot of all biological samples in the experiment. The internal standard is then run on every gel in the experiment. This means that every protein spot from all samples will be represented in the internal standard. This in turn allows more accurate quantification and spot statistics between gels. Based on the literature, it may be possible to profile up to 1,000 protein spots on properly prepared samples that provide well resolved 2D gels. An additional advantage of this system is the ability to detect many post-translational modifications, such as phosphorylation, which often play a key role in modulating protein function and which cannot generally be detected by ICAT-MS based protein profiling.

(d) Isotope coded affinity tag (icat)-based protein profiling- While both MALDI-MS based peptide/protein disease biomarker discovery and DIGE analyses comparatively profile the naturally occurring forms of peptides and proteins, ICAT analysis profiles the relative amounts of cysteine-containing peptides derived from tryptic digests of protein extracts. Recognizing that only a single tryptic peptide is needed to quantify the expression of the corresponding parent protein, the ICAT reagent was designed to affinity isolate and quantifies via the use of a stable isotope the relative concentrations of cysteine-containing tryptic peptides obtained from digests of control versus experimental samples. A nice feature of the ICAT approach is that the *in vitro* incorporation of a stable isotope into one of the two samples being compared. This removes the need to analyze by mass spectrometry the control and experimental samples separately.

The alkyl linker in the ICAT reagent is connected to a (cleavable) biotin group which allows rapid affinity isolation of cysteine-containing tryptic peptides. While a tryptic digest of a whole cell human protein extract might produce 550,000 peptides, less than 100,000 of these might be expected to contain cysteine.

The resulting ICAT data, which is analogous to that obtained via the use of two different fluorescent dyes in DNA microarray analysis of mRNA or DIGE analysis of protein expression, provides the corresponding ratio for the level of expression of the parent protein in the control versus experimental sample.

Two dimensional chromatofocusing/non-porous reverse phase HPLC protein separation (2D-LC)-In this approach the chromatofocusing first dimension separates proteins based on isoelectric point (pI), and subsequent fractions are sequentially and automatically separated by surface hydrophobicity on a reverse-phase HPLC column. Proteins are detected at 214 nm and collected for tryptic digestion and MS/MS analysis.

II. PROTEIN MICROARRAY

Proteins were once thought to be too delicate for long-term storage on microarrays, but that notion has been changed. Arrays are printed on high-quality glass substrates using specialized printing buffers that stabilize proteins on the surface and attached to the glass covalently through appropriate surface chemistries. For example, arrays that employ Array It's SuperAldehyde chemistry enjoy a shelf life of at least one year on par with the stability of DNA microarrays.

For the past 10 years there has been considerable interest in the creation of high-density arrays of nucleic acids for use in genomics and 'transcriptomics'. Since the first model experiments in academic laboratories, the number of spots on each array has steadily grown whilst the overall size of the arrays has shrunk. Today's DNA microarray is the size of a thumbnail and can contain over 10,000 different oligonucleotides. The pages of scientific journals are peppered with advertisements touting robots for probing DNA microarrays, machines for imaging DNA microarrays, software for analyzing DNA microarrays and, of course, the DNA microarrays themselves; the estimated annual market for the technology is in excess of $527 million . The application of DNA microarrays is equally widespread, ranging from DNA sequencing to the genotyping of disease genes

It has long been the goal of molecular biologists to develop a technology that can quantify, in a reliable and reproducible manner, the expression level of every individual protein in a tissue sample. DNA microarrays have been used extensively for the analysis of RNA in cellular extracts, from which the expression of individual proteins can be inferred by assessing the levels of their corresponding mRNAs. Changes at the mRNA level, however, are not necessarily proportional to changes at the protein level because of differences in rates of protein translation and degradation. Furthermore, nucleotide screens are unable to provide information on the post-translational modifications of a protein, which may be critical for a protein's function. After all, it is the protein and not the mRNA that provides cellular function, whether it is for communication, metabolism or building cellular architecture.

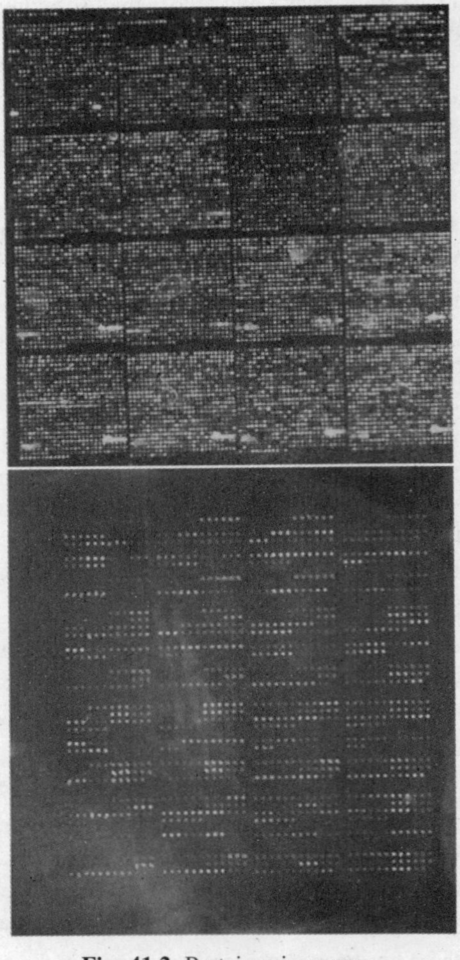

Fig. 41.2. Protein microarray

Although two-dimensional pulse-field gel electrophoresis (2D PFGE) can be used to analyze the proteins expressed in different cellular extracts directly, it requires both a high degree of technical skill and sophisticated computational analysis to identify protein spots that are present on the gel or blot from one extract but not on the other. Even then, it is still necessary to determine the identity of the corresponding protein. The Holy Grail for such proteomic analyses is a highly sensitive protein array screen, in which the strength of the signal at each point on the array provides readout of the expression level of each protein in the human body.

The Brown group was among the pioneers of DNA microarrays (Fig. 41.2) developing many of the robotic systems and protocols that are currently used for screening at the nucleic-acid level. They have now applied the same techniques to the creation and screening of protein microarrays, consisting of either spotted antibodies or spotted protein antigens (Fig. 41.2). As each of the antibody-antigen pairs is mutually exclusive - that is, each antibody only binds to a single cognate antigen and vice versa (a feature that makes antibodies particularly well suited to the task of protein profiling) - a single microarray comprising 115 antibodies can be used to assay the presence of different target antigens in a complex mixture. Although the concept of using antibody arrays to profile protein expression is not new, the screen described by Haab achieves, for the first time, the kind of sensitivity and reproducibility that would be required to detect the vast majority of proteins present in a human cell lysate. Furthermore, the protocol is centered on well-established technology that could easily be reproduced by any researcher well versed in the art of DNA arrays. Indeed, the same glass slides, robotic gridders, dye labels, slide readers and image analysis software have simply been adapted for use with proteins rather than with DNA.

Over the past three years there has been considerable excitement about the impending availability of protein arrays. Of course, protein is quite different from DNA. Whereas the immobilization of DNA to a solid support requires a simple charge interaction, a protein must be attached in such a way as to maintain its folded (and thereby functional) conformation. Although the nucleotide sequence of one piece of DNA differs from the next, they generally have similar chemical and physical properties. Two proteins, on the other hand, may be quite dissimilar, having different sizes, charges, stabilities and solvent solubilities. The binding of one protein to another is an even more complicated affair. The strength of the interaction between a single strand of DNA and its complementary strand is mainly dictated by the length of contiguous base pairing, which on a DNA array can be controlled by laying down different oligonucleotides of the same length. Each interaction that occurs on the surface of a DNA array therefore has a similar (high) affinity. By contrast, different protein-protein interactions may have affinities ranging from femtomolar to micromolar - 12 orders of magnitude difference.

Finally, a truly high-throughput array requires a wide range of binding molecules, each of which has a unique specificity for a given target present in the cell lysate. For DNA arrays, oligonucleotides can be synthesized that correspond to specific nucleotide sequences known to be present at the RNA level. For each protein in the lysate, however, a similar binder must be first isolated and then tested to confirm its contextual specificity. Thus, if a human cell lysate is to be probed, each binder in the array must specifically recognize a single protein in that lysate and not cross-react with any others. This is where antibodies come into their own. Highly specific monoclonal antibodies can be generated either *in vivo* by mouse immunization or *in vitro* using techniques such as phage display. Ideally, the antibodies are coupled or adsorbed to a solid support and then probed with a labeled lysate. Several model systems have demonstrated the feasibility of this approach using a handful of antibodies. Traditionally, CCD detectors employ a white light source and filter sets to read fluorophores of interest. Here, instead of white light, fluorophores are excited specifically at the wavelengths they were designed for. This dramatically reduced the cost of instruments without sacrificing sensitivity and dynamic range.

Qiagen's MSXpress Protein Chip kits provide sensitive high-throughput analysis of proteins using MALDI-MS. Qiagen claims up to 100-fold improvement in sensitivity compared with conventional targets for identifying low-abundance species. MSXpress 192 Protein Chips contain 192 precisely defined sinapinic acid matrix spots, which concentrate the sample protein within the spot volume. After the sample dries, it is crystallized and presented to the sample stage of a MALDI-MS.

III. PROTEOMICS

The identification and qualification of the portion that a whole organism expresses under certain conditions is main goods of high-throughput proteomics. Consequently two main goals of proteomics research are to identify proteins derived from complex mixtures extracted from cells and to quantify expression levels of these identified proteins. The concept of proteome was first proposed by Wilkins in 1995 and refers to the protein equivalent of the genome. Proteomics is a systematic analysis of proteins expressed by a genome. Genome now only represents the first step in the complexity of understanding biological functions. Gene portion relationship may not be monogamous. A single gene can encode for more than one protein species. The recent compilation of human genome database has given rise to a number of predicted proteins, which require primary structural and functional identifies. Hence the gene sequence information and pattern of gene activity inside the cell do not provide a complete picture. The link between gene and the protein has to be the established. Measurement of gene transcript (mRNA levels) cannot give complete information on cellular regulations, as gene expression is regulated post – transcriptionally and by the time many of the post –translational modifications are considered, the number of protein expressed in a cell is many times more than the coding potential of the organism. Hence it is necessary to determine the protein expression levels directly the proteome of the cell will reflect immediate environment of the protein.

One distinct advantage of proteomics in the potential of detecting changes that occur after messenger RNA step thus giving quantitative analysis of protein expression profiles. Therefore, the definition of proteomics should be the protein complement of a given cell at a given time including the set of all protein isoforms and modifications.

(a) Sample preparation - Sample preparation is very important in proteome analysis and it affects reproducibly of the results. Pre-treatment of samples for 2D electrophoresis involve solubilisation, denaturation and reduction to completely break-up the interaction between the proteins. Mass spectrophotometry is used to analyse samples containing thousands of proteins that can differ by many orders of magnitude in concentration. Because of this, the sample preparation plays a critical role. Sample fractionation can be carried out on cellular organelles, protein complexes, whole proteins or peptides to reduce the mixture complexity. Generally, 1D or 2D gel electrophoresis is used to separate proteins, which is based on their molecular weight and isometric point. Recent approaches involve the uses of liquid chromatography separation by size exclusion, strong cation/ anion exchange or hydrophobic interaction (reverse phase chromatography). Individual proteins can also be isolated from complex mixtures using immunoprecipitation.

Protein samples are first denatured using heat, urea, guanidine or surfactants such as sodium dodecyl sulphate to disrupt the 3Dprotein structure and enable proteases to access all of the available cleavage sites. Disulfide bonds also affect cleavage site availability and are usually reduced and alkylated before digestion.

Proteins are typically enzymatically digested into peptides before ms analysing (Fig.41.3). Trypsin is most commonly used because it produces small peptides that are workable to electro spray ionisation: trypsin cleaves C-terminal to basic amino acids, which ensures that there are two possible basic sites for ionisation (N –terminus and C-terminus, Lys or Arg). Other enzymes are also used depending on the desired application but the resulting peptide might not contain any or too many internal basic resides that can affect peptide identification following fragmentation.

After digestion, post-translationally modified peptides can be subjected to enrichment strategies. For example, phosphopeptides can be purified using immobilized metal affinity chromatography, whereas glycopeptides can be purified using lectins. Before mass spectrometric analysis, peptide mixtures are sometimes purified using desalting columns to remove denaturants used in the digestion.

Peptide mixtures are usually separated using online reverse phase liquid chromatography and analysed using mass- spectrophotometry (LC-MS) with tandem mass spectrophotometry (LC-MS-MS). Peptide mixtures, obtained by digestion of whole cell lysate, are more complex than the maximum capacity of both the liquid chromatographic separation and the mass spectrophotometer. Mixture complexity is the main problem associated with high-throughput proteomics experiments and leads to a reduced detection of peptides. Peptide mixture can be separated, offline, using strong cation exchange before the reverse phase separation. Multi-dimensional peptide identification technology, which combines strong cation exchange and reverse phase separations in an online LC-MS-MS experiment, is also common. The use of two dimensional peptide separations results in greater chromatographic peak capacities and protein identifications but requires more analysis time and results is greater numbers of MS-MS spectra (Fig. 41.4).

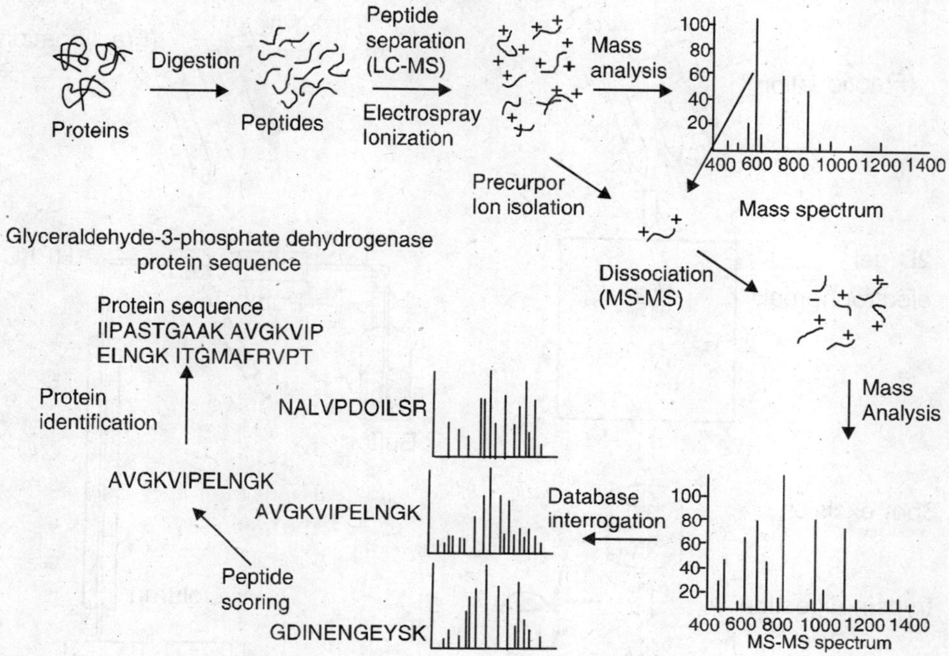

Fig. 41.3. Steps involved in proteome analysis (Peptide separation, mass analysis, protein identification, protein sequence)

(b) 2D – PAGE -A critical requirement of proteomics research in high quality separation of cellular proteins. The 2 D – PAGE (two dimensional poly acrylamide gel electrophoresis) technique developed by O' farrel (1975) led to the development of a classical proteomics tool. The proteins are resolved in two dimensions ; the first dimensions separates protein in a pH gradient according to their isoelectric point, whilst in the second dimension the proteins are separated according to their molecular weight by SDS-PAGE. Combination of these two separation techniques resolves protein to a spot on the two-dimensional map that is fixed by the co-ordinates as determined by its isoelectric point and molecular weight (Fig. 41.5). This map can be considered as the protein finger print of the sample. Two such fingerprints derived from two different cellular stages of an organism can be compared to see the up/down regulation (or appearance/disappearance of protein spots).

The first dimension of electrophoresis involves denaturing isolectric focusing (IEF) using 3 mm wide immobilized pH gradient gels (IPG). IPG strips have a gradient of charge imbedded in acrylamide significantly improving the reproductively and reliability (Fig. 41.6) Samples are applied either by cup loading or by in gel dehydration. In cup loading method, the strips aree re-hydrated with rehydration buffer and the sample are applied into the loading cup at either acidic or basic end. While in gel rehydration, the sample in lysis buffer is diluted with the rehydration buffer. Rehydration of a strip occurs in an individual strip holder. The dry gel matrix takes up the fluid together with the proteins. IEF is carried out on a first dimension electrophoreses unit consisting of five phases of stepped voltage from 500 to 3500 (multiphor) or 500 to 8000 V (IPGPhor). Immobilized dry strips come in a variety of lengths (from 7 to 24 cm) and pH ranges.

The proteins visible on the 2D gel may not give complete information about all the proteins or proteome of the cell. Most proteins seen on the gel are those present in abundance or "house keeping" proteins. Proteins which are in low abundance may be visible in the 2D-gel.

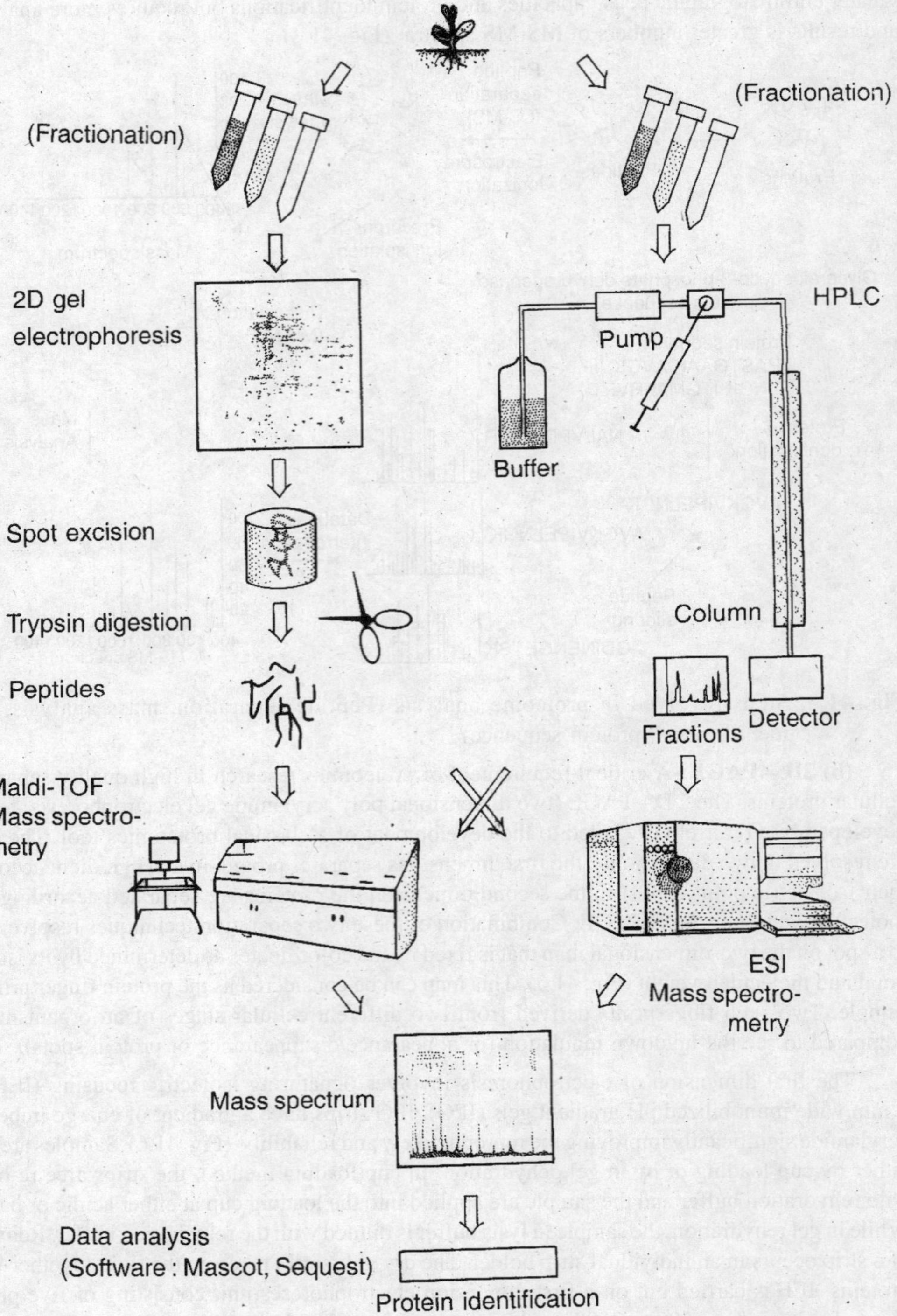

Fig. 41.4. Schematic presentation of proteome analysis

Visualization of proteins following 2D-gel electrophoresis is usually done by staining with either Commassie brilliant blue (detection > 100ng protein) or silver staining (detect 2-5 ng protein) Fig. 41.7. However, silver staining interferes with mass spectrophotometry.

(c) Labelling approaches -This technology is used to compare two samples rather than identification of a particular protein in a sample. Because mass spectrophotometers make measurements of mass to charge this adds additional complexity to proteomics experiments. To differentiate between identified peptides from two samples simultaneously, one sample is labelled with heavy isotopes (^2H, ^{13}C, ^{15}N, or ^{18}O) to produce a mass shift. A mass shift of at least 3 Daltons is desired to prevent isotope overlaps between the two samples, which would affect the quantitation. Samples can be grown on isotope-labelled medium so that the resulting isotopes are incorporated into every amino acid in a protein. Bacterial culture can be grown on ^{14}N-^{15}N media whereas mammalian cells can be cultured using stable isotope – labelled amino acids.

Protein and peptides can be labelled before or after digestion. Protein can be labelled with isotope coded tages (ICAT). ICAT reagents only label cysteine residues but cysteine containing peptides are enriched from the sample using the affinity tag. Proteins can also be digested in ^{18}O-labelled water, which results in incorporation of two labelled oxygen atom. After protein digestion, specific amino acids can be modified using deuterium, ^{13}C – labelled reagents or other reagents. It is preferable to use ^{13}C, ^{15}H or ^{18}O because no chromatographic shifting occurs.

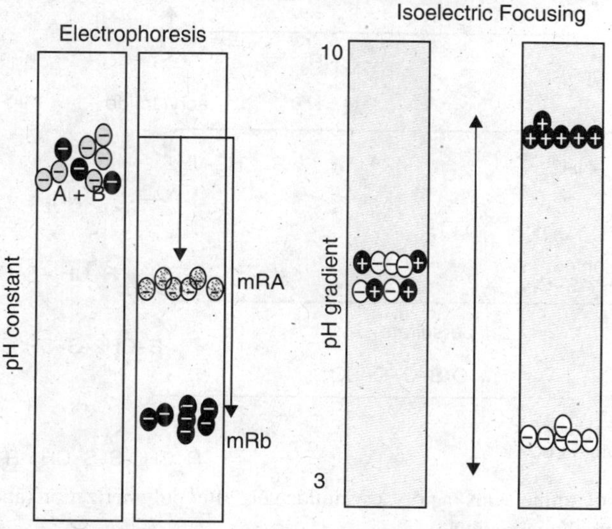

Fig. 41.5. Schematic presentation of principle for electrophoresis and isoelectric focusing

(d) Data pre-processing - Proteomics studies produce large amounts of MS-MS spectra per experiment (up to 8000 per hour). Although each MS-MS spectrum correspondence to only one peptide ion charge state, redundant database searches of multiple charge states are typically conducted because many mass spectrometers lack sufficient isotopic resolution for the correct charge state determination of multiply charged ions. This problem coupled with the ever-increasing size of protein sequences database, creates a substantial data processing difficulty. To overcome this difficulty, several approaches that differentiate between singly versus multiply charged spectra, and doubly versus triply charged spectra have been applied to reduce the number of MS-MS spectra to be searched. The other approach is to avoid comparing poor quality spectra. The new software like 'Charge Czar' and 'CHASTE' can reduce the burden of searching redundant data by ~ 90% without any information loss.

LC-MS

Mass spectrophotometry is used in the analysis of small molecules, peptides and proteins. Mass spectrometers measures ions and make measurements of mass-to-charge. The two main ionisation methods are – matrix assisted laser desorption ionization (MALDI) and electrospray ionisation (ESI). MALDI typically produces singly charged ion whereas ESI produces singly and multiply charged ions whereas ESI produces singly and multiply charged ions depending on the number of available basic sites in the peptide. There are many different types of mass spectrophotometers, each capable of obtaining different levels of isotopic resolution. Most instruments can perform tandem mass spectrometry (MS-MS) to obtain structural information by the fragmentation of peptide ions. Ion-trap instrument can perform multiple stages of MS-MS (MS^n) analysis to provide additional structural information. The most popular high-throughput (HTP) proteomics experiments aimed at protein identification and expression involve online nano-flow liquid chromatography (LC) mass spectrometry (LC-MS) with tandem mass spectrometer (LC-MS-MS). LC-MS-MS experiments use ESI because it is easily coupled to the chromatographic separation. MALDI can also be used and has advantage over ESI (e.g. the samples can be reanalysed from the target plates) but MS-MS spectra of MALDI – generated ions are usually less informative that ESI-generated ions because multiply changed ions typically provide more fragmentation information than singly charged ions.

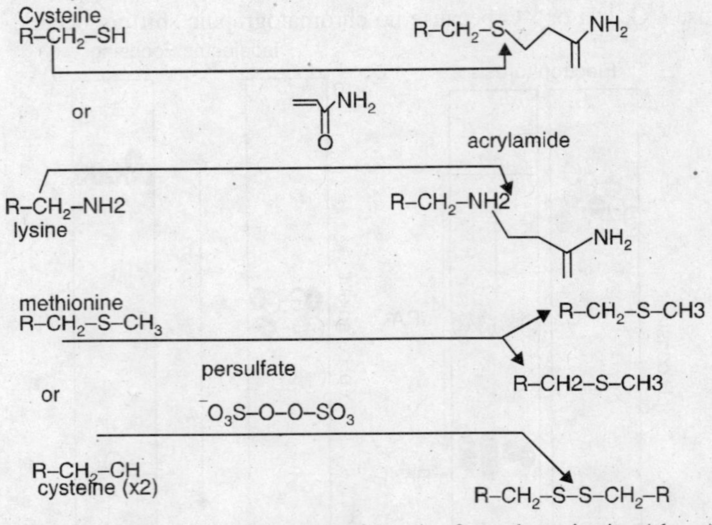

Fig. 41.6. Modifications of amino acids in polyacrylamide gel : after polymerization (above) or by persulphate treatment after polymerization

A typical proteomics experiment involves measuring intact peptide masses following chromatographic separation and ionisation (LC-MS). Individual peptide ions are isolated from the mixture and subjected to dissociation (MS-MS) followed by mass analysis of the resulting fragment ions. Both the intact peptide mass and peptide fragment ions masses are used for peptide identification through database searching because a peptide mass measurement alone is not sufficient for identification when dealing with complex mixtures. LC-MS-MS methodologies are robust but reproducible experiments depend greatly on the variability of chromatographic separation.

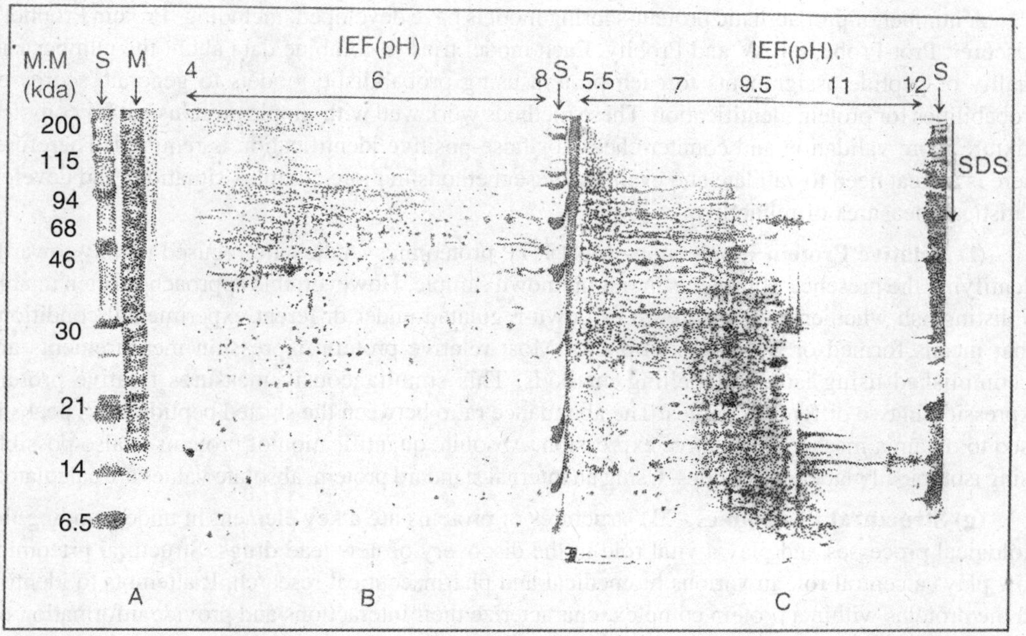

Fig. 41.7. Comparison between SDS-PAGE (A, line M) and two dimensional (B, C) electrophoresis. Line 's' shows standard proteins (kda). For second dimension of electrophoresis, two different pH gradients were used resulting in differential separation.

(e) Peptide identification – The desired and point of any proteomics expression would be to identify and characterize the proteins. One of the most common methods of identifying proteins is through peptide-mass finger printing.

The proteins are digested with a proteolytic enzyme such as trypsin to produce a set of tryptic fragments unique to each protein (Fig. 41.8). The masses of these peptides are then determined by mass spectrometry. It has been possible to determine the molecular structure of proteins that show up as spot on 2D-gels using mass spectrometry. This novel technology determines the mass to charge ratio (m/z) of individual molecules in the gas phase by observing their flight in electric and/ or magnetic fields. Once the ions are formed, they can be separated according to their m/z and detected. The most common system used is matrix-assisted laser desorption/ ionisation time-of-flight mass spectrometer (MALDI-TOF). MALDI allows the analysis of high molecular weight compounds with high sensitivity and, soft ionisation with little or no fragmentation is observed. It is also tolerant to salts in millimolar concentration. The non-volatile matrix plays an important role by absorbing the laser radiation resulting in vaporization of matrix and sample embedded in matrix. Once in gas phase, the analyte molecules are ionised, these are then directed towards the mass analysis, calculating the time of flight. The molecular weight values of the trypsinized peptides obtained by MALDI-TOF are then used to identify the predicted proteins using web-based search engines such as MASCOT, SEQUEST and PROFOUND. The other software available are X' Tandem and OMSSA.

A main goal of HTP proteomics is to identify proteins from complex mixtures. However, the main thrust of research has been towards accurate peptide identification, which does not necessarily translate into accurate identification primarily because of two facts. First, peptides do not necessarily uniquely correspond to a single protein. Second, even small error rates in peptide identification can result in a large number of protein identification errors because of the enormous amount of identified peptides.

A number of probabilistic protein-scoring models have developed, including 'Protein Prophet', 'Qscore', Prot-Probe, OLAV and Probity. Each model tries to combine data about the number and quality of peptide assignments to each protein using probabilistic models to generate scores or probabilities for protein identification. These methods work well with simple proteins but for complex mixture more validation and counter-check for false-positive identification is required. Therefore, there is a great need to validate and evaluate these methods in more complex situations and develop statistical measures of reliability.

(f) Relative Protein Expression- Most HTP proteomics studies are focused mainly towards identifying the presence of proteins in our unknown sample. However, this approach is often unable to distinguish whether proteins are up or down-regulated under different experimental conditions (that means formed or formation stopped). Most relative protein expression measurements are accomplished using isotope-labelling methods. This simultaneously measures relative protein expression in two different samples. The abundance ratio between the shifted peptide mass peaks is used to obtain a measure of relative expression. Absolute quantification of proteins is also possible using isotopically labelled peptides. Using an internal standard protein, absolute values are calculated.

(g) Structural Proteomics - 3D structures of proteins are a key element in understanding the biological processes and play a vital role in the discovery of new lead drugs. Structural prteomics now plays a central role in various biomedical and pharmaceutical research. It attempts to identify all the proteins within a protein complex, characterize their interactions and provide information on protein signalling, disease mechanisms or protein-drug interaction. Since the 3D structure of a protein is more conserved than sequence, these initiatives also open up the possibility of biochemical or biophysical functional characterization via structure. Understanding protein function at molecular level and its mechanism at genetic level provides information about protein – DNA interaction. Unique contributions have been made on the mechanisms for control of genetic information, packaging, repair and other DNA-protein interactions both by x-ray crystallography and nuclear magnetic resonance (NMR) spectroscopy. X-ray crystallography has been used effectively in generating the first structure-based drugs, derived from structure-function studies of neuraminidase inhibitors such as HIV-protease inhibitors amprenavir. Although x-ray crystallography has the advantage of defining ligand-binding sites with more certainty, the ability of NMR to measure proteins in their native state is an important distinction. NMR is also recognized as an important tool in upstream part of drug discovery process. NMR can also be used as a tool for determination of structure and dynamics of protein and protein-ligand complex. It can identify and characterize interactions between protein-protein and/or protein-DNA. The recent sequencing of human genome suggests that there are about 600-3000 genes that can code for potential drug targets for human disease. With human proteome being much larger and diverse than genome, there can be many more potential drug targets. Determination of 3D structure provides an insight to its binding site and biological function. This technology was first used in characterizing the binding of penicillin to serum albumin in a NMR relaxation study.

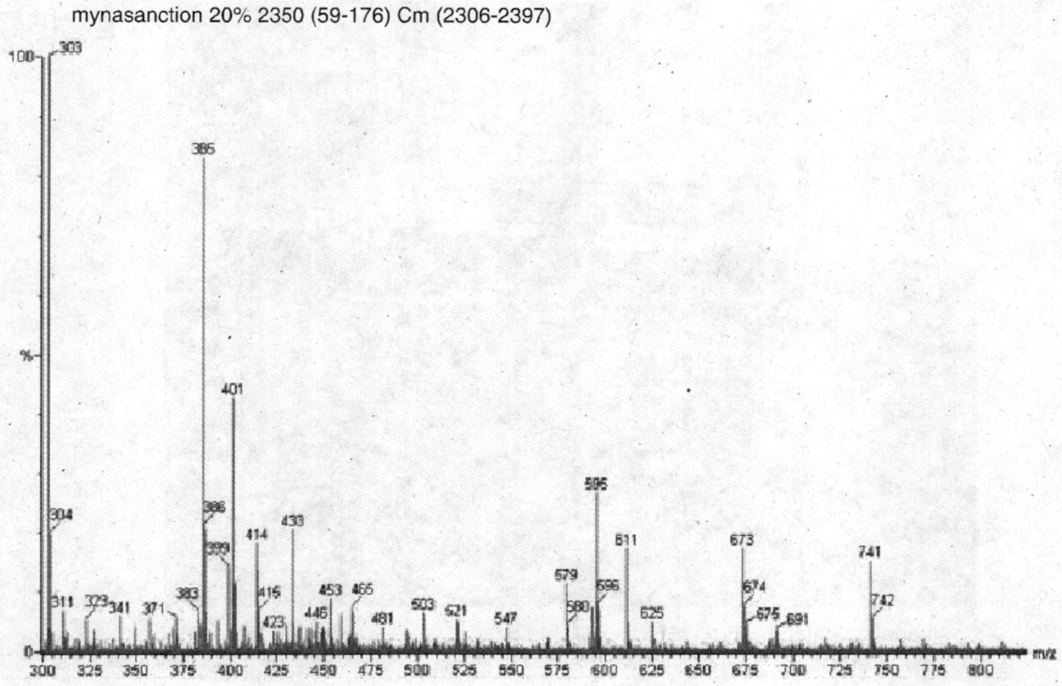

Fig. 41.8. Mass spectrum of a protein

IV. MALDI

The term matrix-assisted laser desorption ionization (MALDI) was coined in 1985 by Franz Hillenkamp, Michael Karas and their colleagues.

(a) Matrix-assisted laser desorption/ionization (MALDI) is a soft ionization technique used in mass spectrometry, allowing the analysis of biomolecules (biopolymers such as proteins, peptides and sugars) and large organic molecules (such as polymers, dendrimers and other macromolecules), which tend to be fragile and fragment when ionized by more conventional ionization methods. It is most similar in character to electrospray ionization both in relative softness and the ions produced (although it causes much fewer multiply charged ions).The ionization is triggered by a laser beam (normally a nitrogen laser). A matrix is used to protect the biomolecule from being destroyed by direct laser beam and to facilitate vaporization and ionization. (Fig. 41.9 a, b).

(*b*) **Matrix**

The matrix consists of crystallized molecules, of which the three most commonly used are *3,5-dimethoxy-4-hydroxycinnamic acid* (sinapinic acid), á-cyano-4-hydroxycinnamic acid (alpha-cyano or alpha-matrix) and 2,5-dihydroxybenzoic acid (DHB). A solution of one of these molecules is made, often in a mixture of highly purified water and an organic solvent (normally acetonitrile (ACN) or ethanol). Trifluoroacetic acid (TFA) may also be added. A good example of a matrix-solution would be 20 mg/mL sinapinic acid in ACN:water:TFA (50:50:0.1).

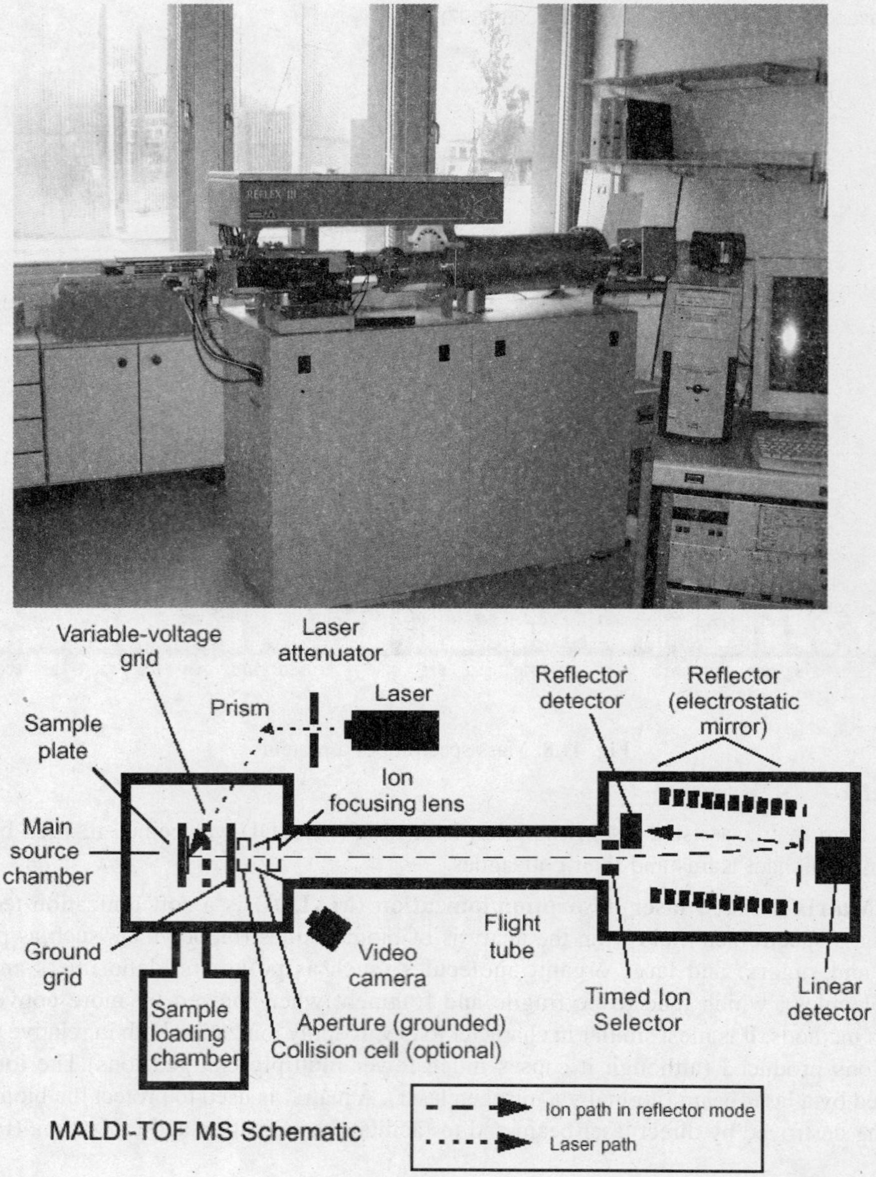

Fig. 41.9. (*a*) MALDI-TOF Instrument (*b*) Schematic presentation of MALDI-TOF MS principle and working.

The matrix solution is mixed with the analyte (e.g. protein-sample). The organic solvent allows hydrophobic molecules to dissolve into the solution, while the water allows for water-soluble (hydrophilic) molecules to do the same. This solution is spotted onto a MALDI plate (usually a metal plate designed for this purpose). The solvents vaporize, leaving only the recrystallized matrix, but now with analyte molecules spread throughout the crystals. The matrix and the analyte are said to be co-crystallized in a MALDI spot.

(*c*) **Laser**

The laser is fired at the crystals in the MALDI spot. The matrix absorbs the laser energy and it is thought that primarily the matrix is ionized by this event. The matrix is then thought to transfer part of its charge to the analyte molecules (e.g. protein), thus ionizing them while still protecting them from the disruptive energy of the laser. Ions observed after this process consist of a neutral molecule [M] and an added or removed ion. Together, they form a quasimolecular ion, for example $[M+H]^+$ in the case of an added proton, $[M+Na]^+$ in the case of an added sodium ion, or $[M-H]^-$ in the case of a removed proton. MALDI generally produces singly-charged ions, but multiply charged ions ($[M+nH]^{n+}$) can also be observed, usually as a function of the matrix, the laser intensity and/or the voltage used. Note that these are all even-electron species. Ion signals of radical cations can be observed eg. in case of matrix molecules and other stable molecules

V. AP-MALDI

Atmospheric pressure (AP) matrix-assisted laser desorption/ionization (MALDI) is an ionization technique (ion source) that in contrast to vacuum MALDI operates at normal atmospheric environment. The main difference between vacuum MALDI and AP-MALDI is the pressure in which the ions are created. In vacuum MALDI, ions are typically produced at 10 mTorr or less while in AP-MALDI ions are formed in atmospheric pressure. Disadvantage of the AP MALDI source is the limited sensitivity observed and the limited mass range.

AP-MALDI is used in mass spectrometry (MS) in a variety of applications ranging from proteomics to drug discovery fields. Popular topics that are addressed by AP-MALDI mass spectrometry include: proteomics, DNA/RNA/PNA, lipids, oligosaccharides, phosphopeptides, bacteria, small molecules and synthetic polymers, similar applications as available also for vacuum MALDI instruments.

The AP-MALDI ion source is easily coupled to an ion trap mass spectrometer[12] or any other MS system equipped with ESI (electrospray ionization) or nanoESI source.

(*a*) Mass spectrometer

The type of a mass spectrometer most widely used with MALDI is the *TOF* (time-of-flight mass spectrometer), mainly due to its large mass range. The TOF measurement procedure is also ideally suited to the MALDI ionization process since the pulsed laser takes individual 'shots' rather than working in continuous operation. MALDI-TOF instruments are typically equipped with an "ion mirror", deflecting ions with an electric field, thereby doubling the ion flight path and increasing the resolution. Commercial reflectron TOF instruments reach today a resolving power m/Äm of well above 20'000 FWHM (full-width half-maximum, Äm defined as the peak width at 50% of peak height).

(*b*) Use

In Biochemistry

In proteomics, MALDI is used for the identification of proteins isolated through gel electrophoresis: SDS-PAGE and two-dimensional gel electrophoresis. One method used is peptide mass fingerprinting by MALDI-MS, or with post ionisation decay or collision-induced dissociation (further use see mass spectrometry).

In Organic Chemistry

Some synthetic macromolecules, such as catenanes and rotaxanes, dendrimers and hyperbranched polymers, and other assemblies, have molecular weights extending into the thousands or tens of thousands, where most ionization techniques have difficulty producing molecular ions. MALDI is a simple and rapid analytical method that can allow chemists to analyze the results of such syntheses and verify their results.

Reproducibility and performance

The sample preparation for MALDI is important for the result. Inorganic salts which are also part of protein extracts interfere with the ionization process. The salts are removed by solid phase extraction or washing the final target spots with water. Both methods can also remove other substances from the sample. The matrix protein mixture is not homogenous because the polarity difference leads to a separation of the two substances during crystallization. The spot diameter of the target is much larger than that of the laser, which makes it necessary to do several laser shots at different places of the target, to get the statistical average of the substance concentration within the target spot. The matrix composition, the addition of trifluoroacetic acid and formic acid, delay between laser pulses, delay time of the acceleration power, laser wavelength, energy density of the laser and the impact angle of the laser on the target are among others the critical values for the quality and reproducibility of the method.

QUESTIONS

1. Define proteomics ? Give various steps involved in proteome analysis.
2. Discuss the role of trypsin, SDS, pH gradient, isoelectricfocusing and MS in proteomics.
3. How proteomics helps in genomics ?
4. Define and compare protein profiling with proteomics analysis.
5. Write short notes :
   (*a*) MALDI
   (*b*) 2D-PAGE
   (*c*) Protein microarray
   (*d*) Differential fluorescence gel electrophoresis
   (*e*) ICAT
   (*f*) Protein Profiling.

CHAPTER 42

MASS SPECTROMETRY

Mass spectrometry is an analytical tool used for measuring the **molecular mass** of a sample.

For large samples such as **biomolecules,** molecular masses can be measured to within an accuracy of **0.01%** of the total molecular mass of the sample *i.e.* within a 4 Daltons (Da) or atomic mass units (amu) error for a sample of 40,000 Da. This is sufficient to allow minor mass changes to be detected, *e.g.* the substitution of one amino acid for another, or a post-translational modification.

For small **organic molecules** the molecular mass can be measured within an accuracy of **5 ppm** or less, which is often sufficient to confirm the molecular formula of a compound. Molecular mass of water is presented in Fig. 42.1 for easy understand of data generated by mass spectrometer.

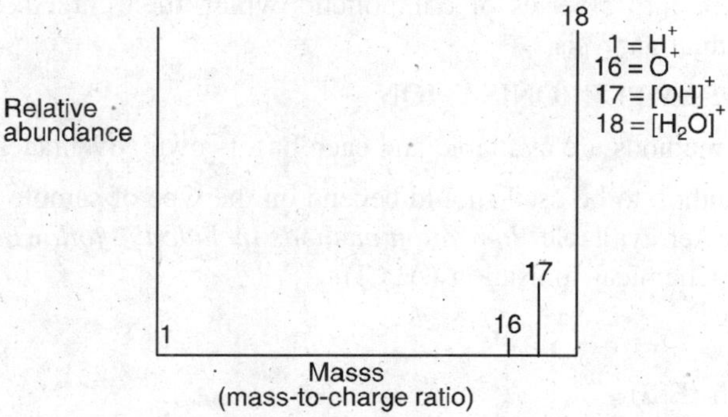

Fig. 42.1. Mass spectrum of water

Structural information can be generated using certain types of mass spectrometers, usually those with multiple analysers which are known as **tandem mass spectrometers.** This is achieved by fragmenting the sample inside the instrument and analysing the products generated. This procedure is useful for the structural elucidation of **organic compounds** and for **peptide** or **oligonucleotide** sequencing.

I. APPLICATIONS

Mass spectrometers are used in industry and academia for both routine and research purposes. The following list is just a brief summary of the major mass spectrometric applications:

- **Biotechnology:** *the analysis of proteins, peptides, oligonucleotides*
- **Pharmaceutical:** *drug discovery, combinatorial chemistry, pharmacokinetics, drug metabolism*
- **Clinical:** *neonatal screening, haemoglobin analysis, drug testing*
- **Environmental:** *PAHs, PCBs, water quality, food contamination*
- **Geological:** *oil composition*

II. WORKING

Mass spectrometers can be divided into three fundamental parts, namely the ionisation source, the analyser , and the detector.

The sample has to be introduced into the ionisation source of the instrument. Once inside the ionisation source, the sample molecules are ionised, because ions are easier to manipulate than neutral molecules. These ions are extracted into the analyser region of the mass spectrometer where they are separated according to their **mass (m) -to-charge (z) ratios (m/z)**. The separated ions are detected and this signal sent to a data system where the m/z ratios are stored together with their relative abundance for presentation in the format of a **m/z spectrum**.

The analyser and detector of the mass spectrometer, and often the ionisation source too, are maintained under high vacuum to give the ions a reasonable chance of travelling from one end of the instrument to the other without any hindrance from air molecules. The entire operation of the mass spectrometer, and often the sample introduction process also, is under complete **data system** control on modern mass spectrometers.

Sample introduction – The method of sample introduction to the ionisation source often depends on the ionisation method being used, as well as the type and complexity of the sample.

The sample can be inserted directly into the ionisation source, or can undergo some type of chromatography *en route* to the ionisation source. This latter method of sample introduction usually involves the mass spectrometer being coupled directly to a high pressure liquid chromatography (HPLC), gas chromatography (GC) or capillary electrophoresis (CE) separation column, and hence the sample is separated into a series of components which then enter the mass spectrometer sequentially for individual analysis.

III. METHODS OF SAMPLE IONISATION

Many ionisation methods are available and each has its own advantages and disadvantages.

The ionisation method to be used should depend on the type of sample under investigation and the mass spectrometer available. *Ionisation methods include the following:*

Atmospheric Pressure Chemical Ionisation (APCI)
Chemical Ionisation (CI)
Electron Impact (EI)
Electrospray Ionisation (ESI)
Fast Atom Bombardment (FAB)
Field Desorption / Field Ionisation (FD/FI)
Matrix Assisted Laser Desorption Ionisation (MALDI)
Thermospray Ionisation (TSP)

The ionisation methods used for the majority of biochemical analyses are **Electrospray Ionisation (ESI) and Matrix Assisted Laser Desorption Ionisation (MALDI).** With most ionisation methods there is the possibility of creating both positively and negatively charged sample ions, depending on the proton affinity of the sample.

Analysis and Separation of Sample Ions – The main function of the **mass analyser** is to **separate,** or **resolve,** the ions formed in the ionisation source of the mass spectrometer according to their **mass-to-charge (m/z)** ratios. There are a number of mass analysers currently available, the better known of which include **quadrupoles, time-of-flight (TOF)** analysers, **magnetic sectors** , and both **Fourier transform** and **quadrupole ion traps .**

Tandem (MS-MS) mass spectrometers are instruments that have more than one analyser and so can be used for structural and sequencing studies. Two, three and four analysers have all been incorporated into commercially available tandem instruments. More popular tandem mass spectrometers include those of the **quadrupole-quadrupole, magnetic sector-quadrupole** , and more recently, the **quadrupole-time-of-flight** geometries.

Detection and recording of sample ions – The **detector** monitors the ion current, amplifies it and the signal is then transmitted to the data system where it is recorded in the form of **mass spectra**. The **m/z** values of the ions are plotted against their **intensities** to show the **number of components** in the sample, the **molecular mass** of each component, and the **relative abundance** of the various components in the sample. The common detectors are the **photomultiplier**, the **electron multiplier** and the **micro-channel plate** detectors.

IV. ELECTROSPRAY IONISATION

Electrospray ionization- **Electrospray Ionisation (ESI)** is one of the **Atmospheric Pressure Ionisation (API)** techniques and is well-suited to the analysis of polar molecules ranging from less than 100 Da to more than 1,000,000 Da in molecular mass. (Fig. 42.2).

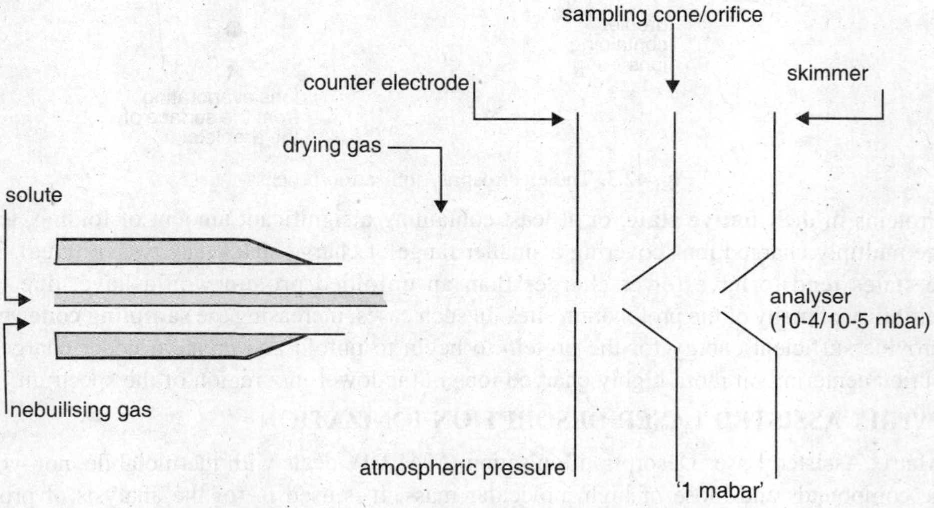

Fig. 42.2. Standard electrospray ionisation source (Platform II)

During standard electrospray ionisation the sample is dissolved in a polar, volatile solvent and pumped through a narrow, stainless steel capillary (75 - 150 micrometers i.d.) at a flow rate of between 1 μL/min and 1 mL/min. A high voltage of 3 or 4 kV is applied to the tip of the capillary, which is situated within the ionisation source of the mass spectrometer, and as a consequence of this strong electric field, the sample emerging from the tip is dispersed into an aerosol of highly charged droplets, a process that is aided by a co-axially introduced nebulising gas flowing around the outside of the capillary Fig. 42.3. This gas, usually nitrogen, helps to direct the spray emerging from the capillary tip towards the mass spectrometer. The charged droplets diminish in size by solvent evaporation, assisted by a warm flow of nitrogen known as the **drying gas** which passes across the front of the ionisation source. Eventually charged **sample ions**, free from solvent, are released from the droplets, some of which pass through **a sampling cone** or orifice into an **intermediate vacuum** region, and from there through a small aperture into the analyser of the mass spectrometer, which is held under **high vacuum**. The lens voltages are optimised individually for each sample.

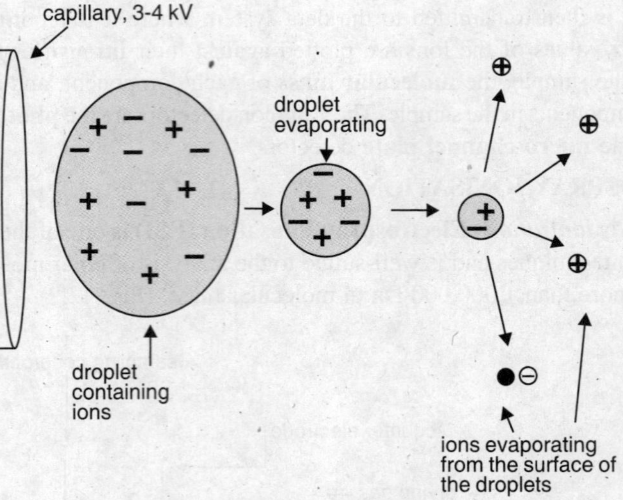

Fig. 42.3. The electrospray ionisation process

Proteins in their **native state**, or at least containing a significant amount of folding, tend to produce multiply charged ions covering a smaller range of charge states (say two or three). These charge states tend to have fewer charges than an unfolded protein would have, due to the inaccessibility of many of the protonation sites. In such cases, increasing the **sampling cone voltage** may provide sufficient energy for the protein to begin to unfold and create a wider charge state distribution centering on more highly charged ions in the lower m/z region of the spectrum.

V. MATRIX ASSISTED LASER DESORPTION IONIZATION

Matrix Assisted Laser Desorption Ionisation (MALDI) deals with thermolabile, non-volatile organic compounds and those of high molecular mass. It is used in for the analysis of proteins, peptides, glycoproteins, oligosaccharides, and oligonucleotides.

MALDI is based on the usage of matrix complexed with a given sample molecule that is bombarded with a laser in order for the sample molecule to form a sample ionization Fig. 42.4. Most commercially available MALDI mass spectrometers are now a pulsed nitrogen laser of wavelength 337 nm. This method is normally coupled with a Time of Flight detector in order to obtain proper charge-mass ratios and calculate a mass spectrum.

MALDI is used successfully in biochemical areas for the analysis of **proteins, peptides, glycoproteins, oligosaccharides, and oligonucleotides**. The mass accuracy depends on the type and performance of the analyser of the mass spectrometer and capable of measuring masses within 0.01% of the molecular mass of the sample, at least up to ca. 40,000 Da.

MALDI is based on the **bombardment** of sample molecules with a **laser** light to bring about **sample ionisation.** The sample is pre-mixed with a highly absorbing **matrix** compound and a low concentration of sample to matrix works best. The matrix transforms the laser energy into **excitation energy** for the sample, which leads to sputtering of analyte and matrix ions from the surface of the mixture. In this way energy transfer is efficient and also the analyte molecules are spared excessive direct energy that may otherwise cause decomposition.

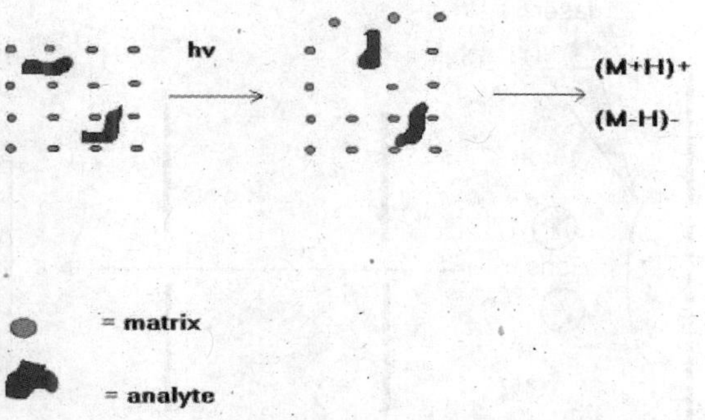

Fig. 42.4. Matrix assisted laser desorption ionisation (MALDI)

The sample to be analysed is dissolved in an appropriate volatile solvent, usually with a trace of trifluoroacetic acid if positive ionisation is being used, at a concentration of ca. **10 pmol/μL** and an aliquot (**1-2 μL**) of this removed and mixed with an equal volume of a solution containing a vast excess of a matrix. A range of compounds is suitable for use as matrices: **sinapinic acid** is a common one for **protein** analysis while **alpha-cyano-4-hydroxycinnamic acid** is often used for **peptide** analysis. An aliquot (1-2 μL) of the final solution is applied to the sample target which is allowed to dry prior to insertion into the high vacuum of the mass spectrometer. The laser is fired, the energy arriving at the sample/matrix surface optimised, and data accumulated until a m/z spectrum of reasonable intensity has been amassed. The time-of-flight analyser separates ions according to their **mass(m)-to-charge(z) (m/z)** ratios by measuring the time it takes for ions to travel through a field free region known as the flight, or drift, tube. The heavier ions are slower than the lighter ones Fig. 42.5 a, b).

The m/z scale of the mass spectrometer is **calibrated** with a known sample that can either be analysed independently (external calibration) or pre-mixed with the sample and matrix (internal calibration).

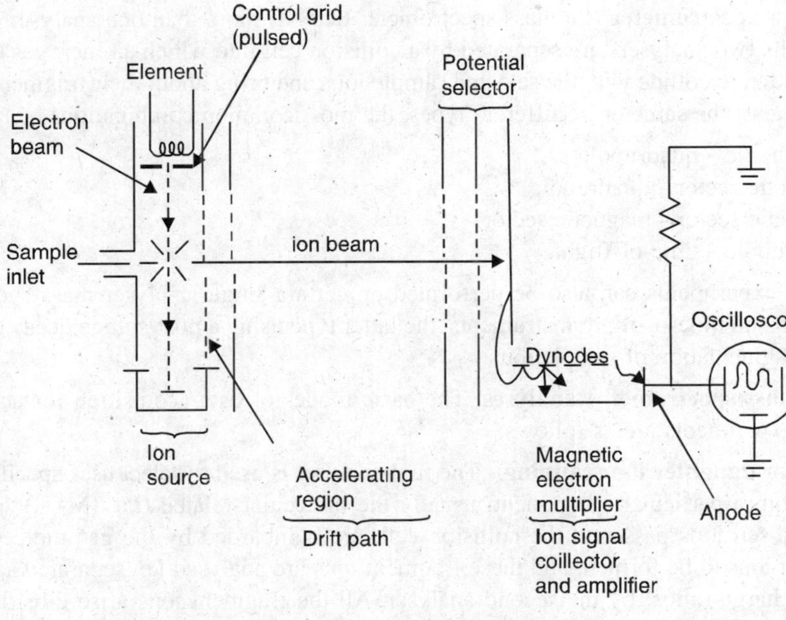

Fig. 42.5(a). Schematic diagram of a time-of-flight mass spectrometer

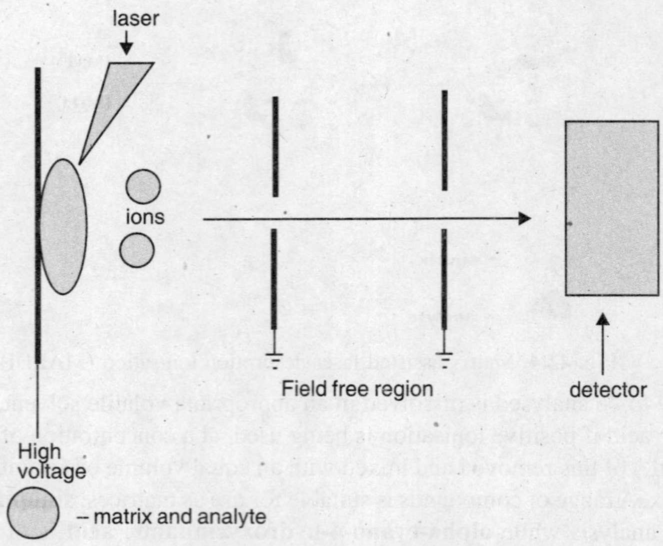

Fig. 42.5b. Simplified Schematic of MALDI-TOF mass spectrometry (linear mode).

Positive ionisation is used in general for **protein** and **peptide** analyses. Negative ionisation can be used for the analysis of **oligonucleotides** and **oligosaccharides.**

VI. TANDEM MASS SPECTROMETRY (MS-MS): STRUCTURAL AND SEQUENCE INFORMATION FROM MASS SPECTROMETRY.

Tandem mass spectrometry – Tandem mass spectrometry (MS-MS) is used to produce structural information about a compound by fragmenting specific sample ions inside the mass spectrometer and identifying the resulting fragment ions. This information can then be pieced together to generate structural information regarding the intact molecule. Tandem mass spectrometry also enables specific compounds to be detected in complex mixtures on account of their specific and characteristic fragmentation patterns.

A tandem mass spectrometer is a mass spectrometer that has more than one analyser, in practice usually two. The two analysers are separated by a collision cell into which an inert gas (e.g. argon, xenon) is admitted to collide with the selected sample ions and bring about their fragmentation. The analysers can be of the same or of different types, the most common combinations being:

- quadrupole - quadrupole
- magnetic sector - quadrupole
- magnetic sector - magnetic sector
- quadrupole - time-of-flight.

Fragmentation experiments can also be performed on certain single analyser mass spectrometers such as ion trap and time-of-flight instruments, the latter type using a post-source decay experiment to effect the fragmentation of sample ions.

(a) Tandem mass spectrometry analyses. The basic modes of data acquisition for tandem mass spectrometry experiments are as follows:

Product or daughter ion scanning:- The first analyser is used to select user-specified sample ions arising from a particular component; usually the molecular-related (i.e. $(M+H)^+$ or $(M-H)^-$) ions. These chosen ions pass into the collision cell, are bombarded by the gas molecules which cause fragment ions to be formed, and these fragment ions are analysed i.e. separated according to their mass to charge ratios, by the second analyser. All the fragment ions arise directly from the precursor ions specified in the experiment, and thus produce a fingerprint pattern specific to the compound under investigation.

(b) Peptide Sequencing by Tandem Mass Spectrometry

The most common usage of MS-MS in biochemical areas is the product or daughter ion scanning experiment which is particularly successful for peptide and nucleotide sequencing.

Peptide sequencing: H_2N-CH(R')-CO-NH-CH(R")-CO_2H

Peptides fragment in a reasonably well-documented manner. The protonated molecules fragment along the peptide backbone and also show some side-chain fragmentation with certain instruments.

There are three different types of bonds that can fragment along the amino acid backbone: the NH-CH, CH-CO, and CO-NH bonds. Each bond breakage gives rise to two species, one neutral and the other one charged, and only the charged species is monitored by the mass spectrometer (Fig. 42.6). The charge can stay on either of the two fragments depending on the chemistry and relative proton affinity of the two species. Hence there are six possible fragment ions for each amino acid residue and these are labelled as in the diagram, with the a, b, and c" ions having the charge retained on the N-terminal fragment, and the x, y", and z ions having the charge retained on the C-terminal fragment. The most common cleavage sites are at the CO-NH bonds which give rise to the b and/or the y" ions. The mass difference between two adjacent b ions, or y"; ions, is indicative of a particular amino acid residue (see Table of amino acid residues at the end).

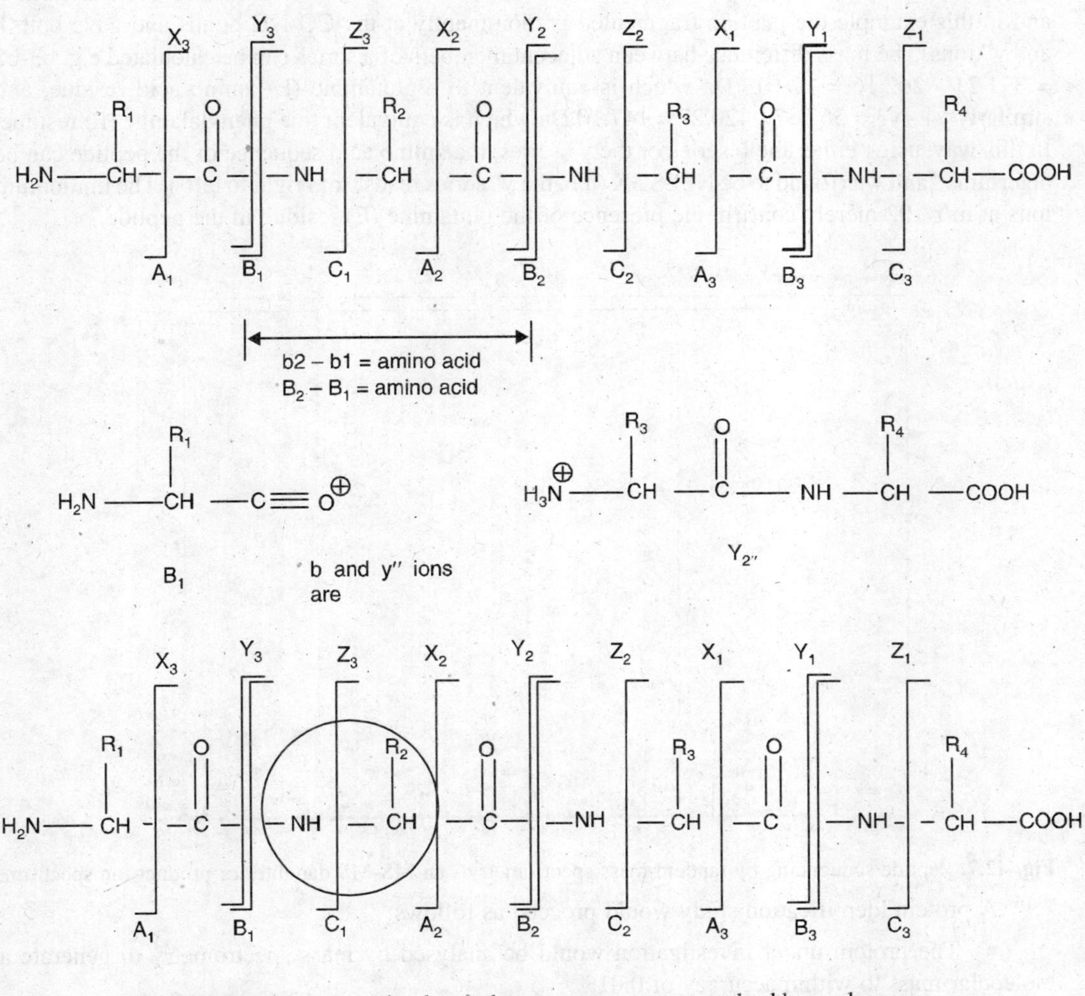

Fig. 42.6. Peptide sequencing by tandem mass spectrometry - backbone cleavages

The extent of side-chain fragmentation detected depends on the type of analysers used in the mass spectrometer. A magnetic sector - magnetic sector instrument will give rise to high energy collisions resulting in many different types of side-chain cleavages. Quadrupole - quadrupole and quadrupole - time-of-flight mass spectrometers generate low energy fragmentations with fewer types of side-chain fragmentations.

Immonium ions (labelled "i") appear in the very low m/z range of the MS-MS spectrum. Each amino acid residue leads to a diagnostic immonium ion, with the exception of the two pairs leucine (L) and iso-leucine (I), and lysine (K) and glutamine (Q), which produce immonium ions with the same m/z ratio, i.e. m/z 86 for I and L, m/z 101 for K and Q. The immonium ions are useful for detecting and confirming many of the amino acid residues in a peptide, although no information regarding the position of these amino acid residues in the peptide sequence can be ascertained from the immonium ions.

An example of an **MS/MS daughter or product ion spectrum** is illustrated below (Fig. 42.7). The molecular mass of the peptide was measured using standard mass spectrometric techniques and found to be 680.4 Da, the dominant ions in the MS spectrum being the protonated molecular ions ($M+H^+$) at m/z 681.4. These ions were selected for transmission through the first analyser, then fragmented in the collision cell and their fragments analysed by the second analyser to produce the following MS/MS spectrum. The sequence (amino acid backbone) ions have been identified, and in this example the peptide fragmented predominantly at the CO-NH bonds and gave both b and y" ions. The mass difference between adjacent members of a series can be calculated e.g. b3-b2 = 391.21 - 262.16 = 129.05 Da which is equivalent to a glutamine (E) amino acid residue; and similarly y4 - y3 = 567.37 - 420.27 = 147.10 Da which is equivalent to a phenylalanine (F) residue. In this way, using either the b series or the y" series, the amino acid sequence of the peptide can be determined and was found to be NFESGK (n.b. the y" series reads from right to left!). The immonium ions at m/z 102 merely confirm the presence of the glutamine (E) residue in the peptide.

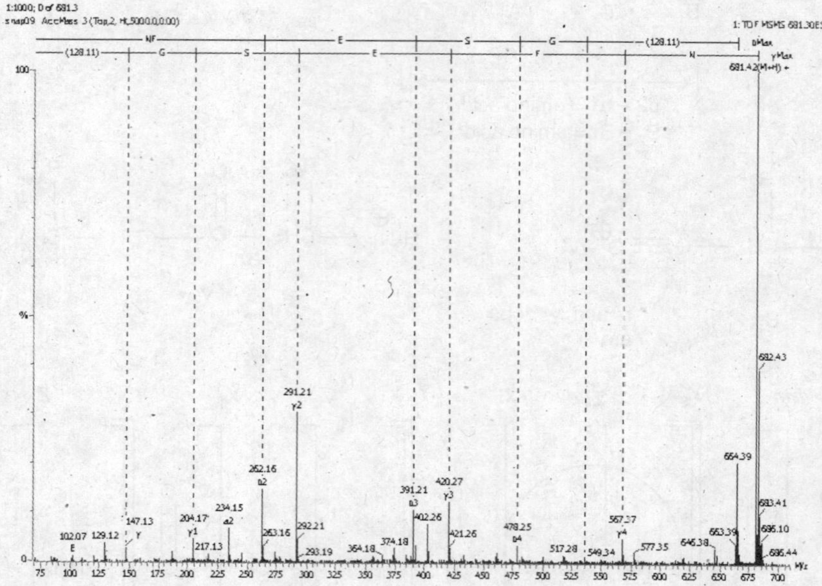

Fig. 42.7. Peptide sequencing by tandem mass spectrometry - an MS-MS daughter or product ion spectrum.

A protein identification study would proceed as follows:

- The protein under investigation would be analysed by mass spectrometry to generate a molecular mass to within accuracy of 0.01%.

- The protein would then be digested with a suitable enzyme. Trypsin is useful for mass spectrometric studies because each proteolytic fragment contains a basic arginine (R) or lysine (K) amino acid residue, and thus is eminently suitable for positive ionisation mass spectrometric analysis. The digest mixture is analysed - without prior separation or clean-up - by mass spectrometry to produce a rather complex spectrum from which the molecular weights of all of the proteolytic fragments can be read. This spectrum, with its molecular weight information, is called a peptide map. (If the protein already exists on a database, then the peptide map is often sufficient to confirm the protein.)

For these experiments the mass spectrometer would be operated in the "MS" mode, whereby the sample is sprayed and ionised from the nanospray needle and the ions pass through the sampling cone, skimmer lenses, Rf hexapole focusing system, and the first (quadrupole) analyser. The quadrupole in this instance is not used as an analyser, merely as a lens to focus the ion beam into the second (time-of-flight) analyser which separates the ions according to their mass-to-charge ratio.

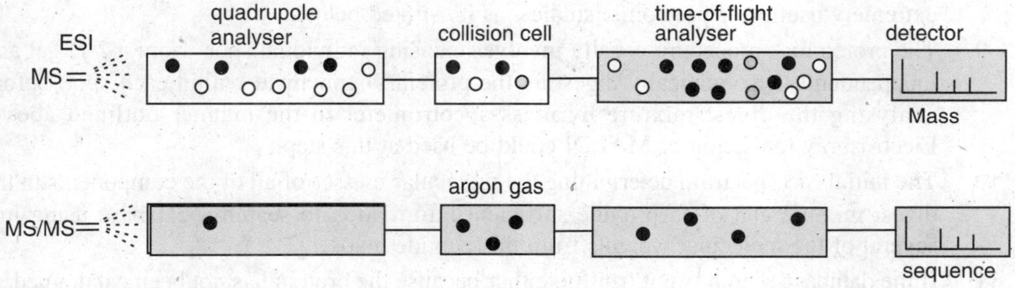

Fig. 42.8. Q-TOF mass spectrometer operating in MS (upper) and MS/MS mode (lower) modes.

- With the digest mixture still spraying into the mass spectrometer, the Q-Tof mass spectrometer is switched into **"MS/MS" mode**. The protonated molecular ions of each of the digest fragments can be independently selected and transmitted through the quadrupole analyser, which is now used as an analyser to transmit solely the ions of interest into the collision cell which lies in between the first and second analysers. An inert gas such as argon is introduced into the collision cell and the sample ions are bombarded by the collision gas molecules which cause them to fragment. The optimum collision cell conditions vary from peptide to peptide and must be optimised for each one. The fragment (or daughter or product) ions are then analysed by the second (time-of-flight) analyser. In this way an MS/MS spectrum is produced showing all the fragment ions that arise directly from the chosen parent or precursor ions for a given peptide component (Fig. 42.8).

An MS/MS daughter (or fragment, or product) ion spectrum is produced for each of the components identified in the proteolytic digest. Varying amounts of sequence information can be gathered from each fragmentation spectrum, and the spectra need to be interpreted carefully. Some of the processing can be automated, but in general the processing and interpretation of spectra will take longer than the data acquisition if accurate and reliable data are to be generated.

The amount of sequence information generated will vary from one peptide to another, Some peptide sequences will be confirmed totally, other may produce a partial sequence of, say, 4 or 5 amino acid residues. Often sequence "tag" of 4 or 5 residues is sufficient to search a protein database and confirm the identity of the protein.

VII. Peptide sequencing in summary:

1. Peptides fragment along the amino acid backbone to give sequence information.
2. Peptides ca. 2500 Da or less produce the most useful data.

3. The amount of sequence information varies from one peptide to another. Some peptides can generate sufficient information for a full sequence to be determined; others may generate a partial sequence of 4 or 5 amino acids.
4. A protein digest can be analysed as an entire reaction mix, without any separation of the products, from which individual peptides are selected and analysed by the mass spectrometer to generate sequence information.
5. About 4 µL of solution is required for the analysis of the digest mixture, with a concentration based on the original protein of ca. 1-10 pmol/µL. MS/MS sequencing is a sensitive technique consuming little sample.
6. Sometimes the full protein sequence can be verified; some proteins generate sufficient information to cover only part of the sequence. 70 - 80% coverage is reasonable.
7. Often a sequence "tag" of 4/5 amino acids from a single proteolytic peptide is sufficient to identify the protein from a database.
8. The final point in this summary means that mass spectrometers have been found to be extremely useful for proteomic studies, as illustrated below.
9. The proteomics procedure usually involves excising individual spots from a 2-D gel and independently enzymatically digesting the protein(s) contained within each spot, before analysing the digest mixture by mass spectrometer in the manner outlined above. Electrospray ionisation or MALDI could be used at this step.
10. The initial MS spectrum determining the molecular masses of all of the components in the digest mixture can often provide sufficient information to search a database using just several of the molecular weights from this peptide map.
11. If the database search is not fruitful, either because the protein has not been catalogued, is previously uncharacterised, or the data are not accurate or comprehensive enough to distinguish between several entries in the database, then further information is required.
12. This can be achieved by sample clean-up and then MS/MS studies to determine the amino acid sequences of the individual proteolytic peptides contained in the digest mixture, with which further database searching can be carried out.

VIII. Table 42.1. Amino acid residues and their mass

Symbol	Structure	Mass (Da)
Ala A	$-NH.CH.(CH_3).CO-$	71.0
Arg R	$-NH.CH.[(CH_2)_3.NH.C(NH).NH_2].CO-$	156.1
Asn N	$-NH.CH.(CH_2CONH_2).CO-$	114.0
Asp D	$-NH.CH.(CH_2COOH).CO-$	115.0
Cys C	$-NH.CH.(CH_2SH).CO-$	103.0
Gln Q	$-NH.CH.(CH_2CH_2CONH_2).CO-$	**128.1**
Glu E	$-NH.CH.(CH_2CH_2COOH).CO-$	129.0
Gly G	$-NH.CH_2.CO-$	57.0
His H	$-NH.CH.(CH_2C_3H_3N_2).CO-$	137.1
Ile I	$-NH.CH.[CH.(CH_3)CH_2.CH_3].CO-$	**113.1**
Leu	$-NH.CH.[CH_2CH(CH_3)_2].CO-$	**113.1**
Lys K	$-NH.CH.[(CH_2)_4NH_2].CO-$	**128.1**
Met M	$-NH.CH.[(CH_2)_2.SCH_3].CO-$	131.0
Phe F	$-NH.CH.(CH_2Ph).CO-$	147.1
Pro P	$-NH.(CH_2)_3.CH.CO-$	97.1
Ser S	$-NH.CH.(CH_2OH).CO-$	87.0

Thr T	-NH.CH.[CH(OH)CH_3).CO-	101.0
Trp W	-NH.CH.[CH_2.C_8H_6N].CO-	186.1
Tyr Y	-NH.CH.[(CH_2).C_6H_4.OH].CO-	163.1
Val V	-NH.CH.[CH(CH_3)$_2$].CO-	99.1

IX. SELDI-TOF

Surface Enhanced Laser Desorption Ionization is a modification of the procedure used in MALDI. Instead of mixing the UV sensitive matrix with the protein sample, the protein sample is spotted on a plate which has some surface binding characteristics such as a chromatographic array. The spots are then washed to remove impurities and weakly bound proteins. The UV matrix is then added to the spot and allowed to co-crystallize. After the ionization with the UV laser, the ions are analyzed using a TOF mass analyzer, in the same manner as MALDI.

SELDI provides on-chip separation as well as the capability to perform enzymatic reactions directly on the chip. However, there are concerns about the reproducibility of SELDI-TOF mass spectra, especially when normal post processing techniques frequently used with MALDI such as baseline correction are applied. Environmental sources of variation such as humidity can also play a large role.

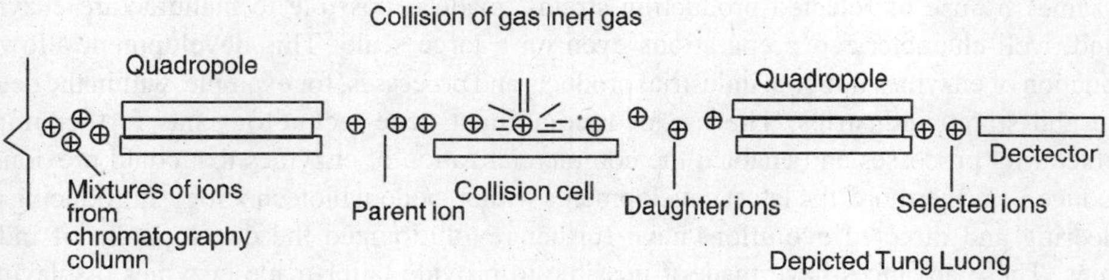

Fig. 42.9. Schematic SELDI-TOF.

QUESTIONS

1. How mass spectophotometry is helpful in solving biological problems? Describe protein profiling by MS-MS.
2. Write short notes :
   (*a*) Mass spectrometry
   (*b*) MALDI-TOF
   (*c*) SELDI-TOF
   (*d*) Electro ionization in MS
   (*e*) Protein peptide sequencing.

ENZYME TECHNOLOGY

CHAPTER 43

I. INTRODUCTION

The enzyme industry as we know it today is the result of a rapid development seen primarily over the past four decades thanks to the evolution of modern biotechnology. Enzymes found in nature have been used since ancient times in the production of food products, such as cheese, sourdough, beer, wine and vinegar, and in the manufacture of commodities such as leather, indigo and linen. All of these processes relied on either enzymes produced by spontaneously growing microorganisms or enzymes present in added preparations such as calves' rumen or papaya fruit. The enzymes were, accordingly, not used in any pure or well-characterized form. The development of fermentation processes during the later part of the last century, aimed specifically at the production of enzymes by use of selected production strains, made it possible to manufacture enzymes as purified, well-characterized preparations even on a large scale. This development allowed the introduction of enzymes into true industrial products and processes, for example, within the detergent, textile and starch industries. The use of recombinant gene technology has further improved manufacturing processes and enabled the commercialization of enzymes that could previously not be produced. Furthermore, the latest developments within modern biotechnology, introducing protein engineering and directed evolution, have further revolutionized the development of industrial enzymes. These advances have made it possible to provide tailor-made enzymes displaying new activities and adapted to new process conditions, enabling a further expansion of their industrial use. The majority of currently used industrial enzymes are hydrolytic in action, being used for the degradation of various natural substances. Proteases remain the dominant enzyme type, because of their extensive use in the detergent and dairy industries. Various carbohydrases, primarily amylases and cellulases, used in industries such as the starch, textile, detergent and baking industries, represent the second largest group (Fig.43.1)

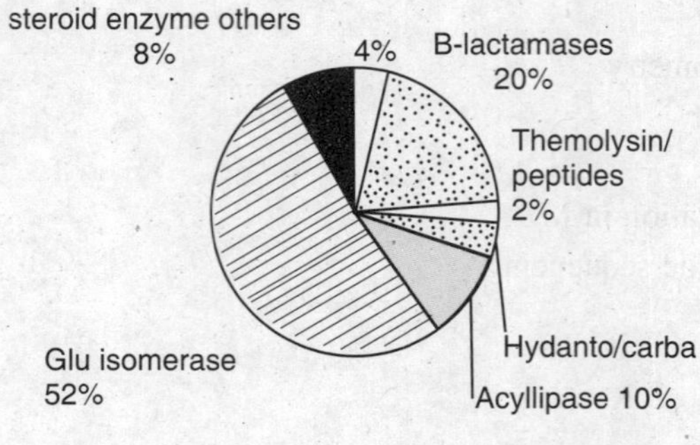

Fig. 43.1. Share of different enzymes in industrial processes.

Enzymes catalyze chemical reactions with great specificity and rate enhancements. These reactions are the basis of the metabolism of all living organisms, and provide tremendous opportunities for industry to carry out elegant, efficient and economical biocatalytic conversions. In the broadest sense, enzyme technology can be seen as the application of free enzymes and whole cell biocatalysts in the production of goods and services. A more narrow definition limits enzyme technology to the technological concepts that allow the use of enzymes in competitive large-scale bioprocesses. Enzyme technology is an interdisciplinary field, recognized by the Organization for Economic Cooperation and Development (OECD) as an important component of sustainable industrial development. Its applications range from straightforward industrial processes to pharmaceutical discovery and development. Table 1 gives a few recent examples of the broad and growing impact of enzyme applications in different industry sectors. At the same time, rapid developments in genetic engineering, high-throughput screening, and other established or emerging technologies, improve the economics of existing processes, and open up new application fields (Table 43.1).

Table 43.1. Impact of emerging and established technologies on enzyme technology

Technology	Key developments
Functional genomics	Enzyme discovery
	Genome link with enzyme activities
	Genomics for enzyme-based drug discovery
Protein structure determination	Structural genomics initiative (www.jcs-org)
	(www.structuralgenomics.org)
High-throughput methods	Enzyme discovery and improvement
	Sol-gel immobilized enzyme arrays
	High-throughput screening methods
	Global analysis of protein activities
Enzyme engineering	Directed evolution to improve enantioselectivity
	Directed evolution and biocatalysis
	New enzyme activities by chemical modification
	Artificial enzymes
Combinatorial biocatalysis	Combinatorial biocatalysis review
	Enzymatic polymerization
	Glycosylation of bioactive compounds
Bioelectro-catalysis	Biosensors, bioreactors and biofuel cells
Metabolic engineering	Metabolities as industrial chemicals
Bioprocess engineering	Dynamic kinetic resolution/ioric liquids
	Industrial biocatalysis
	Enzyme immobilization on EUPERGIT
	Enzymic protection group techniques
Protein production and downstream processing	Protein refolding for industrial processes
	Continuous solid-state fementation
	Field-assisted protein purification
	New developents in bioseparation
Regulatory aspects	Single isomer pharmaceuticals
	Genetically modified organisms
	Pollution and waste reduction

II. TRADITIONAL ENZYME MARKETS

The food, feed, agriculture, paper, leather and textile industries are well suited for enzyme technology because products as well as raw materials consist of biomolecules, which can be produced, degraded or modified by enzymatic processes. Many enzymes are commercially available, and numerous industrial applications have been described. An illustrative overview of new developments in the food sector was recently written by the editor of *Food Technology.* In the pulp and paper industries, enzymes are increasingly used for cleaner production processes, resulting in 1994 enzyme sales in the United States and Europe of about $25 million. The availability of cost-effective and clean enzymatic or biological alternatives to chemical procedures strongly promotes technological improvements. Hence, the resulting market value of biotechnology for clean production in the paper industry ($31–62 billion) is very significant indeed . Enzymes have a similar role in textiles production: the use of laccases for bleaching or catalase to remove hydrogen peroxide cause a significant reduction in the use of raw materials and production of waste.

III. IMPACT OF ENZYMES AS CATALYSTS IN THE CHEMICAL INDUSTRY

The chemical industry represents one of the biggest economic sectors worldwide and European companies play a dominant role. The European chemical industry is highly developed, but introduction of new technologies continues, including biocatalysis. This allows access to new market segments and products in analytical and synthetic applications. Enzyme applications, which are in different development stages, are found mainly in the organics, drugs, and cleaners segments. We expect rapid developments in the polymer field (e.g. in the production of 1,3-propanediol) culminating in applications in all segments of the chemical industry, excluding only bulk products with prices below $1/kg. Today, the market for basic, intermediate, fine and speciality chemicals and polymers made by bioprocesses accounts for only 2% of the chemical market or $25 billions in revenue, excluding revenues from chemicals like ethanol made by traditional fermentation.

IV. Bioprocesses and the environment- The sustainability of a process relates to energy and raw material use, waste production, process stability/safety, and product quality. These factors often translate into a reduction of production costs and then contribute to improved competitiveness, especially in highly regulated countries. In a recent report, the organization for economic cooperation and development (OECD) illustrates this with various examples and case studies (Table 43.2).

Table 43.2. Bioprocesses increasing the sustainability of the chemical industry

Product	Catalyst*	Comparison with conventional process		Company
		Category	Remarks	
Ammonium acrylate	Nitrilase	†	High yield	Ciba
		†	Easy quality control of product	
		‡	No emission of toxic vapour	
Polyester	Lipase	§	High-quality of product	Baxenden
		†	No alternative conventional process	
(*S*)-Chloropropionic acid	(*R*)-Specific dehalogenase	†	Simple one-step process	Avecia
		§	High-quality of product	
		‡	No involvement of toxic raw materials	
7-ACA	D-Amino acid oxidase, glutaryl amidase	‡	No involvement of toxic raw materials	Biochemie
		†	Mild reaction conditions	
		‡	Tenfold reduction of waste	
SO_4^{2-}-removal	Sulfate-reducing microbe	‡	SO_4^{2-} and F^- load in waste water is very low	Budel Zink
		‡	Reduction of gypsum in waste water from 18 tons/day to essentially 0 tons/day	
Removal of hydrogen peroxide from textiles	Catalase	§	High-quality of product	Windel
		†	Simple process	
		‡	Decrease of waste water	
Removal of fatty acid esters from oil	Phospholipase	†	Simple process	Cereol
		‡	Tenfold reduction of waste water	
		‡	Eightfold reduction of sludge	
		†	Reduction of raw materials	

*Origin of enzymes: nitrilase, Rhodococous sp.; lipase, Candida antarctica; (R) specific dehalogenase, Pseudomonas sp.; D-amino acid oxidase, not mentioned; glutaryl amidase, E. coli, catalase, not mentioned; phospholipase, Hyphozymas sp. Process efficiency, environmental sustainability; sporoduct quality, Order of category is based on the importance in each process. Abbreviation; 7-ACA, 7-aminocephalosporanic acid.

Process improvements such as increase in yield and reductions in raw material demand, emissions, and waste result in process cost savings and can give bioprocesses advantages over traditional chemical routes. This was shown by Ciba (UK) for the synthesis of acrylic acid on pilot-scale using nitrilase. Similarly, Budel Zink (NL) are using sulfate reducing bacteria to treat 30 m3 of wash tower acid per hour and recover 8.5 tons/day of ZnS, an impressive application in the industry sector of inorganics. Today, bioprocesses based on genetically modified organisms (GMOs) are economically feasible. This was possible due to harmonized European Community guidelines and quick decisions of authorities. A successful example is the setup of a new plant (on stream since early 2001) for the enzymatic production of 7-aminocephalosporanic acid (7-ACA) by Biochemie (Frankfurt, Germany).

The bioprocess uses no toxic raw materials, operates under mild reaction conditions, and yields significantly less waste. Another example is the one-step manufacture of riboflavin from glucose by Hoffmann La-Roche (Grenzach, Germany) using a mutant of *Bacillus subtilis*. Thanks to a 50% reduction of production costs and fast approval through legislation, a new plant with a capacity of 2000 tons/year started production in May 2000. Sustainability is achieved via the reduction of emissions by 33% and 50% (volatile organic compounds and waste water, respectively). In this case, an existing process was replaced on the basis of new research findings (research-push).

V. COMMERCIALIZED APPLICATIONS OF ENZYMES IN THE CHEMICAL INDUSTRY

Today, key reactions in bioprocesses running on production scale are mostly hydrolytic or the processes are based on fermentations of cheap starting materials like glucose (Table 43.3). Mechanistic exceptions are the hydroxylations of trimethylaminobutyrate and heteroaromatic compounds by Lonza, using oxygen derived from water. DSM is using ammonia lyase for the production of L-aspartic acid and is implementing a process for the enantioselective production of cyanohydrins. A series of processes operate on the multi-hundred or thousand tons/year scale, which illustrates the technological feasibility and increasing acceptance of biocatalysis for industrial organic synthesis (Table 43.3).

Table 43.3. Commercialized bioprocesses

Company	Strategy	Product(s)	Substrate(s)	Catalyst(s)	Remarks
Avecia (ICI)	Kineticresolution	(S)-2-Chloro-propionic acid	Racemic 2-Chloro-propionic acid	Whole cells,(S)-specific dehalogenase	Knock-out of(R)-specificdehalogenase. Scale:several 1000tons/year
BASF	Fermentation	L-Lysine (R)-isobutyl lactate	Glucose Glucose	*C.glutamicum* Microorganism	Scale:>100tons/year Precursor for (S)-chloropropionic acid.Scale:several 100tons/year
	Kinetic resolution	Enatiopure alcohols	Racemic alcohols	Lipases	Enantiospecific acylation, ChiPros™Scale:several 100tons/year Lipases
		Chiral amines	Racemic sec-amines	Lipases	Enantiospecific acylation,ChiPros^{T} Scale:up to 1000tons/year
Chirotech	Kinetic resolution	Various α-amino acids	Lactams, N-protected racemic a-amino acidesters	Lactamases	Complementary stereoisomers viacomplementary lactamases.Scale:kg to tons/year
		Various D-amino acids	Racemic N-acylated amino acids	D-Aminoacylase	
		Various L-amino acids	Racemic N-acetyl amino acids	N-Acetyl-L-amino-acid amidohydrolase (aminoacylase)	Immobilized enzyme in packed-bed reactor.Scale:up to several kg on demand.

	Dynamic resolution	Various (S)-ester amides	Racemic aralactones	Immobilized Triacylglycerol acylhydrolase (triacylglycerol lipase)	The reaction is performed in organic media, spontaneous racemization of the substrate.
Degussa	Fermentation	L-Threonine	Glucose	Whole cells	REXIM (subsidiary in France). Scale:multi 1000 tons/year
		Enantiopure D-amino acids	Racemic hydantoins	Hydantoinases, decarbamylases Racemase	In vivo process,three enzymes cloned in E.coli.Process for L-amino acids under development.
	Enantio-Selective production	L-tert-Leucine	Trimethyl pyruvic acid	Leucine dehydrogenase	NADH-regeneration with formate dehydrogenase in an EMR.Scale:tons/year
		Enantiopure L-amino acids	Racemic amino acidderivatives	Amidases,esteras -es, proteases	Scale:few to several 100 tons/year
	Enantio-selective synthesis	L-aspartic acid	Fumaric acid	Ammonia lyase	Scale;1000 tons/year
		Aspartame	N-Protected L-aspartic acid, D/L-phenylalanine methylester	Thermolysine	Selective coupling.Scale: 1000tons/year
	Enzymatic hydrolysis/ synthesis	6-APA,7-ADCA	Penicillin G/V	Penicillin acylase	Scale:1000 tons/year
		β-Lactam antibiotics	7-ADCA or 6-APA and acid derivatives	Acylases	Enzymatic coupling In aqueous solution. Scale:few to several 100 tons/year
Lonza	Hydrolysts	Niacinamide	Nicotinonitrile whole cells, Nitrile hydratase	Immobilized	Scale:>3000 tons/year
	Kinetic resolution	(S)-Pipecolic (R) and (S)-piperazinic acid	Corresponding racemic amides	Whole cells,(R)- and acid, (S)-amidase	Substrates derived chemically from products of previous entry.
	Oxidation	L-Camitine	4-Trimethylamino butyrate	Whole cells, desaturase and hydratase	Large scale
		Ring-hydroxy-lated pyridine and pyrazine derivative	Pyridine and pyrazine derivatives	Wholecells, Various nitrilases and dehydro-genases	e.g. 6- Hydroxy-nicotinic acid,5-Hydroxypyrazinic acid,6-Hydroxy-(S)-nicotine.Scale:up to a few tons/year
Novartis	Amino-oxidation and hydrolysis	7-ACA	Cephalosporin C	D-Amino acid oxidase and glutaryl amidase	See Table 43.2
Roche	Fermentation	Riboflavin	Glucose	Genetically engineered B.subtilis	Scale:2000 tons/year

Abbreviations:6-APA,6-aminopenicilinic acid;7-ACA,7-aminocephalosporanic acid;7-ADCA,7-aminodesacetoxycephalosporanic enzyme-membrane reactor.

Recently Described Enzyme Reactions of Industrial Relevance

Nature, but also strain collections, harbour a large pool of still unused but industrially interesting enzymes. The diversity of natural compounds suggests the existence of a, yet undiscovered, potential for functional group biocatalysis. Next to selective enrichments and isolation of desired enzymes, the identification of new reactions can be based on high-throughput approaches, which need efficient assays for detecting desired activities. Robust activity assays are also an important requirement for the practical application of powerful enzyme engineering techniques.

The characterization and application of new enzymes to catalyze reactions with commercial potential will significantly broaden the spectrum of industrial biocatalysis. The flavoenzyme tryptophan 7-halogenase has been evaluated for selective chlorination of tryptophan and indole derivatives. An interesting carbon–carbon bond formation reaction was reported for the preparation of (*R*)-phenylacetylcarbinol by carboligation of pyruvate and benzaldehyde using various pyruvate decarboxylases. Continuous production of (*R*)-phenylacetylcarbinol from acetaldehyde and benzaldehyde could be achieved using a mutant of pyruvate decarboxylase from *Zymomonas mobilis* in an enzyme membrane reactor (EMR) with space-time yields of 81g /L/d. Widespread application of enzymes in the chemical industry will depend on the ability to couple enzymatic and chemical steps. Chemoenzymatic reaction sequences profit from the high technical development level of both chemical and enzymatic reactions.

VI. ENZYMES IN OTHER INDUSTRIAL PROCESSES

Enzymes find wide applications in diverse industries including fuel, baking, animal feed, beverages, paper and leather. In these processes different enzymes such as cellulose, amylase, proteases, xylanase, phytase, and lipase are utilized to obtain desired product in bulk quantities (Table 43.4).

(a) The detergent industry- Their use as detergent additives still represents the largest application of industrial enzymes, both in terms of volume and value. The major component is proteases, but other and very different hydrolases are introduced to provide various benefits, such as the efficient removal of specific stains (Table 43.4). Constantly, new and improved engineered versions of the 'traditional' detergent enzymes, proteases and amylases, are developed. These new second and third-generation enzymes are optimized to meet the requirements for performance in detergents, the composition of which is also constantly developed. In particular, the compatibility of enzymes with detergent components (*i.e.* typically stability properties) is addressed, but their ability to function at lower temperatures has also been amongst the recently reported improvements. To save energy, the temperature used in household laundering and automated dishwashers has been reduced in recent years. This often results in problems with efficient cleaning and stains removal that enzyme technology can help overcome. Recent examples of second-generation detergent enzymes include the development of novel amylases that have enhanced activity at lower temperatures and alkaline pH, while maintaining the necessary stability under detergent conditions. These enzymes were developed by the combined use of microbial screening and rational protein engineering. Proteases displaying activity at low temperatures have been isolated from nature, but have also been evolved in the laboratory by a directed evolution approach Segmentation of the industrial enzyme market. In the year 2000, the enzyme market totalled $1.5 billion. The technical industries segment comprises the detergent, starch, textile, fuel alcohol, leather, and pulp and paper industries.

(b) Enzymes for starch conversion. The enzymatic conversion of starch to high fructose corn syrup is a well-established process and provides a beautiful example of a bioprocess in which the consecutive use of several enzymes is necessary. The enzymes utilized in the starch industry are also subjected to constant improvements.

The first step in the process is the conversion of starch to oligomaltodextrins by the action of α-amylase. The concomitant injection of steam puts extreme demands on the thermostability of the

Table 43.4. Enzymes used in various industrial segments and therir applications.

Industry	Enzyme class	Application
Detergent (laundry and dish wash)	Protease	Protein stain removal
	Amylase	Starch stain removal
	Lipase	Lipid stain removal
	Cellulase	Cleaning, color clarification, anti-redeposition (cotton)
	Mannanase	Mannanan stain removal (reappearing stains)
Starch and fuel	Amylase	Starch liquefaction and saccharification
	Amyloglucosidase	Saccharification
	Pullulanase	Saccharification
	Glucose isomerase	Glucose to fructose conversion
	Cyclodextrin-glycosyltransferase	Cyclodextrin production
	Xylanase	Viscosity reduction (fuel and starch)
	Protease	Protease (yeast nutrition – fuel)
Food (including dairy)	Protease	Milk clotting, infant formulas (low allergenic), flavor
	Lipase	Cheese flavor
	Lactase	Lactose removal (milk)
	Pectin methyl esterase	Firming fruit-based products
	Pectinase	Fruit-based products
	Transglutaminase	Modify visco-elastic properties
Baking	Amylase	Bread softness and volume, flour adjustment
	Xylanase	Dough conditioning
	Lipase	Dough stability and conditioning (*in situ* emulsifier)
	Phospholipase	Dough stability and conditioning (*in situ* emulsifier)
	Glucose oxidase	Dough strengthening
	Lipoxygenase	Dough strengthening, bread whitening
	Protease	Biscuits, cookies
	Transglutaminase	Laminated dough strengths
Animal feed	Phytase	Phytate digestibility – phosphorus release
	Xylanase	Digestibility
	β-Glucanase	Digestibility
Beverage	Pectinase	De-pectinization, mashing
	Amylase	Juice treatment, low calorie beer
	β-Glucanase	Mashing
	Acetolactate decarboxylase	Maturation (beer)
	Laccase	Clarification (juice), flavor (beer), cork stopper treatment
Textile	Cellulase	Denim finishing, cotton softening
	Amylase	De-sizing
	Pectate lyase	Scouring
	Catalase	Bleach termination
	Laccase	Bleaching
	Peroxidase	Excess dye removal
Pulp and paper	Lipase	Pitch control, contaminant control
	Protease	Biofilm removal
	Amylase	Starch-coating, de-inking, drainage improvement
	Xylanase	Bleach boosting
	Cellulase	De-inking, drainage improvement, fiber modification
Fats and oils	Lipase	Transesterification
	Phospholipase	De-gumming, lyso-lecithin production
Organic synthesis	Lipase	Resolution of chiral alcohols and amides
	Acylase	Synthesis of semisynthetic penicillin
	Nitrilase	Synthesis of enantiopure carboxylic acids
Leather	Protease	Unhearing, bating
	Lipase	De-pickling
Personal care	Amyloglucosidase	Antimicrobial (combined with glucose oxidase)
	Glucose oxidase	Bleaching, antimicrobial
	Peroxidase	Antimicrobial

enzyme. Using traditional α-amylases, the pH has to be adjusted to an undesirable high level and calcium must be added to stabilize the enzyme. New α-amylases with optimized properties, such as enhanced thermal stability, acid tolerance, and ability to function without the addition of calcium, have recently been developed offering obvious benefits to the industry. Engineering efforts have also been undertaken to develop improved versions of the enzymes used later in the process (i.e. glucoamylase and glucose isomerase.

(c) Fuel alcohol production. In the alcohol industry, the use of enzymes for the production of fermentable sugars from starch is also well established. Over the past decade, there has been an increasing interest in fuel alcohol as a result of increased environmental concern, higher crude oil prices and, more acutely, by the ban in certain regions of the gasoline additive methyl *tert*-butyl ether (MTBE), which can be interchanged directly with ethanol.

(d) Textile applications. In the textile industry a completely new enzymatic activity has recently been introduced. This industry is under considerable environmental pressure owing to its large energy and water consumption and subsequent environmental pollution. One of the most energy- and water-consuming steps in the processing of cotton is the scouring step, the removal of various remaining cell-wall components on the cellulose fibers performed at high temperature and under strong alkaline conditions. An alternative, enzyme-based process performed at much lower temperatures and using less water has now been developed based on a pectate lyase.

(e) Enzymes for the feed industry. The use of enzymes as feed additives is also well established. For example, xylanases and α-glucanases have been used throughout the past decade in cereal-based feed for monogastric animals which, contrary to ruminants, are unable to fully degrade and utilize plant-based feeds containing high amounts of cellulose and hemicellulose. During recent years focus has been on the utilization of natural phosphorus bound in phytic acid in cereal-based feed for monogastrics. Better utilization of total plant phosphorus, of which 85–90% is bound in phytic acid, is only obtained by adding the enzyme phytase to the feed. The attention on this enzyme has increased dramatically during recent years owing to the bovine spongiform encephalopathy (BSE)-related bone-meal ban in many countries, which has effectively removed the main traditional source for inorganic phosphorus. Also, several western countries with intensive animal production have adopted standards for the release of phosphorus into the environment. As the addition of phytase to feeds results in a significant reduction of the phosphorus outlet from monogastrics, the phytases have grown to become the largest enzyme segment in the feed industry. Besides the direct effects on phosphorus uptake and secretion, indirect effects on the uptake of other nutrients are also being recognized.

(f) Enzymes for the food industry. As indicated in Table 43.4, applications of enzymes in the food industry are many and diverse, ranging from texturizing to flavoring. Common to more or less all food applications, the enzymes are applied to processed food products as processing agents upstream from the final product. Several advances have been made in the optimization of enzymes for existing applications and in the use of recombinant protein production to provide efficient mono-component enzymes that do not have potential detrimental side-effects.

Recently, much work has been carried out on the application of transglutaminase as a texturing agent in the processing of, for example, sausages, noodles and yoghurt, where cross-linking of proteins provides improved viscoelastic properties of the products. Within the baking industry there is an increasing focus on lipolytic enzymes. Recent findings suggest that (phospho) lipases can be used to substitute or supplement traditional emulsifiers, as the enzymes degrade polar wheat lipids to produce emulsifying lipids *in situ*.

(g) Brewing and cheese making. These industries are one of the oldest industries and make use of enzymes present in living cells for fermentation and processing of cheese. This is described in detail in general microbiology. Biotechnology is helping in developing more efficient strains of these organisms, particularly those of yeast and fungi used for alcohol making by fermentation and cheese processing, respectively.

(h) Meat tenderizing- Toughness in meat is due primarily to collagen and elastin and also to actomyosin. Lower quality cuts of meat are just as nutritious as prime cuts. Thus to make maximum use of caracass, efforts have been made to tenderize the lower quality cuts. Also the flavour of meat (known as marination) is to some extent related to the concentration of peptides and amino acids present. Proteases are used both I tenderization and marination. Papain is the traditional enzyme used, but others such as bromelain, trypsin, chymotrypsin and microbial proteases from *Aspergillus* are also used

(i) Enzyme applications in agriculture- Enzymes are also used in a wide range of agrobiotechnological processes, such as enzyme-assisted silage fermentation, bioprocessing of crops and crop residues, fibre processing and production of feed supplements to improve feed efficiency. Especially the latter application, which includes the use of phytases to improve the efficiency of nutrient utilization and to reduce waste, is a rapidly growing sector. The feed enzyme market now amounts to $150 million. In fact, 65% of poultry and 10% of swine feed already contain enzymes such as carbohydrases or phytase.

Several developments have started to tie in the agricultural sector with the chemical and pharmaceutical industries. Plants are being modified by genetic engineering for the production of polymers and pharmaceuticals such as antibodies or for improved nutritional value, for example, by increasing lysine or carotenoid content. Only recently, ProdiGene announced the scale-up of trypsin production in recombinant plants (www.prodigene.com), while other enzymes such as laccase may soon be produced in plants as well. Agricultural waste biomass may soon turn into a valuable resource for the production of chemicals and fuels. Biobased renewables have many advantages, such as reduced CO_2 production, flexibility, and self-reliance. This was also recognized by the chemical industry. For example, the Royal Dutch/Shell group estimated that by the year 2050 renewable resources could supply 30% of the worldwide chemical and fuel needs, resulting in a biomass market of $150 billion. Here, the price of cellulase needed to convert cellulosic biomass to fermentable sugars is a major factor. Therefore, the US Department of Energy awarded $32 million to Genencor and Novozymes to reduce the price of cellulose by a factor of ten, which could make bioethanol production and many other sugar-based fermentations economically viable.

(j) Processing of fats and oils- In the fat and oil industries, several new enzyme-based processes have recently been introduced. Even though the use of immobilized lipases in the inter-esterification of triglycerides was first escribed in the 1980s, the process has not been sufficiently cost-effective to be introduced in true large-scale applications, for example, in the production of margarine. Although enzyme production has become much more efficient, the cost of immobilization has remained an obstacle. Recent developments have, however, changed this picture. A new process for immobilizing lipases based on the granulation of silica has dramatically lowered process costs, and procedures based on this new material are now being implemented for the production of commodity fats and oils with no content of *trans*-fatty acids. Another recently introduced process is the removal of phospholipids in vegetable oils ('de-gumming'), using a highly selective microbial phospholipase.

QUESTIONS

1. How enzyme technology is useful in industrial product production ?
2. Discuss the role of enzyme technology on chemical industry.
3. What are new technologies which play role in enzyme technology for industrial production?
4. Enlist a few commercial processes involving enzymes. Discuss future of such industry.
5. What is enzyme technology ? Describe various applications of enzyme technology in industry.
6. Discuss role of enzyme technology in detergent, textile, food and brewing industries.
7. Write notes on :

   (*a*) Enzymes in chemical industry

   (*b*) Traditional enzymes

   (*c*) Enzymes in environment protection

   (*d*) Enzymes in brewing industry.

PROTEIN-PROTEIN INTERACTIONS

CHAPTER 44

Imagine a cell in which, suddenly, the specific interactions between proteins would disappear. This unfortunate cell would become deaf and blind, paralytic and finally would disintegrate, because specific interactions are involved in almost any physiological process. Sensing extracellular signals is a matter of receptor to adaptor interactions and the shape of the cell is maintained by an intricate network of structural protein interactions. Finding interactions between proteins involved in common cellular functions is a way to get a broader view of how they work cooperatively in a cell. Protein-protein interactions are operative at almost every level of cell function, in the structure of sub-cellular organelles, the transport machinery across the various biological membranes, packaging of chromatin, the network of sub-membrane filaments, muscle contraction, and signal transduction, regulation of gene expression, to name a few. Aberrant protein-protein interactions have implicated in a number of neurological disorders such as Creutzfeld-Jacob and Alzheimer's disease. Because of their importance in development and disease, these systems have been the object of intense research for many years. It has emerged from these studies that nature has employed in many instances a strategy of mixing and matching of domains that specify particular classes of protein-protein interactions, modifying the amino acid sequence in order to confer specificity for particular target proteins. The regulation of cell function brought about by the interactions of these proteins is delicately balanced by the relative affinities of the various protein partners and the modulation of these affinities by the binding of ligands, other proteins, nucleic acids, ions such as Ca2+, and covalent modification, such as specific phosphorlyation or acetylation reactions. Specificity and the strength of signal transduction is encoded by the exact amino acid sequence of the domain, and it is this relationship between sequence, structure, dynamics, energetics and function that constitutes the fundamental issue for the biophysics of protein-protein interactions. This endeavor therefore requires a structural characterization of the domains and if possible their dispositions within the complete protein and their complexes with their specific protein partners.

I. DEFINITION

Specific interactions between two or more proteins.

Examples- Enzyme-inhibitor complex; antibody-antigen complex; receptor-ligand interactions, multiprotein complexes such as ribosomes or RNA polymerases.

II. GENOMICS

Bioinformatics helps scientists to handle the vast amount of data generated by different genome-wide approaches. Interesting conclusions about genome evolution and the relationships between genes and function may be drawn by **comparative genomics**. Predictions of function, based on genome sequence analysis, will become more and more accurate as the number and diversity of sequenced genomes steadily increases. **Systematic deletions** (physical or by RNA interference analysis) and **systematic localization** studies are first steps in functional characterization of genes and may be of great help in improving the quality of protein interaction maps. For example, a two-hybrid interaction between a nuclear and a mitochondrial protein is not likely to be biologically significant. Clustering of genes by mRNA expression profiles seem to be an effective way to predict protein function. Protein quantitation and differences in protein expression under different experimental conditions may be analysed by 2D-electrophoresis but the ICAT technique that makes extensive use of mass spectrometry might be more effective in proteome-wide projects. The equivalent of DNA chips in the protein world and the projects of constructing libraries of specific antibodies raised against every single protein of a cell. **Biochemical genomics**, where the entire ORFeome of an organism is cloned, expressed and individually purified, allow the discovery of proteins when a biochemical assay for their function has already been developed. By a different approach, **structural genomics** is trying to decipher biochemical protein function by structural similarity to known proteins; however this method is limited to predicting the cellular function of a given protein (Fig. 44.1).

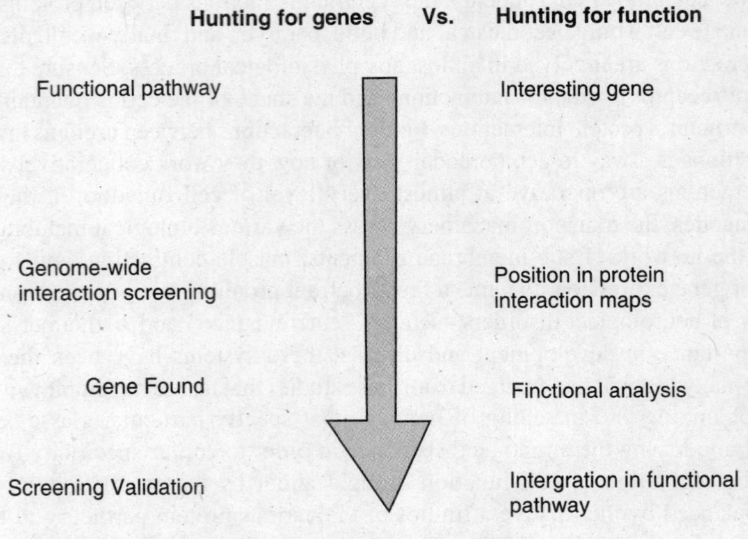

Fig. 44.1. Alternative ways of studying protein function

III. CHARACTERISTICS

Classification: Protein-protein interactions can be arbitrarily classified based on the proteins involved (structural or functional groups) or based on their physical properties (weak and transient, "non-obligate" vs. strong and permanent). Protein interactions are usually mediated by defined domains; hence interactions can also be classified based on the underlying domains.

Universality: All of molecular biology is about protein-protein interactions. Protein-protein interactions affect all processes in a cell: structural proteins need to interact in order to shape organelles and the whole cell, molecular machines such as ribosomes or RNA polymerases are hold together by protein-protein interactions, and the same is true for multi-subunit channels or receptors in membranes.

Specificity distinguishes such interactions from random collisions that happen by Brownian motion in the aqueous solutions inside and outside of cells. Note that many proteins are known to interact although it remains unclear whether certain interactions have any physiological relevance.

Number of interactions: It is estimated that even simple single-celled organisms such as yeast have their roughly 6000 proteins interact by at least 3 interactions per protein, i.e. a total of 20,000 interactions or more. By extrapolation, there may be on the order of ~100,000 interactions in the human body.

Protein-protein interactions and protein complexes: Most protein-protein interactions are detected as interacting pairs or as components of protein complexes. Such complexes may contain dozens or even hundreds of protein subunits (ribosomes, spliceosomes etc.). It has even been proposed that all proteins in a given cell are connected in a huge network in which certain protein interactions are forming and dissociating constantly.

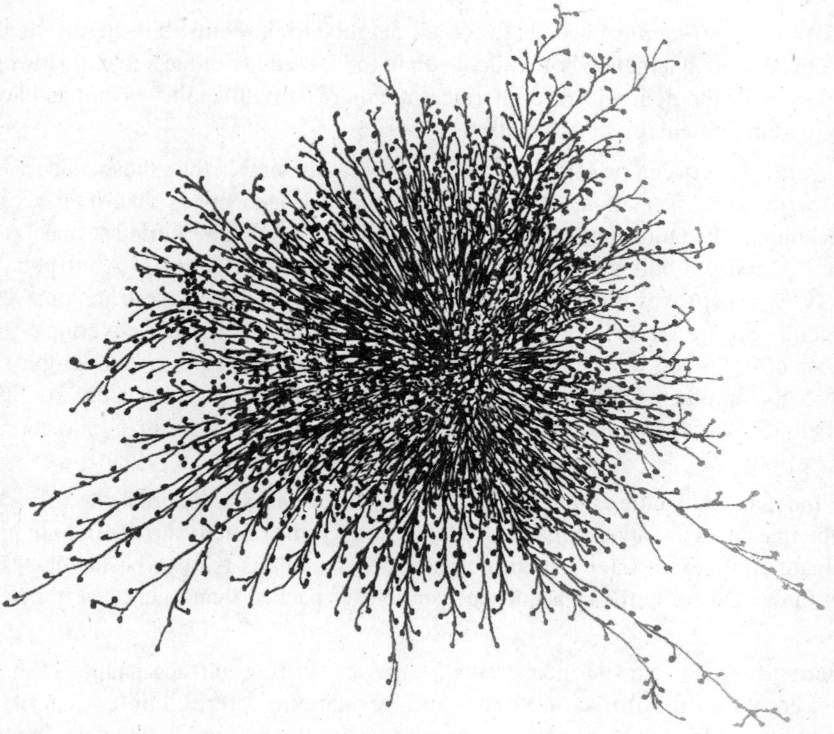

Fig. 44.2. The protein-protein interaction network in yeast.

An interaction map of the yeast proteome assembled from published interactions. The map contains 1,548 proteins (boxes) and 2,358 interactions (connecting lines) (Fig. 44.2).

Structural features of protein-interaction sites

Hundreds of protein complexes have been analyzed by X-Ray crystallography and other methods. Data about the structures of proteins and complexes are available from the Protein Databank (PDB, http://www.rcsb.org/). The following statements about the geometry and energetic of protein interactions have been drawn from the analysis of several dozens to about a 100 protein pairs and complexes that have been crystallized.

The contact area between two proteins is almost always bigger than 1100 $Å^2$ with each of the interacting partners contributing at least 550 $Å^2$ of complementary surface. On average each partner loses about 800 $Å^2$ of solvent-accessible surface upon contact, contributed by some 20 amino acid residues of each partner, i.e. the average interface residue covers some 40 $Å^2$. On average, dimers contribute 12% of their accessible surface area to the contact interface, trimers 17.4% and tetramers 20.9%. However, variations are large and total interface areas range from 6% in inorganic pyrophospatase homodimers to 29% in Trp repressor homodimers. For data sets of 10 enzyme-inhibitor complexes and 6 antibody-antigen complexes the mean interface area was about 780 $Å^2$.

Forces that mediate protein-protein interactions include electrostatic interactions, hydrogen bonds, the van der Waals attraction and hydrophobic effects. The average protein-protein interface is not less polar or more hydrophobic than the surface remaining in contact with the solvent. Water is usually excluded from the contact region. Non-obligate complexes tend to be more hydrophilic in comparison, as each component has to exist independently in the cell. It has been proposed that hydrophobic forces drive protein-protein interactions and hydrogen bonds and salt bridges confer specificity.

Van der Waals interactions occur between all neighbouring atoms, but these interactions at the interface are no more energetically favourable than those made with the solvent. However, they are more numerous, as the tightly packed interfaces are more dense than the solvent and hence they contribute to the binding energy of association.

Hydrogen bonds between protein molecules are more favourable than those made with water. Interfaces in permanent associations tend to have fewer hydrogen bonds than interfaces in non-obligate associations. The number of hydrogen bonds is about 1per 170 $Å^2$ buried surface. A standard size interface (~ 1600 $Å^2$) buries about 900 $Å^2$ of the non-polar surface, 700 $Å^2$ of polar surface, and contains 10 (± 5) hydrogen bonds. In a set of reasonably stable dimers there are, on average, 0.9 to 1.4 hydrogen bonds per 100 $Å^2$ of contact area buried (interfaces usually covering > 1000 $Å^2$), but the number of hydrogen bonds varies from 0 in some complexes (e.g. uteroglobin) to 46 in variant surface glycoprotein. Side-chain hydrogen bonds represent approximately 76-78% of the ineractions. Only 56% of homodimers were found to possess *salt bridges* (many having none, and at the most five).

Shape: Independent studies showed that 83-84% of interfaces are more or less flat. With few exceptions, the interfaces are approximately circular areas on the protein surface in both permanent and non-obligate complexes. Interfaces in permanent associations tend to be larger, less planar, more highly segmented (in terms of sequence), and closer packed than interfaces in non-obligate associations.

Complementarities: can be measured in terms of "fitting surface shape". Interfaces in homodimers, enzyme-inhibitor complexes, and permanent heterocomplexes are the most complementary, whilst the antibody-antigen complexes and the non-obligate heterocomplexes are the least complementary.

Secondary structure: In one study the loop interactions contributed, on average, 40% of the interface contacts. In another study (involving 28 homodimers), 53% of the interface residues were ?-helical, 22% beta sheets, and 12% αβ, with the rest being coils.

Amino acid composition: Interfaces have been shown to be more hydrophobic than the exterior but less hydrophobic than the interior of a protein. In one study, 47% of interface residues were hydrophobic, 31% polar and 22% charged. Permanent complexes have interfaces that contain hydrophobic residues, whilst the interfaces in 5 non-obligate complexes favour the more polar residues. Site-directed mutagenesis showed that in many cases a large majority (i.e. > 50%) of interface residues can be mutated to alanine with little effect on Kd: i.e. the functional epitope is a subset of the structural epitope.

IV. CLASSES OF PROTEIN-PROTEIN INTERACTIONS

(a) Homo-oligomerization

An enormous number of enzymes, carrier proteins, scaffolding proteins, transcriptional regulatory factors, etc. function as homo-oligomers. Incorporation of noncovalent interactions at the level of protein quaternary structure provides a number of advantages in terms of regulatory possibilities that are not afforded in cases where the functional unit is comprised of a single polypeptide chain. First, energy can be stored at the subunit interface that can serve to bind ligand, or modify the protein conformation in response to regulatory ligands. Modulation of subunit affinity in such a manner need not compromise the folded structure of the protein, yet provides a considerable energetic margin for modulation of activity. Another significant advantage to linking regulatory properties to subunit interactions lies in the simple fact that over the appropriate range. the activity of the protein becomes protein concentration dependent. Such control can function either positively or negatively, with the oligomer or the monomer exhibiting the highest activity. Thus, function in oligomers can be very finely tuned by ligand concentration (including ions, substrate, allosteric ligands, protons, etc.) and by protein concentration (expression levels or degradation rates).

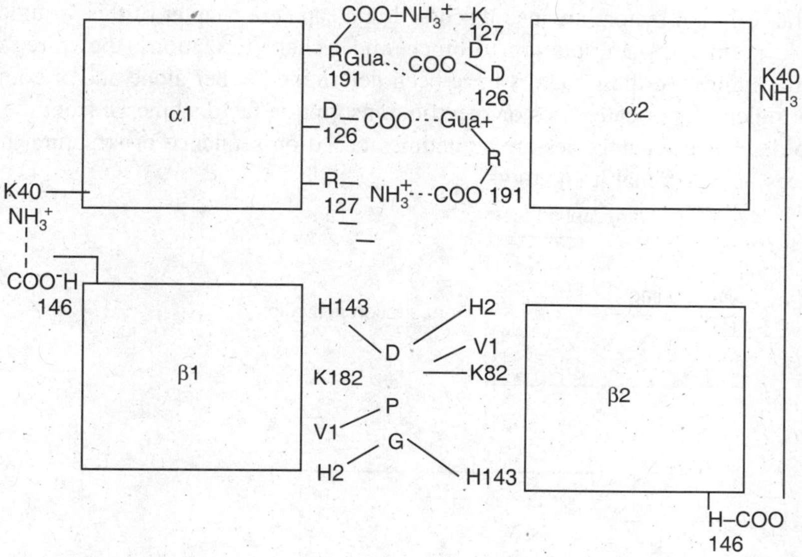

Fig. 44.3. Diagram of the salt bridges at the subunit interfaces of the hemoglobin tetramer.

Hemoglobin represents the best studied oligomeric protein (Fig. 44.3) to date and the structural and energetic linkages between ligation and subunit affinity are known in exquisite detail. Literally thousands of studies have contributed to the detailed understanding of the relationships between sequence, structure, dynamics, energetics and function in this model system.

(b) Interactions between heterologous proteins

Communication at the level of the organism or the cell requires the translation of physical or chemical information signals (e.g., the presence of a particular chemical factor, a change in luminosity, the chemical modification of existing substances, a change in pH or ion concentration) from one compartment to another. In large part, this communication relies on the specific interaction between particular heterologous proteins, in response to particular chemical or physical signals. Clearly all of the mechanisms have not been elucidated that govern the specificity of the interactions between the multitude of protein partners that control cell function. Nonetheless, the resolution of the three dimensional structures of a number of the protein interaction domains by x-ray diffraction or NMR techniques has provided a great deal of insight and the basis for the development of therapeutic strategies designed to modulate some of these interactions known to play a role in various diseases. This section of this chapter is dedicated to a overview of some of the structural motifs involved in these heterologous protein-protein complexes. Protein domains involved in transcriptional regulation A key juncture in the control of cell differentiation and growth is found at the level of transcription of DNA to mRNA. This process is mediated by a number of protein-protein and protein-DNA complexes, in addition to that between the RNA polymerase and the gene and transcript. Protein factors that serve to modulate gene transcription have been termed transcription factors. They fall into three large groups, the general transcription factors involved in the general multi-protein transcriptional machinery, the transcription associated or bridging factors, and specific factors that recognize control sequences in the non-coding regions of the DNA. Modulation of the expression of genes by these factors occurs through their ability to enhance (activation) or disfavor (repression) recruitment of the proteins involved in the transcription. This process therefore involves specific, high affinity proteinprotein interactions, and understanding the structural and energetic basis for the affinity and specificity of these interactions is paramount to the understanding of transcriptional control. Biochemists have identified a very large number of prokaryotic and eukaryotic transcription factors, which cannot all be presented in the context of this introduction to protein-protein interactions. Moreover, transcription factors are also discussed in a separate chapter of this volume concerning protein DNA interactions. In addition to biochemical characterizations, the three dimensional structures of a number of these factors have been determined either alone and/or complexed with DNA. A few subclasses or categories of specific transcription factors emerge from the list of those identified to date. These categories are sometimes based on sequence or structural homology, in other instances on functional homology.

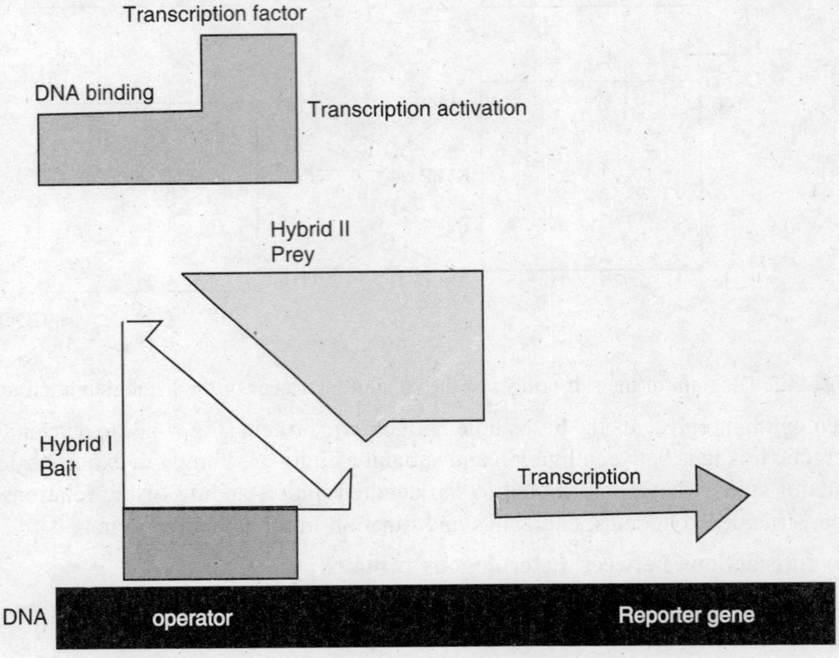

Fig. 44.4. Regulation by hybrid transcription Factor

V. PROTEIN DOMAINS INVOLVED IN SIGNAL TRANSDUCTION

Src homology domains 2 and 3, commonly referred to as SH2 and SH3 domains are protein interaction domains found in proteins involved in signal transduction. The SH2 domains recognize tyrosine phosphorylated proteins, in particular autophosphorylated growth factor receptors. They are found in growth factor receptor binding proteins that intervene in signal transduction downstream of the receptors, but upstream of Ras. They have 7 also been found in docking proteins. A ribbon diagram of the SH2 domain of the Tlymphocyte specific tyrosine kinase, P561ck, is shown below complexed with a carboxymethyl-phenylalanine containing peptide, designed to mimick the the natural phosphphotyrosine containing targets of this domain. The global architecture of SH2 domains consists of a three-stranded twisted beta sheet sandwiched between two alpha-helices. The SH2 domain exhibits a deep binding pocket that presents a high affinity for phosphotyrosine, but not for either phosphoserine or phosphothreonine. Residues carboxyterminal to the phosphotyrosine of the target protein serve to confer the specificity of the interaction. The second pocket of the SH2 domain recognizes these Cterminal specificity determinants. Different classes of SH2 domains recognize different types of C-terminal sequences. For example the class I SH2 domains recognize C-terminal tripeptides P-P-H, where P and H stand for polar and hydrophobic. Class III SH2 domains recognize H-X-H tripeptides. The specificity determinants for the growth factor receptor binding protein 2 (Grb2) resides in the second residue C-terminal to the phosphotyrosine, which must be an asparagine. The three-dimensional structures of a number of SH2 domains have been resolved. There is a very strong structural homology between these which is evident upon comparing the structure of the SH2 domain from p561ck to that show in Figure 5 of the complex between the C-terminal SH2 domain of the Syk tyrosine kinase and a phosphopeptide from the gamma chain of the high affinity immunoglobulin G receptor as determined by NMR. Protein domains involved in transcriptional regulation A key juncture in the control of cell differentiation and growth is found at the level of transcription of DNA to mRNA. This process is mediated by a number of protein-protein and protein-DNA complexes, in addition to that between the RNA polymerase and the gene and transcript. Protein factors that serve to modulate gene transcription have been termed transcription factors. They fall into three large groups, the general transcription factors involved in the general multi-protein transcriptional machinery, the transcription associated or bridging factors, and specific factors that recognize control sequences in the non-coding regions of the DNA. Modulation of the expression of genes by these factors occurs through their ability to enhance (activation) or disfavor (repression) recruitment of the proteins involved in the transcription. This process therefore involves specific, high affinity protein-protein interactions, and understanding the structural and energetic basis for the affinity and specificity of these interactions is paramount to the understanding of transcriptional control.

VI. METHODS TO STUDY PROTEIN-PROTEIN INTERACTIONS

Purification of protein complexes and their separation by 2-dimensional gel electrophoresis or liquid chromatography. After separation the components of a protein complex can be identified by the staining of protein bands by various dyes such as Coomassie Blue. Furthermore, the protein bands can be analyzed by mass spectrometry which can also be used to sequence proteins.

In-vitro binding experiments involve purified proteins whose interaction is detected by retention of one partner by a second protein that has been immobilized on some sort of matrix (such as agarose or glass).

In vivo methods include various **two-hybrid systems**, **FRET**, and other systems that usually require that proteins to be studied are bound to some other detectable protein or chemical compound. Dissociation constants in the micromolar range can be determined by **equilibrium ultracentrifugation**, **microcalorimetry** and other methods. When Kd is in the nanomolar range, radioisotope labelling or ELISA techniques can be used.

Site directed mutagenesis is used to identify the role of individual amino acids.

Structural methods such as **NMR** or **X-ray crystallography.**

Regulation of protein-protein interactions

Interactions between proteins are tightly regulated. The most important ways of regulation involve the following mechanisms:

Expression: Only proteins that are expressed in the same place can interact. RNA and protein expression (i.e. transcription and translation) are therefore regulating such interactions in a time- and space-dependent manner. For example, various Fibroblast Growth Factors (FGFs) are expressed in different tissues such as limbs, brain, kidney etc. while their cognate receptors are similarly expressed in particular tissues where their interaction activates signalling pathways that in turn lead to the expression of certain target genes. FGFs are also expressed in a time-dependent manner, e.g. FGF4 and 8 are only expressed in embryos while others are mostly expressed in adult tissues. The same is true for FGF receptors.

Modifications: Covalent modification of proteins can prevent or induce binding of others. Examples are proteins that contain the so-called bromo domain that binds only acetylated but not unmodified histones.

Ligands: Small molecules or other proteins can regulate protein interactions. For example, binding of GTP to trimeric G proteins promotes their dissociation while binding of GDP favors their association.

Subcellular Localization: Several transcription factors such as NFκB are present in the cytoplasm or even in membranes (such as Notch). They can be activated by translocation to the nucleus where they interact with other transcription factors in order to activate gene transcription. Transport is also important for proteins that are exported from cells such as peptide hormones.

Stability: As for expression, proteins can only interact when they are present. Regulated degradation of proteins prevents them from further interactions.

Prediction of protein-protein interactions

Prediction of interaction sites in proteins of known structure usually focuses on the location of hydrophobic surface clusters on proteins. In one study, this method predicted the correct interaction site in 25 out of 29 cases. Other methods include solvation potential, residue interface propensity, hydrophobicity, planarity, protrusion, and accessible surface area. Among a test set of 28 homodimers, the known interface site was found to be amongst the most planar, protruding, and accessible patches, and amongst the patches with highest interface propensity. Nevertheless, one of the algorithms (PATCH) that uses multiple parameters predicted the location of interface sites in known complexes only for 66% of the structures.

Prediction of interacting protein pairs from genome sequences

Several attempts have been made to predict protein-protein interactions from genome sequences. The major methods are:

Rosetta stone proteins: Some protein sequences have been found to be split into two independent proteins in other organisms. From such sequences it has been concluded that the two independent proteins form a complex, based on the (covalent) association in the former organism. Such fusion proteins are called Rosetta stone proteins. Supposedly they predict interactions among related proteins. Example: human succinyl CoA transferase is split in E. coli into acetate CoA transferases α and β subunits.

Phylogenetic profiles: Some protein pairs are evolutionarily maintained together in many different organisms. It has been concluded that such "co-evolving" proteins are associated either functionally or even physically, i.e. by a protein-protein contact. While the latter is not always true, it has been found to be true in a number of cases, such as the yeast proteins Hog1 and Fus3.

Clinical relevance and applications of protein-protein interaction analysis

Biologically active proteins such as peptide hormones or antibodies act by interacting with other proteins such as receptors or antigens, respectively. Knowing their interaction sites allow the modification of the activity of such proteins or changing their specificity. In addition, small molecules may be designed that block interactions

such as the binding of virus coat proteins to their cellular receptors, thereby blocking infection. Proteins and their interactions are therefore potential drug targets. Sometimes, protein-protein interactions are disadvantageous, such as in insulin that tends to form dimers and hexamers which are less active than onomers. Genetically engineered insulin molecules retain biological activity without oligomerizing.

VII. CHARACTERIZATION OF PROTEIN-PROTEIN INTERACTIONS

Examples of the groups of highly homologous protein interaction domain families presented in the preceding section are found in large numbers of systems involved in the complex control of cellular function. Thus although within each family, the structural and sequence homology are quite high, enough differences are present within each sequence and structure to preclude inappropriate protein pairs from forming. This implies that the relative affinities between these various protein partners have been tuned throughout evolution such that protein pairing occurs only between particular partners, and only under the appropriate conditions. Likewise in homo-oligomeric proteins, the strength of the subunit interactions has evolved to be coupled to substrate or effector binding to just the right levels, such that enzyme activity proceeds at the necessary rate. While biochemists have successfully identified these protein interaction domains and demonstrated interaction preferences and by mutational studies, certain of the necessary sequence determinants of the interactions, in most cases, thorough quantitative thermodynamic and kinetic studies of protein interactions in these systems remain to be carried out. One reason for this lack of information is that priority has been given over the past few years to the identification of these important proteins and to the determination of their three-dimensional structure, both of which are prerequisites to a full understanding of function. Nonetheless, understanding the mechanisms underlying the function of these proteins likewise requires the characterization of their energetic and dynamic properties. We must bear in mind that the study of the energetics and kinetics of protein subunit interactions presents certain practical experimental difficulties that will be discussed below. The good news here is that there are still plenty of interesting and important studies to be undertaken that should keep biophysicists busy for many years to come.

The first major stumbling block encountered by the experimentalist interested in characterizing protein-protein interactions is that of having access to enough purified stable protein, and if possible some interesting constructs, homologous family members or functional mutants in order to carry out a thorough comparative structure/function study. In cases where the NMR or crystal structure of the protein has been determined, this first obstacle can likely be overcome, since the production procedure has been worked out in such cases. The second issue that must be addressed is that of the observable: Given either a homo or hetero-oligomer, how will its degree of oligomerization state be assessed under different conditions? In this section a number of approaches will be discussed briefly, touching on their advantages and disadvantages. These include analytical ultracentrifugation coupled to some sort of optical detector, dynamic light scattering, fluorescence (anisotropy or intensity), surface plasmon resonance and iso-thermal titration calorimetry. Many of these techniques are presented in much more detail in specific volumes of this on-line text dedicated to particular techniques. As a general rule, the characterization will be all the more solid, the more complementary techniques are employed, since there are perturbations and uncertainties associated with all of them. The experimentalist is limited, moreover, by the concentration range of the association, relative to the sensitivity of the various techniques. Characterization of protein interactions for homo-oligomers requires a slightly different experimental approach that the study of hetero-oligomerization. It is rare that only one of 24 these approaches can suffice to provide all of the necessary information, such that it is a good idea to become familiar with several of these techniques. Below is given a very short overview of their usefulness and some of their limitations. The reader is encouraged to consult the appropriate accompanying chapters and volumes of this on-line text for an in-depth discussion.

VIII. BIOPHYSICAL TECHNIQUES USED TO STUDY PROTEIN-PROTEIN INTERACTIONS

(a) Analytical ultracentrifugation (AUC)- AUC is becoming an increasingly popular tool in the study of protein complexes. After having long been relegated to a very few laboratories with the good fortune of possessing the expertise (mechanical as well as theoretical), necessary for its function, AUC is making a strong comeback. This revival is based primarily on the introduction of a much more user friendly analytical ultracentrifuge, the XLA, by Beckman instruments to replace the venerable Model E. The introduction of the XLA coincided with growing interest in examining protein interactions due to the identification, purification and structural characterization of protein interaction domains involved in key aspects of a large number of human diseases. Thus, studies of protein interactions by AUC are becoming almost routine. For an in-depth discussion of this technique as applied to interactions between bio-molecules in general, the reader is referred to the corresponding chapter of this volume. Suffice it to say here that this technique can be applied in sedimentation equilibrium or sedimentation velocity mode, and that the tools for analysis of complex equilibria are much improved. Sedimentation velocity is a much more rapid experiment, whereas equilibrium sedimentation is more accurate. In sedimentation velocity, the speed of sedimentation is measured by determining the rate of movement of the sedimenting boundary. Usually, either the integral distribution of s, the sedimentation coefficient, or its differential distribution is calculated rather than relying on a single boundary point, to take into consideration boundary spreading due to diffusion, complex systems and non-idealities. In sedimentation equilibrium studies the speed of the centrifuge is low, such that the transport by sedimentation is balanced by diffusion. In this case an equilibrium concentration distribution is established along the cell, i.e., the flux of concentration at any point n the cell must be zero. Then the gradient of concentration *v/s* tube distance is used to calculate the molecular weight, since the concentration distribution is exponentially dependent upon the buoyant mass of the protein. Profiles obtained using mixtures of oligomeric states at equilibrium can be analyzed for the equilibrium constant over the appropriate concentration range. Detection is based on absorption, schlieren refractive index optics or on fluorescence, with the relative sensitivity increasing in that order. The major drawback of AUC lies in its limited sensitivity. Protein-protein interactions span from the upper micromolar range to picomolar, and AUC is really only applicable down to around 0.1 μM. With protein oligomers labeled with dyes absorbing (or fluorescing) at different wavelengths, hetero-oligomers can be distinguished from homo-oligomers. Readers interested in pursuing the use of analytical ultracentrifugation should consult the web site for the National Analytical Ultracentrifugation Facility in Storrs Connecticut at the following address: hpp://gopher.uconn.edu

(b) Light scattering- Static and dynamic light scattering represent another approach to studying protein complexes. In static light scattering, the scattering intensity is related to the molecular weight of the protein, in addition to its concentration, the scattering angle, and the wavelength which in principle are known. Dynamic light scattering is based on the auto-correlation of the time dependent fluctuations of scattered light intensity, which in turn depends upon the diffusion constant. This auto-correlation decays more slowly for slowly diffusing particles and thus, the diffusion constant is extracted from the value of the relaxation time of this function. In the case of ideal spherical particles, this provides a measure of the molecular weight. Light scattering, like AUC, is limited principally by sensitivity, with best results around 1 mg/ml, depending upon the size of the protein or the complex. While sensitivity limits preclude the determination of affinities and association or dissociation rate constants in most cases, light scattering is quite useful in characterizing the stoichiometry of complexes at high concentration. This is very important information for the analysis of data obtained by more sensitive techniques. In fact, it should be pointed out that in many cases, biochemical methods provide the information that protein A interacts with protein B, but the stoichiometry of this interaction is often elusive. Even when crystal structures are available, the stoichiometry of the complex in the crystal may not correspond to that observed in solution under various conditions. Thus, determination of the end point stoichiometry, deduced from the molecular weight determination by light scattering is a good idea.

(c) Fluorescence Spectroscopy- There are a number of approaches based on fluorescence that can be used in the study of protein-protein interactions. Again, this on-line text includes an entire volume on spectroscopic methods, including fluorescence, to which the reader is referred for a detailed discussion of the theory and practice. The most widely used approach in the study of protein-protein interactions is that based on the polarization or anisotropy of fluorescence. This value represents the degree to which an excited fluorophore retains the orientation of the exciting light in its emission Small molecules rotate rapidly and thus retain little of this orientation or polarization, whereas large molecules rotate more slowly in solution, and thus retain a higher polarization. Fluorescence can be carried out in static or in time-resolved modes. In the static mode, the polarization measured reflects loss of orientation by global rotation of the protein molecule to which the fluorophore is attached, in addition to loss of orientation by local motions of the fluorophore. Thus, the absolute value of the polarization measured in the static mode is not particularly informative. However, in the case of an equilibrium between associating protein subunits, this value decreases upon dilution of a homo-oligomer or increases upon titration of the labeled protein A with its protein partner B if this results in formation of a hetero-oligomer. In this manner the equilibrium binding constant, and also the kinetics of the interaction can be studied. The sensitivity of fluorescence is quite high, the limit being in the low nanomolar/high picomolar range, depending upon the probe used, such that interactions of very high affinity can be studied. The drawback is that only in cases of low affinity can the intrinsic tryptophan of the protein 26 be used, since it can only be detected down to about 100 nM in the best of cases. Thus, for high affinity interactions, the protein must be labeled with an extrinsic fluorescence probe. Care must be taken to control for function after labeling since the covalently bound probe can conceivably perturb the interaction under study. Time resolved fluorescence of the intrinsic tryptophan residues or of an external probe can separate the global from the local motions and provide information about the stoichiometry of the complex at micromolar concentrations that can then be used to analyze titration curves carried out at lower concentrations. An additional advantage of fluorescence techniques is that they are readily adapted to kinetics experiments (stopped flow or simple mixing). In addition to fluorescence polarization, fluorescence resonance energy transfer between probes on different subunits can be used to study their interactions. The biochemistry involved in these experiments is somewhat more complex (1:1 labeling required for 2 probes), and the sensitivity somewhat more limited. Fluorescence energy transfer is.based on the dipolar interaction of an excited donor fluorophore that will transfer its excited state energy to an acceptor fluorophore provided that the distance between donor and acceptor is not too large and also provided a strong spectral overlap between the emission of the donor and the absorption of the acceptor. Since this is a dipolar interaction, the probability of transfers goes as 1/R6, where R is the distance between the two fluorophores. The transfer efficiency is also related to the lifetime of the donor. The longer the donor remains in the excited state, the higher the probability that transfer will occur. The spectral overlap, the relative orientation of the two probes, the refractive index of the medium and the quantum yield of the donor in absence of the acceptor all contribute to the value of the characteristic transfer distance Ro, the distance at which 50% transfer is achieved. Typical Ro values range from 10 to 50 , allowing detection of transfer at inter-probe distances up to about 80 Å. Recent applications using long lived luminescing lanthanide probes as donors allow for energy transfer at significantly longer distances. Fluorescence resonance energy transfer or FRET involves therefore a large number of parameters, including the elusive orientation factor (typically, but surely not appropriately assumed to correspond to free rotation of both donor and acceptor). Despite the apparent complexity of FRET, it has proven to be a surprisingly robust technique in biophysical applications, even to the point of very accurate distance measurements in a significant number of biomolecules, later confirmed from crystallographic coordinates. Moreover, if FRET is used to detect interactions, rather than to measure accurate distances (i.e., for structural information about complexes) then much of the complexity of the technique does not come into play. The most serious limitations of the technique lie in the necessity of labeling both interacting macromolecules, and in the size of the complex that can be monitored given the distance constraints based on overlap and lifetime. We also note here that fluorescence correlation spectroscopy (FCS) (based on principles similar to that of dynamic light scattering) is also now being used to determine the diffusion constant (and thus the size) of single protein molecules, and to study dissociation of oligomers. This technique requires relatively more sophisticated equipment (pulsed laser (preferably 2-photon) excitation, microscope), than does steady state fluorescence anisotropy 27 or steady-state FRET. The advantage is that such interactions may eventually be studied *in situ*.

(e) Surface Plasmon Resonance- A technique that been the object of a great deal of interest and activity for studying biomolecular interactions is that of surface plasmon resonance (SPR). This technique is based on the excitation of surface plasmons of a gold layer by the light waves of a laser beam impinging on the gold surface. In fact, the reflectivity of the laser beam drops to near zero at a particular angle, termed the surface plasmon resonance angle, due to this phenomenon. The intensity of the optical fields in these surface waves decays exponentially with distance from the surface over about 100 nm, and the value of the SPR angle is therefore very sensitive to the refractive index of the medium adjacent to the surface. SPR gold chips are first derivatized and then one partner of a protein pair is covalently linked to the surface. Buffer solution bearing the other partner flows over the chip and the changes in the SPR angle are monitored. If the protein in the flow solution binds to the immobilized partner protein, the refractive index near the surface changes, leading to a change in the SPR angle. This technique has proven very useful in demonstrating interactions between particular partners.

(e) Calorimetry. Iso-thermal titration calorimetry represents a very powerful tool for the characterization of protein-protein interactions. This technique is based on the relatively straightforward measurement of the heat absorbed or released upon titration of a solution of 28 molecule A with that of molecule B. If the interactions are in the appropriate concentration range, then ITC experiments yield both the enthalpy and the free energy changes of the reaction. Moreover, using the Gibbs function, one can calculate the entropy change. Since protein-protein interaction free energies are usually the result of strong enthalpy-entropy compensations, a thorough understanding of the energetics truly requires the knowledge of all of these thermodynamic parameters. The experiment is done by sequential additions of the titrant to a cell containing a solution of reactant. The heat change associated with each successive addition of titrant is measured using thermopiles whose signals are proportional to the rate of heat transfer from the titration cell to the heat sink, which integrated over time yields the total heat of reaction. The ITC is carried out at constant temperature. However, if the experiment is repeated at multiple temperatures one can obtain the change in the enthalpy change of reaction with respect to the temperature. This value corresponds to the change in heat capacity that occurs upon complexation. Upon complex formation between large biomolecules, a significant amount of hydrophobic surface area is removed from contact with water, which in turn usually results in a large decrease in the heat capacity. Thus, determination of the heat capacity change can provide information about the change in exposed surface area upon complex formation. The only drawback to ITC is its relatively low sensitivity. The lower limit of dissociation constants that can reasonably be determined is about 10 nM, and this under the best of conditions (large heats of reaction, etc.). Nonetheless, one can always obtain values for the enthalpy change of the reaction, as well as the heat capacity change at concentrations much higher than the dissociation constant. If this latter can be obtained under equilibrium binding conditions (i.e. [reaction] << K]) using another more sensitive technique such as fluorescence anisotropy, then all of the thermodynamic parameters can be obtained. One must take care to ensure that the stochiometry of the interaction under the concentration conditions of the ITC experiments is identical to that observed under equilibrium binding conditions. The interested reader is referred to the volume of this on-line text specifically dedicated to calorimetric methods for further information. A national center for biocalorimetry exists at the Johns Hopkins University with the web site address.

IX. PROTEIN-PROTEIN INTERACTION DATABASES

- HPRD Human Protein Reference Database a (manually) curated database of human protein information with visualization tools.
- IntAct Interaction Database is a public repository for manually curated molecular interaction data from literature.

- Database of Interacting Proteins, a manual and automatic catalog of experimentally determined interactions between proteins.
- MINT Molecular INTeraction Database focuses on experimentally verified protein interactions mined from the scientific literature by expert curators.
- The MIPS Mammalian Protein-Protein Interaction Database

X. INTERACTION NETWORK SOFTWARE

- Osprey
- VisANT
- NAViGaTOR
- Cytoscape is an open source bioinformatics software platform for visualizing molecular interaction networks and integrating these interactions with gene expression profiles and other state data.
- yEd, graph editor.
- APID Agile Protein Interaction DataAnalyzer is an interactive bioinformatics web tool to explore and analyze in a unified and comparative platform, mainly for currently known information about protein–protein interactions.
- NetPro Is a comprehensive fully hand-curated knowledge base of Protein-Protein, Protein-Small molecules DNA and RNA interactions.

QUESTIONS

1. What is protein –protein interaction? Why understanding the interaction is important?
2. Define protein interaction. What are characteristics classes and methods of study protein interactions?
3. Write short notes :

   (*a*) Yeast as a model for protein interaction

   (*b*) Hetrologous proteins.

BIOTECHNOLOGY IN BIODIVERSITY CONSERVATION

CHAPTER 45

I. INTRODUCTION

Since man learnt to domesticate plants for using them as a whole, or their products, more and more plants have been taken for cultivation or are being collected from the wilderness for use. This has resulted in the unequal distribution of species in different regions, receiving anthropogenic pressure of differing intensity. More frequent use of certain plant species, through collection from natural vegetation, is causing the shrinkage of their germless and overall reduction of biodiversity of the concerned region. Similarly, more importance on the high yielding varieties of important cultivated species has caused the shrinkage of the shrinkage of the existence and distribution of the traditional land races and the wild species. Now, with the development in the biotechnology, its application for protection and improvement of biodiversity has become a new area of research.

Many modern technologies, including recombinant DNA technologies, genetic engineering, transgenic, and cell and tissue culture protocols, are employed to obtain newer and newer benefits for human beings. The relationship between biotechnology and biodiversity is multidirectional (*i*) biotechnology provides very powerful tools for the critical assessment of biodiversity-genetics, species, and ecosystems, and consequently, the identification of potential bioresearches, themselves, (*ii*) It gives newer methods and guidelines for the conservation of biodiversity, all aiming at utilizing the conserved resources in a sustainable manner for the benefit of mankind, (*iii*) It enhances the wise and efficient utilization of bioresources, both as a genetic resource for production and in the remediation of altered/degraded ecosystems. In other words, the increasing application of biotechnology and molecular tools to biodiversity, has greatly enhanced the value and availability of bioresources, bioprocesses, and the generated products, for providing food and health care to man and his domesticated animals. This is particularly true with reference to: (*i*) increased availability of food, feed and other renewable raw materials; (*ii*) improved human health and hygiene; (*ii*) greater protection of the environment; and (*iv*) enhancement of biosafety and environment friendly technologies.

II. BIORESOURCE ASSESSMENT

Bioresources are defined as materials of biological origin, that are used for the economic betterment of human society through their trade as food, feed or pharmaceuticals. Biotechnological tools enable critical analysis of the diversity of such bioresources (organic compounds) present in living organisms. The latter contributed to a wide variety of environmental services, such as regulation of the gaseous composition of the atmosphere, protection of coastal zones, regulation of the hydrological cycle and the breakdown of wastes, and absorption of pollutants. Many of these services have neither been recognized, nor properly valued in economic terms. Of an estimated number of 1.75 million described species (Table 45.1), their utilization is restricted to very few species and many have not yet been applied for economic development. The progress in biotechnology has provided the technical knowledge to get the maximum benefit from the existing biodiversity, as well as to improve the quality of the resources.

Table 45.1 : Estimated number of described species on the earth

Kingdom	No. of described species (× 1000)
Bacteria	4
Protists (algae, protozoa, etc.)	80
Animals: vertebrates	52
Animals: Invertebrates	1272
Fungi	72
Plants (Excluding Algae)	270
Total described species	1750
Possible no. of species existing on the earth	14000

The development of very sensitive and highly specific bioassays to defect even picogram quantities of potentially useful organic molecules and compounds and automated screening technology, allow screening of thousands of plant, animals and microbial samples at a very quick pace, and efficient selection of those with value in pharmaceutical, agrochemical, food, cosmetic and other industries. We also now possess the expertise to analyze the diversity available in the DNA molecule, that controls the production of novel compounds. DNA sequences of several organisms are being unraveled and specialized methodologies have been developed to analyze and identify specific gene products. The genetic diversity and its analyse and identify specific gene products. The genetic diversity and its potential have been studied in several species to date and some species of critical importance have been fully sequenced to use the information for the development and better use of the same as well as the related species. We are now better equipped to delineate one species from another and accurately estimate the diversity with in and between species. This has given us the opportunity and technical know how to produce hybrids between highly unrelated species (in terms of sexual incompatibility), which could not be possible by conventional breeding methods. Recent data have indicated that genetic variations and hence, biodiversity should be best assessed only on genetic grounds and not exclusively on phenotypic characteristics. Taxonomic systems based on DNA and RNA sequence data or on the basis of variations in gene products should hereafter, form the main basis for biodiversity characterisation and assessment.

Biotechnological tools have enabled identification of desirable and elite genes locked up in wild relatives of domesticated plants, animals and microbes. They have further made possible the specific transfer of such genes to domesticated taxa, to produce transgenics that have increased performance, productivity and resistance to abiotic and biotic stresses of all kinds. For example, medium chain fatty acids, such as lauric acid, form the major source of dietary and industrial oils in the tropics. The production of this fatty acid in seeds is controlled by a specific protein called acyl carrier protein-BTE. Cloning the DNA of the gene coding for this protein from *Umbellularia californica* and its introduction into other oil-seed plants, can substantially increase the level of laurate. Other notable examples of useful genes identified and mapped using biotechnology are: Genes responsible for the production of useful variants in corn identified through the use of quantitative trait loci (QTL) and evolutionary history, hitherto undiscovered genes from wild relatives of potato and tomato (Tanksley and McGough 1997), and several new useful genes in rice and corn and their wild relatives (McCouch 1998). The most notable achievement is the development of virus resistance in the cultivated maize by transfer of a family of genes from the Mexican perennial wild maize *Zea diploperennis*.

III. BIOTECHNOLOGY IN EX SITU CONSERVATION OF BIODIVERSITY

Biotechnology can also play an important role in culturing and conserving biodiversity.In ex situ situations, it provides newer opportunities. A new class of non-living collections of living organisms, in the form of DNA libraries (of naked DNA and genomic and cDNA as well as isolated chromosomes) and sequence databases has been made possible to keep alive the hope of their regeneration in future. The attempt to regenerate the extinct Tasmaian tiger is a major progress in the direction of regeneration of the lost biodiversity. These profoundly increase the option values of genetic resource collections, since the latter are the bases of agricultural development, as well as for providing health care. Such ex situ collections are invariably desirable from plants and animals, especially wild relatives of domesticated plants/microbes and land races, threatened to different extents (critically endangered, endangered and vulnerable). Such collections can be attempted after genetically characterising the wild relatives/land races through RFLP, RAPD (and its modifications) and allozyme techniques and ascertaining their option values for possible future use in improving domesticated taxa. Although, DNA banks are not substitutes for other conservation methods, they have the distinct advantage of requiring minimum space.

The other biotechnological methodologies exploited for *in situ* conservation of plants include rapid multiplication in a very short time using micropropagation, cryopreservation of vegetative or sexual propagules, synthetic sed technology, etc. Under *in situ* situation, application of biotechnology provides data critical for the best management solutions for conservation of the target species. These data allow assessment of optimal or minimal population sizes for maintaining diversity, so that augmentation can be done to increase the population size through transfer from wild populations or through intense breeding programmes.

(a) Field Gene Banks

This is one of the major methods of *ex situ* maintenance of taxa that must be conserved. A field gene bank is an area of land where collections of growing plant species needing conservation are assembled. The assemblage contains as many individuals of the target species as possible, in order to sustain genetic diversity. These field gene banks make available plant material for breeding, reintroduction, research, especially in population genetics, physiology, microbiology, biochemistry, nutrition and processing technology, and for several other purpose.

Field gene banks are particularly useful in the conservation of perennial species and therefore, are of greatest importance in forestry. They have also been established in agriculture/plantation sectors providing the required germplasm of such tropical crops, as cocoa, coconut, rubber, mango, cassava, coffee, banana, sweet potato and yam. The IBPGR has initially established/promoted 23 field gene banks for crops. Several of these also contain germplasms of wild relatives of domesticated crops.

The characterisation and documentation of germplasm resources in a field gene bank consists of five operational procedures, as follows (IBPGR 1991).

(*i*) **Establishing the origin of the plant material:** This is called Passport data. Passport data includes accession data, if an accession has been received from a breeder, another institution, etc., along with details on site of origin, ancestor or pedigree. It will also include collection data if the accession is directly derived from a field collection, along with a description of the collection site.

(*ii*) **Characterisation:** This includes recording all highly heritable and easily identifiable characters of the plant, especially from a fully mature plant. These characters, often called Descriptors, must be those that are expressed in all environments.

(*iii*) **Preliminary evaluation:** It includes details on plants development and recording of characters expressed by the growing plant.

(*iv*) **Further evaluation:** This includes recording all the reactions of plants to physical stresses and to pathogens and predators.

(*v*) **Management data:** 'This includes handling of the genetic resource, its distribution/exchange, regeneration and maintenance, etc.

Over the years, the global network of base collections has swelled to about 50 institutions, all of which have accepted IBPGR's invitation to participate actively in this network. Now, more than 100 species are being conserved in field gene banks throughout the world. The institutions include some of the International Centres of Agricultural Research (IARCs) and some national centres. The total number of accessions of PGRFA conserved in field banks is about a million at present. The advantage of field gene banks is their provision of materials readily accessible for utilisation and evaluation. A major disadvantage is possible reduction in genetic diversity of the material, thereby increasing its susceptibility to pests and diseases. Furthermore, field gene banks involve large areas of land, which limits the genetic range of the material that can be held.

(b) Seed Banks

The seeds are stopped in a seed bank for considerably long period, without being used, in a viable condition. The aim is to preserve a large amount of species, in a viable state, for posterity. This has emerged as a very good device for conservation of plants of the habitats that are undergoing rapid changes and when the existence of the plant in its natural habitat is threatened. Conservation is both the safest and the cheapest if life processes are reduced to a low level, and seed preservation in seed banks fulfils this requirement admirably. Conservation through seed storage aims to preserve as much as possible the genetic integrity of individual or populations of a species.

A seed bank is one of the most efficient methods of *ex-situ* conservation for sexually reproducing species. Seeds suitable for long-term storage are termed orthodox, conventional or desiccation-tolerant. Such seeds can be stored for a long time without substantial loss of vitality and without much genetic changes. Seeds, which cannot be stored in seed banks under the given conditions, are termed recalcitrant. Jackfruit, citrus, avocado, coffee, tea, maple, cinnamon, nutmeg oak, chestnut, mango, cocoa, coconut and rubber belong to this category. It is difficult to conserve recalcitrant seeds. It is estimated that 20% of the world's total plants produce recalcitrant seeds, that do not survive in low temperature and/or dehydrated conditions. They have no natural dormancy and die if no it allowed to germinate immediately. It has also been found that the majority of orthodox seeds are either from temperate regions, where dormancy is enforced by cold winters, or from arid regions where dormancy is enforced by hot and dry climates. Rain forest seeds are often recalcitrant.

Among orthodox seeds, much importance is often given to small seeds, which occupy less space. The FAO constituted a panel of experts on plant exploration and introduction, which formulated certain standards and procedures for storage installations meant for long-term seed conservation. The recommended conditions for seed storage are -18°C or less temperature, use of airtight sealed containers, and a seed moisture content of 5-2%. The seed bank needs proper and continuous power supply. Medium-term storage can be effected at 6 &10°C. Storage at the temperature of liquid nitrogen has given encouraging and problem-free results. It is also less costly than the conventional seed storage. However, the moisture content needs to be strictly controlled. Orthodox seeds can be kept under storage for very long periods at sub-zero degree centigrade, if previously dried to about 5 to 8% moisture content. Although longevity varies from taxon to taxon, orthodox seeds can be stored for 5 to 25 years at 0 to 50°C (medium-term storage) or for up to 100 years, if stored at- 100°C (long-term storage). Drop in every one percent below 8%, invariably doubles the storage life of seeds, but drying to the level below 4% moisture level, causes biochemical damage of seeds and reduces their viability. For every 50°C drop in temperature, the life of the seed is doubled. For example, onion seeds with 10% moisture are viable for 16 weeks at 35°C but will live for 78 years at 0°C. Dropping the temperature further to – 15°C would enhance seed longevity to 624 years. Keeping these in mind, when both temperature and water content are judiciously reduced, these two factors multiply the life span of a seed.

Long-term seed storage requires regular regeneration tests in stored seeds to monitor their viability. The frequency of their germination /regeneration depends on their initial viability and rate of loss of viability due to storage, and storage conditions. The minimum requirement for each regeneration test is 400 seeds. Some seed banks recommend 5% of the seeds stored to be subjected to regeneration tests. In large seed banks, regeneration tests are done continuously, on the seeds of one species or the other. Two types of seed storage are normally practiced in a seed bank. As per the first, base collections are stored under optimum conditions and are not interfered with, until reduced viability is noticed through periodic germination tests. When reduced viability is observed, new seed regeneration is carried out. In the second type of seed storage, active collection of seeds is done, from which subsamples can be taken periodically for experimentation, exchange, evaluation and display. The active collection need not necessarily be stored under optimum seed storage conditions and can be multiplied periodically, by growing a new generation of plants and by reharvesting. This collection acts as a source of genetic material for farmers and plants breeders for their utilisatioin.

While some seed banks specialise in the storage of seeds of taxa in specific geographic areas, others specialise in seeds of a particular taxonomic group. Some banks concentrate on the seeds of forest trees and some on specific crops. No where is there any attempt to store the seeds of the weeds and the weedy relatives of the cultivated species, through it has been well understood that many weeds contain numerous valuable genes that can be used for crop improvement.

It should be kept in mind that a germplasm conservation through seed banks is not an immediate option for biodiversity conservation because the process does not give any importance to the habitat in which the species grow. It only comes as an alternative method when it is realised that there is no other way to conserve the germplasm *in situ*. In addition, it cannot be practiced. for all seeds. It is also criticised because it freezes evolution of plants; seed germplasms held in banks can no longer continuously adapt to changes in the environment, such as exposure to new races of pests or climatic stresses/changes.

(c) Pollen Banks

To make the germplasm conservation economical and to conserve many taxa in a limited space, as limited as a deep freezer, pollen storage can be a good option. Since it is now possible, at least in a few taxa, to raise whole plants as haploids from pollen grains, pollen banks have assumed additional importance. The pollens remain in their normal viable state and can be directly made available for use, for artificial fertilisation and breeding works, especially in breeding tree taxa. The major disadvantage of a pollen bank is that only paternal material can be conserved and regenerated.

Pollen preservation can be made for prolonged periods through cryopreservation. If the moisture content of pollen is suitably reduced, according to the requirements of a particular species, pollen can be kept at cryogenic temperatures for a periods ranging up to 6&12years, but in the majority of species, mostly well below one year. Although the exact reasons for the loss of pollen viability/fertility, in a very short time under apparently good storage conditions, are not known, it is believed that the moisture content of pollen plays a very critical role. Therefore, cryopreservation for long-term pollen storage is unlikely to be within reach for general application, until the causes of pollen breakdown under storage are well understood. However, freeze- and vacuum-drying have both been successfully applied to storing pollen of a number of taxa, even up to 12 years, usually at a storage temperature of +5 to -18°C.

(d) DNA Banks

A DNA bank may be defined as a gene library in which samples of DNA extract are stored. This provides a new option for accession of plant germplasm (Adams et al. 1994). The DNA samples are of three kinds: (*i*) total genomic DNA, (*ii*) DNA libraries, and (*iii*) individual cloned DNA fragments, including RFLP probes, mini- and microsatellites, etc. Samples of the first type are made with DNA, isolated directly from plant tissues. The DNA library preparation requires another step, i.e. fragmenting the isolated DNA with a suitable restriction enzyme and packaging the diverse mixture of fragments into a suitable cloning vector. The purpose of a DNA library is to retain each fragment from the original DNA extract, so that the whole genomic DNA is represented in the mixture. The third type of DNA samples, as individual cloned fragments, are fixed and genetically pure since each vector molecule in the sample is host to the same DNA fragment.

Stored DNA samples ideally, serve two contrasting purposes: i) they are very convenient experimental materials, that can be shipped immediately without quarantine problems and put to use immediately for further analysis and manipulation at the molecular level, and ii) they are ideally suited to a time capsule approach to conservation. In other words, they are frozen genetic resources and potentially, the most stable form of preserved germplasm. They do not require recurrent regeneration to retain their future utility indefinitely. Concomitantly, DNA banks are afflicted with almost all the problems faced by seed banks, but more acutely. Proper, reliable documentation and labelling of DNA samples are very crucial to their use. In addition, the despatch, receipt and use of these samples require a high level of technical skill. Moreover, establishing ownership and control of DNA samples is a complex 'delicate' problem. These factors limit the contribution banks could make to the conservation of biodiversity.

In spite of the above problems, an international network of DNA books (DNA Bank-Net) for the conservation of genomic DNA has been vigorously promoted (Adams 1993). There are three primary data bases that collect and distribute DNA material: Gene Bank (Bethesda, Maryland USA), EMBL (Cambridge, UK) and the DNA Data Bank of Japan (Mishima, Japan). The main aim of such a network is to make biological resources more widely available for human benefits, through increased research work and use. However, it should be understood that it is only an adjunct and not an alternative methodology in conservation. So, a DNA bank can not replace a field gene bank, despite being less costly than the latter.

IV. IN VITRO CONSERVATION METHODS

This form of conservation is very important and is followed throughout the world by many institutions. Mostof the *in-vitro* methods have several advantages, such as greater safety from viral attack, lower cost and accommodation of larger numbers. All the methods involve storage of plant germless under in vitro conditions. Meristem tips and buds are usually stored; such storage is especially followed for those taxa whose seeds are recalcitrant. Very often, the meristems or buds from intact plants or the adventive 'embroyos' produced directly from *in-vitro* grown explants or calli, are coated with materials such as alginate or neutral gums and stored.

V. BIOTECHNOLOGY VS BIODIVERSITY

Interest in research on the potential effects, of the release of genetically modified organisms, on biodiversity is accelerating. Introduction of genetically modified organisms (GMOs) into natural and homeostatic ecosystems is likely to cause the loss of species and habitat diversity; at least there is a strong theoretical possibility for this. Adverse biological effects on non-target populations and ecological and evolutionary disruption may be, either the direct result of the introduced transgene(s) or alternatively the indirect result of the changing socioeconomic conditions, related to the application of recombinant technologies.

Several direct non-target effects on beneficial and native organisms by GMOs have been reported, An example is the transgenic Bt cotton plant (raised to resist the cotton bollworm), which affects a wide array of non-target insects, such as butterflies, moths and beetles. Some GM crops have been shown to affect soil ecosystems by decreasing the rate of decomposition of organic wastes, affecting carbon and nitrogen levels and decreasing the diversity of soil microbial populations. Another possible direct impact of GMOs, raised for conferring viral resistance is the likely emergence of new viruses with new biological characteristics through recombination and capsid modification.

GMOs can cause harm to biodiversity, due to the properties of the transgenes per se or their transfer and expression in non-target organisms. Adverse impacts on biodiversity through the introduction of GMOs may also result from disturbance of the dynamic population equilibrium of ecosystems. Enhanced ability of GMOs to invade natural habitats of native species may lead to reduction of the population size of native taxa and, in extreme cases, to the total elimination of the populations of certain species, especially threatened ones. In view of this, the effects of introducing transgenic plants with conferred disease resistance, pest resistance, herbicide/pesticide/fungicide tolerance, environmental stress tolerance, etc. would have to be very seriously considered as a part of risk assessment and biosafety measures, before field trials and commercialisation of transgenic crops areattempted.

Risk assessment, consequent to the release of a GMO, is not easy and involves several sequential steps: (*i*) a priori identification of the potential/ documented hazards, (*ii*) estimation of their relative importance, and (*iii*) computation of the probability of adverse impacts. The impacts probability and its quantification are hardly feasible in the case of biodiversity. Based on the results of risk assessment, GMOs are classified as low, intermediate or high risk organisms. Certain important measures have been suggested for minimising the direct impacts of GMOs For transgenic crop plants, the measures suggested are: (i) elimination or disarmament of the biological vectors, once transgenics are produced, to arrest the potential for further vector-mediated gene transfer; (*ii*) deletion of the pathogenic sequences contained in some vectors such as *Agrobacterium tumefaciens* Ti plasmid to preclude the possibility of their survival and transmission, and (*iii*) control of dispersal of the transgenic crop, beyond the test site through reproductive isolation. (UNEP 1995). For microbial GMOs , the measures suggested for minimising their direct impacts differ, because the possibility of horizontal gene transfer cannot always be excluded. The measures suggested relate mainly to minimising dispersal of the transgene.

Another direct impact of biotechnology could be episodic genetic erosion, which would threaten the genetic diversity on which this technology depends. For instance, micropropagation and the consequent production of identical cones discourage perpetuation of genetic diversity through evolutionary adaptations. GMOs can cause direct impacts on target organisms and indirect impacts on non-target organisms. Pesticidal proteins produced by GMOs may have effects indirectly through bioaccumulations; when predators consume prey items that contain pesticidal proteins, they ingest the latter.

Indirect impacts of biotechnology are immense and of very great relevance to people, in developing countries, who rely directly on biodiversity for their daily sustenance. These are however, predominantly socioeconomic ones, operated through human economic and social systems. The impacts themselves, are the results of human responses to the changes in relative cost and prices of biotechnologically derived items. Knowledge about the use of different species that exist mostly in the wild would raise the value of that material, resulting in increased collection pressures on that plant, which in turn would lead to overexploitation and species loss. Value increases of biotechnologically derived items can increase the interest and conflict over ownership of genetic resources, the rates at which these resources should be utilises, IPRs, and formulation of laws relating to access to the resources.

VI. GERMPLASM PRESERVATION: CRYOBIOLOGY

In India, Botanical Survey of India has released three volumes (Red Data Book) enlisting the threatened plants of Indian subcontinent. There are several other organisations, which are engaged in the protection and preservation of such plants. The organisations are as Indian Council of Agricultural Research, New Delhi; International Bureau of Plant Genetic Resources, Nottingham, U.K.; National Bureau of Plant Genetic Resources, Pusa Campus, New Delhi; Council of Scientific and Industrial Research, New Delhi and Botany Departments of various Universities. Attempts have been made to conserve the germplasm by preserving the genetic material.

Cryopreservation [preservation in the frozen state(cryo means extreme cold, derived from Latin word kruos = 'frost')] is based on the reduction and subsequent arrest of metabolic functions of biological material by imposition of ultra-low temperature. At the temperature of liquid nitrogen (-196°C) almost all the metabolic activities of cells are ceased and the sample can then be preserved in such state for extended periods. However, only few biological materials, in their natural state, can be frozen to sub-freezing temperatures without adversely affecting the cell viability. Knowledge about the chemicals having cryopreservative properties such as glycerol and dimethyl sulfoxide facilitated the development of effective cryopreservation technique..

One of the earliest records on use or cryopreservation technique is of Sun (1958), who attempted to preserve desiccated seedlings of *Pisum sativum*. With the rapid development in plant cell, tissue and organ culture and regeneration through various explants, emphasis has laid to cryopreserve the genetic resources for their use in the future.

(a) Need for cryopreservation

The main application and objective of developing cryopreservation technology is preservation of valuable genetic resources, especially of vegetatively propagated and also of those species which have short lived seeds. Therefore, it is imperative that use of cryopreservation technology requires efficient regeneration protocols through tissue culture of the species. Cryopreservation of endangered species is one of the most important objectives of NBPGR, New Delhi. Endangered species for which micropropagation techniques have been developed in India are listed in Table 45.2. It has been recommended that the regeneration methods using apical meristem is advantageous as these are simple and there are no chances of genetic variation. It is essential that tissue culture methods do not create genetic variability.

Table 45.2. Threatened species for which micropropagation methods have been developed.

Boswellia serrata	*Commiphora wightii*
Calligonum polygonoides	*Coptis teeta*
Caralluma spp	*Tecomella undulata*
Ceropegia bulbosa	*Nepenthes khasiana*
Chlorophytum borivilianum	*Podophyllum hexandrum*

In contrast to above, the cases where genetic variability has been induced in the cultures, this variability need preservation for use in the future. Therefore, cryopreservation technique is equally good for preserving the genetic resources of existing genotypes and also of new variants.

(b) Procedure of Cryoprotection and pretreatment

Some form of cryoprotection is necessary for cryopreservation of plant material unless they are naturally dehydrated, as in the case of dormant vegetative buds in the winter, or artificially cold acclimated. Several chemicals such as dimethyl sulphoxide(DMSO), glycerol, various sugars and sugar alcohols protect living cells against damage during freezing and thawing (means to become unfrozen or warm after preservation at ultra low temperatures) these compounds lower the temperature at which freezing first occurs and can alter the crystal habit of ice when it separates. The colligative properties of the cryoprotectants minimize the harmful action of electrolyte concentration resulting from conversion of water into ice. High solubility in aqueous phase and low toxicity to the cells are the two essential characteristics for cryoprotectants. Cryoprotectants are categorised as –

1. permeating, e.g., DMSO, methanol, glycerol and
2. non-permeating, e.g., sugars, sugar alcohols, high molecular weight polymers as dextran, polyvinyl pyrrolidone, hydroxy ethyl starch.

The cells require different pretreatment periods with different compounds for proper cryoprotection. DMSO enters more rapidly than glycerol, in permeating types, and therefore requires shorter period for treatment. Most of the cryorpotectants exhibit varying degree of cytotoxicity at higher concentrations. Generally, DMSO at 5 to 10% and glycerol at 10 to 20% are used as cryoprotectants. Sometime, a mixture of cryoprotectants can also be used to improve the efficacy. Addition of osmotically active compounds in the culture medium such as mannitol, sorbitol, sucrose and proline increase the freezing resistance of the cells. These compounds mainly act by their dehydration effect. Mannitol has been found to be beneficial in reducing the mean cell volume of cells of *Acer pseudoplatanus* and *Capsicum annum* and increase their post-freezing survival. Therefore, all such compounds also possess cryoprotective properties besides that they are osmotically active. Sorbitol has been successfully used as an osmotic agent as well as cryoprotectant for *Glycine max, Datura innoxia, Brassica napus,* and *Daucus carota.*

In a detailed study using nuclear magnetic resonance (NMR) and differential thermal analysis (DTA), it was recorded that incorporation of DMSO or sorbitol into the liquid medium had a significant effect in the temperature range for initiation to completion of ice crystallization and the amount of liquid water present in the cells at various sub zero temperatures. A close association between the percent-unfrozen water at –40°C and the percent cell survival after freezing in liquid nitrogen was observed. These results explain the cryoprotective properties of DMSO and sorbitol and showed that a combination of DMSO and sorbitol was more effective than use of single compound as cryoprotectant. The beneficial effect of using a mixture of cryoprotectants involving PEG, glucose and DMSO (10:8:10%) has been demonstrated in sugarcane callus.

Cold hardening or abscisic acid (ABA) treatment of wheat suspension culture increased the freezing resistance and survival following cryopreservation. Similarly, cold hardening of *Panax* cell suspension cultures at 2°C in a medium supplemented with 3 to 25% sucrose increased the survival of cells after storage in liquid nitrogen. Incorporation of amino acids (asparagine, alanine, proline, serine) in the preculture medium also enhanced the survival rate of *Dioscorea deltoidea* cells after storage at low temperature.

It is evident from above cited examples that the use of pretreatment or culture conditions prior to freezing is beneficial for post-freezing survival of the cells.

(c) Freezing methods

Methods used by different workers for cryopreservation of various plant materials can be categorized as – (*i*) slow freezing, (*ii*) rapid freezing and (*iii*) droplet-freezing.

(*i*) **Slow freezing** – Methods of freezing biological specimens slowly are based on the physicochemical events occurring during the freezing. When plant cells are cooled progressively, ice crystal formation is usually initiated extra- cellularly. It is presumed that plasma membrane acts as a barrier that prevents the ice crystal formation in the cytoplasm. In absence of ice crystals, cytoplasm remains super cooled. On further lowering in temperature, the concentration of extra-cellular liquid increases as more water is converted to ice. Since the vapour pressure of the frozen solution is lower than the same concentration of the super cooled liquid, the vapour pressure of slowly cooled cells reach equilibrium with external ice by efflux of water. Thus, slow freezing prevents the intra-cellular ice formation and consequently freezing injury is prevented. It is believed that slow freezing increase the concentration of cytoplasm and increased dehydration increases the survival of cells. There are different methods of obtaining protective dehydration, including slow cooling at constant or varying cooling rates, or keeping the samples at one or more intermediate subzero temperatures.

The development of efficient slow freezing method depends upon several factors like cooling rates, pretreatment and cryoprotection, type and physiological state of the material, and the temperature prior to immersion in liquid nitrogen. Slow cooling method used for different species is given in Table 45.3. The most commonly used methods for the cryopreservation of plant cells generally involves regulated slow cooling at a constant rate of 0.5 to 2°C/min. to terminal temperature between –30°C to –40°C followed by storage in liquid nitrogen.

(ii) **Rapid freezing** – Rapid freezing is unsuitable for the cryopreservation of cell cultures, it is employed to cryopreserve shoot tips. It has been used for preservation of shoot tips of carnation, potato, strawberry and several others (Table 45.4) and somatic embryos of oil palms.

Rapid freezing is accomplished by direct immersion of the cryoprotectant-treated specimens in liquid nitrogen. The cooling rate in this method is very high, usually several hundred degrees per minute. At such high cooling rates, the intracellular fluids do not have sufficient time to equilibrate with the external ice with the possibility of intracellular ice formation, which is considered to be lethal for cells.

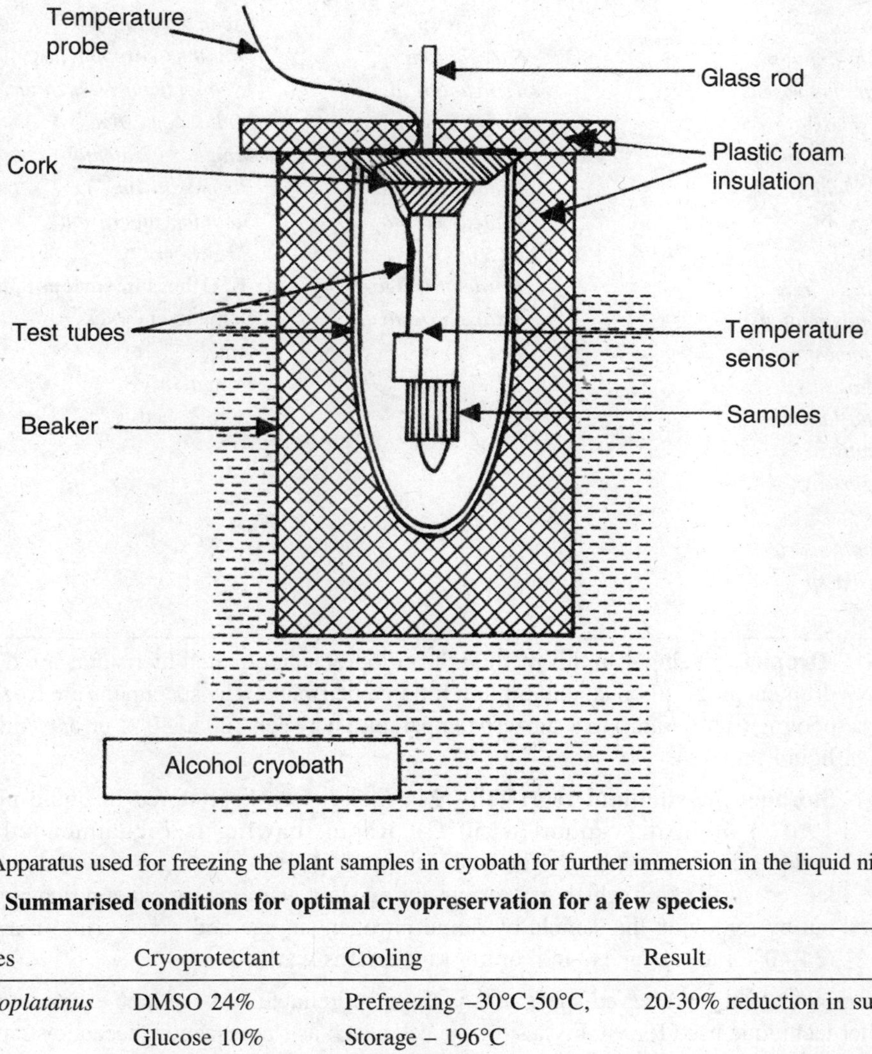

Fig. 45.1. Apparatus used for freezing the plant samples in cryobath for further immersion in the liquid nitrogen

Table 45.3. Summarised conditions for optimal cryopreservation for a few species.

Plant species	Cryoprotectant	Cooling	Result
Acer pseudoplatanus	DMSO 24%	Prefreezing –30°C-50°C,	20-30% reduction in survival
	Glucose 10%	Storage – 196°C	
	Glycerol 10%	Prefreezing 1°C Storage –196°C	28% cell survival
Atropa belladonna	DMSO 5%	Prefreezing 2°C/min. Storage –196°C	40% cell survival
Datura stramonium	DMSO 5%	Prefreezing 1-2°C/min. till-100°C Storage –196°C	40% cell survival
Daucus carota	DMSO 10% or Glycerol 5%	Prefreezing 2-4°C/min Storage –196°C	65% cell survival after 60 days storage, plant regeneration

Ipomoea sp.	DMSO 5% DMSO 2.5% Glycerol 2.5% Sucrose 6.5%	Pre-freezing 2-4°C/min. Storage –196°C	5.2% survival, high growth in few days.

Table 45.4. Plant species in which callus, cells, protoplasts and Meristem/shoot tips have been used for cryopreservation.

A. Cell culture	B. Callus cultures	D.Meristem/shoot tips
Acer pseudoplatanus	*Lavandula vera*	*Arachis hypogaea*
Atropa balladonna	*Medicago sativa*	*Brassica napus*
Brassica napus	*Populus sp.*	*Cicer aerietinum*
Capsicum annuum	*Sacccharum sp.*	*Dianthus caryophyllus*
Catharanthus roseus	*Triticum aestivum*	*Lycopersicon esculentum*
Datura innoxia	*Ulmus americana*	*Malus domestica*
Dioscorea deltoidea	**C. Protoplast**	*Manihot esculenta*
Digitalis lanata	*Datura innoxia*	*Pisum sativum*
Daucus carota	*Daucus carota*	*Solanum tuberosum*
Glycine max	*Glycine max*	*Strawberry*
Hordeum vulgare	*Hordeum vulgare*	**E. Others in various spp.**
Hyosyamus muticus	*Triticum aestivum*	Somatic embryos
Malus domestica	*Zea mays*	Anther
Medicago sativa		Pollen
Nicotiana tabacum		Whole seeds
Oryza sativa		
Panax ginseng		
Populus sp.		
Rose Pauls scarlet		
Saccharum sp.		
Zea mays		

(*iii*) **Droplet freezing** – In this method the cryoprotactant treated meristems are dispensed in droplets of 2-3 μl on an alluminium foil in a petriplate. The specimens are frozen by slow cooling (0.5°C/min.) to a subzero temperature between –20 to 40°C prior to immersion in liquid nitrogen.

(*iv*) **Storage, thawing and regrowth** - Material can be kept stored in liquid nitrogen (–196°C) or in its vapour (–150°C). Rapid thawing is recommended for most cryopreservation methods. The basis of applying rapid thawing is to avoid the damaging ice recrystalization which may occur during slow warming. In general thawing is carried out by removing the sample by liquid nitrogen storage and transferring in a water bath (34-40°C) for about 1-2 min or until material is warmed up.

Regrowth of cryopreserved specimens is the most reliable and accurate estimate of viability. The other technique used for vitality test of the cells such as fluorescence diacetate stain, triphenyl tetrazolium chloride test etc. may provide a quicker method of testing cell viability. The further growth and regeneration of the cryopreserved cells after thawing will be like normal cells. However, care should be taken in handling, plating and subculture of such cells.

(d) Freezing Apparatus. Various types of cryostate and freezing units are available by which different rates of cooling can be easily regulated (Fig. 45.1).

The culture subjected to ultra cooling can be stored at –196°C in the liquid nitrogen container for various lengths of time, and can be taken out and thawed when required.

VII. Applications of cryopreservation

1. Conservation of genetic uniformity.
2. Preservation of rare genomes
3. Freeze storage of cell cultures and cell lines
4. Maintenance of disease free material
5. Cold acclimation and frost resistance
6. Retention of morphogenetic potential in long-term cultures
7. Slow metabolism and aging.

In spite of the information and technology available for *in-vitro* storage of several crop species, *in vitro* active gene banks exist so far only for potato at International Potato Centre, Lima, Peru and casava at Centro International de Agricultura Tropical (CIAT), Cali, Colombia. Recently, efforts have been made under the auspices of International Board for Plant Genetic Resources (IBPGR) to exploit the full potential of *in vitro* storage methods employing meristem and shoot tip cultures. The initiation of a pilot project by collaborative efforts of IBPGR and CIAT, using cassava as a model system, for assessing the potential and feasibility of establishing and running an in-vitro active gene bank is a first significant step. Under the project more than 4000 clones of cassava are being maintained *in vitro* under minimal growth conditions. In India, storage facility has been created at National Bureau of Plant Genetic Resources (NBPGR), New Delhi for germplasm or preservation using *in vitro* methods.

While it is true that biotechnological exploitation of biodiversity has both positive and negative impacts, an optimistic consideration is to make best use of the knowledge of biotechnology for the conservation of the biodiversity. In general, the data suggest that despite a variety of initiatives, biodiversity continues to decline. Most examples of successful conservation action are those, where particular attention, and considerable financial resources have been focused on individual species or localised areas. Many threats to biodiversity, such as habitat modification and natural invasion or artificial introduction, continue to intensify. With the rising interest in biotechnology and the progress made in the field, the gained knowledge can be used to make better utilisation, as well as conservation of bioresources

QUESTIONS

1. Describe the methods of preservation for germplasm.
2. What is cryopreservation ? Describe the procedure and applications of cryopreservation for storage of germplams.
3. Write short notes:

   (*a*) Germplasm

   (*b*) Cryopreservation

   (*c*) *In situ* preservation

   (*d*) *Ex situ* preservation

   (*e*) Thawing.

BIOREMEDIATION

CHAPTER 46

In the past few years, as landfills (specific pits filled with waste material/garbage) have become more and more scarce and concomitantly more and more cost prohibitive, interest in biological methods to treat organic wastes has increased. One area, in particular, that has received increased attention is the biological treatment of petroleum contaminated soils.

The term bioremediation has been given to describe the process by which the use of living organisms (in conjunction with or independent from other technologies) is employed to effectively decontaminate a polluted system. In most cases, the organisms employed are bacteria; however, work is being conducted using fungi and plants. Water hyacinths have been utilized in water systems to effectively remove trace organics and trace metals.

There are two techniques for utilizing bacteria to degrade petroleum in the soil. One method uses the bacteria that can already be found in the soil. These bacteria are stimulated to grow by introducing nutrients into the soil and thereby enhancing the biodegradation process. This process is known as biostimulation. The other method involves culturing the bacteria independently and adding them to the site. This process is known as bioaugmentation.

One advantage of bioremediation is that the process can be done on site with a minimum amount of space and equipment. By treating a site, costs and liability are greatly reduced while extending the life of our current landfills by reducing the amount of waste they would normally receive. On site treatment may involve excavation of the contaminated soil and construction of a lined treatment cell. If excavating is impractical, the treatment may be conducted without disturbing the contaminated site by using a recirculating injection well system. This process is considered *in situ* treatment. Both on site and in situ treatment have their advantages and disadvantages and the decision to use one method of treatment or the other is often dictated by various factors at the site.

I. ON SITE VERSUS IN SITU TREATMENT

On site treatment whereby the contaminated soil is excavated and placed into a lined treatment cell has some distinct advantages. It allows for better control of the system by enabling the engineering firm to dictate the depth of soil as well as the exposed surface area. By controlling the depth and exposed surface area of the soil, one is able to better control the temperature nutrient concentration moisture content and oxygen availability. The presence of the liner is an added benefit, since the liner prevents the migration of the contaminants there is no possibility of contaminating the groundwater. After treatment, the liner is picked up and properly disposed of generally by incineration.

On site treatment has an added benefit in that it is much easier to demonstrate the site is clean than in an in situ cleanup. By isolating the contaminated soil in the treatment cell, it is possible to sample the site in a more thorough, and therefore, representative manner. This may prove a necessity if the regulating agency or the customer desire to optimize the reliability of sampling and analysis.

The excavation of the contaminated soil adds to the cost of a bioremediation project as does the liner and the land farming equipment. In addition to these costs, it is necessary to find enough space to treat the excavated soil on site. In some states, areas are now being set aside to provide the needed space to treat these soils.

In situ treatment is advantageous in instances where the excavation of the contaminated soil is cost prohibitive or impossible. The method of in situ treatment generally involves establishing a hydrostatic gradient through the area of contamination. Water is placed on the site so that it will flow through the area of contamination, carrying nutrients and possibly organisms to the contaminants. Once the water has passed through the site, it is pumped up through wells, and returned to the beginning of the system. This continuous recirculation is carried on until the site has been determined to be clean.

Recovery of the percolating water is the most difficult aspect of this treatment method. Sites may contain a natural clay or rock barrier which collects the percolating water, in which case extraction wells can be placed in this collection zone. Other sites may require the construction of collection trenches or numerous recovery wells at the bottom of the contaminated soil horizon. Given the various geologic/hydraulic conditions that exist at a site, the application of this technology may be limited and would depend on whether regulatory agencies would consider this to be an appropriate and feasible alternative.

The most effective means of implementing these principles depends on the geology/hydrology of the subsurface area, the extent of the contaminated area, and the nature (type) of the contamination. In general, this method is effective only when the subsurface soils are highly permeable, the soil horizon to be treated is within 20-30 feet of grade, and shallow ground water is present, i.e., at 30 feet or less below grade.

As was briefly mentioned above, determining whether or not an in situ remediation process is complete can be a difficult task. If the recirculating water is monitored to check if contaminant has disappeared, then it becomes necessary to somehow correlate the recirculating water to that of the contaminated soil. If monitoring wells are to be used to assess the site, then a preponderance of wells may be necessary to satisfy that the entire site is clean. Due to the poor mixing in these types of systems, it becomes necessary to treat for very long periods of time to ensure that all the pockets of contamination have been treated.

The average time frame for an on site bioremediation project is from sixty to ninety days depending on contamination levels. The average time frame for an in situ bioremediation project can be on the order of twelve to twenty-four months depending on contamination levels and depth of contamination.

The depth of contamination plays an important role in determining whether or not an in situ bioremediation project should be employed. If the contamination is near the groundwater, but the groundwater is not yet contaminated, then it would be unwise to set up a hydrostatic system and further the migration of the contaminant. It would be safer to excavate the soil and treat away from the groundwater by using an on site method of treatment.

II. BIOSTIMULATION VS. BIOAUGMENTATION

Along with deciding whether or not a site should be remediated using on site treatment or in situ treatment, it is necessary to decide how one is going to bioremediate the site. As stated above, there are two methods of employing microorganisms to bioremediate a site. Biostimulation involves the stimulation of indigenous microorganisms to degrade the contaminant. Bioaugmentation involves adding preselected organisms to the site to degrade the contaminant.

A biostimulation project requires that adjustments be made to the soil to enhance the microbial populations already present. These include adding a nitrogen source, a phosphorous source, and a myriad of trace minerals and making appropriate pH adjustments. For an on site treatment, the nutrients are spread over the site and worked into the soil. For an in situ treatment, the nutrients are added to the water upstream in the hydrostatic gradient.

Biostimulation assumes that every organism needed to accomplish the desired treatment results are, in fact, present. Therefore, all that is required to achieve effective biodegradation is to provide (or enhance) an ideal environment for these ubiquitous microorganisms to live and work.

There are numerous shortcomings with this hypothesis. For example, how can we be certain that those organisms present are the most suitable to degrade all materials present? Secondly, what if the only organisms stimulated are those that eliminate the primary substrate, but do not co-metabolize the specifically targeted substrates? At any given site, many of the problem substrates may not be able to be biodegraded directly. If they are the only food source available, the microbes may not be able to degrade these targeted organics, since they do not serve as primary food sources on which the microbes feed.

To ensure that the necessary organisms are present, it is generally necessary to conduct a feasibility study on the soil from the site before any biostimulation project is undertaken. The cost of such a study can range from $5000 to $40,000 depending on the extent of contamination and the characteristics of the contaminants.

Bioaugmentation is the controlled addition of specially formulated biocultures to assist those found naturally in the soil. It is done in conjunction with the development and monitoring of an ideal growth environment in which these selected bacteria can live and work.

In most cases, the targeted organic contaminants either serve as the food source or are co-metabolized. Essential elements are added to the "food source" to provide the required nutrient levels, and water provides the media in which the bacteria function.

The mere addition of bacteria will not, in itself, solve the problem. Studies conducted in 1979 by Dibble and Bartha clearly demonstrated that sewage sludge actually inhibited hydrocarbon biodegradation in soil, and the use of yeast extract had no effects whatsoever(2). The selected microorganisms must be carefully matched to the waste contamination present in the soil, as well as the metabolites formed. They must favorably compete with the ubiquitous organisms found in the expected environmental conditions.

Bioaugmentation allows one to control the nature of the biomass. It provides an element, heretofore not available, that of predictability. Bioaugmentation ensures that the proper team of microorganisms is present in the soil in sufficient type, number, and comparability to effectively and efficiently attack the waste constituents and break them down into their most basic compounds.

One objection to bioremediation has been that it takes an inordinate amount of time for the process to work. In the case of biostimulation, this is true. However, the addition of specially selected microbial consortia allows one to control the biomass of the contaminated site. The additional control of the biomass enables one to increase the kinetic rates of removal from the contaminated site by selecting a more efficient consortia of microorganisms than might be present at the site.

By increasing the kinetic rates, it has been possible to remediate sites in sixty to ninety days using the addition of a selected consortia of microorganisms.

By selecting the microbial consortia beforehand, it is possible to select for organisms that will not produce nuisance odors such as hydrogen sulfide. Petroleum degradation can create anaerobic conditions within the soil. Once anaerobic conditions are present, it becomes possible to generate phytotoxic compounds such as hydrogen sulfide (1). If one augments the soil with organisms that do not possess the ability to generate these phytotoxic compounds, a potential hazard to on site petroleum degradation can be averted.

The cost of the selected microorganisms has been mentioned as a disadvantage in treating contaminated soils. However, if one considers the cost of a feasibility study to ensure that a biostimulation project will work, the cost is considerably less for the bioaugmentation products.

III. PHYTOREMEDIATION

The word's etymology comes from the Greek « phyto » = plant, and Latin « remedium » = restoring balance, or remediating. Phytoremediation consists in depolluting contaminated soils, water or air with plants able to contain, degrade or eliminate metals, pesticides, solvents, explosives, crude oil and its derivatives, and various other contaminants, from the mediums that contain them. It is clean, efficient, inexpensive and non-environmentally disruptive, as opposed to processes that require excavation of soil. Thus phytoremediation is the use of certain plants to clean up soil, sediment, and water contaminated with metals and/or organic contaminants. It is a name for the expansion of an old process that occurs naturally in ecosystems as both inorganic and organic constituents cycle through plants. Plant physiology, agronomy, microbiology, hydrogeology, and engineering are combined to select the proper plant and conditions for a specific site. Phytoremediation is an aesthetically pleasing mechanism that can reduce remedial costs, restore habitat, and clean up contamination in place rather than entombing it in place or transporting the problem to another site.

IV. VARIOUS PHYTOREMEDIATION PROCESSES

Phytoremediation can be used to clean up contamination in several ways (Table 46.1)

- **Phytovolatilization:** Plants take up water and organic contaminants through the roots, transport them to the leaves, and release the contaminants as a reduced of detoxified vapor into the atmosphere.
- **Microorganism stimulation:** Plants excrete and provide enzymes and organic substances from their roots that stimulate growth of microorganisms such as fungi and bacteria. The microorganisms in the root zone then metabolize the organic contaminants.
- **Phytostabilization:** Plants prevent contaminants from migrating by reducing runoff, surface erosion, and ground-water flow rates. "Hydraulic pumping" can occur when tree roots reach ground water, take up large amounts of water, control the hydraulic gradient, and prevent lateral migration of contaminants within a ground water zone.
- **Phytoaccumulation/extraction:** Plant roots can remove metals from contaminated sites and transport them to leaves and stems for harvesting and disposal or metal recovery through smelting processes.
- **Phytodegradation by plants:** Organic contaminants are absorbed inside the plant and metabolized (broken down) to non-toxic molecules by natural chemical processes within the plant.

Table 46.1. The following list gives the media, contaminants and typical plants for the types of phytoremediation listed above.

Application	Media	Contaminants	Typical Plants
1.Phytovolatization	Soil, groundwater, Landfill leachate, land application of wastewater	Herbicides (atrazine, alachlor); Aromatics (BTEX); Chlorinated aliphatics(TCE); Nutrients; Ammunition wastes(TNT,RDX)	Phreatophyte trees (poplar, willow, cottonwood, aspen); Grasses(rye, Bermuda, sorghum, fescue); Legumes (clover, alfalfa, cowpeas)
2.Microorganism stimulation	Soil, sediments, Land application of waste water	Organic contaminants (pesticides aromatic, and polynuclear aromatic hydrocarbons	Phenolics releasers(mulberry, apple,osage orange); Grasses with fribous roots (rye, fescue, bermuda); Aquatic plants for sediments

3.Phytostabilization	Soil, sediments	Metals(Pb,Cd,Zn As,Cu,Cr,Se,U Hydrophobic organics (PAH, PCB,DDT, dieldrin)	Phreatophyte trees to transpire large, amounts of water (hydraulic control); Grasses to stabilize soil erosion; Dense root systems are needed to sorb/bind contaminants
4.Phytoaccumulation/ extraction	Soil, Brownfields, sediments	Metals(Pb,Cd,Zn, As, Cu, Cr,Se,U) with EDTA addition for Pb Selenium	Sunflowers; Indian Mustard; Rape seed plants; Barle, Hops; Crucifers; Serpentine plants; Nettles, dandelions
5.Degradation	Soil, groundwater, Landfill leachate, land application of wastewater	Herbicides (atrazine, alachlor); Aromatics (BTEX); Chlorinated aliphatics (TCE); Nutrients; Ammunition wastes (TNT, RDX)	Phreatophyte trees (poplar, willow, cottonwood, aspen); Grasses (rye, Bermuda, sorghum, fescue); Legumes (clover, alfalfa, cowpeas)

Advantages and Disadvantages of Phytoremediation

When using phytoremediation there are many positive and negative aspects to consider. The advantages and disadvantages are listed below (Table 46.2).

Table 46.2. Advantages and disadvantages of phytoremediation

Advantages	Disadvantages
Works on a variety on organic and inorganic compounds	May take several years to remediate
Can be either *In Situ/ Ex Situ*	May depend on climatic conditions
Easy to implement and maintain	Restricted to sites with shallow contamination within rooting zone
Low-cost compared to other treatment methods	Harvested biomass from phytoextraction may be classified as a RCRA hazardous waste
Environmentally Friendly and aesthetically] pleasing to the public	Consumption of contaminated plant tissue is also a concern.
Reduces the amount wastes to be landfilled	Possible effect on the food chain

A major advantage that is listed above is the low cost. For example, the cost of cleaning up one acre of sandy loam soil at a depth of 50cm with plants is estimated at \$60,000-\$100,000 compared to \$400,000 for the conventional excavation and disposal method. One reason for this low cost is phytoremediation may not require expensive equipment or highly specialized personnel, and can be relatively easy to implement.

One major concern with phytoremediation is the possible affects on the food chain. For example vegetation is used that absorbs toxic or heavy metals and moles or voles eat the metal contaminated plants. The predators of the moles or voles then become victims of intoxication. All though the possibilities of such scenarios are being looked at, more fieldwork and analysis is necessary to understand the possible effects phytoremediation can have.

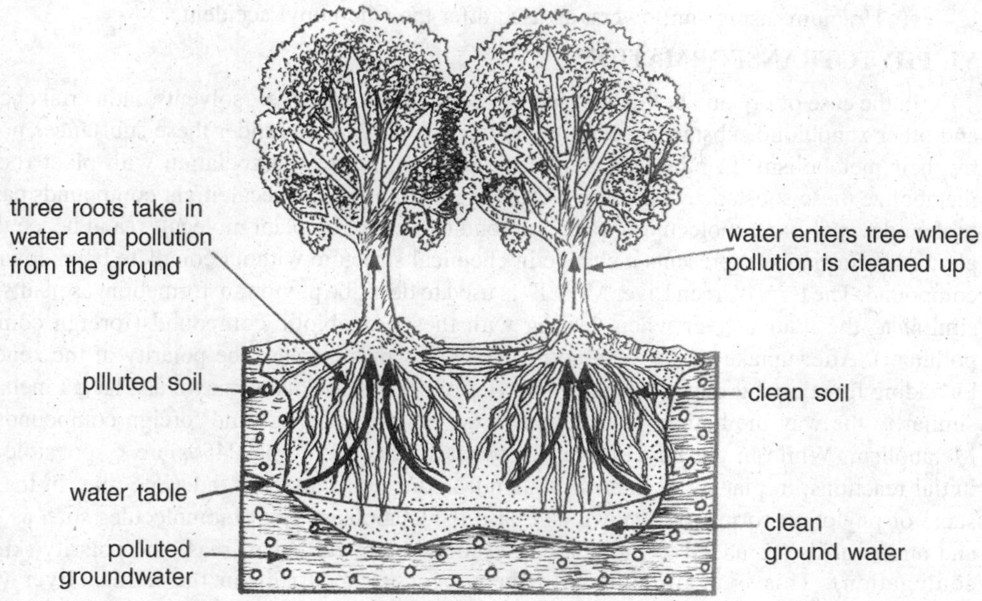

Fig.46.1. cleaning of soil and underground water by plants.

PHYTOEXTRACTION

Phytoextraction (or *phytoaccumulation*) uses plants to remove contaminants from soils, sediments or water into harvestable plant biomass. Phytoextraction has been growing rapidly in popularity world-wide for the last twenty years or so. Generally this process has been tried more often for extracting heavy metals than for organics. At the time of disposal contaminants are typically concentrated in the much smaller volume of the plant matter than in the initially contaminated soil or sediment. 'Mining with plants', or **phytomining**, is also being experimented with.

The plants absorb contaminants through the root system and store them in the root biomass and/or transport them up into the stems and/or leaves. A living plant may continue to absorb contaminants until it is harvested. After harvest a lower level of the contaminant will remain in the soil, so the growth/harvest cycle must usually be repeated through several crops to achieve a significant cleanup. After the process, the cleaned soil can support other vegetation.

Two versions of phytoextraction:

- **natural hyper-accumulation**, where plants naturally take up the contaminants in soil unassisted, and
- **induced or assisted hyper-accumulation**, in which a conditioning fluid containing a chelator or another agent is added to soil to increase metal solubility or mobilization so that the plants can absorb them more easily.

Examples of phytoextraction from soils (see also 'Table of hyperaccumulators'):

- Arsenic, using the Sunflower (*Helianthus annus*), or the Chinese Brake fern ("Pteris spp"], a hyperaccumulator. Chinese Brake fern stores arsenic in its leaves.
- Cadmium and zinc, using alpine pennycress (*Thlaspi caerulescens*), a hyperaccumulator of these metals at levels that would be toxic to many plants. On the other hand, the presence of copper seems to impair its growth (see table for reference).
- Lead, using Indian Mustard (*Brassica juncea*), Ragweed (*Ambrosia artemisiifolia*), Hemp Dogbane (*Apocynum cannabinum*), or Poplar trees, which sequester lead in its biomass.
- Salt-tolerant (moderately halophytic) barley and/or sugar beets are commonly used for the extraction of Sodium chloride (common salt) to reclaim fields that were previously flooded by sea water.

- Uranium, using sunflowers, as used after the Chernobyl accident.

VI. PHYTOTRANSFORMATION

In the case of organic pollutants, such as pesticides, explosives, solvents, industrial chemicals, and other xenobiotic substances, certain plants, such as Cannas, render these substances non-toxic by their metabolism. In other cases, microorganisms living in association with plant roots may metabolize these substances in soil or water. These complex and recalcitrant compounds cannot be broken down to basic molecules (water, carbondioxide etc) by plant molecules, and hence the term phytotransformation represents a change in chemical structure without complete breakdown of the compound. The term "Green Liver Model" is used to describe phytotransformation, as plants behave similar to the human liver when dealing with these xenobiotic compounds(foreign compound/ pollutant). After uptake of the xenobiotics, plant enzymes increase the polarity of the xenobiotics by adding functional groups such as hydroxyl groups (-OH). This is known as Phase I metabolism, similar to the way the human liver increases the polarity of drugs and foreign compounds (Drug Metabolism. While in the human liver, enzymes like Cytochrome P450s are responsible for the initial reactions, in plants enzymes such as nitroreductases carry out the same role. In the second stage of phytotransformation, known as Phase II metabolism, plant biomolecules such as glucose and amino acids are added to the polarized xenobiotic to further increase the polarity (known as conjugation). This is again similar to the processes occurring in the human liver wherein glucuronidation (addition of glucose molecules by the UGT (e.g. UGT1A1) class of enzymes) and glutathione addition reactions occur on reactive centers of the xenobiotic. Phase I and II reactions serve to increase the polarity and reduce the toxicity of the compounds, although many exceptions to the rule are seen at least in the case of the human liver. The increased polarity also allows for easy transport of the xenobiotic along aqueous channels. In the final stage of phytotransformation (Phase III metabolism), a sequestration of the xenobiotic occurs within the plant. The xenobiotics polymerize in a lignin-like manner and get a complex structure which is sequestered in the plant. This ensures that the xenobiotic is safely stored in the plant, and does not affect the functioning of the plant. However, preliminary studies have shown that these plants can be toxic to small animals (such as snails) and hence plants involved in phytotransformation may need to be maintained in a closed enclosure. The human liver differs from plants in Phase III metabolism, since the liver can transport the xenobiotics into the bile for eventual excretion. Since plants have no excretory mechanisms, they sequester the modified xenobiotics. Hence, the plants reduce toxicity (with exceptions) and sequester the xenobiotics in phytotransformation. Trinitrotoluene phytotransformation has been extensively researched a transformation pathway has been proposed .

VII. THE ROLE OF GENETICS

Breeding programs and genetic engineering are powerful methods for enhancing natural phytoremediation capabilities, or for introducing new capabilities into plants. Genes for phytoremediation may originate from a micro-organism or may be transferred from one plant to another variety better adapted to the environmental conditions at the cleanup site. For example, genes encoding a nitroreductase from a bacteria were inserted into tobacco and showed faster removal of TNT and enhanced resistance to the toxic effects of TNT.

Advantages and limitations

- **Advantages:**
  - the cost of the phytoremediation is lower than that of traditional processes both *insitu* and *exsitu*
  - the plants can be easily monitored
  - the possibility of the recovery and re-use of valuable metals (by companies specializing in "phytomining")

- it is the least harmful method because it uses naturally occurring organisms and preserves the natural state of the environment.
- **Limitations**:
- phytoremediation is limited to the surface area and depth occupied by the roots.
- slow growth and low biomass require a long-term commitment
- with plant-based systems of remediation, it is not possible to completely to prevent the leaching of contaminants into the groundwater (without the complete removal of the contaminated ground which in itself does not resolve the problem of contamination)
- the survival of the plants is affected by the toxicity of the contaminated land and the general condition of the soil.
- possible bio-accumulation of contaminants which then pass into the food chain, from primary level consumers upwards.

VIII. HYPERACCUMULATORS AND BIOTIC INTERACTIONS

This section is for the first four points (Protection, Interferences, Mutualism, and Commensalism.

A plant is said to be a hyperaccumulator if it can concentrate the pollutants in a minimum percentage which varies according to the pollutant involved (for example: more than 1000 mg/kg of dry weight for nickel, copper, cobalt, chromium or lead; or more than 10,000 mg/kg for zinc or manganese. 145 hyperaccumulators of nickel (around 300 Ni accumulators are known; 26 of cobalt, 24 of copper, 14 of zinc, four of Lead, and two of Chromium 46.3. This capacity for accumulation is due to hypertolerance, or phytotolerance: the result of adaptative evolution from the plants to hostile environments along multiple generations. Boyd and Martens list 4 biotic interactions that may be affected by metal hyperaccumulation, to which can be added the biofilm as a particular aspect of micorrhizae:

1. protection
2. Interferences with neighbour plants of different species
3. Mutualism (Mycorrhizal associations or micorrhizae, and Pollen and seed dispersal)
4. Commensalism
5. The biofilm

Table 46.3. Hyperaccumulators plant species

Plant species	**Metal**	**Organism(s) affected**
Silene vulgaris (Moench) Garke	Cu (400 mg g-¹)	*Hadena cucubalis Schiff. (Lepidoptera: Noctuidae)*
Streptanthus polygaloides Gray	Ni	*Xanthomonas campestris* (Gram-negative bacterium)
Streptanthus polygaloides Gray	Ni	*Alternaria brassicicola* (Imperfect fungus)
Streptanthus polygaloides Gray	Ni	*Erisyphe polygoni* (Powdery mildew)
Streptanthus polygaloides	Ni	*Pieris rapae L.* (Lepidoptera: Pieridae)
Thlaspi montanum L. var. *montanum*	Ni	*Pieris rapae*
Thlaspi caerulescens J. and C. Presl.	Zn	*Schistocerca gregaria (Forsk.) (Orthoptera: Acrididae)*
Thlaspi caerulescens J. and C. Presl.	Zn	*Deroceras carvanae (Pollonera) (Pulmonata: Limacidae)*
Thlaspi caerulescens J. and C. Presl.	Zn	*Pieris brassicae L. (Lepidoptera: Pieridae)*

IX. MUTUALISM

Two types of mutualism are considered here, mycorrhizal associations or *mycorrhizae*, and animal-mediated pollen or seed dispersal.

1 - **Mycorrhizal associations** or *mycorrhizae*

There are two types of mycorrhizal fungi: ectomycorrhizae and endomycorrhizae. Ectomycorrhizae form sheaths around plant roots, endomycorrhizae enter cortex cells in the roots.

Mycorrhizae are the symbiotic relationship between a soil-borne fungus and the roots of a plant. Some hyperaccumulators may form mycorrhizae and, in some cases, the latter may have a role in metal treatment. In soils with low metal levels, vesicular arbuscular mycorrhizae enhance metal uptake of non-hyperaccumulating species. On the other hand, some mycorrhizae increase metal tolerance by decreasing metal uptake in some low-accumulating species. Mycorrhizae thus assists Calluna in avoiding Cu and Zn toxicity. Most roots need about 100 times the amount of carbon than do the hyphae of its associated ectomycorrhizae in order to develop across the same amount of soil. It is therefore easier for hyphae to acquire elements that have a low mobility than it is for plant roots. Cesium-137 and strontium-90 both have low mobilities.

Mycorrhizal fungi depend on host plants for carbon, while enabling host plants to absorb the soil's nutrients and water with more efficiency. In mycorrhizae, nutrient uptake is enhanced for the plants while they provide energy-rich organic compounds to the fungus. Although certain plant species that are normally symbiotic with mycorrhizal fungi can exist without the fungal association, the fungus greatly enhances the plant's growth. Hosting mycorrhizae is much more energy effective to the plant than producing plant roots.

2 - **Pollen and seed dispersal**

Some animals obtain food from the plant (nectar, pollen, or fruit pulp - Howe & Westley 1988). Animals feeding from hyperaccumulors high in metal content must either be metal-tolerant or dilute it with a mixed diet. Alternatively hyperaccumulators may rely on abiotic vectors or non-mutualistic animal vectors for pollen or seed transport, but we lack information on seed and pollen dispersal mechanisms for hyperaccumulating plants.

X. COMMENSALISM

This is an interaction benefiting one organism while being of neutral value to another. The most likely one with hyperaccumulators would be epiphytism. But this is most noticeable in humid habitats, whereas only a few detailed field studies of hyperaccumulators have been conducted in such habitats, and those studies (mostly to do with humid tropical forests on serpentine soils) pay little or no attention to that point. Proctor *et al.* (1988) studied the tree *Shorea tenuiramulosa*, which can accumulate up to 1000 mg Ni/kg dry weight in leaf material. They estimated covers of epiphytes on the boles of trees in Malaysia, but did not report values for individual species. Boyd *et al.* (1999) studied the occurrence of epiphytes on leaves of the Ni hyperaccumulating tropical shrub *Psychotria douarrei* (Beauvis.). Epiphyte load increased significantly with increasing leaf age, up to 62% for the oldest leaves. An epiphyte sample of leafy liverworts removed from *P. douarrei*, was found to contain 400 mg Ni /kg dry weight (far less than the host plant, whose oldest and most heavily epiphytized leaves contained a mean value of 32,000 mg Ni/kg dry weight). High doses of Ni therefore do not prevent colonization of *Psychotria douarrei* by epiphytes.

In theory another commensal interaction could exist, if the high metal content of the soil under hyperaccumulator plants was needed for another plant species to establish itself. No evidence is known showing such effect.

XI. BIOCATALYSTS

Advances in both chemical catalysis and biocatalysis are determinants in reducing the environmental pollutants of chemical processes and petroleum-based technologies. In the field of chemical catalysis (*i.e.* using catalysts of nonbiological origin) the principle of environmental catalysis is well established and has been accepted for decades: in the 60s and 70s it flourished, producing excellent research on how to decrease, catalytically, the amounts of contaminants in the fuels derived from petroleum, and in the 80s and 90s, increasing interest about catalysis and the environment was documented in many scientific scenarios.

(a) Biocatalysis platform for green processes

'Green chemistry' is defined as the design, development and application of chemical processes and products to reduce or eliminate the use and generation of substances hazardous to human health and the environment. Biocatalysts (either enzymes or whole-cells) constitute a greener alternative to traditional organic synthesis that offers appropriate tools for the industrial transformation of natural or synthetic materials under mild reaction conditions, low energy requirements and minimizing the problems of isomerization and rearrangement

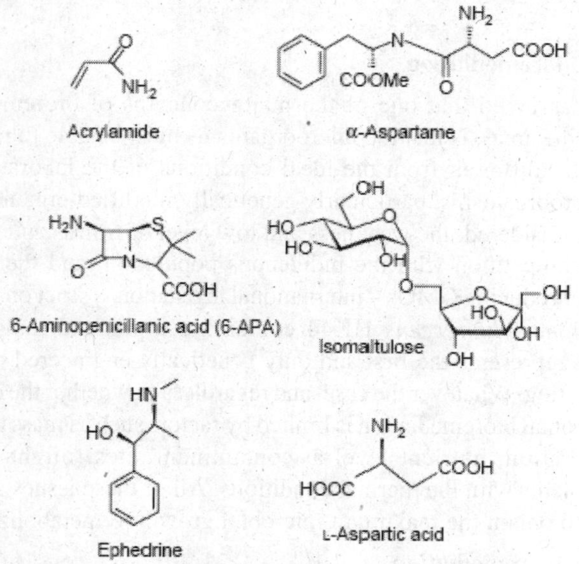

Fig. 46.2. Products of green chemistry

In addition, biocatalysts are biodegradable and can display chemo-, regio- and stereo-selectivity, resulting in decreased by-product formation and avoiding the need for functional-group activation, protection or deprotection.

Large-scale industrial applications of biocatalysts include, for example, the thermolysin-catalyzed synthesis of the low calorie sweetener aspartame, the production of acrylamide and nicotinamide (assisted by nitrile hydratases), the synthesis of the non-cariogenic sweetener isomaltulose by sucrose mutases and, more recently, the production of biopolymers such as polylactic acid (Fig. 46.2). Good examples of the replacement of traditional organic processes by a greener biocatalytic alternative include the industrial synthesis of semisynthetic penicillins and cephalosporins, the transformation of natural and synthetic fibers, the pulp kraft-bleaching and recycling of paper, and the multi-step synthesis of polyketide and glycopeptide antibiotics. To overcome the problem of competing chemical reactions, the current frontiers for biocatalysis are protein activation and stabilization, and reaction specificity. For example, by using protein engineering and high-throughput screening it is currently possible to create stereoselective substrate specificities, which are useful in organic synthesis; stable hydroxynitrile hydrates for the synthesis of for example, substituted R-mandelic acids; and hydrolases capable of enantioselective carbon–carbon bond formation or selective oxidation processes.

(*b*) Bioremediation of persistent contaminants

Removal of widely dispersed, anthropogenic, organic pollutants is considered as one of the main concerns for the reasonable and sustainable development of planet Earth in the 21st century.

Compared with traditional physico-chemical methods, bioremediation is generally the safest, least disruptive and most cost-effective treatment. This biotechnological tool uses whole microorganisms, naturally occurring or introduced, or isolated enzymes to degrade persistent contaminants into non- or less-toxic compounds and, in many cases, can be combined with complementary physical, chemical or mechanical processes to improve the reliability and effectiveness of detoxification. For example, since November 2002, microbial bioremediation using naturally occurring microbes in combination with mechanical approaches are currently being used as the major mechanism of removing low and high molecular weight, polycyclic aromatic hydrocarbons (PAHs) from the Prestige ship spill on the north coast of Spain. In fact, there are many possibilities in this area of biotechnology, either by in situ bio-stimulation (e.g. bioventing and natural attenuation) or ex situ technologies (e.g. biocells and land farming).

(*c*) Microbial bioremediation

It is generally assumed that one of the major concerns of bioremediation with microbes, particularly in maritime areas, is that the microorganisms must be able to resist a variety of adverse conditions that are far different from the ideal conditions of the laboratory. Additionally, when using inoculated microorganisms, particularly genetically modified organisms (GMOs), two major problems should be considered: the weakness and low level of fitness and growth of the inoculated microorganisms in competition with the indigenous population, and the possibility of altering a given ecosystem by introducing GMOs – international legislation is strict on this point. The controlled release of *Pseudomonas fluorescens* HK44 containing a plasmid that encodes a naphthalene catabolizing enzyme represents the first and only genetically engineered microorganism approved for bioremediation testing. Whatever the case, and regardless of whether the organisms are genetically modified or not, microbial bioremediation is limited by factors such as mass transfer (low contaminant bio-availability), aeration, nutrient level at contaminant sites (might require bio-stimulation techniques) and problems with the thermal conditions. All of these issues should be addressed on a case-by- case basis to obtain the maximum microbial growth to metabolize the toxicants.

(*d*) Enzymatic bioremediation

The enzymatic bioremediation has become an attractive alternative to further support the bio-treatment techniques currently available – enzymes provide simpler systems than a whole organism. Most xenobiotics can be submitted to enzymatic bioremediation, for example, polycyclic aromatic hydrocarbons (PAHs), polynitrated aromatic compounds, pesticides such as organochlorine insecticides, bleach-plant effluents, synthetic dyes, polymers and wood preservatives (creosote, pentachlorophenol) (Fig. 46.3). It is just a matter of searching for the microorganisms capable of feeding on a particular pollutant, and then focusing the effort on identifying the enzyme(s) responsible. Historically, the most studied enzymes in bioremediation are bacterial mono- or di-oxygenases, reductases, dehalogenases, cytochrome P450 monoxygenases, enzymes involved in ligninmetabolism (such as laccases, lignin- and manganese peroxidases) from white-rot fungi, and bacterial phosphotriesterases. Moreover, new developments in the design and application of enzymatic 'cocktails' for the biotreatment of wastewaters have recently emerged owing to the effort of many companies and administrations such as American Industry Enzyme. Technologies and Australian Orica. From an environmental point of view, the use of enzymes instead of chemicals or microorganisms undoubtedly presents some advantages:

1. the biotransformation does not generate toxic side products as is often the case with chemical and some microbiological processes, and the enzymes are digested, *in situ*, by the indigenous microorganisms after the treatment;

2. the requirement to enhance bio-availability by the introduction of organic co-solvents or surfactants is much more feasible from an enzymatic point of view than using whole cells; and

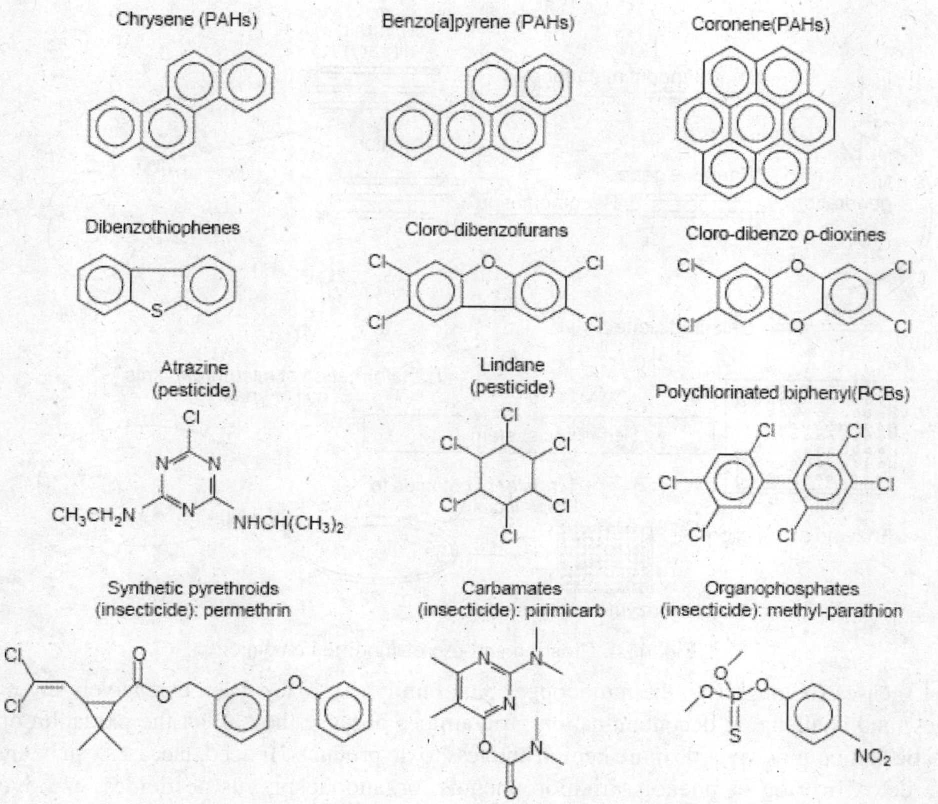

Fig.46.3. Compounds for which enzyme is used for bioremediation.

3. the potential to produce enzymes at a higher scale, with enhanced stability and/or activity and at a lower cost by using recombinant-DNA technology.

When performing enzymatic bioremediation, it is imperative that the optimal conditions for the enzyme are maintained during operational conditions. This requires cheaply produced enzymes (i.e. heterologous expression), with high substrate affinity (Km in the micromolar range), supporting thousands of product turnovers. At the same time, enzymes should display robustness under an array of external factors and low dependency on expensive redox cofactors (i.e. NAD(P)H), which might prove prohibitive in a commercial setting. Many of these shortcomings have been successfully resolved by using directed enzyme evolution (Fig. 46.4) or by semi-rational approaches (*i.e.* combinatorial saturation mutagenesis of several hot-spot residues). Obviously, a convenient method of assessing bioremediation applications should be the improvement of several enzymatic properties simultaneously, such as stability and activity. By directed (or forced) evolution, new features can be conferred to enzymes that, somehow, do not demonstrate affinity against novel or poorly degraded xenobiotics such as insecticides, herbicides, fungicides and mycotoxins. One example of this is atrazine (2- chloro-4-ethylamino-6-isopropylamino-1,3,5-s-triazine), a class of herbicide that first appeared in the 50s. In 2001 Raillard and co-workers applied directed evolution to shuffle two highly homologous triazine hydrolases, and the mutants used triazines that were originally not substrates for parent types. Directed evolution (two rounds of DNA shuffling) was also used to tailor a highly efficient phosphotriesterase to degrade organophosphates (OP) such as methyl parathion, a highly toxic neurotoxin used in insecticides and chemical warfare agents.

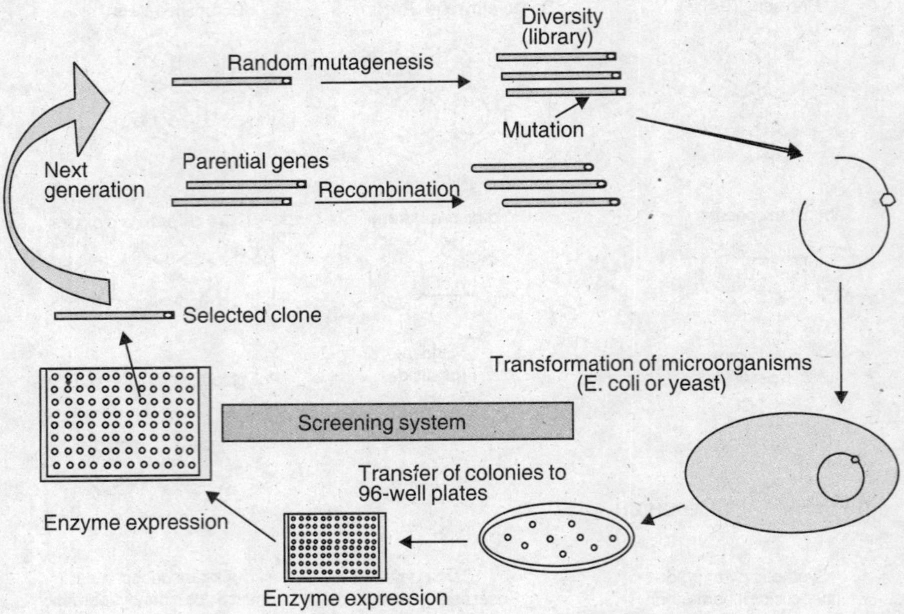

Fig. 46.4. Creation and use of modified enzyme.

Laccases (belonging to the multicopper blue family) have also been extensively investigated for new and challenging decontamination programmes because they affect the oxidation of many aromatic compounds towards more benign and less toxic products. Indeed, laccases can be involved in the detoxification of phenols, trichlorophenols, organophosphorus pesticides, azo dyes and, interestingly, PAHs such as benzo[*a*]pyrene, a class of highly mutagenic and carcinogenic xenobiotics, which are widely distributed in terrestrial and aquatic environments.

(*e*) **Enzymes for clean energy production**

Nowadays, exciting new opportunities for biocatalysis in the production of renewable and clean energy sources, such as biodiesel, bioethanol and biohydrogen, are rapidly emerging. Based on the premise that these alternatives can contribute to a cleaner environment, particularly when using renewable agricultural products, the demand for these energy sources is increasing. Today, bioenergy accounts for approximately 15% of the energy consumption of the world, and it is believed that in the next 20 to 40 years we will be able to convert biomass into our transportation fuels. According to the enzymatic platform and the biomass used, we can distinguish three main groups of clean energies.

(*f*) **Biodiesel.** The conversion of vegetable oils to methyl- or other short chain esters in a single transesterification reaction using lipases has led to the production of high-grade biodiesel. This technology overcomes the disadvantages of chemical transformations based on acid- or base-catalysts because it reduces the consumption of energy and the need for separation of the catalyst from the reaction mixture, which is costly and chemically wasteful. However, there are pros and cons to the use of biodiesel. Biodiesel is renewable, has low emissions per volume and is exempt from diesel tax, through special legislation, in several European countries, making processes involving biocatalysis more competitive. Moreover, the efficient, solventfree synthesis of oleic acid short-chain alcohol esters have been achieved with immobilized lipases such as those from *Pseudomonas cepacia*, *Rhizomocur miehei* and *Candida antarctica*. The limitations include the relatively high production cost, moderate reaction yields and the difficulties found during purification of the unreacted substrates, which obviously will require new future advances.

(*g*) **Bioethanol.** Before the discovery of petroleum, natural carbohydrates were used for the production of food, clothing and energy. Ethanol fuels can be derived from renewable resources, including agricultural crops, such as corn, sugar cane and sugar beet, or from agricultural byproducts, such as whey (from cheese making) and potato processing waste streams. Ethanol can be used either as a 100% replacement for petroleum fuels or as an extender: it can replace the toxic oxygenate methyl tert-butyl ether. The best currently available technology for conversion is the acid hydrolysis of the biomass into sugars; however, alternative technologies, using enzymes such as a-amylases, glucoamylases, invertases, lactases, cellulases and hemicellulases, to hydrolyze starch, sucrose, lactose, cellulose or hemicellulose into fermentable sugars are in development. These sugars can be further fermented with bacteria, yeasts and fungi to produce ethanol, avoiding the use of strong acids and resulting in a cleaner stream of sugars for fermentation with fewer by-products. Again, the environmental benefits stem from the greater utilization of natural, renewable resources, safer factory working conditions, reduced harmful automobile emissions, and the consumer benefits from a safer alternative to the existing supply of liquid fuels such as gasoline.

(*h*) **Biohydrogen and biofuel cells.** The possibility of using molecular hydrogen as a renewable, efficient and pollution-free energy source is also gaining attention. Hydrogen is colorless, odorless, tasteless, non-toxic and, on combustion, it produces water as the only by-product, making it different from every other common fuel we use today. Hydrogen obtained from biomass has the potential to compete with hydrogen produced by other methods such as from natural gas, which requires the catalytic conversion of hydrocarbons or electrochemical or photochemical water splitting. Although certain microbes, such as *Rhodobacter spheroids*, have been used, successfully, in the production of hydrogen from fruit and vegetable waste, the process is currently still at the laboratory stage, and work needs to be done on increasing cost efficiency and applications. For these reasons, most research has concentrated on the use of hydrogenases for the production of hydrogen, for example, by fermentation of sugars or, more attractively, from waste. However, at present, typical production ranges are only between 0.37 and 3.3 moles hydrogen per mole of glucose, and these low yields, in addition to the cost, might explain why hydrogen is not our primary fuel. This has prompted the search for new hydrogenases using genome database mining and metagenomics, although the latter has not been used for enzymatic screening so far.

Robust biocatalysis methods now exist for chemical synthesis and transformation, bioremediation of contaminants and clean-energy production, confirming and reinforcing the potential of this technology for environmental purposes. The main advantages of using enzymes in these processes are their favorable and unique properties, their biodegradability and their high chemo-, region and stereo-selectivity, resulting in low by-product formation, all of which enables their progressive implementation.

However, a more extensive effort is required to overcome several bottlenecks: high enzyme cost, low activity and/or stability under given conditions, low reaction yields and the low biodiversity in those screened so far (it has been estimated that more than 1030 microbial species are still unexplored). The relatively recent introduction and development of novel recombinant DNA technologies, such as (meta)genomics and directed evolution, have had and will continue to have a profound positive effect on the expression and production of greater and greater amounts of recombinant proteins (from gram to kilogram), which means more competitive prices, by introducing new- or tailored catalytic activities. Results obtained to date suggest that the engineering of virtually any enzyme, old or new, for a commercially acceptable price and catalytically optimal properties might be achieved, suggesting a promising future for environmental biocatalysis.

There are several specialized companies doing bioremediation work in the world, especially for oil spills in soil and water (sea) and for mining works

QUESTIONS

1. What is bioremediation and how it is used to reduce soil and water pollution ?
2. How biocatalysts are useful in remediation of pollutants ?
3. What is phytoremediation? How plants can be used in cleaning soil and water pollution ?
4. Write short notes :

   (*a*) biofuels

   (*b*) biodiesel

   (*c*) enzyme bioremediation

   (*d*) accumulator plants

   (*e*)biostimulation

   (*f*) bioaugmentation

   (*g*) phytoextraction.

ANNEXURE – 46A

BIOABSORPTION OF METALS

Bioabsorption is a process that utilizes inexpensive dead biomass, to sequester toxic heavy metals. Bioabsorbents are prepared from naturally abundant or waste biomass of algae, fungi, moss or dead bacteria. The biosorption process has been recently evaluated, since conservation methods like precipitation, oxidation/reduction, ion exchange, filtration, membranes and evaporation are either extremely expensive, or inefficient for metal extraction. Although biosorption is promising, its mechanism is not well elucidated. Microbial cells can accumulate heavy metals. Radionuclides and organmetalloid compounds by a variety of both, physicochemical and biological processes. Metabolisms' independent binding or adsorption to living or dead cells, extracellular polysaccharides, capsules and slime layer is frequently rapid. Bacterial cell walls and envelopes, and walls of algae, fungi and yeast are efficient metal biosorbents, with binding to charged groups, frequently being followed by an increased amount of inorganic metal deposit. Sorption and/ or complexation of dissolved metals based on the chemical activity of microbial biomass are the basis of the new biosorption technology for metal removal and recovery. Because of their wide diversity of habitats, some surprising, therefore, that specific resistance mechanisms have evolved, to cope with toxic heavy metals. In addition to exposure from naturally occurring ores and during bio-geochemical cycling of metals, anthropogenic release of metals has recently increased bacterial exposure to high levels of metals in some environments.

Biosorption is possible by both living and nonliving biomass. Bioaccumulation is a growth dependent process, and biosorption involve mechanisms like ion exchange, chelation, and complexation. As a result of diffusion through cell walls and membranes, inorganic precipitation may occur by hydrolysis, while inorganic deposition occurs via adsorption by physical forces and ion entrapment, in inter- and intra fibrillar capillaries and space of the structural polysaccharide network. Several active groups of cell constituents like acetamido group of chitin, structural polysaccharides of fungi amine, sulphahydral and carboxyl group in protein, phosphodiester (teichoic acid), phosphate, hydroxyl and polysaccharides participate in biosorption.

To develop an efficient biosorbent and its resuse by subsequent desorption processes, knowledge of the mechanisms of metal binding is thus very important. Metal binding appears to be at least a two-step process, where the first step involves stoichiometric interaction between the metal and the reactive chemical groups in the cell wall, and the second step in an inorganic deposition of increased amounts of metal. All the metal ions come across the cell wall, before gaining access to the plasma membrane and cell cytoplasm. The cell wall consists of a variety of polysaccharides and protein and hence offers a number of active sites capable of binding metal ions. It is regarded as a complex ion exchanger similar to a commercial resin. Difference in cell wall composition among different groups of microorganisms viz. aglae, bacteria, cynobacteria and fungi, and the intra group difference can thus cause significant difference in the type and amount of metal ion binding to them. The resistance mechanisms usually involve detoxification (reduction of mercuric ions to elemental mercury, for example) or the energy dependent export of the metal cation (e.g. for cadmium) or oxyanion (e.g. for arsenite). The resistance determinants are frequently found on extrachromosomal elements (plasmids) and the encoded proteins often function independently of other cellular processes, apart from those required to maintain the membrane potential or to provide ATP or NADPH.

HEAVY METALS

There are no general definition of heavy metals. There are 53 metals with a density above 4-5 g/cm reported as heavy metals, but it is better consider them form their physiological effects and toxicity. At high concentrations, metals form unspecific complex compounds in the cell, which lead to various toxic effects, depending on the metal and the micro-organism considered.

Fe, Mo, Mn	Low toxicity
Zn, Ni, Cu, V, Co, W, Cr	Average toxicity
As, Ag, Sb, Cd, Hg, Pb, U	High toxicity

MICROORGANISMS AND METAL ABSORPTION

Bioabsorption by fungi- Among microorganism, fungal biomass offers the advantage of having a high percentage of cell material, which shows excellent metal binding properties. Many fungi and yeast have shown an excellent potential of metal biosorption particularly the genera *Rhizopus, Aspergillus, Streptovertcillum and Saccharomyces.*

Polysaccharides, in associationwith lipids and proteins, represent the main constituent of fungal cell wall. In filamentous fungi, outer cell wall layers mainly contain neutral polysaccharides (glucans and mannans). While the inner layers contain more of glucosamines (chitin and chitosan) in a microfibrillar structure, ligands within these matrices include carboxylate, amine phosphate, hydroxyl, sulphydral and other functional groups. Proteins are also found to be associated with metal binding.

Rhizopus arrhizus has been used for accumulation of uranium and thorium, a common anion exchange resin, used for accumulation of uranium. The study of uranyl ion, lanthanum and other divalent transition metal ion by *Rhizopus arrhizus* uptake by *Aspergillus niger* and *Saccharomyces cerevisiae* have indicated involvement of carboxylic group.

Biosorption by Algae and Moss- Photoautotrophs marine algae have bulk availability of their biomass from water bodies. *Sargassum natans* and *Ascophyllum nodosum* in this group have shown very high biosorptive capacities for various metals. Besides marine algae, there are reports on binding of heavy metal to green algae, viz. Chorella sp. and cyanobacteria. Special polysaccharides are present in the algae cell wall, and contain potential metal ion binding sites. The number and kind of binding sites depends on the chemical composition of the cell wall. In Pheophycean membranes, algin is present and contributes significantly to metal binding. It has been suggested that the polysaccharides of the cell wall can provide not only the amino and carboxyl group, but also the sulphate. The amino, carboxyl group and the nitrogen and oxygen based moieties could also be electrostatically bonded to unprotonaed carboxyl oxygen. Sulphate convalent bonding between divalent cation and algae cell wall proteins has also been reported.

Mechanisms such as entrapment of metal both in the form of insoluble microdeposits in the inter and intra fibrillar capillaries and paracrystalline regions of polysaccharides and the binding to other biopolymer RNA, Polyphosphare can contribute to the metal binding. The photoautotrophs, eukaryotic algae cell wall are mainly cellulosic, and potential metal binding groups are carboxylate, amine, imidazole, phosphate sulfhydryl, sulfate and hydroxyl. Of these, amine and imidazoles are positively charged when protonated, and build negatively charged metal complexes, the amino and carboxyl groups and nitrogen and oxygen of the peptide bonds are also available for coordination bonding with metal ions such as lead(II), copper(II) and chromium. Such bond formation could be accompanied by displacement of protons and is dependent in part on the extent of protonation which is determined by the pH.

Biosorption by Bacteria- Bacteria may carry determinants of resistance to a number of heavy metals, including Ag, Bi, Cd, Co, Cu, Hg, Pb, Ni, Tl, or Zn cation and the oxylation of As, Cr, Sb, or W. Bacterial resistance to heavy metals is conferred by specific resistance determinants, which are often carried on plasmids or transposons. Resistance is specific to one or a few metals and the mechanisms of resistance include efflux of the metal, modification of the speciation of the metal, sequestration of the metal, or a combination of these mechanisms. The proteins conferring metal resistance, the genetic regulatory mechanisms, and the organisms conferring resistance to heavy metals are now beginning to be exploited in bioremediation and biomonitoring strategies for environments, which are contaminated with heavy metals. Cell walls of bacteria and cynobacteria are principally composed of peptidoglycans, N acetylglucasamine, â 1-4 acetyl mucross linked. They have an outer membrane, which is composed of an outer layer of lipopolysaccharides (LPS), phospholipids and proteins. Gram-negative bacteria are more widerspread in metal contaminated soils than Gram-negative bacteria. The anionic nature of bacterial surface enables them to bind metal cations through of Gram positive bacteria has a high capacity for metal binging. It has been reported that peptidoglycan is the major cell wall component responsible for metal binding by *Bacillus subtilis*. In contrast *B. licheniformis*, metal binding is predominately due to teichuronic acid. Among bacteria, Bacillus sp. has been identified as having a high potential for metal sequestration and has been used in commercial biosorbent preparation. Beside, there are reports on the biosorption of metal using *Pseudomonas* sp. *Zoogle*a and *Streptomyces* sp. Teichoic acid In B.subtilis, and teichoic acid and teichouronic acid in B. licheniformis, were found to be prime sites for the metal binding.

MECHANISMS OF BACTERIAL METAL RESISTANCE

There are four mechanisms of bacterial metal resistance.

1. Keeping the toxic ion out of the cell (reduced uptake).
2. Highly specific efflux pumping (*i.e.* removed toxic ions that entered the cell by means of transport system evolved for nutrient cation or anions). Efflux pumps can be either ATPases or cheiosmotic driven. ATPases are enzymes that use the chemical energy from cleavage of the high energy phosphodiester bond of ATP to drive the formation of concentration gradients.
3. Intra- or extracellular sequestration by specific mineral-ion binding components (e.g. metallothioneins) and/ or segregation into complex compounds.
4. Enzymatic detoxification, (oxydoreductions) which converts a more toxic ion to a less toxic one.

The first two mechanisms can be grouped under the term avoidance, whereas the last two mechanisms are known as sequestration mechanisms. It is important to notice that mechanisms of metal tolerance might be adaptation of the processes of normal homeostasis. Homeostasis describes the fact that cells have processes to monitor and maintain intracellular concentrations of metals. So, metal homeostasis must involve uptake of sufficient essential metals, while providing protection against their toxicity.

Mechanism of Uptake

The first barrier to penetration is the wall, which provides some protection for the cytoplasmic membrane. Cell walls, especially those of fungi, can be used as biosorbents. Thus, many works have focused on the description of Langmuir or Freundlich isotherms to describe binding. However, walls cannot act as a perfect barrier to entry of some ions that are essential trace elements for micro-organisms. So, the metal ion is first transported into the cytoplasm despite its high concentration, which is the first reason why metal ions are toxic.

Mechanism of Efflux

Efflux pumps reduce the intracellular concentration of metals by means of transport systems, without any exzymatic transformation. This mechanism is more widespread than enzymatic detoxification. Uptake and efflux mechanism can be classified in 8 protein familier approximately:

the most important are the ABC family (ATP Binding Cassette), the P-and A-type ATPases family, the RND family (Resistance, Nodulation and cell Division) and the MIT family (Metal Inorganic Transport)

Plasmid Resistance

Bacterial plasmids contain gene that provide extra functions to the cells, one of which is resistance to toxic metals. Plasmids are small circular DNA molecules that can move from one cell to another. Thus, the transfer of toxic metal resistance system are found on these plasmids, but some systems are determined by chromosomal genes in other organisms. Plasmid-determined resistance systems are very specific.

Metallothioneins and Binding

Bacterial metallothioneins (MT) are commonly grouped in 3 classes. Class I and II are gene-encoded, where class III is not. Metallothioneins are low molecular mass, cysteine-rich metal-binding proteins, homologous to mammalian metal binding proteins. They have not been widely found (only on cyanobacteria at present), but bind divalent cation, such as Ca^{2+}, Cu^{2+}, and Zn^{2+} so tightly, that it is well protected (sequestration). Here, it is important to make the difference between biosorption (which can be effected by nonviable biomass and bioaccumulation /biosequestration (adsorptive and metabolism-dependent mechanisms).

Genetic Engineering for Specific Proteins

The highly specific metal absorption proteins is the result of a clearly designed genetic circuit that is tightly under the control of gene MerR and ArsR which are regulatory proteins used for controlling the expression of enzymes responsible for mercury and arsenic detoxification. The high affinity and selectivity of MerR towards mercury has been exploited for the construction of microbial biosorbent specific for mercury removal. Presence of surface exposed MerR on an engineered strain enabled 6 fold higher Mg^{2+} binding. Similarly, cell overexpressing, ArsR accumulated 5- and 60 fold higher level of arsenate and arsenite.

BIOREACTORS AND BIOABSORPTION

The use of freely suspended microbial biomass has disadvantages that include small particle size, low mechanical strength and difficulty in separating biomass and effluent. Therefore, process parameters are optimized for bioabsorption of heavy metals. Various modifications are made by using immobilized biomass particle in the bioreactors, and based on this, bioreactors are divided.

Packed Bed Reactor- Immobilized, living biomass has primarily taken the form of biofilms on supports prepard from a range of inert materials. Living cell biofilms may provide additional capacity for the removal of pollutants, including hydrocarbons, pesticides and nitrates. In addition to the use of biofilms, living or dead biomass of all microbial groups has been immobilized by encapsulation or cross linking. Supports include agar, cellulose,alginate, cross-linked ethyl acrylate, ethylene glycol dimethylacrylate, polyacrylamide, silica gel and the cross linking reagents toluene diisocyanate and glutaraldehyde. The biomass may be used in its 'natural state' or modified, for example, by alkali treatment, to improve biosorption efficiency. Small scale systems may be adequate for low volume waste, containing valuable elements such as gold.

Fluidized Bed Reactors- Fluidized beds of alginate and polyacrylamide immobilized algae; for example *Chlorella vulgaris* and *Spirulina platensis* have been used to remove a variety of metals, including Cu2+, Pb2+, Zn2+ and Au3+, from mixtures, with several schemes for selective recovery. Alginate and polyacrylamide provide good resistance to hydrostatic pressure and mechanical degradation, although it is thought that polyacrylamide is not strong enough for commercial applications. The localization some of the processes, especially enzymes and redox reaction is uncertain and may vary between groups, strains and species of both prokaryotic and eukaryotic organisms.

Rotating Disc Reactor- A large scale commercial process treats effluent from metal mining and milling using rotating disc biofilm contacting units for simultaneous degradation of cyanide, thiocyanate, and ammonia, as well as metal removal by biosorption.

Single Blanket Reactor- Hydrogen sulphide is produced by sulphate reducing bacteria (SKBs), such as Desulphovibrio and/or Desulphotomaculum sp. the solubility products of most metal sulphides are extremely low and they are readily precipitated as sulphides. Sulphate- reducing activity can occur as a result of anaerobic biomass decay in biosorption systems, and thus act as a useful auxiliary metal removing mechanism. A purpose designed 9m 3 stainless steel, sludge-blanket reactor, using sulphate reducing bacteria has been piloted. This plant successfully removes using toxic metals and sulphate from contaminated groundwater at a long standing smelter site by precipitation of the metal as metal sulphides.

Sequential Reactor- In sequential bioreactors, two or more bioreactors may be connected to increase the efficiency of the metal absorption. Sometimes, reactor type can be mixed i.e. the use of a biofilm reactor in conjuction with a tandem stirred tank reactor. Living cell in its 'natural state', or modified is used to improve biosorption efficiency. Small scale systems may be adequate for low volume waste containing valuable elements such as gold.

Possible use of 'Metallophores' to aid Phytoextraction of Soil Metals

Poaceae species secrete mugineic acid family phytosiderophores (chelating agents) to solubilize soil, Fe and accumulate the intact chelate into root cells, Transgenic plants could be developed to secrete metal selective ligands into the rhizosphere which could specifically solublize elements of phytoremediation interest. Although this approach holds promise, phytosiderophores obtain their specificity not by chelation specifically only of Fe in soils, but from their uptake of nearly only Fe-phytosiderophores by a membrane carrier. The biosynthetic molecules with selective chelation ability plants can make and secrete into the rhizosphere at adequate concentrations. It is mediated by creating a selective transport protein for the metal chelate. It seems difficult, but worth examination to develop unique phytoremediation tools. Regulatory control phytosiderophore secretion in barley was induced by Fe-deficiency, but not Mn, Zn, or Cu deficiency in contrast with other reports, which indicated that Zn deficiency, also induced the biosynthesis and secretion of phytosiderophores.

Lastly extensive progress has recently been achieved in identifying genes and proteins involve in uptake of Fe by yeast and plants; high affinity Zn and Cu membrane transporter have also been found in yeast. A fundamental understanding of both uptake and translocation processes in normal plants and metal hyperaccumulators, regulatory control of these activities, and the use of tissue specific promoters offers great promise that uses of molecules biology tools can be used to develop effective and economic phytoremediation plants for soil metals.

MECHANISM OF BIOABSORPTION AND BIOACCUMULATION IN SPECIFIC METALS

More recent studies have documented the ability of a number of Desulfo vibrio strains to reduce toxic metals such as uranium and chromium, a process that results in the production of less water soluble species. The modification of solublility properties brought about by changing the redox state of the metal presents itself as a potential avenue for bioremediation of contaminated ground waters or soils. The genetically accessible strain of *Desulfo vibrio* mediate this reduction. Thus studies for the elucidation of the electron path way, the metal reductase (s), their regulation and their specificity deserve special attention. Microarray analyses will prove most valuable in sorting out the effects of toxic metals and the bioenergetics of these microbes.

BIOFERTILISERS AND BIOPESTCIDES

CHAPTER 47

Biofertilisers are considered to be an important alternative source of plant *nutrition*. They are biologically active products, including bacteria, algae or fungi, with the ability to provide plants with nutrients. Most biofertilisers belong to one or two categories: nitrogen fixing and phosphate solubilising. Nitrogen fixing biofertilisers fix atmospheric nitrogen into forms, which are readily useable by plants. These include *Rhizobium, Azatobacter, Azospirillum,* blue green algae (BGA) and *Azolla*. While *Rhizobium* requires symbiotic association with the root nodules of legumes to fix nitrogen, others can fix nitrogen independently. *Phosphate* solubilising microorganisms (PSM) secrete organic acids, which enhance the uptake of phosphorus by plants by dissolving rock phosphate and tricalcium phosphates. PSMs are particularly valuable as they are not crop specific and can benefit all crops. Biofertilisers are a boon for farmers, because they help in increasing the land fertility. Their present consumption in India is about 1200 tonnes per annum.

Biofertilisers may be divided into natural fertilisers and efficient fertilisers, depending upon origin and mode of actions. Natural fertilisers involve the *preparation* of efficient strains of microorganisms capable of converting atmospheric nitrogen into nitrate or solubilising phosphorus from the fixed form. Efficient fertiliser includes BGA, *Rhizohium, Azospirillum* and phosphate solubilising microorganisms (PSM). Rhizobium can fix 50-200 kg N per ha. per year and 40-80 kg nitrogen is left over in land, useful for subsequent crop. It increases the crop yield by 25-30%. Blue green algae, used for paddy fields, increases the yield by 20-30% by fixing nitrogen and producing growth promoting compounds. *Azatobactor* and *Azospirillim,* used for cereal crops can fix 20-30 kg of atmospheric nitrogen. The yield can be increased by 10-15%. Biofertilisers are cheaper, renew-able and pollution free. Such a combination can help in reducing the evils of chemical fertilisers on the environment. Besides being non-toxic, biofertilisers have some distinct advantages over chemical fertilisers. Biofertilisers like blue green algae are capable of providing more nitrogen-an important nutrient for plants-than all chemical fertilisers put together. Moreover, biofertilisers ensure ecological stability to the soil. On the other hand, chemical fertilisers are toxic and pollute the water and disturb the ecological balance. They are also getting costlier as a result of ever increasing price of petroleum products. This, coupled with the low purchasing power of small and marginal farmers continues to be the major concern in our agricultural production process. But biofertilisers are *comparatively* cheaper. Production of chemical fertilisers requires the setting up of big plants involving huge investment, but a concrete tank is sufficient for the produc-tion of blue green algal biofertiliser. It can provide employment to a large number of unemployed graduates and a source of livelihood to a fairly good number of people. Unlike chemical fertilisers, biofertilisers, especially algal, have the *capacity* to hold water. They are also capable of concentrating nutrients such as nitrogen and scavenge sodium salt-affected soils.

Biofertilisers are, however, still not popular due to various factors, such as, lack of proper education, marketing and distribution channels to the farmers and lack of quality of guaranteed products. It seems that the day is not far off when biofertilisers will replace chemical fertilisers for an era of prosperity and clean environment.

I. BIOFERTILISERS IN AGROECOSYSTEM

Azolla-Anabaena. *Azolla-Anabaena* is an excellent example of a mutual relationship between microorganisms and plants, that results into nitrogen fixation. *Azolla* grows on the surface of quite water in tropical and sub-tropical zones. On the lower surface of the *Azolla,* leaves are mucilage and contain cavities that con-tain *Anabaena.* Within the cavities, the *Anabaena* fix large quantities of atmospheric nitrogen.

Azospirillum. These bacteria are capable of increasing the yield o*f agriculturally* important crops in a variety of different soils. Various strains, *A. brasilense* andA. *lipoferum,* are used successfully in crop production with around 70 percent success.

Rhizobia-legume. The nitrogen fixing (diazotrophic) associations of rhizobia with leguminous plants are of great importance in global nitrogen cycling and crop production. The three genera, *Rhizobium, Azorhizobium* and *Bradyrhizobium*, have been investigated, although rhizobia occurs as a free living heterotroph in soil, and does not fix atmospheric nitrogen in this state. However, rhizobia can invade root hairs, initiate the formation of a nodule and develop nitrogen fixing activities. Flavonoids and isoflavonoids secreted by the host plants induce the *expression* of a number of nodulation (nod) genes in rhizobial bacteria. The products of nod genes are enzymes involved in the biosysnthesis of species-specific, *substituted* lipo-oligosaccharides, called nod factor, which elicit the curling of plant root hair and division of meristamatic cells, eventually leading to the formation of root nodules. Within the infected tissue, rhizobia multiply, forming unusually shaped and enlarged cells called bacteroids. The bacteroids within the nodule carry out the fixation of atmospheric nitrogen. The plant-rhizobia require various organic and inorganic compounds for nitrogen fixation. The trace element molybdenum forms an important part of the nitrogenase enzyme and plays an active role in the fixation of nitrogen.

Mycorrhizal Fungi. Mycorrhizae are a symbiotic association between soil fungi and higher plants. The fungi derive nutritional benefits from the plant roots and contribute to plant nutrition. There are two basic types of mycorrhizal *associations*; ectomycorrhizae and endomycorrhirae. The vascular-arbuscular type is a morphologically distinct and widespread form of endomycorrhizae. Mycorrhizae enhance the uptake of mineral nutrients and enable plants to grow in habitats that otherwise might not provide sufficient nutrients.

II. PESTICIDES

The increasing pressure for food production to meet current demand requires *protection* of crops from pests. The use of pesticides is inevitable and constitutes an integral part of modern crop management practices. Pesticides include chemicals to protect crops from insects (insecticides), weeds (herbicides), disease causing microorganisms (fungicides and bactericides), mites (acaricides), nematodes (nematicides), and rodents (rodenticides). Pesticides also include chemicals that control pets directly hazardous to animals and humans, particularly vectors, such as, mosquitoes, flies, fleas, ticks and lice that transmit diseases. "Natural" or "*organic*" pesticide controls a pest. Organic pesticides are usually those pesticides that come from natural sources.

These natural sources are usually plants, as is the case with pyrethrum (pyrethins), rotenone or ryania (botanical insecticides), or minerals, such as boric acid, cryolite, or diatomaceous earth. Organic pesticides are largely insecticides. Even if a product is considered to be organic, it is still a pesticide. One has to be careful while using any pesticide, as they may be toxic despite being organic and natural. Some organic pesticides are more toxic than many synthetic chemical pesticides. Organic pesticides have specific modes of action, in the same manner as synthetic pesticides. While some organic pesticides may be nontoxic or slightly toxic to human beings, they could be very toxic to other animals. For instance, the organic pesticide ryania, which is relatively safe for human beings, is toxic to fish. Also, some organic pesticides may be toxic to beneficial insects, such as *honeybees*, if they are combined with other materials, such as combining pyrethrins with rotenone.

III. BIOPESTICIDES

Biopesticides are pest management tools that are based on beneficial *microorganisms* (bacteria, viruses, fungi and protozoa), beneficial nematodes or other safe, biologically based active ingredients. The benefits of biopesticides include effective control of insects, plant diseases and weeds, as well as human and *environmental* safety. Biopesticides also play an important role in providing pest *management* tools in areas where pesticide resistance, niche markets and environmental concerns limit the use of chemical pesticide products. Biopesticides include *naturally* occurring substances that control pests (biochemical pesticides), *microorganisms* that control pests (microbial pesticides), and pesticidal substances produced by plants containing added genetic PIP (Plant-Incorporated Protectants) materials.

The efficacy of many of the biopesticides can equal that of conventional chemical pesticides. However, the mode of action will be different. With many of the biopesticides, the time from exposure to morbidity and death of the target *insect* may be 2 to 10 days. Understanding the fundamental differences in the mode of action of biopesticides and traditional pesticides is important, since the usage patterns of a biopesticide may be different from traditional pesticides to control a particular pest species. The use of an Integrated Pest Management Program (IPM) is important to ensure success. Biopesticides are an important group of *pesticides* that can reduce risk of pesticides.

(a) Advantages of Biopesticides

1. Biopesticides are inherently less harmful in comparison to conventional pesticides.
2. Biopesticides are designed to affect only one specific pest or, in some cases, a few organisms, in contrast to conventional pesticides, that may affect organisms as different as birds, insects, and mammals, across a broad *spectrum*.
3. Biopesticides are effective in very small quantities and often decompose quickly, thereby resulting in lower exposures and largely avoiding the *pollution* problems caused by conventional pesticides.
4. When used as a component of Integrated Pest Management (IPM) programmes, biopesticides can greatly decrease the use of conventional pesticides, while crop yields remain high.

However, to use biopesticides effectively, the users need to know a great deal about managing the use and mode of action of biopesticides. The world has woken up to the adverse effects of synthetic pesticides all over the world and is rapidly shifting to non-synthetic safer pesticides and "biopesticides". The success of biological control has been spectacular. In many countries, waterways that were once unusable have now returned to full use and the agents have reduced the weeds to relatively insignificant population. Aquatic weeds can bring rivers and lakes to a standstill, destroying their ecology and also the livelihoods of communities who depend on them. Two of the most notorious are salvinia, *Salvinia molesta,* and water hyacinth, *Eichhornia crassipes.* Peptide and protein toxins have a variety of potential uses in protecting crops against insects. In particular, toxins can be delivered directly to the pest insects by expression in genetically engineered crops. Very few genes are known to encode orally active insecticidal proteins and the Bt proteins currently available affect only a limited range of pests. The main focus is on bacteria that are symbiotically associated with insect pathogenic nematodes. The toxin proteins are being tested for insecticidal activity against a range of pest species.

Functional genomics (the determination of biological function of genes) has particular promise in agricultural biotechnology for identifying new targets for pest control and for identifying gene/ enzyme pathways involved in processes such as insect-pathogen and insect-plant interactions. It is dependent on the availability of relevant genomic or Expressed Sequence Tag (EST) databases and sophisticated bioinformatics tools for database mining. At present, opportunities for functional genomics analysis of pest insects are limited by lack of relevant EST databases for gene identification.

(b) Genetic Transformation

A strong platform capability in target validation is being established around insect genetic transformation and interfering RNA technologies with the aim of identifying new pesticide target genes. Genetic transformation, or the introduction of a novel gene into an organism, is an extremely useful tool in the study of gene structure and function. Genome sequencing projects are identifying new genes from pest insects at an accelerating rate and using genetic transformation to identify and validate those genes that have potential as new targets for pest control. In addition, RNA interference as a means of knocking out activity of specific genes to gain insights into gene function are reported by several workers. This technique has facilitated efforts in finding genes useful for the control of insect pests.

One technique being developed is the ability to render an insect sterile. The production of sterile transgenic invertebrates would alleviate concerns about the spread of genetically modified organisms in the environment. Efforts have been made to produce sterile aquaculture species in aquaculture farms or local waters. This technology would thereby prevent the transgenic animals from spreading and potentially causing environmental damage. Similar technologies are being developed for the production of sterile male insects for use in sterile insect release programmes.

IV. MICROBIAL AND ANTIMICROBIAL PESTICIDES

These are two separate and distinct types of pesticides registered by EPA.

Microbial Pesticides are microbes, including bacteria, that help to control insects and weeds, as well as fungi and bacteria that cause plant diseases. These pesticides belong to one category.

Antimicrobial Pesticides are pesticides that control unwanted microbes on inani-mate objects, in water and on selected foods under certain circumstances. These pesticides are almost always chemicals and they act by killing or inactivating microbes that are pests. Antimicrobial pesticides include the disinfectants used in swimming pools, drinking water supplies and in hospitals to control microbes that can cause disease. These are another category of pesticides.

CHARACTERISTICS OF PESTICIDES

Biopesticides have the following characteristics:

(*a*) They have a narrow target range.

(*b*) They have a specific mode of action.

(*c*) They are slow acting and have relatively critical application times.

(*d*) They can suppress, rather than eliminate a pest population.

(*e*) They have limited field persistence and a short shelf life.

(*f*) They are safer to humans and the environment than conventional pesticides.

(*g*) They have a short residual effect.

Biopesticides fall into the following two major categories:

Microbial pesticides are those pesticides, which contain a microorganism like a fungus, virus, protozoan or bacterium as the active ingredient. Microbial insecticides are another kind of biopesticide. They come from naturally-occurring or genetically altered bacteria, fungi, algae, viruses or protozoans. They suppress pests by producing a toxin specific to the pest, causing a disease and preventing establishment of other microorganisms through competition or other modes of action. (Table 47.1)

(b) Bacteria as Pesticides

Bacteria species of the genus *Bacillus* have been successful as biocontrol agent. *B. thuringiensis.cis, B. popillae, B. sphaericus* have been widely used as bio-insecticides. The most widely known microbial pesticides are varieties of the bacterium *Bacillus thuringiensis,* or Bt, which can control certain insects in cabbage, potatoes, and other crops. Bt produces a protein that is harmful to specific insect pests. Certain other microbial pesticides act by out-doing pest organisms. Microbial *pesticides* need to be continuously monitored to ensure so that they do not harm *nontarget* organisms, including humans. Microbial pesticides can control many different kinds of pests, although each separate active ingredient is relatively specific to its target pest[s]. For example, there are fungi that control certain weeds, and other fungi that kill specific insects.

The most widely used microbial pesticides are subspecies and strains of *Bacillus thuringiensis,* or Bt, which is naturally occurring soil bacteria that is toxic to the larvae of several species of insects but nontoxic to non-target organisms. Each strain of this bacterium produces a different mix of proteins and specifically kills one or a few related species of insect larvae. While some Bt's control moth larvae found on plants, other Bt's are specific for larvae of flies and mosquitoes. The target insect species are determined by whether the particular Bt produces a protein that can hind to a larval gut receptor, thereby causing the insect larvae to starve. *Bacillus thuringiensis* can he applied to plant foliage or incorporated into the genetic material ut'crops. *Bacillus thuringiensis,* as discovered, is toxic to the caterpillars (larvae) of moths and butterflies. Several strains of Bt have been developed and now strains are available that control fly larvae. These can be used in controlling mosquitoes and black flies.

(b) Virus as Pesticides

Viruses of group baculovirus are the most potential biocontrol agent. They are highly specific and can quickly infect large populations of insects. Sometime larvae of the viruses feed on the crop after being infected.

(c) Fungi as Biopesticide

Fungi possess a number of characteristics that make them potential biocontrol agents. *Ti ichoderma* species have long been used to control the root-pathogenic fungus *Rhizoctonia solani*. Insect specific fungal diseases are important in the natural control of insects. *Beauveria* and *Metarhizium* fungi cause white and green muscardine disease of insects. Both these fungi infect insects through their cuticle or mouth parts. They produce toxins. *Beauveria* produces beauveracin that acts on membranes by changing their permeability and eventually causing death. *Metarhizium* produces destruxin and cytochalasins. Mycoinsecticide products are now registered in Australia and elsewhere and offer substantial advantages over chemical pesticides in terms of reduced hazards to users and the environment. The challenge has been to develop products that are cost-effective and provide an acceptable level of pest control. BioCaneO, a registered product for control of greyback canegrubs, is now being widely used with a consumption of over 40 tonnes in 2001 and a second product, Greenguard®, was used operationally by the Australian Plague Locust Commission for the first time in the 2000/1 season. It was successfully used to control locusts on over 20,000 ha at. I cost comparable with that of chemical pesticides. Improvements in the production and formulation of the products have recently enabled much lower doses to be effective in the field, thus pushing the existing costs to a great extent. Australia is now recognised as a World leader and has been approached by countries such as China, Indonesia, Madagascar and Mexico to help develop the use of biopesticides for locust control in those countries. The Food and Agriculture Organisation (FAO) of the United Nations is considering recommending the use of Green Guard® in other Countries particularly Africa.

The advances in strain identification, selection, mass production and formulation made in developing these products are now being extended into new research on mycoinsecticides for pecan borers, citrus thrips, whiteflies and the Taro beetle in Papua New Guinea. Environmental sensitivities often prevent the use of *herbicides*. Agricultural/environmental biotechnology has shifted focus and resembles a pharmaceutical model of mechanism-based high-throughput screens for lead identification. The discovery process including target identification and *validation*, screen design and development and sources of novel chemical diversity are significant components of bio-insecticides. The research is directed at key *invertebrate* pests of crops and livestock, principally lepidopteran insects and nematodes. Several potential physiological targets for new insecticides have been identified as the sites of action of insect-selective arthropod neurotoxins and strategies for *cloning* the relevant genes from key pest species have been developed. Novel receptor target has been identified for evaluation of the mode of action of biopesticides.

Fungi that can control nematodes are divided into nematode trapping fungi, endoparasitic fungi and fungi that parasitise nematode eggs, *Paecilomyces lilacines* has been widely used for control of nematodes. Nematode-destroying fungi are applied in the soil together with a large carbon source to enhance the growth of fungi.

(d) Plant Pesticides

Plant pesticides are those pesticides which have been derived from genetic *material* produced by the plants that have pesticidal properties. Pesticidal substances that plants produce from genetic material are added to the plants. For example, scientists can take the gene of the Bt pesticidal protein and introduce the gene into the genetic material of plants. Then the plant instead of the Bt bacterium, *manufactures* the substance that destroys the pests. Plant-Incorporated-Protectants (PIPs) are pesticidal substances that plants produce from genetic material, that has been added to them.

(e) Biochemical Pesticides

Biochemical pesticides are those pesticides that are derived from naturally *occurring* substances and control pests by non-toxic mechanisms. Conventional pesticides, by contrast, are synthetic materials that usually kill or inactivate the pest. Biochemical pesticides include substances that interfere with growth or *mating*, such as plant growth regulators, or substances that repel or attract pests, such as pheromones. It is sometimes difficult to determine whether a natural pesticide controls the pest by a non-toxic mode of action. Biochemical pesticides may have a similar structure to, and function like, naturally occurring chemicals, and have nontoxic modes of *action*. Insect pheromones, for example, are naturally-occurring chemicals that *insects* use to locate mates. Man-made pheromones are used to disrupt insect mating by creating confusion during the search for mates, or can be used to attract male insects to traps. Pheromones are often used to detect or monitor insect *populations*, or in some caes, to control them.

(f) Natural Pesticides

In nature, plant protects themselves against insect attack mainly by mechanical and chemical defences. Mechanical defence includes formation of thick cuticle which prevents insect from picking for food or laying eggs. Chemical defences are highly developed in plants in the form of secondary metabolites. These secondary metabolites are product of primary metabolites and referred to as insecticides. Insecticidal properties of over 2000 plant species and their traditional uses were established from the ancient time. However, up to 1989, insecticidal compounds from only 62 plant species have been isolated and their chemical structures were elucidated. The chemical nature of these insecticidal compounds is widely varying. These vary from simple alkaloids (nicotine, anabasin), complex esters (pyrethrins), heterocyclic aromatic compounds (rotenoids), steroids (phytoecdysones) etc.

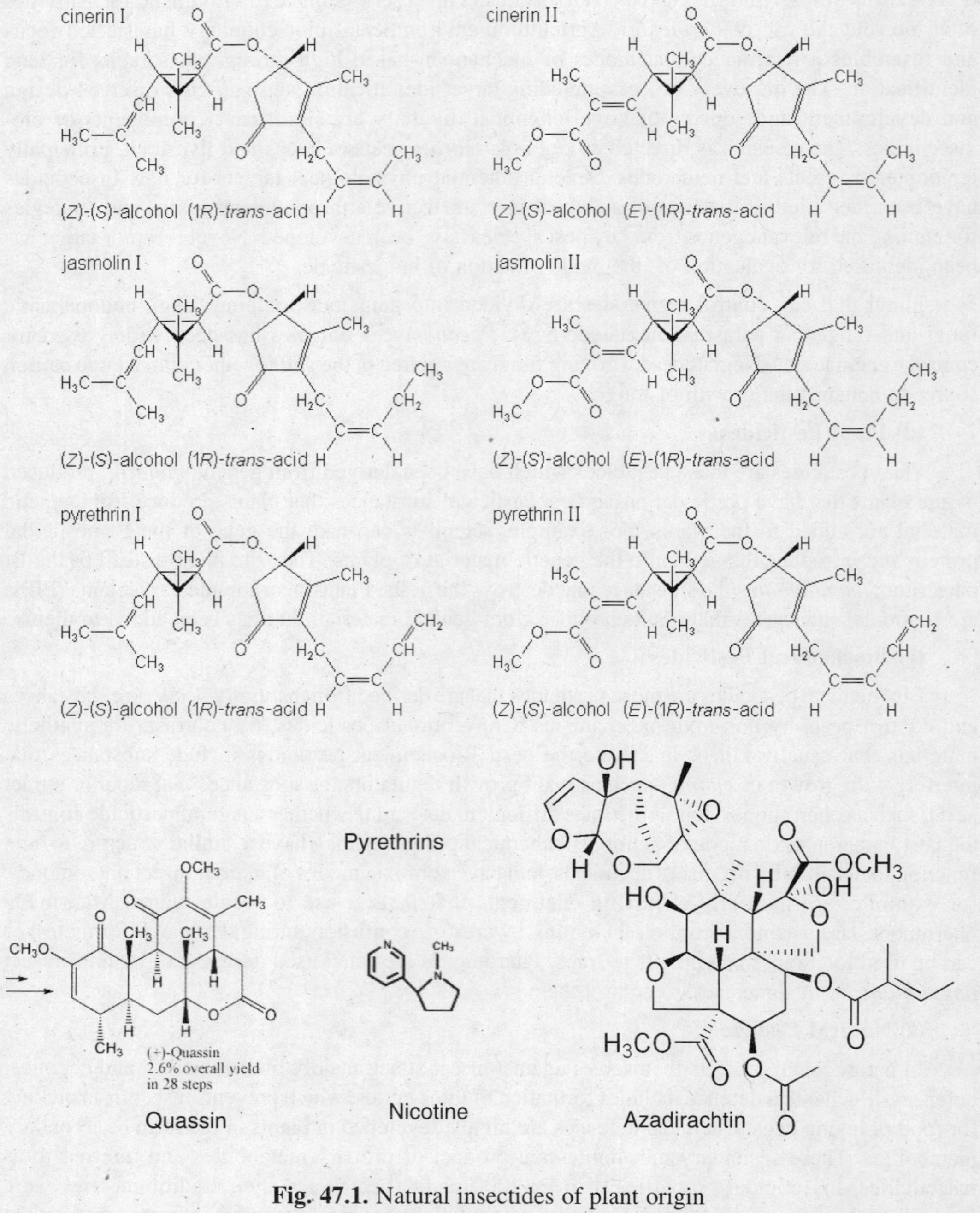

Fig. 47.1. Natural insectides of plant origin

Commercial insecticides of plant origin include pyrethrins, rotenoids, nicotine, quassin etc. Fig. 47.1. The natural insecticides of plant origin are effective against a wide range of insects, many of which can not be successfully controlled by synthetic insecticides. In contrast to the synthetic insecticides, natural compounds are relatively non-hazardous to human beings and to the other mammals. Plant insecticides generally exert their insecticidal effect, by interfering with the physiology of insects, affecting their nervous system, hampering normal development of insects and also act as antifeedants. Some plants contain compounds which are analogs of ecdysteroids, a group of insect hormones, viz. 20-hydroxyecdysone and makisterone, those initiate the cycles of ecdysis in insect development.

Phytoecdysones : Phytoecdysones are found in plants and they can disturb the growth cycles of insects and can result in the formation of abnormal adults. Ecdysterone, a widely occurring phytoecdysterone, has been reported in *Polypodium vulgare, P. virginianum, Achyranthes aspera, Trianthema portulacastrum, Sida carpinifolia, Sesuvium portulacastrum* and *Gomphrene celosiodes*. Azadiractin is obtained from *Azadirachta indica* and *Melia azedarach*. Limonoids are also insecticidal compounds found mainly amongst the members of Family Meliaceae and Rutaceae.

Rotenoids : Rotenoids are used as insecticides and also as fish poison in tropical countries. Rotenoids are found in *Derris, Tephrosia, Lonchocarpus, Millettia* and *Mundulea*. The mean rotenoid content of roots of *D. elliptica* varies from 5-9%, whereas, *D. maccensis* contains 0-4% and *L. utilis* contains 8-11% rotenone. Six insecticidal rotenoids are known to occur naturally, viz., rotenone, elliptone, summatrol, malacol, a-toxicarol and degulin. Occasionally, an oxidative product of degulin is present, viz. tephrosia.

Pyrethrins : Pyrethrins are most economically important natural insecticides of plant origin. They have lethal effect against broad range of insect species, have low mammalian toxicity and without any residual effect. This has high combining capacity with synergists, like, carbon tetrachloride, ethylene dichloride etc. A non-inflammable preparation of pyrethrins is used as sprays in aerosol against insect vectors to prevent transmission of insect borne diseases.

Pyrethrins have been detected in several plant species viz. *Chrysanthemum cinerariaefolium, C. coccinum, Tagetes erecta, T. minuta, Calendula officinalis, Demorphotheca sinuata, Zinnia elegans* and *Z. linnearis*. Pyrethrins consist of six structurally related insecticidal esters formed by the combination of two acids (chrysanthemic acid and pyrethric acid) and three alcohols (pyrethrolone, cinerolone and jasmolone). The esters of pyrethric acid are called pyrethrin I, cinerin I and jasmoline I, respectively and collectively referred as pyrethrins I (P I). The esters of pyrethric acid are called pyrethrin II, cinerin II and jasmoline II and together known as pyrethrins II (P II). P I are responsible for killing insects, whereas, P II provides much of the knock down action against flying insects.

Nicotine and Anabasine : Crude tobacco extracts was used as an insecticide as early as 1763, but the nicotine alkaloid was not isolated until 1828. Subsequently, pure nicotine and nicotine sulphate became the commercial source in the early part of the twentieth century. Preparations of nicotine were very popular until 1945. However, due to the high production cost, its disagreeable odor and toxicity to mammals, use of nicotine has become very restricted in the recent past.

Quassin : Wood extracts of *Quassia amara* and *Q. africana* were used in early part of this century as an insecticide, and lately, its use as a horticultural and agricultural insecticide has increased considerably. Quassin has been approved by the Soil Association as a biocide and has been used for insect control in organic orchards. Quassin and Quassia extracts have been tested on more than 100 insect species, and appear to be selectively effective against sawfly, larvae, aphids and locusts. Commercial Quassin representing approx. 0.2% of the dry wood, and the content varies depending on the source of plant.

Natural pesticides are environment friendly, non-toxic, completely biodegradable natural products with comprehensive insect-pest control properties, effective on more than 300 types of insect-pest based on Neem *(Azadirachta Indica)*. Neemhit is for diverse biological effects such as insecticidal, antifeedant, repellant, nematicidal, antibacteria, antifungal, antiviral, oviposition and deterrent.

Neemaura is a environmental friendly pesticide, which has the natural bitterness of Neem. It is non-toxic and has a safe biodegradable urea coating agent containing Neem Triterpenes, which inhibits the growth of nitrifying bacteria, resulting into the delayed transformation of ammoniacal nitrogen into nitrite nitrogen. This ensures slow and continuous availability of nitrogen, matching the requirements of crop plants during their life cycle and effectively retards the nitrification of urea. Neemaura coated urea mineralises much slower than plain urea, at least two to three times under soil conditions by controlling the multiplication process of soil-borne bacteria like, *Nitrosomanas* and *Nitrobacter* which are responsible for nitrification. Neemaura formulation contains neem bitters and sul-phurous compounds, which are mainly responsible for retarding the process of bacterial action and protects urea from leaching, volatilisation and also protects crop from insect pests, thus resulting in higher yields.

Kuber is a natural photosynthesis enhancer. Kuber is a highly effective natural plant growth promoter, derived from agricultural raw material based on Tricontanol. Technically, it is a mixture of long chain, high molecular weight, fatty alcohols with carbon atoms and sterols. It helps in supplementing the potency of various fertilisers and micronutrients embedded in soil. Kuber helps unlock nature's storehouse of nutrients and stimulates the following growth response in different crops.

I. NEMATODE BIOPESTICIDES

The infective stages of insect-killing or entomopathogenic nematodes (ENs) are primitive microscopic round worms that can seek out and kill a wide range of insect pests but are harmless to other animals, plants and the environment. ENs currently account for most of the world's biopesticide market, in comparison to all other organisms apart from Bt. However, further expansion of this market is lim-ited by expense of production, short shelf life, under-developed formulation and application technology and limited strain collection and evaluation.

Table 47.1. Microorganisms for biocontrol of pests.

Type of Control	Weed Control	Examples
Insect Control	Fungi	*Bacillus thuringiensis, B. sphaericus, B. popilliae, Serratia entomophila, Bacillus cereus, Entomophaga grylli*
Bacteria	Bacteria	
Viruses	**Plant Disease Control**	*Anagrapha falcifera, Nuclear Polyhedrosis, Virus Baculovirus, Autographa californica*
	Agrobacterium tumefaciens	
Fungi	Soil pathogens-*Fusarium*,	*Beauveria, Metarhizium, Entomophaga, Zoopthora, Paecilomyces, Normuraea,. Beauveria bassiana*
	Phythium, some nematodes	
Protozoa	Powdery mildew	*Nosema, Thelohania, Vairimorpha*
Entomopathogenic	*Rhizoctonia, Fusarium,*	*Steinernema, Heterorhabditis,*
nematodes	*Alternaria* & *Aspergillus*, that	*Romanomermis*
Others	cause root rots & seedling diseases,	Pheromones, parasitoids, predators, microbial byproducts
	Postharvest pathogens-*Botrytis*,	
	Penicillium Nematode trapping	*Alternaria,*
	fungi, Competitive inoculants	*Colletotrichum*
	Composts, soil Inoculants	*Xanthomonas*
		Agrobacterium radfobacter
		Ampelomyces quisqualis
		Bacillus subtilis Candida oleophila Arthrobotrys

(*h*) **Biopesticide Chemicals**

Also in commercial or near-commercial use are biopesticides, including biofungicides (e.g. kasugamycin, polyoxins), bioinsecticides (nikkomycin, spinosyns), bioherbicides (bialaphos), antihelminthics (avermectin), coccidiostats, ruminant-growth promoters (monensin, lasalocid, salinomycin), plant-growth regulators (gibberellins), immunosuppressants for organ transplants (cyclosporin A, FK-506, rapamycin), anabolic agents in farm animals (zearelanone), uterocontractants (ergot alkaloids) and antitumour agents -(doxorubicin, daunorubicin, mitomycin, bleomycin).

QUESTIONS

1. What are biofertilizers? How these are helpful in crop growth.
2. What are biopesticides? Describe various types of biopesticides and discuss their merits and demerits.
3. Write short notes:
   (*a*) Biopesticides. (*b*) Fungi as biopesticide (*c*) Plant pesticides.

BIOFUEL

CHAPTER 48

I. BIOMASS ENERGY

The biomass is defined as the biological degradable fraction of products, wastes and residues from agriculture (including animal and vegetable materials), forestry and biological degradable fraction of industrial and household wastes. Biomass is considered as renewable source of energy as it is CO_2 neutral with in its entire life cycle. During their growth plants absorb CO_2 from the air and convert it by photosynthesis into complex biochemical compounds like cellulose and lignin. Sunlight provides the energy for this process. In fact, solar energy is stored by means of biomass. At a later stage this energy can be released and used. During this conversion process the same amount of CO_2 is released. In this way there is no net increase of CO_2 and therefore we can speak of a closed cycle in which the total amount of CO_2 is conserved.

The various types of biomass fuels are classified in Table 48.1. The physical and chemical properties of biomass fuels are shown in Table 48.2 & 48.3 respectively. Various conversion technologies for biomass resources are shown in Figure 48.1. Plants are the major source of biomass because their cell wall is constituted by components such as cellulose, hemicellulose, lignin water-soluble sugars, amino acids and aliphatic acids, ether and alcohol-soluble constituents (fats, waxes, resin, pigments) and proteins.

Table 48.1: Types of biomass fuels.

Woody biomass	Non-woody biomass	Processed Waste	Processed fuels
Trees	Energy crops such as sugarcane	Cereal husks and cobs	Charcoal (wood and residues)
Shrubs	Cereal straw	Wastes from pineapple and other fruits	Briquette/densified biomass
Bushes such as coffee and tea	Cotton, cassava, tobacco stems	Nut shells	Methanol/Ethanol (wood alcohol)
Sweepings from forest floor	Grass	Plant oil cake	Plant oils from palm, sunflower, rape etc.
Bamboo	Soft stems such as pulses and potatoes	Black liquor from pulp mills	Producer gas
Palms	Swamp and water plants	Municipal waste	Biogas

Table 48.2: Physical properties of biomass fuels.

Parameter	Significance
Moisture content	Storage, dry-matter losses
Volatiles	Thermal decomposition, combustion technology
Ash content	Dust emission, ash manipulation
Fixed carbon	Combustion technology
Calorific/heating value	Fuel utilization, plant design
Ash melting	Safety, process control
Fungi	Health risks
Bulk density	Logistics
Particle density, heat capacity, conductivity	Thermal decomposition
Dimension, shape	Conveying, drying, bridging, combustion technology

Table 48.3: Chemical properties of biomass fuels.

Hydrogen H	Heating value
Oxygen O	Heating value
Nitrogen N	NOx, N_2O emissions
Chlorine Cl	HCl, PCDD/F emissions, corrosion
Sulphur S	Sox emissions, corrosion
Fluorine F	HF emissions, corrosion
Potassium K	Corrosion, ash melting
Sodium Na	Corrosion, ash melting
Magnesium Mg	Ash melting, utilization
Calcium Ca	Ash melting, utilization
Phosphorous P	Ash utilization
Heavy metals	Emission, ash melting

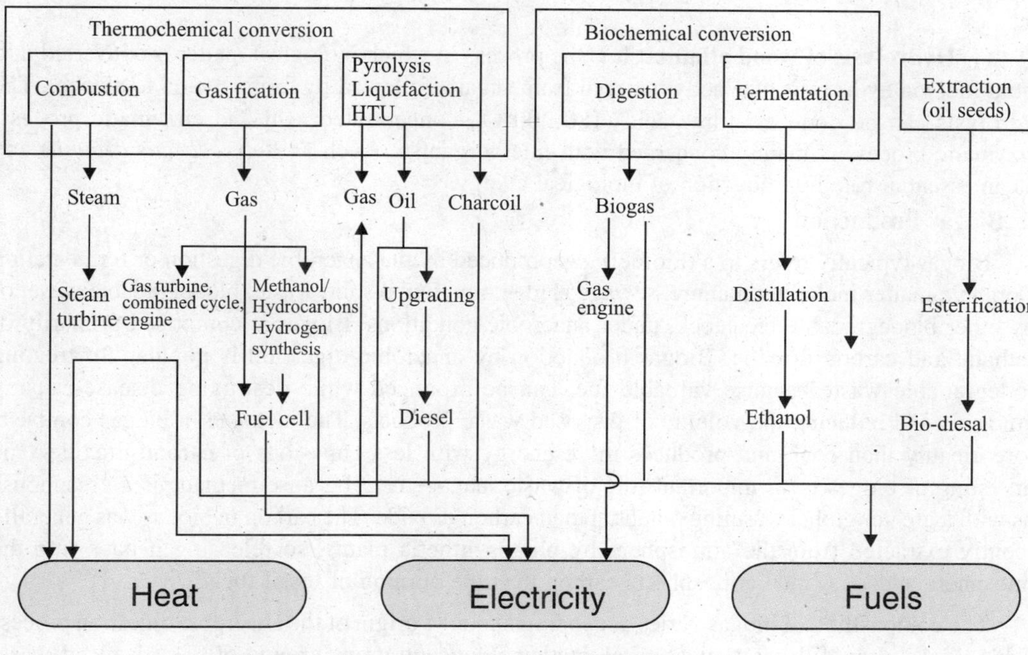

Fig. 48.1. Conversion technologies for biomass resources.

Various technologies involved in biofuel production are presented in figure 48.1. It is combination of biology and engineering technologies.

(*i*) **Pyrolysis:** It is defined as the destructive distillation or decomposition of organic matter, for example solid residues, wastes (saw dust, wood chips, wood pieces) in an oxygen-deficient atmosphere or in absence of oxygen at high temperature (200-500°C or rarely 900°C). Products of pyrolysis are gases, organic liquids and chars, depending on the pyrolysis process and temperature of reaction. The condensable liquids separate into aqueous (pyroligneous acid), oil and tar fraction. The composition of gas is carbon monoxide (28-33%), methane (3-18%), higher hydrocarbons (1-3%) and hydrogen (1-3%). During pyrolysis, hydrogen content of gas increases with the temperature.

(*ii*) **Gasification:** It is a process of thermal degradation of carbonaceous material under controlled amount of air or pure oxygen and high temperature up to around 1000°C. As a result of gasification, high amount of gases are produced. When biomass reacts with steam and oxygen it produces BTU gas, which mainly consist of carbon monoxide and hydrogen. This gas can be burnt as a fuel or can be used as an intermediate for synthesis of fuels or energy intensive chemicals such as methanol, intermediate gasoline, hydrogen and ammonia.

(*iii*) **Liquefaction:** It involves production of oils for energy from wood or agriculture and carbon residues by reacting them with carbon monoxide and water/steam at high pressure (4000 lb/in^2) and temperature (350-400°C) in presence of catalysts. By this method 40-50 % oil is obtained from wood and serves as a good source of fuel.

(*iv*) **Bioconversion:** It involves the conversion of organic materials into energy, fertilizer, food and chemicals through biological agent. Enzymatic digestion involves conversion of cellulosic and lignocellulosic materials into alcohols, acids and animal feeds by using microbial enzymes such as cellulase, hemicellulase, amylase, pectinase etc. An anerobic digestion is a partial conversion by microorganisms of organic substrates into gases in absence of air. The gases produced are collectively known as biogas. Aerobic digestion involves the conversion of organic substrates by microorganisms into utilizable forms in presence of air to produce gases, single cell protein, fertilizers etc.

(*v*) **Hydrolysis of wood ethanol:** It is the process in which biological matter is converted into liquid fuel via hydrolysis of woody material from sugars, followed by fermentation to alcohol. The hydrolysis can be done by dilute acid (180-200°C), concentrated acid and enzymatic process. Enzymatic process is mainly concerned with microorganism (such as *Botryococcus braunii*) and has an essential role in production of biological energy.

II. Biogas Production

Biogas typically refers to a (biofuel) gas produced by the anaerobic digestion or fermentation of organic matter including manure, sewage sludge, municipal solid waste, biodegradable waste or any other biodegradable feedstock, under anaerobic conditions. Biogas is comprised primarily of methane and carbon dioxide. Biogas production by anaerobic digestion is popular for treating biodegradable waste because valuable fuel can be produced while destroying disease-causing pathogens and reducing the volume of disposed waste products. The methane in biogas combusts more cleanly than coal, and produces more energy with less emissions of carbon dioxide. The harvesting of biogas is an important role of waste management because methane is a greenhouse gas with a greater global warming potential than carbon dioxide. The carbon in biogas was generally recently extracted from the atmosphere by photosynthetic plants, so releasing it back into the atmosphere adds less total atmospheric carbon than the burning of fossil fuels.

The composition of biogas varies depending upon the origin of the anaerobic digestion process (Table 48.4). Landfill gas typically has methane concentrations around 50%. Advanced waste treatment technologies can produce biogas with 50-75% CH_4.

Table 48.4. Biogas Composition.

Matter	%
Methane, CH_4	50-75
Carbon dioxide, CO_2	25-50
Nitrogen, N_2	0-10*
Hydrogen, H_2	0-1
Hydrogen sulphide, H_2S	0-3
Oxygen, O_2	0-2*

*often 5 % of air is introduced for microbiological desulphurisation

Three basic points about this process are:

(*i*) that most of the important bacteria involved in biogas production process are anaerobes and slow growing;

(*ii*) that a greater degree of metabolic specialization is observed in these anaerobic microorganisms; and

(*iii*) that most of the free energy present in the substrate is found in the terminal product methane. Since less energy is available for the growth of organism, less microbial biomass is produced and, consequently, disposal of sludge after the digestion may not be a major problem.

Complex polymers are broken down to soluble products by enzymes produced by fermentative bacteria (Fig. 48.2 Group 1), which ferment the substrate to short-chain fatty acids, hydrogen and carbon dioxide. Fatty acids longer than acetate are metabolized to acetate by obligate hydrogen-producing acetogenic bacteria (Fig. 48.2, Group 2). The major products after digestion of the substrate by these two groups are hydrogen, carbon dioxide, and acetate. Hydrogen and carbon dioxide can be converted to acetate by hydrogen-oxidizing acetogens (Fig. 48.2, Group 3) or methane by carbon-dioxide-reducing, hydrogen-oxidizing methanogens (Fig. 48.2, Group 4). Acetate is also converted to methane by aceticlastic methanogens (Fig. 48.2, Group 5). Nearly seventy per cent of methane from biogas digesters fed with cattle dung is derived from acetate.

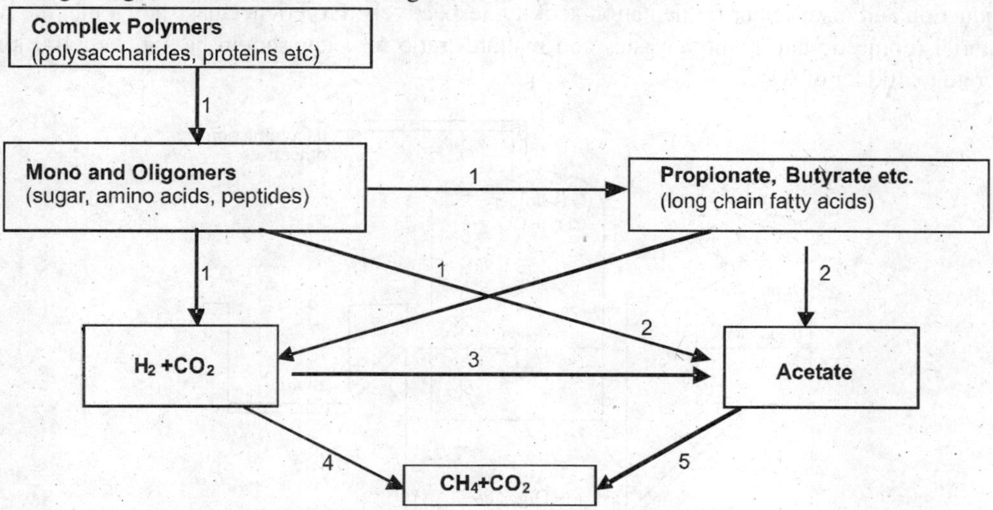

Fig. 48.2. Microbial conversion of organic matter to methane.

Biogas system is a mechanism in which any type of organic waste can be used for the production of biogas and enriched manure in terms of N-P-K contents which can supplement use of chemical fertilizer. This combustible gas can be effectively used for cooking, lightning and motive power generation. The biogas plant consists of two components: a digester (or fermentation tank) and a gas holder (Fig. 48.3). The digester is a cube-shaped or cylindrical waterproof container with an inlet into which the fermentable mixture is introduced in the form of liquid slurry. The gas holder is normally an airproof steel container that, by floating like a ball on the fermentation mix, cuts off air to the digester (anaerobiosis) and collects the gas generated. In one of the most widely used designs, the gas holder is equipped with a gas outlet, while the digester is provided with an overflow pipe to lead the sludge out into a drainage pit. For biogas plant construction, important criteria are the amount of gas required for a specific use or uses, and the amount of waste material available for processing. Digester reactors are constructed from brick, cement, concrete, and steel. Clay bricks have successfully replaced cement blocks and concrete in some countries like Indonesia. In areas where the cost is high, the "sausage" or bag digester appears to be ideal. The digester is constructed of 0.55 mm thick Hypalon laminated with Neoprene and reinforced with nylon. The bag is fitted with an inlet and an outlet made from PVC. Even if imported from the United States, the cost of the digester and the gas holder (both combined in one bag) is only 10 per cent of that for a concrete-steel digester. Another advantage is that it can be mass produced and is easily mailed. In rural areas, the whole installation is completed in a matter of minutes. A hole in the ground accommodates the bag, which is filled two-thirds full with waste water. Gas production fully inflates the bag, which is weighted down and fitted with a compressor to increase gas pressure.

Raw materials may be obtained from a variety of sources - livestock and poultry wastes, night soil, crop residues, food-processing and paper wastes, and materials such as aquatic weeds, water hyacinth, filamentous algae, and seaweed. Different problems are encountered with each of these wastes with regard to collection, transportation, processing, storage, residue utilization, and ultimate use. Residues from the agricultural sector such as spent straw, hay, cane trash, corn and plant stubble, and bagasse need to be shredded in order to facilitate their flow into the digester reactor as well as to increase the efficiency of bacterial action. Succulent plant material yields more gas than dried matter does, and hence materials like brush and weeds need semi-drying. The storage of raw materials in a damp, confined space for over ten days initiates anaerobic bacterial action that, though causing some gas loss, reduces the time for the digester to become operational. Production of biogas is inefficient if fermentation materials are too dilute or too concentrated, resulting in, low biogas production and insufficient fermentation activity, respectively. Experience has shown that the raw-material (domestic and poultry wastes and manure) ratio to water should be 1:1, i.e., 100 kg of excrete to 100 kg of water.

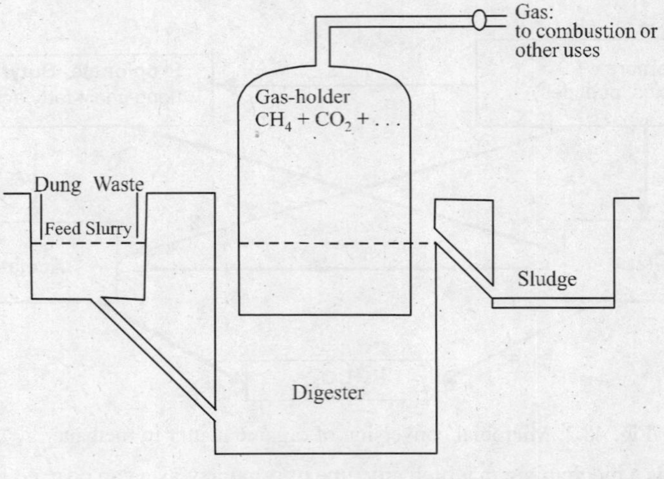

Fig. 48.3. A simple biogas plant.

In the slurry, this corresponds to a total solids concentration of 8 - 11 per cent by weight. The size of the digester depends upon the loading, which is determined by the influent solids content, retention time, and the digester temperature. Optimum loading rates vary with different digesters and their sites of location. Higher loading rates have been used when the ambient temperature is high. In general, the literature is filled with a variety of conflicting loading rates. In practice, the loading rate should be an expression of either (*a*) the weight of total volatile solids (TVS) added per day per unit volume of the digester, or (*b*) the weight of TVS added per day per unit weight of TVS in the digester. The latter principle is normally used for smooth operation of the digester. Common practice involves seeding with an adequate population of both the acid-forming and methanogenic bacteria. Actively digesting sludge from a sewage plant constitutes ideal "seed" material. As a general guideline, the seed material should be twice the volume of the fresh manure slurry during the start-up phase, with a gradual decrease in amount added over a three-week period. If the digester accumulates volatile acids as a result of overloading, the situation can be remedied by reseeding, or by the addition of lime or other alkali. Low pH inhibits the growth of the methanogenic bacteria and gas generation and is often the result of overloading. A successful pH range for anaerobic digestion is 6.0 - 8.0; efficient digestion occurs at a pH near neutrality. A slightly alkaline state is an indication that pH fluctuations are not too drastic. Low pH may be remedied by dilution or by the addition of lime. With a mesophilic flora, digestion proceeds best at 30 - 40° C; with thermophiles, the optimum range is 50 - 60° C. The choice of the temperature to be used is influenced by climatic considerations In general, there is no rule of thumb, but for optimum process stability, the temperature should be carefully regulated within a narrow range of the operating temperature. In warm climates, with no freezing temperatures, digesters may be operated without added heat. As a safety measure, it is common practice either to bury the digesters in the ground on account of the advantageous insulating properties of the soil, or to use a greenhouse covering. Heating requirements and, consequently, costs, can be minimized through the use of natural materials such as leaves, sawdust, straw, etc., which are composted in batches in a separate compartment around the digester. The maintenance of optimum microbiological activity in the digester is crucial to gas generation and consequently is related to nutrient availability. Two of the most important nutrients are carbon and nitrogen and a critical factor for raw material choice is the overall C/N ratio.

Domestic sewage and animal and poultry wastes are examples of N-rich materials that provide nutrients for the growth and multiplication of the anaerobic organisms. On the other hand, N-poor materials like green grass, corn stubble, etc., are rich in carbohydrate substances that are essential for gas production. Excess availability of nitrogen leads to the formation of NH_3, the concentration of which inhibits further growth. Ammonia toxicity can be remedied by low loading or by dilution. In practice, it is important to maintain, by weight, a C/N ratio close to 30:1 for achieving an optimum rate of digestion. The C/N ratio can be judiciously manipulated by combining materials low in carbon with those that are high in nitrogen, and vice versa. Wastes and biodegradable residue are often accompanied by a variety of pollutants that could inhibit anaerobic digestion. Potential toxicity due to ammonia can be corrected by remedying the C/N ratio of manure through the addition of shredded bagasse or straw, or by dilution. Common toxic substances are the soluble salts of copper, zinc, nickel, mercury, and chromium. On the other hand, salts of sodium, potassium, calcium, and magnesium may be stimulatory or toxic in action, both manifestations being associated with the cation rather than the anionic portion of the salt. Pesticides and synthetic detergents may also be troublesome to the process. When solid materials not well shredded are present in the digester, gas generation may be impeded by the formation of a scum that is comprised of these low-density solids that are enmeshed in a filamentous matrix. In time the scum hardens, disrupting the digestion process and causing stratification. Agitation can be done either mechanically with a plunger or by means of rotational spraying of fresh influent. Agitation, normally required for bath digesters, ensures exposure of new surfaces to bacterial action, prevents viscid stratification and slow-down of bacterial activity, and promotes uniform dispersion of the influent materials throughout the fermentation liquor, thereby accelerating digestion.

Microbial diversity in biogas digesters is as great as that of rumen wherein seventeen fermentative bacterial species have been reported to play important role for production of biogas. Furthermore, it is the nature of the substrate that determines the type and extent of the fermentative bacteria present in the digester. Higher presence of proteolytic organisms in cows dung-fed digesters and other animals waste-fed digesters had been reported. It has been observed that while cow dung-fed digesters support higher amylolytic microorganisms, poultry waste-fed digesters show higher proteolytic population. Among fermentative organisms, *Bacteroides* succinogens, *Butyrivibrio fibrisolvens*, *Clostridium cellobioparum*, *Ruminococcus albus* and *Clostridium* sp. are predominant. A clear differentiation exists in the type of cellulolytic bacterial distribution in rumen and biogas digester. Whereas in rumen, *Ruminococcus* sp. alone accounted for 60 per cent of the total population, in the biogas digester the predominant species belonged to the genera Bacteroides and Clostridium rather than the genus *Ruminococcus*. *Ruminococcus flavefaciens*, *Eubacterium cellulosolvens*,

Clostridium cellulosolvens, Clostridium cellulovorans, Clostridium thermocellum, Bacteroides cellulosolvens and *Acetivibrio cellulolyticus* are some of the other predominant fermentative bacteria present in cattle dung-fed digesters. Various factors such as biogas potential of feedstock, design of digester, inoculum, nature of substrate, pH, temperature, loading rate, hydraulic retention time (HRT), C: N ratio, volatile fatty acids (VFA), etc. influence the biogas production.

Several kinetic models have been developed to describe the anaerobic fermentation process. Monod showed a hyperbolic relationship between the exponential microbial growth rate and substrate concentration. In this model, the two kinetic parameters, namely, microorganisms growth rate and half velocity constant are deterministic in nature, and these predict the conditions of timing of maximum biological activity and its cessation. This model can be used to determine the rate of substrate utilization (r_S) by the equation:

$$r_s = q_{max} \cdot S_x/K + S,$$

where S is limiting substrate concentration, K is half constant, x is concentration of bacterial cells, and q_{max} is maximum substrate utilization rate.The above equation is applicable for low substrate concentration. However for high substrate concentration, the equation is re-written as:

$$r_S = q_{max} \cdot X$$

The Monod model suffers from the drawback that one set of kinetic parameters are not sufficient to describe biological process both for short- and long-retention times, and that kinetic parameters cannot be obtained for some complex substrates. To alleviate limitations of the Monod model while retaining its advantages, Hashimoto developed an alternative equation, which attempts to describe kinetics of methane fermentation in terms of several parameters. According to this equation, given below, for a given loading rate S_0/q daily volume of methane per volume of digester depended on the biodegradability of the material (B_0) and kinetic parameters μm and K.

$$r_V = (B_0 \cdot S_0/q) \cdot \{1-(K/q \;\; \mu m - 1 + K)\}$$

where,

r_v is volumetric methane production rate, l CH_4 l^{-1},digester d^{-1}

S_0 is influent total volatile solids (VS) concentration, gl^{-1}

B_0 is ultimate methane yield, l CH_4 g^{-1} VS added as q ® ¥

q is hydraulic retention time d^{-1}

μm is maximum specific growth of microorganism d^{-1}

K is kinetic parameter, dimensionless.

III. ETHANOL PRODUCTION FOR ENERGY

Ethanol fuel is an alternative to gasoline. It can be combined with gasoline in any concentration up to pure ethanol. Anhydrous ethanol, that is, ethanol with at most 1% water, can be blended with gasoline in varying quantities to reduce consumption of petroleum fuels and in attempts to reduce air pollution. Worldwide automotive ethanol capabilities vary widely and most spark-ignited gasoline style engines will operate well with mixtures of 10% ethanol.

In Brazil, ethanol-powered and flexible-fuel vehicles are manufactured to be capable of operation by burning hydrated ethanol, an azeotrope of ethanol (93% v/v) and water (7%). Hydrated ethanol may also be mixed with gasoline in flexible fuel vehicles but a minimum amount of ethanol (granted by legally regulated gasoline type C) is required to avoid problems with the mixture. A few flexible-fuel systems, like Hi-Flex, used by Renault Clio and Fiat Siena, can also run with pure gasoline.

Ethanol is increasingly used as an oxygenate additive for standard gasoline, as a replacement for methyl t-butyl ether (MTBE), the latter chemical being difficult to retrieve from groundwater and soil contamination. At a 10% mixture, ethanol reduces the likelihood of engine knock, by raising the octane rating. The use of 10% ethanol gasoline is mandated in some cities where the possibility of harmful levels of auto emissions are possible, especially during the winter months. Ethanol can be used to power fuel cells, and also as a feed chemical in the transesterification process for biodiesel.

Ethanol can be mass-produced by fermentation of sugar or by hydration of ethylene from petroleum and other sources. Current interest in ethanol lies in production derived from crops (bio-ethanol), and there's discussion about whether it is a sustainable energy resource that may offer environmental and long-term economic advantages over fossil fuels, like gasoline or diesel. It is readily obtained from the starch or sugar in a wide variety of crops (Fig. 48.4). Ethanol fuel production depends on availability of land area, soil, water, and sunlight.

Bioethanol is obtained from the conversion of carbon based feedstock. Agricultural feedstocks are considered renewable because they get energy from the sun using photosynthesis. Ethanol can be produced from a variety of feedstocks such as sugar cane, bagasse, miscanthus, sugar beet, sorghum, grain sorghum, switchgrass, barley, hemp, kenaf, potatoes, sweet potatoes, cassava, sunflower, fruit, molasses, whey or skim milk, corn, stover, grain, wheat, wood, paper, straw, cotton, other biomass, as well as many types of cellulose waste. The production is primarily from sugarcane, maize (corn) and sugar beets - and also as of today technology does not exist that makes it economically competitive to produce ethanol from cellulosic feedstock. Four countries have developed bioethanol fuel programs: Brazil, Colombia, China and the United States. For the ethanol to be usable as a fuel, water must be removed. Most of the water is removed by distillation, but the purity is limited to 95-96% due to the formation of a low-boiling water-ethanol azeotrope. The 96% m/m (93% v/v) ethanol, 4% m/m (7% v/v) water mixture may be used as a fuel, and it's called hydrated ethyl alcohol fuel. In 2006/2007, an estimated 17 billion liters (4,5 billion gallons) of hydrated ethyl alcohol fuel will be produced, to be used in ethanol powered vehicles.

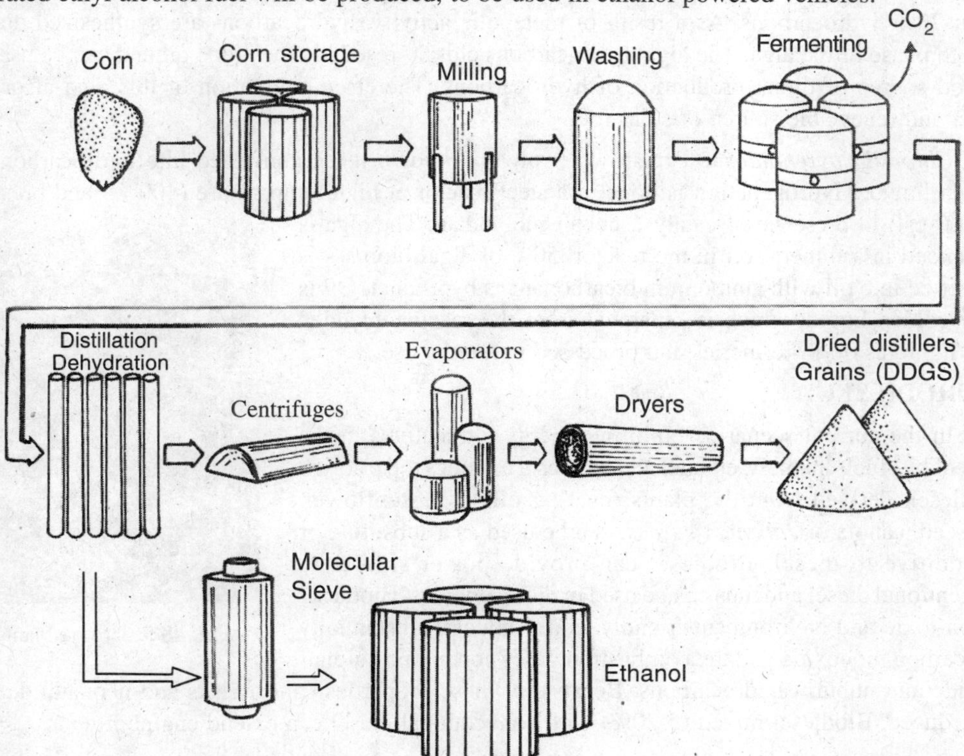

Fig. 48.4. Ethanol production from corn.

IV. HYDROGEN AS ENERGY SOURCE

Hydrogen is also a renewable fuel because the raw material in the form is abundant and solar energy used for decomposition of water to produce hydrogen, will be available for millions of year. Hydrogen production is emerging as an important area of biotechnology research with respect to environment friendly non-conventional and renewable source of energy. In achieving this objective, the principle used in photosynthetic machinery by green plants is used. Enzymes hydrogenase and nitrogenase can be used to convert hydrogen ions (H^+) into hydrogen gas (H_2). Hydrogen is made mainly from natural gas. Most of the hydrogen produced industrial derives from fossil fuels, although some is produced by electrolysis of water.

Some microbes possess the enzyme hydrogenase, which helps the two electrons join two hydrogen ions to produce one molecule of hydrogen gas. These include a number of algae (*Scenedesmus*, *Chlamydomonas*, *Dunaliella*, *Porphyridium*, *Oscillatoria*) and some photosynthetic bacteria (*Rhodospirillium*, *Chromatium*, *Thiocapsa*), which can be gainfully employed for hydrogen production. In cyanobacteria (*Anabanea*, *Nostoc*, *Rivularia*) the nitrogenase enzyme reduces nitrogen to ammonia with associated release of hydrogen gas and this depends on light.

V. HYDROCARBON PRODUCTION

Hydrocarbons produced by some microorganisms may also serve as liquid or gaseous fuels. For instance, microbial production of methane gas from polymers like carbohydrates, proteins, lipids etc. involves: (*i*) breakdown of polymers into organic acids and alcohols by hydrolytic and fermentation bacteria; (*ii*) conversion of acids and alcohols into acetates, H_2 and CO_2 by hydrogen producing acetogenic bacteria and (*iii*) conversion of acetate and $H_2 + CO_2$ to methane by methanogenic bacteria.

Hydrocarbons are also produced by unicellular alga *Botryococcus braunii*, which contains about 70% hydrocarbons. As a result of metabolic activity, hydrocarbons are synthesized during growth phase of the alga. The algal hydrocarbons closely resemble the crude oil and can be used as a good source of direct production of hydrocarbons. Therefore cultivation of this alga affords a direct and renewable source of solar fuel.

Chlorella pyrenoidosa, a fresh water alga, is known to be converted into hydrocarbons as golden liquid. Hydrogenation is done in a steel reactor at high temperature (400°C) and pressure (12000 psi) in presence of catalyst, cobalt molybdate. The alga is suspended in a mineral oil in the reactor. 50% of algal biomass is converted into oil with ammonium bicarbonate as byproduct. Oil is clear golden liquid, which is separated from the reactor, blended with light gas oil in refineries and processed before its use.

VI. BIODIESEL

In the current scenario, exploring energy alternatives in the form of bio-fuels, namely, ethanol and biodiesel assumes top priority. Biodiesel, derived from the plants (bearing oils) like sunflower, rapeseed, canola or *Jatropha curcas*, can be used as a substitute or an additive to diesel. Biodiesel can provide power similar to conventional diesel and thus can be used in diesel engines. Biodiesel is non-toxic and environment friendly as it produces substantially less carbon monoxide and the combustion gases contain no sulphur dioxide and unburnt hydrocarbons. Because of these properties cancer risks and neonatal defects are reduced. Biodiesel mixed (5-20%) with conventional diesel can extend engine life.

Fig. 48.5. Jatropa Plant

Considering all the options in India, *Jatropha curcas* has been identified as the most suitable source due to its lower gestation period. While the seeds are used for oil extraction, other parts of the plant, that is, leaves, bark and the like, can be used for organic dyes, medicines, biogas and the like. Originating in the Caribbean islands, jatropha has spread as a valuable hedge plant to Africa, Asia and India. It grows in a variety of agro-climatic conditions. Thus it ensures a reasonable production of seeds with little inputs (Fig. 48.5, 48.6).

Intercropping of jatropha with lemon grass, turmeric, ashwagandha and the like has been found to be an economically viable option when compared with monocropping. Since jatropha is not suitable for forage, fence farming is also an attractive option. Plantation of saplings from the seeds rather than vegetative propagation is a better option for the dry areas. The plantation of saplings (4 months old) should be carried out before the rains. As the growth and canopy coverage are more in fertile soil, the plants occupy more area when compared with dry land cultivation. Oil from jatropha seeds (oil content: 35% on average) can be extracted by using a screw press for further processing into biodiesel by the transesterification route. Transesterification is the process of converting the raw vegetable oil into biodiesel. The process involves displacement of alcohol from an ester by another alcohol (methanol is used commonly). Biodiesel can safely be stored for up to 6 months. The existing storage facilities and infrastructure for petrol and diesel can be used for the biodiesel with minor alterations. Production of bio-fuel from plant materials is a major step toward harnessing one of the world's most-prevalent, yet least-utilized renewable energy resources.

Fig. 48.6. Biodiesel Plan

QUESTIONS

1. Describe the alternative sources of energy production with reference to biogas production.
2. Discuss plant based fuel production such as biodiesel and ethanol.
3. Discuss role of biotechnology in generating alternative source of energy.

BIOSAFETY SCIENCE

CHAPTER 49

Work in laboratory involves hazardous chemicals, carcinogens, inflammable solvents, radioactive compounds, live and infectious microorganisms, proper disposal of contaminated/dead animal materials and genetically modified crops. Safety of the personnel working in the laboratory involves various measures to prevent accidental damage to the workers and spread of contamination. The term "containment" is used in describing safe methods for managing infectious materials in the laboratory environment where they are being handled or maintained. The purpose of containment is to reduce or eliminate exposure of laboratory workers, other persons, and the outside environment to potentially hazardous agents.

Primary containment, the protection of personnel and the immediate laboratory environment from exposure to infectious agents, is provided by both good microbiological technique and the use of appropriate safety equipment. The use of vaccines may provide an increased level of personal protection. Secondary containment, the protection of the environment external to the laboratory from exposure to infectious materials, is provided by a combination of facility design and operational practices. Therefore, the three elements of containment include laboratory practice and technique, safety equipment, and facility design. The risk assessment of the work to be done with a specific agent will determine the appropriate combination of these elements.

I. LABORATORY PRACTICE AND TECHNIQUE

The most important element of containment is strict adherence to standard microbiological practices and techniques. Persons working with infectious agents or potentially infected materials must be aware of potential hazards, and must be trained and proficient in the practices and techniques required to handle such material safely. The director or person in charge of the laboratory is responsible for providing or arranging the appropriate training of personnel.

Each laboratory should develop or adopt a biosafety or operations manual that identifies the hazards that will or may be encountered, and that specifies practices and procedures designed to minimize or eliminate exposures to these hazards. Personnel should be advised of special hazards and should be required to read and follow the required practices and procedures. A scientist trained and knowledgeable in appropriate laboratory techniques, safety procedures, and hazards associated with handling infectious agents must be responsible for the conduct of work with any infectious agents or material. This individual should consult with biosafety or other health and safety professionals with regard to risk assessment.

Laboratory personnel, safety practices, and techniques must be supplemented by appropriate facility design and engineering features, safety equipment, and management practices.

II. SAFETY EQUIPMENT (PRIMARY BARRIERS)

Safety equipment includes biological safety cabinets (BSCs), enclosed containers, and other engineering controls designed to remove or minimize exposures to hazardous biological materials. The biological safety cabinet (BSC) is the principal device used to provide containment of infectious splashes or aerosols generated by many microbiological procedures. Three types of biological safety cabinets (Class I, II, III) used in microbiological laboratories are described and illustrated in Appendix A. Open-fronted Class I and Class II biological safety cabinets are primary barriers which offer significant levels of protection to laboratory personnel and to the environment when used with good microbiological techniques. The Class II biological safety cabinet also provides protection from external contamination of the materials (e.g., cell cultures, microbiological stocks) being manipulated inside the cabinet. The gas-tight Class III biological safety cabinet provides the highest attainable level of protection to personnel and the environment.

An example of another primary barrier is the safety centrifuge cup, an enclosed container designed to prevent aerosols from being released during centrifugation. To minimize this hazard, containment controls such as BSCs or centrifuge cups must be used when handling infectious agents that can be transmitted through the aerosol route of exposure.

Safety equipment also may include items for personal protection, such as gloves, coats, gowns, shoe covers, boots, respirators, face shields, safety glasses, or goggles. Personal protective equipment is often used in combination with biological safety cabinets and other devices that contain the agents, animals, or materials being handled. In some situations in which it is impractical to work in biological safety cabinets, personal protective equipment may form the primary barrier between personnel and the infectious materials. Examples include certain animal studies, animal necropsy, agent production activities, and activities relating to maintenance, service, or support of the laboratory facility.

III. FACILITY DESIGN AND CONSTRUCTION (SECONDARY BARRIERS)

The design and construction of the facility contributes to the laboratory workers' protection, provides a barrier to protect persons outside the laboratory, and protects persons or animals in the community from infectious agents which may be accidentally released from the laboratory. Laboratory management is responsible for providing facilities commensurate with the laboratory's function and the recommended biosafety level for the agents being manipulated.

The recommended secondary barrier(s) will depend on the risk of transmission of specific agents. For example, the exposure risks for most laboratory work in Biosafety Level 1 and 2 facilities will be direct contact with the agents, or inadvertent contact exposures through contaminated work environments. Secondary barriers in these laboratories may include separation of the laboratory work area from public access, availability of a decontamination facility (e.g., autoclave), and handwashing facilities.

When the risk of infection by exposure to an infectious aerosol is present, higher levels of primary containment and multiple secondary barriers may become necessary to prevent infectious agents from escaping into the environment. Such design features include specialized ventilation systems to ensure directional air flow, air treatment systems to decontaminate or remove agents from exhaust air, controlled access zones, airlocks as laboratory entrances, or separate buildings or modules to isolate the laboratory.

IV. BIOSAFETY LEVELS

Four biosafety levels (BSLs) are described in Section III, which consist of combinations of laboratory practices and techniques, safety equipment, and laboratory facilities. Each combination is specifically appropriate for the operations performed, the documented or suspected routes of transmission of the infectious agents, and the laboratory function or activity.

The recommended biosafety level(s) for the organisms in Section VII (Agent Summary Statements) represent those conditions under which the agent ordinarily can be safely handled. The laboratory director is specifically and primarily responsible for assessing the risks and appropriately applying the recommended biosafety levels. Generally, work with known agents should be conducted at the biosafety level recommended in Section VII. When specific information is available to suggest that virulence, pathogenicity, antibiotic resistance patterns, vaccine and treatment availability, or other factors are significantly altered, more (or less) stringent practices may be specified.

(*a*) **Biosafety Level 1** practices, safety equipment, and facility design and construction are appropriate for undergraduate and secondary educational training and teaching laboratories, and for other laboratories in which work is done with defined and characterized strains of viable microorganisms not known to consistently cause disease in healthy adult humans. *Bacillus subtilis, Naegleria gruberi,* infectious canine hepatitis virus, and exempt organisms under the NIH Recombinant DNA Guidelines are representative of microorganisms meeting these criteria. Many agents not ordinarily associated with disease processes in humans are, however, opportunistic pathogens and may cause infection in the young, the aged, and immunodeficient or immunosuppressed individuals. Vaccine strains that have undergone multiple *in vivo* passages should not be considered avirulent simply because they are vaccine strains.

Biosafety Level 1 represents a basic level of containment that relies on standard microbiological practices with no special primary or secondary barriers recommended, other than a sink for handwashing.

(*b*) **Biosafety Level 2** practices, equipment, and facility design and construction are applicable to clinical, diagnostic, teaching, and other laboratories in which work is done with the broad spectrum of indigenous moderate-risk agents that are present in the community and associated with human disease of varying severity. With good microbiological techniques, these agents can be used safely in activities conducted on the open bench, provided the potential for producing splashes or aerosols is low. Hepatitis B virus, HIV, the salmonellae, and Toxoplasma spp. are representative of microorganisms assigned to this containment level. Biosafety Level 2 is appropriate when work is done with any human-derived blood, body fluids, tissues, or primary human cell lines where the presence of an infectious agent may be unknown. (Laboratory personnel working with human-derived materials should refer to the OSHA Bloodborne Pathogen Standard[(2)] for specific required precautions.)

Primary hazards to personnel working with these agents relate to accidental percutaneous or mucous membrane exposures, or ingestion of infectious materials. Extreme caution should be taken with contaminated needles or sharp instruments. Even though organisms routinely manipulated at Biosafety Level 2 are not known to be transmissible by the aerosol route, procedures with aerosol or high splash potential that may increase the risk of such personnel exposure must be conducted in primary containment equipment, or in devices such as a BSC or safety centrifuge cups. Other primary barriers should be used as appropriate, such as splash shields, face protection, gowns, and gloves.

Secondary barriers such as handwashing sinks and waste decontamination facilities must be available to reduce potential environmental contamination.

(c) Biosafety Level 3 practices, safety equipment, and facility design and construction are applicable to clinical, diagnostic, teaching, research, or production facilities in which work is done with indigenous or exotic agents with a potential for respiratory transmission, and which may cause serious and potentially lethal infection. *Mycobacterium tuberculosis*, St. Louis encephalitis virus, and *Coxiella burnetii* are representative of the microorganisms assigned to this level. Primary hazards to personnel working with these agents relate to autoinoculation, ingestion, and exposure to infectious aerosols.

(d) Biosafety Level 4 practices, safety equipment, and facility design and construction are applicable for work with dangerous and exotic agents that pose a high individual risk of life-threatening disease, which may be transmitted via the aerosol route and for which there is no available vaccine or therapy. Agents with a close or identical antigenic relationship to Biosafety Level 4 agents also should be handled at this level. When sufficient data are obtained, work with these agents may continue at this level or at a lower level. Viruses such as Marburg or Congo-Crimean hemorrhagic fever are manipulated at Biosafety Level 4.

(e) Animal Facilities

Four biosafety levels are also described for activities involving infectious disease work with experimental animals. These four combinations of practices, safety equipment, and **facilities are designated** Animal Biosafety Levels 1, 2, 3, and 4, **and provide increasing levels of protection to personnel and the environment.**

V. CLINICAL LABORATORIES

Except in extraordinary circumstances (e.g., suspected hemorrhagic fever), the initial processing of clinical specimens and serological identification of isolates can be done safely at Biosafety Level 2, the recommended level for work with blood borne pathogens such as hepatitis B virus and HIV. The containment elements described in Biosafety Level 2 are consistent with the OSHA standard, *"Occupational Exposure to Bloodborne Pathogens"* from the Occupational Safety and Health Administration. This requires the use of specific precautions with **all** clinical specimens of blood or other potentially infectious material (Universal or Standard Precautions). Additionally, other recommendations specific for clinical laboratories may be obtained from the National Committee for Clinical Laboratory Standards.

Biosafety Level 2 recommendations and OSHA requirements focus on the prevention of percutaneous and mucous membrane exposures to clinical material. Primary barriers such as biological safety cabinets (Class I or II) should be used when performing procedures that might cause splashing, spraying, or splattering of droplets. Biological safety cabinets also should be used for the initial processing of clinical specimens when the nature of the test requested or other information suggests the likely presence of an agent readily transmissible by infectious aerosols (e.g., *M. tuberculosis*), or when the use of a biological safety cabinet (Class II) is indicated to protect the integrity of the specimen.

VI. IMPORTATION AND INTERSTATE SHIPMENT OF CERTAIN BIOMEDICAL MATERIALS

The importation of etiologic agents and vectors of human diseases is subject to the requirements of the Public Health Service Foreign Quarantine regulations. Companion regulations of the Public Health Service and the Department of Transportation specify packaging, labeling, and shipping requirements for etiologic agents and diagnostic specimens shipped in interstate commerce (see Appendix C).

The U. S. Department of Agriculture regulates the importation and interstate shipment of animal pathogens and prohibits the importation, possession, or use of certain exotic animal disease agents which pose a serious disease threat to domestic livestock and poultry.

VII. HAZARDS AND HAZARDOUS MATERIALS

Numerous hazardous materials, including radioactive materials, volatile organic compounds, acids, solvents, and petroleum products, are used within LBNL in laboratory activities and/or facility operations (such as maintenance). The transportation, use, storage, treatment, and disposal of these materials can expose individuals or the environment to health and/or environmental hazards. LBNL complies with applicable federal, state, and local laws and regulations.

(*a*) **Hazardous Materials**

The term "hazardous material" is defined as any material that, because of quantity, concentration, or physical or chemical characteristics poses a significant present or potential hazard to human health and safety or to the environment. 2 Hazardous materials are grouped into the following four categories, based on their properties: toxic (causes human health effects), ignitable (has the ability to burn), corrosive (causes severe burns or damage to materials), and reactive (causes explosions or generates toxic gases).3 Hazardous materials are commonly used in research laboratories and commercial, agricultural, and industrial applications, as well as in residential areas to a limited extent.

(*b*) **Biosafety Standards**

The US Federal and state laws establish standards for working with biohazardous materials. A hazardous biological material is any potentially harmful biological material (including infectious agents, oncogenic viruses, and recombinant DNA) or any material contaminated with a potentially harmful biological material.

The U.S. Public Health Service, the National Institutes of Health, and the Centers for Disease Control and Prevention operate under the U.S. Department of Health and Human Services. These agencies establish standards for working with biohazardous materials.

(*c*) **Microorganisms**

Projects incorporating **microorganisms** (including bacteria, viruses, viroids, prions, rickettsia, fungi, and parasites), **recombinant DNA (rDNA) technologies** or **human or animal fresh tissues, blood, or body fluids** may involve working with potentially hazardous biological agents. Students are permitted to do research projects with potentially hazardous biological agents as long as every effort is made to ensure that they work safely and that the projects meet the conditions and rules described below. The following rules were developed to protect students and to help them adhere to federal and international biosafety regulations and guidelines.

When dealing with potentially hazardous biological agents it is the responsibility of the student and all of the adults involved in a research project to conduct a **risk assessment**. A risk assessment defines the potential level of harm, injury or disease to **plants, animals** and **humans** that may occur when working with biological agents. The end result of a risk assessment is the assignment of a **final biosafety level** which then determines the laboratory facilities, equipment, training, and supervision required for the research project to proceed.

(*d*) **Additional Rules for Projects Involving Unknown Microorganisms**

Studies involving unknown microorganisms present a challenge because the presence, concentration and pathogenicity of possible agents are unknown. In science fair projects these studies typically involve the collection and culturing of microorganisms from the environment (e.g. soil, household surfaces, skin, etc.) (1) Research with unknown microorganisms can be treated as a

BSL-1 study under the following conditions (a) Organism **is cultured** in a plastic Petri dish (or other standard non-breakable container) **and sealed**. Other acceptable containment include petri film and doubled heavy-duty (2-ply) sealed bags. (b) Experiment involves only procedures in which the Petri dish remains sealed throughout the experiment (i.e. counting presence of organisms or colonies). (c) The sealed Petri dish is disposed of in the appropriate matter under the supervision of the Designated Supervisor. (2) If a culture is opened for identification, sub-culturing or isolation, it must be treated as a BSL-2 study and involve BSL-2 laboratory procedures.

(*e*) **Additional Rules for Projects Involving**

Recombinant DNA (rDNA) Technologies

Studies involving rDNA technologies in which microorganisms have been genetically modified require close review to assess risk level assignment. There are a few rDNA studies that can be safely conducted in a BSL-1 high school laboratory with prior review by a knowledgeable SRC. (1) All rDNA technology studies involving BSL-1 organisms and BSL-1 host vector systems may be conducted in a BSL-1 laboratory under the supervision of a Qualified Scientist or trained Designated Supervisor and must be approved by the SRC prior to experimentation. Examples include cloning of DNA in *E. coli K12*, *S. cerevesiae*, and *B. subtilis* host-vector systems. (2) All rDNA technology studies using the following DNA insert molecules may be conducted in a BSL-1 laboratory under the supervision of a Qualified Scientist or trained Designated Supervisor and must be approved by the SRC prior to experimentation: (a) DNA molecules that are not in the DNA of organisms or viruses, (b) DNA from single nonchromosomal or non-viral sources and (c) DNA that is entirely from a prokaryotic host, including its indigenous plasmids or viruses when propagated only in the host. (3) A rDNA technology study that involves BSL-1 agents that may convert to BSL-2 agents during the course of experimentation must be conducted entirely in a BSL-2 facility. (4) All rDNA technology studies involving BSL-2 organisms and/ or BSL-2 host vector systems must be conducted in a regulated research institution and approved by the IBC prior to experimentation. (5) Propagation of recombinants containing DNA coding for oncogenes or other human, plant or animal toxins (including viruses) are prohibited.

VIII. CLASSIFICATION OF BIOLOGICAL AGENTS

Risk Groups

Biological agents, plant or animal, are classified according to biosafety level risk groups. These classifications presume ordinary circumstances in the research laboratory, or growth of agents in small volumes for diagnostic and experimental purposes.

BSL-1 risk group contains biological agents that pose low risk to personnel and the environment. These agents are highly unlikely to cause disease in healthy laboratory workers, animals or plants. The agents require Biosafety Level 1 containment. Examples of BSL-1 organisms are: *Aspergillus niger, Bacillus thuringiensis, Escherichia coli strain K12, Lactobacillus acidophilus, Micrococcus leuteus, Neurospora crassa, Pseudomonas fluorescens, Serratia marcescens.*

BSL-2 risk group contains biological agents that pose moderate risk to personnel and the environment. If exposure occurs in a laboratory situation, the risk of spread is limited and it rarely would cause infection that would lead to serious disease. Effective treatment and preventive measures are available in the event that an infection occurs. The agents require Biosafety Level 2 containment. Examples of BSL-2 organisms are: *Mycobacterium, Streptococcus pneumonia, Salmonella choleraesuis*.

BSL-3 risk group contains biological agents that usually cause serious disease (human, animal or plant) or that can result in serious economic consequences. These agents are usually not spread by casual contact. The agents require Biosafety Level 3 containment. **(Prohibited).**

BSL-4 risk group contains biological agents that usually produce very serious disease (human, animal or plant) that is often untreatable. These agents are usually easily transmitted from one individual to another, from animal to human or vice-versa, either directly or indirectly, or by casual contact. The agents require Biosafety Level 4 contaminant (Prohibited).

Following guideline need to be followed.

(*a*) Experimentation involving the following types of biological agents will require Institutional Biosafety Committee (IBC) and Institutional Animal Care and Use Committee (IACUC) authorization:

1. All microorganisms.
2. Materials of human origin including blood, fluids, tissues, primary and established human cell lines.
3. Recombinant DNA activities described in the NIH Guidelines for Research Involving Recombinant DNA Molecules.
4. Generation of rodents in which the animal's genome has been altered by stable introduction of recombinant DNA, or DNA derived there from, into the germ-line (transgenic animals).

(*b*) Access to the animal room must be limited to the fewest number of authorized individuals possible. Personnel who must enter the room for program or service purposes when work is in progress are advised of the potential hazard (training).

(*c*) Work Practices

1. Food or drink is prohibited in areas where animals are housed or manipulated.
2. All procedures must be carefully performed to minimize the creation of aerosols or splatters.
3. Biological safety cabinets, other physical containment devices must be used whenever conducting procedures with a high potential for creating aerosols. These include necropsy of infected animals, harvesting of tissues or fluids from infected animals or eggs, or intranasal inoculation of animals. In certain cases where physical containment devices are unavailable, combinations of personal protective equipment (e.g. respirators and face shields) and work practices may provide satisfactory protection. Such instances must be defined in the IACUC protocol application and approved by the IACUC prior to initiating experiments.
4. All personnel must wash hands after handling animals or animal tissues and prior to leaving animal care and use areas, even if gloves are used. If a sink is unavailable for hand washing, an alcohol-based hand sanitizer may be used but hands should be washed with soap and running water as soon as feasible.
5. Equipment and work surfaces must be decontaminated with an effective disinfectant (e.g. 10% bleach) after work with the infectious agent, and after any spill or splash of viable materials.
6. Spills and accidents that result in overt exposure to infectious materials must be immediately reported to the LAS and/or individual supervisor. Supervisors are responsible for reporting incidents to the Biosafety Program at 956-3197.
7. All infectious samples must be collected, labeled, properly transported, and processed in a manner that contains and prevents transmission of the agent(s). The outer surfaces of potentially contaminated containers must be disinfected prior to moving the material.

(*d*) Waste Disposal

1. Animal carcasses that have been intentionally infected with organisms likely to be pathogenic to healthy humans or animals for purposes of research, in vivo testing, production of biological materials, or for any other reason, must be placed into a red plastic bag.
2. All wastes must be properly contained and properly decontaminated.
3. Contaminated bedding and caging from infected animals must be autoclaved prior to cage cleaning. All infectious wastes from the animal room (including animal tissues, carcasses, contaminated bedding, unused feed, sharps, and other refuse) must be transported in a leak-proof, covered container in a manner that contains and prevents transmission of the agents. The outer surfaces of the potentially contaminated containers must be disinfected prior to moving the material.
4. Sharp items (e.g. syringes, needles, scalpels, razors, etc.) must be disposed in appropriately designated puncture-resistant sharps containers.

IX. WORKING WITH GENETICALLY MODIFIED ORGANISMS (GMO)

Genetic Engineering has made a rapid entry into agriculture in the United States, Argentina, Canada, China, Brazil and South Africa, with these countries being responsible for 99 percent of the genetically modified (GM) crops grown globally. These countries grow crops with transgenic resistance to certain herbicides, insects, or diseases supposedly to overcome the productivity constraints of conventional breeding. These transgenic crops, and products such as kernels of maize and Soya beans, are known as genetically modified organisms (GMOs).

Genetic engineering brings with it a wide range of biosafety concerns and broader socio-economic impacts. It requires the acceptance of intellectual property rights on living organisms, the privatisation of public research, and expensive research and development at the expense of farmer-based innovation. Genetic engineering and GMOs impact on several fundamental human rights, including the right to nutritious, safe and culturally acceptable food, the right to informed choice, the right to democratic participation, the right to save and exchange seeds, and the right to a safe and healthy environment. It also raises far-reaching ethical concerns for those that adhere to value systems.

X. CARTAGENA PROTOCOL ON BIOSAFETY

The origin of the Biosafety Protocol is Article 19(3) of the CBD, which in turn, originates from a proposal made by the government of Malaysia during the negotiation of the CBD in November 1991. The Malaysian proposal contained the core element of 'prior informed consent' or 'advanced informed consent' of importing countries that prevailed throughout the process of the Biosafety Protocol negotiation. Article 19(3) of the CBD obliges the Parties to the CBD to consider the need for and modalities of a Protocol setting out appropriate procedures. In particular this includes, 'advance informed agreement', in the field of safe handling and use of any living modified organism that may have adverse effects on the conservation and sustainable use of biodiversity. The Biosafety Protocol negotiations, however, only commenced in 1995 when the 'Jakarta Mandate' was adopted at the second Conference of the Parties of the CBD in Jakarta, Indonesia. The Jakarta Mandate sanctioned the establishment of an Open-ended *Ad Hoc* Working Group on Biosafety, to elaborate a Protocol on Biosafety specifically focusing on the transboundary movement of living modified organisms (LMOs). The Biosafety Protocol negotiations have been described as 'one of the most difficult and complex negotiations between trade and environment'. The negotiations spanned across a period of five years, and collapsed once at the final round of negotiations in Cartagena, Colombia in February 1999 before it was finally adopted in Montreal, Canada in January 2002. The Biosafety Protocol came into effect on 11 September 2003, and to date, 107 countries have ratified the Biosafety Protocol. The central regulatory element of the Biosafety Protocol is the Advanced Informed Agreement procedure, which applies to the first transboundary movement of GMOs for intentional introduction into the environment. The procedure seeks to ensure that importing countries have the opportunity to assess the environmental and human health risks associated with a GMO and take a decision based on the precautionary principle, before agreeing to its import. It obliges exporters to notify importers in advance of the first shipment and to supply certain prescribed information concerning the GMO. Receipt of this information needs to be acknowledged within 90 days. Within 270 days the importing party must communicate its final decision with regard to the future status of the GMO. This decision is to be based on a risk assessment and may either approve or prohibit the import of the GMO, request further information, or extent the deadline by a defined period of time. In each case reasons for the decision need to be stated. Both the importing and exporting Parties may, at any time, initiate a review and change of the decision in the light of new scientific information.

XI. THE WORLD TRADE ORGANISATION

The provisions of the Law, which are trade-related, are compatible with the relevant agreements under the World Trade Organisation (WTO) because these are justified by a rational policy purpose such as, for example, the protection of human life or health. Trade-related measures that are relevant to biosafety fall within the purview of three agreements of the WTO, namely, the Agreement on the Application of Sanitary and Phytosanitary Measures (SPS), the Agreement on Technical Barriers to Trade (TBT) and the General Agreement on Tariffs and Trade (GATT) 1994. These agreements allow biosafety measures to be taken, including import bans, use restrictions such as risk assessments and risk management measures, and traceability, identification and labeling requirements. The SPS agreement regulates sanitary and phytosanitary measures that will affect trade provided it is with respect to protecting plant, animal and human health. Biosafety measures aimed at the regulation of imported GMOs to protect plant, animal and human health would be covered under the SPS agreement. The SPS agreement allows countries to set their own standards, as long as these are based on science, are applied only to the extent necessary to protect human, animal or plant health, and do not arbitrarily or unjustifiably discriminate between countries where identical or similar conditions prevail. Countries are allowed to set higher standards based on appropriate risk assessment, as long as the approach is not arbitrary.

XII. HORIZONTAL GENE TRANSFER

There are also safety concerns over the Genetic Engineering (GE) process itself, which greatly enhances the scope and probability of horizontal gene transfer and recombination. This is the main way to the creation of viruses and bacteria that cause diseases. Destabilising genes and genomes through GE can thus be hazardous. From genetically modified crops and pharmaceutical drugs to health genomics, the hazards are often not known. However, where something can cause irreversible harm, it is right and proper for society, and scientists in particular, to seek evidence that it is safe beyond reasonable doubt. Hence the precautionary principle or approach is crucial.

Horizontal gene transfer may spread transgenes to the entire biosphere

Horizontal gene transfer is the transfer of genetic material between cells or genomes belonging to unrelated species, by processes other than usual reproduction. In the usual process of reproduction, genes are transferred vertically from parent to offspring; and such a process can occur only within a species or between closely related species. Bacteria have been known to exchange genes across species barriers in nature. There are three ways in which this is accomplished. In conjugation, genetic material is passed between cells in contact; in transduction, genetic material is carried from one cell to another by infectious viruses; and in transformation, the genetic material is taken up directly by the cell from its environment. For horizontal gene transfer to be successful, the foreign genetic material must become integrated into the cell's genome, or become stably maintained in the recipient cell in some other form. In most cases, foreign genetic material that enters a cell by accident, especially if it is from another species, will be broken down before it can incorporate into the genome. Under certain ecological conditions which are still poorly understood, foreign genetic material escapes being broken down and become incorporated in the genome. For example, heat shock and pollutants such as heavy metals can favour horizontal gene transfer; and the presence of antibiotics can increase the frequency of horizontal gene transfer 10 to 10 000 folds.

What are the hazards of horizontal gene transfer ?

Most artificial vectors are either derived from viruses or have viral genes in them, and are designed to cross species barriers and invade genomes. They have the potential to recombine with the genetic material of other viruses to generate new infectious viruses that cross species barriers. Such viruses have been appearing at alarming frequencies. The antibiotic resistance genes carried by artificial vectors can also spread to bacterial pathogens. Has the growth of commercial-scale genetic engineering biotechnology contributed to the resurgence of drug and antibiotic infectious diseases within the past 25 years ? There is already overwhelming evidence that horizontal gene transfer and recombination have been responsible for creating new viral and bacterial pathogens and for spreading drug and antibiotic resistance among the pathogens. One way that new viral pathogens may be created is through recombination with dormant, inactive or inactivated viral genetic materials that are in all genomes, plants and animals without exception. Recombination between external and resident, dormant viruses have been implicated in many animal cancers.

Both the artificial vectors used in genetic engineering and the genes transferred to make transgenic organisms are predominantly from viruses and bacteria associated with diseases, and these are being brought together in combinations that have never existed in billions of years of evolution. Genes are never transferred alone. They are transferred in unit-constructs, known as an 'expression cassettes' Transgenic DNA was found to persist in the soil for up to two years after the transgenic crop was planted. Though they did not comment on it, the data showed that the proportion of kanamycin resistant bacteria in the soil increased significantly between 1.5 and 2 years. Could it be due to horizontal transfer of antibiotic resistance marker gene in the transgenic DNA? Although none of 4000 colonies of soil bacteria isolated – a rather small number – was found to have taken up transgenic DNA by the probes available, two out of seven samples of total bacterial DNA yielded positive results after 18 months. This suggests that horizontal gene transfer may have taken place, but the specific bacteria which have taken up the transgenic DNA cannot be isolated as colonies.

Many developing countries are now in the process of developing their national biosafety frameworks. As they do so, the politics and controversies surrounding genetic engineering continue. Numerous developing countries are facing intense pressure to formulate national biosafety frameworks that facilitate biotechnology development and acceptance, without the necessary biosafety considerations in place. Developing countries are also concerned to ensure that their biosafety regulations are trade consistent, so that they would not face the threat of a World Trade Organisation (WTO) complaint, as has happened to the EU.

XIII. THE BIOSAFETY REGULATORY FRAMEWORK IN INDIA

Genetically modified organisms are regulated in India under the purview of the 1986 Indian Environment (Protection) Act (henceforth, the EP Act). The broad objective of the EP Act is the protection and improvement of the environment. To meet this objective, the Act calls for regulation of "environmental pollutants" which are defined as "any solid, liquid or gaseous substance present in such concentration as may be, or tend to be, injurious to the environment" (MOEF 1986). The broad definition of "environmental pollutant" was used by the Ministry of Environment and Forests in 1989 to issue rules to govern use of genetically engineered organisms under the EP Act. The 1989 "Rules for the Manufacture, Use, Import, Export and Storage of Hazardous Microorganisms, Genetically Engineered Organisms or Cells" (henceforth 1989 Rules) constitute the legally binding regulatory framework for GMOs in India. As evident from the title, GMOs are placed here in the same category as hazardous microorganisms and their regulation under the EP Act is justified by their alleged potential to be hazardous substances or environmental pollutants. Since the 1989 Rules call for guidelines to be developed to give them effect, biosafety guidelines were issued by the department of Biotechnology under the Ministry of Science and Technology in 1990, and have been consistently revised and expanded in the last decade. Development of this regulatory framework dating from the early 1990s has been in response to the need for regulatory oversight of the growing domestic community engaged in biotechnology research in both the agricultural and pharmaceutical sectors. It was also in response to debates within the European Union about safe use of genetically modified organisms, during negotiation of regional directives on contained use and deliberate release of GMOs. The possibility that India might be a future importer of GMOs from OECD countries thus also spurred development of a domestic regulatory framework.

XIV. THE SCOPE OF THE BIOSAFETY REGULATORY FRAMEWORK

The biosafety regulatory framework consists of the 1989 Rules issued by the Ministry of Environment and Forests, as elaborated and revised by the 1990, 1994 and 1998 Guidelines issued by the Department of Biotechnology. These cover the entire spectrum of activities relating to genetically modified organisms. This includes "research involving genetically modified organisms, as well as genetic transformations of green plants, rDNA technology in vaccine development, and large-scale production and deliberate/accidental release into the environment of organisms, plants, animals and products derived from rDNA technology" (DBT 1990: 1). Production facilities such as distilleries and tanneries which use genetically modified organisms are also covered (MOEF 1989: Art. 4a-d, p. 437-438). The 1990 "Recombinant DNA Safety Guidelines" and 1994 "Revised Guidelines for Safety in Biotechnology" provide detailed guidance on containment and safe laboratory practices for GMOs in the agricultural and pharmaceutical sectors. They also, however, contain an important change from the 1989 Rules in their treatment of deliberate release of GMOs. While the 1989 Rules effectively banned such releases (permitting them only under special circumstances), the 1990 Guidelines permit them, with a shift in focus to assessing and managing ecological and health risks that might result11. This is similar to the evolution of GMO regulation in OECD countries such as the United States, where self-regulation by scientists in the early 1970s prohibited deliberate release and focused on ensuring safe use of GMOs in contained conditions, yet where the prohibition on release was revoked in a relatively short period of time The 1998 "Revised Guidelines for Research in Transgenic Plants and Guidelines for Toxicity and Allergenicity Evaluation of Transgenic Seeds, Plants and Plant Parts" apply only to transgenics for use in agriculture (in contrast to the 1990 and 1994 rules which covered both the agricultural and pharmaceutical sectors). The latest rules add to the regulatory architecture by calling for toxicity and allergenicity data on transgenic plants and plant parts. They also clarify how the procedures governing import of genetically modified organisms for research are distinct from imports for release or commercialization. However, notwithstanding this comprehensive coverage, GMO use in India to date has occurred largely in contained conditions, with some deliberate releases in the form of experimental field trials. Approvals for commercialization of GMOs or food safety assessments of processed materials have not yet been required. Furthermore, no transgenic products have knowingly been imported into the country for commercial use. Thus, while the biosafety regulatory framework in India is broad in its coverage, its functioning in practice remains in a state of evolution and development.

XV. CONSENT–GIVERS IN THE BIOSAFETY REGULATORY FRAMEWORK

An important component of national biosafety governance, the consent-givers, is examined next. This is left relatively unproblematized in transnational rule-making, where the state is seen as the legitimate national-level consent giver. However, often the consenting authority as well as the *process* of giving consent are the most contentious elements of national risk regulation. In India, much debate has entered around the legitimacy of consent givers for GMO use in agriculture. As per the 1989 Biosafety Rules, the authority to regulate GMOs is divided between the Ministry of Science and Technology and the Ministry of Environment and Forests. Research on GMOs is to be overseen by the Review Committee on Genetic Manipulation (RCGM) under the Ministry of Science and Technology's Department of Biotechnology. Deliberate release and Commercialization of GMOs is to be overseen by the Genetic Engineering Approval Committee (GEAC) under the Ministry of Environment and Forests. Although the functions of these two national biosafety committees now appear to be clearly delineated, the division of responsibility between them has been a source of much controversy. Disputes have centered around where the boundary between research and deliberate release lies and, in particular, whether field trials constitute a research activity or a deliberate release of GMOs into the environment, since, if the former, they would be regulated by the Department of Biotechnology, and if the latter, by the Ministry of Environment. In addition to these national-level regulatory committees, every institution engaged in genetic engineering research in India is required to establish an Institutional Biosafety Committee. Furthermore, State Biotechnology Coordination Committees and District-Level Committees are to be set up to facilitate information exchange between the center and the states. The most recent addition to this institutional framework is a Monitoring and Evaluation Committee to oversee the agronomic evaluation of the transgenic crop during field tests and to monitor biosafety data generation. Finally, a Recombinant DNA Advisory Committee is to meet occasionally to review national and international developments in biotechnology and recommend appropriate biosafety regulations for India (DBT 1990, 1998). While this is an elaborate and participatory decision-making structure on paper, its functioning remains far from smooth, as discussed further below.

Useful sites for more information:

1. American Biological Safety Association: ABSA Risk Group Classification – list of organisms http://www.absa.org
   American Type Culture Collection
   (703) 365-2700; 1(800) 638-6597 (US, Canada, & PR) http://www.atcc.org
2. Bergey's Manual of Systematic Bacteriology website : http://www.bergeys.org
3. Biosafety in Microbiological and Biomedical Laboratories (BMBL) - 4th Edition. Published by CDC-NIH,
   http://www.cdc.gov/od/ohs/biosfty/biosfty.htm
4. World Health Organization Laboratory Safety Manual-3rd Edition
   http://www.who.int/csr/resources/publications/biosafety/Biosafety7.pdf
5. Canada – Agency of Public Health – list of non-pathogenic organisms
   http://www.phac-aspc.gc.ca/ols-bsl/pathogen/organism_e.html
6. Microorganisms for Education Website – list of organisms
   http://www.science-projects.com/safemicrobes.htm
7. NIH Guidelines for Research Involving Recombinant DNAMolecules.
   http://www4.od.nih.gov/oba/
   OSHA – Occupational Health and Safety Administration
   http://www.osha.gov

QUESTIONS

1. What are safety issues working in a laboratory? Discuss various levels of biosafety issues.
2. What are hazardous chemicals and microorganisms? How to handle and transport them.
3. What are rules regarding use of GMO crops? What are rules in India for such crops?

INDUSTRIAL FERMENTATION

CHAPTER 50

I. INTRODUCTION

Fermentation, technically defined, as the chemical transformation of organic compounds with the help of enzymes is an old process. The ability of yeast to make alcohol-beer-was known to the Summerians and the Babylonians before 6000 B.C. Much later, by about 4000 B.C., the Egyptians discovered that the carbon dioxide generated by the action of brewer's yeast could leaven bread. References to wine, another product of fermentation, can be found in the Book of Genesis, where it is noted that the Noal consumed a bit too much of the beverage. Reference has also been made to wine as "Soma" in Indian mythology as the liquid frequently drunk on several occasions. Religious Indian script like "Ramayan" also indicates that alcohol was consumed by "Asuras" millions of years ago.

By the 14th century A.D. the distillation of alcoholic spirits from fermented grain, a practice thought to have originated in China or the Middle East, was common in many parts of the world. Similarly some other fermentation processes such as production of vinegar (and cultivation of acetic acid bacteria) as well as yogurt and cheese were also known. Indian literature is full of examples of 'curd" being utilized by people since millennia.

Although there are reports of microorganism providing food and drink for more than 8000 years. Yet their existence was recognized only in the 17th century by Dutch Microscopist Anton van Leeuwenhoek. It was however not until the second half of the 19th century that Louis Pasteur of France and John Tyndall of Britain demolished by concept of spontaneous generation and proved that existing microbial life comes from pre-existing life.

II. FERMENTATION TECHNOLOGY

The technology utilized in fermentation processes are designed to obtain maximum growth of an organism under the optimum physical conditions in a specific medium for the production of a desired end products. Fermentation is typically either batch operated or continuous in operation mode.

(i) Fermenters: The heart of the fermentation process is the fermenter. A fermenter is a vessel in which, is maintained an environment favorable to the operation of a desired biological process. A fermenter is frequently confused with a bioreactor. However, the major difference between the two is the type of cells being grown. A fermenter grows a prokaryotic cells such as bacteria, fungi etc. While a bioreactor grows eukaryotic cells such as insects, mammalian cells etc. Fig. 50.1). A typical fermenter consists of three major parts-culture vessels, associated supply and environmental systems, its measurement and control systems.

Broadly the fermenters could be grouped under two classes: submerged fermenters and surface fermenters. In suspended growth systems (submerged fermenters), the organisms are immersed in and dispersed throughout, there in nutrient medium and their movement follows that of the nutrient liquid. In the case of supported growth systems (surface fermenters) organisms grow as a layer or a film on a surface in contact with nutrient medium. Practically, the suspended growth systems have a film of the organisms on the surface of the container, and supported growth systems usually have organisms dispersed in the nutrient media.

During fermentation it is necessary to regulate many factors within predetermined values: oxygen and carbon dioxide, pH, temperature, media etc. It is also essential to maintain high degree of sterility within fermenter. The fermenter should be made of stainless steel of copper because such fermenters are resistant to steam sterilization. The details of the instruments and their working is given in the chapter on bioreactor chapter 19.

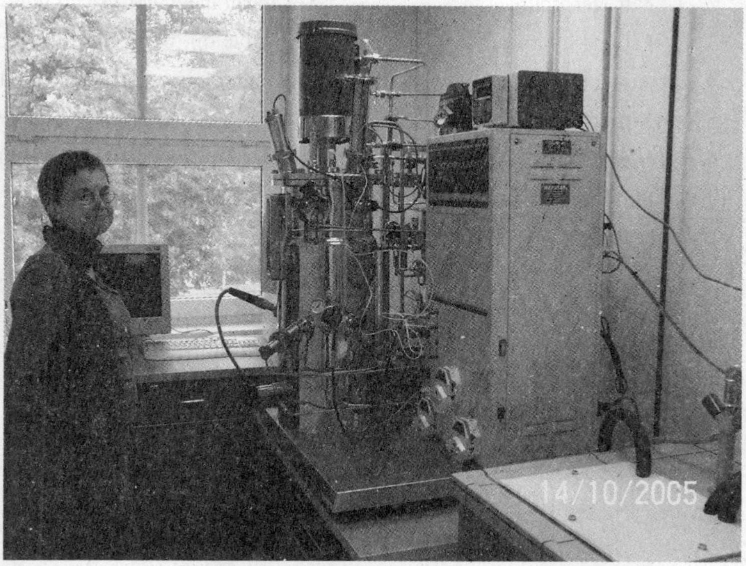

Fig. 50.1. A pilot-scale 50L fermentor

(ii) Selection of microbes: The choice of microorganisms for use in the fermentation industry begins with screening so as to find the right microorganism of the many species of microorganisms, relatively few possess the genetic information needed to produce economically useful products. The selection procedures employed in industry are designed to separate microorganisms that are potentially valuable in producing a commercially useful product from rest. Selection of microbes that possess the potential for producing industrially important substances includes both naturally occurring and genetic variants.

Genetic engineering has opened several new directions for employing microbes to produce economically important substances. Mutation and selection approach is hit or miss, whereas the use of recombinant DNA technology permits the purposeful manipulation of genetic information to engineer a microorganism, which can produce high yields of a variety of products.

(iii) Fermentation medium: The fermentation medium includes nutrients to support the growth of microorganisms, and formation of desired end products. It provides water, sources of carbon, nitrogen, inorganic nutrients, vitamins, amino acids, precursors, inducers, buffers, antifoams etc. Table 50.1 presents various nutrients sources used in industrial fermentation. Good yields of fermentation products occur only if large quantities of carbon and nitrogen are channeled through the metabolic pathways of microorganisms.

Precursors are the substances generally serve to improve the yield or quality of product that are incorporated without any major change into the molecule of the fermentation product; e.g. is phenyl acetic acid for penicillin G and Cobalt for vitamin B12. Regulation by induction is of vital for certain secondary metabolite production; e.g. is α-mannosidase a key enzyme for streptomycin biosynthesis is inducers in the presence of yeast manner.

Antifoam agents (*i.e.* alcohols and fatty acids) included allow the maintenance of nutritive balance of the medium. Phosphates were used as buffering agents although external pH control is normally employed. Inorganic nutrients can be met by potassium phosphates, magnesium sulphate and ammonium sulphate or phosphate. Calcium carbonate is sometimes included either to neutralize the acid produced during fermentation or to prevent rapid decline in pH that occur during growth.

The growth requirements of the process organisms are provided by the crude constituents and some times the additional media sources included with cell hydrolyrates or autolystates as growth factors.

Table 50.1. Nutrient sources used for industrial fermentations.

S.No.	Nutrient	Raw Material
1.	Carbon source	Corn sugar, molasses, starch, glucose
2.	Nitrogen source	Soyabean meal, corn steep liquor, yeast extract
3.	Fat and hydrocarbons	Vegetable oil, petroleum fractions
4.	Ammonia	Pure ammonia or ammonium salts
5.	Nitrate	Nitrate salts
6.	Nitrogen	Air (from nitrogen fixing organisms)
7.	Phosphorous source	Phosphate salts

III. PENICILLIN PRODUCTION

The term penicillin is the generic name for a family of related substances which differ in the nature of side chain, attached to the basic fused β-lactam thiazolidine ring systems (Figure 50.2). From the fermentation viewpoint penicillin G and penicillin V are important. Penicillin G is the starting material for the manufacture of 6-amino penicillanic acid (6-APA), the key intermediate in the production of a wide range of semi synthetic penicillin's.

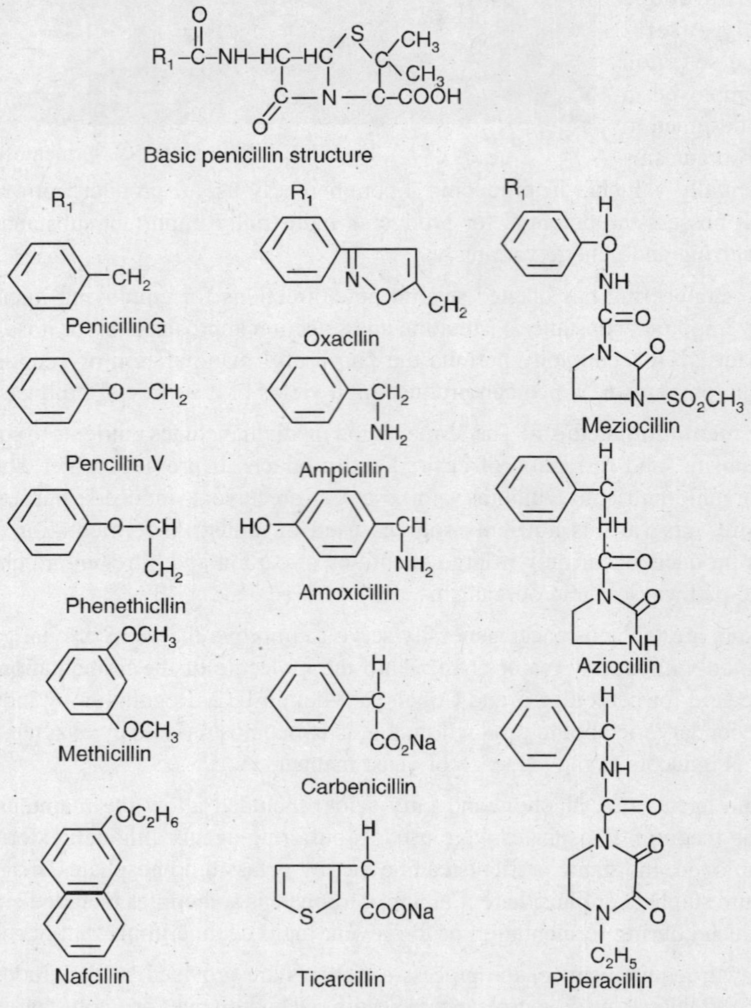

Fig. 50.2. Chemical structures of the penicillin family of antimicrobial agents

The penicillin production follows various stages in the process are described below and in Figure 50.3.

(i) Mold: Several strains of *Penicillium chrysogenum* mold have been employed in past for the production of penicillin but a carefully selected strain of *Penicillium chrysogenum* Thom. is considered to be best. High yielding strains have been developed either by mutation or by recombination over a long period of time. Yield has been increased from less than 10 μg/ml to 30 μg/ml in some of the present high procedures.

(ii) Stock culture: The high-yielding strains of *Penicillium chrysogenum* are maintained as stock culture. The stocks for these organisms are maintained by one of three methods: a well-sporulated agar slant is frozen, soil stocks are prepared and stored at 2 to 4°C or the spores are lyophilized.

(iii) Media: A typical medium for penicillin production described by Jackson (1958) contains:

Cornsteep liquor solids	3.5%
Lactose	3.5%
Glucose	1.0%
Calcium carbonate	1.0%
Potassium dihydrogen phosphate	0.4%
Edible oil	0.25%
Penicillin precursor	0.005%
pH after sterilization is	5.5 to 6.0

A production medium as described by Sylvester and Coghill (1954) is composed of:

Corn steep liquor	30 g/l
Lactose	30 g/l
Glucose	5.0 g/l
Sodium nitrate	3.0 g/l
Magnesium sulphate	0.25 g/l
Zinc sulphate	0.044 g/l
Calcium carbonate	3.0 g/l
Precursor	0.05 g/l

Inoculum media are similar to production media, except that lactose and precursor are not included. Glucose or sucrose serves as the carbon source. Because, during the development of the inoculum, growth and not penicillin production is desired.

Precursors are used to increase the yield of penicillin by the fermentation process. Phenylacetic acid, phenoxy acetic acid and hydroxy phenyl acetic acid are used as precursors for penicillin G, penicillin V and penicillin X respectively. Amino acid such as L-cystine, L-cysteine and valine are important in the synthesis of β-lactam thiazolidine ring system of penicillin.

(iv) Inoculum preparation: For inoculum production, spores from heavily sporulated working stocks (agar sporulation media) are suspended in water or in dilute solution of a nontoxic wetting agent, such as 1:10,000 sodium lauryl sulfonate. These spores are then added to flasks or bottles of wheat bran plus nutrient solution and are incubated at 24°C for 5-7 days. The resulting spores are used directly to inoculate inoculum tanks. The inoculum tanks are incubated 24 to 48 hours with agitation and aeration in order to provide heavy mycelial growth and the resulting inoculum is used for a production tank.

(v) Production in fermenters: The production tanks are inoculated by employing air pressure to force 10% by volume of inoculum into the tank. The actual penicillin fermentation is incubated at 25 to 26°C for 3-5 days. During production, periodic samples are removed for determination of penicillin yields and for contamination checks. During the incubation period, the medium is aerated with sterile air and agitated.

A temperature of 25°C appears to by optimum. In general, the pH remains constant at the start of the fermentation while the glucose, cornsteep liquor, carbon compounds and ammonia are being utilized. Ammonia is liberated by deamination of the amino acids of the cornsteep liquor and the pH rises to 7-7.5, the optimum pH for penicillin production. The penicillin is sensitive to low pH values as well as pH values above 7.5, therefore the pH may be controlled by adding either sodium hydroxide or sulphuric acid. At the end of the fermentation the pH rises to 8 or higher, because depletion of lactose brings on autolysis of the mycelium. However, usually the penicillin is harvested before this phase.

Lipid nutrients are also utilized by the fungus during penicillin production, and fatty oils such as lard oil, soyabean oil, linseed oil and fatty acids of greater than 14 carbon chains and their esters are effective. Some of the oil is added as antifoam and rest is purposely added directly to the medium. These nutrients increase both the amounts of mycelium and yields, but high levels can be deleterious in both early and late stages of the fermentation..

(vi) Recovery and purification: The mycelium is separated from the broth on a rotary vacuum filter, under aseptic conditions. Penicillin in the acid (anion) form is solvent extracted. Phosphoric or sulphuric acids are added to lower the pH to 2-2.5, in order to convert the penicillin to the anionic form. The broth is immediately extracted in a Podbielniak countercurrent solvent extractor, with an organic solvent such as amyl acetate, methyl isobutyl ketone, or butyl acetate. The penicillin is then back-extracted into water from the organic solvent by adding enough potassium or sodium hydroxide to form a salt of the penicillin, and the resulting aqueous solution is again acidified and re-extracted with methyl isobutyl ketone. These shifts between water and solvent aid in the purification of the penicillin. The resulting crystalline penicillin then is washed and dried.

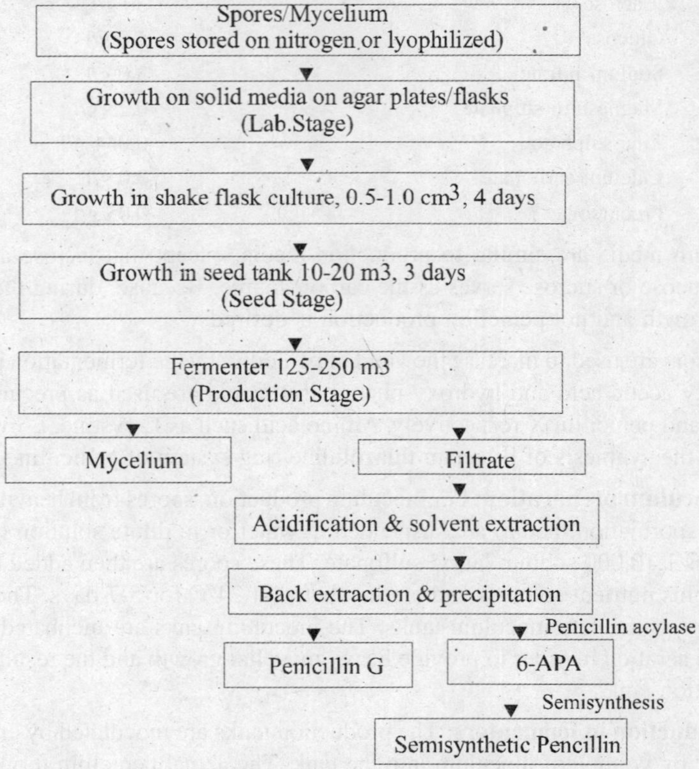

Fig. 50.3. Pencillin production

Semi-synthetic Penicillin's: All the semi-synthetic penicillin's contain 6-APA nucleus. 6-APA is prepared on large scale by the removal of phenyl acetate side chain from penicillin G. This can be accomplished by chemical means or by using acylases of microbial origin (Fig. 50.3).

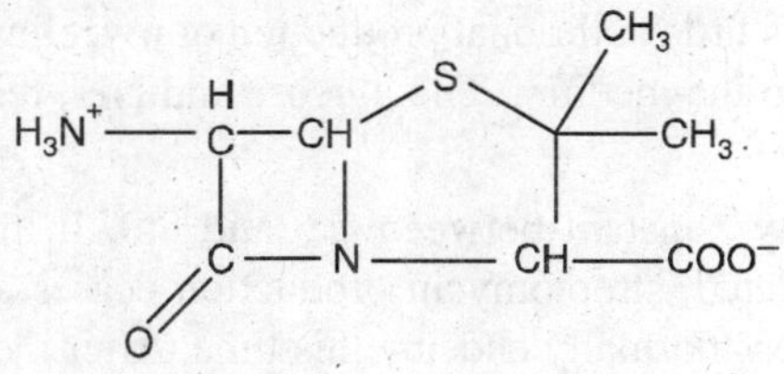

Fig. 50.4. 6-aminopenecillanic acid (6-APA).

At present the majority of the bulk producers employ an enzymatic process. Certain microorganisms produce enzyme, penicillin amidases (acylases), which catalyze the removal of the side chain from benzyl penicillin. *E. coli* and *Alcaligenes* possessing a membrane bound enzyme, with high specificity for penicillin G, have shown acylase activity. The yield of 6-APA (Fig. 50.4) in enzymatic process is usually lower than that of chemical process.

New pencillins can be produced by 6-APA chemical acylation with appropriate acid chlorides. This has led to the production of penicillin's such as amoxicillin, ampicillin, methicillin, piperacillin, azocillin, mezlocillin etc.

IV. STREPTOMYCIN PRODUCTION

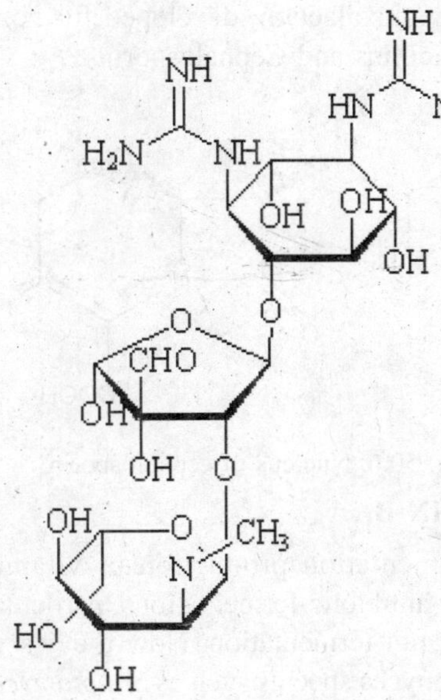

Fig. 50.5. Structure of streptomycin

Streptomycin (Fig. 50.5) is an antibiotic drug, the first of a class of drugs called aminoglycosides to be discovered, and was the first antibiotic remedy for tuberculosis. It is derived from the actinobacterium *Streptomyces griseus*. Streptomycin and various other antibiotics are produced using strains of *Streptomyces griseus*. Spores of this actinomycete are inoculated into a medium to establish a culture with a high mycelial biomass for introduction into an inoculum tank, with subsequent use of the mycelial inoculum to initiate the fermentation process in the production tank.

The medium contains soybean meal (N- source), glucose (C- source) and NaCl. Process is carried out at 28C and the maximum production achieved at pH range of 7.6-8.0. High agitation and aeration are needed. Process lasts for about 10 days. The classic fermentation process involves three phases. During the first phase there is rapid growth of the microbe with production of mycelial biomass. Proteolytic activity of the microbe releases NH_3 to the medium from the soybean meal, causing a rise in pH. During this initial fermentation phase there is little production of streptomycin. During second phase there is little additional production of mycelium, but the secondary metabolite, streptomycin accumulates in the medium. The glucose and NH_3 released are consumed during this phase.

The pH remains fairly constant-between 7.6 and 8.0. In the third and final phase, when carbohydrates become depleted, streptomycin production ceases and the microbial cells begin to lyse pH increases and process normally ends by this time. After the process is complete, mycelium is separated from the broth by filtration and the antibiotic recovered. In one method of recovery and purification, streptomycin is adsorbed onto activated charcol and eluted with acid alcohol. It is then precipitated with acetone and further purified by use of column chromatography.

V. CEPHALOSPORINS PRODUCTION

Cephalosporin C is made as the fermentation product of *Cephalosporium acremonium* Fig. 50.6. However, this form is not potent for clinical use. Its molecule can be transformed by removal of an α-aminoadipic acid side chain to form 7-α aminocephalosporanic acid (7-ACA), which can be further modified by adding side chains to form clinically useful broad-spectrum antimicrobials.

Various side chains can be added to as well as removed from both 6-APA and 7-ACA to produce antibiotics with varying spectra of activities and varying degrees of resistance to inactivation by enzymes produced by pathogenic microbes. Thus we have entered into the so-called third generation cephalosporins, such as moxalactam, developed for control of bacteria that produce enzymes capable of degrading penicillins and Cephalosporins.

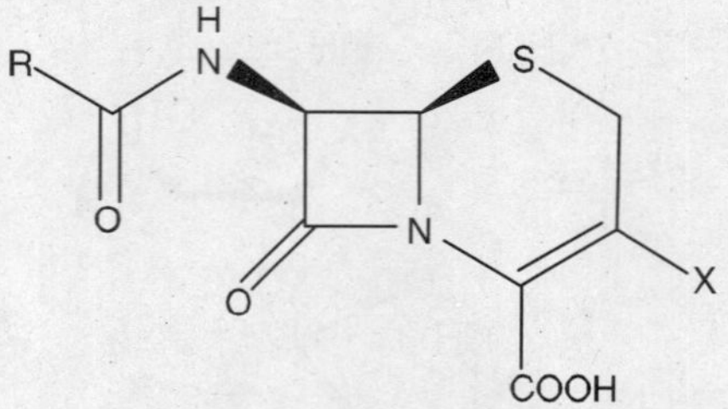

Fig. 50.6. Nucleus of cephalosporin.

VI. PRODUCTION OF VITAMIN B_{12}

The first commercial fermentation production of vitamin B_{12} (Fig. 50.7) was as a by-product of various *Streptomyces* antibiotic fermentations, particularly that for streptomycin and as a by-product of the acetone-butanol fermentation. Now a day's vitamin B_{12} is produced by a direct fermentation utilizing streptomyces species such as *Streptomyces olivaceus*. This organism is grown with aeration at 27°C in a nutritionally rich crude medium with glucose as a major carbon source. Cobalt chloride is added in the medium as a precursor (2 to 10 ppm). The duration of the fermentation is 3-4 days. Vitamin B_{12} remains within the microbial cells until autolysis sets in; hence the recovery of the vitamin from fermentation broth is done by harvesting before autolysis. The solids and mycelium are filtered or centrifuged to separate them from the fermentation broth and then dried for use as B_{12} enriched supplements. The other alternate procedure involves release of vitamin from the cells by procedures such as acidification, heat, alcohol treatment etc. The residual solids and spent mycelium are then separated by centrifugation and the fluid portion is evaporated under vacuum. The vitamin B12 is purified and crystallized.

Vitamin B12 has also been commercially produced by a fermentation using *Bacillus megaterium* and the fermentative recovery procedures are similar to those for *Streptomyces* organisms.

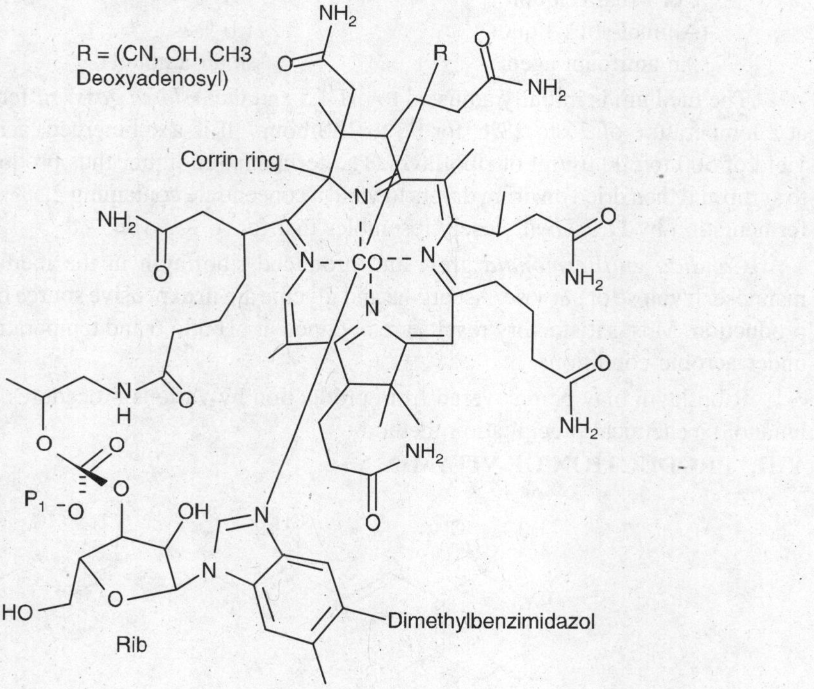

Fig. 50.7. Structure of Cyanocobalamine.

VII. PRODUCTION OF RIBOFLAVIN

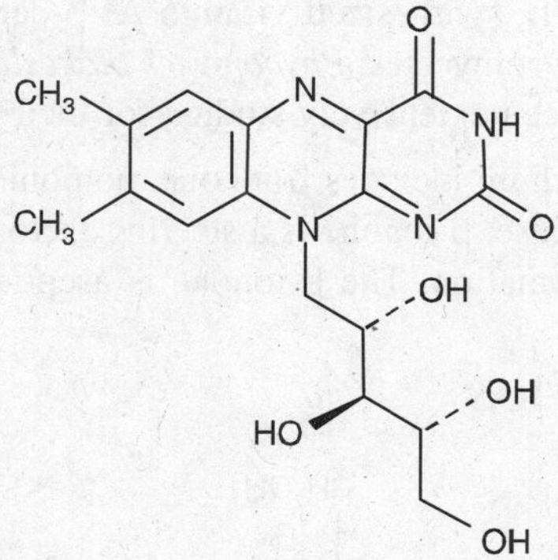

Fig. 50.8. Structure of riboflavin

Riboflavin (Fig. 50.8) is a by-product of acetone-butanol fermentation carried out by microorganisms such as *Clostridium butylicum*, *Clostridium acetobutylicum* and related species. It is produced commercially by direct fermentation utilizing the *Ascomycetes*, *Eremothecium ashbyii* and *Ashbya gossypii*. A gossypii has the greater stability of riboflavin-producing capacity since this organism does not degenerate as readily as does *E. ashbyii* to forms lacking the ability to accumulate riboflavin. Various candida species such as *Candida guilliermondia* and *Candida flareri* also produce riboflavin satisfactorily.

The *Ascomycete* riboflavin fermentation utilizes media containing a semi-purified sugar such as glucose with additional crude organic nutrients. A recommended production media composition is:

Glucose	2.0 %
Corn-steep liquor	1.8 %
Animal-stick liquor	1.0 %
An antifoam agent	small amount

The medium is initially adjusted to pH 4.5 and the *Ashbya gossypii* fermentation is conducted at a temperature of 26 to 28°C for 96 to 120 hours. It is a submerged, aerated fermentation with yield of 500 to 600 µg/ml of riboflavin. The fermentation liquor thus produced may be evaporated to syrup and then dried on drum dryers to yield a concentrate containing 2.5% of riboflavin. Riboflavin fermentation by *E. ashbyii* closely resembles that for *A. gossypii*.

Candida guilliermondia grew and produced riboflavin in the media containing dextrose, mannose, levulose or sucrose. Aspargine and glycine are in expensive source of nitrogen for riboflavin production. Most satisfactory results are obtained at pH 5 to 6 and temperature at 30°C for 6-7 days under aerobic conditions.

Riboflavin may be recovered from production by various procedures such as extracting with butanol or chemical precipitation method.

VIII. PRODUCTION OF VITAMIN A

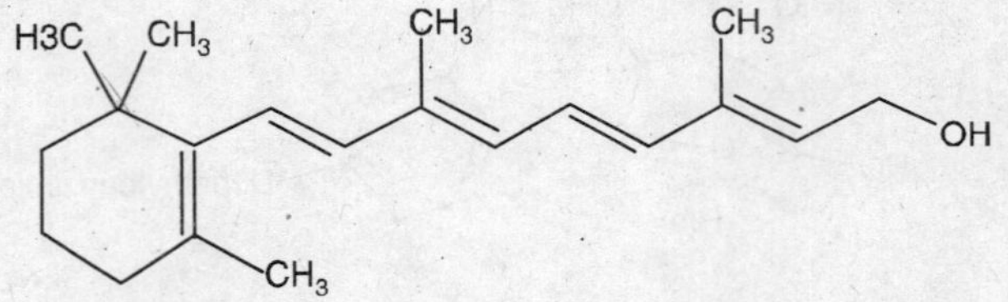

Fig. 50.9. Structure of vitamin A.

β-Carotene also known as provitamin A when ingested is converted in the animal body to vitamin A (Fig. 50.9). β-carotene can be produced by microbial fermentation but yield is relatively low as compared to chemically synthesized vitamin A. β-carotene is produced by members of choanephoraceae family of phycomycetes. *Phycomyces blakesleeanus*, *Choanephora cucurbitarum* and *Blakeslea trispora* have been extensively studied for their ability to produce β-carotene.

The fermentation medium includes β-ionone, nonionic detergent and vegetable oils, fats, fatty acids, waxes or white grease. β-ionone is a steering factor associated for activity of enzymes responsible for β-carotene formation. The β-ionone is aseptically added to the fermentation 48 hours after inoculation.

IX. VITAMIN C PRODUCTION

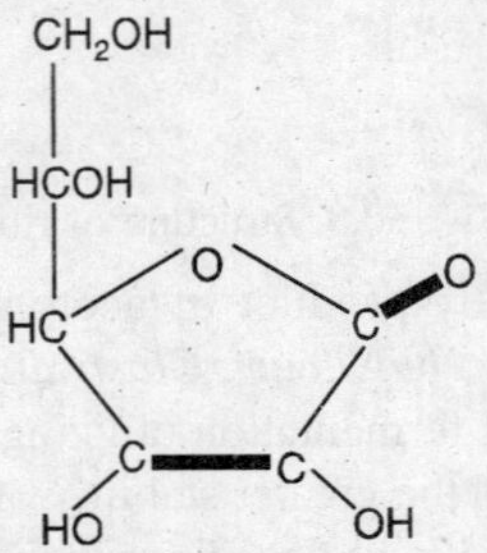

Fig. 50.10. Structure of ascorbic acid

Vitamin C was first recognized when, in 1747, Scottish naval surgeon James Lind discovered that something in Citrus foods prevented scurvy. In 1928 Albert Szent-Gyorgyi, a much admired biochemist, was the first to isolate vitamin C (ascorbic acid) in Hungarian paprika. He later won the 1937 Nobel Prize in Physiology and Medicine for other work. Vitamin C (Fig. 50.10) was the first vitamin to be synthesized artificially in a process invented by Dr. Tadeusz Reichstein, of the Swiss Institute of Technology in Zurich in 1939.

Industrial production of vitamin C is based chiefly on Reichstein procedure (Fig. 50.11). In this method D-glucose is hydrogenated chemically to produce D-sorbitol. The bacterial (*Acetobactor suboxydans*) oxidation of D-sorbitol yield L-sorbose. L-sorbose is then isolated and condensed acetone to form Diacetone-L-sorbose. The diacetone-L-sorbose is then oxidized to diacetone 2-keto-L-gluconic acid, which after hydrolysis, enolization and lactonization yields L-ascorbic acid. The direct oxidation of D-sorbitol or L-sorbose using acetobactor or pseudomonas spp. has led to poor yield. A significant yield of sorose can be obtained by employing a strain of *Gluconobacter oxydans*.

The process with oxidation of D-glucose to 2-keto-L-gluconic acid (KGA) can be performed either by L-idonic acid pathway or the 2,5-diketo-D-gluconic acid pathway. The technique of L-iodonic acid pathway starting with oxidation of glucose is not successful on industrial scale. The 2-keto-L-gluconic acid can be produced in double fermentation by 2,5-diketo-D-gluconic acid pathway that eliminates all the chemical stages involving only two stages. The first stage consist of fermentation for Ca-2,5-diketo-D-gluconate utilizing *Eswinia* Sp. (ATCC 31626), followed by second stage fermentation of Ca-2,5-diketo-D-gluconate by *Cornebacterium* sp. (ATCC 31090) to produce Ca-2-keto-gluonate.

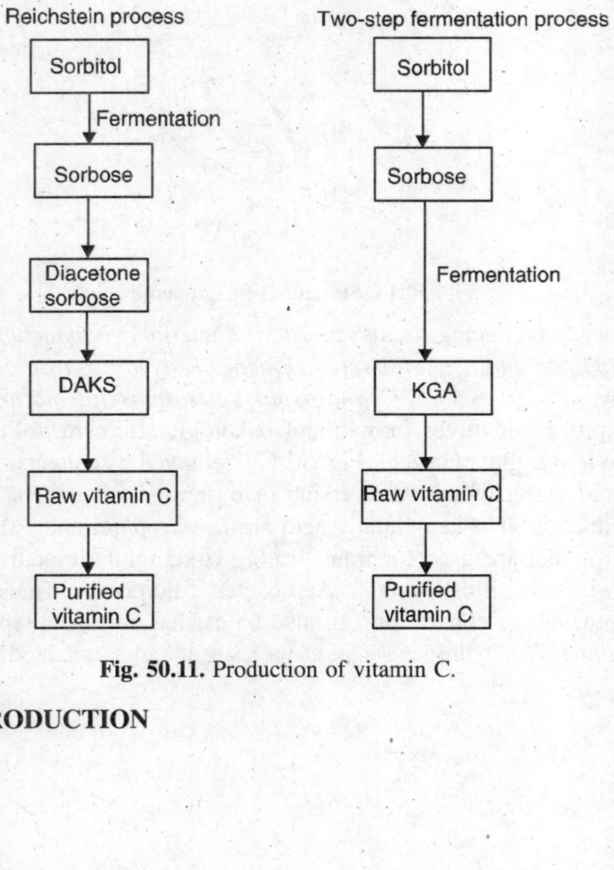

Fig. 50.11. Production of vitamin C.

X. AMINO ACID PRODUCTION

(i) L-Lysine:

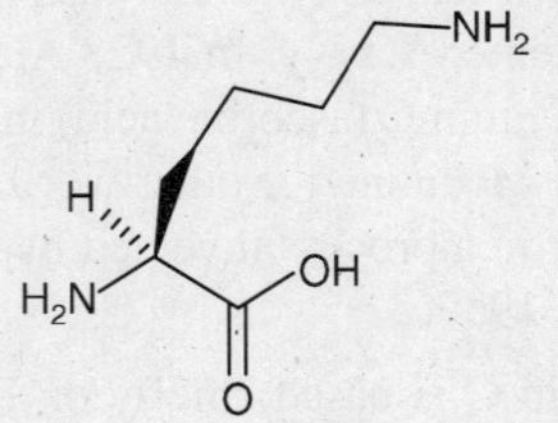

Fig. 50.12. Structure of L-lysine.

The commercial production of lysine (Fig. 50.12) is based on a two step process, that is generally followed at industrial scale.

- Diaminopimelic acid production utilizing corynebacterium diphtheriae and *Mycobacterium tuberculosum.*
- Decarboxylation of diaminopimelic acid to L-lysine utilizing enzyme diaminopimelic acid decaboxylase, produced by strains of *Aerobacter aerogenes* (A.T.C.C. 12409).

For recovery of lysine, the reaction mixture is filtered and the lysine is absorbed on a cation-exchange resin such as Amberlite IR-120 and eluted with dilute alkali by passing it through a weak cation-exchange resin such as Amberlite IRC-50. The eluates are pooled and dried to crystallize lysine, which may be recrystallized for purification.

Further more the mutant strains of *Corynebacterium glutamicum* and *Bravibacterium flavum* has been also utilized for production of lysine.

(ii) L-Glutamic acid:

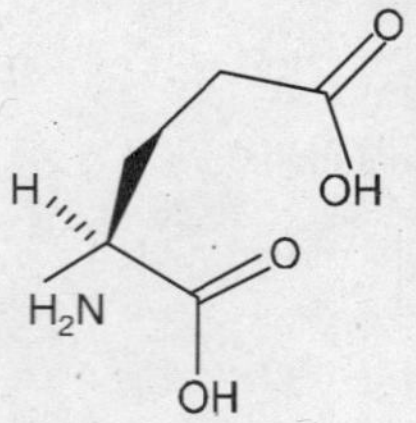

Fig. 50.13. Structure of L-glutamic acid.

Various high yield producing strains are *Corynebacterium glutamicum, Bacerium succinicum, B. natto, E. coli, E. freundii; Aerobacter cloacae, Kluyvera citrophila* var., *Pseudomonas fluorescens, Serratia marcescens, Cephalosporium salmosynnematum, C. dospyri,* and *C. acremorium.* The process involves formationof α-ketoglutaric acid, followed by conversion of α-ketoglutaric acid into L-glutamic acid (Fig. 50.13), followed by conversion of á-ketoglutaric acid into L-glutamic acid via enzymatic conversion or by icroorganism. The most widely used two organisms for production of α-ketoglutaric acid are *K. citrophila* and *Pseudomonas fluorescens.* The microorganisms that are used for transforming α-ketoglutaric acid to L-glutamic acid are Pseudomonas, Erwinia, Bacillus, Serratia, Aerobactes, Micrococus, Hansenula, Mycotorula etc. (Fig. 50.14). The enzymatic conversion can also be carried out; in this process hydrogen donor oxidizes the coenzyme of dehydrogenase and as a result the glutamic acid is formed.

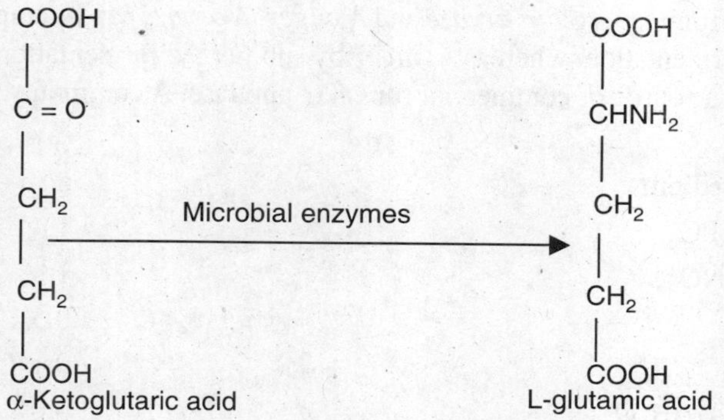

Fig. 50.14. Conversion of a -ketoglutaric acid to L-glutamic acid.

XI. AMYLASES PRODUCTION

Amylases are enzymes that breakdown starch or glycogen. Bacteria and fungi secrete amylases to the outside of their cells to carry out extra cellular digestion. Amylases are classified based on how they breakdown starch molecules.

α-amylase: Reduces the viscosity of starch by breaking down the bonds at random, therefore producing varied sized chains of glucose.

α-amylase: Breaks the glucose-glucose bonds down by removing two glucose units at a turn, thereby producing maltose.

Amyloglucosidase (AMG): Breaks successive bonds from the non-reducing end of the straight chain, producing glucose.

(i) Production of α-amylase: It is an amylolytic enzyme obtained from fungi such as Aspergillus oryzae or from a non-pathogenic variant of bacteria such as Bacillus subtilis. It has the specific activity of converting starch into dextrin and maltose.

(a) Bacterial α-amylase: The main producer organisms of bacterial á-amylases are *Bacillus subtilis*, *Bacillus lincheniformis*, *Bacillus amyloliquifaciens*, *Bacillus stearothermophilus*, *Pseudomonas saccharophiba* and *Clostridium* spp. High yields of bacterial α-amylase can be obtained by *B. subtilis* in stationary culture. The composition of production medium for bacterial α-amylase is:

Ingredient	**g/l**
Bacteriological Peptone	6.0
$MgSO4.7H_2O$	0.5
KCl	0.5
Starch	1

Careful balance of carbohydrate and nitrogen ingredients of the medium is vital. Higher yield can be obtained with glucose as carbon source. A high carbohydrate level in the medium stimulates protease production and depresses amylase production. The medium contains cereal meal, starch and an organic source of nitrogen. The pH of the medium has to be maintained near neutrality during fermentation. The incubation at 25-35°C has to be carried for 3-4 days. The *B. subtilis* in stationary culture produces surface-pellicle growth associated with high amylase yield. At harvest, the culture can either be filtered or centrifuged and bacterial α-amylase can be precipitated from the aqueous solution by addition of acetone, ethanol, isopropanol or ammonium sulphate.

(b) Fungal α-amylases: The fungal α-amylases differ from the bacterial amylases in their lower deactivation temperature, high saccharifying action and lower optimum pH value (4-5). These are mainly produced from *Aspergillus oryaze* and *A. niger*. *A oryaze* can be employed on commercial scale by solid state fermentation, whereas *A. niger* by submerged fermentation technique. *A. oryaze* is commonly used as a source of commercial fungal α-amylase. A commonly employed production medium is:

Ingredients	**g/l**
$KH_2 PO_4$	1.4
NH_4NO_3	10
KCl	0.5
$MgSO_4.7H_2O$	0.1
$FeSO_4.7H_2O$	0.01
Soluble Starch	20
pH adjusted to	6.5

Wheat bran serves as a basic component in solid-state fermentation and is spread in relatively thin layers in trays. It is inoculated with spores after sterilization. The amylase can be extracted by water from the wheat bran culture and precipitated by the addition of alcohol and dried at 50°C or less.

(ii) Production of β-amylases: The α-amylases are mainly obtained from *Bacillus polymyxa*, *B. circulans* and *B. cereus*. The α -amylase genes from *B. cereus* have been cloned in *E. coli*. Recombinant DNA techniques were used to obtain the gene for enzyme on a plasmid suitable for cloning in *B. subtilis* using *E. Coli* as intermediate host. A transformed *B. subtilis* strain can be grown in shake flasks at 30°C using starch, soya meal, and sodium phosphate in medium to increase the yield by 10-20 folds.

XII. PRODUCTION OF PROTEASES

Protease refers to a group of enzymes whose catalytic function is to hydrolyze (breakdown) peptide bonds of proteins. They are also called proteolytic enzymes. Proteases differ in their ability to hydrolyze various peptide bonds. Each type of protease has a specific kind of peptide bonds it breaks. Firstly proteases are divided into two broad categories on the basis of type of attack on the protein; they are exo-and-endo. The two groups are further subdivided on the basis of the mechanism of action at the active site. Thus proteases are divided into five groups. Serine, cysteine (thiol), aspartic (acid), metallo and mixed depending on the principal amino acid participating in catalysis.

The peptidases are classified on the action pattern, thus aminopeptidase cleaves amino acids from the amino end; carboxypeptidase cleaves amino acids from the carboxyl end; dipeptidyl peptidase cleaves two amino acids; dipeptidase splits a dipeptide and tripeptidase cleaves an amino acid from tripeptide. Most proteases are small from 21,000 to 45,000 daltons. Some examples include:

Serine proteinases (alkaline): trypsin, chymotrypsin (animal pancreas), subtilisim (*Bacillus subtilis*)

Cysteine or tiol proteinases: bromelain (pineapple), papain (papaya latex), ficin (*Ficus* latex).

Aspartic (acid proteinases): pepsin (gastric stomach), carboxyl proteinase yeast (*Aspergillus*).

Metallo-proteinase (neutral): collagenase (*Clostridia*), neutral proteinase (*Bacillus subtilis*), keratinase, collagenase (animal tissues).

Aminopeptidase: alanyl aminopeptidase, lysine aminopeptidase.

Diatripeptidase: lysosomal peptidase, bacterial peptidases.

Carboxy peptidase: prolyl carboxypeptidase.

Peptidyl peptidase: cathepsin B.

(a) Bacterial proteases: The main producing organisms are strains of *Bacillus lincheniformis*, *B. amyloliquefaciens* and *B. stearothermophilus*. High yields can be obtained from *B. subtilis* grown in stationary culture and fermentation conditions are similar to those for amylase production by this organism. However, the *Bacillus subtilis* strains are specially selected for high protease activity and not for amylase activity. High carbohydrate content in the medium is utilized to stimulate protease activity and depress amylase production. Media used contains ground basely as carbon source; starch level is limited. Protein hydrolystate or sodium glutamate is used as nitrogen source. The fermentation is incubated 3-5 days at 37^{0}C in pans containing shallow layer of fermentation medium; pH of the medium is within neutral range.

The bulk of protease employed in detergent industry is produced from *B. licheniformis*, which is cultivated in a medium rich in protein and protein hydrolysates at neutral pH between 30 and 40°C.

The harvest procedure is similar to that for bacterial amylase, except that the concentrations of the broth is carried out at reduced pressure and at temperatures of less than 40°C in order to protect the enzyme from denaturation.

(b) Fungal proteases: Various fungi produce enzyme is good yield and they are mainly produced from *Aspergillus flavus*, *Aspergillus oryzae*, *Aspergillus wentii, Mucor delemar*, *Mucormiehei* and *Amylomyces sourii*. The fungus is grown on wheat bran under fermentation conditions similar to those for amylase production. As sporulation, the various fungal proteolytic enzymes are present in the medium and proteases are recovered by procedures similar to those for mold amylases. Submerged fermentation procedures for fungal protease production have been studied and are commercially feasible.

QUESTIONS

1. What is fermentation? Discuss role of bioreactors in fermentation technology.
2. Discuss industrial production of antibiotics.
3. Discuss production of ascorbic acid by fermentation technology.
4. What are applications of amylase and protease enzyme? Describe their production.

INTELLECTUAL PROPERTY RIGHTS AND BIOTECHNOLOGY

CHAPTER 51

Intellectual property refers to creations of the mind: inventions, literary and artistic works, and symbols, names, images, and designs used in commerce. Intellectual property is divided into two categories: Industrial property, which includes inventions (patents), trademarks, industrial designs, and geographic indications of source; and Copyright, which includes literary and artistic works such as novels, poems and plays, films, musical works, artistic works such as drawings, paintings, photographs and sculptures, and architectural designs. Rights related to copyright include those of performing artists in their performances, producers of phonograms in their recordings, and those of broadcasters in their radio and television programs.

In recent years, instruments enforcing intellectual property rights (IPRs), such as patents and trade secrets, have received attention as mechanisms by which biodiversity resources may be maintained while promoting sustainable development and a more equitable distribution of the resulting benefits among nations. Most of the world's biodiversity-rich countries are underdeveloped and lack the necessary technologies to transform biological resources into products yielding significant measurable benefits. With little or insignificant *in situ* market value, biodiver sityrich wild lands may be expected to succumb to pressure from development activities (e.g. conversion to cropland, inundation of forest lands due to hydroelectric and flood-control projects, etc.).

One way to prevent the destruction of wild lands (and, in turn, biodiversity loss) is to promote biodiversity prospecting which creates new markets for biological resources and generates incentives for their conservation. However, biodiversity prospectors generally are multinational corporations from developed countries. These corporations are reluctant to invest in bio-technologies discovered in developing countries due to poorly defined and enforced intellectual property laws. Several scientists currently are addressing this deficiency in IPR protection. Nations which have become signatories of two major inter-national agreements in recent years: the 1992 Con-vention on Biological Diversity (CBD) (UNEP, 1992) and the 1993 Trade-Related Intellectual Property Rights (TRIPS) (UN, 1993). These agreements call for establishing a set of suitable intellec-tual property laws in each nation, depending on the type of intellectual material in question and the economic and technological background of the nation itself.

CBD establishes a formal framework for the reciprocal transfer of biological resources and knowledge (technology) between nations. The convention promotes the idea of biodiversity as a global common heritage, which, therefore, requires biodiversity-rich countries to allow access to biological resources to other countries on 'mutually agreed terms' (UN, 1993). CBD requires technology-rich nations-generally, developed nations to encourage transfer of technology to biodiversity-rich, underdeveloped countries. Thus, the Convention promotes the exchange of biological resources for technology to facilitate biodiversity prospecting, which benefits all nations in the world.

Even though the resource-technology reciprocity and IPR provisions of these agreements have strong economic justification, their use at the global level raises issues concerning the transfer of wealth and respect for national and cultural sovereignty. These issues may become pressing in situations where the resource consumption associated with a patented biotechnology comes in direct conflict with traditional uses of a region's biological resources. Conflict may take different forms. First, the developer of a new biotechnology who normally has greater financial strength-might out-compete traditional users for raw biological material in the input market by paying higher prices. However, the products produced by the new commercial user and traditional users may not compete with each other in the output market. Second, the commercial product developed may have intellectual similarity with traditional products. Then, conferring IPR to a new product/technology may limit or prevent traditional consumers from continuing their use of biological resources. Third, the new commercial product may become a substitute for a traditional product and available at a cheaper price. The lower output market price may drive the traditional producers out of business. Such conflicts can alter the underlying market incentives of competing resource users, with likely adverse implications for biodiversity.

I. INTELLECTUAL PROPERTY PROTECTION OF PLANT BIOTECHNOLOGY INVENTIONS

It is becoming more and more difficult to obtain broad claims in patents and the strength of broad claims in issued patents is weakening. A recent example of this is a case in the USA, in which claims to a diagnostic assay, to differentiate one genus of bacteria from other were invalidated

(*a*) TRIPS

The standard for intellectual property rights is outlined in the global intellectual property treaty agreement of Trade Related Aspects of Intellectual Property Rights (TRIPS) (http://www.wto.org/English/docs - e/legal- e/27-trips. pelf). Member countries that have signed this agreement must ensure that the requirements stated in TRIPS are met in their own legislation. TRIPS states that 'patents shall be available for any inventions, whether products or processes, in all fields of technology, provided that they are new, involve an inventive step and are capable of industrial application'. Although seemingly simple and consistent with the requirements for obtaining a patent in most developed countries, this statement is at the heart of most of the controversy relating to biotechnology patenting. An additional paragraph in TRIPS permits several grounds for exclusion in granting patent protection, including exclusion on moral grounds; or diagnostic, therapeutic and surgical methods for the treatment of humans or animals, life forms other than microorganisms and processes for the production of plants or animals. However, these exclusions are optional and vary from country to country. For example, the European Patent Convention (EPC) provides limited moral grounds for exclusion (http://www.european-patent-office.org/ legal/epc/ e/ar53.html), yet no such grounds are defined in the Patent Acts in Canada, Australia, the USA or Japan. TRIPS also states that if protection of plants is not available by patent, then member countries need to provide protection in some other way. A standard method for such alternative protection is plant variety protection, as set out under Union International pour la Protection des Obtentions Vegetables, known more simply as UPOV.

(*b*) Plant variety protection

UPOV is a global agreement setting out a minimum standard for the protection of plant varieties, similar to that of TRIPS. Member states that have signed the UPOV Convention must ensure that these standards are met within their own legislation. The two versions of UPOV that are currently in force are set out in the UPOV Conventions of 1978 and 1991 and are similar in that they provide protection to a plant variety that is distinct from existing known varieties and that is uniform, stable and novel. However, there are several significant changes in the 1991 Act. For example, The definition of propagating material has been tightened and provisions relating to farmers' rights (or privilege) - the rights that permit a farmer to replant seed for personal use - have been defined. Although plant variety protection meeting the standard set out under UPOV is accepted in 50 countries, such protection has not been uniformly accepted and many countries with strong histories of farmers' privilege have yet to accede to the convention (http://allafi:-ica.com/stories/ 200205090709.html).

A rigorous debate also continues over the effect of plant variety protection and associate Material Transfer Agreements, on sharing and developing new germplasm.

Table. 51.1. Summary of four international conventions with implications for IPR linked to living species and materials of relevance to propagation of living species.

• Convention of Bio-Diversity (CBD)	Proposed in 1992, it came into force by the end of 1993 ratified by more than 169 countries. The conservation of biological diversity, the sustainable use of its components and fair and equitable sharing of benefits arising out of the utilization of genetic resources are the key features. Bio-diversity and its economic benefits were redefined from the "common heritage of mankind" to "national goods" that nations could protect and trade as commodities. It gives a great deal of emphasis to the rights of developing countries to exploit, and reap benefits from, discoveries made in their own territory.
• Union of Protection of Plant Varieties (UPOV) and Plant Breeders Rights	UPOV Convention signed in 1961, came into force in 1968 with revisions in 1972, 1978 & 1991. The 13 essential features of UPOV address the conditions for entitlements to protection to a breeder of a new plant variety provided the variety is distinct from existing, commonly known varieties, is sufficiently homogeneous, stable and new. The plant breeders rights include the requirement of prior authorization of the breeder for the production for the purposes of commercial marketing of the propagating material of the new plant variety, offering for sale of the propagating material, marketing of such material and repeated use of the new plant variety for commercial production of another new variety. For reasons of public interest, the Member States are free to impose restrictions subject to ensuring that breeders receive "equitable remuneration".
• Farmers Rights (UN Food and Agriculture Organization - FAO)	Starting as a concept for debate in 1979 in F AO, it found its way into three F AO conference resolutions 4/89, 5/89 & 3/91. These were negotiated by the Commission on Plant Genetic Resources (PGR) and unanimously adopted by over 160 countries in 1989 & 1991. This defines Farmer's Rights as "rights arising from the past, present, future contribution of farmers in conserving, improving and making available plant genetic resources particularly those in centres of origin/ diversity, These rights are vested in the International Community, as trustees for present and future generations of farmers, for the purpose of ensuring full benefits to farmers for supporting the continuation of

their contributions". Intentionally acceptable mechanisms for implementation of Farmers Rights are yet to be arrived at.

- Trade Related Intellectual Property Rights (TRIPS)

Article 27.1, 27.2 & 27.3 are most relevant to a discussion on patents in the field of biology and biotechnology. Article 27.1 requires members to grant patents for inventions in all fields of technology provided they are new, involve an inventive step and are capable of industrial application. Article 27.2 states "members may exclude from patentability inventions, prevention within their territory of commercial exploitation of which is necessary to protect ordre public or morality, including to protect human, animal or plant life or health or to avoid serious prejudice to the environment, provided that such exclusion is made not merely because the exploitation is prohibited by domestic law", Article 27.3 allows members to exclude from patentability inventions related to diagnostic, therapeutic and surgical methods for treatment of humans or animals. It also allows exclusion of plants and animals other than microorganisms, and essentially biological processes for production of plants or animals other than non-biological and microbiological processes. However members are required to provide for the protection of plant varieties, either by patents or an effective slligeneris system or by a combination thereof.

UPOV 1978 provides an exclusive right to produce and offer for sale propagating materials of a plant variety but not the harvested end product, for example a fruit. Furthermore, the right pertains only to a commercial end product and not to non-commercial uses. As a result, replanting seed is implicitly allowed under UPOV 1978, leading to farmers' privilege. Also provided in the 1978 convention was a breeders' exemption permitting the use of protected varieties as a germplasm source to develop new plant varieties. The 1991 Act introduced several significant amendments. The number of plant genera and species that could be protected under UPOV was increased from the select list of plants in UPOV 1978 to all plants. Furthermore, the 1991 Act provides the option to protect all aspects of the production and reproduction of a plant variety, thereby removing farmers' privilege. However, the application of this provision is discretionary for each member state of UPOV and a country can provide an exemption in its laws to permit farmers' privilege, if desired. Another important change pertains to providing protection to plant varieties that are 'essentially derived' from a protected variety. An 'essentially derived' plant is one that comprises the properties of the protected variety along with only a minor change. The introduction of a gene using recombinant techniques into a protected plant variety might not be sufficient to exceed the 'essentially derived' criteria unless the gene alters the variety in a significant manner (http://www.upov.int/eng/convntns/19911pd£'act1991.pdt).

On 3 November 2001, the International Treaty on Plant Genetic Resources (see http://www.fao.org/ag/ cgrfa/news.htm) was adopted by 116 countries; there were two abstentions (the USA and Japan). Before the Treaty comes into effect 40 countries must ratify it. This Treaty pertains to ensuring that the raw materials used to develop new crop varieties remain publicly available. In so doing, the Treaty promotes the conservation of plant genetic resources for food and agriculture. The aim of the Treaty is to ensure farmers' privilege and to develop a multilateral system comprising an aggregate of genetic material from the member countries, so that, after paying a fee, members can have access to the genetic material. The preface to the Treaty indicates that 'nothing in this Treaty shall be interpreted as implying in any way a change in the rights and obligations of the Contracting Parties under other international agreements'. However, there is an active debate as to whether the Treaty will remain subordinate to TRIPS and UPOV. The USA abstained from signing this treaty partly because of the lack of clarity in the intellectual property provisions.

(*c*) **GATT (General Agreement of Tariffs and Trade)**

During Urugway conference, WTO (World Trade Organization) was created. General Agreement of Tariffs and Trade (GATT) was framed by WTO in 1948 and was meant to be a temporary arranged to settle amicably, among countries, disruptes regarding who gets what share of world trade. This is achieved by determining both tariff rates and quantitative restrictions on imports and exports globally. In 1994, about 100 countries signed this agreement including the then president of USA, Mr. Bill Clinton. This was to be effective from 1-1-95 in phases. Its new quarter is in Geneva, Switzerland. Although GATT has made the world a better place to do business by allowing more free and fruitful flow of goods and services, this benefit has unfortunately gone mainly to developed countries, to the disadvantage of the countries in the third world.

II. US PLANT PATENTS

Another way to protect plant-related subject matter includes a 'plant patent', a unique form of protection offered in the USA. A US plant patent is available for a plant that reproduces through asexual reproduction but it does not include a tuber-propagated plant (Section 161, http:// www.uspto.gov /web / offices /pac/mpep/consolidated_laws.pdf). Although not a common form of plant protection, it is used to protect ornamental and fruit-producing trees, roses, poinsettias, strawberries and other plants that reproduce asexually. A plant patent is different from a regular utility patent.

In the autumn of 2000, the USPTO began rejecting plant patents with a UPOV-based certificate that had been issued before filing for the corresponding plant patent application if the UPOV-based application had been >1 year before the plant patent application had been filed. This interpretation of a UPOV-based disclosure had not been made previously because it was not considered 'enabling', that is, the disclosure of a plant variety within a Plant Breeder Right's certificate did not provide enough information to enable someone- 'one of skill in the art' - to produce the plant variety. The position taken by the USPTO is in direct opposition to that decided in re LeGrice but the USPTO argued that rejection on these grounds is consistent with *Ex parte Thomson*. However, it should be noted that the only public disclosure made in *re LeGrice* was a notice in a publication, which is arguably a non-enabling disclosure, whereas in *ex parte Thomson*, seeds were made publicly available for >1 year before application for a plant patent, clearly placing one of skill in the art in possession of the invention.

The applicant of a US plant patent is provided with a 1 year period of grace. Strong pressure from the industry resulted in a review of USPTO's position and the issue of a preliminary statement recanting its position (http://www.uspto.gov/inappright.html) and, in May 2002, an amendment to the US Patent Act was proposed (it is still under discussion), providing a 10 year period of grace. Even so, the question as to whether a public, non-enabling disclosure of a plant is sufficient to permit one of skill in the art to be in possession of the invention, as is the case in *re LeGrice*, was not addressed.

III. UTILITY PATENTS

Plants can also be protected using a regular (utility) patent in countries that permit patenting of plant or higher life forms (HLFs). This is a more common method for protecting whole novel plants, plant genes, methods for creating novel plants and novel applications for an existing plant. However, the costs are greater and the process more involved than plant variety protection. Many major jurisdictions permit the patenting of non-human HLFs, including Europe, the USA, Japan and Australia.

In the USA, patents have been granted to HLFs since the 1980 landmark decision in Diamond v/s Chakrabarty (http:// people.bu.edu/ ebortman/ index/chakrabarty.html). The recent Supreme Court decision in the case of Pioneer Hi-Bred International, Inc., v/s J.E.M. AG Supply. Inc. et.al. further established that such protection is valid for plants, even if protection of a plant is available through either plant variety protection or plant patent protection. This case also confirmed that plants are a composition of matter, as ruled earlier by the US Patent Board of Appeal.

The scope of protection offered by a utility patent is broader than that available under plant variety protection. As noted above, a farmer saving and replanting seed, and a breeder producing a new variety, can do so without infringing a plant variety certificate. However, if a utility patent, the patent owner, protects the plant or licensee has the right to exclude the making, using or selling of the plant or seed, making a user buy seed every year.

IV. GENE PATENTING

Although patents have been granted on nucleotide sequences for >30 years, there has been much recent controversy surrounding the patenting of genes. Genome sequencing initiatives, coupled with improved techniques for identifying and sequencing genes, has resulted in an exponential increase in the number of gene patents in the last decade. As a result, the obscure world of gene patenting is now being scrutinized closely in many different sectors, not least because the effect of these patents is felt in everyday life, especially healthcare. For example, in Europe, a European Parliament resolution regarding the patenting of BRCA 1 and BRCA 2 (breast cancer associated) genes was passed calling on the EPO to ensure that all patent applications in Europe do not violate the principle of non-patentability of humans, their genes or cells in their natural environment. The resolution identified two European patents related to BRCA 1 and BRCA 2 and asked that an official objection be filed against these patents. The importance of intellectual property in India is well established at all levels- statutory, administrative and judicial. India ratified the agreement establishing the World Trade Organisation (WTO). This Agreement, inter-alia, contains an Agreement on Trade Related Aspects of Intellectual Property Rights (TRIPS) which came into force from 1st January 1995.

V. COPYRIGHTS

India's copyright law, laid down in the Indian Copyright Act, 1957 as amended by Copyright (Amendment) Act, 1999, fully reflects the Berne Convention on Copyrights, to which India is a party. Additionally, India is party to the Geneva Convention for the Protection of rights of Producers of Phonograms and to the Universal Copyright Convention. India is also an active member of the World Intellectual Property Organisation (WIPO), Geneva and UNESCO.

The copyright law has been amended periodically to keep pace with changing requirements. The recent amendment to the copyright law, which came into force in May 1995, has ushered in comprehensive changes and brought the copyright law in line with the developments in satellite broadcasting, computer software and digital technology. The amended law has made provisions for the first time, to protect performer's rights as envisaged in the Rome Convention

VI. TRADE SECRETS

Trade secrets, often include private proprietary information that allows a definite advantage to the owner. This can be illustrated by the popular example of Coca-Cola brand syrup formula which is not known publically under trade-secret. Trade secrets in the area of biotechnology may include material like (*i*) hybridization conditions (*ii*) cell lines (*iii*) corporate merchandising plan or (*iv*) customer lists.

Unlike patents, trade secrets have an unlimited duration and therefore may not be required to satisfy the more difficult conditions laid down for patent applications. Disclosure of a trade secret and its unauthorized use can be punished by the court and the owner may be allowed compensation. However if a trade secret becomes public knowledge by independent discovering or other means, it is no longer protectable.

QUESTIONS

1. What is IPR? Give an account of various international conventions for the protection of IPR.
2. Discuss the IPR in comparison to developed and developing countries.
3. Write short notes :
   (*a*) IPR
   (*b*) TRIPS
   (*c*) GATT
   (*d*) UPNOV
   (*e*) Patent.

BIOINFORMATICS

CHAPTER 52

The terms bioinformatics and computational biology are often used interchangeably. However, bioinformatics more properly refers to the creation and advancement of algorithms, computational and statistical techniques, and theory to solve formal and practical problems arising from the management and analysis of biological data. Computational biology, on the other hand, refers to hypothesis-driven investigation of a specific biological problem using computers, carried out with experimental or simulated data, with the primary goal of discovery and the advancement of biological knowledge. A similar distinction is made by National Institutes of Health in their working definitions of Bioinformatics and Computational Biology, where it is further emphasized that there is a tight coupling of developments and knowledge between the more hypothesis-driven research in computational biology and technique-driven research in bioinformatics. Bioinformatics is also often specified as an applied subfield of the more general discipline of Biomedical informatics. Major research efforts in the field include sequence alignment, gene finding, genome assembly, protein structure alignment, protein structure prediction, prediction of gene expression and protein-protein interactions, and the modeling of evolution. A common thread in projects in bioinformatics and computational biology is the use of mathematical tools to extract useful information from data produced by high-throughput biological techniques such as genome sequencing. A representative problem in bioinformatics is the assembly of high-quality genome sequences from fragmentary "shotgun" DNA sequencing. Other common problems include the study of gene regulation using data from microarrays or mass spectrometry.

Bioinformatics at the basic level deals with biological information like data collection and storage, data searching and retrieval, analysis and predicting patterns. Fredj Tekaia suggested the following defination of bioinformatics: " The mathematical, statistical and computing methods that aim to solve biological problems using DNA and amino acid sequences and related information".

I. OVERVIEW

Bioinformatics has moved just beyond data analysis and includes (1) DNA microarray- There are new technologies designed to measure the relative number of copies of a genetic message (levels of gene expression) at different stages in development or diseases or in different tissues. (2) Functional genomics-Large scale ways of identifying gene functions and associations (e.g. two-hybrid methods). (3) Structural genomics- Attempts to crystallize and /or predict the structure of all proteins. (4) Comparative genomics- Study of multiple whole genome for understanding the differences and similarities between all the genes of multiple species. From such studies we can draw particular conclusions about species and their evolution. (5) Medical informatics- The management of biomedical experimental data associated with particular molecules-from mass spectroscopy to in vitro assays to clinical side effects.

As a result of massive surge in the data and its complexity, many of the challenges in biology have become challenges in computing.

II. SEQUENCE ANALYSIS

Since the Phage Φ-X174 was sequenced in 1977, the DNA sequences of hundreds of organisms have been decoded and stored in databases. The information is analyzed to determine genes that encode polypeptides, as well as regulatory sequences. A comparison of genes within a species or between different species can show similarities between protein functions, or relations between species (the use of molecular systematics to construct phylogenetic trees). With the growing amount of data, it long ago became impractical to analyze DNA sequences manually. Today, computer programs are used to search the genome of thousands of organisms, containing billions of nucleotides. The human genome project has sequenced more than 100,000 genes encoded by 3×10^9 bases, necessitating the use of good computer based system for the analysis of such enormous data. Assembly of the human genome is one of the greatest achievements of bioinformatics. These programs would compensate for mutations (exchanged, deleted or inserted bases) in the DNA sequence, in order to identify sequences that are related, but not identical. A variant of this sequence alignment is used in the sequencing process itself. The so-called shotgun sequencing technique (which was used, for example, by The Institute for Genomic Research to sequence the first bacterial genome, Haemophilus influenzae) does not give a sequential list of nucleotides, but instead the sequences of thousands of small DNA fragments (each about 600-800 nucleotides long). The ends of these fragments overlap and, when aligned in the right way, make up the complete genome. Shotgun sequencing yields sequence data quickly, but the task of assembling the fragments can be quite complicated for larger genomes. In the case of the Human Genome Project, it took several months of CPU time (on a circa-2000 vintage DEC Alpha computer) to assemble the fragments. Shotgun sequencing is the method of choice for virtually all genomes sequenced today, and genome assembly algorithms are a critical area of bioinformatics research.

Another aspect of bioinformatics in sequence analysis is the automatic search for genes and regulatory sequences within a genome. Not all of the nucleotides within a genome are genes. Within the genome of higher organisms, large parts of the DNA do not serve any obvious purpose. This so-called junk DNA may, however, contain unrecognized functional elements. Bioinformatics helps to bridge the gap between genome and proteome projects-for example, in the use of DNA sequences for protein identification.

III. GENOME ANNOTATION

In the context of genomics, annotation is the process of marking the genes and other biological features in a DNA sequence. The first genome annotation software system was designed in 1995 by Dr. Owen White, who was part of the team that sequenced and analyzed the first genome of a free-living organism to be decoded, the bacterium Haemophilus influenzae. Dr. White built a software system to find the genes (places in the DNA sequence that encode a protein), the transfer RNA, and other features, and to make initial assignments of function to those genes. Most current genome annotation systems work similarly, but the programs available for analysis of genomic DNA are constantly changing and improving.

IV. COMPUTATIONAL EVOLUTIONARY BIOLOGY

Evolutionary biology is the study of the origin and descent of species, as well as their change over time. Informatics has assisted evolutionary biologists in several key ways; it has enabled researchers to (1) trace the evolution of a large number of organisms by measuring changes in their DNA, rather than through physical taxonomy or physiological observations alone (2) more recently, compare entire genomes, which permits the study of more complex evolutionary events, such as gene duplication, lateral gene transfer, and the prediction of factors important in bacterial speciation (3) build complex computational models of populations to predict the outcome of the system over time (4) track and share information on an increasingly large number of species and organisms

V. MEASURING BIODIVERSITY

Biodiversity of an ecosystem might be defined as the total genomic complement of a particular environment, from all of the species present, whether it is a biofilm in an abandoned mine, a drop of sea water, a scoop of soil, or the entire biosphere of the planet Earth. Databases are used to collect the species names, descriptions, distributions, genetic information, status and size of populations, habitat needs, and how each organism interacts with other species. Specialized software programs are used to find, visualize, and analyze the information, and most importantly, communicate it to other people. Computer simulations model such things as population dynamics, or calculate the cumulative genetic health of a breeding pool (in agriculture) or endangered population (in conservation). One very exciting potential of this field is that entire DNA sequences, or genomes of endangered species can be preserved, allowing the results of Nature's genetic experiment to be remembered in silico, and possibly reused in the future, even if that species is eventually lost.

VI. ANALYSIS OF GENE EXPRESSION

The expression of many genes can be determined by measuring mRNA levels with multiple techniques including microarrays, expressed cDNA sequence tag (EST) sequencing, serial analysis of gene expression (SAGE) tag sequencing, massively parallel signature sequencing (MPSS), or various applications of multiplexed in-situ hybridization. All of these techniques are extremely noise-prone and/or subject to bias in the biological measurement, and a major research area in computational biology involves developing statistical tools to separate signal from noise in high-throughput gene expression studies. Such studies are often used to determine the genes implicated in a disorder: one might compare microarray data from cancerous epithelial cells to data from non-cancerous cells to determine the transcripts that are up-regulated and down-regulated in a particular population of cancer cells.

VII. ANALYSIS OF REGULATION

Regulation is the complex orchestration of events starting with an extracellular signal such as a hormone and leading to an increase or decrease in the activity of one or more proteins. Bioinformatics techniques have been applied to explore various steps in this process. For example, promoter analysis involves the identification and study of sequence motifs in the DNA surrounding the coding region of a gene. These motifs influence the extent to which that region is transcribed into mRNA. Expression data can be used to infer gene regulation: one might compare microarray data from a wide variety of states of an organism to form hypotheses about the genes involved in each state. In a single-cell organism, one might compare stages of the cell cycle, along with various stress conditions (heat shock, starvation, etc.). One can then apply clustering algorithms to that expression data to determine which genes are co-expressed. For example, the upstream regions (promoters) of co-expressed genes can be searched for over-represented regulatory elements.

VIII. ANALYSIS OF PROTEIN EXPRESSION

Protein microarrays and high throughput (HT) mass spectrometry (MS) can provide a snapshot of the proteins present in a biological sample. Bioinformatics is very much involved in making sense of protein microarray and HT MS data; the former approach faces similar problems as with microarrays targeted at mRNA, the latter involves the problem of matching large amounts of mass data against predicted masses from protein sequence databases, and the complicated statistical analysis of samples where multiple, but incomplete peptides from each protein are detected.

IX. ANALYSIS OF MUTATIONS IN CANCER

In cancer, the genomes of affected cells are rearranged in complex or even unpredictable ways. Massive sequencing efforts are used to identify previously unknown point mutations in a variety of genes in cancer. Bioinformaticians continue to produce specialized automated systems to manage the sheer volume of sequence data produced, and they create new algorithms and software to compare the sequencing results to the growing collection of human genome sequences and germline polymorphisms. New physical detection technology are employed, such as oligonucleotide microarrays to identify chromosomal gains and losses (called comparative genomic hybridization), and single nucleotide polymorphism arrays to detect known point mutations. These detection methods simultaneously measure several hundred thousand sites throughout the genome, and when used in high-throughput to measure thousands of samples, generate terabytes of data per experiment. Again the massive amounts and new types of data generate new opportunities for bioinformaticians. The data is often found to contain considerable variability, or noise, and thus Hidden Markov model and change-point analysis methods are being developed to infer real copy number changes. Another type of data that requires novel informatics development is the analysis of lesions found to be recurrent across many tumors .

X. PREDICTION OF PROTEIN STRUCTURE

Protein structure prediction is another important application of bioinformatics. The amino acid sequence of a protein, the so-called primary structure, can be easily determined from the sequence on the gene that codes for it. In the vast majority of cases, this primary structure uniquely determines a structure in its native environment. (Of course, there are exceptions, such as the bovine spongiform encephalopathy - aka Mad Cow Disease - prion.) Knowledge of this structure is vital in understanding the function of the protein. For lack of better terms, structural information is usually classified as one of secondary, tertiary and quaternary structure. A viable general solution to such predictions remains an open problem. As of now, most efforts have been directed towards heuristics that work most of the time.

One of the key ideas in bioinformatics is the notion of homology. In the genomic branch of bioinformatics, homology is used to predict the function of a gene: if the sequence of gene A, whose function is known, is homologous to the sequence of gene B, whose function is unknown, one could infer that B may share A's function. In the structural branch of bioinformatics, homology is used to determine which parts of a protein are important in structure formation and interaction with other proteins. In a technique called homology modeling, this information is used to predict the structure of a protein once the structure of a homologous protein is known. This currently remains the only way to predict protein structures reliably. One example of this is the similar protein homology between hemoglobin in humans and the hemoglobin in legumes (leghemoglobin). Both serve the same purpose of transporting oxygen in the organism. Though both of these proteins have completely different amino acid sequences, their protein structures are virtually identical, which reflects their near identical purposes. Other techniques for predicting protein structure include protein threading and de novo (from scratch) physics-based modeling.

XI. COMPARATIVE GENOMICS

The core of comparative genome analysis is the establishment of the correspondence between genes (orthology analysis) or other genomic features in different organisms. It is these intergenomic maps that make it possible to trace the evolutionary processes responsible for the divergence of two genomes. A multitude of evolutionary events acting at various organizational levels shape genome evolution. At the lowest level, point mutations affect individual nucleotides. At a higher level, large chromosomal segments undergo duplication, lateral transfer, inversion, transposition, deletion and insertion. Ultimately, whole genomes are involved in processes of hybridization, polyploidization and endosymbiosis, often leading to rapid speciation. The complexity of genome evolution poses many exciting challenges to developers of mathematical models and algorithms, who have recourse to a spectra of algorithmic, statistical and mathematical techniques, ranging from exact, heuristics, fixed parameter and approximation algorithms for problems based on parsimony models to Markov Chain Monte Carlo algorithms for Bayesian analysis of problems based on probabilistic models. Many of these studies are based on the homology detection and protein families computation.

XII. SOFTWARE TOOLS

Software tools for bioinformatics range from simple command-line tools, to more complex graphical programs and standalone web-services. The computational biology tool best-known among biologists is probably BLAST, an algorithm for determining the similarity of arbitrary sequences against other sequences, possibly from curated databases of protein or DNA sequences. The NCBI provides a popular web-based implementation that searches their databases.

SOAP-based interfaces have been developed for a wide variety of bioinformatics applications allowing an application running on one computer in one part of the world to use algorithms, data and computing resources on servers in other parts of the world. The availability of these SOAP-based bioinformatics web services through systems such as the BioMoby service register demonstrate the applicability of web based bioinformatics solutions. These tools range from a collection of standalone tools with a common data format under a single, standalone or web-based interface, to integrative and extensible bioinformatics workflow management systems.

(*a*) Biological Data

Biological data that is avialable is emormous and can be used for bioinformatics to predict the function of actual gene products. The data can be broadly divided into three categories : (A) Raw DNA/protein sequence data-Computational analysis can help to reveal important biological information from nucleotide sequences of DNA and amino acid sequences of proteins. These provide a clue to the structure and functions of genes and proteins. Following applications have been conceived of

(1) DNA and protein sequence database: Enormous databases have been generated which have been deposited in public databases like Genbank, EMBL Databank, SWISS-PROT, NRDB etc. The Genbank repository of nucleic acid sequences holds more than a total of 9.5 billion bases in 8.2 million entries. At the next level are protein sequences comprising of strings of 20 amino acid letters, which have generated about 300,000 known protein sequences. Macromolecular structural data represents a more complex form of information.

(2) Sequence similarity search: Genes or proteins that are sufficiently similar are likely to be rel;ated and are therefore said to be homologous to each other. BLAST and FASTA are some widely used programmes that help in rapid sequence similarity searches against huge databases.

(3) Sequence alignment: Comparison of sequences is done to determine evolutionary relationships or judging their functions based on structure. The retardness of proteins is used to trace the family trees of different molecules through evolutionary time.

(4) Diversity in size: There are invariably more sequence based data than structural data because of the relative ease with which they can be produced. This is partly related to the greater compelxity and information content of individual structures compared to individual sequences. While more biological information can be derived from a single structure than a protein sequence, the problem is overcome in the latter by analysing larger quantities of data.

(5) Predicting genes in genomic sequences: A concept that underpins most research methods in bioinformatics is that much of the data can be grouped together based on biologically meaningful similarities. The real challenge lies in the prediction of new genes in genomic sequences, which can help in designing experiments to determine their functions.

(B) Protein structure data-Proteins are undoubtedly the most complex biological molecules, which are integral constituents of cell structure, help in signal transduction and cellular commmunication, participate in immunologiacal functions, and function as enzymes to catalyse biochemical reactions. In order ot leverage this valuable genetic inforamtion, it is important fo classify new protein sequences into existing superfamilies and new genes into existing phylogenies.

(1) Protein sequence databases: Protein sequence databases are categorised as primary and composite or secondary. Primary databases contain over 300,000 protein sequences and function as repository for the raw data. Two major protein sequence databases are SWISS-

PROT and PIR international which are different from the nucleotide databases in that they are both curated.

(2) Protein structure databases: Structural databases are databases of macromolecular structures. The PDB is the main primary database for 3D structures of proteins determined by X-ray crystallography and NMR.

(3) Structure visualization: Protein structure data is stored as collections of x,y,z coordinates, but proteins cannot be visualized simply by plotting those points. The connectivity between atoms in proteins has to be taken into account, and for the visualization to be effective, a virtual 3D environment which provides the illusion of depth, needs to be created.

(4) Identification of protein family: One of the most popular databases is PROSITE, which is a database of short sequence patterns and profiles that characterise biologically significant sites in proteins.It is a database of protein families and domains.

(5) Prediction of protein structure: The sequence of a protein can be easily determined, but structure prediction is not possible experimentally. It would be interesting to predict function from sequence, to identify functional sites in uncharacterised 3D structures, and eventually build designed preoteins-molecular machines that do whatever we want them to do.

(6) Drug design: Proteins are the molecules of biological specificity. Faulty proteins would often result in diseases. The behaviour of proteins can be controlled by drug molecules which may provide a tool for curing diseases. Hence, knowledge of preotein structure can be exploited to design drug molecules that can bind to target proteins to enhance or inhibit their activity.

(C) Data on gene expression DNA microarrays are miniaturised laboratories for the study of gene expression. Each DNA microarray or gene chip contains a deliberately designed array of probe molecules that can bind specific pieces of DNA or mRNA. Labelling the DNA or RNA with fluorescent tags allows the level of expression of any gene in a cellular preparation to be measured quantitatively. Microarrays also have other applicatins in molecular biology, but their use in studying gene expression has opened up a new way of measuring genome function.

(*b*) BLAST

In bioinformatics, Basic Local Alignment Search Tool, or BLAST, is an algorithm for comparing primary biological sequence information, such as the amino-acid sequences of different proteins or the nucleotides of DNA sequences. The BLAST algorithm and the computer program that implements it were developed by Stephen Altschul, Warren Gish, David Lipman at the U.S. National Center for Biotechnology Information (NCBI), Webb Miller at The Pennsylvania State University, and Gene Myers at the University of Arizona. A BLAST search enables a researcher to compare a query sequence with a library or database of sequences, and identify library sequences that resemble the query sequence above a certain threshold. For example, following the discovery of a previously unknown gene in the mouse, a scientist will typically perform a BLAST search of the human genome to see if humans carry a similar gene; BLAST will identify sequences in the human genome that resemble the mouse gene based on similarity of sequence. BLAST is one of the most widely used bioinformatics programs, because it addresses a fundamental problem and the algorithm emphasizes speed over sensitivity. This emphasis on speed is vital to making the algorithm practical on the huge genome databases currently available, although subsequent algorithms can be even faster. BLAST is also often used as part of other algorithms that require approximate sequence matching. Alternative implementations are available at WU-BLAST and FSA-BLAST.

To run, BLAST requires two sequences as input: a query sequence (also called the target sequence) and a sequence database. BLAST will find subsequences in the query that are similar to subsequences in the database. In typical usage, the query sequence is much smaller than the database, e.g., the query may be one thousand nucleotides while the database is several billion nucleotides.

BLAST searches for high scoring sequence alignments between the query sequence and sequences in the database using a heuristic approach that approximates the Smith-Waterman algorithm. The exhaustive Smith-Waterman approach is too slow for searching large genomic databases such as GenBank. Therefore, the BLAST algorithm uses a heuristic approach that is slightly less accurate than Smith-Waterman but over 50 times faster. The speed and relatively good accuracy of BLAST are the key technical innovation of the BLAST programs.

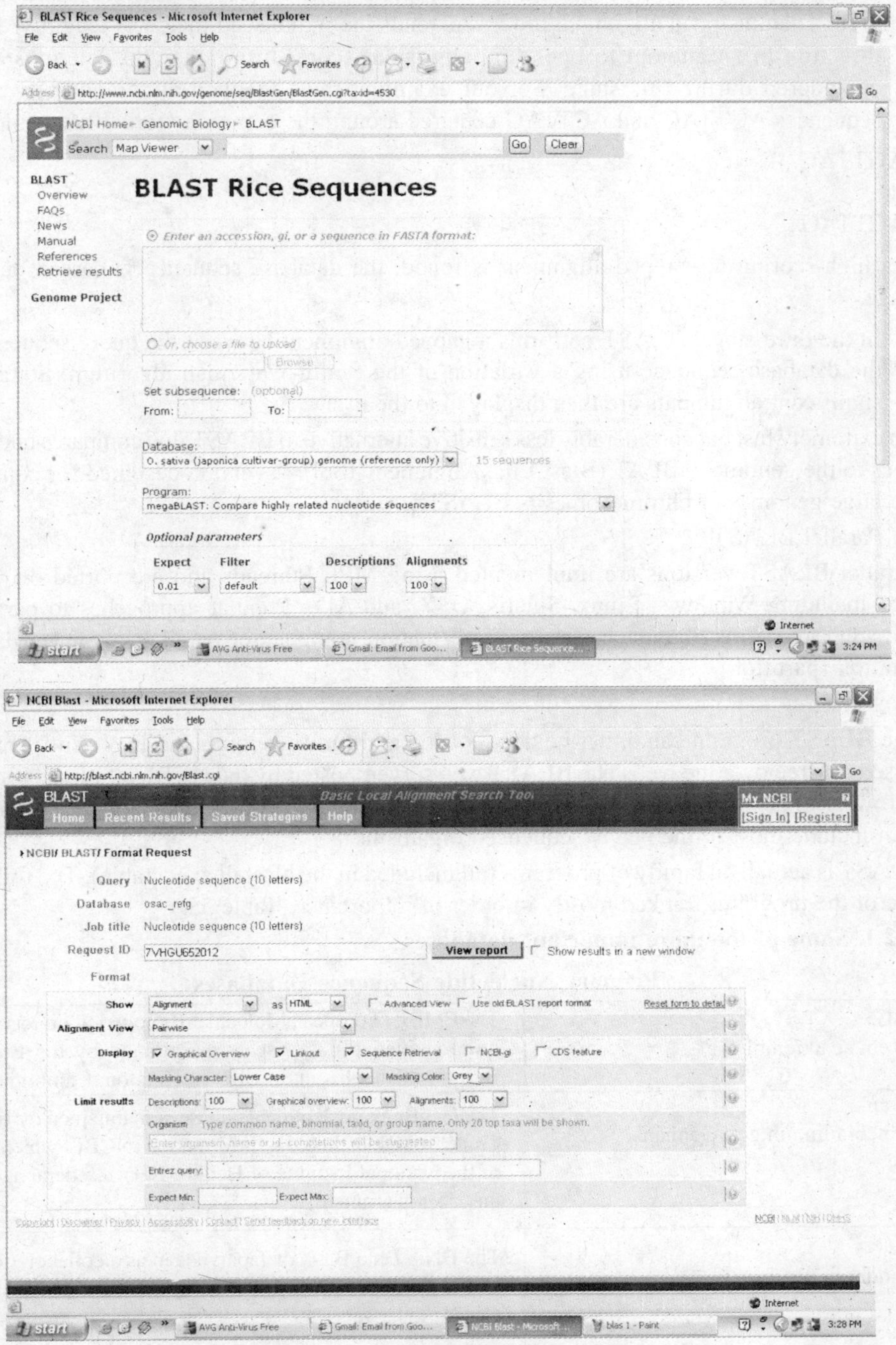

Fig. 52.1. (a) & (b) BLAST window in PC

The BLAST algorithm can be conceptually divided into three stages.

(1) In the first stage, BLAST searches for exact matches of a small fixed length W between the query and sequences in the database. For example, given the sequences AGTTAC and ACTTAG and a word length W = 3, BLAST would identify the matching substring TTA that is common to both sequences. By default, W = 11 for nucleic seeds.

(2) In the second stage, BLAST tries to extend the match in both directions, starting at the seed. The ungapped alignment process extends the initial seed match of length W in each direction in an attempt to boost the alignment score. Insertions and deletions are not considered during this stage. For our example, the ungapped alignment between the sequences AGTTAC and ACTTAG centered around the common word TTA would be:

```
..AGTTAC..
  | |||
..ACTTAG..
```

If a high-scoring un-gapped alignment is found, the database sequence is passed on to the third stage.

(3) In the third stage, BLAST performs a gapped alignment between the query sequence and the database sequence using a variation of the Smith-Waterman algorithm. Statistically significant alignments are then displayed to the user.

An extremely fast but considerably less sensitive alternative to BLAST that compares nucleotide sequences to the genome is BLAT (Blast Like Alignment Tool). A version designed for comparing multiple large genomes or chromosomes is BLASTZ.

(*c*) Parallel BLAST

Parallel BLAST versions are implemented using MPI, Pthreads and are ported on various platforms including Windows, Linux, Solaris, OSX, and AIX. Popular approaches to parallelize BLAST include query distribution, hash table segmentation, computation parallelization, and database segmentation (partition).

Program

The BLAST program can either be downloaded and run as a command-line utility "blastall" or accessed for free over the web. The BLAST web server, hosted by the NCBI, allows anyone with a web browser to perform similarity searches against constantly updated databases of proteins and DNA that include most of the newly sequenced organisms.

BLAST is actually a family of programs (all included in the blastall executable). The following are some of the programs, ranked mostly in order of importance. Table 52.1.

Table 52.1. Some of the more important databases

Primary Nucleotide Sequence Databases

EMBL www.ebi.ac.uk/embl/	The EMBL (European Molecular Biology Laboratory) nucleotide sequence database is maintained by the European Bioinformatics Institute (EBI) in Hinxton, Cambridge, UK
GenBank www.ncbi.nlm.nih.gov/Genbank/:	The GenBank nucleotide database is maintained by National Center for Biotechnology Information (NCBI), which is parrt of the National Institute of Health (NH), a federal agency of the US government;
DDBJ www.ddbj.nig.ac.jp;	The DNA Data Bank of Japan began as a collaboration with EMBL and GenBank. It is run by the National Institue of Genetics, Japan;

Other Nucleotide Sequence Databases	
UniGene www.ncbi.nlm.nih.gov/uniGene/	GenBank sequence data is converted into a non-redundant set of gene-oriented clusters. EAch UniGene cluster contains sequences that represent a unique gene
SGD Genome www.stanford.edu/saccharmoyces/:	The Saccharomyces Genome Database (SGD) is a scientific database of the molecular biology and genetics of the yeast saccharomyces cerevisiae.;
EBI Genomes www.ebi.ac.uk/genomes/	This web site provides access and statistics for the completed genomes, and information about ongoing projects
Genome Biology www.ncbi.nlm.nih.gov/Genomes/	The Genome Biology site at NCBI contains information about the available complete genomes.
Ensembl www.ensembl.org	Ensembl is a joint project between EMBL-EBI and the Sanger Centre to develop a software system which produces and maintains automatic annotation of eukaryotic genomes
Protein Sequence Databases	
SWISS-PROT, TrEMBL www.expasy.ch/sprot/;	SWISS-PROT is a protein sequence database which strives to provide a high level of annotations (such as the description of the function of a protein, its domains structure, post-translational modifications, variants, etc.), a minimal level of redundancy and high level of integration with other databases
PIR pir.georgetown.edu;	The Protein Information Resource (PIR) is a division of the National Biomedical Research Foundation (NBRF) in the US.;
Sequence Motif Databases	
Pfam www.sanger.ac.uk/Software/Pfam/, www.cgr.ki/Pfam/;	Pfam is a database of protein families defined as domains. For each domain, it contains a multiple alignment of a set of defining sequences (the seeds) and the other sequences in SWISS-PROT and TrEMBL., that can be matched to that alignment
PROSITE www.expasy.ch/prosite/	PROSITE is a database of protein families and domains. It consists of biologically significant sites, patterns and profiles collected from multiple alignments of known sequences of the same function.
Macromolecular 3D Structure Databases	
PDB www.rcsb.org/pdb/	The PCB is the main primary database for 3D structures of biological macromolecules determined by X-ray crystallography and NMR
SCOP scop.mrc-lmb.cam.ac.uk/scop/	The SCOP (Structural Classification of Proteins) database was started by Alexey Murzin in 1994 (Lab of Molecular Biology, MRC, Cambridge, UK). Its purpose is to classify protein 3D strucutres in a hierarchical scheme of structural classes
CATH www.biochem.ucl.ac.uk/bsm/cath/	The CATH database (Class, Architecure, Topology, Homologous superfamily) is a hierarchical classification of protein domain structures, which clusters proteins at four major structural levels

Other Databases KEGG www.genome.ad.jp/kegg/;	The Kyoto Encyclopedia of Genes and Genomes (KEGG) is an effort to computerize current knowledge of molecular and cellular biology in terms of the information pathways that consist of interacting molecules or genes and to provide links from the gene catalogs produced by genome sequencing projects.

(1) **Nucleotide-nucleotide BLAST (blastn)** -This program, given a DNA query, returns the most similar DNA sequences from the DNA database that the user specifies.

(2) **Protein-protein BLAST (blastp)** - This program, given a protein query, returns the most similar protein sequences from the protein database that the user specifies.

(3) **Position-Specific Iterative BLAST (PSI-BLAST)** -One of the more recent BLAST programs, this program is used for finding distant relatives of a protein. First, a list of all closely related proteins is created. Then these proteins are combined into a "profile" that is a sort of average sequence. A query against the protein database is then run using this profile, and a larger group of proteins found. This larger group is used to construct another profile, and the process is repeated. By including related proteins in the search, PSI-BLAST is much more sensitive in picking up distant evolutionary relationships than the standard protein-protein BLAST.

(4) **Nucleotide 6-frame translation-protein (blastx)** -This program compares the six-frame conceptual translation products of a nucleotide query sequence (both strands) against a protein sequence database.

(5) **Nucleotide 6-frame translation-nucleotide 6-frame translation (tblastx)**- This program is the slowest of the BLAST family. It translates the query nucleotide sequence in all six possible frames and compares it against the six-frame translations of a nucleotide sequence database. The purpose of tblastx is to find very distant relationships between nucleotide sequences.

(6) **Protein-nucleotide 6-frame translation (tblastn)** -This program compares a protein query against the six-frame translations of a nucleotide sequence database.

(7) **Large numbers of query sequences (megablast)**-When comparing large numbers of input sequences via the command-line BLAST, "megablast" is much faster than running BLAST multiple times. It basically concatenates many input sequences together to form a large sequence before searching the BLAST database, then post-analyze the search results to glean individual alignments and statistical values.

XIII. EUROPEAN BIOINFORMATICS INSTITUTE (EBI)

As the amount of biological data grows, so does the need for biologists to store and access this information in central repositories in a free and unambiguous manner. The European Bioinformatics Institute (EBI) hosts six core databases, which store information on DNA sequences (EMBL-Bank), protein sequences (SWISS-PROT and Tremble), protein structure (MSD), whole genomes (Ensembl) and gene expression (ArrayExpress). But just as a cell would be useless if it couldn't transcribe DNA or translate RNA, our resources would be compromised if each existed in isolation.

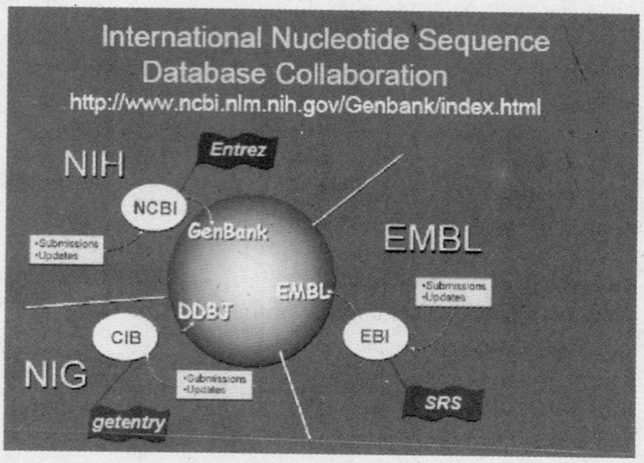

Fig. 52.2. NCBI site for bioinformation

We have therefore developed a range of tools that not only facilitate the deposition and retrieval of biological information, but also allow users to carry out searches that reflect the interconnectedness of biological information. The EBI's databases and tools are all available on our website at www.ebi.ac.uk.

XIV. NATIONAL CENTRE FOR BIOTECHNOLOGY INFORMATION (NCBI, USA)

NCBI has a multi-disciplinary research group composed of computer scientists, molecular biologists, mathematicians, biochemists, research physicians, and structural biologists concentrating on basic and applied research in computational molecular biology. These investigators not only make important contributions to basic science but also serve as a wellspring of new methods for applied research activities. Together they are studying fundamental biomedical problems at the molecular level using mathematical and computational methods. These problems include gene organization, sequence analysis, and structure prediction. A sampling of current research projects includes: detection and analysis of gene organization, repeating sequence patterns, protein domains and structural elements, creation of a gene map of the human genome, mathematical modeling of the kinetics of HIV infection, analysis of effects of sequencing errors for database searching, development of new algorithms for database searching and multiple sequence alignment, construction of non-redundant sequence databases, mathematical models for estimation of statistical significance of sequence similarity, and vector models for text retrieval. Additionally, NCBI investigators maintain ongoing collaborations with several institutes within the NIH and also with numerous academic and government research laboratories.

QUESTIONS

1. What do you understand by bioinfomatics? Discuss its application in biotechnology.

2. Write short notes :

(*a*) BLAST

(*b*) Bioinformatics

(*c*) Sequence matching

(*d*) Protein structure prediction

(*e*) European bioinformatics institute

(*f*) National centre for bioinformatic information(NCBI, USA).

MULTIPLE CHOICE QUESTIONS

S.No. *Question*

1. Which one is known as imino group ? (a)NH_4 (b)NH_3 (c)NH_2 (d) NH
2. DNA is not found in
   (a) Cauliflower mosaic virus (b) Bacteriophage
   (c) TMV (d) Variola major
3. Soilless cultivation is also known as-
   (a) Jhoom cultivation, (b)crop rotation
   (c)Hydroponics (d) Green house
4. Which one of the followings is not a simple protein ?
   (a) albumin (b) globulin
   (c) RNA polymerase (d) glycoprotein
5. In amphoteric compounds such as amino acids, there are two ionisable groups. If there is simultaneous ionization at both the groups, there will be two opposite charges on the ion, such ion is called
   (a) isoelectric ion (b) Zwitter-ion
   (c) cation (d) do not exist
6. A sex linked trait disease is
   (a) color blindness (b) night blindness
   (c) beri beri (d) deafness
7. Leaf abscission takes place
   (a) At internodes (b) At node
   (c) At the base of petiole (d) At leaf margin
8. The word 'inteins' is related to
   (a) sugar (b) protein
   (c) lipid (d) nucleic acids
9. Prions are made of
   (a) RNA (b) DNA
   (c) Protein (d) Polysaccharides
10. A micelle is an association of
    (a) cellulose chains (b) pectin chains
    (c) protein chains (d) polynucleotides
11. Which one is an insecticide of plant origin ?
    (a) Pyrethrin (b) Jasmine
    (c) lycopene (d) ascorbic acid
12. The four elements making 99% of living system are
    (a) C,H,O,S (b) C,H,O,P
    (c) C,H,O,N (d) C,N,O,P
13. Cot curve is related to
    (a) DNA denaturation (b)Protein denaturation
    (c) Starch hydrolysis (d) Lipid burning

14. 'Omega three' gene is important for quality improvement of
    (a) Wheat (b) Brassica (c) Banana (d) Potato
15. An important character of plasmid that separates it from a DNA fragment is
    (a) ability to replicate (b) nucleotide pattern
    (c) insect resistance (d) acid resistance
16. Synthetic seeds are
    (a) chemical seed (b)encapsulated somatic embryos
    (c) clone (d) None above
17. The biosynthetic route for terpenes is
    (a) TCA cycle (b) MVA acetyl co-enzyme
    (c) Shikimic acid (d) CAM pathway
18. Particle gun is an instrument related to
    (a) Drug delivery (b) gene delivery
    (c) insulin delivery (d) insecticide injection
19. Molecular farming is production of bioactive molecules in
    (a) Bacteria (b) Yeast (c) Plants (d) Bioreactor
20. Sephadex column is used to separate
    (a) complex polysaccharides (b) Proteins
    (c) Lipids (d) perfumes
21. DNA polymerase having exo- and endo-nuclease activity, means it can cut DNA:
    (a) X & Y axis (b) X-X axis (c) Y-Y axis (d) None above
22. cDNA microarray chips are used to display differences in
    (a) gene expression (b) nucleus
    (c) protein profile (d) chromosomes
23. Ribozyme is an
    (a) Nucleic acid (b) protein
    (c) another name for zymase (d) none above
24. Golden rice is a new variety of rice having gene for
    (a) Gold (b) insect resistance
    (c) provitamin-A (d) virus resistance
25. Terminator gene related to crops in
    (a) embryo formation (b) terminate insects
    (c) terminate weeds (d) terminate termites
26. The size of lambda DNA molecule is
    (a) 10 kb (b) 20 kb
    (c) 50 kb (d) 100 kb
27. Full form of EDTA is
    (a)ethylene diamine tetra acetic acid (b) ethylene diamide tri acetic acid
    (c) ethylene diamine tri acetic acid (d) ethane diimido tetracetic acid
28. TATA Box is a
    (a)Promoter Region (b)Gene Box
    (c)Protein Block (d)Polypeptide chain

29. Prophage is a
(a) stage of mitosis (b) stage of meiosis
(c) a virus (d) mycoplasma
30. In plasmid pBR322, BR stands for
(a)experiment number (b) plasmid number
(c) name of scientists (d) name of animal
31. Genomic library is a library of
(a) gene related books (b) DNA fragments of a genome
(c) lab working on genes (d) place where genes are stored
32. Chromosome walking is a technique of
(a) chromosome movement (b) gene mapping
(c) jumping genes (d) gene transfer
33. Oligonucleotides are
(a)nucleotide chain (b)polypeptide chain
(c)amino acid sequence (d) none of the above
34. Secondary structure of protein is
(a)alpha-helix (b) beta-pleated
(c)beta-helix (d)both a & b type
35. Glucose is stored as glycogen in
(a) pancreas (b) bone
(c) kidney (d) liver
36. What amongst the following is unique to DNA alone ?
(a) denaturation and renaturation (b) polymer complex
(c) replication (d) resistance to temperature changes
37. An allosteric enzyme has
(a) one active site (b)one active site and one allosteric site
(c) active sites and two types of allosteric sites (d) two types of active sites
38. Inorganic cofactor is often called
(a) coenzyme (b) prosthetic group
(c) modulator (d) activator
39. Plants derive energy from solar radiation, which travels in the form of
(a)heat (b)energy (c)waves (d)water vapour
40. Phytotron is meant for
(a)controlled irradiation (b)induction of mutations
(c)a bioreactor (d)growing plants under controlled environment
41. Phenylketone urea is inherited disease that is characterized by
(a) elimination of gentisic acid in urine (b) increased phenylalanine in blood
(c) elimination of sugar in urine (d) decreased in phenylalanine in blood
42. The conversion of fat into carbohydrate is an example of
(a)glycolysis (b)TCA (c)gluconeogenesis (d)all of the above
43. Which one of the following is a sulphur containing amino acid ?
(a) cysteine (b) leucine (c) valine (d) alanine

44. The nonprotein part of an enzyme is
(a) holoenzyme (b) apoenzyme (c) vitamin (d) prosthetic group
45. Semiconservative replication using ^{15}N was demonstrated by
(a)Taylor (b) Messelson and Stahl (c) Hershey and Chase (d) Messelson
46. Ligase is an enzyme required for
(a) breaking of DNA (b) joining DNA fragments
(c) renaturation of DNA (d) proof reading
47. The process of translation is
(a) ribosome synthesis (b) protein synthesis
(c) DNA synthesis (d) RNA synthesis
48. Triplet codon refers to
(a)mRNA (b) t RNA (c) r RNA (d) all of the above
49. Circular DNA is present in
(a) chloroplast (b) mitochondria (c) plasmid (d) all the above
50. Which enzyme is used to degrade cell wall ?
(a)cellulase (b)amylase (c)pectinase (d)cellulose
51. Best source of renewable source of energy is
(a) cattle (b) petroleum (c) coal (d) trees
52. Change is one nucleotide in DNA results in change in one amino acid in a protein. Such changes are called
(a)Transitions (b) Transversion (c) missense mutation (d) non-sense muations
53. Red data book provides data on
(a) biota of Red sea (b) effect of red light on photosynthesis (c) extinct species (d) threatened species
54. Cryopreservation is carried out at
(a) -10°C (b) -50°C (c) -100°C (d) -196°C
55. Germplasm is contained in
(a) somatic cells (b) growing points
(c) reproductive cells (d) both a and b
56. In a gene bank genetic material is stored as
(a) genes (b) seeds (c) tissue culture plants (d) both b and c
57. Chloroplast of an algae lacks
(a) quantasome (b) lamellae (c) pigments (d) grana
58. Spindle fibre are made of
(a) proteins (b) lipids (c) cellulose (d) polysacchraides
59. DNA replication takes place during
(a) G1-phase (b) G2-phase (c) S-phase (d) prophase
60. ELISA is used for
(a) separate viral RNA (b) purify proteins
(c) isolate DNA (d) identify specific proteins
61. An angiosperm without stomata is
(a)Xerophyte (b) floating hydrophyte
(c) submerged hydrophyte (d) mesophyte

62. One strand of the DNA double helix forming RNA is called
(a) coding strand (b) non-coding strand
(c) primer (d) probe
63. Cocaine is obtained from
(a) *Theobroma cacao* (b) *Erythroxylon coca*
(c) *Cannabis sativus* (d) *Papaver somniferum*
64. Enzyme complex involved in alcoholic fermentation is
(a) lipase (b) invertase (c) amylase (d) zymase
65. Coenzyme is often a
(a) carbohydrate (b) protein (c) vitamin (d) fatty acid
66. Application of which hormone increases the number of female flowers
(a) ethylene (b) ABA (c) auxin (d) GA
67. Enzyme required for hydrolysing the reserve food in castor seed is
(a) amylase (b) protease (c) diastase (d) lipase
68. Maximum number of ATP is synthesised during oxidation of
(a) Amino acids (b) glucose (c) protein (d) palmitic acid
69. The habit of cabbage plant can be changed by application of
(a) GA (b) IAA (c) ABA (d) kinetin
70. Formation of an offspring without fertilization
(a) Parthenocarpy (b) polyembryony
(c) adventitious embryony (d) parthenogenesis
71. Auxin synthesis occurs from following amino acids
(a) tryptophan (b) glycine
(c) methionine (d) cystiene
72. Tunica corpus theory is related to
(a) meristems of all kind (b) apical meristem only
(c) root hair (d) root cap
73. Which of the following is weedicide
(a)IAA (b)IBA (c)NAA (d)2,4-D
74. Sprouting of stored potatoes can be prevented by
(a)IAA (b)GA (c)Kinetin (d)Maleic hydrazide
75. Artificial ripening of fruit is carried out by treatment with
(a)IAA (b)ethrel
(c)zeatin (d)ABA
76. Cellular totipotency refers to capacity of cells to produce
(a) shoots (b) plant (c) embryo (d) roots
77. Papain is obtained from latex of
(a)Calotropis (b)Hevea (c)Papaver (d)Carica
78. Cell division is blocked by
(a)colchicine (b)cucurbitacin (c)carotene (d)morphine
79. To overcome dormancy which one is helpful (a) auxin (b) GA (c) cytokinin (d) ABA
80. Beta -galactosidase is another name for (a) galactase (b) lactase (c) maltase (d) cellulase

81. Colchicine is an alkaloid used to double the number of chromosomes. It is obtained from-
(a) Rhizome of *Colchium autumnale* (b) Corms of *C. autumnale*
(c) Bulbs of *C. autumnale* (d) Root tubers of *C. autumnale*

82. Greater is the distance between the two genes on a chromosome
(a)greater is the linkage strength (b)lesser is the linkage strength
(c)no change in linkage (d)no relationship between the two

83.. Number of chromatids present in a metaphasic chromosome are
(a)1 (b)2 (c)3 (d)4

84. 5' to 3' direction in a DNA helix is determined by carbon atom of
(a)sugar (b)phosphate
(c)base (d)all of the above

85. PCR was invented by
(a) Kary Mullins (b) Carl Fisher (c) Hunt Morgan (d) Carl Zeiss

86. A sheep named Dolly became famous because of
(a) Embryo rescue (b) in vitro fertilization (c) cloning (d) transgenic animal

87. Artificial chromosomes are
(a) chromosome produced by chemical synthesis
(b) Chromosomes produced by plastic
(c) chromosome like bodies having foreign gene and centromere
(d) plasmids

88. Cell fractionation is based on
(a) density centrifugation (b) electric charge
(c) porosity (d) chemical nature

89. FISH is a technique for localization of
(a) Nucleus (b) Nucleolus
(c) genes (d) Sequences on chromosomes

90. AFLP stands for
(a) Arbitrary file length polymorphism (b) Amplified factor late product
(c) Amplified fragment length polymorphism (d) Amplified fragment length polyp

91. Cloning in trees means
(a) individuals derived from bulked seeds (b) plants produced vegetatively
(c) plants produced by anther culture (d) plants produced by transgene

92. Morphogenesis is defined as
(a) origin of forms (b) origin of shoots
(c) origin of roots (d) embryo formation

93. DNA chip is
(a) a computer chip (b) a silicon chip
(c) glass slide with DNA arrays (d) Glass slide used in computers for AIDS

94. Who wrote the work 'origin of species'
(a) Lamarck (b) Darwin (c) De vries (d) Mendel

95. The plant used in social forestry is
(a) Teak (b) sandalwood
(c) subabool (d) rose

96. The best technique to separate isoenzymes is
(a) paper chromatography (b) electrophoresis
(c) microscopy (d) Thin layer chromatography
97. Chloroplast contains maximum quantity of
a) RuBP carboxylase (b) Pyruvic carboxylase
(c)hexokinase (d) oxidase
98. Embryo rescue is
(a) somatic embryo growth (b) somatic embryo culture
(c) hybrid's embryo culture (d) animal embryo transfer
99. 'Shadow casting' is used in microscopy
(a) TEM (b) SEM
(c) phase contrast (d) fluorescence
100. Virus free plants are obtained by
(a) meristem culture (b) embryo culture
(c) protoplast culture (d) axillary buds
101. Single nucleotide polymorphism can be detected by
(a) RAPD (b) Gel electrophoresis (c) PCR (d) HPLC
102. Apoptosis is a phenomenon of
(a) programmed cell death (b) cancerous cell formation
(c) movement of cells (d) movement of cytoplasm
103. Aquaporins are
(a) large pores on membranes (b) membrane proteins
(c) fluidity of membranes (d) membrane fatty acids
104. Bacterial DNA is
(a)SS DNA circular (b) DS DNA circular
(c) DS DNA open (d) SS DNA open
105. Extrachromosomal DNA found in several type of bacteria is called
(a) f factor (b) plasmid
(c)genome (d) all of these
106. Pilli of bacteria are actually
(a) Growth on cell wall (b) extension of cell membrane
(c) extension of protoplast (d) capsule
107. Which of the following is a protein rich animal fee
(a) *Spirulina* (b) *Nostoc* (c) *Oscillatoria* (d) *Azolla*
108. Tomato fruit becomes red due to
(a) anthocynin (b) lycopene (c) carotin (d) xanthophyll
109. Artificial intelligence is –(a) Use of satellite (b) use of computers for biological parameters (c) use of computers for data comparison (d) use of biochips for human brain
110. Biosensors are –(a) sensing by biological organs (b) sensing by cells (c) cell signalling (d) use of biological reactions with electronics
111. Nanoparticles created by nanotechnology are of
(a) orange coloured (b) of millimetre size
(c) of nm size (d) created in space

112. Patch clamp technique is related to measuring
(a) Electrical pulse (b) osmosis
(c) photosynthesis (d) respiration

113. The gas used in supercritical fluid extraction is
(a) carbon dioxide (b) nitrogen (c) oxygen (d) argon

114. Cytokinin was discovered by
(a) Murashige (b) Skoog (c) Steward (d) Thiman

115. Haploids through anther culture were discovered by
(a) Guha and Maheshwari (b) Gupta and Maheshwari
(c) Vasil (d) Rangaswamy

116. Numerical aperture is related to
(a) black hole (b) digital data
(c) resolving power of microscope (d) size of sample for TEM

117. Best separation of essential oil can be achieved by
(a) HPLC (b) TLC (c) GLC (d) Electrophoresis

118. Commonly used gel for cell immobilization is
(a) agar (b) alginate (c) sodium silicate (d) sephadex

119. Protoplasts are isolated by the treatment of
(a) cellulase (b) lipase (c) amylase (d) beta-galactosidase

120. Electoporation is a technique of
(a) membrane rupturing by electrical pulse (b) electron transfer
(c) electroplating (d) none

121. Synchronous cultures are those which
(a) grow in phytotron (b) divide at the same time
(c) communicate to each other (d) a type of callus culture

122 Paclitaxel is another name for
(a) taxol (b) diacetyl baccatin
(c) taxotere (d) none

123. Xenobiotics are chemicals that
(a) kills plant (b) kills animal
(c) resistant to degradation (d) produced by cell culture

124. Xenotransplantation means
(a) chemical patch implant
(b) transplantation of organs from animal to humans
(c) biosensors (d) biochip

125. X-ray crystallography is used to elucidate the following
(a) crystal structure (b) graphical presentation of X-rays
(c) X-ray photography of human body (d) X-ray treatment of plants

126. Haploids can be obtained from
(a)anther wall and pollen (b) pollen tetrad
(c) filament and petals (d) apical meristem

127. An antitumor agent

(a) vincristine (b) taxol
(c) camptothecin (d) all

128. Biological control means
(a) Spray of chemical weedicide (b) use of living organisms to control pests
(c) chemical control of fungi (d) eliminating birds from airport

129. Bioreactor is a
(a) atomic reactor (b) large vessel to grow cells
(c) BOD incubator (d) COD incubator

130. Functional foods are
(a)materials which can function as food (b) alternative food
(c) Office time food (d) food having medicinal value

131. Chemically, insulin is a
(a) lipid (b) steroid (c) peptide (d) oligonucleotide

132. Phytoremediation is a process of
(a) cleaning plants (b) sanitization of plants
(c) use of plants to clean pollution (d) medicines derived from plants

133. Polymerase chain reaction is used to produce a polymer of
(a) sugar (b) fatty acids
(c) nucleotides (d) amino acids

134. Tm represents
(a) temperature of evaporation (b) evaporation of essential oil
(c) temperature of DNA denaturation (d) Trade-mark of products

135. Molecular markers are
(a) phenotypic trait in a crop (b) differences in essential oil content
(c) molecular marks on a plant (d) parameters of molecular biology

136. Transformation in an animal system means
(a) transgenic cell lines
(b) conversion of a normal cell into a tumour cell
(c) a cloned animal (d) metamorphosis in insects

137. Hybridomas are produced by fusion of
(a) selected lymphocytes (b) lymphocytes and tumour cells
(c) tumour cells and HeLa cells (d) HeLa cells and plant cells

138. Sanger is famous for his work on
(a) polypeptide sequencing (b) fatty acid sequencing
(c) polynucleotide sequencing (d)YAC

139. Hairy roots are produced by
(a) *A. tumefaciens* (b) *A. rhizogenes* (c) *E. coli* (d) *Lactobacillus*

140. Hairy root phenotype means
(a) growth of monocots roots (b) negative geotropic roots
(c) prop roots (d) stilt roots

KEY

1.	D	26.	C	51.	D	76.	B	101	C	126	B
2.	C	27.	A	52.	C	77.	D	102.	A	127.	D
3.	C	28.	A	53.	D	78.	A	103.	B	128.	B
4.	D	29.	C	54.	D	79.	B	104.	B	129.	B
5.	B	30.	C	55.	D	80.	B	105.	B	130.	D
6.	A	31.	B	56.	D	81.	B	106.	B	131.	C
7.	C	32.	B	57.	D	82.	B	107.	A	132.	C
8.	B	33.	A	58.	A	83.	B	108.	B	133.	C
9.	C	34.	D	59.	C	84.	A	109.	B	134.	C
10.	A	35.	D	60.	D	85.	A	110.	D	135.	D
11.	A	36.	A	61.	C	86.	C	111.	C	136.	B
12.	C	37.	C	62.	B	87.	C	112.	A	137.	B
13.	A	38.	D	63.	B	88.	A	113.	A	138.	C
14.	B	39.	C	64.	D	89.	D	114.	B	139.	B
15.	A	40.	D	65.	C	90.	C	115.	A	140.	B
16.	B	41.	B	66.	C	91.	B	116.	C		
17.	B	42.	D	67.	D	92.	A	117.	C		
18.	B	43.	A	68.	D	93.	C	118.	B		
19.	C	44.	D	69.	A	94.	B	119.	A		
20.	B	45.	B	70.	D	95.	C	120.	A		
21.	A	46.	B	71.	A	96.	B	121.	B		
22.	A	47.	B	72.	B	97.	A	122.	A		
23.	A	48.	A	73.	D	98.	C	123.	C		
24.	C	49.	D	74.	D	99.	B	124.	B		
25.	A	50.	A	75.	B	100.	A	125.	A		

Glossary

Acclimatization- Adaptation of an organism to a new environment.

Acidophilus milk - Milk seeded with living *Lactobacillus acidophilus* bacteria used to improve the balance of beneficial bacteria in the intestines of the person who drinks the milk.

Action letter- An official FDA communication that informs an NDA or BLA sponsor of a decision by the agency. An approval letter allows commercial marketing of the product.

Adventitious, adventive - Developing from unusual points of origin, such as shoots or roots arising from a leaf or stem tissues other than the axils or apex; often dependent on close physical or temporal association with organized or semi-organized tissues or cells. Compare to "*de-novo*"

Aerobic- Needing oxygen for growth.

Agrobacterium tumefaciens- A common soil bacterium used as a vector to create transgenic plants.

Allele- Any of several alternative forms of a gene.

Alzheimer's disease -A disease characterized by, among other things, progressive loss of memory. The development of Alzheimer's disease is thought to be associated, in part, with possessing certain alleles of the gene that encodes apolipoprotein E.

Amino acids - Long chains of amino acids make up proteins. About 20 are known. Some amino acids are made in the body; those that are not are called essential amino acids and must be supplied in food.

Amplification- The process of increasing the number of copies of a particular gene or chromosomal sequence.

Anaerobic- Growing in the absence of oxygen.

Androgenesis - Development of haploid plants from the male gametophyte following a developmental pattern resembling embryogenesis, resulting from the culture of anthers or microspores.

Anther culture - The aseptic culture of anthers for the production of haploid callus or plants from the microspores, see "androgenesis".

Antibiotic- Chemical substance formed as a metabolic byproduct in bacteria or fungi and used to treat bacterial infections. Antibiotics can be produced naturally, using microorganisms, or synthetically.

Antibody - Protein produced in body tissues in response to the presence of a specific antigen. See antigen.

Anticodon -Triplet of nucleotide bases (codon) in transfer RNA that pairs with (is complementary to) a triplet in messenger RNA. For example, if the codon is UCG, the anticodon is AGC. See also Base; Base pair; Complementarity.

Antigen-A substance that, when introduced into the body, induces an immune response by a specific antibody.

Antisense -A piece of DNA producing a mirror image ("antisense") messenger RNA that is opposite in sequence to one directing protein synthesis. Antisense technology is used to selectively turn off production of certain proteins.

Antiserum- Blood serum containing specific antibodies against an antigen. Antisera are used to confer passive immunity to many diseases.

Anti-tumour agents - Compounds active against or inhibit neoplasm or cancerous growth.

Aseptic - The absence of contaminating fungi, bacteria, viruses, mycoplasma and other competing microorganisms in cultures of eukaryotic cells or tissues.

Assay - Technique for measuring a biological response; a test.

Attenuated -Weakened; with reference to vaccines, made from pathogenic organisms that have been treated so as to render them avirulent.

Autoimmune disease- A disease in which the body produces antibodies against its own tissues.

Autosome- Any chromosome other than a sex chromosome.

Avirulent -Unable to cause disease.

Axenic - A pure culture of single organism that is not contaminated by symbionts or parasites.

B lymphocytes (B-cells) A class of lymphocytes, released from the bone marrow, that produce antibodies.

Bacillus - Singular for a rod-shaped bacterium (plural, bacilli). Also used as the name of a genus of bacteria, including the species *Bacillus thuringiensis* .

Bacillus subtilis- A bacterium commonly used as a host in recombinant DNA experiments. Important because of its ability to secrete proteins.

Bacillus thuringiensis (Bt) - Bacterium that produces a protein called Bt toxin, a biological insecticide. Bt toxin is used to control insect pests by dusting the crop with Bt bacteria. When ingested, Bt toxin kills certain insect larvae, but is regarded as harmless to humans, pets and most beneficial insects such as bees. Inserting a copy of the Bt gene into plants enables them to produce Bt toxin protein. Such plants can resist some insect pests. See biological control, microbial insecticide.

Bacteriophage -Virus that lives in and kills bacteria. Also called phage.

Bacterium - Class of single-cell organisms (plural, bacteria). One member, *E. coli*, is commonly used in recombinant DNA technology for producing proteins and other chemicals.

Base - On the DNA molecule, one of the four chemical units that are linked in a series to make a strand of DNA. The four DNA bases are: adenine (A), cytosine (C), guanine (G), and thymine (T). In RNA, uracil (U) substitutes for thymine. See DNA finger printing, nucleotide.

Base pair -Two nucleotide bases on different strands of the nucleic acid molecule that bond together. The bases can pair in only one way: adenine with thymine (DNA), or uracil (RNA) and guanine with cytosine.

Batch culture - A suspension culture in which cells grow in a finite volume of nutrient medium and follow a sigmoidal pattern of growth between subcultures.

Bioassay- Determination of the effectiveness of a compound by measuring its effect on animals, tissues or organisms in comparison with a standard preparation.

Bioaugmentation- Increasing the activity of bacteria that break down pollutants by adding more of their kind. A technique used in bioremediation.

Biocatalyst -In bioprocessing, an enzyme that activates or speeds up a biochemical reaction.

Biochip- An electronic device that uses organic molecules to form a semiconductor.

Bioconversion- Chemical restructuring of raw materials by using a biocatalyst.

Biodegradable- Capable of being reduced to water and carbon dioxide by the action of microorganisms.

Bioenrichment- A bioremediation strategy that involves adding nutrients or oxygen, thereby bolstering the activity of microbes as they break down pollutants.

Bioinformatics -The science of informatics as applied to biological research. Informatics is the management and analysis of data using advanced computing techniques. Bioinformatics is particularly important as an adjunct to genomics research, because of the large amount of complex data this research generates.

Biolistic device- A device that shoots microscopic DNA-coated particles into target cells.

Biologic -A therapeutic or prophylactic derived from a living source (human, animal or unicellular). Most biologics are complex mixtures that are not easily identified or characterized, and many are manufactured using biotechnology. Biological products often represent the cutting-edge of biomedical research and are sometimes the most effective way to prevent or treat a disease.

Biological control (biocontrol) - Managing pest populations by purposefully manipulating beneficial natural enemies - predatory or parasitic insects that kill pest insects, or microbes that cause insect diseases. See Bacillus thuringiensis (Bt).

Biomass- The totality of biological matter in a given area. As commonly used in biotechnology, refers to the use of cellulose, a renewable resource, for the production of chemicals that can be used to generate energy or as alternative feed stocks for the chemical industry to reduce dependence on nonrenewable fossil fuels, callus or cells biomass in culture.

Biomaterials- Biological molecules, such as proteins and complex sugars, used to make medical devices, including structural elements used in reconstructive surgery.

Bioprocess A process in which living cells, or components thereof, are used to produce a desired product.

Bioreactor - A vessel, such as a fermenter, used for fermentation or other biological processes.

Bioremediation - Using organisms to remove toxins from the environment. Examples: a fungus to detoxify a wood preservative that contaminates soil at sawmills; a bacterium to help clean up oil spills.

Biosynthesis -Production of a chemical by a living organism.

Biotechnology – Development of products by a biological process requiring engineering technologies such as fermentation or controlled environments, or utilizing biochemical or genetic technologies such as recombinant DNA techniques for the modification and improvement of biological systems.

Biotransformation- (biochemical transformation or Bioconversion) – The use of cultured cells to convert substrates into desired organic compounds by virtue of an endogenous enzyme which catalyzes the reaction.

Blastocyst (Blastula)- The 4- to 5-day-old ball of undifferentiated cells from which a prospective embryo develops. In mammals it consists of two distinct parts: the inner cell mass and the trophoblast.

Bovine - Formal word for any animal in the cattle family.

Bovine growth hormone/bovine somatotropin (BGH/BST) - Two names for one protein hormone produced in the pituitary gland of cattle. BGH/BST can also be produced by inserting a copy of the gene for BGH/BST into laboratory bacteria. Such recombinant BGH/BST is also referred to as rBGH/rBST. Purified from pituitary glands or from bacterial cultures, BGH/BST injected into dairy cows can increase milk production up to 20 percent. Compare to porcine somatotropin (PST).

Bovine somatotropin (BST)- A hormone secreted by the bovine pituitary gland. It is used to increase milk production by improving the feed efficiency in dairy cattle milk. Also called bovine growth hormone.

BRCA1 and BRCA2 -(Breast Cancer genes 1 and 2) Two genes that normally help to restrain cell growth, but which can contain certain genetic mutations associated with the development of breast and ovarian cancer. Note, however, that inherited BRCA1 and BRCA2 mutations are thought to account for less than 10 percent of all breast and ovarian cancers. Recent evidence suggests that somatic cell genetic mutations (i.e., noninherited genetic mutations) in these two genes may also play a role in the development of cancer.

Callus (plural, calli) – (1) An undifferentiated growth of plant cells *in vitro* on a culture medium, (2) A wound response *in vivo* resulting in a tumorous growth.

Cancerous growth – Uncontrolled and abnormal cell division of cells leading to tumour formation.

Carbohydrate- A type of biological molecule composed of simple sugars such as glucose. Common examples include starch and cellulose.

Carcinogen- Cancer-causing agent.

Catalyst- An agent (such as an enzyme or a metallic complex) that facilitates a reaction but is not itself changed during the reaction.

Cell - Smallest unit of living matter able to grow and reproduce independently. Cells contain DNA for storing information, ribosomes for making proteins, and mechanisms for converting energy.

Cell culture - The growing of cells *in vitro*, including the culture of single cells or small aggregates of cells in a liquid medium.

Cell differentiation- The process by which descendants of a common parental cell achieve specialized structure and function

Cell fusion- See Fusion.

Cell line - Derived from a primary culture and selected at the time of the first subculture, or derived from a cell population with particular attributes.

Cell -The smallest structural unit of a living organism that can grow and reproduce independently.

Chemical genomics- Using structural and functional genomic information about biological molecules, especially proteins, to identify useful small molecules and alter their structure to improve their efficacy.

Chemically defined medium – A nutritive solution for culturing cells in which each component is of known chemical structure.

Chimera -The individual (animal or lower organism) produced by grafting an embryonic part of one individual onto an embryo of either the same or a different species.

Chromosomes- Threadlike components in the cell that contain DNA and proteins. Genes are carried on the chromosomes.

Chymosin - Enzyme, also called rennet, used in making cheese. Chymosin can be extracted from the stomach of veal calves, or from genetically engineered bacteria that have the gene for chymosin.

Clinical studies- Human studies that are designed to measure the efficacy of a new drug or biologic. Clinical studies routinely involve the use of a control group of patients that is given an inactive substance (placebo) that looks like the test product.

Clonal propagation - A sexual reproduction or vegetative propagation of plants considered to be physiologically and/or genetically uniform and which originated from a single individual or explant.

Clone - *as noun* (1) Population of plants derived from a single individual by vegetative propagation or by regeneration of plants from *in vitro* cultures. (2) Population of cells mitotically derived from a single cell. (3) Population or recombinant DNA molecules carrying the same inserted sequence. *as verb*- (1) To propagate a clone. (2) To insert a particular DNA sequence into a vector. Researchers have achieved laboratory cloning using genetic material from adult animals of several species, including mice, pigs and sheep.

Cloning - Technique of creating a group of genetically identical cells or DNA molecules from a single ancestor. In horticulture, cloned plants are reproduced asexually from a single parent.

Co-cultivation - In transformation experiments, the incubation of host cells or tissues with *Agrobacterium* or another organism acting as a vehicle for a genetic vector.

Codon- A sequence of three nucleotide bases that specifies an amino acid or represents a signal to stop or start a function.

Co-enzyme- An organic compound that is necessary for the functioning of an enzyme. Co-enzymes are smaller than the enzymes themselves and sometimes separable from them.

Co-factor- A nonprotein substance required for certain enzymes to function. Co-factors can be co-enzymes or metallic ions.

Combinatorial chemistry- A product discovery technique that uses robotics and parallel synthesis to generate and screen quickly as many as several million molecules with similar structure in order to find chemical molecules with desired properties.

Complementarity- The relationship of the nucleotide bases on two different strands of DNA or RNA. When the bases are paired properly (adenine with thymine [DNA] or uracil [RNA]; guanine with cytosine), the strands are complementary.

Complementary DNA (cDNA)- DNA synthesized from a messenger RNA rather than from a DNA template. This type of DNA is used for cloning or as a DNA probe for locating specific genes in DNA hybridization studies.

Computational biology- A subdiscipline within bioinformatics concerned with computation-based research devoted to understanding basic biological processes.

Conjugation- Sexual reproduction of bacterial cells in which there is a one-way exchange of genetic material between the cells in contact.

Continuous culture - A suspension culture held at constant volume and continuously supplied with nutrients by the inflow of fresh medium.

Cosmid - A plasmid vector which contains the *cos* site of phage and one or more selectable markers such as an antibiotic resistance gene.

Culture - Cultivate cells or living organisms in a prepared medium under laboratory conditions. "Culture" is both the process and the growing cells.

Culture medium -Any nutrient system for the artificial cultivation of bacteria or other cells; usually a complex mixture of organic and inorganic materials.

Cybrid - A cytoplasmically hybrid cell with organelles from both parental sources, e.g., achieved by fusion of a cytoplast with a whole cell. During subsequent cell divisions random or nonrandom segregation of organelles may occur, resulting in variable parental combinations in cybrid cell lines.

Cystic fibrosis - Disease of mucous glands throughout the body that usually develops during childhood, and makes breathing increasingly difficult. If a child receives two copies of the defective gene called the CF gene - one copy from each parent - then the child will develop the disease.

Cytotoxic- Able to cause cell death.

De-novo **-** Literally to "arise anew", with reference to plant regeneration or developmental processes, arising from unorganized cells or tissues or from undetermined cells. Compare to "adventitious".

Deoxyribonucleic acid (DNA)- The molecule that carries the genetic information for most living systems. The DNA molecule consists of four bases (adenine, cytosine, guanine and thymine) and a sugar-phosphate backbone, arranged in two connected strands to form a double helix. See also Complementary DNA; Double helix; Recombinant DNA.

Diagnostic- A product used for the diagnosis of disease or medical condition. Both monoclonal antibodies and DNA probes are useful diagnostic products.

Differentiation - The process of biochemical and structural changes by which cells become specialized in form and function.

Diploid- A cell with two complete sets of chromosomes. Compare Haploid.

DNA analysis - See polymerase chain reaction (PCR) and RFLP mapping. Both PCR and RFLP analysis can be used in DNA fingerprinting for genealogical studies and forensics. See next entry.

DNA chip- A small piece of glass or silicon that has small pieces of DNA arrayed on its surface.

DNA fingerprinting - Detecting patterns in DNA that indicate the presence of a gene for a trait. The pattern resembles a bar code printed on a commercial product so computers can scan the price. Forensics experts can use this distinct pattern to link or clear an Individual suspected of being involved in a crime, like they compare fingerprints. Breeders can use these patterns to find and select breeding stock with traits such as disease resistance.

DNA hybridization- The formation of a double-stranded nucleic acid molecule from two separate strands. The term also applies to a molecular technique that uses one nucleic acid strand to locate another.

DNA library- A collection of cloned DNA fragments that collectively represent the genome of an organism.

DNA polymerase- An enzyme that replicates DNA. DNA polymerase is the basis of PCR—the polymerase chain reaction.

DNA probe- A small piece of nucleic acid that has been labeled with a radioactive isotope, dye or enzyme and is used to locate a particular nucleotide sequence or gene on a DNA molecule.

DNA repair enzymes- Proteins that recognize and repair certain abnormalities in DNA.

DNA sequence- The order of nucleotide bases in the DNA molecule.

DNA vaccines- Pieces of foreign DNA that are injected into an organism to trigger an immune response.

Double haploid – A plant with two sets of haploid chromosomes.

Double helix- A term often used to describe the configuration of the DNA molecule. The helix consists of two spiraling strands of nucleotides (a sugar, phosphate and base) joined crosswise by specific pairing of the bases. See also Deoxyribonucleic acid; Base; Base pair.

Drug delivery- The process by which a formulated drug is administered to the patient. Traditional routes have been oral or intravenous perfusion. New methods deliver through the skin with a transdermal patch or across the nasal membrane with an aerosol spray.

Electrofusion - The application of electroporation techniques for protoplast fusion.

Electrophoresis - Technique for analyzing and separating molecules based on the movement of charged particles in an electric field.

Electroporation- The creation of reversible small holes in a cell wall or membrane through which foreign DNA can pass. This DNA can then integrate into the cell's genome.

Enzyme-linked immunosorbent assay (ELISA) A technique for detecting specific proteins by using antibodies linked to enzymes.

Embryo - Early stages of an animal's development that results when a sperm fertilizes an egg. Embryo cloning produces many calves from one embryo, for example.

Embryo rescue - Application of embryo culture techniques to obtain interspecific hybrid plants prior to abortion of the embryos due to post-fertilization hybridization barriers.

Embryogenesis (embryogeny) – Initiation and development of embryos from zygotes, or of bipolar structures from somatic cells that parallel the developmental path of zygotic embryos. In the later case, such embryos may be either adventitious or *de novo* in origin.

Embryonic stem cells- Cells that can give rise to any type of differentiated cell. They can be derived from two sources: the inner cell mass from a blastocyst or the primordial germ cells (eggs and sperm) of an older embryo.

Endostatin- An endogenous protein that blocks the proliferation of blood vessels.

Enzyme - A protein catalyst that facilitates specific chemical or metabolic reactions necessary for cell growth and reproduction.

Escherichia coli (E. coli) - Common bacterium found in human and mammalian digestive tracts. Some strains of E. coli are used in recombinant DNA work because they have been genetically well-characterized and are easily grown in laboratory fermenters.

Eukaryote- A cell or organism containing a true nucleus, with a well-defined membrane surrounding the nucleus. All organisms except bacteria, viruses and cyanobacteria are eukaryotic. Compare Prokaryote.

Exon- In eukaryotic cells, that part of the gene that is transcribed into messenger RNA and encodes a protein. See also Intron; Splicing.

Explant - The tissue taken from a plant or seed and transferred to a culture medium to establish a tissue culture system or regenerate a plant.

Expression - In genetics, manifestation of a characteristic specified by a gene. In industrial biotechnology, production of a specific protein by inserting a gene into a new host organism.

Extremophiles- Microorganisms that live at extreme levels of pH, temperature, pressure and salinity.

Frmentation - Chemical reaction induced by a living agent - yeast, bacterium or mold - that splits complex organic compounds to simple ones. For example, yeast converts sugar to alcohol and carbon dioxide. In biotechnology, the process of growing microbes to produce chemical or pharmaceutical compounds. Also referred to as classical biotechnology, traceable back 6,000 years.

Functional foods-Foods containing compounds with beneficial health effects beyond those provided by the basic nutrients, minerals and vitamins. Also called nutraceuticals.

Functional genomics-A field of research that aims to understand what each gene does, how it is regulated and how it interacts with other genes.

Fusion - Joining the membrane of two cells of different origin to create a cell that contains the parent cells' nuclear material. Used in monoclonal antibody technology to make hybridomas

- fusing an immortal cell (one that divides continuously) and an anti body-producing cell. See monoclonal antibody.

Gametoclonal variation - Phenotypic variation apart from normal segregation occurring during gamete formation, both epigenetic and genetic, arising from the culture of gametophytic cells or tissues, such as in anther culture or unfertilized ovule culture. Related to "somaclonal variation".

Gel electrophoresis -A process for separating molecules by forcing them to migrate through a gel under the influence of an electric field.

Gene - For genetic engineering, the smallest portion of a chromosome that contains the hereditary information for the production of a protein.

Gene amplification- The increase, within a cell, of the number of copies of a given gene.

Gene gun - A particle acceleration device for physically delivering recombinant DNA, typically precipitated onto a microprojectile, into a cell using an explosive propellant.

Gene knockout-The replacement of a normal gene with a mutated form of the gene by using homologous recombination. Used to study gene function.

Gene machine-A computerized device for synthesizing genes by combining combing nucleotides (bases) in the proper order.

Gene mapping-Determination of the relative locations of genes on a chromosome.

Gene sequencing- Determination of the sequence of nucleotide bases in a strand of DNA. See also Sequencing.

Gne splicing - Inserting new genetic information into a chromosome using recombinant DNA techniques.

Gene therapy- The replacement of a defective gene in an organism suffering from a genetic disease. Recombinant DNA techniques are used to isolate the functioning gene and insert it into cells. More than 300 single-gene genetic disorders have been identified in humans. A significant percentage of these may be amenable to gene therapy.

Genetic analysis - Studying how traits and genes for traits are passed from generation to generation, and how genes and the environment interact to result in traits.

Genetic code - Information coded within nucleotide sequences of RNA and DNA that specifies the amino acid sequence in protein synthesis and on which heredity is based.

Genetic counseling - Providing current or prospective parents with information on the probabilities of inherited diseases occurring in their children, and on diagnosis and treatment of such diseases.

Genetic engineering - Using recombinant DNA techniques and related methods to move one or several genes from one organism to another, to rearrange one or several genes within a cell, or to alter gene-controlled processes. Transferring a DNA segment from one organism and inserting it into the DNA of another organism to modify, amplify, transform and express genetic information. The two organisms can be totally unrelated. See recombinant DNA.

Genetic modification- A number of techniques, such as selective breeding, mutagenesis, transposon insertions and recombinant DNA technology, that are used to alter the genetic material of cells in order to make them capable of producing new substances, performing new functions or blocking the production of substances.

Genome -The total hereditary material of a cell, comprising the entire chromosomal set found in each nucleus of a given species.

Genomics- The study of genes and their function. Recent advances in genomics are bringing about a revolution in our understanding of the molecular mechanisms of disease, including the complex interplay of genetic and environmental factors. Genomics is also stimulating the discovery of breakthrough health-care products by revealing thousands of new biological targets for the development of drugs and by giving scientists innovative ways to design new drugs, vaccines and DNA diagnostics. Genomic-based therapeutics may include "traditional" small chemical drugs, as well as protein drugs and gene therapy.

Genotype- Genetic makeup of an individual or group. Compare Phenotype.

Germ cell - Reproductive cell (sperm or egg). Also called gamete or sex cell.

Germplasm - The total genetic variability, represented by germ cells or seeds, available to a particular population of organisms.

Glycoprotein- A protein conjugated with a carbohydrate group.

Granulocyte-macrophage colony stimulating factor (GMCSF) - A natural hormone that stimulates white blood cell production, particularly that of granulocytes and monocytes (the precursors of macrophages).

Growth factors - Naturally occurring proteins that stimulate the growth and reproduction of specific cell types. Growth factors are essential to regenerative medicine and tissue engineering.

Growth hormone- A protein produced by the pituitary gland that is involved in cell growth. Human growth hormone is used clinically to treat dwarfism. Various animal growth hormones can be used to improve milk production as well as produce a leaner variety of meat.

Haploid- A cell with half the usual number of chromosomes, or only one chromosome set. Sex cells are haploid. Compare Diploid.

Heredity- T-ransfer of genetic information from parent cells to progeny.

Homologous- Corresponding or alike in structure, position or origin.

Homozygous - Having the same allele of a gene in homologous chromosomes.

Hormone - Chemical that acts as a messenger relaying instructions to start certain physiological activities. Hormones are synthesized in one type of cell, and released to direct the function of other cell types.

Host- A cell or organism used for growth of a virus, plasmid or other form of foreign DNA, or for the production of cloned substances.

Host-vector system- Combination of DNA-receiving cells (host) and DNA-transporting substance (vector) used for introducing foreign DNA into a cell.

Human Genome Project- An international research effort aimed at discovering the full sequence of bases in the human genome. Led in the United States by the National Institutes of Health and the Department of Energy.

Human growth hormone - Human somatotropin, a protein hormone made in the pituitary gland. Lab-grown (HGH) bacteria given a copy of the gene for human growth hormone can then produce large amounts of HGH, which can be purified and used to treat certain kinds of human dwarfism or pathological condition of growth arrested by various causes.

Human immunodeficiency virus (HIV) - The virus that causes acquired immune deficiency syndrome (AIDS).

Hybridization- Production of offspring, or hybrids, from genetically dissimilar parents. The process can be used to produce hybrid plants (by crossbreeding two different varieties) or hybridomas (hybrid cells formed by fusing two unlike cells, used in producing monoclonal antibodies). See DNA hybridization.

Hybridoma- The cell produced by fusing two cells of different origin. In monoclonal antibody technology, hybridomas are formed by fusing an immortal cell (one that divides continuously) and an antibody-producing cell. See also Monoclonal antibody; Myeloma.

Immobilization - Entrapment /attachment of cells or protoplasts into/onto a matrix.

Immunoassay- Technique for identifying substances based on the use of antibodies.

Immunodiagnostic- The use of specific antibodies to measure a substance. This tool is useful in diagnosing infectious diseases and the presence of foreign substances in a variety of human and animal fluids (blood, urine, etc.). The approach is currently being investigated as a way of locating tumor cells in the body.

Immunofluorescence- Technique for identifying antigenic material that uses an antibody labeled with fluorescent material. Specific binding of the antibody and antigen can be seen under a microscope by applying ultraviolet light rays and noting the visible light that is produced.

Immunogen- Any substance that can elicit an immune response.

Immunoglobulin - General name for proteins that function as antibodies. These proteins differ somewhat in structure and are grouped into five categories on the basis of these differences; immunoglobulin G (IgG), IgM, IgA, IgE and IgD.

Immunology- Study of all phenomena related to the body's response to antigenic challenge (i.e., immunity, sensitivity and allergy).

Immunomodulators- A diverse class of proteins that boost the immune system. Many are cell growth factors that accelerate the production of specific cells that are important in mounting an immune response in the body. These proteins are being investigated for use in possible treatments for cancer.

Immunotoxins- Specific monoclonal antibodies that have a protein toxin molecule attached. The monoclonal antibody is targeted against a tumor cell, and the toxin is designed to kill that cell when the antibody binds to it.

In situ- In its original or natural place or position.

***In vitro* -** Any process carried out in sterile cultures, or the measurement of biological processes outside the intact organism such as enzyme reactions. Literally, "in glass".

In vivo- In a living organism.

Inducer- A molecule or substance that increases the rate of enzyme synthesis, usually by blocking the action of the corresponding repressor.

Insulin - Protein hormone that regulates blood sugar, made in cells of the pancreas. In the laboratory, microbes given a copy of the gene for human insulin can make insulin to treat diabetes mellitus, a shortage of insulin.

Interferon- A class of lymphokine proteins important in the immune response. There are three major types of interferon: alpha (leukocyte), beta (fibroblast) and gamma (immune). Interferons inhibit viral infections and may have anticancer properties.

Interleukin- A type of lymphokine that regulates the growth and development of white blood cells. Twelve interleukins (IL-1 through IL-12) have been identified to date.

Intron- In eukaryotic cells, a sequence of DNA that is contained in the gene but does not encode for protein. The presence of introns "splits" the coding region of the gene into segments called exons. See also Exon; Splicing.

Islet cells- Pancreatic cells that are the source of insulin and two other hormones involved in regulating glucose metabolism and absorption.

Isoenzyme- One of the several forms that a given enzyme can take. The forms may differ in certain physical properties, but function similarly as biocatalysts.

Isogenic- Of the same genotype.

Kidney plasminogen activator- A precursor to the enzyme urokinase, which has blood-clotting properties.

Lactase - Enzyme in certain yeasts and mammalian intestinal tracts that catalyzes converting lactose to glucose and galactose. Lactase tablets remove lactose from milk for people who are lactose intolerant. See next entry.

Lactose - Milk sugar; also white crystal sugar made from whey used in baby food, baked goods, candies and pharmaceuticals.

Library- A set of cloned DNA fragments that taken collectively contain the entire genome of an organism. Also called a DNA library.

Ligase- An enzyme used to join DNA or RNA segments together.

Linkage- The tendency for certain genes to be inherited together due to their physical proximity on the chromosome.

Linker-A fragment of DNA with a restriction site that can be used to join DNA strands.

Lipoproteins- A class of serum proteins that transport lipids and cholesterol in the bloodstream. Abnormalities in lipoprotein metabolism have been implicated in certain heart diseases.

Lymphocyte- A type of leukocyte found in lymphatic tissue in the blood, lymph nodes and organs. Lymphocytes are continuously made in the bone marrow and mature into antibody-forming cells. See also B lymphocytes; T lymphocytes.

Lymphokine- A class of soluble proteins produced by white blood cells that play a role, as yet not fully understood, in the immune response. See also Interferon; Interleukin.

Lymphoma- Form of cancer that affects the lymph tissue.

Macrophage- A type of white blood cell produced in blood vessels and loose connective tissues that can ingest dead tissues and cells and is involved in producing interleukin-1. When exposed to the lymphokine macrophage-activating factor, macrophages also kill tumor cells. See also Phagocyte.

Macrophage colony stimulating factor (M-CSF)-A natural hormone that stimulates the production of white blood cells, particularly monocytes (the precursors of macrophages).

Marker gene - Gene that is easy to find or observe. Attaching a marker gene to another gene that is hard to find, such as an antisense gene, is like putting a reflective collar on a dog so you can see the collar at night - even though you cannot see the dog, you know where it is. See antisense.

Mastitis - Inflammation of milk-producing glands in a cow's udder, caused by bacteria. Mastitis can be treated with antibiotics.

Medium- A substance containing nutrients needed for cell growth.

Meiosis- Process of cell reproduction whereby the daughter cells have half the chromosome number of the parent cells. Sex cells are formed by meiosis. Compare Mitosis.

Messenger RNA (mRNA)- Nucleic acid that carries instructions to a ribosome for the synthesis of a particular protein.

Metabolism- All biochemical activities carried out by an organism to maintain life.

Microbiology- Study of living organisms that can be seen only under a microscope.

Microinjection-The injection of DNA using a very fine needle into a cell.

Microorganism- Any organism that can be seen only with the aid of a microscope. Also called microbe.

Mitosis- Process of cell reproduction whereby the daughter cells are identical in chromosome number to the parent cells. Compare Meiosis.

Molecular genetics- Study of how genes function to control cellular activities.

Monoclonal antibody (MAb) – Highly specific, purified antibody that is derived from only one clone of cells and recognizes only one antigen. See also Hybridoma; Myeloma.

Morphogenesis - (1) The ontological development of differentiated structures. (2) The development of a structure from an unorganized state to a differentiated and organized state.

Multigenic- Of hereditary characteristics, one that is specified by several genes.

Mutagen - Agent that causes biological mutation. Examples include chemicals, radioactive elements and ultraviolet light. See pentachlorophenol.

Mutant - A cell that manifests new characteristics due to a change in its DNA.

Mutation - A heritable alteration in a gene; the process by which such changes occur.

Myeloma- A type of cancer cell (plasma cell) that is used in monoclonal antibody technology to form hybridomas.

Natural active immunity - Immunity that is established after the occurrence of a disease.

Noncoding DNA- DNA that does not encode any product (RNA or protein). The majority of the DNA in plants and animals is noncoding.

Nuclease- An enzyme that, by cleaving chemical bonds, breaks down nucleic acids into their constituent nucleotides.

Nucleic acids- Large molecules, generally found in the cell's nucleus and/or cytoplasm, that are made up of nucleotides. The two most common nucleic acids are DNA and RNA.

Nucleotide - Organic compound composed of a sugar, a phosphate and a base. Two nucleotides joined crosswise by specific pairings of the bases make one rung in the DNA molecule's double helix "ladder."

Oligonucleotide- A polymer consisting of a small number (about two to 10) of nucleotides.

Oncogene- Gene thought to be capable of producing cancer.

Oncogenic- Cancer causing.

Operator gene- A region of the chromosome, adjacent to the operon, where a repressor protein binds to prevent transcription of the operon.

Operon- Sequence of genes responsible for synthesizing the enzymes needed for biosynthesis of a molecule. An operon is controlled by an operator gene and a repressor gene.

Optimization - Manipulation of chemical and physical environment of cells to obtain optimal growth of the cells or their contents.

Organ culture - The maintenance or growth of organ primordia or whole parts of an organ *in vitro*, in a way that allows differentiation and/or preservation of its structure and/or function.

Organic compound- A compound containing carbon.

Organogenesis - The process of initiation and development of a structure which shows natural organ form and/or function, the initiation of which is temporally separated from the initiation of other organs. Such organs may be adventive or *de novo* in origin.

Packed cell volume - A measurement that estimates the total cell mass in a given volume of cell culture by centrifugation of the cells to a constant volume.

Parafilm - A stretchable film based on paraffin wax which is used to seal tubes and petri dishes.

Passive immunity- Immunity acquired from receiving preformed antibodies.

Pathogen - Organism able to cause disease in a certain host.

Peptide- Two or more amino acids joined by a linkage called a peptide bond.

Phagocyte- A type of white blood cell that can ingest invading microorganisms and other foreign material. See also Macrophage.

Phenotype- Observable characteristics resulting from interaction between an organism's genetic makeup and the environment. Compare Genotype.

Phytoremediation- The use of plants to clean up pollution. .

Plant regeneration - The process of producing plantlets from *in vitro* cultures through organogenic or embryogenic modes of development, either by adventitious or *de novo* origination. This term should not be used with reference to micropropagation by enhanced axillary branching.

Plasmid- A small circular form of DNA that carries certain genes and is capable of replicating independently in a host cell.

Pluripotent cells- Having the capacity to become any kind of cell or tissue in the body. Embryonic stem cells and cells of the inner cell mass are pluripotent. Adult stem cells are multipotent. The mammalian embryo (blastocyst trophoblast plus inner cell mass) is totipotent because it can become an entire organism. Fully differentiated cells from many plants are totipotent.

Polyclonal- Derived from different types of cells.

Polymer-A long molecule of repeated subunits.

Polymerase chain reaction (PCR) - Multiplying a particular DNA segment in repeated cycles. The "copies" made in a previous cycle are used as "originals" or templates in the next cycle. For example, PCR enables forensics experts to do DNA testing on very small blood samples.

Polypeptide- Long chain of amino acids joined by peptide bonds.

Primary products – Compound produced by primary metabolism and directly involved in cell metabolism/ cell building processes.

Prokaryote- An organism (e.g., bacterium, virus, cyanobacterium) whose DNA is not enclosed within a nuclear membrane. Compare Eukaryote.

Promoter- A DNA sequence that is located in front of a gene and controls gene expression. Promoters are required for binding of RNA polymerase to initiate transcription.

Propagule - A plant part serving for propagation.

Prophage- Phage nucleic acid that is incorporated into the host's chromosome but does not cause cell lysis.

Protein - Composed of amino acids, proteins are a key component in many cell structures. In addition, many hormones and enzymes that regulate cells are proteins.

Protein A- A protein produced by the bacterium Staphylococcus aureus that specifically binds antibodies. It is useful in the purification of monoclonal antibodies.

Proteomics- Each cell produces thousands of proteins, each with a specific function. This collection of proteins in a cell is known as the proteome, and. unlike the genome, which is constant irrespective of cell type, the proteome varies from one cell type to the next. The science of proteomics attempts to identify the protein profile of each cell type, assess protein differences between healthy and diseased cells, and uncover not only each protein's specific function but also how it interacts with other proteins.

Protoplast- The cellular material that remains after the cell wall has been removed from plant and fungal cells.

Pure culture- In vitro growth of only one type of microorganism.

Radioimmunoassay (RIA)- A test combining radioisotopes and immunology to detect trace substances. Such tests are useful for studying antibody interactions with cell receptors, and can be developed into clinical diagnostics.

Rational drug design- Using the known three-dimensional structure of a molecule, usually a protein, to design a drug molecule that will bind to it. Usually viewed as an alternative to drug discovery through screening many molecules for biological activity.

Reagent- Substance used in a chemical reaction.

Recombinant DNA (rDNA) - Technique of isolating DNA molecules and inserting them into the DNA of a cell. This technique includes taking copies of genes from one organism and inserting them in another organism. The two organisms can be totally unrelated. Recombinant DNA has a variety of uses, such as studying how genes work, and producing medicines such as human insulin and other commercial products. See genetic engineering.

Recombinant DNA (rDNA)- The DNA formed by combining segments of DNA from two different sources.

Regeneration- Laboratory technique for forming a new plant from a clump of plant cells.

Regulatory gene- A gene that acts to control the protein-synthesizing activity of other genes.

Replication- Reproduction or duplication, as of an exact copy of a strand of DNA.

Replicon- A segment of DNA (e.g., chromosome or plasmid) that can replicate independently.

Repressor- A protein that binds to an operator adjacent to a structural gene, inhibiting transcription of that gene.

Restriction enzyme- An enzyme that breaks DNA in highly specific locations, creating gaps into which new genes can be inserted.

Restriction fragment length polymorphism (RFLP)- The variation in the length of DNA fragments produced by a restriction endonuclease that cuts at a polymorphic locus. This is a key tool in DNA fingerprinting and is based on the presence of different alleles in an individual. RFLP mapping is also used in plant breeding to see if a key trait such as disease resistance is inherited.

Retrovirus- A virus that contains the enzyme reverse transcriptase. This enzyme converts the viral RNA into DNA, which can combine with the DNA of the host cell and produce more viral particles.

RFLP (restriction fragment length polymorphism) - RFLP mapping detects patterns in DNA that can indicate the presence of a gene for a trait. Both RFLP and polymerase chain reaction (PCR) analysis can be used in DNA fingerprinting for genealogical studies and forensics.

Rheology- Study of the flow of matter such as fermentation liquids.

Rhizobium- A class of microorganisms that converts atmospheric nitrogen into a form that plants can utilize for growth. Species of this microorganism grow symbiotically on the roots of certain legumes, such as peas, beans and alfalfa.

Ribonucleic acid (RNA) - A molecule similar to DNA that delivers DNA's genetic message to the cytoplasm of a cell where proteins are made.

Ribosome - Site of protein synthesis in the cytoplasm. selection and breeding - Manipulating microbes, plants or animals, and choosing desirable individuals or populations as breeding stock for new generations.

RNA (ribonucleic acid) - Molecule similar to DNA that functions primarily to decode the instructions that genes carry for protein synthesis.

RNA interference- A natural process used by organisms to block protein production.

Scale-up - Transition from small-scale production to production of large industrial quantities.

Secondary products - Metabolites produced as end products of primary metabolism and not involved directly in metabolic activity.

Selection, selective medium - Culture conditions and/or nutrient material constituted such that they will support the growth of specific organisms or genotypes while inhibiting the growth of others.

Selective medium- Nutrient material constituted such that it will support the growth of specific organisms while inhibiting the growth of others.

Sepsis- The presence in the blood or other tissues of pathogenic microorganisms or their toxins; the condition associated with such presence.

Sequencing- Decoding a strand of DNA or gene into the specific order of its nucleotides: adenine, cytosine, guanine and thymine. This analysis can be done manually or with automated equipment. Sequencing a gene requires analyzing an average of 40,000 nucleotides.

Serology- Study of blood serum and reactions between the antibodies and antigens therein.

Single-cell protein- Cells or protein extracts from microorganisms, grown in large quantities for use as protein supplements.

Somaclonal variation - Variation which occurs in cultures of cells and tissues that may be either genetic or epigenetic.

Somatic cell gene therapy- Somatic cell gene therapy involves the insertion of genes into cells for therapeutic purposes; for example, to induce the treated cells to produce a protein that the body is missing. It does not affect genetic makeup of a patient's offspring and generally does not change all, or even most, cells in the recipient. Somatic cell gene therapy is only one way of applying the science of genomics to improve health care.

Somatic cell nuclear transfer- The transfer of a nucleus from a fully differentiated cell into an egg that has had its nucleus removed.

Splicing - The removal of introns and joining of exons to form a continuous coding sequence in RNA.

Somatic cells - Cells other than sex or germ cells. High somatic cell counts found in milk indicate mastitis in dairy cows. Low somatic cell counts indicate better quality milk and net a premium price from dairy processors.

Species - Group of organisms with common or similar characteristics and capable of inter breeding.

Stop codon- One of three codons in messenger RNA that signal the end of the amino acid chain in protein synthesis.

Structural gene- A gene that codes for a protein, such as an enzyme.

Substrate- Material acted on by an enzyme.

Suicide gene- A gene that codes for an antibiotic that can kill the host bacterial cell. It is genetically modified into the bacterium along with a molecular switch that is controlled by a nutrient in the environment. When the nutrient disappears, the suicide gene is switched on and the bacterium dies.

Suppressor gene - A gene that can reverse the effect of a mutation in other genes.

Synchronous culture - A culture in which a specified proportion of the cells are at the same indicated phase of the cell cycle, enter the same indicated phase of the cell cycle simultaneously, and/or exhibit the same approximate duration of cell cycle.

Synthesis - The process whereby separate elements are combined to form a new complex product, synthetic chemical compound or material.

Systems biology- A hypothesis-driven field of research that creates predictive mathematical models of complex biological processes or organ systems.

T lymphocytes (T-cells)- White blood cells that are produced in the bone marrow but mature in the thymus. They are important in the body's defense against certain bacteria and fungi, help B lymphocytes make antibodies and help in the recognition and rejection of foreign tissues. T lymphocytes may also be important in the body's defense against cancers.

Taxol(TM) paclitaxel - Natural product approved by the U.S. Food and Drug Administration in 1992 to treat ovarian cancer after failure of first-line chemotherapy; also used experimentally to treat breast, lung and other cancers. Manufactured by Bristol-Myers Squibb U.S. Pharmaceutical Group from bark of old-growth Pacific yew (cedar) trees. Because these forests are also home to endangered spotted owls, researchers are working to synthesize paclitaxel from twigs and needles, leaving the tree standing.

Technology transfer- The process of transferring discoveries made by basic research institutions, such as universities and government laboratories, to the commercial sector for development into useful products and services.

Template- A molecule that serves as the pattern for synthesizing another molecule.

Terminator- Sequence of DNA bases that tells the RNA polymerase to stop synthesizing RNA.

Tertiary structure - The total three-dimensional shape of a protein that is essential to protein function.

Therapeutics- Compounds that are used to treat specific diseases or medical conditions.

Tissue culture - Growing plant or animal tissues or cells in test tubes or other laboratory glassware, without other contaminating organisms, for propagation, chemical production and medical research.

Totipotency - The property of normal cells that they have the genetic potential to give rise to a complete individual. Exceptions include terminally specialized cells.

Toxin- A poisonous substance produced by certain microorganisms or plants.

Transcription- Synthesis of messenger (or any other) RNA on a DNA template.

Transdifferentiation- The process whereby a specialized cell de-differentiates and re-differentiates into a different cell type; or the process whereby an adult stem cell from a specific tissue type becomes a cell type from a very different tissue (for example a nerve stem cell differentiates into a kidney cell).

Transduction- Transfer of genetic material from one cell to another by means of a virus or phage vector.

Transfection- Infection of a cell with nucleic acid from a virus, resulting in replication of the complete virus.

Transfer RNA (tRNA)- RNA molecules that carry amino acids to sites on ribosomes where proteins are synthesized.

Transformation - The transfer and incorporation of DNA, especially recombinant DNA, into a cell. Cells, plants, or progeny resulting from this process are said to be transformed upon demonstration of the expression in the recipient organism of unique marker genes carried by the transferred DNA.

Transgenic - Refers to animals or plants in which novel DNA has been incorporated into the germ line.

Transgenic organism- An organism formed by the insertion of foreign genetic material into the germ line cells of organisms. Recombinant DNA techniques are commonly used to produce transgenic organisms.

Translation- Process by which the information on a messenger RNA molecule is used to direct the synthesis of a protein.

Transposon- A segment of DNA that can move around and be inserted at several sites in bacterial DNA or in a phage, thus alerting the host's DNA.

Tumor necrosis factors (TNFs)- Rare proteins of the immune system that appear to destroy some types of tumor cells without affecting healthy cells.

Vaccine- A preparation that contains an antigen, consisting of whole disease-causing organisms (killed or weakened) or parts of such organisms, that is used to confer immunity against the disease that the organisms cause. Vaccine preparations can be natural, synthetic or derived by recombinant DNA technology.

Vector- The agent (e.g., plasmid or virus) used to carry new DNA into a cell.

Vegetative propagation - Reproduction of plants using a nonsexual process, involving the culture of plant parts such as stem and leaf cuttings. Synonymous with "asexual propagation".

Virion -An elementary viral particle consisting of genetic material and a protein covering.

Virology - Study of viruses.

Virulence- Ability to infect or cause disease.

Virus -A submicroscopic organism that contains genetic information but cannot reproduce itself. To replicate, it must invade another cell and use parts of that cell's reproductive machinery.

Wild type- The form of an organism that occurs most frequently in nature.

Xenobiotics- Synthetic chemicals believed to be resistant to environmental degradation. A branch of biotechnology called bioremediation is seeking to develop biological methods to degrade such compounds.

Xenotransplantation- The transplantation of living organs, cells or tissues from animals into humans.

X-ray crystallography- An essential technique for determining the three-dimensional structure of biological molecules. This information aids in the discovery of products that will interact with the biological molecule.

Yeast- A general term for single-celled fungi that reproduce by budding. Some yeasts can ferment carbohydrates (starches and sugars) and thus are important in brewing and baking.

Bibliography

Akashi, K., Kondo, M., Cheshier, S., Shizuru, J., Gandy, K., Domen, J., Mebius, R., Traver, D., and Weissman, I.L. (1999). Lymphoid development from stem cells and the common lymphocyte progenitors. Cold Spring Harb. Symp. Quant. Biol. *64*, 1-12.

Akashi, K., Traver, D., Kondo, M., and Weissman, I.L. (1999). Lymphoid development from hematopoietic stem cells. Int. J. Hematol. *69*, 217-226.

Akashi, K., Traver, D., Miyamoto, T., and Weissman, I.L. (2000). A clonogenic common myeloid progenitor that gives rise to all myeloid lineages. Nature. *404*, 193-197.

Alison, M.R., Poulsom, R., Jeffery, R., Dhillon, A.P., Quaglia, A., Jacob, J., Novelli, M., Prentice, G., Williamson, J., and Wright, N.A. (2000). Hepatocytes from non-hepatic adult stem cells. Nature. *406*, 257.

Altman, J. and Das, G.D. (1965). Autoradiographic and histological evidence of postnatal hippocampal neurogenesis in rats. J. Comp. Neurol. *124*, 319-335.

Atherly A.G., Girton J.R. McDonald J.F. (1999); The Science of Genetics, Saunders College Pub, Fort Woth, Philadelphia, USA.

Axtell, J.D. 1981. Breeding for improved feeding quality, in: Frey K.J. (Ed) Plant Breeding II, Iowa State Univ. Press. pp 365–414.

Banga, S.K. (1996). Breeding for oil and meal quality, in: Chopra V.L., Prakash S. (Eds). Oilseed and Vegetable Brassicas: Indian Perspective, Chapter 11, pp. 234–249.

Barnes, D., and Sato, G., 1980a, Methods for growth of cultured cells in serum-free medium, *Anal. Biochem.* **102:**255¨C270.

Barnes, D., and Sato, G., 1980b, Serum-free cell culture: A unifying approach, *Cell* **22:**649¨C655.

Bisaria V.S., A.K. Panda, Large scale plant cell culture methods, applications and products, Curr. Opin. Biotechnol. 2 (1991) 370–374.

Bongso, A., Fong, C.Y., Ng, S.C., and Ratnam, S. (1994). Isolation and culture of inner cell mass cells from human blastocysts. Hum. Reprod. *9*, 2110-2117.

Bongso, A., Fong, C.Y., Ng, S.C., and Ratnam, S.S. (1994). Blastocyst transfer in human in vitro: fertilization; the use of embryo co-culture. Cell Biol. Int. *18,* 1181-1189.

Bongso, A., Fong, C.Y., Ng, S.C., and Ratnam, S.S. (1995). Co-culture techniques for blastocyst transfer and embryonic stem cell production. Asst. Reprod. Rev. *5,* 106-114.

Bottenstein, J., Hayashi, I., Hutchings, S. H., Masui, H., Mather, J., McClure, D. B., Chasa, S., Rizzino, A., Sato, G., Serrero, G., Wolfe, R., and Wu, R., 1979, The growth of cells in serum free hormone supplemented media, *Methods Enzymol.* **58:**94¨C109.

Bounassisi, V., Sato, G., and Cohen, A., 1962, Hormone-producing cultures of adrenal and pituitary tumor origin, *Proc. Natl. Acad. Sci. USA* **48:**1184¨C1190.

Bradley, A., Evans, M., Kaufman, M.H., and Robertson, E. (1984). Formation of germ-line chimaeras from embryoderived teratocarcinoma cell lines. Nature. *309,* 255-256.

Brown TA (2007) Genomes 3, Garland Science, Taylor and Francis Group, NY.

DBT, Plant Tissue Culture: From Research to Commercialisation, A Decade of Support, Department of Biotechnology, New Delhi, 2000, pp. 1–224.

De Vos, A. and Van Steirteghem, A. (2000). Zona hardening, zona drilling and assisted hatching: new achievements in assisted reproduction. Cells Tissues Organs. *166,* 220-227.

DiCosmo F, Misawa M, eds. (1996) Plant Cell culture, Secondary Metabolism, CRC Press, Boca Raton.

DiCosmo F, P.J. Facchini, M.M. Kraml, Cultured plant cells—the chemical factory within Chemistry in Britain 25 (1989) 1001–1004.

Dorak MT, ed. (2006) Real Time PCR, Taylor and Francis, NY, USA.

Downey, R.K. 1990. *Brassica* oilseed breeding—achievements and opportunities. Plant Breeding Abstracts. 60(10): 1165–1170.

El-Lakhany M.H., Rapid propagation of fast growing tree species in developing countries: It potentials, constraints and future developments, in: F.W.G Baker (Ed) Rapid Propagation of Fast Growing Woody Species, CAB International, (1992) pp.102–108.

Evans, M.J. and Kaufman, M.H. (1981). Establishment in culture of pluripotential cells from mouse embryos. Nature. *292,* 154-156.

Flores H.E., W.R. Curtis, Approaches to understanding and manipulating the biosynthetic potential of plant roots, Ann. New York Acad. Sci. 665 (1992) 188–209.

Fong, C.Y., Bongso, A., Ng, S.C., Kumar, J., Trounson, A., and Ratnam, S. (1998). Blastocyst transfer after enzymatic treatment of the zona pellucida: improving in-vitro fertilization and understanding implantation. Hum. Reprod. *13,* 2926-2932.

Freshney RI (2005) Culture of Animal Cells, A manual of basic techniques, John Wiley & Sons Inc., Hoboken, NJ, USA.

Friedrich, T.D., Regenass, U., and Stevens, L.C. (1983). Mouse genital ridges in organ culture: the effects of temperature on maturation and experimental induction of teratocarcinogenesis. Differentiation. *24,* 60-64.

Gardner EJ, Simmons MJ, Snustad DP (1991) Principles of Genetics, John Wiley & Sons, Inc. NY.

Gardner, D.K. and Schoolcraft, W.B. (1999). Culture and transfer of human blastocysts. Curr. Opin. Obstet. Gynecol. *11,* 307-311.

George EF, Hall MA, Klerk GJ (2008) Plant Propagation by Tissue Culture Vol 1-The background, Springer, Dordrecht, The Netherlands.

Gibson G, Muse SV (2004) A Primer of Genome Science, Sinauer Associates Inc. Pub. Sunderland, MS, USA.

Glick BR, Pasternak JJ (1998) Molecular Biotechnology, Principles and applications of recombinant DNA, ASM Press, Washington, USA.

Gopalan, C.D., Krishnamurthi, D., Shenolikar, I.S. and Krishnamachari, K.A.V.R. 1974. Myocardial changes in monkey fed on mustard oil. Nutr. Metab. 16: 352–365.

Hammer, R.E. et al. (1985) Production of transgenic rabbits, sheep and pigs by microinjection. Nature 315, 680–683

Ho RJY, Gibaldi MG (2003) Biotechnology and Biopharmaceuticals, transforming proteins and genes into drugs, Wiley–Liss, Hoboken, NJ, USA.

Jain S.M., K. Ishii, Recent Advances in somatic embryogenesis in forest trees, in: Recent Advances in Biotechnology for Tree Conservation and Management, Proc. of IFS Workshop, Florianoplis, Brazil, 15–19 September 1997, International Foundation for Science. 1998, pp. 214–231

Jain S.M., P.K. Gupta, R.J. Newton (eds.) Somatic Embryogenesis in Woody Plants, Vol. 1–3, Kluwer Academic Publishers, Netherlands, 1995.

Jolicoeur M., C. Chavarre, P.J. Carreau, J. Archambault, Development of helical-ribbon impeller bioreactor for high density plant cell suspension culture, Biotechnol. Bioeng. 39 (1992) 511–521.

Karp G (1999) Cell and Molecular Biology: Concepts and Experiments, John Wiley & Sons Inc. NY.

Khuspe S.S., P.K. Gupta, D.K. Kulkarni, U.J. Mehta, A.F. Mascarenhas, Increased biomass production by tissue culture of *Eucalyptus*, Can. J. For. 17(1987) 1361–1363.

Kieran P.M, P.F. MacLoughlin, D.M. Malone, Plant cell suspension cultures: some engineering considerations, J. Biotechnol. 59 (1997) 39–52.

Klug WS, Cummings MR (2003) Concepts of Genetics, Pearson education Inc., New Delhi, India.

Leckie F., A.H. Scragg, K.C. Cliffe, (1991) An investigation into the role of initial *kLa* on the growth and alkaloid accumulation by cultures of *Catharanthus roseus*, Biotechnol. Bioeng. 37; 364–370.

Lewin B (2008) Genes IX, Jones and Bartlett Publishers, Boston, Toronto.

Li, R. H., Gao, W.-Q., and Mather, J. P., 1996, Multiple factors control the proliferation and differentiation of rat early embryonic (day 9) euroepithelial cells, *Endocrine* **5:**205¨C217.

Li, R., Phillips, D. M., and Mather, J. P., 1995, Activin promotes ovarian follicle development *in vitro, Endocrinology* 136:849¨C856.

Loo, D., Rawson, C., Helmrich, A., and Barnes, D., 1989, Serum-free mouse embryo cells: Growth responses *in vitro, J. Cell. Physiol.* 139:484¨C491.

Mascarenhas A.F., E.M. Muralidharan, Clonal Forestry with Tropical Hardwoods, in: M.R. Ahuja, W.J. Libby (Eds), Clonal Forestry II: Conservation and Application, Springer-Verlag. (1993) pp 169–187.

Mascarenhas A.F., E.M. Muralidharan, Tissue culture of forest trees in India, Curr. Sci. 58 (1989) 606– 613.

Mather, J. P., and Sato, G. H., 1979, The use of hormone-supplemented serum-free media in primary cultures, *Exp. Cell Res.* 124:215¨C221.

Meyers RA, ed. (1995) Molecular Biology and Biotechnology, VCH Publishers, Inc. NY.

Mullins K.V., D.J. Llewellyn, V.J. Hartney, S. Strauss, E.S. Dennis, Regeneration and transformation of *Eucalyptus camaldulensis*, Plant Cell Rep. 16 (1997) 787–791.

Muralidharan E.M., A.F. Mascarenhas, Somatic embryogenesis in Eucalyptus, in: S.M. Jain, P.K., Gupta, R.J. Newton (Eds) Somatic Embryogenesis in Woody Plants, Vol. 2, Kluwer Academic Publishers, Netherlands, (1995), pp 23–40.

Murray, J.D. et al. (1989). Production of transgenic merino sheep by microinjection of ovine metallothionein-ovine growth hormone fusion genes. Reprod. Fertil. Dev. 1, 147–155

Nijkamp HJJ, VanderPlas LHW, Aartrijk JV (1990) Progress in plant cellular and molecular biology, Kluwer Acad. Pub., Dordrechet, The Netherlands.

Oksman-Caldantey KM, Barz WH (2002) Plant Biotechnology and Transgenic plants, Marcel Dekers Inc. NY.

Panda A.K, S. Mishra, V.S. Bisaria, S.S. Bhojwani, Plant cell reactors—a perspective, Enz. Microbial Technol. 11 (1989) 386–397.

Palmiter, R.D. et al. (1982). Dramtic growth of mice that develop from eggs microinjected with metallothionein-growth hormone fusion genes. Nature 300, 611–615

Payne G.F, M.L. Shuler, P. Brodelius, Plant cell culture, in: B. K. Lydensen (Ed), Large Scale Cell Culture Technology, Hanser Publishers, New York, 1987, pp. 193–229.

Prakash, S., Kumar, P.R., Sethi, M., Singh, C., Tandon, R.K. (2000). Mustard Oil: The Ultimate Edible Oil. The Botanica 50: 94–101.

Primose SB, Twyman RM (1996) Principles of Gene Manipulation and Genomics, Blackwell Publishing,Victoria, Australia.

Puck, T., 1972, *The Mammalian Cell as a Microorganism: Genetic and Biochemical Studies in Vitro,* Holden-Day, San Francisco.

Ramawat KG, Merillon JM, eds. (2007) Biotechnology, Secondary Metabolites: Plant and Microbes, Sci Pub,NJ, USA.

Reinert J, Bajaj YPS, eds. (1977) Plant Cell, Tissue and Organ Culture, Spinger –Verlag, Berlin , Heildelberg.

Renarid, S., McGregor, L. 1976. Antithrombogenic effects of erucic acid poor rapeseed oils in the rats. Rev. Fr. Crops. 23: 393–396. Cited in: Chopra P.K., Kirti P.B. 1996. Biotechnology in: Chopra V.L., Prakash S. (Eds). Oilseed and Vegetable Brassicas: Indian Perspective, Chapter 12, pp. 250–269.

Sauer, F.D., Kramer, J.K.G. (1983) The problems associated with the feeding of high erucic acid rapeseed oils and some fish oils to experimental animals, in: High and Low Erucic Acid Rapeseed Oils (Kramer J.K.G., Sauer F.D., Figden W.J. (Eds), Toronto, Canada Academic Press. pp 254–292.

Scragg A.H., The problems associated with high biomass levels in plant cell suspensions, Plant Cell Tissue Organ Cult. 43 (1995) 163–170.

Stahl E (1969) Thin Layer Chromatography, Springer-Verlag, Berlin, Heidelberg.

Tiwari KK, Singhal DS (1997) Plant Molecular Biology and Biotechnology, Narosa Pub House New Delhi, India

Vasil IK, Thorpe TA (1994) Plant Cell and Tissue Culture, Kluwer Acad. Pub., Dordrechet, The Netherlands.

Verpoorte R, Alfermann AW, eds. (2000) Metabolic Engineering of Plant Secondary Metabolism, Kluwer Acad. Pub., Dordrechet, The Netherlands.

Wall, R.J. et al. (2005) Genetically enhanced cows resist intramammary Staphylococcus aureus infection. Nat. Biotechnol. 23, 445–451

Watson JD et al., (2004) Molecular Biology of the Genes, Pearson education Inc., New Delhi, India.

Wink M, ed. (2006) An Introduction to Molecular Biotechnology, Molecular fundamentals methods and applications in Modern Biotechnology, Wiley VCH Verlag GmbH and Co. KGa, Weinheim, Germany.